Field Manual of Techniques in Invertebrate Pathology

Field Manual of Techniques in Invertebrate Pathology

Application and Evaluation of Pathogens for Control of Insects and other Invertebrate Pests

Edited by

Lawrence A. Lacey

USDA-ARS Wapato, WA, USA

and

Harry K. Kaya

University of California Davis, CA, USA

Springer

A C.I.P. Catalogue record for this book is available from the Library of Congress.

ISBN 978-1-4020-5932-2 (PB)
ISBN 978-1-4020-5931-5 (HB)
ISBN 978-1-4020-5933-9 (e-book)

Published by Springer,
P.O. Box 17, 3300 AA Dordrecht, The Netherlands.

www.springer.com

The credits for cover photos:

Upper left corner, application of codling moth granulovirus. photograph by H. Headrick; Upper right, cabbage looper infected with the nucleopolyhedrovirus of *Autographa californica.* Photo courtesy of D. Hoffmann. Lower left, resting spores of the fungus *Entomophaga maimaiga.* Photo courtesy of A. Hajek. Lower right, adult banana weevil infected with *Beauveria bassiana.* Photo courtesy of R. Duncan and J. Peña

Printed on acid-free paper

Dedication

We dedicate the *Field Manual* to Dr. H. Denis Burges in recognition of his pioneering and visionary research in insect pathology, and in particular, for his contributions with developing quality standards for bacterial pathogens of insects. His scientific career has spanned more than 56 years, and he continues to be active in the field as evidenced by his contribution to this book. Dr. Burges' many scientific achievements include elucidation of the biology and mode of action of *Bacillus thuringiensis* and its standardization, formulation of microbial insecticides, applied microbial control, and many other contributions. These have not only enriched our profession, but have also contributed to the increased use of microbial insecticides around the world. As a long term member and president of the Society for Invertebrate Pathology, he helped to create a home for invertebrate pathologists where intellectual stimulation, collaboration, information sharing, and scientific broad-mindedness were the emphasis. For these achievements alone, he deserves to be honored, but Denis has made other contributions to the field of insect pathology as well – contributions that are at least as unique and important as his impressive research record. As an articulate and intelligent spokesperson for insect pathology and microbial control, he brings integrity, wisdom and humor to the discussion, and in so doing has raised the profile of our profession within, and beyond the scientific community. And he has helped our profession grow by serving as an inspiring mentor to many insect pathologists. By taking young scientists seriously, and by sharing his vast knowledge and enthusiasm with them, he has helped us all.

Contents

SECTION IX – Resistance

SECTION X – Non-target Organisms

Preface

The *Field Manual of Techniques in Invertebrate Pathology* is designed to provide background and instruction on a broad spectrum of techniques and their use in the evaluation of entomopathogens in the field. The second edition of the *Field Manual* provides updated information and includes two additional chapters and 12 new contributors. The intended audience includes researchers, graduate students, practitioners of integrated pest management (IPM), regulators and those conducting environmental impact studies of entomopathogens. Although it can function as a stand alone reference, the *Field Manual* is complementary to the laboratory oriented *Manual of Techniques in Insect Pathology* and to comprehensive texts in insect pathology.

We have structured the 40 chapters of the *Field Manual* into 10 sections to provide the tools required for planning experiments with entomopathogens and their implementation in the field. The basic tools include chapters on the theory and practice of application of microbial control agents (MCAs) (Section I), statistical considerations in the design of experiments (Section II), and three chapters on application equipment and strategies (Section III). Section IV includes individual chapters on the major pathogen groups (virus, bacteria, microsporidia, fungi, and nematodes) and special considerations for their evaluation under field conditions. This section sets the stage for subsequent chapters on the impact of naturally occurring and introduced exotic pathogens and inundative application of MCAs. Twenty-three chapters on the application and evaluation of MCAs in a wide variety of agricultural, forest, domestic and aquatic habitats comprise Section VII of the *Field Manual*. In addition to insect pests, the inclusion of mites and slugs broadens the scope of the book. Most of the chapters in this section include step by step instructions on handling of inoculum, design of field experiments and experimental plots and application and assessment of efficacy of dozens of MCAs. Several of these chapters include supplementary techniques and media for conducting follow up laboratory studies for confirmation of infection, determination of persistence, etc. The three final chapters include: special consideration for evaluation of *Bt* transgenic plants (Section VIII); resistance to insect pathogens and strategies to manage resistance (Section IX); and guidelines for evaluating effects of MCAs on nontarget organisms (Section X).

Due to uncertainty regarding the future availability of organophosphate and other conventional chemical insecticides, MCAs will play increasingly important roles in IPM. The *Field Manual* will provide researchers and IPM practitioners with techniques and practical guidance for the study and optimal use of MCAs in a variety of settings.

We are especially indebted to all 81 of the contributing authors. Their expertise represents virtually every aspect of invertebrate pathology and microbial control from an international perspective. We also thank our colleagues who furnished information, photographs, and reviews of the chapters. We gratefully appreciate the secretarial assistance provided to the various authors for individual chapters and Heather Headrick, Hanayo Arimoto, Belinda Bishop and Joanne Kaya for help with the second edition of the book. Special thanks to Zuzana Bernhart and Springer Scientific Publishers for the invitation to produce an updated and affordable second edition. We are especially appreciative of the patience and understanding of our spouses, Cindy and Joanne, during the process of producing the original and second edition of the *Field Manual*.

Lawrence A. Lacey
Harry K. Kaya

Contributors

Dr. Steven P. Arthurs
Yakima Agricultural Research Laboratory
USDA-ARS
5230 Konnowac Pass Road
Wapato, WA 98951, USA
e-mail: sarthurs@yarl.ars.usda.gov

Dr. Roy P. Bateman
International Pesticide Application Research Centre (IPARC)
Imperial College, Silwood Park
Ascot, Berkshire SL5 7PY, UK
e-mail: r.bateman@imperial.ac.uk

Dr. Leah S. Bauer
USDA Forest Service
North Central Research Station
1407 S. Harrison Road
East Lansing, MI 48823, USA
e-mail: lbauer@fs.fed.us

Dr. James J. Becnel
USDA-ARS-CMAVE
1600 SW 23rd Drive
Gainesville, FL 32608, USA
e-mail: jbecnel@gainesville.usda.ufl.edu

Dr. Ralph E. Berry
Emeritus Professor of Entomology
Oregon State University
18160 Cottonwood Road, PMB #798
Sunriver, OR 97707, USA
e-mail: berryr@onid.orst.edu

Dr. Steven R. Booth
2523 B 44th Ave N.W.
Olympia, WA 98502, USA
e-mail: boothswa@comcast.net

Dr. Denny J. Bruck
Horticultural Crops Research Laboratory
USDA-ARS
3420 N.W. Orchard Avenue
Corvallis, OR 97330, USA
e-mail: bruckd@onid.orst.edu

Dr. John Burand
Department of Plant, Soil and Insect Science
Fernald Hall
University of Massachusetts
Amherst MA 01003-0002, USA
e-mail: jburand@microbio.umass.edu

Dr. H. Denis Burges (retired)
21 Withdean Ave.
Goring-by-Sea, Worthing
West Sussex, BN12 4XD, UK

Dr. James F. Campbell
Grain Marketing & Production Research Center
Biological Research Unit
USDA-ARS
1515 College Avenue
Manhattan, KS 66502, USA
e-mail: james.campbell@gmprc.ksu.edu

Dr. Michael A. Caprio
Department of Entomology and Plant Pathology
Box 9775
Mississippi State University
Mississippi State, MS 39762-9775, USA
e-mail: caprio@ra.msstate.edu

Dr. Andrew C. Chapple
Modelling, Metabolism and Environmental Fate
Bayer CropScience AG
Gebaude 6990, Zm. 5
50 Alfred Nobel Strasse
D-40789 Monheim, Germany
e-mail: andrew.chapple@bayercropscience.com

Dr. Ho Yul Choo
Department of Agricultural Biology
Gyeongsang National University
Gyeongnam Province, Chinju, 660–701
Republic of Korea
e-mail: hychoo@nongae.gsnu.ac.kr

Dr. Jenny S. Cory
Department of Biology
Algoma University College
1520 Queen Street East
Sault Ste. Marie, ON P6A 2G4, Canada
e-mail: cory@algomau.ca

Dr. Jack D. DeAngelis
College of Agricultural Sciences
138 Strand Hall
Oregon State University
Corvallis, OR 97331-2907, USA
e-mail: deangelj@bcc.orst.edu

Dr. Italo Delalibera, Jr.
Departamento de Entomologia, Fitopatologia e Zoologia Agrícola
ESALQ-USP
Av. Pádua Dias 11 Caixa Postal 9
Piracicaba-SP 13418-900 Brazil
e-mail: italo@esalq.usp.br

Dr. Patrick F. Dowd
Crop Bioprotection Research Unit
USDA-ARS-NCAUR
1815 N. University Street
Peoria, IL 61604, USA
e-mail: DOWDPF@ncaur.usda.gov

Dr. Roger A. Downer
Lab for Pest Control Application Technology
OARDC, 1680 Madison Avenue
Wooster, OH 44691-4096, USA
e-mail: downer.2@osu.edu

Dr. Frank A. Drummond
Department of Biological Sciences
University of Maine
Orono, ME 04469, USA
frank.drummond@umit.maine.edu

Dr. Normand R. Dubois [deceased]
USDA Forest Service
Northeastern Research Station
Center for Forest Health Research
Hamden, CT 06514, USA

Dr. Larry W. Duncan
Citrus Research & Education Center
University of Florida, IFAS
700 Experiment Station Road
Lake Alfred, FL 33850, USA
e-mail: lwdn@crec.ifas.ufl.edu

Dr. Joseph S. Elkinton
Department of Plant, Soil and Insect Sciences
Fernald Hall
University of Massachusetts
Amherst, MA 01003, USA
e-mail: elkinton@ent.umass.edu

Dr. Martin A. Erlandson
Agriculture & Agri-Food Canada
Saskatoon Research Centre
107 Science Place
Saskatoon, SK S7N OX2, Canada
e-mail: erlandsonm@em.agr.ca

Dr. Hugh F. Evans
Entomology Branch
Forest Research
Alice Holt Lodge - Wrecclesham
Farnham Surrey GU10 4LH, UK
e-mail: h.evans@forestry.gov.uk

Dr. Stephen F. Garczynski
Yakima Agricultural Research Laboratory
USDA-ARS
5230 Konnowac Pass Road
Wapato, WA 98951, USA
e-mail: sgarczynski@yarl.ars.usda.gov

Dr. Mark S. Goettel
Agriculture & Agri-Food Canada
Lethbridge Research Centre
P.O. Box 3000
Lethbridge, AB T1J 4B1, Canada
e-mail: goettelM@agr.gc.ca

Dr. Dawn H. Gouge
University of Arizona
Maricopa Agricultural Center
37860 Smith-Enke Road
Maricopa AZ 85239-3010, USA
e-mail: DHGouge@Ag.Arizona.edu

Dr. Parwinder S. Grewal
Department of Entomology
Ohio State University, OARDC
Wooster, OH 44691-4096, USA
e-mail: grewal.4@osu.edu

Dr. Eleanor Groden
Department of Biological Sciences
University of Maine
Orono, ME 04469, USA
e-mail: eleanor.groden@umit.maine.edu

Dr. Ann E. Hajek
Department of Entomology
Cornell University
Ithaca, NY 14853-0901, USA
e-mail: aeh4@cornell.edu

Dr. Frank R. Hall
Department of Entomology
Lab for Pest Control Application Technology
Ohio State University, OARDC
Wooster, OH 44691, USA
e-mail: hall.1@osu.edu

Dr. Jürg Huber
Institut für Biologischen Pflanzenschutz
Heinrichstrasse 243
64287 Darmstadt, Germany
e-mail: J.Huber @bba.de

Dr. G. Douglas Inglis
Agriculture & Agri-Food Canada
Lethbridge Research Centre
P.O. Box 3000
Lethbridge, AB T1J 4B1, Canada
e-mail: InglisD@AGR.GC.CA

Dr. D. Michael Jackson
U. S. Vegetable Laboratory
USDA-ARS
2875 Savannah Highway
Charleston, SC 29414, USA
e-mail: mjackson@awod.com

Dr. Jan J. Jackson
Pioneer Hi-Bred International
South Dakota Research Center
216 West 2nd St. South, Suite A
Brookings SD 57006, USA
e-mail: jan.jackson@pioneer.com

Dr. Trevor A. Jackson
AgResearch
Canterbury Agricultural & Science Centre
P. O. Box 24
Lincoln, New Zealand
e-mail: jacksont@agresearch.cri.nz

Dr. Harry K. Kaya
Department of Nematology
University of California
One Shields Avenue
Davis, CA 95616-8668, USA
e-mail: hkkaya@ucdavis.edu

Dr. James Kerwin
Manager, Analytic Core
School of Pharmacy
University of Southern California
Los Angeles, CA 90089-9121, USA
e-mail: jlkerwin@hotmail.com

Dr. Michael G. Klein (retired)
Department of Entomology
Ohio State University, OARDC
Wooster OH 44691, USA
e-mail: Klein.10@osu.edu

Dr. Alan L. Knight
Yakima Agricultural Research Laboratory
USDA-ARS
5230 Konnowac Pass Road
Wapato, WA 98951, USA
e-mail: aknight@yarl.ars.usda.gov

Dr. Albrecht M. Koppenhöfer
Department of Entomology
Rutgers University
New Brunswick, NJ 08901, USA
e-mail: koppenhofer@aesop.rutgers.edu

Dr. Lawrence A. Lacey
Yakima Agricultural Research Laboratory
USDA-ARS
5230 Konnowac Pass Road
Wapato, WA 98951, USA
e-mail: llacey@yarl.ars.usda.gov

Dr. Jarrod E. Leland
Novozymes Biologicals Inc.
5400 Corporate Circle
Salem, VA 24153, USA
e-mail: JRRL@novozymes.com

Dr. Leslie C. Lewis
Corn Insects & Crop Genetics Research Unit
USDA-ARS
Genetics Laboratory
Iowa State University
Ames, IA 50011, USA
e-mail: leslewis@iastate.edu

Dr. Jeffrey C. Lord
Grain Marketing & Production Research Center
Biological Research Unit
USDA-ARS
1515 College Avenue
Manhattan, KS 66502, USA
e-mail: jeffrey.lord@gmprc.ksu.edu

Dr. Graham A. Matthews
International Pesticide Application Research Centre (IPARC)
Imperial College, Silwood Park
Ascot, Berkshire SL5 7PY, UK
e-mail: g.matthews@ic.ac.uk

Dr. Clayton W. McCoy (retired)
Citrus Research & Education Center
University of Florida, IFAS
700 Experiment Station Road
Lake Alfred, FL 33850, USA
e-mail: facelkm@aol.com

Dr. Michael L. McManus
USDA Forest Service
Northeastern Research Station
Center for Forest Health Research
51 Mill Pond Road
Hamden, CT 06514, USA
e-mail: mmcmanus/ne_ha@fs.fed.us

Dr. Karl Mierzejewski
Droplet Technologies
937-1 West Whitehall Road
State College, PA 16801-2906, USA
e-mail: karlem@comcast.net

Dr. Richard J. Milner (retired)
CSIRO, Division of Entomology
GPO Box 1700
Canberra ACT 2601, Australia
e-mail: richardm@ento.csiro.au

Dr. Flávio Moscardi
Embrapa Soja, C.P. 231
Londrina, PR 86001-970, Brazil
e-mail: moscardi@cnpso.embrapa.br

Dr. Bradley A. Mullens
Department of Entomology
University of California
Riverside, CA 92521, USA
e-mail: bradley.mullens@ucr.edu

Dr. Judith K. Pell
Department of Entomology & Nematology
IACR, Rothamsted Research
Harpenden, Hertfordshire, AL5 2JQ, UK
e-mail: judith.pell@bbsrc.ac.uk

Dr. Roberto M. Pereira
Entomology & Nematology Dept.
University of Florida
Gainesville, FL 32611-0620, USA
e-mail: rpereira@ufl.edu

Dr. Dudley E. Pinnock
Microbial Products Pty. Ltd.
P.O. Box 494
Kent Town, Adelaide
South Australia 5071
Australia
e-mail: dpinnock@bigpond.com

Dr. Tadeusz J. Poprawski [deceased]
USDA-ARS-SPA
Subtropical Agricultural Research Center
Weslaco, TX 78596, USA

Dr. Richard C. Reardon
USDA Forest Service
Forest Health Technology
Enterprise Team
180 Canfield Street
Morgantown, WV 26505, USA
e-mail: rreardon@fs.fed.us

Dr. William C. Rice
USDA-ARS
Renewable Energy and Manure Management Research Unit
2300 Experiment Station Road
Bushland, TX 79012, USA
e-mail: wrice@cprl.ars.usda.gov

Dr. Richard T. Roush
Department of Entomology
University of California
One Shields Avenue
Davis, CA 95616-8668, USA
e-mail: rtroush@ucdavis.edu

Dr. John D. Sedlacek
Department of Entomology
Kentucky State University
Frankfurt, KY 40601, USA
e-mail: john.sedlacek@kysu.edu

Dr. David I. Shapiro-Ilan
SE Fruit and Tree Nut Research Unit
USDA-ARS, SAA
21 Dunbar Road
Byron, GA 31008, USA
e-mail: david.shapiro@ars.usda.gov

Dr. Anthony M. Shelton
Department of Entomology
Cornell University
New York State Agricultural Experiment Station
416 Barton Lab
Geneva NY 14456, USA
e-mail: ams5@cornell.edu

Dr. Joel P. Siegel
San Joaquin Valley Agricultural Sciences Center
USDA-ARS
9611 South Riverbend Avenue
Parlier, CA 93648, USA
e-mail: jsiegel@fresno.ars.usda.gov

Dr. Ole Skovmand
118 Chemin des Alouettes
34170 Castelnau le Lez, France
e-mail: Ole.skovmand@insectcontrol.net

Dr. Leellen F. Solter
Illinois Natural History Survey
607 E. Peabody Drive
Champaign, IL 61820, USA
e-mail: lsolter@uiuc.edu

Dr. Daniel R. Sosa-Gómez
Embrapa Soja, C.P. 231
Londrina, PR 86001-970, Brazil
e-mail: drsg@cnpso.embrapa.br

Dr. Marc Sporleder
Centro Internacional de la Papa
Apartado 1558
Lima, Peru
e-mail: m.sporleder@cgiar.org

Dr. Donald C. Steinkraus
Department of Entomology
University of Arkansas
Fayetteville, AR 72701, USA
e-mail: steinkr@comp.uark.edu

Dr. Robin J. Stuart
Citrus Research & Education Center
University of Florida, IFAS
700 Experiment Station Road
Lake Alfred, FL 33850, USA
e-mail: rstuart@ufl.edu

Dr. Douglas V. Sumerford
Corn Insects & Crop Genetics Research Unit
USDA-ARS
Genetics Laboratory
Iowa State University
Ames, IA 50011, USA
e-mail: sumerford@iastate.edu

Dr. Harold Thistle
USDA Forest Service
Forest Health Technology Enterprise Team
180 Canfield Street
Morgantown, WV 26505, USA
e-mail: hthistle@fs.fed.us

Dr. Patrick V. Vail [deceased]
San Joaquin Valley Agricultural Sciences Center
USDA, ARS
9611 South Riverbend Avenue
Parlier, CA 93648, USA

Dr. Kees van Frankenhuyzen
Great Lakes Forestry Centre
Natural Resources Canada
Canadian Forestry Service
P. O. Box 490
Sault Ste. Marie, ON P6A 5M7, Canada
e-mail: kvanfran@NRCan.gc.ca

Dr. John D. Vandenberg
Plant, Soil & Nutrition Laboratory
USDA-ARS
Tower Road
Ithaca, NY 14853, USA
e-mail: JDV3@cornell.edu

Dr. Fernando E. Vega
Insect Biocontrol Laboratory
USDA-ARS
10300 Baltimore Boulevard
Bldg. 011A, Rm. 214,
BARC-West
Beltsville, MD 20705-2350, USA
e-mail: VegaF@ba.ars.usda.gov

Dr. Ping Wang
Department of Entomology
Cornell University
619 Barton Lab
New York State Agricultural
Experiment Station
Geneva, NY 14456, USA
e-mail: pw15@cornell.edu

Dr. David K. Weaver
Department of Land Resources
and Environmental Sciences
Montana State University
Bozeman, MT 59717, USA
e-mail: weaver@montana.edu

Dr. Michael J. Wilson
Department of Plant & Soil Sciences
University of Aberdeen
Aberdeen AB24 3UU, Scotland, UK
e-mail: m.j.wilson@abdn.ac.uk

Dr. Stephen P. Wraight
Plant, Soil & Nutrition Laboratory
USDA-ARS
Tower Road
Ithaca, NY 14853, USA
e-mail: spw4@cornell.edu

Dr. Seth Y. Young (retired)
Department of Entomology
University of Arkansas
Fayetteville, AR 72701, USA
e-mail: syoung@comp.uark.edu

Dr. Jian-Zhou Zhao
Department of Entomology
Cornell University
417 Barton Lab
New York State Agricultural Experiment Station
Geneva, NY 14456, USA
e-mail: jz49@cornell.edu

SECTION I

INTRODUCTION

Chapter I-1

Introduction to microbial control

Harry K. Kaya

Department of Nematology
University of California
Davis, CA 95616, USA

Lawrence A. Lacey

USDA-ARS
Yakima Agricultural Research Laboratory
Wapato, WA 98951, USA

1 Introduction

Pathogens, including viruses, bacteria, rickettsia, fungi, protozoans, and nematodes, are commonly isolated from insect and other invertebrate hosts. Their natural occurrence in invertebrate populations contributes to the regulation of injurious pests of humans and their crops, households and domestic animals. These entomopathogens have potential for biological (*i.e.*, microbial) control programs (Steinhaus, 1956), and many of them have been exploited for insect pest control through inoculative, inundative and augmentative releases (Lacey *et al.*, 2001). Spectacular successes have been reported for a few entomopathogens as classical biological control agents such as the *Oryctes* virus against the coconut palm rhinoceros beetle in the South Pacific, (Huger, 2005), a nucleopolyhedrovirus against the European pine sawfly in Canada (Bird, 1955), and the nematode, *Beddingia* (=*Deladenus*) *siricidicola* against siricid woodwasps in Australia (Bedding, 1993) and several countries in South America and in South Africa (Bedding and Iede, 2005). However, many classical biological control introductions of pathogens have resulted in establishment, but not necessarily in pest control (Hajek *et al.*, 1996). In some instances, fortuitous or accidental introduction of a pathogen has also resulted in excellent biological control of a pest. These include a nucleopolyhedrovirus against the European spruce sawfly (Bird and Burk, 1961), the fungus, *Entomophaga maimaiga*, against the gypsy moth (Hajek *et al.*, 1995), and the microsporidium, *Nosema* (=*Perezia*) *pyrausta*, against the European corn borer (Kramer, 1959).

Aside from natural infections and their introduction as classical biological control agents, entomopathogens have been commonly used as inundative agents for the suppression of pests. A number of them have been registered and/or are commercially available for use against pest species. The first commercial entomopathogens were the milky disease bacteria, *Paenibacillus* (=*Bacillus*) *popilliae* and *P. lentimorbus*, that were registered in 1948 and used primarily in augmentative releases to suppress populations of the Japanese beetle in the USA (Klein, 1981). Microbial control took a significant step forward in the USA with the registration and commercialization of the bacterium, *Bacillus thuringiensis* (*Bt*), in 1961 (Glare and O'Callaghan, 2000). Large scale production of *Bt* on artificial media, application with conventional sprayers, safety, and selectivity for lepidopterous pests were its major positive attributes (Glare and O'Callaghan, 2000). Despite these advantages, the initial *Bt* isolate (*e.g.*, subsp. *thuringiensis*)

L.A. Lacey and H.K. Kaya (eds.), *Field Manual of Techniques in Invertebrate Pathology*, 3–7.

used in commercial products was not very efficacious against the targeted lepidopterous pests. Fortunately, *Bt* subsp. *kurstaki* commercialized in the early 1970s, followed by other *Bt* subspecies (*e.g.*, *aizawai*, *tenebrionis*, and *israelensis*), showed marked improvement in potency and host range to increase their utility in microbial control programs. Although the major share of the microbial control market goes to *Bt*-based products, a variety of other microbial control agents (MCAs) are commercially produced. Thus, viruses, other bacteria, fungi, and nematodes have one or more species available for microbial control of insect pests and other invertebrates in many different countries.

MCAs (*i.e.*, biopesticides) are produced by both large and small commercial companies. Some like baculoviruses and *P. popilliae* are obligate pathogens and are produced *in vivo*, whereas others such as many fungi, entomopathogenic nematodes and *Bt* can be produced *in vitro* in large bioreactors. These MCAs are formulated and sold for pest suppression or, if not yet registered, are made available to researchers to test in the field. In many cases, if the MCA has not generated the interest of a commercial company or is still in the early developmental stage, the researcher will produce the MCA in the laboratory for field testing.

To test these microbial commercial products or if registration of an insect pathogen is contemplated, a number of issues need to be addressed. In this book, we bring together experts in the field who have the experience and knowledge about pathogens and their use against insect and other invertebrate pests. Our purpose is to provide sufficient background in spray equipment, experimental design and statistical analyses, the biology and the natural occurrence of pathogens, cropping systems, pests, environmental effects on the pathogen, resistance management, non-targets effects, etc. so that a researcher can design and conduct proper field experiments. We cover various agroecosystems and other environments including row crops, orchards, forestry, greenhouses, mushroom production, stored products, nursery and landscape, aquatic, livestock and poultry, and urban habitats. Moreover, naturally occurring entomopathogens can regulate insect populations in agricultural and forestry ecosystems (see Chapters V-1 and V-2), and have been introduced as classical biological control agents that have resulted in significant pest population reduction (see Chapter VI-1).

2 Advantages and disadvantages of microbial agents

The comparison of biopesticides with conventional chemical pesticides is usually solely from the perspective of their efficacy and cost. However, taking into account environmental benefits including (1) safety for humans and other non-target organisms [see Chapter X-1; recently, however, *Beauveria bassiana* (Henke *et al.*, 2002) and the bacterium, *Photorhabdus asymbiotica* associated with a *Heterorhabditis* sp. (Gerrard *et al.*, 2006) have been implicated in human health issues], (2) reduction of pesticide residues in the aquatic and terrestrial environments, (3) increased activity of most other natural enemies, and (4) increased biodiversity in aquatic ecosystems, their advantages are numerous (Lacey *et al.*, 2001). Besides these four advantages of MCAs, others listed by Tanada and Kaya (1993) include:

- specificity to the target organism or to a limited number of host species,
- little or no development of resistance by the target organism [however, resistance has developed in some target insects to *Bt*, *Bacillus sphaericus* and baculoviruses (see Chapter IX-1)],
- no secondary pest outbreak,
- compatibility with other biological control agents,
- possibility of long term control,
- ease of application,
- no pre-harvest interval, and
- adaptable to genetic modification through biotechnology.

Although the advantages of MCAs are many, some of them are also disadvantages. For example, host specificity is a double-edged sword and is an advantage and a disadvantage. If more than one pest species occurs in the ecosystem, additional control measures are needed. The disadvantages of MCAs compared with chemical pesticides include the following:

- specificity only to target organism,
- strict timing of application for maximal effect,
- long period of lethal infection (*i.e.*, little or no "knock-down" effect),
- inactivation by environmental factors (*e.g.*, ultraviolet light, desiccation, temperature extremes, etc.) and therefore, short field persistence,
- difficult to produce obligate pathogens, and/or difficult to formulate,
- short shelf life,
- development of resistance by target organisms, especially to *Bt* and *B. sphaericus*,
- uneconomical except for niche markets, and
- risks or non-acceptability by consumers associated with genetically-modified organisms.

Chemical pesticides have a number of advantages over MCAs including broad host range, faster "knockdown" and kill, longer persistence, ease of formulation, longer shelf-life (*i.e.*, stability of product), ease of application, and lower cost, but there are also many disadvantages. In fact, some of these advantages have become major impediments to their use such as broad host range and long persistence in the environment. Chemical pesticides may cause secondary pest outbreaks, can result in pest resistance to the chemicals, and as alluded to above, have adverse effects on the environment (*e.g.*, ground water contamination, accumulation in the soil, etc.) and human health. Indeed, to minimize impact of these chemicals to the environment and humans (and other non-target vertebrates), a number of laws have been passed, the most recent of which has been the Food Quality Protection Act (FQPA) of 1996 in the USA. Other countries have passed similar initiatives to protect the biosphere from these chemicals.

As the older, broad-spectrum chemical pesticides (*i.e.*, organophosphates, organochlorines and carbamates) are being phased out through legislative actions or by the chemical companies themselves, newer chemical compounds (*e.g.*, halofenozide, neonicotinoid, and phenylpyrazole) and some others that are byproducts of microorganisms (*e.g.*, spinosad and avermectins), are on the market (Georgis, 1997). These compounds are affecting the use of MCAs because they tend to be user friendly, have a narrower host spectrum, are less toxic to humans and non-target vertebrates, and have less environmental impact. However, they are more expensive than the conventional chemical insecticides, have a broader host range than most MCAs, and have a potential for eliciting resistance in the target pest. Transgenic plants, ironically led by the incorporation of *Bt* genes, are also major factors affecting the use of MCAs.

3 Role of entomopathogens in pest suppression

MCAs can play an important role in pest suppression. In some systems, entomopathogens are the significant or dominant control agents. In forestry, *Bt* is used for spruce budworm control (see Chapter VII-9), whereas in potatoes and crucifer crops, *Bt* is used for pest control because of chemical resistance occurring in the target pests (see Chapters VII-1 and VII-2, respectively). Although *Bt* resistance occurs in the diamondback moth in crucifers, the discovery and commercial development of different *Bt* isolates or subspecies have proven to be effective against this pest. Furthermore, *Bt* has the potential to control the sheep louse attacking sheep in Australia (see Chapter VII-21). In aquatic systems, *Bt* subsp. *israelensis* have been important in mosquito and black fly larval control (see Chapter VII-22). In citrus, entomopathogenic nematodes are used for Diaprepes weevil control because they are efficacious and cost effective and there are no other alternatives available for the larval stage of this soil pest (Georgis *et al.*, 2006; see Chapter VII-13). In other cropping systems, the application of a nucleopolyhedrovirus has played a significant role in suppressing a lepidopterous pest in soybeans in Brazil (see Chapter VII-5), and the use of entomopathogenic fungi have played a key role against sucking insects in greenhouses (see Chapter VII-8). These are but a few examples of the important roles that MCAs have contributed towards pest suppression.

Even where *Bt* transgenic plants are used, some acreages are set aside for non-transgenic plants to minimize resistance developing in the target insect (see Chapters VIII-1 and IX-1). Acreage planted with non-transgenic plants cannot be treated with *Bt*, but it does open the door for other entomopathogens, especially baculoviruses and fungi.

MCAs are not restricted to entomopathogens. Major pests of agriculture, nurseries, and homeowners are slugs. Molluscidal nematodes are commercially available in Europe and offer an alternative to chemical pesticides (see Chapter VII-23).

In many niche markets, the use of standard chemical insecticides is being phased out by FQPA. Many of the newer chemistry insecticides will replace the older, more toxic chemicals, but MCAs will take part of that share. In addition, there is a continued pressure to use entomopathogens in an integrated approach. Because MCAs have many advantages alluded to earlier, they are ideal candidates for integrated pest management programs.

4 Conclusions

It is critical that we know how to evaluate entomopathogens in the different ecosystems where pests occur. Entomopathogens, whether they are locally produced or produced by larger companies, will have to be applied and evaluated against the target pest. We need to know the pest, the cropping system, the application equipment, plot design, and the entomopathogens to properly evaluate their effectiveness. We need to evaluate non-target effects and determine the effects of the naturally occurring pathogens. In some systems, the entomopathogens are proven control agents and the protocol is well established. In other cases, entomopathogens are potential control agents against pests. Although the chapters in this book do not address all pests or every potential use of entomopathogens, sufficient examples of protocols are available that can be adapted to almost any situation. This book, plus its companion "Manual of Techniques in Insect Pathology" edited by Lacey (1997), should be useful for anyone interested in evaluating MCAs against a variety of pests in various cropping systems, veterinary situations, or natural habitats.

5 References

Bedding, R. A. 1993. Biological control of *Sirex noctilio* using the nematode *Deladenus siricidicola*. *In* "Nematodes and the Biological Control of Insect Pests" (R. Bedding, R. Akhurst and H. Kaya, Eds.), pp. 11–20. CSIRO Publications, East Melbourne, Australia.

Bedding, R. A. and Iede, E. T. 2005. Application of *Beddingia siricidicola* for Sirex woodwasp control. *In* "Nematodes as Biocontrol Agents" (P. S. Grewal, R.-U. Ehlers, and D. I. Shapiro-Ilan, Eds.), pp. 385–399. CABI Publishing, Wallingford, UK.

Bird, F. T. 1955. Virus disease of sawflies. *Can Entomol.* **87**, 437–446.

Bird, F. T. and Burk, J. M. 1961. Artificially disseminated virus as a factor controlling the European spruce sawfly, *Diprion hercyniae* (Htg.) in the absence of introduced parasites. *Can. Entomol.* **93**, 228–238.

Georgis, R. 1997. Commercial prospects of microbial insecticides in agriculture. *In* "Microbial Insecticides: Novelty or Necessity?" (H. F. Evans, chair). *Proc. British Crop Prot. Coun. Symp.* **68**, 243–252.

Georgis, R., Koppenhöfer, A. M., Lacey, L. A., Bélair, G., Duncan, L. W., Grewal, P. S., Samish, M., Tan, L., Torr, P., and van Tol, R. W. H. M. 2006. Successes and failures in the use of parasitic nematodes for pest control. *Biol. Control* **38**, 103–123.

Gerrard, J. G., Joyce, S. A., Clarke, D. J., ffrench-Constant, R. H., Nimmo, G. R., Looke, D. F. M., Feil, E. J., Pearce, L., and Waterfield, N. R. 2006. Nematode Symbiont for *Photorhabdus asymbiotica*. *Emerging Infectious Dis.* 12,

Henke, M. O., de Hoog, G. S., Gross, U., Zimmerman, G., Kraemer, D., and Weig, M. 2002. Human deep tissue infection with an entomopathogenic *Beauveria* species. *J. Clinical Microbiol.* **40**, 2698–2702.

Glare, T. R. and O'Callaghan, M. 2000. *Bacillus thuringiensis*: Biology, Ecology and Safety. J. Wiley & Sons, Ltd. Chichester, UK.

Hajek, A. E., Elkinton, J. S., and Witcossky, J. J. 1996. Introduction and spread of the fungal pathogen *Entomophaga maimaiga* (Zygomycetes: Entomophthorales) along the leading edge of gypsy moth (Lepidoptera: Lymantriidae) spread. *Environ. Entomol.* **25**, 1235–1247.

Hajek, A. E., Humber, R. A., and Elkinton, J. S. 1995. The mysterious origin of *Entomophaga maimaiga* in North America. *Am. Entomol.* **41**, 31–42.

Huger, A. 2005. The *Oryctes* virus: its detection, identification, and implementation in biological control of the coconut palm rhinoceros beetle, *Oryctes rhinoceros* (Coleoptera: Scarabaeidae). *J. Invertebr. Pathol.* **89**, 78–84.

Klein, M. G. 1981. Advances in the use of *Bacillus popilliae*. In "Microbial Control of Pests and Plant

Disease 1970–1980." (H. D. Burges, Ed.), pp. 183–192. Academic Press, New York.

Kramer, J. P. 1959. Observations on the seasonal incidence of microsporidian in European corn borer populations in Illinois. *Entomophaga* **4**, 37–42.

Lacey, L. A. 1997. "Manual of Techniques in Insect Pathology." Academic Press, London.

Lacey, L. A. Frutos, R., Kaya, H. K., and Vail, P. 2001. Insect pathogens as biological control agents: do they have a future? *Biol. Control* **21**, 230–248.

Steinhaus, E. A. 1956. Microbial control – an emergence of an idea. *Hilgardia* **26**, 107–160.

Tanada, Y. and Kaya, H. K. 1993. "Insect Pathology." Academic Press, San Diego, CA.

Chapter I-2

Theory and practice of microbial insecticide application

Andrew C. Chapple
Modelling, Metabolism and Environmental Fate, Bayer CropScience AG
Alfred Nobel Str. 50
D-40789, Monheim, Germany

Roger A. Downer
Laboratory for Pest Control Application Technology, OARDC
Wooster, OH 44691, USA

Roy P. Bateman
IPARC, Imperial College at Silwood Park
Sunninghill, Ascot
Berkshire, SL5 7PY, UK

1 Introduction

Spray application takes input from a wide range of disciplines, from the traditional crop protection disciplines of entomology, weed science, and plant pathology, formulation chemistry, and agricultural engineering, to those of fluid dynamics, meteorology, and others. In general, spray application is either largely ignored–one just sprays a crop with a microbial control agent (MCA)–or it is seen as a panacea for all ills, increasing the efficacy and effectiveness of an MCA to the point of commercial viability. Spray application is neither. It is an important and much neglected component of the process from discovery to commercial success of any active ingredient, microbial or conventional agrochemical, but it goes hand in hand with other components, especially formulation, strain selection and improvement, and an understanding of the target/MCA interaction.

Figure 1 shows the steps from the formulation being added to the spray tank to the eventual biological result for an MCA, a process known as "dose transfer". All the steps are interlinked (see Section 3. **A**.4). Aspects of this process have been investigated in some depth; for example the effects on atomization of various adjuvants, surfactants, and formulants (see Hall *et al.*, 1993 for a review), and the effect of deposit structure on biological effect (*e.g.*, Bryant and Yendol, 1988; Maczuga and Mierzejewski, 1995; Ebert *et al.*, 1999a,b). However, there is a "black box" within which almost nothing is known (Figure 1), *i.e.*, the relationship between the in-flight characteristics of the spray cloud (droplet size spectra, velocity, surfactant effects within the droplets while in flight, *etc.*) and the eventual deposit in the crop canopy. It is this critical information gap which impedes the progress of spray application research in general, and severely limits the usefulness of spray application to researchers attempting to improve the efficacy of any MCA or conventional active ingredient (AI). The principal assumption made is that changes in formulation, atomization, or other spraying parameters give rise to changes

L.A. Lacey and H.K. Kaya (eds.), Field Manual of Techniques in Invertebrate Pathology, 9–34.

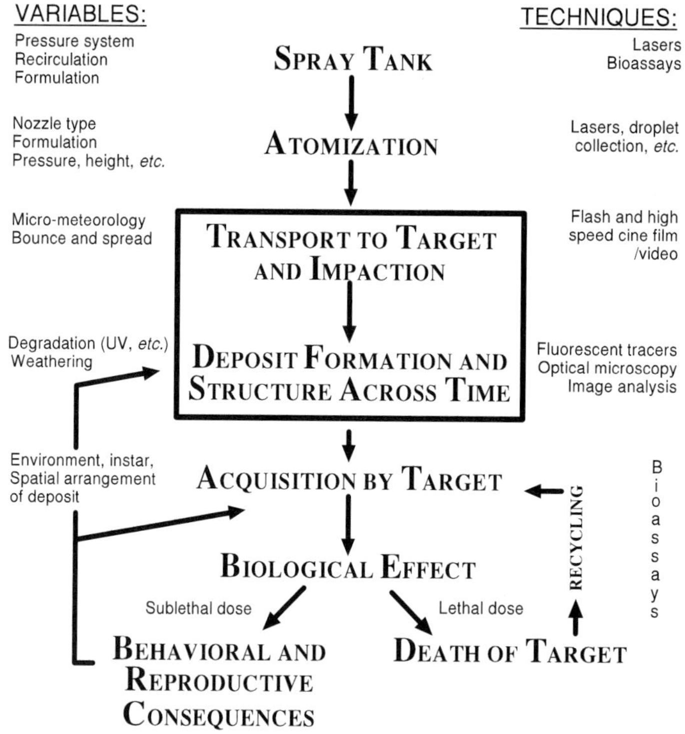

Figure 1. Schematic of dose transfer, adapted from Young (1986)

in deposit structure in the canopy which in turn are reflected in changes in biological performance and effect, *and that these are in some way linearly related.* Clearly they are not, but because of our almost total lack of knowledge concerning the eventual fate of spray deposits, we are left with this crude approximation. [For simplicity, in this chapter, "active ingredient" (AI) means the biologically active component of a formulation, whether a microbial insecticide or conventional agrochemical.]

Therefore, this chapter is intended to be pragmatic in outlook, a guide to what is possible in the field for researchers attempting to take microbial control of insect pests out of the laboratory or glasshouse and into the field, and who wish to optimize application. Hence, we have focused on those aspects of spray application that are relevant to the development of field entomopathogens. Our experience is taken from not only the general field of spray application research, but in particular from the development of *Metarhizium anisopliae* var. *acridum* (=*flavoviride*) against locusts in hot, dry African environments and, in contrast, the commercial development of *Ampelomyces quisqualis*, the hyper-parasite of powdery mildew in field vines and glasshouses, commercially available as AQ10® from Intrachem Bio International SA [Geneva, Switzerland]. For a review of its development, see Hofstein and Chapple (1999). Also, as there are already several excellent texts on spray application, *e.g.*, Matthews (2000), we will not attempt to cover the subject *per se* but rather to focus on those aspects of spray application that we have found to be relevant to microbial pesticides.

According to the *The Manual of Biocontrol Agents* (formerly the *Biopesticide Manual*: Copping, 2004), there are currently 106 living systems available to the grower for microbial control of crop pests, other than through the use of natural products, pheromones, or insect predators. Of these 106, about a quarter are for control of insect pests, in which group, there are insect viruses, entomopathogenic nematodes, fungi [based on spores (= "conidia" which is used in other chapters)], bacteria, and microsporida. A number of products are available based on *Bacillus thuringiensis* (*Bt*) subspecies as the natural strains or with genetic modification, and *B. sphaericus*. Other than the *Bts*, none are used on a wide-area basis comparable with the conventional agrochemicals used in the same crops against similar pests. However, what most have in common is that they must be delivered inundatively into the target crop: *i.e.* they must be applied.

Unlike agrochemicals which are tested in the field very early in their development, MCAs tend to be developed in a laboratory and/or glasshouse environment over a long period of time. This is not surprising considering the budgets usually available to researchers developing MCAs. Field trials are a substantial component of the total development costs of a conventional agrochemical, which can exceed \$30 million. Microbial insecticides in particular are almost always developed for pests of glasshouse crops, although *Bts* are an important exception. This approach has the advantage of being directly related to the glasshouse industry where almost all microbial insecticides have started their commercial, as opposed to research, life, but has the distinct disadvantage of being no closer to field crops than the original laboratory. It is at this point, the attempt to take a microbial pesticide from the glasshouse to the field, that researchers come up against the "lab to field performance" problem and where most researchers become aware of "application" and its limitations.

Therefore, we start with the "real world" – the use of microbial pesticides in broad-acre crops – and return to theory later in the chapter. By broad-acre agriculture, we mean field crops: cotton, soya, cereals, rice, maize, small grains, top fruit, vines, orchards, large scale vegetable production, but *not* protected crops, nurseries, extensively irrigated crops, or hydroponics. This is for three reasons. The first is that much application theory remains untested and should be considered as a pointer to where thinking about application can take research, and in this case, data from field trials and commercial experience have greater value.

Second, the glasshouse industry is a well researched and well documented field. A large and extensive literature exists covering all aspects of microbial control in the glasshouse, including application. Microbials, including the biorationals (*e.g.*, the botanicals), currently account for less than 1% of pesticide use worldwide. Of this 1%, the *Bts* take approx. 80%, nematodes another 13%, and "all others" approx. 7%: an unpromising picture (Lisansky and Coombs, 1994). Of this latter and tiny proportion, most *commercially* available microbial insecticides are used in the glasshouse, ornamental, or protected crops industries. For almost all microbials, broad-acre agricultural use is either nil or minute. (The principal exception in terms of area treated is viruses in forestry, *e.g.*, *Lymantria dispar* nucleopolyhedrovirus, commercialized as Gypcheck, for gypsy moth control.)

Third, of the 106 organisms used for microbial control of insects listed in the *The Manual of Biocontrol Agents* (Copping, 2004), the majority of these are applied by spraying, including the most widely used *Bt* formulations. Of the remainder, 19 are typically applied as drenches, are watered on or incorporated into soil irrigation systems (*e.g.*, many entomopathogenic nematodes, *Bacillus sphaericus, Serratia entomophilia*) or applied as dry granules or baits (*e.g., Beauvaria brongniarti, Paranosema* (= *Nosema*) *locustae*), or applied as seed treatments (*e.g., Bacillus subtilis*). Some 25 other organisms are typically applied by other means, including: granules, seed treatments, dips, baits, water treatments, paint-on herbicides, or take advantage of natural dispersal. For products that are not sprayed, biological efficacy is largely application independent, *i.e.*, spray parameters (droplet size, concentration, etc.) are not relevant.

The other area of application not covered is forestry. Agroforestry is a specialized case (see Chapter III-2), usually requiring aerial application at very low volumes to allow for commercially viable work rates. Forestry is the best, if not sole, example of the successful implementation of spray application knowledge to pest control, particularly *Bt* and lately viruses. However, the logistical lessons learnt from forestry cannot be applied to broad-acre field crops: the crop and application systems are simply too different. What is clear, however, is that the basic principles applied in forestry (*i.e.*, knowing where the spray is deposited in the canopy and understanding the toxicological and behavioral relationships between pest and AI) must be emulated in field crops. If these can be determined, even for a specific crop, significant increases in application efficiency can be made.

Researchers attempting to make the step from glasshouse to field soon run into the problem of much reduced efficacy, sometimes no efficacy, when comparing field results with those obtained in the glasshouse. Microbials are in good company: this problem is the focus of much interest in the agrochemical industry and much can be learnt from the attempts to improve the link between laboratory and field. In other respects, the agrochemical and biologicals industries have little in common (Waage, 1997). Discovery of new AIs, development, formulation, even the end use in agriculture are all very different, and from our experience in both industries, it is fair to say that there is little to be learnt from agrochemicals that applies to microbials (and *vice versa*). However, the lab to field performance problem is one area where there is extensive overlap. For example, Pittis and Russell (1994) have addressed the problem of imitating the field plant pathogen environment in the glasshouse, a situation with lessons for the development of field entomopathogenic fungi. They argue that "…it is virtually impossible to reproduce field effects in the glasshouse." Another area of overlap is application, and lessons drawn from agrochemical experience can be very relevant to microbials.

2 Application systems

Many attempts have been made to replace the application system for delivering AIs into crops, and almost without exception, these systems have failed to replace the standard flat fan hydraulic nozzle as the principal delivery system. Yet it is well recognized that the hydraulic nozzle is very inefficient. Estimates of the inefficiency vary from 90% upwards. From a purely theoretical viewpoint, Hall and Adams (1990) have shown the efficiency of application of permethrin onto cabbage for diamondback moth control to be worse than 90% by > 5 orders of magnitude. This has been confirmed with *Bt* (Chapple, unpublished data, LPCAT). It can be argued that inefficiency of application applies more to microbial pesticides than agrochemicals as microbials tend to be far less forgiving of application errors such as poor tank mixing. Loss of efficacy of an MCA can be attributed to many factors, amongst which are:

- poor spray retention leading to losses to the ground.
- poor or uneven spray distribution (*i.e.*, concentration at specific sites such as leaves, and poor coverage on stems and petioles).
- environmental and microbial degradation, inter-microflora competition, and inactivation within a target organism.
- mismatching the position of the deposit relative to target pest, etc.
- drift and interception by non-target foliage.

The hydraulic nozzle is also largely uncontrollable. For a given nozzle type, there is very little that can be done to alter droplet spectra (the frequency distribution of droplets produced by the nozzle) other than changing nozzle orifice size, altering pressure, or adding adjuvants (Hall *et al.*, 1993). [Throughout, nozzles will be described either by their exact trade description (*e.g.*, TeeJet XR8003VS flat fan) or by the more general description covering types of nozzles, as described by Southcombe *et al.* (1997), *e.g.*, fine, coarse.]

There are currently two principal application systems used to deliver AI into broad-acre crops. These are:

- the hydraulic nozzle (principally the flat fan) used on large vehicle mounted booms up to and exceeding 36 m in length, on large numbers of knapsack sprayers, and for aerial application. Other hydraulic nozzles include the hollow cone nozzle (commonly found on knapsack sprayers, on smaller vehicle mounted booms, and for aerial application), anvil, and "drift reducing" nozzles.
- the air-shear sprayer (or motorized mist blower), used almost exclusively in tree crops, vines, etc., and not to be confused with air-assisted (or air-blast) orchard sprayers (hydraulic nozzles spraying into an airstream: see Chapter III-1).

Other minor use sprayers are spinning discs and electrohydrodynamic sprayers (also known collectively as controlled droplet application or CDA), electrostatically charged sprays, foggers, rotary cages, etc. The development of CDA systems was largely based on theoretical considerations concerning "optimum droplet size" (Potts, 1946, but more usually referenced to

Himel, 1969). These devices have a long history. For example, electrohydrodynamic devices are based on research from the last century (Rayleigh, 1882), but have been developed more recently for agriculture as the Electrodyn® system produced by ICI (now Syngenta, Basel, Switzerland). Although electrohydrodynamic devices have relied on formulations inimical to microbials, use of vegetable oils as the carrier liquid can allow microbials to be applied using such technology. The main use for spinning discs and rotary cages is forestry (very low volume aircraft application), migratory pests (ultra-low volumes, often using aircraft), and protected crops. However, their use relative to the main sprayer types is negligible, and these "minor use" sprayers have had virtually no effect on spraying practices outside protected crops and certain special circumstances (*e.g.*, insecticide application in cotton).

It is important to note that even where such minor-use sprayers are used successfully [*e.g.*, for locust control in Africa (Bateman, 1997)], the choice of sprayer has often been forced upon the researcher by the local equipment used for that pest. Requiring Sahelian farmers to use knapsack sprayers and water at 200 liters/ha or even 20 liters/ha volume application rate for locust control would cause any project to fail. What is less obvious is that requiring a Canadian wheat farmer to use spinning-discs is equally unrealistic. [Volume application rate or VAR is an IPARC/NRI agreed term for the volume of liquid used as the carrier for an AI.]

There is a consensus of opinion amongst researchers in spray application that when developing a new AI, unless there are exceptional circumstances, the only available spray application system is **that system in widespread use in the target crop in the geographical area of interest or local market**. We cannot stress this enough: if the efficacy of a novel MCA requires application through a system different from that used by the target grower, then the chances of that MCA being used on anything but the smallest scale is very slight.

Therefore, this chapter will move from what it is possible to change in terms of field application systems, and only then to the relevance of current application theory.

A Possible changes to the application system

Taking a microbial control agent from the glasshouse to the field involves changing many parameters. These include, to name but a few:

- changing method of production (scale-up), a major undertaking even for agrochemicals.
- exposing the MCA to a far wider range of variability in the pest population, not only in the insects' ability to resist the MCA but also in the population structure.
- differing micro- and macro-climate.
- a far harsher environment, (*e.g.*, UV-light and competition from other microflora).
- different carrier liquid volumes and application systems. Reduced application volumes can be a severe restriction. For example, entomopathogenic nematodes are often applied to protected crops as "drenches", up to a liter of water per 50 cm pot. This translates to a VAR per hectare of around 50,000 liters, 250 times more than the 200 liters/ha more commonly used for field application of insecticides.

With respect to application, there is one major difference from which almost all other changes arise: work rates have to be much higher as field crops are typically worth much less per unit area than protected/glasshouse crops. Areas to be treated tend to be far larger, so the weather limits application windows, etc. This leads to:

- much reduced application volumes with consequent effects on the concentration of AI and any adjuvants.
- a requirement for versatile, easy-to-use formulations.
- direct competition with chemical alternatives.

Volume application rates in the glasshouse are typically far higher than the open field: 2000 liters/ha or more is typical for the glasshouse. In broad-acre crops, the tendency is to spray as little carrier liquid as possible. This has lead not only to the use of CDA on boom sprayers (now rare), but to modern developments in herbicide spraying in cereals in Canada and the USA (*e.g.*, coarse nozzles applying relatively low volumes per ha at high travel speeds: 80 liters/ha at > 25 kph). CDA systems have failed to replace the hydraulic nozzle due to issues of complexity of use, poor canopy penetration with AIs where this factor is critical (*i.e.*, some important cereal fungicides and post emergence herbicides), and failure under some conditions where a pest has escaped control. There is also the failure to fulfill the often perceived rather than real promise of a reduction in the overall requirements of AI for pest control [see Arnold *et al.* (1984a, b, c) for the definitive work on this subject]. Reduced AI requirements has been argued as the critical parameter for successful introduction of novel application systems into broad-acre agriculture (Chapple *et al.*, 1997).

There is an increasing trend for modern agrochemical AIs to be less intrusive to the environment and to have very low mammalian toxicology (if not the AI, then the formulations). This has eroded many of the advantages of MCAs such as environmental effects and operator safety. Hence, MCAs must compete on a cost basis as the "green premium" has more to do with what a farmer can be asked to do rather than pay. Microbials require farmers to use the knowledge and understanding they have of their own farming systems as opposed to what is traditionally required when using agrochemicals. To put it another way, farmers will expend intellectual capital but as businessmen, are unlikely to expend hard cash without good reason. This can be exploited, for example, in the requirement of scouting systems for IPM. There is also some scope for changing application practices.

1 Hydraulic nozzles

When asking farmers to make changes to their existing hydraulic application systems or methods, it is realistic to ask them to:

- change spray quality by changing nozzle (fine, medium, or coarse).
- change volume application rate by altering travel speed, nozzle, or hydraulic pressure, although the limits are from approx. 100 liters/ha to 500 liters/ha.
- use tank adjuvants, but these *must* be easy to use and commonly available.
- change application timing (*e.g.*, am. *vs.* pm. or even night spraying).

Modern booms are often equipped with turret-type nozzle selectors: changing nozzle can be very simple. However, farmers will consistently use the least amount of water possible in order to increase work rates and decrease application costs. Attempting to translate glasshouse conditions to the field, *e.g.*, applying very high volumes of water to aid "dew" period, goes against the commercial interests farmers. Adjuvants that require excessive tank cleaning or special solvents after use should be avoided. Finally, hydraulic nozzles are *not* all alike: there are substantial difference between conventional flat fans, hollow cones, anvils, and venturi (or drift reduction) nozzles (Matthews, 2000).

2 Spinning disc applicators

Rotary atomization (the use of spinning discs, cups, cages, etc.) is the most reliable way of producing sprays with a narrow droplet size spectrum at low flow rates. Although often presented as a paradigm for efficient dose transfer (*e.g.*, Bals, 1969), in practical terms, spinning disc sprayers simply provide an effective means of applying pesticides at ultra-low volume application rates (ULV: < 5 liters/ha), and thus dramatically improving work rates.

Bals' original concept was to spray non-evaporative (usually oil-based) ULV formulations (ULs) where no mixing of pesticide is required by farmers and in this respect, one hazard in the spray operation has been reduced. If the formulation is relatively non-evaporative, wide track spacing (up to 10 m with hand-held equipment, *i.e.*, "drift" spraying) is possible, achieving high work rates especially when the plants are small. Attempts have been made to lower costs by reducing volume application rates to 1 liter/ha and atomizing ULs into even smaller droplets and widening swaths, but yields have sometimes proved less reliable, especially for control of sucking pests. MCAs are necessarily suspensions and the oil-based equivalents of ULV formulations are called ULV suspensions (SUs) or oil-miscible flowable concentrates (OFs; see Chapter III-1).

Several workers have noted that, at a given rate of AI/ha, ULs were considerably more expensive than equivalent quantities of more conventional formulations. The possibility of lowering the cost of CDA by rotary atomization of aqueous mixtures containing cheaper formulations such as wettable powders was first examined and found to be effective in Malawian cotton (Mowlam *et al.*, 1975). This technique was initially called water-based ULV or WULV, but it has subsequently come to be known as Very Low Volume (VLV: *e.g.*, spraying up to 16 liters/ha but more typically 10 liters/ha) of spray mixtures to fully grown cotton. The technique uses narrower swaths and is effectively "placement" rather than "drift" spraying. Volume application rates of a "bucket-full of water per smallholding" are negligible compared with conventional knapsack spraying, but battery consumption may be more than doubled when compared with conventional ULV/CDA.

Typical CDA operations may therefore involve: **a**. ULV with volume application rates of 1 liter/ha (*e.g.*, locust control) where work rate is crucial, or **b**. the use of the water-based VLV technique at approximately 10 liters/ha where this is more acceptable to farmers (Bateman, 1997). Figure 2 shows the droplet size spectra obtained from a Micron Ulva+, spraying two formulations of *M. anisopliae*, under these operating conditions.

3 *Air shear systems*

Air shear application systems are typically operated at 30 to 250 liters/ha volume application rates (in this chapter we have chosen the higher application volume). Their principal use is in crops such as vines or orchards, and they are used in tree crops to lift AI as far as 10 m up into canopies, *e.g.*, cacao. The main distinction between a volume applied through an air shear system and a hydraulic nozzle is the far smaller size and vaster numbers of droplets produced by the air shear system (Table 4).

B *Formulations and adjuvants*

MCAs and surfactants tend to be incompatible. Solvent systems are worse, although this is not the case for *Bt*s and the botanicals. Hence, formulation of MCAs is limited to either: a very narrow range of surfactants (*e.g.*, some of the Tween series); oils; and polymers; or granules and powders based on the fermentation product itself. However, effective spray application with hydraulic systems usually requires

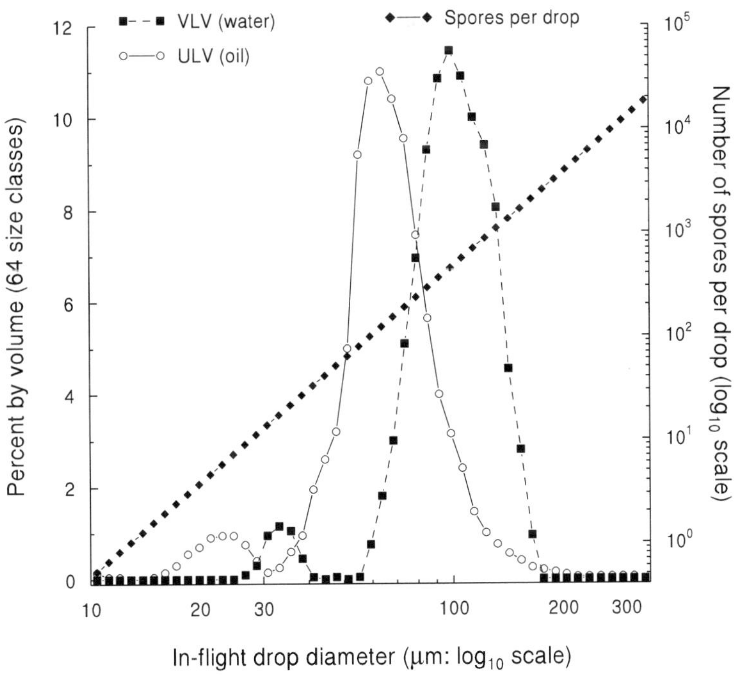

Figure 2. A comparison of drop spectra for two CDA formulations, oil *v.* water (NB. log_{10} scales). Oil based: 8000 rpm, 60 ml/min, 1 l/ha, 10 m swath, 5×10^{12} conidia/ha; and water based:, 6000 rpm, 77 ml/min, 6.4 l/ha, 2 m swath, 7.8×10^{11} conidia/ha. Also shown, the relationship between drop size and spores per drop for a formulation. [VLV = very low volume; ULV = ultra low volume]

surfactants, if only as wetting agents. In developing *A. quisqualis*, Ecogen [Langhorne, PA, USA] screened a very large number of surfactants in glasshouse and laboratory tests for tank-mixing compatibility (Hofstein and Chapple, 1999). Almost all affected the efficacy of the hyperparasite, most to the extent of killing it. Those that did not were considered unusable by the grower. A surfactant had to be used in the field trial program, so a commonly used non-ionic surfactant was used at 0.05%, despite being toxic to *A. quisqualis* in laboratory and glasshouse tests. In the field, across extensive trials in four countries, no deleterious effect of the surfactant could be detected. In short, there was no reasonable correlation between results in the glasshouse and those observed in the field: the glasshouse gave too many false-negative results.

Oils are typically used to enhance an MCA's performance. Their main contribution to efficacy appears to be the reduction, if not complete eradication, of any necessary "dew" period or requirement for long periods of unrealistically high relative humidity. However, oils are by definition non-miscible with water and need to be emulsified into the spray volume if they are to be dispersed uniformly over the period of application. Emulsions require energy for their formation, and surfactants are used to reduce this energy. Although oils can be used that have intrinsic surfactant properties, surfactants are often still needed. Either way, the MCA is exposed to surfactants.

The outlook for the practical use of application research seems at first glance quite constrained. Formulations are limited to the very few surfactants that can be shown to be non-toxic to the MCA, and these can only be found through extensive screening, which itself may not reflect detrimental activity in the field. The sprayer type is dictated by the user, whether a hydraulic flat fan in European cereals or a spinning disc in African cotton (*e.g.*, Egypt, Francophone West Africa). There would appear to be little that can be taken from application research and applied to MCAs. Fortunately, this is not the case, although many of the assumptions made in most application research do not apply to MCAs. The following sections deal with how application technology applies to the various categories of MCA and considers the useful information available to the MCA researcher.

3 Droplet spectra

A Terminology

Terminology is covered in Lefevbre's (1989) and Matthew's (2000) excellent books, the latter being the classic text covering agricultural spray application. Other related texts are Bache and Johnstone's (1992) book on spray deposition and microclimate (for the mathematically inclined), two books covering aspects of biorationals and biologicals (Hall and Barry, 1995; Hall and Menn, 1999) and two symposia proceedings that address relevant issues (Hewitt *et al.*, 1994; Evans, 1997). The following is an overview of the more commonly used terminology.

1 Droplet spectra statistics (VMD, NMD, relative span, etc.)

It should be noted that there are large differences in droplet spectra parameters for the same spray cloud depending on system of measurement and methodology. In general, it is a rule of thumb that if the different sized droplets in a spray cloud have very different velocities or a wide range of velocities within a droplet size, then a temporal measurement system [*e.g.*, the PMS (Particle Measuring Systems Inc., Boulder, CO, USA) or PDPA (phase-Doppler particle analyzer, Aerometrics Inc., Sunnyvale, CA, USA])] is more reliable than a spatial measurement system [*e.g.*, the Malvern (Malvern Instrument Ltd., Worcs. UK)]. When dealing with spinning discs or comparisons between formulations using the same nozzle, then either spatial or temporal sampling systems are applicable.

Various terms are used to describe the droplet spectra of spray clouds (Lefebvre, 1989). Those most widely mentioned in papers involving spray application in the context of agriculture include:

- volume median diameter (VMD, $D_{V0.5}$, or MMD, mass median diameter) and the 10% and 90% points ($D_{V0.1}$, $D_{V0.9}$), *i.e.*, the droplet diameters such that 50, 10, and 90% of the volume of liquid sprayed is in droplets of smaller diameter, respectively.

- the D_{10}, D_{30}, and D_{32} [length, volume, and Sauter (*i.e.*, area) mean diameters respectively: see Lefebvre (1989), pp. 90–99].
- the number median diameter (NMD), *i.e.*, the droplet diameter such that 50% of the droplets by number are of smaller diameter.
- various fractions of the spray cloud, *e.g.*, % volume of the spray cloud or % number of droplets less than or greater than a given diameter of interest.

As a generalization and provided the atomization systems are similar (*e.g.*, comparing the effects of adjuvants on a particular nozzle or comparing within types of nozzle), VMD and NMD give information about the large and small droplets in a spray cloud, respectively. The volume *mean* diameter (D_{30}, sometimes referred to as the volume average diameter to avoid confusion with VMD) is particularly useful: the spray volume per ha divided by the D_{30} gives the number of droplets/ha for a given atomization system.

Two other terms regularly published are relative span (or just "Span") and "*R*" (the VMD:NMD ratio). These are both measures of the spread or "quality" of the droplet sizes present in the spray cloud. Span is defined as the ($D_{V0.9}$ minus the $D_{V0.1}$) divided by the VMD ($D_{V0.5}$). "*R*" is the VMD divided by the NMD.

There is no single descriptor that describes a spray cloud adequately for all situations. None of the above single descriptors (*e.g.*, NMD, VMD) takes into account the variability in droplet sizes in a spray cloud. With rare exceptions, spray clouds are often described in terms of their VMD and NMD for size, occasionally in terms of fractions (% volume and % number) below a certain droplet size, and rarely as relative span or "*R*" for quality (Bateman, 1993). AI per droplet is usually assumed to be directly proportional to the droplet volume (but see Section 3. **B**.3 below). Unfortunately, because droplet volume is a cube function of diameter, small changes in droplet size (or descriptor) can translate to very large changes in droplet volume. For example, water sensitive paper (WSP) is a widely used guide to in-flight droplet size, and a general assumption is that droplets spread by a factor of 2. In fact, in-flight diameter is *not* directly linearly related to WSP stain size. One equation fit (data taken from that provided by Ciba-Geigy with the WSPs) used at LPCAT is:

$$\text{Spread factor on WSP} = \log[\text{in-flight diameter } (\mu\text{m}) - 1]^{0.73037} \quad (1)$$

A theoretical example of the size of errors that can occur is given in Table 1 for a mycoinsecticide applied at 1×10^{11} spores/ha in water to glasshouse cabbage at either 6.2 liters/ha (*e.g.*, spinning disc) or 200 liters/ha (*e.g.*, hydraulic nozzle). Table 1 shows a range of in-flight-droplet sizes, their actual volume, the back-calculated in-flight diameter from the deposit on WSP using a spread factor of 2, the same using the above equation, the difference in the number of spores per deposit using the two approaches, and the effect of the different spread factor obtained for cabbage - approx. 85% of that for WSP. The back calculation used (from deposit diameter to in-flight droplet diameter) is the polynomial: $8.382 + (0.573 \times DD) + (-1.61 \times 10^{-4} \times DD^2) + (6.54 \times 10^{-8} \times DD^3)$, where DD = WSP deposit diameter. For droplet sizes > 30 μm, this has a correlation (r) of > 0.999. It can be seen that in terms of spores per droplet, the errors are not large for either system. However, if a *Bt* or virus were considered, then the AI delivered is proportional to the droplet volume (see Section 3. **B**.2 for a clarification), and the errors can be substantial.

Relative span is rarely reported. However, VMD and NMD are usually reported together, and these can be used to assess the variability of droplet sizes in the spray described (see Figure 3 and Table 2 for a comparison of three application systems). Representative droplet spectra for three types of commonly used sprayer are given in Figure 3: a conventional flat fan [XR8003VS, 40 psi (276 kPa)]; a Micron Ulva+ (12 volts, red restrictor); and an air-shear sprayer. The differences are notable, principally the difference within sprayers between distribution by number and by volume. The smaller the relative span (or VMD:NMD ratio), the smaller the difference between volume and number distributions (*c.f.*, Figure 3 and Table 2).

There is a commonly-held assumption that CDA means "uniform" or "near-uniform" droplet sizes. However, this is not the case (see Figures 2 and 3). Rather, these devices attempt to control the droplet spectrum created, normally

Table 1. Errors associated with changes in deposit diameter as measured on water sensitive paper: 10^{11} spores/ha, 6.2 liters/ha (spinning disc) or 200 liters/ha (hydraulic nozzle)

Parameters	Sprayer type						
	Spinning disc				Hydraulic nozzle		
True in-flight drop diameter (μm)	50	100	150	200	200	400	600
True volume (pl)	65	524	1,767	4,189	4,189	33,510	113,097
Actual spores per drop (deposit)	1	8	29	68 : 2	2	17	57
Measured deposit diam. on WSP (μm)	73	166	264	367	367	804	1,265
Back calculated drop size, assuming spread factor = **2** (μm)	37	83	132	184	184	402	633
Back calculated volume (pl)	26	297	1,210	3,244	3,244	34,017	132,566
Underestimate on volume calculation (%)	61	43	32	23	22.6	−1.5	−17
Back calculated spores/deposit	0	5	20	52 : 2	2	17	66
Back calculated drop size using eqn. (μm)	50	99	150	200	200.4	399	608
Deposit diameter on glasshouse grown cabbage (μm)	62	141	225	312	312	683	1,075

by creating a narrower spread of droplet sizes relative to the hydraulic nozzle. Figure 4 shows the effect of changing VMD:NMD ratio (and relative span) on the consumption of cabbage treated with *Bt* by a modeled lepidopteran (Chapple *et al.*, 1994). As droplet spectra widens (*i.e.*, moves further away from a mono-dispersed distribution where all droplets are the same size: VMD/NMD = 1.00), efficacy of the *Bt* deposit lessens (*i.e.*, more cabbage is eaten). The arrow (**C↓/C↑**) indicates the VMD:NMD ratio or Span for the spinning disc sprayer given as an example in Table 2 and Figures 3 and 4. Although the CDA sprayer has the narrowest VMD:NMD ratio and relative span of the three sprayer types, this statistic cannot be used alone to judge its suitability as a sprayer for applying an MCA. Also, the 250 liters/ha volume application rate

Table 2. Spray descriptors for three types of application system: hydraulic nozzle [XR8003VS, 40 psi (276 kPa)]; airshear sprayer; and CDA spinning disc (Micron Ulva+, red restrictor)

Parameters ↓ : Sprayer →	Hydraulic nozzle	Airshear	Spinning disc
Volume/ha	170 liters/ha	250 liters/ha	6.25 liters/ha
Arithmetic mean diameter: D_{10}μm	93	37	50
Volume mean diameter: D_{30}μm	162	56	57
10% Point: $D_{V0.1}$μm	145	46	44
90% Point: $D_{V0.9}$μm	835	170	95
Volume mean diameter: VMD μm	321	105	67
Number mean diameter: NMD μm	57	26	47
% Volume < 100 μm	4.6	12.5	17.1
% Volume < 150 μm	10.7	45.9	93.1
% Number < 100 μm	73.3	80.3	56.2
% Number < 150 μm	84.6	96	99.1
VMD/NMD ratio	5.6	4	1.4
Relative Span	2.15	1.18	0.76

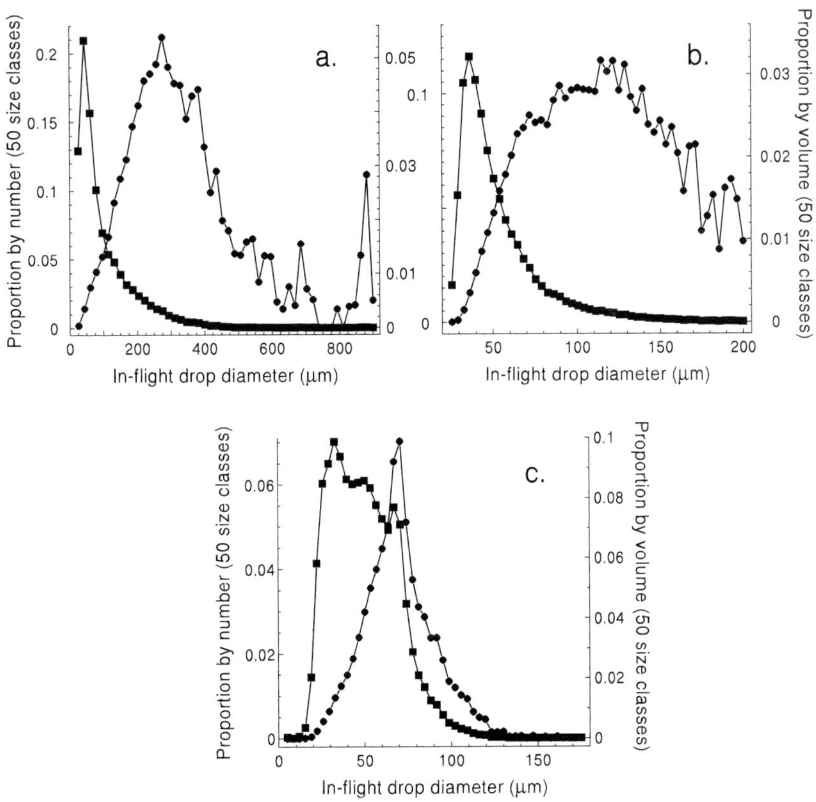

Figure 3. Drop spectra for a falt fan hydraulic nozzle (**a.**), and airblast sprayer (**b.**), and a spinning disc sprayer **c.**). ■ : proportion by number: •: proportion by volume

for the air-shear sprayer is a top limit for usual practice: these sprayers can be used at volume application rates as low as 30 liters/ha.

2 *Cover*

Cover is that proportion of the leaf or crop covered by deposits. Once again, it must be stressed that the broad assumptions made concerning cover are either extrapolations from droplet spectra (sometimes including spread factors for the different sized droplets) or from mass balance studies (*e.g.*, Cooke *et al.*, 1985) where the *amount* of active deposited is determined for different parts of a crop canopy. Neither take into account the actual spatial variation in deposit structure in a canopy. Extrapolating from droplet spectra ignores differential sampling by a crop canopy and assumes that all the droplets produced have the same chance of impaction or retention. It also assumes that the larger droplets do not shatter and redistribute further into the crop canopy, which of course they do. Droplets of water as small as 60 μm can bounce several times before finally settling on a difficult-to-wet surface such as cabbage. Mass balance studies ignore both the size *and* distribution of the deposits. Clearly, the amount of AI contained in one hundred 50 μm droplets distributed over a given area of canopy may be of much greater value than one 232 μm droplet, the equivalent volume. Conversely, Hislop (1987) suggests that there is no reliable correlation in the field between good deposition (*i.e.*, cover) and biological result, laboratory experiments notwithstanding (see Ebert *et al.*, 1999a,b). Unfortunately, cover falls into the "black box" shown in Figure 1.

3 *Particle size distributions and pumping systems*

Field trials more closely resemble true field practices than the glasshouse trials that precede them. The latter attempt to obtain the maximum efficacy from an MCA, often using commercially unacceptable formulations (*e.g.*, simple

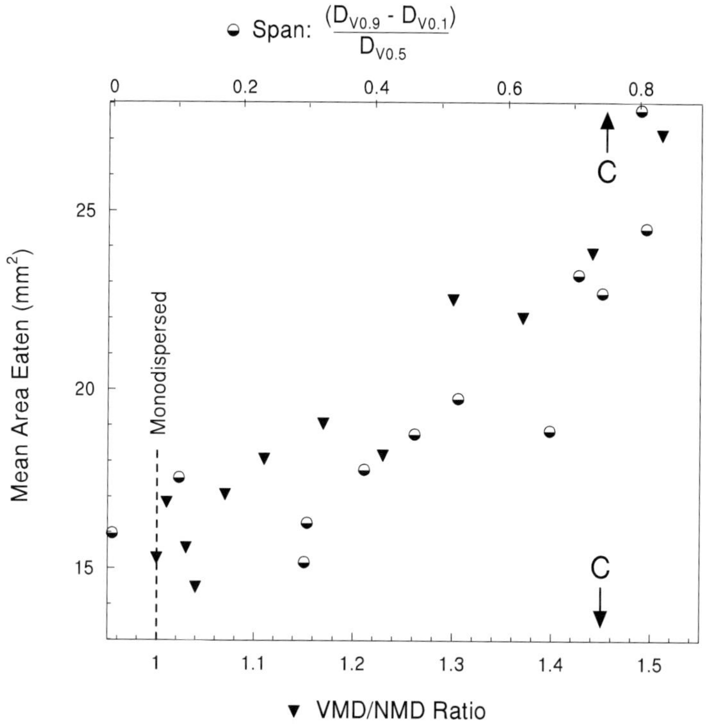

Figure 4. Effect of changing variability of drop sizes in drop spectra (as measured by VMD/NMD ratio and relative span for a theoretical normal distribution) on the consumption of cabbage by the diamond back moth, as modelled. The "AL" is *Bt*. (From Chapple *et al.*, 1994)

suspensions of the micro- organism under study) or excessively high volume application rates to simulate dew periods and raise relative humidity. However, even the traditional application techniques for field trials (*i.e.*, small, compressed-air driven precision sprayers) do not reflect the application system likely to be used in the field. The principal differences are the absence of a pump and re-circulation system, and the tendency to use application volumes higher than those commonly used in the field.

Two commonly-made assumptions are that the distribution of an AI through a spray tank is uniform (*i.e.*, random) and that the distribution of AI in the spray tank is similar to that seen in the laboratory. Chapple and Bateman (1997) have shown that the frequency distribution of *A. quisqualis* spores in the spray tank changes after passage through a pump. Figure 5 shows the distribution of spores: in a glass beaker **B**, *e.g.*, the distribution present in a small hand-held mist-blower typical of a laboratory test; in a larger volume **T**, in this case, 50 liters, comparable to the volumes used in larger scale glasshouse trials with air-pressurized containers; after recycling through a pump **R**, representative of a field sprayer tank with a pump, either an air-shear or conventional hydraulic system; and after recycling through the pump and spraying through a flat fan nozzle at 40 psi (276 kPa), **S**, *i.e.*, adding the shear stresses of atomization to those of the pump. The work described considered a "worst case" scenario: the formulation of *A. quisqualis* used was not allowed time to hydrate. [Other data (Chapple, unpublished) suggest that this particular formulation breaks up into near mono-dispersed particles after hydrating properly.] *A. quisqualis* spores have dimensions of approx. 3 by 5 μm, and a large proportion of particles in the spray tank comprise single spores. Clumping of spores is much higher if the spray suspension is not sheared by the pump. An even distribution of viable spores across target foliage is important for efficiency: "clumping" to the extent seen in the beaker or tank must be detrimental (see Section 3. **B**.3 below).

The oils used in either formulations or as tank adjuvants must be delivered with the AI.

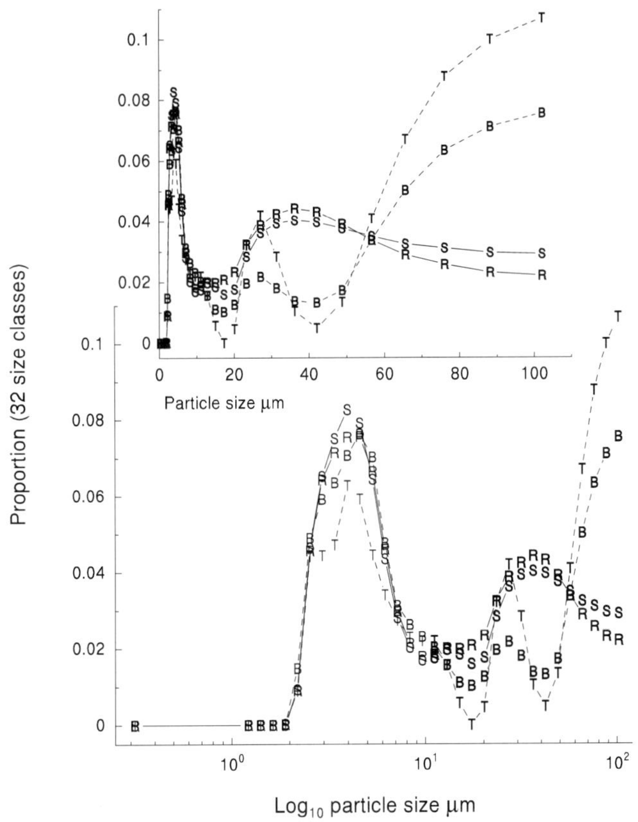

Figure 5. Frequency distribution of particle sizes (linear and $\log_{10}$ axes) of *A. quisqualis* (formulation) when: stirred in a beaker, **B**; stirred in 50 l. of water in a spray tank, **T**; after 5 minutes recycling through a diapragm pump, **R**; and after recycling and spraying throuugh a flat fan hydraulic nozzlwe, **S** (see text for details)

For example, effective field use of *A. quisqualis* requires an adjuvant oil (approx. 4 to 8 liters/ha of a proprietary oil: ADDQ®, Ecogen Inc., Langhorne, PA, USA). From Figure 6 which shows the same sort of data as Figure 5, it can be seen that without the energy imparted by the pump, the particle sizes of the oil emulsion are large relative to the size of the *A. quisqualis* spores. After passage through the pump, either recycling or spraying, the emulsion produced is far closer in size distribution to the spores, and when sprayed, droplets containing spores are much more likely to carry oil with them.

Three conclusions can be derived from this work. The first and most important is that field trial equipment must resemble as closely as possible the equipment that is eventually going to be used in the true field situation. The differences are not trivial. In one set of glasshouse trials with *A. quisqualis*, the biological effects of the adjuvant oil could not be detected, despite field trials data to the contrary. As an explanation, it was noted that the majority of the oil remained on the surface of the liquid in the pressure containers, despite vigorous shaking, and any oil that was emulsified was in such large particle sizes that very little was delivered to the foliage with the spores it was supposed to enhance.

Second, pump systems must be considered as part of formulation research. Pumps can damage MCAs (Klein and Georgis, 1994; Nilson and Gripwall, 1999), and can seriously affect the physico-chemical properties of polymer-based formulations and tank adjuvants. They should therefore be part of the routine investigation of formulations for microbials.

Third, the equipment encountered in the field, especially systems imparting shear

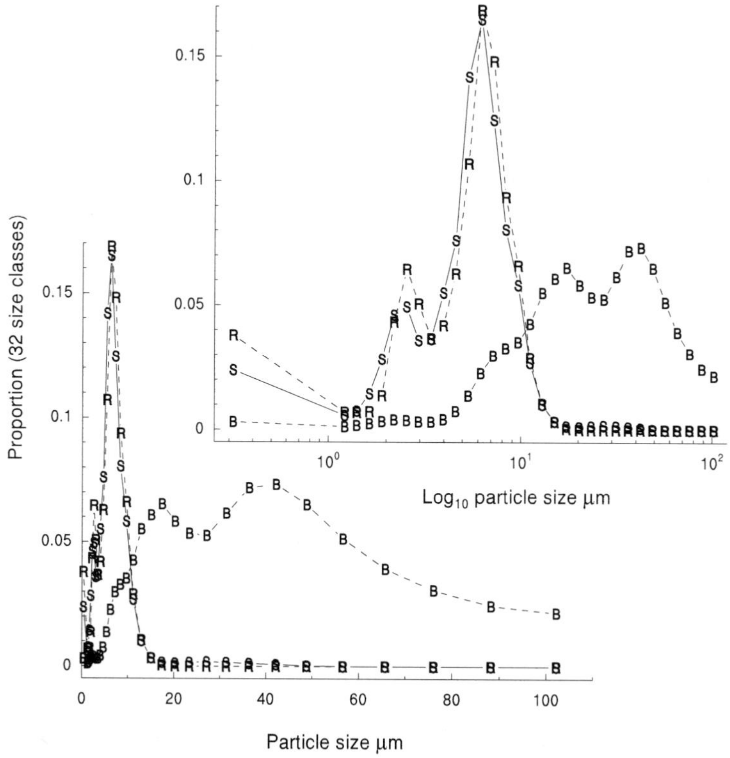

Figure 6. Frequency distribution of oil particles of ADDQ® an oil enhancement agent for *A. quisqualis*, when: stirred in a beaker **B**; after 5 minutes recycling though a diaphragm pump, **R**; and after recycling and spraying through a flat hydraulic nozzle **S**

stresses, can be used to alleviate some of the problems normally solved through formulation. For example, a grower can be legitimately asked to fill the spray tank a quarter full, add the MCA formulation and any oil adjuvants, start the recycling system, and then add the remainder of the carrier liquid, thus emulsifying a formulation without the need for excessive surfactants.

4 Interaction of application parameters

Five interlinked parameters that ultimately affect application efficiency can be controlled to a greater or lesser extent by the grower:

- volume application rate per hectare.
- forward speed.
- pressure.
- droplet spectra.
- AI (especially spores)/ha.

Volume application rate can be altered by changing nozzle, pressure, and forward speed. Droplet spectra can be altered by changing nozzle or pressure. AI per ha can be altered by changing AI concentration, volume application rate, pressure, or forward speed. All are interlinked. For a given AI concentration, change forward speed, and one has altered volume application rate per ha and therefore AI per ha. Alter pressure, and volume application rate changes, droplet spectra changes, and AI per ha changes. *All* parameters must be considered: attempting to alter one parameter in isolation is self-defeating (see Ebert *et al.*, 1999a,b).

B "Optimum droplet size": three classifications

With respect to microbials, "optimum droplet size" depends on the MCA's size and the frequency distribution of sizes of the particle(s) constituting the MCA. When considering application, microbials can be broken down into three classifications:

- Very large particles, *e.g.*, entomopathogenic nematodes (*Heterorhabditis* and *Steinernema* spp.).
- Very small particles, *e.g.*, viruses (granuloviruses, nucleopolyhedroviruses) or small particles with a non-uniform size distribution, *e.g., Bt*s.
- Spores of bacteria, fungi, or protista, *e.g., Erwinia carotovora, Beauvaria bassiana*, of approx. 1 to 10 μm minimum dimension, and applied at approximately the same number per ha as there are droplets in the delivery spray cloud (within two orders of magnitude).

1 Large particulate microbials

For very large particulate MCAs, there is no "optimum" droplet size. For example, the application of entomopathogenic nematodes can largely ignore droplet size: in general, nematodes are large and overcome the atomization process. As a nematode reaches the edge of the sheet of liquid that is being broken up into droplets, either the nematode will be included in any large-enough droplet that is being formed or the nematode itself will be a focus for droplet formation. If a small droplet was likely to be formed at that moment, then the nematode will cause a much larger droplet to be formed, containing the nematode and carrying with it a quantity of formulation components and spray tank adjuvants, if any. Consequently, nematodes are deposited where the volume of applied liquid is found (see Figure 7). However, if 500,000 infective juveniles/m^2 (or 5×10^9/ha) are applied in 242 liters/ha and assuming that a nematode will "fit" into a minimum diameter droplet of 300 μm, then only approx. 29% of the spray volume will be associated with the nematodes (for calculation, see Box 1). Or, put another way, approx. 70% of the formulation additives or tank adjuvants are applied in droplets *not* containing nematodes. As volume application rates increase, this wasted proportion of adjuvants and carrier liquid will also increase. Therefore, where tank adjuvants might be critical to efficacy, for example foliar application, and especially where they are expensive relative to the AI, lower carrier liquid volumes should be considered, and perhaps CDA.

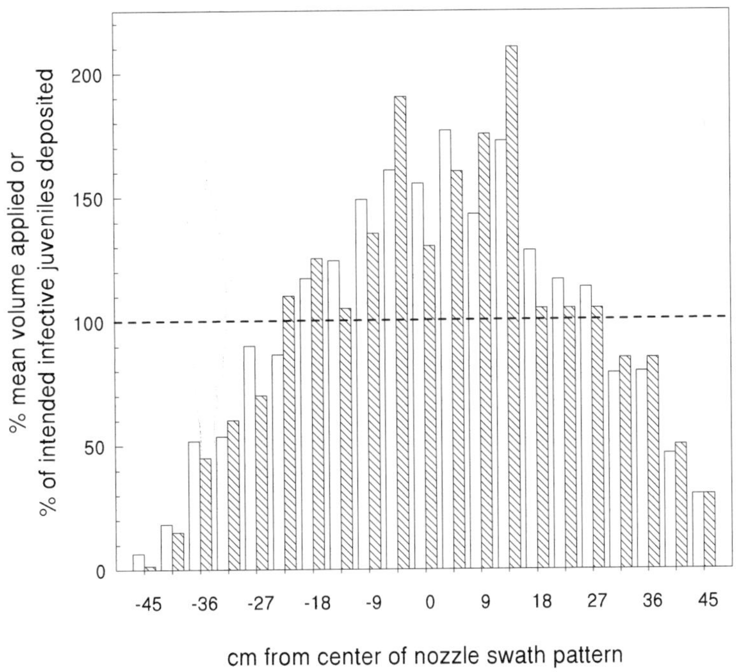

Figure 7. The distribution of nematodes (hatched bars) and carrier liquid (clear bars), applied with an XR8003VS flat fan nozzle, 45 psi (310 kPa), using a CO_2 – pressurized cylinder, at 1 m/s walk speed, 45 cm above the collecting petri dishes. Data obtained at a Cost 819 (Co-operation of Science and Technology) workshop, held at Ecogen Europe, Pantalla, Umbria, Italy, 1996, and reported by Chapple, 1996)

Box 1

How to calculate the percentage of spray volume containing large MCAs

To calculate the amount of the volume application rate atomized *without* large particulate biologicals, for example, nematodes (*i.e.*, the percent of formulation components or spray tank adjuvants not associated with nematodes/infective juveniles):

$$\frac{ApplVol - (\min\ DVol.NemHa)}{AppVol} \times 100$$

where:

- AppVol = volume application rate in m^3 (*e.g.*, 242 liters/ha = $0.242\,m^3/ha$);
- NemHa = number of nematodes or infective juveniles per hectare (*e.g.*, $500,000/m^2 = 5 \times 10^9/ha$); and
- minDVol = volume (in m^3) of minimum droplet size necessary to contain one infective juvenile.

2 *Small or non-uniformly distributed particles*

Much of the research into alternative application devices such as CDA and electrostatics has been based on work with conventional insecticides which has shown that for a given amount of AI, efficiency of utilization of insecticide is inversely proportional to droplet size - *i.e.*, small droplets work better for the same amount of AI (Adams *et al.*, 1990). The same is true for MCAs such as *Bt* (Bryant and Yendol, 1988; Maczuga and Mierzejewski, 1995). Hence, when applying very large numbers per ha of very small MCA particles (*e.g.*, viruses), or emulsions (*e.g.*, botanicals such as neem), or particles with a non-uniform distribution (*e.g.*, *Bt*), the general principles derived from conventional agrochemical insecticides apply. In brief, these are:

- Cover is important, but application to run-off is wasteful and must be avoided.
- The droplet is the dose – *i.e.*, the amount of AI present in any droplet is directly proportional to its size.
- Efficiency of application is indirectly proportional to droplet size (*i.e.*, the smaller the droplet, the more efficient the use of AI), down to a droplet size which contains less than an LD_{95} or similar critical factor.

Cover is always important. However, there are many situations where deposition of AI in that part of the canopy containing the pest is more important than an even distribution (*e.g.*, cotton). In the final analysis, cover needs to be considered in the light of the target/deposition interaction sought (Hislop, 1987).

Efficiency of application is demonstrated in Figure 8, from modeling the diamond back moth on cabbage using the Pesticide Dose Simulation (PDS) model, a stochastic cellular automaton model, which models the insect behavior when presented with deposits of *Bt*, and lately fipronil (Chapple *et al.*, 1994; Ebert *et al.*, 1999a,b; Hall *et al.*, 1995). The general trends elicited with the PDS have been confirmed against data for *Bt* (*e.g.*, Bryant and Yendol, 1988; Maczuga and Mierzejewski, 1995) and the general principles derived from agrochemicals (see Adams *et al.*, 1990). Efficiency of application, in this case inversely proportional to the time alive of the third instar diamondback moth, increases with decreasing droplet diameter, most steeply for low concentrations of AI.

Ascertaining the "optimum droplet size" is critical and can be determined in the laboratory, using a variety of devices from piezo-electric mono-dispersed droplet generators to "flickers" to air-shear systems. However, when the liquid being atomized is either viscous or has inclusions (*e.g.*, emulsions or particles such as *Bt*), then these devices often fail to deliver a uniform droplet size or to function at all. Although simple, the spinning disc and moveable slit device (*e.g.*, Frick, 1970) can handle particles and viscous fluids and, with care, can be used to produce droplets below 50 μm diameter. Experiments can then be done to determine the optimum relationship between droplet size, concentration of AI, and concentration of adjuvant. From these data, changes that might improve application efficacy can be determined [see Ebert *et al.* (1999a,b) for an experimental design]. Again, forestry is the classic example of the successful extrapolation of such data to the field.

Once an optimum droplet size is determined, then modeling the interaction between insect behavior and the pickup of AI can lead to other

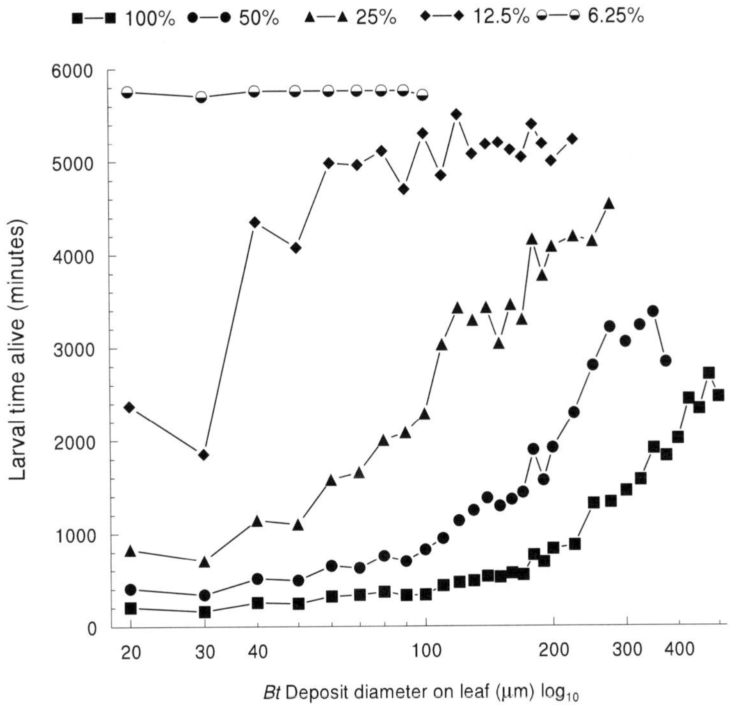

Figure 8. Relationship between survival time of diamondback month 3rd instar larvae exposed to *Bt* as a range of mono-dispersed drop sizes, with dose constant application rate, for approx. field application rate (100%) and various fractions (50, 25, 12.5, and 6.25%). From modeling of diamondback moth on cabbage, Chapple *et al.*, (1994)

insights which may have use in optimizing application efficiency (*e.g*., Hall *et al*., 1995; Ebert *et al*., 1999a,b).

3 *Fungal and bacterial spores*

In the following discussion, the reader is cautioned to make a careful distinction between hydraulic nozzle and CDA application systems. The latter have unique characteristics that can be exploited in very different ways to the conventional hydraulic nozzle. The reasons for using the two application systems are based on different assumptions, and this must be borne in mind when assessing the following.

For hydraulic nozzle application, fungal and bacterial spores have a distinct advantage over the above two categories of MCA and agrochemicals in general: the AI is dispersed through the spray tank in mono-dispersed, unit doses (although see Section 3. A.3. above). Mono-dispersed distributions of MCA particles in a spray tank permit some simplifying assumptions which allow the spray cloud to be assessed differently to agrochemicals.

- There is often no dose response in the agrochemical sense and within the context of a deposit: the AI (a spore) is either viable or it is not, and only viable spores are relevant. It has been clearly demonstrated for some entomopathogens that a number of infective spores/insect are required for mortality. However, with regards to application, this is a moot point. Application systems and parameters can be manipulated to allow for any number of spores per droplet or deposit and to derive measures of efficiency of application. Also, multiple secondary pickup of spores may offset the need to deliver a single deposit containing a lethal number of spores. Modeling pickup, the interaction between an insect's behavior and the distribution of AI, can be important, and can be used to link the dose response (number of spores/insect for a % mortality for a given population) to application variables (spores/deposit, probability of pickup, etc.).
- For foliar application, efficiency of application can be defined as how effectively the application system delivers spores across the target canopy. For biofungicides and mycoherbicides, maximum efficacy would be each spore being delivered *alone*

per deposit. For microbial insecticides, clearly other definitions may apply.

- For biofungicides and mycoherbicides, "optimum droplet size" can be exactly defined: the volume application rate (liter/ha) divided by the numbers of spores/ha. Therefore, if the spray volume could be divided up into mono-dispersed droplets (and it should be remembered that there is no system that can do this outside the laboratory apart from the Electrodyn® system previously mentioned), then application efficiency would be 100%. This calculation can also be applied to any oils used provided the distribution of the oil emulsion is very narrow (*e.g.*, Figure 6.) This does not exclude the possibility of having more droplets than spores, so that although the spores may be delivered individually (100% efficiency), a proportion of the droplets will contain no spores but will contain water soluble adjuvants, surfactants, salts, etc., and these will therefore be wasted. Hence, there is a second efficiency parameter describing the efficiency of use of adjuvants: the proportion of adjuvant delivered with spore(s), *i.e.*, adjuvant efficiency.

An even deposition of single spores through a canopy is the most efficient use of a mycoherbicide or fungal biofungicide. However, this may not be the case for other MCAs, especially microbial insecticides. Laboratory work can demonstrate the inter-action between deposit contents (number of spores) and optimum pickup. As a very simple theoretical example, it might be shown in the laboratory and glasshouse that for a particular insect pest, 20 fungal spores is the optimum dose. Modeling different encounter scenarios might show that distribution of deposits across the leaf surface should not fall below $5/cm^2$ and that at this density, the insect will on average encounter four deposits in a critical time period. Therefore, each deposit should hold five spores. Application can then be modified to move the eventual deposit structure in the canopy towards an optimum, and then be tested in field trials. This has been done very rarely, the work on *M. anisopliae* being an instructive exception. Therefore, for microbial insecticides, there is no "optimum droplet size", rather a "critical droplet size" for a given target/canopy/MCA combination. To complicate matters further, this may not be a single size, but could be a range of sizes or even a bimodal distribution. Salt and Ford (1996) have demonstrated through modeling that for conventional pyrethroid insecticides, a "composite spray" consisting of large dilute-concentration deposits combined with small high-concentration deposits can extend the window of control for sedentary pests. However, the following discussion of the relationship between numbers of spores per droplet and critical droplet size will assume that the number of spores determining the critical droplet size is one, the same as for biofungicides and mycoherbicides. Considerations of more than one spore per droplet as being "critical" or bimodal distributions are too complex to be discussed in this chapter. Besides, the relationships are most easily seen for a critical droplet size containing one spore, and readers are left to extrapolate to their particular pest/MCA combination.

The use of *M. anisopliae* for locust control contrasts substantially with the hydraulic systems described above. Oil-based SU formulations for CDA are highly concentrated, with 5×10^{12} conidia/liter recommended for a one liter/ha volume application rate. Figure 2 shows that the volumes of most droplets in the principal mode range from approximately 45–100 μm (or 50–520 pl). On the assumption that particle distributions in individual droplets will be normally distributed at high concentrations, locusts would be expected to encounter deposits containing 250–2600 conidia. Secondary pick-up of this contact insecticide is known to be more important than direct impaction of spray droplets on the insects themselves (Bateman, 1997); in the short-term at least, efficacy is known to be dose dependent (*e.g.*, Bateman *et al.*, 1993). Dosage levels in the region of 2 to 5×10^{12} *Metarhizium* conidia/ha are required in order to achieve mortality of many locust and grasshopper species in a reasonable length of time (say < 3 weeks).

Conidia are normally the method of fungal propagation in nature and likely to be the most infective and robust propagules for any fungal agent. Unfortunately, it has been (and still is) possible to obtain mycopesticide formulations containing large (> 100 μm) particles that block filters and nozzles, which makes products unattractive to growers and may cause variable efficacy in the field. Such problems were especially pronounced and unacceptable in ULV locust control, so a device called the 'MycoHarvester' was designed to harvest conidia safely and efficiently from a solid substrate as single conidia. Devices such as these have enabled the production of high quality preparations of powdery, lipophilic conidia of genera such as *Metarhizium* and *Beauveria* spp., and more recently with various mycofungicide species of *Trichoderma* that have more hydrophilic cell-walled conidia. Well separated conidial preparations, in turn, enable the development of biologically and physically stable formulations that are also more "user friendly".

Returning to the hydraulic nozzle, if the spray cloud is no longer considered as an *application* system but rather as a *delivery* system, then the droplets created by the atomization process can be considered as a sampling system: different sized droplets sample different volumes of liquid from the spray tank and have different probabilities of containing a spore or spores. Taking the simplest case, a critical droplet size containing just one spore, then various parameters describing a particular application scenario and its efficiency can be calculated *if* one assumes that the spores are distributed through the spray tank in a uniform manner. Clearly, they are not. The distribution is random and the actual sampling regime, considering the different sized spray droplets as a sampling system for the spores in the spray tank, is therefore Poisson. However, the errors caused by the former assumption are not large. Table 3 shows the breadth of data available from calculations of droplet spectra and efficiency.

Within the context of medium to high volume application, three measures of efficiency are proposed:

- Relative Efficiency (the percentage of the AI delivered with the "critical" number of spores per droplet).
- The converse, Relative Inefficiency, the % AI delivered in droplets containing more than the critical number of spores per droplet.
- Adjuvant Efficiency, the % of the volume applied [VAR] containing water soluble adjuvants associated with spores, irrespective of the number of spores per droplet.

In this simplest example, Relative Efficiency is the percentage of spores applied per ha that are delivered singly (here, one per droplet), and one would aim to raise this. Relative Inefficiency is the percentage of spores delivered per droplet in pairs or greater: having more than the "critical" number of spores per deposit can be considered a waste of AI. As "critical" droplet diameter doubles, the number of spores per droplet increases eight fold. Therefore, for the example nozzle in Table 3, the "critical" droplet size is 148 μm and contains 1 spore, a 290 μm droplet contains 8 spores of which 7 are wasted, a 580 μm droplet contains 64 spores of which 63 are wasted. It should be noted that the VMD (Table 3: 50% of the volume applied is in droplets larger) is more than double the "critical" droplet size (the % volume applied with 2 or more spores per

Table 3. Efficiency of application for a medium nozzle (XR8003VS) applying 1×10^{11} spores/ha, in 170 liters water per ha, at 8 kph, 40 psi (276 kPa) (1.13 liters/min), assuming that the "critical" number of spores per deposit is 1 (therefore the "critical droplet size" is 148 μm)

Parameter	Value	Parameter	Value	Parameter	Value
"Critical" drop size (μm)	148	% Droplets with no spores	71	% Droplets with >/= 2 spores	15.5
Droplets/ha	9.22×10^{10}	Number of droplets with no spores	6.54×10^{10}	No. droplets with >/= 2 spores	1.44×10^{10}
VMD (μm)	321	% Volume with no spores	6.2	% Volume with >/= 2 spores	88
NMD (μm)	57	liters/ha with no spores	10.6	liters/ha with >/= 2 spores	149
D_{30} (μm)	162	% Droplets with 1 spore	13.5	Volume> 300(μm)	55
Relative Span	2.15	Number of droplets with 1 spore	1.24×10^{10}	Relative Efficiency	12.5
VMD/NMD ratio	5.62	% Volume with 1 spore	6.3	Relative Inefficiency	87.5
		liters/ha with 1 spore	10.7	Adjuvants Efficiency	94

droplet is 94%, although this is only 29% by *number*.) Relative Inefficiency should be lowered, but has other implications. Large droplets (*e.g.*, > 300 μm, 55% of the spray volume in the example given in Table 3) containing many spores impact poorly, shatter, and perhaps redistribute through a canopy. However, their momentum is just as likely to carry them through a canopy to the soil, where the AI is not just inefficiently applied, it is lost completely. Adjuvant Efficiency should be raised, especially where the adjuvants used are expensive and critical to performance. Each must be balanced against other considerations (see Section 2. **A**.1).

Table 4 compares four very different sprayers for the simplest example of a "critical" droplet size containing 1 spore: the hydraulic nozzle described in Table 3, a motorized mist-blower, applying 250 liters/ha (a high volume application rate for this type of sprayer), a CDA spinning disc applying 6.3 liters/ha (water-based formulation), and a "theoretical electrohydrodynamic" sprayer. Efficiency of application of the CDA sprayer is not very different to that of the hydraulic nozzle, especially when compared with the efficiency of the airshear sprayer (89%). However, the gains in efficiency with the airshear sprayer must be offset against the 63% waste of the adjuvants applied with the spores. It is here that the advantages of CDA appear. The "hypothetical electrohydrodynamic" sprayer is included as an extreme example of CDA, and demonstrates the risks of extrapolating conclusions from a medium or high volume application system to a CDA system.

Clearly, a "critical" droplet size containing one spore is an unrealistic scenario for many MCA/insect pest combinations. If the critical droplet size were determined by, say, 20 spores per droplet, then the droplet size required from an hydraulic system would be either unobtainable or rare: CDA is much more flexible. However, hydraulic nozzles are often the only realistically available application system (see Section 2). Considering just hydraulic nozzles, Table 5 shows the effect on application efficiency of changing nozzle type from fine to coarse and the impact of low concentrations of an adjuvant polymer. Larger orifice nozzles produce larger droplets but for the same application speed, apply higher volumes. The number of droplets produced is a balance between volume and droplet spectra (Table 5). As orifice

Table 4. Efficiency of application for a medium nozzle [XR8003VS, 40 psi (276 kPa), 8 kph], an air-shear system (knapsack motorized mistblower operating at near maximum volume application rate), a CDA system (Ulva+, red restrictor, 6.3 liters/ha), and a "hypothetical electrohydrodynamic" (ED) system, all applying 1×10^{11} spores/ha in water and all assuming that the "critical" number of spores per deposit is 1 (therefore defining the "critical droplet size" as the volume application rate divided by the spores per hectare)

Nozzle and parameters	Medium-high volume		CDA	
	XR8003VS	Airshear	Ulva+, red	"ED"
Volume/ha	170 liters/ha	250 liters/ha	6.3 liters/ha	1 liter/ha
"Critical" drop size (μm)	148	168	49	27
VMD (μm)	321	105	67	45
NMD (μm)	57	26	46.5	44.5
D_{30} (μm)	162	56	57	43
VMD/NMD	5.6	4.0	1.4	1.01
Relative Span	2.15	1.18	0.76	0.25
Droplets per ha	9.2×10^{10}	2.7×10^{12}	6.55×10^{10}	2.4×10^{10}
% Droplets with 0 spores	70.9	96.4	30.5	0
% Volume with 0 spores	6.2	63	5.9	0
% Droplets with just 1 spore	13	3.3	24	0
No. droplets with just 1 spore	1.2×10^{10}	8.9×10^{10}	1.6×10^{10}	0
% volume with just 1 spore	6.3	27	10	0
l/ha with just 1 spore	10.7	67	0.6	100
% droplets with 2 or more spores	15.7	0.3	45.2	100
Relative Efficiency	12.5	89.5	16	0
Relative Inefficiency	87.5	10.5	84	100
Adjuvant Efficiency	94	37	94	100

Table 5. The effect on efficiency of application of: changing nozzle from fine to medium to coarse, and the effect of adding a polymer at various concentrations, all at 1×10^{11} spores/ha, forward speed 8 kph, 40 psi (276 kPa) pressure, and all assuming that the "critical" number of spores per deposit is 1 (therefore defining the "critical droplet size" as the volume application rate divided by the spores per hectare)

Nozzle type:	XR8001VS	XR8002VS	XR8003VS	XR8004VS	XR8003VS	XR8003VS	XR8003VS
Polymer concentration	–	–	–	–	0.001%	0.01%	0.1%
| VAR[1] (liters/ha)	56	114	170	228	170	170	170
Optimum drop size (μm)	102	130	148	163	148	148	148
VMD (μm)	172	245	321	395	267	303	465
NMD (μm)	50	50	57	58	55	53	46
D30 (μm)	105	128	162	185	144	152.3	200.1
VMD/NMD ratio	3.4	4.9	5.6	6.8	4.9	5.8	10
Relative Span	1.09	1.48	2.15	1.67	1.22	1.21	0.9
Droplets/ha	9.2×10^{10}	1.1×10^{11}	9.2×10^{10}	8.1×10^{10}	1.1×10^{11}	9.2×10^{10}	4.1×110^{10}
% Droplets 0 spores[2]	62.2	72.1	70.9	72.4	71.8	72.8	73.5
% Volume 0 spores	7.9	8.3	6.2	4.9	6.85	5.65	2.06
Relative Efficiency[3]	18.1	16.6	12.4	9.5	14	11.3	4
Relative Inefficiency	81.9	83.4	87.6	90.5	85.9	88.6	96
Adjuvant Efficiency	92.1	91.7	93.8	95.1	93.1	94.3	97.9

[1] Volume Application Rate.
[2] *et seq.*: % droplets and %VAR containing no spores.
[3] *et seq.*: "efficiency" as defined in Section 3. **B**.3.

size increases, Relative Efficiency declines markedly whereas adjuvant Efficiency remains relatively stable.

Perhaps of more importance is the effect on Relative Efficiency of adjuvant polymers, which are becoming important as MCA formulation components. In this case, low concentrations of polymer had only a small effect on Relative Efficiency. However, above 0.01% of this particular polymer, Relative Efficiency declined rapidly to a minimum of 0.03% at 0.1% concentration at 97 liters/ha (Table 5). If polymers are to be used in the formulation of a microbial, then the eventual concentration of the polymer in the spray tank must be considered, and this may set a lower limit on volume application rates.

Finally, and taking all the above into account, two figures (Figures 9 and 10) demonstrate the Relative Efficiency calculated for four different hydraulic nozzles at four pressures [15, 20, 40, and 60 psi (103, 138, 276, and 414 kPa, respectively)], applying 10^9 to 10^{13} spores/ha at four forward speeds (8, 10, 12, and 14 kph), again for a "critical" droplet size of one spore per droplet. Relative Efficiency has been calculated for each scenario, and in Figure 9 (data broken down by nozzle), the highlighted points show the effect on the efficiency of one nozzle, an XR8003VS applying 10^{10} spores/ha at the four pressures. A subset of the data is expanded, showing the effect of changing forward speed for one pressure. The overall picture shows that manipulating "critical" droplet size is a simple way of improving application efficiency. Even within a small range of "critical" droplet sizes [here, 237, 249, 279, and 298 μm for the 15, 20, 40, and 60 psi (103, 138, 276, and 414 kPa), respectively], efficiency is improved from approx. 15% to 75%. Within one scenario (XR8003VS, 10^{10} spores/ha, and 40 psi (276 kPa)), altering AI concentration by changing volume application rate by changing forward speed also has a significant effect, although not as great as changing nozzle. From Figure 10 (points broken down by spores/ha), for AI application rates below 10^{10} and above 10^{12} spores/ha, application efficiency is little affected by application parameters. The highlighted points show the lowest and highest efficiency within a spore/ha rate. The differences can be substantial, as much as two orders of magnitude. In general, within a nozzle, low volume application rates combined with low nozzle pressures lead to poor application efficiency, and *vice versa*. The widest spread of efficiency is within small nozzles, but the overall conclusion holds for the four nozzles described.

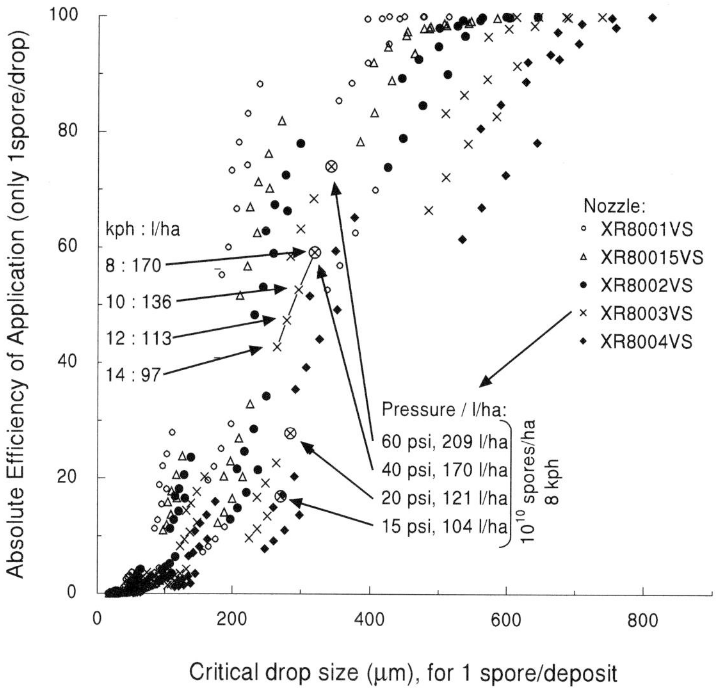

Figure 9. The effect on Relative Efficiency of application ("critical" drop size assumes 1 spore per drop: therefore, 100% efficiency would be each spore delivered in a separate drop) for five hydraulic nozzles spraying at 4 pressures (15, 20, 40, and 60 psi), applying 10^9 to 10^{13} spores/ha at four forward speeds (8, 10, 12 and 14 kph, translated to different volume application rates and AI concentrations). Data points broken down by nozzle type (see text for more details)

4 Conclusions

When developing a microbial control agent (MCA) for use in broad-acre agriculture, unless there are exceptional circumstances, the only available spray application system is that system in widespread use in the target crop in the geographical area of interest or local market. This is especially the case where an MCA is being included with other pesticides as part of an IPM system: the principal application system is the system that must be used for an MCA. Alternative application systems (*e.g.*, CDA, air shear systems, foggers) often have substantial advantages over the conventional system and should not be ignored, especially for niche markets where either no application system exists or where other application systems are available.

It is important to match field trial equipment to that used in the true field situation, especially nozzle type, volume application rate, and pump system. In particular, pump systems can be used to alleviate formulation problems by breaking up "clumped" formulations and by imparting energy to oil adjuvants and emulsifying them, thus making them much finer and therefore better distributed through the spray volume. Pump systems can alleviate some formulation problems.

When AI efficiency improvements are sought through altering application parameters, *e.g.*, droplet size, number, concentration of AI and adjuvant, it is important to work within the constraints set by the grower or farmer. A farmer can be asked to change nozzle, volume application rate, and other parameters such as nozzle pressure and forward speed, but within limits which must be determined. With microbials, it is possible to ask more of a farmer than would normally be asked when applying conventional agrochemicals. However, there are limits, and these must be explored properly before making label recommendations.

Application of large microbials (*e.g.*, entomopathogenic nematodes) is independent of droplet size. Essentially, large particle microbials will be delivered where the volume of liquid is applied. However, any water soluble or emulsified adjuvants will be associated with the spray liquid directly. The waste of

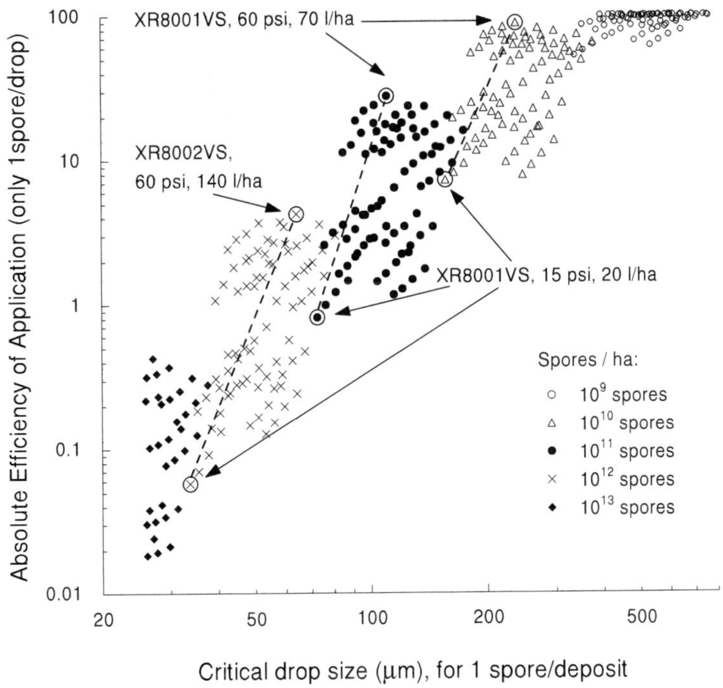

Figure 10. The effect of changing optimum drop size on Relative Efficiency of application ("critical" drop size assumes 1 spore per drop: therefore, 100% efficiency would be each spore delivered in a separate drop) for five hydraulic nozzles spraying at 4 pressures (15, 20, 40, and 60 psi), applying 10^9 to 10^{13} spores/ha at four forward speeds (8, 10, 12, and 14 kph, translated to different volume application rates and AI concentrations). Data points broken down by spores/ha (NB. log10 scale: see text for more details)

adjuvant contained in droplets containing no MCA should be borne in mind when altering application practices.

Small particulate microbials applied in very large numbers (*e.g.*, viruses), biorationals (*e.g.*, the botanicals), and small particles with a non-uniform distribution (*e.g.*, *Bt*s) can be considered as conventional agrochemicals, *i.e.*, the corpus of conventional spray application knowledge applies directly to such MCAs. In short, efficiency of application should be indirectly proportional to droplet size: the smaller the droplet, the more efficient the use of the AI. However, there is a trade-off between small droplets and canopy penetration, cover, and the behavior of the insect target, and these must also be considered as important parts of the whole equation of dose transfer.

Particles of a uniform size distribution but approximately 1 to 10 μm in diameter (*e.g.*, fungal or bacterial spores) can be considered differently from viruses, *Bt*s or biorationals. Because the volume of the droplet is no longer directly related to the dose of AI delivered by that droplet, it is possible to come to different generalizations than for microbials with non-uniform size distributions or of very small size. From theoretical considerations of spray clouds as sampling systems and the interaction of spray cloud droplet spectra and parameters (such as the number of droplets and spores applied per ha, the volume application rate, and the concentration of any adjuvants present), the following generalizations can be made. On the simplifying assumption that application efficiency is related to even dispersal of, say, spores through a canopy and that 100% efficient application is each spore delivered separately, then:

- "Critical droplet size" is the important parameter to consider, not the generalization that small droplets are more efficient.
- At AI application rates above 10^{12} spores per ha, the generalizations that hold for conventional agrochemicals apply (small droplets = better efficiency), whereas at AI application rates below 10^{10} spores/ha, manipulation of spray application

practices (changing nozzles [*i.e.*, droplet spectra], volume application rate (*i.e.*, liter/ha and therefore AI concentration and droplets per ha) etc.) have little or no effect on application efficiency as defined here.

- Within the AI application rate of 10^{10} to 10^{12} spores/ha inclusive, a smaller "critical" droplet size, a shift to a smaller droplet spectra, and higher volume application rates results in greater efficiency of application.
- The above statements cannot be applied to adjuvants, which being either water soluble or broken up into fine emulsions, must be considered in the same way as conventional agrochemicals. In general, a smaller "critical" droplet size means reduced waste of adjuvant, *i.e.*, a greater proportion of the adjuvant is delivered with the MCA.

5 Caveats

It must be stressed that the discussion of "critical droplet size" and the generalizations drawn from such work are theoretical. In particular, the use of the simplest example of "critical" droplet size (1 spore per droplet) is clearly *not* a typical scenario for microbial insecticides. However, we offer these guidelines more as an example of what can be done rather than what should be done when considering the application of microbials. This work is ongoing at LPCAT and Silwood Park, and the generalizations made are based on unpublished data (A. C. Chapple, R. A. J. Taylor, LPCAT, and R. P. Bateman, Imperial College). Research on more complex definitions of "efficiency" continues. In particular, the parameters governing typical hydraulic application are very different from those of oil-based ULV systems.

Any discussion of the importance of application variables in improving AI efficiency rests on the assumption that what has been measured prior to deposition in the canopy bears a direct relationship to the eventual biological result, whether the measurement is the spray cloud droplet spectra or the elongational viscosity of the liquid in the spray tank. However, the "black box" referred to in the Introduction (Figure 1) is very much an unknown quantity. Large changes in the process of dose transfer prior to deposition in the canopy often have little or no effect when tested in the field. The consequence of this lack of fundamental information is that field trial data is to be trusted over theory. Theory such as that presented in this chapter can only be considered as a guideline.

6 Acknowledgments

The authors would like to acknowledge Ecogen Inc. (Langhorne, PA, USA) for allowing the publication of the data and insights obtained whilst Dr. Chapple worked for Ecogen Europe.

7 References

Adams, A. J., Chapple, A. C., and Hall, F. R. 1990. Droplet spectra for some agricultural fan nozzles, with respect to drift and biological efficiency. *In* "Pesticide Formulations and Application Systems: Tenth Symposium." ASTM STP 1078. (E. Bode, J.L Hazen, and D. G. Chasin eds.), American Society for Testing and Materials, Philadelphia, USA, pp. 156–169.

Arnold, A. J., Cayley, G. R., Dunne, Y., Etheridge, P., Greenaway, A. R., Griffiths, D. C., Phillips, F. T., Pye, B. J., Rawlinson, C. J., and Scott, G. C. 1984a. Biological effectiveness of electrostatically charged rotary atomisers. III. Trials on arable crops other than cereals, 1982. *Ann. Appl. Biol.* **105**, 369–377.

Arnold, A. J., Cayley, G. R., Dunne, Y., Etheridge, P., Griffiths, D. C., Jenkyn, J. F., Phillips, F. T., Pye, B. J., Scott, G. C., and Woodcock, C. M. 1984b. Biological effectiveness of electrostatically charged rotary atomisers. II. Trials with cereals, 1982. *Ann. Appl. Biol.* **105**, 361–367.

Arnold, A. J., Cayley, G. R., Dunne, Y., Etheridge, P., Griffiths, D. C., Phillips, F. T., Pye, B. J., Scott, G. C., and Vojvodic, P. R. 1984c. Biological effectiveness of electrostatically charged rotary atomisers. I. Trials on field beans and barley. *Ann. Appl. Biol.* **105**, 353–359.

Bache, D. H. and Johnstone, R. J. 1992. "Microclimate and Spray Dispersion." Ellis Horwood, Chichester, UK. 238 pp.

Bals, E. J. 1969. The principles of and new developments in Ultra Low Volume spraying. *Proc. of the 5th. British Insecticide and Fungicide Conf.*, pp.189–193.

Bateman, R. P. 1993. Simple, standardised methods for recording droplet measurements and estimation of deposits from controlled droplet applications. *Crop Prot.* **12**, 201–206.

Bateman, R., Carey, M., Moore, D., and Prior, C. 1993. The enhanced infectivity of M. flavoviride in oil formulations to desert locusts at low humidities. *Ann. Appl. Biol.* **122**, 145–152.

Bateman, R. P. 1997. Methods of application of microbial pesticide formulations for the control of grasshoppers and locusts. *Mem. Entomol. Soc. Canada*, 171, 67–79.

Bryant, J. E. and Yendol, W. G. 1988. Evaluation of the influence of droplet size and density of Bacillus thuringiensis against gypsy moth larvae (Lepidoptera: Lymantriidae). *J. Econ. Entomol.* **81**, 130–134.

Chapple, A. C. 1996. Application of Biological Control Agents: Some Theoretical Considerations of Dispersal. *In*: "Proc. 5th European Meeting of the International Organisation of Biological Control, West and East Palearctic Regions: Microbial Control of Pests." Poznan, Poland. pp. 24–28.

Chapple, A. C., Taylor, R. A. J., Hall, F. R, and Downer, R. A. 1994. Understanding the dose transfer process as a prerequisite to a more predictable lab-to-field performance correlation. *In*: "Comparing Laboratory and Field Pesticide Performance II." *BCPC Monograph No. 59*: Proceedings of an International Symposium held at the University of Kent 12–14, April, 1994 pp. 267–272.

Chapple, A. C. and Bateman, R. P. 1997. Application systems for microbial pesticides: necessity not novelty. *In*: "Microbial Insecticides: Novelty or Necessity?" H. F. Evans, Ed., *BCPC Monograph No. 68* pp. 181–190.

Chapple, A. C., Downer, R. A., Hall, F. R., and Wolf, T. M. 1997. Application of biological pesticides: limitations and a practical solution. *Entomophaga* **41**: 465–474.

Cooke, B. K., Hislop, E. C., Herrington, P. J., Western, N. M., Jones, K. G., Woodley, S. E., and Chapple, A. C. 1985. The physical, chemical, and biological appraisal of alternative spray techniques in cereals. *Crop Prot.* **5**, 155–164.

Copping L G 2004 (Ed.) The Manual of Biocontrol Agents (3rd Edition, The BioPesticide Manual). BCPC, Alton, Hants. 702 pp.

Ebert, T. A., Taylor, R. A. J., Downer, R. A., and Hall, F. R. 1999a. Deposit structures and efficacy of pesticide application: 1 Interaction between deposit size, toxicant concentration, and deposit number. *Pestic. Sci.* **55**, 783–792.

Ebert, T. A., Taylor, R. A. J., Downer, R. A., and Hall, F. R. 1999b. Deposit structures and efficacy of pesticide application: 2. Trichoplusioa ni control on cabbage with fipronil. *Pestic. Sci.* **55**, 793–798.

Evans, H. F. 1997. (Ed.) "Microbial Insecticides: Novelty or Necessity?" *Symposium Proceedings 68, British Crop Protection Council*. BCPC, Farnham, UK. 301 pp.

Frick, E. L. 1970. The effects of volume, droplet size, and concentration, and their interaction on the control of apple powdery mildew by Dinocap. *British Crop Protection Council Monogram 2*, pp. 23–33.

Hall, F. R. and Adams, A. J. 1990. Microdroplet application for determination of comparative topical and residual efficacy of formulated permethrin to two populations of diamondback moth (Plutella xylostella L.). *Pestic. Sci.* **28**, 337–343.

Hall, F. R., Chapple, A. C., Downer, R. A., Kirchner, L. M, and Thacker, J. R. M. 1993. Pesticide application as affected by spray modifiers. *Pestic. Sci.* **38**, 123–133.

Hall, F. R. and Barry, J. W. (Eds.) 1994. "Biorational Pest Control Agents: Formulation and Delivery." ACS Symposium Series 595, American Chemical Society, Washington, DC, USA.

Hall, F. R., Chapple, A. C., Taylor, R. A. J., and Downer, R. A. 1995. Modelling the dose acquisition process of Bacillus thuringiensis: the influence of feeding pattern on survival. *In*: "Biorational Pest Control Agents: Formulation and Delivery." Eds. Hall F. R. and Barry J. ACS Symposium Series 595, American Chemical Society, Washington, DC, USA. pp. 68–78.

Hall, F. R. and Menn, J. 1999. "Biopesticides: Use and Delivery." Humana Press Inc. NJ, USA. 626 pp.

Hewitt, H. G., Casely, J., Copping, L. G., Grayson, B. T., and Tyson, D. (Eds.) 1994. "Comparing Laboratory and Field Pesticide Performance II." *BCPC Monograph No. 59* Proceedings of an International Symposium held at the University of Kent 12–14, April, 1994. 323 pp.

Himel, C. M. 1969. The optimum droplet size for insecticide spray droplets. *J. Econ. Entomol.* **62**, 919–925.

Hislop, E. C. 1987. Can we define and achieve optimum pesticide deposits? *Asp. Appl. Biol.* **14**, 153–165.

Hofstein, R. and Chapple, A. C. 1999 Commercial Development of Biofungicides. *In* "Biopesticides: Use and Delivery." pp. 77–102 Eds. J. Menn and F. R. Hall. Humana Press Inc. NJ, USA. 626 pp.

Klein, M. G. and Georgis, R. 1994. Application techniques for entomopathogenic nematodes. Proc. IV Int. Coll. on Invertebrate Pathology and Microbial Control, Montpelier, France, pp. 483–484.

LeFebvre, A. H. 1989. "Atomization and Sprays." Hemisphere Pub. Co. NY., London, UK. 421 pp.

Lisansky, S. G. and Coombs, J. 1994. Developments in the market for biopesticides. *Brighton Crop Protection Conference – Pests and Diseases – 1994* Vol.3. pp. 1049–1060.

Maczuga, S. A. and Mierzejewski, K. J. 1995. Droplet size and density effects of Bacillus thuringiensis kurstaki on gypsy moth (Lepidoptera: Lymantriidae) larvae. *J. Econ. Entomol.* **88**, 1376–1379.

Matthews, G. A. 2000. "Pesticide Application Methods." Blackwell Science, Oxford, UK, 432 pages.

Mowlam, M. D., Nyirenda, G. K. C. and Tunstall, J. P. 1975. Ultra low volume application of water-based formulations of insecticides to cotton. *Cotton Growers Rev.*, **52**, 360–370.

Nilsson, U. and Gripwall, E. 1999. Influence of application technique on the viability of the biological control agents Verticillium lecannii and Steinernema feltiae. *Crop Prot.* **18**, 53–59.

Pittis, J. E. and Russell, P. E. 1994. Factors affecting disease expression by plant pathogens. *In*: "Comparing Glasshouse and Field Pesticide Performance II." *BCPC Monograph No. 59*, pp. 57–66.

Potts, S. F. 1946 Particle size of insecticides and its relation to application, distribution, and deposit. *J. Econ. Entomol.* **39**, 716–720.

Rayleigh, Lord 1882. On the equilibrium of liquid conducting masses charged with electricity. *Phil. Mag. Series 5* **14**, 184–186.

Salt, D. W. and Ford, M. G. 1996. Deterministic models to simulate the movement of pesticide from discrete deposits and to predict optimum deposit characteristics on leaf surfaces for control of sedentary crop pests. *Pestic. Sci.* **48**, 77–87.

Southcombe, E. S. E, Miller, P. C. H., Ganzelmeier, H., Zande, J. C. van de, Miralles, A., and Hewitt, A. J. 1997. The international (BCPC) spray classification system including a drift potential factor. *Proc. Brighton Crop Prot. Conf. – Weeds 1997*, Vol. 1, pp. 371–380.

Waage, J. K. 1997. Biopesticides at the crossroads: IPM products or chemical clones? *In*: "Microbial Insecticides: Novelty or Necessity?" H. F. Evans, Ed., *BCPC Monograph No. 68*, pp. 11–19.

Young, B. W. 1986. The need for greater understanding in the application of pesticides. *Outlook on Agriculture* **15**, 80–87.

SECTION II

STATISTICAL CONSIDERATIONS

Chapter II-1

Experimental design: statistical considerations and analysis

James F. Campbell
USDA-ARS Grain Marketing and Production Research Center
Biological Research Unit
Manhattan, KS 66502, USA

Stephen P. Wraight
USDA-ARS, Plant, Soil & Nutrition Laboratory
Ithaca, NY 14853, USA

1 Introduction

In this chapter, information on how field experiments in invertebrate pathology are designed and the data collected, analyzed, and interpreted is presented. The approach will be to present this information in a step-by-step fashion that, hopefully, will emphasize the logical framework for designing and analyzing experiments. The practical and statistical issues that need to be considered along the way and the rationale and assumptions behind different designs or procedures will be given, rather than the nuts-and-bolts of specific types of analysis. We want to emphasize that we are not statisticians by training and strongly recommend consulting a statistician during the planning stages of any experiment.

Writing a chapter on this topic presents a number of difficulties. The first is the incredible breadth of the topic. Second, the wide range of statistical expertise of the target audience of this book means that there will be a correspondingly wide range of expectations over what should be included in this chapter. Third, making general recommendations is complicated by the many factors that need to be considered in selecting an experimental design and statistical analysis and the fact that there is often disagreement over the best statistical approaches to use for particular situations. We have geared this chapter for researchers with very limited statistical and experimental design experience. The roadmap provided here is intended to assist beginning field researchers to make decisions about designing and analyzing experiments, dealing with some of the real world problems that arise, finding sources of additional in-depth information, and, most importantly, communicating with statisticians.

There is a large body of information available on the design and analysis of agricultural and ecological experiments that is applicable to invertebrate pathology field experiments. Experimental design and analysis principles are similar in the laboratory and the field, but the experimental approaches taken for invertebrate pathology field experiments differ from those in the laboratory in a number of significant ways. Many laboratory studies focus on individuals, and their analyses emphasize detecting trends in susceptibility using such statistical tools as probit, survival or failure time, or logistic regression. In the field, researchers typically measure population changes and have to deal with more confounding factors (*e.g.*, variation within the field site, insect movement, inability to sample all individuals exposed to a treatment) and interactions than in the laboratory.

L.A. Lacey and H.K. Kaya (eds.), Field Manual of Techniques in Invertebrate Pathology, 37–69.

2 Types of experiments and analyses

A *Experiments*

Experiments in all areas of science serve as important tools to test hypotheses. However, the experiment must be planned, performed, analyzed and interpreted in such a way that misleading interpretations resulting from faults in design and execution are minimized. One of the most important aspects of experimental design is determining the specific objectives of the experiment. A clear statement of the objectives will keep the experimental design focused and provide a means of assessing if the experiment or a portion thereof, is worth performing. The statement can be in the form of questions to be answered, hypotheses to be tested, or effects to be estimated. The population about which generalizations from the results of the experiment are going to be made needs to be clearly defined and included in this statement. In addition, each treatment should be clearly defined and its role in reaching the objectives of the experiment determined. To provide useful results, experiments should have the following characteristics (Little and Hills, 1978): (1) the experimental design should be as simple as possible and still meet the objectives of the experiment; (2) the experiment should be able to measure differences with a degree of precision that is desired by the experimenter (*e.g.*, by using an appropriate experimental design and level of replication); (3) the experiment should be planned to minimize the chance of systematic error (*e.g.*, by appropriate blocking of experimental treatments); (4) the range of validity of the conclusions from an experiment should be as wide as feasible (*e.g.*, by repeating the experiment in space and time or by performing a factorial experiment); and (5) the degree of uncertainty in the experiment should be determined.

Experiments can be divided into two broad categories: mensurative and manipulative (Hurlbert, 1984). Mensurative experiments involve measurements of some parameter taken at one or more points in time or space and, typically, the experimenter does not manipulate or perturb the system. Statistical comparisons may or may not be necessary for these types of experiments. In mensurative experiments, analysis is typically performed to determine how a population conforms to a predicted value (goodness-of-fit), if there are differences among systems, or how variables are correlated.

Manipulative experiments are those where treatments are assigned to experimental units or plots and these assignments can be randomized. Manipulative experiments are discussed extensively in most books on experimental design and are probably the most common experiments in invertebrate pathology. Manipulative experiments have an advantage over mensurative experiments in that confounding variables can be better controlled. However, both types of experiments are useful and have important roles in the development of theory.

B *Hypothesis testing*

Hypothesis testing is central to the scientific method, but there are two types of hypothesis testing that need to be considered: experimental hypothesis testing and statistical hypothesis testing. Experimental hypotheses involve the system that the researcher is investigating. For example, an experimental hypothesis might be that a particular pathogen will reduce host populations under a certain set of environmental conditions. Statistical hypothesis tests are used as tools to test experimental hypotheses. In testing hypotheses, statistical procedures are used to draw inferences about a test population by examining samples from that population. Commonly, inferences are drawn with regard to population means, and the first step in hypothesis testing is to formulate a precise statement or hypothesis about the mean. Based on the results of the experiment, the hypothesis is either falsified (the prediction is not met) or not falsified. If the hypothesis is falsified, the set of assumptions is altered and the theory revised.

The term assumption can also have multiple meanings. First, explanatory assumptions are those made about the universe and involve the theory that generated the hypothesis that the experiment is testing. Second, simplifying assumptions are made for analytical convenience. These assumptions have either been tested by previous experiments or are robust, based on independent experiments or analyses. Third, statistical assumptions underlie the statistical procedure used. Statistical assumptions are the basis for the stated probability that the deviation of the observed parameter is significantly different from the predicted. Meeting these statistical assumptions is critical if the statistical procedure is to provide a correct conclusion, but these assumptions are often not fully understood or tested by researchers.

After establishing the specific question or hypothesis about what will happen under a certain set of experimental conditions, it is necessary to test it. This involves creating the conditions required by the hypothesis and observing the results. In all statistical hypothesis testing, the nature of the hypothesis is an important consideration. Proving that a hypothesis is true is difficult, because, based on inductive reasoning, the results for every possible circumstance need to be observed or inferred. This requirement is usually impossible to meet, because there is always the potential that an additional experiment could disprove the hypothesis (Underwood, 1990). Consequently, a falsification procedure is employed with the objective being to disprove the hypothesis. This is more straightforward because once disproved, additional experimentation would not alter the conclusion. Accordingly, a hypothesis that is actually the opposite of the one we want to test is proposed and is termed the null hypothesis (abbreviated H_0 and called null because it comprises a statement of no difference, *e.g.*, no difference between two means). Experiments then attempt to disprove the null hypothesis. If a statistical test leads to a conclusion to reject the null hypothesis, then an alternative hypothesis is needed. This alternative hypothesis (H_A) will usually not specify a single value, but, rather, is formulated to account for all possible outcomes not stated by the null hypothesis. Thus, H_0 and H_A together account for all possible outcomes, and rejection of H_0 leads to acceptance of H_A. See Sokal and Rohlf (1995) and Zar (1999) for more extensive discussions of hypothesis testing. Experiments should be designed so that there are only two possible outcomes, either the null hypothesis or the alternative hypothesis is accepted.

C Type I and type II errors

Two types of error, type I and II, can arise over the decision to reject a null hypothesis being tested. A type I error occurs when the null hypothesis is false when it is actually true (*i.e.*, a false positive). A type II error is the failure to falsify the null hypothesis when it is actually false. Statistical procedures are designed to indicate the likelihood or probability of these errors. The ultimate goal of the experiment needs to be considered in setting type I and type II error rates and interpreting results (Scheiner, 1993).

Most researchers are familiar with the probability of committing a type I error; this is indicated by the α (threshold level determined before analysis) and P values (estimated from the analysis) that are typically chosen by the experimenter and reported with statistical tests. However, the power of a test is also important to consider when designing an experiment and interpreting the results. The power of an experiment is the probability of correctly rejecting a false null hypothesis and is one minus the probability of committing a type II error ($1-\beta$). At some point, the acceptable probability of committing a type I error became conventionalized at 0.05. While it is important that some sort of objective criterion be used to avoid altering expectations to meet results, and being 95% confident is a reasonable standard, the actual probability used should be based on the biology of the system and the requirements of the investigator. For example, in the medical field a 99% confidence level is often used, but in field experiments the replication needed to obtain that level of confidence may not be possible and that level of confidence unnecessary. Prior to starting an experiment, it is desirable to balance the probability of committing a type I error, the power of the test, and the number of replicates. The objectives of the experiment and the biology of the system also need to be kept in mind when designing statistical hypothesis tests. See Toft and Shea (1983), Young and Young (1991), and Shrader-Frechette and McCoy (1992) for more discussion of this issue.

When a value of α is set for a statistical test, the accuracy of that value depends upon the validity of the statistical assumptions. When the assumptions do not hold, the true alpha level may be higher or lower than the nominal value set by the researcher. When the actual alpha is higher than nominal, the test is described as liberal. A liberal test declares more significant differences than it should (rejects a true null hypothesis too frequently). When alpha is lower than nominal, the test is conservative and declares fewer differences than it should. Conservative tests are also described as having low power to detect real differences.

3 Choosing a statistical analysis

Choosing the type of analysis to perform can be a difficult problem and should be considered during the planning stages of an experiment. There are a large number of approaches, and selection of the best test depends on the type of data collected, the objective of the experiment, and the assumptions of the analysis. Statistical tests can be divided into two categories: parametric and nonparametric. Parametric tests are based on the assumption that the data are sampled from a normal or Gaussian distribution (*e.g.*, t-test, Analysis of Variance). Nonparametric, or distribution free, tests do not make assumptions about any specific population distribution and typically involve analyzing the ranking of data (*e.g.*, Wilcoxon test, Kruskal-Wallis test). Nonparametric tests may be based on an assumption that the samples being compared came from populations with similar shapes and dispersions. However, these tests are often applied to data with heterogeneous variances as they are less affected by differences in population dispersions than ANOVA (Sokal and Rohlf, 1995; Zar, 1999).

Nonparametric tests tend to be less powerful than their parametric counterparts if the data are normally distributed or even approximately normally distributed, and it is therefore best to use parametric approaches where justified. If a large number of data points are collected, it may be possible to determine if the distribution is normal by plotting the data or performing a test such as the Komogorov-Smirnoff test (Zar, 1999). If data from a previous experiment are available, then these can also be used to help determine if the population has a normal distribution. The decision of whether to use parametric or nonparametric tests is difficult for small data sets because the nature of the distribution cannot be determined; ultimately, the parametric test may not be robust and the nonparametric test may not be powerful. It is sometimes suggested that nonparametric tests are the only option for testing of ordinal-scale (ranked) data, but this is incorrect. Parametric tests may be applied to such data provided the usual assumptions hold (Zar, 1999).

It is noteworthy that ratio data are not normally distributed, and in some cases, nonparametric analyses of ratios may be markedly more powerful than parametric analyses. Of considerable relevance to the present topic are ratios derived from two independent normally distributed variables. Such data comprise a heavy-tailed distribution known as the Cauchy distribution. Prominent examples of Cauchy-distributed data are LC_{50} values, which are generated from regression analyses essentially as the intercept divided by the slope. Outliers (aberrant data points) tend to occur frequently in heavy-tailed distributions, and in the presence of outliers, application of traditional analysis of variance (ANOVA) results in a conservative test. Nonparametric tests, in this case, exhibit substantially greater power than standard ANOVA, and are strongly recommended (Anderson and Lydic, 1977; Randles and Wolfe, 1979; Zimmerman, 2001). More information on nonparametric approaches can be obtained in Krauth (1988), Siegel and Castellan (1988), Daniel (1990), and Conover (1999).

A Comparison of two treatments

Many statistical comparisons involve only two treatments. These types of comparisons are relatively simple to perform, and the analyses are well covered in most general statistics texts (*e.g.*, Sokal and Rohlf, 1995; Zar, 1999). Different approaches are used for paired versus unpaired data and normal versus nonnormal population distributions.

1 Comparing two unpaired groups

When comparing two unpaired groups, an unpaired t-test can be used for situations where a parametric analysis is appropriate and a Mann-Whitney test where a nonparametric analysis is appropriate. The t-test produces results mathematically identical to analysis of variance (ANOVA), and the statistical assumptions are the same for both tests; these are discussed in section B. Motulsky (1995) describes a procedure for estimating the power of a t-test. The Mann-Whitney test does not assume a normal distribution, but does assume that the samples are selected randomly from a larger population, the measurements were obtained independently, and that the distributions of the two populations are similar. When variances or standard deviations are not equal, Day and Quinn (1989) recommend the Welch (1938) t-test, generalized with Satterthwaite's degrees of freedom (Winer, 1971), as a parametric procedure and the Fligner-Policello test (Fligner and Policello, 1981) for nonparametric comparisons.

Resampling statistics (*e.g.*, Monte Carlo methods such as randomization and bootstrapping) are a nonparametric approach that is becoming more commonly used in agricultural and ecological experimental analysis. Resampling approaches use the entire set of data that has been collected to produce new samples of simulated data and then compare the actual results to the simulated data set. A resampling test can be constructed for almost any statistical inference and they are computationally easier and have fewer assumptions than most traditional tests. These approaches are not typically covered in general biostatistics texts, but there is a growing number of statistical texts dealing with this type of analysis (*e.g.*, Dixon, 1993; Efron and Tibshirani, 1993; Manly, 1997).

2 *Comparing two paired groups*

Paired tests can be used whenever the value of one replicate in the first group is expected to be more similar to a particular replicate in the second group than to a randomly selected replicate in the second group. This occurs when measurements are taken from a replicate before and after the treatment is applied, when replicates are matched for certain variables, and when an experiment is run multiple times with a control and treated replicate performed in parallel. Paired statistical tests such as the paired t-test and the Wilcoxon paired-sample test can be used under these circumstances (*e.g.*, Motulsky, 1995; Sokal and Rohlf, 1995; Zar, 1999). Paired statistical tests have an advantage over unpaired tests because they distinguish variation among replicates from variation due to differences between treatments. For a parametric approach, the paired t-test can be used. Assumptions of this test include that the pairs are selected randomly from a larger population, the samples are paired or matched, each pair is selected independently of the others, and the differences in the population are normally distributed. A nonparametric approach that can be used is the Wilcoxon paired-sample test. This nonparametric test has the same assumptions as the paired t-test, with the exception of the normal distribution.

3 *Comparing two proportions*

Many measurements collected in invertebrate pathology experiments can be expressed as proportions, which do not have normal distributions. For example, experimental treatments are applied and after a certain exposure period the number of individuals live or dead is counted and the result expressed as a proportion or percentage. Two commonly used tests for comparing the proportion responding to different treatments are Fisher's exact test and chi-square test (Motulsky, 1995; Zar, 1999). The chi-square test is easier to calculate than Fisher's test, but is not as accurate. Both tests use contingency tables to estimate the probability of obtaining the observed results. Contingency tables typically consist of rows that represent exposure to treatments (*e.g.*, pathogen dosage) and columns that represent alternative outcomes (*e.g.*, number of individuals live or dead). The rows must be mutually exclusive and the columns must be mutually exclusive. Each cell of the table contains the number of subjects matching the combination of row and column categories. Both tests have the following assumptions: random sampling of the data, the data must form a contingency table (*i.e.*, values must be number of subjects observed and the categories defining the rows and columns must be mutually exclusive), the subjects are independent, and the measurements are independent. The chi-square test should be used for large samples and if the total number of subjects is more than 20 and no expected value is less than five (but see section B.2.). For small to moderate samples, Fisher's exact test is recommended; however, the method generally requires a computer. If a computer program is not available, the chi-square test can be used with Yates' correction (Zar, 1999). There are several approaches that can be used to calculate the power of comparisons of proportions (Cohen, 1988; Motulsky, 1995).

McNemar's test can be used for comparing paired samples when the collected data are dichotomous and two treatments are compared (*e.g.*, the number of live and dead insects in response to application of a pathogen compared to the number of live and dead among controls) (Zar, 1999). It can also be used for comparisons of measurements made before and after some event (*e.g.*, the number of live and dead insects in a population before and after application of a pathogen). The McNemar's test uses a table similar to a contingency table, but the rows and columns are not independent. This test calculates a chi-square value and, like the chi-square test, should not be used for small samples.

Unfortunately, chi-square and related tests are among the most misused in biometrics (see Hurlbert, 1984). The most common misapplication in microbial control research involves sampling from unreplicated treatment versus control plots. Samples from a single field plot to which a treatment has been applied are correlated pseudoreplicates or subsamples, not true replicates, and generally should not form the basis of a statistical analysis. This approach is acceptable only in cases where no treatment has been applied (*e.g.*, to compare rates of natural disease prevalence at two different sites).

Misapplication of chi-square testing is also common in cases where treatments have been replicated, but where replicates comprise groups of individuals. In pathogen efficacy testing, insects are commonly treated and maintained in groups (*e.g.*, groups of insects in Petri dishes, or groups of insects infesting field plots). Subsequently, these groups, or samples from these groups, might be pooled and subjected to chi-square analysis. However, as in the previous example, this approach violates the assumption of independence (that each insect represents an independent replicate). Nevertheless, this assumption is often overlooked in cases where absence of significant unexplained error can be demonstrated; data are pooled, but only if the responses among the similarly treated replicates are homogeneous. This is readily determined by a preliminary chi-square test for heterogeneity (described in most standard biometry texts). Pooling in spite of significant heterogeneity puts any subsequent chi-square analysis at risk of declaring significant differences that may be the result of factors other than the applied treatment (*e.g.*, lack of proper randomization). It can certainly be argued that treatment of any pseudoreplicates as true replicates carries similar risk, regardless of the degree of heterogeneity; however, heterogeneity testing is considered an adequate safeguard by many statisticians, as long as the data are carefully examined for systematic errors. An analogous example, is the test for heterogeneity (goodness-of-fit) in probit analysis. If significant heterogeneity is not detected, individual test subjects that have been treated in groups are accepted as independent replicates, and a large-sample t value is used for calculation of confidence intervals (Finney, 1971). Sokal and Rohlf (1995) describe a procedure for analysis of replicated tests of goodness of fit and discuss consequences of pooling data without testing for heterogeneity.

The reader is alerted to the many seemingly routine applications of pooling for chi-square analysis, such as that employed for detection of synergism or antagonism (chi-square test of independence) between control agents (Robertson and Priesler, 1992; Sokal and Rohlf, 1995). Pooling may be questionable whenever subjects have been treated in groups and especially in the absence of heterogeneity testing. One should obviously avoid chi-square analysis of pseudoreplicated data whenever possible. ANOVA based on the truly replicated groups (*e.g.*, % mortality among individuals in each group) is a sound alternative following application of any necessary transformation.

B Comparing more than two groups

Many statistical analyses require the comparison of more than two treatments or groups;

particularly in field experiments. The problem with using multiple two-treatment comparisons is that the probability of obtaining a significant P value by chance (*i.e.*, probability of committing a type I error) increases with each comparison. The best approach is to compare all the groups at once using analysis of variance or an equivalent type of approach.

1 Analysis of variance

Analysis of Variance (ANOVA) is a general test that can make many types of comparisons and, as the name implies, analyzes the variance among values. In depth coverage of ANOVA is beyond the scope of this chapter. More detailed discussions of ANOVA can be obtained by reading Scheffé (1959), Sokal and Rohlf (1995), Underwood (1997), Zar (1999) and many other general statistics books. The emphasis taken here will be on the underlying assumptions of ANOVA and how they influence experimental design and analysis. The ANOVA procedure has several assumptions that need to be taken into account in the design and analysis. Failure to meet the assumptions of a statistical test can affect the level of significance of the test. Some of the common violations of ANOVA assumptions in agricultural and ecological experiments are discussed by Gomez and Gomez (1983) and Underwood (1997), respectively. The major assumptions of ANOVA are discussed below.

a Normally distributed data

Fortunately, the results of an ANOVA are generally not greatly affected by violations of this assumption. This is particularly true when there are either many treatments, many replicates per treatment, or the experiment is balanced. Tests for normality are not very useful unless the number of replicates is quite large, but plotting the relationship between the sample means and the variances can often enable the researcher to determine how normal is the distribution of the data. Outliers in the data can affect the normality of the distribution and the data should be checked for presence of outliers. If the data are highly skewed, the log transformation can make the data more normally distributed. When this is not possible, nonparametric tests should be used instead of ANOVA.

b Homogeneity of variances

If the variances within treatments are different among treatments, pooling them is not justified and the validity of the ANOVA is called into question. Heterogeneity of variance can affect both the type I and type II error rates. Sampling from populations with unequal variances means that there is an increased probability that populations will differ statistically than if they were drawn from a population of equal variances. There are a number of tests for the heterogeneity of variances; the most useful of which is probably Cochran's (1951) test (Underwood, 1997). If the Cochran's test is significant, it indicates the potential for serious problems with the ANOVA.

If the variances are not equal, it is important to determine why this has occurred, because the experimental units were assumed to represent samples from a single population. Differences in variation may indicate that (1) the treatments are affecting the variance differently among treatments (which would be something interesting to investigate further), (2) the experimental units were not from the same population, (3) test subjects were not randomly assigned to the treatments, or (4) errors in data collection may have skewed the distribution of the data in some treatments. If the variances are not equal, several approaches can be used to try and meet the assumption of homogeneity of variances: data could be separated into groups with similar variances, the data could be transformed, or a procedure of weighting means according to their variances could be performed (Underwood, 1997).

c Additivity of the main effects

In two-way or multi-way ANOVA, the effects of the factor levels must be additive, meaning that the levels of each factor must differ, on average, by a constant value. Violations of this assumption are generally reflected in highly significant interactions and, notably, may lead to false conclusions in analysis of synergism/antagonism. A common failure to meet this assumption occurs when factor effects are multiplicative rather than additive, and this problem is most readily corrected by log transformation. A test devised by Tukey (1949) and detailed by Sokal and Rohlf (1995) is useful for determining if a significant interaction is the result of nonadditivity.

d Independence of data within and among treatments

Independence indicates that there is no relationship between the size of the error terms and the experimental grouping. This is a reason to avoid having all plots receiving a given treatment occupying adjacent positions in the field. Adjacent portions of a field are more closely related to each other than randomly scattered plots. Lack of independence in sampling is perhaps the greatest problem faced in biological experiments (Hurlbert, 1984). However, independence of the data can be achieved with an understanding of the biology of the system and proper planning and effort by the experimenter. Some examples of the biological basis for nonindependence of data and how it impacts analysis are provided by Underwood (1997).

The best way to deal with non-independence is to give careful consideration to the biology of the system when designing the experiment. Pilot studies and previous research can also be helpful in assessing the independence of sampled data. However, some violations of the independence of data are not necessarily biologically based. A positive correlation between means and variances can be encountered when there is a wide range of sample means. For example, if comparing insect densities with means ranging from 5 to 500, the variation around the average of 5 would be expected to be less than that around the average of 500. This would violate the assumption of the ANOVA. If this occurs, the data frequently can be transformed so that the assumption of independence is supported.

2 Alternatives to analysis of variance

There are a number of alternatives to ANOVA that can be tried if the assumptions of the ANOVA are not supported and problems cannot be corrected using transformation. Day and Quinn (1989) list four parametric alternatives to ANOVA, but recommend the W test (Welch, 1951), which can be used when the assumption of homogeneous variances is questioned. These alternatives work for one-way ANOVA designs when variances are heterogeneous, but there are no robust alternatives for nested or factorial ANOVAs (Day and Quinn, 1989; Weerahandi, 1995). A non-parametric procedure for one-way designs that is widely used is the Kruskal-Wallis test. This test is appropriate if the distribution of the data is not normal but the variances of the data (or transformed data) are equal. A robust version of the Kruskal-Wallis test that requires only that the variances be symmetrical, not necessarily equal, is provided by Rust and Fligner (1984). Nonparametric analysis of two-way designs without interaction (randomized blocks) can be achieved using the Friedman test.

The above-mentioned nonparametric alternatives are valid only for one-way ANOVA or two-way ANOVA without interaction. There are, unfortunately, few options for nonparametric analysis of factorial tests with interaction. An extension of the Kruskal-Wallis test in which data are simply transformed to single-array ranks and subjected to standard ANOVA (Iman, 1974; Conover and Iman, 1976; Scheirer *et al.*, 1976), has been recommended in the past (*e.g.*, see Sokal and Rohlf, 1995); however, a theoretical study by Thompson (1991) revealed that it is not reliable. Conover (1999) recommends use of aligned-rank procedures; however, these tests are not strictly distribution free (nonparametric), and Mansouri and Chang (1995) found they performed poorly with respect to interaction testing for Cauchy-distributed data. Conover (1999) also recommends analysis by standard ANOVA of both the unranked and rank-transformed data and comparison of the results; if the results produced by the two procedures are nearly identical, the analysis of the unranked data can be considered valid.

Proportions can be transformed (by arcsine) to normalize the distribution and analyzed using ANOVA. However, chi-square analysis can also be used when more than two groups of proportional data need to be compared. The chi-square test can be used for contingency tables with more than two rows or columns if the following assumptions are met (Motulsky, 1995; Zar, 1999): data are randomly selected from populations, the data form a contingency table, and the observations are independent (see previous discussion of pseudoreplication problems in chi-square analysis). The use of the chi-square, as discussed above, has several simplifying assumptions that make it valid only for large data sets. A general rule is that 80% of the expected values must be greater or equal to five and all must be greater or equal to 2 (greater than or equal to 1 if the table has more than 30 df) (Motulsky, 1995). However, Zar, (1999) and Conover, (1999) suggest guidelines that are less conservative. If these rules are violated it may be possible to combine two or more rows or columns to increase the expected values. Three-dimensional contingency tables (*e.g.*, species, location, and disease) can also be used to analyze factorial experiments (Zar, 1999).

3 Multiple comparison tests

After rejecting the null hypothesis of a multifactor ANOVA, it may be necessary to determine which means are actually different from each other. There are many procedures for determining which of the many alternatives to the null hypothesis is best supported. However, selecting a procedure can be confusing because of considerable controversy about the use of various techniques (Day and Quinn, 1989; Underwood, 1997).

An essential point when considering multiple comparisons is to determine if specific comparisons can be determined prior to the start of an experiment (*i.e.*, planned or *a priori*). This means that the potential outcome needs to be specified before the experiment is conducted and should be based on the knowledge of the researcher and the particular question being asked. These comparisons can be performed regardless of the results of the ANOVA but should not have been suggested by the results and should test completely separate hypotheses (*i.e.*, planned orthogonal comparisons) (Sokal and Rohlf, 1995). For example, in an experiment where the effects of a formulation technique on pathogen persistence are being studied, a specific hypothesis to be tested could be that survival is greater in treatments with the formulation than in the control. In this case, we have established before the start of the experiment what we expect our alternative(s) to the null hypothesis to be. Thus, fewer than all possible comparisons will be performed and this reduces the likelihood of making a type I error.

The specific number of planned comparisons depends on the hypotheses about the data, but the number should be limited. It is not proper to consider the decision to compare all of the means with every other mean as a planned comparison. The number of comparisons should not exceed the total number of treatments or means – 1 (Sokal and Rohlf, 1995). *A priori* comparisons have many advantages and are generally more powerful than unplanned comparisons (Winer, 1971; Sokal and Rohlf, 1995). Researchers should attempt to ask focused questions that can be analyzed using planned comparisons when possible to take advantage of the benefits of this type of analysis.

Planned comparisons use the per-comparison error rate and can be pairwise comparisons, where two means or contrasts (*i.e.*, combinations of means) are compared. As long as each comparison addresses a separate biological question, a per-comparison error rate is appropriate and the statistical approaches listed for the comparison of two treatments can be used. The least significant difference (LSD) test can be used for a limited number of *a priori* comparisons, but should not be used for *a posteriori* comparisons because the experiment-wise type I error rate is uncontrolled (Day and Quinn, 1989).

If the experiment does not have a clear, expected pattern to the results or the question being asked is more general or exploratory, *a posteriori* or unplanned comparisons are needed for multiple comparisons. For example, in an experiment comparing a variety of species and or strains of pathogens to control a certain pest, a researcher may not have any specific predictions and is just interested in how the different pathogens rank. This area of statistics has been quite controversial and there is considerable disagreement over the best approaches and even if these tests should be performed at all. Typically, unplanned comparisons are analyzed by comparing all the possible pairs of means using a multiple comparison procedure. Unplanned comparisons should be run as a two stage or protected procedure: only when the null hypothesis is rejected using ANOVA are the multiple comparison procedures performed.

It is important to assess if multiple comparison procedures are necessary and that the assumptions of the analyses are met by the data. Often experimenters do not have a specific rationale for using a certain procedure, but the different mean comparison procedures have different levels of robustness to violations of assumptions and different probabilities of committing type I and type II errors. The assumption of equal variances is especially important. Multiple comparisons are also often performed in situations where they are unnecessary. For example, in situations where different doses or distances are being compared, the relationship between levels of treatment may be explained using regression analysis. Under these situations the trend may be more important than whether successive treatments are different.

The different types of analyses fall into two categories: repeated pairwise comparisons or multivariate approaches. Concerns about using many pairwise comparisons arise because of the potential problems of excessive type I errors. If repeated analyses at a pre-set probability of a type I error (α) are performed, the set of tests has a greater probability of error than each individual test (Ryan, 1959). This problem is even greater when there are multiple factors in an experiment or multiple experiments are compared because of the introduction of additional sources of error. The Bonferroni or Dunn-Sidak methods can be used to correct the significance level of multiple pairwise tests but tend to be very conservative (Rice, 1989; Sokal and Rohlf, 1995). The problem is to determine the appropriate error for a given experiment.

A number of *a posteriori* tests are commonly used for unplanned comparisons of means in invertebrate pathology field experiments. These tests all have different levels of power, control of the type I error rate, and underlying assumptions that need to be taken into account. Two tests that have been used traditionally as *a posteriori* tests, but which are no longer considered valid because of problems with the control of the type I error rate are the LSD test and Duncan's (1955) multiple range test (Scheffé, 1959; Day and Quinn, 1989). The Student-Newman-Keuls (SNK) test can also have problems controlling the type I error under certain situations and is not recommended by Day and Quinn 1989), but is recommended by Underwood (1997) for many situations because of its level of power. Several approaches that are generally considered to be valid are Tukey's honestly significant difference (HSD) method or Scheffe's test for equal sample sizes and the Tukey-Kramer method for unequal sample sizes (Sokal and Rohlf, 1995). Day and Quinn (1989) recommend what they call the Ryan's Q test (Einot and Gabriel, 1975) for parametric unplanned multiple comparisons because it is powerful, controls for the experiment error rate, and is easy to use. Day and Quinn (1989) also recommend the Joint-Rank Ryan test with treatments ordered by their joint-rank sums (Campbell *et al.*, 1985) as a nonparametric procedure for unplanned multiple comparisons. The Dunnett's test used in a stepwise manner (Miller, 1981) is a useful technique when the researcher is just interested in comparing experimental manipulations to a control. See Day and Quinn (1989), Sokal and Rohlf (1995), and Underwood (1997) for more details on the specific tests mentioned above, among other procedures, and under what conditions different tests are appropriate.

C *Analysis of trends*

1 *Correlation and regression*

Often a researcher is interested in the relationship among variables, not just if their effects differ significantly from one another. Regression analysis can be used to address questions concerning the functional relationship between one variable and another [*i.e.*, the dependence of a variable (y) on an independent variable (x)]. The purpose of this type of analysis is to evaluate the impact of a variable on a particular outcome. For example, regression analysis could be used to determine how increasing pathogen density impacts host mortality. Correlation analysis can be used for determining the amount of association between variables without the assumption of causation (*i.e.*, to determine the amount of interdependence between variables x and y). The purpose of this type of analysis is usually to measure the strength of relationships between two continuous variables, and is good for developing hypotheses rather than testing causation.

Correlation analysis could be used to investigate the relationships between disease prevalence in an insect population and characteristics of the insect's host plant such as age. Most general statistics books cover the techniques of correlation and linear regression in considerable detail (*e.g.*, Snedecor and Cochran, 1989; Sokal and Rohlf, 1995; Zar, 1999).

Correlation analysis is commonly used in mensurative experiments when relationships need to be determined between measured variables. Correlation analysis does not assume a causal relationship between the variables (*i.e.*, the relationship does not change, regardless of which variables are on the x or y axis). Correlation should be used when the causal relationship is not known or assumed or if the two variables are both effects of the same cause (*e.g.*, relationship between insect weight and length). Correlation analysis should also be used when neither of the variables is under the control of the investigator (*i.e.*, when the variables are random rather than fixed). The assumptions for correlation analysis are that: (1) the replicates are randomly selected from a larger population, (2) each replicate has both x and y values, (3) each replicate is independent, (4) the x and y values are measured independently, (5) x values are measured and not controlled, (6) x and y are each sampled from a normal distribution, and (7) covariation is linear. If the assumption that x and y are from a population with a normal distribution are not supported, there are nonparametric alternatives to calculating the correlation coefficient. The Spearman rank correlation is one of the most widely used nonparametric correlation analyses, and it has the same assumptions as above, except for the normality assumption (Sokal and Rohlf, 1995; Zar, 1999).

For manipulative experiments, the regression analysis is generally more appropriate because the researcher has set the values of x (*e.g.*, dose). Using regression analysis, the relationship between two variables is explained using a linear or more complex model that relates the value of one variable (dependent) as a function of the other (independent). Regression analysis can be used for a number of applications, but care needs to be taken to insure that the independent variable is measured with minimal error. Sokal and Rohlf (1995) discuss a number of uses for regression analysis; a couple of which are mentioned here. One of the most common is the study of causation, where the variation in y is caused by changes in another variable (x). Regression analysis can also be used to estimate functional relationships between variables. There are a number of statistical methods for analyzing regressions, including ANOVA. There are also many types of regression, including linear regression, multiple regression, logistic regression and nonlinear regression.

Linear regression is one of the most commonly used regression approaches, and it involves the calculation of the slope and intercept of the relationship. Usually, if the researcher thinks that there is a linear functional relationship between two variables, the analysis will address one of two hypotheses: (1) there is some slope (β) to the relationship that is different from zero ($\beta > 0$ or $\beta < 0$) or (2) there is a particular relationship ($\beta = z$). In the first case, the null hypothesis is that $\beta = 0$ and in the second case, the null hypothesis is that $\beta = z$. The null hypothesis tests for regressions are similar to those used for means.

Linear regression has the following assumptions: (1) x and y are asymmetrical, (2) the relationship between x and y is linear, (3) the variability of values around the line follows a normal distribution, (4) the standard deviation of the variability is the same regardless of the value of x (*i.e.*, homoscedasticity), (5) x is fixed and measured without error, or the imprecision in the measurement of x is small compared to that of y, and (6) each xy pair is randomly and independently sampled from the population. Some of these assumptions are under the control of the researcher and can be addressed using good experimental design. In field applications especially, it is unlikely that x is fixed because of the imprecision inherent in applying treatments, but in most field trials the imprecision in treatment application (under investigator control) is small relative to the imprecision in the measurement of the response to the treatment (y). The independence assumption is frequently violated, particularly when using repeated sampling of an experimental unit. Dealing with the assumptions of linear regression in ecological experiments is discussed in Underwood (1997).

2 Repeated measures

For some experiments, the researcher is interested in the impact of a pathogen application on an insect population over time and it may be desirable or necessary to make repeated measures from the same experimental units. However, repeated sampling of an experimental unit raises a number of statistical concerns, especially the lack of independence among sampling times. Repeated measures designs are quite common in laboratory experiments that measure survival of insects after exposure to a pathogen, and the use of survival analysis is often appropriate (Kalbfeisch and Prentice, 1980). Repeated measures can be analyzed using ANOVA (Snedecor and Cochran, 1989; Crowder and Hand, 1990; Winer *et al.*, 1991), and the experimental design is similar to that for a split-plot design, which is discussed in more detail in a later section. However, there are a number of problems with this approach. One potential problem is the assumption that there is no interaction between sampling times and experimental units within a treatment. If an interaction exists, then there is no overall interpretation of the interactions between sampling time and treatment using repeated measures ANOVA (Underwood, 1997). Another underlying assumption with split-plot and repeated measures ANOVA is equality of the variances of the differences for all pairs of levels of the repeated measures factor (Stevens, 1996). This requirement is called sphericity, and violation of this assumption is common in repeated measures analyses due to correlations among the data. Many software packages calculate corrections for violation of sphericity (*e.g.*, the Greenhouse-Geisser and Huynh-Feldt estimators). These corrections reduce the degrees of freedom for the F test to hold alpha near nominal (Girden, 1992).

There are a number of alternative approaches for analysis of time series data that may be more appropriate than ANOVA. The first is to avoid repeated sampling altogether and set up enough experimental units so that separate and independent samples can be taken at each time. This would be a two-factor design (*e.g.*, treatment and sampling time) and it is straightforward to analyze, but the number of experimental units needed may not be feasible. Second, if experimental units are repeatedly sampled, each sampling time can be analyzed separately. Thus, at each sampling time the data conform to a single factor analysis of variance. The probabilities of committing type I errors should be adjusted because of the series of comparisons that will be performed. If the number of sampling times is small, this adjustment could be made using the Bonferroni procedure (also called the Dunn test), where the acceptable alpha value (*e.g.*, 0.05) is divided by the number of sampling times. Power of this test is rapidly lost as the number of comparisons increases.

A third approach can be used if the researcher is just interested in differences in the temporal trend among treatments. In this case, each experimental unit is an independent measure of the temporal trend and can be analyzed using techniques such as linear regression or nonlinear curve fitting. Repeated measures data can also be analyzed using multivariate techniques such as MANOVA (multivariate analysis of variance). Using this approach each measurement (*e.g.*, sampling time) for an experimental unit is treated as a different dependent variable. MANOVA is also a useful approach if more than one variable is measured at each sampling time. A strong advantage of MANOVA is that it does not depend on the sphericity assumption; however, it does have limitations due to constraints on the number of levels of sampling (*i.e.*, within-subject sampling) and problems with low power when sample sizes are small. Stevens (1996) recommends application of MANOVA in addition to sphericity-corrected ANOVA (with α for each test set at 0.025) to substantiate test results. Issues regarding the use of multivariate vs. univariate techniques for repeated measures analysis are also discussed in von Ende (1993) and Underwood (1997).

The best approach for dealing with repeated measures data is probably that of Mixed Linear Models, which is a generalization of the mixed-model ANOVA (Crowder and Hand, 1990; Littell *et al.*, 1996). This statistically sophisticated approach is useful because it can more accurately handle the inclusion of random and fixed effects in the same model; typically, experimental units are considered to be random effects and sampling date as a fixed effect. This technique can also handle missing data and many levels of repeated factors that can cause problems for some of the other approaches. However, mixed linear models can be complicated to perform because different forms of the variance-covariance matrix can provide different results. Therefore, the goodness of fit measure for the different matrices needs to be determined and the best one selected for analysis.

D *Analysis of covariance*

Analysis of covariance (ANCOVA) encompasses a large number of statistical methodologies. The principle behind ANCOVA is to use the information about the relationship of the variable (y) to the covariate (x) to estimate the values of the variable in each treatment if all measurements had been performed on the same value of x. This allows a test of the null hypothesis of no differences among y, having removed any scatter due to x (Underwood, 1997). ANCOVA is particularly useful for situations where the initial allocation of experimental units to treatments is not equally representative. For example, if experimental units differ in some variable other than the treatment that may contribute to observed differences among treatments. If these variables can be determined in advance, they can potentially be controlled when designing the experiment, but these variables may not be known in advance or it may not be possible to control them. ANCOVA can be used to control for this variation after the experiment is performed. Another use for analysis of covariance is when we know in advance that the variable we are interested in is correlated with another variable. For example, growth rate measurements will often depend on the original size of the animal. If we estimate the relationship between the two, we can use this information to improve the precision of the estimate of the differences between treatments.

In ANCOVA, the regression relationship between the variable (y) and the covariate (x) is determined. It is also possible to use a series of covariates. This relationship is used to adjust the data to a chosen value of the covariate. This involves fitting a series of three regression models and then comparing the deviations from each of the models. In the first model, a regression is constructed for each treatment. In the second model a common regression is made for each treatment by constructing a regression model with the smallest squared deviations over all the treatments. In the third model, all the data are combined and a total regression, including all of the treatments, is constructed. The deviations from this third model should be larger than in the second model. If the deviation is similar between models 2 and 3, there is no difference among treatments. If there is sufficient similarity of the slopes among the treatments so that a common regression can be fitted to the data, then the data can be adjusted. The adjustment is to move the mean value of the variable in every treatment from its value at the mean of the covariate in its treatment to its predicted value at the mean of the covariates in all treatments. Adjusted means can be compared as if there were no influence of the covariate.

ANCOVA cannot, or need not, be performed under certain circumstances. If there are no differences among the treatments in the mean values of the covariates, there is no reason to perform the analysis of covariance. A preliminary ANOVA on the covariates from the different treatments can be performed and ANCOVA only performed if there is a significant difference among treatments. If there is heterogeneity of slopes among the treatments, the means of the treatments cannot be compared using a uniform adjustment procedure because there is no common slope (Underwood, 1997). However, this variation in slopes may be of biological interest even if it complicates statistical analysis. Thus, if the relationship between pathogen dose and host population density differed among habitats, this would be an interesting result and require additional experiments to further study the phenomenon.

If the regressions are similar and there are significant differences among adjusted means of the treatments, multiple comparison procedures are needed to determine which alternative hypothesis is supported. The procedures, like those following ANOVA, may be *a priori* or *a posteriori*. A useful *a priori* test is the Dunn-Bonferroni test and a useful *a posteriori* test is the Bryant, Paulson, and Tukey test. These and other procedures are described by Huitema (1980).

There are many assumptions behind ANCOVA because it combines the assumptions of both regression and analysis of variance. In addition to the assumptions of ANOVA and regression, there are also some assumptions specific to ANCOVA. These include the homogeneity of regressions (*i.e.*, the slopes are parallel) and that the treatment and covariate are independent. These assumptions are discussed in more detail in Underwood (1997). The ANCOVA approach can still be used if there is more than one covariate and/or if non-linear regression is used to determine the relationship between variable and covariate. These more complex approaches are discussed in Huitema (1980).

4 Experimental treatments, material and units

The experimental units and the treatments that they receive are two fundamental components of an experimental design, and there are theoretical and practical considerations involved in making decisions about both components (Mead and Curnow, 1983). Selection of these components of the design has an impact on what experimental design will be used and how the data will be analyzed and interpreted. Determining the structure of the experimental units involves identifying the units and describing the patterns of variation among the units. This variation is taken into account in experimental design by replication, blocking, and randomization. Determining the structure of the treatments involves choosing the different treatments to be included in the experiment. Parameters such as the type of analysis to be performed, the factorial structure of the design, whether the treatments are qualitative or quantitative, and the nature of suitable control treatments need to be taken into account in choosing treatments.

The reason for doing a manipulative experiment is to examine the effects of two or more treatments. Treatment in its broadest sense includes applied treatments such as different concentrations of a control agent, and inherent treatments such as strains or varieties of crops, and controls. To compare two or more treatments, the effect of the treatment on experimental units needs to be observed. These experimental units may be a field, section of a field, group of plants or insects, or even individual plants or insects. Because the responses of these experimental units to a treatment differ, we need two or more experimental units per treatment. Some large-scale studies cannot be truly replicated, but may be analyzed using methods discussed later.

A Selection of treatments

The selection of treatments is often not given the level of thought that it should receive. This is unfortunate because this step has many influences on the subsequent steps of experimental design and analysis. Keep in mind that the objectives of the experiment can help with treatment selection and enable the use of a more powerful statistical test. The more focused the objectives of the experiment, the fewer treatments needed, and the experiment may also be easier to design and analyze. However, many experimental situations exist where large numbers of treatments are needed (*e.g.*, screening large numbers of pathogen strains or large factorial experiments). The influence of the number of treatments on block design will be discussed later. In treatment selection, it is also important to keep in mind that not all treatments need to have equal status in the experiment. The treatments can be compared with different levels of precision and this can influence the number of replicates of each treatment and how the experiment is designed.

Manipulative experiments need one or more controls. The control is the baseline to which the other treatments are compared. The control may be either experimental units with no applied treatment, application of a sham treatment, or the use of standard control tactic technique to which other techniques are being compared. Sham treatments receive everything applied to the other treatments except the ingredient being tested. Common sham treatments include application of so-called spray carriers or formulation blanks. When using sham treatments it is useful, when possible, to also include controls that receive no applications to determine the impact of the other ingredients. The use of standard grower practices as a control is also recommended when testing material on a grower's field or orchard. This type of control has two uses; it provides an additional realistic control treatment and, if the grower complains that the experimental treatment has caused harm to his crop, it may provide evidence that the damage was due to factors other than the experimental treatments (Gomez and Gomez, 1983).

Because before and after (*i.e.*, paired) comparisons are statistically powerful, it is often desirable to take measurements of all experimental units (both control and experimental treatments) before and after applying experimental materials. This type of approach is particularly relevant when studying the impact of a pathogen on a host population, but the biology of the target invertebrate and the pathogen need to be considered so that before and after measurements can be meaningfully compared. This approach is appropriate when a fast acting agent is used and the same generation of hosts is measured at both sampling times, or the post-treatment population comprises the direct offspring of the treated population. This approach is not valid if there is significant insect movement into or out of the test plots, as may occur, *e.g.*, if a pathogen is slow-acting and considerable time elapses between samples. Generally, as the length of time between measurements increases, the two measurements become less related and this statistical approach less applicable. Obviously, these types of problems will influence the interpretation of the results of unpaired comparisons as well. Henderson and Tilton (1955) developed a modification of Abbott's formula as a method for calculating the percent control due to treatment based on before-and-after counts.

It is usually desirable to use treatments that are 'realistic.' For applied experiments, this means using (1) experimental materials, especially for controls, that are in common usage, (2) application techniques that are appropriate for a particular crop, or (3) application rates that are economically feasible. In more basic experiments, realistic treatments might mean using pathogen densities that are within the range of naturally occurring populations. However, attempts to use realistic treatments should not stand in the way of using treatments that are appropriate for addressing the hypothesis being tested or that will provide insight into the biology of the system.

B Selection of experimental material

Many of the parameters associated with selecting the experimental material are context specific and are made based on the experimenter's experience with the system and the pathogen. Three major considerations when selecting experimental material are:

1. Experimental material should be consistent with the population or system about which generalizations are to be made. Ideally the experimental material should be drawn from the same population about which generalizations based on the results of the experiment are to be made. For example, if making recommendations about commercial products, the products and application rates that a grower would actually use should be selected as the experimental materials.
2. The experimental material should be homogeneous across all experimental units. The key points for maintaining homogeneity are: (1) making sure that the sample of material applied to all experimental units in a treatment is drawn from the same population; (2) if the material comes from several production batches or sources, then it should be mixed well before applying; (3) material should be applied in a consistent manner among the treatments; and (4) some assessment of the variation in the experimental material (*e.g.*, viability, infectivity, etc.) should be made before application in the field.
3. The experimental material should be applied uniformly across all experimental units. Application rates of microbial control agents are often reported as viable propagules, infectious units, or colony-forming units per unit area. These dosages are typically based on an applied amount of pathogen suspension, and the concentration of the active ingredient in the suspension is routinely determined using hemacytometer or Petroff-Hausser counting chambers. In making counts, a common error is to consider the numbers from the individual counting units (etched squares) in a single chamber sample as true replicates. These counts are, in fact, pseudoreplicates that do not take into account the potentially considerable error associated with extraction of a minute sample ($< 10\,\mu l$) from a suspension and loading

it into a counting chamber. Valid estimates of concentrations and standard deviations can only be obtained from counts of multiple samples, preferably using multiple counting chambers to account for chamber- or coverslip-related errors. Independently prepared and quantified suspensions should be applied to replicate plots, and if possible, application rates should be sampled in the field. For example, slides, coverslips, or Petri dishes containing selective media can be positioned in the field to confirm uniformity of spray depositions or characterize the level of variability potentially resulting from numerous factors, especially variable weather conditions, or spray equipment malfunctions.

C Selection of experimental unit

Experiments involve the taking of measurements from an experimental unit. The experimental unit is the smallest division to which experimental material is applied and usually receives different treatments and is replicated. Replicates are experimental units that have received the same treatment. In field experiments, an experimental unit can be a section of land that contains plants and insects, a plant or group of plants or an insect aggregation or nest. An experimental unit can also be an individual insect, but this is rarely used in field experiments. The experimental unit may not be the unit that is actually measured during the experiment. For example, subsamples may be taken from an experimental unit to calculate an estimate of the impact of a treatment on the experimental unit. Failure to accurately define what the experimental unit is can lead to problems of pseudoreplication (Hurlbert, 1984).

1 Size and shape of experimental unit

The size and shape of an experimental unit influences the precision of the experiment; variability between experimental units generally decreases with increase in plot size, but variability within units can increase. Above a certain size, the rate of decrease in variability among units is reduced and there is a diminishing return to increasing unit size. How the experimental unit is sampled will also influence how the increase in experimental unit size influences precision. Increasing the number of objects being sampled (*e.g.*, insects) per experimental unit increases precision. As the experimental unit increases in size it becomes increasingly more difficult to sample it accurately. Also, more than one type of measurement is often made from a sampling unit (*i.e.*, insect density, disease prevalence, plant damage, etc.), and the influence of experimental unit size on all of the measured variables needs to be considered.

The optimum size and shape of an experimental unit depends on a number of parameters, many of which are practical or biological in nature. Cultural practices and application equipment can impact the decision on what size and shape of experimental unit to use. For example, aerial or tractor applications of treatment material can limit how small an experimental unit can be, whereas hand applications can limit how large an experimental unit can be. The limitations on the amount of sampling effort, space, treatment material, and other parameters also influence the size of experimental units.

The nature of the experimental material (*i.e.*, the pathogen) and the target organism should both be considered when determining the size and shape of an experimental unit. The spatial distribution and movement patterns of arthropods should be considered because the efficiency of sampling techniques and the variance of the means depend on the density within an experimental unit. If the density of arthropods is too low per experimental unit, the data may not be normal in distribution and many sampled units may not have insects. If the density is too high, sub-sampling of the experimental unit may be necessary. The feasibility of accurately sub-sampling an experimental unit also limits the size of the unit. The experimental unit should be large enough to contain the normal movement patterns of the arthropods that are being measured; otherwise the affects of the treatments may be diluted and inaccurate conclusions may be drawn from the analysis. How the application of a treatment will impact insect movement should also be considered. If these factors cannot be addressed by the experimental design, then they should be acknowledged when interpreting the results of an experiment.

The influence of the shape of the experimental unit is likely to increase as size increases. For small-scale field experiments, shape of the plot may have little influence. For larger size experimental units, considerations discussed later for determining the shape of experimental blocks should be taken into account. Factors such as the distribution and movement patterns of the target organism or gradients of environmental factors such as soil type can have implications for the best shape experimental unit. Again, biological considerations should be taken into account, but practical constraints may limit the alternatives available in experimental unit shape.

Statistical considerations also play a role in determining experimental unit size. In some cases, such as incomplete block experimental designs, the number of treatments and replications is rigidly set. More commonly, constraints on space or resources will limit experimental unit size. In blocking experiments, the size of the blocks coupled with the number of treatments constrains the size of the experimental units. A number of methods have been proposed for determining an optimal size of an experimental unit (Federer, 1955), but because practical concerns and biology are often more important in this determination, these methods are not particularly useful.

2 *Borders*

Sampling units should be independent from each other but, with very mobile insects for example, this independence can be difficult or impossible to obtain. In invertebrate pathology, two concerns need to be addressed. First, the application of the treatment needs to be confined to each experimental unit. For instance, spray applications can drift between sampling units. Second, treatment effects need to be confined to the experimental unit. For example, insects, after application of a control agent, may move among sampling units and immigration and emigration from the experiment may occur. Attempts to control these effects include using an appropriate size and shape experimental unit and by introduction of a border area between sampling units that is not sampled. In field trials with small experimental units, physical barriers to prevent movement of arthropods among experimental units (*e.g.*, screen cages to confine insect movement, barren soil strips to inhibit walking insects) or to prevent movement of the experimental material among experimental units (*e.g.*, metal barriers to impede nematode movement in the soil) may be used. In field trials with larger experimental units or where barriers are impractical or undesirable, then border areas are often used. Whether border areas are treated or left untreated can influence the sampling unit. Edge effects in field trials are often underappreciated and it may be desirable to analyze data with and without samples taken near the edges. In most cases, border areas should be planted and treated the same way as the experimental units because of the microclimatic effects that they can have on plants and animals in the targeted areas.

3 *Subsampling experimental units*

Frequently in field experiments it is difficult to measure all of the possible results of a particular treatment within an experimental unit. As the size of the experimental unit increases, the ability to sample all of the subjects becomes more difficult. The method of sampling an experimental unit should yield a sample value that is as close as possible to the value that would be obtained if the whole sampling unit was measured. The difference between these two measurements is the sampling error. The smaller the sampling error is, the better the sampling technique. To develop a plot sampling technique, the following parameters need to be specified: sampling unit, number of samples, and sampling design (Gomez and Gomez, 1983).

a *Number and size of subsamples*

The unit on which the actual measurement is made is the sampling unit. A sampling unit could be a leaf, a whole plant, a volume of soil, or a certain area of the experimental unit. To facilitate sampling, the sampling units should be easy to identify. The size of the unit should be appropriate for the type of measurement being taken. For example, if performing a total count of all insects present, the size of the sampling unit will depend on the distribution of the insects in the experimental units and the constraints on the number of insects that can be counted in a reasonable period of time. The precision of the estimate and the cost of generating that measurement need to be balanced.

The number of samples is determined by the amount of variability among sampling units within the same experimental unit (sampling variance) and the degree of precision wanted by the researcher. The number of samples required from an experimental unit can be estimated for a certain degree of precision expressed as the margin of error of the plot mean or treatment mean. Gomez and Gomez (1983) present the details of these techniques. Many of the techniques discussed below for estimating the number of replicates can be used for estimating the number of samples. Krebs (1989) also lists a number of approaches and a stepwise empirical approach for estimating the number of samples when repeatedly measuring from a site.

b Sampling design

Sampling design determines how the sampling units are distributed within an experimental unit. Taking a representative sample is not necessarily the same as taking a random sample. Random samples do avoid biases introduced by the experimenter selecting the sampling units directly. However, when additional information is available, other sampling techniques may be more representative of the experimental unit. Four sampling designs are presented here and further information on these and other techniques can be found in Gomez and Gomez (1983), Krebs (1989), and Underwood (1997). These same techniques also apply to developing sampling schemes in large scale unreplicated or mensurative experiments.

1. *Random sampling*. This is the simplest and most widely used sampling design. In this design, there is only one type of sampling unit and all of the sampling units are known. Sampling units to be measured can be selected using probability sampling (Krebs, 1989). In many situations, *e.g.*, in field or forest sampling, it is rarely possible to enumerate all sample units and select a truly random sample. A common approach in these cases is to devise a sampling scheme that directs the sampler to a random location where the sample is then taken. For example, the sampler might be instructed to walk a randomly determined number of meters or paces down a randomly selected crop row or transect line to locate a specific plant to be sampled. If the sample unit does not comprise the whole plant, the sampler might be given additional instructions for random selection (or at least unbiased selection) of a specific sample unit from the plant. Random samples are on average representative, but depending on the sample size may not be representative of a particular plot.
2. *Stratified random sampling*. This method of sampling is similar to blocking experimental units and is a powerful tool in sampling design. In stratified sampling, the experimental unit is divided into non-overlapping subpopulations and each of the subpopulations is sampled using a random sampling technique. The rationale behind dividing the experimental unit into strata is that some pattern of variation within the unit, typically population density, exists. If done properly, stratification can increase the precision of the estimate. The number of samples per strata can be proportional (equal percentage of samples from each strata) or optimal (number of samples per strata varies and depends on prior information) (Krebs, 1989). In optimal sampling, more samples can be taken if a stratum is larger, more variable, or cheaper to sample than another stratum [see Krebs (1989) for details].
3. *Multistage sampling*. In this process sampling units are selected randomly and elements within these units are selected randomly and measurements are taken from these elements. For example, several trees may be selected randomly from an experimental unit, a number of leaves are selected from random locations on that tree, and insects are counted on each of the leaves. Additional subsampling can also be performed to make even more subdivisions.
4. *Systematic sampling*. This is a useful technique because it is simple and can sample evenly across an experimental unit (Krebs, 1989). A common technique of systematic sampling is the centric systematic area-sample. In this technique, the experimental unit is divided into N sampling units and a sample is taken from the center of each sampling unit. As the individual sampling units are not identified for random selection, special care is required to avoid bias in the final selection of a sample. If, for example, insects or insect damage is prominently visible to the sampler, a method must be devised to insure blind (unbiased) selection of the sample (unless the protocol specifically

calls for sampling of infested or damaged sample units). Regardless of the sampling method used, all treatments within an experimental block should be sampled by a single individual to account for sampler bias. Randomization is generally preferable to systematic sampling when feasible.

c *Pseudoreplication*

When setting up an experiment with subsampling or analyzing the results, it is important to differentiate between true replicates (experimental units) and subsamples. As related earlier in the discussion of chi-square testing, pseudoreplication involves subsampling from a single experimental unit. These subsamples are not true replicates and are not statistically independent (Hurlbert, 1984). Subsamples taken from a single experimental unit thus do not increase the degrees of freedom available for statistical tests. The simplest approach is to use only the mean of the subsamples from an experimental unit in statistical tests. An alternative approach is to use an analysis that can take into account the levels of independence such as nested analysis of variance, but this does not increase the power to find differences among treatments. Pseudoreplication is of special concern when using large-scale mensurative experiments. For example, spraying a section of forest with a pathogen and then taking samples of disease prevalence at several locations is not true replication. In order for a treatment to be truly replicated, each replicate plot must be independently treated with an independently prepared pathogen preparation (the applied treatment must be replicated). Strict replication may not be feasible in every field test situation, depending on such factors as the type and size of spray equipment used, the numbers of treatments called for within a given time frame, and the amount of pathogen preparation available. In such cases, true replication may be achieved by repeating the experiment over time. Considering the degree to which pathogen efficacy is affected by weather conditions, repetition over multiple field seasons is highly recommended regardless of pseudoreplication problems. Whenever tests involve subsampling, this should be clearly reported in the methods of the research report. The classic paper by Hurlbert (1984) should be consulted for excellent coverage of these issues.

5 Select an experimental design

Experimental design is simply the logical structure of the experiment (Fisher, 1951), and a large number of different designs have been developed over the years to aid in the process of assigning treatments to experimental units in manipulative experiments. In single factor experiments, one factor is varied and all others are held constant (*e.g.*, comparing the efficacy of several species of pathogen with that of a chemical pesticide standard). In more complex experimental designs, more than one factor is varied (*e.g.*, comparing how irrigation rate influences the efficacy of several species of pathogen). The factors being compared can be either fixed or random. Fixed factors are generally selected by the experimenter or all levels of a factor are included in the experiment. Random factors in the experiment are randomly selected from the set of all possible factors. A number of statistical approaches can be used to analyze simple experimental designs such as the completely randomized design, but ANOVA is the most useful analysis for more complicated designs and will be the approach that is emphasized in this section. More detailed information on the setup and analysis of experimental designs can be found in Federer (1955), Cochran and Cox (1957), Little and Hills (1978), Gomez and Gomez (1983), Mead and Curnow (1983), and Pearce (1983).

When designing an experiment, it is important to consider the nature of the site where the experiment is to be performed. Field sites often have permanent features that may influence the response of the experimental units to a treatment. These features include soil depth, moisture, air movement, border areas, insect density, etc. Some features of the site can be identified and their effects controlled in the experimental design, but others are less apparent and/or difficult to control. Often more than one feature will vary at a field site and the relative importance of a feature will differ. This makes it difficult to develop an experimental design that controls for all features of the site. It takes experience with the system to make decisions about how to handle these sources of variation. Reviewing the results of previous experiments can provide insights into the relative importance of these various features on the system of interest. Understanding the variance present in the field site will influence what experimental design is appropriate and how experimental units and treatments are allocated.

A Single factor experimental designs

1 Completely randomized design

The completely randomized design is the simplest type of experimental design. Treatments are randomly assigned to each experimental unit so each unit has the same chance of receiving a treatment. This design is appropriate where the experimental area is virtually uniform or where variation is suspected but the variation has no pattern or its pattern is unknown to the experimenter. The completely randomized design is most useful for small-scale field trials in relatively uniform environments. The advantages of this design are that (1) it is simple to perform, (2) the placement of the experimental units is flexible, and (3) the degrees of freedom for estimating experimental error is maximized and the F value in an ANOVA required for determining statistical significance is minimized. A principal disadvantage of this design is that the process of randomization can assign most or all of the replicates of a particular treatment into experimental units that are different in some way from others in the experiment. Thus, randomization can introduce a systematic bias, especially when the number of replicates is low. For this design, differences among experimental units are included in the experimental error. Using more complex experimental designs, the sources of variation among the experimental units can be identified and incorporated into the analysis and smaller statistically significant differences among treatments can be detected.

The number of experimental units will be the number of treatments multiplied by the number of replicates. Before assigning treatments, each experimental unit is given a number. Treatments are assigned to the experimental units by selecting random numbers from a random numbers table or using some other source of random numbers. There are two sources of variation among the experimental units in a completely randomized design; treatment variation and experimental error. The relative size of the two is used to determine if differences among treatments are due to chance (*i.e.*, the difference is significant if the treatment variation is sufficiently larger than the experimental error).

2 Randomized complete block design

In the randomized complete block design, the treatments are assigned randomly to a group of experimental units that are termed a block. The objective of this approach is to minimize the probability of placing most or all replicates in locations that are unique in some way other than treatment by minimizing the variability among experimental units within a block and maximizing the variability among blocks. This is one of the most widely used experimental designs in agricultural research. The advantage of this approach is that experimental error is reduced because the block variability is removed from the experimental error term in an ANOVA. This design becomes more efficient, relative to the completely randomized design, at detecting differences among treatments as the variability among blocks increases. If there are no differences among blocks, this design will not contribute to the precision in detection of treatment differences. Therefore, it is most effective when there is a predictable pattern of variability.

Two important parameters in blocking are the selection of a source or sources of variability to use for blocking and selection of block shape and orientation. To ensure that a block is as uniform as possible, the experimental units should be grouped on the basis of some parameter that will provide uniformity. Blocks should be kept compact, because as size increases so does the probability of within-block variability. The fewer the number of treatments the more compact the blocks can be and still have at least one replicate of each treatment in each block. Generally, blocks that are long and narrow in shape perpendicular to the direction of the gradient and that are close together are best when variability is in one direction. Blocks that are square are better when variability is not predictable. Keeping blocks the same size will also reduce within-block variability. The treatments are assigned randomly to each experimental unit in a block and a separate randomization is performed for each block. For the analysis of variance, there are three sources of variability in this design; treatment, block, and experimental error.

3 Latin square design

The Latin square design is not used very often in insect pathology field research but is applicable when there are two sources of variability that vary in different directions. Treatments are randomized into columns as well as in rows. Experimental error from both of the sources of variation can therefore be removed. A limitation of this design is the requirement that the number of replicates equal the number of treatments, thus space rapidly becomes a limiting factor. The number of treatments that can be accommodated will depend on a variety of factors such as the space available and the experimental unit size, but experiments with more than eight to twelve treatments are generally not practical. For ANOVA there are four sources of variation; row, column, treatment, and experimental error.

4 Incomplete block designs

When a large number of treatments are needed in an experiment, an incomplete block design may be appropriate. As the number of treatments increase, it becomes difficult to use a complete block design. This is because each block must contain at least one replicate of each treatment and to accommodate additional treatments, the block size has to increase. As block size increases the homogeneity of the block decreases and the experimental error increases. Incomplete block designs do not have all of the treatments represented in each block and this enables the size of the blocks to remain small even with a large number of treatments. These types of designs do have some significant disadvantages: (1) the number of treatments and replicates is relatively inflexible, (2) there is unequal precision in the comparison of treatment means, and (3) the data analysis is more complex. However, the availability of sophisticated computer software capable of handling these analyses (general linear models) has made the latter issue less important. There are many incomplete block designs, with the balanced lattice design being the most commonly used, and more information on these designs can be found in Cochran and Cox (1957).

B Multiple factor designs

Factorial experimental designs are used for situations where two or more variable factors and their interactions need to be studied. The results of a single factor experiment are technically only applicable to the conditions present during the experiment. However, the way that an organism responds to an experimental treatment is often influenced by other factors. It is often desirable to determine how the response to a variable is influenced by one or more other factors. For example, it may be of interest to determine how the efficacy of an entomopathogenic nematode is influenced by the level of fertilizer applied to the soil. The two factors (frequently termed primary and secondary factor) interact if the effect of one factor changes with the level of another factor. When there is no interaction among factors, the simple or direct effect of the factor is the same as the average effect of the primary factor across the levels of the secondary factor (*i.e.*, main effect) and generalizations can be made. If there is an interaction, no generalizations can be made across levels of the secondary factor. Although interactions can cause statistical problems they are often the most biologically interesting results of an experiment.

A complete factorial experiment is one in which all combinations of the selected levels in two or more factors are present. For example, if three species of entomopathogenic nematode and three levels of fertilizer were used in a factorial experiment, it would be a 3×3 factorial experiment with nine different treatments representing each possible combination of factors. The number of treatments increases rapidly with the addition of factors or levels of factors. Factorial experiments can become quite large and costly to perform, and this can limit their applicability. The running of a single factor experiment first and then asking selected questions with a factorial experiment can be a preferred approach.

The term factorial refers to the method of determining the treatments, not to a specific experimental design. The completely randomized and randomized complete block designs can be used for factorial experiments. The analysis of variance does differ for factorial experiments, because the treatment sum of squares are partitioned into the main effects of the individual factors and their interactions. Because of the large number of treatments that can result from a factorial experiment, complete block designs may be too large to be used efficiently. Incomplete block designs like the balanced lattice design are not appropriate for factorial experiments, but there are comparable designs that can be used with factorial experiments. The most commonly used experimental design for factorial experiments is the split-plot design.

1 Split-plot design

The basic split-plot design involves assigning the treatments of one factor to main plots (main plot factor) and the treatments of a second factor (subplot factor) to subplots within each main plot. This design sacrifices precision in estimating the average effects of the treatments assigned to the main plots, but it often improves the precision of comparing the average effects of treatments in the subplots. When interactions exist, the precision is increased for comparisons of subplot treatments for a given main plot treatment. The experimental error for main plots is usually larger than the experimental error used to compare subplot treatments. The error term for subplot treatments is smaller than would be obtained if all treatment combinations were arranged in a randomized complete block design. Because of this variation in precision, the selection of which factors to assign to the main plot or subplot is very important. Several parameters are important in making this decision. Two parameters are statistical and deal with the degree of precision desired for either factor and the expected size of the main effects of each factor. The factor of greater interest or with the smaller expected main effect size should be the subplot factor. Management practices also influence the decision of which factors should be main plot or subplot factors. Using our above example, fertilizer levels would be a likely main plot factor and nematode species a subplot factor.

There are two steps to the randomization of a split plot design. Main plot factors can be arranged using a completely random or randomized complete block design. The randomization procedure of the main plot depends on the design that was selected. Subplot factors are randomized within each main plot, with separate randomizations carried out within each main plot. The analysis of variance of a split-plot design is divided into two analyses; the main-plot and the subplot analysis.

2 Designs with more than two factors

The number of factors included in a factorial experiment can be increased beyond two factors, but there is a rapid increase in the number of treatments and in the number and types of interactions that can occur. Sometimes these interactions are of interest, but often the complexity and cost of a large factorial experiment are prohibitive. There are a number of experimental designs that can be used for three-or-more factor experiments; including randomized complete block design, split-split-plot design, and fractional factorial design. Generally, it is preferable that simpler experimental designs than multi-factorial designs be used. Information on these more complex designs can be obtained from the references cited in the beginning of this section.

6 Improving the precision of an experiment

The level of precision in sampling refers to the size of the confidence interval that with a high probability contains the true mean. The smaller the confidence interval, the greater the precision and the smaller the detectable difference between treatments. The level of precision is influenced by three things: the sample size, variation in the population, and the probability used to construct the confidence interval (Underwood, 1997). Methods that increase precision are designed to lower unaccountable variability per plot and to increase the effective number of replicates.

Precision can be improved by increased replication, careful selection of treatments, refinement of technique, selection of experimental materials, selection of experimental units, taking additional measurements, and planned grouping of experimental units. The goal in experimental design is to maximize precision within the constraints of cost (*e.g.*, labor and time availability, space and material limitations).

A Replication

Before starting an experiment, the number of experimental units receiving the same treatment (replication or sample size) needs to be determined. How many replicates to perform is perhaps one of the most frequently asked questions concerning experimental design and, unfortunately, it is a difficult one to answer. Determining the number of replicates involves balancing a number of factors, only some of which are statistical. Careful consideration is given to replicate number because it influences the power and precision of the experiment. Enough replication is needed to obtain valid results, but not so much as to make the experiment impractical. The amount of replication is often determined by estimation by the experimenter, but with experience these estimates can be quite accurate. In addition, the amount of replication is constrained by space availability and economics. Some experimental designs, as already discussed, can influence decisions on the number of replicates.

The precision of an experiment can be increased by adding additional replication, but the degree of improvement falls off rapidly as the number of replicates increases. In general, field research with four to eight replicates will provide reasonable precision. However, estimating the amount of replication needed based on the difference in magnitude the experimenter is interested in detecting or the desired power of the experiment is a preferable approach when feasible. There are a number of statistical approaches to determine sample size, but they all require some estimation of the expected variation in the data based on previous data or pilot studies. The experimenter also needs to have some idea of the size of the difference between treatments that must be detectable. The calculated replicate number is still only an estimate because it is based on estimates and arbitrarily set values and should not necessarily be followed blindly. If sufficient resources are lacking to perform the experiment at the level of precision needed, it is best to determine this before conducting the experiment.

1 Estimating the number of replicates needed

There are different ways of estimating the number of replicates needed for an experiment. Simple formulae are available for calculating the appropriate sample size for comparing two means or two proportions (Motulsky, 1995; Zar, 1999). To estimate the optimal number of replicates prior to performing an ANOVA, we can use the following equation and the process of iteration (Zar, 1999).

$$\phi = \sqrt{\frac{n\delta^2}{2ks^2}} \tag{1}$$

In equation (1) the variables are: ϕ, a value related to the noncentrality parameter; k, the number of treatments used in the ANOVA; s^2, the error MS from an ANOVA of a similar experiment; δ, the minimum detectable difference that the experimenter wants to be able to detect; n, the number of replicates. This approach requires the results of an ANOVA performed on a similar system. The process of determining the number of replicates involves making an initial guess and then repeatedly refining that estimate. To do this, select a power for the experiment, *e.g.*, 80%, and an initial estimate of the number of replicates and solve for ϕ. The parameter ϕ is then plotted on the graphs prepared by Pearson and Hartley (1951; also included in Zar, 1999) to determine the power of the test. Each graph is for a different v_1 (group df) and the position where ϕ (on the x-axis) intersects a curve for a given v_2 (error df) provides a value for power (y axis) (Zar, 1999). If the resulting power is above or below 80%, the number of replicates can be adjusted appropriately and a new value of ϕ calculated. This process is repeated until the required power is achieved and this value of n is used for the number of replicates.

2 Different amounts of replication among treatments

In most experiments, the design uses an equal number of replicates for all treatments. In some situations, not all comparisons of treatments are of equal interest, and this can influence the number of replicates that are needed for each treatment. For example, frequently there are two comparisons of interest: one among experimental treatments and one between each of these treatments and the control treatment. In this situation it may be desirable to have two replicates of the control and one for each of the other treatments in a block. This will result in different power levels for each type of comparison. This consideration is also of importance when there is not sufficient experimental material or experimental units for equal replication.

B Statistical power

Prior to running a field experiment, the power of the proposed experiment should be determined. The results of this determination can help in deciding if an experimental design needs to be modified or even if the experiment should be run at all. If the power is not adequate, a number of parameters associated with an experiment can be modified to increase the power of the test (*e.g.*, increase sample size, increase difference among population means, decrease the number of groups, decrease the variability within populations, or use a larger value of α).

A number of procedures have been described for estimating the power, required sample size, or detectable difference among means for ANOVA, and most biostatistics books present methods for determining the power of an experimental test. The technique described above for estimating replication can also be modified to estimate power. Here techniques described in Zar (1999) that use the graphs of Pearson and Hartley (1951) are described. The term ϕ can be calculated for an ANOVA in several ways. If the mean squares of an ANOVA from a similar experiment are available, then the following formula can be used.

$$\phi = \sqrt{\frac{(k-1)(groupsMS)}{ks^2}} \qquad (2)$$

The variables in formula (2) are: ϕ, a value related to the noncentrality parameter; k, the number of treatments used in the ANOVA; groups MS, the mean squares from an ANOVA for a similar experiment; and s^2, the error MS from an ANOVA for a similar experiment. The estimated power can then be determined by converting ϕ to a power measurement using the figure listed above.

If the variability among populations is expressed in terms of deviations of the k population means, μ_i, from the overall mean of all populations, μ, then the following calculation can be used.

$$\phi = \frac{n\sum_{i=1}^{k}(\mu_i - \mu)^2}{ks^2} \qquad (3)$$

A third approach to determining the power of an experiment is to specify the smallest difference between two populations that we want to detect. This term is the minimum detectable difference (δ). We can then calculate ϕ using the following formula that is similar to equation (1).

$$\phi = \sqrt{\frac{n\delta^2}{4ks^2}} \qquad (4)$$

The variables in formula (4) and the method of calculating power are the same as in equation (1). This formula can also be used to estimate the power of comparing two means by substituting another error term for the error MS.

7 Issues associated with data analysis

A Correction for control mortality

In many experiments it is desirable to correct the mortality in the experimental treatments by the mortality that occurs in the control treatment. When the treatments consist of a graded series of treatments, the response can be corrected for control mortality using probit analysis (Finney, 1971). When there is a small number of treatments or treatments are not related in a series, correction for control mortality has traditionally involved the use of Abbott's formula (Abbott, 1925). However, Abbott's formula is not a complete correction for control mortality because it does not include an estimate of the variance in the control mortality (Rosenheim and Hoy, 1989). It is common practice to retain the estimate of variance for the uncorrected treatment mortality and use it for the corrected treatment mortality, but this can lead to misleading conclusions. Rosenheim and Hoy (1989) presented a technique to correct for control mortality that provides an approximate confidence interval for the corrected treatment mortality. An alternative, but more complex, approach that provides a better approximation was pointed out by Koopman (1994). This approach derives the point and confidence intervals directly from likelihood theory for binomial data (Finney, 1971; Gart and Nam, 1988). Another approach is to use a resampling technique such as bootstrapping to calculate the variance estimate of the index (Manly, 1997). Any of these approaches to variance estimation should be used rather than the original Abbott's formula alone when correcting for control mortality.

B Calculating the power of the ANOVA

It is useful to determine the power of an ANOVA, or any other statistical procedure, after it has been performed. This is especially useful if the null hypothesis is not rejected because we would like to know how likely the test was to detect a true difference among the population means. This can be done in a manner similar to that described above for estimating power before running an ANOVA. The following equation, which is similar to equation (2) can be used.

$$\phi = \sqrt{\frac{(k-1)(groupsMS - s^2)}{ks^2}} \tag{5}$$

C Data transformation

Data transformation is a common method used to overcome violations of the assumptions of a statistical test. During data transformation, the original data are converted to a new scale that is expected to be more consistent with the assumptions of the analysis. Because all data are treated the same, comparisons among treatments remain valid. However, data transformations are performed only so that the analysis will be valid, not so that more desirable results are obtained. Moreover, it is important to recognize that application of a transformation to data that already satisfy the assumptions of an analysis will produce data that violate the assumptions and may lead to erroneous conclusions.

After a valid transformation has been selected, all analyses are conducted in the transformed scale. For presentation purposes, results are most easily comprehended if converted back to the original scale; however, the back-transformation is not straightforward in all respects. Means can be directly back-transformed to yield correctly weighted means in the original scale; however, this is not the case with standard errors, as they are symmetrical (expressible as the familiar ± values) only in the transformed scale. Instead, confidence limits must be calculated in the transformed scale and then converted back to the original scale, a process that generates asymmetrical confidence intervals. Any consideration of means from the ANOVA as being "correctly" weighted, of course depends on the reliability of all the usual ANOVA assumptions, and in reporting results, it is desirable, space permitting, to present both the weighted and unweighted means. The standard errors of the unweighted means (the original data) are especially useful as indicators of potential problems (heterogeneity of variances, outliers, etc).

Some common data transformations to deal with ANOVA assumption violations are listed below. More detail on data transformations is available in many general statistics books (*e.g.*, Gomez and Gomez, 1983; Pearce, 1983; Sokal and Rohlf, 1995; Zar, 1999). After the data are transformed, the assumptions of the analysis should be tested again to make sure that the adjustment is adequate. Alternative statistical approaches that have assumptions that are not violated should also be considered as an alternative to transformation.

1 The log transformation

The log transformation is useful for data where the means and variances are not independent and/or the effects are multiplicative not additive.

This generally occurs with data that are ratios of two variables or are whole numbers that vary over a wide range of values (*e.g.*, the number of insects per sampling unit). Data are transformed by taking the logarithm of each data point. Logarithms of any base can be used. Data with negative values or zeros cannot be transformed this way. If there are zeros or values less than 10 in the data set, a positive value, such as 1.0 or 0.1, should be added to all of the data before transforming. If most of the data are large, adding a large constant such as 1 makes little difference. However, if some samples are small (< 0.1) with some zeros and others are large (> 1.0), adding 1.0 can make a big difference in their relative magnitudes. In this situation, a small number such as 0.1 should be added. More on the influence of adding a constant to log transformations can be found in McArdle and Gaston (1992).

2 The square-root transformation

The square root transformation is useful when the data are not normally distributed but follow a Poisson distribution. In this case, the relationship between mean and variance (mean = variance) violates the homogeneity of variance assumption. This type of distribution occurs frequently when the data are based on counts and there is a low probability of an event occurring for any individual sample. This type of distribution is typical of insect counts per plant, quadrant, or net sweep. Causes other than Poisson distributions can also generate data where the variance is proportional to the mean. This transformation can also be used for percentage data if the values are between 0 and 30% or between 70 and 100%. Plotting the variances against the means can be useful for determining if a transformation is necessary. Adding 0.5 or 1.0 to the data before taking the square root can reduce the heterogeneity, especially when the data are close to zero (*i.e.*, < 10).

3 The arcsine transformation

The arcsine transformation is useful for data based on percentages or proportions derived from counts (*e.g.*, the proportion of insects that are infected). This type of data is binomial in distribution rather than normal. In this type of distribution, variances tend to be small at values close to 0% or 100% and large near 50%, and this can lead to heterogeneity of the variances. To make the distribution more normal, the arcsine of the square root of the data is taken. In this transformation, the proportion is used, not the percentage. If the majority of the percentage data lie between 30 and 70%, the data probably do not need to be transformed. The arcsine transformation is not as effective for data near the extreme values of 0% or 100%. If the actual proportions are known, the transformation is improved by replacing 0/n with 1/4n and n/n with 1–1/4n (Bartlett, 1947). Even greater improvements are realized with a number of other transformations (see Sokal and Rohlf, 1995; Zar, 1999).

4 The Box-Cox transformation

A common approach for transforming to normality involves raising the data to some power. Box and Cox (1964) developed a procedure to identify the best transformation from a defined family of power transformations. This is accomplished through an iterative process, and for practical application, requires a computer. The Box-Cox algorithm is included in many statistical software packages. If not available, Sokal and Rohlf (1995) recommend trying the series $1/\sqrt{Y}$, $\sqrt{Y}$, ln Y, and 1/Y for distributions skewed to the right and Y^2, $Y^3, \ldots$ for distributions skewed to the left. In regression, the reciprocal transformation, 1/Y or $1/(Y+1)$ allowing for zero values, is often effective for linearization of the hyperbolic curves that characterize many rate phenomena such as eggs produced/female/day (Sokal and Rohlf, 1995). It is possible to apply the Box-Cox transformation to either the independent or dependent variable or to both variables simultaneously in regression analysis.

5 No transformation is possible

Sometimes the violations of ANOVA assumptions, such as heterogeneous variances, cannot be corrected with data transformation. However, if the data are balanced and there are a large number of treatments and replicates ($n > 5$), then ANOVA is relatively robust to departures from the assumptions, and in some cases it may be worthwhile to proceed with the ANOVA. If violation of an assumption is known to produce a liberal test (*e.g.*, analysis of data with unequal variances), it is reasonable to accept an ANOVA finding of no significant difference. Similarly, if a violation creates a conservative test (*e.g.*, analysis of Cauchy-distributed data), it is reasonable to accept a result of significant difference. In each case; however, the alternative result must be viewed with considerable caution. Uses of nonparametric tests as alternatives to ANOVA in cases where the underlying assumptions cannot be met were discussed previously.

D Missing data, outliers, and other 'mistakes'

Mistakes happen in field experiments and unlike in many laboratory experiments, it may not be possible to repeat the experiment. Every effort should be made to avoid losing data, but clearly, some things are out of an experimenter's control. Missing data occur when an experimental unit is being excluded from analysis for reasons that have nothing to do with the experimental treatments. Missing data can result from a number of different situations, but care needs to be taken in deciding when data should not be used in an analysis. Discussed below are some of the common causes of missing data, along with ways to deal with them.

1. In many cases the decision on what experimental units should be excluded is clear-cut. For example, if a treatment was not properly applied (*e.g.*, no application or application of an incorrect concentration) to an experimental unit, that unit should be excluded. An exception would be if the treatment was applied improperly to all replicates and the researcher wanted to retain the modified treatment. Another example would be data or samples that are lost after they are collected in the field. These data can be excluded from analysis.
2. Careful thought should be given to the decision to exclude data when the cause is less clear cut because an incorrect choice could lead to inaccurate conclusions being drawn from the experiment. For example, if all or most of the crop plants are lost from an experimental unit, it needs to be determined if this loss is in any way treatment-related. If it is not treatment related, the replicate could be declared missing data, but if this cannot be determined clearly, the data should be included.
3. In some cases, experimental units are mixed (*e.g.*, the samples from two experimental units are not labeled and their origin cannot be determined). Common sense may be all that is needed to determine if an experimental unit can be assigned to a particular treatment with reasonable certainty or if it should be thrown out. Some mathematical tests for dealing with this problem are provided by Pearce (1983).
4. If so many replicates are lost that a treatment has only one replicate remaining, the treatment should be excluded. Similarly, a block should be excluded if all replicates of all treatments but one are lost.
5. After the data have been recorded, some data points may appear illogical (*e.g.*, have values that are beyond the normal range for a particular material). Performing an objective check for outliers in the data can be useful. However, illogical data should not be excluded just because it does not meet expectations. Only illogical data that result from some identifiable type of error should be manipulated or excluded from analysis.
6. If detected early enough, some errors in the data (*e.g.*, misread observations, improper use of equipment, transcription errors) can be corrected by repeating the measurement or adjusting the data. Check the data immediately after taking measurements so that errors may be detected and corrected before the opportunity is lost.
7. If there is a good reason to think that a data point is wrong, but it is not different from the rest of the data, it should be left in the analysis.
8. If there are objective reasons to think that an extreme data point is flawed, there are a number of approaches to assess if it is an outlier and should be excluded (Dunn and Clark, 1987; Hoaglin *et al.*, 1983). However, any data set is likely to have some outliers (*i.e.*, data points that are not typical of the other data collected), and it is therefore important to have some objective criteria for exclusion (Underwood, 1981). Often, examination of outliers can provide useful insights into some interesting aspects of the biology of the system. It is this variation that is often of greatest interest when trying to understand how a system works.

9. If an experimental unit or replicate is to be excluded from analysis, there are a variety of procedures to deal with the missing data [see Pearce (1983), Gomez and Gomez (1983), or Smith (1981) for more detailed information]. One approach is to analyze the data without the excluded replicates and use an analysis that is appropriate for unbalanced data (*e.g.*, General Linear Models procedure). This is by far the most common approach and is the least contentious. A second approach is to estimate missing data values to fill the gaps left by the missing data. A third approach is to replace the missing values with approximate values and use the analysis of covariance to make adjustments. The later two approaches are more useful for complex experimental designs that have partitioning of treatment effects.

8 Interpretation and presentation of results

A *Biological versus statistical significance*

Statistical analyses provide P values, but often a researcher is interested in determining if the obtained P value is statistically significant by using statistical hypothesis testing. This decision is based on comparing the P value calculated from the data to a predetermined value of alpha (typically 0.05) that is based on the consequences of type I and type II errors. Values of P less than alpha are considered statistically significant and the null hypothesis is rejected (a difference is declared). The determination of statistical significance has its advantages in situations where a dichotomous decision needs to be made. However, the disadvantage is that this conclusion can be misinterpreted (Jones and Matloff, 1986). If the results of an experiment are statistically significant, this means that one of three possibilities is true (Motulsky, 1995): (1) the null hypothesis is actually true and the P value indicates the probability of this fact; (2) the null hypothesis is false and the populations are different in some biologically meaningful way; and (3) the null hypothesis is false, but the populations are not different in any meaningful biological way. With large sample sizes, even very small differences between populations will be significantly different. The scientific importance of the conclusion depends on the size of the difference between populations and the significance of this difference in the biology of the organism or the requirements of the researcher. The fact that the results of a statistical analysis are significant does not necessarily mean that the results of the experiment are biologically or economically meaningful. On the other hand, with small or even moderate sample sizes, small differences may not be declared significant that are, in fact, important. Thus, the fact that the results of a statistical analysis are not significant does not necessarily mean that the results are not biologically or economically significant. A common example is statistical analysis of crop yields. Small differences in yield may not be detectable even in a well-designed experiment; yet the difference could be of considerable economic importance to a grower. Multiple repetitions of a test may be required to obtain sufficient statistical power to detect small but important differences.

In biological experiments, it is not always necessary to reach a sharp decision from each P value; instead the P value itself can be used as an indication of the degree of confidence in the result. However, there is disagreement over how to use P values. Many researchers and statisticians believe that how close the P value is to alpha does not matter. If it is below the threshold it is significant and if it is above, it is not significant. Others think that differences in P values do provide important information and that more confidence can be placed in the results of an experiment with $P = 0.001$ than in an experiment with $P = 0.049$. Sometimes results of statistical analyses are presented with a scale of significance levels: $P < 0.05$ being significant (*), $P < 0.01$ being highly significant (**), and $P < 0.001$ being extremely significant (***). Regardless of the approach that a researcher chooses, differences of questionable biological relevance and P values near the value of alpha may warrant further experimentation before reaching scientific conclusions. Calculating the power of the analysis can also help with the interpretation of nonsignificant results.

B Presentation of results

In publications, the details needed to repeat the experimental procedure are usually reported, but details needed to understand the experimental design and statistical analysis are often not as clearly stated. Not reporting procedures adequately is one of the most common statistical problems (Fowler, 1990). Some things that are important to state when reporting the results of an experiment are listed below.

1. Clearly state the objectives of the experiment to enable the reader to determine the relationship between the logical structure, the data collected, and the conclusions reached.
2. Report clearly what statistical methods were used and why they were chosen so that readers can draw their own conclusions.
3. The means, sample sizes, and standard errors should be the minimum amount of summary information reported.
4. State how experimental units were selected and sampled so that the reader can determine if pseudoreplication may have occurred.
5. If questionable data (outliers) are adjusted or eliminated from analysis, these manipulations should be described in the research report.
6. State the assumptions of the tests and how this influenced the selection of transformation procedures and selection of statistical procedures.
7. Cite the software package that performed the analysis if a computer was used or the reference for the procedure if done by hand.
8. Describe completely the experimental material and how it was handled (*e.g.*, commercial source, generations in laboratory culture, storage time and conditions).
9. Distinguish between statistical and biological or economical significance when presenting the results of an experiment. Performing power analysis before running an experiment can help with this problem.

9 Special cases

A Large scale unreplicated trials

Some important questions can only be addressed on large scales that are difficult or impossible to replicate and randomize. These types of experiments may be natural experiments (*e.g.*, perturbations of the system that result from unexpected events such as natural movement of a pathogen into a particular area) or experimenter manipulations (*e.g.*, classical biological control introductions, large-scale applications). Smaller scale experiments that can be replicated and randomized may be used, but these may not provide accurate predictions of large-scale phenomena. Some large-scale studies can be analyzed using techniques developed for time series data. However, even on large scales, replication is preferred if at all possible.

Most of these types of experiments have a before and after component to them; measurements are made of the system before and after an intervention. Time series data are obtained by repeated subsampling from the same experimental unit through time. Time series analysis techniques can be used to determine if an abnormal change followed a manipulation of the system compared to the normal variation in the experimental unit. Determining the amount of normal variation through time is more difficult than determining the experimental error in a replicated and randomized experiment. The different samples that are taken through time are not independent from each other and are not the same as replicates. Good design and analysis of unreplicated experiments are difficult and time consuming. Rasmussen *et al.* (1993) provide an excellent introduction to this type of analysis. More detailed information can be obtained from Box and Jenkins (1976) and other references cited in Rasmussen *et al.* (1993).

These designs are prone to having random events influence interpretation. Some ways around this are to have multiple interventions or to switch back and forth between treatment and control. The probability of the response being coincidental is reduced with each observed response. Using paired units, where one unit receives a treatment and the other does not, is also a useful approach (Stewart-Oaten *et al.*, 1986). In this design, the control unit provides support that the response in the perturbed unit is not due to random changes. Two important factors need to be considered in designing this type of experiment: (1) what are the experimental units and (2) how many and how often should the experimental units be sampled. Treatment and control experimental units should be similar to each other and they should represent the systems to which the results will be generalized. The samples should be evenly spaced through time with the time interval determined by the rate of change in the population being studied and the specific questions being asked. For example, insect populations with large fluctuations in density will need to be studied over longer periods of time than insects with more stable population densities. Generally, the greater the number of sampling times the stronger the statistical analysis will be (usually 50 or more samples are needed). The number of samples taken from an experimental unit at a particular point in time will depend on the characteristics of the experimental units and the question of interest.

There are many different statistical approaches for the analysis of time-ordered sequences of observations. Rasmussen *et al.* (1993) describes the autoregressive integrated moving average (ARIMA) models, which are a subset of the available statistical approaches. These models describe a wide range of processes and are appropriate for evenly spaced intervals. The analysis of time series data from unreplicated experiments can be more subjective than other more traditional statistical approaches and caution must be used in analysis and interpretation. Consultation with a statistician is strongly recommended.

B Meta-analysis

The ability to generalize and summarize is an essential part of statistical analysis and in science in general. If a number of independent experiments have been performed, it may be desirable to have a statistical synthesis of this research: this type of analysis is termed meta-analysis. This approach is somewhat controversial, but has been used in fields such as medicine and the social sciences and is becoming more frequently used in ecology (Gurevitch and Hedges, 1993). Progress in science depends on the ability to reach general conclusions from a body of research. Meta-analysis provides a way to reach general conclusions based on quantitative techniques: *e.g.*, how large is the effect, how frequently does it occur, what is the difference in the magnitude of the effect among different studies.

Counting up the number of statistically significant results from various studies to gain insight into the importance of an effect (*i.e.*, vote counting) is commonly done but is subject to serious flaws. This is because the significance level of a study is the result of the magnitude of the effect and the sample size. Small studies are less likely to produce significant results (lower power), and therefore, vote counting is strongly biased toward finding no effect (Gurevitch and Hedges, 1993). Even review articles that summarize results of previous studies can be subject to the bias associated with vote counting because these qualitative summaries are based on the significance of the outcomes of previous studies without considering sample size and statistical power.

Meta-analysis begins by representing the outcome of each experiment by a quantitative index of the effect size. Effect size is chosen to reflect differences between experimental and control treatments in a way that is independent of sample size and scale of measurement. An effect size can be determined by dividing the difference between the treatment and the control by the pooled standard deviation. The average magnitude of the effect across all studies is determined and whether that effect is significantly different from zero is tested. This procedure is not subject to the problems associated with vote counting, but some specificity and fine detail are lost in the analysis.

The calculations involved in meta-analysis are relatively simple, but the gathering and handling of data can be complex. Cooper (1989), Cooper and Hedges (1993), and Light and Pillemer (1984) deal with some of the issues associated with handling data for meta-analysis. The problems associated with publication bias (*e.g.*, not publishing negative results) and ways to deal with these issues have also been discussed (Hedges and Olkin, 1985; Cooper and Hedges, 1993). Gurevitch and Hedges (1993) provide a good introduction to the use of meta-analysis and some the issues related to its use for ecological studies which also have relevance to many agricultural situations.

10 References

Abbott, W. S. 1925. A method of computing the effectiveness of an insecticide. *J. Econ. Entomol.* **18**, 265–267.

Anderson, D. E. and Lydic, R. 1977. On the effect of using ratios in the analysis of variance. *Biobehav. Rev.* **1**, 225–229.

Bartlett, M. S. 1947. The use of transformations. *Biometrics* **3**, 39–52.

Box, G. E. P. and Cox, D. R. 1964. An analysis of transformations. *J. R. Stat. Soc., Ser. B.* **26**, 211–243.

Box, G. E. P. and Jenkins, G. M. 1976. "Time Series Analysis: Forecasting and Control," Holden-Day, San Francisco.

Campbell, G., and Skillings, J. H. 1985. Nonparametric stepwise multiple comparison procedures. *J. Am. Stat. Assoc.* **80**, 998–1003.

Cochran, W. G. 1951. Testing a linear relation among variances. *Biometrics* **7**, 17–32.

Cochran, W. G. and Cox, G. M. 1957. "Experimental Designs". John Wiley & Sons, New York.

Cohen, J. 1988. "Statistical Power Analysis for the Behavioral Sciences". Lawrence Erlbaum, Hillsdale.

Connor, E. F. and Simberloff, D. 1986. Competition, scientific method and null models in ecology. *Am. Sci.* **75**, 155–162.

Conover, W. J. 1999. "Practical Nonparametric Statistics, Third Edition." John Wiley and Sons, New York.

Conover, W. J. and Iman, R. L. 1976. On some alternative procedures using ranks for the analysis of experimental designs. *Commun. Stat. Theory and Methods* **A5**, 1349–1368.

Cooper, H. M. 1989. "Integrating Research: A Guide for Literature Reviews." Sage Publications, Newbury Park, CA.

Cooper, H. M., and Hedges, L. V. 1993. "Handbook of Research Synthesis." Russell Sage Foundation, New York.

Crowder, M. J., and Hand, D. J. 1990. "Analysis of Repeated Measures." Chapman and Hall, London.

Daniel, W. W. 1990. "Applied Nonparametric Statistics, Second Edition." PWS-KENT Publishing Co., Boston.

Day, R. W. and Quinn, G. P. 1989. Comparisons of treatments after an analysis of variance in ecology. *Ecol. Monogr.* **59**, 433–463.

Dixon, P. M. 1993. The bootstrap and the jackknife: describing the precision of ecological indices. *In* "Design and Analysis of Ecological Experiments" (S. M. Scheiner and J. Gurevitch, Eds.), pp. 290–318. Chapman and Hall, New York.

Duncan, D. B. 1955. Multiple range and multiple F tests. Biometrics **11**, 1–42.

Dunn, O. J. and Clark, V. A. 1987. "Applied statistics: analysis of variance and regression." John Wiley & Sons, New York.

Efron, B. and Tibshirani, R. 1993. "Introduction to the Bootstrap." Chapman & Hall, London.

Einot, I. and Gabriel, K. R. 1975. A study of the powers of several methods of multiple comparisons. *J. Am. Stat. Assoc.* **70**, 574–583.

Federer, W. T. 1955. "Experimental Design." MacMillan Company, New York.

Finney, D. J. 1971. "Probit Analysis." Cambridge University Press, Cambridge.

Fisher, R. A. 1951. "Design of Experiments." Oliver and Boyd, Edinburgh.

Fligner, M. A. and Policello, G. E. 1981. Robust rank procedures for the Behrens-Fisher problem. *J. Am. Stat. Assoc.* **76**, 162–168.

Fowler, N. 1990. The 10 most common statistical errors. *Bull. Ecol. Soc. Am.* **71**, 161–164.

Gart, J. J. and Nam, J. 1988. Approximate interval estimation of the ratio of binomial parameters: A review and corrections for skewness. *Biometrics* **44**, 323–338.

Girden, E. R. 1992. "ANOVA: Repeated Measures." Sage publications, Newbury Park, CA.

Gomez, K. A. and Gomez, A. A. 1983. "Statistical Procedures for Agricultural Research." John Wiley & Sons, New York.

Gurevitch, J. and Hedges, L. V. 1993. Meta-analysis: Combining the results of independent experiments. *In* "Design and Analysis of Ecological Experiments" (S. M. Scheiner and J. Gurevitch, Eds.), pp. 378–398. Chapman and Hall, New York.

Hedges, L. V. and Olkin, I. 1985. "Statistical Methods for Meta-Analysis." Academic Press, New York.

Henderson, C. F. and Tilton, E. W. 1955. Tests with acaricides against the brown wheat mite. *J. Econ. Entomol.* **48**, 157–161.

Hoaglin, D. C., Mosteller, F. and Tukey, J. W. 1983. "Understanding Robust and Exploratory Data Analysis." John Wiley and Sons, New York.

Hollander, M., and Wolfe, D. A. 1973. "Nonparametric Statistical Methods." John Wiley & Sons, New York.

Huitema, B. E. 1980. 'The Analysis of Covariance and Alternatives.' Wiley Interscience, New York.

Hurlbert, S. J. 1984. Pseudoreplication and the design of ecological field experiments. *Ecol. Monogr.* **54**, 187–211.

Iman, R. L. 1974. A power study of a rank transform for the two-way classification model when interaction may be present. *Can. J. Stat. Section C Applications* **2**, 227–239.

Jones, D., and Matloff, N. 1986. Statistical hypothesis testing in biology: a contradiction in terms. *J. Econ. Entomol.* **79**, 1156–1160.

Kalbfeisch, J. D. and Prentice, R. L. 1980. "The statistical analysis of failure time data". John Wiley and Sons, New York.

Koopman, P. A. R. 1994. Confidence intervals for the Abbott's formula correction of bioassay data for control response. *J. Econ. Entomol.* **87**, 833.

Krauth, J. 1988. "Distribution-free Statistics: An Application-oriented Approach." Elsevier, Amsterdam.

Krebs, C. J. 1989. "Ecological Methodology." Harper & Row, New York.

Light, R. J. and Pillemer, D. B. 1984. "Summing Up: The Science of Reviewing Research." Harvard University Press, Cambridge, MA.

Littell, R. C., Milliken, G. A., Stroup, W. W., and Wolfinger, R. D. 1996. "SAS System for Mixed Models." SAS Institute Inc., Cary NC.

Little, T. M. and Hills, F. J. 1978. "Agricultural Experimentation." John Wiley & Sons, New York.

Manly, B. F. J. 1997. "Randomization, Bootstrap and Monte Carlo Methods in Biology." CRC Press, Boca Raton.

Mansouri, H. and Chang, G.-H. 1995. A comparative study of some rank tests for interaction. *Comput. Stat. Data Anal.* **19**, 85–96.

McArdle, B. H. and Gaston, K. J. 1992. Comparing population variabilities. *Oikos* **64**, 610–612.

Mead, R. and Curnow, R. N. 1983. "Statistical Methods in Agricultural and Experimental Biology." Chapman and Hall, London.

Miller, R. G. 1981. "Simultaneous Statistical Inference." Springer-Verlag, Berlin.

Motulsky, H. 1995. "Intuitive Biostatistics." Oxford University Press, New York.

Pearce, S. C. 1983. "The Agricultural Field Experiment." John Wiley and Sons, New York.

Pearson, E. S., and Hartley, H. O. 1951. Charts for the power function for analysis of variance tests, derived from the non-central F-distribution. *Biometrika* **38**, 112–130.

Randles, R. H. and Wolfe, D. A. 1979. "Introduction to the Theory of Nonparametric Statistics." John Wiley and Sons, New York.

Rasmussen, P. W., Heisey, D. M., Nordheim, E. V., and Frost, T. M. 1993. Time-series intervention analysis: unreplicated large-scale experiments. In: "Design and Analysis of Ecological Experiments" (S. M. Scheiner and J. Gurevitch, Eds.), pp. 138–158. Chapman and Hall, New York.

Rice, W. R. 1989. Analyzing tables of statistical tests. *Evolution* **43**, 223–225.

Robertson, J. L. and Priesler, H. K. 1992. "Pesticide Bioassays with Arthropods." CRC Press, Boca Raton, FL.

Rosenheim, J. A. and Hoy, M. A. 1989. Confidence intervals for the Abbott's formula correction of bioassay data for control response. *J. Econ. Entomol.* **82**, 331–335.

Rust, S. W., and Fligner, M. A. 1984. A modification of the Kruskal-Wallis statistic for the generalized Behrens-Fisher problem. *Communications in Statistics. Part A – Theory and Methods* **13**, 2013–2027.

Ryan, T. A. 1959. Multiple comparisons in psychological research. *Psych. Bull.* **56**, 26–47.

Scheffé, H. 1959. "The Analysis of Variance." John Wiley and Sons, New York.

Scheiner, S. M., 1993. Introduction: Theories, hypotheses, and statistics. *In*: "Design and Analysis of Ecological Experiments" (S. M. Scheiner and J. Gurevitch, Eds.), pp. 1–13. Chapman and Hall, New York.

Scheirer, C. J., Ray, W. S., and Hare, N. 1976. The analysis of ranked data derived from completely randomized factorial designs. *Biometrics* **32**, 429–434.

Shrader-Frechette, K. S. and McCoy, E. D. 1992. Statistics, costs and rationality in ecological inference. *Trends Ecol. Evol.* **7**, 96–99.

Siegel, S. and Castellan, N. J., Jr. 1988. "Nonparametric Statistics for the Behavioral Sciences." McGraw-Hill, New York.

Smith, P. L. 1981. The use of analysis of covariance to analyze data from designed experiments with missing or mixed-up values. *Appl. Stat.* **30**, 1–8.

Snedecor, G. W. and Cochran, W. G. 1989. "Statistical Methods." University of Iowa Press, Ames, IA.

Sokal, R. R. and Rohlf, F. J. 1995. "Biometry: the Principles and Practice of Statistics in Biological Research, Third Edition." W. H. Freeman, San Francisco.

Stevens, J. 1996. "Applied Multivariate Statistics for the Social Sciences, Third Edition." Lawrence Erlbaum Associates, Mahwah, NJ.

Stewart-Oaten, A., Murdoch, W. W., and Parker, K. R. 1986. Environmental impact assessment: "Pseudoreplication" in time. *Ecology* **67**, 929–940.

Thompson, G. L. 1991. A note on the rank transform for interactions. *Biometrika* **78**, 697–701.

Toft, C. A. and Shea, P. J. 1983. Detecting community-wide patterns: estimating power strengthens statistical inference. *Am. Nat.* **122**, 618–625.

Tukey, J. W. 1949. One degree of freedom for non-additivity. *Biometrics* 5, 232–242.

Underwood, A. J. 1981. Techniques of analysis of variance in experimental marine biology and ecology. *Ocean. Marine Biol. Annu. Rev.* **19**, 513–605.

Underwood, A. J. 1990. Experiments in ecology and management: their logics, functions and interpretations. *Aust. J. Ecol* **15**, 365–389.

Underwood, A. J. 1997. "Experiments in ecology: their Logical Design and Interpretation using Analysis of Variance." Cambridge University Press, Cambridge, UK.

von Ende, C. 1993. Repeated-measures analysis: growth and other time-dependent measures. *In*: "Design and Analysis of Ecological Experiments" (S. M. Scheiner and J. Gurevitch, Eds.), pp. 113–137. Chapman and Hall, New York.

Weerahandi, S. 1995. Anova under unequal error variances. *Biometrics* **51**, 589–599.

Welch, B. L. 1938. The significance of the difference between two means when the population variances are unequal. *Biometrika* **29**, 28–35.

Welch, B. L. 1951. On the comparison of several mean values: an alternative approach. *Biometrika* **38**, 330–336.

Winer, B. J. 1971. "Statistical Principles in Experimental Design." McGraw-Hill, New York.

Winer, B. J., Brown, D. R., and Michel, K. M. 1991. "Statistical Principals in Experimental Design." McGraw-Hill, New York.

Young, L. J. and Young, J. H. 1991. Alternative view of statistical hypothesis testing. *Environ. Entomol.* **20**, 1241–1245.

Zar, J. H. 1999. "Biostatistical Analysis, Fourth Edition." Prentice-Hall, Upper Saddle River, NJ.

Zimmerman, D. W. 2001. Increasing the power of the ANOVA F test for outlier-prone distributions by modified ranking methods. *J. Gen. Psychol.* 122, 83–94.

SECTION III

APPLICATION EQUIPMENT

Chapter III-1

Ground-based application equipment

R. P. Bateman and G. A. Matthews
International Pesticide Application Research Centre (IPARC)
Imperial College at Silwood Park
Sunninghill
Ascot Berks, SL5 7PY, UK

F. R. Hall
Department of Entomology
LPCAT, OARDC
Wooster, OH 44691, USA

1 Introduction

Biopesticides reached prominence due to severe criticisms of chemical pesticides for both environmental and operational reasons (*e.g.*, insecticide resistance). After the identification of a promising pathogen, its successful implementation will depend on the development of an effective "delivery system" that includes logistics of supply, formulation and application (Bateman, 1999). In practice this often means minimizing changes to procedures set-up for the chemical pesticides it aspires to replace, because as pointed out in Chapter I-2, it is usually unrealistic to expect users to change (often expensive) application equipment. This problem can be ameliorated by good formulation: an issue which is linked to application technology and is discussed in detail in Burges (1998).

As biopesticides are generally slower acting than modern chemical pesticides, their potential markets have been somewhat limited; promising roles for biopesticides and the consequences for their application are listed in Table 1. The key areas of current use are in glass-houses (where environmental conditions can to some extent be controlled) and in tree crops, (both orchards and forests) in which the local environment is more stable than in arable farming systems. However, as Gelernter (2005) has pointed out, there have now been 3 phases of biopesticide development and concludes that (a) they have been as much about values as commerce and (b) product development needs to be driven by highly motivated scientists and research groups. Biopesticides are important also in special circumstances, such as locust control, to minimize the impact of control measures on non-target organisms in natural habitats. For the purposes of this chapter, we will focus on the application of true biopesticides: now represented by > 100 living micro-organisms listed in Copping (2004).

A common assumption is that a biopesticide has to be applied as if it were a chemical pesticide. In consequence when promising activity has been established under laboratory conditions, the usual route to the field has been to mix the biopesticide with a surfactant and apply it with a large volume of water through a hydraulic sprayer. The carrier liquid will be formulated with additives to enhance the survival of the biopesticide. However droplet size in hydraulic nozzles can be heavily affected by the

L.A. Lacey and H.K. Kaya (eds.), *Field Manual of Techniques in Invertebrate Pathology*, 73–98.

Table 1. Some markets for biopesticides and the consequences for their application

Some existing and potential markets for biopesticides	Possible consequences for application
Large scale pest control in natural and semi-natural habitats (forests, range-land, wetlands *etc.*)	Ultra-low volume (ULV) application (ground-based, aerial) or baits to achieve very high work rates (Bateman, 1997)
High value crops in glass-houses (IPM to preserve bees and natural enemies)	Special sprayers possible (*e.g.*, air assist); humidity control possible (Helyer *et al.*, 1992)
Where chemicals are unacceptable (residue problem, organic market *etc.*)	These situations include many arable crop applications where conventional medium – high volume hydraulic
Where chemicals must be replaced (*e.g.*, after pesticide resistance, need for IPM strategy with low impact on other natural enemies)	Spraying will continue to predominate for the foreseeable future (see Chapter I-2).

use of polymeric adjuvants and emulsified oils (see Chapter I-2; Butler Ellis *et al.*, 1997).

Traditionally (since the discovery of Bordeaux mixture to protect vines) pesticides have been applied at high volume application rates (VARs) to "run-off", but this wastes a very high proportion of the chemical that drips from foliage. While a high volume treatment may be effective against soil pests, foliar treatments with this technique are grossly inefficient even if effective against the pest.

Many biopesticides have been more expensive to manufacture than their chemical alternatives, so it is vital that the application method is as efficient as possible. Clearly attention is needed to determine the priorities for application of biopesticides. In addition to identifying the position within the crop where the most susceptible stage of the pest is located, it is important to know the extent of coverage required at the target site, and how quickly an area has to be treated to ensure delivery is accurately timed. Climatic conditions at the time of treatment, for example temperature and humidity, are often crucial and can affect choice of delivery system. In certain circumstances, formulation can play a crucial role: *e.g.*, the use of oils to apply *Metarhizium* against locusts in arid conditions, or humectants to ensure adequate moisture for spore germination of mycoherbicides (Burges, 1998). Such formulations can place severe restrictions on the choice of application equipment. All formulation names and definitions used in this chapter conform to the CropLife International (CropLife, 2002) standards.

One of the key issues relating to biopesticide application is that, unlike chemicals which can be dissolved in a solvent (as in emulsifiable formulations), fungal spores, viruses and entomopathogenic nematodes (EPNs) are all particulate suspensions. The particle size spectrum and concentration of the suspension have important consequences for application:

1. Depending on their specific gravity and suspensibility in the carrier liquid, there is a definite settling of the biopesticide if the liquid is left standing for a prolonged period. This may occur in the formulation prior to preparation of the spray or at any stage during application if the liquid is not agitated.
2. The particulates can be retained on filters used to prevent blockages of nozzles, unless the aperture size is carefully selected and the flow route allows passage to the nozzle without stagnant areas in which particles can aggregate within the sprayer.
3. Care may be needed to avoid stress which could affect the physical characteristics of the particles and reduce viability of the biopesticide. Thus shearing effects through nozzle orifices and in pumps could adversely affect some biopesticides.
4. Large droplets are likely to be needed for nematodes simply because of their size.
5. Attention must be paid to the numbers of infective particles that can be expected to occur in each size class of a spray droplet distribution. This can be done by assuming a random distribution in the spray tank and that the numbers of particles contained in

a droplet are related to its volume (*e.g.*, Chapter I-2; Bateman, 1999). These assumptions, coupled with data on the properties of droplets as shown in Table 2, can give an insight into the proportion of pathogen leaving the sprayer that is likely to come into contact with the target pest.

Conventional hydraulic nozzles produce a very wide spectrum of droplet sizes so the distribution of particles within individual droplets will depend on their dispersion within the carrier liquid and their size. Clearly very small droplets cannot contain larger discrete particles; thus when applying nematodes, a high proportion of droplets contain no organisms. At the other extreme, ULV applications of *Metarhizium* to control locusts emphasized the relationship between spore concentration, volume applied and optimum droplet size to ensure an adequate delivery of spores per droplet (*e.g.*, Bateman, 1993). In this case, it was crucial to develop a cost effective device for separation of pure (mostly single) conidia from their production substrate (Cherry *et al.*, 1999), and a commercial version of which is now available for small-medium scale extraction of experimental samples (see www.mycoharvester.info; Chapter I-2).

Biopesticide propagules are almost invariably very delicate: even the *Bacillus thuringiensis* crystal is less robust than many chemicals. Application methods should take into consideration their live nature and the multitude of factors affecting host-pathogen relationships (*e.g.*, a pathogen's capacity for secondary cycling – horizontal transmission – where direct impact with sprays and their residues may be less important). However to use a more "chemical" abstraction, biological agents are rarely "systemic": unlike some of their chemical counterparts. High "coverage" rates (the probability that a pest will encounter a pesticide) will be very important for biological agents that: act after ingestion, are used against sucking pests or in other situations such as hyperparasitic fungicides where high contact rate is required. Achieving good under-leaf coverage can also enhance pathogen survival by protecting against sunlight. We have therefore

Table 2. Ranges of droplet sizes, with their volumes and properties

Diameter range (μm)	volume (maximum) in pico liters or l^{-12}	Properties/function
< 10	0.52	Potentially hazardous very fine aerosols or particles, with a progressively increasing risk of inhalation by operators as the size diminishes (greatest risk at ≈ 1–3 μm)
Note:		"Fogs"
< 15	1.8	thermal fog
< 30	14	cold fog
< 50	65	Aerosols (appropriate for direct contact with small insects)
50–100	524	Mists appropriate for oil-based ULV spraying
75–150	1,767	Maximizes coverage with water-based insecticide and fungicide sprays
150–300	14,137	Maximizes coverage for herbicide sprays, avoiding drift especially at low (< 2 m.s^{-1}) wind speeds.
300–500	65,450	Coarse spray: maximum avoidance of drift; at > 500 μm droplets become drops and progressively less efficient at covering ground, unless total volume applied is substantially increased.

included certain potentially useful application techniques (including some fairly new products) that have yet to be widely used with biological agents.

2 Nozzles and biological agents

A *Optimum droplet sizes*

Delivery of droplets to the target will be affected significantly by their size. The high terminal velocity of large droplets ensures a predominantly vertical trajectory and a high proportion of droplets will be deposited by sedimentation on horizontal surfaces. This is satisfactory for pesticides applied to the soil, but can result in poor distribution on foliage, especially as large droplets tend to be retained poorly on waxy leaf surfaces, unless the formulation contains sufficient surfactant to avoid droplets bouncing off leaf surfaces. Bergeron *et al.* (1998) describe the ways in which dynamic surface tension and elongational (dynamic, non-Newtonian) viscosity of the spray formulation can affect spray retention, and how these effects can be modified with adjuvants.

Coverage of leaves is commonly improved if small droplets are applied, but if too small, the droplets remain airborne and are most liable to drift with the wind out of the treated area. Impaction of droplets within foliage depends on air turbulence and can be generally improved by using an air-assisted sprayer. The optimum droplet size for individual particular types of biopesticide application has yet to be established, and we have to rely on known characteristics and optima (see Table 2) for chemicals (*e.g.*, Himel, 1969, Matthews, 2000).

B *Hydraulic atomizers*

1 *Features*

Hydraulic nozzles have a shaped orifice through which the pesticide liquid is forced under pressure. For most applications pressures of 1–3 bar (100–300 kPa; 15–44 psi) are considered adequate, but much higher pressures (up to 20 bar) have been used to apply some pesticides. This is the longest established and most widely used mode of atomization and it encompasses a very wide range of application systems; for example Wraight *et al.* (Chapter VII-1) describe a remarkable variety of methods that have been used for just one crop: potatoes.

An elliptical orifice will allow a thin fan-shaped spray sheet to be formed which disintegrates as it is stretched. A fan-shaped pattern is also formed if liquid is emitted through a circular orifice and is deflected immediately by a flat surface; these are referred to as deflector nozzles, although previously the terms impact or anvil type nozzles were used (Figure 1). If the spray is emitted through a circular orifice without deflection a cone pattern is obtained. Circular orifices are less likely to be blocked by particulates, which can accumulate at the edges of the fan-shaped orifice and affect output and spray pattern.

Droplets can be formed from the edge of a sheet of liquid by 'rim disintegration', but most of the droplets are formed as holes develop

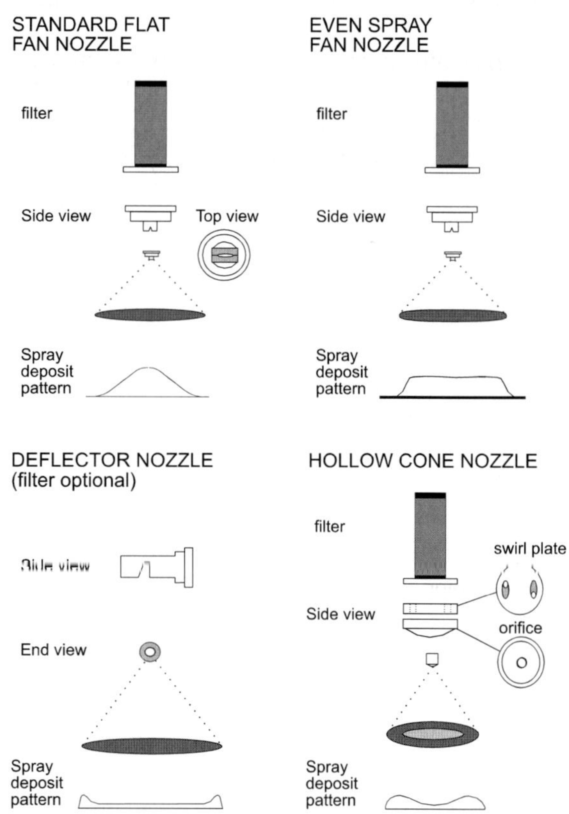

Figure 1. Conventional hydraulic nozzles and their "footprints"

within the sheet to form threads of liquid that are unstable and break up to form the individual droplets. The distance from the orifices at which the threads form will partly depend on the physical characteristics of the liquid, thus the addition of surfactant tends to stretch the sheet further and form smaller droplets, whereas a more viscous liquid may form larger droplets. Air turbulence due to the forward passage of the nozzle, or an air flow at the nozzle on air-assisted sprayers, will also affect the stability of the liquid sheet and affect droplet formation. Increasing pressure reduces the average size and increasing flow rate (with a larger nozzle orifice size) creates larger droplets. In all cases a very wide range of droplet sizes is produced. Hydraulic nozzles were traditionally made in brass (or sometimes stainless steel), but hard-wearing engineering plastics are now most commonly used. If nozzle wear from abrasive particles is a special problem, some manufacturers supply ceramic tips inserted into a plastic mounting. The use of plastics has enabled color coding of hydraulic nozzles, according to their flow rates (Table 3).

Nozzle blockages are always a problem with particulate preparations, and these will generally be reduced by avoiding narrow orifice sizes. A number of variations of hydraulic nozzle have been manufactured for chemical applications (Figure 2). These include pre-orifice nozzles in which a circular opening upstream within the nozzle reduces the pressure at the final orifice and nozzles that suck in air by a Venturi effect to produce droplets that contain bubbles of air. These newer nozzles have been used by farmers primarily to decrease the risk of downwind drift by having fewer droplets smaller than 100 μm diameter. This is achieved by effectively enlarging the VMDs; as with more conventional hydraulic nozzles, the droplet size spectra remain broad (large relative spans of $\gg 1$*). The 'Turbo Teejet' – effectively a downward spraying deflector nozzle – has the added capacity to produce sprays at relatively low flow-rates, without the use of very narrow orifices.

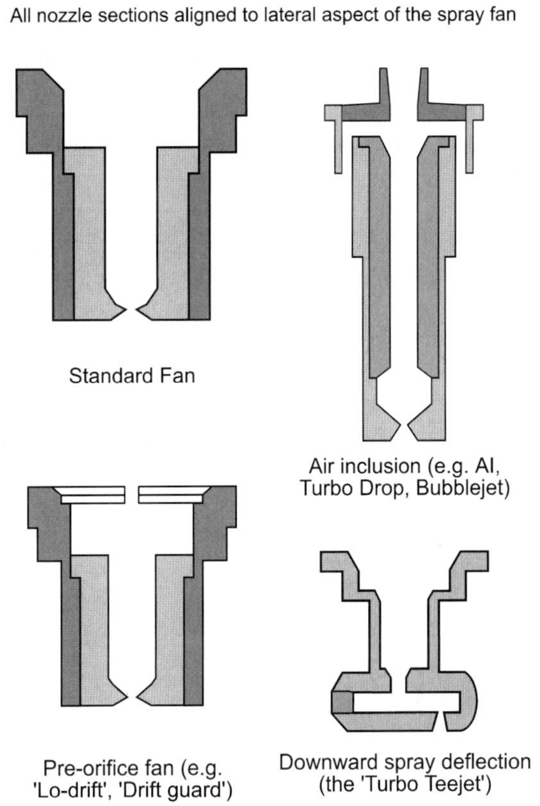

Figure 2. Variants on the hydraulic fan nozzle

Table 3. International (ISO 10625: 1996) color codes for hydraulic nozzles

Color	Flow rate (liters/min. at 300 kPa)
Orange	0.4
Green	0.6
Yellow	0.8
Blue	1.2
Red	1.6
Brown	2.0
Grey	2.4
White	3.2

Hydraulic nozzles remain the most widely used method of spraying chemical pesticides. They are fitted to a wide range of spraying systems ranging from hand-held "trombone sprayers", side-lever knapsack sprayers (used by the majority of small-holder farmers and for small-scale field trials), compression sprayers (designed for use in vector control operations), to the wide range

* Relative span $\{(D_{[v,0.9]} - D_{[v,0.1]}) \div D_{[v,0.5]}\}$ is a dimensionless statistic describing the uniformity of spray droplet spectra (see Chapter 1)

of tractor mounted equipment (see Matthews, 2000). Figure 3 shows two forms of side-lever knapsack sprayer; the tank mixture is pumped using a diaphragm or a piston mechanism, both requiring two valves. "Pulsation" (variations in pressure with pumping) is minimized with a pressure chamber that is mounted either internally or externally to the main tank, and certain sprayers have a pressure control mechanism mounted either in the tank or on the spray lance ("wand"). Filters adjacent to the tank lid are usually fairly coarse and unlikely to retain material from any but the crudest biopesticide formulations; problems are much more likely to occur with the filters next to the nozzle tips (Figure 1).

Distribution of spray deposited with a lance depends very much on the skill of the operator in keeping a steady pumping and walking speed and directing the nozzle to the key target areas of foliage. Too often the nozzle is directed downwards over a crop so under-leaf coverage is exceedingly poor. Adaptations with vertical and horizontal booms have been used to increase work rate (by increasing the number of nozzles) and spray distribution. When mounted to the rear of the spray tank in the form of a vertical tail-boom (Figure 4), operator contamination is reduced and excellent under-leaf cover is achieved. Multi-nozzle booms require a pump with sufficient output. Hollow cone nozzles with low (200 ml/min) flow-rates are prone to blockages with particulate formulations.

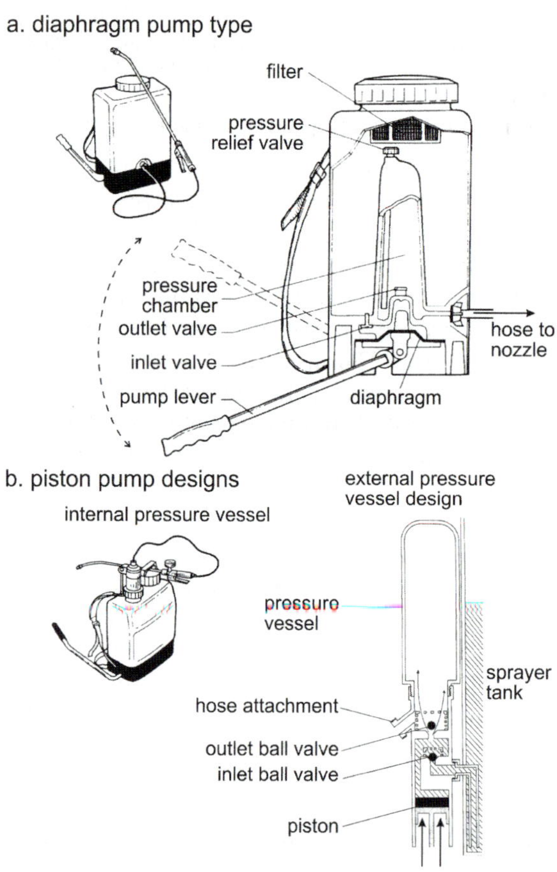

Figure 3. Side lever knapsack sprayers and their pumps

2 Nozzle classification

Development of laser/computer systems of measuring the size of droplets has enabled rapid characterization of droplet spectra. The spray from hydraulic nozzles has been classified in relation to the spectra from selected reference nozzles that demarcate the boundaries between categories, referred-to by the British Crop Production Council (BCPC) as very fine, fine, medium, coarse and very coarse sprays (Southcombe *et al.*, 1997). An International (ISO) Standard is in preparation to define droplet size spectrum classes and specify a means for relative nozzle comparisons based on droplet size only. Inside glass-houses or storage sheds where small droplets are used in space sprays the very fine spray is subdivided into mists (with no more than 5% by volume below 30 μm droplet diameter) and fogs (with > 90% by volume below 30 μm), thus the latter are the most hazardous spray in relation to inhalation of pesticides (see Table 2). Unlike other size classes, Matthews and Bateman (2004) suggested that a 52 mm rotary nozzle, operated at 17,000 RPM and a flow rate of 35 ml/min, be used to differentiate between the spray spectra produced by mist and fog application equipment.

3 Use with biopesticides

How appropriate are hydraulic nozzles for biopesticides? Clearly, they are the most widely-used means of atomization, and where the particle size is very small and equivalent to a highly micronized dispersible/wettable powder formulation, application of the biopesticide should be possible through a complete range of

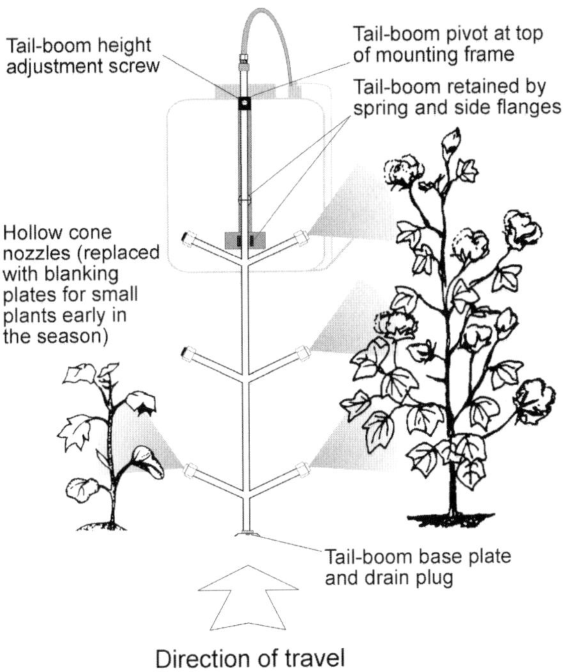

Figure 4. A side-lever knapsack sprayer with tail-boom

hydraulic nozzles. Conversely, larger particles will need larger orifices to avoid blockages, but this inevitably increases the total volume of spray applied. Fife *et al.* (2004) investigated the potential effects of hydraulic spraying equipment on biopesticides such as EPN and concluded that operating conditions should never exceed 2000 kPa (290 psi) (with lower pressures for species such as *Heterorhabditis megidis*). Energy dissipation rates (assumed to represent shear forces acting on micro-organism particles) were found to be higher in the exit orifices of flat fan tips than in equivalent hollow-cone nozzles, thus it was inferred that the latter may be more suitable for more delicate organisms.

Spodoptera littoralis nucleopolyhedrovirus (NPV) was applied with hydraulic nozzles in Egypt, but as the early instars feed on the undersurface of leaves, positioning of the nozzles to direct spray up through the foliage was required (Burges, 1998). In trials a 'tail-boom' (see above) enables the spray to be applied simultaneously to the different strata of foliage. Increasing under-leaf coverage reduces the impact of ultraviolet light and improves the persistence of viral deposits. Baculoviruses have been applied in this way through cone nozzles with an output of 200 ml/min at 3 bar (300 kPa; 44 psi) pressure.

The number of EPN infective juveniles (IJs) deposited on leaves was shown to be greatest with high output nozzles *e.g.*, 80° cone nozzle applying 2.6 liters/minute at 3 bar pressure, but high mortality of *Plutella xylostella* was obtained with similar nozzles applying 1.6 liters/minute at 2 bar pressure (Lello *et al.*, 1996).

C Twin fluid nozzles

1 Features

In some sprayers, the hydraulic nozzle is replaced by an air-shear mechanism to break up the liquid pesticide and produce a spray; the air stream generated may also be used to assist projection of the droplets to the target. Spray volume application rates (VARs) are reduced, and in consequence the concentration of the pesticide in the tank mix is higher.

Some twin-fluid nozzles have an internal mixing of air with the spray liquid, while others have an air flow over a liquid jet. The nozzle may have a straight jet of air at high velocity, but in nozzles used to create 'cold fogs' a vortex of air is used to create droplets of $< 30\,\mu m$. As with hydraulic nozzles, a wide range of droplet sizes is usual.

Motorized knapsack mist-blowers (or air blast sprayers: see Figure 9) have many uses, although prototypes were originally developed for obtaining good droplet coverage in trees and bushes. Nozzles are usually of the air-shear type, although rotary atomizers are also available (see 3.C.2). Various designs of air-shear nozzle are produced, which frequently aim to introduce thin layers of liquid into the air stream and thus produce finer sprays. Apart from the ULV versions, there are usually no very fine restrictions in the liquid delivery system, so clogging is unlikely: making mistblowers versatile and appropriate for use with biopesticides.

2 Use with biopesticides

Cold fogging equipment (Micro-Gen) was used experimentally to apply *B. thuringiensis*, NPVs and *Paecilomyces farinosus* to a range of field crops (Falcon and Sorensen, 1976). The aim was to increase coverage of pathogen particles within the crop canopy, however such small droplets (> 30 μm) are not readily deposited and are intended primarily as a space spray.

Deflector nozzles such as the 'Airtec' illustrated in Figure 5 can produce fine sprays, which are delivered to the crop with air assistance. Being modified flood jet tips, the orifice size is large and therefore has a reduced risk of blockage. Since a compressor is required for the air supply, these nozzles are suitable only for tractor booms.

D Thermal energy nozzles

1 Features

ULV thermal foggers have a pulse jet engine, and are available as hand-held or vehicle-mounted equipment. To start small manually carried machines, a manual or electrically operated pump is used to pressurize a petrol tank so the fuel is delivered via a control valve to an open-ended combustion chamber with a spark plug. Once the engine has started, further pumping is not necessary as the hot exhaust gases ignite the incoming fuel and air mix. The rate of pulsation is determined by resonance and is related the length of the chamber. A high temperature is used to vaporize the spray liquid that then condenses to form a dense 'fog' consisting usually of droplets that are mostly < 15μm in diameter. Fogs have traditionally been used for space treatments inside buildings but have also been used for outdoor mosquito control when inversion conditions exist.

2 Use with biopesticides

Any pathogen formulation applied with this equipment must be able to withstand at least transient exposure to high temperatures. Formulations are vaporized by temperatures generally in excess of 500 °C, so some degradation of the active ingredient can be expected. There are certain types of thermal fogging machines that operate at lower temperatures; however they were also found to be unsuitable for applying *B. thuringiensis* spores and toxin crystals, since formulation was exposed to the hot combustion chamber over a longer period (Burges and Jarrett, 1979). It is known that blastospores of *Lecanicillium (Verticillium) lecanii* are killed by

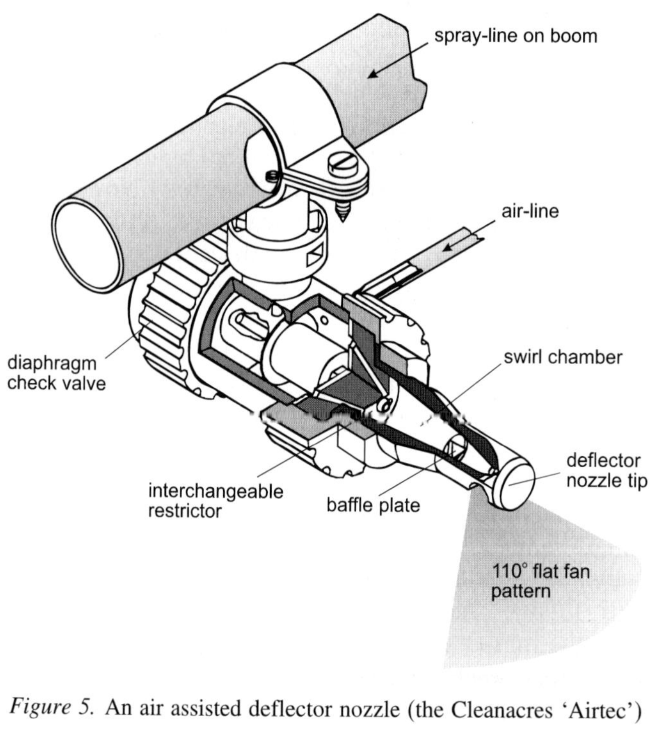

Figure 5. An air assisted deflector nozzle (the Cleanacres 'Airtec')

certain thermal fogging machines, but apparently are unharmed by cold foggers (Jarrett and Burges, 1982).

In contrast to ULV cold foggers, chemical pesticides are normally formulated in oil, but water-based mixtures can be applied with a thermal fogger, in which case an adjuvant is usually included to limit droplet evaporation and ensure that the fog is visible. Burges and Jarrett (1979) substituted water for oils, since the latter were phytotoxic to glass-house crops, at the high volumes required for *B. thuringiensis* formulations. They showed that very brief exposure to a high temperature had not affected the viability of spores collected from leaf surfaces, but 95% of the spray was deposited on the upper surfaces of leaves. Virtually no thermal fogging machines are equipped with an agitation mechanism in the spray tank, and we can conclude that this equipment has only a very limited use with biopesticides.

E CDA systems (mostly rotary atomizers)

1 Features

The essential feature of controlled droplet application (CDA) is the production of sprays with a relatively narrow range of droplet sizes (typically with a relative span of < 0.9). In most cases this requires a rotary atomizer, although narrow droplet spectra can also be produced using electrodynamic energy ('Electrodyn' or ED nozzles). The latter have not been used commercially with biopesticides, partly because the polar solvents (that adjust the electrical conductivity of the special formulations required) kill many pathogens. Smits *et al.* (1988) successfully applied an experimental ED formulation of NPV to *Spodoptera exigua* on chrysanthemums; they obtained comparable results to high and low volume applications at the same rate of virus application with short plants (but inferior control with tall crops).

With rotary atomizers, spray liquid fed on to a spinning disc or cage is thrown from the periphery as single droplets, at low flow rates. For insecticides, higher disc speeds are used which result in the generation of ligaments (individual threads of liquid) that break up into smaller droplets. Care is needed to avoid flooding a rotary nozzle as this produces a sheet of liquid in much the same manner as a hydraulic nozzle. Within operational limits an increase in flow rate will increase droplet size, but the main control of droplet size is by adjusting the rotational speed; thus increasing speed reduces droplet size. Rotary nozzle design aims to ensure that the liquid is fed uniformly to the edge of the rotating surface. Relatively large restrictor orifices can be used to apply low volumes as liquid is fed by gravity or low pressure to the nozzle.

Rotary atomizers are most commonly used in the form of hand-held Spinning disc sprayers, although they have also been fitted to motorized mistblowers, tractor booms and other vehicle mounted equipment (see 3.C.3). One version of spinning disc sprayer – the 'Motax' – has a small axial fan mounted on a knapsack frame (Figure 6). This was designed to provide CDA air-assisted sprays on small bushes such as coffee and has been used with a Colombian *Beauveria bassiana* formulation.

Spinning disc sprayers have been used for a wide range of laboratory and "pre-field" tests, for which they have several advantages including:

1. The production of droplets that are a better simulation of field applications than spray residues created with standard equipment such as the Potter tower (Matthews, 1997)
2. They enable easy spray application of small quantities of experimental products (that may block narrow Potter tower or 'air-brush' nozzles)
3. They produce a narrow droplet size spectrum with a VMD that easily can be adjusted by regulating the applied voltage. The residues produced therefore consist of more succinct "doses" that simplify the process of pathogen acquisition by the target pest.

2 Use with biopesticides

Rotary atomizers have been used with both manually-carried and vehicle-mounted equipment to apply *Metarhizium* for locust and grasshopper control as the droplet size can be very effectively optimized and the mycopesticide applied at ultra-low volume in an oil formulation (Bateman, 1997). Most authorities agree that, where resources for control campaigns of migratory pests are limited, Ultra Low Volume (ULV) techniques are the only economical and practical means of applying pesticides over large

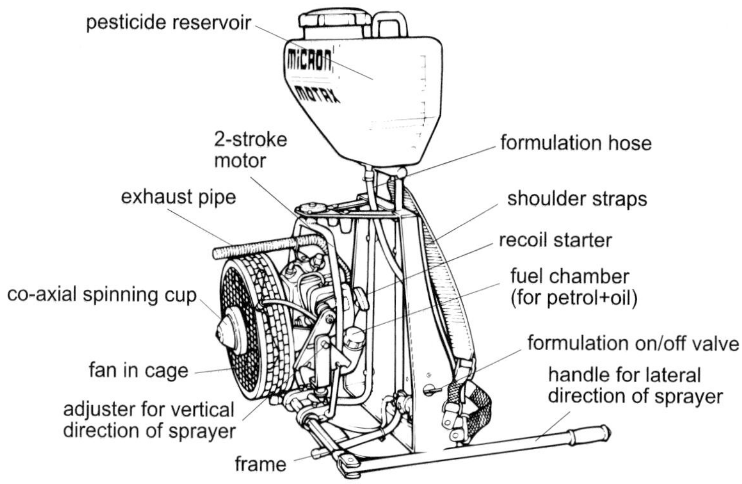

Figure 6. The Micron 'Motax'

areas of often difficult terrain, where water supplies are poor. Table 4 summarizes estimated costs for formulation and application in Senegal in 1986 (U.S. Congress, 1990).

Rotary atomizers have also been used to apply viruses, for example in forest trials with *Panolis flammea* in Scotland. Parnell *et al.* (1999) reported that spinning disc sprayers were superior to other application methods, such as motorized mistblowers, for applying NPV formulations to *Helicoverpa armigera* on cotton. Preliminary work with EPNs showed a high level of mortality with a very low deposit of IJs. However, subsequent studies have suggested that the distribution of large EPNs on a grooved cup-shaped rotor can be affected by centrifuging the liquid carrier from the IJs (Lello *et al.*, 1996). Some equipment (such as the Berthoud 'C5') has a smooth internal surface on the rotor, which may lessen this problem.

Table 4. Estimated relative formulation and application costs for chemical pesticides used in Senegal (1986)

Application method	Cost (US$/ha)	Ha treated/hour
Dusts Baits	18.00–26.00 8.00–12.00	0.5 (distributed by hand)
ULV spraying (ground)	6.00–8.50	8–12
Aerial ULV spraying (depending on aircraft type)	2.20 upwards	450–470

3 Main Types of Ground-Based Spraying Equipment for Biopesticides and their Calibration

A *Special considerations for biopesticides*

The size of spray tanks varies from < 1 liter on some manually carried spinning disc sprayers to > 2000 liters on trailed orchard sprayers. In some tanks agitation is not provided except for any movement caused by the operator, while other tanks have either a mechanically operated paddle or re-circulation system through nozzles that flush liquid over the bottom of the tank. If no agitation is provided, the operator will need to keep the biopesticide in suspension by periodically shaking the container. On some knapsack sprayers, especially those fitted with a diaphragm pump, agitation can be provided only by movement of the whole tank. Where there is re-circulation, a proportion of the spray will be circulated through the pump several times before reaching a nozzle; thus particles could be damaged by any shearing effects with their passage within the pump and through filters.

B *Calibration: general*

The ultimate level and uniformity of dosage is dependent on several factors that are discussed in a number of standard texts (*e.g.*, Matthews, 2000; Bache and Johnstone, 1992). There are essentially four variables (besides changing nozzles) that operators can adjust in order to adjust mass application rate and VAR (see Figure 7); alterations to flow rate (via operational pressure with hydraulic nozzles) will also affect droplet size and spray quality. Cory and Evans (see Chapter IV-1) further discusses the complex set of factors that must be considered in order to achieve a "control window" for viral control agents.

A satisfactory and practical way to obtain an outlook of where spray actually lands at the target site is to use tracers (Cooke and Hislop, 1993). The most commonly used method is to add fluorescent pigments to the spray tank mix and assess droplets with an ultra violet lamp (Staniland, 1959). Various colored pigments and dyes are available, but concentrated (*e.g.*, ULV) formulations containing *Metarhizium* conidia are dark green and opaque, and may mask many of them. Very bright yellow pigments such as 'Lumogen' added to the spray mixtures are visible at the rate of 0.5–1% w/v in the formulation; however substantially cheaper alternatives such as 'Solar yellow' are also sufficiently apparent for most purposes. Droplet coverage can be estimated rapidly using standard cards (Bateman, 1993) or accurate measurements of total deposit can be made using fluorimetry (*e.g.*, Pickin in Heinrichs *et al.*, 1981).

Calibration requires measurement of the nozzle output, forward speed and selection of track spacing. The general formula for calculating volume application rate (VAR) is:

$$\textbf{VAR in liters/hectare (V)} = \frac{\mathbf{k \times F}}{\mathbf{T \times S}} \qquad (1)$$

where:

F (liter or ml/min):	flow rate
T (m):	width of track spacing
S (m/min, *etc.*):	forward speed
k:	constant

Different values of '**k**' can be selected according to the way in which the flow rate and forward speed is measured. Other values can be used to adjust for other local units of measurement, and an example is given here (d) for use with vehicles with speedometers reading in statute miles per hour.

value of **k** when: **S** is measured in: And **F** is measured in:

	ml/min	liters/min
a) meters/min.	**10**	**10,000**
b) seconds for 100m	**600**	**600,000**
c) km/hour	**0.6**	**600**
d) miles/hour	**0.3726**	**372.6**

Whether spraying on foot or using a vehicle, it can be easier to simply keep **S** as the time taken to spray a fixed length of field (b):

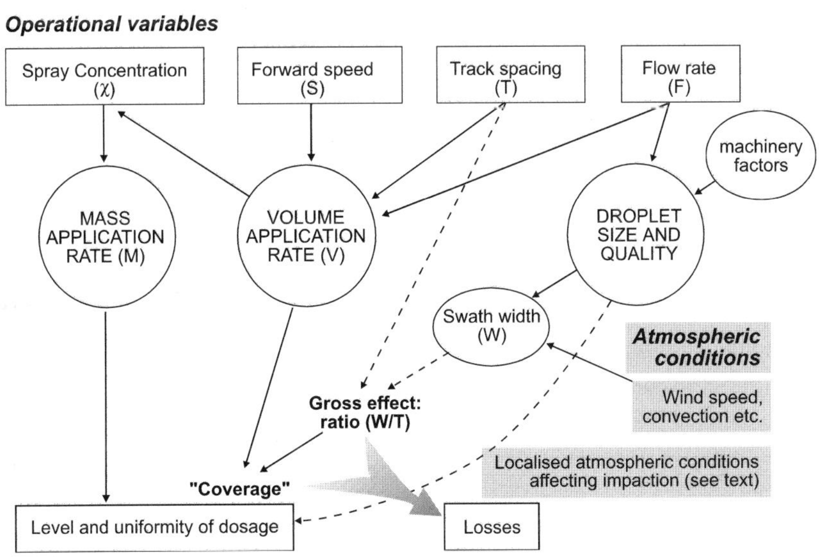

Figure 7. Factors affecting coverage and dosage

1. Measure out and mark a **100 m** track through the terrain to be sprayed.
2. Maintain a **sustainable** walking speed for the operator within the crop, or choose an appropriate vehicle gear ratio and accelerator position for the terrain.
3. Time how long it takes (**t**: in seconds) to travel between the markers.

For applications using vehicle mounted sprayers, speedometers are often inaccurate (and may not even register at low speeds). For locust control operations a speed of approximately 10 km/hour is appropriate. The forward speed in km/hour is **360/t**, when calculated from the method above.

It can be important to estimate the work rate (hectares treated per hour) for both trials and operations. The actual spraying time for a **1 ha** plot (where **T** is measured in m):

$$\text{Spraying time (minutes/ha)} = \frac{\mathbf{k}}{(\mathbf{60 \times T \times S})} \tag{2}$$

The total work rate depends on several additional factors, including the times taken for:

1. turning time between swaths ($\approx$ **width of field (m)/(60** $\times$ **S)** minutes);
2. mixing and measuring formulation then refilling the reservoir/sprayer tank (multiplied by the number of reservoirs to treat a hectare);
3. distance to and from the re-filling site;
4. resting (for arduous conditions) and washing.

Items 2 and 3 become increasingly significant at higher volume application rates; and can be reduced by enlarging the effective reservoir capacity (*e.g.*, using back-packs with spinning disc sprayers).

To select a suitable flow rate (*e.g.*, choosing a nozzle or restrictor), rearrange formula **(1)** as follows:

$$\mathbf{F} = \frac{\mathbf{V \times T \times S}}{\mathbf{k}} \tag{3}$$

To adjust the track spacing with a given flow rate and speed, for desired volume application rate:

$$\mathbf{T} = \frac{\mathbf{F \times k}}{\mathbf{V \times S}} \tag{4}$$

A computer program, 'APLICALC' is available from www.dropdata.org for rapid calculation of these values from any combination of the other factors. This web site also gives further information on nozzles, sprayer maintenance and calibration, *etc.*

Note: The calculations above all refer to a single nozzle spray release. If more than one nozzle is used **F** refers to the total flow rate (the sum of each calibrated nozzle output). Likewise, **T** refers to the total sprayed width from a tractor boom, or true track spacing in a targeted "band" application.

C Manually carried sprayers

1 Manually carried hydraulic sprayers

The most commonly used equipment on small scale farms and in experimental plots is the lever-operated knapsack sprayer. These sprayers have a 10–20 liter tank, 15 liter being the most common, with either a diaphragm or piston pump, the latter being generally better for higher pressures. Piston pump sprayers with an internal air pressure chamber often have a paddle attached to the pump mechanism. Spray is delivered to a lance with a trigger valve and is usually applied through a single nozzle, often of poor quality. To avoid the drudgery of manual pumping, hydraulic knapsack sprayers with a small 2-stroke engine or electrically operated pump are available. The latter has a rechargeable battery to allow spraying for about two hours before recharging is required. Quality of manufacture and price of equipment vary considerably, so FAO has introduced specifications in an attempt to achieve minimum standards and reduce operator contamination (FAO, 1998).

In situations where it is difficult to direct a nozzle carefully and pump at the same time, the lever operated sprayer is replaced by a compression sprayer. A 5–10 liter container is filled about 3/4 full with spray and then an air pump in the lid is used to pressurize the contents. The pressure decreases as the tank empties so the operator periodically has to stop spraying and re-pressurize the tank.

With both lever-operated and compression sprayers fluctuations in pressure alter the output at the nozzle and the droplet spectrum. This problem could be alleviated by fitting pressure control valves, but until now these have not been widely accepted. Useful devices (such as the Hardy 'Kali bottle': see Figure 8) have been developed by several companies to enable calculation-free measurement of VARs after simulated treatment of a small (*e.g.*, 25 m^2) area of crop.

2 Motorized knapsack mistblowers

Another type of manually carried sprayer is the motorized knapsack mistblower (Figure 9). Because they produce relatively small droplets (Chapter I-2), mistblowers are usually operated at lower VARs than with the other types of hydraulic sprayer. Atomization occurs either conventionally with an air-shear nozzle, or with a rotary atomizer supplied separately or (more economically) from the mistblower manufacturer. They are typically used to apply water based mixtures at 20–100 liters/ha, but low flow rate ULV adapters are available, achieving VARs of as little as 2 liters/ha with oil-based formulations.

An air-cooled, 2-stroke engine drives a centrifugal fan to provide a high velocity ($\approx 40\,m/s$) air jet to the nozzle. Either the 10–12 liter tank is slightly pressurized or a small pump is used to deliver spray liquid via a restrictor to the nozzle. Although versatile and effective, they are fairly expensive (US$ 300–600) and their engines require a substantial commitment to maintenance and repair in comparison with other manual equipment. Sutherland (1980) gives a useful guide on the use and maintenance of mistblowers and Bateman and Alves (2000) reviewed their performance with a view to applying microbial agents such as *Metarhizium anisopliae* and *Trichoderma* spp. to cocoa.

The main (and original) use of motorized mistblowers is to project pesticide up into trees, but they are also used to treat a wide swath across crops such as rice, which are less easy to walk through. The vertical and horizontal throw (distance that spray can be delivered) depends primarily on the power of the motor and size of the fan. The maximum horizontal throw is approximately 15m although the usual working track spacing is 5–10m. This equipment can be adapted for granule application (see below).

Collection of the spray in an air stream is extremely impractical; simply measuring the flow of liquid in the formulation line past the restrictor will always give a substantial (often > 30%) under-estimate of operational flow since there is a strong "suction effect" in the twin-fluid nozzle.

Accurate calibration involves the following procedure:

1. Place the sprayer on a firm horizontal surface and mark a level near the top of the pesticide tank (but not full);

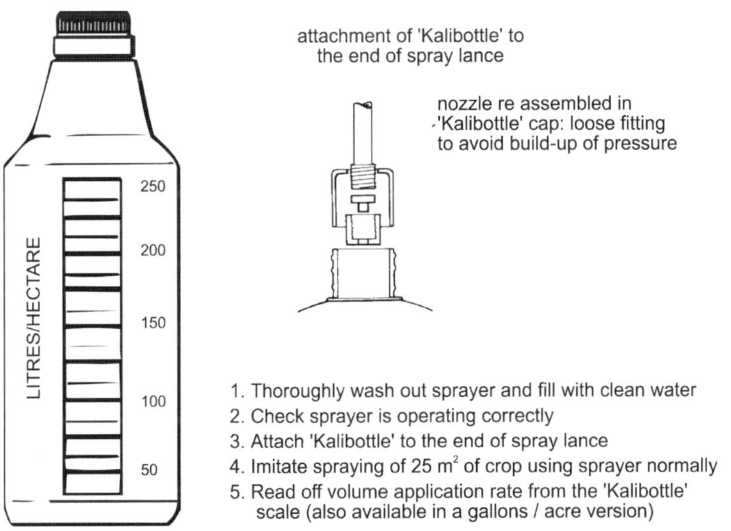

Figure 8. The 'Kali bottle': a device for calibration of hydraulic nozzles

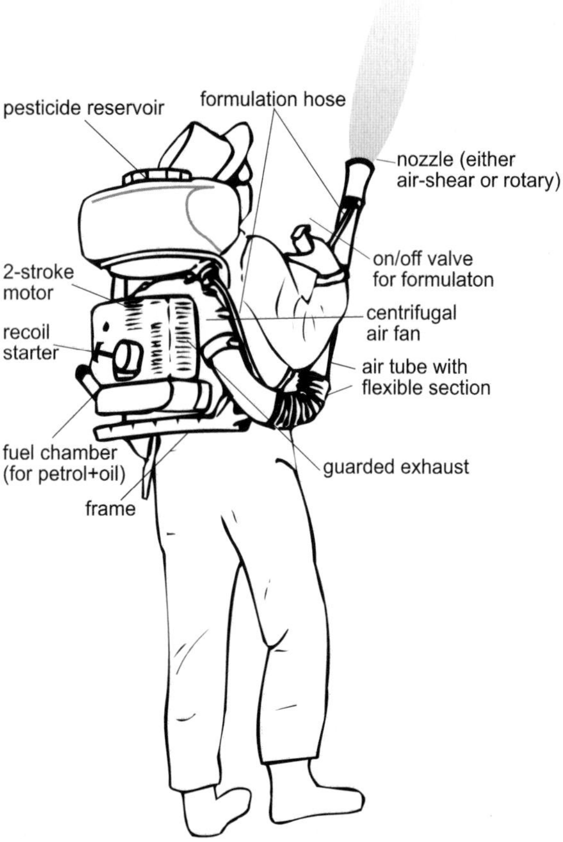

Figure 9. A motorized knapsack mistblower in use spraying trees

2. pour in water (or blank formulation) to the mark on the tank;
3. run the engine at normal operating speed (normally full throttle) for a set time (say 2 minutes);
4. turn off the engine and measure the volume of water/formulation that is needed to the fill the tank to the mark;
5. calculate **F** = volume/time.

A further complication is that flow rate can vary substantially with the angle at which the nozzle tube is directed (*e.g.*, spraying upwards into trees vs. horizontally into crops). Jollands (1991) noted that tank pressurization was often inadequate for consistent formulation flow, and recommended that sprayers should be selected with an independent pump. Certain models such as the Jacto PL50bv have a centrifugal pump fitted, connected with a belt drive to the main motor; others, such as the Stihl machines have "booster pumps" available as expensive (approximately $100) optional extras.

3 Spinning disc sprayers

Rotary atomization is the most practical means of achieving controlled droplet application of biological pesticides: it maximizes droplet production within appropriate ranges for the biological target (see Table 2). There are essentially two types of hand carried battery-operated spinning disc sprayer. One designed for insecticides and fungicides, has a long handle in which batteries are carried to operate a small DC motor to drive the spinning disc. Spray in a small (1 liter) container is fed by gravity through a restrictor to the rotating disc. The disc is held above the crop and relatively small droplets are carried downwind and into the crop by air turbulence and gravity (Figure 10). Two use strategies for spinning disc sprayers have been described in Chapter I-2: conventional ULV spraying (< 5 liters/ha) with oil-based formulations and very low volume (VLV) application with water-based mixtures at 5–30 liters per hectare. As the swath will vary depending on wind velocity, the passes through a crop (track or lane separation) are decided with reference to crop structure and should allow for low wind speeds. Successful applications of this type typically achieve 20–50 droplets/cm^2 on foliage.

The second type of spinning disc sprayer designed for herbicide application, has either a governed motor or alternative system to ensure a slow disc speed for droplets of > 200 μm that are deposited rapidly by gravity. The disc is held just above the weeds so downwind displacement of droplets is minimal. Spray deposits of 5–10 droplets/cm^2 on the ground or foliage are typical.

Checking flow rate is a simple process with modern, single-disc sprayers (Figure 11) and the latest atomizers (such as the Ulva+) have convenient holes at the base of the atomizer cup, so it need not be removed for calibration. The viscosity of certain oil formulations (particularly those with a high vegetable oil content) can change substantially with temperature (they have a low viscosity index), and must be checked under normal working conditions. Consequently, it is often necessary to change the restrictor and/or adjust the track spacing to achieve

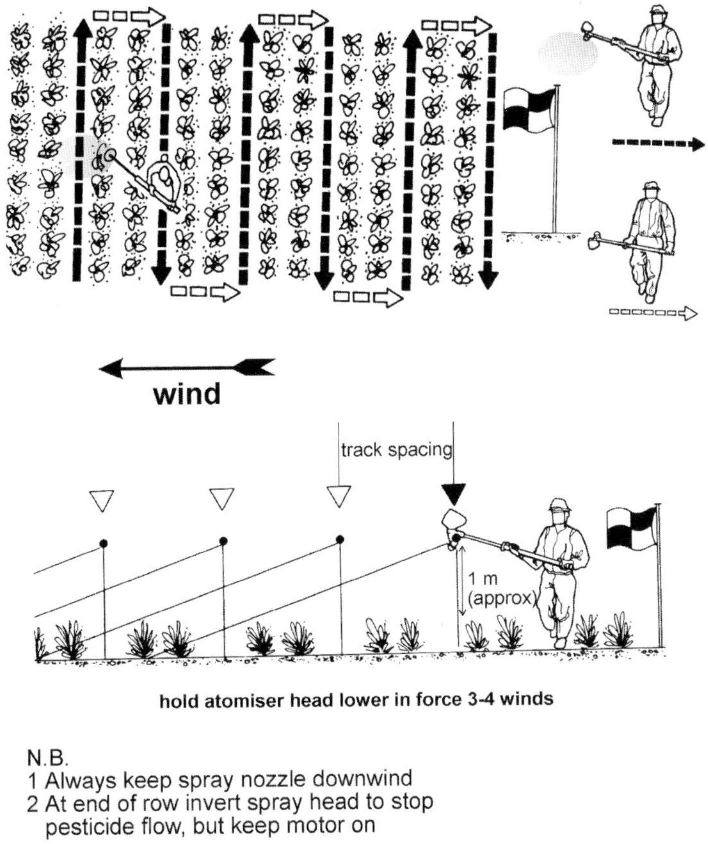

Figure 10. Spraying rows of a crop with a hand-held spinning disc sprayer

a given VAR, using formulae (3) and (4). Flow rate calibration with vehicle mounted sprayers involves adjustment of a needle valve or (preferably) changing a fixed orifice restrictor. The latest version of the 'Ulvamast' (version III) has an electronically controlled positive displacement gear pump, so fluid flow can simply be pre-set with a dial.

Particulate suspension formulations are often viscous and have non-Newtonian properties: they include suspension concentrates (SC) and oil-miscible flowable concentrates (OF) both of which must be diluted before use. Figure 12 shows the flow rates of serial dilutions of a *B. thuringiensis* SC product ('Biobit XL') in water, through an 'Ulva+' (fitted with a black restrictor). The flow rates for mixtures containing up to 30% SC are little different from those of water, but then decrease with the increase in viscosity accompanying higher concentration. The resulting curve for flow of active ingredient shows a peak at approximately 50% dilution, although there is little change over the 40–70% range. In practical terms, only the flow rate (affecting the volume application rate, coverage and work rate) would change. Paradoxically, the preparation of higher concentration mixtures (in the case illustrated, $>$ 70% product) substantially reduces the flow of active ingredient (a.i.).

D Tractor and other vehicle mounted sprayers

1 Conventional boom

Large arable areas are treated by a series of hydraulic nozzles mounted across a horizontal boom, up to 36m in width. Booms may be attached: directly to tractors (or alternative low ground pressure vehicles), to trailed sprayers, or to large, dedicated self-propelled spraying vehicles. A review of large-scale ground-based equipment is given by Robinson (1993). The spray tank may be up to 1000 liter capacity mounted on the tractor, but larger tanks are

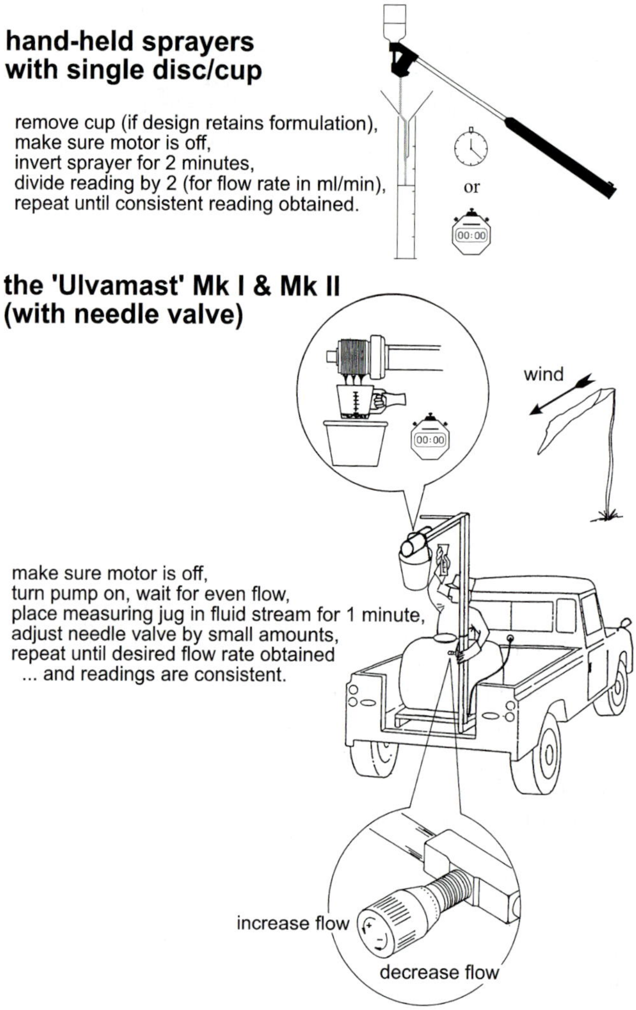

Figure 11. Sprayer calibration: rotary atomizers

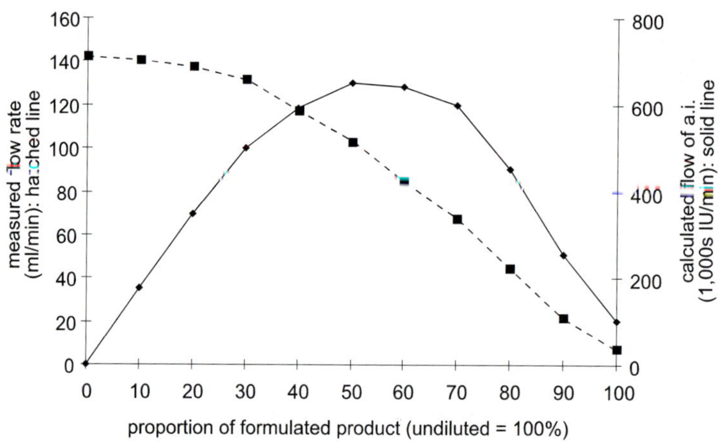

Figure 12. Flow characteristics through an 'Ulva+' of dilutions of a *B. thuringiensis*. SC formulation (see text)

trailed. Diaphragm or piston pumps may be used with sufficient capacity to supply the maximum number of high output nozzles on the boom plus a surplus for re-circulation to the tank to provide agitation. Alternatively it is also possible to use a roller vane pump. From the pump, spray liquid is fed to a pressure regulator with a return flow to the tank.

Calibration is essentially the same as for manual equipment, but it is important to check all the nozzles periodically to check for:

1. uniformity of flow rate across the boom, and
2. erosion of nozzle orifices, causing an increase in output.

2 Air-assisted boom

Some tractor-mounted boom sprayers are equipped with a sleeve through which air is ducted to provide a curtain of air alongside the hydraulic nozzles and project the spray into the crop (Figure 13). This air entrainment improves deposition within a crop canopy and reduces downwind drift of small droplets, provided there is sufficient canopy to collect the spray (Taylor and Andersen, 1991). Fans are often the axial type, but some sprayers have a centrifugal fan to produce a higher air velocity through a smaller duct and also to use air shear twin fluid nozzles.

The vortices created by air assistance help to improve coverage on the under-sides of leaves (which is frequently essential for biopesticides). Another approach with vehicle mounted booms is to attach drop legs that pass in between crop rows; these can either be fitted with diagonally upwardly pointing nozzles – as in the tail boom (Figure 4) – or by an air shear nozzle (Figure 14). As with the 'Airtec' nozzle (see section 2.C.2), a compressor is required and supplied with the 'Dropspray' rig.

3 Orchard sprayers

Deciduous tree orchards provide a wide diversity of targets due to the variation in tree height row spacing and density of the crop canopy. Opinions differ about the volume of spray required, but often trees have been sprayed to runoff using in excess of 2000 liters/ha. Invariably spray has been directed upwards to achieve adequate coverage of the tops of trees, but much of this spray, particularly the larger droplets, falls to the ground and the smaller droplets projected over the top of the crop have led to downwind drift over considerable distances. Optimization of coverage therefore requires careful matching of forward speed, output and the direction of droplet laden air. Tractor mounted or trailed air assisted sprayers are most commonly used in conventional orchards; axial fans (Figure 15) are fitted to many orchard sprayers and provide the high air volumes (relative to forward speed) needed

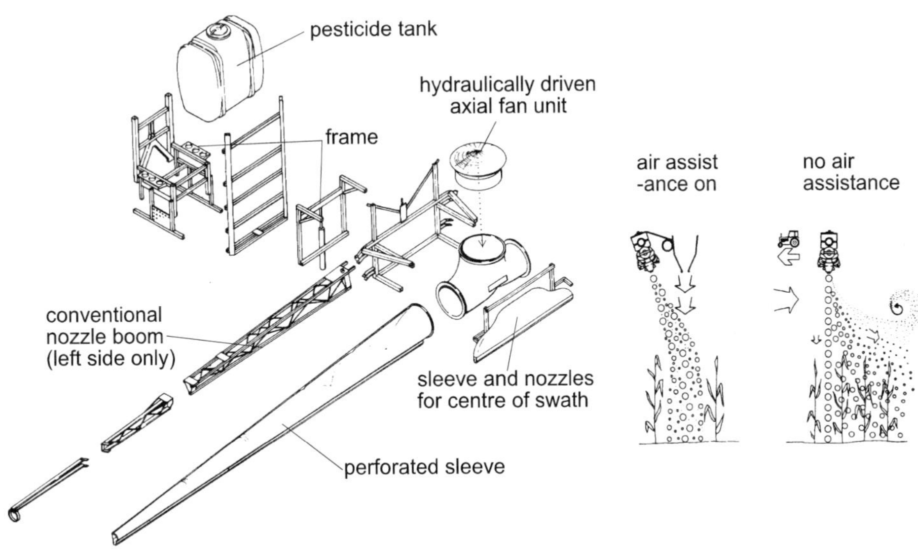

Figure 13. A perforated air sleeve tractor boom sprayer (Hardi 'Twin')

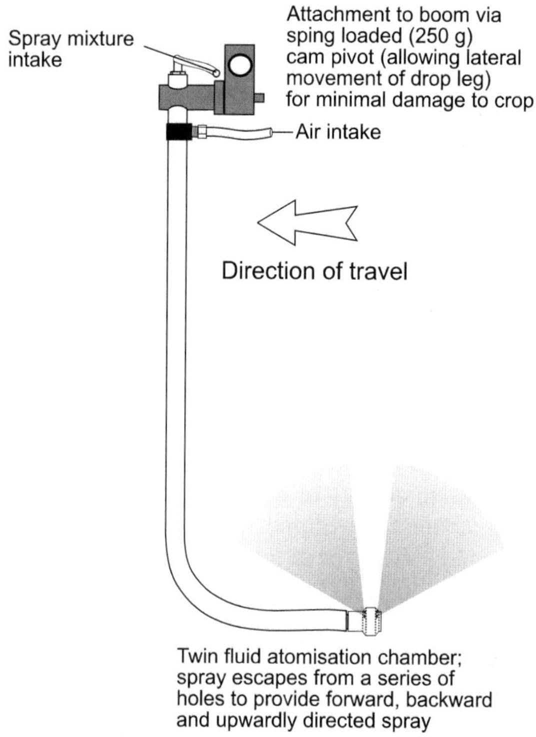

Figure 14. The tractor boom mounted Micron 'Dropspray' (formerly the Benest 'Dropleg')

to optimize spray distribution within the crop (Randall, 1971).

Recently the trend has been to growing dwarf rootstocks (apples) and multiple row of trees in beds. This has allowed use of vertical ducts to project spray horizontally at the trees, reducing drift (an issue for all types of pesticide in N. America) and the distance from nozzle to target. Some of these sprayers utilize cross-flow fans rather than the axial type. Dwarfing rootstocks has also led to the development of tunnel sprayers (Cross and Berrie, 1995) and other equipment that recaptures spray projected through a tree to recycle it. In one version, the tunnel sprayer is designed for automatic travel through the orchard. Experimental sprayers have been developed to assess the value of adding an electrostatic charge to orchard sprayers to minimize drift. Work is also in progress to assess newer types of nozzle, including air induction nozzles; one problem is that the larger droplet size spectra produced are less easily conveyed in the air flow.

With the basic axial fan type sprayer, manual adjustment of vanes, and operation of specific nozzles allows adjustments to suit the different tree heights in each orchard being treated. This requires the operator to adjust for the geometry of each orchard. Multiple nozzle (turret) units facilitate changes in VAR (or hydraulic nozzles). On/off solenoid switches also ease these adjustments and are especially useful in hilly orchards. Advances in sensor technologies allow automatic adjustments by the machine to allow for missing trees, replants, and different heights within a row, saving considerable amounts of pesticide.

VARs and forward speeds may be adjusted during the season as the crop leaf index increases with foliar growth and maturation of the crop. Audits of: active ingredient distribution, spray coverage and total deposits per unit emission show decreased retention in the spring with increased retention towards harvest. As tree row distances and tree width are decreased, there appears to be more opportunity to increase both forward speed of the sprayer as well as air velocities so as to utilize air turbulence and reduce drift. Alternate row spraying with increased tree rows/hectare is also a technique advanced for reducing pesticide load, enhancing IPM and improving the economics of application; this is especially useful for small trees early in the season.

Tank sizes vary considerably, some having a capacity of over 2000 liters on trailed equipment. Traditionally the tank was filled through the opening at the top, but Health and Safety authorities are increasingly demanding closed transfer and other systems of filling to reduce operator exposure to the undiluted pesticide. Various systems have been proposed to calculate the dosage required. Raisigl and Felber (1991) describe a kit developed for calibrating air-assisted, tractor-drawn mistblowers. The tree-row-volume (TRV) relationship (Sutton and Unrath, 1984) has not been widely used but relates VAR with the cubic capacity of the tree canopy. Furness *et al.* (1998) describe the Australian unit canopy row system (UCR) where tree and vine sprayer calibration is based on canopy size and length of row. In the UK, a system called Pesticide dosage Adjustment to the Crop Environment (PACE) was developed to change fungicide dosage according to canopy density optimizes pesticide application on tree crops (Cross *et al.*, 2004) and would also

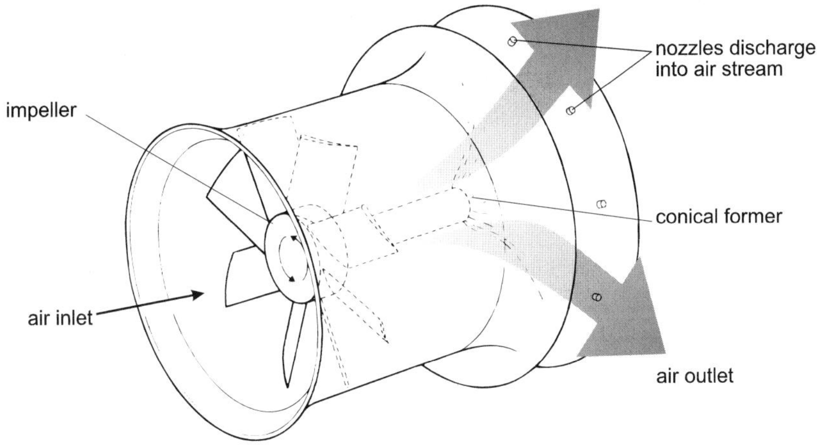

Figure 15. Axial flow fan of an orchard sprayer

be expected to improve application of biopesticide. Low single rows of trees and vines can be sprayed using tunnel sprayers to minimize wind affecting spray distribution (Holownicki *et al.*, 1996)

Thus the tree height, width of the canopy crown, row spacing and length of row need to be known. An assessment of the density of foliage in the canopy is also needed. For example:

If rows are 6 m apart, trees 4 m. high, and 3 m. wide, then (with 10, 000 m^2/ha):

- the length of row is 10, 000/6 = 1667m;
- $1667 \times 4 \times 3 = 20,000\,m^3$ of foliage per hectare;
- the spray volume applied to this volume of foliage will vary from 10 to 100 liters per $1000\,m^3$;
- so if we selected 50 liters, then VAR $- 20,000\,m^3 \times 50/1000 = 1000$ liters/ha.

In assessing the application of (often more expensive) biological agents, not only is the tree canopy important, but also the locality and distribution of the target pest within the canopy. The target may be confined to new foliage on the outer canopy, or it may be that the fruit (rather than foliage) requires protection. Hall (1993) obtained a 6 fold increase in coverage efficiency as the canopy volume decreased from approximately $33\,m^3$ with pyramid canopies to $17.8\,m^3$ per tree on a trellis planting. Doruchowski *et al.* (1996) also showed that in similar plantings the dosage could be reduced ca 50% by improved matching of VAR with crop canopy volume. Applications of entomopathogenic nematodes have utilized very high VARs without an air blast to enable the tree trunk to be wetted in early autumn.

Relatively few studies have examined the optimization of droplet size on tree crops due to the wide diversity of target surfaces within a crop canopy. Smaller droplets are more suitable for transport in air flows and provide better coverage with minimal VARs, but are more prone to drift. Most orchard sprayers are fitted with hydraulic nozzles, while some use rotary nozzles. In some tests 140 μm droplets were the most efficient for mildew and scab control with chemical fungicides. Smaller droplets are less likely to be suitable for particulate biological insecticides, while the larger droplets will be more exposed to the influence of gravity and (being larger) will require larger volumes to achieve adequate coverage. At present, the optimum delivery system (nozzles and air volumes) for maximum coverage remain unclear. With the application of biopesticides, which are usually more selective than chemicals, the need for good coverage will require careful adjustment of equipment to optimize delivery and minimize wastage of spray on the orchard ground.

4 Vehicle mounted ULV sprayers

An early vehicle mounted sprayer for locust control used the exhaust gases from the vehicle engine to atomize ULV formulations. Its use has been largely replaced by CDA sprayers, but the exhaust gas nozzle sprayer (ENS) is still available in many countries. The application of conidia of *Metarhizium* with the ENS has been evaluated in the laboratory by Griffiths and Bateman (1997) who showed that spore survival may be sufficient to achieve adequate field efficacy.

Vehicle mounted CDA sprayers either rely on the natural wind to disperse the spray or initially have an air-blast to project spray upwards. Downwind displacement of the spray with both types is very much dependent on the wind speed and amount of turbulence in relation to the extent and type of vegetation present. Provision of the air-blast requires a separate 4-stoke engine to drive the fan. The atomizer consists of multiple spinning discs or rotary cages to accommodate the higher flow rates needed for the vehicle speed. An electrically operated pump is provided and on the latest version an electronic control box facilitates calibration. None of these machines is equipped with any agitation in the tank.

In some cases orchard sprayers have been adapted to fit on the flat bed of a truck. Other vehicle mounted air-assisted sprayers are those with vortical nozzles to produce cold fogs used in mosquito control. Fogging droplets are extremely small ($< 30\,\mu m$), and are too small for most biopesticides except *B. thuringiensis* and viruses.

E Baits, granules and dusts

1 Features

Bait, granule and dust formulations have the disadvantage that the inert filler is usually consists of particles of irregular shape and size. Clays such as attapulgite or dried waste products such as citrus peel and maize corn grits are used (Burges, 1998). Dusts are extremely small particles, liable to be inhaled and prone to be carried by the wind. Dusts are seldom used with conventional pesticides except for treating grain in stores, although they may also have a role for small scale grasshopper control, where application equipment is unavailable (but see 2.E.2).

Granules are larger particles used mainly for soil treatments, but when small micro-granules are applied to foliage they can be caught in leaf axils. Irregularly shaped particles are difficult to dispense at a uniform rate and variation in size may result in winnowing, so that the pesticide is deposited unevenly. The desirable properties and use of granules of chemical pesticides have been reviewed by Walker (1976).

Granules can be coated with a polymer to form a spherical particle to facilitate flow and more accurate metering. Release of the pesticide is extended over a longer period by including a polymer, but this adds to the cost of formulation, which is inherently high due to the low proportion of active ingredient. Baits may be formulated with a water repellent to improve their stability under wet conditions. When controlling pests such as leaf-cutting ants, a slow-acting insecticide is often preferred in baits (a trait favoring microbial products) so that more of the bait is carried back to the nests.

Granules are usually formulated so that the product falls through a particular size of screen, but is retained by a smaller size screen (see Table 5); thus a product may have 30/60 granules, where the sieves give a product of 246–540μm particles. Application rate is also affected by the bulk density so equipment needs to be calibrated for each batch of granules. Machines with a rotary metering system are preferred over those with a static aperture

Table 5. Granule sizes and estimated numbers per unit area (from Matthews, 2000)

Mesh size: wires per inch (= 25.4 mm)	Particle size (μm)	Calculated number* of granules per m^2 @1 kg/ha
8/15	2360–1080	32
15/30	1080–540	253
20/40	830–400	817
30/60	540–246	2712
80/120 (micro-granules)	200–80	78, 125

* Example based on an attapulgite base: the number of granules per kilogram will depend on whether they are dried or calcined. The number of granules per plant will depend on the crop spacing.

through which the granules have to pass. Bridging of particles over a fixed opening may occur resulting in intermittent application. Granules generally fall by gravity from the metering system through a tube to the dispenser, but some machines have forced air to provide distribution over a wider swath.

In their original simple form, dusts were relatively cheap to formulate, the major cost being transportation of the bulky carrier. Fine dusts with a particle size of < 50 μm are subject to wind drift, there is often a poor (< 20%) recovery on foliage, and the active ingredient may be winnowed from its carrier. There may also be objections to the widespread use of dusts on human safety grounds: with a particle size in the range 2–9 μm, many anamorphic fungal conidia can be inhaled efficiently and deposited in the alveoli (Clay and Clarke, 1987). Inhalation studies must therefore be an essential part of product safety testing. In an effort to ameliorate these effects, 80–200 μm micro-granules have been developed (especially for rice in Asia) but these are expensive to formulate (Matthews, 2000).

Unless they are very sticky, baits can be applied using the same equipment as granular fertilizers and pesticides. This ranges from simple hand held equipment such as "horn seeders" and "pepper-pots" (Figure 16) to large vehicle mounted applicators (Matthews, 2000). Machines should contain a device for metering the material - typically using a positive displacement rotor. Soft baits can clog this mechanism and Johnson and Goettel (1993) used a hydraulically driven rubber belt mechanism at the base of the hopper. Particles are discharged at speed either by use of a spinner or a blower unit. Small to medium sized field trials can be carried out using motorized mistblowers, most of which can be converted for dust or granule application.

2 *Use with biopesticides*

Feng *et al.* (1994) review the use of *Beauveria bassiana* preparations in China (and elsewhere); dusts have been tested often for field crops (including rice, soy beans, and vegetables) and forest trees. Moscardi and Sosa Gómez describe dust applications of entomopathogenic fungi to hemipteran pests of soybeans (see Chapter VII-5). In early field demonstrations of grasshopper population reduction using *B. bassiana*, conidia were mixed with a wheat bran carrier and applied with a motor driven blower (Johnson and Goettel, 1993). Subsequent examination of residues indicated that the conidia had separated from the bait and were effectively applied as a dust.

Granular formulations of *B. bassiana* developed in France include 'Ostrinil' – for use against corn borers (*Ostrinia nubilalis*); since it has a low mammalian toxicity, it can be hand distributed into the central whorls of

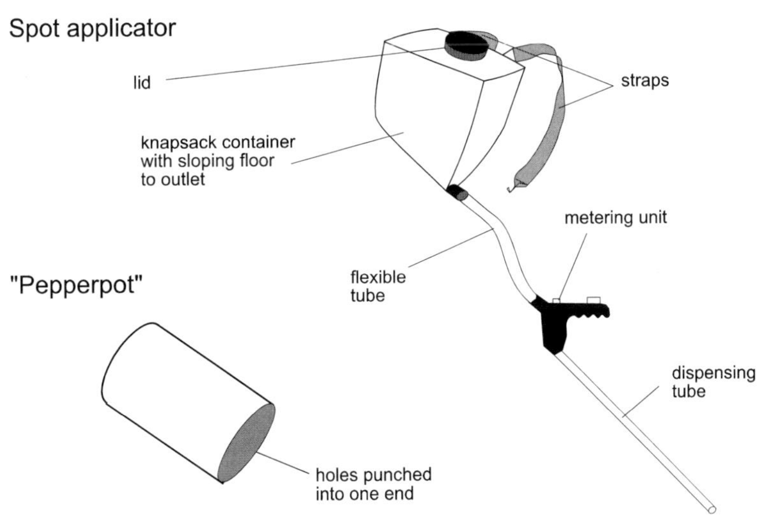

Figure 16. Simple hand-held equipment for distributing granules or granular baits

maize plants where application equipment is unavailable.

The development of bait formulations to "lure and kill" grasshoppers is discussed by Moore and Caudwell (1997). Zimmermann (1993) has reviewed the development of *Metarhizium* products, including an interesting granular formulation of *M. anisopliae* ('Bio-1020') for black vine weevil (*Otiorhynchus sulcatus*) control. Dried pellets were prepared directly from the fermentation product for inclusion in soil, where the granules produced infective conidia. Unfortunately, its marketing was halted for commercial reasons.

F Specialized application systems: soil drenches, etc.

1 Soil application

Copping (2004) lists at least 19 micro-organisms that are usually applied to soil, in high volumes of water by watering-on, drenching or through a drip irrigation system. Whether the dilution involved is beneficial has not been established and it may be that granules should be used (Burges, 1998), or the large volume of water should be applied separately to minimize dilution of formulation additives. Thus a dual boom system could be envisaged with an optimized application of the biopesticide followed by the high volume "irrigation" to disperse the agent into the soil profile.

Soil drenches have been used mainly with plant symbionts, plant disease antagonists and EPNs, which need the water to move through the soil. If soil is too dry, organisms are likely to desiccate or have only local effect. Steinkraus *et al.* (Chapter VII-6) describes the use of various irrigation and sprinkler systems (as well as sprays) for the application of nematodes in American cotton. The formulation of organisms "with a power of search" has been described by Georgis and Kaya in the comprehensive review of microbial biopesticide formulations by Burges (1998).

The application of soluble chemicals through drip irrigation systems has become a common practice in modern irrigated agriculture (Papadopoulos, 1996). Drip irrigation lines can be utilized as an application tool for EPN (see Chapter VII-13). There is however a drop off in EPN output towards the end of each line, the size of which will be determined by the specifications of the irrigation model (Brown A. P., personal communication). This drop off in EPN output with increased distance needs to be considered by a grower when designing an application regime. Application of EPN via drip irrigation systems is a practice used by growers in the UK as a method of applying the control agents to soil dwelling pests which might otherwise be hard to target (*e.g.*, infested soil is covered by plastic sheeting). This method of application has potential to allow highly targeted application. Utilizing the already installed drip irrigation system as an application tool adds value to the expensive investment such a system costs a grower.

Drenches have also been used for plant pathogenic nematode control. The fungus *Paecilomyces lilacinus* was originally developed in Australia and the Philippines, by adding a growth medium that feeds both the fungus and the plant. Later produced in S. Africa as 'PL Plus' it is a combined formulation of wettable powder and soluble nutrients. The fungal agent is applied in a water suspension to plant holes; seeds or tubers at planting, the roots of plantlets, or the soil at the base of established plants. Favorable environmental conditions are essential for the success of the product; these include adequate soil moisture content at time of application and planting of the crop and a well-prepared seedbed.

The application of soluble chemicals through drip irrigation systems has become a common practice in modern irrigated agriculture. Drip irrigation lines can be utilized as an application tool for EPN. There is however a drop off in EPN output towards the end of each line, the size of which will be determined by the specifications of the irrigation model (Brown A. P., personal communication). This drop off in EPN output with increased distance needs to be considered by a grower when designing an application regime. Application of EPNs via drip irrigation systems is a practice used by growers in the UK as a method of applying the control agents to soil dwelling pests which might otherwise be hard to target (*e.g.*, infested soil is covered by plastic sheeting). This highly targeted method of appli cation utilizes already installed drip irrigation equipment and thus adds value to the expensive investments incurred by growers.

The efficacy of biopesticides for use against soil pests can be enhanced substantially by drilling below the soil surface. A good example is the use of modified seed drills for soil injection of a liquid ferment product of *Serratia entomophila* ('Invade') against grass scarab grubs (*Costelytra zealandica*) in New Zealand.

Pinnock and Mullens (Chapter VII-21) refer to jetting equipment, which is widely used for applying control agents against sheep ectoparasites as an alternative to plunge dipping. It has a small hydraulic pump with re-circulating bypass supplying a crook-shaped hand wand which usually carries four nozzles.

4 Conclusions

In this chapter we have reviewed not only the application techniques that have already been used for the application of biopesticides, but also techniques that are undergoing field testing. We have assumed that efforts to optimize dose transfer and work rate will be rewarded by improved efficacy and cost effectiveness. We especially draw readers' attention to methods that improve "coverage" and thus the likelihood of contact with the target pest (especially if the pest is small and/or sedentary).

A Application equipment at the field trial stage

Clearly the first hurdle to overcome is whether the biopesticide will work under field conditions, so the use of commercial application techniques is often disregarded at this stage. Assuming that the target is foliage, the next step is to assess how much of the biopesticide is needed at prime sites (such as the under-surfaces of leaves), with reference to droplets size, spray concentration and volume criteria.

Initial field trials can be on quite small plots so field trials require portable applicators (knapsack, manual or motorized) that achieve appropriate deposition on target. Using relatively small equipment allows manageable assessment of the actual quantity delivered, and can enable several variations of an application (different nozzles, volume application rates, *etc.*) to be replicated. When adequate control of the pest has been demonstrated, the translation of these application criteria can be focused on which practical application technique will deliver the biopesticide most economically.

Selection of full scale application equipment should therefore begin at an early stage in field testing; some points to consider are:

1. If you are not familiar with the application equipment, try it out using water or an appropriate blank formulation. *Time spent in reconnaissance is never wasted.*
2. Remember to calibrate equipment appropriately before the trial under realistic conditions (*i.e.*, a <u>sustainable</u> walking pace, a realistic VAR, *etc*).
3. Always consider the use of tracers to find out what is really happening to spray droplets.
4. Check carefully the amount of biopesticide retained inside the tank (does it settle out?) or is it collected on filters/nozzles.
5. With prototype biopesticide formulations it is important to check for viability after passage through spray equipment.
6. Trials must be staged to appropriately translate screen-house and other "pre-field trial" methods (including application) into large scale field use strategies.
7. Consider the eventual cost of labor (*e.g.*, work rate vs. maximizing coverage).
8. It is important to consider formulation and application together as a complete "delivery system" at an early stage.
9. True biopesticides are living organisms capable of reproduction and, whereas the "chemical model" can be a useful aid to understanding the dose transfer mechanism, they may also have profound long-term effects that are less influenced by application techniques.

It will be important to assess whether any pest control failure is due to:

1. insufficient biopesticide being deposited on target,
2. problems due to rapid loss of active residue after deposition due to UV light, rain washing off, *etc.*
3. lack of biopesticide transfer to the pest (*e.g.*, low mobility, no penetration of a leaf cuticle).

Combination with an additive or use of an unconventional application may be needed to ensure availability of the biopesticide where it will function. Unconventional applications may include a twin application: *e.g.*, a surfactant or some other initial treatment to alter the leaf or insect cuticle to allow greater contact with the biopesticide. Alternatively attraction of the pest to a pheromone trap may be needed to infect part of a population which is then used as the delivery system to distribute the biopesticide on their bodies when the trapped insects are released. Retention of sufficient moisture on a leaf surface may be another crucial factor that will decide whether the biopesticide can survive until the pest is in contact with the biopesticide.

B Application equipment for operations

If the use of biopesticides had been as straight forward as applying a chemical pesticide, they would be more widely used. There is an inherent problem that biopesticides are more specific and thus an application must be more closely related to the biology of the pest, and the most susceptible stage of the pest often needs to be targeted.

We have repeatedly stated the need to conform to standard application equipment whenever possible, but is there a better way? Once the biopesticide has worked in the field under small plot conditions, the translation to larger equipment may require different approaches depending on how unique the situation is for optimizing biopesticide application. The more similar it is to applying a conventional chemical pesticide, the easier it will be to scale up the process, but some aspects of the design of equipment must be carefully thought out. Will there be sufficient agitation to keep a particulate formulation in suspension (especially with larger tanks)? If it is necessary to make alterations to equipment, how much modification will be acceptable to farmers and operators? Only time and experience will tell, but if microbial agents are to be utilized more extensively, more scientists must take an interest in dose transfer mechanisms and the practicalities associated with them.

5 Acknowledgments

We much appreciate the comments of Dr. H. Denis Burges and the late Dr. Chris Lomer in the preparation of this chapter.

6 References

Bache, D. H. and Johnstone, D. R. 1992. "Microclimate and spray dispersion". Ellis Horwood, Chichester, England. 238 pages.

Bateman, R. P. 1993. Simple, standardised methods for recording droplet measurements and estimation of deposits from controlled droplet applications. *Crop Prot.* **12**, 201–206.

Bateman, R. P. 1997. Methods of application of microbial pesticide formulations for the control of grasshoppers and locusts. *Mem. Entomol. Soc. Can.* **171**, 67–79.

Bateman, R. P. 1999. Delivery systems and Protocols for Biopesticides. In "Biopesticides: Use and Delivery" Eds. F. R. Hall and J. Menn, pp. 509–528. Humana Press, Totowa, NJ, USA.

Bateman, R. P. and Alves, R. T. 2000. Delivery systems for mycoinsecticides using oil-based formulations. *Aspects Appl. Biol.*. **57**, 163–170.

Bergeron, V., Martin, J. Y. and Vovelle, L. 1998. Interaction of droplets with a surface: impact and adhesion, *Proc. Fifth Intern. Symp. Adjuvants for Agrochemicals: Memphis, TN, USA*, **Vol. 1**, 377–382.

Burges, H. D. (ed.) 1998. "Formulation of Microbial Biopesticides, beneficial microorganisms, nematodes and seed treatments". Publ. Kluwer Academic, Dordrecht, The Netherlands.

Burges, H. D. and Jarrett, P. 1979. Application and distribution *of Bacillus thuringiensis* for control of tomato moth in glasshouses. *Proc. 1979 British Crop Prot. Conf.* 433–39.

Butler Ellis, M. C., Tuck, C. R. and Miller, P. C. H. 1997. The effect of some adjuvants on sprays produced by agricultural flat fan nozzles. *Crop Prot.* **16**, 41–50.

Cherry, A., Jenkins, N., Heviefo, G., Bateman, R. and Lomer, C. 1999 A West African pilot scale production plant for aerial conidia of *Metarhizium* spp. for use as a mycoinsecticide against locusts and grasshoppers. *Biocontrol Sci. Technol.* **9**, 35–51.

Clay, M. M. and Clarke, S. W. 1987. Effect of nebulised aerosol size on lung deposition in patients with mild asthma. *Thorax* **42**, 190–194.

Cooke, B. K. and Hislop, E. C. 1993. Spray tracing techniques. Ch. 5. In: *Application Technology for Crop Protection*, G. A. Matthews and E. C. Hislop Eds. pp. 85–100. CAB International, Wallingford, UK.

Copping L G (Ed.) 2004. *The Manual of Biocontrol Agents* (3rd Edition of *The BioPesticide Manual*). BCPC, Alton, Hants, UK.

CropLife 2002. Catalogue of pesticide formulation types and international coding system 5th Edition. *Croplife Technical Monograph No. 2*, CropLife International (formerly GIFAP and GCPF), Avenue Louise 143, 1050 Brussels, Belgium. 9 pp available on: http://www.croplife.org/monographs.aspx.

Cross, J. and Berrie, A. 1995. Field evaluation of a tunnel sprayer and effects of spray volume at constant drop size on spray deposits and efficacy of disease control on apple. *Ann. Appl. Biol*, **127**, 521–532.

Cross, J. V., Murray, R. A., Walklate, P. J., and Richardson, G. M. 2004. Pesticide dose Adjustment to the Crop Environment (PACE): Efficacy evaluations in UK apple orchards 2002–2003. *Aspects Appl. Biol.* **71**, 287–294.

Doruchowski, G., Svensson, S. and Mordmork, L. 1996. Spray deposit with apple trees of differing sizes and geometry at low, medium and high spray volumes. *IOBC/WPRS Bull.* **19**, 289–294.

Falcon, L. A. and Sorensen, A. A. 1976. Insect Pathogen – U.l.v. combination for crop pest control. *PANS*, **22** 322–326.

Feng, M. G., Poprawski, T. J. and Khachatourians, G. G. 1994. Production, formulation and application of the entomopathogenic fungus *Beauveria bassiana* for insect control: current status. *Biocontrol Sci. Technol.* **4**, 3–34.

Fife J P, Ozkan, H E, Derksen R C. 2004. Physical effects of conventional spray equipment on a biological pesticide. *Aspects Appl. Biol.* **71**, 495–502.

Food and Agriculture Organisation (FAO) of the United Nations 1998. "Agricultural Pesticide Sprayers; Volume 2": FAO technical standards: sprayer specifications and test procedures, *FAO-AGSE*, 63 pp.

Furness, G. O., Magarey, P. A., Miller, P. H. and Drew, H. J. 1998. Fruit tree and vine sprayer calibration based on canopy size and length of row: unit canopy row (UCR) method. *Crop Prot.* **17**, 639–644.

Gelernter, W. D. 2005. Biological control products in a changing landscape. *Br. Crop Production Council (BCPC) Conference - Pests and Diseases*, 293–300.

Griffiths, J. and Bateman, R. P. 1997. Evaluation of the Francome Mk.II Exhaust Nozzle Sprayer to apply oil-based formulations of *Metarhizium flavoviride* for locust control. *Pestic. Sci.*, **51**, 176–184.

Hall, F. R. 1993. Application to plantation crops. In Matthews, G. A. and Hislop, E. C. (Eds.) *Application Technology for Crop Protection* CAB International, Wallingford. 187–213.

Heinrichs, E. A., Chelliah, S., Valencia, S. L., Arceo, M. B., Fabellar, L. T., Aquino, G. B. and Pickin, S. R. 1981. *Manual for testing insecticides on rice*, Ch. 5 (pp. 52–66), IRRI, Los Baños, Philippines.

Helyer, N., Gill, G. and Bywater, A. 1992. Elevated humidities for control of chrysanthemum pests with *Verticillium lecanii. Pestic. Sci*, **36**, 373–378.

Himel, C. M. 1969. The optimum drop size for insecticide spray droplets. *J. Econ. Entomol.* **62**, 919–925.

Holownicki, R., Goszcynski, W., Doruchowski, G., and Nowacka H. 1996. Comparison of apple scab control with traditional and tunnel sprayers at full and reduced chemical rates. *IOBC/WPRS Bull.* **19** 383–384.

Johnson, D. L. and Goettel, M. S. 1993. Reduction of grasshopper populations following field application of the fungus *Beauveria bassiana. Biocontrol Sci. Technol.* **3**, 165–175.

Jarrett, P. and Burges, H. D. 1982. Use of fogs to disseminate pathogens. *Proc. IIIrd Int. Colloq. Invertebr. Pathol.* pp. 49–54.

Jollands, P. 1991. Evaluation of knapsack mistblowers for the control of coffee leaf rust in Papua New Guinea, *BCPC Monograph* **46**, 177–184.

Lello, E. R., Patel, M. N., Matthews, G. A. and Wright D. J. 1996. Application technology for entomopathogenic nematodes. *Crop Prot.* **15**, 567–574.

Matthews, G. A. 1997. Techniques to evaluate insecticide efficacy. *In* "Methods in Ecological & Agricultural Entomology" Ch. 10 pp 243–269. Ed. D. R. Dent and M. P. Walton, CAB International, Wallingford, UK.

Matthews G..A. 2000. "Pesticide Application Methods" 3rd Edition. Blackwell Science, Oxford, UK, 432 pp.

Matthews G..A. and Bateman, R. P. 2004. Classification criteria for Fog and Mist application of pesticides. *Aspects App. Biol.*. **71**, 55–60.

Moore, D. and Caudwell, R. W. 1997. Formulation of entomopathogens for the control of grasshoppers and locusts. *Mem. Ent. Soc. Can.* **171**, 49–67.

Parnell, M. A., King, W. J., Jones, K. A., Ketunuti, U. and Wetchakit, D. 1999. A comparison of motorised knapsack mistblower, medium volume application, and spinning disk, very low volume application of *Helicoverpa armigera* nuclear polyhedrosis virus on cotton in Thailand. *Crop Prot.* **18**, 259–265

Raisigl, U. and Felber, H. 1991. Caliset: safety and ecology in a holdall. *BCPC Monograph*, **46**, 269–270.

Randall, J. M. 1971. The relationship between air volume and pressure on spray distribution in fruit trees. *J. Agric. Eng. Res.* **16**, 1–31.

Robinson, T. H. 1993. Large-scale ground-based application techniques. *In* "Application Technology for Crop Protection" (Matthews, G. A. and Hislop, E. C., Eds.) pp. 163–186. CAB International, Wallingford.

Smits, P. H., Rietstra, I. P. and Vlak, J. M. 1988. Influence of application techniques on the control of beet armyworm larvae (Lepidoptera: Noctuidae) with nuclear polyhedrosis virus. *J. Econ. Entomol*, **81**, 470–475.

Southcombe, E. S. E., Miller, P. C. H., Ganzelmeier, H., van de Zande, J. C., Miralles, A. and Hewitt, A. J. 1997. The international (BCPC) spray classification system including a drift potential factor. *Proc. Brighton Crop Prot. Conf. – Weeds*, **Vol. 1**, 371–380.

Staniland, L. N. 1959. Fluorescent tracer techniques for the study of spray and dust deposits. *J. Agric. Eng. Res.* **4**, 42–81.

Sutton,T. and Unrath, C. 1984. Evaluation of tee-row volume concept with density adjustments in relation to spray deposits in apple trees. *Plant. Dis.* **68**, 480–484.

Sutherland, J. A. 1980. "Mistblowers". Centre for Overseas Pest Research (now NRI), London. 57 pp.

Taylor, W. A. and Andersen, P. G. 1991. Enhancing conventional hydraulic nozzle use with the twin spray system. *BCPC Monograph* **46**, 125–136.

U.S. Congress, Office of Technology Assessment 1990. A plague of locusts – special report, *Washington DC: U.S. Government Printing Office* **OTA-F-450**, 129 pp.

Walker, P. T. 1976. Pesticide granules: development overseas and opportunities for the future. *Brit. Crop Prot. Council Monograph* **18**, 115–121.

Zimmermann, G. 1993. The entomopathogenic fungus *Metarhizium anisopliae* and its potential as a biocontrol agent. *Pestic. Sci.* **37**, 375–379.

Chapter III-2

Conventional application equipment: aerial application

Karl Mierzejewski
Droplet Technologies, State College PA 16801, USA

Richard C. Reardon and Harold Thistle
USDA Forest Service
Forest Health Technology Enterprise Team
Morgantown, WV 26505, USA

Normand R. Dubois
Deceased, formerly with the USDA Forest Service, Northeastern Research Station
Center for Forest Health Research
Hamden, CT 06514, USA

1 Introduction to aerial application

Chemical control of insects, diseases and weeds began in earnest with the introduction of organic pesticides in the 1940s. Chemical pesticides offer a powerful control method and, until recently, their use had been overemphasized to the detriment of research and development of cultural and biological control methods. Impacts on non-target organisms and emergence of resistance to chemical pesticides prompted the development of biopesticides as well as the implementation of integrated pest management, where pesticides are used compatibly as one of many control methods. Even though spray application of biopesticides is based on many of the same principles and technologies as the application of chemical pesticides, their unique biological characteristics and the need to target use within narrow application windows require a better understanding of the application process as well as of the biological and behavioral aspects of the targeted pest. This chapter covers the aerial application of biopesticides and the associated technology.

A Use of aircraft in pesticide application

The popular press occasionally still refers to aerial applicators as "crop dusters". The name dates back to 1922 when the first biplanes were used to apply lead arsenate dusts to control lepidopteran pests in cotton. The aerial application industry began to expand before World War II. This expansion continued after the war, when aerial spraying began to be widely adopted as many larger surplus military trainer aircraft became available and were easily modified by replacing the front seat with a tank for holding pesticides. In the 1970s, the first purpose-designed agricultural aircraft began to appear. Most of these aircraft were monoplanes with piston engines and a high cockpit allowing the pilot a good view of the ground. Today, new agricultural airplanes are mostly fitted with turbine engines, resulting in application speeds

L.A. Lacey and H.K. Kaya (eds.), Field Manual of Techniques in Invertebrate Pathology, 99–126.

of 240 km/h, that enable high productivity rates measured in hundreds of hectares per hour. New microprocessor-based and global navigation-based systems which monitor, measure and even control flow rates and position, are now commonly installed in cockpits.

B Overview of factors influencing the aerial spray process

Successful and efficient application and use of biopesticides are influenced not only by the insecticidal agent and the delivery system but also by the physical characteristics of droplets and formulations, and by dose acquisition and processing by the target pest.

The process of releasing a cloud of droplets (which may range from 30–2,000 μm in diameter) from a fast moving aircraft with the intention of delivering them to a known specific stationary target is fraught with many uncertainties that will limit the accuracy of delivery and deposition. However, many of the factors influencing the formation and delivery of droplets to the target are either controllable or selectable by users. The primary factors influencing deposition on a given target in aerial application are: (1) Droplet size distribution, smaller droplets give better coverage but increase the potential for drift, larger droplets spend less time exposed to the vagaries of the atmosphere but may not give adequate coverage and canopy penetration, (2) Release height, the higher a droplet is released the more time it is available for off target transport and evaporation before deposition, (3) Wind speed, the higher the wind speed, the greater the lateral displacement of the droplet from the flight line, and (4) Humidity, if the formulation has significant volatility, droplets released into low humidity environments can become much smaller before reaching the target and more prone to drift as they evaporate. A further factor that can determine the success of an aerial application is the timing. The timing should match the most favorable phenology of the target pest or target vegetation. In practice, limitations of equipment availability and performance, and narrow target windows serve to limit the range of possible application scenarios. Nevertheless, a complete understanding of these factors and how they affect droplet deposition can greatly influence the success of a biopesticide application.

2 Aerial spray delivery

A Aircraft

1 Aircraft characteristics

a Types

When project planners begin the process of selecting an aircraft for a project, they may encounter many possible aircraft types within the two broad aircraft categories of fixed-wing and rotary-wing. In practice, however, each aerial applicator will generally operate only one category of aircraft and a limited number of specific types of aircraft within that category. Which of these aircraft types is selected to perform the project is a compromise among numerous factors (*e.g.*, maneuverability, payload, proximity to an airport, swath width). An excellent reference on specific aircraft types and their associated parameters is the Aerial Application Equipment Guide 2003 (Kilroy *et al.*, 2003). This reference discusses a large number of aircraft, ranging from purpose-designed agricultural aircraft to modified civil and military aircraft. Rotary-wing aircraft (helicopters) are more expensive to operate than fixed-wing aircraft (airplanes) but do offer some distinct advantages. Helicopters can operate from remote areas using very small landing zones (LZs) close to the area being sprayed, thereby minimizing time and cost that would be spent ferrying an airplane from an airport or airstrip to the spray block. Helicopters can turn more rapidly and fly more slowly than an airplane, enabling smaller areas to be more accurately sprayed. They can maintain a more constant height above ground over steeper slopes in mountainous terrain.

Contrary to commonly held belief, there is no "fan" effect of the rotor blowing a spray cloud downward into a crop or forest canopy for better penetration, or minimization of drift when the helicopter is operated at a normal application speed (90-160 km/h for turbine helicopters).

However, it is possible to operate a helicopter at slower speeds, and the rotor wash may interact with the target foliage and/or the ground (*e.g.*, application of herbicides to maintain a clear area along rights-of-way). This flight regime is unsafe as the helicopter is not able to make a gentle landing in the event of engine malfunction, and productivity is low because of its slow ground speed.

The disadvantages of helicopters are higher operating costs and smaller hopper capacities compared with airplanes of similar size and weight. Fixed-wing aircraft offer the advantages of greater productivity by covering more area per unit of time and carrying a larger load on each flight. Also, purpose-designed agricultural fixed-wing aircraft are designed for pilot safety.

b Productivity

The productivity of an aircraft (area sprayed per unit time) is influenced by several aircraft parameters: the capacity of the spray hopper, application speed, endurance (the time an aircraft can stay airborne without refueling), and lane separation (the width of the 'stripe' of ground covered with each pass – often called swath width or effective swath width). As the size of the aircraft increases, all of these parameters increase as well.

There is generally a close link between the size of the spray project (both in total area and size of individual spray blocks) and the size of the aircraft used to apply the biopesticide. A representative from an aerial application company can provide initial input on the types of aircraft that are suitable for the project. Kilroy *et al.* (2003) is also a good source of information. Use the modified Baltin-Amsden formula (Ekblad *et al.*, 1988, Curbishley *et al.*, 1993) as well as a computer program entitled Computer Assisted Spray Productivity and Efficiency Routine (CASPER) (USDA-Forest Service, Asheville, NC) to compare cost and productivity of several types of aircraft. This subject is covered later in this chapter.

A large aircraft with high productivity will finish a spray project in a shorter period of time than a smaller aircraft. However, larger aircraft, especially converted transports, cannot fly as low as smaller aircraft. Larger aircraft are more apt to cause the spray to drift past the intended site and have a greater turning radius that results in lost spraying time. If the spray blocks are small and oddly shaped, large aircraft will therefore not be able to spray them efficiently. In such a situation, the only solution may be to use several smaller aircraft that together can match the productivity of the larger aircraft, but are able to spray the small blocks more effectively.

Table 1 shows productivity parameters for some commonly used types of spray aircraft within several broad categories of fixed-wing and rotary-wing aircraft. The numbers in this table were gathered from operational data dealing with the application of *Bacillus thuringiensis* var. *kurstaki* (*Btk*) applied at droplet sizes of 100–200 μm volume median diameter (VMD) in broadleaved and coniferous forests. The table should be used only as a broad guideline.

c Number and type of engines (turbine or piston)

The adoption of turbine engines in agricultural aircraft has resulted in the development of fast single-engine aircraft capable of carrying up to 3,800 liters (1,000 gallons) of spray material, although 2,000 liter and 2,700 liter aircraft are more typical. Such aircraft are reliable and productive when compared with older piston-engine types.

Agricultural aircraft are registered in a restricted category as defined by the US Federal Aviation Administration (FAA). This often precludes single-engine aircraft flying over large congested areas, for safety reasons. In such a situation, multi-engine aircraft such as older converted airliners like the DC-3 can be used. Also, projects requiring the rapid spraying of tens of thousands of contiguous hectares may require the use of larger multi-engine aircraft (*e.g.*, DC-4 and DC-6) which are capable of carrying loads of 20,000 liters.

2 Safety

Safety is a major topic that cannot be adequately covered within a few paragraphs. Aircraft that operate close to personnel and equipment during refueling and reloading of biopesticides and fly low present major risks to support personnel

Table 1. Productivity parameters for some commonly used spray aircraft

Aircraft category and type	Hopper capacity (liters)	Lane separation range	
		m	ft
Single-engine Fixed-Wing Aircraft			
Piper Pawnee	550	20–30	65–100
Piper Brave	750	23	75
Cessna Ag Truck, Ag Wagon, Ag Husky	750	23–30	75–100
Ag Cat Model B	1,500	30–40	100–130
Ayres Thrush SR2 - Turbine	1,900	45	150
Ayres Thrush SR2 - Piston	1,900	45	150
PZL M-18 Dromader	1,900	45–53	150–175
Air Tractor 400 Piston	1,700	45	150
Air Tractor 502 Turbine	1,900	53	175
Air Tractor 802 Turbine	3,000	60	200
Multi-engine Fixed-Wing Aircraft			
DC-3	$\leq$ 8,000	75	225
DC-4, DC-6, DC-7	$\leq$ 20,000	120	400
C-130	$\leq$ 20,000	120	400
Beech 18	$\leq$ 2,000	65	200
Rotary-Wing Aircraft			
Bell 47G	300	23	75
Hiller 12E	300	23	75
Hughes/MD 500	450	23–27	75–90
Bell 47G Soloy	700	30	100
Hiller 12E Soloy	600	30	100
Bell 206 Jetranger, Long Ranger	700	30–36	100–120
Bell 204/205/212/412/UH-1	1,300	45	150

who are not familiar with the potential hazards. Therefore, the use of a full service contract will require the contractor to provide well-maintained ground support equipment and trained personnel along with the aircraft and pilot.

The following guidelines will help prevent accident or injury when working around aircraft (USDA, 1996). General guidelines for working safely around both categories of aircraft as well as differences in safety procedures between fixed-wing and rotary-wing aircraft are outlined here.

a General guidelines for working safely around aircraft (Figures 1 & 2)

- Wear personal protective equipment (*e.g.*, safety eyewear, ear plugs, fire-resistant clothing).
- Keep all unauthorized personnel away from the aircraft at all times.
- If the engines are running, stay at least 30 meters away.
- Never smoke within 20 meters of the aircraft.
- Never fuel the aircraft while engines are running.
- Never load the aircraft while engines are running.
- Never work on the aircraft while engines are running.
- Stay away from aircraft propellers and helicopter blades.
- Tell the pilot of any obstacles that may be encountered when flying to a spray block as well as any located within the spray block.

b Guidelines for working safely around fixed-wing aircraft (Figure 1)

- If the aircraft's engine is running, never walk in front of it.
- If you need to approach an aircraft while the engine is running, approach only from the side or rear.
- Always maintain a safe distance from the aircraft's engine–even when it's not running.
- Never put your hand on the propeller. The slightest movement of the propeller could cause the engine to start.

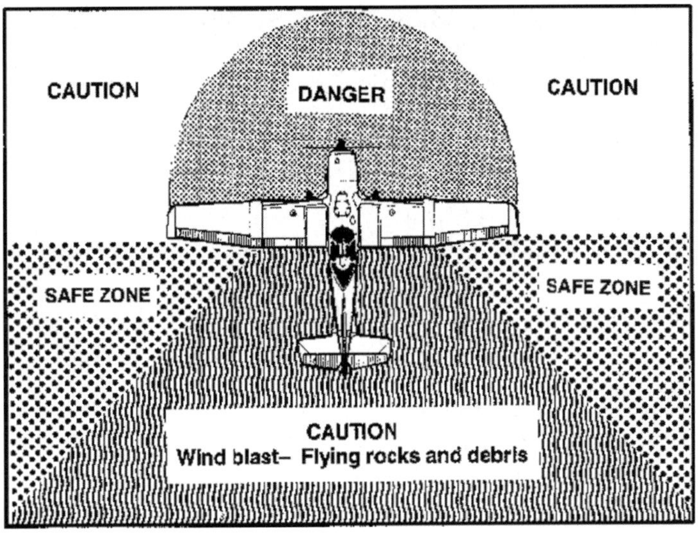

Figure 1. Safe and danger zones when working with single-engine, fixed-wing airplanes (USDA, 1996)

c Guidelines for working safely around rotary-wing aircraft (Figure 2)

- If the helicopter's engine is running, never walk behind it.
- Approach a helicopter only from the front, and then only if the pilot can see you and motions you to proceed.
- Leave a helicopter only from the front or side.
- On sloping ground, never approach the helicopter from high ground.

B Factors affecting droplet formation

The means by which a formulation is atomized and the chemical and physical properties of that formulation are crucial to how it behaves from release to deposition, and how it interacts with physical and biological targets. The selection of the atomizer is the key determinant of the nature of the generated droplet spectrum. Atomizers affect the size of the droplet produced, can influence the width of the droplet spectrum, and can offer a method of adjusting it as conditions change during a spray project. The Droplet Size Distribution (DSD) is the primary determinant in predicting the landing position of an aerially released droplet. The point of deposition of a large droplet is much easier to determine than that of a small droplet.

1 Atomizers

The terms "atomizer" and "nozzle" are often used interchangeably. The term "atomizer" is used to describe a piece of equipment which has moving parts used to apply insecticide, such as a rotary atomizer. The term "nozzle" generally refers to a hydraulic nozzle, a device with no moving parts, not very different from that found at the end of a garden hose. In this chapter, the term "atomizer" is used to describe any kind of device used for effecting break-up of a fluid, and the term "nozzle" solely for hydraulic nozzles.

In order to overcome the surface tension and viscosity forces of a liquid and break it up into droplets, atomizers use different methods: (1) hydraulic nozzles use pressure, (2) gaseous (twin fluid) nozzles use shear between the air and formulation, and (3) rotary atomizers use centrifugal force. Other physical processes can also be harnessed to aid delivery. For example, electrostatic spraying uses hydraulic nozzles and high electrical voltages to charge droplets so that they can get to their targets efficiently.

When selecting an atomizer for an aircraft, it is important to consider the speed of the relative wind at the device and air interface. The air shear forces encountered by slow moving droplets being ejected from an atomizer or nozzle into a rapidly moving aircraft's airstream produces

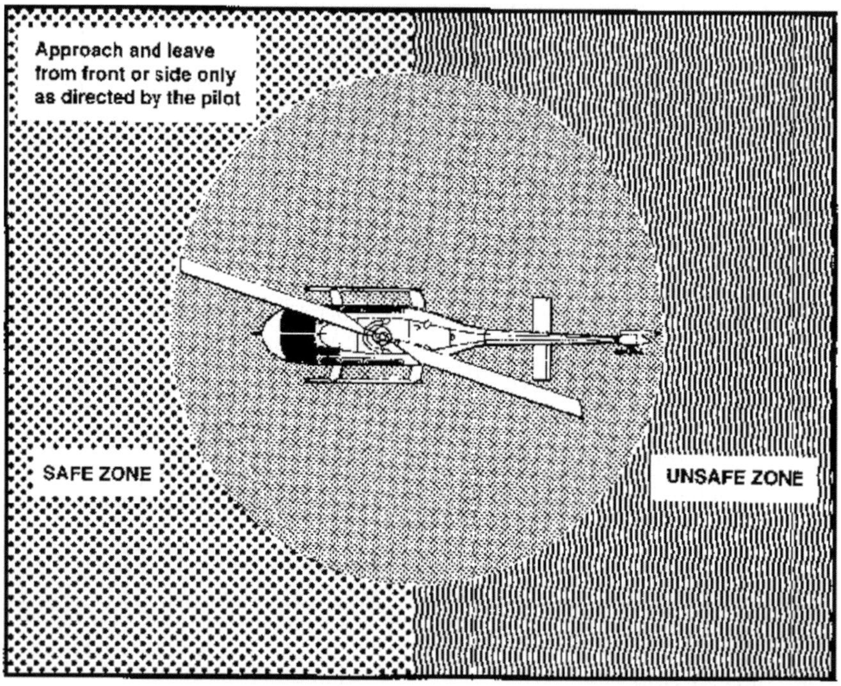

Figure 2. Safe and danger zones when working with helicopters (USDA, 1996). Note that every project should have a detailed safety plan that is made available to all project personnel. Agencies such as the USDA-Forest Service have detailed requirements for the contents of safety plans. Further, in the United States, it is currently a strong recommendation that all projects follow State and Federal Guidelines to insure operational security. These requirements vary by agency (for instance, the USDA-Forest Service has specific lock down and security requirements for parked aircraft during spray projects). Incident reporting is also critical in this context

a blast-shattering effect which results in the formation of small droplets. This shearing can often change a given atomizer DSD from coarse when released from a nozzle on a stationary sprayer to fine when released from the same nozzle at typical aircraft speeds. This effect can be greatly reduced by releasing the biopesticide formulation parallel to the slip stream of the aircraft.

Relative span is a measure of "kurtosis" or narrowness of a droplet size distribution. This number varies over commercial nozzles. None of the commercially available atomizing devices are capable of producing a very narrow (technically referred to as mono-disperse) droplet spectrum. This should not necessarily be regarded as being entirely disadvantageous, as any biological target demonstrates some variability in size, shape or location. Hydraulic nozzles and rotary atomizers are the most commonly used types of atomizing devices used in aerial application.

a Hydraulic nozzles

A very wide range of hydraulic nozzles is available, using pressure to force the formulation through a small opening with sufficient velocity to spread out as a thin sheet. A common characteristic of all hydraulic nozzles is the wide droplet spectrum that each type produces. Manufacturers' improvements have focused on making designs for quick-release from the boom, "dialable" DSD, adopting new hard materials to limit tip wear, and color-coding the tip mounts to make identification easier.

There are four main types of hydraulic nozzles used in aerial application: (1) flat fan, (2) cone (hollow and solid), (3) impact, and (4) needle or solid-stream. When using hydraulic nozzles, major changes in the flow rate can be accomplished by using different sized tips and minor changes made by adjusting the pressure (the flow rate increases in proportion to the square root of the pressure).

Flat fan and cone nozzles use the same one-piece body, check valve (anti-drip device) and filter or strainer components. It is only the orifice and swirl plate that need to be changed to give the nozzle its different characteristic. Figure 3 shows the components of a hydraulic nozzle in cross-section. The tip may be made up of one or more parts. The strainer can be made up of a wire mesh or of a slotted solid metal construction. The nozzle body is usually mounted on a valve seat with a check valve, which opens and closes at 40–70 kPa (6–10 psi) allowing a positive opening and shut-off as pressure builds up and drops in the boom.

Atomizing devices are mounted on a hollow spray boom, which can extend to the length of the aircraft's wing span or rotor span. Devices should not be mounted beyond 75% of the wing or rotor span as those further out contribute strongly to drift (Teske *et al.*, 1998). The boom may be either round or streamlined; therefore, the appropriate mounting brackets are needed for rotary atomizers. Booms should be fitted with bleed lines joining the end of the boom to one of the nozzles to prevent air locks forming at the boom ends.

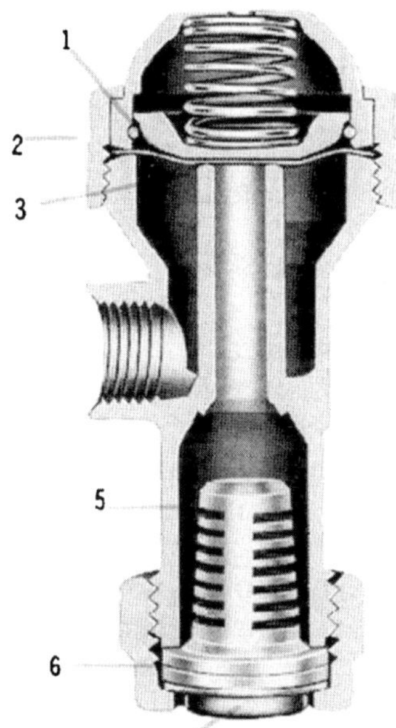

Figure 3. A typical hydraulic nozzle in cross-section. (From spraying systems Co., Wheaton, IL)

1. Diaphragm (synthetic rubber).
2. Nozzle body is usually made of brass.
3. Valve seat.
4. Spring clip (not shown) which secures the cap holding the check valve.
5. Slotted strainer (can also be wire mesh).
6. Nozzle tip (*e.g.*, Core Disc) can be replaced with several different size nozzle tips.
7. Orifice Disc

Flat fan: Principle: Two streams of liquid strike each other at an angle inside the nozzle body creating a flat sheet of liquid in the shape of a fan.

Characteristics:

- Flow rate at a fixed pressure usually 275 kPa (40 psi) varies between 0.38 and 6.8 liters (0.1 and 1.5 gal)/nozzle/minute.
- Tip designations are based on fan angle and flow rate at a pressure of 275 kPa. For example, a Spraying Systems Co. (Wheaton, IL) tip which has 80° fan angle and flow rate of 1.16 liters/min (0.3 US gal/min) is called an 8003 tip.
- The nozzle bodies are usually made of brass and the tips of polymer or stainless steel, both of which show good wear life. Brass tips were once common but are rarely used now because of poor wear characteristics.
- Larger orifices produce larger droplets for decreased drift.
- Widely used in agriculture and forestry as the tips generally show very little tendency to block with most formulations.

Cone *(hollow and solid)*: Principle: Liquid is forced through a swirl plate into a chamber. An air core is formed as the liquid passes with high rotational velocity from the chamber through a circular orifice disc and swirl plate (core). The rotating sheet emerges in the pattern of a hollow cone. A variation passes liquid centrally through the nozzle to produce a solid core. This version generally gives a narrower spray angle and larger droplets. It is common to have the swirl plate separate from the orifice, although combinations are available.

Characteristics:

- Different angle and size of swirl plate alters the tangential speed of the liquid which affects droplet size (faster tangential speeds generate smaller droplets).
- Spraying Systems Co. uses a numbering system based on the orifice size in 0.40 mm (1/64 inch) increments and a core number. D8-46 is an 8/64th inch orifice disc and a 46 swirl plate or core size.

- A wide range of materials is used for cores, with brass being the most used. Orifice disks tend to be very prone to wear, therefore, are made of hardened stainless steel, ceramic or polymer.

Impact: A liquid stream impinges at high velocity on a smooth surface set at an angle to the stream, thereby causing it to break up. Impact nozzles (brand name Floodjet® by Spraying Systems Co. and CP nozzles by CP Products Co., Mesa, AZ) produce large droplets and are generally used where high flow rates are required or when clogging is a problem.

CP nozzles offer flexibility with respect to how much liquid flows through each nozzle, and whether the droplet spectrum is skewed towards larger or smaller droplets. These options are achieved by having a range (usually four) of orifice sizes and a range of angles (three on plastic bodied and two on stainless steel bodied nozzles) on the 'anvil' part of the nozzle. Some CP nozzles also enable a direct stream to be produced, without any impact on an anvil. This flexibility, which enables aerial applicators to use a single set of nozzles for a range of applications, though they use an atomization method that results in a droplet spectrum range that is similar to other types of hydraulic nozzles.

Needle or Solid Stream: Needle nozzles is a generic name for nozzles using fine hollow needles to produce large droplets, typically several millimeters in diameter. Their main use is to apply herbicides from slow-moving helicopters, although they are used on fixed-wing aircraft as well; they are rarely used in biopesticide application. However, if the need arises for a material to be applied in large raindrop-sized droplets, the Thru-Valve Boom (Waldrum Specialties, Inc, Doylestown, PA) or Accu-Flow Nozzles (Bishop Engineering, Hatfield, PA) nozzles would be suitable.

b Rotary atomizers

There are two principal ways that rotational energy is used to break up a liquid: ligament formation (where liquid leaves the edge of the atomizer in curved threads or ligaments that breakdown to form droplets), and sheet or stream break-up. Atomizers using ligament formation utilize spinning discs that produce a narrow range of droplet sizes. However, the fragility, low flow rates and tendency to clog easily have precluded the large-scale commercial development of atomizers based on this principle for the aerial application of biopesticides.

Most commercially available rotary atomizers use sheet or stream break-up where sheets of liquid produced by a feed device in the center of the unit are mechanically broken by impinging on a rotating cage. Although they produce droplet spectra similar in range to hydraulic atomizers, the ability to modify the rotational speed of the units enables adjustment of the droplet sizes. Such effects on droplet sizing are largely independent of the flow rate through the atomizer. The ability to "uncouple" flow rate and droplet spectrum gives these atomizers considerable advantage over hydraulic nozzles. The two most widely used atomizers are Micronair and Beecomist (covered below in further detail). Both of these atomizers can output larger volumes per unit than hydraulic nozzles, and consequently fewer units are needed on an aircraft. Depending upon size, it is common to mount no more than 8 atomizers on a typical single-engine agricultural aircraft.

Bioinsecticides that are applied using ultra low volume (ULV) principles– generally taken to mean application rates lower than 5 liters/ha require fine atomization to ensure effective distribution of the product. Therefore, manufacturers of products such as *Btk* frequently recommend the use of rotary atomizers.

Micronair atomizers: Micronair atomizers (Micron Sprayers Limited, Bromyard, Herefordshire, UK) are high-volume, wind driven atomizers that use a cylindrical spinning cage as the primary method of liquid atomization (Figure 4). They are made to rotate by air flow over fan blades, which are attached radially around the main rotating axis. The speed of rotation can be adjusted by adjusting the blade angle and/or shape and size of the blades.

Micronairs come in different sizes, with different flow rate ranges (Table 2). Although the reported flow rate range for each model is wide, in practice the upper range is somewhat lower than the published figures, as the cages can become overloaded resulting in coarse atomization. The AU3000 model, which was

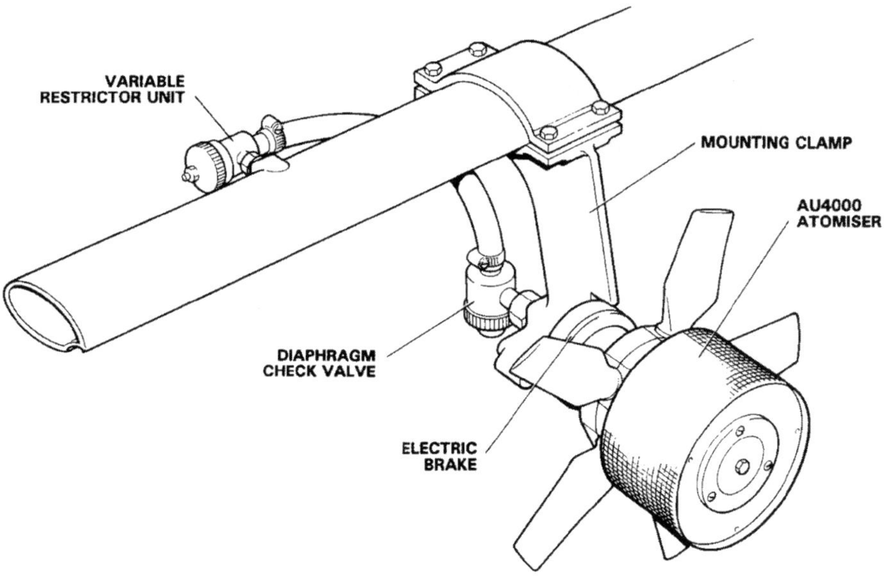

Figure 4. A Micronair AU4000 rotary atomizer, with mount and flow controller (variable restrictor unit) (Source Micronair literature)

Table 2. Micronair atomizers used for aerial application[1]

Model	Flow rate range liters/min (gal/min)	Unit weight kg (lb)
AU3000	0.36–24.4 (0.1–6.5)	3.9 (with brake) (8.6)
AU4000	0.34–23.8 (0.09–6.2)	2.8 (6.4)
AU5000	0.29–19.1 (0.07–5.1)	1.8 (4.0)
AU7000	0.27–12.8 (0.07–3.4)	1.5 (3.3)

[1] Micronair literature.

the first unit manufactured in the 1950s is little used now, and the AU7000 is typically used for very light and slow flying aircraft, such as small helicopters and ultralight airplanes. The various models use identical variable restrictor units (VRUs) to control the flow rate.

Micronair atomizers were developed for use on piston-engine aircraft with maximum application speeds of up to 190 km/h. As fast, turbine-powered agricultural aircraft became more common, the atomizers needed to be modified to limit their rotation speed to 10,000 rpm, the maximum allowable sustained rpm (*e.g.*, the rpm's of individual atomizers can be monitored using tachometers). Micron Sprayers has produced short blades and deflector rings, which limit high air velocities around the cage and serve to limit the effect of wind shear on droplet size.

Beecomist atomizers: Beecomist atomizers (Beecomist Systems Inc., Telford, PA) are compact electrically or hydraulically driven units that use spinning interchangeable perforated or porous sleeves to produce droplets of 20 to 150 μm (Figure 5). The units are designed to turn at high speeds (10,000 rpm) for ULV applications but do not provide a means of speed regulation. The porous sleeves are made to be used with pure liquids, whereas the perforated sleeves can be used with particulates such as aqueous flowable formulations. Porous sleeve assemblies are available in 20, 40, and 60 μm sizes made of high density polyethylene plastic (HDPE), and perforated stainless steel sleeve assemblies in 0.75 mm hole size (producing approximately 80–100 μm sized droplets) and 1.5 mm hole size (producing approximately 100–150 μm sized droplets). Before selecting them as the atomizer of choice, users should check with the manufacturer to ensure that their product can be applied with Beecomist atomizers.

These atomizers are powered by electric (12V or 24V D.C.) or hydraulic motors that match the power source in the aircraft or ground vehicle. The maximum flow rates are 4 liters/min for the 24V system, and 2 liters/min for the 12V system.

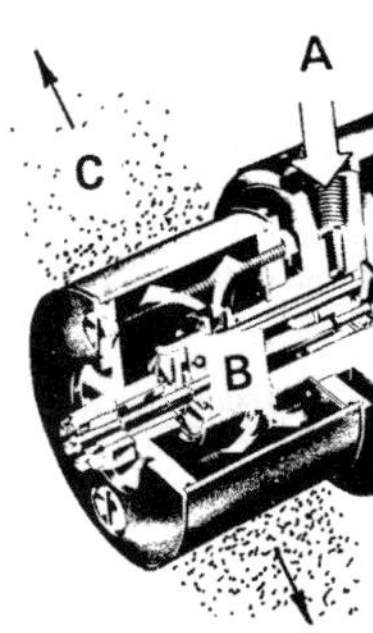

Figure 5. Beecomist rotary atomizer. A. Liquid entry point B. Disperser body (outer sleeve is interchangeable) C. Droplet production D. Power source (electric or hydraulic motor)

c Strainers

Strainers in spray systems are designed and installed to prohibit particles from blocking the system. In-line strainers are used to capture or break up particles that would plug the pump or orifices found in the nozzle tips. The most common mesh size found in aircraft in-line screen type strainers is 50 mesh (openings in a 50 mesh screen are 0.28 mm (0.011 inches) across). Although an undiluted *Btk* formulation will initially pass through a 50 mesh screen, the product solids will eventually build on the screen. As foreign matter collects on the screens, the buildup will occur more rapidly and will eventually cause blocking of the screen.

Table 3 lists some of the mesh sizes in screen and equivalent mesh size slotted type strainers for in-line strainers, and compares them with the orifice sizes found in various atomizers. The table shows that the common, smallest nozzle openings (D-3, 8003 and Micronair VRU setting 1 or 3) are significantly larger than pore size of a 25 slotted or 30 mesh in-line screen. *Btk* manufacturers therefore recommend the use of 30 mesh screens in in-line strainers and 25 mesh equivalent slotted strainers in the nozzles to avoid blockages at the nozzle tips.

Table 3. Pore sizes of small atomizers and nozzle screens

Item	Orifice diameter (mm)
50 mesh screen in-line strainer	0.28
30 mesh screen in-line strainer	0.53
25 mesh equiv. slotted strainer	0.51
Flat Fan 02 series	0.91
Flat Fan 03 series (8003)	1.09
Hollow Cone D-2 orifice	1.04
Hollow Cone D-3 orifice	1.19
Micronair VRU 1	0.76
Micronair VRU 3	1.17

d Atomizer selection

The choice of an atomizing device is largely determined by the required droplet spectrum. Biopesticides such as *Btk* are typically delivered in droplet sizes of 75–200 μm volume median diameter (usually abbreviated to VMD or $D_{v0.5}$). At airspeeds below 190 km/h, rotary atomizers such as the Micronair can better produce much smaller droplets than conventional hydraulic nozzles. An additional advantage of rotary atomizers is the capability to alter droplet size independently of aircraft boom pressure or airspeed by adjusting atomizer rotational speed. At higher airspeeds, small drop diameter ranges are possible with standard hydraulic nozzles through the action of high pressure and wind shear (*i.e.*, directing the nozzle forward at an angle of 45° into the flight direction). Such airspeeds are typically obtained with single-engine turbine agricultural airplanes and multi-engine airplanes. In large *Btk* spray projects requiring large aircraft (frequently converted four-engine passenger airplanes) excellent results have been obtained with standard hydraulic nozzles fitted with large tips.

e Wind tunnel profile

The use of equipment that is capable of sizing moving droplets in wind tunnels has provided a system for studying droplet spectra. Various *Btk* formulations have been evaluated in wind tunnel studies and the droplet spectra are available as inputs to swath pattern simulation and drift models.

There has been some controversy over the quality of the data generated in wind tunnel studies. The discrepancy deals with two droplet sampling techniques, spatial and flux (also called temporal). Typically, a spray from an atomizer measured using the spatial technique will record drops smaller on average than the same spray measured using the flux technique (Schick, 1998). In applications requiring accurate spray deposition such as agriculture, the flux sampling technique is considered by the industry to be more appropriate. This issue is discussed in detail in Fife (2006).

2 *Formulations*

Most active ingredients need to be combined with numerous inerts and adjuvants to produce a formulation or tank mix that can be used in the field. Companies do not disclose the formulation ingredients, making the behavior of a spray formulation in the field difficult to predict. Study of product literature (label, Material Safety Data Sheet [MSDS]), as well as experimentation in the laboratory on the target plant and pest with the commercial formulation will provide some insight into how the product is likely to atomize and evaporate during its deposition phase.

a Physical properties

When aerially applying a biopesticide, various physical characteristics of the formulation or tank mix need to be considered. Some of these characteristics can be obtained from the product literature, whereas others can only be obtained directly from the manufacturer.

Unlike the majority of chemical pesticides, biopesticide formulations cannot be diluted below a given concentration because of their particulate nature. The solids can settle out in these suspension formulations. Discrete particulate suspensions may constrain the amount of active ingredient that can be contained in a given droplet, particularly in very small droplets where the volume may be insufficient to contain large quantities of a pathogen (assume a random distribution in the spray tank and that the number of particles contained in a droplet are related to its volume).

One of the most important characteristics of a formulation is whether it is formulated for undiluted (neat) application or needs to be mixed with water or another inert carrier. Undiluted application is preferred as aircraft have limited payloads. High application rates made with very diluted sprays can greatly increase the cost of an application. When applied in undiluted form, most formulations exhibit a higher viscosity than water. When applying formulations containing living organisms, it is necessary to pay attention to the temperature of the material. Storage of these formulations in small volumes (100 liters or less) in cool conditions (below 5 °C as can occur overnight) can markedly increase their viscosity. The flow rate through previously calibrated spray equipment will be reduced and equipment will need to be recalibrated. Oil-formulations, which have become contaminated by small amounts of water (1% of total volume), can become very viscous as invert emulsions are formed and the material will not flow.

The water content of aqueous formulations can evaporate from droplets making them smaller. Check the product label to determine whether evaporation retardants are included in the formulation. Although such additives may limit some evaporation, care should be taken if the conditions are very warm and dry and if the droplets have to travel considerable distances to reach their target (such as occurs in forestry spraying).

Some *Btk* formulations use vegetable or mineral oil to limit evaporation during spraying, and to aid in spreading on the target plant surface. Do not add small amounts of water to these formulations as invert emulsions (mayonnaise-like mixtures) can result. It is best to flush out any water from loading and spray systems prior to using an oil formulation. If necessary, an oil-miscible, organic solvent (diesel, kerosene, crop oil) can be used to evacuate any remaining water. Trace amounts of water do not adversely affect the spray systems. If it is necessary to dilute the oil formulation before spraying, *the formulation should be added to the water in the mix tank*, never the other way round.

Between 67% and 95% of the droplet spectrum size characteristics of an atomized formulation can be accounted for by the three physical parameters of a spray liquid (Esterly, 1998): (1) maximum extension viscosity (η_{ext}), (2) shear viscosity (η_{shr}), and (3) dynamic surface tension 20 msec after formation at the surface of the final formulation (γ_{dyn}). These physical parameters are not widely published (and little understood by anyone except specialists), it is possible that in the future manufacturers will include these statistics on the labels of products as droplet spectrum simulation software becomes available

(Hermansky, 1998). Statistical models exist that calculate DSD from nozzle type and orientation, aircraft speed and formulation (Kirk 2001, Kirk 2002, Teske *et al.*, 2005).

b Toxicity of active ingredients

Because of their biological origin, biopesticide formulations are frequently aqueous in nature and represent minimal corrosive risk to equipment or safety risk to personnel. The product label details the standard safety precautions that should be taken when handling the material. If questions about safety remain unanswered, call the National Pesticide Telecommunications Network (1-800-858-7378) in the USA, or contact the manufacturer.

c Adjuvants

The term "adjuvants" is used to cover many different additives (stickers, sunscreens, antievaporants, encapsulating materials) which are either included in a biopesticide formulation or can be mixed in the spray tank before application. These are inert ingredients (EPA's List 4) included to protect often delicate biological active ingredients. If manufacturers formulate their products for undiluted application, the adjuvants are carefully balanced so that there is no adverse interaction and their effect on the living organism is minimal. The label often contains warnings against the addition of specific adjuvants, as breakdown of the suspension formulation could occur. Generally speaking, all inert ingredients are under increased regulatory scrutiny and many compounds are under going regulatory review.

Viscoelastic anti-drift additives, which serve to skew the droplet spectrum towards larger droplets, have become popular in agricultural applications due to the increased concerns about drift. In general, such anti-drift agents should not be used with biopesticides as they may adversely affect the stability and palatability of the formulation. Doubts have emerged concerning the effectiveness of such viscoelastic additives. Studies performed at New Mexico State University (Hewitt *et al.*, 1994) showed that their effectiveness greatly diminished after multiple exposures to a high shear rate, such as is typically found inside a pump.

d Accidental spills

Commonly used biopesticides such as *Btk* and nucleopolyhedroviruses (NPVs) usually present a low risk (*e.g.*, low mammalian toxicity) if spilled at aircraft loading sites. Observe the label instructions in dealing with spills. If you have any questions, call the National Pesticide Telecommunications Network (1-800-858-7378) in the USA, or contact the manufacturer.

C Droplet dispersal and deposition

1 Overview of factors affecting dispersal

Once droplets are atomized and released from the aircraft, final destination is influenced by their properties (size, density, volatility, surface tension, viscosity) and the atmosphere through which they move. The temperature and relative humidity of the air have a major effect on the behavior of aqueous formulations, as does the speed and turbulence of the air mass. Once close to a surface, the aspect, texture, size and shape of the surface influence deposition.

2 Droplet characteristics

Invariably, selection of a suitable droplet spectrum is the result of a compromise between different desirable characteristics. For example, creating 30 μm droplets of *Btk* to control spruce budworm in coniferous forests results in an excellent distribution of suitable sized droplets for capture by needles and the silken thread web woven by the insect. However, such droplets may contain sublethal doses, and some may not contain any doses at all, which could lead to ingestion but little larval mortality (see Chapter VII-9, Forest Defoliators). Such doses may protect the larvae by inhibiting feeding and preventing continued ingestion of *Btk* deposits for a few days until recovery. Additionally, small droplets will increase the potential for off-target drift. Nevertheless, making the droplets larger so that they would each contain several toxic doses of *Btk* could result in poor distribution in the forest canopy and consequently reduce the chance that a larva would contact a droplet.

a Droplet size

The most widely used parameter to describe droplet size is volume median diameter (VMD or $D_{V0.5}$). The VMD of a droplet spectrum is the droplet diameter which has one half of the total spray volume in smaller diameter droplets and the other half in larger diameter droplets. Doubling the diameter of a droplet increases its volume by a factor of 8 (*i.e.*, volume increases as a cube of the diameter). Due to this fact of geometry, there are many more droplets smaller than the VMD than there are larger. When people speak about 'spraying an application consisting of 200 μm sized droplets', they are generally referring to VMD.

The number median diameter (NMD or $D_{N0.5}$) of a spectrum is the droplet diameter which has half of the total number of droplets with smaller diameters than itself and half of droplets with larger diameter. Neither the VMD nor the NMD give any indication of the range of droplet sizes, but the ratio (VMD/NMD) can provide an indication of droplet spectrum range. In a spectrum of identically sized droplets, the ratio would be equal to one. In general, a narrow spectrum would have a typical ratio of 2 to 4 and broad spectrum of up to 10.

An additional physical process that should be considered is the deposition or impaction of the droplets on their target. A final target that is a flat, more-or-less horizontal surface, is suitable for bigger droplets with a large vertical speed component. Such droplets have large impaction energy and are likely to deposit on the first surface they encounter, similar to raindrops. On the other hand, thin, vertically oriented targets will more effectively capture smaller droplets which are moving with a strong horizontal component. These droplets may encounter many surfaces in their trajectory after release before depositing. Thus larger droplets have higher momentum and the collection efficiency of a given surface to these droplets is said to be high. Smaller droplets have lower impaction energy and typically collection efficiencies of smaller droplets are lower. Wind speed is also important. As wind speed increases, the deposition on flat surfaces will decrease, while the "capture efficiency" of vertical surfaces will increase. In practice, biological targets rarely have one aspect and droplet spectra are fairly broad, so the aerial application process can be viewed as a "shotgun" approach where a broad range of targets is hit by a droplet spectrum with a broad range of droplet sizes. The main consideration is choosing a suitable spectrum for the intended target.

Product literature and labels often give instructions or recommendations on droplet size to use and the proper atomizer or nozzle to achieve the desired DSD. These instructions should be followed, but can be adjusted for specific conditions. In fact, under some conditions it is common to adjust the droplet spectrum in the course of a morning's spraying. If an aqueous formulation of *Btk* is being applied to a forest during a dry, sunny day, many applicators who use Micronair atomizers reduce the speed of the units as the temperature increases in order to produce a droplet spectrum that contains larger droplets. This reduces the evaporative loss of water from the *Btk*.

b BCPC nozzle classification system

The British Crop Protection Council (BCPC) has developed an easy-to-understand method of classifying spray nozzles that is in the process of being adopted by the U.S. Environmental Protection Agency (EPA) and the American Society of Agricultural and Biological Engineers (ASABE). The original BCPC system included five size classes, to which the ASABE added one additional size class (modified BCPC system). Droplet spectra are classified under the modified BCPC system as very fine, fine, medium, coarse, very coarse and extremely coarse. Although the names seem somewhat arbitrary, they represent tightly defined boundary parameters that are used to define the output of atomizers or nozzles for the benefit of users. It is not just the atomizer or nozzle that is characterized by this system, but an atomizer or nozzle and formulation combination.

In the future, product labels will include recommendations on which atomizer or nozzles to use in order to produce droplet spectra in the different size classes. The symbols and color code that have been adopted worldwide by

Table 4. Classification categories of the modified BCPC atomizer and formulation system

Classification category	Symbol	Color code
Very Fine	VF	Red
Fine	F	Orange
Medium	M	Yellow
Coarse	C	Blue
Very Coarse	VC	Green
Extremely Coarse	XC	White

the agricultural industry are shown in Table 4. Nozzle manufacturers such as Spraying Systems Co. have adopted the modified BCPC classification and are color-coding the appropriate portions of each table in their literature.

c Composition (susceptibility to evaporation)

As a droplet gets smaller the ratio of surface area to diameter increases geometrically according to a square relationship. Because evaporation takes place on the surface of a droplet, the proportional rate at which a droplet loses its water content increases with decreasing droplet size.

Undiluted formulations contain adjuvants to retard evaporation. Experimental data show that under extreme conditions, once the water content of a droplet has evaporated, there is no further loss of its components because they are non-volatile. Consequently, the droplet diameter does not change after the water has evaporated. With modern formulations, the dehydrated droplet is still liquid, and will adhere to surfaces on impact, although subsequent spread after impaction is minimal (*i.e.*, beading of droplets on foliage).

Biopesticides formulated with low-volatile vegetable and mineral oils can withstand a broader range of temperature and low humidity conditions.

3 Aircraft wake

The wake of an aircraft contains air currents of complex form and velocity. Two vortices, one associated with each wing tip and the slipstream from the propeller, interact with crosswinds to influence the path of spray droplets. Small droplets released close to the point of generation of the vortices may be entrained and carried outside the target area. The slipstream may also entrain droplets which are deflected to the left of the flight path.

The spray simulation models covered in Section 5B accurately calculate the degree of entrainment by the wing-tip vortices of the different sized droplets emitted by the aircraft's spray system.

4 Weather

The weather has a major impact on the aerial application process. Wind, temperature, humidity and their interaction affect how droplets are deposited on the desired targets. Weather effects do not end after deposition; both the ultraviolet rays of the sun and the solubilizing action of rain on foliage can reduce the active biological ingredient component of biopesticides as well as the feeding activity of the targeted insect.

a Wind speed and direction

The primary meteorological variable in the determination of landing position of a spray droplet is wind speed. Away from aircraft wake effects, the droplet trajectory is determined by the vector resultant of settling velocity and mean wind speed. For example, a 10 μm water droplet has a settling velocity of .003 ms^{-1} while a 100 μm water droplet has a settling velocity of 0.25 ms^{-1}. In a 3 ms^{-1} wind, the smaller drop will be displaced laterally 10000 m over a fall distance of 10 m, while the larger drop will be displaced 120 m over the same distance. This displacement will increase or decrease linearly with wind speed, so at 1.5 ms^{-1} the displacement distances for the same droplets will be 5000 m and 60 m, respectively. Wind direction determines the direction the droplets will move so that droplet displacement is downwind.

b Atmospheric turbulence and stability

Air is a low density fluid that is capable of mixing rapidly. As a spray cloud moves with the wind, its dispersal depends on the size and speed of the air motions that mix the air containing the droplets with the surrounding air. The scale of these turbulent movements (eddies) can vary from the order of a few centimeters to hundreds of meters. It is the atmospheric turbulence that disperses the spray cloud. The degree of dispersal or mixing varies according to the *stability* of the atmosphere (the rate of temperature change with height), the *wind speed* and the *surface roughness* of the ground over which the air mass is moving.

Cold air is denser than warmer air. During the portion of the diurnal cycle that the sun is not in the sky, the Earth's surface will lose heat to space if the sky is clear. This radiative cooling causes the surface to be cold and the near surface atmosphere to be colder than that above. This condition is known as a stable atmosphere or an "inversion." In a stable atmosphere, turbulent mixing is suppressed and a cloud of fine droplets will remain relatively concentrated (Thistle, 2000). During the day, when the sun is in the sky, the earth's surface heats and the hot air near the surface wants to rise through the overlying cooler air. This is known as an unstable atmosphere and in this situation turbulent mixing is enhanced and fine droplets can mix through a deep volume of atmosphere. Either strong winds or cloud cover will prevent the atmosphere from becoming either strongly stable or unstable so stability is primarily a problem in clear, low wind speed conditions. Thus, on a still clear morning, just after dawn when much forestry spraying commences, the atmosphere is often stable (inversion conditions) and fine droplets hang in the air and don't mix into the canopy. This situation has the potential for these airborne droplets to move off site in a concentrated cloud and cause off target damage. Conversely, in the afternoon on a clear, relatively calm day, the deep convective mixing associated with the warm ground surface will cause fine droplets to loft into the atmosphere and miss the target area. In this case, the droplets are so disperse that off-target damage is not typically a problem but the desired efficacy is not attained on the target.

Two considerations are essential when aerially applying material to an agricultural crop or forest – maximizing deposit in the plant canopy while minimizing drift. Table 5 summarizes the effects of different wind speeds at different times of the day. The best spraying conditions are during light to moderate wind with neutral stability, such as occurs during cloudy days.

Table 5. Summary of stability scenarios

Night to early morning, clear, light wind *Atmosphere near the surface is stable. Fine droplets will hang in the air as turbulent mixing is damped. Hanging droplet cloud can move off-target and remain concentrated in low mixing environment. Classic scenario for off-target damage. Large droplets deposit on or near target as wind speed is low. Canopy penetration is poor because turbulence is damped.*

Transition (early to mid morning), clear, light wind *Air layer near surface begins to warm as the sun heats the ground. Warm air begins rising and well mixed layer deepens from surface upwards. Spray may hang above 'convective' layer. Turbulence will develop in near surface convective layer so mixing improves. Difficult application environment as stability changes abruptly at or below the height of the application aircraft.*

Mid-day to dusk, clear, light wind *Sun has warmed surface and warm air near surface rises through cooler air above. This convective layer deepens so that it includes the layer the plane is working in and can reach up thousands of meters into the atmosphere under favorable conditions. This is the scenario that produces 'thermals.' Fine drops sprayed in this scenario will mix through the entire convective layer. It is difficult to get small drops down onto the target so application may be inefficient but lost material mixes into a large volume of atmosphere so is very dilute, making off-target damage unlikely.*

High wind (anytime) *High winds will mix atmospheric layers together. When the wind is strong, temperature gradients are lessened as the air layers mix together and stability effects are diminished. Moderate winds may help mix spray material into a forest canopy. High winds (as defined on many pesticide labels) will cause material to be blown off-site.*

Cloudy, low wind (anytime) *Cloud cover reduces the surface heating effects of the sun and prevents radiative cooling after sunset. In this way, clouds prevent temperature gradients from forming so stability effects are lessened. Cloudy, low wind conditions can represent good application scenarios especially in terms of off-target drift.*

This table considers the near surface layer where application aircraft usually work. Pressure (altitude) effects (lapse rates) on temperature are ignored in this relatively shallow layer.

As droplets get bigger, their settling velocity in still air greatly increases. Smaller droplets get entrained in turbulence, whereas larger droplets tend to be less affected by the same turbulent eddies (Thistle *et al.*, 2004).

c Diurnal variation

Diurnal variation is the change in temperature from day to night brought about by the daily rotation of the earth. During the 24 h diurnal cycle, there may be dramatic changes in the structure of the atmosphere that greatly affect spray behavior. A clear day with low to moderate winds, during which aerial applications most commonly take place, will exhibit stable characteristics overnight and in the early morning, which will rapidly change to unstable conditions as the sun heats the ground. The ground in turn transfers heat to the adjacent air layer, causing the heated air to become buoyant and rise up to the top of the mixing layer. Later during the day, as the sun gets lower in the sky, the vertical air movement subsides as it no longer has the degree of heating it had during the day.

The absence of clouds typically found during fair weather will also remove any barrier that may exist to the surface losing heat. As night falls, the warmed ground radiates its heat out to space, and cools rapidly. The cooling ground chills the air in contact with it, producing a temperature inversion. The surface loses heat quickly through radiation, when the wind speed is low, the higher layers of the atmosphere (situated sometimes just a few meters above the ground), do not get the same degree of cooling as the layer next to the ground. Therefore the temperature inversion layer may be quite shallow.

d Temperature and humidity

The water vapor content of the atmosphere can be described in different ways. Two commonly used terms are relative humidity and dew point. The relative humidity describes the ratio between the actual water vapor present and that which would be present if the air was fully saturated (*i.e.* holding all the water that it physically can at that temperature). The dew point is the temperature to which air must be cooled in order to become saturated at its existing water vapor content.

Biopesticides are usually manufactured as aqueous formulations, making the droplets which form a spray cloud particularly sensitive to conditions conducive to evaporation. Such conditions are determined by a combination of temperature and humidity, and are usually expressed as a difference between wet and dry bulb temperatures (or dew point depression). Relative humidity by itself is not a valid parameter for determining whether such conditions exist. Cool air can be very dry, but because of its low temperature, is not able to hold much moisture and does not substantially affect the evaporation of the water content of droplets.

Manufacturers formulate their aqueous products with anti-evaporants to resist dehydration. In general, this makes them suitable for aerial application in warm and dry conditions as may be encountered during a warm spell in temperate climates. However, these products would not be suitable for application in hot dry regions. For these extreme conditions, manufacturers prepare oil-based formulations that are less affected by high temperatures or low humidities, and can be sprayed undiluted under very dry conditions. Figure 6 shows how temperature and relative humidity interact in altering the application conditions for an aqueous *Btk* formulation.

As well as affecting droplet behavior, the air temperature also affects aircraft performance. Warm air is less dense than cold air and adversely affects both the aerodynamics of an aircraft (its ability to take off from short runways and carry a heavy load) and the performance of piston engines. A similar decrease in aircraft performance occurs as the air gets less dense at high elevation. The term density altitude is used to describe pressure altitude (*i.e.*, a theoretical air density which exists under standard conditions at a given altitude) corrected for air temperature and humidity. As density altitude increases, air thins out. All pilots are trained to be aware of high density altitudes, which reduce engine performance and lift by a wing or helicopter rotor. Density altitude forces both fixed wing and rotary-wing aircraft to reduce hopper load on each flight.

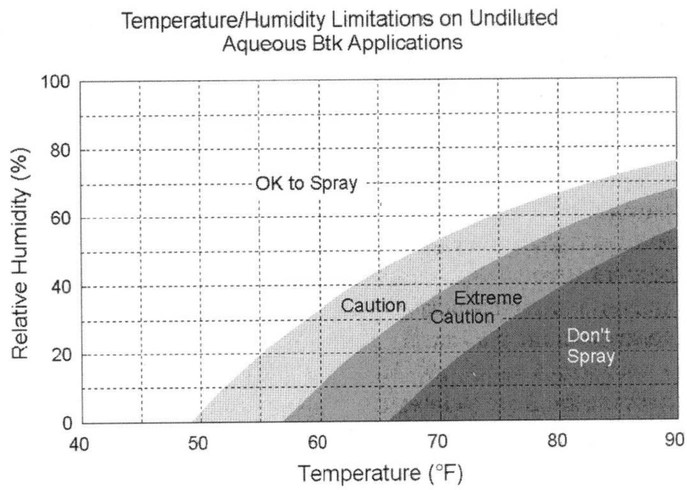

Figure 6. Effect of temperature and relative humidity on spraying a generic undiluted aqueous *Btk* formulation. The evaporative condition experienced by droplets is a function of a combination of the relative humidity and temperature (Source: Abbott Laboratory Forestry Technical Manual)

e Rain and dew

Spray deposits of aqueous formulations on exposed surfaces such as foliage are prone to be diluted by water in the form of rain or dew. Manufacturers add adjuvants to the formulations to aid droplet adhesion to surfaces; however, these generally become effective only once the deposit has dried on the surface. Manufacturers do not recommend spraying products on wet foliage or when rain is imminent. If sufficient moisture or rain is present, it is likely that much of the product will not be able to dry on the target surface, will mix with the surface water, and find its way to the ground. For example, a minimum of a 6 h period free of precipitation following *Btk* application is recommended to allow the *Btk* spray deposit to dry and adhere to the foliage.

f Solar UV radiation

Living organisms and products derived from organisms are frequently sensitive to ultraviolet (UV) radiation. Manufacturers attempt to reduce the damaging effects of UV by adding absorbing compounds that screen the harmful radiation from deposits. The short half-life of biopesticides is often the result of their sensitivity to UV radiation. The likely exposure of spray deposits to sunlight should always be a consideration when applying biopesticides. Manufacturers rarely document the longevity of their products, and often it is up to the user to check the scientific literature to determine how sensitive the specific products are to environmental conditions, especially their expected activity following deposition on the targeted plant.

5 Target characteristics

When taking into account how a dose of a biopesticide will be acquired by a living organism, consideration has to be given to how droplets get captured by the desired target objects and how the dose is transferred to the organism. The deposition of droplets in a canopy can be viewed as a conditional probability. The density of the canopy elements in space determines the probability that a given droplet with a given trajectory will encounter a foliar element. This is termed an encounter probability. The ratio of the number of drops that impact on the collecting surface of the target to the total number of drops approaching the surface is called the "collection efficiency" of the target foliage. Multiplying these two probabilities will yield the likelihood a given droplet will be deposited in a given target canopy layer. Collection efficiency is determined by both the shape (flat, cylindrical, etc.) and by the surface characteristics. Both of these properties vary over an almost continuous spectrum (see Hickey and King, (2002) for instance for an extensive, diagrammatic table of plant surface types) making collection efficiency quite variable among plant types. Theoretical considerations such as collection efficiency of droplets by target elements (*e.g.*, vertically oriented conifer needles capture smaller droplets more efficiently, whereas larger droplets deposit more efficiently in broadleaved forests) should be evaluated, but the complexity of plant and forest canopy structures make empirical studies essential.

6 *Drift*

Recent large studies have focused on the off-target drift of pesticide (Hewitt *et al.*, 2002, Bird *et al.*, 2002). These studies have contributed to existing work by the USDA-Forest Service (Teske *et al.* 1996) as well as work in Canada (Crabbe and McCooye, 1990). Detailed studies of drift have led to the development of computer models that simulate aerial spraying of pesticides. The original model known as AGDISP (Bilanin *et al.*, 1989) has been updated (Teske *et al.*, 2003) and is in the public domain and available for applicator use. A close relative of AGDISP is AgDrift which is a proprietary model distributed by the chemical manufacturers (both models use the same core calculation). These models allow assessment of spray deposition and drift and are discussed later.

a. *Definitions of spray drift and buffer zone.*

Pesticide Drift: The physical movement of pesticide through the air at the time of pesticide application or soon thereafter from the target site to any non- or off-target site. Pesticide drift shall not include the movement of pesticides to non- or off-target sites caused by erosion, migration, volatility, or windblown soil particles that occurs after application unless specifically addressed on the pesticide label with respect to drift control requirements.

Buffer Zone: A buffer zone is an area where pesticide is not directly applied, thereby providing protection to a defined area. Buffer zone designations are contingent upon: state regulation, pesticide product labels, prevailing weather conditions, and sensitive or protected areas. Buffer zones are usually adjacent to a protected area.

b. *Drift awareness.* Whenever biopesticides are being applied aerially, users should be aware of the possibility of drift. The amount of spray drift from a given application depends on many factors. These can be broken down into factors related to the material properties of the sprayed material, factors related to the application mechanism and method, and factors related to the ambient environment, including both the state of the atmosphere and nature of the target. The degree of concern should reflect the nature of the material being applied and the specific areas of risk in the immediate vicinity. The primary meteorological factors that affect spray drift are wind speed and direction (towards a sensitive area), relative humidity through its influence on drop size, and atmospheric stability. During the planning phase of the operation, managers and applicators should become familiar with product literature including recommendations on reducing drift and maintaining an efficacious droplet spectrum for the treatment. They should determine any likely sensitive areas in the vicinity of their spray operation. If circumstances dictate, a pre-spray investigation using AGDISP simulation may indicate what meteorological conditions could be risky, and what atomizer and setting to adopt.

As the use of satellite navigation (DGPS-based) recording equipment becomes standard on aircraft, it is worthwhile maintaining records of flights. Although not automatically admissible as evidence in a court of law (there exists the possibility that programmers could edit files after the flight), they serve as an initial defense against chemical trespass lawsuits.

3 Aircraft calibration and spray pattern assessment

A *Overview*

When a specific type of aircraft is selected for a spray operation, it is reasonable to expect that the aerial applicator has set up the aircraft based on contract specifications or previous successful performance for similar projects. If a unique spray project is being conducted or the client desires to verify the setup of the aircraft, it is worthwhile to perform field testing. These tests may include calibration of the aircraft (assuring that a known flow rate is achieved with even distribution among the atomizers) and an analysis of the swath pattern for effective swath width and droplet spectrum size. Whenever possible, the actual formulation that will be applied should be used for the calibration as many of the manufacturers' nozzle flow rates were determined with water. The widespread adoption of accurate electronic aircraft spray-system flow meters and flow controllers has reduced the need to periodically check the calibration but the system should still be initially manually calibrated and checked for leaks.

B *Aircraft calibration*

Proper calibration and spray atomization is paramount for maximizing efficacy. Several methods exist for calibrating flow rates, based on the kind of equipment that is mounted on the aircraft. In all cases, some basic calculations must first be performed to establish the flow rate required by the spray system and the flow rate through each atomizer or nozzle.

(1) Determine the spray system flow rate:
The formulae for determining the *system flow rate* for US units and metric units are:
US Units

$$\text{Flow rate (gal/min)} = \frac{\text{Airspeed (mph)} \times \text{Swath (ft)} \times \text{Application Rate (gal/acre)}}{495}$$

Metric Units

$$\text{Flow rate (liters/min)} = \frac{\text{Airspeed (km/h)} \times \text{Swath (m)} \times \text{Application Rate (liters/ha)}}{600}$$

(2) Choose the atomizer type and number:
The type of atomizer selected for use is based on the desired droplet spectrum. Manufacturer flow charts for various types of atomizers are used to determine the most accurate combination of pressure and flow setting (or nozzle orifice size) to deliver the desired flow per minute through each atomizer.
The determination of the correct number of atomizers varies somewhat according to whether hydraulic nozzles or atomizers that have a wide range of flow rates (such as Micronair) will be used.

- Hydraulic nozzles have narrow pressure and flow rate ranges for any particular droplet size. The number of nozzles must therefore be calculated by dividing the system flow rate by the flow rate per nozzle for the type selected (see manufacturers' tables for this value).

$$\text{Number of nozzles} = \frac{\text{System Flow Rate}}{\text{Flow Rate per Nozzle}}$$

  The aircraft spray boom must be inspected to see whether it can accommodate this number of nozzles. If the number of required nozzles is equal to or less than the number of attachment points, the selection process is complete. If the nozzle number slightly exceeds the attachment point number, a small increase in boom pressure will usually enable the system flow rate to be achieved or inverted Y-shaped adapters can be attached to one or several openings in the boom for multiple nozzles. Changing the pressure has limited effect on the flow rate (*e.g.*, doubling the spray pressure increases flow rate by about 40%). Therefore, adjustments to flow rate are best achieved by changing the number and size of nozzles. If many more attachment points are required, then the process should be repeated using a nozzle tip with a higher flow rate. Pressure gauges are usually mounted on one of the booms.
- Rotary atomizers offer a wide range of flow rates and can be adjusted to produce different droplet sizes independent of flow rate. Aircraft using rotary atomizers are configured by the applicator with a fixed number of atomizers so that an even spray pattern can be attained. The atomizer selection process therefore verifies that the flow rate per atomizer is within the atomizer manufacturers' specifications.

To calculate the flow rate per atomizer for either US or metric units, divide the system flow rate by the number of atomizers that will be fitted to the aircraft.

$$\text{Flow/Atomizer/Minute} = \frac{\text{System Flow Rate}}{\text{Number of Atomizers}}$$

Example: (metric units) If airspeed is 170 km/h, and expected swath is 70 m, what is the calibrated flow rate through each rotary atomizer if six Micronair AU5000 units will be used and the application rate is 5liters/ha?

Flow rate (liters/min)

$$= \frac{\text{Airspeed (170)} \times \text{Swath (70)} \times \text{ApplicationRate (5)}}{600}$$

$$= 99.2 \text{ liters/min}$$

$$\text{Liters/Min/Atomizer} = \frac{99.2}{6} = 16.5$$

The next step in the calibration process will depend upon the type of equipment mounted on the aircraft. If the spray system is powered by an engine driven pump (hydraulic or electric), and rotary atomizers are fitted, the aircraft can be calibrated by running the spray system and catching and measuring the output of the atomizers. If there are many nozzles or if the system pump is wind-driven, then ground calibration becomes impractical, and an airborne method is required.

1 Ground calibration for aircraft with hydraulic or electrical pumps

(1) Load sufficient product into the aircraft hopper both for priming the spray system and the required number of tests.
(2) Place collectors under each atomizer or nozzle and operate spray system for one or more minutes, so that a measurable volume is produced. Measure volume output per minute from each atomizer or nozzle and compare to the calculated rate. Check total output.
(3) Adjust system pressure and/or atomizer setting, or change nozzle orifice size, to raise or lower output as needed. Retest system as per step (2).

2 Airborne calibration for aircraft with wind driven pumps

***Note**: It is often possible to get sufficient wind pump pressure by applying power while stationary. Consult the pilot for the standard operating procedure. In such cases follow the calibration procedure as detailed above.*

(1) Load product into the hopper as described in step (1) on the previous page, with the exception that the system be primed in flight.
(2) After system is primed, position the aircraft on a level surface and mark the spot. Add a measured volume of product to the spray tank and note the level either through the sight window or by measuring the distance from the top or bottom of the tank to the fluid surface of the product.
(3) Fly the aircraft as in a normal application and operate the spray system for a set amount of time, *e.g.*, 1 min. using a stopwatch.
(4) Return aircraft to the exact spot on the ground as marked in step (3) above, and measure the volume of product needed to refill to the original level. This volume can then be used to calculate output per minute.
(5) Make adjustments to the spray system, if necessary, to change output.

3 Aircraft with electronic flow meters

Electronic flow monitors such as the Crophawk (Onboard Systems, Portland, OR) and those manufactured by Micron Sprayers Ltd. greatly facilitate aircraft calibration and enable in-flight adjustments. However, flow monitors and application computers should be pre-calibrated with the formulation or tank mix prior to operational use. Refer to equipment manufacturers' directions for volumetrically calibrating flow meters with fluids other than water. Also, flow monitors that have interchangeable cartridges of different flow range sensitivities should have the correct cartridge or flow turbine installed. For example, for general ULV use, the Onboard Systems "Crophawk Series 4100" uses cartridge #2, which has a listed range of accuracy between 5 and 100 liters/min (1.5 and 30 gal/min). Consult other manufacturers' directions for other flow monitors.

Flow Meter Calibration Procedure: *Bacillus thuringiensis (Bt)* formulations are optimized so that their viscosity is as low as possible.

(1) Assume that the *Bt* formulation of choice will behave like water and use the appropriate calibration factor in the flow meter.
(2) Fill up the aircraft hopper with a known volume of formulation using ground loading equipment fitted with an accurate positive displacement type flow meter (most are). Use the (previously calibrated) site gauge on the aircraft hopper if you are not confident about the ground rig. You may want to fly this first load as part of the operation, if you are confident that the application rate will be within ±10% of the target rate.

(3) Make an adjustment to your flow meter calibration constant if the total volume sprayed on your flow meter is different from the amount that was pumped into the hopper. Typically this adjustment is:

$$\text{New Calibration Constant} = \text{Old Calibration Constant} \times \frac{\text{Volume Applied}}{\text{Volume Indicated}}$$

With the new calibration constant, adjust the pressure of the spray system until the desired flow rate appears. This step may have to be repeated once or twice to arrive at the correct figure.

C Spray pattern assessment

In many spray projects, it is common practice to examine the spray pattern of aircraft systems before spraying, to ensure effective swath width, drop size spectrum, and droplet density. Because many bioinsecticides have to be ingested to be efficacious, accurate application is essential to ensure that an adequate dose is deposited in the sprayed area. By contrast, chemical insecticides can often exert their contact effects at doses below those recommended on the label, giving the applicator more flexibility at the edges of the swath, where dose is typically low. Swath pattern testing should duplicate the field situation with regards to aircraft height, airspeed, spray pressure and atomizer or nozzle settings and locations. The movement of atomizers or nozzles to fill in gaps or minimize heavy deposited areas is a trial and error procedure. Several investigative methods are available for spray swath pattern analysis.

Although established methods using microscopes to analyze spray deposits on cards have been used for many years, the 1990s have seen the development of rapid and powerful machine measurement methods. Two examples are the Swath Kit manufactured by Droplet Technologies (State College, PA – www.droptech.com) and Stainalysis available from REMSpC Spray Consulting, Ayr, ON (www.remspc.com). These systems use image analysis to assess and analyze aircraft spray deposits. They permit aerial applicators or spray program managers to evaluate the range of droplet sizes in the spectrum, spray deposit patterns and if necessary, to make corrective actions to ensure a uniform spray pattern and a suitable swath width. Figure 7 shows some typical output of the Swath Kit.

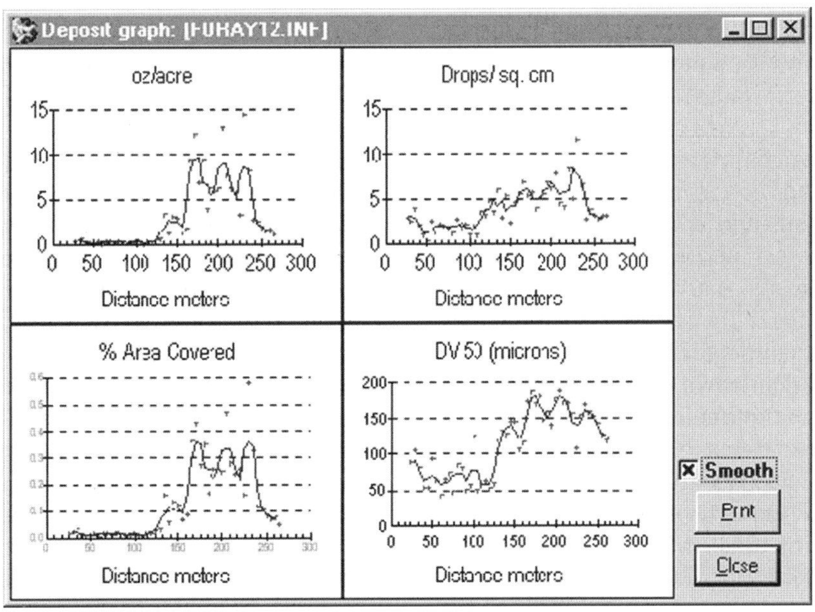

Figure 7. Swath Kit display showing summary swath pattern data for a single pattern: application rate, droplet density, percentage coverage and VMD of droplets (using a spread factor to convert stain size into droplet size)

When image analytical techniques are used for spray analysis, it is necessary to convert the diameter of the measured stain size on the collector to that of the droplet which caused the stain. A spread factor coefficient is used for this calculation.

Swath pattern analysis equipment is most commonly used to establish the offset distance between parallel tracks flown by a spray aircraft, commonly known as the *lane separation* or the *effective swath width*. When an aircraft is accurately flown under suitable weather conditions, there will be no significant over- or under-application if this lane separation spacing is maintained.

The lane separation of aircraft varies depending upon the aircraft and spray system parameters (release height, atomizer setting, aircraft speed), the biopesticide formulation, as well as factors such as meteorology and forest canopy architecture. Table 1 shows lane separations that are used in forestry for a range of agricultural aircraft applying *Btk*.

Wind direction has little effect on increasing the effective part of the swath width droplet spectra typically used for biopesticide application. Although fine droplets may drift long distances, they do not form part of the effective swath, as there is very little activity in this portion of the droplet spectrum. For convenience, lane separation determinations are normally performed by flying the aircraft into wind, so that any lateral drift of deposit is then a function of the aircraft's wake, rather than wind-borne distribution.

The technique most often used to measure the swath width assesses deposits of dyed droplets on collectors, usually flat cards. White, coated card stock (commercially known as Kromekote) has been the most popular collector. Historically, a droplet density of between 5 and 20 spots per square centimeter (range dependent on product potency) has been commonly held as an effective deposit for *Btk* products for field use. This standard is rarely used now as the number of fine (biologically insignificant) droplets that are caught by flat cards is greatly influenced by the wind speed and can give wildly varying results under a small range of wind speeds. When image analysis pattern testing equipment is used, it is more common to measure swath patterns in application rate units of liters/ha or oz/acre. A system made byWRK Inc. (Manhattan, KS), widely used by state aerial applicator associations in the USA, uses a suspended string to capture droplets for deposit analysis. It is of limited use however, as it only provides qualitative data describing the relative deposition across the swath pattern.

Measuring spray deposit on the target foliage instead of artificial collectors would certainly provide more meaningful data concerning swath patterns and deposition available to the target insect although this requires more sophisticated measurement techniques and is generally not practical. These techniques involve the use of fluorescent tracer additives to the spray formulation and analysis of fluorescing deposits on leaves collected after spraying (see ChapterVII-9, Forest Defoliators). When working in forests, twigs with attached leaves can be collected by tree climbers the use of pole pruners or by using shotguns.

4 Operational equipment

A Mixing tanks

It is important to thoroughly flush all tanks, pumps, pump lines and aircraft systems with clean water or oil (depending on the type of formulation), followed by complete draining, before addition of any formulation. Any in-line strainers should be cleaned and inspected for holes or gaps. It is sensible to use strainers (between 16 and 50 size mesh) as formulation clumping may have occurred during storage. A 30-mesh strainer is a good general recommendation. For nozzle strainers, follow the equipment manufacturer's recommendations.

Biopesticides such as *Btk* are formulated to be applied undiluted, therefore, require minimum preparation before application. Most manufacturers recommend the recirculation of the contents of drums, minibulks (plastic or metal containers containing around 1,000 liters of liquid) or tankers prior to loading the aircraft to ensure homogeneity in the formulation.

Biopesticide formulations that need to be diluted prior to application require mixing tanks. The mixing and water tanks are usually adjacent to each other and interconnected through a series of hoses and valves which are actuated externally. Water used in mixing should be clean (filtered of any coarse suspended matter), of the proper pH and free from chemicals that could inactivate the active ingredient. The preferred mixing sequence is to add the product to water.

If many hopper-loads are to be transferred from storage or mixing tank to the aircraft, pumps with a 7.5 cm (3 inches) suction inlet are recommended. They should be powerful enough to transfer a minimum of 400 liters/min (100 gal/min). If 5 cm (2 inches) suction pumps are used with bulk tankers, it is better to use a 7.5 cm suction hose from the tanker to the pump and then reduce from 7.5 cm to 5 cm at the pump. At the start of the season, the calibration of transfer pumps should be checked by pumping material into a previously calibrated container such as an aircraft hopper, and comparing the pump readings to the actual volume transferred. Back-up pumps are essential.

Bulk tank shipments are made in standard U.S./Canadian tank trucks equipped with 7.5 cm (3 in.) male Camlock outlet fittings. Most bulk tankers are equipped with two valves to avoid accidental discharge. Both valves (internal and external) must be open to allow discharge of contents. External valve controls should be secured during non-use periods to prohibit unauthorized operation. The lid of the tanker manhole must always be open when pumping to prevent the collapse of the tanker walls. The lid should be vented prior to opening the tanker to release any built-up pressure. If the tanker has not been completely emptied, the lid must be closed in order to prevent rain from contaminating undiluted product.

If the tanker is being dropped, it must be placed on solid ground with the front support dollies on solid planks or timbers (10–15 cm thick). The ground should be solid and level (or slightly inclined towards outlet) and the trailer wheels should be locked and chocked. If the tanker unloads from the back, the rear of the tanker should be lower than the front. If it unloads from the center, it should be level. If loading into a compartmentalized tanker that has been spotted on its dolly legs, load into the middle compartment first, the wheel end, and then the dolly end. Reverse the procedure when unloading from a compartmentalized tanker. No specific unloading procedure is necessary for compartmentalized tankers that are connected to a tractor. Whether unloading a tanker into another tanker or holding tank, the storage tanker or tank must be flushed and cleaned with clear water and completely drained prior to transferring undiluted product.

B Meters

Meters are usually integrated with the transfer pumps that are used to load aircraft hoppers. Meter accuracies will vary with the slippage of the liquid past the meter vanes and by the amount of entrapped air in the product. Meters should be pre-calibrated using the actual formulation. If meters are calibrated with water, some biopesticide aqueous formulations such as *Btk* with a specific gravity of approximately 1.0 produce meter readings which are typically 3 to 4% higher than the actual amount of non-aerated product delivered. Therefore, a meter calibrated with water reading 100 liters will have actually only delivered 96 or 97 liters of the product. See Section 3.B.3 dealing with Aircraft Calibration for a discussion of aircraft electronic flow meters.

C Boom timers

Some government agencies fit external boom timers on contract aircraft. The timer's switch is fitted to the boom and is pressure-activated. The timer is only on for the period that the boom pressure reaches a certain threshold.

More typically, the boom pressure sensor (or a spray handle position microswitch) is connected to the aircraft's satellite (DGPS) tracking guidance and recording system. The spray record then shows where the aircraft was during its flight and the status of the spray system at each point in the flight.

D Pumps

Pumps used in aerial spraying generally fall into two broad categories: positive displacement and centrifugal. In a positive displacement pump, the liquid in the working chamber of the pump is physically replaced by a solid object, therefore making the output of this type of pump independent of pressure and proportional to the speed. Slippage due to increased viscosity is avoided, as any resistance to flow would tend to slow the pump down rather than affect the volume delivered by each stroke.

Positive displacement pumps are mostly used in ground equipment, especially where accurate metering is required, such as in aircraft loading. Piston pumps are the most widely used type of positive displacement pump, but diaphragm and roller vane pumps also produce a flow rate that is proportional to their speed.

Centrifugal pumps are by far the most commonly used aircraft pumps. The pump has an impeller, which rotates at high speed in a round case, and liquid is drawn in through the center, and flung centrifugally to a channel situated on the periphery. The pump is capable of handling high flow rates at pressures up to 2 or 3 bars, but when a strong backpressure develops, the output rapidly drops off. A big advantage of this type of pump is that the outlet can be shutoff while the pump is spinning and the increase in pressure and resulting slippage causes no damage to the pump.

Some aqueous *Btk* formulations have caused the seals of centrifugal pumps to leak. Replacement of the inexpensive carbon-ceramic seals that are standard with most centrifugal pumps with tungsten-carbide or silicon-carbide typically alleviates the leak problem.

5 Technological development

A Guidance systems

In the last 15 years, major advancements have been made in accurate navigational guidance equipment for use during aerial application. A Global Positioning System (GPS) receiver using simultaneous signals from a number of satellites and differential corrections (DGPS) from a base station can determine the location of a spray aircraft and provide guidance information to within a sub-meter accuracy. Prior to this, the pilot of an agricultural aircraft used external physical markers (usually provided by a person waving a flag) in crop agriculture, and used block markers when working in hilly, forested areas where complicated shaped spray blocks were marked by helium-filled balloons or large (many square meters) pieces of orange paper or cloth. In the agricultural aviation industry, the immediate cost savings in personnel and reduction in human exposure to pesticides make this technology very attractive.

GPS was developed by the US Department of Defense (DOD) for military use. The first satellite was launched in 1978. Since 1995, when the system became fully operational, the DOD has maintained 24 satellites in the GPS constellation. GPS receivers need to see only four satellites to compute a three dimensional position. Ionospheric and tropospheric effects, poor satellite geometry as well as reflections from nearby objects (multipath) all serve to degrade the accuracy of the service.

The civilian community can overcome these natural errors with differential correction to the GPS signal. A ground station must be established at a known position. This station receives the GPS satellite signals and determines its exact position by averaging the uncorrected signal over time. Once its position is established, any difference between that exact position and the computed position is the error that can be used to correct other computed positions at other locations. These corrections are broadcast to other GPS receivers in the local area. However, it is now more common to use a series of continent-wide base stations to integrate a series of correction equations, and relay these via a geosynchronous satellite and broadcast over a wide area, allowing other receivers to use the part of the correction which corresponds to their locations. The delay in the entire process (including the back and forth trip to the satellite) typically takes around 6 seconds. The correction is broadcast once every 2 sec.

Equipment for the agricultural aviation market has become fairly standardized among the

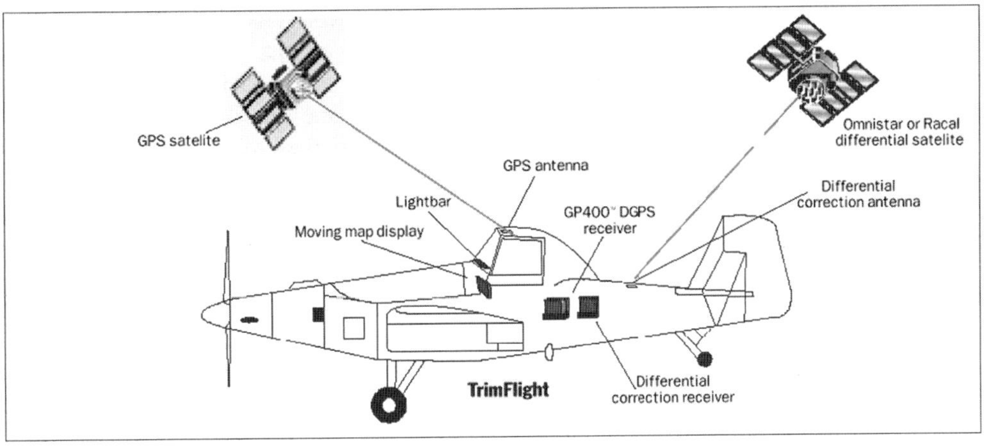

Figure 8. DGPS installation in an aircraft showing the different elements of a DGPS system. The lightbar shows the pilot the deviation from the desired track. The moving map display provides a visual feedback of the pilot's location in the spray area, and the quality of the coverage performed. The separate antennae and receivers for the GPS and differential signals have been combined into single units for the most modern equipment. (Source: Trimble Precision Agricultural Systems)

different manufacturers. The GPS is capable of calculating the aircraft's position at 10 times/sec. Linked to the computer is a method of data entry, a Course Deviation Indicator (CDI) normally in the form of a lightbar, showing the pilot the deviation from the desired track, and in the premium systems, a moving map display showing a map-like view of the aircraft in relation to the spray area, and other landmarks. A recording system logs the exact position, height and speed of the spray aircraft and whether the spray is on or off, to a data file. Additional recording channels are available to log data from any other sensors mounted on the aircraft. Figure 8 shows a schematic of a typical aircraft DGPS installation.

B Spray simulation models

1 Background

Growing awareness of the need for a healthy environment has led to an increased demand for pesticide-use accountability from the agriculture industry. This need has been accentuated through the involvement of the EPA in regulatory affairs concerning drift. Numerous research projects have been performed in order to detect and quantify pesticide deposition over target areas and determine off-target drift (Section C.6). Such projects enabled empirical data to be collected and used to test simulation models that were appearing on mainframe and minicomputers. In the 1990s, the availability of powerful desktop computers made these models available to most users. The model now widely used by applicators and sanctioned by EPA is the AGDISP model developed by the USDA Forest Service and being supported by a consortium of government agencies and private sector partners. As mentioned previously, this model is in the public domain and is computationally very similar to the AgDrift model which is proprietary.

The AGDISP near-wake model solves a Lagrangian system of equations for the position and position variance of spray material in the form of a range of droplet size classes released into the atmosphere from specified atomizer locations on an aircraft. It utilizes a line source approach and interfaces with a Gaussian model to calculate deposition beyond downwind distances of 1 km (Teske and Thistle, 2004).

The USDA Forest Service and US EPA are working together to bring the AGDISP spray modeling results onto a Geographical Information System (GIS) platform. Two models are being developed. The first is SprayAdvisor which is available for beta testing from the authors as this chapter goes to press. This model brings the model deposition results onto a real map that can either be used in planning mode or in post-spray analytical mode. The model will operate using ESRI based GIS software and will allow various GIS functionalities such as integrated deposition to real water bodies, numerical deposition at any point on the map (a selected residence for instance) and many other features. This software will accept actual flight lines recorded with aircraft DGPS systems for detailed post-spray analysis. The basic AGDISP model includes an evaporation model for volatile spray components, a canopy penetration model for forest canopy interception, an accountancy model to recover environmental fate of the released material and various other specialized modules and outputs. SprayAdvisor will contain all the capabilities of AGDISP. The second model being developed will be known as SprayTrans and this model will allow deposition calculations to be made at 10 to 20 km distance from the release area. SprayTrans will use the AGDISP model to calculate deposition in the near field, and will then hand the fine fraction that remains aloft at some distance (250–500 m) to a puff model (a modified version of CalPuff). The puff model will allow effects of varying terrain and varying meteorology to be incorporated in the deposition calculation. This model should be available for validation in the near future.

The AGDISP model has been extensively validated with existing sets of aerial spray data representing a wide range of test site geography, spray systems, spray aircraft, and types of canopies. It is now recognized as the industry standard spray dispersion and deposition model, and is currently used by government agencies and private industry in the United States, Canada and New Zealand.

2 *Using the models*

The simulation models have paid dividends to operational programs. Canadian researchers have been able to optimize strategies for maximizing deposition in spray blocks. Empirical work (Mickle and Rousseau, 1998) confirmed the results that were obtained with model simulations. The researchers noted that the small droplets that are used in spruce budworm suppression projects often are able to drift out of a forest spray block, especially if the blocks are small. Additionally, because several spray runs typically overlap on any particular area of forest being sprayed, the traditional method of applying swaths of spray over the block with a slight upwind offset would result in poor deposition on the upwind part of the block, and considerable escape of material from the downwind side of the block.

Optimization strategies resulted in multiple passes (up to six) on a single line displaced up to one half swath upwind of the block boundary. Swath displacement and multiple passes increased with increasing wind speed. Measured deposits within the optimized block showed increases of 13%–185% over the traditionally sprayed block, with an increase in deposit uniformity.

C *Meteorological atmospheric profile analysis equipment*

Large-scale forest spray projects, where several aircraft spray large tracts of forest, benefit from the availability of comprehensive meteorological information about the airmass in which the application is taking place. Often above-canopy winds are very different than those occurring at the airport, especially early in the morning. Recently, Aventech Research Inc. (Concord, ON, Canada) developed an aircraft-mounted meteorological system. It uses state-of-the art instruments, along with DGPS geo-referencing to collect wind, temperature and humidity data for onboard storage or for immediate transmission to a ground station for real-time analysis. The self-contained package can be mounted on a variety of light aircraft. This instrument package has now been combined with DGPS, GIS, a flow controller and AGDISP to yield real time swath offsets and plot them on an in-cockpit screen. This allows the pilot to use the meteorological information in real time to determine the swath offset.

D *Electrostatic spraying*

Electrical charging of sprays has always shown promise for improving spray deposition in closed environments, and the first patent on electrostatic spraying was issued in 1932. The paint spraying industry recognized the benefits of electrostatic spraying for paint finishing and for several decades all automobile manufacturers worldwide have been using various types of electrostatic sprayers for all their automotive painting. Electrostatic spraying in agriculture has been more technically evasive to such solutions because of the conductive nature and wide variety of the spray materials used. Additionally, as aircraft are airborne in an insulated medium (air), there have been difficulties in avoiding the build-up of the opposite charge on the aircraft to that being applied to the droplets.

The objective of electrostatic spraying is to utilize electrical forces to control the behavior of spray droplets. Charge added to the surface of airborne spray droplets helps to change the droplet trajectory around the target being sprayed. It is possible for droplets to reverse their flight path and coat areas of the target that would not otherwise be coated with uncharged spray. Such attraction can only occur if electrostatic forces on the droplets are dominant over gravity, inertia, and the force of air currents. Typically that means that droplets less than 100 μm in size, carrying a high charge in low winds give best results. Electrical fields around plants are not strong at typical release heights of aerial application, so the risk for off-target drift remains as in typical (uncharged) application. Companies which make air-assisted electrostatic spray systems state that aqueous sprays do not present an obstacle to the chargingprocess.

Recent work by Carlton *et al.* (1995) using a prototype aerial electrostatic spray charging system revealed that a dual-polarity charging protocol gave superior depositional results when used with chemical insecticides. A four-fold increase was obtained from spray deposited on cotton leaves over that of uncharged spray. The spray plume's electric field was sufficient to enhance underleaf deposit in the upper canopy. A significant amount of spray was driven through the upper plant canopy and onto the upper leaf surface at mid canopy. Preliminary results indicate the electrostatic system at 10 liters/ha provided whitefly control equivalent to conventional aerial systems at 50 liters/ha.

There have been no reported studies done on electrostatically charged biopesticides. However, the potential of spraying small droplets that are attracted to areas of plants (such as the underside of leaves) which are not normally accessible to conventional sprays remains attractive.

6 Summary

Pesticide spraying, whether performed by equipment working on the ground or in the air is an inherently rudimentary process. Only miniscule proportions of an applied material are actually used to control the pest. For example, a toxin like *Btk* has to be sprayed on as much of the foliage area of a forest, in as even a manner as possible, to control a defoliator, because the insects' exact locations and movements are unknown. Therefore application efficiency is measured by how completely and uniformly target vegetation is covered rather than how accurately the pesticide is targeted to the insect.

In the last decade, aerial application has made many advances that improve the targeting of biopesticides. Very accurate, universally available tracking guidance using satellite navigation, electronic flow meters, flow controllers and a greatly expanded understanding of the physical processes involved in the production, transport and deposition of deposits and simulation models have all served to achieve this improvement. Better behaved formulations ensure that more of the active ingredient reaches its destination, and remains viable for longer periods. Real time use of GIS and deposition modeling results in the cockpit are giving the applicator more information to aid in decision making.

Ultimately, it is the combined responsibility of equipment manufacturers, biopesticide manufacturers, applicators and program managers to develop and utilize technology and information in a way that allows aerial spraying of biopesticides to remain an efficacious and publicly acceptable approach to forest pest control.

7 References

Bilanin, A.J., Teske, M.E., Barry J.W., and Eckblad, R.B. 1989. AGDISP: the aircraft spray dispersion model, code development and experimental evaluation. *Trans. Am. Soc. Agric. Engin.* **32**, 327–334.

Bird, S.L., Perry, S.G., Ray, S.L., and Teske, M.E. 2002. Evaluation of the AGDISP aerial spray algorithms in the AgDrift Model. *Environ. Toxicol. Chem.* **21**, 672–681.

Carlton, J.B., Kirk, I.W., and Latheef, M.A. 1995. Cotton pesticide deposition from aerial electrostatic sprays. American Society of Agricultural Engineers Paper No. 95-007, St. Joseph, MI.

Crabbe, R.S. and McCooye, M. 1990. Wind drift from ultra-low-volume aerial insecticide applications in stable and turbulent conditions. Proceedings of Associate Committee on Agriculture and Forestry Aviation Symposium and Trade Show, Winnipeg, Manitoba 1990, pp. 158–168.

Curbishley, T.B., Teske, M.E., and Barry, J.W. 1993. A Productivity Model for Aerial Application. American Society of Agricultural Engineers Paper No.931072, St. Joseph, MI.

Ekblad, R.B., Teske, M.E., Banaugh, R., and Barry, J.W. 1988. Computer Assisted Spray Productivity Routine (CASPR). American Society of Agricultural Engineers Paper No.AA88-009, St. Joseph, MI.

Esterly, D.M. 1998. Neural network analysis of spray drift task force atomization. Dropkick™ II. Paper 981014, Proceedings of the American Society of Agricultural Engineers Annual International Meeting, Orlando, FL; St. Joseph, MI.

Fife, J. 2006. Evaluation of Spatial Droplet Sizing Methodology for Mass Closure. Contract Ref. # OP34809. Battelle-Columbus, Columbus, OH.

Hermansky, C.G., 1998. A regression model for estimating spray quality from nozzle, application and physical property data. Institute of Liquid Atomiztion and Spray Systems Americas. Proceedings of the 11th Annual Conference on Liquid Atomization and Spray Systems, Sacramento, CA.

Hewitt, A.J., Sanderson, R., Huddleston, E.W., and Ross, J.B. 1994. Polymer and invert emulsifying oil effects upon droplet size spectra of aerial sprays. *J. Environ. Sci. Health, Part B* **29**, 815–830.

Hewitt, A.J., Johnson, D.R., Fish, J.D., Hermansky, C.G., and Valcore, D.L. 2002. Development of the spray drift task force database for aerial applications. *Environ. Toxicol. Chem.* **21**, 648–658.

Hickey, M. and King, C. 2002. "The Cambridge Illustrated Glossary of Botanical Terms." Cambridge University Press, New York, NY.

Kilroy, B., Karsky, D., and Thistle, H. 2003. Aerial Application Equipment Guide 2003. FHTET-2003, USDA-FS Morgantown, WV.

Kirk, I.W. 2001. Droplet Spectra Classification for Fixed-Wing Aircraft Spray Nozzles. Technical Paper No.011082,American Society of Agricultural Engineers, St. Joseph, MI.

Kirk, I.W. 2002. Measurement and prediction of helicopter spray nozzle atomization. *Trans. Am. Soc. Agric. Engin.* **45**, 27–37.

Mickle, R.E., and Rousseau, G. 1998. Optimization Trials for Insecticide Spraying into Small Blocks – Murdochville Results. Spray Efficacy Research Group (SERG) 1998/04 Publication.

Schick, R.L. 1998. An Engineer's Practical Guide to Drop Size. Spraying Systems Co., Wheaton, IL.

Teske, M.E., Thistle, H.W., Hewitt, A.J., Kirk, I.W., Dexter, R.W., and Ghent, J.H. 2005. Rotary atomizer drop size distribution database. *Trans. Am. Soc. Agric. Engin.* **48**, 917–921.

Teske, M.E. and Thistle, H.W. 2004. Aerial application model extension into the far field. *Biosys. Engin.* 89, 29–36.

Teske, M.E., Thistle, H.W., and Ice, G.G. 2003. Technical advances in modeling aerially applied sprays. *Trans. Am. Soc. Agric. Engin.* 46, 985–996.

Teske, M.E., Thistle, H.W., Barry, J.W. and Eav, B. 1998. A simulation of boom length effects for drift minimization. *Trans. Am. Soc. Agric. Engin.*- pp. 545–551 May/June.

Teske, M.E., Thistle, H.W., and Barry, J.W. 1996. Topics in aerial spray drift modeling. In "Environmental Modeling Vol. 3: Computer Methods and Software for Simulating Environmental Pollution and Its Adverse Effects." (P. Zannetti, Ed.) pp. 17–54. Southhampton: Computational Mechanics Publications.

Thistle, H.W. 2000. The role of stability in fine pesticide droplet dispersion in the atmosphere: a review of chysical Concepts. *Trans. Am. Soc. Agric. Engin.* **43**, 1409–1413.

Thistle, H.W., Teske, M.E., and Richardson, B. 2004. Spray Droplet Trajectory Length in Ambient Vortices. American Society of Agricultural Engineers Paper No.041088, St. Joseph, MI.

US Department of Agriculture. 1996. Aerial Application Manual. Washington, D. C.

Chapter III-3

Dissemination of beneficial microbial agents by insects

Fernando E. Vega
Insect Biocontrol Laboratory
USDA-ARS
Beltsville, MD 20705, USA

Patrick F. Dowd
Crop Bioprotection Research Unit
National Center for Agricultural Utilization
Research, USDA-ARS
Peoria, IL 61604, USA

Lawrence A. Lacey
Yakima Agricultural Research Laboratory
USDA-ARS
Wapato, WA 98951, USA

Judith K. Pell
Department of Entomology and Nematology
IACR-Rothamsted, Harpenden
Hertfordshire, AL5 2JQ, UK

D. Michael Jackson
U. S. Vegetable Laboratory
USDA-ARS
Charleston, SC 29414, USA

Michael G. Klein
Department of Entomology, OARDC
Wooster, OH 44691, USA

1 Introduction

Recent trends in pest management show an increased shift from insecticidal sprays towards the use of transgenic crops with insecticidal properties. This shift is partly due to the development of insect resistance to conventional insecticides and the large and increasing costs of developing and registering new chemicals. However, the long-term effectiveness of transgenic crops remains a hotly-debated issue. This scenario confronts and challenges entomologists to develop and evaluate other methods of pest management.

Of the alternatives, the use of entomopathogens has long been recognized to be a possible alternative for pest management. The current approach is to use entomopathogens as if they were chemical insecticides, *i.e.*, using the same equipment and application strategies. This has

L.A. Lacey and H.K. Kaya (eds.), Field Manual of Techniques in Invertebrate Pathology, 127–146.

led to problems with obtaining effective formulations that allow the different entomopathogens to withstand the adverse environmental conditions faced in the field, and the need for a long shelf life at room temperature. A different approach would be to selectively disseminate entomopathogens among target pest populations. Even though pathogen dispersal occurs naturally, *e.g.*, via parasitoids (Levin *et al.*, 1983; Young and Yearian, 1990), predators (Pell *et al.*, 1997; Roy *et al.*, 1998), insect feces (Vago *et al.*, 1966; Forschler and Young, 1993), bird and mammal feces (Entwistle *et al.*, 1978; Lautenschlager and Podgwaite, 1979), wind and rain (Wilding, 1970; Hukuhara, 1973), transovarial transmission (Bird, 1961; Doane, 1970), and surface contamination by infected insects (Neilson and Elgee, 1968; Kellen and Hoffmann, 1987; Scholte *et al.* 2004), it would be advantageous to manipulate such dispersal by using devices that attract insect pests into a focus of entomopathogens, from which the pathogen can be disseminated to other members of the target pest population. This insect dispersal of a pathogen to members of its own population is known as autodissemination (Soper, 1978; Hunter-Fujita *et al.*, 1998), and as assisted-autodissemination when devices are used to promote this dispersal (Hunter-Fujita *et al.*, 1998). The devices used to serve as a focus of infection are known as autoinoculators; examples of pathogens tested with autoinoculators are given in Table 1. Rather than using the term "assisted-autodissemination," we will use autodissemination throughout the chapter. The objective of this chapter is to introduce the concept of autodissemination by presenting four case studies involving the dissemination of entomopathogenic fungi and viruses and plant disease antagonists.

2 Case study #1. Autodissemination of entomopathogenic fungi for management of the diamondback moth

Cruciferous vegetables are economically important crops worldwide; 2.2 million ha are grown with 50% of production coming from Asia where crucifers are an essential dietary component and an important market crop for resource-poor farmers. The most destructive and persistent pest of crucifers for the last 40 years has been the diamondback moth, *Plutella xylostella* (Lepidoptera: Plutellidae), with an estimated annual management cost of US$1 billion (Shelton *et al.*, 1997). The reasons for the pest status of this insect are its high reproductive potential (up to 20 generations a year), the disruption or lack of natural enemies (caused by excessive insecticide use against co-occurring pests), and its impressive ability to develop resistance to all currently available insecticides including toxins of the microbial agent, *Bacillus thuringiensis*. This demonstrates the problems associated with relying on a single component pest management approach, *i.e.* chemical and biological pesticides. Integrated strategies are required which incorporate the use of natural enemies and sound cultural practices with existing and novel management approaches of the whole crop. One such novel strategy is the combined use of fungal natural enemies and synthetic female sex pheromones. Such approaches are environmentally friendly and sustainable and should extend the useful lifetime of the synthetic insecticides that are still available (Talekar and Shelton, 1993; Shelton *et al.*, 1997).

The entomopathogenic fungus *Zoophthora radicans* (Zygomycetes: Entomophthorales) is a widespread and important member of the natural enemy complex attacking the diamondback moth and has the potential to be encouraged as one component of an integrated management strategy. It commonly produces epizootics in larval populations that can, locally, be eliminated (*e.g.*, Ooi, 1981; Riethmacher *et al.*, 1992). However, epizootics are unpredictable and often only occur in large host populations too late in the crop season to restrict damage below the economic threshold. To maximize the potential of *Z. radicans* for diamondback moth management, we must ensure that epizootics develop early in the season when pest populations are low. We hypothesize that this can be achieved by manipulating moth behavior to facilitate autodissemination of the fungus to susceptible conspecifics on the crop.

The hypothesis behind the strategy is that male moths would be attracted into a specially designed, fast entry, slow exit trap in response to synthetic female sex pheromone. While

Table 1. Examples of manipulated dissemination of entomopathogens by insects using autoinoculator devices

Pathogen	Target	Reference
BACTERIA		
Paenibacillus (=Bacillus) popilliae	*Popillia japonica*	Skadeland, 1981
BACULOVIRUSES[1]		
*Ac*MNPV	*Heliothis virescens*	Nordin *et al.*, 1990, 1991
		Jackson *et al.*, 1992
*Hz*SNPV	*Helicoverpa zea*	Gard, 1975
		Jackson *et al.*, 1993
		Gross *et al.*, 1994a, b
*Pi*GV	*Plodia interpunctella*	Kellen and Hoffmann 1987; Vail *et al.*, 1993
*Se*NPV	*Spodoptera exigua*	Yu, 1996; Yu and Brown, 1997
*Cp*GV	*Cydia pomonella*	Pultar *et al.*, 2000;
		Winstanley *et al.*, 2005
FUNGI		
Beauveria bassiana	*Carpophilus lugubris*	Vega *et al.*, 1995
	Ips typographus	Vaupel and Zimmermann, 1996
	Cosmopolites sordidus	Tinzaara *et al.*, 2004
	Cylas formicarius	Yasuda, 1999
Metarhizium anisopliae	*Popillia japonica*	Klein and Lacey, 1999
	Glossina spp.	Maniania, 1998, 2002; Maniania *et al.*, 2006
Zoophthora radicans	*Plutella xylostella*	Pell *et al.*, 1993a, b
		Furlong *et al.*, 1995
PROTOZOA		
Mattesia trogodermae	*Trogoderma glabrum*	Shapas *et al.*, 1977

[1] Definitions for acronyms used: *Autographa californica* multiply-embedded nucleopolyhedrovirus (*Ac*MNPV); *Helicoverpa zea* singly-embedded nucleopolyhedrovirus (*Hz*SNPV); *Plodia interpunctella* granulosis virus (*Pi*GV); *Spodoptera exigua* nucleopolyhedrovirus (*Se*NPV); *Cydia pomonella* granulovirus (*Cp*GV)

inside the trap they would become contaminated with infectious conidia from a sporulating source of *Z. radicans*. Contaminated moths would leave the trap on habituation to the pheromone and return to the crop disseminating the entomopathogen among their own populations. The benefits of this system over the conventional use of mycoinsecticides are threefold: (1) use of a specific sex pheromone targets the inoculum to the diamondback moth as this is the only insect entering the trap; (2) only small quantities of fungal inoculum and pheromone are required thereby limiting production costs; and (3) while inside the autoinoculator, the fungus is in an environment that favors sporulation and infection and is protected from the damaging effects of UV radiation.

Autoinoculators that have been evaluated in the laboratory and the field to test this hypothesis will be described here. Complementary laboratory studies have selected fungal isolates with high virulence against the diamondback moth (Pell *et al.*, 1993a) and low virulence against non-target organisms (Furlong and Pell, 1996). These isolates are, therefore, suitable in integrated pest management strategies using insect natural enemies. In fact, certain insect natural enemies can improve the impact of the fungus (Furlong and Pell, 1996).

A Design and evaluation

The autoinoculator was designed to allow rapid entry of the insects to a central inoculation arena but to make departure from the arena more difficult, though not impossible, thereby ensuring heavy contamination of moths with infectious conidia. The autoinoculator is constructed from clear plexiglass and comprises two 300 mm^2

horizontal base plates 55 mm apart, separated by a series of baffles that surround the central arena and delay exit; moths enter and leave the central arena through the baffles (Figure 1, Pell *et al.*, 1993b). The central arena contains the sporulating fungus in an inverted Petri dish and a polythene vial impregnated with the synthetic pheromone lure (Pell *et al.*, 1993b); both are protected from rain and UV radiation by a plastic pyramid placed above them. A water reservoir, located beneath the central arena (connected by a wick) ensures that humidity remains high within the autoinoculator. During evaluation, the autoinoculator was either placed at the height of the cabbage crop (ca. 0.5 m) or on the soil between rows.

In field evaluations in the Cameron Highlands, Malaysia, moths responded to the synthetic sex pheromone and entered the autoinoculator at all times of the day and night. From observations of moths entering the autoinoculator, they spent a mean of 1.5 min in the central inoculation arena. Parallel laboratory studies showed that when a sporulating source of *Z. radicans* was held above the pheromone lure in the trap, the LT_{50} for a single visit by male moths to the inoculation arena was ca. 5 min. When the inoculum was above and below the lure, this time was reduced to < 60 sec (Furlong *et al.*, 1995; Furlong and Pell, 1997a). Complementary studies demonstrated the potential for using a conidial powder of the fungus *Beauveria bassiana* (Hypocreales, ARSEF 2729, isolated from the diamondback moth) in the inoculation arena. Moths received a lethal concentration when they spent 1 sec in an inoculation arena containing only 25 μg of *B. bassiana* conidia/cm^2 (Furlong *et al.*, unpublished data). Therefore, the duration of the visits observed in the field would guarantee infection if appropriate quantities of either inoculum were placed in the trap (Furlong *et al.*, 1995). Under field conditions, contaminated moths succumbed to *Z. radicans* within 4 days of inoculation regardless of concentration. This was within the active lifetime of healthy moths which was vital as *Z. radicans* only develops within a living host and cannot grow and sporulate on an insect that has died before development is complete (Furlong *et al.*, 1995). Time to death after inoculation with *B. bassiana* under simulated field temperatures was concentration related, but was usually within 3–5 days of inoculation (Furlong *et al.*, unpublished data).

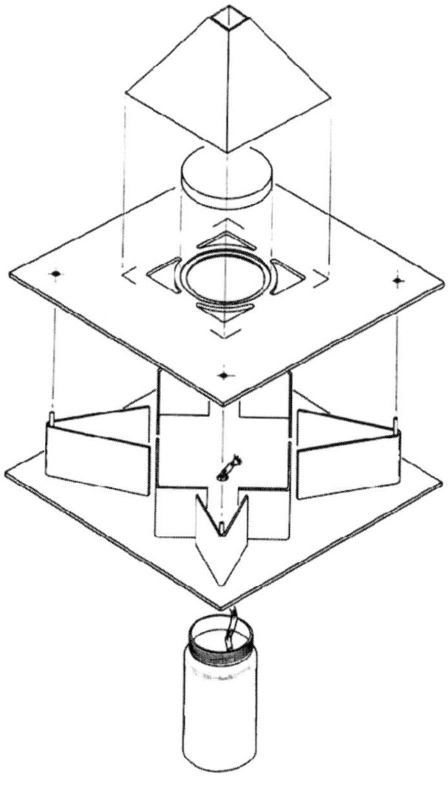

Figure 1. Diagram of autoinoculator used for autodissemination of fungal entomopathogens by *Plutella xylostella* (from Pell *et al.* 1993b. Printed with permission from Carfax Publishing Ltd., P. O. Box 25, Abingdon, Oxfordshire OX14 3UE, UK.)

The autoinoculator was both complex in design and expensive to construct and, therefore, impractical for use by resource-poor farmers. Simpler traps have now been designed based either on the standard Delta® trap (Oecos, Kimpton, UK), or materials that are readily available everywhere, such as soda cans. A Delta® trap with a Petri dish insert (inoculation arena) and the pheromone lure (= Delta-Petri trap, Figure 2) was more effective at attracting moths (16.2 ± 4.2 moths entered in 15 min) than the prototype trap (7.9 ± 2.8 moths entered in 15 min) when tested in kale fields in Kenya (Furlong and Pell, unpublished data). The duration of visits to the central inoculation arena of the Delta-Petri trap (79.9 ± 19.2 sec), was less than to the same region of the prototype trap (122.2 ± 19.4 sec) but still within the time necessary to ensure infection of the moth (Furlong and Pell, unpublished data). Simple traps also required

Figure 2. Photograph of Delta-Petri trap used for autodissemination of fungal pathogens by *Plutella xylostella*

modifications to compensate for the removal of the water reservoir in the prototype trap. To maintain a high humidity within the inoculation chambers, the inner surfaces of the traps were coated with a layer of 1% tap water agar. Sporulation rates of *Z. radicans* inside Delta-Petri traps or soda can traps were assessed in a wind tunnel. Traps were either placed with their "entrance" at 90° or 180° to the airflow (1.5 m/sec). Sporulation was only maintained for longer than 2 h in the Delta-Petri traps placed with their "entrance" at 90° to the air flow. Sporulation rates of control mycelium maintained at 100% RH during the experiment remained high (Furlong *et al.*, unpublished data). Although the position of the trap in relation to wind direction can be achieved using a fin on the trap as in the autoinoculator of Vega *et al.* (1995), more work is required to ensure a continual high humidity in these simple traps.

B Pathogen dispersal

Once effective autoinoculators were designed that reliably contaminated insects with fatal doses of inoculum, it was essential to quantify dispersal and transmission to the targeted population. In a preliminary field trial, a fluorescent marker was placed in the inoculation arena. Moths became contaminated and were observed depositing the marker on foliage up to 5 m away from the trap demonstrating that dispersal would occur over at least that distance and perhaps much further (Pell *et al.*, 1993b).

Dispersal of inoculum from contaminated moths to conspecifics can occur in two ways: (1) passive transfer from contaminated males to other adults and larvae on the crop; and, (2) on death of the contaminated moth, transfer of conidia from the resulting cadaver to larvae on the crop. In laboratory studies, although the dose of either *Z. radicans* or *B. bassiana* conidia acquired in the inoculation arena was sufficient to kill moths, passive mechanical vectoring of *Z. radicans* to other adults and larvae was extremely limited whereas *B. bassiana* conidia were readily vectored in this way. For example, a mean of 70% and 15% of larvae on plants in contact with *B. bassiana*-contaminated or *Z. radicans*-

contaminated moths became infected, respectively (Furlong and Pell, 1997b). Insects killed by *Z. radicans* were characteristically attached to the leaf surface by rhizoids and the sticky conidia (covered in pre-formed mucus) were actively discharged forming a halo around the dead insect. Conversely, insects that had been killed by *B. bassiana* were not attached to the plant and the conidia (not covered in pre-formed mucus) remained associated with the body of the host. These biological characteristics of the 2 species contributed to a difference in the level of transmission achieved from cadavers to larvae. Cadavers that produced *Z. radicans* conidia caused disease in a greater proportion of the surrounding larval population than cadavers that produced *B. bassiana* conidia (Furlong *et al.*, unpublished data). As there was little interference between *Z. radicans* and *B. bassiana* (Furlong and Pell, 1997b and unpublished data), combined use of both pathogens in the trap is likely to have greatest potential for population suppression.

Passive vectoring of *B. bassiana* conidia from contaminated males to females during copulation would remove females from the population. However, if an epizootic is to establish, the contaminated male should ultimately succumb to *Z. radicans* infection and produce *Z. radicans* conidia. In addition to ensuring death of the male moths entering the trap with the resulting potential to initiate epizootics, *Z. radicans* also has detrimental impacts on its host prior to death. Infected larvae consume up to 45% less foliage and infected adults lay fewer eggs and are less able to produce and respond to sex pheromone (Furlong *et al.*, 1997a,b; Reddy *et al.*, 1998). These effects also contribute to a reduction in crop damage.

C Conclusion

Autodissemination of entomopathogenic fungi has a role to play in integrated pest management, and particularly in resistance management. This strategy is compatible with insect natural enemies and could be used to disseminate other microbial organisms for use against co-occurring insect pests and plant diseases. Large field tests designed to evaluate this strategy will be needed. The number of traps required per ha and the position and timing of their use requires careful evaluation before this technology can be exploited practically. Autoinoculators can target control agents to particular pests or diseases, can be constructed cheaply from locally available materials, are environmentally friendly and will reduce dependence on chemical pesticides. Their potential is undeniable.

3 Case study #2: Autodissemination of fungal pathogens of the Japanese beetle

A Design and evaluation

Adults and larvae of many scarab beetles are serious pests of turf, ornamentals and crops throughout the world (Jackson, 1992). The Japanese beetle, *Popillia japonica* (Coleoptera: Scarabaeidae), is one of the most serious pests of turfgrass in the eastern USA (Ahmad *et al.*, 1983; Tashiro, 1987). Additionally, adult beetles feed on a wide variety of host plants.

A rich diversity of entomopathogens is reported for scarab species including viruses, fungi, nematodes, bacteria, rickettsia and protozoa (Jackson and Glare, 1992). However, only *Paenibacillus popilliae* and entomopathogenic nematodes have been commercially developed for control of the Japanese beetle (Klein, 1992; Klein and Georgis, 1992). These agents are generally applied to larval habitats where grub damage is severe and the cost of treatment is economically justified. Heavily infested lawns and golf courses are the most commonly treated habitats. Economic considerations limit the feasibility of treating large expanses of larval habitats, especially pastures and roadside grasses where damage to turf may not be as noticeable or serious as in the aforementioned habitats. In these situations, it may be more practical to lure and infect adult beetles with entomopathogens followed by subsequent dissemination to larval habitats by infected adults. Autodissemination of virus and fungal pathogens by adult beetles into the larval habitats of two other scarab species has been successfully employed by Zelazny *et al.* (1992), Keller *et al.* (1989), Huger (2005) and others.

Fungi and nematodes are infectious for both adult and larval *P. japonica* and were investigated for their potential for autodissemination (Lacey *et al.*, 1994a). Nematodes in the families Steinernematidae and Heterorhabditidae readily infect adult beetles but subsequent dissemination is limited due to the rapid debilitation of the host insect (Lacey *et al.*, 1993, 1994a). The fungi, *Beauveria* spp. and *Metarhizium anisopliae*, infect adults but do not rapidly kill the insects (Lacey *et al.*, 1994b). The delay in mortality observed in adult beetles allows dispersal up to and over 1 km from release sites (Lacey *et al.*, 1995).

The combination of floral lures and attractant traps has been thoroughly developed for monitoring the Japanese beetle (Ladd *et al.*, 1976; Klein and Edwards, 1989). The combination of this technology with *M. anisopliae* appeared to offer potential for the autodissemination of entomopathogenic fungi by adult beetles. Combining the standard Japanese beetle trap (Catch Can™, Trécé, Inc., Salinas, CA) (Klein and Edwards, 1989) containing floral lure (phenethyl propionate and eugenol [3:7]) (Ladd *et al.*, 1976), and an autoinoculating chamber containing fungal conidia provided a means of attracting and infecting large numbers of male and female beetles (Klein and Lacey, 1999). The autoinoculating chamber consists of a rectangular wooden box (40.5 × 11.5 × 4.5 cm) made with a threaded receptacle for attachment of the Trécé trap near the end on the top side of the box (Figure 3). A 10-cm section in the middle of the autoinoculating chamber between the entrance hole in the roof and the exit hole in the floor allows space for a Petri dish containing a conidia:talc mixture (1:2.5 or 1:5), which is placed into the chamber through a door in the side of the unit (Figure 3A). Beetles attracted to the trap fall through a hole in the autoinoculation chamber and are subsequently attracted to light passing through the plexiglass window (Figure 3D). Passing through the chamber, the beetles walk or fly into the conidial preparation in a Petri plate in the center of the chamber. A funnel attachment secured beneath a hole in the floor at the opposite end of the autoinoculation chamber allows contaminated beetles to leave (Figure 3A and 3C).

The trap was tested in the field (Figure 4) on Terceira Island in the Azorean Archipelago with *M. anisopliae*, using a 2.5:1 or 5:1 mixture of talc and *M. anisopliae* conidia placed in the autoinoculation chamber. Beetles emerging from the autodissemination device in the field were captured and returned to the laboratory where the presence of conidia and the ultimate mortality to adult beetles from the fungus were confirmed. Contaminated beetles were held in cages described by Lacey *et al.* (1994b) and

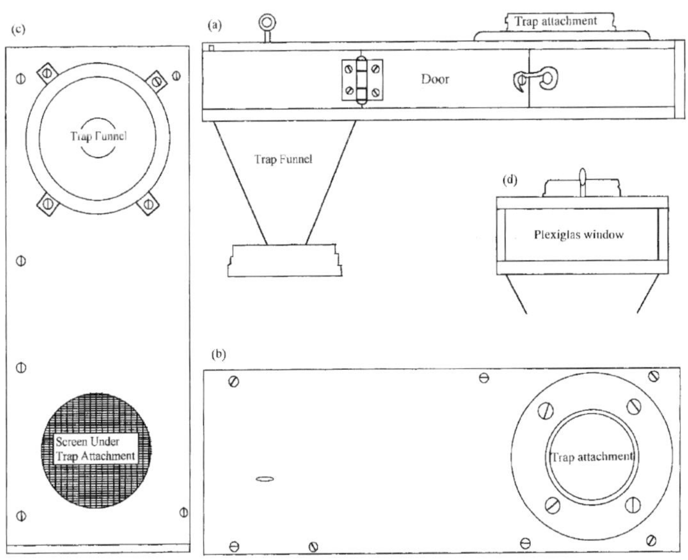

Figure 3. Schematic drawing of autoinoculation chamber for contamination of Japanese beetles with fungi. (A) Side view showing door, trap funnel (left), and trap attachment (right). (B) Top view, showing trap attachment (right) and eye-hook (left) to hang and stabilize the trap. (C) Bottom view, to show screen under trap attachment and exit funnel. (D) End view of trap showing plexiglass window

Figure 4. Photograph of the Japanese beetle autodissemination trap in the field

supplied with blackberry leaves and a source of water. Mortality was assessed daily for 10 days. Over 95% of beetles that passed through the traps during a one-week period in the field died within the 10 day observation period while control mortality during the same period was 8%.

The Japanese beetle autoinoculator can facilitate augmentation of endemic entomopathogenic fungi or the introduction of exotic fungi. Foreign exploration on Hokkaido Island, Japan, the proposed center of origin of the Japanese beetle, yielded isolates of *Beauveria brongniartii* from *P. japonica* larvae (Lacey and Klein, unpublished). This fungus is not reported from the beetle in the United States or Terceira Island and may offer potential as an inoculated biological control agent.

In the United States, J. Grundler (IPM Coordinator, Missouri Department of Agriculture, unpublished data) has successfully used this autoinoculator in combination with *M. anisopliae* to reduce population levels of Japanese beetles in a Missouri state park.

B Conclusion

Concomitant with host factors, pathogen virulence, infectivity and persistence, the capacity to disperse is a key factor in the ability of entomopathogens to produce epizootics (Tanada, 1963). Using adult beetles to disperse entomopathogenic fungi through the soil barrier could provide an inexpensive means of introducing entomopathogens into larval habitats that are difficult or expensive to treat.

4 Case study #3: Autodissemination of baculovirus by *Heliothis virescens* and *Helicoverpa zea*

Noctuidae is the largest and most destructive family of Lepidoptera, and several species in the *Heliothis/Helicoverpa* spp. complex are especially damaging to field crops (Reed, 1981; Fitt, 1989). Although many different control options have been employed for management of *Heliothis/Helicoverpa* (Reed, 1981), there is a need for control techniques that are less reliant on chemical pesticides with their well-documented problems (King and Coleman, 1989). Baculoviruses, especially nucleopolyhedrovirus (NPV), are effective microbial control agents against many lepidopteran pests, including several species of Noctuidae (Huber, 1986; Hamm, 1994), but problems with delivery systems, slow viral activity, cost effectiveness, efficacy, and environmental degradation have prevented their widespread use and commercialization (Falcon, 1976; Tinsley, 1979). Chapter IV-I in this book presents other information relevant to baculoviruses.

This case study will discuss an autodissemination approach for deploying baculoviruses for two noctuid pests: the tobacco budworm, *Heliothis virescens* in tobacco, *Nicotiana tabacum* (Solanaceae) (Nordin *et al.*, 1990, 1991; Jackson *et al.*, 1992); and the corn earworm, *Helicoverpa zea* in sweet maize, *Zea mays* (Gramineae) (Jackson *et al.*, 1993). An effective autodissemination method for baculoviruses could (1) solve delivery problems, (2) deposit virus in a protected environment, (3) use a smaller amount of virus thereby reducing costs, (4) eliminate plant coverage considerations, (5) avoid timing problems by using the appropriate insect stage as the disseminating vehicle, and (6) increase efficacy by concentrating virus at the most susceptible life stage, neonatal larvae.

A Heliothis virescens-AcMNPV-tobacco system

The first evaluation of an autodissemination approach for managing noctuid pests with an NPV was for *H. virescens* in North Carolina and Kentucky (Jackson *et al.*, 1992). Briefly, sex pheromone-baited traps were used to direct male moths into autoinoculators where they were forced to crawl through a powder formulation of NPV. After they were contaminated with NPV, males escaped into the field and initiated the vertical transmission of the virus. When males mated with wild females, they transferred some of the NPV powder to them, which in turn surface-contaminated their eggs during oviposition. When larvae chewed through their egg chorion, some of them ingested enough viral polyhedra to become lethally infected.

Initial experiments determined the transovum transmission efficiency and field persistence of powder formulations of two NPVs with known activity toward larvae of *H. virescens* (Nordin *et al.*, 1990, 1991). The viruses evaluated were *Autographa californica* multiply-embedded NPV (*Ac*MNPV) (derived from a plaque-purified isolate from E. M. Dougherty, USDA, Beltsville, MD) and *Helicoverpa zea* singly-embedded NPV (*Hz*SNPV) (derived from Elcar, a commercial *Hz*SNPV product; Sandoz Crop Protection Corp., Des Plaines, IL) (Nordin *et al.*, 1990, 1991). The *in vivo* production of these viruses (Boucias and Nordin, 1977; Nordin *et al.*, 1990) was through a healthy *H. virescens* colony reared on a pinto bean-based medium (Shorey and Hale, 1965). The dried, purified virus was formulated as a dust with walnut shell flour at 4.4×10^{10} polyhedral occlusion bodies (OBs)/gram of product.

H. virescens moths were capable of causing substantial transovum transmission of baculoviruses to F_1 progeny when one partner of a mating pair was surface-contaminated with the viral dust formulation. Scanning electron microscopy (SEM) revealed that many NPV OBs adhere to the wings and body scales of surface-contaminated *H. virescens* adults, and that NPV OBs were concentrated on the upper hemisphere of virus-contaminated eggs near the rosette, where they are most likely to be ingested during eclosion (Nordin *et al.*, 1990). In laboratory experiments, the viral-induced larval mortality for progeny arising from mating pairs in which only male moths were surface-contaminated with virus powder was 69% for the *Ac*MNPV formulation and 53% for the *Hz*SNPV formulation (Nordin *et al.*, 1990). *Ac*MNPV induced greater mortality to F_1 progeny than did *Hz*SNPV in both laboratory and field experiments. Studies in small field cages containing tobacco plants indicated that *Ac*MNPV persisted longer on foliage than *Hz*SNPV, and that enough OBs remained after 3 days to initiate lethal infections in early instars of *H. virescens* (Nordin *et al.*, 1991). Therefore, *Ac*MNPV was selected for field evaluations of the autodissemination approach described here (Jackson *et al.*, 1992).

The autodissemination technique was tested during 1988–1990 by tandem field experiments that were replicated 3 times at each location (North Carolina and Kentucky). A modified wire cone trap was used, in which Plexiglass contamination stations were attached at the top, replacing the collection baskets (Jackson *et al.*, 1994) (Figure 5). This combination of the wire cone trap with the contamination station will be referred to as "autoinoculator."

Treatments consisted of (1) autoinoculators containing viral contamination stations with

Figure 5. Photograph of plexiglass contamination station attached to a pheromone-baited wire-cone trap for autodissemination of nucleopolyhedroviruses

virus and marking powder; (2) autoinoculators containing contamination station without virus but with marking powder; and (3) control plots. Wire-cone traps (Hartstack *et al.*, 1979) baited with Virelure (Scentry, Buckeye, AZ), a synthetic sex attractant, were placed in the corners of all fields to monitor dispersal of insects emerging from the autoinoculators. Different colors of fluorescent marking powders (Day-Glo Color Corp., Cleveland, OH) for each treatment were added to the powder formulation so that the proportion of the male population passing through the autoinoculators could be estimated via recapture in the monitoring traps.

Male moths entered the bottom of the autoinoculator and moved upward through the wire cone to eventually enter the contamination station. The contamination stations contained three baffles (Figure 6) that forced the moths to crawl through an internal reservoir containing the NPV powder before they exited. Contamination stations were re-charged with 5 g of *Ac*MNPV formulation or control powder every 2 weeks. All moths emerging from the contamination station carried the powder.

Evaluation of this autodissemination technique required measurements of contamination rates of eggs, larvae, and male moths. Monitoring traps showed that up to 30% of the males captured in treatment fields were marked with the correct color of fluorescent marking powder. Examination by SEM showed that an average of 6–8% of the eggs collected from *Ac*MNPV-treated fields had OBs clustered on the upper hemisphere near the micropyle. Bioassays showed that *Ac*MNPV-induced larval mortality peaked at ca. 25% at the Kentucky location. Although this autodissemination technique functioned, it was not economically effective because *Ac*MNPV transmission and larval mortality were limited. In part, this may have been because the dispersal area of *H. virescens* moths was large relative to the small experimental plot size and the short distance between fields, resulting in migration of moths and dilution of *Ac*MNPV-induced mortality.

B Helicoverpa zea-Hz*SNPV-sweet maize system*

A *H. zea*-sweet maize system was selected in an attempt to induce a higher rate of viral infection. This system was chosen for several reasons: (1) ovipositing *H. zea* typically concentrate their eggs in the maize silks (Johnson *et al.*, 1975) where hatching larvae have a higher probability of being contaminated by infected cohorts; (2) both sexes of *H. zea* moths are attracted to lights (Latheef *et al.*, 1991), which could be used as the lure; (3) *Hz*SNPV is highly effective against *H. zea* larvae (Ignoffo *et al.*, 1965), and the commercial product Elcar was available; (4) like tobacco, sweet maize is a high-value crop that can bear higher input costs than other crops affected by *H. zea*; and (5) an autodissemination system for NPVs of *H. zea* using a light trap as the lure has already been developed (Gard, 1975).

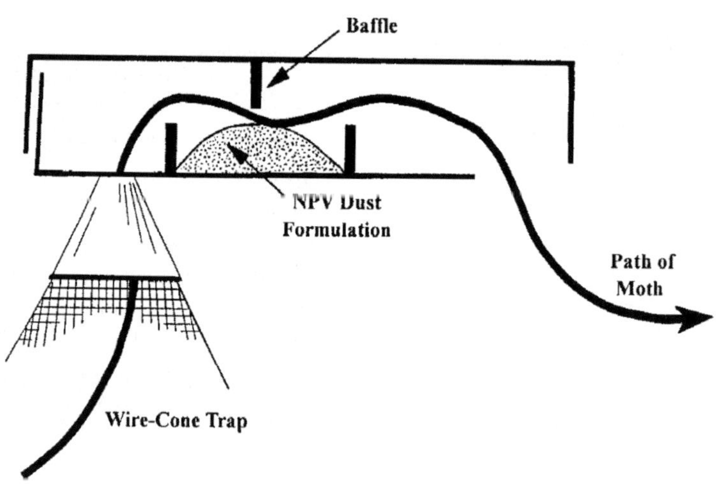

Figure 6. Diagram of plexiglass contamination station for autodissemination of entomopathogens

After a preliminary study in 1991, the autodissemination system was modified for experiments conducted in 1992 and 1993. Rather than using pheromones, a battery-operated blacklight was used as the lure, although male *H. zea* moths were still monitored by pheromone-baited wire-cone traps on each corner of the plots. Because the blacklight traps directed the insects down to the collection basket (Hollingsworth and Hartstack, 1972), a new aluminum contamination station based on Gard (1975) was developed; each contamination station (10 × 25 × 50 cm) was attached to the bottom of a blacklight trap (Figure 7). The last 8 cm of the bottom portion was bent upwards at 45° to form a reservoir that contained the *Hz*SNPV dust formulation. Directly below the entrance into the contamination station from the light trap funnel was a 15-cm diameter hole through the bottom of the trap. This hole was covered with window screen, allowing rain to pass through while retaining

Figure 7. Photograph of aluminum contamination station attached to the bottom of a battery-powered blacklight trap for autodissemination of nucleopolyhedroviruses

H. zea moths. There was a set of downward-directed baffles from the top of the contamination station that forced the captured moths to crawl through the viral dust.

One blacklight trap with a contamination station (= autoinoculator) was placed on the center line of the field at the interfaces of the first two plantings of sweet maize. After the first planting had matured, the autoinoculator was moved to the center line of the field at the interfaces of the second and third plantings. An interval-timer device was used to turn the blacklights on and off in 30 min cycles, which accentuated dispersal of moths away from the contamination station. Photocell switches were also utilized so that the blacklights only operated at night.

In 1992 and 1993 in North Carolina, up to 10% of the *H. zea* male moths captured in the monitor traps bore the fluorescent marker powder. Even though SEM showed that up to 17% of *H. zea* eggs had OBs on their chorions, there was little viral-induced mortality in bioassays of eggs collected from the virus-treated plots. There might have been a problem with the bioassay system, because damage assessment of the sweet maize plots in North Carolina had up to 50% less ear damage and 50% fewer larvae in the *Hz*SNPV-treated plots than in the control plots, showing that the autodissemination technique worked. The reduction in damage to sweet maize ears in the NPV-treated plots was lower at the Kentucky site, probably due to much higher pest pressures.

C Conclusion

The two case studies with *Heliothis/Helicoverpa* demonstrate that vertical transmission of baculoviruses can be achieved through an autodissemination approach. Although in neither example was larval damage reduced below acceptable pest management requirements, these autodissemination systems could be refined to achieve higher rates of viral infection. Indications that these systems could be improved were revealed by a computer simulation model of the autodissemination system for tobacco (Brown *et al.*, 1989). A sensitivity analysis of that model revealed that the system was most sensitive to the trapping efficiency. Other factors influencing vertical transmission of virus include weather factors, virulence of the viral formulation, lure attractiveness, pest population size, and local movement of moths. This model also pointed out the importance of horizontal transmission of baculoviruses for enhancing control of pest populations. Egg clustering, cannibalism, and dissemination by non-target species positively affect horizontal transmission of baculoviruses.

5 Case study #4: Dissemination of *Bacillus subtilis* and *Beauveria bassiana* by sap beetles

Sap beetles (Coleoptera: Nitidulidae) are small insects (typically < 5 mm) that feed on a variety of fresh and fermenting plant materials (Hinton, 1945). Several species can be pests of fresh fruits or vegetables, while a few species are also pests of stored dried fruits or grains. Sap beetles are also known to vector a variety of plant pathogens to fruit, vegetables, grain crops and trees (Dowd, 1991, 1995). In some cases, the fungi that these beetles carry will produce toxins that are carcinogenic (*e.g.*, aflatoxins produced by *Aspergillus flavus*); these can be a problem in peanuts (ground nuts), cotton seed, figs, and maize (Dowd, 1998).

Sap beetles respond strongly to many volatile compounds that are produced by plants or microorganisms (Dowd, 1991, 1995). Maize damaged by caterpillars is more attractive to sap beetles than maize that has only received mechanical damage (Sanford and Luckmann, 1963). Males in several species of sap beetles (*e.g.*, *Carpophilus* spp.) produce aggregation pheromones that are attractive to both sexes (Bartelt, 1997). The pheromones and host volatiles interact in a highly synergistic manner (Bartelt, 1997). Methods for effectively releasing both the pheromones (Bartelt, 1997) and host volatiles (Bartelt *et al.*, 1992) are available; thus, they can be used to attract and trap sap beetles.

Several different traps baited with pheromones and host-derived attractants have been developed to capture sap beetles (Dowd *et al.*, 1992; Williams *et al.*, 1993). A component in one of these traps (Dowd *et al.*, 1992) has been modified and the entire device is now referred to as an autoinoculator (Figure 8) (Vega *et al.*, 1995; Dowd and Vega, 1995). Sap beetles that are attracted to the modified trap fall into the autoinoculator chamber containing a formulated microbial agent in powder form and leave the chamber with the material covering their body parts (Vega *et al.*, 1995). The development of this device has led to studies to determine if sap beetles will carry a biological control agent active against a mycotoxin-producing fungus that occurs in maize, as well as to autodisseminate an insect fungal pathogen to selectively increase mortality of sap beetles during overwintering.

A *Dispersal of* Bacillus subtilis *for reduction of aflatoxin in maize*

The dusky sap beetle (*Carpophilus lugubris*) is among the species that are attracted to damaged maize in the milk stage (Connell, 1956; Sanford and Luckmann, 1963). Because of the ability of these beetles to readily locate damaged kernels, which may ultimately become infected with beetle-vectored fungi, it should be possible to use the beetles to carry a biological control organism that acts against mycotoxigenic fungi such as *A. flavus*. A powder formulation of a strain of *Bacillus subtilis* designed as a seed protectant against fungi such as *A. flavus* became commercially available in the USA in 1993 as Kodiak™ (Gustafson Inc., Plano, TX) (Gustafson Inc., 1993). When applied in the field directly to damaged maize kernels it reduced colonization by *A. flavus* and subsequent aflatoxin production (Cuero *et al.*, 1991).

Laboratory studies in a model system indicated that dusky sap beetles could deliver *B. subtilis* to purposely damaged maize ears and inhibit colonization by *A. flavus* (Dowd *et al.*, 1998). Field cage studies indicated that dusky sap beetles could disperse *B. subtilis* from autoinoculators to purposely damaged maize, thereby inhibiting colonization by *A. flavus* (Dowd *et al.*, 1998). *B. subtilis* was readily re-isolated from the damaged ears and sap beetle adults and larvae collected from the damaged ears inside the field cage. No *B. subtilis* was isolated from ears outside the cage. Studies in maize fields in the first year also indicated that these beetles could disperse *B. subtilis* from autoinoculators to purposely damaged ears and thereby inhibit the colonization by *A. flavus* (Dowd *et al.*, 1998).

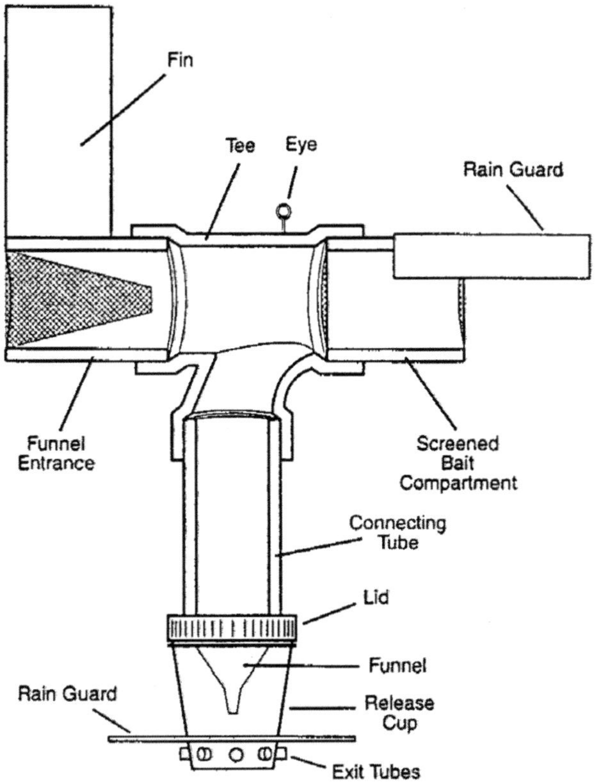

Figure 8. Diagram of the autodissemination device used to disseminate bioactive agents by sap beetles

When beetles were excluded by screening from damaged maize ears, 92% of the ears had visibly sporulating *A. flavus*. When maize ears were accessible to sap beetles, the presence of *A. flavus* could not be visually detected.

In another field study, a single autoinoculator containing *B. subtilis* was placed adjacent to a maize field immediately prior to pollen shed. The objective was to allow sap beetles to pick up the bacteria and disperse it to pollen that accumulated in leaf axils, where sap beetles feed prior to ear fill. This would allow for additional foci of the bacteria to develop and potentially to be dispersed. *B. subtilis* was reisolated from 13% of sampled leaf axils sampled, from 25% of embryonic ears damaged by vertebrates such as deer and raccoons, and from all sap beetles collected from vertebrate damaged ears immediately before and during the ear-wounding experiment (Dowd *et al.*, 1998). *B. subtilis* was also reisolated from beetles captured in most traps several hundred meters away after the experiment was completed. Overall, visible *A. flavus* was rare in ears that were accessible to sap beetles, but common in ears protected from sap beetles. Over 80% of ears protected from sap beetles had aflatoxin levels over 20 ppb, but less than 20% of the ears that were accessible to sap beetles had aflatoxin levels over 20 ppb. As would be expected, ears that were accessible to sap beetles for 7 days had significantly less aflatoxin that those accessible for only 3 days (Dowd *et al.*, 1998).

B *Sap beetles autodissemination of* Beauveria bassiana *to overwintering sites*

Sap beetles, because of their generally secondary pest status, do not receive the attention that other pest species such as lepidopteran larvae receive, although they are controlled along with caterpillars when insecticides are used in crops (James *et al.*, 1995; Dowd, 1998). Previous reports (Pree, 1968; Foott and Timmins, 1971) and surveys by (Dowd and Vega 2003) indicated *B. bassiana* was frequently recorded from sap beetles, but typically < 2% of the population was infected. Field collected strains of *B. bassiana* were relatively poor pathogens for the dusky sap beetle, even in ideal situations (Dowd, unpublished data). However, when the *B. bassiana* strain AF-4 (produced by Mycotech Corp., Butte, MT), and provided by S. Ferguson, (CIBA, Vero Beach, FL) was tested in the laboratory, it killed most of the adult dusky sap beetles within 2 days (Vega, unpublished data). Thus, AF-4 was a more suitable strain to test in the field for control of dusky sap beetles, using rapid kill as the criterion.

Aggregation pheromones of the dusky sap beetle are effective in the field until it becomes too cold for the beetles to fly (Dowd and Nelsen, 1994). Preliminary observations indicated some sap beetle species overwintered in groups, making them a good target for late season autodissemination of *B. bassiana* into places unlikely to be frequented by beneficial insects. Initially, autoinoculators containing a powder formulation of *B. bassiana* strain AF-4 were placed in the field in early November. The following spring, trap monitoring indicated little or no increase in *B. bassiana* prevalence over background levels noted in the past. Overwintering traps were developed because it was possible that the sap beetles were being killed in overwintering sites and not emerging to be detected (Dowd, unpublished data). The first of these was placed so that sap beetles would be able to enter the duff/leaf litter/soil to overwinter. While these traps were very effective for *Coleopterus truncatus* [a sap beetle that can vector *Ceratocystis* tree pathogens (Hinds, 1972)], very few of the target dusky sap beetles were captured. Due to subsequent hanging trap captures that suggested the dusky sap beetles were overwintering in hollow trees, a tree stump-like overwintering trap was constructed, which has yielded much larger numbers of this beetle relative to the ground traps (Dowd, unpublished data).

As in studies with *B. subtilis*, timing of placement of the autoinoculator in the field was found to be important. Initial studies encountered relatively early winters, and relatively low prevalence of sap beetles infected with *B. bassiana* were observed the following spring. The prevalence of *B. bassiana* in dusky sap beetles collected from both overwintering and hanging traps (see Dowd *et al.*, 1992, for hanging trap design) has greatly increased with placement of the autoinoculator in the field just prior to maize harvest in early autumn (Dowd, unpublished data). It was not uncommon to recover sap beetles from overwintering traps that had > 20% of the individuals infected with *B. bassiana*. Hanging-trap capture rates have generally exceeded 10% infection rates with *B. bassiana*, with much higher levels occurring for some sample dates in recent years. No economically important beneficial insects have ever been found in the overwintering traps, indicating the selective delivery of the *B. bassiana*. Molecular methods will be used in future studies to assess whether field collected strains of *B. bassiana* match the released strain (AF-4).

C Conclusion

Overall, this methodology has several advantages. The beetles can carry the *B. subtilis* directly to damaged maize. This targeted vectoring is presumably less costly and more effective than conventional methods of application due to the difficulty of reaching the cryptically damaged ears. Because a natural control agent can be used, this method should be more environmentally benign than use of conventional chemical fungicides, which are typically of low efficacy for *A. flavus*. If suitable biological control agents are available, this strategy should be successful for use against other toxin producing fungi, provided sufficient populations of appropriate sap beetles are available.

Autoinoculators can also be efficient in disseminating an insect pathogen in field situations. The technique allows for specific and targeted delivery of the bioactive agent. which would not be possible using conventional application methods. However, the timing and placement of the autoinoculators in the field is very important. Combination of different bioactive agents in the same autoinoculator might be desirable and could potentially solve different insect pest- originated problems.

6 Other examples

An early example of autodissemination, although without the use of an autoinoculator device, is that of augmented "natural" dissemination of the *Oryctes* virus by the palm rhinoceros beetle, *Oryctes rhinoceros*, in the South Pacific (Zelazny, 1976; Zelazny *et al.*, 1992; Huger,2005).

Other examples using autoinoculators involve the use of inoculum dispensers placed on beehives or on colony boxes of bumble bees to inoculate bees (*Apis mellifera*) and bumble bees (*Bombus* spp.) with microbial agents that antagonize plant pathogens [*e.g., Gliocladium roseum* against *Botrytis cinerea* (Peng *et al.*, 1992; Yu and Sutton, 1997) and P*seudomonas fluorescens* against *Erwinia amylovora* (Johnson *et al.*, 1993)]. Bees have also been used to disseminate insect pathogens. Gross *et al.* (1994a, b) adapted a device to a conventional beehive to have honey bees disseminate *Heliothis*-NPV to crimson clover (*Trifolium incarnatum*) as a means of managing *H. zea* larvae. Similarly, Butt *et al.* (1998) used bees to disseminate the fungal entomopathogen *M. anisopliae* to buds and flowers of oilseed rape (*Brassica napus* and *B. rapa*) and successfully reduced populations of pollen beetles (*Meligethes* spp.). Their finding that *M. anisopliae* did not adversely affect honey bees is encouraging, although an important aspect that has not been examined in these bee dissemination studies is whether the bioactive agent contaminates the honey, which may present problems for human consumption.

Autoinoculator devices have also been developed for use against sweetpotato weevils (*Cylas formicarius* (F.); Yasuda, 1999), the cabbage root fly (*Delia radicum*; Meadow *et al.*, 1998), the spruce bark beetle (*Ips typographus*; Vaupel and Zimmermann, 1996), banana weevils (*Cosmopolites sordidus* (Germar); Tinzaara *et al.*, 2004), codling moth (*Cydia pomonella*; Pultar *et al.*, 2000; Winstanley *et al.*, 2005) and against insects of medical importance such as *Glossina* spp. (Maniania, 1998, 2002; Maniania *et al.*, 2006).

7 Conclusions

The adaptability of autoinoculator devices in commercial agriculture will depend, in part, on growers, extension service personnel, and the public being aware of the existence of this alternative method of pest management. The availability of pheromones and other lures for the most important agricultural pests make this method an alternative that deserves further study by specialists in different agroecosystems. To successfully use these devices, additional studies will be necessary to provide guidelines on the number of autoinoculator devices needed to adequately protect a certain area.

Several patents have been issued to cover autoinoculator devices (*e.g.*, Skadeland 1981; Gunner *et al.*, 1991a, b; Miller *et al.*, 1993; Chang and Gehret, 1993; Gross *et al.*, 1994b; Jackson *et al.*, 1994; Dowd and Vega, 1995). Only one of these inventions (Gunner *et al.*, 1991a, b; Miller *et al.*, 1993), a cockroach control chamber using fungal entomopathogens and marketed as BioPath™ by EcoScience Corp. (East Brunswick, NJ) has reached the market. The existence of these patents should serve as encouragement for industry to license the invention and to make it available to the public.

One of the most important advantages of the autodissemination technique is the targeted delivery of a pest control agent; consequently, there are likely to be limited detrimental effects to other organisms or the environment. This targeted delivery can be directed at concealed pests (Jackson and O'Callaghan, 1998) that cannot be reached via conventional methods (*i.e.*, insecticidal sprays). The technique can be adapted for different insect species and bioactive agents, and may also be suitable for disseminating weed biological control agents using an insect that also attacks the target weed.

8 Acknowledgments

Thanks to M. Greenstone (USDA) for reviewing the manuscript. J. K. Pell's work was supported by Shell Research International, The Leverhulme Trust, Department for International Development (Crop Protection Programme, Project Number R6615), Ministry for Agriculture Fisheries and Food, UK. IACR-Rothamsted receives grant-aided support from the Biotechnology and Biological Sciences Research Council of the UK. The mention of firm names or trade products does not imply that they are endorsed or recommended by the U. S. Department of Agriculture (USDA) over other firms of similar products not mentioned. Names are necessary to report factually on data; however, the USDA neither guarantees nor warrants the standard of the product, and the use of the name by USDA implies no approval of the product to the exclusion of others that may be suitable.

9 References

Ahmad, S., Streu, H. T., and Vasvary, L. M. 1983. The Japanese beetle: a major pest of turfgrass. *Am. Lawn Appl.* **4**, 2–10.

Bartelt, R. J. 1997. Aggregation pheromones of *Carpophilus* spp. (Coleoptera: Nitidulidae): Review of chemistry and biology. *Rec. Res. Devel. Entomol.* **1**, 115–129.

Bartelt, R. J., Dowd, P. F., Vetter, R. S., Shorey, H. H., and Baker, T. C. 1992. Responses of *Carpophilus hemipterus* (Coleoptera: Nitidulidae) and other sap beetles to the pheromone of *C. hemipterus* and host-related co-attractants in California field tests. *Environ. Entomol.* **21**, 1143–1153.

Bird, F. T. 1961. Transmission of insect viruses with particular reference to ovarial transmission and its importance in the development of epizootics. *J. Insect Pathol.* **3**, 352–380.

Boucias, D. S., and Nordin, G. L. 1977. Inter-instar susceptibility of the fall webworm, *Hyphantria cunea*, to its nucleopolyhedrosis and granulosis viruses. *J. Invertebr. Pathol.* **30**, 68–75.

Brown, G. C., Nordin, G. L., Jackson, D. M., and Johnson, D. W. 1989. Autodissemination of a baculovirus for suppression of *Heliothis virescens* on tobacco: A computer simulation model. 22nd Annu. Meet., Soc. Invertebr. Pathol., Abstract No. 113.

Butt, T. M., Carreck, N. L., Ibrahim, L., and Williams, I. H. 1998. Honey-bee mediated infection of pollen beetle (*Meligethes aeneus* spp.) by the insect-pathogenic fungus, *Metarhizium anisopliae. Biocontrol Sci. Technol.* **8**, 533–538.

Chang, F. N., and Gehret, M. J. 1993. Insect bait station. U. S. Patent No. 5,238,681.

Connell, W. A. 1956. Nitidulidae of Delaware. Univ. Del. Agric. Exp. Sta. Bull. No. 318, 67 pp.

Cuero, R. G., Duffus, E., Osuji, G., and Pettit, R. 1991. Aflatoxin control in preharvest maize: effects of chitosan and two microbial control agents. *J. Agric. Sci.* **117**, 165–169.

Doane, C. C. 1970. Primary pathogens and their role in the development of an epizootic in the gypsy moth. *J. Invertebr. Pathol.* **15**, 21–33.

Dowd, P. F. 1991. Nitidulids as vectors of mycotoxin-producing fungi. *In* "Aflatoxin in Corn: New Perspectives"(O. L. Shotwell and C. R. Hurburgh, Jr., Eds.), pp. 335–342. North Central Regional Publication #329, Iowa State University, Ames, IA.

Dowd, P. F. 1995. Sap beetles and mycotoxins in maize. *Food Add. Contam.* **12**, 497–508.

Dowd, P. F. 1998. Involvement of arthropods in the establishment of mycotoxigenic fungi under field conditions. *In* "Mycotoxins in Agriculture and Food Safety" (K. K. Singh and D. Bhatnagar, Eds.), pp. 307–350. Marcel Dekker, NY.

Dowd, P. F., Bartelt, R. J., and Wicklow, D. T. 1992. Novel insect trap useful in capturing sap beetles (Coleoptera: Nitidulidae) and other flying insects. *J. Econ. Entomol.* **85**, 772–778.

Dowd, P. F., and Nelsen, T. C. 1994. Seasonal variation of sap beetle (Coleoptera: Nitidulidae) populations in central Illinois cornfield-oak woodland habitat and potential influence of weather patterns. *Environ. Entomol.* **23**, 1215–1223.

Dowd, P. F., and Vega, F. E. 1995. Autoinoculating device for contaminating insects with active agents. U. S. Patent No. 5,452,540.

Dowd, P. F., Vega, F. E., Nelsen, T. C., and Richard, J. L. 1998. Dusky sap beetle mediated dispersal of *Bacillus subtilis* to inhibit *Aspergillus flavus* and aflatoxin production in maize *Zea mays. Biocontrol Sci. Technol.* **8**, 233–247.

Dowd, P. F., and Vega, F. E. 2003. Autodissemination of *Beauveria bassiana* by sap beetles (Coleoptera: Nitiulidae) to overwintering sites. *Biocontrol Sci. Technol.* **13**, 65–75.

Entwistle, P. F., Adams, P. H. W., and Evans, H. F. 1978. Epizootiology of a nuclear polyhedrosis virus in European spruce sawlfy (*Gilpinia hercyniae*): the rate of passage of infective virus through the gut of birds during cage tests. *J. Invertebr. Pathol.* **31**, 307–312.

Falcon, L. A. 1976. Problems associated with the use of arthropod viruses in pest control. *Annu. Rev. Entomol.* **21**, 305–324.

Fitt, G. P. 1989. The ecology of *Heliothis* species in relation to agroecosystems. *Annu. Rev. Entomol.* **34**, 17–52.

Foott, W. H., and Timmins, P. R. 1971. Importance of field corn as a reproductive site for *Glischrochilus quadrisignatus* (Say) (Coleoptera: Nitidulidae). *Proc. Entomol. Soc. Ontario* **101**, 73–75.

Forschler, B. T., and Young, S. Y. 1993. Southern corn rootworm adults (Coleoptera: Chrysomelidae) may act as a nonhost disseminator of nuclear polyhedrosis virus. *J. Invertebr. Pathol.* **61**, 313–314.

Furlong, M. J., Pell, J. K., Ong, P. C. and Syed, A. R. 1995. Field and laboratory evaluation of a sex pheromone trap for the autodissemination of the fungal entomopathogen *Zoophthora radicans* (Entomophthorales) by the diamondback moth, *Plutella xylostella* (Lepidoptera: Yponomeutidae). *Bull. Entomol. Res.* **85**, 331–337.

Furlong, M. J., and Pell, J. K. 1996. Interactions between the fungal entomopathogen *Zoophthora radicans* Brefeld (Entomophthorales) and two hymenopteran parasitoids attacking the diamondback moth, *Plutella xylostella. J. Invertebr. Pathol.* **68**, 15–21.

Furlong, M. J., and Pell, J. K. 1997a. Integration of *Zoophthora radicans* and synthetic female sex pheromone for the control of diamondback moth. *In* "Proceedings of the Third International Workshop on the Management of Diamondback Moth and other Crucifer Pests" (A. Sivapragasm, W. H. Loke, A. K. Hussan and G. S. Lim, Eds.), pp. 123–129. Kuala Lumpur, Malaysia.

Furlong, M. J., and Pell, J. K. 1997b. The relative merits of the fungal entomopathogens *Zoophthora radicans* (Entomophthorales) and *Beauveria bassiana* (Deuteromycetes) for use in autodissemination techniques. Proceedings of the 6th European meeting of the IOBC Working Group "Insect Pathogens and Insect Parasitic Nematodes", Copenhagen.

Furlong, M. J., Pell, J. K. and Reddy, G. V. P. 1997a. Premortality effects of *Zoophthora radicans* infection in *Plutella xylostella. J. Invertebr. Pathol.* **70**, 214–220.

Furlong, M. J., Reddy, G. V. P., Pell, J. K. and Poppy, G. M. 1997b. Premortality effects of *Zoophthora radicans* infection in the diamondback moth. *In* "Proceedings of the Third International Workshop on the Management of Diamondback Moth and other Crucifer Pests" (A. Sivapragasm, W. H. Loke, A. K. Hussan, and G. S. Lim, Eds.), pp. 116–122. Kuala Lumpur, Malaysia.

Gard, Jr., I. E. 1975. Utilization of light traps to disseminate insect viruses for pest control. Ph.D. Dissertation, University of California, Berkeley, 174 pp.

Gross, H. R., Hamm, J. J., and Carpenter, J. E. 1994a. Design and application of a hive-mounted device that used honey bees (Hymenoptera: Apidae) to disseminate *Heliothis* nuclear polyhedrosis virus. *Environ. Entomol.* **23**, 492–501.

Gross, H. R., Johnson, R., and Walters, J. C. 1994b. Beehive-mounted device for utilizing honeybees (Hymenoptera: Apidae) in the dissemination of biocontrol agents. U. S. Patent No. 5,348,511.

Gunner, H. B., Agudelo-Silva, F., and Miller, D. W. 1991a. Method and device for the biological control of cockroaches. U. S. Patent No. 5,057,315.

Gunner, H. B., Agudelo-Silva, F., and Miller, D. W. 1991b. Method and device for the biological control of insects. U. S. Patent No. 5,057,316

Gustafson, Inc. 1993. Kodiak registration sheet; Kodiak formulations. Gustafson Corp., Plano, Texas, 7 pp.

Hamm, J. J. 1994. Nuclear polyhedrosis viruses for insect control. *In* "Natural and Engineered Pest Management Agents", (P. A. Hedin, J. J. Menn, and R. M. Hollingworth, Eds.), ACS Symp. Ser. No. 551, pp. 324–335. Am. Chem. Soc., Wash., DC

Hartstack, A. W., Witz, J. A., and Buck, D. R. 1979. Moth traps for the tobacco budworm. *J. Econ. Entomol.* **72**, 519–522.

Hinds, T. E. 1972. Insect transmission of *Ceratocystis* species associated with aspen cankers. *Phytopathology* **62**, 221–225.

Hinton, H. E. 1945. A Monograph of Beetles Associated with Stored Products. Jarrold and Sons, Norwich, England.

Hollingsworth, D. E. and Hartstack, A. W. 1972. Effect of components on insect light trap performance. *Trans. Am. Soc. Agric. Engineer.* **5**, 924–927.

Huber, J. 1986. Use of baculoviruses in pest management programs. *In* "The Biology of Baculoviruses, Vol. 2" (R. R. Granados and B. A. Federici, Eds.), pp. 181–202. CRC Press, Boca Raton, FL.

Huger, A. M. 2005. The Oryctes virus: its detection, identification, and implementation in biological control of the coconut palm rhinoceros beetle, *Oryctes rhinoceros* (Coleoptera: Scarabaeidae). *J. Invertebr. Pathol.* 89, 78–84.

Hukuhara, T. 1973. Further studies on the distribution of a nuclear-polyhedrosis virus of the fall webworm, *Hyphantria cunea*, in soil. *J. Invertebr. Pathol.* **22**, 345–350.

Hunter-Fujita, F. R., Entwistle, P. F., Evans, H. F., and Crook, N. E. 1998. "Insect Viruses and Pest Management". J. Wiley and Sons, Chichester, England. 620 pp.

Ignoffo, C. M., Chapman, A. J., and Martin, D. F. 1965. The nuclear polyhedrosis virus of *Heliothis zea* (Boddie) and *Heliothis virescens* (Fabricius). III. Effectiveness of the virus against field populations of *Heliothis* on cotton, corn and grain sorghum. *J. Invertebr. Pathol.* **7**, 227–235.

Jackson, T. A. 1992. Scarabs - pests of the past or the future? *In* "Use of Pathogens in Scarab Pest Management" (T. A. Jackson and T. R. Glare, Eds.), pp. 1–10. Intercept Ltd., Andover, UK.

Jackson, T. A., and Glare, T. R. (Eds.) 1992. *Use of Pathogens in Scarab Pest Management.* Intercept Ltd., Andover, UK, 298 pp.

Jackson, T. A., and M. O'Callaghan. 1998. Lure, infect and disperse – a strategy for control of concealed pests (M. O'Callaghan and T. A. Jackson, Eds.), pp. 125–130. Proceedings of the 4th International Workshop on Microbial Control of Soil Dwelling Pests, AgResearch, Lincoln, New Zealand.

Jackson, D. M., Brown, G. C., Nordin, G. L., and Johnson, D. W. 1992. Autodissemination of a baculovirus for management of tobacco budworms (Lepidoptera: Noctuidae) on tobacco. *J. Econ. Entomol.* **85**, 710–719.

Jackson, D. M., Nordin, G. L., Brown, G. C., and Johnson, D. W. 1993. Autodissemination of a baculovirus for the suppression of corn earworm larvae in sweet corn. 26th Annu. Meet., Soc. Invertebr. Pathol., Abstract No. 140.

Jackson, D. M., G. Brown, and G. L. Nordin. 1994. Method and apparatus for autodissemination of insect pathogens. U. S. Patent No. 5,359,807.

James, D. G., Faulder, R. J., and Bartelt, R. J. 1995. Fauna and seasonal abundance of *Carpophilus* spp. (Coleoptera: Nitidulidae) in four stone fruit growing regions of southeastern Australia as determined by pheromone trapping. *J. Aust. Entomol. Soc.* **34**, 327–333.

Johnson, K. B., Stockwell, V. O., Burgett, D. M., Sugar, D., and Loper, J. E. 1993. Dispersal of *Erwinia amylovora* and *Pseudomonas fluorescens* by honey bees from hives to apple and pear blossoms. *Phytopathology* **83**, 478–484.

Johnson, M. W., Stinner, R. E., and Rabb, R. L. 1975. Ovipositional response of *Heliothis zea* (Boddie) to its major hosts in North Carolina. *Environ. Entomol.* **4**, 291–297.

Kellen, W. R., and Hoffmann, D. F. 1987. Laboratory studies on the dissemination of a granulosis virus by healthy adults of the Indianmeal moth, *Plodia interpunctella* (Lepidoptera: Pyralidae.) *Environ. Entomol.* **16**, 1231–1234.

Keller, S., Keller, E., Schweizer, C., Auden, J. A. L., and Smith, A. 1989. Two large field trials to control the cockchafer *Melolontha melolontha* with the fungus *Beauveria brongniartii* (Sacc.) Petch. British Crop Protection Council Monograph **43**, 183–190.

King, E. G., and Coleman, R. J. 1989. Potential for biological control of *Heliothis* species. *Annu. Rev. Entomol.* **34**, 53–76.

Klein, M. G. 1992. Use of *Bacillus popilliae* in Japanese beetle control. *In* "Use of Pathogens in Scarab Pest Management" (T. R. Glare and T. A. Jackson, Eds.), pp. 179–90. Intercept. Andover, UK.

Klein, M. G., and Edwards, D. C. 1989. Captures of *Popillia lewisi* (Coleoptera: Scarabaeidae) and other scarabs on Okinawa with Japanese beetle lures. *J. Econ. Entomol.* **82**, 101–103.

Klein, M. G., and Georgis, R. 1992. Persistence of control of Japanese beetle (Coleoptera: Scarabaeidae) larvae with steinernematid and heterorhabditid nematodes. *J. Econ. Entomol.* **85**, 727–730.

Klein, M. G., and Lacey, L. A. 1999. An attractant trap for the autodissemination of entomopathogenic fungi into populations of the Japanese beetle, *Popillia japonica* (Coleoptera: Scarabaeidae). *Biocontrol Sci. Technol.* **9**, 151–158.

Lacey, L. A., Bettencourt, R., Garrett, F. J., Simões, N. J., and Gaugler, R. H. 1993. Factors influencing parasitism of adult Japanese beetles, *Popillia japonica* (Coleoptera: Scarabaeidae) by entomopathogenic nematodes. *Entomophaga* **38**, 501–509.

Lacey, L. A., Amaral, J. J., Klein, M. G., Simões, N. J., Martins, A., and Mendes, C. 1994a. Microbial control of *Popillia japonica* (Coleoptera: Scarabaeidae) on Terceira Island (Azores Portugal): the role of operational research. Proc. VIth Int. Colloq. Invertebr. Pathol. Microb. Cont. pp. 409–415.

Lacey, L. A., Martins, A., and Ribeiro, C. 1994b. The pathogenicity of *Metarhizium anisopliae* and *Beauveria bassiana* for adults of the Japanese beetle, *Popillia japonica* (Coleoptera: Scarabaeidae). *Euro. J. Entomol.* **91**, 313–319.

Lacey, L. A., Amaral, J. J., Coupland, J., Klein, M. G., and Simões, A. M. 1995. Flight activity of *Popillia japonica* (Coleoptera: Scarabaeidae) after treatment with *Metarhizium anisopliae. Biol. Control* **5**, 167–172.

Ladd, T. L., Jr., McGovern, T. P., Beroza, M., Buriff, C. R., and Klein, M. G. 1976. Japanese beetles: attractancy of phenethyl propionate+eugenol (3:7) and synthetic eugenol. *J. Econ. Entomol.* **69**, 468–470.

Latheef, M. A., Lopez, Jr., J. D., and Witz, J. A. 1991. Reproductive condition of female corn earworm (Lepidoptera: Noctuidae) moths from sweep nest and blacklight trap collections in corn. *Environ. Entomol.* **20**, 736–741.

Lautenschlager, R. A., and Podgwaite, J. D. 1979. Passage of the nucleopolyhedrosis virus by avian and mammalian predators of the gypsy moth, *Lymantria dispar. Environ. Entomol.* **8**, 210–214.

Levin, D. B., Laing, J. E., Jaques, R. P., and Corrigan, J. E. 1983. Transmission of the granulosis virus of *Pieris rapae* (Lepidoptera: Pieridae) by the parasitoid *Apanteles glomeratus* (Hymenoptera: Braconidae). *Environ. Entomol.* **12**, 166–170.

Maniania, N. K. 1998. A device for infecting adult tsetse flies, *Glossina* spp., with an entomopathogenic fungus in the field. *Biol. Control* **11**, 248–254.

Maniania, N. K. 2002. A low-cost contamination device for infecting adult tsetse, *Glossina* spp., with the entomopathogenic fungus *Metarhizium anisopliae* in the field. *Biocontrol Sci. Technol.* **12**, 59–66.

Maniania, N. K., Ekesi, S., Odulaja, A., Okech, M. A., and Nadel, D. J. 2006. Prospects of a fungus-contamination device for the control of tsetse fly *Glossina fuscipes fuscipes. Biocontrol Sci. Technol.* **16**, 129–139.

Meadow, R., Shelton, A. M., and Vandenberg, J. D. 1998. Autodissemination of *Beauveria bassiana* for microbial control of the cabbage root fly in sustainable agriculture. *IOBC Bull.* **21**, 37–40.

Miller, D. W., Agudelo-Silva, F., Gunner, H. B., LeFevre, V. F., and Broza, M. 1993. Method and device for the biological control of insects. U. S. Patent No. 5,189,831.

Neilson, M. M., and Elgee, D. E. 1968. The method and role of vertical transmission of a nucleopolyhedrosis virus in the European spruce sawfly *Diprion hercyniae. J. Invertebr. Pathol.* **12**, 132–139.

Nordin, G. L., Brown, G. C., and Jackson, D. M. 1990. Vertical transmission of two baculoviruses infectious to the tobacco budworm, *Heliothis virescens* (F.) (Lepidoptera: Noctuidae) using an autodissemination technique. *J. Kans. Entomol. Soc.* **63**, 393–398.

Nordin, G. L., Brown, G. C., and Jackson, D. M. 1991. Transovum transmission of two nuclear polyhedrosis viruses (Baculoviridae) by adult tobacco budworm and viral persistence on tobacco foliage. *Trans. Ky. Acad. Sci.* **52**, 33–39.

Ooi, P. A. C. 1981. Microbial control of the diamondback moth in Cameron Highlands, Malaysia. *Malaysian Appl. Biol.* **10**, 49–56.

Pell, J. K., Wilding, N., Player, A. L., and Clark, S. J. 1993a. Selection of an isolate of *Zoophthora radicans* (Zygomycetes: Entomophthorales) for biocontrol of the diamondback moth, *Plutella xylostella* (Lepidoptera: Yponomeutidae). *J. Invertebr. Pathol.* **61**, 75–80.

Pell, J. K., Macaulay, E. D. M., and Wilding, N. 1993b. A pheromone trap for dispersal of the pathogen *Zoophthora* radicans Brefeld. (Zygomycetes: Entomophthorales) amongst populations of the diamondback moth, *Plutella xylostella* L. (Lepidoptera: Yponeumatidae.) *Biocontrol Sci. Technol.* **3**, 315–320.

Pell, J. K., Pluke, R., Clark, S. J., Kenward, M. G., and Alderson, P. G. 1997. Interactions between two aphid natural enemies, the entomopathogenic fungus *Erynia neoaphidis* Remaudière & Hennebert (Zygomycetes: Entomophthorales) and the predatory beetle *Coccinella septempunctata* L. (Coleoptera: Coccinellidae). *J. Invertebr. Pathol.* **69**, 261–268.

Peng, G., Sutton, J. C., and Kevan, P. G. 1992. Effectiveness of honey bees for applying the biocontrol agent *Gliocladium roseum* to strawberry flowers to suppress *Botrytis cinerea*. *Can. J. Plant Pathol.* **14**, 117–129.

Pree, D. J. 1968. Control of *Glischrochilus quadrisignatus* (Say) (Coleoptera: Nitidulidae), a pest of fruit and vegetables in southwestern Ontario. *Proc. Entomol. Soc. Ont.* **99**, 60–64.

Pultar, O., Kocourek, F., Berankova, J., Stara, J., Kuldova, J., and Hrdy, I. 2000. Codling moth management by means of pheromone stations with *Cydia pomonella* granulosis virus. *Acta Horticult.* **525**, 477–480.

Reed, W. 1981. International Workshop on *Heliothis* Management. 15–20 November 1981, International Crops Research Institute for the Semi-Arid Tropics (ICRISAT) Center, Patancheru, Andhra Pradesh, India. ICRISAT.

Reddy, G. V. P., Furlong, M. J., Pell, J. K., and Poppy, G. M. 1998. *Zoophthora radicans* infection inhibits the response to and production of sex pheromone in the diamondback moth. *J. Invertebr. Pathol.* **72**, 167–169.

Riethmacher, G. W., Rombach, M. C., and Kranz, J. 1992. Epizootics of *Pandora blunkii* and *Zoophthora radicans* (Entomophthoraceae: Zygomycotina) in diamondback moth populations in the Philippines. *In* "Proceedings of the Second International Workshop on Diamondback Moth and other Crucifer Pests" (N. S. Talekar, Ed.), pp. 193–199. Tainan, Taiwan.

Roy, H. E., Pell, J. K., Clark, S. J., and Alderson, P. G. 1998. Implications of predator foraging on aphid pathogen dynamics. *J. Invertebr. Pathol.* **71**, 236–247.

Sanford, J. W., and Luckmann, W. H. 1963. Observations on the biology and control of the dusky sap beetles in Illinois. *Proc. North Central Branch Entomol. Soc. Am.* **18**, 39–43.

Scholte, E.-J., Knols, B. G. J., and Takken, W. 2004. Autodissemination of the entomopathogenic fungus *Metarhizium anisopliae* amongst adults of the malaria vector *Anopheles gambiae*. *Malaria J.* **3**, 45.

Shapas, T. J., Burkholder, W. E., and Boush, G. M. 1977. Population suppression of *Trogoderma glabrum* by using pheromone luring for protozoan pathogen dissemination. *J. Econ. Entomol.* **70**, 469–474.

Shelton, A. M., Perez, C. J., Tang, J. D., and Vandenberg, J. D. 1997. Prospects for novel approaches towards management of the diamondback moth. *In* "Proceedings of the Third International Workshop on the Management of Diamondback Moth and other Crucifer Pests" (A. Sivapragasm, W. H. Loke, A. K. Hussan, and G. S. Lim, Eds.), pp. 17–20. Kuala Lumpur, Malaysia.

Shorey, H. H., and Hale, R. L. 1965. Mass-rearing of the larvae of nine noctuid species on a simple artificial medium. *J. Econ. Entomol.* **58**, 522–524.

Skadeland, D. A. 1981. Dispersal of pathogenic material for pest control. U. S. Patent No. 4,301,147.

Soper, R. S. 1978. Autodissemination of entomopathogens: fungi. *In* "Microbial control of insect pests: future strategies in pest management systems" (G. E. Allen, C. M. Ignoffo, and R. P. Jaques, Eds.), pp. 63–65, NSF-USDA-University of Florida Workshop. Gainesville, FL.

Talekar, N. S., and Shelton, A. M. 1993. Biology, ecology and management of the diamondback moth. *Annu. Rev. Entomol.* **38**, 275–302.

Tanada, Y. 1963. Epizootiology of infectious diseases. *In* "Insect Pathology: An Advanced Treatise. Vol. 2." (E. A. Steinhaus, Ed.), pp. 423–475. Academic Press, NY.

Tashiro, H. 1987. "Turfgrass Insects of the United States and Canada". Cornell University Press. Ithaca, NY 391 pp.

Tinsley, T. W. 1979. The potential of insect pathogenic viruses as pesticidal agents. *Annu. Rev. Entomol.* **24**, 63–88.

Tinzaara, W., Gold, C. S., Nankinga, C., Dicke, M., van Huis, A., Ragama, P. E., and Kagezi, G. H. 2004. Integration of pheromones and the entomopathogenic fungus for the management of the banana weevil. *Uganda J. Agric. Sci.* **9**, 621–629.

Vago, C., Fosset, J., and Bergoin, M. 1966. Dissémination des virus de polyédries par les éphippigères prédateur d'insectes. *Entomophaga* **11**, 177–182.

Vail, P. V., Hoffmann, D. F., and Steven Tebbets, J. 1993. Autodissemination of *Plodia interpunctella* (Hübner) (Lepidoptera: Pyralidae) granulosis virus by healthy adults. *J. Stored Prod. Res.* **29**, 71–74.

Vaupel, V. O., and Zimmermann, G. 1996. Preliminary trials on the combination of pheromone traps with the entomopathogenic fungus *Beauveria bassiana* (Bals.) Vuill. against the bark beetle species *Ips typographus* L. (Col., Scolytidae). *Anz. Schädlingskde., Pflanzenschutz, Umweltschutz* 69, 175–179. (In German).

Vega, F. E., Dowd, P. F., and Bartelt, R. J. 1995. Dissemination of microbial agents using an anutoinoculating device and several insect species as vectors. *Biol. Control* **5**, 545–552.

Wilding, N. 1970. *Entomophthora* conidia in the air-spora. *J. Gen. Microbiol.* **62**, 149–157.

Williams, R. N., Fickle, D. S., Bartelt, R. J., and Dowd, P. F. 1993. Responses by nitidulid beetles to synthesized aggregation pheromones, a coattractant, and effects of trap design and placement. *Eur. J. Entomol.***90**, 287–294.

Winstanley, D., Cross, J. V., Naish, N., Keane, G., Hilton, S., Gajek, D., and van Wezel, R. 2005. Semiochemical driven auto-dissemination of viruses for the control of orchard pests. Abstract. Ann. Meeting, Society for Invertebrate Pathology, Anchorage, AK, Aug. 7–11, 2005.

Yasuda, K. 1999. Auto-infection system for the sweet potato weevil, *Cylas formicarius* (Fabricius) (Coleoptera: Curculionidae) with entomopathogenic fungi, *Beauveria bassiana* using a modified sex pheromone trap in the field. *Appl. Entomol. Zool.* **34**, 501–505.

Young, S. Y., and Yearian, W. C. 1990. Transmission of nuclear polyhedrosis virus by the parasitoid *Microplitis croceipes* (Hymenoptera: Braconidae) to *Heliothis virescens* (Lepidoptera: Noctuidae) on soybean. *Environ. Entomol.* **19**, 251–256.

Yu, Z. 1996. Autodissemination of a baculovirus for suppression of beet armyworm, *Spodoptera exigua* Hübner (Lepidoptera: Noctuidae). Ph.D. Thesis, University of Kentucky, Lexington.

Yu, Z., and Brown, G. C. 1997. Autodissemination of beet armyworm (Lepidoptera: Noctuidae) baculovirus under laboratory conditions. *J. Econ. Entomol.* **90**, 1187–1194.

Yu, H., and Sutton, J. C. 1997. Effectiveness of bumblebees and honeybees for delivering inoculum of *Gliocladium roseum* to raspberry flowers to control *Botrytis cinerea. Biol. Control* **10**, 113–122.

Zelazny, B. 1976. Transmission of a baculovirus in populations of *Oryctes rhinoceros. J. Invertebr. Pathol.* **27**, 221–227.

Zelazny, B., Lolong, A., and Pattang, B. 1992. *Oryctes rhinoceros* (Coleoptera: Scarabaeidae) populations suppressed by a baculovirus. *J. Invertebr. Pathol.* **59**, 61–68.

SECTION IV

OVERVIEW OF PATHOGEN GROUPS

Chapter IV-1

Viruses

Jenny S. Cory
Department of Biology
Algoma University College
Laurentian University
and Great Lakes Forestry Centre
Sault Ste. Marie, ON
P6A 2G4, Canada

Hugh F. Evans
Forest Research
Alice Holt Lodge, Wrecclesham
Farnham, Surrey, GU10 4LH, UK

1 Introduction

A Current status

Microbial control agents, including bacteria, fungi, nematodes, protozoa and viruses, provide a more environmentally acceptable and sustainable form of insect pest management than chemical insecticides, but are most effective when underpinned by a detailed knowledge of specific host-pathogen interactions. Among the fifteen or more families of viruses of invertebrates, it is mainly those having virus particles (virions) occluded within a proteinaceous matrix, termed an occlusion body (OB), that have been used successfully in microbial control programs. The Baculoviridae (rod-shaped viruses) are by far the most studied and extensively used of the occluded viruses in pest management regimes (Hunter-Fujita *et al.*, 1998; Moscardi, 1999).

Recognition that baculoviruses were the causal agents of disease, especially their association with OBs, has resulted in a long history of use, with particular progress being made during the last 150 years (Steinhaus, 1956). While the prospect of using baculoviruses has excited scientists for many years, their penetration into the world markets for bio-insecticides has been very small (with the exception of soybean in Brazil (Moscardi, 1999)). However, there is growing interest in baculoviruses in relation to their potential for particular markets, such as orchard produce (Cross *et al.*, 1999; Arthurs *et al.*, 2005), greenhouse crops (Bianchi *et al.*, 2000, 2002a,b,c) and agricultural production in developing countries (Parnell *et al.*, 1999; Williams *et al.*, 1999; Cherry *et al.*, 2000), in addition to more traditional areas such as forestry (Moscardi, 1999). There has also been steady interest in the potential to produce recombinant baculoviruses to increase their speed of kill or reduce insect feeding (Miller, 1995; Inceoglu *et al.*, 2001), particularly in crops, such as cotton, where resistance to other forms of insect pest control is high (Smith *et al.*, 2000; Sun *et al.*, 2004).

B Types of insect viruses

Descriptions of the various families of viruses can be found in several publications (Adams, 1991; Evans and Shapiro, 1997; Hunter-Fujita *et al.*, 1998). The main characteristics of the principal families that influence their use as

L.A. Lacey and H.K. Kaya (eds.), Field Manual of Techniques in Invertebrate Pathology, 149–174.

possible microbial insecticides are summarized in Table 1. This confirms that, with the exception of the non-occluded DNA virus of *Oryctes rhinoceros*, all the viruses used in commercial or large-scale field programs produce OBs. This is one of the main characteristics that enables them to be enumerated and handled without recourse to highly specialized equipment, thus offering the potential for use in both developed and developing countries. The descriptions below, therefore, concentrate on the occluded viruses. Other groups of viruses, however, such as the tetraviruses, are starting to receive attention from a pest control perspective (Christian *et al.*, 2005)

1 Baculoviridae

The Baculoviridae (BV) has received greatest attention for development as pest management tools. They are double stranded DNA viruses containing rod-shaped virus particles (virions). Changes in nomenclature have removed the non-occluded DNA virus of *O. rhinoceros* from the Baculoviridae leaving two genera – *Polyhedrovirus* and *Granulovirus*. The representatives of these genera are known as nucleopolyhedroviruses (NPV) and granuloviruses (GV), respectively. More recently, it has been suggested that the nomenclature should be more closely allied to taxonomic groupings and should include four genera: *Alphabaculovirus* (lepidopteran-specific NPV), *Betabaculovirus* (lepidopteran-specific Granuloviruses), *Gammabaculovirus* (hymenopteran-specific NPV) and *Deltabaculovirus* (dipteran-specific NPV) (Jehle *et al.*, 2006), although this scheme has yet to be adopted. As the name NPV implies, replication takes place in the nuclei of infected cells. Within the virogenic stroma produced during virus replication in the cell, virions are gradually occluded within a proteinaceous crystalline matrix to develop well-defined OBs that fill the nucleus and eventually cause the entire cell to disintegrate. Although an initial cycle of replication may take place in the cells of the midgut, there are two principal tissue replication strategies exhibited within the BVs. The NPVs of sawflies (Hymenoptera: Symphyta) and the GV of the moth *Harrisina brillians* (=*metallica*) complete their replication cycles in the midgut cells, whereas infection in all other Lepidoptera takes place in organs other than the gut, producing virtually systemic infection. Growth of virus in gut infections are generally relatively low such that yields of NPV OBs are an average of 70 times greater for systemic infections (Lepidoptera, which can be in excess of 10^9 OBs for late instar caterpillars) compared with gut infections (Hymenoptera).

The unique nature of the structure of BVs and the fact that they have been found only in arthropods provide strong support for their use as safe, specific microbial control agents. For further information see Evans and Shapiro (1997), Miller (1997) and Hunter-Fujita *et al.* (1998).

2 Entomopoxviridae

Entomopoxviruses (EPVs) have large linear molecules of double stranded DNA formed as brick-shaped virions that are occluded in a paracrystalline protein known as spheroidin. The OBs are known as spheroids or spherules. Replication takes place in the cytoplasm of infected cells, mainly confined to the fat body, although systemic infection may also take place. Spindle-shaped occlusions may be present in the EPVs of Lepidoptera and Coleoptera but are absent in Diptera and Orthoptera. One of the most dramatic symptoms of infection is the extreme longevity of infected individuals. There has been some interest in use of EPVs for pest management and safety testing has shown them to be environmentally benign to non-target taxa. For further information see Arif (1984) and Goodwin *et al.* (1991).

3 Reoviridae

Representatives of the Reoviridae that infect arthropods are known as cypoviruses (CPVs) (previously cytoplasmic polyhedrosis viruses). They are double stranded RNA viruses in which the virions have 12 spikes on the icosahedral particles within the OBs. Replication takes place exclusively in the cytoplasm of midgut epithelial cells and this often leads to chronic infection and stunting of individuals, particularly in later

Table 1. The characteristics of the principal families of viruses affecting arthropods

Virus family	Genus	Nucleic acid type	Occlusion bodies	Hosts	Site of replication	Large scale use
Ascoviridae	Ascovirus	ss DNA	–	Lepidoptera, Noctuidae only	Nuclei: fat body, hypodermis, tracheal matrix	No
Baculoviridae	Nucleo-polyhedrovirus (NPV)	ds DNA	Polyhedral	Lepidoptera, Diptera, Hymenoptera	Nuclei: gut or systemic	Yes
	Granulovirus (GV)	ds DNA	Granular			Yes
Iridoviridae	Iridovirus and Chloriri-dovirus	ds DNA	–	Wide range of invertebrate families	Cytoplasm: fat body, hemocytes, hypodermis, sometimes systemic	No
Parvoviridae	Densovirus	ss DNA	–	Diptera, Blattoideae, Lepidoptera, Odonata, Orthoptera	Most tissues except midgut	No
Polydnaviridae	Ichnovirus	ds DNA	–	Hymenoptera, Ichneumonidae	No effects on parasitoids	No
	Bracovirus	ds DNA	–	Hymenoptera, Braconidae		No
Poxviridae	Entomopoxvirus	ds DNA	Spheroid	Coleoptera, Diptera, Hymenoptera, Lepidoptera, Orthoptera	Cytoplasm: mainly fat body but other organs can be infected	Minor
Unclassified	*Oryctes* virus	ds DNA	–	Coleoptera	Nucleus: gut in adults, systemic in larvae	Yes
Birnaviridae	Birnavirus	ds RNA		Diptera, *Drosophila* X virus	No tissue symptoms: adults sensitive to CO_2	No
Caliciviridae	Chronic stunt virus	ss RNA		Lepidoptera,	Pathology poorly understood	No
Nodaviridae	Nodavirus	ss RNA	–	Diptera, Coleoptera, Lepidoptera	Cytoplasm: gut and later systemic	No
Picornaviridae	Picornavirus	ss RNA	–	Diptera, Lepidoptera and other families by injection	Cytoplasm: gut and other organs	No
Reoviridae	Cypovirus (CPV)	ds RNA	Polyhedral	Diptera, Hymenoptera, Lepidoptera	Cytoplasm: gut only	Minor
Rhabdoviridae	Sigma virus	ss RNA	–	Diptera	No tissue symptoms: adults sensitive to CO_2	No
Tetraviridae	Omegatetravirus	ss RNA	–	Lepidoptera	Cytoplasm: chronic infection. Stunt virus	No

instars of the Lepidoptera. Most infections have been recorded in the Lepidoptera with other records from the Diptera and Hymenoptera (Symphyta). Surprisingly, although restricted to the midgut cells, CPVs yield only seven fold less than comparable systemic infections with NPVs. However, much of these differences are removed when yields were expressed in relation to body weight, implying that the mean weights of CPV-infected individuals are generally lower than for NPV-infected larvae (Entwistle and Evans, 1985).

Safety testing has indicated that use of CPVs could be viable within pest management regimes, but there has been relatively little work on this family of viruses, compared with the BVs. Hukuhara and Bonami (1991) have reviewed the Reoviridae, both for their basic characteristics and for their potential as pest control agents.

2 Biology and ecology of viruses

Understanding the biology and ecology of viruses is crucial for optimizing pest control strategies and for meaningful risk assessment of exotic and genetically modified organisms. While this is particularly the case for developing longer term control strategies, it is also important for the success of inundative spray application programs. For example, not only will this information contribute to the design of more successful spray regimes, it can also be used to minimize flying time, reduce the need for diluents in remote areas and utilize the build up of inoculum in more stable ecosystems. For viruses, the ecology of the baculoviruses is by far the best studied (Cory *et al.*, 1997; Cory and Myers, 2003), although many questions still remain in terms of their role in natural insect populations. The key features of virus biology and ecology, all of which are likely to have an impact on the success of pest management programs, are discussed below.

A Environmental persistence

Considering the resources required to produce the proteinaceous matrix of the OB, there must be an adaptive significance that has led to the evolution of this characteristic structure. Each OB contains from one to many virions, singly or multiply enveloped, thereby implying a high level of redundancy to potentially cause infection. The adaptive significance of occlusion appears to be to increase the survival of the virus outside the host, as infection of new hosts is thought primarily to be accomplished via horizontal transmission and ingestion of OBs. In this respect, there is clear evidence that survival of virus is greater within the OB than as free virions in the environment. For example, *Autographa californica* MNPV (*Ac*MNPV) engineered to delete the polyhedrin gene, had a much shorter survival period than polyhedrin positive variants of the virus (Wood *et al.*, 1994). OBs are susceptible to degradation by ultra violet irradiation; however OBs released from infected cadavers will persist for varying periods of time depending on their level of protection from UV. For example, NPV produced in induced epizootics in the pine beauty moth, *Panolis flammea*, persisted on pine foliage until the following year in Scotland (although at low levels) (Carruthers *et al.*, 1988). Occluded virus can persist in soil for even longer, up to several decades for the NPV of some forest insect species (England *et al.*, 1998). Occlusion of multiple virions could also be an aid to retaining diversity within the virus population and allow rapid adaptation to novel situations (see below). In pest management terms, the presence of OBs is critical to success for inundative control approaches; they can be counted using conventional microscopy, are simple to formulate, are able to survive the forces induced during passage through a sprayer, and have an improved level of survival relative to non-occluded viruses. Taking a more ecological approach to insect pest control, the long term environmental persistence of OBs can be exploited in order to build up reservoirs of virus, either within or between generations. Knowledge of the role of persistent virus is particularly important in more stable ecosystem such as forests. In the gypsy moth, *Lymantria dispar*, contamination of egg masses at oviposition with NPV released onto plant surfaces during the preceding virus epizootic, was found to be the main means of introduction of virus into the population (Murray and Elkinton, 1989).

Insect behavior can also be important in initiating virus infection. Young larvae of the tussock moth, *Orgyia antique*, acquired NPV infection by dispersing to the contaminated foliage beneath pine trees and then transmitted infection to their conspecifics by climbing back up the trees (Richards *et al.*, 1999).

B Dispersal capacity

Among the pathological changes that take place during infection, the behavior of the infected host can have a considerable bearing on the site of death and eventual release of the virus. This has major implications for the ability of the virus to spread within and between generations of the host. There have been anecdotal reports for many years of baculovirus-infected larvae climbing up plants before death; this behavior has been named *wipfelkrankheit* (tree top disease) (Hofman, 1891). Quantification of the behavior of different instars of the cabbage moth, *Mamestra brassicae*, demonstrated that larvae infected with NPV moved up to five times more than their healthy counterparts during the middle stages of infection (Vasconcelos *et al.*, 1996b). NPV infection also resulted in larvae altering their behavior from feeding within the hearts of cabbages (*i.e.*, cryptic), to feeding on the upper edges of the outer leaves (exposed) (Evans and Allaway, 1983). These behavioral changes are thought to result in more rapid dissemination of the virus and increased transmission (although an adaptive explanation has yet to be demonstrated (Cory and Myers, 2003)). Overall, behavioral changes in *M. brassicae* as a result of infection led to rapid dispersal of virus infection within the field sites (Evans and Allaway, 1983; Vasconcelos *et al.*, 1996b). Although it is generally accepted that larvae infected with baculoviruses tend to move up their host plant, there have been very few studies that have actually quantified this behavior. A recent study on the winter moth, *Operopthera brumata*, has shown the opposite behavior; larvae infected with NPV moved down their host plants and died on the stems (Raymond *et al.*, 2005). The authors' demonstrated that virus persistence was greater on the stems than on the foliage and speculated that such behavioral changes would enhance between generation persistence of the virus in such a univoltine species (whereas moving upward might be more beneficial for increasing within season transmission in multivoltine species). Insect behavior is a crucial component of virus transmission, both in terms of where an infected insect dies and the likelihood of a healthy insect acquiring an infective virus dose. Further studies are needed in specific target systems in order to optimize the targeting of virus sprays.

Adult insects can also disperse viruses via vertical transmission. There is good evidence for adult dispersal of both sawfly NPVs (Smirnoff, 1972; Oloffson, 1989) and CPVs of Lepidoptera (Sikorowski *et al.*, 1973), reflecting the continuation of gut infection into the adult stage. In these cases, it would appear that the route is transovum transmission as an external contaminant of the deposited eggs. There is considerable evidence that baculoviruses can be transmitted vertically (Kukan, 1999), although the route of infection and its impact on insect-virus population dynamics is not clear (see below).

Host movement is not the only means of virus dispersal; virus OBs can be carried by a range of biotic and abiotic mechanisms. Predators and scavengers can pass baculoviruses in their feces, without appreciable loss of infectivity, in part because their gut conditions are neutral to acid and thus do not break down the OB structure. Longer distance spread can be achieved by birds. Passage of virus can be rapid; NPV OBs appeared in the feces of blue tits within 30 min of being fed infected European spruce sawflies, *Gilpinia hercyniae*, and some of the birds passed infective virus for up to 6 days (Entwistle *et al.*, 1978). Field studies in Scotland showed that up to 77% of birds collected were feeding on NPV infected *P. flammea* larvae and cadavers (Entwistle *et al.*, 1993), clearly indicating that passive virus dispersal by birds could be a significant means of long distance movement. A variety of invertebrate predators also feed on infected larvae. Laboratory tests with several ground beetle (Carabidae) species indicated that they showed no preference between diseased and healthy lepidopteran larvae, and that they excreted viable virus for at least 15 days after feeding (Vasconcelos *et al.*,1996a). Field experiments showed that the predators released sufficient levels of virus to cause low levels of mortality in caterpillar populations (Vasconcelos *et al.*, 1996a). Parasitoids have also been implicated in the spread of baculoviruses as a result of contamination of ovipositors and direct injection of virus into the hemocoel of susceptible hosts during oviposition, although the levels of infection by this route appear to be low (Sait *et al.*, 1996; Matthews *et al.*, 2004). The combined effects of predators (mainly), parasitoids and abiotic mechanisms resulted in a rate of spread of soybean looper, *Anticarsia gemmatalis*, NPV of up to 1.2 m per day during the first year after point release (Fuxa and Richter, 1994).

C *Vertical transmission*

Persistence within the insect host population is crucial to the survival of any pathogen, and viruses have developed a wide range of strategies to deal with periods when transmission to new hosts might be low or unpredictable. In fact, many viruses can be transmitted by both horizontal and vertical modes of transmission and understanding the relative contributions of each mode and the conditions that maintain them is a key issue in pathogen ecology and evolution. Baculoviruses are the best studied group in this respect, although even here, there are many unanswered questions as to the nature and role of vertically transmitted virus infection (Cory and Myers, 2003). Baculoviruses can be transmitted by transovum (external contamination of the egg) and transovarial (within the egg) routes and either can result in overt disease in the next generation or the virus infection can remain sublethal or quiescent. Kukan (1999) has reviewed vertical transmission in NPVs and it appears that levels of transmission of overt disease is specific to particular host-virus systems and can vary from zero to 50% infection in progeny larvae. In the laboratory it is relatively easy to estimate transovum transmission as contaminating virus can be removed by various techniques; however, in the field it can be obscured by environmental contamination of the eggs. Most vertical transmission that results in active infection is likely to result from transovum transmission.

Transovarial transmission is the area that has raised most speculation. There are numerous reports, reaching back many decades, of apparently healthy lepidopteran larvae suddenly succumbing to baculovirus infection. This has lead to the widespread belief that many insect populations support latent virus infections. Only recently, with the advent of sensitive and specific molecular tools such as PCR, has it been possible to address the issue of asymptomatic infections. Initial studies with *M. brassicae* indicated that insects supported low level persistent NPV infections that exhibit some transcriptional activity (rather than truly latent viruses that are inactive) (Hughes *et al.*, 1997.) More recent data have shown that persistent NPV infections are widespread in wild populations of *M. brassicae* in the UK (Burden *et al.*, 2003). Similar results have been obtained from PCR studies on NPV in the African armyworm, *Spodoptera exempta* in Tanzania (L. Vilaplana, K. Wilson, E. Redman and J.S. Cory, *submitted*). These results indicate that persistent baculovirus infections may be much more common in natural populations than previously thought. We know little about how these infections are initiated, although it seems likely that they are the result of surviving a non-lethal virus challenge. Studies on the GV of *Plodia interpunctella* have shown that insects that survive virus exposure pass a persistent infection onto the next generation, and that this can be achieved by both sexes (Burden *et al.*, 2002). More pertinent questions are, how important are these infections in initiating baculovirus epizootics and can this be exploited for strategic pest control? Several studies have demonstrated that persistent infections can be triggered to overt disease by a variety of means including, infection with heterologous baculoviruses, crowding, temperature stress and dietary changes. For example, cross infection of field collected forest tent caterpillar, *Malacosoma disstria*, larvae with the NPV from the western tent caterpillar, *M. californium pluviale*, resulted in the insect dying from *M. disstria* NPV infection (Cooper *et al.*, 2003a). However, while cross-infection seems the most promising method for triggering a persistent baculovirus infection, none of these methods produces consistent results, and so it may well be that triggering is a more stochastic process (Cory and Myers, 2003).

D Sublethal effects

Another feature of baculoviruses that might have been underestimated is sublethal effects. If insects survive virus challenge they often pay a cost in terms of reduced fecundity, egg viability or size and altered development time (Rothman and Myers, 1996; Milks *et al.*, 1998; Matthews *et al.*, 2002). These effects may be a direct result of fighting off infection, such as damage to the gut, or may be a result of persistent virus infection, or a combination of both. Thus baculoviruses can affect insect populations by direct mortality and delayed effects, such as reduced fecundity and these should be factored into longer term control programs.

E Specificity

Knowledge of virus host range is an important component of their environmental evaluation and necessary for registration and commercial exploitation (Laird *et al.*, 1990). It is particularly crucial in relation to ecological risk assessment for the introduction of novel or genetically modified viruses (Cory, 2003). In comparison to synthetic chemical insecticides, the relative specificity of baculoviruses is very high, but the ability to replicate requires a different form of safety assessment compared to the toxicity testing employed for chemical insecticides. Wide ranging host testing that covers species of ecological, as well as commercial interest, is rare. In addition, studies do not always confirm that the virus produced in cross-infection assays is the same as the inoculum; as discussed above, cross-infection can also trigger persistent virus infections (Bourner and Cory, 2004). Bearing these issues in mind, Table 2 shows the recorded host ranges of several baculoviruses. Many baculoviruses appear to be limited to a single or a few host species, although several NPVs have broader host ranges within the Lepidoptera, particularly *Autographa californica* MNPV and *Anagrapha falcifera* MNPV. However, it also needs to be remembered that not all species are equally susceptible to these viruses and host range, as determined in the laboratory, is likely to be much greater than that found in the field.

Narrow host ranges carry both advantages and disadvantages in the development of viral insecticides. A narrow host range is undoubtedly the key characteristic that makes viral agents (or any biological control agent) the most environmentally acceptable form of pest control. Direct effects on non-target organisms will be small or non-existent. There may also be indirect effects on natural enemies because the availability of prey may decline resulting in a decrease in other natural enemies. The interaction between biological control agents therefore needs to be studied carefully, particularly for longer term control strategies. However, such effects are likely to be very small compared to the use of broad-spectrum chemical insecticides that may not discriminate between pest and beneficial targets. The principal disadvantage of a narrow host range is the potential restriction in the commercial viability of the agent, particularly if the host is one of several that have to be controlled on a given crop. The need to produce viruses *in vivo* is a further disadvantage arising

Table 2. Host ranges of some baculoviruses, illustrating the range of families and species affected

Original host	Order/Family	Baculovirus	Families	Species	No. tested
Neodiprion sertifer	Hymenoptera/Diprionidae	NPV	1	3	–
Orgyia antique	Lepidoptera/Lymantriidae	NPV	1	1	23
Euproctis chrysorrhoea	Lepidoptera/Lymantriidae	NPV	1	1	73
Pieris rapae	Lepidoptera/Pieridae	GV	1	3	–
Cydia pomonella	Lepidoptera/Tortricidae	GV	1	8	–
Helicoverpa zea	Lepidoptera/Noctuidae	NPV	1	7	–
Anticarsia gemmatalis	Lepidoptera/Noctuidae	NPV	2	9	–
Mamestra brassicae	Lepidoptera/Noctuidae	NPV	4	32	66
Anagrapha falcifera	Lepidoptera/Noctuidae	NPV	10	30	38
Autographa californica	Lepidoptera/Noctuidae	NPV	15	95	105

from high specificity, because each species of virus has to be produced in a given host, requiring a high degree of duplication of effort and facilities.

F Variation

Variation appears to be ubiquitous in insect viruses (Williams and Cory, 1993; Graham *et al.*, 2006), but it has only been studied in any detail in the baculoviruses. Numerous studies have shown that baculoviruses can exhibit high levels of genetic variation; however, while this has been exploited in *in vitro* cloning studies and the development of recombinant baculoviruses, the importance of genetic variation for baculovirus biology and pest control and the mechanisms that maintain it in baculovirus populations have been largely ignored. Baculoviruses can vary at different spatial scales; isolates collected from the same species at diverse geographic sites are often different (Cory and Myers, 2003). Isolates from individual larvae are frequently genetically distinct and it has also been shown that an individual larva can contain up to 24 distinct genotypic NPV variants (Cory and Myers, 2003; Cory *et al.*, 2005). However, we know little about patterns of genetic variation in nature and how this might change over different spatial scales and as a result of exposure to baculovirus epizootics. An exception to this is a study on *M. calfornicum pluviale* NPV which showed that different NPV isolates (individual larvae) showed hierarchical spatial structure (Cooper *et al.*, 2003b).

A more important issue than demonstrating the existence of genetic variation, is whether the presence of multiple variants alters the biology of the virus population and whether this has implications for insect pest management. Several studies have shown that individual variants, cloned either *in vitro* or *in vivo*, exhibit significant differences in pathogenicity and speed of kill (*e.g.*, Ribeiro *et al.*, 1997; Hodgson *et al.*, 2001). There has thus been a tendency to select the variant that has the highest pathogenicity or the fastest speed of kill, as determined in laboratory bioassays, for genetic modification or field application. However, while this might be promising under laboratory conditions, the situation in the field is often very different and far more variable. A recent study using the NPV of *P. flammea* has shown that inoculum containing more than one variant is more pathogenic than either variant alone (Hodgson *et al.*, 2004). Similar results have been obtained in another system, *S. exempta* and it's NPV (E. M. Redman and J. S. Cory, *unpubl. data*). Other studies that have focussed on parasitic NPV genotypes, *i.e.*, genotypes that contain deletions and are usually unable to be transmitted or infect alone, also suggest that baculovirus population structure could be important. Genotypes with deletions appear to be common in natural baculovirus populations (Simon *et al.*, 2004) and can have both positive and negative effects on pathogenicity (Munoz and Caballero, 2000; Lopez-Ferber *et al.*, 2003). Mixed baculovirus populations are also likely to be able to retain a suite of biological characteristics and to have the capacity to adapt more rapidly to changing conditions. For example, it has already been shown that different baculovirus variants vary in their pathogenicity for different host species (Hitchman *et al.*, 2007; Paul and Cory, *unpubl. data*) and in their interactions with different host plants (Hodgson *et al.*, 2002; Cory and Myers, 2004). It is therefore important to maintain variation in baculovirus preparations for insect pest control, but it is also necessary to gain a better understanding of the interactions and evolution of these genetic mixtures.

G Secondary cycling

A key feature of any pathogen infection is that it results in the pathogen being amplified, thereby potentially producing a secondary wave of infection which may increase the overall effects of the application. Viruses that replicate in the host gut are released into the environment more rapidly than those infecting other body tissues. This is particularly well illustrated by NPV infection in *G. hercyniae*. This insect uses defensive regurgitation to repel predators and parasitoids and, in doing so, includes gut contents in the fluid ejected through the mouth (Entwistle *et al.*, 1983). Within days of infection, this fluid contains OBs that are then available for ingestion by other larvae and can induce rapid infection in susceptible hosts. Thus, infection is passed on well before death of the host and subsequent disintegration of the larval cadaver. Similar routes of infection are observed for CPVs that are transmitted rapidly from host to host. In contrast, in most BV infections of Lepidoptera, virus accumulates within various host tissues including hemocytes, fat bodies, muscles and hypodermis, but not directly in the midgut. Release of secondary inoculum to the environment is dominated by a single massive event arising from rupture of the fragile integument, although there may be some release through the gut late in infection (Vasconcelos, 1996; Fuxa and Richter, 1998). This gives rise to a sudden, locally concentrated augmentation of inoculum that is sufficiently large to kill any susceptible host life stage. The GV of *H. brillians*, where virus infection is restricted to the midgut, is the exception to all other baculovirus infections of Lepidoptera (Federici and Stern, 1990).

Understanding the processes that influence pathogen transmission is therefore crucial to elucidating how insect viruses affect the population dynamics of their host (and thus the success of pest control) both within a season and over the longer term. There is a burgeoning theoretical literature on host-microparasite dynamics (see Cory and Myers, 2003) and manipulative studies on the horizontal transmission of baculoviruses have made a significant contribution to our understanding of disease transmission dynamics in general (*e.g.*, D'Amico *et al.*, 1995; Dwyer *et al.*, 1997, 2000; Beisner and Myers, 1999; Hails *et al.*, 2002). Virus transmission is expected, and usually does, increase with increasing host density and thus has the potential to regulate insect populations. Virus transmission depends on the rate of contact between susceptible larvae and diseased individuals and many theoretical models assume that this transmission coefficient (the likelihood of transmission) is constant. However manipulative studies have shown that this is not the case, which is perhaps not surprising as transmission depends upon the spatial distribution of inoculum, insect behavior and the range of susceptibilities within the insect population (Cory and Myers, 2003). Interestingly, there is also evidence that the susceptibility of lepidopteran larvae varies with density (Goulson and Cory, 1995). In *S. exempta*, the LD_{50} for *S. exempta* NPV is higher in gregarious that solitary forms of the caterpillar (Reeson *et al.*, 1998), and this increased resistance translates into decreased transmission of NPV among the gregarious larvae in the field (Reeson *et al.*, 2000). This difference is thought to be due to an increased investment in immunity in situations where the risk of infection is higher (Reeson *et al.*, 1998; Wilson and Reeson, 1998). Diet can also influence immunity and virus susceptibility; recent studies with *Spodoptera littoralis* and its NPV have shown that a high protein diet after virus challenge can significantly increase larval survival and there are even indications that a caterpillar can alter its diet to promote this (Lee *et al.*, 2006). Other aspects of insect diet and disease are discussed below.

H Tritrophic interactions

Insect viruses that are introduced into pest populations, for example by large scale inundative spray application, need to be ingested to initiate infection, thus the host plant can potentially have a large impact on the outcome of the interaction (Cory and Hoover, 2006). Plant architecture will influence spray deposition and adhesion; it will also influence the longer term persistence of virus OBs or virus OBs in cadavers through its capacity to shade (Raymond *et al.*, 2005) or through plant surface chemistry. Some species of plants, particularly cotton and soybean, produce alkaline exudates that contain basic ions that can inactivate baculoviruses (Duffey *et al.*, 1995). Plant-virus-insect interactions in the midgut can also have a significant impact on resulting mortality through interactions with polyphenols, such as tannins, and peroxidases in particular (Cory and Hoover, 2006). Plant chemistry can alter other aspects of the insect-virus interaction, such as speed of kill, resulting OB yield and even the infectivity of the progeny virus (Ali *et al.*, 2002; Raymond *et al.*, 2002). Plant chemistry also varies between different plant structures and as a result of the induced changes resulting from insect damage (Hunter and Schultz, 1993; Ali *et al.*, 1998). Tritrophic relationships are clearly complex as there will be a trade off between how plant allelochemicals affect the pathogen and their direct effects on the host insect. It is also important to consider the methodology involved in assessing potential plant effects; many bioassays are designed to assess the effect of secondary plant chemicals at the point of ingestion, whereas in the natural situation this will be combined with the effects of both phytochemicals and nutrients on insect growth, feeding behavior and physiology. In reality the high concentrations of baculovirus OBs found in the field may well overwhelm any effects of plant chemistry, but in some cropping systems it will need to be given serious consideration, particularly if incorporated with plant resistance mechanisms which may add another dimension.

3 Ecologically-based control strategies for use of virus

Most examples of the successful use of viruses as pest control agents are based on application of inoculum through spray application onto high density pest populations. However, there is scope for the use of more ecologically-based strategies that increase the level of virus infection in insect populations, thereby reducing pest populations below damage thresholds. In addition, ecological information can be used to maximize the effectiveness and reduce the cost of inundative release programs.

A Classical biological control and augmentation

The closest example to classical biological control using viruses is the introduction of the non-occluded virus of the coconut palm rhinoceros beetle, *O. rhinoceros*. The Oryctes virus affects both larvae and adults of *O. rhinoceros*, causing mortality and reduced fecundity. When it was first released into Samoa it rapidly became established in the population and spread to neighbouring areas, leading to a decline in beetle populations without the need for reintroduction (Huger, 2005). The rapid spread of the virus was explained by the fact that it causes widespread infection in the gut and thus the adults are continually shedding virus that infects any other beetles sharing feeding sites. The adult acts as a vector of disease and in this way infection persists in the beetle population, maintaining the insect below the economic threshold. Since the first trials, the virus has been successfully released in numerous locations in the South Pacific, as well as Papua New Guinea, the Maldives and Mauritius (Huger, 2005). However, it is important to note that this long term suppression will only be achieved if this approach is combined with cultural practices, such as the removal of breeding sites (rotting palm logs) (Huger, 2005).

While examples of classical biological pest control using viruses are rare, augmentation of virus in pest populations at intervals can be an excellent (and relatively cheap) means of reducing pest densities. For this type of approach it is essential to understand the ecology of the insect-virus system and use this to exploit routes of virus transmission or to manipulate the natural insect-virus cycle. Augmentation can be achieved through manipulation of the environment or by direct application of limited quantities of virus to specific parts of the host substrate or through early addition of virus, before the host has appeared. An excellent example of the importance of secondary inoculum in pest management is provided by the dynamics of NPV infection in field control of porina, *Wiseana* spp. (Lepidoptera: Hepialidae), populations in pastures in New Zealand (Fleming *et al.*, 1986). Inoculum arising from NPV infection in the larvae accumulates in undisturbed pastures and careful management practices, such as oversowing rather than ploughing and re-seeding, maintained sufficient inoculum in the soil to control the pest population.

B Autodissemination

Autodissemination is the carriage of virus into susceptible portions of the target population by the action of the pest species itself, such as by vertical transmission via adults, and is essentially a component of the manipulation of virus in pest populations (*e.g.*, *Oryctes* in which the virus is very effectively introduced into the population by the release of infected adults). However, techniques to increase the carriage of virus in pest populations artificially have also been investigated, such as by applying the virus inoculum directly to adult stages that were then released. A more elegant approach is to attract adult male moths to traps baited with female sex-attractant pheromones, where the moths become contaminated with virus. This was attempted by Jackson *et al.* (1992) who employed a powder formulation of *Ac*MNPV in pheromone traps attractive to *Heliothis virescens* males. Although contamination of the males was apparent, inoculum transfer only took place during mating, so that female contamination of eggs was actually only around 8%, with subsequent larval infection of up to 12%. A more successful use of the technique for autodissemination of *P. interpunctella* GV through contaminated pheromone traps resulted in up to 50% infection of progeny (Vail *et al.*, 1993). Also see Chapter III-3 for additional information on autodissemination.

C Combination with theoretical studies

Population cycles in forest insects have long attracted the attention of theoretical ecologists and mathematical modelers aiming to explain the driving forces of these regular population fluctuations. The theoretical mathematical models of Anderson and May (1980, 1981), were the first to indicate the potential of micro-parasites, such as viruses, as the cause of insect population cycles. Since then, numerous mathematical explorations of host-microparasite dynamics have been published. While many of these models do not produce realistic dynamics or infection levels, they are still valuable in exploring the factors that could be important in insect outbreaks and population cycles. Knowledge of natural virus infection cycles can also be invaluable for pest control. For example, the Douglas fir tussock moth, *Orgyia pseudosugata*, is frequently devastated by baculovirus epizootics that occur regularly in high density populations. Spraying NPV onto an increasing insect population, a year before a natural virus epizootic was predicted, allowed Otvos *et al.* (1987) to initiate an early virus epizootic and control the population. Such tactics are only likely to work in a long-lived crop with a relatively high damage threshold. It also appears to be more effective in species where virus is naturally part of their population cycle; applications to other forest insects tend to show less virus persistence (and thus long term suppression) within the population (Cory and Myers, 2003). It is possible that this difference is due to the presence of vertically transmitted virus.

Spatial aspects of disease spread are also important. Manipulative studies on genetically modified baculovirus insecticides have demonstrated that the number of infected cadavers is more important than the quantity of virus that they contain (Hails *et al.*, 2002). Other studies on *O. pseudosugata* showed that virus transmission was lower when the distribution was heterogeneous than when it was uniform. Age structure of the insect population can also be important, with virus distribution having less effect on transmission for the more mobile later instars (Dwyer, 1991). Rate of spread of infection is one of the most relevant topics for pest control. It is well established that virus infection rapidly spreads from the areas where virus has been applied. It is not clear whether this results from the action of the insect itself via adult or larval dispersal, or whether it is the result of passive carriage by predators, but understanding this process will allow us to predict how rapidly infection will spread through a pest population. This provides the opportunity not only to enhance the number of potential vectors, but also to reduce the quantity of inoculum necessary for pest control by spraying in strips and allowing the virus to spread naturally, rather than blanket spraying. Knowledge of the rates of primary spread of *G. hercyniae* NPV has led to the suggestion that application of virus on a lattice arrangement, with centers 200 m apart would give effective control within one year of application, resulting in a potential saving in inoculum of several orders of magnitude (Hunter-Fujita *et al.*, 1998).

Rather than searching for general concepts through analytical models, other modelling studies attempt to produce more detailed simulations that explain the dynamics of specific insect-virus-crop systems. This approach requires many more parameters to be estimated but can be valuable for designing pest control programs and making short-term predictions as to the efficacy of different control options. Greenhouse crops can be particularly amenable to such approaches; for example, Bianchi *et al.* (2000, 2002a,b,c) used this method to assess the potential for using baculoviruses to control *S. exigua* on chrysanthemum. The authors estimated parameters such as, insect development rate, feeding behavior, virus transmission, speed of action and rate of virus inactivation. The model generally was able to predict the behavior of both AcMNPV and SeMNPV against *S. exigua* larvae. Differences between the model and recorded results can also indicate where more data are needed; time-mortality relationships predicted by the model were generally lower than those found in experiments. This result was thought to be explained by the fact that larvae found a refuge on the underside of the leaves where their exposure to NPV was less. Overall, this proved to be a good method of assessing the role of viruses with varying biological properties, different virus concentrations, times of application and changing pest population structure without the need to run innumerable spray trials.

4 Design of field application systems

Successful field application of viral agents has been demonstrated for many pests, and has resulted in a number of useful and practical publications on the subject (Granados and Federici, 1986a,b; Hunter-Fujita *et al.*, 1998). In common with other microbial agents, the key to success in using viral pesticides is to gain a detailed understanding of the biology and ecology of both the target host species and of the pathogen being utilized for pest management (Fuxa and Tanada, 1987).

The conceptual basis for design of a pest management regime can be summarized within a *Control Window* that can be adapted easily to take account of the particular target pest (Evans, 1994b). A typical *Control Window* is illustrated in Figure 1 and identifies the primary components of the system and the factors that interact with them. These are described in more detail below as a basis for quantification of the main parameters of a given pest management regime involving viruses.

A Target pest species

Although the target host is clearly the key component within a viral application system, it is often given less attention than the pathogen and formulation aspects of the program. However, application of virus can be targeted

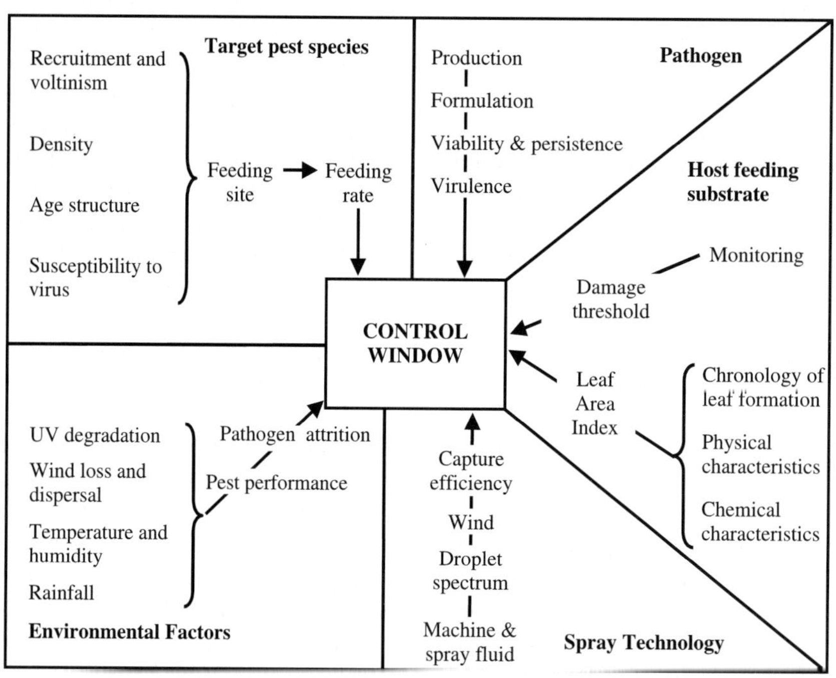

Figure 1. The *Control Window* schematic model for development of field application of viral control agents in pest management (from Evans, 1994b)

more accurately and cost-effectively by taking account of the biology of the pest, particularly recruitment, voltinism, feeding site, larval development rates and rates of food consumption. Indeed, such information is central to *a priori* design of a spray regime and can reduce the numbers of trials required to refine the use of the pathogen and also to identify limitations within the system.

At the most simplistic level, the fact that an insect species has been designated a pest implies that a damage threshold has been exceeded and that some form of control is desirable. Damage thresholds are based on the amount of feeding carried out by the pest and a necessary characteristic of a pest control strategy is to reduce total food consumption to a level that is economically acceptable. It is, therefore, important to have a detailed knowledge of sites and rates of feeding by the pest, both to determine thresholds and, more significantly, to aid the design of a viral application regime. The majority of agriculture, horticulture and forestry pests are foliage feeders and, thus, it is a relatively simple matter to gather information on the following key characteristics of the target pest species.

1 Recruitment and voltinism

The life cycle of a pest species determines periods of damage to the crop and also highlights the opportunities to target the pest for maximum impact. Information on the life cycle should be gained from field observation and, where possible, development of predictive models to determine in advance the likely appearance of the pest in the field. Examples of such models are those for *L. dispar*, in North America, (Thorpe *et al.*, 1992), for cotton pests in Australia (Room, 1979) and for European corn borer, *Ostrinia nubilalis*, in Europe (Labatte *et al.*, 1996). Such models use data on insect biology combined with meteorological information to predict the phenological development of the host, especially egg hatch and larval development. They provide guidance on the likely appearance of the damaging stages of the pest and the distribution of instars in the field once the pests have appeared. This is critical when using viruses for field application in view of the rapid decreases in susceptibility with increasing larval age.

Once egg hatch has commenced, the distribution of larval instars will also be dependent on temperature and on suitability of the food source for development. Typically, there will be several instars present, again spanning a wide range of susceptibility. Therefore, decisions on whether to target the most susceptible stage present must be balanced against the need to control the larger, less susceptible stages that cause disproportionately more damage to the crop. Such decisions are made more difficult if the pest has multiple, overlapping generations such that the range of larval instars present at any one time may span the full spectrum of host susceptibility to virus. In such situations, it may not be possible to consider targeting the least susceptible stages because damage arising from larger larvae would be too great to tolerate.

2 Host feeding rate

While the most obvious aspect of host feeding is to provide information on the amount of damage caused to the crop, of greater importance is the quantification of food consumption in relation to acquisition of virus applied to that food source. The average food consumption rate per instar is, thus, a fundamental input for the design of a spray regime using insect viruses. This information can only be gained using an appropriate food source, in most cases the different aged leaves from a host plant species, and by providing similar temperature and humidity conditions to those typically encountered in the field. The period of feeding will ultimately be determined by the rate of attrition of virus in the field by factors like UV radiation and, to a lesser extent, by the rate of development of the target host stage. For the younger instars of a pest in temperate regions, a feeding period of 24–48 h is practical, during which time the larvae will have consumed a measurable area of foliage but will not have caused major damage. Targeting the early instars also avoids the disproportionately greater feeding rates as larvae age. For example, cabbage white butterfly, *Pieris rapae*, larvae consumed about 70% of their total food in the fifth instar, compared with less than 5% in the first instar stage (Tatchell, 1981). The dose of GV required to control the larvae was also reduced [LD_{50} of 5 capsules for the first instar compared with 662 capsules for the fourth instar (Payne *et al.*, 1981)]. Similarly, Harper (1973) showed that larvae of cabbage looper moth, *Trichoplusia ni*, infected with NPV during the first instar stage ate less than 2% of potential food consumption.

The feeding period of pests that develop rapidly, such as those in the tropics or subtropics, will need to be considerably shorter than for temperate pests. Recent research into the potential for use of the baculovirus of teak defoliator moth, *Hyblaea puera*, in India has employed a feeding period of 6 h to determine the areas of teak leaves consumed by the various larval stages of the moth (H.F. Evans, K.S.S. Nair and V.V. Sudheendra Kumar, unpublished data). This period was determined by the rapid rate of larval development (2 days/instar) and the likely rate of attrition of virus in the field.

Data on host feeding rate are used to calculate the required dosage of virus per unit area of foliage, such that a known amount of virus is consumed in the pre-determined time period. This dosage will ideally be in excess of that required to kill 90% of the target larval stages, allowing for attrition of virus once applied to the foliage. If the pest has a non-random spatial distribution on the feeding substrate, then account must be taken of that distribution pattern in determining the way in which virus is applied. Changes in age and density of the pest must also be taken into account. Patterns of spatial distribution have been discussed by many authors (*e.g.*, Taylor, 1984) and these studies emphasize that the spatial distribution can change dramatically with changes in host density. Such changes can profoundly affect the target locations for the placement of virus in the field and, provided that targeting can be tailored accordingly, offers prospects for greater precision in field use of microbial and other pesticides.

B *Susceptibility to virus: dosage-mortality relationships*

Knowledge of the relationships between virus application rate and host response is essential to optimize field use and is one of the few parameters that can be quantified accurately in the laboratory, but caution is required in interpolating such data to field conditions (Evans, 1994a). Although basic LD_{50} data are relatively easy to calculate, it is the slope of the line between log dose and mortality that is used to determine field rates. Knowledge of the slope relationship also provides a quantitative basis for adjusting field rates if the calculated quantity of virus is unrealistically high. The majority of studies cited in the literature indicate that larval susceptibility decreases, often dramatically, with increasing larval age. This is particularly well documented for the baculoviruses and has been summarized by Evans (1986). Much of the enormous variability in susceptibility with age was explained by the increase in body weight during development and this factor can be used to aid prediction of field responses if the weights of each larval stage are known. Dosages required to kill larvae in the first instar can be extremely low, for example LD_{50} is less than 10 OBs for most sawflies and for many Lepidoptera (Entwistle and Evans, 1985).

There can be a great deal of variability in the slopes of dosage-mortality lines for different virus-host interactions, irrespective of the method of analysis used [see Ridout *et al.* (1993) for a useful discussion on the merits of different methods for analyzing mortality data for insect viruses]. Knowledge of the slope is important for interpreting potential field responses and, particularly, for refining application parameters to optimize the use of viral inoculum. This is illustrated in Figure 2, in which dosage-mortality lines with a common LD_{90} value, but with different slopes, are analyzed in relation to impacts on predicted field responses.

A steep slope provides a proportionately higher increase in mortality rate for a given rise in dose so that a relatively small increase in dosage provides a significant increase in response. However, in many cases the dosage necessary to achieve more than 90% mortality may be very high and difficult to achieve in practice. In such situations, a flatter slope may actually be of greater value in relation to responses achieved at lower dosages. Although high mortality may be much more difficult to achieve, shallower slopes have the advantage that responses do not decrease significantly as dosages drop. Thus, if

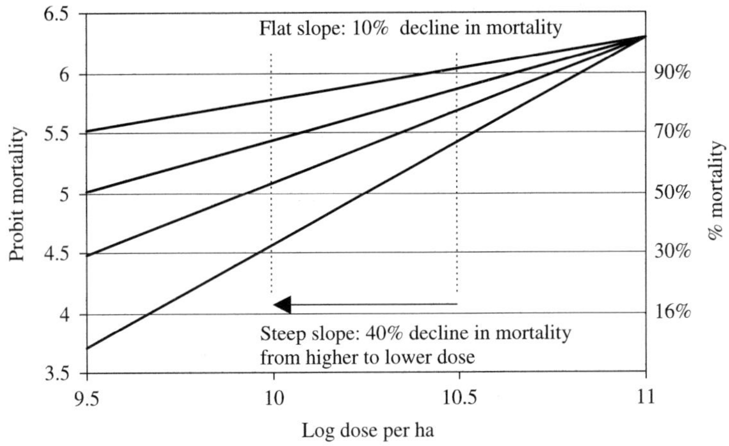

Figure 2. The effects of slope value of dosage-mortality relationships on predicted field mortality from applied virus

it is accepted that it is not possible to kill all target pest stages with the initial dose application, then it may be more cost effective to use lower dosages, which will still give acceptably high mortality. In addition, if the inoculum produced from these initial moralities is available to recycle to remaining larvae, then the final response could be higher than expected from the applied dosage. Data on slope values can, therefore, be used to estimate trade-offs between cost (*i.e.*, the quantity of inoculum that can be delivered to the host feeding site at acceptable cost) and desired pest mortality response for a given virus-host system. Once a slope value has been estimated, it is a simple matter to extrapolate back from the ideal 90+% predicted mortality to a lower mortality based on reduced dosage. Thus, taking the hypothetical example in Figure 2, a decrease in dosage of approximately 0.5 logarithms gives reductions in predicted percentage mortalities of 40% and 10% for the steepest and the flattest slopes, respectively. Such slope values are, in fact, typical of the range that would be expected for BVs, based on the theory of one-hit infection events within the gut of host insects (Ridout *et al.*, 1993).

C *Virus/environmental factors: retention of activity in the field*

Viruses are subject to environmental degradation from a number of causes, although the principal factor is UV radiation. Losses can be high and so it is important to compensate for this in determining the initial tank concentration of virus and the efficacy of any formulation for enhanced survival.

1 *Ultraviolet irradiation*

The photo-destructive effects of sunlight are caused by UV irradiation (290–380 nm) which accounts for about 1% of total solar radiation. Viruses applied as sprays are more vulnerable to degradation by UV than virus produced in cadavers as the latter has some protection afforded by insect tissue, as well as being found in much greater concentrations. The activity of virus sprays without UV protectants can decline significantly in a matter of hours after exposure to sunlight. For example, *Baculovirus heliothis* on soybean lost over 80% of its activity after 10 hour exposure to sunlight (Ignoffo *et al.*, 1997).

It is important to distinguish between the highly adverse and extremely rapid effects of laboratory generated UV-C using germicidal lamps (wavelength approximately 250 nm) and the effects of UV-A and UV-B in natural sunlight. Nevertheless, comparative studies have shown that viruses and a range of other microbial organisms have intrinsic differences in ability to withstand UV degradation, irrespective of the type of UV. Thus, Ignoffo *et al.* (1977) showed, under laboratory conditions, that the relative stability of various microbial control agents when exposed to artificial UV light was Bt > *Nomuraea rileyi* (fungus)>entomopoxvirus >

NPV = CPV > *Vairimorpha necatrix* > GV. It is interesting to note that EPV was more stable than both NPV and CPV, possibly reflecting the greater size of OBs in this group of viruses.

In the field, the total quantity (flux) of UV light varies both geographically and seasonally (Barker, 1968). Models are available to predict average UV fluxes for exposed surfaces at different locations and times of the year, thus providing a useful measure of assessing whether viruses are likely to be degraded rapidly once applied in the field (Cutchins, 1982). However, the fact that application inevitably results in droplets containing virus impacting on both top and bottom surfaces of leaves and other surfaces, means that the effects of shading may provide significant protection from direct UV. For example, Killick and Warden (1991) showed that survival of the NPV of *P. flammea*, was greater in more shaded parts of a pine canopy. A shaded position *per se* does not guarantee protection from UV because of the effects of Rayleigh scattering of light by reaction with nitrogen, oxygen and other molecules in the air or by Mie scattering arising from presence of particulate or aerosol entities in the air (Leckner, 1978). Nevertheless, significantly greater survival can be expected in shaded positions, possibly making the critical difference between success and failure in field use of virus.

2 Rainfall

Virus deposits in the field could be affected by rainfall. This is most likely to affect virus preparations applied in aqueous formulations, particularly if the foliage is wet at the time of application, but information on the specific effects of rainfall is scant and is difficult to separate from the more serious effects of UV degradation. There is some indication that adhesion of baculovirus OBs to foliage is in fact quite strong, with little active ingredient being lost by exposure to rainfall after application (Ignoffo *et al.*, 1997); however, this is likely to change with the target host plants and their surface properties, as well as the formulants used (Tamez Guerra *et al.*, 2000a). Rainfall is also a key factor in local dispersal of virus liberated after death of infected individuals, particularly in the case of the baculoviruses that cause systemic infections and release large quantities of inoculum when the insect's fragile integument is ruptured. For example, with *L. dispar* NPV, rainfall clearly moved virus OBs in the cadavers from branch to branch, increasing virus spread and the likelihood of transmission to new susceptible insects via secondary cycling (D'Amico and Elkinton, 1995). The obvious tendency will be for inoculum to be carried downwards and to end up ultimately in the soil where persistence can be prolonged; however, rainfall can reintroduce viable virus OBs from the soil to plants by rainsplash, albeit at low levels (Fuxa and Richter, 2001).

3 Temperature

Occluded viruses are quite tolerant of temperature extremes and survive well under frozen conditions (*e.g.*, storage at −70°C for several years without loss of activity) and at most field temperatures. Studies on heat inactivation indicate that temperatures higher than 40 °C can give rise to rapid inactivation, particularly in aqueous suspensions. At 60 °C most viruses lose up to 50% activity within 10 min, although differences in responses of singly enveloped NPVs (SNPVs) compared with multiply enveloped NPVs (MNPVs) were observed by Martignoni and Iwai (1977). Under the majority of conditions likely to be encountered in the field, high temperature can be disregarded as a significant attritional factor in use of occluded viruses as control agents. However, there is some indication that storage of virus preparations can vary depending on the formulants used (Tamez-Guerra *et al.*, 2000b).

4 Physical losses

Retention of viruses in the habitat of the host may also be influenced by the ability of the virus to remain attached to the host feeding site. Presence of insect remains can act as a sticker when virus is released following infection of larvae; a number of examples of long-term persistence on foliage have already been cited. Studies by Small *et al.* (1986) showed that OBs of NPVs were attached firmly to plant surfaces by short range forces arising from chemical bonding between the OB membrane and the plant surface. Indeed, it is difficult to dislodge, by washing, OBs that are attached to leaf surfaces by such forces.

D Virus/spray technology: formulation of the spray mix

The net effects of the factors discussed above result in reduction of the quantity of virus at the target field site. It is, therefore, important to assess whether these losses in field use of the virus will be unacceptably high, possibly indicating the need to formulate the viral preparation before it is used on a large scale. Assessment of the literature indicates that considerable effort has been put into development of formulations for practical field use, but it is not always apparent that the need for, or efficacy of, such formulations have been adequately quantified. Jones *et al.* (1997) discussed this point in assessing whether formulation is "an excuse for poor application." They concluded that formulation is not always essential for success, but that quality control, in terms of shelf life, ease of handling and more predictable field use, can be enhanced by formulation. Although the theory of efficient spray application is well developed for both ground and aerial application (Picot and Kristmanson, 1997; Matthews, 1992; see also Chapters III-1 and III-2), the reality is that application is often based on machinery designed for general agricultural pesticide usage that may not be appropriate for delivery of microbial agents (Chapple and Bateman, 1997).

Formulation must, therefore, be used in the full knowledge of why it is being employed and with due consideration to the costs and benefits of such an approach. Specific aspects of formulation are discussed in various chapters in this book, but for the *Control Window* approach advocated for invertebrate viruses; a number of key points must be taken into account:

1. Is loss over the defined host feeding period so high that UV protection is necessary?
2. Does the host feeding site confer some shade protection from UV, thus defining the scale of UV protection necessary?
3. How efficient is the match between droplet delivery and host feeding site?
4. Will the formulation affect the ability to deliver droplets to the required host feeding site?
5. How does the slope of the dosage-mortality relationship for the virus-pest system influence the predicted field responses and, hence, the need for formulation?
6. Do formulation components discourage larval feeding?

These are discussed further in section 4F2 in relation to optimization of applied virus through the use of *Control Window* equations. Formulation *per se* is a large subject that has been discussed in a number of publications, the precise details of which are beyond the scope of this chapter (Bryant, 1994; Shapiro, 1995; Burges, 1998; Hunter-Fujita *et al.*, 1998). Perhaps the most significant recent advance in formulation for insect viruses is the discovery that optical (fluorescent) brighteners can both provide UV protection and intrinsically enhance the biological activity of baculoviruses (Shapiro, 1992; Webb *et al.*, 1994, 1996). For example, the stilbene disulfonic acid optical brightener, Blankophor BBH, significantly increased the efficacy of both applied and naturally occurring NPV of *L. dispar* in tests in woodlands in the USA (Webb *et al.*, 1994). Comparisons of the efficacies of different optical brighteners indicated that the brighter (more fluorescent) the preparation, the greater the enhancement of infection and protection from UV (Argauer and Shapiro, 1997).

Recent research has provided an indication that a stilbene based optical brightener, designated M2R (Calcofluor white M2R = Tinopal LPW = fluorescent brightener 28), improved the effectiveness of *Ac*MNPV in tests against larvae of *T. ni* and *H. virescens* by reducing the normal sloughing of infected midgut epithelial cells and, therefore, encouraging cell-to-cell infection processes (Washburn *et al.*, 1998). Such fundamental information may provide the basis for improvement to the design of efficacy enhancing chemicals. This is particularly the case for those situations in which it has been shown that the addition of optical brighteners can actually increase host range as well as intrinsic efficacy. A good example was the change from resistance to susceptibility of *L. dispar* when challenged by either *Ac*MNPV or *Amsacta* EPV in the presence of Phorwite AR (Shapiro and Dougherty, 1994).

E *Spray technology: calibration of the spray system*

The theory and practice of droplet generation and application in the field for a range of spray equipment are covered in Chapters I-2, III-1, and III-2. Knowledge of the generated droplet spectra for the spray fluid and equipment likely to be used in the field must be known, particularly in relation to the Volume Median Diameter (VMD) and span of the spray. Ideally, this would be pre-determined for a range of equipment rotation speeds (in revolutions per minute [rpm]) and flow rates and taking account of the ways in which formulation may affect both droplet generation and the retention of virus within those droplets in flight. In this way, the relative loading of virus OBs within droplets can be determined accurately and decisions made on the best compromise between droplet sizes, total spray volumes and the distribution of OBs between droplets. In the majority of cases, the high cost of virus production and the vulnerability of inoculum to UV degradation, requires that wastage of virus is kept to the absolute minimum. The cubic relationship between droplet diameter and droplet volume is an important consideration in deciding both on VMD and the range (span) of droplet sizes to be used in the field. Doubling droplet diameter increases the number of OBs present by 8-fold. Considering that inefficient spray machinery may have a very wide span of droplet sizes, perhaps ranging up to 5 fold or more in diameter from the smallest to the largest droplets, then the range of droplet loadings could vary by 125-fold. Clearly, this points to large potential wastage of inoculum and inefficient pest control if droplet parameters are not taken into account. In addition, there is some evidence that particulate matter can be lost from droplets between their generation at the spray head and their arrival at the target surface. This was discussed by Hunter-Fujita *et al.* (1998), who summarized a number of instances of loss of particulate matter in spray equipment (*Bt*, fluorescent particles, GV OBs, NPV OBs) ranging up to 97% apparent loss of *Bt* from Micronair atomizers (Morris, 1977). Although there appears to be no known strategy to combat such losses within the spray generation process, it is important to be aware that they can occur and to take account of the losses in calculating field spray parameters.

Overall, within a given combination of spray equipment parameters, there is considerable scope for generating a spray cloud that gives the optimal setting for emitted volume and droplet size range, after allowing for any constraints inherent in the system. As will be discussed in Section 4F, the key to success in using viral agents for pest management is to ensure that the host encounters a lethal dose of virus in the shortest time, taking account of losses arising from shortcomings in spray deposition in the field. This requires careful measurement of droplet generation and deposition for the selected spray equipment, but it is vital that this is done on the basis of a *Control Window* style of approach, rather than *ad hoc* experimentation. In reality, it is only those droplets that impact at the feeding sites of the target host stage that contribute to pest management and, conversely, all other droplets are lost to the system. This loss must be accounted for in calculating field dosage rates.

F Control Window*: calculation of tank mix based on known virus and host parameters*

Knowledge of the parameters discussed earlier can be brought together to calculate spray parameters and predicted field mortality rates in relation to practical limits such as cost and damage threshold. This is discussed in relation to a theoretical model for design of a spray regime, based on Evans (1998). The process requires that droplets landing at the feeding sites of target host stages should contain sufficient virus to exceed the LD_{90}, in relation to host feeding rates over a given time and of environmental losses likely to be encountered during that time. This process can be broken down into a series of equations in which parameters, calculated from both laboratory and field observations, are entered. The equations can be incorporated into a spreadsheet model so that sensitivity analysis can be carried out, particularly on those parameters under the direct control of the operator. This analysis can be used to optimize spray characteristics to increase efficacy and/or reduce costs.

1 Calculation of spray parameters

Calculations of tank mix can be carried out using simple relationships among virus, host and droplet parameters (Evans, 1998). Although there are many other physical and biological parameters that could be included, experience has demonstrated that predictions from the equations below provide a reliable first approximation to the necessary field dosage. The process is illustrated by a hypothetical example in which values are entered in a simple spreadsheet model (Table 3).

By using the *name* option (in the case of Microsoft Excel) or equivalent for the main parameter values within the spreadsheet, the formulae can be set up using precisely the same expressions as the equations described below; the Microsoft Excel spreadsheet example in Table 3 employs this terminology.

Step 1: In relation to droplet emission, measure:

N = Numbers of droplets emitted by the atomizer per liter of formulated spray fluid.

As discussed above, it is important that information on the range of droplet sizes is known, but in practice it is usually the VMD that is used for this calculation.

CE = Capture Efficiency.

This is the number of droplets generated to ensure that at least one droplet is captured per host feeding area. The parameter is expressed in terms of droplets per unit ground area, based on Leaf Area Index (LAI), loss to ground, etc., thus allowing comparison between sites on a common basis.

Step 2: Determine from laboratory and field experimentation and observation:

Host feeding rate (*fr*) as unit area in a pre-determined time, typically 6 h (tropics) to 48 h (temperate regions).

LD_{90} (d), based on laboratory bioassay studies, including the slope value for later optimization. Ideally, the value should be taken to the LD_{95} level, but experience has shown that it is rarely possible to reach such a high mortality in practice and it is, therefore, more realistic to start with the lower target value. However, where virus supply is not a problem and dosage mortality relationships indicate high susceptibility [*e.g.*, *N. sertifer* NPV, (Entwistle *et al.*, 1990)], then it will be possible to base calculations on a higher LD value.

Virus activity loss (a), based on studies of UV inactivation and rainfall attrition.

The loss rate is used as a multiplier to give the initial dose D_i expressed as OBs/mm^2. For

Table 3. Example of the spreadsheet version of the *Control Window* model to determine optimal field application parameters for application of arthropod pathogenic viruses in the field (based on Evans, 1998)

Enter information on the host, virus, etc. for reference purposes.

Hypothetical example of lepidopteran forest defoliator with virus OBs sprayed onto the upper canopy.

Enter the parameters below:

Droplet diameter (μm)		100		
Feeding rate of larvae (mm^2)	*fr*	18		
Area of ha in mm^2	*area*	1.00E+10		
Leaf Area Index (ratio to ground area)	*LAI*	4		
Loss of spray fluid to ground area (one/proportion)	*s*	0.6		
Virus attrition rate (proportion of original)	*a*	0.5		
LD_{90} for target larval stage	LD_{90}	1000		

Results are shown below:

Capture efficiency = (area)*(LAI*(1/(s*fr))	*CE*	3.70E+09	Equation *i*	=required droplets per ha
Number of droplets per liter:	*N*	1.91E+09		
Initial dose (OBs/mm^2) = LD_{90}/a	D_i	2000		
Theoretical Volume = CE/N liters per ha	*V*	1.94	Equation *ii*	
Dose per liter	D_l	3.82E+12	Equation *iii*	
Dose per ha	D_{ha}	7.41E+12	Equation *iv*	

example, 50% loss will double D_i in relation to the calculated value for d.

Step 3: Enter values in *Control Window* model equations (Table 3):

The parameters from Steps 1 and 2 are then entered in a series of equations to calculate the dose per ha and the concentration of virus in the tank mix:

$$CE = (1 \times 10^{10}) LAI \frac{1}{s \times fr} \text{ droplets per ha} \quad (1)$$

where 1×10^{10} = area of 1 ha in mm^2
LAI = Leaf Area Index, a multiplier to express surface area of foliage in units of ground area, determined by the physical structure (height, leaf number, leaf area) of the canopy of the target host plant.
s = Loss of spray fluid to non-target area, determined by comparison of the actual numbers of droplets captured at the designated host feeding site (area determined by fr) and the known number of droplets emitted.

Theoretical minimum volume,

$$V = \frac{CE}{N} \text{ liters per ha} \quad (2)$$

Dose per liter

$$D_l = N \times D_i \text{ expressed in OBs per liter} \quad (3)$$

Dose per ha,

$$D_{ha} = CE \times D_i \text{ expressed in OBs per ha} \quad (4)$$

2 *Optimization of spray parameters*

It is a simple matter to change some of the physical parameters in the spreadsheet model in Table 3 and to assess their effects on both dosage of virus and the volume of spray delivered per ha. This process will, of course, depend on the dosage-mortality relationships of the particular virus-host interaction and on the attrition rate of applied virus in the field. The necessity for optimization will depend on whether the original predictions indicate that efficacy and/or costs are unacceptable for management of the pest. Parameters having the greatest potential for producing reductions in virus dosages or increases in efficacy are:

a. The slope of the dosage-mortality relationship and its effect on predicted mortality if the field dosage is reduced. A relatively flat slope implies that field mortality will not decline significantly with reductions in field dosage, even though mortality will not reach the original target level. In respect of induced mortality, it is more important to assess the effects on damage reduction than on population reduction *per se*. Thus, efficacy should be measured as a combination of both damage and host population reduction.
b. The distribution of susceptibility within the target host population. As a general rule, susceptibility is greatest, and the amount of damage induced is lowest, in younger larval stages, indicating that less virus would be required to target such stages. However, the time course of recruitment of larvae to the host population will, inevitably, increase the range of susceptibility as a whole and would require greater quantities of virus to kill all the larvae present at the time of virus application. One aspect that can be considered is to improve the monitoring of host recruitment and so to allow better timing of the spray application to target the greatest proportion of highly susceptible individuals. Predictive phenological models, supplemented by scouting for appearance of larval stages, can aid this process.
c. Reduce the rate of attrition of applied virus. In section 4C1, the rapid losses that can be experienced from the effects of UV light on unformulated virus in the field were considered. The rate of loss (a), can be reduced considerably by appropriate formulation. In the tropics, a simpler approach can also be adopted by delaying spray application until late afternoon so that the uptake of virus within the timescale of the feeding rate (fr) takes place in darkness.
d. Improve the degree of coverage of droplets in the target spray area. Coincidence of droplet capture and host feeding rate, fr, is essential to ensure an early encounter between the host and virus, *i.e.*, to improve the capture efficiency (CE). If there is a high rate of loss of droplets to non-target areas (s), then improvements to the spray regime should be sought. There is certainly a strong case to consider controlled droplet application and smaller diameter droplets to both reduce the range of droplet sizes (particularly avoiding large diameter droplets that waste a disproportionate quantity of inoculum) and to increase coverage per unit area. Such an

approach was adopted in control of pine beauty moth, *P. flammea*, using NPV in Scotland in which it was shown that egg and early larval distribution was concentrated in the upper third of the pine canopy. Use of rotating disk atomizers and droplets around 50 μm diameter sprayed incrementally at wind speeds not less than 5 m/sec resulted in more than 90% of the emitted droplets impacting on the critical upper region of the canopy (Entwistle *et al.*, 1990). It was also demonstrated that increased efficacy could be achieved by apportioning fewer NPV OBs per droplet, but increasing the numbers of droplets generated.

The above generalities provide some guidance on the scope for spray application of viruses within a *Control Window* approach, and show the value of obtaining and using detailed information on key parameters in designing and optimizing a virus control strategy. There is considerable scope for further improvement, particularly in integrating some of the genetically modified viruses that are now approaching commercial availability.

5 Conclusions

Viral agents have attributes that, in many respects, make them ideal for pest management purposes. Specificity, persistence ability to replicate and dispersal capacity are key factors that contribute to the position of viruses as the most environmentally acceptable components of direct and integrated management regimes. However, care must be taken to avoid regarding viruses as merely analogues of chemical pesticides. By paying due regard to the range of characteristics described in this chapter, particularly within the *Control Window* concept, virus performance can be optimized for both efficacy and cost.

Further research is required to improve knowledge of the components of the *Control Window* that are under the direct control of the operator. Some of these components, particularly spray technology, are discussed elsewhere in this volume. More can be made of the biology and ecology of viruses to optimize these aspects, especially where current performance is erratic or where the cost of virus production is prohibitive. A valuable and increasingly important element of such an approach is provision of the tools necessary to carry out risk assessments for release of genetically modified viral agents (Cory *et al.*, 1997; Cory, 2000). This is one of the growth areas in pest management, but must be handled extremely carefully in bringing the new "designer" viruses to the marketplace. At the moment, the science of risk assessment lags behind the science of genetic modification but, in reality, the tools are already available to carry out the necessary quantitative risk assessment.

In conclusion, the future for viral agents will depend on recognition of their particular characteristics and on integrating those attributes in Integrated Crop Management programs. This requires investment of both research and development time but, as indicated in the world survey included in Hunter-Fujita *et al.* (1998), there is global interest in use of viral agents. It is to be hoped that this interest continues and expands to make full use of these environmentally benign pest management agents.

6 Acknowledgments

HFE would like to thank Dr. Richard Jinks, Forest Research, for his help in improving the layout and clarity of the paper. We are grateful to anonymous referees for their helpful suggestions.

7 References

Adams, J.R. 1991. Introduction and classification of viruses of invertebrates. *In* "Atlas of Invertebrate Viruses" (Adams, J.R. and Bonami, J.R., Eds.) pp. 1–8. CRC Press, Boca Raton, FL.

Ali, M.I., Felton, G.W., Meade, T. and Young, S.Y. 1998. Influence of interspecific and intraspecific host plant variation on the susceptibility of heliothines to a baculovirus. *Biol. Control* **12**, 42–49.

Ali, M.I, Young, S.Y., Felton, G.W and New, R.W. 2002. Influence of host plant on occluded virus productivity and lethal infectivity of a baculovirus. *J. Invertebr. Pathol.* **81**, 158–165.

Anderson, R.M. and May, R.M. 1980. Infectious disease and population cycles of forest insects. *Science* **210**, 658–661.

Anderson, R.M. and May, R.M. 1981. The population dynamics of microparasites and their invertebrate hosts. *Phil. Trans. R. Soc. London. Ser. B* **291**, 451–524.

Argauer, R. and Shapiro, M. 1997. Fluorescence and relative activities of stilbene optical brighteners as enhancers for the gypsy moth (Lepidoptera: Lymantriidae) baculovirus. *J. Econ. Entomol.* **90**, 416–420.

Arif, B.M. 1984. The Entomopoxviruses. *Adv.Vir.Res.* **29**, 195–213.

Arthurs, S.P., Lacey, L.A. and Fritts, R. 2005. Optimizing use of codling moth granulovirus: Effects of application rate and spraying frequency on control of codling moth larvae in Pacific Northwest apple orchards. *J. Econ. Entomol.* **98**, 1459–1468.

Barker, R.E. 1968. The availability of solar radiation below 290 nm and its importance in photomodification of polymers. *Photochem. Photobiol.* **7**, 275–295.

Beisner, B.E. and Myers, J.H. 1999. Population density and transmission of virus in experimental populations of the western tent caterpillar (Lepidoptera: Lasiocampidae). *Environ. Entomol.* **28**, 1107–1113.

Bianchi, F.F.J.A, Joosten, N. N., Vlak, J. M. and van der Werf, W. 2000. Greenhouse evaluation of dose- and time-mortality relationships of two nucleopolyhedroviruses for the control of beet armyworm, *Spodoptera exigua,* on chrysanthemum. *Biol.. Control* **19**, 252–258.

Bianchi, F.F.J.A, van der Werf, W and Vlak, J.M. 2002a. Validation of a comprehensive process-based model for the biological control of beet armyworm, *Spodoptera exigua*, with baculoviruses in greenhouses. *Biol. Control* **23**, 47–55.

Bianchi, F.F.J.A, Vlak, J.M., Rabbinge, R. and van der Werf, W. 2002b. Biological control of beet armyworm, *Spodoptera exigua*, with baculoviruses in greenhouses: Development of a comprehensive process-based model. *Biol. Control* **23**, 35–46.

Bianchi, F.F.J.A, van der Werf, W and Vlak, J.M. 2002c. Evaluation of the control of beet armyworm, *Spodoptera exigua*, with baculoviruses in greenhouses using a process-based simulation model. *Biol.. Control* **24**, 277–284.

Bourner, T.C. and Cory, J.S. 2004. Host range of an NPV and a GV isolated from the common cutworm, *Agrotis segetum*: pathogenicity within the cutworm complex. *Biol. Control* **31**, 372–379.

Bryant, J.E. 1994. Commercial production and formulation of *Bacillus thuringiensis. Agric. Ecosys. Environ.* **49**, 31–35.

Burden, J.P., Griffiths, C.M., Cory, J.S., Smith, P. and Sait, S.M. 2002. Vertical transmission of sublethal granulovirus infection in the Indian meal moth, *Plodia interpunctella. Mol. Ecol.* **11**: 547–555.

Burden, J.P., Nixon, C.P., Hodgkinson, A.E., Possee, R.P. Sait, S.M., King, L.A., and Hails, R.S. 2003. Covert infections as a mechanism for long-term persistence of baculoviruses. *Ecol. Letts.* **6**, 524–531.

Burges, H.D. (ed.). 1998. "Formulation of Microbial Biopesticides, Beneficial Micro-organisms and Nematodes." Kluwer Academic. Dordrecht, The Netherlands.

Carruthers, W.R., Cory, J.S. and Entwistle, P.F. 1988. Recovery of pine beauty moth (*Panolis flammea*) nuclear polyhedrosis virus from pine foliage. *J. Invertbr. Pathol.* **52**, 27–32.

Chapple, A.C. and Bateman, R.P. 1997. Application systems for microbial pesticides: Necessity not novelty. *In* "Microbial Insecticides: Novelty or Necessity?" (Evans, H.F., Ed.) pp. 181–190. British Crop Protection Council, Farnham.

Cherry, A.J., Rabindra, R.A., Parnell, M.A., Geetha, N., Kennedy, J.S. and Grzywacz, D. 2000. Field evaluation of Helicoverpa armigera nucleopolyhedrovirus formulations for control of the chickpea pod-borer, *H-armigera* (Hubn.), on chickpea (*Cicer arietinum* var. *Shoba*) in southern India. *Crop Protect.* **19**, 51–60.

Christian, P.D. Murray, D., Powell, R., Hopkinson, J., Gibb, N.N. and Hanzlik, T.N. 2005. Effective control of a field population of *Helicoverpa armigera* by using the small RNA virus *Helicoverpa armigera* stunt virus (Tetraviridae: Omegatetravirus). *J. Econ. Entomol.* **98**, 1839–1847.

Cooper, D., Cory, J.S. and Myers, J.H. 2003b. Hierarchical spatial structure of genetically variable nucleopolyhedroviruses infecting cyclic populations of western tent caterpillars. *Mol. Ecol.* **12**, 881–890.

Cooper, D., Cory, J.S., Theilmann, D.A. and Myers, J.H. 2003a. Nucleopolyhedroviruses of forest and western tent caterpillars: cross-infectivity and evidence for activation of latent virus in high density field populations. *Ecol. Entomol.* **28**, 41–50.

Cory J.S. 2000. Assessing the risks of releasing genetically modified virus insecticides; progress to date. *Crop Protect.* **19**, 779–785.

Cory J.S. 2003. Ecological impacts of virus insecticides: host range and non-target organisms. *In "Environmental impacts of microbial insecticides*". (H. Hokkanen and A. Hajek, Eds.), pp. 73–91. Kluwer Academic Publishers, Drodrecht.

Cory, J.S., Green, B.M., Paul, R.K. and Hunter-Fujita, F.R. 2005. Genotypic and phenotypic diversity of a baculovirus population within an individual insect host. *J. Invertebr. Pathol.* **89**, 101–111.

Cory, J.S., Hails, R.S. and Sait, S.M. 1997. Baculovirus ecology. In "*The Baculoviridae*" (L. K. Miller, Ed.), pp. 301–339, Plenum Press, New York.

Cory, J.S. and Hoover, K. 2006 Plant-mediated effects in insect-pathogen interactions. *Trends Ecol. Evol.* **21**, 278–286.

Cory, J.S. and Myers, J.H. 2003. The ecology and evolution of insect baculoviruses. *Annu. Rev. Ecol. Evol.Syst.* **34**, 239–272.

Cory, J.S. and Myers, J.H. 2004. Adaptation in an insect host-plant pathogen interaction. *Ecol. Letts.* **7**, 632–639.

Cross J.V., Solomon M.G., Chandler D., Jarrett P., Richardson P.N., Winstanley D., Bathon H., Huber J., Keller B., Langenbruch G.A., Zimmerman G. 1999.

Biocontrol of pests of apples and pears in northern and central Europe: 1. Microbial agents and nematodes. *Biocontr. Sci. Tech.* **9**, 125–149.

Cutchins, P. 1982. A formula for comparing annual damaging ultra-violet (DUV) radiation doses at tropical and mid-latitude sites. *In* "NATO Conference Series No. IV: The Role of Solar Ultra-Violet Radiation in Marine Ecosystems" (Calkins, J., Ed.) pp. 213–228. Plenum Press, New York.

D'Amico, V. and Elkinton, J.S. 1995. Rainfall effects of transmission of gypsy moth (Lepidoptera Lymantriidae) nuclear polyhedrosis virus. *Environ. Entomol.* **24**, 1144–1149.

D'Amico, V., Elkinton, J.S., Dwyer, G., Burand, J.P. and Buonaccorsi, J.P. 1995. Virus transmission in gypsy moths is not a simple mass action process. *Ecology* **77**, 201–206.

Duffey, S.S, Hoover, K., Bonning, B.C. and Hammock, B.D. 1995. The impact of host-plant on the efficacy of baculoviruses. In "*Reviews in Pesticide Toxicology*" (Roe, M. and Kuhr, R., ed.) pp. 137–275. CTI Toxicology Communications.

Dwyer, G. 1991. The roles of density, stage and patchiness in the transmission of an insect virus. *Ecology* **72**, 559–574.

Dwyer, G., Dushoff, J., Elkinton, J.S and Levin, S.A 2000. Pathogen-driven outbreaks in forest defoliators revisited: building models from experimental data. *Am. Nat.* **156**, 105–120.

Dwyer, G., Elkinton, J. and Buonaccorsi, J.P. 1997. Host heterogeneity in susceptibility and disease dynamics: tests of a mathematical model. *Am. Nat.* **150**, 685–707.

England, L.T., Holmes, S.B. and Trevors, J.T. 1998. Review: persistence of viruses and DNA in soil. *World J. Micro. Biotech.* **14**, 163–169.

Entwistle, P.F., Adams, P.H.W. and Evans, H.F. 1978. Epizootiology of a nuclear polyhedrosis virus in European spruce sawfly (*Gilpinia hercyniae*): the rate of passage of infective virus through the gut of birds during cage tests. *J. Invertebr. Pathol.* **31**, 307–312.

Entwistle, P.F., Adams, P.H.W. and Evans, H.F. 1983. Employing defensive regurgitation to estimate levels of nuclear polyhedrosis virus infection in *Gilpinia hercyniae* (Hymenoptera: Diprionidae) larvae. *J.Invertebr. Pathol.* **41**, 262–264.

Entwistle, P.F. and Evans, H.F. 1985. Viral control. *In* "Comprehensive insect physiology, biochemistry, and pharmacology" (Kerkut, I. and Gilbert, L.I., Eds.) pp. 347–412. Pergamon Press, Oxford.

Entwistle, P.F., Evans, H.F., Cory, J.S. and Doyle, C.J. 1990. Questions on the aerial application of microbial pesticides to forests. *Proc. Vth Int. Coll. Invertebr. Pathol.* 159–163.

Entwistle, P.F., Forkner, A.C., Green, B.M. and Cory, J.S. 1993. Avian dispersal of nuclear polyhedrosis viruses after induced epizootics in the pine beauty moth, *Panolis flammea* (Lepidoptera: Noctuidae). *Biol. Control* **3**, 61–69.

Evans, H.F. 1994a. Laboratory and field results with viruses for the control of insects. *In* "BCPC Monograph No 59: Comparing Glasshouse and Field Pesticide Performance II" (Hewitt, H.G., Caseley, J., Copping, L.G., Grayson, B.T. and Tyson, D., Eds.) pp. 285–296. BCPC, Farnham.

Evans, H.F. 1994b. The Control Window: A conceptual approach to using baculoviruses for forest pest control. *Proc. VI Int. Colloq. Invertbr. Pathol. Microb. Control* 380–384.

Evans, H.F. 1986. Ecology and epizootiology of baculoviruses. *In* "Biology of Baculoviruses Vol.2. Practical Application for Insect Pest Control" (Granados, R.R. and Federici, B.A., Eds.) pp. 89–132. CRC Press, Boca Raton, FL.

Evans, H.F. 1998. Principles of dose acquisition for bioinsecticides. *In* "Biopesticides: Use and Delivery" (Hall, F.R. and Menn, J., Eds.) pp. 553–573. Humana Press, Totowa, NJ.

Evans, H.F. and Allaway, G.P. 1983. Dynamics of baculovirus growth and dispersal in *Mamestra brassicae* L. (Lepidoptera: Noctuidae) larval populations introduced into small cabbage plots. *Appl. Environ.Microbiol.* **45**, 493–501.

Evans, H.F. and Shapiro, M. 1997. Viruses. *In* "Manual of Techniques in Insect Pathology" (Lacey, L.A., Ed.) pp. 17–53. Academic Press, London.

Federici, B. A. and Stern, V. M. 1990. Replication and occlusion of a granulosis virus in the larval and adult midgut epithelium of the western grapeleaf skeletonizer, *Harrisina brillians*. *J. Invertebr. Pathol.* **56**, 401–414.

Fleming, S.B., Kalmakoff, J., Archibald, R.D. and Stewart, K.M. 1986. Density-dependent virus mortality in populations of *Wiseana* (Lepidoptera: Hepialidae). *J. Invertebr. Pathol.* **48**, 193–198.

Fuxa, J.R. and Richter, A.R. 1994. Distance and rate of spread of *Anticarsia gemmatalis* (Lepidoptera: Noctuidae) nuclear polyhedrosis virus released into soybean. *Environ. Entomol.* **23**, 1308–1316.

Fuxa, J.R. and Richter, A.R. 1998. Host insect survival time and disintegration in relation to population density and dispersion of recombinant and wild-type nucleopolyhedroviruses. *Biol. Control* **12**, 143–150.

Fuxa, J.R. and Richter, A.R. 2001. Quantification of soil-to-plant transport of recombinant nucleopolyhedrovirus: effect of soil type and moisture, air currents and precipitation. *Appl. Environ. Micro.*, **67**, 5166–5170.

Fuxa, J.R. and Tanada, Y. 1987. "Epizootiology of Insect Diseases." John Wiley & Sons Inc., New York.

Georgis, R. 1997. Commercial prospects of microbial insecticides in agriculture. *In* "Microbial Insecticides: Novelty or Necessity?" (Evans, H.F., Ed.) pp. 243–252. British Crop Protection Council, Farnham.

Goodwin, R.H., Milner, R.J. and Beaton, C.D. 1991. Entomopoxvirinae. *In* "Atlas of Invertebrate Viruses" (Adams, J.R. and Bonami, J.R., Eds.) pp. 259–285. CRC Press, Boca Raton, FL.

Goulson, D. and Cory, J.S. 1995. Responses of *Mamestra brassicae* (Lepidoptera: Noctuidae) to crowding: interactions with disease resistance, colour phase and growth. *Oecologia* **104**, 416–423.

Graham, R.I., Rao, S.J., Possee, R.D., Sait, S.M., Mertens, P..P.C. and Hails, R.S. 2006. Detection and characterisation of three novel species of reovirus (Reoviridae), isolated from geographically separate populations of the winter moth *Operophtera brumata* (Lepidoptera: Geometridae) on Orkney. *J. Invertebr. Pathol*, **91**, 79–87.

Granados, R.R. and Federici, B.A. 1986a. "The Biology of Baculoviruses: Volume 1 – Biological Properties and Molecular Biology." CRC Press, Boca Raton, FL.

Granados, R.R. and Federici, B.A. 1986b. "The Biology of Baculoviruses: Volume 2 – Practical Application for Insect Control." CRC Press, Boca Raton, FL.

Hails, R.S., Hernández-Crespo, P., Sait, S.M., Donnelly, C.A., Green, B.M. and Cory, J.S. 2002. Transmission patterns of natural and recombinant baculoviruses. *Ecology* **83**, 906–916.

Harper, J.D. 1973. Food consumption by cabbage loopers infected with nuclear polyhedrosis virus. *J.Invertebr. Pathol.* **21**, 191–197.

Hitchman, R.B., Hodgson, D.J., King, L.A., Hails, R.S., Cory, J.S. and Possee, R.D. 2007. Host mediated selection of pathogen genotypes as a mechanism for the maintenance of baculovirus diversity in the field. *Journal of Invertebrate Pathology* **94**, 153–162.

Hodgson D.J., Hitchman R.B., Vanbergen A.J., Hails R.S., Possee R.D. and Cory J.S. 2004. Host ecology determines the relative fitness of virus genotypes in mixed-genotype nucleopolyhedrovirus infections. *J. Evol. Biol.* **17**, 1018–1025.

Hodgson, D.J., Vanbergen, A.J., Hartley, S.E., Hails, R.S. and Cory J.S. 2002. Differential selection of baculovirus genotypes mediated by different species of host food plant. *Ecol. Lett.* **5**, 512–518.

Hodgson, D.J., Vanbergen, A.J., Watt, A.D., Hails, R.S. and Cory, J.S. 2001. Phenotypic variation between coexisting genotypic strains of a lepidopteran baculovirus. *Evol. Ecol. Res.* **3**, 687–701.

Hofman, O. 1891. Die Schlaffsucht (Flacherie) der Nonne (*Liparis monacha*) nebst einem Anhang. *In* "Insektentotende Pilze mit besonderer Berucksichtigung der Nonne" pp. 31–31. P. Weber, Frankfurt.

Huger, A.M. 2005. The *Oryctes* virus: its detection, identification, and implementation in biological control of the coconut palm rhinoceros beetle, *Oryctes rhinoceros* (Coleoptera:Scarabidae). *J. Invertebr. Pathol.* **89**, 78–84.

Hughes, D.S., Possee, R.P. and King, L.A. 1997. Evidence for the presence of a low-level persistent baculovirus infection of *Mamestra brassicae* insects. *J. Gen. Virol.* **78**, 1801–1805.

Hukuhara, T. and Bonami, J.R. 1991. Reoviridae. *In* "Atlas of Invertebrate Viruses" (Adams, J.R. and Bonami, J.R., Eds.) pp. 393–434. CRC Press, Boca Raton, FL.

Hunter, M.D. and Schultz, J.C. 1993. Induced plant defenses breached? Phytochemical induction protects an herbivore from disease. *Oecologia* **94**, 195–203.

Hunter-Fujita, F.R., Entwistle, P.F., Evans, H.F. and Crook, N.E. 1998. "Insect Viruses and Pest Management." John Wiley & Sons Ltd, Chichester.

Ignoffo, C.M., Garcia, C. and Saathoff, S.G. 1997. Sunlight stability and rainfastness of formulations of Baculovirus heliothis. *Environ. Entomol.* **26**, 1470–1474.

Ignoffo, C.M., Hostetter, D.L., Sikorowski, P.P., Sutter, G. and Brooks, W.M. 1977. Inactivation of representative species of entomopathogenic viruses, a bacterium, fungus, and protozoan by an ultraviolet light source. *Environ. Entomol.* **6**, 411–415.

Inceoglu, A.B., Kamita, S.G., Hinton, A.C., Huang, Q.H., Severson, T.F., Kang, K.D. and Hammock, B.D. 2001. Recombinant baculoviruses for insect control. *Pest Manage. Sci.* **57**, 981–987.

Jackson, D.M., Brown, G.C., Nordin, G.L. and Johnson, D.W. 1992. Autodissemination of a baculovirus for management of tobacco budworms (Lepidoptera: Noctuidae) on tobacco. *J. Econ. Entomol.* **85**, 710–719.

Jehle, J.A, Blissard, G. W., Bonning B. C., Cory, J. S., Herniou, E. A, Rohrmann, G. F., Theilmann, D. A., Thiem, S. A. and Vlak,. J. M. 2006. On the classification and nomenclature of baculoviruses: a proposal for revision. *Arch. Virol.* **151**, 1257–1266.

Jones, K.A., Cherry, A.J., Grzywacz, D. and Burges, H.D. 1997. Formulation: Is it an excuse for poor application? *In* "Microbial Insecticides: Novelty or Necessity?" (Evans, H.F., Ed.) pp. 173–180. British Crop Protection Council, Farnham.

Killick, H.J. and Warden, S.J. 1991. Ultraviolet penetration of pine trees and insect virus survival. *Entomophaga* **36**, 87–94.

Kukan, B. 1999. Vertical transmission of nucelopolyhedrovirus in insects. *J. Invertbr. Pathol.* **74**, 103–111.

Labatte, J.M., Meusnier, S., Migeon, A., Chaufaux, J., Couteaudier, Y., Riba, G. and Got, B. 1996. Field evaluation of and modeling the impact of three control methods on the larval dynamics of *Ostrinia nubilalis* (Lepidoptera: Pyralidae). *J. Econ. Entomol.* **89**, 852–862.

Laird, M., Lacey, L.A. and Davidson, E.W. (*Eds.*) 1990. "Safety of Microbial Insecticides." CRC Press, Boca Raton, FL.

Leckner, B. 1978. The spectral distribution of solar radiation at the Earth's surface - elements of a model. *Solar Energy* **20**, 143–150.

Lee, K.P., Cory, J.S., Wilson, K., Raubenheimer, D. and Simpson, S.J. 2006. Flexible diet choice offsets protein costs of pathogen resistance in a caterpillar. *Proc. R. Soc. London B*, **273**, 823–829

Lopez-Ferber, M., Simon, O., Williams, T. and Caballero, P. 2003. Defective or effective? Mutualistic interactions between virus genotypes. *Proc. R. Soc. B.* **270**, 2249–2255.

Martignoni, M.E. and Iwai, P.J. 1977. Thermal inactivation characteristics of two strains of nucleopolyhedrosis virus (Baculovirus subgroup A) pathogenic for *Orgyia pseudotsugata. J.Invertebr. Pathol.* **30**, 255–262.

Matthews, G.A. 1992. "Pesticide application methods." Longman, London.

Matthews, H.J., Smith, I. and Edwards, J.P. 2002. Lethal and sublethal effects of a granulovirus on the tomato moth *Lacanobia oleracea. J. Invertebr. Pathol.* **80**, 73–80.

Matthews, H.J., Smith, I., Bell, H.A and Edwards, J.P. 2004. Interactions between the parasitoid *Meteorus gyrator* (Hymenoptera: Braconidae) and a granulovirus in *Lacanobia oleracea* (Lepidoptera: Noctuidae). *Environ. Entomol.* **33**, 949–957.

Milks, M.L., Burnstyn, I. and Myers, J.H. 1998. Influence of larval age on the lethal and sublethal effects on the nucleopolyhedrovirus of *Trichoplusia ni* in the cabbage looper. *Biol. Control* **12**, 119–126.

Miller, L.K. 1995. Genetically engineered insect virus pesticides: present and future. *J. Invertebr. Pathol.* **65**, 211–216.

Miller, L.K. 1997. "The Baculoviruses". 447 pp Plenum Publishing Corp., New York.

Morris, O.N. 1977. Relationship between microbial numbers and droplet size in aerial spray application. *Can.Entomol.* **109**, 1319–1323.

Moscardi, F. 1999. Assessment of the application of baculoviruses for the control of Lepidoptera. *Annu. Rev Entomol.* **44**, 257–289.

Munoz, D. and Caballero, P. 2000. Persistence and effects of parasitic genotypes in a mixed population of the *Spodoptera exigua* nucleopolyhedrovirus. *Biol. Control* **19**, 259–264.

Murray, K.D. and Elkinton, J.S. 1989. Environmental contamination of egg masses as a major component of transgenerational transmission of gypsy moth nuclear polyhedrosis virus (LdMNPV). *J. Invertebr. Pathol.* **53**, 324–334.

Olofsson, E. 1989. Transmission of the nuclear polyhedrosis virus of the European pine sawfly from adult to offspring. *J.Invertebr.Pathol.* **54**, 322–330.

Otvos, I.S., Cunningham, J.C. and Alfaro, R.I. 1987. Aerial application of nuclear polyhedrosis virus against Douglas-fir tussock moth *Orgyia pseudosugata* (McDunnough) (Lepidoptera: Lymrantriidae): II. Impact 1 and 2 years after application. *Can. Entomol.* **119**, 707–716.

Parnell, M.A., King, W.J., Jones, K.A., Ketunuti, U. and Wetchakit, D. 1999. A comparison of motorised knapsack mistblower, medium volume application, very low volume application of *Helicoverpa armigera* nuclear polyhedrosis virus on cotton in Thailand. *Crop Protect.* **18**, 259–265.

Payne, C.C., Tatchell, G.M. and Williams, C.F. 1981. The comparative susceptibilities of *Pieris brassicae* and *P. rapae* to a granulosis virus from *P. brassicae. J.Invertebr.Pathol.* **38**, 273–280.

Picot, J.J.C. and Kristmanson, D.D. 1997. "Forestry Pesticide Aerial Spraying: Spray Droplet Generation, Dispersion, and Deposition." Kluwer Academic Publishers, Dordrecht, The Netherlands.

Raymond, B., Hartley, S.E., Cory, J.S. and Hails, R.S. 2002. Host plant species can influence the fitness of herbivore pathogens: the winter moth and its nucleopolyhedrovirus. *Oecologia* **131**, 533–541.

Raymond, B., Hartley, S.E., Cory, J.S. and Hails, R.S. 2005. The role of food plant and pathogen-induced behaviour in the persistence of a nucleopolyhedrovirus. *J. Invertebr. Pathol.* **88**, 49–57.

Reeson, A.F., Wilson, K., Gunn, A., Hails, R.S. and Goulson, D. 1998. Baculovirus resistance in the noctuid *Spodoptera exempta* is phenotypically plastic and responds to population density. *Proc. R. Soc. London. Ser. B* **265**, 1787–1791.

Reeson, A.F., Wilson, K., Cory, J.S., Hankard, P., Weekes, J.M., Goulson, D. and Hails, R.S. 2000. Phenotypic plasticity in a noctuid caterpillar affects pathogen transmission in the field. *Oecologia* **124**, 373–380.

Ribeiro, H.C.T., Pavan, O.H.O. and Muotri, A.R. 1997. Comparative susceptibility of two different hosts to genotypic variants of *Anticarsia gemmatalis* nuclear polyhedrosis virus. *Entomol. Exp. Appl.* **83**, 233–237.

Richards, A.R, Cory, J., Speight, M. and Williams, T. 1999. Foraging in a pathogen reservoir can lead to local host population extinction: a case study of Lepidoptera-virus interaction. *Oecologia.* **118**, 29–38.

Ridout, M.S., Fenlon, J.S. and Hughes, P.R. 1993. A generalized one-hit model for bioassays of insect viruses. *Biometrics* **49**, 1136–1141.

Room, P.M. 1979. A prototype 'on-line' system for management of cotton pests in the Namoi Valley, New South Wales. *Protect. Ecol.* **1**, 245–264.

Rothman, L.D. and Myers, J.H 1996. Debilitating effects of viral diseases on host Lepidoptera. *J.Invertebr.Pathol.* **61**, 1–10.

Sait, S.M, Begon, M., Thompson, D.J. and Harvey, J.A 1996. Parasitism of baculovirus-infected *Plodia interpunctella* by *Venturia canescens* and subsequent virus transmission. *Funct. Ecol.* **10**, 586–591.

Shapiro, M. 1992. Use of optical brighteners as radiation protectants for gypsy moth (Lepidoptera: Lymantriidae) nuclear polyhedrosis virus. *J. Econ. Entomol.* **85**, 1682–1686.

Shapiro, M. 1995. Radiation protection and activity enhancement of viruses. *In* "Biorational Pest Control Agents: Formulation and Delivery" (Hall, F.R. and

Barry, J.W., Eds.) pp. 153–164. American Chemical Society Symposium 595, Washington, DC.

Shapiro, M. and Dougherty, E.M. 1994. Enhancement in activity of homologous and heterologous viruses against the gypsy moth (Lepidoptera, Lymantriidae) by an optical brightener. *J. Econ. Entomol.* **87**, 361–365.

Sikorowski, P.P., Andrews, G.L. and Broome, J.R. 1973. Transovum transmission of a cytoplasmic polyhedrosis virus of *Heliothis virescens* (Lepidoptera: Noctuidae). *J.Invertebr.Pathol.* **21**, 41–45.

Simon, O., Williams, T., Lopez-Ferber, M and Caballero, P. 2004. Genetic structure of a *Spodoptera frugiperda* nuclopolyhedrovirs population: high prevalence of deletion genotypes. *Appl. Environ. Microbiol.* **70**, 5579–5588.

Small, D.A., Moore, N.F. and Entwistle, P.F. 1986. Hydrophobic interactions involved in attachment of a baculovirus to hydrophobic surfaces. *Environ. Microbiol.* **52**, 220–223.

Smirnoff, W.A. 1972. Promoting virus epizootics in populations of the Swaine Jack Pine sawfly by infected adults. *BioScience* **22**, 662–663.

Smith C.R., Heinz K.M., Sansone C.G. and Flexner J.L. 2000. Impact of recombinant Baculovirus applications on target heliothines and nontarget predators in cotton. *Biol. Control* **19**, 201–214.

Steinhaus, E.A. 1956. Microbial control. The emergence of an idea. A brief history of insect pathology through the nineteenth century. *Hilgardia* **26**, 107–160.

Sun X.L, Wang H.L., Sun X.C., Chen X.W., Peng C.M., Pan D.M., Jehle J.A., van der Werf W., Vlak J.M. and Hua Z. 2004. Biological activity and field efficacy of a genetically modified Helicoverpa armigera single-nucleocapsid nucleopolyhedrovirus expressing an insect-selective toxin from a chimeric promoter. *Biol. Control* **29**, 124–137.

Tamez-Guerra, P., McGuire, M.R., Behle, R.W., Hamm, J.J., Sumner, H.R. and Shasha, B.S. 2000a. Sunlight persistence and rainfastness of spray-dried formulations of baculovirus isolated from *Anagrapha falcifera* (Lepidoptera: Noctuidae). *J. Econ. Entomol.* **93**, 210–218.

Tamez-Guerra, P., McGuire, M.R., Behle, R.W., Shasha, B.S. and Pingel, R.L. 2000b. Storage stability of *Anagrapha falcifera* nucleopolyhedrovirus in spray-dried formulations. *J. Invertebr. Pathol.* **79**, 7–16.

Tatchell, G.M. 1981. The effects of a granulosis virus infection and temperature on the food consumption of *Pieris rapae* (Lep.: Pieridae). *Entomophaga* **26**, 291–299.

Taylor, L.R. 1984. Assessing and interpreting the spatial distributions of insect populations. *Annu. Rev.Entomol.* **29**, 321–357.

Thorpe, K.W., Ridgway, R.L. and Webb, R.E. 1992. A computerized data management and decision support system for gypsy moth management in suburban parks. *Computers Electronics Agric.* **6**, 333–345.

Vail, P.V., Hoffmann, D.F. and Tebbets, J.S. 1993. Autodissemination of *Plodia interpunctella* (Hubner) (Lepidoptera, Pyralidae) granulosis virus by healthy adults. *J. Stored Prod. Res.* **29**, 71–74.

Vasconcelos, S.D.V. 1996. Alternative routes for the horizontal transmission of a nucleopolyhedrovirus. *J.Invertebr. Pathol.* **68**, 269–274.

Vasconcelos S.D.V., Cory J.S., Wilson K.R., Sait S.M. and Hails R.S. 1996b. The effect of baculovirus infection on the mobility of *Mamestra brassicae* L. (Lepidoptera: Noctuidae) larvae at different developmental stages. *Biol. Control* **7**, 299–306.

Vasconcelos S.D.V., Williams T., Hails R.S. and Cory J.S. 1996a. Prey selection and baculovirus dissemination by carabid predators of Lepidoptera. *Ecol. Entomol.* **21**, 98–104.

Vilaplana, L., Wilson, K., Redman, E.M. and Cory, J.S. High levels of persistent, vertically-transmitted baculovirus infection in field populations of the migratory African armyworm, *Spodoptera exempta*. Mol. Ecol. *submitted.*

Washburn, J.O., Kirkpatrick, B.A., Haas-Stapleton, E. and Volkman, L.E. 1998. Evidence that the stilbene-derived optical brightener M2R enhances *Autographa californica* M nucleopolyhedrovirus infection of *Trichoplusia ni* and *Heliothis virescens* by preventing sloughing of infected midgut epithelial cells. *Biol. Control* **11**, 58–69.

Webb, R.E., Dill, N.H., McLaughlin, J.M., Kershaw, L.S., Podgwaite, J.D., Cook, S.P., Thorpe, K.W., Farrar, R.R., Ridgway, R.L., Fuester, R.W., Shapiro, M., Argauer, R.J., Venables, L. and White, G.B. 1996. Blankophor BBH as an enhancer of nuclear polyhedrosis virus in arborist treatments against the gypsy moth (Lepidoptera: Lymantriidae). *J. Econ. Entomol.* **89**, 957–962.

Webb, R.E., Shapiro, M., Podgwaite, J.D., Ridgway, R.L., Venables, L., White, G.B., Argauer, R.J., Cohen, D.L., Witcosky, J., Kester, K.M. and Thorpe, K.W. 1994. Effect of optical brighteners on the efficacy of gypsy moth (Lepidoptera, Lymantriidae) nuclear polyhedrosis virus in forest plots with high or low levels of natural virus. *J. Econ. Entomol.* **87**, 134–143.

Williams T. and Cory J. S. 1993. DNA restriction fragment polymorphism in iridovirus isolates from individual blackflies (Diptera: Simuliidae). *Med. Vet. Entomol.* **7**, 199–201.

Williams, T. Goulson, D., Caballero, P., Cisneros, J., Martinez, A.M., Chapman, J.W., Roman, D.X. and Cave, R.D. 1999. Evaluation of a baculovirus bioinsecticide for small-scale maize growers in Latin America *Biol. Control* **14**, 67–75.

Wilson, K. and Reeson, A.F. 1998. Density-dependent prophylaxis: evidence from Lepidoptera-baculovirus interactions. *Ecol. Entomol.* **23**, 100–101.

Wood, H.A., Hughes, P.R. and Shelton, A. 1994. Field studies of the co-occlusion strategy with a genetically altered isolate of the *Autographa californica* nuclear polyhedrosis virus. *Environ. Entomol.* **23**, 211–219.

Chapter IV-2

Bacteria

Stephen F. Garczynski
USDA-ARS
Yakima Agricultural Research Laboratory
Wapato, WA 98951, USA

Joel P. Siegel
USDA-ARS
San Joaquin Valley Agricultural
Sciences Center
Parlier, CA 93648, USA

1 Introduction

This chapter will focus on the major species of bacteria used as biopesticides to control insect pests. These insect pathogens mainly occur in the families Bacillaceae, Pseudomonadaceae, Enterobacteriaceae, Streptococcaceae, and Micrococcaceae (Tanada and Kaya, 1993). Today, the principal microbial insecticides utilize spore-forming bacteria (Bacillaceae) or toxins produced by these bacteria as their active ingredient, either in formulations or by incorporation of toxin genes with insecticidal properties into transgenic plants. With the exception of *Serratia entomophila* and *Paenibacillus* spp., the entomopathogenic bacteria most readily available in commercial formulations belong to the genus, *Bacillus*. Although *Paenibacillus* (= *Bacillus) popilliae* was the first microbial insecticide registered in the USA in 1948 for use against the Japanese beetle, *Popillia japonica* (Klein, 1992), its importance was eclipsed by the commercialization of *Bacillus thuringiensis (Bt)*, whose host spectrum includes species in the orders Coleoptera, Diptera, Hymenoptera and Lepidoptera. *Bacillus sphaericus* (VectoLex®) was commercialized in the late 1990s by Abbott Laboratories (North Chicago, IL) and is currently being marketed by Valent BioSciences Corporation (Libertyville, IL) for use in mosquito control programs.

The relative commercial importance of *Bt* compared to the other species is illustrated by examining the number of bacterial pesticides registered with the United States Environmental Protection Agency in 1995. There were 249 products registered, of which 182 (73%) contained *Bt* spores and toxins or only *Bt* toxins, two products contained *P. popilliae* or *P. lentimorbus* and one product contained *B. sphaericus*. Currently (as of 2006), there are 361 products that contain Bt spores and toxins or toxins alone, three products containing *B. sphaericus* and only one containing *P. popilliae*. The relative importance of each species can be further exemplified by performing Internet searches. "Google" searches for each species yielded over 2,100,000 hits for *Bt*, 678,000 hits for *B. sphaericus*, and 32,900 hits for *Paenibacillus* or *Bacillus popilliae*. In 1993, *Bt* insecticides produced in North America and Western Europe were valued at \$100 million to \$150 million/annum (Bernhard and Utz, 1993); the value of transgenic crops incorporating *Bt* toxins is substantially greater and as of 2004 there were 22.5 million hectares of insect resistant crops planted worldwide (James, 2004). The primary focus of this chapter will be on *Bt*,

L.A. Lacey and H.K. Kaya (eds.), Field Manual of Techniques in Invertebrate Pathology, 175–197.

which is also covered extensively in several chapters in Section VII and transgenic plants are discussed in Chapter VIII-1. Further information on *P. popilliae* and *B. sphaericus*, respectively, is found in Chapters VII-18 and VII-22.

2 *Bacillus thuringiensis*

A *Identification and taxonomy*

In 1901, S. Ishiwata isolated a bacillus from infected *Bombyx mori* larvae. Ishiwata described the pathology of his isolated bacillus in *B. mori* and suggested that the disease was caused by a toxin because insect death occurred prior to bacterial multiplication. Ten years later, E. Berliner described a bacillus isolated from infected *Anagasta kuehniella* larvae and named the bacterium *Bacillus thuringiensis*. Berliner is credited for naming *B. thuringiensis* because Ishiwata did not provide an adequate description of the bacterium he isolated (reviewed in Beegle and Yamamoto, 1992; Federici 2005). *Bacillus thuringiensis* (*Bt*) is a Gram-positive, rod-shaped, aerobic, endospore-forming bacterium closely related to *B. cereus*. Aside from the production of parasporal crystals, it is difficult to distinguish *Bt* from *B. cereus*, and their distinction as two separate species has been questioned (Gordon *et al.*, 1973; Rasko *et al.*, 2005).

Beegle and Yamamoto (1992) summarized the history of commercial development of *Bt*, which began with Sporeine, produced in France during the late 1930s by Laboratoire Libec, and ending with the numerous *Bt* products available today. These products are primarily based on four subspecies and their trade names include Bactimos®, Biobit®, Bitayon®, Condor®, Dipcl®, Foil®, Foray®, Javelin®, Novodor®, Teknar®, Thuricide®, Vectobac®, and XenTari®. Heimpel and Angus (1958) introduced a classification scheme to identify *Bt* isolates based on morphological and biochemical characterization, but this scheme was superceded by the serological analysis of flagellar (H) antigens (de Barjac and Bonnefoi, 1962). This system served as the basis for designating serovars or subspecies, and by 1996, the collection of the Pasteur Institute contained 63 *Bt* serovars. Thiery and Frachon (1997) summarized the protocol for flagellar typing and also presented an overview of other bacteriological techniques used for characterizing entomopathogenic bacteria. A current listing of the 69 *Bt* subspecies and 82 serovars contained in the Pasteur Institute collection can be found in Lecadet *et al.* (1999).

It was Berliner who first noted the production of crystals by *Bt* (Figure 1). Hannay (1953) reported that pathogenic strains of *Bt* produced diamond-shaped crystals during spore formation and proposed that the alkali soluble crystals play some role in pathogenicity. Less than a year later, the parasporal crystal was shown to be the agent responsible for the toxicity of *Bt* toward *B. mori* larvae, and toxicity varied with crystal count (Angus, 1954). The parasporal crystals of different strains of most *Bt* subspecies contain varying combinations of insecticidal crystal proteins (ICPs). For example, some isolates of *Bt* subsp. *morrisoni* produce crystals containing ICPs toxic to Coleoptera while other isolates produce ICPs toxic to Diptera or Lepidoptera. Crystals can take various shapes (bipyramidal, cuboidal, flat rhomboid, round, amorphous) depending on their ICP composition. The crystal depicted in Figure 1 is round. Some researchers reported that crystal morphology is

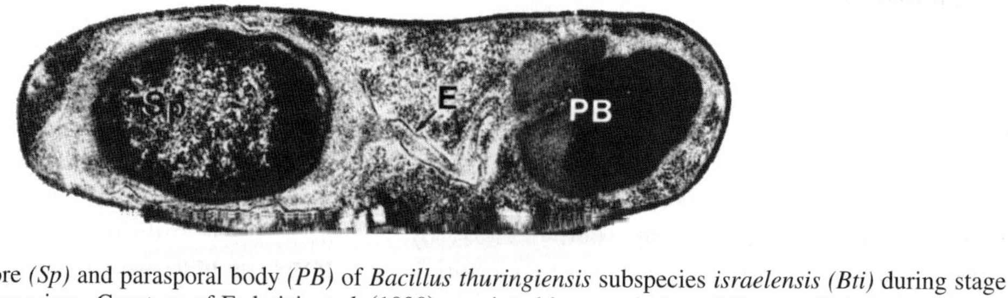

Figure 1. Developing spore *(Sp)* and parasporal body *(PB)* of *Bacillus thuringiensis* subspecies *israelensis (Bti)* during stage 5 of sporulation; (*E*), exosporium. Courtesy of Federici *et al.* (1990), reprinted by permission of Rutgers University Press

correlated with activity against target insects (Bulla *et al.*, 1977; Höfte and Whiteley, 1989), but as new subspecies were discovered this correlation broke down.

Genes encoding ICPs were identified in the early 1980s and are usually located on large, transmissible plasmids that can be cured and exchanged during conjugation (Gonzalez and Carlton, 1980; Gonzalez *et al.*, 1981). The ICP gene sequences enabled researchers to construct gene-specific probes to screen established *Bt* strains by hybridization analysis for the presence of known nucleotide sequences and to characterize novel ICPs from new and existing *Bt* isolates (Adang *et al.*, 1985; Kronstad and Whiteley, 1986; Höfte *et al.*, 1988; Visser *et al.*, 1990). By 1989, the nucleotide sequences of 42 crystal protein genes had been reported (Höfte and Whiteley, 1989).

Based on structural similarities and insecticidal spectra, a classification system for crystal protein genes cloned from *Bt* was developed (Höfte and Whiteley 1989). From their analysis, 14 distinct genes falling into two families were found among the 42 reported nucleotide sequences. The *cry* (crystal protein) genes encode a family of Cry proteins which were put into four classes and several subclasses. The CryI proteins are 130–140 kDa, sharing > 50% amino acid identity and are toxic to lepidopteran larvae; CryII proteins are 65 kDa and show insecticidal activity to both lepidopteran and dipteran larvae; CryIII proteins are 72 kDa and have coleopteran-specific activity; CryIV proteins are toxic to dipteran larvae. The *cyt* (cytolytic crystal protein) gene family had one member encoding the 27 kDa CytA protein displaying dipteran and general cytolytic activity (Höfte and Whiteley, 1989). Mycogen Corporation (now Dow Agro Sciences) patented a *Bt* isolate that produced ICPs toxic to nematodes (Narva *et al.*, 1991), which expanded the target range of Cry proteins beyond insects (Cry VI).

The nomenclature of Cry genes was revised in 1995 in order to accommodate the ever growing list of *cry* genes. Crickmore *et al.* (1998) proposed a revision of the nomenclature for *B. thuringiensis* Cry proteins based solely on sequence similarities, foregoing pesticidal activity. This new classification system is designed to handle the increase of reported gene sequences that encode *B. thuringiensis* crystal proteins. In the proposed revision, Cry and Cyt are kept as mnemonic roots, Arabic numerals are used for a primary ranking (replacing the Roman numerals above), and letters are assigned to indicate the degrees of phylogenetic divergence.

Possession of Cry genes is not unique to *Bt*. Several other bacterial species including *B. sphaericus, P. popilliae, P. lentimorbus* and *Clostridium bifermentans* produce Cry proteins (Barloy *et al.*, 1998; Zhang *et al.*, 1997; Yokoyama *et al.*, 2004). Currently, 145 holotype crystal protein genes have been assigned to 50 sets at the primary rank (*cry*1–50) and 9 *cyt* genes have two primary ranks (*cyt*1–2). In addition to the ICPs, 21 holotype vegetative insecticidal protein (*vip*) genes with three primary ranks (*vip*1–3) have been classified. A list of the holotype toxins using the revised nomenclature is presented in Table 1.

Some *Bt* subspecies produce other types of toxins in addition to Cry proteins. The toxin that has been most studied is the β-exotoxin, which is a heat stable nucleotide composed of adenine, glucose, and allaric acid (Farkas *et al.*, 1976). β-exotoxin inhibits RNA polymerase enzymes by competing with ATP; consequently, it is toxic to almost all insect species as well as vertebrates. *Bt* subspecies containing strains that produce β-exotoxin include *darmstadiensis, galleriae, morrisoni, thuringiensis* and *tolworthi*; its presence in *Bt* products is banned in the USA (Cantwell *et al.*, 1964; Mohd-Salleh *et al.*, 1980). Additional toxins produced by some *Bt* subspecies include α-toxin, which is a phospholipase C and affects cell membrane phospholipids, γ-toxin which is heat labile and affects Tenthredinidae (sawflies), a water soluble toxin which paralyzes Lepidoptera and a "mouse factor" toxin that affects mice and Lepidoptera (Heimpel, 1967; Fast, 1971; Krieg, 1971).

B Cry toxin mode of action

The Cry proteins that compose the ICPs are responsible for the insecticidal activity of *Bt* formulations. Commercial insecticides contain ICPs or ICP-spore complexes; the spore wall may contain Cry toxins and in some insects,

Table 1. The current 154 Holotype crystal toxins, 9 cytolytic toxins, and 21 vegetative insecticidal proteins of *Bacillus thuringiensis, B. sphaericus, B. cereus, Brevibacillus laterosporus, Clostridium bifermentans, Paenibacillus popilliae*, and *P. lentimorbus*[1]

CRYSTAL PROTEIN TOXINS
Lepidoptera Activity:
Cry1A(a-i), Cry1B(a-g), Cry1C(a-b), Cry1D(a-b), Cry1E(a-b), Cry1F(a-b), Cry1G(a-c), Cry1H(a-b), Cry1I(a-f), Cry1J(a-d), Cry1Ka, Cry1La, Cry2A(a-e), Cry9Aa, Cry9B(a-b), Cry9Ca, Cry9E(a-d), Cry15Aa, Cry26Aa, Cry28Aa, Cry32Aa

Diptera Activity:
Cry1Ca, Cry2A(a-e), Cry4Aa, Cry4Ba, Cry10Aa, Cry11Aa, Cry11B(a-b), Cry16Aa, Cry17Aa, Cry19Aa, Cry19Ba, Cry24Aa, Cry24Ba, Cry26Aa, Cry29Aa, Cry30Aa, Cry39Aa, Cry40Aa, Cry40Ba, Cry44A, Cry47Aa, Cry48Aa, Cry49Aa, Cry50Aa

Coleoptera Activity:
Cry1I(a-f), Cry3Aa, Cry3B(a-b), Cry3Ca, Cry7A(a-b), Cry8Aa, Cry8B(a-c), Cry8Ca, Cry8Da, Cry8Ea, Cry8Fa, Cry8Ga, Cry9D(a-b), Cry18Aa, Cry18Ba, Cry18Ca, Cry22Ab, Cry22Ba, Cry23Aa, Cry34Aa/Cry35Aa, Cry34Ab/Cry35Ab, Cry34Ac/Cry35Ac, Cry34Ba/Cry35Ba, Cry36Aa, Cry 37Aa, Cry38Aa, Cry43Aa, Cry43Ba

Hymenoptera Activity:
Cry22Aa

Nematoda Activity:
Cry5A(a-c), Cry5Ba, Cry6Aa, Cry6Ba, Cry12Aa, Cry13Aa, Cry14Aa, Cry21Aa, Cry21Ba

Human Cancer Cell Line Activity:
Cry31A(a-b), Cry41A(a-b), Cry42Aa, Cry45Aa, Cry46A(a-b)

CYTOLYTIC TOXINS:
Cyt1Aa, Cyt1Ab, Cyt1Ba, Cyt1Ca, Cyt2Aa, Cyt2Ba, Cyt2Bb, Cyt2Bc, Cyt2Ca

VEGETATIVE INSECTICIDAL PROTEINS:
Vip1A(a-b), Vip1B(a-b), Vip1Ca, Vip1Da, Vip2A(a-d), Vip2B(a-b), Vip3A(a-g), Vip3B(a-b)

[1] Crickmore, N., Zeigler, D. R., Feitelson, J., Schnepf, E., Van Rie, J., Lereclus, D., Baum, J. and Dean, D. H. "*Bacillus thuringiensis* toxin nomenclature" (2005). Accessed 5/8/06. http://www.lifesci.sussex.ac.uk/Home/Neil_Crickmore/Bt/

spores synergize the activity of the ICPs (Du and Nickerson, 1996). The site of action for *Bt* Cry proteins is the insect midgut (Heimpel and Angus, 1959); therefore, the ICPs must be ingested by the target insect pest in order to be effective. Once ingested, the midgut provides an environment suitable for solubilizing the ICPs releasing Cry protoxins. Serine proteases present in the insect midgut cleave the protoxin to form a stable toxin core. This activated toxin core passes through the peritrophic membrane, then binds to sites located on the apical membrane of the columnar epithelial cells lining the insect midgut. Once bound, the toxin inserts into the cell membrane forming pores that cause ion imbalances leading to osmotic lysis of the midgut epithelial cells (reviewed in English and Slatin, 1992; Gill *et al*., 1992; Rajamohan *et al*., 1998; Schnepf *et al*., 1998). Damage to the midgut epithelium is associated with the cessation of feeding and gut paralysis observed in many insect species. Midgut lesions caused by the toxins enable the hemolymph and gut contents to mix, which in turn results in a septicemia that may contribute to or cause the death of the insect (Angus, 1954). The mixing of the hemolymph with the gut contents provides an environment suitable for spore germination. The Cry and Cyt toxins of *Bti* are among the most potent of the *Bt* toxins on a ng/ng basis; behavioral changes can be observed in susceptible mosquito larvae within 5 min after ingesting these toxins (Federici *et al*., 1990).

1 Cry toxin structure

Resolution of the three-dimensional structures for the Cry3Aa (Li *et al*., 1991) and Cry1Aa (Grochulski *et al*., 1995) toxins shows that these globular proteins are made of three distinct domains. Domain I is composed of seven anti-parallel α-helices, with the hydrophobic helix α5 surrounded by the other six amphipathic helices. Domain I extends from the amino (N)-terminus to amino acid 290 for the Cry3A toxin (Li *et al*., 1991) and for Cry1Aa extends from amino acids 33–253 (Grochulski *et al.*, 1995). Domain II is composed of three anti-parallel β-sheets with the first two sheets formed by four strands each, connected by the "Greek key" topology (Grochulski *et al.*, 1995). For Cry3A, domain II stretches from amino acids 291–500, and residues 265–461 for Cry1Aa. Domain III is two anti-parallel β-sheets that are highly twisted forming a β-sandwich. This domain is composed of amino acids 501–644 for Cry3A, and residues 463–609 for Cry1Aa.

Höfte and Whiteley (1989) analyzed the deduced amino acid sequences from the cloned genes encoding Cry proteins, and they identified five conserved amino acid blocks for crystal proteins encoded by 10 of the then known 13 *cry* genes. Schnepf *et al.* (1998) extended the comparison of deduced amino acid sequences and found that the five conserved amino acid blocks are present in Cry1, Cry3, Cry4, Cry7, Cry8, Cry9, Cry10, Cry16, Cry17, Cry19 and Cry20 proteins. The five conserved amino acid blocks are either centrally located within domains or at the junctions between domains. The location of the conserved regions implied that homologous Cry proteins would fold in a similar manner (Li *et al.*, 1991). Indeed this was the case for Cry1Aa, and it is proposed that Cry proteins with the five conserved amino acid blocks will have similar structure (Rajamohan *et al.*, 1998; Schnepf *et al.*, 1998).

Proposed functions for each of the structural domains determined for Cry3A were deduced by Li *et al.* (1991). Domain I is proposed to function in membrane insertion and pore formation, domain II in receptor binding, and domain III providing stability to the overall structure (Li *et al.*, 1991). Functions for each of the domains have been assessed by measuring binding and pore formation after introducing mutations in different regions of Cry toxins. Recent reviews discuss the literature on Cry toxin mutagenesis and domain swapping to probe the function of each domain (Dean *et al.*, 1996, Rajamohan *et al.*, 1998, Schnepf *et al.*, 1998; de Maagd *et al.*, 2001). Domain I is still proposed to be involved with membrane insertion and pore formation, domain II with receptor binding, and domain III with ion channel function, receptor binding and membrane insertion (de Maagd *et al.*, 2003).

2 Cry toxin binding

Once ingested and activated, Cry toxins bind to receptor sites located on the surface of the midgut epithelial cells. The term receptor has been used to refer to a molecule (or molecules) on the surface of an insect cell to which a Cry toxin binds (Knowles and Dow, 1993). This is a rather vague definition and does not imply any physiological function associated with these to cell surface molecules or indicate an active role in Cry toxin action. Receptors for Cry proteins have been identified in insects and nematodes as proteins or glycolipids and recently reviewed (Griffitts and Aroian, 2005).

Cry1 toxins bind to the apical membrane of columnar epithelial cells in the insect midgut. Preparations of brush border membrane vesicles (BBMV) from the midguts of lepidopteran larvae provide an *in vitro* representation of the apical microvillar membrane (Wolfersberger *et al.*, 1987). The use of BBMV has facilitated study of the molecular interactions between Cry1 toxins and insect midgut epithelial cells. Hofmann *et al.* (1988a,b) were first to use BBMV in competitive binding assays to characterize the interaction of Cry1 toxins with its target site from the apical membrane. They determined that Cry1 toxins bind saturably and with high affinity to BBMV isolated from the midguts of *Pieris brassicae* and *Manduca sexta* larvae, and toxin binding was correlated with insecticidal activity (Hofmann *et al.*, 1988a, b). Binding studies have been done to assess the interaction of several Cry1 toxins with BBMV prepared from a variety of lepidopteran larvae, and specific binding sites with affinity constants in the nM range have been reported (Van Rie *et al.*, 1989; Van Rie *et al.*, 1990a; Wolfersberger, 1990; Ferré *et al.*, 1991; Garczynski *et al.*, 1991; Denolf *et al.*, 1993; Sanchis and Ellar, 1993). Cry1 toxin binding to high-affinity sites present on BBMV has been positively correlated with insecticidal activity (Hofmann *et al.*, 1988a, b; Van Rie *et al.*, 1989; Van Rie *et al.*, 1990a; Garczynski *et al.*, 1991; Denolf *et al.*, 1993). However, the presence of binding sites does not ensure *in vivo* toxicity (Wolfersberger, 1990; Ferré *et al.*, 1991; Garczynski *et al.*, 1991; Sanchis and Ellar, 1993; Luo *et al.*, 1999), suggesting that post-binding events (membrane insertion and pore formation; see section 3 below) are important for Cry1 toxicity. For example, Cry1Ac is not toxic to *Spodoptera frugiperda* larvae yet ^{125}I-Cry1Ac bound to BBMV prepared from the midguts isolated from this insect (Garczynski *et al.*, 1991; Luo *et al.*, 1999).

There are two phases associated with Cry1 binding to BBMV, consisting of reversible (Hofmann *et al.*, 1988a, b) and irreversible steps (Van Rie *et al.*, 1989; Ihara *et al.*, 1993; Chen *et al.*, 1995; Liang *et al.*, 1995 Rajamohan *et al.*, 1995). Ihara *et al.* (1993) were first to correlate irreversible binding with the insecticidal activity of Cry1A toxins. This correlation is consistently seen with Cry1A toxins (Chen *et al.*, 1995; Liang *et al.*, 1995; Rajamohan *et al.*, 1995). It should be noted that the decrease in irreversible binding of mutant Cry toxins is not correlated with changes in high-affinity binding (Chen *et al.*, 1995; Liang *et al.*, 1995), possibly suggesting two separate classes of receptor molecules. Reduced pore function was observed for mutant toxins with decreased irreversible binding (Chen *et al.*, 1995), which may be due to a "severely disturbed membrane insertion process."

3 *Cry toxin-induced pores*

Knowles and Ellar (1987) showed that Cry toxins affect membrane permeability in insect cell lines, and proposed that Cry toxins kill cells by forming non-specific pores leading to cell death by colloid osmotic lysis. Models for membrane insertion and disruption of ion gradients and osmotic balances of the midgut epithelial cells have been reviewed (English and Slatin, 1992; Knowles and Dow, 1993; Knowles, 1994). Cry1 toxin pores have been characterized in planar phospholipid bilayers (Slatin *et al.*, 1990; Schwartz *et al.*, 1993; English *et al.*, 1995), insect cell lines (Schwartz *et al.*, 1991), hybrid phospholipid/BBMV (English *et al.*, 1991; Wolfersberger, 1995), and BBMV (Hendrickx *et al.*, 1990; Carroll and Ellar, 1993; Wolfersberger, 1995; Escriche *et al.*, 1998). Mutagenesis experiments assessing the role of the different domains of Cry proteins in membrane insertion and pore formation function have been reviewed (Rajamohan *et al.*, 1998; Schnepf *et al.*, 1998).

Cry1 toxins insert into artificial phospholipid bilayers at high doses ($>1\,\mu g/ml$), creating pores that are permeable to ions and small molecules (Haider and Ellar, 1989; Slatin *et al.*, 1990; English *et al.*, 1991). The concentration of Cry1Ac needed to create pores is drastically reduced when BBMV are incorporated into phospholipid bilayers (English *et al.*, 1991; Martin and Wolfersberger, 1995). Carroll and Ellar (1993) used a scattered light assay to show the specificity of Cry1Ac pore formation in BBMV isolated from the midguts of *M. sexta* larvae. Pore formation is also correlated with Cry1 toxicity in *Spodoptera exigua* and *S. frugiperda* (Luo *et al.*, 1999). Because pore formation has been correlated with toxicity, assays monitoring this post-binding event have been used to determine functionality of Cry1 toxin receptors.

4 *Insect resistance to* Bt *formulations*

Resistance to a spore-crystal formulation of *Bt* was first reported for the Indianmeal moth, *Plodia interpunctella* (McGaughey, 1985). It was determined that the mechanism of this resistance was correlated with a reduction of toxin affinity for its binding site present on BBMV prepared from susceptible and resistant laboratory colonies of *P. interpunctella* (Van Rie *et al.*, 1990b). In most cases, resistance has been associated with a recessive or partially recessive trait(s) and appears to be linked to a single gene (Boucias and Pendland, 1998).

The importance of binding sites in the insect midgut is best illustrated in the diamondback moth, *P. xylostella*, which has acquired resistance to Cry1 toxins. *P. xylostella* was the first insect to develop field resistance to *Bt* toxins. A reduced number binding sites was found in a Cry1A resistant population of *P. xylostella* (Ferré *et al.*, 1991). In further studies with a different population of *P. xylostella* larvae resistant to the Cry1A toxins, Tabashnik *et al.* (1994) found no binding of Cry1Ac and that the resistance was reversible. The reversal of resistance was correlated with the return of Cry1 binding sites. Cry1Ac binds with high-affinity to BBMV isolated from susceptible and revertant strains of *P. xylostella*, while no binding was detected for resistant larvae (Tabashnik *et al.*, 1994). The Cry1Ac binding protein in *P. xylostella* was identified as aminopeptidase N (Luo *et al.*, 1997).

Analysis of BBMV isolated from susceptible and resistant strains of *P. xylostella* revealed that aminopeptidase N activity was the same, and there was no difference in Cry1Ac binding on ligand blots (Luo *et al.*, 1997). These results do not support aminopeptidase as a determinant of toxicity, and possible explanations as to how *P. xylostella* larvae could attain Cry1Ac resistance with a receptor present in BBMV were offered (Luo *et al.*, 1997). Potential mechanisms have been proposed for the presence of binding proteins on ligand blots, but not in soluble assays. These include epitope masking (Lee *et al.*, 1995) and loss of binding based on GPI anchorage (Luo *et al.* 1997). In epitope masking, toxin binding to receptor proteins may be blocked by other midgut molecules that are not recognized by Cry toxins (Lee *et al.*, 1995). Luo *et al.*, (1997) suggest that it is also possible that an endogenous phosphoinositol phospholipase C releases aminopeptidase from the apical membrane in resistant strains prior to Cry1Ac toxin pore formation. One obvious possibility that was not addressed is that other BBMV molecules may also serve as toxin binding determinants. Other mechanisms of invertebrate resistance to *Bt* toxins has recently been reviewed (Ferré and Van Rie, 2002; Griffitts and Aroian, 2005)

C *Natural occurrence of* Bt

Bt is widespread in nature and readily persists in soil and sediment. Early research concentrated on its persistence in soil. DeLucca *et al.* (1981) reported that *Bt* represented between 0.5 and 0.005% of *Bacillus* species isolated from soil samples in the USA. Martin and Travers (1989) recovered *Bt* from soils in the USA and from 29 other countries. Meadows (1993) isolated *Bt* from 785 of 1,115 soil samples that were collected globally and the percentage of samples containing *Bt* ranged from 56% to 94% in samples from Asia, and central and southern Africa. Ohba and Aizawa (1986) isolated *Bt* from 136 of 189 soil samples in Japan.

In the 1990s, *Bt* was found to be an inhabitant of the phylloplane (Smith and Couche, 1991). Numerous *Bt* subspecies have been recovered from coniferous and deciduous trees as well as other plants. *Bt* isolates recovered from these habitats possess diverse Cry toxins (Hansen *et al.*, 1998).

Many *Bt* subspecies have also been isolated from dead or moribund insect larvae, and as many as 9.2×10^7 spores/cadaver have been recovered from the Egyptian cotton leafworm *(Spodoptera littoralis)* (Jarrett and Stephenson, 1990; Meadows, 1993). Upon disintegration of cadavers, spores and ICPs are released into the environment (Prasertphon *et al.*, 1974; Aly, 1985; Akiba, 1986). Infected larvae can also excrete spores in their frass, and *Bt* spores have also been associated with insect eggs in the laboratory and in the field (Lynch *et al.*, 1976; Burges and Hurst, 1977; Siegel *et al.*, 2000). One of the mysteries surrounding *Bt* is the ability of researchers to isolate this organism from soils in habitats devoid of insects, such as high mountains in the Himalayas (Martin, 1994). It is possible that *Bt* is ubiquitous because its spores are distributed in the soil and phylloplane by living insects and spores are airborne by wind and distributed throughout the environment.

There is little *Bt* multiplication while the insect is alive but considerable proliferation can occur in insect cadavers as noted above. Laboratory studies have demonstrated that *Bt* subspecies can recycle to varying degrees (Knell *et al.*, 1998) and *Bt* can cause problems in laboratory colonies (Itoua-Apoyolo *et al.*, 1995; Federici, 1999; Siegel *et al.*, 2000). *Bt* has been isolated from stored grain products and epizootics have been observed in grain silos (Burges and Hurst, 1977; Vankova and Purrini, 1979; Meadows *et al.* 1992). The best-documented *Bt* epizootic in the field led to the discovery of *Bti* (Goldberg and Margalit, 1977).

D *Industrial production of* Bt *and* B. sphaericus

All commercial *Bt* and *B. sphaericus* productions originate with a "mother" culture that is used to start production batches of bacteria in liquid media in a submerged fermentation, each ending in harvest and storage. The mother culture usually is maintained as a group of lyophilized spores in vials or powders stored under controlled conditions. Registration of a product is based on this material. Periodically, a single vial is used to inoculate a number of starter agar slants or agar plates, that are incubated and after sufficient bacterial multiplication has occurred, they are then refrigerated for a limited period of time. These are the actual starters used to initiate fermentation runs. A typical run begins when a 500 ml flask of medium is inoculated with bacteria from a starter; cell densities as great as 10^{10} cells/ml may be achieved (Bernard and Utz, 1993). This flask is used to inoculate a larger amount of medium, typically in a 20-liter fermentor. After 6–8 h this small fermentor-load is in the late log-phase and is transferred to a 200-liter fermentor. Growth in the fermentor is monitored, and this product is used to inoculate still larger fermentors; fermentors as large as 100,000 liters have been used to produce *Bt* and *B. sphaericus* (Bernard and Utz, 1993).

Paddle blades on a vertical rotating shaft and static baffles on the inner walls of the fermentor agitate the culture medium. Large engines are used to turn the paddles. The viscosity of the culture medium gradually increases because of sporulation and evaporation; this determines the size of engine necessary to turn the paddles. Aeration is achieved using filter-sterilized air delivered through perforated pipes. Various steps are taken to avoid foaming during fermentation. Oxygen concentration and pH are monitored constantly using probes; entry of the bacteria into the stationary growth phase is marked by a gradual increase in oxygen level and pH (Bernard and Utz, 1993). The details on the composition of media used (protein sources include cotton seed flour, soy bean powder, pig kidneys and molasses may be used as a source of sugar), step-up fermentation sequence, temperature maintained during fermentation, oxygen and pH levels that trigger the decision to terminate a fermentation run, and the length of the fermentation run are proprietary.

At the end of the fermentation run, the medium contains a mixture of crystals, spores, medium solids, and water. The medium is concentrated by centrifugation, or filtration through bags and the slurry that results contains approximately 80% water (Burges and Jones, 1998). This technical fermentor slurry may be stabilized as a suspension concentrate or by spray drying. This technical material is bioassayed and its potency is standardized before it is formulated and marketed.

E Product standardization

In the early years, standardization of commercial products was problematic because it was based on spore counts; there is only a poor relationship between the number of spores in a product and insect toxicity. According to Beegle and Yamamoto (1992), the person most responsible for standardization in Europe was Dr. H. D. Burges and in North America it was Drs. H. Dulmage and A. Heimpel. Currently, the final formulation of a *Bt* product is bioassayed against an accepted standard using a specific test insect (Dulmage *et al.*, 1981). Typically, the manufacturer uses an internal standard that is periodically calibrated against an International Standard that has an assigned potency. The product potency is expressed in International Toxicity Units (ITU)/mg product. It is crucial to preserve the toxicity of whatever the International or internal standard that is used to determine the potency of a production lot. This is illustrated by the fate of the 1980 USA Standard (11–114-BD) for *Bt* subsp. *kurstaki (Btk).* After its production, the *Btk* standard was shipped to several locations before a final storage location was designated and during this process the standard was not kept refrigerated. This led to a loss of 2,000 units of potency (Beegle and Yamamoto, 1992). Some in the *Bt* industry nicknamed this standard the "vagabond" standard, because it wandered to several depots over a period of months before reaching its final destination. This nickname was prophetic because researchers subsequently discovered that in the mother culture, one of the complement of four Cry genes was lost prior to the fermentation of the standard (Wilcox *et al.*, 1986). Not only was this standard less potent than its label stated because of mishandling, but it also had a reduced spectrum of susceptible insect species.

Considerable effort has also been spent standardizing bioassay procedures because both intra-laboratory and inter-laboratory variation can affect the determination of potency (Skovmand *et al.*, 1998). In the USA, a collaborative effort between the United States Department of Agriculture (USDA) and *Bt* producers resulted in a standardized lepidopteran assay with a designated target insect species. There are two standardized assays for assessing dipteran activity using *Aedes aegypti*; one assay was developed by the World Health Organization (WHO) and the other is the USA standard bioassay (Beegle and Yamamoto, 1992). Knowledge of test insect feeding preferences, target instar, assay incubation temperature, length of assay, crowding limits of the test insect and stability of the test preparation is essential in order to conduct valid bioassays. For example, bioassays of *Bt* conducted with black fly larvae pose special challenges because the larvae require running water.

To determine the LD_{50}, a range-finding assay must be conducted first. Once the range is determined, six-seven dilutions should be assayed. Ideally, two dilutions above and below the LD_{50} should be used. At a minimum, three replicates conducted on different days and using different cohorts of larvae are necessary; the number of replicates may need to be increased if there is considerable control mortality or variation. There are a number of software programs available that utilize probit analysis to analyze mortality data, such as POLO, SAS, and SPSS. The issues described above are more extensively discussed and procedures for conducting *Bt* bioassays for Diptera and Lepidoptera are summarized by Lacey (1997) and McGuire *et al.* (1997).

F *Environmental fate of* Bt *and factors affecting persistence*

Both environmental factors and the site of application influence the distribution, transport, and degradation of *Bt* formulations. The diverse habitats to which commercial products must be applied pose different challenges with regard to field stability, and reapplication at varying intervals may be necessary to achieve insect control (Bulla *et al.*, 1985; Andrews *et al.*, 1987; Mulla, 1990). Regardless of habitat, it is important to distinguish between changes in the toxicity of ICPs and in the numbers of viable *Bt* spores over time when assessing the environmental degradation of a *Bt* insecticide. Formulations for coleopteran and lepidopteran pests are used in agriculture and forestry and are typically directed toward plant surfaces, whereas formulations for dipteran pests (mosquitoes and black flies) are applied to aquatic habitats. Approximately half of the current *Bt* larvicides on the market are oil-in-water emulsions (Burges and Jones, 1998). Formulations that are applied in terrestrial habitats have ranged from ultra-low volume oil to high-volume, wettable powder aqueous suspensions and water dispersible granules. *Bti* and *B. sphaericus* formulations applied to water include suspension concentrates, wettable powders, slow release briquette and corncob grit formulations (Burges and Jones, 1998; Lacey, 2007).

1 *Delivery*

Many physical factors can affect the stability of the ICPs of *Bt* products before they reach their target. According to Burges and Jones (1998), tank mixed *Bt* must not stand for long periods because the ICPs can deteriorate or settle in the bottom of the tank. Various additives, as well as continuously stirring the tank, can slow settling. Cold and pulse-jet thermal foggers can also be used to dispense *Bt*. Whatever the method, droplet size is critical because smaller droplets may partially or completely evaporate after they leave the sprayer. If droplet size is reduced too much, there is increased spray drift as well as other problems; Burges and Jones (1998) state that the optimal droplet size is 40–100 μm. The rate of evaporation may be slowed by adding humectants and anti-evaporants to the mixture, but some solvents may damage ICPs or cause clumping that can speed the degradation of the ICPs due to increased exposure to UV radiation. The nature of the carrier can affect deposition of the ICPs. For example, oil droplets impinge onto leaves better than water droplets, possibly because partial evaporation makes water droplets more prone to bounce off leaf surfaces. Finally, ICPs can be lost during droplet formation or destroyed by shear forces during atomization.

2 *Solar radiation*

Solar radiation is the most devastating environmental factor that influences the stability of *Bt* ICPs and spores on plant surfaces. The tryptophan residues of ICPs are damaged by radiation in the 300–380 nm range and spores are rapidly inactivated by UV radiation (Pinnock *et al.*, 1977; Griego and Spence, 1978; Pusztai *et al.*, 1991). The impact of solar radiation is extensively reviewed in Burges and Jones (1998). The combined effect of UV, leaf temperature and vapor pressure deficit contribute more to decreased biological activity than any single factor (Leong *et al.*, 1980). Ignoffo (1992) summarized data for the reduction of ICP activity and spore viability on leaves of various plants in sunlight. ICP and spore viability were reduced 8–80% in one day, but β-exotoxin activity declined more slowly. The use of UV protectants can extend residual activity out to 2 weeks post treatment (Hostetter *et al.*, 1975; Burges and Jones, 1998). Unformulated *Bt* preparations on foliage can have a half-life of only a few hours, but formulated products can have a half-life of 10 days (Dent, 1993).

3 *Leaf surface*

Spore survival can be affected by the surface to which the material is applied; *Bt* spores on the leaves of California live oak (*Quercus agrifolia*) have a 3.9 day half-life compared to a 0.63 days half-life on western redbud (*Cercis occidentalis*) (Pinnock *et al.*, 1974). Lüthy (1986) demonstrated that tannins could inactivate ICPs and that cotton leaf extracts reduced ICP activity more than spruce needle extracts. Additionally, both highly acidic (pH 1) and strongly basic (pH 11) plant surfaces can inactivate ICPs (Burges and Jones, 1998). In contrast, there have also been studies documenting that *Bt* persisted on leaves at low levels for considerable lengths of time after formulated products were sprayed. Reardon and Hassig (1983) reported *Btk* spores were still present at unquantified levels on balsam fir (*Abies balsamea*) one year after application for control of spruce budworm, *Choristoneura fumiferana*.

4 *Spore survival in soil*

Bt applied as a foliar insecticide can enter the soil because of overspray or physical removal by rain. Within the soil, several abiotic and biotic factors play a role in its persistence, movement, and degradation. Once *Bt* spores and crystals reach the soil they are relatively immobile. Martin and Reichelderfer (1989) found no vertical movement of spores below a 6 cm deep zone in soil and less than 10 m lateral movement, even along drainage courses. Akiba (1991) reported that there was no translocation of *Bt* spores down to a depth of 10 cm. In this same study, in artificially irrigated soils receiving the equivalent of 45 cm rainfall, *Bt* spores could be detected at a depth of 3–6 cm. *Bt* and its ICPs have varying affinities for different soil types. Venkateswerlu and Stotsky (1992) reported that adsorption and binding of *Bt* ICPs were greater on montmorillonite than on kaolinite clays. Maximum adsorption occurred within 30 min and temperatures between 7° and 50 °C did not significantly affect adsorption.

There are conflicting studies concerning the length of time that *Bt* vegetative cells, spores, and ICPs remain viable in soil. Petras and Casida (1985) reported that spore counts (commercial and laboratory cultures) decreased by one log unit in the first week after application, then remained constant for 8 months. However, spores produced in the soil survived less than 2 weeks. Saleh *et al.* (1970) reported that spores remained viable for several months and could germinate when soil conditions favored bacterial growth. West *et al.* (1984a, b) reported that vegetative cells of *Bt* subspecies *aizawai* (*Bta*) in soil disappeared rapidly at an exponential rate, ICPs disappeared non-exponentially and more slowly, and spore numbers remained unaltered with no detectable germination through 91 days at an incubation temperature of 25 °C. The half-life of the ICPs varied from 3 to 6 days. West *et al.* (1984b) reported that the loss in activity of *Bta* was greater in nonautoclaved than autoclaved soils. The initial decrease was rapid and then stabilized at approximately 10% of the original inoculum level after 250 days incubation. No activity was lost in autoclaved soil. These data suggest that the environmental persistence of *Bt* spores in soils depends on their failure to germinate.

There is no evidence to suggest that when *Bt* is applied as a larvicide, it multiplies in the environment. Bernier *et al.* (1990) conducted an environmental fate study in which a *Btk* product was applied aerially at 100 times the concentration used for operational programs. There was no significant spore increase in either organic or mineral layers of soil over an 11 month period and spores were not present in running water.

5 Spore survival in aquatic habitats

Laboratory studies have reported that *Btk* spores applied to fresh and salt water survived for more than 70 and 40 days, respectively. A higher percentage of *Btk* survived for extended periods in lake water than in tap and distilled water, presumably because of the presence of more nutrients in lake water, implying germination and growth. Spores of *Bti* remained viable for shorter periods when suspended in moving water versus static bottles, possibly indicating that laboratory trials may overestimate the longevity of spores in the field (Menon and DeMestral, 1985; Yousten *et al*., 1992).

Bti for the control of mosquitoes and black flies is purposefully applied to water or containers holding water. Rapid sedimentation is an important limitation on the efficacy of these insecticides. Sheeran and Fisher (1992) demonstrated that the sedimentation of *Bti* is facilitated by sorption onto particulate material. Contact of *Bti* with mud can immediately affect larvicidal activity due to sorption of ICPs by soil particles. This inactivation can be reversed by agitation, presumably because of reversible sorption. Adsorption to soil does not affect spore viability. Within sediments, *Bt* spores may persist for as long as 22 days, and the spores may be liberated from these sediments and resuspended during floods (Ohana *et al*., 1987). *Bti* in water is also subject to the UV and bacterial degradation mentioned above. In aquatic systems, *Bti* may be distributed by fish, but there is no evidence to suggest that *Bti* accumulates or multiplies in fish (Snarski, 1990).

G Non-target effects of Bt

There have been thousands of research papers published on *Bt*, but there are relatively few published studies on vertebrate safety of *Bt* because these data are proprietary. In depth reviews of these studies are presented in Lacey and Siegel (2000), Glare and O'Callaghan (2000), Siegel (2001), and Lacey and Merritt (2003). A discussion of the philosophy behind mammalian safety testing as well as a useful definition of the term "infection" is presented in Siegel (1997). Chapter X-1 in this book presents guidelines for evaluating the effects of entomopathogens on non-target organisms (NTO). It is worth emphasizing that when assessing the impact of *Bt* larvicides on vertebrate and invertebrate NTO, it is essential that the effect of applying *Bt* is weighed against the impact of widespread forest defoliation if the target pest is uncontrolled, or the deleterious effects caused by the broad spectrum chemical insecticides used in its place.

Initially, one of the main issues raised about the safety of *Bt* was its close relationship to *Bacillus anthracis*. Some feared that it would somehow mutate and become a human pathogen, although Steinhaus (1959) eloquently rebutted these concerns. More recent questions have centered on its relationship to *B. cereus* because *B. cereus* has been recognized as the causal agent of an increasing number of cases of food poisoning and as a source of ocular infections (Drobniewski, 1993; Jackson *et al*., 1995). However, no evidence of mammalian toxicity or infectivity has been found in the numerous laboratory safety studies conducted on *Bt* insecticides; many of these tests were designed to determine if toxicity occurred after ingestion (Ignoffo, 1973; Hadley *et al*., 1987; Siegel, 2001). In the past few years the question of *Bt* infectivity has been complicated by semantics. Some researchers equate simple recovery of *Bt* from a human as an infection, even when there is no evidence of illness or tissue damage, and these reports can cause confusion. The following examples consist of cases where *Bt* was isolated from humans but there was no evidence of illness. Jensen *et al*. (2002) isolated *Bt* from a fecal sample obtained from workers at a greenhouse where *Bt* insecticides were used. Valadres de Amorim *et al*. (2001) and Pearce *et al*. (2002), as part of a Canadian epidemiology study following a large-scale spray campaign, isolated *Bt* from humans using nasal swabs. All three studies concluded that there was no evidence of illness, yet if a definition of infection is used that equates bacterial recovery with infection, then these studies can be cited as evidence that *Bt* is infectious to humans. A different problem of semantics and taxonomy is illustrated by the study of Helgason *et al*. (2000). These authors did not distinguish between *B. cereus* and *Bt* and the title of their paper seemingly linked *Bt* with periodontitis and other human infections although none of their human isolates were in fact *Bt*.

There are three commonly cited reports associating human infection with *Bt*. In the first case, a farmer developed an ocular ulcer after being splashed in the face by a larvicide and *Btk* was recovered 10 days after this exposure (Samples and Buettner, 1983). However, there was no evidence of multiplication and viable spores may have been recovered when the eye was swabbed. In the second case, a laboratory worker stuck a finger with a needle used to resuspend a cell spore crystal pellet of *Bti* and *Acinetobacter calcoaceticus* var. *anitratus* resulting in pain and rapid swelling (Warren *et al.*, 1984). It is impossible to conclude that *Bt* alone caused the infection. In the final case, a French soldier stepped on a land mine and injured his leg. Multiple abscesses developed within 24 h after the blast and *Bt* subsp. *konkukian* was recovered (Hernandez *et al.*, 1998). Land mine blasts are notoriously difficult to deride and other species of bacteria may have contributed to this injury or *Bt* subsp. *konkukian* may have grown saprophytically in damaged tissue. It is also worth noting that *Bt* subsp. *konkukian* is not used commercially.

There have been several epidemiology studies following large-scale spraying of *Bt* in Canada, New Zealand, and the USA. These studies concluded that there were no adverse effects associated with *Bt* (Glare and O'Callaghan, 2000; Siegel, 2001). Petrie *et al.* (2003) reported a contrary finding. The authors administered a questionnaire to residents of West Auckland, New Zealand following a large-scale spray operation. Some residents reported significant increases in subjective symptoms such as sleep problems, stomach discomfort, throat irritation, dizziness, gas discomfort, extra heartbeats, and difficulty in concentrating. Despite these self reported symptoms, there was no increase in clinic visits. This study had serious problems with statistical methodology and was fatally flawed because of the failure to include a Control group. The findings of this study were contradicted by a larger New Zealand government-sponsored study that used appropriate methodology and found no increased illness associated with the use of *Bt* (Anonymous, 2002).

In the USA, as part of the testing necessary for registration, *Bt* products were administered orally to two species of birds (mallard ducks and northern bobwhite quail). Three species of fish were also tested during the registration process (sheepshead minnow, steelhead trout and bluegill sunfish). There was no evidence of pathogenicity or infectivity (bacterial recovery 100 times the administered dose to these organisms) in these studies. Data published in refereed journals on direct effects of *Bt* insecticides on non-target organisms support the conclusions of these industry studies (Smirnoff and MacLeod, 1961; Merritt *et al.*, 1989; Wipfli and Merrit, 1994).

Numerous published studies of the indirect effects of *Bt* on small mammals and birds concluded that any effects were minor. In a study reported by Innes and Bendell (1989), *Btk*, fenitrothion and aminocarb were applied aerially for control of jack pine budworm, and there were no significant differences in abundance of small mammal populations that could be attributed to *Bt*. However, fenitrothion altered the abundance of shrews, primarily *Sorex cinereus*. Nagy and Smith (1997) studied the effect of *Btk* aerial application on hooded warblers (*Wilsonia citrina*), Rodenhouse and Holmes (1992) studied the impact of food reduction due to aerial spraying in populations of black-throated blue warblers (*Dendroica caerulescens*), and Holmes (1998) studied the reproduction and behavior of Tennessee warblers (*Vermivora peregrina*) in forests treated with *Btk* and tebufenozide. All studies reported that overall, reduction in lepidopteran larvae due to spraying with *Btk* had minimal effect on bird populations.

The direct and indirect effects of *Bt* on non-target invertebrates is reviewed in Glare and O'Callaghan (2000), Lacey and Siegel (2000) and Lacey and Merritt (2003). In the case of parasitoids, the effects of *Bt* treatment of host insects on survival and percentage parasitism depends on the host, timing of applications and dosage of *Bt*. The effects of host removal on the survival of predators depend on the specificity of the predator and availability of other prey. Concern over the impact of *Bt* on non-target Lepidoptera is predominantly focused on indigenous species found mainly in forest habitats. These concerns may be alleviated by timing the application to avoid the most sensitive life stage of the NTO. Few long-term effects of repeated applications of *Bt* have been reported. The impact of *Bti* on aquatic community structure and diversity is reviewed by Lacey and Merritt (2003).

3 *Bacillus sphaericus*

A B. sphaericus *identity and toxin identification*

Bacillus sphaericus is a heterogeneous grouping that can be divided into 6 DNA homology groups; all isolates with mosquito activity belong to DNA homology group IIA, and mosquitoes are the only insects affected by this bacterium (Krych *et al.*, 1980; Woodburn *et al.*, 1995). Mosquitocidal isolates of *B. sphaericus* were first reported by Kellen *et al.* (1965) and Singer (1973), but their larvicidal activity was very low in comparison with conventional chemical insecticides. Subsequently, isolates with elevated larval activity were discovered and those with the highest activity belong to serotypes 5a5b and 25 (de Barjac *et al.*, 1980; de Barjac *et al.*, 1985). These high-toxicity strains are characterized by the production of a parasporal crystal at sporulation; this crystal is a binary protein toxin. In addition to the binary toxin, many high toxicity isolates also contain a cell wall-associated toxin coded by the *Mtx* chromosomal gene (Boucias and Pendland, 1998). The toxins of *B. sphaericus* must be ingested in order to be effective but, unlike *Bti*, the host spectrum is narrower. The toxins are almost inactive against *Ae. aegypti*, but are extremely effective against *Culex* and *Psorophora* species in waters with high organic content, as well as members of the genus *Ochlerotatus* such as *Oc. triseriatus* (Siegel and Novak, 1997). *B. sphaericus* can persist in organically enriched habitats and there is some evidence that it can recycle under certain conditions (Mulligan *et al.*, 1980; Davidson *et al.*, 1984; Karch and Coz, 1986; Nicolas *et al.*, 1987; Mulla *et al.*, 1988; Karch *et al.*, 1990; Siegel and Novak, 1997). In addition to acute mortality in larvae following ingestion of *B. sphaericus*, a range of sublethal effects has also been reported. These include prolongation of the pupal stage and reduced lifespan and energy reserves in adults (Lacey *et al.*, 1987).

B *Non-target effects of* B. sphaericus

Mammalian safety tests of *B. sphaericus* included inhalation, oral and intraperitoneal exposure to at least 10^6 colony forming units (cfu) per test animal, as well as intraocular and intracerebral injection. Non-target vertebrate studies included fish and birds. There was no evidence of infectivity or pathogenicity in these studies. Many of these tests emphasized intracerebral injection (as many as 10^7 cfu) because there were reports in the literature associating strains of *B. sphaericus* with fatal human central nervous system infections. It is noteworthy that in all cases, when these human isolates were injected in experimental animals, they were uninfectious (Shadduck *et al.*, 1980; Siegel and Shadduck, 1990; Lacey and Siegel, 2000). The most well documented human isolate of *B. sphaericus* was in fact misidentified and is a different species (Siegel *et al.*, 1997). The direct impact of *B. sphaericus* on non-target invertebrates is for the most part nonexistent due to its specificity, and some NTO may even assist in the dispersal of this bacterium (Mulla *et al.*, 1984; Aly and Mulla, 1987; Lacey and Mulla, 1990; Yousten *et al.*, 1991; Yousten *et al.*, 1992; Lacey and Merritt, 2003; Lacey, 2007). One exception is a predatory mosquito, *Toxorhynchites rutilus*, which are killed after consumption of *Ae. aegypti* larvae that have ingested *B. sphaericus* (Lacey, 1983; Lacey *et al.*, 1988). Few long-term effects of repeated applications of *B. sphaericus* on aquatic community structure and diversity have been reported. Mulla *et al.* (1984) and Lacey and Mulla (1990) reported no noticeable adverse effects on invertebrate fauna after season-long control of *Culex* spp. with *B. sphaericus*. Merritt *et al.* (2005) conducted a 3-year study in Wisconsin to assess the effects of *B. sphaericus* applied for mosquito control on non-target wetland invertebrates. No detrimental effects to non-target organisms could be attributed to routine application of *B. sphaericus*.

C B. sphaericus *use and persistence*

Commercial development of *B. sphaericus* has lagged behind that of *Bti*. Several experimental formulations have been developed and evaluated (Lacey *et al.*, 1984; Lacey *et al.*, 1988).

A commercial corncob granule formulation based on isolate 2362, serotype 5a5b, is currently available from Valent Biosciences and is marketed under the tradename VectoLex®. This formulation is targeted for mosquito control in water with high organic content and is also suitable for treating containers that hold water (Mulla *et al.*, 1997; Siegel and Novak, 1997, 1999).

The most important biotic and abiotic environmental factors that influence the larvicidal activity of *B. sphaericus* are: species of mosquito, age and feeding behavior of larvae, water quality and depth, pH, water temperature, vegetative cover, and solar radiation (Lacey, 1985; Lacey and Undeen, 1986). Extremely high levels of resistance in certain populations of *Cx quinquefasciatus* to *B. sphaericus* toxin have been reported within the past few years (Rao *et al.*, 1995; Nielsen-Leroux *et al.*, 1995, 1997).

4 *Paenibacillus popilliae*

A P. popilliae *biological characteristics*

Paenibacillus popilliae (Pettersson *et al.*, 1999) is the causal agent of "milky disease" in the Scarabaeidae. It is an obligate pathogen and has been only associated with scarab larvae belonging to the subfamilies Melolonthinae, Rutleinae, and Dynastinae (Dutky, 1963; Klein, 1981; Obenchain and Ellis, 1990, Boucias and Pendland, 1998). This bacillus must be ingested to kill, and unlike *Bt* and *B. sphaericus*, it was believed that mortality is caused solely by bacteremia. The traditional pathway of infection is thought to be spore germination in the gut of grubs and subsequent penetration of the epithelial, basal lamina, and capsular barriers of the host and ultimately entry into the hemolymph by vegetative rods. Proliferation of refractive spores and parasporal bodies during the final phase of infection gives the hemolymph a milky-white appearance for which the disease is named. However, there is also evidence that the parasporal bodies may also contribute to mortality. Significant toxicity was observed when intact or solubilized parasporal bodies were injected into the hemolymph of *Popillia japonica* (Weiner, 1978). Cry toxin genes have recently been found in the parasporal crystals of *P. popilliae* subsp. *melolonthae* H1, and the gene coding for this protein shares about 40% sequence identity with some of the *Bt* Cry toxins (Zhang *et al.*, 1997). This toxin is designated Cry18Aa (Table 1).

B *Commercial production*

P. popilliae is extremely fastidious and unlike *Bt* and *B. sphaericus*, it must be grown *in vivo*. The inability to produce spores *in vitro* has limited commercial production of this species. Two formulations were sold commercially under the trade names Doom and Japademic, and a spore-talc powder was developed and produced by the USDA Japanese Beetle Laboratory (Klein, 1992). These products were made by macerating infected grubs and were formulated on talc; they contained 10^8 spores/g powder (Redmond and Potter, 1995). Considerable effort was made by the USDA to produce *P. popilliae in vitro*, but these efforts failed to produce spores at the level necessary for a commercial product. Currently, *P. popilliae* is produced commercially by St. Gabriel Laboratories (Orange, VA, USA) and marketed under the name Milky Spore.

At one time there was a *P. popilliae* product (Grub Attack), marketed commercially by Reuter (and then Ringer), based on *in vitro*-production, but this product did not work (Redmond and Potter, 1995). Subsequent investigation revealed that this product, in fact, was *B. polymyxa*, a bacillus with no insecticidal activity (Stahly and Klein, 1992). *B. polymyxa* was most likely introduced into the starter cultures when a fermentor belonging to another company was used for production. Interestingly, other cultures deposited with the American Type Culture Collection by Reuter as part of the patent process contained *B. amylolyticus*, another bacillus with no coleopteran activity (Redmond and Potter, 1995).

C *Efficacy of* P. popilliae

In several locations in the USA, naturally occurring *P. popilliae* caused epizootics and persistently recycled in the Japanese beetle, *Popillia japonica* (Klein, 1992). However, other researchers have reported control failures using commercially produced *P. popilliae* (Lacey *et al.*,

1994; Klein and Kaya, 1995; Redmond and Potter, 1995). In the Azorean archipelago, despite treatment rates that were several multiples higher than those used in the USA for inoculative treatment of grassland, patently infected larvae were extremely rare (Lacey *et al.*, 1994). This may have been because of relatively low soil temperatures, the lack of crowding stress, or uneven distribution of larvae in the soil. Unlike *Bt* and *B. sphaericus*, where the insect control can be similar to that obtained by chemicals, *P. popilliae* appears to provide partial but long term control and the interaction between this pathogen and environmental conditions needs further study in order to improve efficacy.

D Non-target (NTO) effects

The mammalian safety studies and NTO studies conducted on *P. popilliae* are summarized by Obenchain and Ellis (1990) and Lacey and Siegel (2000). Test animals included mice, rats, guinea pigs, rabbits, monkeys, starlings, and chickens. Doses as high as 10^8 spores were used in these studies, and there was no evidence of infectivity or pathogenicity and the NTO effects were minimal. In one notable instance, Dr. G. Langford ate a "heaping spoonful of spore dust" to demonstrate its safety after a farmer claimed that *P. popilliae* had killed her chickens (Heimpel, 1971).

5 *Serratia entomophila*

A S. entomophila *biological characteristics*

S. entomophila, the causative agent of amber disease in the grass grub, *Costelytra zealandica*, is a Gram-negative bacterium that attaches to the cuticle surface of the foregut and colonizes the surface of the alimentary tract; vegetative cells are especially numerous around the cardiac valve (Klein and Jackson, 1992). It was first isolated from field populations of *C. zealandica* in 1981 (Jackson *et al.*, 1992). This bacterium produces fimbriae, which have been proposed to mediate adhesion to the foregut. Within several days infected insects cease to feed. This cessation in feeding is correlated with a reduction in trypsin activity. Infected grubs undergo a period of starvation that may last from 1 to 3 months, and during this time the insect takes on a characteristic amber appearance. Bacteria continue to grow in the foregut and may also colonize the external cuticle, the crop, cardiac valve domain and the hindgut (Boucias and Pendland, 1998). In the final stage of the disease (more than 1 month after ingestion) the bacterium penetrates the hemocoel. *S. entomophila* strains capable of causing amber disease possess a large 105 kilobase plasmid that contains genes responsible for the antifeedant activity and the amber characteristic (Boucias and Pendland, 1998). A commercial formulation, marketed under the name Invade®, is currently registered in New Zealand. Since *S. entomophila* does not produce spores, this product must be kept refrigerated and has a shelf life of approximately 3 months (Jackson *et al.*, 1992).

6 Acknowledgments

We thank both the International Program on Chemical Safety and the Special Program for Research and Training in Tropical Disease, World Health Organization, for their roles in sponsoring research and discussions that contributed to this chapter. We also thank Brian Melin, Robert Cibulsky, and Robert Smith, Abbott Laboratories, as well as John Shadduck and H. Denis Burges, for sharing their insights on *Bacillus thuringiensis* and *Bacillus sphaericus* over the years. We thank Lcrry Lacey and Juan Luis Jurat-Fuentes for their review and helpful comments on the revised chapter.

7 References

Adang, M. J., Staver, M. J., Rocheleau, T. A., Leighton, J., Barker, R. F., and Thompson, D. V. 1985. Characterized full-length and truncated plasmid clones of the crystal protein of *Bacillus thuringiensis* subspecies *kurstaki* HD-73 and their toxicity to *Manduca sexta*. *Gene* **36**, 289–300.

Akiba, Y. 1986. Microbial ecology of *Bacillus thuringiensis* VII. Fate of *Bacillus thuringiensis* in larvae of the silkworm, *Bombyx mori*, and the fall webworm, *Hyphantria cunea*. *Jpn. J. Appl. Entomol. Zool.* **30**, 99–105.

Akiba, Y. 1991. Assessment of rainwater-mediated dispersion of field-sprayed *Bacillus thuringiensis* in soil. *Appl. Entomol. Zool.* **26**, 477–483.

Aly, C. 1985. Germination of *Bacillus thuringiensis* variety *israelensis* spores in the gut of *Aedes* larvae (Diptera: Culicidae). *J. Invertebr. Pathol.* **45**, 1–8.

Aly, C. and Mulla, M. S. 1987. Effect of two microbial insecticides on aquatic predators of mosquitoes. *J. Appl. Entomol.* **103**, 113–118.

Andrews, R. E., Jr., Faust, R. M., Wabiko, H., Raymond, K. C., and Bulla, L. A., Jr. 1987. The biotechnology of *Bacillus thuringiensis. CRC Crit. Rev. Biotechnol.* **6**, 163–232.

Angus, T. A. 1954. A bacterial toxin paralysing silkworm larvae. *Nature* **173**, 545.

Anonymous. 2002. Public Health Service, Auckland District Health Board, health risk assessment of the 2002 aerial spray eradication programme for the painted apple moth in some western suburbs of Auckland. Wellington: Ministry of Agriculture and Forestry.

Barloy, F., Lecadet, M. M., and Delecluse, A. 1998. Distribution of clostridial cry-like genes among *Bacillus thuringiensis* and *Clostridium* strains. *Curr. Microbiol.* **36**, 232–237.

Beegle, C. C. and Yamamoto, T. 1992. History of *Bacillus thuringiensis* Berliner research and development. *Can. Entomol.* **124**, 587–616.

Bernhard, K. and Utz, R. 1993. Production of *Bacillus thuringiensis* insecticides for experimental and commercial uses. *In* "*Bacillus thuringiensis*: an Environmental Biopesticide: Theory and Practice" (P. F. Entwhistle, J. S. Cory, M. J. Bailey, and S. Higgs, Eds.), pp. 255–267. Wiley and Sons, Chichester, UK.

Bernier, R. L., Jr., Gannon, D. J., Moser, G. P., Mazzarello, M., Griffiths, M. M., and Guest, P. J. 1990. Development of a novel Bt strain for the control of forestry pests. Brighton Crop Protection Conference-Pests and Diseases. pp. 245–252, Brighton, UK.

Boucias, D. and Pendland, J. C. 1998. *Bacillus thuringiensis. In* "Principles of Insect Pathology" pp. 217–257. Kluwer Academic Publishers, Boston, MA.

Boucias, D. and Pendland, J. C. 1998. Insect Pathogenic Bacteria. *In* "Principles of Insect Pathology" pp. 178–216. Kluwer Academic Publishers, Boston, MA.

Bulla, L. A. Jr., Faust, R. M., Andrews, R., and Goodman, N. 1985. Insecticidal bacilli. *In* "The Molecular Biology of the Bacilli, vol.2" (D. A. Dubnau, Ed.), pp. 185–209. Academic Press Inc., New York.

Bulla, L. A., Jr., Kramer, K. J., and Davidson, L. I. 1977. Characterization of the entomocidal parasporal crystal of *Bacillus thuringiensis. J. Bacteriol.* **130**, 375–383.

Burges, H. D. and Hurst, J. A. 1977. Ecology of *Bacillus thuringiensis* in storage moths. *J. Invertebr. Pathol.* **30**, 131–139.

Burges, H. D. and Jones, K. A. 1998. Formulation of bacteria, viruses, and protozoa to control insects. *In* "Formulation of Microbial Biopesticides" (H. D. Burges, Ed.), pp. 34–127, Kluwer Academic Publishers, Dordrecht, The Netherlands.

Cantwell, G. E., Heimpel, A. M., and Thompson, M. J. 1964. The production of an exotoxin by various crystal-forming bacteria related to *Bacillus thuringiensis* variety *thuringiensis* Berliner. *J. Insect Pathol.* **6**, 466–480.

Carroll, J., and Ellar, D. J. 1993. An analysis of *Bacillus thuringiensis* δ-endotoxin action on insect-midgut-membrane permeability using a light-scattering assay. *Eur. J. Biochem.* **214**, 771–778.

Chen, X. J., Curtiss, A., Alcantara, E. and Dean, D. H. 1995. Mutations in Domain I of *Bacillus thuringiensis* δ-endotoxin CryIAb reduce the irreversible binding of toxin to *Manduca sexta* brush border membrane vesicles. *J. Biol. Chem.* **270**, 6412–6419.

Crickmore, N., Zeigler, D. R., Feitelson, J., Schnepf, E., Van Rie, J., Lereclus, D., Baum, J., and Dean, D. H. 1998. Revision of the nomenclature for the *Bacillus thuringiensis* pesticidal crystal proteins. *Microbiol. Molec. Biol. Rev.* **62**, 807–813.

Davidson, E. W., Urbina, M., Payne, J., Mulla, M. S., Darwazeh, H., Dulmage, H. T., and Correa, J. A. 1984. Fate of *Bacillus sphaericus* 1593 and 2362 spores used as larvicides in the aquatic environment. *Appl. Environ. Microbiol.* **47**, 125–129.

de Barjac H. and Bonnefoi, A. 1962. Essai de classification biochimique et serologique de 24 souche de *Bacillus* de type *thuringiensis. Entomophaga* **7**, 5–31.

de Barjac, H., Larget-Thiery, I., Dumanoir, V. C., and Ripoutearu, H. 1985. Serological classification of *Bacillus sphaericus* strains on the basis of toxicity to mosquito larvae. *Appl. Microbiol. Biotechnol.* **21**, 85–90.

de Barjac H., Veron, M., and Cosmao Dumanoir, V. 1980. Characterisation biochimique et serologique de souches de *Bacillus sphaericus* pathogens ou non pour les moustiques. *Ann. Microbiol.* **131B**, 191–201.

de Maagd, R. A., Bravo, A. and Crickmore, N. 2001. How *Bacillus thuringiensis* has evolved specific toxins to colonize the insect world. *TRENDS in Genetics.* **17**, 193–199.

de Maagd, R. A., Bravo, A., Berry, C., Crickmore, N. and Schnepf, H. E. 2003. Structure, diversity, and evolution of protein toxins from spore-forming entomopathogenic bacteria. *Annu. Rev. Genet.* **37**, 409–433.

Dean, D. H. 1984. Biochemical genetics of the bacterial insect-control agent *Bacillus thuringiensis*: Basic principles and prospects for genetic engineering. *Biotechnol. Genet. Eng. Rev.* **2**, 341–363.

Dean, D. H., Rajamohan, F., Lee, M. K., Wu, S.-J., Chen, X. J, .Alcantara, E., and Hussain, S. R. 1996. Probing the mechanism of action of *Bacillus thuringiensis* insecticidal proteins by site-directed mutagenesis – a minireview. *Gene* **179**, 111–117.

DeLucca, A. J. II, Simonson, J. G., and Larson, A. D. 1981. *Bacillus thuringiensis* distribution in soils of the United States. *Can. J. Microbiol.* **27**, 865–870.

Denolf, P., Jansens, S., Peferoen, M., Degheele, D., and Van Rie, J. 1993. Two different *Bacillus thuringiensis* delta-endotoxin receptors in the midgut brush border membrane of the european corn borer, *Ostrinia nubilalis* (Hubner) (Lepidoptera: Pyralidae). *Appl. Environ. Microbiol.* **59**, 1828–1837.

Dent, D. R. 1993. The use of *Bacillus thuringiensis* as an insecticide. In "Exploitation of Microorganisms" (D. G. Jones, Ed.), pp. 19–44, Chapman and Hall, London.

Drobniewski, F. A. 1993. *Bacillus cereus* and related species. *Clin. Microbiol. Rev.* **6**, 324–338.

Du, C. and Nickerson, K. W. 1996. *Bacillus thuringiensis* HD-73 spores have surface-localized Cry1Ac toxin: physiological and pathogenic consequences. *Appl. Environ. Microbiol.* **62**, 3722–3726.

Dulmage, H. T. and Cooperators. 1981. Insecticidal activity of isolates of *Bacillus thuringiensis* and their potential for pest control: 1. Background of the International Cooperative Program. *In* "Microbial Control of Pests and Plant Diseases 1970–1980" (H. D. Burges, Ed.), pp 130–200. Academic Press Inc., London.

Dutky, S. R. 1963. The milky diseases. In "Insect Pathology: An Advanced Treatise. Vol. 2." (E. A. Steinhaus, Ed.) pp. 75–115. Academic Press, New York.

English, L., Readdy, T. L., and Bastian, A. E. 1991. Delta-endotoxin-induced leakage of $^{86}Rb^{+}$–K^{+} and H_2O from phospholipid vesicles is catalyzed by reconstituted midgut membrane. *Insect Biochem.* **21**, 177–184.

English, L. and Slatin, S. L. 1992. Mode of action of delta-endotoxins from *Bacillus thuringiensis*: A comparison with other bacterial toxins. *Insect Biochem. Mol. Biol. 22*, 1–7.

English, L., Walters, F., Von Tersch, M. A., and Slatin, S. 1995. Modulation of δ-endotoxin ion channels, p. 302–307. *In* J. Marshall Clark (ed.), Molecular action of insecticides on ion channels. American Chemical Society, Washington, D.C.

Escriche, B., De Decker, N., Van Rie, J., Jansens, S., and Van Kerkhove, E. 1998. Changes in permeabilty of brush border membrane vesicles from *Spodoptera littoralis* midgut induced by insecticidal crystal proteins from *Bacillus thuringiensis. Appl. Environ. Microbiol.* **64**, 1563–1565.

Farkas, J., Sebesta, K., Horska, K., Samek, Z., Dolejs, L., and Sorm, F. 1976. Structure of thuringiensin, the thermostable exotoxin from *Bacillus thuringiensis. Coll. Czechoslov. Chem. Commun.* **42**, 909–929.

Fast, P. G. 1971. Isolation of a water-soluble toxin from a commercial microbial insecticide based on *Bacillus thuringiensis. J. Invertebr. Pathol.* **17**, 301.

Federici, B. A. 2005. Insecticidal bacteria: An overwhelming success for invertebrate pathology. *J. Invertebr. Pathol.* **89**, 30–38.

Federici, B. A. 1999. *Bacillus thuringiensis* in Biological Control. *In* "Handbook of Biological Control" (T. W. Fisher, T. S. Bellows, L. E. Caltagirone, D. L. Dahlsten, C. Huffaker and G. Gordh, Eds.), pp. 575–592. Academic Press, San Diego, CA.

Federici, B. A., Lüthy, P., and Ibarra, J. E. 1990. Parasporal body of *Bacillus thuringiensis*: structure, protein composition, and toxicity. *In* "Bacterial Control of Mosquitoes and Black Flies: Biochemistry, Genetics, and Applications of *Bacillus thuringiensis israelensis* and *Bacillus sphaericus*"(H. de Barjac and D. Sutherland, Eds.) pp. 16–44, Rutgers University Press, New Brunswick, NJ.

Ferré, J., Real, M. D., van Rie, J., Jansens, S., and Peferon, M. 1991. Resistance to the *Bacillus thuringiensis* bioinsecticide in a field population of *Plutella xylostella* is due to a change in a midgut membrane receptor. *Proc. Natl. Acad. Sci. USA* **88**, 5119–5123.

Ferré, J. and Van Rie, J. 2002. Biochemistry and genetics of insect resistance to *Bacillus thuringiensis. Annu. Rev. Entomol.* **47**, 501–533.

Garczynski, S. F., Crim, J. W., and Adang, M. J. 1991. Identification of a putative insect brush border membrane-binding molecules specific to *Bacillus thuringiensis* δ-endotoxin by protein blot analysis. *Appl. Environ. Microbiol.* **57**, 2816–2820.

Gill, S., Cowles, E. A. and Pietrantonio, P. V. 1992. The mode of action of *Bacillus thuringiensis* endotoxins. *Annu. Rev. Entomol.* **37**, 615–636.

Glare, T. R. and O'Calaghan, M. 2000. *Bacillus thuringiensis*: Biology, Ecology, and Safety". 350 pp. John Wiley & Sons Ltd, Chichester, UK.

Goldberg, L. J. and Margalit, J. 1977. A bacterial spore demonstrating rapid larvicidal activity against *Anopheles sergentii, Uranotaenia unguiculata, Culex univiattus, Aedes aegypti*, and *Culex pipiens. Mosq. News* **37**, 355–358.

Gonzalez, J. M., Jr. and Carlton, B. C. 1980. Patterns of plasmid DNA in crystalliferous and acrystalliferous strains of *Bacillus thuringiensis. Plasmid* **3**, 92–98.

Gonzalez, J. M., Jr., Dulmage, H. T., and Carlton, B. C. 1981. Correlation between specific plasmids and (delta)-endotoxin production of *Bacillus thuringiensis. Plasmid* **5**, 351–365.

Gordon, R. E., W. C. Haynes, and C. H.-N. Pang. 1973. The genus *Bacillus*. Agricultural handbook 427. United States Department of Agriculture, Washington, D. C.

Griego, V. M. and Spence, K. D. 1978. Inactivation of *Bacillus thuringiensis* spores byultraviolet and visible light. *Appl. Environ. Microbiol.* **35**: 906–910.

Griffitts, J. S. and Aroian, R. V. 2005. Many roads to resistance: how invertebrates adapt to Bt toxins. *BioEssays.* **27**, 614–624.

Grochulski, P., Masson, L., Borisova, S., Pusztai-Carey, M., Schwartz, J. L., Brousseau, R.and Cygler, M. 1995. *Bacillus thuringiensis* CryIA(a) insecticidal toxin: crystal structure and channel formation. *J. Mol. Biol.* **254**, 1–18.

Hadley, W. M., Burchiel, S. W., McDowell, T. D., Thilsted, J. P., Hibbs, C. M., Whorton, J. A., Day, P. W., Friedman, M. B., and Stoll, R. E. 1987. Five-month oral (diet) toxicity/infectivity study of *Bacillus thuringiensis* insecticides in sheep. *Fund. Appl. Toxicol.* **8**, 236–242.

Haider, M. Z. and Ellar, D. J. 1989. Mechanism of action of *Bacillus thuringiensis* δ-endotoxin: interaction with phospholipid vesicles. *Biochim. Biophys. Acta* **978**, 216–222.

Hannay, C. L. 1953. Crystalline inclusions in aerobic spore-forming bacteria. *Nature* **172**, 1004.

Hansen, B. M., Damgaard, P. H., Eilenberg, J., and Pedersen, J. C. 1998. Molecular and phenotypic characterization of *Bacillus thuringiensis* isolated from leaves and insects. *J. Invertebr. Pathol.* **71**, 106–114.

Heimpel, A. M. 1967. A critical review of *Bacillus thuringiensis* variety *thuringiensis* Berliner and other crystalliferous bacteria. *Annu. Rev. Entomol.* **12**, 287–322.

Heimpel, A. M. 1971. Safety of insect pathogens for man and vertebrates. "Microbial Control of Insects and Mites" (H. D. Burges and N. W. Hussey, Eds.), pp 469–489. Academic Press, New York.

Heimpel, A. M. and Angus, T. A. 1958. The taxonomy of insect pathogens related to *Bacillus cereus. Can. J. Microbiol.* **4**, 531–541.

Heimpel, A. M. and Angus, T. A. 1959. The site of action of crystalliferous bacteria in lepidopteran larvae. *J. Insect Pathol.* **1**, 152–170.

Helgason, E., Caugant, D., Olsen, I., and Kolsto, A. B. 2000. Genetic structure of population of *Bacillus cereus* and *B. thuringiensis* isolates associated with periodontitis and other human infections. *J. Clin. Microbiol.* **38**, 1615–1622.

Hendrickx, K., De Loof, A., and Van Mellaert, H. 1990. Effects of *Bacillus thuringiensis* delta-endotoxin on the permeability of brush border membrane vesicles from tobacco hornworm (*Manduca sexta*) midgut. *Comp. Biochem. Physiol.* **95C**, 241–245.

Hernandez, E., Ramisse, F., Ducoureau, J. P., Cruel, T., and Cavallo, J. D. 1998. *Bacillus thuringiensis* subsp. *konkukian* (Serotype H34) superinfection: case report and experimental evidence of pathogenicity in immunosuppressed mice. *J. Clin. Microbiol.* **36**, 2138–2139.

Hofmann, C., Vanderbruggen, H., Höfte, H., Van Rie, J., Jansens, S., and Van Mellaert, H. 1988a. Specificity of *Bacillus thuringiensis* delta-endotoxins is correlated with the presence of high-affinity binding sites in the brush border membrane of target insect midguts. *Proc. Natl. Acad. Sci. USA* **85**, 7844–7848.

Hofmann, C., Lüthy, P., Hutter, R., and Pliska, V. 1988b. Binding of the delta-endotoxin from *Bacillus thuringiensis* to brush-border membrane vesicles of the cabbage butterfly *(Pieris brassicae). Eur. J. Biochem.* **173**, 85–91.

Höfte, H., Van Rie, J., Jansen, S., Van Houtven, A., Vanderbruggen, H., and Vaeck, M. 1988. Monoclonal antibody analysis and insecticidal spectrum of three types of lepidopteran-specific insecticidal crystal proteins of *Bacillus thuringiensis. Appl. Environ. Microbiol.* **54**, 2010–2017.

Höfte, H. and Whiteley, H. R. 1989. Insecticidal crystal proteins of *Bacillus thuringiensis. Microbiol. Rev.* **53**, 242–255.

Holmes, S. B. 1998. Reproduction and nest behavior of Tennessee warblers *Vermivora peregrina* in forests treated with Lepidoptera-specific insecticides. *J. Appl. Ecol.* **35**, 185–194.

Hostetter, D. L., Ignoffo, C. M., and Kearby, W. H. 1975. Persistence of formulations of *Bacillus thuringiensis* spores and crystals on eastern red cedar foliage in Missouri. *J. Kan. Entomol. Soc.* **48**, 189–193.

Ignoffo, C. M. 1973. Effects of entomopathogens on vertebrates. *Ann. NY Acad. Sci.* **217**, 141–164.

Ignoffo, C. M. 1992. Environmental factors affecting persistence of entomopathogens. *Fla. Entomol.* **75**, 516–525.

Ihara, H., Kuroda, E., Wadano, A., and Himeo, M. 1993. Specific toxicity of δ-endotoxins from *Bacillus thuringiensis* to *Bombyx mori. Biosci. Biotechnol. Biochem.* **57**, 200–204.

Innes, D. G. and Bendell, J. F. 1989. The effects of small-mammal populations of aerial applications of *Bacillus thuringiensis*, fenitrothion, and Matacil® used against jack pine budworm in Ontario. *Can. J. Zool.* **67**, 1318–1323.

Itoua-Apoyolo, C., Drif, L., Vassal, J. M., de Barjac, H., Bossy, J. P., Leclant, F., and Frutos, R. 1995. Isolation of multiple subspecies of *Bacillus thuringiensis* from a population of the European Sunflower moth, *Homoeosoma nebulella. Appl. Environ. Entomol.* **61**, 4343–4347.

Jackson, S. G., Goodbrand, R. B., Ahmed, R., and Kasatiya, S. 1995. *Bacillus cereus* and *Bacillus thuringiensis* isolated in a gastroenteritis outbreak investigation. *Letters Appl. Microbiol.* **21**, 103–105.

Jackson, T. A., Pearson, J. F., O'Callaghan, M., Mahanty, H. K., and Willocks, M. J. 1992. "Pathogen to product-Development of *Serratia entomophila* (Enterobacteriaceae) as a commercial biological control agent for the New Zealand grass grub (*Costelytra zealandica*)." *In* "Use of Pathogens in Scarab Pest Management" (T. A. Jackson and T. R. Glare, Eds.), pp. 191–198. Intercept Limited, Hampshire, UK.

James, C. 2004. Global Review of Commercialized Transgenic Crops: 2002. ISAAA Briefs No. 32, ISAAA: Ithaca, NY.

Jarrett, P. and Stephenson, M. 1990. Plasmid transfer between strains of *Bacillus thuringiensis* infecting *Galleria mellonella* and *Spodoptera littoralis. Appl. Environ. Microbiol.* **56**, 1608–1614.

Jensen, G. B., Larsen, P., Jacobsen, B. L., Madsen, B., Wilcks, A., Smidet, L., and Andrup, L. 2002. Isolation and characterization of *Bacillus cereus*-like bacteria from faecal samples from greenhouse workers who are using *Bacillus thuringiensis*-based insecticides. *Int. Arch. Occup. Environ. Health* **75**, 191–196.

Karch, S. and Coz, J. 1986. Recyclage de *Bacillus sphaericus* chez les larves mortes de *Culex pipiens* (Diptera: Culicidae). *Cah. ORSTOM ser. Entomol. Med. Parasit.* **24**, 41–44.

Karch, S., Monteny, N., Jullien, J. L., Sinegre, G., and Coz, J. 1990. Control of *Culex pipiens* by *Bacillus sphaericus* and role of nontarget arthropods in its recycling. *J. Am. Mosq. Control Assoc.* **6**, 47–54.

Kellen, W. R., Clark, T. B., Lindegren, J. E., Ho, B., Rogoff, C., and Singer, S. 1965. *Bacillus sphaericus* Neide as a pathogen of mosquitoes. *J. Invertebr. Pathol.* **7**, 442–448.

Klein, M. G. 1981. Advances in the use of *Bacillus popilliae* for pest control. *In* "Microbial Control of Pests and Plant Diseases 1970–1980" (H. D. Burges, Ed.) pp. 183–192. Academic Press, London.

Klein, M. G. 1992. Use of *Bacillus popilliae* in Japanese Beetle Control. *In* "Use of Pathogens in Scarab Pest Management" (T. A. Jackson and T. R. Glare, Eds.), pp. 179–189. Intercept Limited, Hampshire, UK.

Klein, M. G. and Jackson, T. A. 1992. Bacterial diseases of scarabs. *In* "Use of Pathogens in Scarab Pest Management" (T. A. Jackson and T. R. Glare, Eds.), pp 43–61. Intercept Limited, Hampshire, UK.

Klein, M. G. and Kaya, H. K. 1995. *Bacillus* and *Serratia* species for scarab control. *Mem. Inst. Oswaldo Cruz* **90**, 87–95.

Knell, R. J., Begon, M., and Thompson, D. J. 1998. Host-pathogen population dynamics, basic reproductive rates and threshold densities. *Oikos*, **81**, 299–308.

Knowles, B. H. and Ellar, D. J. 1987. Colloid-osmotic lysis is a general feature of the mechanisms of action of *Bacillus thuringiensis* (delta)–endotoxins with different insect specificity. *Biochim. Biophys. Acta* **924**, 509–518.

Knowles, B. H. and Dow J. A. T. 1993. The crystal delta-endotoxins of *Bacillus thuringiensis*: models for their mechanism of action on the insect gut. *Bioessays* **15**, 469–476.

Knowles, B. H. 1994. Mechanism of action of *Bacillus thuringiensis* insecticidal δ-endotoxins. *Adv. Insect Physiol.* **24**, 275–308.

Krieg, A. 1971. Concerning (alpha)-exotoxin produced by vegetative cells of *Bacillus thuringiensis* and *Bacillus cereus*. *J. Invertebr. Pathol.* **17**, 134–135.

Kronstad, J. W. and Whiteley, H. R. 1986. Three classes of homologous *Bacillus thuringiensis* crystal protein genes. *Gene* **43**, 29–40.

Krych, V. K., Johnson, J. L., and Yousten, A. A. 1980. Deoxyribonucleic acid homologies among strains of *Bacillus sphaericus*. *Int. J. Syst. Bacteriol.* **30**, 476–484.

Lacey, L. A. 1983. Larvicidal activity of *Bacillus* pathogens against *Toxorhynchites* mosquitoes (Diptera: Culicidae). *J. Med. Entomol.* **20**, 620–624.

Lacey, L. A. 1985. The effects of pH and storage temperature on spore viability and mosquito larvicidal activity of *Bacillus sphaericus*. *Bull. Soc. Vect. Ecol.* **10**, 102–106.

Lacey, L. A. 1997. Bacteria: Laboratory bioassay of bacteria against aquatic insects with emphasis on larvae of mosquitoes and black flies. *In* "Manual of Techniques in Insect Pathology"(L. A. Lacey, Ed.), pp. 79–90. Academic Press, San Diego, CA.

Lacey, L. A. 2007. *Bacillus thuringiensis* serovariety *israelensis* and *Bacillus sphaericus* for mosquito control. In "Biorational Control of Mosquitoes" *Am. Mosq. Control Assoc. Bull.7, in press.*

Lacey, L. A., Amaral, J. J., Klein, M. G., Simoes, N. J., Martins, A., and Mendes, C. 1994. Microbial control of *Popillia japonica* (Coleoptera: Scarabaeidae) on Terceira Island (Azores, Portugal): the role of operational research. Proc. VIth Int. Colloq. Invertebr. Pathol. Microb. Control pp 409–415.

Lacey, L. A., Day, J., and Heitzman, C. M. 1987. Long-term effects of *Bacillus sphaericus* on *Culex quinquefasciatus*. *J. Invertebr. Pathol.* **49**, 116–123.

Lacey, L. A. and Merritt, R. W. 2003. The safety of bacterial microbial agents used for black fly and mosquito control in aquatic environments. *In* "Environmental Impacts of Microbial Insececticides: Need and Methods for Risk Assessment"(H. M. T. Hokkanen and A. E. Hajek, Eds.), pp. 151–168. Kluwer Academic Publishers, Dordrecht.

Lacey, L. A. and Mulla, M. S. 1990. Safety of *Bacillus thuringiensis* (H-14) and *Bacillus sphaericus* to non-target organisms in the aquatic environment. *In* "Safety of Microbial Insecticides" (M. Laird, L. A. Lacey, and E. W. Davidson, Eds.)pp. 169–188. CRC Press, Boca Raton, FL.

Lacey, L. A., Ross, D. H., Lacey, C. M., Inman, A., and Dulmage, H. T. 1988. Experimental formulations of *Bacillus sphaericus* for the control of anopheline and culicine larvae. *J. Indus. Microbiol.* **3**, 39–47.

Lacey, L. A. and Siegel, J. P. 2000. Safety and ecotoxicology of entomopathogenic bacteria. *In* "Entomopathogenic Bacteria: From Laboratory to Field Application" (J. F. Charles, A. Delecluse, and C. Nielsen-LeRoux, Eds.), Kluwer Academic Publishers, Dordrecht, The Netherlands.

Lacey, L. A. and Undeen, A. H. 1986. Microbial control of black flies and mosquitoes. *Annu. Rev. Entomol.* **31**, 265–296.

Lacey, L. A., Urbina, M. J., and Heitzman, C. M. 1984. Sustained release formulations of *Bacillus sphaericus* and *Bacillus thuringiensis* (H-14) for control of container-breeding *Culex quinquefasciatus*. *Mosq. News* **44**, 26–32.

Lacey, L. A., Lacey, C. M., Peacock, B., and I. Thiery. 1988. Mosquito host range and field activity of *Bacillus sphaericus* isolate 2297 (serotype 25). *J. Amer. Mosq. Control Assoc.* **4**, 51–56.

Lecadet, M. M., Frachon, E., Dumanoir, V. C., Ripouteau, H., Hamon, S., Laurent, P., and Thiery, I. 1999. Updating the H-antigen classification of *Bacillus thuringiensis. J. Appl. Microbiol.* **86**, 660–672.

Lee, M. K., Rajamohan, F., Gould, F., Dean, D. H. 1995. Resistance to *Bacillus thuringiensis* CryIA delta-endotoxins in a laboratory-selected *Heliothis virescens* strain is related to receptor alteration. *Appl Environ Microbiol.* **61**, 3836–3842.

Leong, K. L. H., Cano, R. J., and Kubinski, A. M. 1980. Factors affecting *Bacillus thuringiensis* total field persistence. *Environ. Entomol.* **9**, 593–599.

Li, J., Carroll, J., and Ellar, D. J. 1991. Crystal structure of insecticidal delta-endotoxin from *Bacillus thuringiensis* at 2.5 angstrom resolution. *Nature* **353**, 815–821.

Liang, Y., Patel, S. S., and Dean, D. H. 1995. Irreversible binding kinetics of *Bacillus thuringiensis* CryIAa δ-endotoxins to gypsy moth brush border membrane vesicles is directly correlated to toxicity. *J. Biol. Chem.* **270**, 24719–24724.

Luo, K., Tabashnik, B. E. and Adang, M. J. 1997. Binding of *Bacillus thuringiensis* Cry1Ac toxin to aminopeptidase in susceptible and resistant diamondback moths *(Plutella xylostella). Appl. Environ. Microbiol.* **63**, 1024–1027.

Luo, K., Banks, D., and Adang, M. J. 1999. Toxicity, binding, and permeability analyses of four *Bacillus thuringiensis* Cry1 δ-endotoxins using brush border membrane vesicles of *Spodoptera exigua* and *Spodoptera frugiperda. Appl. Environ. Microbiol.* **65**, 457–464.

Lüthy, P. 1986. Insect pathogenic bacteria as pest control agents. *Fortschr. Zool.* **32**, 201–216.

Lynch, R. E., Lewis, L. C., and Brindley, T. A. 1976. Bacteria associated with eggs and 1st-instar larvae of the European corn borer: Isolation techniques and pathogenicity. *J. Invertebr. Pathol.* **27**, 325–331.

Martin, F. G. and Wolfersberger, M. G. 1995. *Bacillus thuringiensis* delta-endotoxin and larval *Manduca sexta* midgut brush-border membrane vesicles act synergistically to cause very large increases in the conductance of planar lipid bilayers. *J. Exp Biol.* **198(Pt 1)**, 91–6

Martin, P. A. W. 1994. An iconoclastic view of *Bacillus thuringiensis* ecology. *Am. Entomol.* **40**, 85–90.

Martin, P. A. W. and Travers, R. S. 1989. Worldwide abundance and distribution of *Bacillus thuringiensis* isolates. *Appl. Environ. Microbiol.* **55**, 2437–2442.

Martin, W. F. and Reichelderfer, C. F. 1989. *Bacillus thuringiensis*: Persistence and movement in field crops. Proceedings and Abstracts Society for Invertebrate Pathology XXIInd Annual Meeting, p 25. University of Maryland, USA.

McGaughey, W. H. 1985. Insect resistance to the biological insecticide *Bacillus thuringiensis. Science* **229**, 193–195

McGuire, M. R., Galan-Wong, L. J., and Tamez-Guerra, P. 1997. Bacteria: bioassay of *Bacillus thuringiensis* against lepidopteran larvae. *In* "Manual of Techniques in Insect Pathology" (L. A. Lacey, Ed.), pp. 91–100. Academic Press, San Diego, CA.

Meadows, M. P. 1993. *Bacillus thuringiensis* in the environment: ecology and risk assessment. *In* "*Bacillus thuringiensis*, an Environmental Biopesticide: Theory and Practice" (P. F. Entwhistle, J. S. Corey, M. J. Bailey, and S. Higgs) pp. 193–220, Wiley and Sons, Chichester, UK.

Meadows, M. P., Ellis, D. J. Butt, J., Jarrett, P., and Burges, H. D. 1992. Distribution, frequency, and diversity of *Bacillus thuringiensis* in an animal feed mill. *Appl. Environ. Microbiol.* **58**, 1344–1350.

Menon, A. S., and DeMestral, J. 1985. Survival of *Bacillus thuringiensis* var. *kurstaki* in waters. *Water, Air Soil Pollut.* **25**, 265–274.

Merritt, R. W., E. D. Walker, M. A. Wilzbach, K. W. Cummins, and W. T. Morgan. 1989. A broad evaluation of B.t.i. for black fly (Diptera: Simuliidae) control in a Michigan river: efficacy, carry and nontarget effects on invertebrates and fish. *J. Am. Mosq. Control Assoc.* **5**, 397–415.

Mohd-Salleh, M. B., Beegle, C. C., and Lewis, L. C. 1980. Fermentation media and production of exotoxin by three varieties of *Bacillus thuringiensis. J. Invertebr. Pathol.* **35**, 75–83.

Mulla, M. S. 1990. Activity, field efficacy, and use of *Bacillus thuringiensis israelensis* against mosquitoes. *In* "Bacterial Control of Mosquitoes and Black Flies: Biochemistry, Genetics and Applications of *Bacillus thuringiensis israelensis* and *Bacillus sphaericus*" (H. de Barjac and D. J. Sutherland, Eds.) pp. 134–160. Rutgers University Press, New Brunswick, NJ.

Mulla, M. S., Axelrod, H., Darwazeh, H. A., and Matanmi, B. A. 1988. Efficacy and longevity of *Bacillus sphaericus* 2362 formulations for control of mosquito larvae in dairy wastewater lagoons. *J Am. Mosq. Control Assoc.* **4**, 448–452.

Mulla, M. S., Darwazeh, H. A., Davidson, E. W., Dulmage, H. T., and Singer, S. 1984. Larvicidal activity and field efficacy of *Bacillus sphaericus* strains against mosquito larvae and their safety to nontarget organisms. *Mosq. News* **44**, 336–342.

Mulla, M. S., Rodcharoen, J., Ngamsuk, W., Tawatsin, A., Pan-Urai, P., and Thavara, U. 1997. Field trials with *Bacillus sphaericus* formulations against polluted water mosquitoes in a suburban area of Bangkok, Thailand. *J. Am. Mosq. Control Assoc.* **13**, 297–304.

Mulligan, F. S. III, Schaefer, C. H., and Wilder, W. H. 1980. Efficacy and persistence of *Bacillus sphaericus* and *B. thuringiensis* H-14 against mosquitoes under laboratory and field conditions. *J. Econ. Entomol.* **73**, 684–688.

Nagy, L. R. and Smith, K. G. 1997. Effects of insecticide-induced reduction in lepidopteran larvae on reproductive success of hooded warblers. *The Auk* **114**, 619–627.

Narva, K. E., Payne, J. M., Schwab, G. E., Hickle, L. A., Galasan, T., and Sick, A. J. 1991. Novel *Bacillus thuringiensis* microbes against nematodes, and genes encoding novel nematode-active toxins cloned from *Bacillus thuringiensis* isolates. European Patent Application Number 91305047.2, filing date 04.06.91. Publication number: **0 462 721 A2**

Nicolas, L., Dossou-Yovo, J., and Hougard, J. M. 1987. Persistence and recycling of *Bacillus sphaericus* 2362 spores in *Culex quinquefasciatus* breeding sites in west Africa. *Appl. Microbiol. Biotech.* **25**, 341–345.

Nielsen-LeRoux, C., Charles, J. F., Thiery, I., and Georghiou, G. P. 1995. Resistance in a laboratory population of *Culex quinquefasciatus* (Diptera: Culicidae) to *Bacillus sphaericus* binary toxin is due to a change in the receptor on midgut brush-border membranes. *Euro. J. Biochem.* **228**, 206–210.

Nielsen-LeRoux, C., Pasquier, F., Charles, J. F., Sinegre, G., Gaven, B., and Pasteur, N. 1997. Resistance to *Bacillus sphaericus* involves different mechanisms in *Culex pipiens* (Diptera: Culicidae) larvae. *J. Med. Entomol.* **34**, 321–327.

Obenchain, F. D. and Ellis, B. J. 1990. Safety considerations in the use *of Bacillus popillae*, the milky disease pathogen of Scarabeidae. *In* "Safety of Microbial Insecticides" (M. Laird, L. A. Lacey, and E. W. Davidson, Eds.) pp. 189–201. CRC Press, Boca Raton, FL.

Ohana, B., Margalit, J., and Barak, Z. 1987. Fate of *Bacillus thuringiensis* subsp. *israelensis* under simulated field conditions. *Appl. Environ. Microbiol.* **53**, 828–831.

Ohba, M. and Aizawa, K. 1986. Distribution of *Bacillus thuringiensis* in soils of Japan. *J. Invertebr. Pathol.* **47**, 277–282.

Ohba, M., Yu, Y. M., and Aizawa, K. 1988. Occurrence of noninsecticidal *Bacillus thuringiensis* flagellar serotype 14 in the soil of Japan. *Appl. Microbiol.* **11**, 85–89.

Pearce, M., Habbick, B., Williams, J., Eastmen, M., and Newman, M. 2002. The effects of aerial spraying with *Bacillus thuringiensis kurstaki* on children with asthma. *Can. J. Public Health.* **93**, 21–25.

Petras, S. F. and Casida, L. E. Jr. 1985. Survival of *Bacillus thuringiensis* spores in soil. *Appl. Environ. Microbiol.* **50**, 1496–1501.

Petrie, K., Thomas, M., and Broadbent, E. 2003. Symptom complaints following aerial spraying with biological insecticide Foray 48B. *NZMJ* **116**, 1–7.

Pettersson, B., Rippere, K. E., Yousten, A. A., and Priest, F. G. 1999. Transfer of *Bacillus lentimorbus* and *Bacillus popilliae* to the genus *Paenibacillus* with emended descriptions of *Paenibacillus lentimorbus* comb. nov, and *Paenibacillus popilliae* comb. nov. *Int. J. Syst. Bacteriol.* **49**, 531–540.

Pinnock, D. E., Brand, R. J., Jackson, K. L., and Milstead, J. E. 1974. The field persistence of *Bacillus thuringiensis* spores on *Cercis occidentalis* leaves. *J. Invertebr. Pathol.* **23**, 341–346.

Pinnock, D. E., Milstead, J. E., Kirby, M. E., and Nelson, B. J. 1977. Stability of entomopathogenic bacteria. *Misc. Publ. Entomol. Soc. Am.* **10**, 77–97.

Prasertphon, S., Areehul, P., and Tanada, Y. 1974. Sporulation of *Bacillus thuringiensis* in host cadavers. *J. Invertebr. Pathol.* **21**, 205–207.

Pusztai, M., Fast, P., Gringorten, L., Kaplan, H., Lessard, T., and Carey, P. R. 1991. The mechanism of sunlight-mediated inactivation of *Bacillus thuringiensis* crystals. *Biochem. J.* **273**, 43–47.

Rajamohan, F., Alcantara, E., Lee, M. K., Chen, X. J., Curtiss, A. and Dean, D. H. 1995. Single amino acid changes in domain II of *Bacillus thuringiensis* CryIAb δ-endotoxin affect irreversible binding to *Manduca sexta* midgut membrane vesicles. *J. Bacteriol.* **177**, 2276–2282.

Rajamohan, F., Lee, M. K. and Dean, D. H. 1998. *Bacillus thuringiensis* insecticidal proteins: molecular mode of action. *Prog. Nucleic Acids Mol. Biol.* **60**, 1–27.

Rao, D. R., Mani, T. R., Rajendran, R., Joseph, A. S., Gajanana, A. and Reuben, R. 1995. Development of a high level of resistance to *Bacillus sph*aericus in a field population of *Culex quinquefasciatus* from Kochi, India. *J. Am. Mosq. Control Assoc.*, **11**, 1–5.

Rasko, D. A., Altherr, M. R., Han, C. S., and Ravel, J. 2005. Genomics of the *Bacillus cereus* group of organisms. *FEMS Microbiol. Rev.* **29**, 303–329.

Reardon, R. C. and Hassig, K. 1983. Spruce budworm larval populations and field persistence of *Bacillus thuringiensis* after treatment in Wisconsin. *J. Econ. Entomol.* **76**, 1139–1143.

Redmond, C.T and Potter, D. A. 1995. Lack of efficacy of *in vivo* and putatively *in vitro* produced *Bacillus popilliae* against field populations of Japanese beetle (Coleoptera: Scarabaeidae) grubs in Kentucky. *J. Econ. Entomol.* **88**, 846–854.

Rodenhouse, N. L. and Holmes, R. T. 1992. Results of experimental and natural food reductions for breeding black-throated blue warblers. *Ecology* **73**, 357–372.

Saleh, S. M., Harris, R. F., and Allen, O. N. 1970. Fate of *Bacillus thuringiensis* in soil: effect of soil pH and organic amendment. *Can. J. Microbiol.* **16**, 677–680.

Samples, J. R. and Buettner, H. 1983. Ocular infection caused by a biological insecticide. *J. Inf. Dis.* **148**, 613–614.

Sanchis, V. and Ellar, D. J. 1993. Identification and partial purification of a *Bacillus thuringiensis* CryIC δ-endotoxin binding protein from *Spodoptera litoralis* gut membranes. *FEBS Letters* **3**, 264–268.

Schnepf, E., Crickmore, N., Van Rie, J., Lereclus, D., Baum, J., Feitelson, J., Zeigler, D. R., and Dean, D. H. 1998.

Bacillus thuringiensis and its pesticidal crystal proteins. *Microbiol. Mol. Biol. Rev.* **62**,775–806.

Schwartz, J. L., Garneau, L., Masson, L., and Brousseau, R. 1991. Early response of cultured lepidopteran cells to exposure to δ-endotoxin from *Bacillus thuringiensis*: involvement of calcium and anionic channels. *Biochim. Biophys. Acta.* **1065**, 250–260.

Schwartz, J. L., Garneau, L., Savaria, D., Masson, L., Brousseau, R., and Rousseau, E. 1993. Lepidopteran-specific crystal toxins from *Bacillus thuringiensis* form cation- and anion-selective channels in planar lipid bilayers. *J. Membr. Biol.* **132**, 53–62.

Shadduck, J. A., Singer, S., and Lause, S. 1980. Lack of mammalian pathogenicity of entomocidal isolates of *Bacillus sphaericus*. *Environ. Entomol.* **9**, 403–407.

Sheeran, W. and Fisher, S. W. 1992. The effects of agitation, sediment, and competition on the persistence and efficacy of *Bacillus thuringiensis* var. *israelensis* (Bti). *Ecotoxicol. Environ. Safety* **24**, 338–346.

Siegel, J. P. 1997. Testing the pathogenicity and infectivity of entomopathogens to mammals. *In* "Manual of Techniques in Insect Pathology" (L. A. Lacey, Ed.) pp. 325–336, Academic Press, San Diego, CA.

Siegel, J. P. 2001. The mammalian safety of *Bacillus thuringiensis*-based insecticides. *J. Invertebr. Pathol.* **77**, 13–21.

Siegel, J. P., and Novak, R. J. 1997. Field trials of VectoLex CG®, a *Bacillus sphaericus* larvicide, in Illinois waste tires and catch basins. *J. Am. Mosq. Control Assoc.* **13**, 305–310.

Siegel, J. P. and Novak, R. J. 1999. Duration of activity of the microbial larvicide VectoLex CG® (*Bacillus sphaericus*) in Illinois catch basins and waste tires. *J. Am. Mosq. Control Assoc.* **15**, 366–370.

Siegel, J. P. and Shadduck, J. A. 1990. Safety of microbial insecticides to vertebrates-humans. *In* "Safety of Microbial Insecticides" (M. Laird, L. A. Lacey, and E. W. Davidson, Eds.) pp. 101–113. CRC Press, Boca Raton, FL.

Siegel, J. P., Smith, A. R., and Novak, R. J. 1997. Cellular fatty acid analysis of a human isolate alleged to be *Bacillus sphaericus* and *Bacillus sphaericus* isolated from a mosquito larvicide. *J. Appl. Environ. Microbiol.* **63**, 1006–1010.

Siegel, J. P., Tebbets, S. J., and Vail, P. V. 2000. Prevalence and transovum transmission of *Bacillus thuringiensis* Berliner in a navel orangeworm colony. *J. Invertebr. Pathol.* In press.

Singer, S. 1973. Insecticidal activity of recent bacterial isolates and their toxins against mosquito larvae. *Nature* **244**, 110–111.

Skovmand, O., Thiery, I., Benzon, G. L., Sinegre, G., Monteny, N., and Becker, N. 1998. Potency of products based on *Bacillus thuringiensis* var. *israelensis*: inter-laboratory variations. *J. Am. Mosq. Control Assoc.* **14**, 298–304.

Slatin, S. L., Abrams, C. K., and English, L. 1990. Delta-endotoxin forms cation-selective channels in planar lipid bilayers. *Biochem Biophys. Res. Commun.* **169**, 765–772.

Smirnoff, W. A. and MacLeod, C. F. 1961. Study of the survival of *Bacillus thuringiensis* var. *thuringiensis* Berliner in the digestive tracts and in feces of a small mammal and birds. *J. Insect Pathol.* **3**, 266–270.

Smith, R. A. and Couche, G. A. 1991. The phylloplane as a source of *Bacillus thuringiensis* variants. *Appl. Environ. Microbiol.* **57**, 311–315.

Snarski, V. M. 1990. Interactions between *Bacillus thuringiensis* subsp. *israelensis* and fathead minnows, *Pimephales promelas* Rafinesque, under laboratory conditions. *Appl. Environ. Microbiol.* **56**, 2618–2622.

Stahly, D. P. and Klein, M. G. 1992. Problems with *in vitro* production of spores *of Bacillus popilliae* for use in biological control of the Japanese beetle. *J. Invertebr. Pathol.* **60**, 282–291.

Stahly, D. P., Dingman, D. W., Bulla, L. A., Jr., and Aronson, A. I. 1978. Possible origin and function of the parasporal crystals in *Bacillus thuringiensis. Biochem. Biophys. Res. Commun.* **84**, 581–588.

Steinhaus, E. A., 1959. On the improbability of *Bacillus thuringiensis* Berliner mutating to forms pathogenic for vertebrates. *J. Econ. Entomol.* **52**, 506–508.

Tabashnik, B. E., Finson, N., Groeters, F. R., Moar, W. J., Johnson, M. W., Luo, K., and Adang, M. J. 1994. Reversal of resistance to *Bacillus thuringiensis* in *Plutella xylostella. Proc. Natl. Acad. Sci. USA* **91**, 4120–4124.

Tanada, Y. and Kaya, H.K 1993. "Insect Pathology". Academic Press, San Diego, CA.

Thiery, I. and Frachon, E. 1997. Identification, isolation, culture and preservation of entomopathogenic bacteria. *In* "Manual of Techniques in Insect Pathology" (L. A. Lacey, Ed.), pp 55–90. Academic Press, San Diego, CA.

Thomas, W. E. and Ellar, D. E. 1983. *Bacillus thuringiensis* var. *israelensis* crystal delta endotoxins: effects on insect and mammalian cells *in vitro* and *in vivo. J. Cell Sci.* **60**, 181–197.

Valadares de Amorim, G., Whittome, B., Shore, B., and Levin, D. B. 2001. Identification of *Bacillus thuringiensis* subsp. *kurstaki* strain HD1-like bacteria from environmental and human samples after aerial spraying of Victoria, British Columbia, Canada, with Foray 48B. *Appl Environ Microbiol* **67**, 1035–1043.

Vankova, J. and Purrini, K. 1979. Natural epizootics caused by bacilli of the species *Bacillus thuringiensis* and *Bacillus cereus. Z. ang. Entomol.* **88**, 216–221.

Van Rie, J., Jansens, S., Höfte, H., Degheele, D., and Van Mellaert, H. 1989. Specificity of *Bacillus thuringiensis* delta-endotoxins. *Eur. J. Biochem.* **186**, 239–247.

Van Rie, J., Jansens, S., Hofte, H., Degheele, D., and Van Mellaert, H. 1990a. Receptors on the brush border membrane of the insect midgut as determinants of the specificity of *Bacillus thuringiensis* delta-endotoxins *Appl. Environ. Microbiol.* **56**, 1378–1385.

Van Rie, J., McGaughey, W. H., Johnson, D. E., Barnett, B. D., and Van Mellaert, H.. 1990b. Mechanisms of insect resistance to the microbial insecticide *Bacillus thuringiensis. Science* **247**, 72–74.

Venkateswerlu, G. and Stotsky, G. 1992. Binding of the protoxin and toxin proteins of *Bacillus thuringiensis* subspecies *kurstaki* on clay minerals. *Curr. Microbiol.* **25**, 225–233.

Visser, B., Munsterman, E., Stoker, A., and Dirks, W. G. 1990. A novel *Bacillus thuringiensis* gene encoding a *Spodoptera exigua*-specific crystal protein. *J. Bacteriol.* **172**, 6783–6788.

Warren, R. E., Rubenstein, D., Ellar, D. J., Kramer, J. M., and Gilbert, R. J. 1984. *Bacillus thuringiensis* var. *israelensis*: protoxin activation and safety. *Lancet* **8378**, 678–679.

Weiner, B. A. 1978. Isolation and partial characterization of the parasporal body of *Bacillus popilliae. Can. J. Microbiol.* **24**, 1557–1561.

West, A. W., Burges, H. D., White, J. R., and Wyborn, C. H. 1984a. Persistence of *Bacillus thuringiensis* parasporal crystal insecticidal activity in soil. *J. Invertebr. Pathol.* **44**, 128–133.

West, A. W., Burges, H. D., and Wyborn, C. H. 1984b. Effect of incubation in natural and autoclaved soil upon potency and viability of *Bacillus thuringiensis. J. Invertebr. Pathol.* **44**, 121–127.

Wilcox, D. R., Shivakumar, A. G., Melin, B. E., Miller, M. F., Benson, T. A., Schopp, C. W., Casuto, D., Gundling, G. J., Bolling, T. J., Spear, B. B., and Fox, J. L. 1986. Genetic engineering of bioinsecticides. *In* "Protein Engineering: Applications in Science, Medicine, and Industry" (M. Inouye and R. Sarma, Eds.), pp. 395–413. Academic Press, Orlando, FL.

Wipfli, M. S. and Merritt, R. W. 1994. Effects of *Bacillus thuringiensis* var. *israelensis* on nontarget benthic insects through direct and indirect exposure. *J. N. Am. Benthol. Soc.* **13**, 190–205.

Wolfersberger, M. G., Lüthy, P., Maurer, A., Parenti, P., Sacchi, V. F., Giordana, B., and Hanozet, G. M. 1987. Preparation and partial characterization of amino acid transporting brush border membrane vesicles from the larval midgut of the cabbage butterfly (*Pieris brassicae*). *Comp. Biochem. Physiol.* **86A**, 301–308.

Wolfersberger, M. G. 1990. The toxicity of two *Bacillus thuringiensis* δ-endotoxin to gypsy moth larvae is inversely related to the affinity of binding sites on midgut brush border membranes for the toxins. *Experientia.* **46**, 475–477.

Wolfersberger, M. G. 1995. Permeability of *Bacillus thuringiensis* CryI toxin channels. *In* "Molecular Action of Insecticides on Ion Channels" (J. Marshall Clark, Ed.), pp. 294–301. American Chemical Society, Washington, D. C.

Woodburn, M. A., Yousten, A. A., and Hilu, K. H. 1995. Random amplified polymorphic DNA fingerprinting of mosquito-pathogenic and nonpathogenic strains of *Bacillus sphaericus. Int. J.System. Bacteriol.* **45**, 212–217.

Yokoyama, T., Tanaka, M., and Hasegawa, M. 2004. Novel cry gene from *Paenibacillus lentimorbus* strain Semadara inhibits ingestion and promotes insecticidal activity in *Anomala cuprea* larvae. *J. Invertebr. Pathol.* **85**, 25–32

Yousten, A. A., Benfield, E. F., Campbell, R. P., Foss, S. S., and Genthner, F. J. 1991. Fate of *Bacillus sphaericus* 2362 spores following ingestion by nontarget invertebrates. *J. Invertebr. Pathol.* **58**, 427–435.

Yousten, A. A., Genthner, F. J., and Benfield, E. F. 1992. Fate of *Bacillus sphaericus* and *Bacillus thuringiensis* serovar *israelensis* in the aquatic environment. *J. Am. Mosq. Contol. Assoc.* **8**, 143–148.

Zhang, J., Hodgman, T. C., Krieger, L., Schnetter, W., and Schairer, H. U. 1997. Cloning and analysis of the first cry gene from *B. popilliae. J. Bacteriol.* **179**, 4336–4341.

Chapter IV-3

Entomopathogenic microsporidia

Leellen F. Solter
Illinois Natural History Survey
Champaign, IL 61820, USA

James J. Becnel
USDA-ARS, Center for Medical
Agricultural and Veterinary Entomology
Gainesville, FL 32608, USA

1 Introduction

Microsporidia are single-celled eukaryotes that are unique among the entomopathogenic fungi with which they share a genetic relationship but few morphological characteristics. The systematics of many microsporidian groups, as well as the recent taxonomic placement of the microsporidia in the Kingdom Fungi, are incompletely resolved; however, Vossbrinck and DeBrunner-Vossbrinck (2005) evaluated the phylogenetic data in GenBank for nearly all microsporidia that have been deposited and present a phylogenetic tree that is useful to compare relationships among species.

Although many species in several taxa are common pathogens of arthropods, few microsporidia have been used in microbial control programs because certain fundamental characteristics inhibit their use. These include complicated life cycles, some with intermediate hosts, obligate parasitism, and typically chronic rather than acute effects on their hosts. The expense of *in vivo* production, difficulties in long-term storage, and low persistence and enzootic prevalence in host populations add to the obstacles (Brooks, 1988). Nevertheless, microsporidia constitute a significant portion of the primary pathogens of insects and produce considerable impacts on some economically important beneficial and deleterious species. The role of microsporidia in biological control programs will likely continue to be that of naturally occurring and classical biological control agents, and their value in population suppression, particularly in situations where chemical inputs are inappropriate or are not feasible, suggests that further studies of their biology, interactions with their hosts, specificity, and potential for manipulation are warranted.

Concerning the Protozoa, the taxonomic group with which microsporidia were, until recently, aligned, few have been studied to any great extent as biological control agents of insects. Species of entomopathogens in the Phyla Sarcomastigophora (amoebae and flagellates) and Apicomplexa (gregarines and coccidia) have been recently studied in the laboratory (*e.g.*, Kirchhoff and Führer, 1990; Lange and Wittenstein, 1998; Pereira *et al.*, 2002; Lord, 2003; Valles and Pereira, 2003) and surveyed in natural host populations (*e.g.*, Hanula and Andreadis, 1988; Apuya *et al.*, 1994; Wegensteiner *et al.*, 1996). Although the potential exists that species in these groups could be manipulated in the field for biological control purposes, few field studies have been conducted.

2 Microsporidia

A Microsporidia in insect control

Microsporidia are obligately parasitic, single-celled organisms that can only reproduce in

L.A. Lacey and H.K. Kaya (eds.), Field Manual of Techniques in Invertebrate Pathology, 199–221.

living cells. The infective form for all species is an environmentally resistant spore (environmental spore) consisting of a proteinaceous exospore and an inner plasma membrane, the plasmalemma, which surrounds the sporoplasm. In addition, the spore wall of most species possesses an endospore layer formed of a protein-chitin matrix between the exospore and plasmalemma. Tanada and Kaya (1993) described the complex of organelles composing the sporoplasm of the microsporidia. One feature, unique to the microsporidia, is a polar filament that is attached to one end of the spore and coiled within the sporoplasm. The polar filament is everted from the spore to infect cells of the host (Figure 1). Eversion or germination of spores can occur either in the gut lumen of the host if environmental spores have been ingested, or within the cells of the host. Development and replication occurs within the cytoplasm of the targeted cells of the host tissues and different species may be specific to one or more particular tissues of a host. Effects on hosts range from benignity, or nearly so, to acute malignancy, depending on the microsporidian species and the host species.

Life cycles of the microsporidia vary among species and range from quite simple to very complex. Typically, environmental spores are ingested by a host and conditions within the midgut lumen initiate polar filament eversion or germination. The everting polar filament punctures a midgut epithelial or underlying midgut muscle cell and the sporoplasm, containing one or two nuclei as well as other organelles, passes through the hollow polar filament into the host cell. The sporoplasm divides to form amoeba-like but nonmotile meronts. Vegetative reproduction may or may not continue to occur in the midgut tissues but the meronts eventually become committed to sporogony and divide to form sporonts. Sporonts form sporoblasts, the immature forms of spores, sometimes after another division. Finally, one infective, environmental spore develops from each sporoblast (Figure 2).

Avery and Anthony (1983) reported the presence of newly formed spores that germinated within host tissues early in the infection process. Recent findings suggest that for many species, particularly species for which the midgut is not the sole target tissue for production of environmental spores, these early (primary) spores, may be responsible for spreading the parasite within the host by germinating within the tissues and infecting adjacent cells and/or other target tissues, including hemocytes (Iwano and Ishihara, 1991; Fries *et al.*, 1992; Becnel *et al.*, 1989; Johnson *et al.*, 1997; Solter and Maddox, 1998a). Another vegetative cycle occurs in the target tissues and may involve yet another cycle of primary spore formation and germination. This is one of the simplest microsporidian life cycles and is typical of the genus *Nosema*, as well as other genera in terrestrial and aquatic insects.

Many microsporidian life cycles are more complicated and may involve any or all of the following: haplosis, meiosis, sexual cycles, multiple spore types, and intermediate hosts (Andreadis, 1988, 2002; Andreadis and Vossbrinck, 2002; Becnel *et al.*, 1989; Johnson *et al.*, 1997). Vertical transmission of

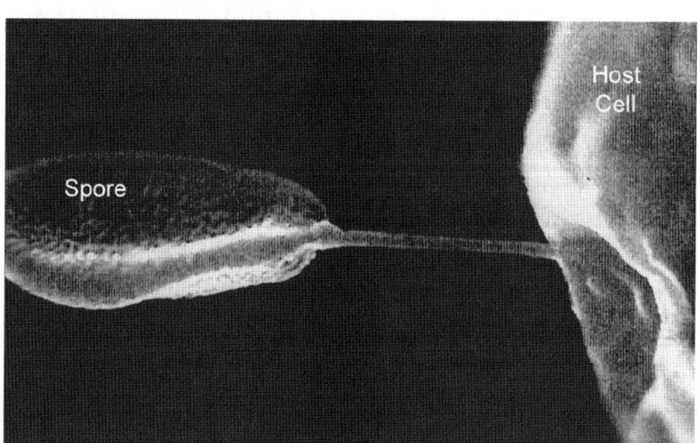

Figure 1. A microsporidian spore germinates by everting the polar filament and injecting the contents of the spore (the sporoplasm) into a host midgut cell. Scanning electron micrograph, 13.200 X. Micrograph courtesy of A. Linde

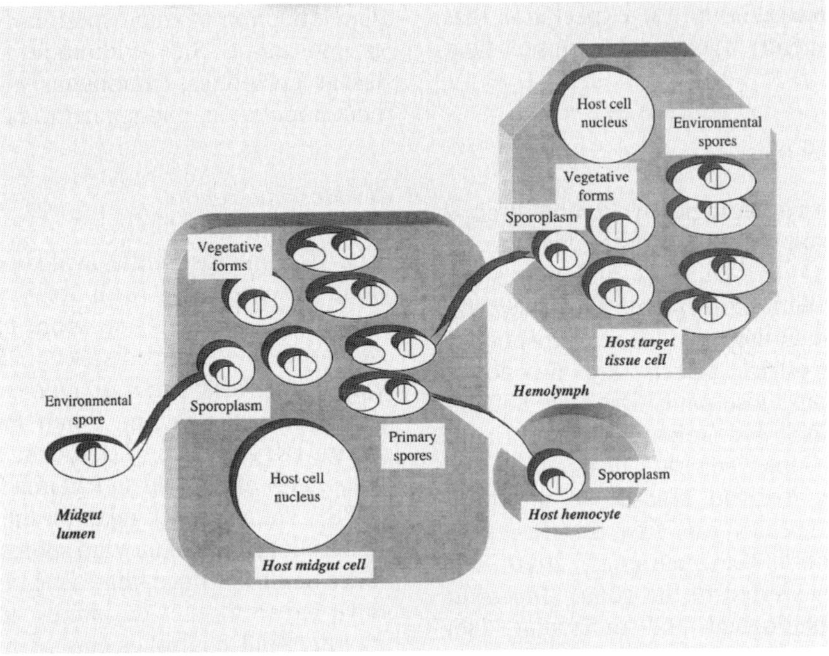

Figure 2. Generalized life cycle of the diplokaryotic *Nosema*-type microsporidia

microsporidia from infected host, usually the female parent, to offspring is common and, for some species, is the major mechanism for transmission (Solter, 2006).

Some species of microsporidia cause high mortality in larval host populations. For example, *Nosema tortricis* produces epizootics in the green tortrix, *Tortrix viridana* (Franz and Huger, 1971); *Cougourdella* sp. devastates populations of its caddisfly host *Glossosoma nigrior* (Kohler and Wiley, 1992); and *Vairimorpha necatrix* induces rapid host mortality when large numbers of spores are ingested (Maddox, 1986a). Most species, however, cause chronic infections (Gaugler and Brooks, 1975), the consequences of which range from little effect to quite significant effects on individual hosts or host populations. Documented deleterious effects on hosts include increased larval mortality, developmental delays and growth reduction, reduced adult longevity and fecundity, and increased susceptibility to environmental stresses (Siegel, 1985; Williams *et al.*, 1999; Henn and Solter, 2000).

Nearly 1,000 species of microsporidia have been described, the majority from insect hosts. They have been recovered from nearly every order of insects that have been intensively sampled for protist infections. The few field populations of insect species that have been monitored for long-term periods exhibit prevalences ranging from 1% to nearly 100% (Wilson, 1973; Maddox, 1986b; Pilarska *et al.*, 1999; Briano 2005). Only one microsporidium, *Paranosema* (= *Nosema*; = *Antonospora*) *locustae* (see Section 4A for genus name change) used for control of rangeland grasshoppers, has been registered as a microbial insecticide (Henry and Oma, 1981). Although microsporidia are probably not suitable for use as microbial insecticides, their potential as classical biological control agents is far more promising (Federici and Maddox, 1996). The majority of current efforts to manipulate microsporidia as biological control agents are within the context of classical or augmentative biological control programs emphasizing inoculative release, or local inundative release into introduced pest populations. In addition, natural epizootics such as *Nosema pyrausta* in European corn borer, *Ostrinia nubilalis*, populations may be monitored for decision-making in control programs. Therefore, the emphasis of this chapter, a necessary departure from others in this manual, will be the evaluation of microsporidia released with the expectation that they will permanently cycle in their insect host populations.

B Key target insects

Insect pests in two ecologically diverse groups have been targeted for control programs, aquatic hosts in the Order Diptera, and terrestrial hosts in the Orders Coleoptera, Hymenoptera, Lepidoptera and Orthoptera. The microsporidia parasitizing terrestrial insects are taxonomically distinct from those parasitizing the aquatic insects (Vossbrinck and DeBrunner-Vossbrinck, 2005), despite the generic designation of the well-known *Nosema algerae*, a mosquito pathogen. This species was first transferred to the genus *Brachiola* (Lowman *et al.*, 2000), but was recently transferred to the genus *Anncaliia* based on morphological and molecular data (Franzen *et al.*, 2005). The recent finding that *A. algerae* was responsible for a fatal myositis in a human patient on immunosuppressant drugs (Coyle *et al.*, 2004) makes it highly unlikely that *A. algerae* could ever be considered for use as a biological control agent for mosquitoes. Likewise, *V. culicis* has also been found to be phylogenetically related to *Trachipleistophora hominis* (Cheney *et al.*, 2000), an opportunistic microsporidian parasite of AIDS patients that is infectious for mosquitoes and can be passively transferred from infected adults during feeding (Weidner *et al.*, 1999).

Table 1 lists species of pests for which microsporidia have been tested as biological control agents, or for which the microsporidia are known to be naturally occurring biological control agents (Maddox, 1986a).

3 Microsporidia for management in aquatic systems

Microsporidia have been described and reported from a variety of medically important Diptera including members of the Ceratopogonidae, Simuliidae, Chironomidae and Culicidae. While microsporidia play a role as natural regulators of many aquatic Diptera, the only attempts to utilize microsporidia as control agents in aquatic systems have been against mosquitoes. The attraction of microsporidia for management of mosquitoes lies with their ability to cause larval epizootics, continuously cycle within a host population, and spread to new habitats.

A Mosquito control

The first microsporidia evaluated for mosquito control were those with relatively simple life cycles that produce one spore type responsible for horizontal transmission. The two species evaluated were *Vavraia culicis* against *Culex pipien fatigans* on the South Pacific island of Nauru (Reynolds, 1972) and *A. algerae* against *Anopheles albimanus* in Panama (Anthony *et al.*, 1978). The approach taken in these releases was to inundate the habitat with spores to effect short-term control. Since that time, two discoveries have led to a renewed interest in microsporidia as microbial control agents of mosquitoes. The first was the involvement of an intermediate copepod host in the life cycle of *Amblyospora* spp. (Andreadis, 1985; Sweeney *et al.*, 1985) and *Parathelohania* spp. (Avery, 1989; Avery and Undeen, 1990), documenting for the first time the mechanism for horizontal transmission in these genera. The second was the identification of a new genus of microsporidia, *Edhazardia*, that is both horizontally and vertically transmitted but does not require an intermediate host (Hembree, 1979; Becnel *et al.*, 1989). Andreadis (1990) suggested that the best approach for utilizing these heterosporous microsporidia is via inoculative releases for long-term establishment and population suppression.

B Environmental considerations for microsporidia in aquatic systems

Introduction of microsporidia into an aquatic habitat depends on the delivery of spores such that they will be encountered and ingested by the target host. Spore viability and dosage, estimated by concentration of spores in the water, are major factors in establishing infections in the target host. Recycling of the pathogen in the host population depends on horizontal transmission within a larval population and/or vertical transmission by infected female adults within and between larval breeding sites.

Table 1. Target insects for microsporidian pathogens, and naturally occurring biological control agents[1]

Microsporidia by host order	Insect host	Common name	Life stage[2]	Application method[3]	Commodity
COLEOPTERA					
Nosema scolyti	Scolytidae	bark beetles		NC	softwood timber
Unikaryon minutum	*Dendroctonus frontalis*	southern pine beetle		NC	pine timber
Nosema sp.	*Pissodes strobe*	white pine weevil		NC	pine timber
Microsporidium sp.	*Otiorhynchus sulcatus*	black vine weevil	L	I	various nursery shrubs
DIPTERA					
Amblyospora spp.	*Aedes* spp. & *Culex* spp.	mosquitoes		NC	animal & human pest
Edhazardia aedis	*Aedes aegypti*	yellow fever mosquito	L	I	animal & human pest
Parathelohania spp.	*Anopheles* spp.	mosquitoes		NC	human & animal pest
Nosema algerae	*Anopheles* spp.	mosquitoes	L	I	human & animal pest
Vavraia culicis	*Culex* spp.	mosquitoes	L	I	human & animal pest
HYMENOPTERA					
Thelohania solenopsis					
Vairimorpha invicta	*Solenopsis invicta*	fire ants	L, Q	IBT	human & food animal pest
LEPIDOPTERA					
Nosema spp.	*Cactoblastis cactorum*	prickly pear moth			native cactus
Nosema fumiferanae	*Choristoneura fumiferana*	spruce budworm		NC	fir trees
Vairimorpha necatrix	*Lacanobia oleracea*	tomato moth	L		tomato & various shrubs
Vairimorpha disparis	*Lymantria dispar*	gypsy moth	L	I & Inoc	forest trees & shrubs
Nosema portugal	*L. dispar*	gypsy moth	L		
Nosema sp. (10 isolates)[a]	*L. dispar*	gypsy moth	L		
Nosema pyrausta	*Ostrinia nubilalis*	European corn borer		NC	corn
Pleistophora oncoperae	*Oncopera alboguttata*			NC	pasture grasses
Vairimorpha sp.	*Plutella xylostella*	diamondback moth	L	I	cabbage, other cole crops
Nosema tortricis	*Tortrix viridana*	green tortrix		NC	oak species
ORTHOPTERA					
Paranosema locustae	various spp.[4]	grasshoppers & locusts	N	I & Inoc	rangeland grasses

[1] Adapted from Maddox (1986a) and Roberts *et al*. (1990);
[2] L = larval stage. N = nymph. Q = queen;
[3] NC = natural control-monitored. I = Inundative release. Inoc = Inoculative release. IBT = Infected brood transfer.
[4] see Brooks (1988) for species lists;
[a] Austria (2 isolates) Bulgaria (1 isolate), Germany (1 isolate), Hungary (1 isolate) Poland (1 isolate), Portugal (1 sp.), Romania (1 isolate), Slovakia (2 isolates).

1 Spore survival

The survival of spores in the aquatic habitat after introduction has not been fully investigated. The aquatic habitat of mosquitoes is a complex system of biotic and abiotic environmental factors, some of which may reduce the probability that the introduced spores will encounter the target host as viable infectious agents. Abiotic factors such as extreme temperatures, water chemistry and sunlight can inactivate spores or cause them to germinate prematurely (Undeen and Vávra, 1997). In addition, most spores from aquatic hosts cannot withstand drying (Undeen and Vávra, 1997). Biotic factors such as secondary hosts, predators, and non-target organisms may remove spores prior to encountering the target host and reduce infectivity levels (Brooks, 1988).

The general approach to establishing infections of microsporidia in a population of mosquitoes has been the inundative application of spores to larval mosquitoes. This assumes that the host population is actively feeding and that mosquito larvae would ingest spores within hours of their introduction. Because of this, inactivation of spores by extreme temperatures and sunlight has not been a major consideration when applying spores to an aquatic environment.

The inactivation of spores of aquatic microsporidia by sunlight has only been investigated for *A. algerae* and *V. culicis*. Purified spores of *A. algerae* (5×10^6 spores in 10 ml water in a 60×15 mm Petri dish) exposed to sunlight for up to 4 h showed no significant decline in the incidence of infection in *An. albimanus*, but the intensity of infection declined significantly (Kelly and Anthony, 1979). Both the prevalence and intensity of infection in *An. albimanus* was determined for *V. culicis* spores exposed to sunlight and the combination of the 2 measurements gave an LT_{50} of 3.1 h and an LT_{90} of 6.1 h (Kelly *et al.*, 1981).

The temperature tolerances of aquatic species of microsporidia vary with species but generally the pathogen will tolerate conditions that support development of the target host. For example, the ecologically relevant temperature range for normal development of larval *Aedes aegypti* is 14–31 °C (Gilpin and McClelland, 1979). Results of field studies conducted at this temperature range have found that spores of *Edhazardia aedis* introduced into containers maintained high levels of infectivity for 3 days, but by day 4 and 5 infectivity levels fell to approximately 50% and by day 10 was about 10% (Becnel and Johnson, 2000). This time frame of survival would provide ample opportunity for viable spores of *E. aedis* to be ingested by larvae of *Ae. aegypti*.

Water quality is an important consideration when applying spores to an aquatic habitat because it has been well established that germination is affected by pH, various ions, and inhibitors (Undeen, 1990). Certain factors in the water could stimulate or inhibit spores to germinate prior to ingestion by larvae, which could influence infectivity. For example, shifts in pH are known to cause spores of *A. algerae* to germinate (Undeen and Avery, 1988), and ammonia and calcium can inhibit germination (Undeen, 1978). In field studies with *V. culicis*, water from one site produces no infections in *C.p. fatigans* because "apparently the water from that site contained some factor inimical to *V. culicis*" (Reynolds, 1972). Laboratory bioassays using field water conducted prior to introduction of spores into the habitat should indicate whether water chemistry would affect the infectivity of spores to the target larval hosts.

Spores introduced into the aquatic system can be lost due to the activity of non-target organisms including secondary mosquito hosts, scavengers, filter feeders, and predators. Laboratory studies have shown that spores of *E. aedis* germinated in each of 17 aquatic scavengers, filter feeders, and predators tested but failed to cause infections (Becnel, 1992). Nine mosquito predators were fed larvae of *Anopheles quadrimaculatus* infected with *A. algerae* and resulted in the infection of only one hemipteran species (Van Essen and Anthony, 1976). In field studies with *A. algerae*, none of the non-target mosquito species examined were infected nor were any of the aquatic Hemiptera, Coleoptera, Diptera, Odonata, Ephemeroptera, and Crustacea (Anthony *et al.*, 1978). Therefore, in aquatic habitats with high levels of non-target aquatic organisms, a certain loss of spores due to the activity of these organisms must be considered in determining the appropriate dose to be applied.

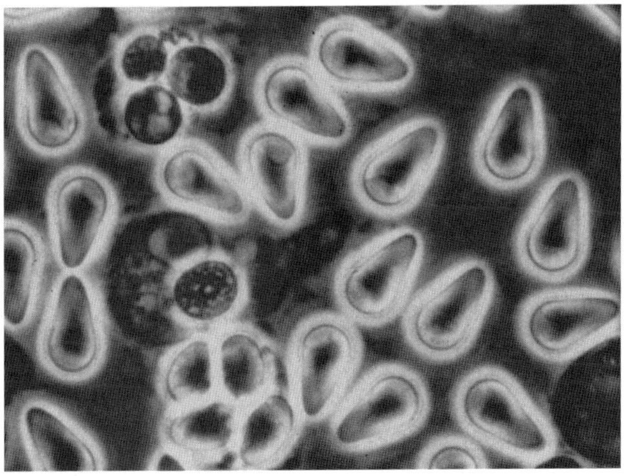

Figure 3. Spores of *Edhazardia aedis* as observed under phase contrast microscopy (Photo courtesy J.J.B. and Society for Invertebrate Pathology)

2 Recycling

In a three-year study on natural ecology, it was demonstrated that there is a well-defined seasonality to transmission cycles and epizootics involving *Amblyospora connecticus* (Andreadis, 1988). Larval epizootics as a result of transovarial transmission occur each fall and routinely produce infection prevalences of 80–100%. The two major events identified as critical to the maintenance of *A. connecticus* are 1) the fall epizootic responsible for significant larval mortality of *Aedes cantator* and, concurrently, the infection of the copepod intermediate host to provide for overwintering of the parasite, and 2) the spring epizootic when spores formed in the copepod are responsible for horizontal transmission to larval *Ae. cantator* and results in infected adults. Awareness of these critical windows in the maintenance of *A. connecticus* suggests a control strategy that precludes larviciding these habitats during the spring and fall as this would be counterproductive and disrupt the natural balance of the disease cycle. Larviciding during the summer is suggested because of the absence of infected mosquitoes as well as copepods. This strategy also targets the healthy part of the mosquito population that had escaped infection by *A. connecticus. A. connecticus* was successfully introduced and maintained in a field population of *Ae. cantator* via infected *Acanthocyclops vernalis* (Andreadis 1989b), demonstrating that microsporidia with complex life cycles involving an intermediate host can be employed as part of a classical biological control project.

C Application methods and effects on microsporidia in aquatic systems

All efforts to utilize microsporidia for mosquito control have been directed at initiating horizontal infections in the larval stages of the mosquito host. Viable spores must be delivered to the habitat of the target host and remain in the feeding zone for a suitable time for ingestion to occur.

1 Dosage

The concentration of spores to apply to the aquatic habitat is an important decision that will be influenced by a variety of factors. The starting point is to determine the spore dosage that will consistently yield 100% infection in the target host in the laboratory. Losses due to the factors discussed above must be determined by bioassay of field-collected mosquitoes in field water (Becnel and Johnson, 2000). Examples of spore application rates and the resulting infection levels found in previous studies are given below.

In field studies with *A. algerae* against *Anopheles albimanus* in Panama, infection levels were directly related to the rate of application (Anthony *et al.*, 1978). The peak infection rates ranged from 16% for 1 application of spores at 1.1×10^7 spores/m^2 to 86% for four applications at 2.2×10^9 spores/m^2. Infection levels of *A. algerae* against *An. stephensi* in Pakistan were also dosage related (Brooks, 1988). In the studies of *Ae. cantator* infected with *A. connecticus* via introduction of infected *A. vernalis*, however, infection rates were no higher in tests where total number of spores introduced were 10-fold higher than dosages for other treatment groups (Andreadis, 1989a).

2 *Delivery*

Introductions of microsporidia into the aquatic habitat has been via: 1) aqueous suspensions of purified spores (Reynolds, 1972; Anthony *et al.*, 1978); 2) mosquito larvae infected with horizontally infectious spores (Becnel and Johnson, 2000); and 3) intermediate copepod hosts infected with horizontally infectious spores (Andreadis, 1989a). Regardless of the type of inoculum used, timing is critical to successful introduction of the pathogen into the target population. Aqueous spore suspensions have been applied with a backpack sprayer with no apparent adverse effect on the viability of the spores (Anthony *et al.*, 1978). Small-scale introductions require estimating the volume of water in each test site and a uniform distribution of spores to give the desired final concentration (Reynolds, 1972).

In addition to the introduction of microsporidia into a larval field population of mosquitoes via the release of the infected copepod intermediate hosts (Andreadis, 1989a), *E. aedis* has also been successfully introduced into larval populations of *Ae. aegypti* through the release of infected larvae (Becnel and Johnson, 2000). This classical biological control approach via inoculative introduction depends on the ability of the microsporidium to persist and spread within the target population and exert long-term population suppression.

3 *Storage and formulations*

The optimal storage conditions for spores from aquatic hosts must be determined on an individual species basis. Generally, purified spores will store for much longer periods of time than spores in suspensions contaminated with host tissues. Spores of aquatic microsporidia cannot tolerate temperatures at or below 0 °C or above 40 °C (Brooks, 1988; Maddox and Solter, 1996). Purified spores of *V. culicis* and *A. algerae* can be stored at ca. 5 °C for extended periods of time. Spores of *V. culicis* maintained their viability for 4 months at 4 °C (Wang, 1982), while spores of *A. algerae* were held for approximately 10 years at 4 °C with little loss of activity (Undeen and Vávra, 1997). Antibiotics have been used to control the growth of adventitious microorganisms that are usually detrimental to spore survival (Pilley, 1978; Undeen and Vávra, 1997).

Little information is available on the storage of spores of heterosporous microsporidia from aquatic hosts. Andreadis (1991) stored whole larvae of *Ae. cantator* infected with *A. connecticus* at 4 °C for up to 7 years. Meiospores extracted from the larvae and assayed in the copepod intermediate host *A. vernalis* revealed a significant decrease in meiospore viability after 5 months of storage in distilled water and a virtual loss of all infectivity after 17 months. Some infections, however, were obtained in copepods after 65 months of storage. Undeen *et al.* (1993) found that purified spores of *E. aedis* were totally inactivated by storage at 0–5 °C for 24 h but survived, with some loss of activity, for 30 days at 10–30 °C.

Non-sporal stages can be "stored" within the host. *E. aedis* is most efficiently stored within the eggs of the mosquito *Ae. aegypti* (Becnel *et al.*, 1989; Sweeney and Becnel, 1991). Sporoplasms, not spores, are found in the eggs of *Ae. aegypti* and can survive for at least 3–6 months and perhaps as long as the eggs remain viable (Becnel *et al.*, 1989; Becnel and Johnson, 2000). Furthermore, the sporoplasms in the egg can survive a 24 h exposure to 5 °C indicating that the developmental stages of *E. aedis* are more tolerant to chilling than spores (Undeen *et al.*, 1993).

The spores of *A. algerae* and *V. culicis* used in field studies have been applied as aqueous suspensions (Anthony *et al.*, 1978; Reynolds, 1972), but spores of these microsporidian species do not float (Anthony *et al.*, 1978; Wang, 1982). A formulation that would keep the spore in the feeding zone would increase the rates of infection and extend the residual activity of spores (Anthony *et al.*, 1978). The spore application rate could also be reduced, and the time between applications could be increased. A floating formulation of *A. algerae* was developed by J.V. Maddox but proved to be unsuccessful in field trials (Brooks, 1988). No further attempts to develop floating formulations for spores from aquatic hosts have been reported.

4 Evaluations

Infection levels in larvae and adults indicate the effectiveness of applications of microsporidia to the aquatic environment for mosquito control. For species such as *A. algerae* and *V. culicis*, larval mortality is a small but important factor, (Undeen and Dame, 1987). Impact on adult longevity and fecundity must also be considered (Anthony *et al.*, 1972). In the case of vectors, reducing longevity to allow for only one blood meal could be effective in preventing transmission of diseases (Anthony *et al.*, 1972). In the case of heterosporous microsporidia, vertical transmission is a critical component of the life cycle. Therefore, the effects of infections on adult longevity, fecundity, and mortality in progeny must be considered when evaluating the effectiveness of the pathogen to suppress mosquito populations (Andreadis, 1990; Becnel *et al.*, 1995). In such cases, ovipositional sampling may be an essential tool to provide important information on the levels of infections in the adult mosquitoes. The general techniques and procedures necessary to identify infections of microsporidia can be found in Undeen and Vávra (1997).

In most cases, the only practical use for microsporidia in mosquito control involves species that can cause long-term suppression of the population. This will usually involve the ability of the pathogen to persist and spread within the population. Studies conducted thus far have not demonstrated long term persistence or spread of the introduced pathogens. The field release of *V. culicis* against *Cx. p. fatigans* initially involved four sites. Three of the sites contained *V. culicis* infected larvae 15 days post-introduction, with infection rates up to 30% (Reynolds, 1972). Infected larvae were found at one site 66 days post-exposure and in a follow-up survey during the period of 18 – 24 months post-exposure, infection levels of < 0.1% and approximately 2% were found at 2 sites. There was no evidence of dispersal to other sites and no apparent reduction in mosquito populations. Field-tests with *A. algerae* in Panama against *An. albimanus* produced peak infection levels of about 86% 12 and 15 days from the first application, but only one infected larva was found on day 70 (Anthony *et al.*, 1978). No follow-up surveys were conducted. These two field studies have demonstrated that aquatic microsporidia such as *Anncalia* and *Vavraia* can be introduced and infect a population of mosquitoes, but long-term control would require augmentative releases.

4 Microsporidia for management in terrestrial systems

Despite the large numbers of species of microsporidia described and reported from terrestrial insects, epizootiological data on microsporidian infections have been documented for very few natural host populations or for introductions of microsporidia into populations of pest insects. Nevertheless, those microsporidia studied had significant deleterious effects on their host populations, including mortality and sublethal effects throughout the life cycle of the insect hosts (Roberts *et al.*, 1990). Microsporidia have been studied for use as biological control agents both as microbial insecticides and as natural enemies. The latter use, where microsporidia are regarded as microbial counterparts to insect parasites and predators, is probably the more feasible approach considering the biology of these pathogens (Roberts *et al.*, 1990; Federici and Maddox, 1996).

A Microsporidia for use as microbial insecticides

Canning (1953, 1962) described and characterized *Nosema locustae*, the only microsporidium that has been extensively studied for use as a microbial insecticide. Originally described from the African migratory locust, it was recently renamed *Paranosema locustae* (Sokolova *et al.* 2003; see also Slamovits *et al.*, 2004 and Sokolova *et al.*, 2005 for discussions on genus designation). *P. locustae* is primarily a fat body parasite, reproducing slowly in the host tissues, but eventually starving the host of energy reserves (Johnson, 1997). This species is also found naturally occurring in North American Great Plains and prairie populations of grasshoppers. In addition to Chapter VII-17 in this manual, a series of excellent reviews of three decades of research on this pathogen as a biological control agent of a variety of grasshopper and locust species was published by Goettel and Johnson (1997). In this collection of papers, various strategies for acridoid pests are considered, including use of *P. locustae* within the scope of an IPM program, formulation of *P. locustae* in baits, current status and prospects for the use of microsporidia and protozoans, and issues associated with augmentative inoculation of grasshopper populations (Bateman, 1997; Johnson, 1997; Johnson and Dolinski, 1997; Prior and Streett, 1997). We refer the reader to these sources and to a review by Lange (2002) for further information on the use of *P. locustae*, however, we will discuss some experimental data from field research on *P. locustae*, because other microsporidia produce pathogenicity and host effects similar to this well-studied organism.

One other microsporidium, *Vairimorpha necatrix*, a highly virulent species infecting various noctuid pests (Maddox *et al.*, 1981), has been field-tested as a microbial insecticide (Mistric and Smith, 1973; Fuxa and Brooks, 1978), but application timing and cost of *in vivo* production has hindered interest in commercial production (Roberts *et al.*, 1990).

B Terrestrial microsporidia as natural enemies

Microsporidia and other protists are most fittingly utilized as natural enemies of insect pests in the following ways: 1) inoculative or inundative augmentation, 2) introduction and establishment, and 3) manipulation as naturally occurring pathogens (Roberts *et al*, 1990). The most intensively studied example of natural control of an insect pest by a microsporidium is *Nosema pyrausta* in the European corn borer, *Ostrinia nubilalis*. Probably introduced to North America shortly after the introduction of the host, this microsporidium is the single most important natural enemy of *O. nubilalis* in many areas of the United States (Andreadis, 1984; Siegel *et al.*, 1986) and adversely affects *O. nubilalis* larval populations even when it is in enzootic phase (Andreadis, 1984). Other naturally occurring microsporidia shown to have detrimental effects on host populations include *Nosema fumiferanae*, which suppresses forest populations of the spruce budworm, *Choristoneura fumiferana* (Wilson, 1973; 1981); *Nosema tortricis*, which greatly reduces populations of *Tortrix viridana* (Franz and Huger, 1971); *N. heliothidis*, a pathogen of *Helicoverpa zea* (Brooks *et al.*, 1978); *Thelohania solenopsae* and *Vairimorpha invictae* in the red imported fire ant (Oi and Williams, 2002; Briano 2005) and the *Nosema lymantriae/Vairimorpha disparis* complex in the gypsy moth (McManus and Solter, 2003).

C Considerations for use of microsporidia in terrestrial systems

1 Taxonomic uncertainties

Microsporidian taxonomy is not well resolved. Morphologically similar species described from different host taxa (including different phyla) have often been placed in the same family or even genus based on characteristics such as the following: life cycle; number of nuclei in different life stages; morphology of the organelles including the polar filament, presence or absence of vesicles of parasite or host origin enclosing "packets" of spores (Figure 4B); and size of the environmental spores. Unfortunately, molecular analyses performed to date have shown that many of these characters may not hold up to molecular scrutiny (Baker *et al.*, 1994, 1995; Vossbrinck and DeBrunner-Vossbrinck, 2005). Many morphological characters are probably ancestral or the result of convergent evolution in these organisms (Baker *et al.*, 1997a,b). No matter how these problems are viewed, it is difficult to study organisms for use in biological control when taxonomic information is questionable. A carefully considered combination of molecular and biological studies is greatly needed to address these taxonomic difficulties.

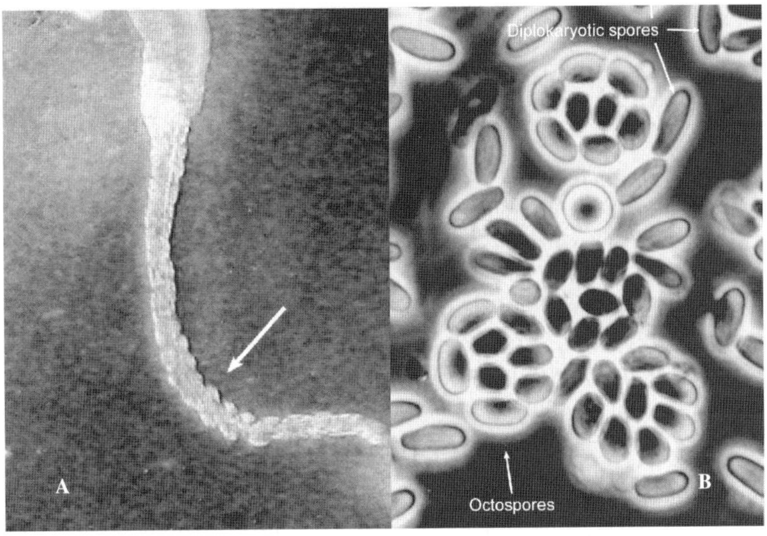

Figure 4. Phase contrast light micrograph (100x) of a lepidopteran silk gland infected with microsporidia. The hypertrophied or swollen cells appear as beads filled with spores (arrow). Released spores appear as cloudy material surrounding the silk gland. The spores are easily identified at 400x. B. Monokaryotic octospores and diplokaryotic free spores of the *Lymantria dispar* microsporidium *Vairimorpha disparis* (1000x phase contrast). Photos by L.F.S

2 *Spore survival*

Despite the relatively tough exospore that encloses the infective environmental spores of terrestrial microsporidia, the spores are sensitive to ultraviolet (UV) radiation and high temperatures (Kaya, 1977; Chu and Jaques, 1981), repeated freezing and thawing (Maddox and Solter, 1996; Undeen and Solter, 1996), and bacterial and fungal degradation (Brooks, 1980).

Consideration must be given to the factors that result in spore degradation and inviability, as well as to the ecology of the host, in order to use microsporidia as insecticides or in augmentative or classical biological control programs. Lewis and Johnson (1982), for example, determined that *N. pyrausta*, sprayed in an aqueous suspension on plants during whorl and pollen-shedding stages, remained infective to *O. nubilalis* larvae for at least 12 days. The addition of Shade®, a UV protectant, did not increase the life of the microsporidian spores because the spores were sufficiently protected from UV radiation in the whorls of the leaves or behind the sheaf collar area. Studies of survival of microsporidian spores in other field situations showed that spores were only viable for a few hours after application if they were applied to leaf surfaces exposed to sunlight (Maddox, 1966). Onstad and Bauer (personal communication) encountered difficulties with spore viability when *Lymantria dispar* egg masses were dipped in suspensions of *Nosema portugal* and stapled to tree trunks. Emerging larvae did not become infected because spore viability decreased before the eggs hatched. Placing egg masses on the north sides of tree trunks to avoid UV light, as well as spraying the egg masses with spore solutions just as the eggs begin to hatch may rectify a difficulty of this kind.

Spores that are sprayed in an aqueous suspension or otherwise are used to inoculate a field population of a target host should be freshly produced and applied as soon as possible after isolation from the laboratory hosts to ensure viability. In addition, the spores should be sprayed at the time the host is actively feeding and is in the appropriate life stage (Henry and Oma, 1974). Otherwise, transovarially infected egg masses or infected larvae should be used to inoculate a population (L. Bauer, personal communication).

3 *Host specificity*

Microsporidia have been variously reported as generalist pathogens and as host specific pathogens. Although there are marked differences between species regarding host range, the results of microsporidian surveys in natural insect populations (Nordin, 1971; Roberts *et al.*, 1977; Andreadis *et al.*, 1983; Siegel *et al.*, 1988) and evaluations of laboratory (physiological) host range (Andreadis, 1989b; 1994; Johnson *et al.*, 1997; Solter and Maddox, 1998b; Solter *et al.*, 2005; Solter, 2006) suggest that the ecological host ranges of microsporidia are probably much more restricted than laboratory bioassays predict. The results of laboratory experiments using techniques that circumvent or overwhelm the natural barriers to infection in non-target hosts may be important when the pathogen is to be used in an inundative approach where large numbers of infectious units are expected to produce mortality (Maddox, 1992). Infections produced by these "unnatural" procedures, however, may not be predictive of the ecology of these pathogens in field situations, nor accurately reflect the consequences of inoculative release (Solter and Maddox, 1998b).

The concept that ecological host range is much narrower than physiological host range was tested by Undeen and Maddox (1973), Andreadis (1989a; 1994), Johnson *et al.* (1997), Solter and Maddox (1998b) and Solter *et al.* (2005), and was found to hold true for the mosquito pathogens *A. algerae, Am. connecticus*, and *E. aedis*, as well as 22 species of lepidopteran microsporidia. Undeen and Maddox (1973) injected *A. algerae* into 16 non-target hosts in eleven orders (four phyla) and all species tested in the seven arthropod orders developed infections using this artificial technique. When the spores were administered *per os*, only one species developed an infection. In other experiments, infections were produced in most lepidopteran non-target hosts when high spore concentrations of lepidopteran microsporidia were fed, but the few microsporidian species that were horizontally transmitted from infected to uninfected conspecific non-target larvae were transmitted at much lower frequencies than between infected and uninfected natural hosts (Solter and Maddox, 1998b).

Because most terrestrial microsporidia will only be useful in classical biological control programs using inoculative release (or inundative release in a small area) against a natural host, it is important to evaluate the host specificity of these pathogens using methods that elucidate ecological host specificity. To base decisions about release of microsporidia on host specificity testing methodologies used for chemical or microbial pesticides risks elimination of potentially useful pathogens that may well have a very narrow host range in the field. Simple testing procedures can be devised that more closely approximate a natural system, for example, the horizontal transmission tests used by Solter and Maddox (1998b) and vertical transmission studies of Andreadis (1989a) and Solter *et al.* (2005). Additionally, ecologically relevant non-target hosts should be tested, and the nature of infections produced should be carefully evaluated (Andreadis, 1989a; Hasan and Delfosse, 1995; Solter and Maddox, 1998b). A pathogen may reproduce in a non-target host but in a suboptimal manner that suggests that transmission between conspecific non-target individuals is not likely to occur. New molecular techniques are being developed that will allow reasonably cost-effective evaluation of the host range of a microsporidium in its aboriginal range, including difficult-to-detect infections in hosts other than natural hosts, and also allow any microsporidia released in a biological control program to be identified in the host and non-target insects. Several research groups have designed primers to unique areas of rDNA for selected microsporidian species, which identify low-level infections that are not detectable under light microscopy (Malone and McIvor, 1995; Hatakeyama *et al.*, 1997; Valles *et al.*, 2002; Sokolova *et al.*, 2004). Nevertheless, moderate to heavy infections are easily detectable using light microscopy at magnifications of 250–400x (best with phase contrast lenses) and the microscope is still the most frequently used tool for exploration (Figure 4A,B).

4 *Persistence in inoculated host populations*

Jeffords *et al.* (1989) determined that microsporidia inoculatively released into *L. dispar* populations in a small woodlot cycled in the host population for at least 3 years, and Lange and De Wysiecki (1996) reported the presence of *P. locustae* in release areas 12 years after application. In naturally infected populations, microsporidia often are maintained in the host population in low prevalences for many years, then respond to increasing host population density with increases in prevalence, sometimes to high levels (Pilarska *et al.*, 1999).

Different microsporidian species utilize different strategies for maintenance in the host population. *N. pyrausta* overwinters in infected fifth instar *O. nubilalis* larvae. Infected female hosts that survive winter conditions and pupation mate and oviposit eggs containing transovarially infected embryos. Infected first-generation larvae may die before pupation, inoculating tunnels in the corn plants with spores that infect the second host generation (Zimmack and Brindley, 1957). *N. pyrausta* and other microsporidian species that are systemic or occur primarily in host gut tissues and/or the Malpighian tubules continually inoculate the environment by release of spores in the feces of the host (Maddox, 1987). Uninfected conspecific larvae feeding in the same environment ingest the spores and become infected. Species in the genus *Vairimorpha* are extremely prolific in the host fat body tissues and large quantities of spores are released into the environment when the host dies. Some microsporidia may be disseminated by predators; *Podisus maculiventris*, the spined soldier bug, excretes large amounts of infective spores after feeding on bright-line brown-eye, *Lacanobia oleracea*, larvae infected with *V. necatrix*. Mortality on plants where the predators were allowed to defecate was over 60% in laboratory trials, suggesting that predators may provide a significant role in use of the microsporidium in biological control applications on tomatoes (Down *et al.*, 2004).

D Application methods for control of terrestrial insects

1 Production and storage

Microsporidia can only be produced in living cells, either in host insects or in tissue culture. Tissue culture has not been developed for mass production of microsporidian spores due to cost of culture media and low spore production (Brooks, 1988). Infective concentrations of spores vary depending on the microsporidian species and host species chosen for production. Usually the natural host is used if it can be mass reared and is of sufficiently large size to produce large quantities of spores (Brooks, 1980). It is important to determine the spore concentration that will be fed to the host and to determine which host stage will be inoculated. Spore concentrations that are too low fail to infect sufficient numbers of hosts or produce sufficient spores for harvest. Concentrations that are too high for the stage chosen often kill the host before maturation of the infective environmental spores. *Vairimorpha necatrix* is highly infective to several species of noctuid hosts, *e.g., Spodoptera exigua*, and *Trichoplusia ni*, at less than 100 spores/mm^2 diet surface fed to third instars. Other species may require spore concentrations 10–100x higher to infect a large percentage of larvae and produce heavy infections. Most *Endoreticulatus* species, for example, require 10^4 or 10^5 spores/mm^2 of diet surface to heavily infect second or third instar *L. dispar* (Solter *et al.*, 1997). Brooks (1988) compiled a comprehensive list of microsporidia and their natural and alternate hosts used for spore production, as well as the yield per host for each species and host combination. Maximum spore yields in natural hosts range from approximately 10^4 spores/host for several microsporidian species in mosquitoes to 10^{10} spores/host for *V. necatrix* produced in *H. zea*.

Terrestrial microsporidia can be isolated, purified, and stored in frozen water (Henry and Oma, 1974), in sterile water at refrigeration temperatures (4–6 °C) (Fuxa, 1979), lyophilized or vacuum-dried (Lewis and Lynch, 1974) and in liquid nitrogen (Maddox and Solter, 1996). They can also be stored for some time in dried cadavers, either at room temperature or frozen (Henry and Oma, 1974). Dried or fresh cadavers can also be suspended in water or water/glycerol and stored in liquid nitrogen. Spores lose viability at temperatures above 35 °C, and upon drying after isolation (Maddox, 1973). Different species of microsporidia vary in their capacity to tolerate various storage methods and some experimentation is needed to determine the best method for each species (Undeen and Vávra, 1997). For maximum retention of viability in terrestrial microsporidian spores, the protocols in Table 2 serve as a starting point. The reader is also referred to Brooks (1988), Undeen and

Table 2. Storage methods for terrestrial microsporidia (see also Brooks, 1988; Undeen and Vávra, 1997; Undeen, 1997)

Storage methods[a]	Time period
In situ: maintenance of infected living hosts in colony	indefinite[b]
Dried host cadavers, −80 to 30 °C	extended periods[c]
Isolated and purified spores, −80 to −20 °C; add s/f[d] and glycerol to suspension	extended periods
Lyophilized	< 2 years
Refrigeration of isolated, purified spores, add s/f, purified distilled or deionized water suspension	3 months– 10 years
Liquid nitrogen: cadavers, dried or fresh	indefinite
Liquid nitrogen: spores in cell culture	indefinite
Liquid nitrogen: isolated spores in 1:1 water/glycerol, s/f	indefinite

[a] No method allows for storage in temperatures over 35 °C (Vávra and Maddox, 1976). Repeated freezing and thawing degrades spores (Maddox and Solter, 1996).
[b] Repeated cycling in hosts or use of alternate hosts may change virulence and/or infectivity of some microsporidian species.
[c] Storage period may decrease as temperature increases.
[d] Approximately 5 μg of 1:1 streptomycin:fungizone (s/f) per ml suspension (Maddox and Solter, 1996). Other antibiotics may be used (Pilley, 1978).

Vávra (1997), and Undeen (1997) for excellent and detailed reviews of storage procedures.

With the exception of *P. locustae*, so few field studies have been performed and evaluated that it is difficult to generalize about the appropriate numbers of spores for application. Nevertheless, some of the factors that should be considered are 1) infectivity and virulence of the microsporidian species used; 2) susceptibility of the target host; 3) stage of the host; 4) life history of the host, *e.g.*, burrowing or mining stages; 5) host plant structure, canopy, and environment; 6) goal of the control attempt, *e.g.*, immediate control or expectations for cycling of the pathogen in the host population; 7) the level of host population suppression necessary to reduce damage below the economic injury level; and 8) effects of high dosages in the microhabitats of non-target organisms. Even more difficult to predict or control is the persistence of infective stages of microsporidia in the environment and the transmission efficiency of the pathogen within the host population.

2 *Dosages, formulation and delivery*

P. locustae has been applied at concentrations ranging from 1×10^8 to 5×10^9 spores/ha (Roberts *et al.*, 1990), but the standard formulation adopted was 2.46×10^9 spores/ha. This application rate produced 50–60% reductions in grasshopper populations and 35–50% infection among survivors (Henry *et al.*, 1973). The spores were formulated as a bait in wheat bran with hydroxymethyl cellulose sticker and applied with Buffalo® turbine equipment mounted on a truck. Ultra low volume (ULV) aqueous sprays were not as successful as the application of wheat bran bait; ULV sprays required much higher concentrations of spores and resulting in less population reduction (Henry *et al.*, 1978).

Limited area inundative releases have also been used to inoculate target host populations with microsporidia expected to function as natural enemies. Using backpack sprayers, Wilson and Kaupp (1975, 1976), applied *N. fumiferanae* spores in 25% molasses and a sun protectant at 2.5×10^{10} to 1.8×10^{11} spores/tree to accelerate an epizootic in a *C. fumiferanae* population. Lewis and Raun (1978) used augmentative aqueous suspension sprays of *N. pyrausta* at an average of 2.3×10^7 spores/plant against *O. nubilalis*, obtaining significantly higher mortality in the first generation and higher prevalence of infection in the second generation. Weiser and Novotny (1987) added wetting agents to aqueous sprays of *N. lymantriae* (=*Vairimorpha disparis* in these experiments; Novotny, personal communication) to gypsy moth, and these methods were repeated in recent field studies using *V. disparis* and *N. lymantriae*, producing infections in the gypsy moth hosts but few or no infections in non-target lepidopteran species (L.F.S., unpublished data). The addition of UV protectants increased the half life of *V. necatrix* spores from a few hours (Maddox, 1977) to about 6 days on tobacco and soybean leaves (Fuxa and Brooks, 1978).

Jeffords *et al.* (1988) and Bauer and Onstad (personal communication) tested the feasibility of inoculating *L. dispar* populations in the U.S. with egg masses from laboratory-reared gypsy moth females dipped in microsporidia suspensions. Jeffords *et al.* (1988) dipped the egg masses in 1.3×10^5 spore/μl suspensions of *Nosema* sp. (*N. portugal*) and 3.3×10^6 spores/μl suspensions of *Vavraia* sp. (=*Endoreticulatus* sp.). Larvae hatched in the laboratory using these spore concentrations were 100% infected. In the field, egg masses were stapled to trees. Infection was overall 21% for *Vavraia* sp. 7 weeks after release, 10% for *N. portugal*, with approximately 23% of trees hosting infected larvae. Bauer and Onstad dipped egg masses in spore suspensions of approximately 1×10^6 spores/ml for 5 minutes. While field releases were not successful because of unusually cold weather and resulting delay in egg hatch, 75–92% of larvae hatching from treated egg masses in the laboratory became infected. When egg masses were soaked for 10 min, infection rates were 86–100%. Odindo (1992) and Odindo *et al.* (1993) did not increase mortality of cereal stem borer, *Chilo partellus*, by inoculating egg masses, but damage in sorghum was reduced up to 92.5% by spraying leaves with an aqueous suspension of *N. marucae* spores. Based on screenhouse trials, the authors suggested application of 4.5×10^{11} spores/ha for sorghum (Odindo *et al.*, 1993).

Solter and McManus (North American Plant Protection Organization Proposal, approved March 2006) have proposed to inoculate gypsy moth, *Lymantria dispar*, populations with microsporidia by releasing gypsy moth larvae that will be infected in the laboratory with a chronic but fatal dosage of the pathogens. This method is intended to circumvent the problems of environmental degradation of the spores, as well as to disseminate the spores over a 2–3 week period in natural populations.

Thelohania solenopsae and *Vairimorpha invictae* were successfully transmitted between colonies to red imported fire ant, *Solenopsis invicta*, queens by the introduction of brood (mixture of eggs, larvae and pupae) from infected *S. invicta* colonies (Williams *et al.*, 1999; Oi *et al.*, 2001; Oi *et al.*, 2005). Introductions in the field were made by opening fire ant mounds with a shovel and adding 5 g of infected brood. After 22 weeks, infected queens were producing infected brood via transovarial transmission. After 48 weeks, the inoculated mounds were still active, but there was a 30% reduction in the average population index for the inoculated colonies as compared to a 5% increase in the controls. *T. solenopsae* spread to other colonies in the test site 48–66 weeks post inoculation (Williams *et al.*, 1999). *T. solenopsae* slowly debilitates the *S. invicta* queen resulting in a decrease in reproductive capacity, and both species of microsporidia deleteriously affect colonies. While the mechanisms involved in the transmission of *T. solenopsae* by introduction of infected brood are not well understood at present, use of the method shows promise for long-term population suppression.

3 *Interactions with other pathogens and pesticides*

With the exception of *P. locustae* studies (Goettel and Johnson, 1997), there are few recent reports on pathogen-pathogen or pathogen-pesticide interactions involving microsporidia. Bauer *et al.* (1998) looked at the effects on *L. dispar* of a mixture of nucleopolyhedrovirus, *Ld*MNPV, and *N. portugal*. They determined that viral infectivity increased ten-fold and time to death was decreased when the larvae were first infected with *N. portugal*. While the production of virus particles was decreased, probably due to the shorter lethal time, increased susceptibility to the virus by infected insects may mean that *L. dispar* populations do not need to be particularly dense for maintenance of both pathogens. Virus and microsporidia interactions differ depending on the host-pathogen-pathogen complex and the reader is encouraged to examine papers by Nordin and Maddox (1972), Fuxa (1979), Cossentine and Lewis (1984), and Moawed *et al.* (1987) for further information. Solter *et al.* (2002) and Pilarska *et al.* 2006) found antagonism between two phylogenetically related microsporidian species, *N. lymantriae* and *V. disparis*, that naturally occur in the fat body tissues of *L. dispar*, but a third species, *Endoreticulatus schubergi*, was compatible with both *N. lymantriae* and *V. disparis*. While *E. schubergi* was occasionally found naturally co-infecting *L. dispar* in the field, usually with the less virulent *N. lymantriae, N. lymantriae* and *V. disparis* have not been found co-occurring in *L. dispar* populations (Solter *et al.*, 2002). Long-term post-release studies of *P. locustae* in grasshopper populations where another microsporidium, *Perezia dichroplusae*, is naturally occurring in grasshopper populations found almost no overlap of the two species in different localities and no mixed infections in individuals (Lange, 2003).

Goertz *et al.* (2004) studied the influence of Dimilin®, a diflubenzuron used for gypsy moth control, on microsporidian infection in the gypsy moth. They found reduced establishment of the pathogen in the host and, when establishment did occur, development of the pathogen was curtailed and spore viability and production was reduced. Microsporidia contain alpha-chitin in the protective thick endospore, and Dimilin is a chitinsynthetase inhibitor. Pierce *et al.* (2001) determined that *Bacillus thuringiensis* LC_{50} values were significantly lower for European corn borers infected with *N. pyrausta* than for uninfected corn borers. Field studies conducted shortly after *Bt* corn was introduced did not show significant reduction of *N. pyrausta* in corn borer populations in *Bt* corn, but no follow-up studies were conducted to determine whether strong susceptibility of corn borers to *Bt* eventually extinguished the *Nosema*. A field monitoring study by Novotny *et al.* (2000) points out the importance of a complex of pathogens in population suppression of outbreaking insects. In seven low-density gypsy moth populations, virus infections were the most common, followed by mixed infections of virus and bacteria (probably virus-induced septicemia), virus and microsporidia, and virus and fungi.

4 *Evaluation*

Evaluation procedures for efficacy of microsporidia as biological control agents logically involve assessments of damage reduction in the commodity of interest. It can also involve assessments of host mortality, generally accomplished by comparing pest populations in test areas with those in similar untreated areas. Wilson and Kaupp (1975, 1976) and Lewis and Raun (1978) acknowledged the difficulties in producing sufficient microsporidian spores to adequately utilize *N. fumiferanae* and *N. pyrausta* as microbial insecticides. These researchers suggested that the major expected outcome of augmentative application would be advancement of epizootics and/or an increase of the prevalence of disease in future generations of pests in the inoculated area.

Unlike studies using predators and parasitoids in biological control programs, there is documented research on relatively few exotic pathogens being introduced into pest populations as classical biological control agents. The only well-known exotic introduction of microsporidia, *N. pyrausta* in *O. nubilalis* in the U.S., was an accidental introduction. In addition to the taxonomic and biological reasons listed in the introduction of this chapter, many other issues probably play a role in this lack of effort (McManus *et al.*, 1989; Hajek *et al.*, Chapter VI-1 of this manual).

Most evaluations of microsporidia as candidates for biological control programs still take place in the laboratory as predictive efforts, such as the host specificity studies discussed earlier. A few limited-scale or contained release programs have also been performed. Jeffords *et al.* (1989) evaluated overwintering success of *N. portugal*, and *Vavraia* sp. (=*Endoreticulatus*) in *L. dispar*. Circular plots with radii of 18 m were established in isolated woodlots. The plots were divided into quadrants with six concentric sampling zones. Larvae were collected from foliage or other resting sites from the ground to approximately 2 m in trees and shrubs, and were returned to the laboratory for dissection and determination of infection. For the collection of late stage larvae, trees were wrapped with burlap at chest height to provide resting sites for the larvae and facilitation of collection. *N. portugal*, which is transovarially transmitted, persisted and occurred in nearly the same prevalences the following year.

5 Sources of spores and information on pathogen host associations

The American Type Culture Collection (ATCC) occasionally has available a small number of insect pathogenic microsporidian species in limited amounts (www.atcc.org/catalogs). This collection should be augmented by microsporidiologists and care should be taken to supply spores that originate from field-collected hosts. Species discovered in laboratory colonies are not a good resource because the closed environment encourages infections in unnatural hosts; thus, the true origin and natural host species of the microsporidia might be unknown. Field-collected microsporidia can, however, be reproduced in the laboratory in any susceptible host in which infective environmental spores are produced. Care should be taken to ensure that no other disease organisms are present before the hosts are inoculated, as well as to avoid cross-contamination of the cultures with other species of microsporidia or entomopathogens.

Spores of *P. locustae* marketed as NoLo-Bait may be obtained from a variety of businesses specializing in the sale of biological control agents. Enter the product name on any internet search engine to locate convenient websites. Note that most of the sites continue to use the original Latin name *Nosema locustae*, and most still refer the pathogen as a protozoan.

There are no commercial sources of other species of microsporidia. Most researchers who study microsporidia will provide inoculum of the species with which they work to other researchers. Transfer between laboratories of microsporidia isolated from exotic hosts usually requires the receiving laboratory to have approved permits. For the U.S.A., obtain permits from the United States Department of Agriculture, Animal and Plant Health Inspection Service, Plant Protection Quarantine (USDA APHIS PPQ Form 526). Forms may currently be found at the USDA APHIS website: http://www.aphis.usda.gov/permits/

An excellent source of information about microsporidian-host associations is the Ecological Database of the World's Insect Pathogens (EDWIP) (Braxton *et al.*, 2003). The website for the database is http://cricket.inhs.uiuc.edu/edwipweb/edwipabout.htm

6 References

Andreadis, T. G. 1984. Epizootiology of *Nosema pyrausta* in field populations of the European corn borer (Lepidoptera: Pyralidae). *Environ. Entomol.* **13**, 882–887.

Andreadis, T. G. 1985. Experimental transmission of a microsporidian pathogen from mosquitoes to an alternate copepod host. *Proc. Natl. Acad. Sci.* **82**, 5574–5577.

Andreadis, T. G. 1988. *Amblyospora connecticus* sp. nov. (Microsporida: Amblyosporidae): Horizontal transmission studies in the mosquito *Aedes cantator* and formal description. *J. Invertebr. Pathol.* **52**, 90–101.

Andreadis T. G. 1989a. Infection of a field population of *Aedes cantator* with a polymorphic microsporidium, *Amblyospora connecticus* via release of the intermediate copepod host, *Acanthocyclops vernalis. J. Am. Mosq. Control Assoc.* **5**, 81–85.

Andreadis, T. G. 1989b. Host specificity of *Amblyospora connecticus*, a polymorphic microsporidian parasite of *Aedes cantator. J. Med. Entomol.* **26**, 140–145.

Andreadis, T. G. 1990. Polymorphic microsporidia of mosquitoes: potential for biological control. *In* "New Directions in Biological Control: Alternatives for Suppressing Agricultural Pests and Diseases" (P. E. Dunn and R.R. Baker, Eds.). pp. 177–188. New York: Alan R. Liss, Inc.

Andreadis, T. G. 1991. Experimental observations on the longevity of meiospores of *Amblyospora connecticus* (Microsporida). *J. Invertebr. Pathol.* **58**, 458–460.

Andreadis, T. G. 1994. Host range tests with *Edhazardia aedis* (Microsporida: Culicosporidae) against northern nearctic mosquitoes. *J. Invertebr. Pathol.* **64**, 46–51.

Andreadis T. G. 2002: Epizootiology of *Hyalinocysta chapmani* (Microsporidia: Thelohaniidae) infections in field populations of *Culiseta melanura* (Diptera: Culicidae) and *Orthocyclops modestus* (Copepoda: Cyclopidae): a three-year investigation. *J. Invertebr. Pathol.* **81**, 114–121.

Andreadis, T. G., Dubois, N. R., Moore, R. E. B., Anderson, J. F., and Lewis, F. B. 1983. Single applications of high concentrations of *Bacillus thuringiensis* for control of gypsy moth (Lepidoptera: Lymantriidae) populations and their impact on parasitism and disease. *J. Econ. Entomol.* **76**, 1417–1422.

Andreadis T. G. and Vossbrinck C. F. 2002. Life cycle, ultrastructure and molecular phylogeny of *Hyalinocysta chapmani* (Microsporidia: Thelohaniidae) a parasite of *Culiseta melanura* (Diptera: Culicidae) and *Orthocyclops modestus* (Copepoda: Cyclopidae). *J. Eukaryot. Microbiol.* **49**, 350–364.

Anthony, D. W., Savage, K. E., Hazard, E. I., Avery, S. W., and Boston, M. D. O. S. W. 1978. Field tests with *Nosema algerae* Vavra and Undeen (Microsporida, Nosematidae) against *Anopheles albimanus* Wiedemann in Panama. *Misc. Publ. Entomol. Soc. Am.* **11**, 17–28.

Anthony, D. W., Savage, K. E., and Weidhaas, D. E. 1972. Nosematosis: its effect on *Anopheles albimanus* Wiedemann, and a population model of its relation to malaria transmission. *Proc. Helminthol. Soc. Washington* pp. 428–433.

Apuya, L. C., Stringham, S. M., Arends, J. J., and Brooks, W. M. 1994. Prevalence of protozoon infections in darkling beetles from poultry houses in North Carolina. *J. Invertebr. Pathol.* **63**, 255–259.

Avery, S. W. 1989. Horizontal transmission of *Parathelohania obesa* (Protozoa: Microsporidia) to *Anopheles quadrimaculatus* (Diptera: Culicidae). *J. Invertebr. Pathol.* **53**, 424–426.

Avery, S. W. and Anthony, D. W. 1983. Ultrastructural study of early development of *Nosema algerae* in *Anopheles albimanus. J. Invertebr. Pathol.* **42**, 87–95.

Avery, S. W. and Undeen, A. H. 1990. Horizontal transmission of *Parathelohania anophelis* to the copepod *Microcyclops varicans*, and the mosquito, *Anopheles quadrimaculatus. J. Invertebr. Pathol.* **56**, 98–105.

Baker, M. D., Vossbrinck, C. R., Becnel, J. J., and Andreadis, T. G. 1997a. Phylogeny of *Amblyospora* (Microsporidia: Amblyosporidae) and related genera based on small subunit ribosomal DNA data: a possible example of host parasite cospeciation. *J. Invertebr. Pathol.* **71**, 199–208.

Baker, M. D., Vossbrinck, C. R., Becnel, J. J., and Maddox, J. V. 1997b. Phylogenetic position of *Amblyospora* Hazzard and Oldacre (Microspora: Amblyosporidae) based on small subunit rRNA data and its implication for the evolution of the microsporidia. *J. Eukaryot. Microbiol.* **44**, 220–225.

Baker, M. D., Vossbrinck, C. R., Didier, E. S., Maddox, J. V., and Shadduck, J. A. 1995. Small subunit ribosomal DNA phylogeny of various microsporidia with emphasis on AIDS related forms. *J. Eukaryot. Microbiol.* **42**, 564–570.

Baker, M. D., Vossbrinck, C. R., Maddox, J. V., and Undeen, A. H. 1994. Relationships among *Vairimorpha* and *Nosema* species (Microspora) based on ribosomal RNA sequence data. *J. Invertebr. Pathol.* **64**, 100–106.

Bateman, R. 1997. Methods of application of microbial pesticide formulations for the control of grasshoppers and locusts. *Memoirs Entomol. Soc. Can.* **171**, 69–81.

Bauer, L. S., Miller, D. L., Maddox, J. V., and McManus, M. L. 1998. Interactions between a *Nosema* sp. (Microspora: Nosematidae) and nuclear polyhedrosis virus infecting the gypsy moth, *Lymantria dispar* (Lepidoptera: Lymantriidae). *J. Invertebr. Pathol.* **74**, 147–153.

Becnel, J. J. 1992. Safety of *Edhazardia aedis* (Microspora, Amblyosporidae) for non-target aquatic organisms. *J. Am. Mosq. Control Assoc.* **8**, 256–260.

Becnel, J. J., Garcia, J. J., and Johnson, M. A. 1995. *Edhazardia aedis* (Microspora: Culicosporidae) effects on the reproductive capacity of *Aedes aegypti* (Diptera: Culicidae). *J. Med. Entomol.* **32**, 549–553.

Becnel, J. J., Sprague, V., Fukuda, T., and Hazard, E. I. 1989. Development of *Edhazardia aedis* (Kudo, 1930) n. g., n. comb. (Microsporida: Amblyosporidae) in the mosquito *Aedes aegypti* (L.) (Diptera: Culicidae). *J. Protozool.* **36**, 119–130.

Becnel, J. J. and Johnson, M. A. 2000. Impact of *Edhazardia aedis* (Microsporidia: Culicosporidae) on a seminatural population of *Aedes aegypti* (Diptera: Culicidae). *Biol. Contr.*, **18**, 39–48.

Braxton, S. M., Onstad, D. W., Docktor, D. E., Giordano, R., Larsson, R. Humber, R. A. 2003. Description and analysis of two internet-based databases of insect pathogens: EDWIP and VIDIL. *J. Invertebr. Pathol.* **83**, 185–195.

Briano, J.A. 2005. Long-term studies of the red imported fire ant, *Solenopsis invicta*, infected with the microsporidia *Vairimorpha invictae* and *Thelohania solenopsae* in Argentina. *Environ. Entomol.* **34**, 124–132.

Brooks, W. M. 1980. Production and efficacy of protozoa. *Biotech. Bioeng.* **22**, 1415–1440.

Brooks, W. M. 1988. Entomogenous Protozoa. *In* "Handbook of Natural Pesticides, Vol. V, Microbial Insecticides" (C. M. Ingnoffo Ed.), pp. 1–150. CRC Press, Inc., Boca Raton, Florida.

Brooks, W. M. and Cranford, J. D. 1978. Host parasite relationships of *Nosema heliothidis* Lutz and Splendor. *Misc. Publ. Entomol. Soc. Am.* **11**, 51–633.

Canning, E. U. 1953. A new microsporidian, *Nosema locustae* n. sp., from the fat body of the African migratory locust, *Locusta migratoria migratorioides* (R. and F.). *Parasitol.* **43**, 287–290.

Canning, E. U. 1962. The life cycle of *Nosema locustae* Canning in *Locusta migratoria migratorioides* Reiche and Fairmaire), and its infectivity to other hosts. *J. Insect Pathol.* **4**, 237–247.

Cossentine, J. E. and Lewis, L. C. 1984. Interactions between *Vairimorpha necatrix, Vairimorpha* sp., and a nuclear polyhedrosis virus from *Rachiplusia ou* in *Agrotis ipsilon* larvae. *J. Invertebr. Pathol.* **44**, 28–35.

Chu, W. H. and Jacque, R. P. 1981. Factors affecting infectivity of *Vairimorpha necatrix* (Microsporidia: Nosematidae) in *Trichoplusia ni* (Lepidoptera: Noctuidae). *Can. Entomol.* **113**, 93–102.

Cheney, S. A., Lafranchi-Tristem, N. J., and Canning, E. U. 2000. Phylogenetic relationships of *Pleistophora*-like microsporidia based on small subunit ribosomal DNA sequences and implications for the source of *Trachipleistophora hominis* infections. *J. Eukaryot. Microbiol.* **47**, 280–287.

Coyle, C. M., Weiss, L. M., Rhodes, L. V., Cali, A., Takvorian, P. M., Broun, D. F., Visvesvara,, G. S., Xiao, L., Naktin, J., Young, E., Gareca, M., Colasante, G., Wittner, M. 2004. Fatal myositis due to the microsporidian *Brachiola algerae*, a mosquito pathogen. *N. Engl. J. Med.* **351**, 42–47.

Down, R. E., Bell, H. A., Matthews, H. J. Kirkbride-Smith, A. E., and Edwards, J. P. 2004. Dissemination of the biocontrol agent *Vairimorpha necatrix* by the spined soldier bug, *Posidus maculinventris*. *Entomol. Exp. Appl.* **110**, 103–114.

Federici, B. A. and Maddox, J. V. 1996. Host specificity in microbe-insect interactions. *Bioscience* **46**, 410–421.

Franz, J. M. and Huger, A. M. 1971. Microsporidia causing the collapse of an outbreak of the green tortrix (*Tortrix viridana* L.) in Germany. *Proc. 4th Intern. Colloq. Insect Pathol.* pp. 48–53

Franzen, C. Nassonova, E. S., Scholmerich, J., and Issi, I. V. 2005. Transfer of the members of the genus *Brachiola* (Microsporidia) to the genus *Anncaliia* based on ultrastructural and molecular data. *J. Eukaryot. Microbiol.* **53**, 26–35.

Fries, I., Granados, R. R., and Morse, R. A. 1992. Intracellular germination of spores of *Nosema apis*. *Z. Apidol.* **23**, 61–70.

Fuxa, J. P. 1979. Interactions of the microsporidium *Vairimorpha necatrix* with a bacterium, virus, and fungus in *Heliothis zea*. *J. Invertebr. Pathol.* **33**, 316–323.

Fuxa, J. P. and Brooks, W. M. 1978. Persistence of spores of *Vairimorpha necatrix*, a microbial control agent for *Heliothis zea*, in tobacco, cotton, and soybean. *J. Econ. Entomol.* **71**, 169–172.

Fuxa, J. P. and Brooks, W. M. 1979. Mass production and storage of *Vairimorpha necatrix* (Protozoa: Microsporida). *J. Invertebr. Pathol.* **33**, 86–94.

Gaugler, R. R. and Brooks, W. M. 1975. Sublethal effects of infection by *Nosema heliothidis* in the corn earworm, *Heliothis zea*. *J. Invertebr. Pathol.* **26**, 57–63.

Gilpin, M. E. and McClelland, G. A. H. 1979. Systems analysis of the yellow fever mosquito *Aedes aegypti*. *Fortschr. Zool.* **25**, 355–388.

Goertz, D., Pilarska, D., Kereselidze, M., Solter, L., and Linde, A. 2004. Studies on the impact of two *Nosema* isolates from Bulgaria on the gypsy moth (*Lymantria dispar* L.) *J. Invertebr. Pathol.* **87**, 105–113.

Goettel, M. S. and Johnson, D. L. (Eds.). "1997. Microbial Control of Grasshoppers and Locusts" *Mem. Entomol. Soc. Can.* **171**, 400 pp.

Hanula, J. L. and Andreadis, T. G. 1988. Parasitic microorganisms of Japanese beetle (Coleoptera: Scarabaeidae) and associated scarab larvae in Connecticut USA soils. *Environ. Entomol.* **17**, 709–714.

Hasan, S. and Delfosse, E. S. 1995. Susceptibility of the Australian native, *Heliotropium crispatum*, to the rust fungus *Uromyces heliotropii* introduced to control common heliotrope, *Heliotropium europeaum*. *Biocontrol Sci. Tech.* **5**, 165–174.

Hatakeyama, Y., Kawakami, Y., Iwano, H., Inoue, T., and Ishihara, R. 1997. Analyses and taxonomic inferences of small subunit ribosomal RNA sequences of five microsporidia pathogenic to the silkworm, *Bombyx mori*. *J. Seric. Sci. Jap.* **66**, 242–252.

Hembree, S. C. 1979. Preliminary report of some mosquito pathogens from Thailand. *Mosq. News* **39**, 575–582.

Henn, M. W. and Solter, L. F. 2000. Food utilization values of gypsy moth, *Lymantria dispar* (Lepidoptera: Lymantriidae) larvae infected with the microsporidium *Vairimorpha* sp. (Microsporidia: Burenellidae). *J. Invertebr. Pathol.* **76**, 263–269.

Henry, J. E. and Oma, E. A. 1974. Effect of prolonged storage of spores on field applications of *Nosema locustae* (Microsporida: Nosematidae) against grasshoppers. *J. Invertebr. Pathol.* **23**, 371–377.

Henry, J. E. and Oma, E. A. 1981. Pest control by *Nosema locustae*, a pathogen of grasshoppers and crickets. *In* "Microbial Control of Pests and Plant Diseases, 1970–1980" (Burges, H. D., Ed.), pp. 573–586. Academic Press, London.

Henry, J. E., Oma, E. A., and Onsager, J. A. 1978. Relative effectiveness of ULV spray applications of spores of *Nosema locustae* against grasshoppers. *J. Econ. Entomol.* **71**, 629–632.

Henry, J. E., Tiahrt, K., and Oma, E. A. 1973. Importance of timing, spore concentrations, and levels of spore carrier in applications of *Nosema locustae* (Microsporida: Nosematidae) for control of grasshoppers. *J. Invertebr. Pathol.* **21**, 263–272.

Iwano, H. and Ishihara, R. 1991. Dimorphism of spores of *Nosema* spp. in cultured cell. *J. Invertebr. Pathol.* **57**, 211–219.

Jeffords, M. R., Maddox, J. V., McManus, M. L., Webb, R. E., and Wieber, A. 1988. Egg contamination as a method for the inoculative release of exotic microsporidia of the gypsy moth. *J. Invertebr. Pathol.* **51**, 190–196.

Jeffords, M. R., Maddox, J. V., McManus, M. L., Webb, R. E., and Wieber, A. 1989. Evaluation of the overwintering success of two European microsporidia inoculatively released into gypsy moth populations in Maryland. *J. Invertebr. Pathol.* **53**, 235–240.

Johnson, D. L. 1997. Nosematidae and other Protozoa as agents for control of grasshoppers and locusts: current status and prospects. *Mem. Ent. Soc. Can.* **171**, 375–389.

Johnson, D. L. and Dolinski, M. G. 1997. Attempts to increase the prevalence and severity of grasshoppers with the entomopathogen *Nosema locustae* Canning (Microsporida: Nosematidae) by repeated field applications. *Mem. Entomol. Soc. Can.* **171**, 391–400.

Johnson, M. A., Becnel, J. J., and Undeen, A. H. 1997. A new sporulation sequence in *Edhazardia aedis* (Microsporidia: Culicosporidae), a parasite of the mosquito *Aedes aegypti* (Diptera: Culicidae). *J. Invertebr. Pathol.* **70**, 69–75.

Kaya, H. K. 1977. Survival of spores of *Vairimorpha* (=*Nosema*) *necatrix* (Microsporidia: Nosematidae) exposed to sunlight, ultraviolet radiation, and high temperatures. *J. Invertebr. Pathol.* **30**, 192–198.

Kelly, J. F. and Anthony, D. W. 1979. Susceptibility of spores of the microsporidian *Nosema algerae* to sunlight and germicidal ultraviolet radiation. *J. Invertebr. Pathol.* **34**, 164–169.

Kelly, J. F., Anthony, D. W., and Dillard, C. R. 1981. A laboratory evaluation of the microsporidian *Vavraia culicis* as an agent for mosquito control. *J. Invertebr. Pathol.* **37**, 117–122.

Kirchhoff, J. F. and Führer, E. 1990. Experimental analysis of infection and developmental cycle of *Malamoeba scholyti* Purrini in *Dryocoetes autographus* Ratz. (Coleoptera: Scholytidae). *Entomophaga* **35**, 537–544.

Kohler, S. L. and Wiley, M. J. 1992. Parasite-induced collapse of populations of a dominant grazer in Michigan streams. *Oikos* **65**, 443–449.

Lange, C. E. 2002. The development of *Nosema locustae* (Protozoa: Microsporidia) for the biological control of grasshoppers (Orthoptera: Acridoidea) and the consequences of its use in Argentina. *Rev. Soc. Entomol. Argent.* **61**, 1–9.

Lange, C. E. 2003. Long-term patterns of occurrence of *Nosema locustae* and *Perezia dichroplusae* (Microsporidia) in grasshoppers (Orthoptera: Acrididae) of the Pampas, Argentina. *Acta Protozool.* **42**, 309–315.

Lange, C. E. and De Wysiecki, M. L. 1996. The fate of *Nosema locustae* (Microsporida: Nosematidae) in Argentine grasshoppers (Orthoptera: Acrididae). *Biol. Control* **7**, 24–29.

Lange, C. E. and Wittenstein, E. 1998. Susceptibility of the locust *Schistocerca cancellata* (Orthoptera: Acrididae) to different entomopathogens. *Rev. Soc. Entomol. Argentina.* **57**, 19–22.

Lewis, L. C. and Lynch, R. E. 1974. Lyophilization, vacuum drying, and subsequent storage of *Nosema pyrausta* spores. *J. Invertebr. Pathol.* **24**, 149–153.

Lewis, L. C. 1982. Persistence of *Nosema pyrausta* and *Vairimorpha necatrix* measured by microsporidiosis in the European corn borer. *J. Econ. Entomol.* **75**, 670–674.

Lewis, L. C. and Lynch, R. E. 1978. Foliar application of *Nosema pyrausta* for suppression of populations of European corn borer. *Entomophaga* **23**, 83–88.

Lord, J. C. 2003. *Mattesia oryzaephili* (Neogregarinorida: Lipotrophidae), a pathogen of stored-grain insects: Virulence, host range and comparison with *Mattesia dispora. Biocontrol Sci. Technol.* **13**, 589–598.

Lowman, P. M., Takvorian, P. M. and Cali, A. 2000. The effects of elevated temperatures and various time-temperature combinations on the development of *Brachiola* (*Nosema*) *algerae* n. comb. in mammalian cell culture. *J. Eukaryot. Microbiol.* **47**, 221–234.

Maddox, J. V. 1966. Generation time of the microsporidian, *Nosema necatrix*, in larvae of the armyworm, *Pseudaletia unipuncta. J. Invertebr. Pathol.* **11**, 90–96.

Maddox, J. V. 1973. The persistence of the microsporidia in the environment. *Misc. Publ. Entomol. Soc. Am.* **9**, 99–104.

Maddox, J. V. 1977. Stability of entomopathogenic Protozoa. *Misc. Publ. Entomol. Soc. Am.* **10**, 3–18.

Maddox, J. V. 1986a. Current status on the use of microsporidia as biocontrol agents. *In* "Fundamental and Applied Aspects of Invertebrate Pathology" (R. A. Samson, J. M. Vlak, and D. Peters, Eds.) Proc. IVth Intl. Coll. Soc. Invertebr. Pathol. pp. 518–522. Veldhoven, The Netherlands.

Maddox, J. V. 1986b. Possibilities for manipulating epizootics caused by Protozoa: a representative case-history of *Nosema pyrausta*. *In* "Fundamental and Applied Aspects of Invertebrate Pathology" (R. A. Samson, J. M. Vlak, and D. Peters, Eds.) Proc. IVth Intl. Coll. Soc. Invertebr. Pathol. pp. 563–566. Veldhoven, The Netherlands.

Maddox, J. V. 1987. Protozoan diseases. *In* "Epizootiology of Insect Diseases" (J. R. Fuxa and Y. Tanada, Eds.) pp. 417–452. J. Wiley & Sons, Inc.

Maddox, J. V. 1992. The effect of regulations on the use of insect pathogens as biological control agents. *In* "Regulations and Guidelines: Critical Issues in Biological Control" (R. Charudattan and J. Browning, Eds.) USDA/CSRS Workshop. pp. 73–81. Vienna, Virginia.

Maddox, J.V., Brooks, W. M. and Fuxa, J. R. 1981. *Vairimorpha necatrix*, a pathogen of agricultural pests: potential for pest control. *In* "Microbial Control of Pests and Plant Diseases" (Burges, H.D., Ed.). pp. 587–594. Academic Press, London.

Maddox, J. V. and Solter L. F. 1996. Long-term storage of viable microsporidian spores in liquid nitrogen. *J. Eukaryot. Microbiol.* **43**, 221–225.

Malone, L. A. and McIvor, C. A. 1995. DNA probes for two microsporidia, *Nosema bombycis* and *Nosema costelytrae. J. Invertebr. Pathol.* **65**, 269–273.

McManus, M. L., Maddox, J. V., Jeffords, M. R., and Webb, R. E. 1989. Evaluation and selection of candidate European microsporidia for introduction into U.S. gypsy moth populations. *In* "Proceedings, Lymantriidae: a Comparison of Features of New and Old World Tussock Moths". USDA Forest Ser. NE For. Expt. Stn. Gen. Tech. Rpt. NE-123. pp. 455–468.

McManus, M. L. and Solter, L. 2003. Microsporidian pathogens in European gypsy moth populations. Proceedings: Ecology, survey, and management of forest insects. USDA Forest Service, Northeast Research Station Gen. Tech. Rep. NE-311, 44–51.

Mistric, W. J. and Smith, F. D. 1973. Tobacco budworm: control on flue-cured tobacco with certain microbial pesticides. *J. Econ. Entomol.* **66**, 979–982.

Moawed, S. M., Saleh, M. R., and Salem, S. A. 1987. Interaction of the microsporidium *Vairimorpha ephestiae* with nuclear polyhedrosis virus and their effect on Egyptian cotton leafworm, *Spodoptera littoralis. Acta Entomol. Bohemoslov.* **84**, 422–430.

Nordin, G. L. 1971. "Studies on a Nuclear Polyhedrosis Virus and Three Species of Microsporidia Pathogenic to the Fall Webworm, *Hyphantria cunea*

(Drury)". Ph.D. thesis, University of Illinois at Urbana-Champaign. 229 pp.

Nordin, G. L. and Maddox, J. V. 1972. Effect of simultaneous virus and microsporidian infection on larvae of *Hyphantria cunea*. *J. Invertebr. Pathol.* **20**, 66–69.

Novotny, J., Pilarska, D, Maddox, J. and McManus, M. 2000. Studies of the frequency of entomopathogens in populations of gypsy moth (*Lymantria dispar* L.) in Slovakia during a latency period. *Acta Zool. Bulg.* **52**, 45–49.

Odindo, M. O. 1992. Screenhouse trials on the potential of *Nosema marucae* for control of *Chilo partellus*. *Entomol. Exp. Appl.* 65, 283–289.

Odindo, M. O., Amutalla, P. A., Opondo-Mbai, M., and Onyango Odero, T. 1993. Production of *Nosema marucae* for biological control of cereal stem borers. *Entomol. Exp. Appl.* **67**, 143–148.

Oi, D. H., Becnel, J. J., and Williams, D. G. 2001. Evidence of intracolony transmission of *Thelohania solenopsae* (Microsporidia: Thelohaniidae) in red imported fire ants (Hymenoptera: Formicidae) and the first report of spores from pupae. *J. Invertebr. Pathol.* **78**, 128–134.

Oi, D. H., Briano, J. A., Valles, S. M. and Williams, D. F. 2005. Transmission of *Vairimorpha invictae* (Microsporidia: Burenellidae) infections between red imported fire ant (Hymenoptera: Formicidae) colonies. *J. Invertebr. Pathol.* **88**, 108–115.

Oi, D. H., and Williams, D. F. 2002. Impact of *Thelohania solenopsae* (Microsporidia: Thelohaniidae) on polygyne colonies of red imported fire ants (Hymenoptera: Formicidae). *J. Econ. Entomol.* **95**, 558–562.

Pereira, R. M., Williams, D. F., Becnel, J. J., and Oi, D. H. 2002. Yellow-head disease caused by a newly discovered *Mattesia* sp. in populations of the red imported fire ant, *Solenopsis invicta*. *J. Invertebr. Pathol.* **81**, 45–48.

Pierce, C. M. F., Solter, L. F., and Weinzierl, R. A. 2001. Interactions between *Nosema pyrausta* (Microsporidia: Nosematidae) and *Bacillus thuringiensis* subsp. *kurstaki* in the European corn borer (Lepidoptera: Pyralidae). *J. Econ. Entomol.* **94**, 1361–1368.

Pilarska, D. K., Solter, L. F., Kereselidze, M., Linde, A., and Hoch, G. Microsporidian infections in *Lymantria dispar* larvae: interactions and effects of multiple species infections on pathogen horizontal transmission. *J. Invertebr. Pathol.* **93**, 105–113.

Pilarska, D. K., Solter, L. F., Maddox, J. V. and M. L. McManus. 1999. Microsporidia from gypsy moth (*Lymantria dispar* L.) populations in Central and Western Bulgaria. *Acta Zool. Bulg.* **50**, 109–113.

Pilley, B. M. 1978. The storage of infective spores of *Vairimorpha necatrix* (Protozoa: Microspora) in antibiotic solution at 4 °C. *J. Invertebr. Pathol.* **31**, 341–344.

Prior, C. and Streett, D. A. 1997. Strategies for the use of entomopathogens in the control of the desert locust and other acridoid pests. *Mem. Entomol. Soc. Can.* **171**, 5–25.

Reynolds, D. G. 1972. Experimental introduction of a microsporidian into a wild population of *Culex pipiens fatigans* Wied. *Bull. W. H. O.* **46**, 807–812.

Roberts, D. W., Fuxa, J. R., Gaugler, R., Goettel, M., Jaques, R., and Maddox, J. 1990. Use of pathogens in insect control. *In* "CRC Handbook of Pest Management in Agriculture" (Pimentel, D., Ed.). pp. 243–278. CRC Press, Boca Raton.

Roberts, S. J., Maddox, J. V., Bartell, D. P., and Armbrust, E. J. 1977. Disease as a larval mortality factor in alfalfa weevil (Coleoptera: Curculionidae) populations in Illinois. *Great Lakes Entomol.* **10**, 199–200.

Siegel, J. P. 1985. "The Epizootiology of *Nosema pyrausta* (Paillot) in the European Corn Borer, *Ostrinia nubilalis*, in Central Illinois". Ph.D. thesis, University of Illinois at Urbana-Champaign. 112 pp.

Siegel, J. P., Maddox, J. V., and Ruesink, W. G. 1986. Lethal and sublethal effects of *Nosema pyrausta* on the European corn borer (*Ostrinia nubilalis*) in central Illinois. *J. Invertebr. Pathol.* **48**, 167–173.

Siegel, J. P., Maddox, J. V., Ruesink, W. G. 1988. Seasonal progress of *Nosema pyrausta* in the European corn borer, *Ostrinia nubilalis*. *J. Invertebr. Pathol.* **52**, 130–136.

Slamovits, C, Williams, B. A. P., and Keeling, P. J. 2004. Transfer of *Nosema locustae* (Microsporidia) to *Antonospora locustae* n. comb. Based on molecular and ultrastructural data. *J. Eukaryot. Microbiol.* **51**, 207–213.

Sokolova, Y. Y., Sokolov, I. M., and Fuxa, J. M. 2004. Identification of microsporidia infections in nature: light microscopy or PCR? *Protistology* **3**, 273–281.

Sololova, Y. Y., Issi, I. V., Morzhina, E. V., Tokarev, Y. S, and Vossbrinck, C. R. 2005. Ultrastructural analysis supports transferring Nosema whitei Weiser 1953 to the genus Paranosema and creation of a new combination, Paranosema whitei. *J. Invertebr. Pathol.* **90**, 122–126.

Sokolova Y. Y., Dolgihk, V. V., Morzhina, E. V., Nassonova, E. S., Issi, I. V., Terry, R. S., Ironside, J. E., Smith, J. E., and Vossbrinck, C. R. 2003. Establishment of the new genus *Paranosema* based on the ultrastructure and molecular phylogeny of the type species *Paranosema grylli* gen. nov., comb nov., from the cricket *Gryllus bimaculatus*. *J. Invertebr. Pathol.* **84**, 159–172.

Solter, L. F. 2006. Transmission as a predictor of ecological host specificity with a focus on vertical transmission of microsporidia. *J. Invertebr. Pathol.* **92**, 132–140.

Solter, L. F. and Maddox J. V. 1998a. Timing of an early sporulation sequence of microsporidia in the genus *Vairimorpha* (Microsporidia: Burenllidae). *J. Invertebr. Pathol.* **72**, 323–329.

Solter, L. F. and Maddox J. V. 1998b. Physiological host specificity of microsporidia as an indicator of ecological host specificity. *J. Invertebr. Pathol.* **71**, 207–216.

Solter, L. F., Maddox, J. V., and McManus M. L. 1997. Host specificity of microsporidia (Protista: Microspora) from European populations of *Lymantria dispar* (Lepidoptera: Lymantriidae) to indigenous North American Lepidoptera. *J. Invertebr. Pathol.* **69**, 135–150.

Solter, L. F., Maddox, J. V., and Vossbrinck, C. R. 2005. Physiological host specificity: a model using the European corn borer, *Ostrinia nubilalis* (Hübner) (Lepidoptera: Crambidae) and microsporidia of row crop and other stalk-boring hosts. *J. Invertebr. Pathol.* **90**, 127–130.

Solter, L. F., Siegel, J. P., Pilarska, D. K. and Higgs, M. C. 2002. The impact of mixed infection of three species of microsporidia isolated from the gypsy moth, *Lymantria dispar* L. (Lepidoptera: Lymantriidae) *J. Invertebr. Pathol.* **81**, 103–113.

Sweeney, A. W., and Becnel, J. J. 1991. Potential of microsporidia for the biological control of mosquitoes. *Parasitol. Today* **7**, 217–220.

Sweeney, A. W., Hazard, E. I., and Graham, M. F. 1985. Intermediate host for an *Amblyospora* sp. (Microspora) infecting the mosquito *Culex annulirostris. J. Invertebr. Pathol.* **46**, 98–102.

Tanada, Y. and Kaya, H. K. 1993. "Insect Pathology". pp. 414–458. Academic Press, Inc., San Diego.

Undeen, A. H. 1978. Spore-hatching processes in some *Nosema* species with particular reference to *Nosema algerae* Vavra and Undeen. *Misc. Publ. Entomol. Soc. Am.*, **11**, 29–49.

Undeen, A. H. 1990. A proposed mechanism for the germination of microsporidian (Protozoa, Microspora) spores. *J. Theor. Biol.*, **142**, 223–235.

Undeen, A. H. 1997. Microsporidia (Protozoa): A Handbook of Biology and Research Techniques. So. Assoc. Agric. Expt. Stn. Dir., So. Coop. Ser. Bull. #387. Website: http://agweb.okstate.edu/pearl/ scsb387/

Undeen, A. H. and Avery, S. W. 1988. Effect of anions on the germination of *Nosema algerae* (Microspora: Nosematidae) spores. *J. Invertebr. Pathol.* **52**, 84–89.

Undeen, A. H. and Dame, D.A. 1987. Measurement of the effect of microsporidian pathogens on mosquito larvae mortality under artificial field conditions. *J. Amer. Mosq. Control Assn.* **3**, 91–93.

Undeen, A. H., Johnson, M. A., and Becnel, J. J. 1993. The effects of temperature on the survival of *Edhazardia aedis* (Microspora, Amblyosporidae), a pathogen of *Aedes aegypti. J. Invertebr. Pathol.* **61**, 303–307.

Undeen, A. H. and Maddox, J. V. 1973. The infection of nonmosquito hosts by injection with spores of the microsporidian *Nosema algerae. J. Invertebr. Pathol.* **22**, 258–265.

Undeen, A. H. and Solter L. F. 1996, The sugar content and density of living and dead microsporidian (Protozoa: Microspora) spores. *J. Invertebr. Pathol.* **67**, 80–91.

Undeen, A. H. and Vávra, J. 1997. Research methods for entomopathogenic Protozoa. *In* "Manual of Techniques in Insect Pathology" (L. A. Lacey, Ed.). pp. 117–151. Academic Press, London.

Valles, S. M., Oi, D. H., Pereira, R. M., and Williams, D. F. 2002. Detection of *Thelohania solenopsae* (Microsporidia: Thelohaniidae) in *Solenopsis invicta* (Hymenoptera: Formicidae) by multiplex PCR. *J. Invertebr. Pathol.* **81**, 196–201.

Valles, S. M. and Pereira, R. M. 2003. Use of ribosomal DNA sequence data to characterize and detect a neogregarines pathogen of *Solenopsis invicta* (Hymenoptera: Formicidae). *J. Invertebr. Pathol.* **84**, 114–118.

Van Essen, F. W. and Anthony, D. W. 1976. Susceptibility of non-target organisms to *Nosema algerae* (Microsporida: Nosematidae), a parasite of mosquitoes. *J. Invertebr. Pathol.* **28**, 77–85.

Vávra, J. and Maddox, J. V. 1976. Methods in microsporidiology. *In* "Comparative Pathobiology, Vol. 1, Biology of the Microsporidia" (L. A. Bulla and T. C. Chang, Eds.) pp. 281–319. Plenum, New York.

Vossbrinck, C. R. and DeBrunner-Vossbrinck, B. A, 2005. Molecular phylogeny of the microsporidia: ecological, ultrastructural and taxonomic considerations. *Folia Parasitol.* **52**, 131–142.

Wang, B. 1982. The pathobiology of the mosquito parasite, *Vavraia culicis* (Weiser). Ph.D. thesis, University of California, Los Angeles. 159 pp.

Weidner E, Canning EU, Rutledge CR, Meek CL. 1999. Mosquito (Diptera:Culicidae) host compatibility and vector competency for the human myositic parasite *Trachipleistophora hominis* (Phylum Microspora). *J. Med. Entomol.* **36**, 522–525.

Wegensteiner, R., Weiser, J., Fuehrer, E. 1996. Observations on the occurrence of pathogens in the bark beetle *Ips typographus* L. (Col., Scolytidae). *J. Appl. Entomol.* **120**, 199–204.

Weiser, J. and Novotny, J. 1987. Field application of *Nosema lymantriae* against the gypsy moth, *Lymantria dispar* L. *J. Appl. Entomol.* **104**, 58–62.

Williams, D. F., Oi, D.H., and Knue, G.J. 1999. Infection of red imported fire ant (Hymenoptera: Formicidae) colonies with the entomopathogen *Thelohania solenopsae* (Microsporidia: Thelohaniidae). *J. Econ. Entomol.* **92**, 830–836.

Wilson, G. G. 1973. Incidence of microsporidia in a field population of spruce budworm. *Can. For. Ser. Bi-mon. Res. Notes* **29**, 35.

Wilson, G. G. 1981. *Nosema fumiferanae*, a natural pathogen of a forest pest: potential for pest management. *In* "Microbial Control of Pests and Plant Diseases" (H. D. Burges, Ed.). pp. 595–602. Academic Press, London.

Wilson, G. G. and Kaupp, W. J. 1975. Application of a microsporidia, *Nosema fumiferanae*, against spruce budworm on Manitoulin Island. *Environ. Can. For. Ser. Inf. Rep*. 1P-X-11.

Wilson, G. G. and Kaupp, W. J. 1976. Application of *Nosema fumiferanae* and *Pleistophora schubergi* (Microsporida) against spruce budworm in Ontario. *Environ. Can. For. Ser. Inf. Rep*. 1P-X-15.

Zimmack, H. L. and Brindley, T. A. 1957. The effects of the protozoan parasite *Perezia pyrausta* Paillot on the European corn borer. *J. Econ. Entomol.* **50**, 637–640.

Chapter IV-4

Fungi

Stephen P. Wraight
USDA-ARS
Plant Soil & Nutrition Laboratory
Ithaca, NY 14853, USA

G. Douglas Inglis and Mark S. Goettel
Agriculture & Agri-Food Canada
Lethbridge Research Centre
Lethbridge, AB T1J 4B1, Canada

1 Introduction

Fungi contain a diverse array of taxa that exhibit a great diversity of properties and requirements, collectively occupying virtually every niche in which arthropods are also found. Consequently, there has been great interest in the use of fungi as microbial biocontrol agents. Numerous species have been developed for invertebrate pest control (Wraight *et al.*, 2001; Copping, 2004; Kabaluk and Gazdik, 2004), but to date, none has gained a significant portion of the overall commercial market for insect control. In this chapter, we restrict our discussion to entomopathogenic fungi that infect their hosts and cause premature death. We thus exclude ectoparasites (*e.g.*, Laboulbeniales) and other insect-associated fungi (*e.g.*, Trichomycetes) whose potential role in microbial control has not been well established. We attempt to provide an overview of the special considerations and methods which would aid the reader in evaluating the efficacy of fungi under field conditions. For more detailed information on entomopathogenic fungi and their role in microbial control, we refer the reader to the reviews of Carruthers and Soper (1987), McCoy *et al.* (1988), Samson *et al.* (1988), Evans (1989), Glare and Milner (1991), Tanada and Kaya (1993), Hajek (1997), Boucias and Pendland (1998), Wraight and Carruthers (1999), Goettel *et al.* (2005) and several chapters in Butt *et al.* (2001), and Upadhyay (2003).

A Entomopathogenic fungi

The fungi, as defined by Kendrick (2000), are eukaryotic, heterotrophic, absorptive organisms that develop a rather diffuse, branched, tubular body (hyphae) and reproduce by means of sexual and/or asexual spores. Organisms that fit this general description are now known to be phylogenetically diverse, and are currently classified in two kingdoms, the Straminipila (=Stramenopila or Straminopila, =Chromista) and the Eumycota (see Alexopoulos *et al.*, 1996; Blackwell and Spatafora, 2004). Though no longer recognized as true fungi, the fungus-like straminipiles resemble the fungi in their morphological, reproductive, nutritional, and ecological characteristics, and their continued study by mycologists is eminently appropriate and practical (Bruns *et al.*, 1991). These microorganisms are now referred to by many systematists as the straminipilous fungi. As pointed out by Kendrick (2000), use of the term fungus in this sense is similar to use of the term alga; the phylogenetic classification of algae also spans multiple kingdoms. A third group traditionally studied by mycologists, the slime molds, are no longer considered fungi, because they do not produce hyphae and

L.A. Lacey and H.K. Kaya (eds.), Field Manual of Techniques in Invertebrate Pathology, 223–248.

feed by phagocytosis rather than by absorption; the somatic phases of these organisms are essentially amoeboid and lack cell wall; slime molds are now classified in the kingdom Protozoa. There are over 700 recognized species of entomopathogenic fungi representing the kingdoms Straminipila and Eumycota. The vast majority of these are eumycotan fungi. There are no entomopathogenic species of slime molds.

1 Straminipilous fungi

The kingdom name Straminipila derives from the Latin word "stramen" meaning straw, which describes the structure of the flagellar hairs produced by these organisms. One of the two flagellae that are usually produced by each zoospore bears numerous tubular hairs and is called a tinsel flagellum (the second flagellum is a smooth whiplash type). Some members of the group have lost the whiplash flagellum, or both flagellae, through evolution. Most straminipiles are algae with chlorophyll c in addition to chlorophyll a (the brown, golden-brown, and yellow-green algae and diatoms), but as indicated above, this group also includes many non-photosynthetic microorganisms that fit the broad definition of a fungus. The best-known straminipilous fungi belong to the divisions Oomycota and Hyphochytriomycota. The Oomycota contains a number of entomopathogens. These microorganisms are aquatic, possess coenoyctic hyphae that are mainly aseptate, have cell walls composed predominantly of a glucan-cellulose, reproduce asexually by zoospores with two flagellae of unequal length (one a whiplash, and the other a tinsel type), and produce sexual spores, termed oospores. Entomopathogenic taxa belong to either the orders Saprolegniales or Lagenidiales. Within the Saprolegniales, three genera (*Aphanomycopsis, Couchia*, and *Leptolegnia*) are entomopathogenic.

The primary entomopathogens within the Lagenidiales belong to the genus, *Lagenidium. Lagenidium giganteum* is an important pathogen of mosquitoes. Its life cycle includes both zoospores and oospores, the latter are thick-walled, formed within the host, and function as resting spores. Zoospores are the infective propagules, they encyst on the larval cuticle and produce germ-tubes that penetrate through the integument. Information regarding the biology and study of this species is presented by Kerwin and Petersen (1997) and in Chapter VII-22 of this volume.

2 Eumycotan fungi

Most of the entomopathogenic fungi are classified in the kingdom Eumycota or true fungi and represent all four of the major divisions that traditionally comprise the kingdom (Chytridiomycota, Zygomycota, Ascomycota, and Basidiomycota). In contrast to the straminipilous fungi, the true fungi are mostly nonmotile and do not produce tinsel flagellae (chytridiomycetes produce motile zoospores and gametes with smooth, whiplash flagellae). These groups are also differentiated by ultrastructural, biochemical and molecular characters too numerous to detail here. Chemical composition of the cell wall has been a prominent feature used to separate the two groups. Chitin is a characteristic component of the cell walls of true fungi but absent from the walls of most straminipiles; cellulose is an important component of straminipile cell walls, but it is rarely found in the walls of true fungi (Alexopoulos *et al.*, 1996).

The division Chytridiomycota contains the single class Chytridiomycetes. Most chytrids inhabit fresh water, but many are also found in soil, which is transformed into an aquatic habitat when flooded. These microorganisms produce zoospores with a single, posteriorly directed, whiplash flagellum. The best-known insect pathogenic species are in the genus *Coelomomyces*; all are obligate parasites of mosquitoes and other Diptera and require crustacean alternate hosts to complete their life cycles.

Entomopathogenic fungi in the Zygomycota, Ascomycota, and Basidiomycota do not form motile spores and are primarily terrestrial fungi. Most entomopathogens within the Zygomycota occur within the order Entomophthorales ("insect destroyers") of the class Zygomycetes (characterized by the formation of a sexual spore, termed a zygospore). Entomophthoralean fungi produce coenocytic hyphae, but some septation occurs, particularly in older hyphae. In addition to the formation of sexual zygospores (formed through conjugation of gametangial cells), some zygomycetes produce azygospores. Azygospores are morphologically similar to zygospores but are produced parthenogenetically. Both are thick walled and function as resting spores capable of surviving periods of unfavorable environmental conditions. This group of fungi also produces uninucleate or multinucleate asexual conidia, and upon germination, conidia may produce secondary or tertiary conidia. Many species have worldwide distributions and are often responsible for wide-scale epizootics. Most are obligate parasites possessing restricted host ranges; however, some species, such as *Zoophthora radicans*, have relatively wide host ranges. The most common genera include *Conidiobolus, Entomophaga, Entomophthora, Erynia, Pandora, Neozygites* and *Zoophthora*.

A few genera of entomopathogenic fungi produce an ascomycetous sexual state (*i.e.*, ascospores produced within an ascus), and are assignable to the division Ascomycota. The most important of these include *Cordyceps* and *Torrubiella* (Pyrenomycetes: Hypocreales) and *Ascosphaera* (Ascosphaeromycetes: Ascosphaerales). Species of *Cordyceps* parasitize a broad range of insects and are the most common entomopathogenic fungi occurring in tropical forests. *Torrrubiella* spp. attack primarily spiders and scale insects. The entomopathogenic species of *Ascosphaera* affect only bees.

Many important species of entomopathogenic fungi appear to have lost most or all of their capacity to produce a sexual state. These include prominent pathogens in the genera *Aspergillus, Aschersonia, Beauveria, Culicinomyces, Fusarium, Gibellula, Hirsutella, Hymenostilbe, Lecanicillium, Metarhizium, Nomuraea, Paecilomyces, Sorosporella*, and *Tolypocladium*. All of the asexual (anamorphic) forms of these fungi produce spores, termed conidia, and some produce chlamydospores. Because the sexual (teleomorphic) states of most species of these genera are unknown or only rarely encountered in close association with their anamorphs, these fungi have traditionally been placed in the form-class Hyphomycetes within the form-division Deuteromycota; however, these are not formal taxonomic categories (*i.e.*, not monophyletic) and are being abandoned as the phylogenetic affinities of so many conidial fungi become known. The current higher classification of nearly all of these fungi is: class, Sordariomycetes, order, Hypocreales, and family, Clavicipitaceae. Although teleomorphs are not known for most species in the genera listed above, representative species from each genus have been found to have ascomycete sexual forms. In most cases, known teleomorphs belong to the clavicipitaceous genera *Cordyceps* or *Torrubiella* (Hodge, 2003). *Aschersonia* spp. are anamorphs of *Hypocrella* spp., also in the Clavicipitaceae. *Fusarium* is an extremely large and diverse group, whose species are anamorphs of hypocrealean fungi in the family *Nectriaceae*. *Aspergillus* spp. are anamorphs of Eurotiomycetes (Eurotiales).

It warrants note that species in the genus *Lecanicillium* were formerly identified as *Verticillium* spp. In a recent revision by Zare and Gams (2001), most of the insect pathogenic species of *Verticillium* were reclassified into the new genus, *Lecanicillium*, based on molecular and morphological differences. In addition, isolates comprising the most important and well-known insect pathogenic species of *Verticillium, V. lecanii*, were separated into four species of *Lecanicillium, L. lecanii, L. muscarium, L. longisporum*, and *L. nodulosum*.

Only a small number of fungi within the division Basidiomycota are true entomopathogens. This group of fungi is characterized by the formation of sexual spores, termed basidiospores, from a specialized dikaryotic and, subsequently, meiosporangial cell termed a basidium. Representatives of the Septobasidiales (Teliomycetes), primarily from two genera, *Septobasidium* and *Uredinella*, are obligate pathogens of armored scale insects (Hemiptera: Diaspididae). The relationship between *Septobasidium* spp. and their scale hosts is mutualistic at the host population level, as the fungi provide shelter for the scales while parasitizing only a small percentage of individuals (see Evans, 1989).

Keys and illustrations to major genera of entomopathogenic fungi are provided by Samson (1981), Samson *et al.* (1988), Humber (1997), and Tzean *et al.* (1997).

B Pathogenesis and epizootiology

Entomopathogenic fungi most commonly invade their hosts through the external cuticle, although some are capable of breaching the alimentary canal (*e.g.*, *Ascosphaera* and *Culicinomyces*). Spores (the term "spores" is used generically to refer to the different types of infective units) attach to the cuticle, germinate, and penetrate the host. The fungus then proliferates in the host hemocoel as cell-walled hyphal bodies (some forms referred to as blastospores) or wall-less amoeboid protoplasts. Host death results from a combination of actions, including depletion of nutrients, physical obstruction or invasion of organs, and toxinosis. Soon after host death, and under favorable conditions, hyphae emerge from the cadaver and produce sporogenous cells; sporulation occurs on the host surface, and the spores are liberated.

The Oomycetes, many species of Zygomycetes, and some ascomycetous fungi produce resting spores within the cadaver body. Production of these spores is regulated by both biotic and abiotic factors (Hajek, 1997). Often, resting spores are thick-walled and pigmented, and are capable of persisting under adverse environmental conditions for many years. Once conditions are favorable, they germinate and release or produce infective spores, or infect the host directly.

In most entomopathogenic fungi, spore dispersal is passive, relying principally on wind and water. However, in the Entomophthorales, spores are forcibly discharged and can land many centimeters from the host or be carried longer distances on air currents. Some insects infected with entomophthoralean fungi climb to the tops of plants just prior to death, where they die firmly clasping the plant (*e.g.*, grasshoppers infected with *Entomophaga grylli*). In other cases, the fungus produces rhizoids with holdfasts which firmly anchor the insect to the substrate (*e.g.*, aphids infected with *Pandora neoaphidis*). Such adaptations help ensure that discharged spores contact potential hosts within and beneath the plant canopy. In species that form motile spores, zoospores selectively attach and encyst on the cuticle of specific hosts (*e.g.*, *Lagenidium giganteum* on mosquito larvae).

Fungi, as a group, have one of the widest host ranges among the pathogens of arthropods. However, host spectra vary widely depending on fungal species. For instance, *E. grylli* infects only acridid grasshoppers, *Aschersonia aleyrodis* infects only whiteflies, and *Nomuraea rileyi* almost exclusively infects lepidopterans. In contrast, species such as *Beauveria bassiana, Metarhizium anisopliae*, and *Z. radicans* have much wider host ranges, spanning numerous orders within the Arthropoda. It is now recognized, however, that many species of common entomopathogenic fungi comprise diverse assemblages of genotypes and likely comprise species complexes. Therefore, it is not surprising that within taxa exhibiting wide host ranges, individual isolates or pathotypes may have substantially restricted host ranges. Most studies on host range are the result of laboratory testing. Therefore, host-range determinations based on "physiological" host-range experimentation in the laboratory do not necessarily represent "ecological" host ranges under field conditions (Jaronski *et al.*, 2003). See Chapter X-1 for more information and a discussion on host ranges.

Epizootics of entomopathogenic fungi are relatively common, and can be important in the natural regulation of insect populations. The most spectacular epizootics are those in which the insect cadavers are conspicuous and occur over large areas (*e.g.*, epizootics of *E. grylli* and *Neozygites fresenii* in grasshopper and aphid populations, respectively). Unfortunately, little is known about host-pathogen-environment interactions and the factors that initiate epizootics of most insect pathogenic fungi. Such an understanding is necessary if we are to use these pathogens successfully in integrated pest management (IPM) programs which attempt to augment natural and inundative biological control.

C Microbial biocontrol potential

Fungi have demonstrated considerable potential in microbial control of arthropods, especially within IPM programs (Lacey and Goettel, 1995). Their restricted host ranges allow for control of pests with limited harm to non-target organisms including predators, parasites, and other pathogens (see Goettel *et al.*, 1990, 2001; Goettel and Hajek, 2001; Vestergaard *et al.*,

2003). In addition, fungi are also compatible with some fungicides and many other types of pesticides.

Commercialization of entomopathogenic fungi has been restricted to those species that are amenable to mass production *in vitro* on economical substrates. Consequently, with the exception of *L. giganteum* for mosquito control, virtually all commercial products developed to date have been based on species within the Hypocreales (Wraight *et al.*, 2001; Copping, 2004; Kabaluk and Gazdik, 2004). Examples include *B. bassiana* for control of a great variety of insects, including pine caterpillar and corn borer in Asia and thrips, whiteflies, aphids, and diamondback moth in North and South America; *Beauveria brongniartii* for control of the cockchafer in Europe; *M. anisopliae* for control of spittle bugs in South America, vine weevils in Europe, and chafers in Australia; *Metarhizium anisopliae* var. *acridum* (initially identified as *Metarhizium flavoviride*) for control of locusts and grasshoppers in Africa and South America; *Nomuraea rileyi* for control of lepidopterous pests in South America; *Paecilomyces fumosoroseus* for control of whiteflies in Europe and North America; and *Lecanicillium* spp. for control of aphids and whiteflies in Europe and South America.

Although fastidious species with narrow host ranges are difficult to commercialize, they should not be overlooked, especially in classical biological control applications. Regulatory agency concerns regarding importation of exotic strains and their safety to non-target organisms, have generally limited this approach (see Chapter X-1). Nevertheless, natural pest control by such species as *Entomophaga maimaiga, N. fresenii, Z. radicans, and E. grylli* underscores the tremendous potential of these fungi for managing insect pests through classical biological control programs.

2 Environmental constraints

As with all entomopathogens, an array of biotic and abiotic factors affect the ability of fungi to survive, propagate, infect, and kill their host. It is highly unlikely that any commercial preparation of an entomopathogenic fungus could be used successfully under all conditions that occur in the field. Consequently, environmental constraints must be well understood in order that windows of opportunity for the effective use of entomopathogenic fungi can be identified. Since these constraints are still relatively poorly understood, every attempt should be made to monitor as many environmental parameters as possible when conducting field trials and to include environmental data in published reports. Only after comparing the specific environmental conditions that occur during successful and unsuccessful applications will it be possible to elucidate which constraints are most critical and find ways to overcome them. Although environmental constraints must ultimately be tested under field conditions, identification of important constraints under controlled conditions is required before the commencement of field tests. The following is a synopsis of the most important environmental constraints, and the methods that are used to measure their importance under controlled conditions. Examples of subsequent field-testing are also included.

A Solar radiation

Fungal propagules are highly susceptible to damage by solar radiation. Radiation of the shortest wavelengths transmitted by the earth's atmosphere, the ultraviolet B (UVB) portion of the solar spectrum (280–320 nm), is most lethal; however, radiation in the UVA range (320–400 nm) is also damaging (Braga *et al.*, 2001a). The persistence of fungal propagules on substrates exposed to direct solar radiation is substantially reduced relative to propagules in protected locations (such as within plant canopies or on the undersides of leaves, which absorb UV radiation). Even within shaded areas, propagules will eventually be killed by reflected radiation (Smits *et al.*, 1996). Despite this high inherent susceptibility, entomopathogenic fungal species and strains within species differ significantly in their susceptibility to solar irradiation (Morley-Davies *et al.*, 1995; Fargues *et al.*, 1996; Braga *et al.*, 2001b).

While the UV components of the solar spectrum are detrimental, irradiation at higher wavelengths (visible light) may be beneficial by stimulating a photoreactivation mechanism that repairs damaged DNA (see Friedberg *et al.*, 1995). The importance of this repair mechanism appears to vary among entomopathogenic fungi.

Chelico *et al.* (2005) reported photoreactivation of *B. bassiana* conidia following DNA damage by UV radiation, but Braga *et al.* (2002) observed no significant levels of photoreactivation in strains of *Lecanicillium sp.* (identified as *V. lecanii), Aphanocladium album*, or *M. anisopliae*.

The influence of sunlight on entomopathogenic fungi is most often studied using artificial sunlight devices. Due to the possibility of a photoreactivation phenomenon, polychromatic light should preferably be used. However, useful information also can be obtained using much simpler and cheaper light sources. Fungal propagules are irradiated on a defined substrate using a defined wavelength and lux for a specified period, and the viability of the propagules is quantitatively compared to propagules shielded from the radiation. Readers are referred to Goettel and Inglis (1997) for details.

Although natural sunlight is variable and unpredictable, much useful information can be gained by studying the effects of natural sunlight on fungal survival and efficacy against arthropods. A number of researchers have measured the persistence of fungal propagules exposed to direct and indirect solar radiation in the field environment (*e.g.*, Inglis *et al.*, 1993; Inglis *et al.*, 1995; Smits *et al.*, 1996). However, a number of potentially confounding variables (*e.g.*, temperature, precipitation, relative humidity, time of year, cloud cover, etc.) greatly complicate assessments of the detrimental impacts of solar radiation in the field. In an attempt to address some of these limitations, Rougier *et al.* (1994) utilized a device that maintained a perpendicular exposure to the sun at constant temperature; the device used water-cooled plates that were rotated and inclined so that their surfaces were kept perpendicular to the sun. To measure global solar radiation, pyranometers (calibrated for daylight spectrum) are primarily used, but quantum sensors are used to measure photosynthetic wavelengths within and outside of plant canopies. For the UVB range, radiometers are used.

Measuring and understanding the direct (*e.g.*, germicidal) and indirect (*e.g.*, host behavior) effects of solar radiation on fungal efficacy against arthropods is difficult. Inglis *et al.* (1997a) utilized a variety of cage environments in an attempt to measure the influence of sunlight on susceptibility of grasshoppers to *B. bassiana* in a field experiment. Conditions of incoming solar radiation (visible range), temperature, relative humidity, and wind speed were recorded. A cage that was protected from the UVB portion of the solar spectrum (< 355 nm) by an absorbing plastic film (1.8 × 3.0 m) attached to a wood frame was used. The front of the frame (facing south) was situated 0.9 m above the soil surface and the back of the frame was 1.2 m above the soil (see Chapter VII-17); this configuration was shown to have no impact on air movement and cages were shaded for most of the day (ca. 0900 to 1700 h). Mortality of field-collected grasshoppers and survival of *B. bassiana* conidia were compared in cages that were UVB-protected or exposed to full-spectrum sunlight. Though conditions of temperature, relative humidity, and visible light were very similar between the two environments, conidial survival was significantly greater, and rates of infection were higher in the UVB-shielded environment (43% vs. 15% infection). Disease development was also more rapid in the protected environment.

B Temperature

Most entomopathogenic fungi have a wide range of temperature tolerances, but temperature optima for infection, growth, and sporulation are usually much more restricted (generally 20–30 °C). Propagules for most species survive well at sub-zero temperatures and can be stored for long periods at −20 to −80 °C or in liquid nitrogen (−196 °C). Spores of some species can tolerate very high temperatures for very short periods (*e.g.*, 150 °C for 30 sec); however, the maximum threshold for long periods is usually close to 40 °C.

The influence of temperature on *in vitro* germination, vegetative growth, and sporulation is generally conducted in controlled-environment chambers using a variety of culture media (Goettel and Inglis, 1997). Bioassays using target hosts is much more involved (see Butt and Goettel, 2000). In designing a bioassay, it is imperative that consideration be given to the temperatures that would be expected in a field environment. A bioassay should not only consider ambient temperature, but also insect behavior that may influence body temperature. For example, grasshoppers elevate their body temperatures higher than ambient by basking; use of bioassay cages fitted with incandescent bulbs allows grasshoppers to thermoregulate (Inglis *et al.*, 1996a; Goettel and Inglis, 1997). To date, most studies on entomopathogen efficacy have focused on the effects of constant temperatures and have largely ignored fluctuating temperatures (*e.g.*, Mohamed *et al.*, 1977; Inglis *et al.*, 1997b; Inglis *et al.*, 1999).

The influence of temperature on field efficacy of entomopathogenic fungi is primarily limited to detailed measurements of macro- and micro-temperatures following application of a fungus or during the course of the growing season(s). The temperature data obtained are then equated with observations of epizootic development. While this strategy can provide valuable insight into the factors affecting epizootic initiation and development and facilitate further experimentation, a number of potentially confounding variables (*e.g.*, solar radiation and moisture) limit the ability of the researcher to make definitive conclusions. Nevertheless, conditions of temperature should be recorded in all field experiments with fungi and may help explain situations where the entomopathogen is ineffective. A number of devices can be used to measure conditions of macro- and micro-climate temperatures (*e.g.*, temperature/relative humidity probes or wire thermistors interfaced with a datalogger), and readers are referred to any number of general references on microclimatic measurements (*e.g.*, Rosenberg, 1974; Unwin, 1981).

To study the impact of temperature on diseases of grasshoppers incited by *B. bassiana* in a field environment, Inglis *et al.* (1997a) utilized different cage environments. In shaded cages (accomplished using a black plastic screen), temperature was as much as 6 °C cooler, light levels were 61–80% lower, and mycosis was 80% greater than in cages exposed to direct sunlight. To determine the relative importance of grasshopper thermoregulation and the deactivation of conidia by UVB radiation on mycosis, a UVB-protected cage was included in the experiment (see section 2A). Using these three environments, it was concluded that the indirect effects of temperature and light on the susceptibility of grasshoppers to a commercial strain of *B. bassiana* (*i.e.*, behavioral thermoregulation) had a greater influence on disease development than did the rapid deactivation of conidia by UVB radiation.

C Moisture

Fungal sporulation and spore germination require high moisture or free water. Many studies have attempted to demonstrate the importance of moisture in the initiation of natural fungal epizootics (see Carruthers and Soper, 1987), and many past failures in microbial control attempts with fungi have been attributed to adverse (especially dry) weather conditions. This has led to the general belief that high-moisture conditions are essential for effective use of fungi in microbial control. Although this is certainly the case in many insect-fungal-pathogen associations, it is not the case for all. For example, *M. anisopliae* var. *acridum* is capable of infecting the desert locust, *Schistocerca gregaria*, at relative humidities as low as 13% (Fargues *et al.*, 1997). This fungus is further adapted to dry conditions by producing spores within the cadaver. The documented capacity of some fungi to operate under dry conditions generally has been attributed to the presence of moisture in the microhabitats within which they are active (*e.g.*, abaxial leaf surfaces or membranous folds of insect cuticle). In some cases, fungi are capable of exploiting cyclical or episodic periods of elevated moisture conditions (especially during overnight periods) to infect hosts under what might generally be described as dry conditions (*e.g.*, see Galaini-Wraight *et al.*, 1991).

Moisture can also have very significant effects on the persistence of fungal inocula. Conidia generally exhibit greatest stability under cool, dry conditions; however, survival of conidia of some fungi (*e.g., M. anisopliae*) is greatest under a combination of moderate-temperature, high-moisture conditions (Clerk and Madelin, 1965; Daoust and Roberts, 1983).

Measuring the effects of moisture on viability, germination, growth, and sporulation is usually carried out using media adjusted to different water activities; details are provided by Goettel and

Inglis (1997). Effects of vapor pressure deficit on host-pathogen relationships are frequently studied in humidity chambers. Since aerial humidities occur at equilibrium only at the solution-air interface, it is important that the air be constantly circulated to obtain a homogeneous environment. To achieve this, air can be circulated over a saturated salt solution in one chamber into a second chamber containing the target hosts (Fargues *et al.*, 1997). An important complication, however, is that the drying capacity of an air mass at any specific vapor pressure deficit is highly dependent upon its movement, and selection of air speeds representative of wind conditions in the field is difficult. Relative humidity (vapor pressure) can be measured in combination with temperature using a wide variety of commercially available probes interfaced with a datalogger (and these data can be used to calculate vapor pressure deficits). Dew instruments and leaf wetness meters are commonly used in the study of foliar fungal pathogens, and these are also available commercially. Soil moisture (water potentials) can be quantified using several methods such as pressure plates, resistance blocks, tensiometers, thermocouple psychrometers, neutron scattering, gamma-ray attenuation, ultrasonic energy, and/or filter paper methods (see Goettel and Inglis, 1997).

D Rainfall

The impact of rain on the persistence of fungal propagules on insects and on foliage has not been extensively studied. Poor recovery of propagules from insects following vigorous washing (Boucias *et al.*, 1988) has led to the general belief that rain does not "wash off" influential numbers of spores from the integument of insects. However, in a laboratory study, simulated rain was found to remove significant numbers of *B. bassiana* conidia from the integuments of Colorado potato beetle larvae (Inglis *et al.*, 2000). Field-testing by Wraight and Ramos (2002) provided evidence that supports this finding, at least with respect to rain occurring soon after fungal applications. Laboratory studies by Inglis *et al.* (2000) and Inyang *et al.* (2000) indicate that formulation can significantly affect conidial retention on foliage exposed to rain; conidia applied in oil or emulsifiable oil formulations were found to persist longer on foliage exposed to simulated rain than conidia in aqueous formulations. Rainfall, especially wind-blown rain, also has been shown to disrupt development of fungal epizootics by removal of fungal inoculum or sources of inoculum (fungal-killed insects) from a crop canopy (Kish and Allen, 1978; Galaini-Wraight *et al.*, 1991), and rainfall could, in this way, interfere with inoculative microbial biocontrol programs, which rely on spore production by the primary inoculum to effect control.

Rainfall simulators are generally used to study the effects of rain on persistence of propagules on leaf surfaces and insect integuments. To study the effects of rain, researchers use both unspecialized (*e.g.*, pesticide spray chambers) and relatively specialized apparatuses (*e.g.*, Tossell *et al.*, 1987; Nord, 1991). Most simulators produce a relatively wide range of droplet sizes, but the low boom heights used produce rain with relatively low droplet velocities. To measure the influence of rain, propagule densities are normally quantified on leaf surfaces before and after exposure to rain (Goettel and Inglis, 1997). Measurements of rain effects on conidial persistence in field environments are extremely difficult due to confounding variables, especially solar radiation. As suggested above, another important variable is the timing of rainfall relative to a biopesticide application. Actual or simulated rain that impacts treated foliage before spray deposits have thoroughly dried is more likely to remove significant amounts of inoculum. A number of devices are used to measure rainfall in field environments. Commercially available electronic measurement devices (*e.g.*, tipping bucket rain gauges) are commonly used, but simple rain gauges may suffice.

3 Inoculum preparation, handling, and characterization

A Preparation of spore formulations

While the propagules of some entomopathogenic fungi are hydrophilic and suspend well in aqueous suspension (*e.g.*, *Lecanicillium* spp.), those of others (*e.g.*, *Beauveria* and *Metarhizium*) possess hydrophobic cell walls that make their suspension in water-based formulations problematic. A number of techniques have been used to facilitate the suspension of hydrophobic propagules; these include the use of surfactants, sonication, mechanical agitation, or a combination of these techniques. Hydrophobic propagules are readily suspended in oils, and many conidia-based biopesticides on the market today are flowable or emulsifiable oils.

To prepare the fungal propagules for application, the number of spores/unit weight (g) or volume (ml) of the inoculum preparation must be determined. Hemacytometers are commonly used to visually quantify numbers of propagules/unit volume or weight; a detailed description of their use is presented by Goettel and Inglis (1997). Dilution plating and the most-probable number methods can also be used to enumerate propagules/unit volume or unit weight (see Goettel and Inglis, 1997). Unlike molecules of a toxin in solution, spores suspended in liquid or mixed with dry carriers are not uniformly distributed, and preparation of dilution series and use of counting chambers such as hemacytometers must take this into account. During sampling for enumeration, spore preparations must be continually mixed or agitated. Even with diligent mixing, small aliquots of liquid transferred between dilution tubes, minute samples ($< 10\,\mu l$) loaded into hemacytometer chambers, and even relatively large samples of powders weighed on a balance can be important sources of error and must be adequately replicated in order to assign meaningful measures of error to estimates of spore concentrations. Incomplete dispersal of spores, leaving large aggregates that are uncountable, or that distribute nonrandomly across the grid of a counting chamber, is another important source of error that is often overlooked. We have found that mechanical wetting of propagules using micropestles provides excellent suspension of conidia in water without causing damage to cells (see Goettel and Inglis, 1997). Sonication in the presence of organosilicone wetting agents is also an effective method of dispersing conidia and blastospores in dried, highly aggregated formulations or technical preparations. It may be necessary to work with subsamples of suspensions prepared for bioassay if harsh treatment is required for dispersal, or if the dispersion of propagules resulting from normal methods of suspension (*e.g.*, according to product label instructions) is desired.

Oil-formulated propagules present a problem in that they settle very slowly in standard hemacytometer chambers (Goettel and Inglis, 1997). Time required for counting of replicate samples can be much reduced by using a Petroff-Hausser bacterial counting slide (Hausser Scientific), which has a chamber depth of only 0.02 mm. Also, these slides are thinner than hemacytometers, which improves optics for use of phase contrast microscopy to identify and count spores in formulations with particulate carrier ingredients.

Once the numbers of propagules/unit weight or volume has been determined, the next step is to determine the viability of the conidia (see section 3D). Doses are then adjusted accordingly and spores are added to the required amount of carrier. When using an emulsifiable oil, the conidial mass is often added directly to the oil and mixed thoroughly, and this mixture is then added to the water carrier while agitating. The volume added by the conidia is negligible and usually disregarded in calculating the total volume of water required. On the other hand, the volume or weight of conidia in the undiluted formulation may be significant, and this should be taken into consideration when preparing carrier-control preparations.

B Preparation of mycelial formulations

Mycelial inoculum is usually fragmented, pelletized or encapsulated, and is often deployed with insect baits and/or with nutrients to facilitate fungal growth. This type of inoculum is generally quantified as weight of formulation/unit area or per plant. The size distribution of the particles and density (particles/unit area) are also important dosage parameters. If incorporated into soil, the weight and number of particles/unit volume of soil should be reported.

Formulations of vegetative hyphae designed to support production of infectious propagules following field application may be characterized with respect to sporulation potential. Such determinations are also made during assessments of storage stability or field persistence. To assess sporulation potential of formulated mycelium, mycelial particles are allowed to sporulate under optimal conditions (generally on agar substrates approaching 100% relative humidity), or if possible, under simulated or actual field conditions (*e.g.*, Wraight *et al.*, 2003). Hydrophobic conidia of many Hypocreales (*e.g.*, *Beauveria, Metarhizium*, etc.) are removed from the particles, and from the surrounding substrate, by washing with various oils or water/buffer (usually with surfactants). The hydrophilic, forcibly ejected conidia produced by mycelial particles of entomophthoralean species have been collected by inverting the inoculated substrate over aqueous collecting solutions. Conidia of *Z. radicans* can be collected in water with Atmos 300 (0.1%) and Tween 80 (0.25%) as wetting agents if preservation of viability is desired. If conidia are collected only for enumeration, a solution containing maleic acid (0.02%) and Triton X-100 (1%) can be used to prevent germination (Soper, 1985).

C *Short-term storage*

Unless long-term effects of formulation ingredients on the viability and virulence of the entomopathogen are known, it is preferable that final formulation and mixing be conducted in the field, immediately prior to application. For instance, it may be possible to mix ingredients directly in equipment fitted with recirculation pumps. If application is not immediately possible, the formulated product should be kept cool, or at least shaded from direct sunlight.

D *Viability assessments*

Measurements of inoculum quality prior to, during, and after application are essential. Assessments should be made on the active unformulated ingredient, on the formulated product, and at the field site, preferably by collecting the inoculum directly from the spray nozzle at intervals during the spray operation.

1 *Germination*

Suspend at least three replicate samples of propagules in an appropriate solution, adjust the concentration to approximately $1–2 \times 10^7$ conidia/ml (see section 3A), and evenly spread a $100\,\mu l$ aliquot of each replicate suspension over the surface of an appropriate medium. A general purpose medium such as potato dextrose agar or Sabouraud dextrose agar amended with an antibacterial agent(s) works well, but the thickness of the medium should be minimized to allow transmission of light for microscopic examinations (*e.g.*, 15 ml of medium/8.5-cm-diam. Petri dish). Unused dishes must be stored in sealed containers to prevent desiccation, as agar with low moisture content can inhibit the speed and ultimate percentage of germination. The cultures are then incubated at an appropriate temperature for 12–24 h, after which time, a drop of liquid (*e.g.*, water, lactic acid, or stain) is placed on the medium followed by a coverslip. The culture is then placed on the stage of a light microscope and examined at approximately 400x under phase contrast. If examination is not immediately possible, a fixative such as lactophenol can be used as the liquid prior to placement of the coverslip. The culture plates can then be stored and examined at a later time.

Propagules are usually considered viable if germ-tube lengths are two times the diameter of the propagule in question. For some taxa (*e.g.*, *B. bassiana*), conspicuous swelling of germinating conidia may be used to indicate viability. Numbers of germinated and non-germinated propagules in arbitrarily-selected fields of view or in parallel transects, usually defined with an ocular micrometer, are counted. It is desirable to determine the viability of propagules at various positions on the same plate. Although assessing ca. 100 propagules per replicate culture is generally sufficient, counting greater numbers can increase accuracy. However, propagules incubated on a single plate are technically pseudoreplicates, and standard errors and statistical comparisons should be based on the number of replicate culture plates rather than the total number of propagules assessed (see Chapter II-1). Viability of conidia in an oil carrier may also be determined as outlined previously. Caution is advised, however, because problems occur when attempting to enumerate propagules of species with small round conidia (*e.g.*, *B. bassiana*) as the oil can form tiny emulsion droplets on the agar surface that are difficult to distinguish from ungerminated conidia. To circumvent this problem, propagules in oil should be stained with a drop of lactofuchsin, which is miscible in oil, prior to placement of the coverslip (see Goettel and Inglis, 1997). Magalhães *et al.* (1997) developed a method for assessing viability of conidia in non-emulsifiable oil formulations that involves placing a small sample (ca. 10 μl) on a block of agar (2%) and applying a coverslip with firm pressure to increase contact between the conidia and the agar substrate.

A major disadvantage of the above-described plating methods is that hyphae from early germinating propagules may obscure late-germinating or non-viable propagules, resulting in inaccurate assessments of viability. The fungicide benomyl affects cell replication in many Hypocreales fungi (*e.g., Beauveria, Metarhizium, Aspergillus*); benomyl binds to β-tubulin and prevents spindle tubule formation, thereby inhibiting mitosis. With low concentrations of benomyl in a medium (see Goettel and Inglis, 1997), propagules are able to form germ tubes, but cell division and hyphal growth are prevented. In our laboratories, we have successfully used this method to measure viability of conidia and blastospores of *Beauveria* and *Metarhizium*.

Recent studies (Moore, *et al.*, 1997) indicate that conidia of *M. anisopliae* var. *acridum* dried for long-term storage are damaged by rapid rehydration such as normally occurs when preparing suspensions for viability checks (*i.e.*, inoculation of dry conidia directly onto agar plates). These researchers recommend that conidia be rehydrated in a high-moisture environment before use. Dried conidia (moisture content 4.5%) exposed to high-humidity conditions for 5–40 min prior to plating on an agar-based substrate exhibited three-fold-greater germination (73–78% vs. 25%) compared to conidia plated without preconditioning. Jin *et al.* (1999) claimed that viability of dry, dormant conidia of *M. anisopliae* was dependent upon the type of surfactant used for rehydration. Further experimentation to address the effects of rapid rehydration on other desiccation-stabilized propagules, including hyphal bodies and mycelia is warranted. In addition, protocols are needed to minimize viability losses when large amounts of material are rehydrated for field-tests and for operational control programs.

2 Vital staining

Although non-fluorescing vital stains may be used, fluorochromes are most commonly used to assess the viability of entomopathogens. Fluorescein diacetate (FDA) and fluorescein isothiocyanate (FITC) are commonly used fluorochromes. The use of FDA alone (FDA enters the propagules and is hydrolyzed by enzymes before fluorescing) or in combination with propidium iodide (PI) is described by Firstencel *et al.* (1990), Schading *et al.* (1995), and Goettel and Inglis (1997). FDA can be used to determine viability in bacteria, fungi and protozoa, but difficulties have been encountered with using these fluorochromes to quantify viability of entomophthoralean resting spores due to autofluorescence (Papierok and Hajek, 1997).

To determine viability of *B. bassiana* conidia, aqueous suspensions of conidia (4 μl) are mixed with equal amounts of freshly prepared working solutions of FDA and PI (optional) in a dark room illuminated with a 40 W photographic safelight (see Goettel and Inglis, 1997). The mixture is stirred with a pipette tip and covered with a cover slip. The slides are then viewed under epifluorescence using a 450–490 nm (blue light) exciter filter and a barrier filter in conjunction with a DM500 dichromatic mirror. Viable conidia fluoresce bright green, and if PI is also used, nuclei of dead conidia fluoresce red. Only PI will fluoresce when green light (515–560 nm) is used. If PI is not used, the green conidia are counted using epifluoresence, and then just enough substage light (phase contrast) is added to identify and count all conidia present within the same field. Viable conidia maintain their fluorescence for only short periods (10 to 30 sec) once normal substage light is applied. Concentrations of the fluorochromes used are critical, and optimization may require experimentation to select the best concentration of stain. The most significant disadvantage of this protocol is that FDA and PI are toxic substances that require special handling and disposal.

4 Application

In most instances, fungal propagules can be applied using the same equipment that has been developed for application of chemical pesticides (see Chapters III-1 and III-2). Many entomopathogenic fungi are readily prepared as dry powders, and during the early years of development of fungi for pest control, applications were often made with crop dusting equipment. However, due to environmental contamination and applicator safety issues, dusting has been almost completely replaced in modern pest control by various spray technologies. Choice of application equipment will very much depend on the formulation (*e.g.*, aqueous or oil) and the target host and habitat. Since many fungi infect primarily through the external integument, optimization of coverage is the most important objective driving development of technologies for microbial control applications. Detailed descriptions of the different types of application equipment used for application of fungi in different target habitats are outlined in several chapters in Sections III and VII of this volume.

A Application equipment and strategies

Efforts to increase the efficacy of mycoinsecticide spray applications have generally focused on maximization of spray coverage by the atomization of liquid suspensions to produce a large number of droplets containing high concentrations of propagules. Coverage, especially of lower (abaxial) leaf surfaces, is further enhanced using air streams, drop-tubes, or other technologies to generate turbulence or direct sprays from below canopy level. Hydraulic spray systems are used to apply water-based formulations (including emulsifiable oils and wettable powders) onto a variety of crops (*e.g.*, row, cereal, and rangeland) at medium (200–600 liters/ha) to high (>600 liters/ha) volumes. Mistblowers and other air-assist technologies are primarily used for low volume applications in fields (50–200 liters/ha) and orchards (200–500 liters/ha), and rotary atomizers are used to apply oil formulations at very low volumes (5–50 liters/ha, usually 5–20 liters/ha) and ultra-low volumes (ULV; 0.5–5 liters/ha). The preceding volume classifications are as defined by Matthews (2000). Hand-held spinning disk applicators can be used for very low and ULV applications, but spinning disk, cup, or cage systems are usually mounted on vehicles or in aircraft for application to large tracts of rangeland or forest (see Bateman, 1997).

Electrostatic charging of spray droplets has been considered a promising technology for application of microbial biocontrol agents due to its potential to increase deposition on leaf undersides. Sopp *et al.* (1989) reported improved coverage of abaxial leaf surfaces and greater efficacy when applying *Lecanicillum* sp. against whiteflies using an electrostatic versus a conventional hydraulic sprayer. Law and Mills (1980) reported similar results from tests of *B. thuringiensis* against caterpillar pests of broccoli. However, electrosatic charging of the spray droplets was not the only variable in these tests (*e.g.*, droplet sizes generated by different sprayers were not considered). In controlled spray deposition studies, electrostatic charging of spray droplets has been found to reduce penetration of dense crop foliage and provide only small improvements in overall coverage. These improvements often failed to translate into significant increases in efficacy (Abdelbagi and Adams, 1987; Sopp and Palmer, 1990; Kirk *et al.*, 2001). Electrostatic spray technologies have not been widely adopted by growers, and Matthews (2000) attributes this to the higher cost of electrostatic equipment versus the relatively small gains in efficacy typically achieved under field conditions.

Although most fungi do not infect insects *per os*, insects can become surface-inoculated with propagules during ingestion of bait substrates with loosely adhered fungal propagules, and use of baits that promote tactile contact may also be an efficacious method of targeting insects with fungi that infect externally (Inglis *et al.*, 1996b; Moore and Caudwell, 1997). Granular formulations (*e.g.*, baits and encapsules) are generally applied dry, although spray application of small granules (*e.g.*, pelletized or encapsulated mycelia) can be achieved with large-orifice nozzles. Commercial-scale technologies for efficient application of large granules are not well developed; this is especially true where precise targeting is required. As a result, field tests have employed such methods as broadcasting with mechanical spreaders, use of modified blower equipment, incorporation into soil by tilling, or simply application to a desired target by hand. Conventional seed drills have been modified for subsurface soil applications of granules and liquid suspensions; the drill described by Jackson *et al.* (1992) for application of a bacterial pathogen could be used for application of fungal spore suspensions.

B Effects of application on inoculum

The effects of the spraying method on viability and virulence of entomopathogenic fungi have not been well studied. However, Nilsson and Gripwall (1999) determined that the viability of both blastospores and conidia of *Lecanicillium* sp. (identified as *V. lecanii*) decreased as the length of the pumping period and pressure in a high-pressure hydraulic sprayer increased. The loss was attributed to mechanical shearing forces, but elevated temperatures generated by the high pressure system could not be ruled out. In contrast, application with a cold fogger or backpack sprayer did not affect viability. Griffiths and Bateman (1997) determined that the viability of oil-formulated conidia of *M. anisopliae* var. *acridum* was reduced by 30%, after passage through an exhaust nozzle sprayer, where conidia were briefly subjected to temperatures exceeding 100 °C. However, there was no detectable difference in virulence between conidia collected from the sprayed and unsprayed formulations.

Due to the dearth of information on the possible effects of spray equipment on fungal propagules, it is of the utmost importance that the quality of the inoculum be closely monitored during the application procedure. During initial testing, samples of the inoculum must be taken from the spray nozzle at the beginning of experimentation and at regular intervals thereafter, until the whole spray procedure has been completed. Viabilities and virulence of the inoculum must then be compared to the original, non-sprayed inoculum through germination tests (see section 3D) and using bioassay methodologies (see Butt and Goettel, 2000).

C Compatibility with agrochemicals

There have been numerous studies of the effects of agrochemicals on entomopathogenic fungi (see McCoy *et al.*, 1988). Most have indicated that entomopathogenic fungi are compatible with many agrochemicals, including most insecticides. Fungicides are the most important exception. However, most studies investigating the influence of fungicides on fungi have been conducted in the laboratory (*e.g.*, effects of different concentrations of fungicides on spore germination and vegetative growth *in vitro*). The vast majority of these studies did not attempt to address the potential inhibition of fungi in field environments.

Results from a recent study on effects of timing of fungicide application on *B. bassiana*-induced mortality in *L. decemlineata* indicate that fungicides, previously determined to be detrimental under laboratory conditions, did not significantly inhibit the direct effect of the mycoinsecticide when applications were made asynchronously (Jaros-Su *et al.*, 1999). However, fungal persistence in the habitat treated with fungicides may be significantly reduced.

Although laboratory tests may provide useful information, especially if the fungus and chemical are to be applied simultaneously (*e.g.*, tank-mix situations), results from such studies must not be used to base conclusions on compatibility in the field. Only through careful field evaluations can compatibility or incompatibility be determined.

D Assessing application rates and targeting

Reporting of application parameters other than just numbers of spores/ha is important in assessing efficacy and for making valid comparisons among field tests. The researcher should include detailed descriptions of crop-plant spacing and architecture, the spray technology employed, specific sprayer configurations, spray volumes, etc. If the actual amounts of inoculum deposited on targeted plants or insects are not measured, reports of dose/unit area based on sprayer calibrations can be misleading and contribute to erroneous conclusions. For example, using a standard spray boom to make a broadcast application of 1×10^{13} conidia to a 1 ha field of melon seedlings planted in broad beds achieves a substantially different concentration/plant than an application of the same amount of material to the same field using a sprayer with fewer nozzles targeting only narrow bands about each row. This so-called banding of pesticide applications is often achieved by using drop tubes with laterally directed nozzles. Banding is a highly effective strategy for improving the efficacy and economics of microbial control. For research purposes, it is important that application rate be expressed both in terms of the total area of the field and the area actually treated.

As indicated previously, most fungi infect insects through the external integument. Since a threshold of inoculum is required to incite disease in insects, quantification of the deposition of propagules at the targeted site is of paramount importance in assessing the efficacy of an entomopathogenic fungus. Although most research has focused on the direct deposition of propagules onto insects, secondary acquisition from foliage, soil, and bait substrates, may be of equal importance. Furthermore, the influence of propagule aggregation (*i.e.*, spatial distribution) on the insect integument on efficacy has not been adequately addressed. The primary methods used to quantify targeting are described below.

1 Spray droplet deposition

Water- or oil-sensitive papers are often used to obtain an indirect measure of fungal propagule deposition rates. For qualitative assessments of size and density of spray droplet impressions, cards can be visually compared to standards. Water-sensitive papers must be treated with care. For example, they cannot be used under conditions of high humidity, or where they may come in contact with free water (*e.g.*, dew on leaves or precipitation). Also, the droplets on oil- and water-sensitive papers are not stable and therefore cannot be archived; they must either be processed soon after use or photographed. Matthews (2000) indicates that stability of deposition patterns on water-sensitive cards can be improved by treating them with ethyl acetate. Spray droplets flatten upon impact, and the diameter of an impression on a spray card is therefore not reflective of actual droplet diameter. However, differences in size between droplet impact impressions and true droplet diameters can be determined and used to calculate a spread factor, which, in turn, can be used to convert stain measurements to actual droplet sizes (Matthews, 2000). For quantitative estimation of droplet density, the number of droplets/unit area (usually 1 cm^2) can be counted with the aid of a magnifying lens or stereomicroscope. In addition to droplet densities, image analysis systems can provide information on droplet size distribution, percent coverage, and spray volume.

Another method which has been used to chemically assess targeting involves the use of tracer dyes that are included in the formulation before application (*e.g.*, Cilgi and Jepson, 1992; see Matthews, 2000).

2 Direct observations

With the aid of light or scanning electron microscopy, direct observation of fungal propagules on foliage, insects, and/or other substrates can be used to provide a measure of the number and spatial distribution of propagules. With light microscopy, differential staining is usually necessary. Fluorescence techniques have been used most extensively; however, for translucent hosts (*e.g.*, thrips, aphids, or whitefly nymphs), normal light microscopy using non-fluorescent dyes, such as acid fuchsin, can be equally effective (*e.g.*, Ugine *et al.*, 2005). Compared to scanning electron microscopy and molecular tagging, conventional microscopy is generally simpler and less costly. Large numbers of samples can be processed relatively quickly, and evaluations can be extended to germination and, in some cases, host penetration (*e.g.*, Wraight *et al.*, 1990). Nevertheless, staining methods are considerably more difficult and time consuming than indirect methods such as counting of colony-forming units (CFU). In some instances, autofluorescence can obscure observations of fungal propagules, and identification of propagules can become problematic. Labeling of propagules with a fluorescent stain (*e.g.*, FITC) before application can also be used to facilitate visualization of spatial distributions and germination rates of conidia on the cuticle of insects (*e.g.*, Sosa-Gomez *et al.*, 1997).

3 Washing

Measures of propagule density/unit area can be obtained directly from foliage. Leaves are sampled from defined positions in the canopy and stored at low temperatures prior to being processed. If a leaf-area meter is available, leaves can be cut into pieces of approximately equal size and shape. If a leaf area meter is unavailable, a cutting device of defined area (*e.g.*, a cork borer) can be used to remove a disk of leaf tissue; this is more time consuming and increases potential variation as a result of the smaller areas obtained from each leaf. If necessary, the leaf pieces or disks can be pooled prior to recovery.

Propagules are detached from the leaf segments by washing, which may include sonication. The Stomacher Lab Blender (Seward Medical Ltd, London, UK) usually provides excellent dislodgement of propagules (see Goettel and Inglis, 1997). Once in suspension, the density of propagules/unit area is quantified using the spread-plate and/or most-probable number methods (see Goettel and Inglis, 1997). The suspensions of propagules are diluted (*e.g.*, 2-, 4-, or 10-fold dilution series), and aliquots from each dilution are spread (usually in duplicate or triplicate) onto an appropriate medium. Colony-forming units are ultimately counted, and propagule densities/unit area is calculated. Efficacy of washing methods should be verified by homogenization and plating or by direct examination of leaf surfaces (*e.g.*, by fluorescence microscopy).

Numbers of fungal spores in wash suspensions are usually estimated by culturing and CFU counting; however, if spores are washed from reasonably clean substrates and can be readily identified (differentiated from debris and spores of other microorganisms), quantification may be possible by direct microscopic observations at the surface of the spread plates (or drop plates) using methods similar to those described for assessing viability. This may enable estimation of total propagule densities (viable plus non-viable spores).

Use of washing methods to quantify doses of fungal propagules on targeted insects must be undertaken with considerable caution. The hydrophobic conidia of many entomopathogenic fungi adhere rapidly and tenaciously to insect host cuticle and may be exceedingly difficult to dislodge (Boucias *et al.*, 1988). Extraordinary care must be taken to verify the effectiveness of the selected procedure.

While the dilution spread-plate method is frequently used to quantify deposition and persistence of fungal propagules, there are a number of potential disadvantages/limitations to this technique. Some of the most important include: 1) the method can only be used for saprotrophic fungi; 2) it only detects viable propagules, and samples must therefore be collected immediately after application (before spores are killed by environmental factors such as solar radiation or desiccation); 3) latent propagules will be determined to be nonviable; 4) allelochemicals washed from the substrate may be inhibitory to fungal germination and growth; and 5) CFU counts virtually always provide a conservative estimate of propagule densities because not every CFU will be derived from a single propagule (given the strong tendency of fungal spores to aggregate). Spore aggregates may be dispersed using sonication or other methods.

4 *Homogenization*

Spray targets such as insects or the leaves of plants may be homogenized (by grinding) and plated on an appropriate medium for CFU counting (a discussion of selective media for *B. bassiana* and *M. anisopliae* is presented in a later section dealing with propagule recovery from soil). Protocols for homogenization of whole insects are detailed by Goettel and Inglis (1997). This procedure circumvents the often difficult problems associated with removing propagules from targeted substrates; however, in addition to the above-outlined problems with plating methods, an important disadvantage with respect to quantifying numbers of propagules inoculated onto insects is the incapacity of these methods to distinguish between propagules recovered from the exterior host cuticle and those from internal tissues or the alimentary tract.

5 *Leaf imprinting*

To obtain a measure of the spatial distribution of propagules on leaves, it may be possible to use a leaf imprint technique. Leaves are uniformly pressed against the surface of an appropriate agar medium, the outline of the leaf is marked, and the position of CFU recorded after incubation of the cultures. The medium used should be selective or semi-selective to prevent growth of other microorganisms that can obscure colonies of the applied entomopathogen. Leaf imprint methods, including their use for propagule quantification, are considered in greater detail under section 5B.

6 *Spore trapping*

A number of other techniques can be used to quantify propagule deposition rates. One of the most common involves use of spore traps (artificial substrates placed in the field to sample spray deposits). Fungal propagules can be recovered from a number of substrates that possess a defined or measurable area, such as glass or plastic coverslips. Round coverslips (approximately 10- to 15-mm diam) are especially suitable because they are relatively robust and readily fit into containers used for washing. One to several coverslips can be attached to a holding platform, such as a plastic Petri dish lid, with double-sided tape or other adhesive. The platforms with attached coverslips can be placed at the base of a spray tower in the laboratory or on the soil surface in the field. If desirable, the height of the coverslips can be varied (*e.g.*, relative to the plant canopy) by attaching the holding platforms to the tops of stakes. Alternatively, cover slips can be placed in specific microhabitats within the plant canopy (*e.g.*, attached by pinning directly to adaxial and/or abaxial surfaces of leaves in the upper, middle, or lower canopy). In field settings, it is recommended that coverslips be placed at defined sites within each plot (*i.e.*, subplots) to obtain a measure of spatial variation. See Chapter VII-1 for further details.

Following the application of propagules, the carrier is allowed to dry, and the coverslips are transported to the laboratory. Propagules can be counted indirectly (*e.g.*, using the wash method) or directly using microscopic methods. If a measure of spore viability is desired, the coverslips should be collected immediately, allowed to dry, and stored at low temperature (ca. 5 °C) until they can be processed. To recover propagules, coverslips are either pooled as necessary or washed individually. The wash suspension is diluted, spread onto an agar medium and the number of CFU counted at the appropriate dilution. For counts of total viable and non-viable spores, samples of the wash suspension can be counted in a hemacytometer or Petroff-Hausser chamber. As would be expected, spray deposits on small spore traps (and even on whole leaves) are extremely variable, and large numbers of samples are required for detecting statistically significant differences among treatments. Wraight *et al.* (1998) described a rapid method for direct enumeration of spores on large numbers of coverslips. Populations are calculated as: (n x dilution factor)/area; where n is the number of propagules at the desired dilution and the area is usually expressed in mm^2 or cm^2. As suggested previously, use of acid fuchsin (1 mg/ml of mounting fluid) can greatly aid differentiation of spores from particulate matter and oil droplets. Oil residues slowly evaporate from stored coverslip samples.

Quantification of conidia production from mycelium formulations of entomophthoralean fungi in the phyllosphere poses a unique problem in that these conidia are forcibly discharged during multiple cycles of rehydration and rapidly germinate to produce secondary conidia that are in turn forcibly discharged. Entomophthoralean conidial depositions (commonly referred to as conidial showers) are usually also too dense for direct enumeration. Wraight *et al.* (2003) developed a conidial trap incorporating the previously described Triton X-100/maleic acid fixative of Soper (1985) to inhibit germination. Conidial traps are fabricated from pieces of pliable plastic (*e.g.*, disposable coverslips). These are bent into an asymmetrical z-shape with a narrow base for pinning to the leaf surface and a large trap surface that is positioned over a mycelium granule without coming in contact with it (the traps can also be positioned over fungus-killed insects for quantification of fungal inoculum in epizootiological studies). Prior to the deployment of the traps, a 10% gelatin solution is prepared and maintained at ca. 45 °C. The ingredients of the fixative are added, and a film of this solution is applied to the trap surface with a fine paintbrush and allowed to dry. The trap is preferably situated over particles on abaxial leaf surfaces so that spore collection is assisted by gravity. The conidia collected on the gelatin are inhibited from germinating and, when dried, can be stored indefinitely.

For enumeration, the dry gelatin film containing the conidial shower is peeled from the conidial trap, immersed in a small amount of water in a test tube, and warmed to 40–45 °C. After the gelatin melts, the tube is vortexed or otherwise agitated to produce a suspension of dispersed conidia, which can then be counted using a hemacytometer. A salient problem with the use of such traps is that they do not catch all of the conidia. Many conidia are discharged onto the leaf substrate surrounding the granule (even in the case of granules or cadavers on leaf undersides). Nevertheless, they are useful for sampling conidial production. They are easily collected and replaced to sample conidia production at desired intervals during a single sporulation episode or over multiple episodes (*e.g.*, over consecutive nights).

7 *Propagule recovery from soil*

Although molecular methods may eventually become the methods of choice for quantification of fungal biomass in soil (*e.g.*, enzyme-linked immunosorbent assay), quantification of inocula (spores and mycelia) applied to soil is currently based primarily on use of semi-selective media (see below). Soil samples (usually 10 g) are collected from desired depths using a soil core sampler. These are added to 100 ml of water or buffer containing a surfactant such as Tween 80 (0.05%), the suspension is homogenized (*e.g.*, using a Waring blender or Stomacher Lab Blender), and the number of CFU enumerated on an appropriate medium using the dilution spread-plate method or most probable number (MPN) methods (see Goettel and Inglis, 1997). Subsamples of soil are used for dry weight and bulk density determinations. Propagule titers are quantified as CFU/g dry soil and as spores/unit volume of soil ($CFU/cm^3 = CFU/g$ x the bulk density). Generally, neither the spread-plate nor MPN method provides an accurate measure of fungal biomass in mycelial inocula, as a small hyphal fragment and a large aggregation of mycelium will both produce a single CFU.

Semi-selective media have been used extensively for recovery of *Beauveria* and *Metarhizium* spp. from soil and various other substrates. Most of these media are based on the differential activity of the fungicide, dodine (N-dodecylguanidine monoacetate). An oatmeal agar medium amended with 650 μg dodine/ml suppresses growth of *Penicillium* spp., *Trichoderma viride*, and Mucorales species, but supports growth of *B. bassiana* and *Metarhizium* spp. (Beilharz *et al.*, 1982). Chase *et al.* (1986) confirmed that an oatmeal medium amended with 600 μg dodine/ml resulted in good isolation of *B. bassiana*, but found that this concentration was inhibitory to *M. anisopliae*. When the concentration was reduced to 500 μg/ml, and 400 μg of the fungicide benomyl (Benlate)/ml was added to the medium, both *B. bassiana* and *M. anisopliae* were effectively isolated from soil. Subsequently, however, Liu *et al.* (1993) observed that even 300 μg dodine/ml inhibited isolation of some isolates of *M. anisopliae*. They found that a low concentration of dodine (10 μg/ml), in combination with 500 μg cyclohexamide/ml, increased recovery of *Metarhizium* from soil. Chase *et al.* (1986) added crystal violet to the medium to enhance the contrast with fungal colonies. These studies indicate that the concentration of dodine is a critical factor in the use of these selective media (especially for quantification of propagule persistence in the field), and the researcher must test the effects of various concentrations on the fungal isolate to be evaluated in the field. Since these media are semi-selective, arbitrarily-selected colonies should be examined microscopically and the identity of the fungus confirmed using the appropriate criteria (see Goettel and Inglis, 1997).

Fungal propagules (*e.g.*, resting spores) also can be recovered from soil using differential centrifugation (*e.g.*, Macdonald and Spokes, 1981; Hajek and Wheeler, 1994). Soil samples are added to water and mixed as described above. The resulting slurry is then washed consecutively through a series of sieves (*e.g.*, 500 to 20 μm) and the slurry retained on the final sieve is resuspended in buffer. After thorough mixing, the suspension is layered on top of a discontinuous density gradient. Although several density gradient methods can be used, Percoll, diluted with 0.15 M NaCl to 1.05, 1.10 and 1.13 g/ml, has been used successfully to extract resting spores of *Conidiobolus obscurus* (Macdonald and Spokes, 1981) and *E. maimaiga* (Hajek and Wheeler, 1994). With Percoll, the sample is centrifuged at 3800 *g* for 10 min, the

1.1 g/ml layer and its interfaces are removed, and the numbers of resting spores are enumerated in a counting chamber.

Whatever method of quantification of propagules within soil is adopted, its accuracy should always be established by adding known quantities of propagules to soil and determining the recovery rate.

5 Assessing persistence

Elucidation of the fate of fungal propagules following field applications poses a difficult challenge, and a variety of approaches and protocols have been developed. These include leaf and soil washes, homogenization, leaf impressions, differential staining, bioassay of field-collected substrates, and monitoring of sentinel insects. Each of these methods has specific advantages and disadvantages, and these techniques are frequently used in combination to obtain a more accurate measure of fungal persistence. Because some isolates of *B. bassiana* are capable of colonizing plants (Bing and Lewis, 1991), it may be necessary to also assess the extent of endophytism after field application of this fungus. Methods for assessing the occurrence of endophytism by *B. bassiana* are presented in Chapter VII-3. It is essential that plants be surface sterilized prior to sampling of internal tissues, because the epidermis in nodal areas, or other niches protected from solar radiation, may be superficially contaminated with spores of the suspected endophyte, especially if the endophyte is also a ubiquitous, aerially dispersed microorganism like *B. bassiana*.

A Washing and homogenization methods

Leaf-, soil-, and insect-washing and homogenization methods have already been described in considerable detail (see section 4D). The principal advantage of these protocols is that the methods are simple and rapid. If carefully applied, they can provide accurate measures of total viable propagules, and in some cases, total viable plus non-viable propagules persisting on target substrates over time. Washing and CFU counting is the only practical method for monitoring persistence of fungi in soil and similar substrates such as animal bedding or crop residues.

B Leaf imprints

Leaf impressions were described previously as a means of determining spatial distribution of spores on leaves. However, this method can also be used to quantify propagule persistence. To achieve this, substrates are pressed onto agar surfaces (usually a selective medium) using a standardized pressure, dislodging the propagules from the substrate. The cultures are incubated under appropriate conditions, and the prevalence of ungerminated and germinated propagules are enumerated after a specified time using a compound microscope (*e.g.*, Fransen, 1995). An alternative method of leaf imprinting involves pressing the leaves against transparent adhesive tape rather than agar (Gouli *et al.*, 2005). The tape with adhered spores can be placed on a nutrient medium for direct assessments of viability.

The principal advantages of leaf imprint methods are that they provide a means for assessing both survival (viability) and physical persistence (densities) of total inoculum (viable plus nonviable spores). Furthermore, specific tissues (*e.g.*, adaxial and abaxial leaf surfaces) can be easily processed.

The primary problem encountered with the leaf-imprint technique is that spore removal efficiency is highly variable, depending upon factors such as substrate topography (especially with respect to leaf pubescence and rugosity), adhesiveness of the propagules, and the pressure with which the substrate comes in contact with the medium. Imprint samples therefore may not be representative of the entire residual spore population, and efficiency of the sampling method must be evaluated, either by homogenization and plating or by direct microscopic observations. Another potential disadvantage is that it may be impossible to visually distinguish propagules of the target entomopathogen from propagules of other phylloplane fungi. Leaves growing under field conditions are commonly colonized by a variety of fungi and may become heavily contaminated with soil particles and other debris. The investigator must take extraordinary precautions to verify that the spores tallied (both germinated and ungerminated) are in fact, the insect pathogen of interest. Identification of nonviable propagules poses a special problem, as these may begin to degrade after several days. If non-viable spores cannot be identified, or if they are weathered from the substrate at a different rate than viable spores, viability will be overestimated. Presses of untreated leaves must be evaluated in the same way as treated leaves, and counts must be corrected for background microorganisms, organic and mineral debris, or oil droplets that could be mistaken for propagules of the subject pathogen. Use of stains can greatly facilitate differentiation between fungal spores and various contaminants.

C Bioassay and sentinel methods

The bioassay method to assess field persistence typically involves exposure of insects in the laboratory to an appropriate substrate (*e.g.*, foliage) collected at various intervals from fungus-treated field plots and recording mortality in the caged insects. It has been used in studies of both entomophthoralean and hypocrealean pathogens (*e.g.*, Brobyn *et al.*, 1985; Vandenberg *et al.*, 1998). The primary advantage is in detecting the presence of infectious inoculum or efficacious residues, rather than simply viable spores. Infectious inoculum may be in the form of germlings or hyphae (possibly endophytic) as well as spores. The procedure does not provide a direct estimate of propagule numbers; however, numbers can be estimated indirectly by comparing recorded levels of mortalities with those from laboratory bioassays with known dosages or concentrations. The obvious disadvantages are that this method, like CFU counting, cannot elucidate mechanisms of inoculum loss, and insects may be extraordinarily susceptible to exceedingly low doses of fungal pathogens under favorable conditions of temperature and moisture in the laboratory.

The use of sentinel insects provides essentially the same information as bioassays and this method is commonly used to elucidate the importance of secondary acquisition of lethal doses from substrates (*e.g.*, foliage) and to measure potential residual activity under more natural conditions. Sentinel insects are often caged on treated plants or in a defined area, and proper interpretation requires a knowledge of the artifactual effects of confinement on susceptibility. In most instances, insects reared in captivity are used as sentinels because their age and physiological status (*e.g.*, pathogen free) can be defined or because they are known to be susceptible to a variety of entomopathogenic fungi (*e.g., Galleria mellonella*). However, field-collected insects are used in some circumstances.

D Molecular methods

The use of molecular methods offers a number of novel possibilities for quantifying the levels of persistence, and for epizootiological studies with fungal entomopathogens. The primary techniques include immunological methods using both polyclonal and monoclonal antibodies, and the quantitative and qualitative use of DNA methodologies [*e.g.*, Southern dot blots and polymerase chain reaction technology (PCR)]. In many instances, a knowledge of the isolate responsible for initiating an epizootic is an important consideration in elucidating the salient factors dictating epizootic development (*i.e.*, was the genotype applied responsible). A variety of PCR-based methods can be used to identify and differentiate specific genotypes (*e.g.*, random amplified polymorphic DNA, amplified polymorphic length polymorphism, and microsatellite), but robust and user-friendly PCR-based techniques for species- or strain-specific detection offer exciting possibilities for monitoring the fate of an applied entomopathogenic fungus, particularly long-term assessments. For example, sequence characterized amplified region (SCAR) markers specific to *B. bassiana* strain GHA, a strain registered in the U.S. for pest control in field and greenhouse crops, have been used to detect presence and persistence of this strain in soil and foliage samples without need for selective culture and fungal isolation (Castrillo *et al.*, 2003). Preliminary screening studies with SCAR markers for *B. bassiana* isolates from GHA-treated fields revealed persistence of this strain for at least four years post-application (L.A. Castrillo, personal communication).

6 Assessing efficacy

The insect pest to be targeted will, to a large degree, dictate the best experimental design and deployment strategy. As with other relatively slow acting entomopathogens (*e.g.*, many bacteria and viruses), it is usually not possible to quantify insects killed by entomopathogenic fungi in a field environment (*e.g.*, due to rapid removal of cadavers by scavengers). As a result, it is necessary to collect insects and maintain them in enclosures and/or to use indirect techniques such as crop damage assessments, changes in population size and species composition, and methods that measure infection and disease progression. While a number of entomopathogens have been shown to exhibit sublethal effects in controlled environment studies, very few studies have attempted to address this possibility in a field setting. Quantification of sublethal effects of entomopathogens on field populations of non-caged insects will require long-term monitoring of populations, including feeding, fecundity, and crop-damage assessments.

A *Controlled-environment cages*

A common method used to assess the efficacy of entomopathogenic fungi involves capture of the targeted hosts at various times post-application, and their subsequent maintenance in cages. Environmental conditions in these cages may be highly controlled (*e.g.*, in an environmental chamber) or relatively variable (*e.g.*, in a greenhouse); however, in each case, conditions of light, temperature, relative humidity, and vapor pressure deficit should be monitored in detail. The researcher must be aware that the environment in which the cages are maintained may have a substantial impact on disease development, and extrapolation to field efficacy must be made with extreme care. For example, Inglis *et al.* (1997a) collected grasshoppers at various times after application of *B. bassiana* in a rangeland agroecosystem. One half of the grasshoppers were maintained in cages in the field, whereas, the other half were maintained in cages in a greenhouse environment. Conditions of temperature, light exposure, and relative humidity differed between the two environments (nighttime temperatures were ca. 8 °C higher and light levels were reduced by ca. 74% in the greenhouse cages). Conidial persistence was substantially enhanced and disease progressed more rapidly and attained a substantially higher prevalence in greenhouse than in field cages. Although less mycosis was observed in grasshoppers collected at later sampling times, mycosis was first observed 3 to 4 days after placement of the nymphs in greenhouse cages regardless of the collection time. This example illustrates the erroneous conclusion that would be made with respect to field efficacy if the researcher relied exclusively on the greenhouse cage results. This test provided valuable evidence that environment, and not targeting, was responsible for the poor efficacy of this entomopathogenic fungus in the field.

The possibility of latent infections in insects treated with fungal pathogens has not been extensively studied, and in conjunction with histopathological, microbiological, and/or molecular techniques, the maintenance of insects collected at various times after application of an entomopathogenic fungus can be used to elucidate the importance of latent infections under sub-optimal environmental conditions.

B *Field cages*

Assessments of efficacy using field cages can provide valuable information. However, the cage can result in an environment which is substantially different from ambient. For example, the mesh screen commonly used on cages can reduce the amount of light in the cage by >50% (Inglis *et al.*, 1997a). As in the controlled environment assessments, insects are collected immediately after application of the entomopathogenic fungus and at regular intervals thereafter. Mortality is generally assessed daily, and compared to insects collected from control plots. If supplemental feeding is required, the diet must be nutritionally adequate and free of pathogens. Despite the inherent problems with cage studies to predict field efficacy, they can play an essential role in assessing efficacy of entomopathogenic fungi. The appropriate environmental conditions should be monitored throughout the experimental period, and ideally the cage design should mimic field conditions (including microclimate) as closely as possible if the goal is to provide a measure of field efficacy.

C *Infection and disease diagnoses of field-collected hosts*

Following field collection, target hosts can be killed using methods that would not harm the pathogen (*e.g.*, CO_2 or freezing), surface sterilized (Goettel and Inglis, 1997), and assessed for disease (see Lacey and Brooks, 1997). To induce external sporulation of fungi on the surface of cadavers, they should be placed in suitable environments (usually high humidity). A high-humidity environment is easily achieved using a moistened cotton batten or piece of filter paper placed in a sterile container such as a Petri dish sealed with parafilm or plastic film. Another frequently used method is to place cadavers on water agar amended with antibacterial agents. Once adequate growth is observed on cadavers, the disease can usually be diagnosed visually (Lacey and Brooks, 1997; Papierok and Hajek, 1997).

Alternatively, disease progression can be monitored through determination of the presence of the pathogen within the hemocoel. Tissues or entire cadavers may be homogenized using a variety of methods and then propagules can be observed microscopically or isolated by centrifugation or filtration or enumerated by plating on a semi-synthetic medium. To homogenize insects, Potter-Elvehjem tubes may be used for larger insects and micropestles in 1.5 ml microcentrifuge tubes for smaller insects. Polytron tissue grinders and blenders may also be used, particularly for insects that possess hard exoskeletons. However, it is more difficult to maintain a sterile environment using Polytron grinders, because they cannot be readily autoclaved. Stomacher blenders have been used to homogenize insects, but bag punctures can be a problem, depending on the insect. To reduce the likelihood of the hemolymph having an adverse affect on the entomopathogen, it is recommended that the homogenization buffer be chilled prior to use and the sample maintained on ice.

Molecular techniques can also be used for species- and strain-specific detection of fungal pathogens within the hemocoel of infected insects. Serological methods such as the enzyme-linked immunosorbent assays (ELISA) were originally used for this purpose (*e.g.*, Guy and Rath, 1990; Hajek *et al.*, 1991), but more sensitive and specific PCR and DNA hybridization methods are now available (Hegedus *et al.*, 1996; Bidochka *et al.*, 1997; Entz, *et al.*, 2005). It is important to note that in the initial stages of disease development, fungal proliferation within the insect may be minimal and difficult to detect. Furthermore, determination of the precise cause of death of a diseased insect is difficult.

D *Host population densities and crop yield*

Since it is not possible to measure the prevalence of cadavers in most circumstances, assessments of the efficacy of entomopathogenic fungi often rely on quantification of temporal changes in insect population densities. The primary methods used to measure insect densities include fixed quadrat, sweepnet sampling, whole-plant or foliage sampling, suction or vacuum sampling, drop cloths, pitfall traps, pheromone traps, etc. The reader is referred to other chapters in Section VII of this volume and the general literature for details on collection methodologies for specific insects.

The primary goal of any management strategy is to prevent economic loss, and this is often overlooked in evaluating efficacy of microbial control agents. Given that many fungi are slow-acting and can have sublethal effects, this is an important consideration in assessing the efficacy of entomopathogenic fungi. The insect pest targeted in a microbial control program will determine the type of parameters utilized. The primary strategies used to measure crop losses include defoliation indices (used particularly for assessments of the antifeedant properties of botanical insecticides), yield and quality of crop, and remote sensing. The reader is referred to other chapters in Section VII for more detailed descriptions of crop loss measurements used to assess the efficacy of entomopathogens.

E *Sublethal effects*

Although most assessments of the efficacy of entomopathogenic fungi have concentrated on mortality of the host, sublethal effects should not be overlooked. Sublethal effects could include decreased vigor, fecundity, and food consumption, altered behavior, and other factors that could decrease the overall damage levels, reproductive rate or survival of a pest population. In particular, sublethal effects may increase susceptibility of affected insects to other natural enemies.

Most attempts at elucidating sublethal effects of entomopathogenic fungi on insect pests have been completed under controlled-environment conditions. One approach to measuring sublethel effects in the field involves use of field cages (*e.g.*, Thomas *et al.*, 1997). Bottomless field cages are often used. These are usually arranged in a randomized block design, and treatments most commonly include various doses of an entomopathogenic fungus and uninoculated and/or spray carrier controls. Insects may be inoculated in the laboratory before introduction into the field cages. An appropriate number of individuals are placed in each cage, and mortality and other pertinent factors (*e.g.*, behavior) are monitored as necessary. Surviving individuals are counted at the end of the experimental period. For food consumption assessments, remaining foliage is collected and oven-dried. Estimates of the quantity of foliage consumed by the insects in each cage are obtained by subtracting the dry weights of foliage remaining in the cages with insects from those not containing insects. Further details on this method are provided in Chapter VII-17. A number of other parameters (*e.g.*, fecundity, behavior, etc.) can be investigated instead of, or in conjunction with, food consumption.

7 Sources of entomopathogenic fungi

There are numerous small collections of entomopathogenic fungi held by researchers at many institutions worldwide. The most important collections housing significant numbers of entomopathogenic fungi are:

1) The United States Department of Agriculture Agricultural Research Service (USDA/ARS) Collection of Entomopathogenic Fungi (ARSEF); Tower Road, Ithaca, NY, 14853-2901; "arsef.fpsnl.cornell.edu/mycology/ARSEF_culture_collection.html";
2) CABI Bioscience, (IMI) (formerly International Mycological Institute), Silwood Park, Ascot, Berkshire SL5 7YA, U.K. "www.cabi.org";
3) Centraalbureau voor Schimmelcultures (CBS), Oosterstraat 1, P.O. Box 273, 3740 AG Baarn, The Netherlands "www.cbs.knaw.nl"
4) American Type Culture Collection (ATCC), 10801 University Boulevard, Manassas, VA 20110-2209; "www.atcc.org/"

Most of these organizations offer an online catalog of accessions available at their web sites. Samples of presently registered fungal control agents are usually available from companies producing these products. These companies are also often willing to provide samples of non-registered products for researchers to study and field test. Many of the companies may also be willing to produce small quantities of novel products on contract.

8 References

Abdelbagi, H. A. and Adams, A. J. 1987. Influence of droplet size, air-assistance and electrostatic charge upon the distribution of ultra-low-volume sprays on tomatoes. *Crop Prot.* **6**, 226–233.

Alexopoulos, C. J., Mims, C. W., and Blackwell, M. 1996. "Introductory Mycology," fourth edition. John Wiley & Sons, New York.

Bateman, R. 1997. Methods of application of microbial pesticide formulations for the control of grasshoppers and locusts. *In* "Microbial Control of Grasshoppers and Locusts" (M. S. Goettel and D. L. Johnson, Eds.), pp. 69–81. *Mem. Entomol. Soc. Can.* Volume 171.

Beilharz, V. C., Parbery, D. G., and Swart, H. J. 1982. Dodine: a selective agent for certain soil fungi. *Trans. Br. Mycol. Soc.* **79**, 507–551.

Bidochka, M. J., Walsh, S. R. A., Ramos, M. E., St. Leger, R. J., Carruthers, R. I., Silver, J. C., and Roberts, D. W. 1997. Cloned DNA probes distinguish endemic and exotic *Entomophaga grylli* fungal pathotype infections in grasshopper life stages. *Mol. Ecol.* **6**, 303–308.

Bing, L.A. and Lewis, L.C. 1991. Suppression of the European corn borer, *Ostrinia nubilalis* (Hübner) (Lepidoptera: Pyralidae), by endophytic *Beauveria bassiana* (Balsamo) Vuillemin. *Environ. Entomol.* **20**, 1207–1211.

Blackwell, M. and Spatafora, J. W. 2004. Fungi and their allies. *In* "Biodiversity of Fungi: Inventory and Monitoring Methods" (G. M. Mueller, G. F. Bills, and M. S. Foster, Eds.), pp. 7–21. Elsevier Academic Press, Amsterdam.

Boucias D.G. and Pendland, J.C. 1998. "Principles of Insect Pathology." Kluwer Academic Publishers, Dordrecht, The Netherlands.

Boucias, D. G., Pendland, J. C., and Latge, J. P. 1988. Nonspecific factors involved in attachment of entomopathogenic Deuteromycetes to host insect cuticle. *Appl. Environ. Microbiol.* **54**, 1795–1805.

Braga, G. U. L., Flint, S. D., Miller, C. D., Anderson, A. J., and Roberts, D. W. 2001a. Both solar UVA and UVB radiation impair conidial curability and delay germination in the entomopathogenic fungus *Metarhizium anisopliae*. *Photochem. Photobiol.* **74**, 734–739.

Braga, G. U. L., Flint, S. D., Miller, C. D., Anderson, A. J., and Roberts, D. W. 2001b. Variability in response to UV-B among species and strains of *Metarhizium* isolated from sites at latitudes from 61 °N to 54 °S. *J. Invertebr. Pathol.* **78**, 98–108.

Braga, G. U. L., Rangel, D. E. N., Flint, S. D., Miller, C. D., Anderson, A. J., and Roberts, D. W. 2002. Damage and recovery from UV-B exposure in conidia of the entomopathogens *Verticillium lecanii* and *Aphanocladium album*. *Mycologia* **94**, 912–920.

Brobyn, P. J., Wilding, N., and Clark, S. 1985. The persistence of infectivity of conidia of the aphid pathogen *Erynia neoaphidis* on leaves in the field. *Ann. Appl. Biol.* **107**, 365–376.

Bruns, T. D., White, T. J., and Taylor, J. W. 1991. Fungal molecular systematics. *Annu. Rev. Ecol. Syst.* **22**, 525–564.

Butt, T. M. and Goettel, M. S. 2000. Bioassays of entomogenous fungi. *In* "Bioassays of Entomopathogenic Microbes and Nematodes" (A. Navon and K. R. S. Ascher, Eds.), pp. 141–195. CABI International Press, Wallingford, UK.

Butt T.M., Jackson C.W., and Magan N. (Eds.) 2001. "Fungi as Biocontrol Agents." CABI Publishing, Wallingford, UK.

Carruthers, R. I. and Soper, R. S. 1987. Fungal diseases. *In* "Epizootiology of Insect Diseases" (J. R. Fuxa and Y. Tanada, Eds.), pp. 357–416. John Wiley and Sons, New York.

Castrillo, L. A., Vandenberg, J. D., and Wraight, S. P. 2003. Strain-specific detection of introduced *Beauveria bassiana* in agricultural fields by use of sequence-characterized amplified region markers. *J. Invertebr. Pathol.* **82**, 75–83.

Chase, A. R., Osborne, L. S., and Ferguson, V. M. 1986. Selective isolation of the entomopathogenic fungi *Beauveria bassiana* and *Metarhizium anisopliae* from an artificial potting medium. *Fla. Entomol.* **69**, 285–292.

Chelico, L., Haughian, J. L., Woytowich, A. E., and Khachatourians, G. G. 2005. Quantification of ultraviolet-C irradiation induced cyclobutane pyrimidine dimers and their removal in *Beauveria bassiana* conidiospore DNA. *Mycologia* **97**, 621–627.

Cilgi, T. and Jepson, P. C. 1992. The use of tracers to estimate the exposure of beneficial insects to direct pesticide spraying in cereals. *Ann. Appl. Biol.* **121**, 239–247.

Clerk, G. C. and Madelin, M. F. 1965. The longevity of conidia of three insect-parasitizing Hyphomycetes. *Trans. Br. Mycol. Soc.* **48**, 193–209.

Copping, L.G. (Ed.) 2004. "The Manual of Biocontrol Agents." Third edition of The Biopesticide Manual. British Crop Protection Council, Alton, UK.

Daoust, R. A. and Roberts, D. W. 1983. Studies on the prolonged storage of *Metarhizium anisopliae* conidia: Effect of temperature and relative humidity on conidial viability and virulence against mosquitoes. *J. Invertebr. Pathol.* **41**, 143–150.

Entz, S. C., Johnson, D. L., and Kawchuk, L. M. 2005. Development of a PCR-based diagnostic assay for the specific detection of the entomopathogenic fungus *Metarhizium anisopliae* var. *acridum*. *Mycol. Res.* **109**, 1302–1312.

Evans, H. C. 1989. Mycopathogens of insects of epigeal and aerial habitats. *In* "Insect-fungus Interactions" (N. Wilding, N. M. Collins, P. M. Hammond and J. F. Webber, Eds.), pp. 205–238. Academic Press, London, UK.

Fargues, J., Goettel, M. S., Smits, N., Ouedraogo, A., Vidal, C., Lacey, L. A., Lomer, C. J., and Rougier, M. 1996. Variability in susceptibility to simulated sunlight of conidia among isolates of entomopathogenic Hyphomycetes. *Mycopathologia* **135**, 171–181.

Fargues, J., Ouedraogo, A., Goettel, M. S., and Lomer, C. J. 1997. Effects of temperature, humidity and inoculation method on susceptibility of *Schistocerca gregaria* to *Metarhizium flavoviride*. *Biocontrol Sci. Technol.* **7**, 345–356.

Firstencel, H., Butt, T. M., and Carruthers, R. I. 1990. A fluorescence microscopy method for determining the viability of entomophthoralean fungal spores. *J. Invertebr. Pathol.* **55**, 258–264.

Fransen, J. J. 1995. Survival of spores of the entomopathogenic fungus *Aschersonia aleyrodis* (Deuteromycotina: Coelomycetes) on leaf surfaces. *J. Invertebr. Pathol.* **65**, 73–75.

Friedberg, E. C., Walker, G. C., and Siede, W. 1995. "DNA repair and Mutagenesis." ASM Press, Washington D.C.

Galaini-Wraight, S., Wraight, S. P., Carruthers, R. I., Magalhães, B. P., Roberts, D. W. 1991. Description of a *Zoophthora radicans* (Zygomycetes: Entomophthoraceae) epizootic in a population of *Empoasca kraemeri* (Homoptera: Cicadellidae) on beans in Central Brazil. *J. Invertebr. Pathol.* **58**, 311–326.

Glare, T. R. and Milner, R. J. 1991. Ecology of entomopathogenic fungi. *In* "Handbook of Applied Mycology, Vol. 2: Humans, Animals, and Insects" (D. K. Arora, L. Ajello and L. G. Mukerji, Eds.), pp. 547–612. Marcel Dekker, New York.

Goettel,. M. S. and Hajek, A.E. 2001. Evaluation of nontarget effects of pathogens used for management of arthropods *in*: "Evaluating Indirect Ecological

Effects of Biological Control" (E. Wajnberg, J.K Scott and P.C. Quimby, Eds.), pp. 81–97. CABI Press, Wallingford, UK.

Goettel, M.S., Hajek, A.E., Siegel, J.P., and Evans, H.C. 2001. Safety of Fungal Biocontrol Agents, *In* "Fungal Biocontrol Agents – Progress, Problems and Potential" Chapter 13, (T. Butt, C. Jackson and N. Magan (Eds.), pp. 347–375. CABI Press, Wallingford, UK.

Goettel, M.S., Eilenberg, J. and Glare, T.R. 2005 Entomopathogenic Fungi and Their Role in Regulation of Insect Populations *In* "Comprehensive Molecular Insect Science" Vol. 6 (L.I. Gilbert, K. Iatrou and S. Gill, Eds.) pp. 361–406. Elsevier Pergamon, Oxford, UK.

Goettel, M. S. and Inglis, D. G. 1997. Fungi: Hyphomycetes. *In* "Manual of Techniques in Insect Pathology" (L. Lacey, Ed.), pp. 213–249. Academic Press, London, UK.

Goettel, M. S., Poprawski, T. J., Vandenberg, J. D. Li, Z., and Roberts, D. W. 1990. Safety to nontarget invertebrates of fungal biocontrol agents. *In* "Safety of Microbial Insecticides" (M. Laird, L. A. Lacey, and E. W. Davidson, Eds), pp. 209–232. CRC Press, Boca Raton, FL.

Gouli, V. V., Gouli, S. Y., Costa, S. D., and Shternshis, M. V. 2005. Comparison of wash, leaf imprint and adhesive tape methods for estimation of fungal spore deposition on leaves. *Mycol. Phytopathol.* **39**, 99–103.

Griffiths, J. and Bateman, R. 1997. Evaluation of the Francome MkII exhaust nozzle sprayer to apply oil-based formulations of *Metarhizium flavoviride* for locust control. *Pestic. Sci.* **51**, 176–184.

Guy, P. L. and Rath, A. C. 1990. Enzyme-linked immunosorbent assays (ELISA) to detect spore surface antigens of *Metarhizium anisopliae*. *J. Invertebr. Pathol.* **55**, 435–436.

Hajek, A. E. 1997. Ecology of terrestrial fungal entomopathogens. *Adv. Microbiol. Ecol.* **15**, 193–249.

Hajek, A. E., Butt, T. M., Strelow, L. I., and Gray, S. M. 1991. Detection of *Entomophaga maimaiga* (Zygomycetes: Entomophthorales) using enzyme-linked immunosorbent assay. *J. Invertebr. Pathol.* **58**, 1–9.

Hajek, A. E. and Wheeler, M. M. 1994. Application of techniques for quantification of soil-borne entomophthoralean resting spores. *J. Invertebr. Pathol.* **64**, 71–73.

Hegedus, D. D. and Khachatourians, G. G. 1996. Detection of the entomopathogenic fungus *Beauveria bassiana* within infected migratory grasshoppers (*Melanoplus sanguinipes*) using polymerase chain reaction and DNA probe. *J. Invertebr. Pathol.* **67**, 21–27.

Hodge, K. T. 2003. Clavicipitaceous anamorphs. In "Clavicipitalean Fungi" (J. F. White, Jr., C. W. Bacon, N. L. Hywel-Jones, and J. S. Spatafora, Eds.), pp. 75–123. Marcel Dekker, New York.

Humber, R. A. 1997. Fungi: Identification. *In* "Manual of Techniques in Insect Pathology" (L. Lacey, Ed.), pp. 153–185. Academic Press, London.

Inglis, G. D., Duke, G. M., Kawchuk L. M., and Goettel, M. S. 1999. Influence of oscillating temperatures on the competitive infection and colonization of the migratory grasshopper by *Beauveria bassiana* and *Metarhizium flavoviride*. *Biol. Control* **14**, 111–120.

Inglis, G. D., Goettel, M. S., and Johnson, D. L. 1993. Persistence of the entomopathogenic fungus, *Beauveria bassiana*, on phylloplanes of crested wheatgrass and alfalfa. *Biol. Control* **3**, 258–270.

Inglis, G. D., Goettel, M. S., and Johnson, D. L. 1995. Influence of ultraviolet light protectants on persistence of the entomopathogenic fungus, *Beauveria bassiana*. *Biol. Control* **5**, 581–590.

Inglis, G. D., Ivie, T. J., Duke, G. M., and Goettel, M. S. 2000. Influence of rain and conidial formulation on persistence of *Beauveria bassiana* on potato leaves and Colorado potato beetle larvae. *Biol. Control* **18**, 55–64.

Inglis, G. D., Johnson, D. L., Cheng, K. -J., and Goettel, M. S. 1997b. Use of pathogen combinations to overcome constraints of temperature on entomopathogenic Hyphomycetes against grasshoppers. *Biol. Control* **8**, 143–152.

Inglis, G. D., Johnson, D. L., and Goettel, M. S. 1996a. Effects of temperature and thermoregulation on mycosis by *Beauveria bassiana*. *Biol. Control* **7**, 131–139.

Inglis, G. D., Johnson, D. L., and Goettel, M. S. 1996b. Effect of bait substrate and formulation on infection of grasshopper nymphs by *Beauveria bassiana*. *Biocontrol Sci. Technol.* **6**, 35–50.

Inglis, G. D., Johnson, D. L., and Goettel, M. S. 1997a. Effects of temperature and sunlight on mycosis (*Beauveria bassiana*) (Hyphomycetes: Sympodulosporae) of grasshoppers under field conditions. *Environ. Entomol.* **26**, 400–409.

Inyang, N., McCartney, H. A., Oyejola, B., Ibrahim, L., Pye, B. J., Archer, S. A., and Butt, T. M. 2000. Effect of formulation, application and rain on the persistence of the entomogenous fungus *Metarhizium anisopliae* on oilseed rape. *Mycol. Res.* **104**, 653–661.

Jackson, T. A., Pearson, J. F., O'Callaghan, M., Mahanty, H. K., and Willocks, M. J. 1992. Pathogen to product – development of *Serratia entomophila* (Enterobacteriaceae) as a commercial biological control agent for the New Zealand grass grub (*Costelytra zealandica*). *In* "Use of Pathogens in Scarab Pest Management" (T. R. Glare and T. A. Jackson, Eds.), pp. 191–198. Intercept Limited, Andover.

Jaronski, S.T., Goettel, M.S., and Lomer, C.J. 2003. Regulatory requirements for ecotoxicological assessments of microbial insecticides – how relevant are they? *In*: "Environmental Impacts of Microbial Insecticides", (H.M.T. Hokkanen and A.E. Hajek, Eds.), pp. 237–260, Kluwer Academic Publishers, Dordrecht, The Netherlands.

Jaros-Su, J., Groden, E., and Zhang, J. 1999. Effects of selected fungicides and the timing of fungicide application in *Beauveria bassiana*-induced mortality of the Colorado potato beetle (Coleoptera: Chrysomelidae). *Biol. Control* **15**, 259–269.

Jin, X., Grigas, K. E., Johnson, C. A., Perry, P., and Miller, D. W. 1999. Method for storing fungal conidia. United States Patent No. 5,989,898.

Kabaluk, T. and Gazdik, K. 2004. Directory of Microbial Pesticides for Agricultural Crops in OECD Countries http://www.agr.gc.ca/env/pest/index_e.php?s 1=pub& page=micro

Kendrick, B. 2000. "The Fifth Kingdom," third edition. Focus Publishing, Newburyport, MA.

Kerwin, J. L. and Petersen, E. E. 1997. Fungi: Oomycetes and Chytridiomycetes. *In* "Manual of Techniques in Insect Pathology" (L. Lacey, Ed.), pp. 251–268. Academic Press, London, UK.

Kirk, I. W., Hoffmann, W. C., and Carlton, J. B. 2001. Aerial electrostatic spray system performance. *Trans. ASAE (Am. Soc. Agric. Eng.)* **44**, 1089–1092.

Kish, L. P. and Allen, G. E. 1978. The biology and ecology of *Nomuraea rileyi* and a program for predicting its incidence on *Anticarsia gemmatalis* in soybean. *Fla. Agric. Exp. Stn. Bull.* 795.

Lacey, L. A. and Brooks, W. M. 1997. Initial handling and diagnosis of diseased insects *In* "Manual of Techniques in Insect Pathology" (L.A. Lacey, Ed.), pp. 1–15. Academic Press, London.

Lacey, L. A. and Goettel, M. S. 1995. Current developments in microbial control of insect pests and prospects for the early 21st century. *Entomophaga* **40**, 3–27.

Law, S. E. and Mills, H. A. 1980. Electrostatic application of low-volume microbial insecticide spray on broccoli plants. *J. Am. Soc. Hortic. Sci.* **105**, 774–777.

Liu, Z. Y., Milner, R. J., McRae, C. F., and Lutton, G. G. 1993. The use of dodine in selective media for isolation of *Metarhizium* spp. from soil. *J. Invertebr. Pathol.* **62**, 248–251.

Macdonald, R. M. and Spokes, J. R. 1981. *Conidiobolus obscurus* in arable soil: a method for extracting and counting azygospores. *Soil Biol. Biochem.* **13**, 551–553.

Magalhães, B., Faria, M., and Frazão, H. 1997. A technique to estimate the conidial viability of *Metarhizium flavoviride* Gams & Rozsypal (Hyphomycetes) formulated in vegetable oil. *An. Soc. Entomol. Bras.* **26**, 569–572.

Matthews, G. A. 2000. "Pesticide Application Methods", third edition. Blackwell Science Ltd., Oxford, UK.

McCoy, C. W., Samson, R. A., and Boucias, D. G. 1988. Entomogenous fungi. *In* "CRC Handbook of Natural Pesticides. Microbial Insecticides, Part A. Entomogenous Protozoa and Fungi, Volume 5" (C.M. Ignoffo, Ed.), pp. 151–236. CRC Press, Boca Raton, FL.

Mohamed, A. K. A., Sikorowski, P. P., and Bell, J. V. 1977. Susceptibility of *Heliothis zea* larvae to *Nomuraea rileyi* at various temperatures. *J. Invertebr. Pathol.* **30**, 414–417.

Moore, D. and Caudwell, R. W. 1997. Formulation of entomopathogens for the control of grasshoppers and locusts. *In* "Microbial Control of Grasshoppers and Locusts" (M. S. Goettel and D. L. Johnson, Eds.), pp. 49–67. *Mem. Entomol. Soc. Can.* Volume 171.

Moore, D., Langewald, J., and Obognon, F. 1997. Effects of rehydration on the conidial viability of *Metarhizium flavoviride* mycopesticide formulations. *Biocontrol Sci. Technol.* **7**, 87–94.

Morley-Davies, J., Moore, D., and Prior, C. 1995. Screening of *Metarhizium* and *Beauveria* spp. conidia with exposure to simulated sunlight and a range of temperatures. *Mycol. Res.* **100**, 31–38.

Nilsson, U. and Gripwall, E. 1999. Influence of application technique on the viability of the biological control agents *Verticillium lecanii* and *Steinernema feltiae. Crop Prot.* **18**, 53–59.

Nord, J. C. 1991. A laboratory rain chamber for testing rainfastness of insecticide deposits. *J. Entomol. Sci.* **26**, 267–270.

Papierok, B. and Hajek, A. E. 1997. Fungi: Entomophthorales. *In* "Manual of Techniques in Insect Pathology" (L. A. Lacey, Ed.), pp. 187–212. Academic Press, London, UK.

Rosenberg, N. J. 1974. "Microclimate: The Biological Environment." J. Wiley and Sons, New York.

Rougier, M., Fargues, J., Goujet, R., Itier, B., and Benateau, S. 1994. Mise au point d'un dispositif d'étude des effets du rayonnement sur la persistance des microorganismes pathogènes. *Agronomie* **14**, 673–681.

Samson, R. A. 1981. Identification: entomopathogenic Deuteromycetes. *In* "Microbial Control of Pests and Plant Diseases 1970–1980" (H. D. Burges, Ed.), pp. 93–106. Academic Press, London.

Samson, R. A., Evans, H. C., and Latgé, J. -P. 1988. "Atlas of Entomopathogenic Fungi." Springer-Verlag, Berlin, Germany.

Schading, R. L., Carruthers, R. I., and Mullin-Schading, B. A. 1995. Rapid determination of conidia viability for entomopathogenic Hyphomycetes using fluorescence microscopy techniques. *Biocontrol Sci. Technol.* **5**, 201–208.

Smits, N., Rougier, M., Fargues, J., Goujet, R., and Bonhomme, R. 1996. Inactivation of *Paeciliomyces fumosoroseus* conidia by diffuse and total solar radiation. *FEMS Microbiol. Ecol.* **21**, 167–173.

Soper, R. S. 1985. *Erynia radicans* as a mycoinsecticide for spruce budworm control. *In* "Proceedings, Symposium: Microbial Control of Spruce Budworms and Gypsy Moths" pp. 69–76. Forest Service, U.S. Department of Agriculture, GTR-NE-100.

Sopp, P. I., Gillespie, A. T.,and Palmer, A. 1989. Application of *Verticillium lecanii* for the control of *Aphis gossypii* by a low-volume electrostatic rotary atomiser and a high-volume hydraulic sprayer. *Entomophaga* **34**, 417–428.

Sopp, P. I. and Palmer, A. 1990. Deposition patterns and biological effectiveness of spray deposits on pot plants applied by the ulvafan and three prototype electrostatic sprayers. *Crop Prot.* **9**, 295–302.

Sosa-Gomez, D. R., Boucias, D. G., and Nation, J. L. 1997. Attachment of *Metarhizium anisopliae* to the southern green stink bug *Nezara viridula* cuticle and fungistatic effect of cuticular lipids and aldehydes. *J. Invertebr. Pathol.* **69**, 31–39.

Tanada, Y. and Kaya, H. K. 1993. "Insect Pathology." Academic Press, San Diego, CA.

Thomas, M. B., Blanford, S., and Lomer, C. J. 1997. Reduction of feeding by the variegated grasshopper, *Zonocerus variegatus*, following infection by the fungal pathogen, *Metarhizium flavoviride. Biocontrol Sci. Technol.* **7**, 327–334.

Tossell, R. W., Dickinson, W. T., Rudra, R. P., and Wall, G. J. 1987. A portable rainfall simulator. *Can. Agric. Eng.* **29**, 155–162.

Tzean, S. S., Hsieh, L. S., and Wu, W. J. 1997. "Atlas of Entomopathogenic Fungi from Taiwan." Council of Agriculture, Executive Yuan, Taiwan.

Ugine, T. A., Wraight, S. P., and Sanderson, J. P. 2005. Acquisition of lethal doses of *Beauveria bassiana* conidia by western flower thrips, *Frankliniella occidentalis*, exposed to foliar spray residues of formulated and unformulated conidia. *J. Invertebr. Pathol.* **90**, 10–23.

Unwin, D. J. 1981. "Microclimate Measurement for Ecologists." Academic Press, New York.

Upadhyay R.K. (Ed.) 2003. "Advances in Microbial Control of Insect Pests." Kluwer Academic Press, New York.

Vandenberg, J. D., Shelton, A. M., Wilsey, W. T., and Ramos, M. 1998. Assessment of *Beauveria bassiana* sprays for control of diamondback moth (Lepidoptera: Plutellidae) on crucifers. *J. Econ. Entomol.* **91**, 624–630.

Vestergaard, S., Cherry, A., Keller, S., and Goettel, M. 2003. Hyphomycete fungi as microbial control agents Chapter 3, *In* "Environmental Impacts of Microbial Insecticides" (H.M.T. Hokkanen and A.E. Hajek, Eds.), pp. 35–62. Kluwer Academic Publishers, Dordrecht, The Netherlands.

Wraight, S. P., Butt, T. M., Galaini-Wraight, S., Allee, L. L., Soper, R.S., and Roberts, D. W. 1990. Germination and infection processes of the entomophthoralean fungus*Erynia radicans* on the potato leafhopper, *Empoasca fabae. J. Invertebr. Pathol.* **56**, 157–174.

Wraight, S. P. and Carruthers, R. I. 1999. Production, delivery, and use of mycoinsecticides for control of insect pests of field crops. *In* "Biopesticides: Use and Delivery" (F. R. Hall and J. J. Menn, Eds.), pp. 233–269. Humana Press, Totowa, NJ.

Wraight, S. P., Carruthers, R. I., Bradley, C. A., Jaronski, S. T., Lacey, L. A., Wood, P., and Galaini-Wraight, S. 1998. Pathogenicity of the entomopathogenic fungi *Paecilomyces* spp. and *Beauveria bassiana* against the silverleaf whitefly, *Bemisia argentifolii. J. Invertebr. Pathol.* **71**, 217–226.

Wraight, S. P., Galaini-Wraight, S., Carruthers, R. I., and Roberts, D. W. 2003. *Zoophthora radicans* (Zygomycetes: Entomophthorales) conidia production from naturally-infected *Empoasca kraemeri* and dry-formulated mycelium under laboratory and field conditions. *Biol. Control* **28**, 60–77.

Wraight, S.P., Jackson, M.A., and De Kock, S.L. 2001. Production, stabilization, and formulation of fungal biocontrol agents. *In* "Fungi as Biocontrol Agents: Progress, Problems and Potential" (T. M. Butt, C.W. Jackson, and N. Magan, Eds.), pp. 253–287. CAB International, Wallingford, UK.

Wraight, S. P. and Ramos, M. E. 2002. Application parameters affecting field efficacy of Beauveria bassiana foliar treatments against Colorado potato beetle *Leptinotarsa decemlineata. Biol. Control* **23**, 164–178.

Zare, R. and Gams, W. 2001. A revision of *Verticillium* section *Prostrata*. IV. The genera *Lecanicillium* and *Simplicillium* gen. nov. *Nova Hedwigia* **73**, 1–50.

Chapter IV-5

Nematodes

Albrecht M. Koppenhöfer
Department of Entomology
Rutgers University
New Brunswick, NJ 08901, USA

1 Introduction

A Nematodes in insect control

Nematodes, non-segmented animals ranging between 0.1 mm and several meters in length, are also referred to as round-, eel-, or threadworms because of their usually cylindrical and elongate body shape. They have digestive, reproductive, muscular, and simple excretory and nervous systems but lack circulatory and respiratory systems. Many nematode species are associated with insects and the types of relationship range from phoresis to parasitism and pathogenesis. Nematodes that have parasitic associations with insects have been described from 23 nematode families. Seven of these families contain species that have potential for biological control of insects: Mermithidae and Tetradonematidae (Order: Stichosomida); Allantonematidae, Phaenopsitylenchidae, and Sphaerulariidae (Order: Tylenchida); Heterorhabditidae and Steinernematidae (Order: Rhabditida). Presently, only the Heterorhabditidae and Steinernematidae are used as microbial insecticides and are produced commercially by various companies around the world. With the exception of the tylenchid, *Deladenus* (= *Beddingia*) *siricidicola*, which was successfully used for inoculative control of a woodwasp species in Australia (Bedding, 1993), the microbial control potential of the other nematode species is rather limited because of problems with their culture and/or limited virulence. Therefore, this chapter will concentrate on the Heterorhabditidae and Steinernematidae. Detailed information on the other groups can be obtained from Poinar (1979), Nickle (1984, 1991), Kaya and Stock (1997), and Kaya and Koppenhöfer (2004). For information on use of mermithid nematodes in rice and of *Phasmarhabditis* for use against molluscs, please refer to Chapters VII-4 and VII-23, respectively, in this book.

Steinernematidae and Heterorhabditidae are obligate pathogens in nature. They have been recovered from soils throughout the world (Hominick *et al.*, 1996; Hominick, 2002) and their distribution may be primarily limited by the availability of susceptible hosts. The term "entomopathogenic nematode" refers to the nematode's ability to quickly kill hosts (1–4 days depending on nematode and host species) that is facilitated by their mutualistic association with bacteria in the genus *Xenorhabdus* for Steinernematidae and *Photorhabdus* for Heterorhabditidae. All known entomopathogenic nematode (EPN) species seem to have a similar biology. The only stage that survives outside of a host is the non-feeding, non-developing third stage infective juvenile (IJ) or dauer juvenile. The IJs carry cells of their bacterial symbiont in their intestines. After locating a suitable host, the IJs invade it through natural openings (mouth, spiracles, anus) or thin areas of the host's cuticle [common in heterorhabditids; in steinernematids observed with tipulid larvae (Peters and Ehlers, 1994)] and penetrate into the host hemocoel. The IJs release their symbiotic bacteria that propagate, kill the host

L.A. Lacey and H.K. Kaya (eds.), *Field Manual of Techniques in Invertebrate Pathology*, 249–264.

by septicemia, and metabolize its tissues. The nematodes start developing and feed on the bacteria and metabolized host tissues and go through 1–3 generations until a new generation of IJs emerges from the depleted host cadaver. One of the major differences between Heterorhabditidae and Steinernematidae is that *Heterorhabditis* adults are hermaphrodites in the first generation inside the cadaver but amphimictic in following generations, whereas *Steinernema* adults are almost always amphimictic. The exception is *Steinernema hermaphroditum* which has hermaphrodites in the first generation (Stock *et al.*, 2004)

EPN infections can be recognized by various signs. Soon after death the cadavers turn flaccid and start changing color. In the case of wax moth, *Galleria mellonella*, larvae and depending on nematode species, steinernematid-killed insects turn to various shades of brown, from ochre to almost black, whereas heterorhabditid-killed insects turn red, brick-red, purple, orange, yellow, and green. The color of the cadaver is attributed to the associated bacterium, especially for *Photorhabdus* that produce pigments. The steinernematid-killed insects stay flaccid throughout nematode development, whereas heterorhabditid-killed insects become less flaccid. If the cuticle is transparent, nematodes are visible inside the cadaver. The cadavers do not putrefy, and even when dissected, do not have a putrid odor. The body contents lose their integrity over time but never liquefy. In heterorhabditid-killed insects, the body contents become ropy.

The range of insects infected by many of the well-studied EPN species (*e.g., S. carpocapsae, S. feltiae*, and *H. bacteriophora*) is broad under laboratory conditions. The range of insects infected by EPN after inundative field releases and especially the natural host range of most species are more restricted due to the ecology of the nematodes and their potential hosts as well as environmental factors (Peters, 1996). The effect of inundative applications of EPN on non-target organisms is negligible, adding to their safety as biological control agents (Akhurst, 1990; Bathon, 1996). Some known species have a restricted host range. For example, *S. scapterisci* is adapted to mole crickets (Parkman and Smart, 1996) and *S. kushidai* (Mamiya, 1989) and *S. scarabaei* (Koppenhöfer and Fuzy, 2003) are adapted to scarab larvae. Because isolation of new nematode strains is usually done using wax moth larvae as bait insects, the host range of known species is likely to be biased towards generalists or Lepidoptera adapted species.

One of the factors restricting nematode host range is the type of foraging behavior exhibited by the IJs. Thus, some species are sit-and-wait strategists or ambushers that tend to stay near the soil surface where their specialized foraging behaviors (nictation, jumping) allow them to efficiently infect mobile host species (*e.g., S. carpocapsae, S. scapterisci*) (Campbell and Gaugler, 1997; Lewis, 2002; Lewis *et al.*, 2006). Other species are widely searching foragers or cruisers that distribute themselves actively throughout the soil profile and are well adapted to infecting sessile or less mobile hosts (*e.g., S. glaseri, H. bacteriophora*). Most known species appear to be situated somewhere along a continuum between these two extremes (*e.g., S. riobrave, S. feltiae*) (Campbell and Gaugler, 1997; Lewis, 2002).

B Key target insects

EPN have been tested against a large number of insect pest species with results varying from no effect to excellent control (Klein, 1990; Begley, 1990; Bedding *et al.*, 1993; Grewal *et al.*, 2005; Georgis *et al.*, 2006; Kaya *et al.*, 2006). Many factors can influence the success of nematode applications. Numerous failures can be attributed to poor understanding of the nematodes' (and pests') ecology. Matching biology and ecology of both the nematode and the target pest is of great importance if nematode applications are to result in significant pest reductions. Among the factors that need to be considered are the foraging behavior and temperature requirements of a nematode species as well as the pest's accessibility in its habitat to nematodes and its suitability as a host.

EPN have been applied against insect pests in soil and cryptic habitats, in manure, in aquatic habitats, and on foliage (Begley, 1990; Grewal *et al.*, 2005). Most applications of EPN have been made against insects in soil (Klein, 1990). Soil is the natural habitat of these nematodes and offers protection from environmental extremes that can affect IJ activity and survival. In addition, many insect species spend at least part of their life cycle in or on the soil. Some of the most consistent and efficacious results with nematodes have been obtained in cryptic habitats because, like soil, they buffer against environmental extremes, but also have less limiting factors than soil and can more easily be manipulated to the nematodes' advantage. Excellent control has been achieved against insects that bore into plants. Manure appears to be another habitat with buffering effects for IJs, but high temperatures in animal rearing facilities and toxic effects of manure contents result in poor IJ survival. Aquatic habitats provide excellent conditions for IJ survival, but the nematodes are not adapted to directed motility in this environment. Finally, the exposed surfaces of foliage and similar habitats are very hostile environments and even adjuvants can only marginally remedy detrimental effects on IJs.

Some insects cannot be infected by nematodes because the usual portals of entry used by IJs are not accessible (Eidt and Thurston, 1995). For example, the mouth may be blocked by oral filters (wireworms) or too narrow (sucking/piercing mouth parts or young instars of insects with chewing mouthparts), the anus may be constricted by muscles or other structures (wireworms), the spiracles may be covered with septa (wireworms) or sieve plates (scarab grubs) or simply too narrow for passage (some Diptera and Lepidoptera). In many insects, intersegmental membranes, fore- and hindgut cuticular linings, or the peritrophic membrane may be too thick or dense to allow for penetration at these sites. Nematode infection may also be hindered by aggressive grooming or evasion behavior (Gaugler *et al.*, 1994) or formation of impenetrable cocoons or soil cells (Eidt and Thurston, 1995). Social insects may be susceptible as individuals but behavioral responses to pathogens (social grooming, isolation/removal of infected siblings, translocation of colony) effectively protect the colonies from significant impacts (Klein, 1990; Gouge, 2005). The immune response of insects to the nematodes as they penetrate into their body cavity also varies with nematode species and insect hosts (Wang *et al.*, 1995).

All the above factors should be considered when selecting nematode species for applications against a given life stage of an insect pest. Table 1 provides a list of some insect pests along with the commodities in which they have been successfully controlled with entomopathogenic nematodes.

2 Environmental considerations for the use of nematodes for insect control

A *Motility and survival*

The performance of EPN applied for insect control is dependent on the motility and persistence of the applied IJs. Active IJ dispersal, although rather limited with up to 90 cm in both horizontal and vertical dispersal within 30 days, gives EPN the ability to actively seek out hosts (Kaya, 1990). Passive nematode dispersal by water, wind, phoresis, infected hosts, human activity, etc. can cover much greater distances and may account for their widespread distribution. Persistence of IJs is also rather limited. When applied onto the soil surface, losses can reach 50% within hours of application due primarily to UV radiation and desiccation (Smits, 1996). Even those IJs that settle in the soil suffer 5–10% losses per day until usually only around 1% of the original inoculum survives after 1–6 weeks. Because of this limited persistence, EPN should only be applied when susceptible stages of the target pest are present. Both motility and persistence of applied IJs are influenced by various intrinsic (*e.g.*, behavioral, physiological, and genetic characteristics) and extrinsic factors. This section concentrates on the effect of extrinsic factors including abiotic factors [extreme temperatures, soil moisture, soil texture, RH, UV radiation (Kaya, 1990; Smits, 1996; Glazer, 2002)] and biotic factors [antibiosis, competition and natural enemies (Kaya and Koppenhöfer, 1996; Kaya, 2002)].

UV light can inactivate and kill nematodes within minutes. Direct exposure to UV light (*i.e.*, sunlight) should be minimized by applying IJs early in the morning or in the evening. For soil applications, timing may not be as important if sufficient amounts of water are used to rinse the IJs into the soil at the time of or directly after application (Selvan *et al.*, 1994). For application

Table 1. Target pests for entomopathogenic nematodes[1]

Pest group	Common name	Life-stage[2]	Application site	Commodity	Nematode sp.[3]
COLEOPTERA					
Curculionidae	Billbugs	L	Turf	Turf	Sc, Hb
	Root weevils	L	Soil	Banana[4], berries, citrus, forest seedlings, hops, mint, ornamentals, sweet potato, sugar beets	Sc, Sk, Hb, Hi Hm, Sr
Chrysomelidae	Flea beetles	L	Soil	Mint, potato, sweet potato, sugar beets, vegetables	Sc
	Rootworms	L	Soil	Corn, peanuts, vegetables	Sc, Sr
Scarabaeidae	White grubs	L	Soil, turf	Berries, field crops, ornamentals, turf	Hb, Sg, Hm
DIPTERA					
Agromyzidae	Leaf miners	L	Foliage	Ornamentals, vegetables	Sc
Ephydridae	Shore flies	L	Soil	Ornamentals, vegetables	Sf
Sciaridae	Sciarid flies	L	Soil	Mushrooms, ornamentals, vegetables	Sf
Tipulidae	Crane flies	L	Turf, soil	Turf, ornamentals	Sc, Hm
Muscidae	Filth flies	A	Baits	Animal rearing facilities	Sf, Hb
LEPIDOPTERA					
Noctuidae	Cutworms, armyworms	L/P	Soil, turf	Corn, cotton, peanuts, turf, vegetables	Sc
Pterophoridae	Plume moths	L	Cryptic	Artichoke	Sc
Pyralidae	Webworms	L	Soil, turf	Cranberries, ornamentals, turf	Sc
Sesiidae	Crown borers	L	Cryptic	Berries	Sc
	Stem borers	L	Cryptic	Cucurbits, ornamentals, shrubs, fruit trees	Sc
Cossidae	Carpenter worms	L	Cryptic	Ornamentals, shrubs	Sc
	Leopard moth	L	Cryptic	Apple, pear	Sc
Carposinidae	Peach borer moth	L	Soil	Apple	Sc
ORTHOPTERA					Sc
Gryllotalpidae	Mole crickets	N, A	Turf, soil	Turf, vegetables	Sc, Ss, Sr
BLATTODEA					
Blattellidae	German cockroach	N, A	Baits	Apartments/structures	Sc
SIPHONAPTERA					
Pulicidae	Cat fleas	L/P	Soil, turf	Pet/vet	Sc
NEMATODA	Plant-parasitic nematodes		Turf	Turf	Sr

[1] After Begley (1990), Klein (1990), Georgis and Manweiler (1994), Georgis *et al.* (1995).
[2]L = larva; P = pupa; N = nymph; A = adult.
[3]Sc = *S. carpocapsae*; Sf = *S. feltiae*; Sk = *S. kraussei*; Sr = *S. riobrave*; Ss = *S. scapterisci*; Hb = *H. bacteriophora*; Hi = *H. indica*; Hm = *H. megidis*.
[4]Applied to residual rhizomes in a cryptic habitat (see text).

to foliage or other exposed habitats, timing is essential, even if UV protectants can be added to the IJ suspension.

Moisture is the most important factor influencing nematode performance. IJs need a water film for effective propulsion. In soil, IJs move through the water film that coats the interstitial spaces. If this film becomes too thin (in dry soil) or the interspaces are completely filled with water (in saturated soil), nematode movement is restricted (Koppenhöfer *et al.*, 1995). IJs can survive desiccation to relatively low moisture levels if water removal is gradual giving them time to adapt to an inactive stage (Womersley, 1990). This is generally the case in natural soils (except near the soil surface in sandy soils low in organic matter), where the RH in the soil pores remains close to 100%. When inactive, IJs may actually persist longer in dry soil (Kaya, 1990); infection, however, will be impeded. Moderate soil moisture is essential for good nematode performance (Georgis and Gaugler, 1991). Similar conditions can be expected in other moisture retaining and cryptic habitats. On foliage and in other exposed habitats, nematode survival is generally a matter of hours unless the RH is close to 100%. To retard desiccation and prolong nematode persistence on foliage, moisture-retaining adjuvants (humectants) can be added to the nematode suspensions.

The effect of temperatures on nematode performance varies with nematode species and strains (Kaya, 1990; Griffin, 1993; Grewal *et al.*, 1994). Generally, IJs become sluggish at low temperatures (< 10–15 °C) and will be inactivated at higher temperatures (> 30–40 °C). Good performance for most commercially available nematode species can be expected between 20 and 30 °C. Exceptions are *S. feltiae* with an efficacious performance between ca. 12 and 25 °C, and *S. riobrave* with exceptional performances between 25 and 35 °C. Extended exposure to temperatures below 0 °C and above 40 °C is lethal to most EPN species, but the effect depends on duration of exposure (Glazer, 2002). In the soil environment, IJs are normally buffered from temperature extremes or usually have enough time to disperse into deeper soil layers where the buffering effect is stronger. For most species, optimal survival temperatures lie between 5 and 15 °C (Georgis, 1990). Higher temperatures will increase metabolic activity and depletion of energy reserves and shorten life span.

Nematode dispersal and survival can differ among different soil types (Kaya, 1990). It has been assumed that dispersal and survival tend to be lower in fine textured soils with the lowest survival in clay soils. The lower survival rate is probably related to lower oxygen levels in the smaller soil pores. Similarly, oxygen may become a limiting factor in water-saturated soils and soils with high contents of organic matter (Kaya, 1990). The pH value of the soil does not have a strong effect on IJ survival. Thus, pH values between 4 and 8 do not vary in their effect on IJs, but at pH 10, IJ survival declines rapidly (Kaya, 1990). However, additional research has shown that different nematode species may be differently affected by various soil parameters (*e.g.*, Koppenhöfer and Fuzy, 2006), which may be at least in part due to differences in behavioral and physiological adaptations among nematode species. In addition, we have to keep in mind that soils occur in an almost unimaginable variety as they may vary in parameters such as soil particle size composition, structure, organic matter content, pH, and nutrient concentration. Due to the conglomeration of these parameters within each soil type, it is difficult to make generalizations on the effect of specific soil parameters on nematode performance.

Biotic factors and their effect on nematode survival have been extensively reviewed (Kaya and Koppenhöfer, 1996; Kaya, 2002). Antibiosis can occur when chemicals with adverse effects on the nematodes are released from roots in the soil affecting IJ host finding or when such chemicals are present in an infected host and affect nematode infection and reproduction. Intraspecific competition may reduce nematode fitness when too many IJs infect one host. Interspecific competition between nematode species may cause local extinction of one competitor. Interspecific competition may also occur with other insect pathogens, especially if they are applied in the same location as the nematodes. The outcome of the competition will depend on the kind of competitor (*e.g.*, entomopathogenic fungi, bacteria, or viruses), the timing of infection, and environmental factors such as temperature or soil moisture. Among the natural enemies of nematodes, nematophagous fungi are the best studied. Other natural enemies include invertebrate predators such as collembolans, mites, tardigrades and predatory nematodes. These natural enemies reduce IJ populations in soil in laboratory experiments, but their impact under field conditions is poorly understood.

B Recycling of nematode populations

The performance of EPN applied for insect control may be additionally improved if recycling in nematode-killed insects produces new generations of IJs. Numerous studies have provided evidence for nematode recycling in the soil environment after inundative releases (Kaya, 1990; Klein, 1993). To some extent, it may also occur in other habitats that provide protection from environmental extremes. Although host cadavers can provide some protection from desiccation (Koppenhöfer *et al.*, 1997; Brown and Gaugler, 1997), conditions in exposed habitats such as on foliage are too detrimental to allow recycling. Nematode recycling is highly desirable because it can provide additional control of a pest and extend it into following pest generations. Although recycling probably is quite common, it is not clear what factors influence its occurrence. Most of the abiotic and biotic factors that influence the persistence, infectivity, and motility of individual IJs, also influence nematode recycling. Some of these factors may even be more stringent for recycling than for persistence and infection. For example, the temperature range for successful reproduction inside a host cadaver is usually narrower than that for infection (Grewal *et al.*, 1994).

As obligate pathogens, natural populations of EPN have to recycle in their hosts, but only few studies have examined the dynamics of persisting nematode populations and what factors allow them to persist in some areas and not in others. Several studies have shown that within-site distribution of nematode populations is patchy (Stuart and Gaugler, 1994; Campbell *et al.*, 1995, 1997; Strong *et al.*, 1996) and may depend on various biotic and abiotic factors including seasonal fluctuations, foraging strategy of the IJs, host population dynamics, alternate hosts, etc. It is likely that EPN populations persist as metapopulations which exhibit a 'shifting mosaic' type of dynamics with asynchronous fluctuations and little migration between patches (Levins, 1970). Patches are highly vulnerable to extinction, and in order to persist as a metapopulation, the founding rate of local populations has to be the same as the extinction rate (Lewis *et al.*, 1998). Stuart *et al.* (2006) provide an excellent overview on EPN population biology.

C *Formulations/storage/temporary storage under field conditions*

The simplest way of storing IJs is in aqueous suspensions (Kaya and Stock, 1997). After harvesting IJs from *in vivo* or *in vitro* laboratory cultures, they can be stored in tissue culture flasks at densities of 300–5,000 IJs/ml, depending on size and activity level of the IJs. To allow for air exchange, the unscrewed flasks have to lay flat with the suspension not deeper than 5–7 mm. Considerably higher concentrations (up to 100,000 IJs/ml) can be stored if the suspension is sufficiently aerated (*e.g.*, with an aquarium pump). A drop of Triton X-100 (wetting agent) may be added to the suspension to prevent sticking of IJs to the container surface and a few drops of sodium bicarbonate solution prevents the formation of rosettes (clumps) in heterorhabditids (Kaya and Stock, 1997). IJs can be stored in aqueous suspension at 4–15 °C (depending on nematode species) without much loss of activity for 6–12 months for *Steinernema* species and 3–6 months for *Heterorhabditis* species. At higher temperatures, storage life is considerably shorter. For experiments, it is preferable to limit the storage time after production to < 1–2 months if cooled (4–15 °C) and < 2 weeks if kept at room temperature.

Many commercial nematode products are still formulated on moist substrates (*e.g.*, sponge, vermiculite, aqueous suspensions) and require continuous refrigeration to maintain nematode quality for extended periods of time. To improve IJ shelf-life and resistance to temperature extremes, a large number of formulations have been developed including alginate, clays, activated charcoals, and polyacrylamide (Georgis, 1990; Georgis and Kaya, 1998; Grewal, 2002; Grewal and Peters, 2005). These formulations reduce IJ metabolism by immobilization or partial desiccation. Presently, the most promising formulation consists of water dispersible granules that combine long nematode shelf-life without refrigeration (9–12, 4–5, and 2 months at 2–10, 22–25, and 30 °C, respectively, for *S. carpocapsae*; less for other species) with ease of handling (Georgis *et al.*, 1995; Georgis and Kaya, 1998; Grewal, 2002). The partially desiccated IJs rehydrate after application to a moist environment such as soil. However, to achieve optimal infectivity the IJs require rehydration for up to 3 days (Baur *et al.*, 1997b). In desiccating environments like foliage, rehydration has to occur before application making this formulation unpractical for these situations (Baur *et al.*, 1997b). Optimal formulations differ for the various nematode species because of their specific requirements for moisture and oxygen.

For field experiments, appropriate methods used for transportation and temporary storage of IJs depend on the nematode quantities used. For small trials, aqueous suspensions (for short periods IJs may be concentrated to 2,000–10,000 IJs/ml) can be transported and stored in tissue culture flasks inside cooler boxes (10–15 °C). For larger quantities, it is easier to transport formulated IJs and prepare the nematode suspension on site. Whether the nematodes are being transported to the application site, stored temporarily under field conditions, or kept in the tank mix before application, the applicator has to be aware of the IJs' sensitivity to extreme temperatures, UV light, and oxygen deficiency (*i.e.*, the tank mix should be prepared immediately before use, should not be unnecessarily exposed to sunlight and heat, and should be agitated for aeration).

3 Application technology and effect on nematodes

A Application strategies

Releases of EPN have almost exclusively used the inundative approach where high numbers of IJs are released in a uniform distribution and control of pest populations is expected to be fast and thorough. However, their limited shelf life, susceptibility to environmental extremes, high price, etc. limit the potential of nematodes, as that of most other biological control agents, in an approach following the chemical pesticide paradigm. Other approaches including inoculative and augmentative releases, and conservation and management of endemic nematode populations need to receive considerably more attention in the future, as they may be more promising and feasible in many pest situations.

Inoculative release of EPN, *i.e.*, the release of relatively small numbers of IJs with the expectation that they establish new populations for long-term pest suppression, has only been attempted a few times and little is known about the optimal approach to this strategy. *Steinernema glaseri*, isolated originally from scarab larvae, was released in a massive inoculative control program in New Jersey from 1939 to 1942 against the Japanese beetle, an introduced pest. Although reisolated from southern New Jersey (Gaugler *et al.*, 1992), the elimination of bacterial symbionts by the use of antimicrobials in the *in vitro* rearing procedure, and possibly poor climatic adaptation of this neotropical nematode limited the success of this program. More recently, *S. scapterisci*, originally isolated from southeast South America, was successfully introduced into Florida for the classical biological control of mole cricket pests (Parkman and Smart, 1996). The nematode established successfully after treatment of 50 m^2 plots in pastures with either IJs (2×10^9/ha) or nematode-infected mole crickets (4 cadavers/m^2) or after release of mole crickets that had been exposed to nematodes in mole cricket sound traps.

For successful inoculative releases of EPN, long-term, multigenerational survival and recycling of the nematode populations are essential. To achieve this goal, several conditions are important including (1) presence of moderately susceptible insect hosts throughout most of the year, (2) high economic threshold level of the target insect pests, and (3) soil conditions favorable for nematode survival (Kaya, 1990). The optimal release method for inoculative releases of EPN may depend on the systems into which they are released, *i.e.*, spatial and temporal distribution and susceptibility of target hosts and potential alternative hosts, seasonal fluctuations in other biotic and abiotic factors etc. Release methods could involve high or low application rates that may be uniformly distributed or applied in points from which colonization of other areas can occur.

Periodic augmentative releases into established nematode populations, and/or management of the susceptibility of the host/pest populations (for example using stressors such as other control agents) are two other approaches that may be used to boost or manage established nematode populations and warrant more attention. The possibilities and requirements for using EPN in a conservation approach of biological control have been extensively discussed by Lewis *et al.* (1998).

B Application method/equipment

The most commonly used application method for EPN is spraying directly onto the soil (or other) surface. This method is simple and quick and provides good coverage. For small plots this may be accomplished with watering cans. However, nematodes can be applied with most commercially available spray equipment including hand or ground sprayers (pressurized or electrostatic), mist blowers, and aerial equipment on helicopters (Georgis *et al.*, 1995; Grewal, 2002; Wright *et al.*, 2005; Shapiro-Ilan *et al.*, 2006) (see also Chapters III-1 and III-2). Filters and sieves should be at least 300 μm wide and nozzle apertures > 500 μm and operating pressure should not exceed 2000 kPa (295 psi) for *S.carpocapsae* and 1380 kPa (204 psi) for *H. bacteriophora* and *H. megidis* (Wright *et al.*, 2005). With some formulations it may be necessary to remove the screens to prevent clogging of the nozzle. Different types of spray nozzles may affect the initial distribution of IJs; *e.g.*, fan nozzles provide a more uniform coverage than hollow cone nozzles (Curran, 1993). Nematodes can also be delivered via irrigation systems including drip, microjet, sprinkler, and furrow irrigation (Georgis *et al.*, 1995; Cabanillas and Raulston, 1996; Wright *et al.*, 2005). When selecting a delivery system for nematodes, the applicator should also consider which system delivers the nematodes most effectively to the target pest. Care should be taken not to expose the IJs to high temperatures during application, be it in the tank mix or in the application equipment.

For soil insects, spray volumes of 750–1890 liters/ha are usually required depending on soil cover (Georgis *et al.*, 1995). To quickly remove the IJs from exposure to UV light and desiccating conditions on the soil surface, post-application and, in the case of dry soil, also pre-application irrigation are recommended. The ideal amount of irrigation will depend on the soil type, temperature, soil cover, the soil moisture before application, and the depth of the target insect in the soil. In addition, continued moderate soil moisture is essential for good nematode performance (Georgis and Gaugler, 1991). When water is limited, subsurface injection of nematodes appears to be an efficient delivery method but does not improve nematode efficacy compared to application with sufficient amounts of water (Klein, 1993). Spray volumes for foliar application vary with the type and morphology of crop to which applications are made (Georgis, 1990).

Different techniques are used to deliver IJs efficiently for the control of boring insects. Larvae of tree borers have been successfully controlled by injecting nematode suspensions directly into the borer holes or blocking the holes with sponges soaked with nematode suspensions (Yang *et al.*, 1993). Adult banana weevils can be successfully controlled by cutting cones out of residual banana rhizomes, filling the holes with nematode preparations, and replacing the cones (Treverrow and Bedding, 1993). The weevils are attracted to the bait holes for 30–40 days, and the ideal microclimate of the holes and intimate contact with the weevils allows the nematodes to kill most of them.

Baits containing IJs can offer a cost-effective way of controlling mobile insects. Only moderate control was achieved against cutworms, grasshoppers, and tawny mole cricket with baits applied open on the ground (Georgis, 1990). But when trap stations were used that ensured intimate IJ-pest contact and protected the IJs from light and desiccation, the baits outperformed standard insecticide baits for housefly adult control in intensive pig units (Renn, 1998) and German cockroach control in apartments (Appel *et al.*, 1993).

Finally, application of IJs in the form of nematode-killed hosts may hold some promise as IJs emerging directly from infected-hosts can exhibit superior performance compared with nematodes applied in aqueous suspensions (Shapiro-Ilan *et al.*, 2003). Technology to facilitate storage and application has been developed to overcome problems with storage and application (Shapiro-Ilan *et al.*, 2001).

C Effects of formulation adjuvants

The detrimental effects of environmental extremes often can be alleviated by the addition of adjuvants to the nematode formulation/suspension (Wright *et al.*, 2005). Because exposure to these extremes is especially high after foliar applications, work with adjuvants has focused on improving nematode performance against foliar-feeding pests. Solar radiation can be filtered with stilbene brighteners, especially Blankophor BBH (Nickle and Shapiro, 1994; Baur *et al.*, 1997a). Effective antidesiccants can be TX7719, Rodspray oil, and Nufilm P (Baur *et al.*, 1997a), Folicote (Glazer *et al.*, 1992), or glycerin (Broadbent and Olthof, 1995). Because adjuvants may have phytotoxic effects, negatively affect the IJs, serve as substrates for fungal or bacterial growth, or vary in effect depending on the plant morphology, adjuvants should be screened for each pest/plant complex (Baur *et al.*, 1997a). Surfactants such as Silwett L-77, Kinetic, or dish detergents may also improve nematode speed of penetration into soil (Schroeder and Sieburth, 1997), but further studies are necessary to determine the mechanism of this interaction and whether these combinations are feasible under field conditions.

D Effects of agrochemicals and other IPM components

EPN are usually applied to systems/substrates that are regularly treated with many other agents, including chemical or botanical pesticides, bioinsecticides, soil amendments, and fertilizers. Depending on the agents, application timing, physico-chemical characters of the system, etc., the nematodes may or may not interact with these other agents, with interactions ranging from antagonistic to synergistic. An extensive summary of such interactions is provided by Koppenhöfer and Grewal (2005).

EPN appear to be compatible with many herbicides, fungicides, acaricides, insecticides, nematicides (Rovesti and Deseö, 1990; Ishibashi, 1993), azadirachtin (Stark, 1996), *Bacillus thuringiensis* products (Kaya *et al.*, 1995), and pesticidal soap (Kaya *et al.*, 1995). However, many other pesticides have limited to strong toxic effect on IJs, *e.g.*, oxamyl, fenamiphos, carbaryl, bendiocarb, diazinon, dodine, paraquat or methomyl (Rovesti and Deseö, 1990; Zimmerman and Cranshaw, 1990; Patel and Wright, 1996). On the other hand, synergistic interaction between EPN and other control agents has been observed for various insecticides such as imidacloprid (Koppenhöfer and Kaya, 1998) or tefluthrin (Nishimatsu and Jackson, 1998) and pathogens such as *Paenibacillus* (= *Bacillus*) *popilliae* (Thurston *et al.*, 1994) or *Bacillus thuringiensis* (Koppenhöfer and Kaya, 1997). In view of the diversity of available chemical insecticides, a generalization on pesticide-nematode compatibility cannot be made.

Inorganic fertilizer may be compatible with nematodes for short-term inundative pest control (Bednarek and Gaugler, 1997). Similarly, composted manure or urea does not have negative effects on nematode virulence but fresh manure does (Shapiro *et al.*, 1997). Natural nematode populations, on the other hand, have been negatively affected by inorganic fertilizers, but positively affected by organic manure (Bednarek and Gaugler, 1997).

4 General methodology for determining performance

A Preparation of inoculum

When preparing applications of EPN, it is advisable to examine IJ quality in subsamples of the batch intended for use. Formulated IJs need to be brought into the aqueous suspension in which they are usually applied before determining viability. If the IJs in the formulated product are partially desiccated (water dispersible granules or clay), it may be necessary to allow some time for rehydration of the IJs. The number of dead and living IJs and their quality should be determined in at least 3 subsamples of > 50 IJs using a microscope at 10–40x. IJ viability can be determined by their movement. Although a large proportion of the IJs in some nematode species may assume an immobile straight or slightly curved posture in storage (Table 2), the majority of the IJs will start moving if the suspension is stirred vigorously. Viable IJs that still are straight should move when touched with a probe. Dead IJs tend to be more transparent than living ones. In an acceptable nematode batch, > 90% of the IJs should be alive. Based on the viability counts, the concentration of live IJs should be adjusted for the target application rate (see Kaya and Stock, 1997).

Vigorous IJs contain abundant energy reserves, *i.e.*, the body except for head and tail region

Table 2. Characteristics of commercially available entomopathogenic nematode species for the confirmation of species identity in commercial shipments

Species	IJ length (Range) (μm)	Host cadaver color[1]	Resting posture[2]
S. carpocapsae	558 (468–650)	Beige	J
S. riobrave	622 (561–701)	Beige	J
S. feltiae	849 (736–950)	Tan/walnut-brown	C
S. glaseri	1130 (864–1448)	Grayish-dark brown	None
S. kraussei	951 (797–1102)	Tan/walnut-brown	C
H. bacteriophora	588 (512–670)	Brick red to dark purple	None
H. indica	528 (479–573)	Dark red	None
H. megidis	768 (736–800)	Orange-brown	None
H. zealandica	685 (570–740)	Pale mint green	None

[1] Nematode-killed insect with a transparent cuticle change to a characteristic color. Color can vary with nematode strains.
[2] IJs in some species can assume an immobile straight (usually with a slight bend in the tail, therefore "J") or slightly curved ("C") posture when stored in water. IJs of species without resting posture are mostly active.

should be dark and non-transparent because it is densely packed with lipids. The infectivity of the IJs should also be checked, preferably before the actual application in an experiment. However, this procedure requires at least 48 h. Bioassays that can be used to determine infectivity include the one-on-one assay, the sand barrier assay, and the Petri dish assay (Kaya and Stock, 1997), the latter being the easiest to perform (see Appendix A). If the IJs are obtained commercially, confirmation of the nematode species should also be considered. Because only a limited number of species is commercially available, this can be done by observing IJ size and color of nematode-infected wax worms (Hominick *et al.*, 1997) (Table 2).

B Efficacy

Numerous tests have been used to assess the efficacy of EPN in greenhouse and field experiments. Because test methods vary considerably with substrate/crop and insect pest against which the nematodes are to be tested, only a few generalizations can be made in this chapter. For more detailed information for various cropping systems, please refer to the chapters in section VII of this book.

When conducting field trials against soil-dwelling insects, it is important to survey the intended test site for the presence of natural nematode populations and other pathogens that may interfere with the outcome of the experiment. Nematode presence can be detected by baiting soil samples with test insects (see section 4 C). A special consideration for nematode field trials is plot spacing. Because of the IJs' ability to actively disperse, nematode-treated plots should be separated from other plots by at least 1 m to prevent contamination. The minimum distance will naturally vary with the characteristics of the cropping and irrigation system, soil type, vegetation, slope, application method, etc. The insertion of barriers such as metal or plastic strips or rings used to restrict lateral movement of insects (*e.g.*, white grubs in turfgrass) will also minimize lateral dispersal of IJs, allowing closer spacing of plots.

Two other special considerations for nematode field trials are the fast speed of kill (usually 1–2 days at $\geq 20\,^\circ C$) and the frailty and rapid disintegration of nematode-killed insects. Depending on the nematode species, the size and type of host, and the temperature, IJ emergence may start as early as 6 to 10 days after infection. At this point, recovery of cadavers becomes increasingly difficult. Even before emergence has started, disturbance of the substrate in which the insect was killed may cause the cadaver to rupture and become mixed or encrusted with substrate particles hindering recovery.

When determining infection rates, the dead insects need to be recovered before they deteriorate. The experimenter has to be aware that in situations where infection of a population is a gradual process (see below), the samples may have to be taken before the maximum mortality has been achieved. This can be remedied by sequential sampling of pest populations. The infection rate is determined by looking for signs of nematode infection in cadavers (see section 1 A). Surviving insects can be incubated for an additional 48 h to determine if they have already been penetrated by nematodes at the time of sampling. If population sampling is difficult to perform, sentinel insects can be placed in or on the substrate and recovered at various times to determine the rate of infection. Mobile sentinel insects may have to be placed inside cages that do not interfere with nematode activity. Caging also facilitates recovery of non-mobile insects such as pupae and protects them from predation.

The efficacy of EPN is most commonly determined through the rate of mortality or rather, because of the above mentioned limitations, the rate of survival of the target pest. This can be done by direct observation of the pest population or by observing rates of emergence of adults when immature stages are treated in a substrate. Where useful and possible, determination of crop damage reduction and yield should also be performed.

The timing of the evaluation is critical and may depend on the specific question(s) addressed in an experiment. In exposed habitats (*e.g.*, on foliage) evaluation can be done within a relatively short time after application of IJs considering the limited window of nematode activity and survival and the rapid mortality of infected insects. In other more complex pest habitats, such as soil, evaluation needs to be delayed for several weeks for two reasons. First, IJs may require a longer time to reach the target because they have to penetrate through layers of organic matter and/or soil. Second, these habitats usually allow for longer IJ survival, and nematode-related pest mortality may accumulate over time. If nematode recycling is possible in the habitat, repeated sampling even into the following growing season may give further information on the performance of the nematodes.

C *Persistence/establishment/recycling*

In environments conducive to IJ survival, prolonged persistence can be important for successful pest management with EPN. For example, more than one generation of pests with short generation time may be controlled by one nematode application provided sufficient nematodes survive in the environment. The control window becomes even wider if recycling and establishment of nematode populations occurs. Because of the difficulty to recover nematode-killed insects, the practical way to determine persistence is by extraction of IJs from the environment. As EPN occur naturally in soil, it is essential to take samples before their application to establish zero-baseline data.

The purpose of the sampling and the system to be sampled will determine how samples should be taken, *i.e.*, size, number, timing. Soil samples generally can be taken with a soil corer, trowel, auger, etc. and should include at least the upper 10–20 cm of the soil horizon (Kaya and Stock, 1997). Because even inundatively released nematode populations tend to assume a patchy distribution within weeks or months after release (Campbell *et al.*, 1998), the variability of the samples will increase over time. To obtain a good estimate with low variability, a rather large number of samples may have to be taken (Duncan *et al.*, 1996). Where appropriate, variability can be reduced by combining subsamples. For example, if persistence in small replicated plots (1–10 m^2) is to be evaluated, 3 subsamples (2.5 cm diam × 15 cm depth) from each replicate could be combined and mixed for further processing.

The samples should be protected from desiccation (plastic bags) and high temperatures (cooler). If not processed within a few days of sampling, the samples should be kept at 8–15 °c. between taking samples that are to be evaluated separately, sampling tools should be wiped with 70% alcohol or 0.5% bleach, or thoroughly washed with water to prevent contamination of subsequent samples.

To allow differentiation between the originally released IJs and IJs originating from recycling in hosts, or native and released IJs, stained IJs can be produced on media containing dyes (0.5% Sudan II, Sudan III, oil soluble blue, 0.4% neutral red) (Kaya and Stock, 1997). The stain can last for several months and effects on IJ persistence, if any, seem to be very limited.

Extraction of IJs can be performed in a number of ways including direct extraction of IJs from the substrate or by using bait insects (Kaya and Stock, 1997). Methods commonly used for the extraction of terrestrial nematodes including Baermann funnels, sieving, elutriation, and centrifugal flotation, can provide good quantitative estimates of IJ densities. However, they tend to be time consuming, expensive, and require a high level of expertise in nematode taxonomy because the methods are not specific to EPN. Generally, methods based on the use of trap insects are simpler to perform. Although these methods were developed for baiting soil samples, modifications for other substrates and pest habitats such as manure or organic litter should be possible.

The majority of known species of EPN including those that are commercially available (except for *S. scapterisci*) can be baited with wax worms or meal worms that are readily available and inexpensive. Nematode density is estimated by determining the mortality of bait insects incubated for an adequate period of time (2–7 days) in soil samples. To improve the efficiency of these methods, more than one baiting round should be performed. More detailed information can be obtained if the nematode-killed bait insects are dissected before progeny can develop and the number of established nematodes counted. Counting can be facilitated by digesting the insect tissues in a Pepsin solution (Kaya and Stock, 1997). A recently developed extraction method (Koppenhöfer *et al.*, 1998) is based on a strong linear relationship between the number of nematode-killed bait insects and nematodes in the bait insects after exhaustive sequential baiting of soil samples (see Appendix B). This method is less time consuming than other baiting methods because it makes bait dissection and nematode counting superfluous, and it has a very high extraction efficiency.

5 Sources of nematodes

Most EPN can be easily cultured in wax worms, mealworms, or other susceptible host insects (Kaya and Stock, 1997). Using this *in vivo* rearing method, up to 1–2 × 10^5 IJs/wax worm can be obtained (*S. carpocapsae, S. riobrave, H. bacteriophora*) providing enough IJs for smaller field trials. The harvest, however, generally decreases with the size of the IJs and only around 10^4 IJs/wax worm can be obtained for species with large IJs (*e.g., S. glaseri*). If larger nematode quantities are needed, *in vitro* rearing methods are more appropriate (Kaya and Stock, 1997). Alternatively, nematodes can be obtained from commercial sources (Hunter, 1997; Shah and Goettel, 1999; see Appendix C). Starter cultures can be requested from most laboratories that regularly work with EPN.

6 Appendix

A *Petri dish assay to determine infectivity*

1. Line lid of a Petri dish (35 × 10 to 100 × 15 mm diam)* with two filter papers or a layer of sand.
2. Add required number of IJs*. Substrate should be moist but not soaked.
3. Add an adequate number of host insects* (preferably late instar wax worms if nematode species infects them efficiently) and cover with lid.
4. Protect dish from desiccation in a plastic bag, allowing air exchange. Incubate at an adequate temperature for at least 48 h (longer at cooler temperatures).
5. Record mortality and verify nematode infection by dissection of cadavers or count number of nematodes established in cadavers within 48 h of death.

B *Estimation of IJ population in soil samples*

1. Place soil sample (100 cm^3) in a Petri dish (100 × 25 mm). Break up soil clumps and adjust soil moisture as needed (*i.e.*, soil should be moist but not saturated).
2. Add ten wax moth larvae onto the soil surface. Do not invert the dish.
3. Incubate samples for 3 days at 20–25 °C.
4. Recover all wax moth larvae.†

* Number of IJs added should be higher in larger dishes; there should not be more than 50 IJs/wax worm. For example, when using a 60 × 15 mm dish, 5 wax worms and 100 IJs may be used.

† To minimize work and the number of wax moth larvae required, the surviving larvae can be left in the soil, and only the dead larvae are replaced with new larvae. However, this increases the number of non-nematode related deaths in subsequent baiting rounds that may be difficult to distinguish from a nematode infection unless larvae are dissected. This approach works well, if the soil samples contain a known nematode species that causes a very distinct coloration in infected larvae.

a. Record number of nematode-infected larvae.
b. Incubate live larvae for an additional 3 days, then record the number of nematode-killed larvae.

5. Repeat steps 2–4 until no more nematode-infected larvae are recorded for two consecutive baiting rounds.
6. Calculate total number of nematode-infected larvae in all baiting rounds and insert in equation: $\mathbf{y = 10^{[-0.34+2.01^{*}\log(x)]}}$ where x = number of nematode-killed bait insects; y = number of nematodes established in bait insects; this equation yields an estimate of IJ number in $100\,cm^3$ of soil.

C Commercial suppliers of entomopathogenic nematodes

Information available on the internet:
http://www.cdpr.ca.gov/docs/ipminov/bscover.html (electronic copy of Hunter 1997)
http://www.nysaes.cornell.edu/ent/biocontrol/pathogens/nematodes.html
http://www2.oardc.ohio-state.edu/nematode
http://www.sipweb.org

D Some commercial suppliers of wax worms in the USA

Vanderhorst Canning Co., POB 37, ST. Mary's, OH 45885, USA; Ph: (419) 394–5236.
Rainbow Mealworms, 126 East Spruce St., Compton, CA 90220, USA; Ph: (800) 777–9676.
Reeve's Cricket Ranch, P. O. Box 2874, La Grande, OR 97850; Ph: (800) 526–4410 or (800) 238–2808
Grubco, P. O. Box 15001, Hamilton, OH 45015; Ph: (800) 222–3563
Fax: (513) 874–5878
Armstrong's Cricket Farm, PO Box 125, West Monroe, LA 71294, (800) 345–8778

7 References

Appel, A. G., Benson, E. P., Ellenberger, J. M. and Manweiler, S. A. 1993. Laboratory and field evaluations of an entomogenous nematode (Nematoda: Steinernematidae) for German cockroach (Dictyoptera: Blatellidae) control. *J. Econ. Entomol.* **86**, 777–784.

Akhurst, R. J. 1990. Safety to nontarget invertebrates of nematodes of economically important pests. *In* "Safety of Microbial Insecticides," (M. Laird, L. A. Lacey, and E. W. Davidson, Eds.), pp. 233–240. CRC Press, Boca Raton, FL.

Bathon, H. 1996. Impact of entomopathogenic nematodes on on-target hosts. *Biocontrol Sci. Technol.* **6**, 421–434.

Baur, M. E., Kaya, H. K., Gaugler, R. and Tabashnik, B. 1997a. Effects of adjuvants on entomopathogenic nematode persistence and efficacy against *Plutella xylostella. Biocontrol Sci. Technol.* **7**, 513–525.

Baur, M. E., Kaya, H. K. and Tabashnik, B. 1997b. Efficacy of a dehydrated steinernematid nematode against black cutworm (Lepidoptera: Noctuidae) and diamondback moth (Lepidoptera: Plutellidae). *J. Econ. Entomol.* **90**, 1200–1206.

Bedding, R. 1993. Biological control of *Sirex noctilio* using the nematode *Deladenus siricidicola. In* "Nematodes and the Biological Control of Insects" (R. Bedding, R. Akhurst and H. Kaya, Eds.), pp. 11–20. CSRIO, East Melbourne, Australia.

Bedding, R., Akhurst, R. and Kaya, H. K. 1993. "Nematodes and the Biological Control of Insects." CSRIO, East Melbourne, Australia.

Bednarek, A. and Gaugler, R. 1997. Compatibility of soil amendments with entomopathogenic nematodes. *J. Nematol.* **29**, 220–227.

Begley, J. W. 1990. Efficacy against insects in habitats other than soil. *In* "Entomopathogenic Nematodes in Biological Control" (R. Gaugler and H. K. Kaya, Eds.), pp. 215–321. CRC Press, Boca Raton, FL.

Broadbent, A. B., and Olthof, Th. H. A. 1995. Foliar application of *Steinernema carpocapsae* (Rhabditida: Steinernematidae) to control *Liriomyza trifolii* (Diptera: Agromyzidae) larvae in chrysanthemums. *Environ. Entomol.* **24**, 431–435.

Brown, I. M. and Gaugler, R. 1997. Temperature and humidity influence emergence and survival of entomopathogenic nematodes. *Nematologica* **43**, 363–375.

Cabanillas, H. E. and Raulston, J. R. 1996. Effects of furrow irrigation on the distribution and infectivity of *Steinernema riobravis* against corn earworm in corn. *Fundam. Appl. Nematol.* **19**, 273–281.

Campbell, J. F., Lewis, E. E., Yoder, F. and Gaugler, R. 1995. Entomopathogenic nematode (Heterorhabditidae and Steinernematidae) seasonal population dynamics and impact on insect populations in turfgrass. *Biol. Control* **5**, 598–606.

Campbell, J. F. and Gaugler, R. 1997. Inter-specific variation in entomopathogenic nematode foraging strategy: dichotomy or variation along a continuum? *Fundam. Appl. Nematol.* **20**, 393–398.

Campbell, J. F., Lewis, E. E., Yoder, F. and Gaugler, R. 1997. Entomopathogenic nematode (Heterorhabditidae and Steinernematidae) spatial distribution in turfgrass. *Parasitology* **113**, 473–482.

Campbell, J. F., Orza, G., Yoder, F., Lewis, E. and Gaugler, R. 1998. Spatial and temporal distribution of endemic and released entomopathogenic nematode populations in turfgrass. *Entomol. Exp. Appl.* **86**, 1–11.

Curran, J. 1993. Post-application biology of entomopathogenic nematodes in soil. *In* "Nematodes and the Biological Control of Insects" (R. Bedding, R. Akhurst and H. Kaya, Eds.), pp. 67–77. CSRIO, East Melbourne, Australia.

Duncan, L. W., McCoy, C. W., and Terranova, A. C. 1996. Estimating sample size and persistence of entomogenous nematodes in sandy soils and their efficacy against the larvae of *Diaprepes abbreviatus* in Florida. *J. Nematol.* **28**, 56–67.

Eidt, D. C. and Thurston, G. S. 1995. Physical deterrents to infection by entomopathogenic nematodes in wire worms (Coleoptera: Elateridae) and other soil insects. *Can. Entomol.* **127**, 423–429.

Gaugler, R., Campbell, J. F., Selvan, S. and Lewis, E. E. 1992. Large-scale inoculative releases of the entomopathogen *Steinernema glaseri*: assessment 50 years later. *Biol. Control* **2**, 181–187.

Gaugler, R., Wang, Y. and Campbell, J. F. 1994. Aggressive and evasive behaviors in *Popillia japonica* (Coleoptera: Scarabaeidae) larvae: defenses against entomopathogenic nematode attack. *J. Invertebr. Pathol.* **64**, 193–199.

Georgis, R. 1990. Formulation and application technology. *In* "Entomopathogenic nematodes in biological control" (R. Gaugler and H. K. Kaya, Eds.), pp. 153–191. CRC Press, Boca Raton, FL.

Georgis, R. and Gaugler, R. 1991. Predictability in biological control using entomopathogenic nematodes. *J. Econ. Entomol.* **84**, 713–720.

Georgis, R. and Manweiler, S. A. 1994. Entomopathogenic nematodes: a developing biological control technology. *Agric. Zool. Rev.* **6**, 63–94.

Georgis, R., Dunlop, D. B., and Grewal, P. S. 1995. Formulation of entomopathogenic nematodes. *In* "Biorational Pest Control Agents: Formulation and Delivery" (F. R. Hall and J. W. Barry, Eds.), pp 197–205. ACS Symposium Series No. 595, Amer. Chem. Soc., Washington, DC.

Georgis, R. and Kaya, H. K. 1998. Formulation of entomopathogenic nematodes. *In* "Formulation of Microbial Biopesticides: Beneficial Microorganisms, Nematodes and Seed Treatments" (H. D. Burges, Ed.), pp 289–308. Kluwer, Dordrecht, The Netherlands.

Georgis, R., Koppenhöfer, A. M., Lacey, L. A., Bélair, G., Duncan, L. W., Grewal, P. S., Samish, M., Tan, L., Torr, P. and Van Tol, R., W. H. M. 2006. Successes and failures in the use of parasitic nematodes for pest control. *Biol. Control* **38**, 103–123.

Glazer, I. 2002. Survival biology. *In* "Entomopathogenic Nematology" (R. Gaugler, Ed.), pp 169–187.

Glazer, I., Klein, M., Navon, A. and Nakache, Y. 1992. Comparison of efficacy of entomopathogenic nematodes combined with antidesiccants applied by canopy sprays against 3 cotton pest (Lepidoptera: Noctuidae). *J. Econ. Entomol.* **85**, 1636–1641.

Gouge, 2005. Applications for social insect control. *In* "Nematodes as Biocontrol Agents." (P. S. Grewal, R.-U. Ehlers, and D. I. Shapiro-Ilan, Eds.), pp. 317–329. CABI Publishing, Wallingford, UK.

Grewal, P. S. 2002. Formulation and application technology. *In* "Entomopathogenic Nematology" (R. Gaugler, Ed.), pp. 265–287. CABI Publishing, Wallingford, UK.

Grewal, P. S., Selvan, S., and Gaugler, R. 1994. Thermal adaptation of entomopathogenic nematodes: niche breadth for infection, establishment, and reproduction. *J. Thermal. Biol.* **19**, 245–253.

Grewal, P. S., Ehlers, R.-U., and Shapiro-Ilan, D. I. 2005. "Nematodes as Biocontrol Agents." CABI Publishing, Wallingford, UK.

Grewal, P. S., and Peters, A. 2005. Formulation and quality. *In* "Nematodes as Biocontrol agents." (P. S. Grewal, R.-U. Ehlers, and D. I. Shapiro-Ilan, Eds.), pp 79–90. CABI Publishing, Wallingford, UK.

Griffin, C. T. 1993. Temperature responses of entomopathogenic nematodes: implications for the success of biological control programmes. *In* "Nematodes and the Biological Control of Insects" (R. Bedding, R. Akhurst and H. Kaya, Eds.), pp. 115–126. CSRIO, East Melbourne, Australia.

Hominick, W. R. 2002. Biogeography. *In* "Entomopathogenic Nematology" (R. Gaugler, Ed.), pp. 115–144. CABI Publishing, Wallingford, UK.

Hominick, W. M., Reid, A. P., Bohan, D. A. and Briscoe, B. R. 1996. Entomopathogenic nematodes: Biodiversity, geographical distribution and the convention on biological diversity. *Biocontrol Sci. Technol.* **6**, 317–331.

Hominick, W. R., Briscoe, B. R., del Pino, F. G., Heng, J., Hunt, D. J., Kozodoy, E., Mracek, Z., Nguyen, K. B., Reid, A. P., Spiridonov, S., Stock, P., Sturhan, D., Waturu, C. and Yoshida, M. 1997. Biosystematics of entomopathogenic nematodes: current status, protocols and definitions. *J. Helminthol.* **71**, 271–298.

Hunter, C. D. 1997. "Suppliers of beneficial organisms in North America," California Environ. Protection Agency - Dept. Pestic. Regulation, Sacramento, CA. 32 pp.

Ishibashi, N. 1993. Integrated control of insect pests by *Steinernema carpocapsae*. *In* "Nematodes and the Biological Control of Insects" (R. Bedding, R. Akhurst and H. Kaya, Eds.), pp. 105–113. CSRIO, East Melbourne, Australia.

Kaya, H. K. 1990. Soil ecology. *In* "Entomopathogenic Nematodes in Biological Control" (R. Gaugler and H. K. Kaya, Eds.), pp. 93–115. CRC Press, Boca Raton, FL.

Kaya, H. K. 2002. Natural enemies and other antagonists *In* "Entomopathogenic Nematology" (R. Gaugler, Ed.), pp. 189–203. CABI Publishing, Wallingford, UK.

Kaya, H. K., Burlando, T. M., Choo, H. Y. and Thurston, G. S. 1995. Integration of entomopathogenic nematodes with *Bacillus thuringiensis* or pesticidal soap for control of insect pests. *Biol. Control* **5**, 432–441.

Kaya, H. K., and Koppenhöfer, A. M. 1996. Effects of microbial and other antagonistic organism and competition on entomopathogenic nematodes. *Biocontrol Sci. Technol.* **6**, 333–345.

Kaya, H. K. and Stock, S. P. 1997. Techniques in insect nematology. *In* "Manual of Techniques in Insect Pathology" (L. Lacey, Ed.), pp. 281–324. Academic Press, London.

Kaya, H. K. and Koppenhöfer, A. M. 2004. Biological control of insects and other invertebrates with nematodes. *In* "Nematology, Advances and Perspectives" (Z. X. Chen, S. Y. Chen and D. W. Dickson, Eds.), pp. 1083–1132. CABI Publishing, Wallingford, UK.

Kaya, H. K., Aguillera, M. M., Alumai, A., Choo, H. Y., de la Torre, M, Fodor, A., Ganguly, S., Hazir, S., Lakatos, T., Pye, A., Wilson, M., Yamanaka, S., Yang, H. and Ehlers. R.-U. 2006. Status of entomopathogenic nematodes and their symbiotic bacteria from selected countries or regions of the world. *Biol. Control* **38**, 134–155.

Klein, M. G. 1990. Efficacy against soil-inhabiting insect pests. *In* "Entomopathogenic Nematodes in Biological Control" (R. Gaugler and H. K. Kaya, Eds.), pp. 195–214. CRC Press, Boca Raton, FL.

Klein, M. G. 1993. Biological control of scarabs with entomopathogenic nematodes. *In* "Nematodes and the Biological Control of Insects" (R. Bedding, R. Akhurst and H. Kaya, Eds.), pp. 49–57. CSRIO, East Melbourne, Australia.

Koppenhöfer, A. M., Kaya, H. K. and Taormino, S. 1995. Infectivity of entomopathogenic nematodes (Rhabditida: Steinernematidae) at different soil depths and moistures. *J. Invertebr. Pathol.* **65**, 193–199.

Koppenhöfer, A. M., Baur, M. E., Stock, S. P., Choo, H. Y., Chinnasri, B. and Kaya, H. K. 1997. Survival of entomopathogenic nematodes within host cadavers in dry soil. *App. Soil Ecol.* **6**, 231–240.

Koppenhöfer, A. M. and Kaya, H. K. 1997. Additive and synergistic interaction between entomopathogenic nematodes and *Bacillus thuringiensis* for scarab grub control. *Biol. Control* **8**, 131–137.

Koppenhöfer, A. M., Campbell, J. F., Kaya, H. K. and Gaugler, R. 1998. Estimation of entomopathogenic nematode population density in soil by correlation between bait insect mortality and nematode penetration. *Fundam. Appl. Nematol.* **21**, 95–102.

Koppenhöfer, A. M. and Kaya, H. K. 1998. Synergism of imidacloprid and entomopathogenic nematodes: A novel approach to white grub control in turfgrass. *J. Econ. Entomol.* **91**, 618–623.

Koppenhöfer, A. M. and Fuzy, E. M. 2003. Ecological characterization of *Steinernema scarabaei*: a natural pathogen of scarab larvae. *J. Invertebr. Pathol.* 83, 139–148.

Koppenhöfer, A. M. and Grewal, P. S. 2005. Compatibility and interactions with agrochemicals and other biocontrol agents. *In* "Nematodes as Biocontrol Agents." CABI Publishing, Wallingford, UK, pp 363–381.

Koppenhöfer, A. M. and Fuzy, E. M. 2006. Effect of soil type on infectivity and persistence of the EPNs *Steinernema scarabaei, Steinernema glaseri, Heterorhabditis zealandica*, and *Heterorhabditis bacteriophora. J. Invertebr. Pathol.* **92**, 11–22.

Levins, R. 1970. Extinction. *In* "Lectures on mathematics in the life sciences," Vol. 2 (M. Gerstenhaber, Ed.), pp. 77–107. Amer. Math. Soc., Providence, RI.

Lewis, E. E. 2002. Behavioural ecology. *In* "Entomopathogenic nematology" (R. Gaugler, Ed.), pp 205–224. CABI Publishing, Wallingford, UK.

Lewis, E. E., Campbell, J. F., and Gaugler, R. 1998. A conservation approach to using entomopathogenic nematodes in turf and landscapes. *In* "Conservation biological control" (P. Barbosa, Ed), pp. 235–254. Academic Press, San Diego, CA.

Lewis, E. E., Campbell, J., Griffin, C., Kaya, H. and Peters, A. 2006. Behavioral ecology of EPNs. *Biol. Control* **38**, 66–79.

Mamiya, Y. 1989. Comparison of infectivity of *Steinernema kushidai* (Nematoda: Steinernematidae) and other steinernematid and heterorhabditid nematodes for three different insects. *Appl. Entomol. Zool.* **24**, 302–308.

Nickle, W. R. 1984. "Plant and Insect Nematodes." Marcel Dekker, New York.

Nickle, W. R. 1991. "Manual of Agricultural Nematology." Marcel Dekker, New York.

Nickle, W. R. and Shapiro, M. 1994. Effects of eight brighteners as solar radiation protectants for *Steinernema carpocapsae*, All strain. *Suppl. J. Nematol.* **26**, 782–784.

Nishimatsu, T. and Jackson, J. J. 1998. Interaction of insecticides, entomopathogenic nematodes, and larvae of the western corn rootworm (Coleoptera; Chrysomelidae). *J. Econ. Entomol.* **91**, 410–418.

Parkman, J. P. and Smart, G. C., Jr. 1996. Entomopathogenic nematodes, a case study: Introduction of *Steinernema scapterisci* in Florida. *Biocontrol Sci. Technol.* **6**, 413–419.

Patel, M. N. and Wright, D. J. 1996. The influence of neuroactive pesticides on the behaviour of entomopathogenic nematodes. *J. Helminthol.* **70**, 53–61.

Peters, A. 1996. The natural host range of *Steinernema* and *Heterorhabditis* spp. and their impact on insect populations. *Biocontrol Sci. Technol.* **6**, 389–402.

Peters, A. and Ehlers, R.-U. 1994. Susceptibility of leatherjackets (*Tipula paludosa* and *Tipula oleracea*; Tipulidae; Nematocera) to the entomopathogenic nematodes *Steinernema feltiae. J. Invertebr. Pathol.* **63**, 163–171.

Poinar, G. O., Jr. 1979. "Nematodes for Biological Control of Insects." CRC Press, Boca Raton, FL.

Renn, N. 1998. The efficacy of entomopathogenic nematodes for controlling housefly infestations of intensive pig units. *Med. Vet. Entomol.* **12**, 46–51.

Rovesti, L. and Deseö, K. V. 1990. Compatibility of chemical pesticides with the entomopathogenic nematodes, *Steinernema carpocapsae* Weiser and *S. feltiae* Filipjev (Nematoda: Steinernematidae). *Nematologica* **36**, 237–245.

Schroeder, W. J. and Sieburth, P. J. 1997. Impact of surfactants on control of the root weevil *Diaprepes abbreviatus* larvae with *Steinernema riobravis. J. Nematol.* **29**, 216–219.

Selvan, S., Grewal, P. S., Gaugler, R. and Tomalek, M. 1994. Evaluation of steinernematid nematodes against *Popillia japonica* (Coleoptera: Scarabaeidae) larvae: species, strains, and rinse after application. *J. Econ. Entomol.* **87**, 605–609.

Shah, P. A. and Goettel, M. S. 1999. "Directory of microbial control products and services." Microbial Control Div., Soc. Invertebr. Pathol., Gainesville, FL.

Shapiro, D. I., Tylka, G. L. and Lewis, L. C. 1997. Effects of fertilizers on virulence of *Steinernema carpocapsae. Appl. Soil Ecol.* **3**, 27–34.

Shapiro-Ilan, D. I., Lewis, E. E., Behle, R. W., and McGuire, M. R. 2001 Formulation of EPN-infected cadavers. *J. Invertebr. Pathol.* **78**, 17–23.

Shapiro-Ilan, D. I., Lewis, E. E., Tedders, W. L., and Son, Y. 2003. Superior efficacy observed in EPNs applied in infected-host cadavers compared with application in aqueous suspension. *J. Invertebr. Pathol.* **83**, 270–272.

Shapiro-Ilan, D. I., Gouge, D. H., Piggott, S. J., Fife, J. P. 2006. Application technology and environmental considerations for use of EPNs in biological control. *Biol. Control* **38**, 124–133.

Smits, P. 1996. Post-application persistence of entomopathogenic nematodes. *Biocontrol Sci. Technol.* **6**, 379–387.

Stark, J. D. 1996. Entomopathogenic nematodes (Rhabditida: Steinernematidae): toxicity of neem. *J. Econ. Entomol.* **89**, 68–73.

Stock, S. P., Griffin, C. T. and Chaerani, R. T. 2004. Morphological and molecular characterisation of *Steinernema hermaphroditum* n. sp. (Nematoda: Steinernematidae), an entomopathogenic nematode from Indonesia, and its phylogenetic relationships with other members of the genus. *Nematology* **6**, 401–412.

Strong, D. R., Kaya, H. K., Whipple, A. V., Child, A. L., Kraig, S., Bondonno, M. and Maron, J. L. 1996. Entomopathogenic nematodes: natural enemies of root-feeding caterpillars on bush lupine. *Oecologia* **108**, 167–173.

Stuart, R. J. and Gaugler, R. 1994. Patchiness in populations of entomopathogenic nematodes. *J. Invertebr. Pathol.* **64**, 39–45.

Stuart, R. J., Barbercheck, M. E., Grewal, P. S., Taylor, R. A. J. and Hoy, C. W. 2006. population biology of EPNs: concepts, issues, and models. *Biol. Control* **38**, 80–102.

Thurston, G. S., Kaya, H. K. and Gaugler, R. 1994. Characterizing the enhanced susceptibility of milky disease-infected scarabaeid grubs to entomopathogenic nematodes. *Biol. Control* **4**, 67–73.

Treverrow, N. L. and Bedding, R. A. 1993. Development of a system for the control of the banana weevil borer, *Cosmopolites sordidus* with entomopathogenic nematodes. *In* "Nematodes and the Biological Control of Insects" (R. Bedding, R. Akhurst and H. Kaya, Eds.), pp. 41–47. CSRIO, East Melbourne, Australia.

Wang, Y., Campbell, J. F. and Gaugler, R. 1995. Infection of entomopathogenic nematodes *Steinernema glaseri* and *Heterorhabditis bacteriophora* against *Popillia japonica* (Coleoptera: Scarabaeidae) larvae. *J. Invertebr. Pathol.* **66**, 178–184.

Womersley, C. Z. 1990. Dehydration survival and anhydrobiotic potential. *In* "Entomopathogenic nematodes in biological control" (R. Gaugler and H. K. Kaya, Eds.), pp. 117–137. CRC Press, Boca Raton, FL.

Wright, D. J., Peters, A., Schroer, S. and Fife, J. P. 2005. Application technology. *In* "Nematodes as Biocontrol Agents." (P. S. Grewal, R.-U. Ehlers, and D. I. Shapiro-Ilan, Eds.), pp. 91–106. CABI Publishing, Wallingford, UK.

Yang H., Zhang, G., Zhang, S. and Jian, H. 1993. Biological control of tree borers (Lepidoptera: Cossidae) in China with the nematode *Steinernema carpocapsae*. *In* "Nematodes and the Biological Control of Insects" (R. Bedding, R. Akhurst and H. Kaya, Eds.), pp. 33–40. CSIRO, East Melbourne, Australia.

Zimmerman, R. J. and Cranshaw, W. S. 1990. Compatibility of three entomogenous nematodes (Rhabditida) in aqueous solutions of pesticides used in turfgrass maintenance, *J. Econ. Entomol.* **83**, 97–100.

SECTION V

NATURALLY OCCURRING PATHOGENS

Chapter V-1

Documentation of naturally occurring pathogens and their impact in agroecosystems

Donald C. Steinkraus
Department of Entomology
University of Arkansas
Fayetteville, AR 72701, USA

1 Introduction

Microbial pathogens play a vital role in the natural regulation of many arthropod populations in agricultural systems. Their importance has been underemphasized compared to parasitoids and predators because fewer specialists work on pathogens, and pathogens are generally less noticeable than parasitic and predaceous arthropods.

Because of the disturbed nature of agroecosystems and the need for rapid control of arthropod pests in crops, many entomologists are of the opinion that naturally occurring pathogens have limited value in integrated pest management (IPM). This assumption is inaccurate. Ecologists have become increasingly aware of the role microorganisms play in population regulation. May (1988) argued that the significance of pathogens has generally been underestimated in ecology. Loehle (1995) stated that the idea that pathogens are important regulators only in overly large, stressed populations is incorrect. Invertebrate pathologists have long been aware that microbial pathogens play a major role in regulation of arthropod populations. The eminent pathologist E. A. Steinhaus stated that a "*. . . knowledge of insect diseases (whether or not it has to do with control) is of fundamental and far-reaching importance in the study of insect ecology*" (Steinhaus, 1949). A worthwhile goal for pest managers is to develop methods to take advantage of the natural control provided by pathogens of pests.

According to Stern *et al.* (1959) in their seminal paper on IPM, *"Chemical control should be used only when the economic threshold is reached and when the natural mortality factors present in the environment are incapable of preventing the pest population from reaching the economic-injury level."* Yet, this central concept of IPM, "natural mortality," especially that provided by pathogens, has rarely been considered in making treatment recommendations in most agroecosystems. Whitten (1992) stated that *"our overriding philosophy is to reduce mankind's dependence on chemical pesticides by identifying other methods which are economic, environmentally sound and lasting in effectiveness. Seeking these alternatives is the exciting challenge."* Natural control of pests by pathogens should play a more prominent role in IPM.

There have been many efforts to predict the prevalence of plant disease (Cook, 1949; Van der Plank, 1963; Waggoner and Horsfall, 1969) and plant pathogen diagnostic laboratories are common, but prior to the work of Kish and Allen (1978), there were no attempts to predict the natural prevalence of disease in insect pest populations. Since then there have been several attempts to develop models or systems to predict the levels of control provided by pathogens to reduce unnecessary pesticide usage. The main factors that need to be determined are: (1) the principal pathogen(s) involved and (2) the effects of these pathogens on pest population density (Kish and Allen, 1978). Ideally such studies

L.A. Lacey and H.K. Kaya (eds.), Field Manual of Techniques in Invertebrate Pathology, 267–281.

will be coordinated with similar studies on the effects of predators and parasitoids on the host population and population models of the pest that take into account oviposition, hatching, pupation, and environmental effects on the pest. The insect disease prediction systems proposed by Kish and Allen (1978), Steinkraus and Hollingsworth (1994) and Hollingsworth *et al.* (1995) are first steps in the incorporation of natural control by pathogens into IPM systems.

Documentation of arthropod pathogens in agroecosystems is important in four ways. First, identification and isolation of pathogens from specific arthropod pests are important initial steps in development of new classical biological control agents or microbial pesticides, as described below in the section on the entomopathogenic nematode *Steinernema riobrave*. Techniques for identification and isolation of pathogens are covered in appropriate chapters in Lacey (1997) and will not be covered further here. Second, our understanding of the population dynamics of arthropod pests is enhanced by documentation of the role played by pathogens. Pathogens are often found to be as important or more important in regulation of arthropod numbers as predators and parasitoids. Examples of this are described below. Third, naturally occurring epizootics are sometimes a valuable natural resource for harvesting pathogens for research or use in biological control (Moscardi, 1999; Steinkraus and Boys, 2005). Fourth, in situations where pathogen-induced epizootics regularly reduce pest populations, it may be possible to develop sampling programs or models to predict pathogen-induced pest declines. Such predictive sampling programs can lead to reduced inputs of chemical pesticides. Two case histories of this approach are described below.

2 General considerations

There are a number of variables that need to be considered when documenting the impact of naturally occurring entomopathogens in agroecosystems. These can be divided into two main interrelated categories: first, factors related to the environment a pathogen encounters and, second, factors integral to the pathogen. The first category provides background information for applied epizootiological studies and may require collection of data on temperature, humidity, rainfall, light intensity and duration, soil properties, crop varieties involved, plant stage and architecture, planting dates, cultural practices, as well as pesticide, herbicide, fertilizer and other chemical inputs. In addition, the arthropod host(s) for a pathogen constitutes a portion of the environment experienced by the pathogen. Therefore, the host's biology and stage(s) of importance need to be well understood by the investigator. The second category involves laboratory studies to determine the optimum conditions for growth and survival of the pathogen. More specific requirements are considered below.

A Pathogen prevalence

A major objective is to determine prevalence of the pathogen(s) in the pest population. Prevalence is the most commonly used measure of pathogen impact on a host population and is defined as the number or proportion of hosts infected by a specific pathogen at a given point in time (Fuxa and Tanada, 1987). To accurately determine prevalence, the investigator must not bias his sample towards or against healthy or diseased individuals; in short, samples must be truly random. However, to do this requires knowledge of the life cycle of the pest species. For example, larvae of the corn earworm, *Helicoverpa zea*, develop in corn ears, but when mature, leave the corn to pupate in the soil. A sampling procedure for nucleopolyhedrovirus (NPV) or microsporidium that sampled corn ears after most healthy earworms had left the plants to pupate would be heavily biased in favor of infected larvae, resulting in an overestimate of prevalence, and therefore, impact of the pathogen(s). Almost always, prevalence determinations require periodic sampling of the host population at intervals determined by the biology of the host.

Accurate prevalence requires collecting live arthropod hosts as well as dead hosts that may have been killed by the pathogen. Living hosts collected in the field must survive in the laboratory long enough for the pathogen to become evident. The hosts require suitable food, temperature, light, and relative humidity. Ideally, live hosts should be held individually to prevent the possibility of transmission of pathogens during the collection and handling procedures, which could increase the apparent prevalence, or contamination by pathogens, resulting in erroneous diagnoses. Excellent examples of methods utilizing these principles are found in Ruano-Rossil *et al.* (2002), Feng *et al.* (2004), and Nielsen and Hajek (2005).

B Identification of pathogens

Host morphology is an important factor in choosing the methods used to diagnose diseased arthropods and identify pathogens. For instance, it is possible to squash entire small arthropods, like tetranychid mites (Carner, 1976; Klubertanz *et al.*, 1991) or aphids (Steinkraus *et al.*, 1991), then directly examine them on a microscope slide for the microscopic signs of a pathogen. But squashing an entire acridid, or mature noctuid larva, would not be feasible. Specific methods for preparing specimens for diagnosis are presented in Lacey and Brooks (1997). Many other sophisticated diagnostic techniques are available and are discussed in chapters in Section IV. Each pathogen group (viruses, bacteria, protozoa, fungi and nematodes) requires somewhat different handling of hosts collected from the field and for pathogen identification. Novices in insect pathology are strongly encouraged to develop cooperative efforts with skilled diagnosticians or taxonomists of the appropriate pathogen groups. It is also advisable to prepare permanent voucher specimens of the pathogens involved and deposit these in suitable museums or other repositories such as the USDA-ARS Collection of Entomopathogenic Fungal Cultures (Ithaca, NY).

3 Control of pests by naturally occurring pathogens

The concept that the activity of naturally occurring pathogens against pests is important is not new. Steinhaus (1949) stated "...one definite fact stands out clear and indisputable: Entomogenous fungi, in nature and without any help from man, cause a regular and tremendous mortality of many insect pests in many parts of the world and do, in fact, constitute an efficient and extremely important natural control factor. Accordingly, entomogenous fungi are of great economic importance in insect control, even though man has not yet learned how to use them artificially in most instances."

There are many examples of the importance of documenting the occurrence of naturally-occurring pathogens in pest populations. Speare (1922) studied the natural control of the citrus mealybug in Florida by *Neozygites fumosa*, and stated that "entomogenous fungi are worth millions of dollars to the citrus industry. Owing to their excellent work, oranges and grapefruit are grown at a profit in many parts of the State where no money whatsoever is spent on artificial remedial measures." In studies on tetranychid spider mites, Klubertanz *et al.* (1991) found that a *Neozygites* sp. reduced a two-spotted spider mite, *Tetranychus urticae*, population 95% over a 6-day period on soybean. Smitley *et al.* (1986) found that *N. floridana* was the most important cause of population declines in *T. urticae* populations in field corn. Similar results have been reported on the importance of *Neozygites* spp. in natural control of mites on cotton (Carner and Canerday, 1968), lima beans (Brandenburg and Kennedy, 1983), and peanuts (Boykin *et al.*, 1984).

Valuable work has been done on the importance of entomopathogenic fungi in regulating aphid populations on potato in Maine. These studies have shown that 5 species of Entomophthorales are the most important natural control agents of several aphid species (Patch, 1907; Shands *et al.*, 1972). An intensive survey of potato-aphid pathogens made between 1952–1962 and 1963–1969 showed that fungal activity was the major cause of aphid declines in most years (Shands *et al.*, 1972).

The importance of naturally occurring pathogens may be revealed during careful studies of the effect of pesticides, such as fungicides, on pest populations. For instance, Ruano-Rossil *et al.* (2002) conducted research on the green peach aphid, *Myzus persicae*, on potato in Minnesota. They found that fungicides disrupted control by naturally occurring entomopathogenic fungi. When fungicides were applied extremely high aphid populations resulted. The three most important fungal species involved were *Pandora neoaphidis* (=*Erynia*), *Entomophthora planchoniana*, and *Conidiobolus obscurus*. Their research clearly indicated the importance of entomopathogenic fungi in controlling early season aphids when the fungi were keeping the aphids below the economic injury level. The Ruano-Rossil *et al.* (2002) study shows that entomopathogens are important in regulating low as well as high pest populations.

Nielsen and Hajek (2005) evaluated the diversity and prevalence of naturally occurring entomopathogenic fungi on the invasive soybean aphid, *Aphis glycines*, in New York State. They found seven species of fungi attacking the aphid, with *P. neoaphidis* the most abundant. In an outbreak of soybean aphids, *P. neoaphidis* caused infection in 84% of the aphids and the population crashed.

Mirid bugs in the genus *Lygus* are important pests of many crops. McGuire (2003) surveyed populations of *L. hesperus* from alfalfa fields in the San Joaquin Valley of California and found that the fungus *Beauveria bassiana* was found infecting *L. hesperus* in every county sampled and was present at almost every sampling date. In one field, the prevalence of *B. bassiana* reached 50%. Steinkraus and Tugwell (1997) and Leland and Snodgrass (2004) also reported on natural *B. bassiana* infections of *Lygus lineolaris*. More information on *B. bassiana* and mirids is presented in Chapter VII-6.

While entomopathogenic fungi often are responsible for dramatic, easily-noticed epizootics, other pathogens, particularly viruses and nematodes also provide control of agricultural pests. For instance, naturally occurring epizootics of *Trichoplusia ni* nucleopolyhedrovirus were recognized by Semel (1956) as important in controlling cabbage looper populations. Two selected examples of the importance of naturally occurring pathogens are discussed in detail below.

A *Natural occurrence of* Steinernema riobrave *in* Helicoverpa zea *in corn*

Noctuids in the genera, *Helicoverpa, Heliothis*, and *Spodoptera*, are serious crop pests worldwide (King, 1994). Raulston *et al.* (1992) isolated a new species of steinernematid nematode, *Steinernema riobrave* (described by Cabanillas *et al.*, 1994), infecting prepupae and pupae of corn earworm, *H. zea*, and fall armyworm, *S. frugiperda*, from cornfields in the Lower Rio Grande Valley in Texas and northern Mexico. The 5-year study found that 34 and 24% of all fields contained infected *H. zea* or *S. frugiperda*, respectively. They further found that *S. riobrave* accounted for 49.4 and 46.1% of the mortality of *H. zea* and *S. frugiperda*, respectively. Not only did this research increase our understanding of the mortality factors acting against two important pests during their pupation stage in the soil, but the new nematode species that was discovered has proven to be useful as a biological control agent and is commercially available. The methods used in this research were as follows.

1 *Characteristics of study area*

The study area was defined in terms of geography, rainfall, irrigation, soil type, crop phenology, and pesticide usage. Such details are important. The *H. zea* mortality caused by *S. riobrave* in Texas and Mexico may not occur in other geographical regions or soil types. For example, in a similar study on pupal mortality of *H. zea* in corn fields in Arkansas, Kring *et al.* (1993) did not find *S. riobrave*.

2 *Sampling procedure and statistical analysis of data*

A sampling procedure was developed for prepupae and pupae of *H. zea* and *S. frugiperda*. Two areas were sampled in each of 90–120 fields/year. The samples were taken 50 and 100 m from the field edges after most larvae had left corn ears to pupate. Each sample was taken by carefully scraping 1 m^2 of the soil surface with a garden trowel to uncover pupal tunnels, then excavating the soil to a depth of 10–15 cm following the tunnels to find the pupal chambers. All extracted insects were placed in individual 20 ml plastic cups. Host species, sex, and developmental stage were recorded, and each pupa that died was examined with a dissecting microscope for the presence of nematodes. Data were analyzed by ANOVA and means separated by computing least square means and testing the hypothesis, Ho: LSM(i) = LSM(j) (Raulston *et al.*, 1992).

The nematodes were preliminarily identified to genus with the works of Poinar (1990) and later described as a new species by Cabanillas *et al.* (1994).

3 Crucial factors in this study

First, the study was replicated in many fields over a 5-year period, providing strong evidence that the mortality caused by *S. riobrave* was not an isolated incident. Second, the scientists understood the life cycle of the hosts. In this case, the fact that *H. zea* and *S. frugiperda* pupate in the soil, the average depth of pupation, and the phenology of larval movement from host plant to soil, were important facts in the experimental design. Third, holding the extracted pupae individually in cups prevented contamination of healthy pupae with nematodes which would have resulted in erroneous prevalence rates. Fourth, each dead pupa was examined for presence of nematodes to confirm their diagnosis. Fifth, an expert nematode systematist (G. O. Poinar, Jr.) was enlisted to assist in the identification and description of *S. riobrave*.

B Natural occurrence of Zoophthora phytonomi in Hypera postica populations in alfalfa

The alfalfa weevil, *Hypera postica*, is an introduced pest of alfalfa in the USA that can cause significant yield losses (Harcourt *et al.*, 1974). Several studies have shown that the entomopathogenic fungus, *Zoophthora phytonomi*, is one of the primary mortality factors operating on the alfalfa weevil. In Ontario, Canada, a 10-year study concluded that the alfalfa weevil had ceased to be an economic problem due to early spring epizootic levels of infection by *Z. phytonomi* (Harcourt *et al.*, 1984). This pathogen is also important in the various parts of the USA, but epizootics may not occur early enough to provide complete control. In Illinois, Morris *et al.* (1996) found that *Z. phytonomi* was responsible for preventing *H. postica* populations in some fields from reaching economic thresholds, but not in others. Methods for documenting the occurrence of *Z. phytonomi* are drawn from DeGooyer *et al.* (1995).

1 Field descriptions and plot design

Four Iowa alfalfa fields were chosen about 40 km apart latitudinally. Field locations, size, management, alfalfa variety, and stand densities were documented. Within each field a study site 50 by 50 m or 25 by 100 m was chosen, and divided into 8 equal 12.5 by 25 m plots.

2 Sampling plan

Larvae were sampled using a 0.1 m^2 sampling frame. Each sample consisted of 32 six-stem sampling units from each site on each sampling date. Fields were sampled twice weekly from 1 April until the first cutting of alfalfa. Larvae were extracted from the stems and placed individually into glass vials, then reared at 24 °C with a photoperiod of 16:8 (LD). Each vial contained a fresh alfalfa terminal that was changed every other day. Larvae were reared until they reached the adult stage or died from *Z. phytonomi* infections.

3 Life table construction

Collected data were used to construct life tables. The authors found that *Z. phytonomi* was the key factor regulating within-generation population trends in 1991, causing high mortality of 3rd and 4th stage larvae. Harcourt *et al.* (1990) developed detailed information on the occurrence of epizootics caused by *Z. phytonomi* in southern Ontario. They found that the first diseased alfalfa weevils occurred 204 degree days (DD) above a base threshold of 9 °C from 1 April. Epizootics began 57 DD later and lasted for 10 to 14 days, killing up to 99% of the larvae. They concluded that *Z. phytonomi* was the principal variable regulating alfalfa weevil populations. This information has obvious implications for alfalfa weevil control in Ontario and for reducing pesticide applications.

4 A model for predicting control of *Anticarsia gemmatalis* populations on soybean by *Nomuraea rileyi*

Soybean, *Glycine max*, is an important crop worldwide as a source of protein and oil for humans and livestock. Severe outbreaks of the velvetbean caterpillar, *Anticarsia gemmatalis*, can defoliate and damage soybean in the southern USA (Funderburk, 1993), and other locations, such as Brazil. In Florida, under certain conditions, *N. rileyi* can decimate *A. gemmatalis* populations with mortality levels approaching 100% Allen *et al.* (1971). Kish and Allen (1978) formulated ideas on predicting prevalence of the entomopathogenic fungus *N. rileyi* on *A. gemmatalis* on soybean. Their aim was to reduce insecticide use by utilizing the natural control provided by *N. rileyi*. They discussed two problems with biological control with naturally occurring pathogens: (1) humankind's lack of control over the natural environmental conditions that favor development of disease, and (2) epizootics may occur after peak pest populations, when crop damage has already resulted (Kish and Allen, 1978). They noted that the consistency of occurrence and high levels of population control provided by *N. rileyi* made it a candidate to be one of the first pathogens to be incorporated into an IPM program. For this approach to work, systems must be identified in which a pathogen naturally provides regular, predictable, epizootics.

Kish and Allen (1978) developed methods to examine and model each factor affecting the host population. Understanding of the environmental effects on pathogen/host dynamics is essential for developing confidence in a predictive model. The sample protocol below is drawn largely from the work of Kish and Allen (1978).

A *Sample protocol to document naturally occurring pathogens and prediction of pest control*

1 *Identify appropriate crop/pest/pathogen system*

The first need is to identify a pest population in a selected crop that has a pathogen acting against it that causes regular epizootics. This can only be achieved by field studies or observations by insect pathologists, or by entomologists working in cooperation with pathologists. It is not uncommon for field trials of pesticides, or ecological studies on predators or parasitoids to fail because of the action of pathogens on the target pest. Field entomologists observing pathogens that interfere with chemical pesticide trials, or other field studies, should consider these interferences as opportunities to conduct research on pathogens.

2 *Identification, isolation, and culture of pathogen*

It is essential that the pathogen(s) under consideration be identified (see Lacey, 1997) and the identification(s) be confirmed by taxonomic experts in the particular pathogen group. Deposit voucher specimens of the pathogen into an appropriate collection. If possible the pathogen(s) should be isolated and cultured *in vitro* or *in vivo* for studies on pathogen biology in the laboratory, such as effects of photoperiod, temperature, and relative humidity.

3 *Determine the pathogen's biology and epizootiology*

Determine the basic epizootiological factors governing transmission of the pathogen to hosts, pathogen overwintering, behavior of infected hosts, incubation period, pathogen propagules produced per infected host, density of pathogen propagules in the agroecosystem, and temporal occurrence of the pathogen in the air, water, soil, or on plant surfaces. It is essential to know when the pathogen first appears in the host population and when it peaks. Experiments to determine these facts must be designed differently for each pathogen group (viruses, bacteria, protozoa, fungi, nematodes). For example, fungi may be dispersed by air movements, but aerial dispersal is less important with protozoa.

4 *Conduct field epizootiological studies*

a *Plot studies*

The crop must be planted in an appropriate experimental design. A randomized complete block design with at least 4 replications is desirable. The crop variety, row spacing, seeding rate, and other factors should be defined and recorded. Often the cooperation of an agronomist is of great value. More realistic studies can be made on commercial plantings, but the researcher loses some control over crop management.

b Sampling methods

The pest population must be sampled during the season using methods appropriate to its biology. Kish and Allen (1978) used the shake cloth method to sample *A. gemmatalis* with 61 row m/0.4 ha sampled bi-weekly.

c Prevalence determination

Sub-samples of larvae collected from the field were held in the laboratory for determination of the prevalence of the pathogen in the host population. Larvae were retained for 5 days, but not longer, to ensure that the infections had originated naturally in the field and were not a consequence of contamination during collection or in the laboratory. This point needs to be emphasized. First, a random sample of the pest population must be collected to ensure that infection rates accurately represent the situation in the field. Infected insects may not behave normally. Healthy insects may move more rapidly than moribund infected insects, and escape more often from a sweep net, beat sheet, or other method, resulting in an overrepresentation of infected specimens in a sample. Conversely, infected insects may fall to the ground and/or be removed by scavengers, and therefore be collected less frequently than healthy individuals, resulting in an underestimation of prevalence. Second, during sampling, individual arthropods may become contaminated with pathogens during close contact with other captured individuals. Therefore, care must be taken to prevent such events. The fate of individual infected *A. gemmatalis* in the field was determined by tagging hosts on plants and observing them over time.

d Effect of agrochemicals

The effect of pesticides, fungicides, and other management options on the pathogen/pest interaction should be determined. For instance, certain fungicides, such as benomyl may suppress development of epizootics of *N. rileyi* (Kish and Allen, 1978).

e Analyze and understand the system

Kish and Allen (1978) showed that the general concept held by many workers that entomopathogenic fungi require wet, rainy conditions to cause epizootics had to be qualified. In fact, they found excessive rain may actually impede fungal epizootic development. Dry, windy conditions promoted dispersal of *N. rileyi* conidia, but also retarded germination and infection unless humid conditions prevailed. Once the fungus was within the host, development of the pathogen was independent of weather conditions. Rain, dew, and relative humidity above 70% promoted conidiophore formation and conidiogenesis, while conidia were washed from host cadavers to the ground by rain. An alternation of wet and dry conditions was most effective in dispersing the pathogen, and excessive free water during early stages of the epizootic retarded development of epizootics.

They then analyzed the results statistically and developed a mathematical model to predict the amount of natural control provided by the pathogen. Pesticide use can be avoided when the pathogen is predicted to reduce the pest population. In practice growers usually follow a variation of the following practices: (1) wait until large, damaging pest populations are observed before spraying a pesticide, or (2) spraying on a schedule, regardless of pest populations. Heavy damage may result from the first approach, and the second, may result in unnecessary control, secondary pest outbreaks, reductions of beneficial arthropods, environmental pollution, and excessive input costs (Kish and Allen, 1978). Neither approach takes into account natural control factors that might suppress the pest population at any given time.

f Implement the model

The model was validated with actual field data. The final step was to incorporate the predictive model into practice. Kish and Allen (1978) believed that the ability to monitor the fungal entomopathogen, *N. rileyi*, and predict mortality levels in *A. gemmatalis* populations on a daily basis could lead to increased accuracy of management decisions and reduce use of pesticides for *A. gemmatalis* on soybean by 60% to 100%.

They planned to implement this model via a computer network with a main centralized computer. Extension inputs would be coordinated by a pest management specialist. This specialist would train extension agents, growers, scouts, and private consultants, in Florida counties. Trained personnel at the farm level would provide the basic data on the pest and beneficial insect populations, insect disease prevalence, and environmental conditions that would be placed in the central computer. The model would then be used to predict the control by *N. rileyi*, and this information provided to county level personnel for rapid and accurate decision-making.

The model of Kish and Allen (1978) does not appear to have been implemented due to decreased planting of soybean in Florida, the low value of soybean crops, and the ease of managing lepidopteran pests on soybean with insecticides (J. E. Funderburk, personal communication).

5 Extension-based sampling service for prediction of natural epizootics of *Neozygites fresenii* in *Aphis gossypii* populations on cotton

A *Background*

The cotton aphid, *Aphis gossypii*, is a serious pest of many crops worldwide. It is a small aphid with a very rapid life cycle; as short as 4 days during the summer (Isely, 1946). Cotton aphids can directly reduce yield and lint quality through photosynthate removal, indirectly reduce photosynthesis due to sooty mold growth on excreted honeydew, and can cause honeydew-contaminated sticky cotton. During the period 1988 through 2006, the aphid has been a serious cotton pest in the USA. The 1996 introduction of transgenic cottons for noctuid pest control in the USA and recent efforts to eradicate the boll weevil, *Anthonomus grandis*, may unpredictably affect the future pest status of cotton aphids.

Our understanding of aphid population levels that cause economic damage in cotton is incomplete. Studies in Oklahoma (Karner *et al.*, 1997) and California (Godfrey *et al.*, 1997; Godfrey and Wood, 1998) showed significant cotton yield losses due to aphids. However, studies in Tennessee (Lentz and Austin, 1998) and Mississippi (Hardee and Adams, 1998) showed no yield losses due to cotton aphids, and no benefit from insecticides applied for aphid control. In 1998, the cotton aphid was ranked the most important pest of cotton in the San Joaquin Valley, California (Williams, 1998); however, its status as a pest varies year by year. An added complication is the resistance aphid populations develop to insecticides (Grafton-Cardwell, 1991; Kerns and Gaylor, 1992; O'Brien *et al.*, 1992; Harris and Furr, 1993). Insecticides may reduce aphid numbers for only a few days, and in some cases may actually result in aphid population increases (Karner *et al.*, 1997). Insecticide application when aphid populations are not causing economic injury, when aphids are resistant, or when natural control will keep aphid numbers beneath the economic threshold, is an unnecessary expenditure of money and effort for cotton growers.

In 1989, cotton entomologists in the midsouthern USA observed major die-offs of cotton aphid populations, presumably caused by a pathogen. In 1990, the causal agent of the aphid declines observed in the midsouth was identified as the entomopathogenic fungus *Neozygites fresenii* (Steinkraus *et al.*, 1991). The importance of *N. fresenii* in the control of the cotton aphid has been subsequently well-documented (Wells *et al.*, 2000; Marti and Olson, 2006). In 1991, reports of epizootics in *A. gossypii* populations on cotton were reported from the USA (Steinkraus *et al.*, 1991) and Africa (Silvie and Papierok, 1991). These epizootics in cotton aphids on cotton may have been a new phenomenon; certainly there were no similar reports prior to 1991. The occurrence of epizootics in a serious pest (*A. gossypii*), that is difficult to control with insecticides, on a valuable crop (cotton) provided the impetus to develop means of utilizing natural control by this pathogen in cotton IPM. The following is an account of the development and implementation of *N. fresenii* into cotton IPM in the USA.

B *Pathogen biology*

Upon discovery that a pathogen is responsible for a high level of natural control, the first step is to conduct research on the biology and occurrence of the pathogen. Preliminary data on the effectiveness of the pathogen may help justify funding from various agencies. Funds are necessary for detailed studies. In this case, funding was initially obtained from the USDA-NRI and USDA Southern IPM competitive grants programs, from the Arkansas Agricultural Experiment Station, and from Cotton Incorporated (a cotton industry supported group). Cotton Incorporated has been instrumental in the success of this program by providing steady funding for 14 years.

Information on the pathogen's biology and interactions with environmental conditions are also needed to manipulate the pathogen, accurately diagnose infected hosts, sample for the pathogen, and develop methods to predict its occurrence. The entomopathogenic fungus, *N. fresenii*, infects many aphid species worldwide. It has been reported from Aphidinae, including at least 9 *Aphis* spp., and *Brevicoryne brassicae, Acyrthosiphum pisum, Myzus persicae, Schizaphis graminum, Rhopalosiphum padi*, and occasionally from certain species of Cinarinae and Chaitophorinae (Thoizon, 1970). This fungus infects only aphids and has no effect on humans, plants, or beneficial insects. Since 1991, many aspects of the biology, epizootiology, and aerobiology of this natural enemy of aphids have been elucidated (Steinkraus and Slaymaker, 1994; Steinkraus *et al.*, 1995; Vingaard *et al.*, 2003).

It is necessary to have viable pathogen material to study its biology. Unfortunately, *N. fresenii* has not yet been cultured *in vitro* in any practical way; therefore, methods were developed to culture the pathogen *in vivo* and store the fungus in desiccated, frozen, aphid mummies (Steinkraus *et al.*, 1993). This provided a source of *N. fresenii* material that permitted determination of the average number of primary conidia produced per host cadaver (3,052 conidia/aphid) and where conidia were dispersed after discharge (77% into the air). It also permitted determination of the time from adherence of capilliconidia to a host aphid till death, and the effect of temperature on this process. The time from host contact with an infective capilliconidium to host death and fungal sporulation was as short as 3 days at 30 °C; one reason why *N. fresenii* produces epizootics quickly.

Further biological studies were made on the pathogen to determine the temporal pattern of sporulation, conidial discharge, and formation of capilliconidia, and the effects of relative humidity and temperature on these processes (Steinkraus and Slaymaker, 1994). These studies indicated that sporulation occurred rapidly within a 5 h period and that temperatures above 35 °C or relative humidities below 85% prevented sporulation. Most primary conidia (93%) germinated to form capilliconidia within 6 h at 25 °C and 100% RH.

This fungus has efficient mechanisms for infecting aphids on plants, and for dispersal within and between fields as airborne conidia. Aerial conidia of *N. fresenii* were collected during epizootics in commercial cotton fields in Arkansas using Rotorod and Burkard spore samplers (Steinkraus *et al.*, 1996). Discharge of primary conidia showed a clear diel periodicity, with most conidia present in the air between 0100 and 0500 h. Therefore, primary conidia are dispersed through a cotton field when it is dark, relatively humid and cool, maximizing conidial longevity and survival. Forty-eight percent of sentinel *A. gossypii* exposed to air in a commercial cotton field in Louisiana during an epizootic became infected after 8 h exposure (Steinkraus *et al.*, 1999). Exposure of sentinel aphids outside the cotton field, at 10 and 100 m downwind, resulted in 34.8% and 24.0% infected aphids, respectively. Aerial primary conidial densities reached 90,437/m^3 at 0015 h on 2 July 1995. The extremely high numbers of conidia in the air and the precise timing of their discharge, play an important role in the rapid development of epizootics in *A. gossypii* populations.

Based on several years of sampling aphid populations throughout the midsouth and southeast, it was determined that aphid populations generally begin a precipitous decline when the prevalence level reaches 15% (Steinkraus and Hollingsworth, 1994; Hollingsworth *et al.*, 1995). Diagnostic procedures developed during these research projects made it possible to diagnose aphid fungus levels from individual fields and predict whether natural declines caused by the fungus will occur. This is practical information for consultants, growers, and extension agents making management decisions. In 1993, an extension-based aphid fungus sampling service was developed to provide Arkansas growers with this information. In the late 1990s the service was expanded to include Alabama, Florida, Georgia, Louisiana, Mississippi, North Carolina, and South Carolina. In 2006, the service completed its 14th year of operation.

C Diagnosis of diseased aphids

Preliminary work resulted in the development of methods for accurately diagnosing aphids for the pathogen. The cotton aphid is tiny and soft-bodied, making it possible to squash an entire aphid in lactophenol on a microscope slide (Steinkraus *et al.*, 1991). The slide is then scanned at 200x magnification with a phase microscope and the stages of *N. fresenii* are recorded. The following categories of infection status are readily observed: capilliconidia attached to the aphid, protoplasts, hyphal bodies or resting spores present within the aphid, and conidiophores and primary conidia present. This research and service has been greatly simplified by the fact that *A. gossypii* is generally the only aphid species regularly encountered in cotton, and *N. fresenii* is the most common pathogen regularly found infecting *A. gossypii* on cotton. The aphid pathogen, *P. neoaphidis*, is occasionally found in low levels in cotton aphids but is usually easily recognized and not involved in major epizootics in this system (Steinkraus, unpublished data). In systems, such as potato, which are attacked by multiple species of aphids that in turn are infected by many species of entomopathogenic fungi, sampling procedures will be more complicated (Shands *et al.*, 1972).

D Geographical and temporal prevalence

In 1992 and 1993, a study was made to determine how widespread *N. fresenii* was in the states of Arkansas, Louisiana, and Mississippi, when the epizootics occurred, and the speed with which the epizootics reduced aphid populations (Steinkraus *et al.*, 1995). This was done by collecting aphids in 32 (1992) and 35 commercial cotton fields (1993) along major north-south highways in the Mississippi flood plain in eastern Arkansas, western Mississippi, and northern Louisiana. Aphid-infested leaves were collected and stored in 70% ethanol. Randomly chosen subsamples of 30 aphids from each field were diagnosed for *N. fresenii*. This research showed that *N. fresenii* was widespread in Arkansas, Louisiana, and Mississippi fields. A similar study was conducted with cooperating entomologists in 10 cotton growing states in 1995 (Steinkraus *et al.*, 1996). Cooperators collected aphids from cotton fields throughout their states from a total of 47 counties. From each field, 50 randomly chosen aphids were diagnosed for *N. fresenii*. This study showed that *N. fresenii* was present in cotton fields in all 10 states, and in 66% of the samples received. These data indicated that epizootics occurred regularly over wide areas of the USA.

E Methods for predicting epizootics

Based on the understanding of the temporal and geographical occurrence of *N. fresenii* epizootics, prevalence was monitored in *A. gossypii* populations for 3 years to develop sampling strategies for predicting aphid population declines due to *N. fresenii* (Hollingsworth *et al.*, 1995). Regression analysis on average aphid densities per leaf and *N. fresenii* prevalence in 6 fields indicated that aphid populations began to decline when prevalence reached 15%, and declined to a low level 5–16 days later. Declines were more rapid in fields with higher aphid densities, and fungus-infected aphids could be detected 10 days before prevalences reached 15%. A sample of only 4–5 leaves was required to detect fungal-infected aphids when prevalence reached 4–5%.

F Sampling service to predict aphid epizootics

An important key to the success of this program has been developing links between research and cooperative extension personnel. In each state, a cotton extension entomologist or IPM specialist has the responsibility of selecting participants for the program and obtaining their addresses, and phone and FAX numbers. Based on the available laboratory resources of space, microscopes, and labor, and the need for participants in most cotton-growing counties in Alabama, Arkansas, Florida, Georgia, Louisiana, Mississippi, North

Carolina, and South Carolina, 20–30 participants per state were chosen. Participants have been county extension agents, private consultants, growers, or researchers who have responsibility for sampling cotton fields and making aphid management decisions. The number of participants varies each year based on how serious a pest cotton aphids were perceived to be the previous year.

In May, each participant is supplied with sampling kits, instructions on how to sample aphids from the fields, and data sheets. The kits consist of 30 ml vials containing 70% ethanol placed inside padded mailing tubes, and pre-addressed return envelopes. Participants sample aphids from their fields when aphids are perceived to be a problem, then mail them via a 2-day express mail service to our laboratory at the University of Arkansas for processing where laboratory technicians have been trained to mount and diagnose aphids for *N. fresenii*. Fifty aphids are randomly selected and diagnosed from each sample. Each aphid is examined for fungal infection at 200x magnification with a phase microscope. Results, expressed as percentage prevalence, are supplied to participants within 48 h by FAX or telephone. In addition, summaries of diagnostic results are faxed to state coordinators weekly for dissemination of the data within their respective states. In 1997, the service developed an Internet website (http://www.uark.edu/misc/aphid) containing all results. Results from each field are uploaded daily onto the Internet site, making them rapidly available to growers and other interested parties in the 8 state area. At the end of the 1997 season, participants were surveyed to determine the value of the service.

G *Results of the sampling service in 1997*

In 1997, 97 participants received kits and 64% of the participants sent samples to our laboratory. Samples were received from 54 counties: 13 in Arkansas, 11 in Louisiana, and 30 in Mississippi. A total of 469 samples were received; 162 from Arkansas, 109 from Louisiana, and 198 from Mississippi. A total of 13,880 ha were sampled; 3,835 in Arkansas, 3,117 in Louisiana, and 6,928 in Mississippi. Each sample took approximately 2 h to process. The estimated final cost per sample, including labor, slides, shipping charges, and kits, but excluding costs of phase and dissecting microscopes, was approximately $40.

No fungus was present in samples collected from any state in 1997 before 7 June. The first samples containing infected aphids were collected in Franklin parish, Louisiana, on 12 June (4%), Attala County, Mississippi, on 25 June (2%) and Chicot County, Arkansas, on 3 July (2%). The first samples containing 15% or more infected aphids were collected in Franklin parish, Louisiana on 23 June, in Leflore and Sunflower Counties, Mississippi, on 3 July, and in Ashley County, Arkansas, on 7 July. Infection levels of 15% are important because they are usually followed within a few days by aphid population declines. Therefore, the fungus provided natural control of cotton aphids in 1997 in early to mid-July. The timing of the epizootics has varied by 1–4 weeks in different years.

The first samples with infected aphids in Mississippi and Arkansas lagged behind those of Louisiana by ca. 2–3 weeks. This lag time in infection levels among states is indicative of a general south-to-north progression of the fungus which has been observed in previous years. Because epizootics occur first in Louisiana, it may be possible to use the sampling service to locate Louisiana fields in which there are ongoing early epizootics, then collect aphid-infested plants from these fields to inoculate northern Arkansas fields in order to initiate earlier aphid epizootics. This approach was attempted in 1999 with some success.

The range of infection levels and numbers of samples in Arkansas are shown in Figure 1. Daily variations in infection levels from field to field within the same county may be due in part to the effect of wind on conidial dispersal between fields. Sampling frequency and infection levels in Arkansas both peaked the week of 20 July. After this date, few samples were received because aphid populations had declined to negligible levels across the state due to *N. fresenii*. These findings demonstrate that *N. fresenii* usually is most prevalent in hot weather, unlike most fungal pathogens which infect aphids in cooler weather during the spring and fall.

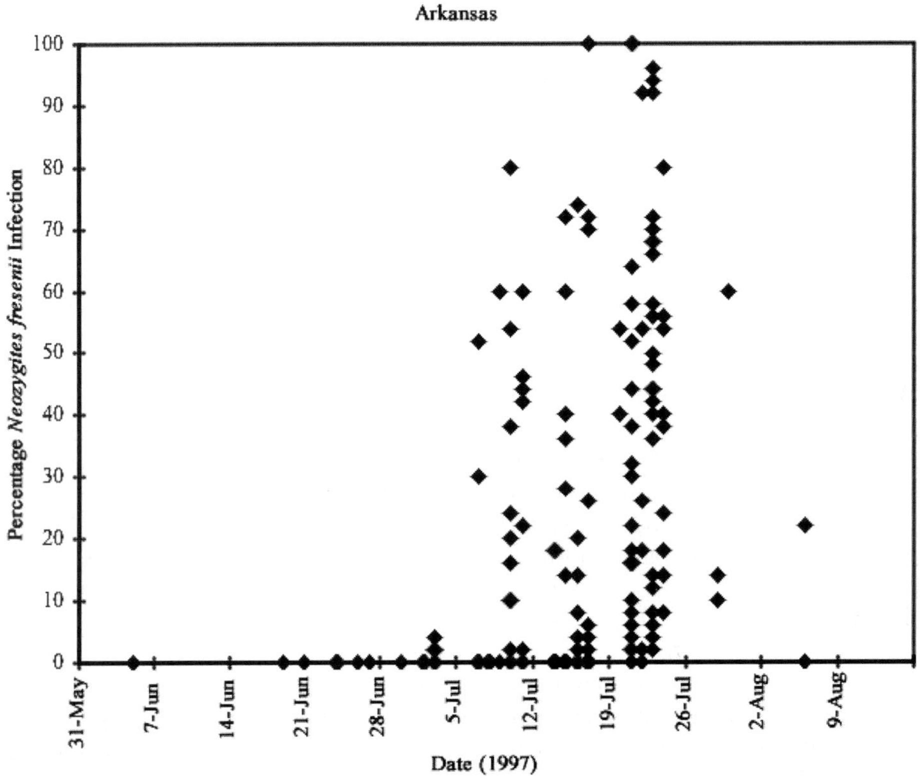

Figure 1. Percentage of aphids infected with *Neozygites fresenii* from Arkansas cotton aphid samples. Further graphs of results of the Cotton Aphid Fungus Sampling Service can be view at the internet website (http://www.uark.edu/misc/aphid)

Because of the time involved in collecting, shipping, and processing samples, it is crucial that samples be shipped in a timely manner in order for diagnostic results to be useful in management decisions, especially if an aphicide treatment is being considered. If there is no fungus present in a field being considered for an insecticide spray for aphids, then a treatment may be in order, though at the risk of reducing beneficial arthropods. On the other hand, if the fungus level in the field is 15% or greater, a treatment may be unnecessary because the aphid populations will decline naturally due to the fungus.

Seventy-one percent of the participants who sent samples in 1997 responded to a follow up survey. Ninety percent said that the service saved money because they had avoided aphid-insecticide treatments when the fungus was present and 98% said that they used service data in making aphid management decisions. Many positive comments were received during the post-season survey on the service. Eight representative comments are listed below .

1. *"This program was very helpful and saved thousands of dollars in insecticide costs."*
2. *"We had 5 growers prepared to spray ca. 2500 acres until the survey revealed that the fungus was present. At $7.50/acre, this was a significant savings."*
3. *"Without it I would have made a follow-up aphid spray. It saved the farmer money."*
4. *"Detection of fungal infestations before they are observed visually is extremely helpful."*
5. *"Speed with which you identified percentage aphids having fungus was excellent."*
6. *"Information gained from the service made decisions easier on many fields not sampled."*
7. *"Gave the grower and myself and other consultants a way to define what was happening in the field instead of just wondering."*
8. *"Participating helped my understanding of how/when the fungal disease works & how we can best fit reliance of this disease into our pest management program."*

6 Summary

Naturally occurring pathogens of arthropod pests in agroecosystems in some cases may be providing important control of arthropod pests. Identification of such situations may permit the development of sampling procedures for determining or predicting the natural control provided. Research in this area can result in increased reliance on natural enemies in the IPM programs on crops, with concomitant reductions in pesticide usage. These are important goals of IPM. Natural control by pathogens is considerable and often underestimated. As Waage (1992) stated, *"It is tempting to see classical biological control as an unusual, one-off event, but this misses an important point: there is no fundamental difference between these successes with exotic natural enemies and the action of indigenous species in our local crops. The striking before and after picture of classical biological control simply isolates a process which is going on around us all the time, and largely undetected. What is revealed is an ecological phenomenon which is relevant to all pest management."*

7 References

Allen, G. E., Greene, G. L., and Whitcomb, W. H. 1971. An epizootic of *Spicaria rileyi* on the velvetbean caterpillar, *Anticarsia gemmatalis*, in Florida. *Fla. Entomol.* **54**, 189–191.

Boykin, L. S., Campbell, W. V., and Beute, M. K. 1984. Effect of pesticides on *Neozygites floridana* (Entomophthorales: Entomophthoraceae) and arthropod predators attacking the twospotted spider mite (Acari: Tetranychidae) in North Carolina peanut fields. *J. Econ. Entomol.* **77**, 969–975.

Brandenburg, R. L. and Kennedy, G. G. 1983. Interactive effects of selected pesticides on the two-spotted spider mite and its fungal pathogen, *Neozygites floridana*. *Entomol. Exp. Appl.* **34**, 240–244.

Cabanillas, H. E., Poinar, G. O., and Raulston, J. R. 1994. *Steinernema riobravis* n. sp. (Rhabditida: Steinernematidae) from Texas. *Fund. Appl. Nematol.* **17**, 123–131.

Carner, G. R. 1976. A description of the life cycle of *Entomophthora* sp. in the twospotted spider mite. *J. Invertebr. Pathol.* **28**, 245–254.

Carner, G. R. and Canerday, T. D. 1968. Field and laboratory investigations with *Entomophthora fresenii*, a pathogen of *Tetranychus* spp. *J. Econ. Entomol.* **61**, 956–959.

Cook, H. T. 1949. Forecasting late blight epiphytotics of potatoes and tomatoes. *J. Agr. Res.* **78**, 545–563.

DeGooyer, T. A., Pedigo, L. P., and Giles, K. L. 1995. Population dynamics of the alfalfa weevil (Coleoptera: Curculionidae) in Central and Southern Iowa. *J. Kansas Entomol. Soc.* **68**, 268–278.

Feng, M. G., Chen, C., and Chen, B. 2004. Wide dispersal of aphid-pathogenic Entomophthorales among aphids relies upon migratory alates. *Environ. Microbiol.* **6**, 510–516.

Funderburk, J. E. 1993. Velvetbean caterpillar. *In* "Handbook of Soybean Insect Pests" (L. G. Higley and D. J. Boethel, Eds.), pp. 95–97. Entomol. Soc. Am., Lanham, Maryland.

Fuxa, J. R. and Tanada, Y. (Eds.) 1987. "Epizootiology of Insect Diseases." Wiley and Sons, New York.

Godfrey, L. D., Fuson, K. J., and Wood, J. P. 1997. Physiological and yield responses of cotton to mid-season cotton aphid infestations in California. *Proc. Beltwide Cotton Conf.*, pp. 1048–1051.

Grafton-Cardwell, E. E. 1991. Geographical and temporal variation in response to insecticides in various life stages of *Aphis gossypii* (Homoptera: Aphididae) infesting cotton in California. *J. Econ. Entomol.* **84**, 741–749.

Harcourt, D. G., Guppy, J. C., and Binns, M. R. 1984. Analysis of numerical change in sub economic populations of the alfalfa weevil, *Hypera postica* (Coleoptera: Curculionidae), in Eastern Ontario. *Environ. Entomol.* **13**, 1627–1633.

Harcourt, D. G., Guppy, J. C., and Tyrrell, D. 1990. Phenology of the fungal pathogen *Zoophthora phytonomi* in Southern Ontario populations of the alfalfa weevil (Coleoptera: Curculionidae). *Environ. Entomol.* **19**, 612–617.

Harcourt, D. G., Mukerji, M. K., and Guppy, J. C. 1974. Estimation of egg populations of the alfalfa weevil, *Hypera postica* (Coleoptera: Curculionidae). *Can. Entomol.* **106**, 337–347.

Hardee, D. D. and Adams, L. C. 1998. Influence of timing of sprays for cotton aphid (Homoptera: Aphididae) on cotton yield. *Proc. Beltwide Cotton Conf.*, pp. 1061–1064.

Harris, F. A. and Furr, R. E., Jr. 1993. Cotton aphicide resistance phenomenon in the Mississippi Delta. *Proc. Beltwide Cotton Conf.*, pp. 961–963.

Hassell, M. P. 1978. "The Dynamics of Arthropod Predator-Prey Systems." Princeton Univ. Press, Princeton, New Jersey.

Hollingsworth, R. G., Steinkraus, D. C., and McNew, R. W. 1995. Sampling to predict fungal epizootics on cotton aphids (Homoptera: Aphididae). *Environ. Entomol.* **24**, 1414–1421.

Isely, D. 1946. The cotton aphid. *Arkansas Agr. Exp. Sta. Bull.* **462**, 4–29.

Karner, M. A., Goodson, J. R., and Payton, M. 1997. Efficacy of various insecticide treatments to control cotton aphids and prevent economic loss in Oklahoma in 1995. *Proc. Beltwide Cotton Conf.*, pp. 1054–1057.

Kerns, D. L. and Gaylor, M. J. 1992. Insecticide resistance in field populations of the cotton aphid (Homoptera: Aphididae). *J. Econ. Entomol.* **85**, 1–8.

King, A. B. S. 1994. *Heliothis/Helicoverpa* (Lepidoptera: Noctuidae). *In* "Insect Pests of Cotton". (G. A. Matthews and J. P. Tunstall, Eds.), pp. 39–106. CAB International, Cambridge.

Kish, L. P. and Allen, G. E. 1978. The biology and ecology of *Nomuraea rileyi* and a program for predicting its incidence on *Anticarsia gemmatalis* in soybean. *Univ. of Fla. Agric. Exp. Stat. Tech. Bull.* **795**.

Klubertanz, T. H., Pedigo, L. P., and Carlson, R. E. 1991. Impact of fungal epizootics on the biology and management of the twospotted spider mite (Acari: Tetranychidae) in soybean. *Environ. Entomol.* **20**, 731–735.

Kring, T. J., Ruberson, J. R., Steinkraus, D. C., and Jacobson, D. A. 1993. Mortality of *Helicoverpa zea* (Lepidoptera: Noctuidae) pupae in ear-stage field corn. *Environ. Entomol.* **22**, 1338–1343.

Lacey, L. A. 1997. "Manual of Techniques in Insect Pathology." Academic Press, London.

Lacey, L. A. and Brooks, W. M. 1997. Initial handling and diagnosis of diseased insects. *In* "Manual of Techniques in Insect Pathology" (L. A. Lacey, Ed.), pp. 1–15. Academic Press, London.

Leland, J. E. and Snodgrass, G. L. 2004. Prevalence of naturally occurring *Beauveria bassiana* in *Lygus lineolaris* (Heteroptera: Miridae) populations from wild host plants of Mississippi. *J. Agri. Urban Entomol.* **23**, 157–163.

Lentz, G. and Austin, N. 1998. Impact of cotton aphid on lint yield. *Proc. Beltwide Cotton Conf.*, pp. 1058.

Loehle, C. 1995. Social barriers to pathogen transmission in wild animal populations. *Ecology* **76**, 326–335.

Marti, O. G. and Olson, D. M. 2006. *Neozygites fresenii* (Entomophthorales: Neozygitaceae) prevalence in *Aphis gossypii* (Homoptera: Aphididae) in three south central Georgia cotton fields. *J. Entomol. Sci.* **41**, 22–32.

May, R. M. 1988. Conservation and disease. *Conservation Biol.* **2**, 28–30.

McGuire, M. R. 2003. Impact of naturally occurring *Beauveria bassiana* on *Lygus hesperus* populations. *Proc. Beltwide Cotton Conf.*, pp. 980–984.

Morris, M. J., Roberts, S. J., Maddox, J. V., and Armbrust, E. J. 1996. Epizootiology of the fungal pathogen, *Zoophthora phytonomi* (Zygomycetes: Entomophthorales) in field populations of alfalfa weevil (Coleoptera: Curculionidae) larvae in Illinois. *Great Lakes Entomol.* **29**, 129–140.

Moscardi, F. 1999. Assessment of the application of baculoviruses for control of Lepidoptera. *Annu. Rev. Entomol.* **44**, 257–289.

Nielsen, C. and Hajek, A. E. 2005. Control of invasive soybean aphid, *Aphis glycines* (Hemiptera: Aphididae), populations by existing natural enemies in New York State, with emphasis on entomopathogenic fungi. *Environ. Entomol.* **34**, 1036–1047.

O'Brien, P. J., Abdel-all, Y. A., Otten, J. A., and Graves, J. B. 1992. Relationship of insecticide resistance to carboxylesterases in *Aphis gossypii* (Homoptera: Aphididae) from midsouth cotton. *J. Econ. Entomol.* **85**, 651–657.

Patch, E. M. 1907. The potato plant louse. *Maine Agr. Exp. Sta. Bull.* **147**, 235–257.

Poinar, G. O., Jr. 1990. Taxonomy and biology of Steinernematidae and Heterorhabditidae. *In* "Entomopathogenic Nematodes in Biological Control" (R. Gaugler and H. K. Kaya, Eds.), pp. 23–61. CRC Press, Boca Raton, FL.

Raulston, J. R., Pair, S. D., Loera, J., and Cabanillas, H. E. 1992. Prepupal and pupal parasitism of *Helicoverpa zea* and *Spodoptera frugiperda* (Lepidoptera: Noctuidae) by *Steinernema* sp. in cornfields in Lower Rio Grande Valley. *J. Econ. Entomol.* **85**, 1666–1670.

Ruano-Rossil, J. M., Radcliffe, E. B., and Ragsdale, D. W. 2002. Disruption of entomopathogenic fungi of green peach aphid, *Myzus persicae* (Sulzer), by fungicides used to control potato late blight. *In* "Aphids in a New Millennium" (J. C. Simon, C. A. Dedryver, C. Rispe, and M. Hullé, Eds.), pp. 365–370. Institut National de la Recherche Agronomique, Paris.

Semel, M. 1956. Polyhedrosis wilt of cabbage looper on Long Island. *J. Econ. Entomol.* **49**, 420–421.

Shands, W. A., Simpson, G. W., Hall, I. M., and Gordon, C. C. 1972. Further evaluation of entomogenous fungi as a biological agent of aphid control in northeastern Maine. *Maine Agr. Exp. Sta. Tech. Bull.* **58**.

Silvie, P. and Papierok, B. 1991. Les ennemis naturels d'insectes du cotonnier au Tschad: premieres donnÈes sur les champignons de l'ordre des Entomophthorales. *Coton Fibres Trop.* **46**, 293–308.

Smitley, D. R., Kennedy, G. G., and Brooks, W. M . 1986. Role of the entomogenous fungus, *Neozygites floridana*, in population declines of the twospotted spider mite, *Tetranychus urticae*, on field corn. *Entomol. Exp. Appl.* **41**, 255–264.

Speare, A. T. 1922. Natural control of the citrus mealy-bug in Florida. *U. S. D. A. Bull.* **117**, 19 pp.

Stern, V. M., Smith, R. F., van den Bosch, R., and Hagen, K. S. 1959. The integrated control concept. *Hilgardia* **29**, 81–101.

Steinhaus, E. A. 1949. "Principles of Insect Pathology." McGraw-Hill, New York.

Steinkraus, D. C., Kring, T. J., and Tugwell, N. P. 1991. *Neozygites fresenii* in *Aphis gossypii* on cotton. *Southwest. Entomol.* **16**, 118–122.

Steinkraus, D. C., Boys, G. O., and Slaymaker, P. H. 1993. Culture, storage, and incubation period of *Neozygites fresenii* (Entomophthorales: Neozygitaceae) a pathogen of the cotton aphid. *Southwest. Entomol.* **18**, 197–202.

Steinkraus, D. C. and Hollingsworth, R. G. 1994. Predicting fungal epizootics on cotton aphids. *Ark. Farm Res.* **43**, 10–11.

Steinkraus, D. C., Hollingsworth, R. G., and Boys, G. O. 1996. Aerial spores of *Neozygites fresenii* (Entomophthorales: Neozygitaceae): density, periodicity, and potential role in cotton aphid (Homoptera: Aphididae) epizootics. *Environ. Entomol.* **25**, 48–57.

Steinkraus, D. C., Howard, M. N., Hollingsworth, R. G., and Boys, G. O. 1999. Infection of sentinel cotton aphids (Homoptera: Aphididae) by aerial conidia of *Neozygites fresenii* (Entomophthorales: Neozygitaceae). *Biol. Control* **14**, 181–185.

Steinkraus, D. C., Hollingsworth, R. G., and Slaymaker, P. H. 1995. Prevalence of *Neozygites fresenii* (Entomophthorales: Neozygitaceae) on cotton aphids (Homoptera: Aphididae) in Arkansas cotton. *Environ. Entomol.* **24**, 465–474.

Steinkraus, D. C. and Slaymaker, P. H. 1994. Effect of temperature and humidity on sporulation and conidia of *Neozygites fresenii* (Zygomycetes: Neozygitaceae) from *Aphis gossypii* (Homoptera: Aphididae). *J. Invertebr. Pathol.* **64**, 130–137.

Steinkraus, D. C. and Tugwell, N. P. 1997. *Beauveria bassiana* (Deuteromycotina: Moniliales) effects on *Lygus lineolaris* (Hemiptera: Miridae). *J. Entomol. Sci.* **32**, 79–90.

Steinkraus, D. C. and Boys, G. O. 2005. Mass harvesting of the entomopathogenic fungus, *Neozygites fresenii*, from natural field epizootics in the cotton aphid, *Aphis gossypii. J. Invertebr. Pathol.* **88**, 212–217.

Thoizon, G. 1970. Spècificité du parasitisme des aphides par les Entomophthorales. *Ann. Soc. Entomol. France*, N. S. **6**, 517–562.

Waage, J. K. 1992. Biological control in the year 2000. *In* "Pest Management and the Environment in 2000" (A. A. S. A. Kadir and H. S. Barlow, Eds.), pp. 329–340. CAB International, Wallingford, UK.

Waggoner, P. E. and Horsfall, J. G. 1969. EPIDEM – A simulator of plant disease written for a computer. *Bull. Conn. Agr. Exp. Sta.* New Haven.

Wells, M. L., McPherson, R. M., Ruberson, J. R., and Herzog, G. A. 2000. Effect of fungicide application on activity of *Neozygites fresenii* (Entomophthorales: Neozygitaceae) and cotton aphid (Homoptera: Aphididae) suppression. *J. Econ. Entomol.* **93**, 1118–1126.

Whitten, M. J. 1992. Pest management in 2000: what we might learn from the Twentieth Century. *In* "Pest Management and the Environment in 2000" (A. A. S. A. Kadir and H. S. Barlow, Eds.), pp. 9–44. CAB. International, Wallingford, UK.

Williams, M. R. 1998. Cotton insect losses - 1997 compiled for National Cotton Council. *Proc. Beltwide Cotton Conf.*, pp. 904–925.

Van der Plank, J. E. 1963. "Plant Diseases: Epidemics and Control." Academic Press, New York.

Vingaard, M. G., Steinkraus, D. C., Boys, G. O., and Eilenberg, J. 2003. Effects of long-term storage at −14 °C on the survival of *Neozygites fresenii* (Entomophthorales: Neozygitaceae) in cotton aphids (Homoptera: Aphididae). *J. Invertebr. Pathol.* **82**, 97–102.

Chapter V-2

Assessing impact of naturally occurring pathogens of forest insects

Joseph S. Elkinton
Division of Entomology in the Department of Plant, Soil & Insect Sciences
University of Massachusetts
Amherst, MA 01003, USA
and
Graduate Program in Organismic and Evolutionary Biology
University of Massachusetts
Amherst, MA 01003, USA

John Burand
Division of Entomology in the Department of Plant, Soil & Insect Sciences
University of Massachusetts
Amherst, MA 01003, USA

1 Introduction

Pathogens have long been known to play a major role in the population dynamics of many important forest insects. For many irruptive species, outbreaks are terminated by baculovirus epizootics that cause dramatic declines of host density. Such epizootics are well known for Lepidoptera such as gypsy moth, *Lymantria dispar*; Douglas fir tussock moth, *Orgyia pseudotsugata*; nun moth, *L. monacha*; pine beauty moth, *Panolis flammea*; forest tent caterpillar, *Malacosoma disstria*; western tent caterpillar, *M. californicum pluviale*, and the European larch budmoth, *Zeiraphera diniana*. Similar epizootics are known in sawflies (Hymenoptera: Diprionidae) including the European pine sawfly, *Neodiprion sertifer*; the European spruce sawfly, *Gilpinia hercyniae* and the red-headed pine sawfly, *Neodiprion lecontei*.

The role of disease organisms in low density populations is less well documented. Indeed, the dynamics of most forest insects are poorly understood. For many forest Lepidoptera, predators or parasitoids are thought to regulate low density populations, but for most species this has not been established conclusively. Even for such well-studied species as the eastern spruce budworm, *Choristoneura fumiferana*, there exists little consensus regarding the factors responsible for causing outbreaks or for maintaining populations at low density for the long periods between outbreaks. For example, in this system, Royama (1984, 1992) discounted the long-held view proposed by Morris (1963) that spruce budworms were maintained at a low density equilibrium by the action of vertebrate predators, and that outbreaks resulted from a combination of host tree effects and favorable weather conditions that caused high larval survival allowing the populations to escape from the equilibrium. Instead, Royama (1984, 1992) presented evidence that changes in budworm density were caused by decade-long oscillations in mortality from a

L.A. Lacey and H.K. Kaya (eds.), *Field Manual of Techniques in Invertebrate Pathology*, 283–296.

complex of natural enemies which include a microsporidium (*Nosema fumiferanae*). Work by Myers (Rothman and Myers, 1994; Myers and Kukan, 1995; Myers, 2000) suggests sublethal effects of viral infection or even latent virus infections (Cooper *et al*., 2003) may influence population dynamics of tent caterpillars. Maddox (1987) summarizes evidence of a gradual, several-year increase in *N. fumiferanae* prevalence associated with declines in spruce budworm populations. Diseases such as *Nosema* produce chronic infections which often do not kill the host. The impact of this agent in combination with other sources of budworm mortality is thus very hard to quantify and, as a result, spruce budworm dynamics remain an enigma.

Applied insect pathology has focused on efforts to develop pathogens as biopesticides. There are relatively few examples of successful classical biological control, where pathogens have been introduced and have caused a measurable long-term change in the density of the target organism. Several exceptions involve accidental introductions of pathogens: The European spruce sawfly *G. hercyniae* was introduced into Canada around 1920 and caused widespread defoliation in Quebec in the 1930s (Bird and Elgee, 1957). Several parasitoid species were introduced, accompanied inadvertently by a baculovirus that became the agent primarily responsible for the decline of this species to densities that have remained low ever since. For gypsy moth, there are two major pathogens that play a critical role. The multiple nucleopolyhedrovirus of *L. dispar* (*Ld*MNPV) was evidently introduced into North America around 1900 from an unknown source and was not previously observed in the years following the introduction of gypsy moth in 1868 (Glaser, 1915). This agent is largely responsible for the collapse of high density populations (Doane, 1970). In 1989, *Entomophaga maimaiga*, a fungal pathogen of the gypsy moth known from Japan, appeared unexpectedly in North America (Andreadis and Weseloh, 1990, Hajek *et al*., 1990) and decimated both low and high density populations. The capacity of this pathogen to cause high mortality in low density populations means that it has the ability to prevent outbreaks from getting started. It is too soon to say what role this pathogen will play in future gypsy moth dynamics in North America, but it has evidently prevented several gypsy moth outbreaks in the northeast.

2 Detection of pathogens

Field studies of insect pathogens depend upon adequate techniques to detect the pathogen either within the host, or more rarely, on substrates outside of the host. For most field studies, the first step of analysis is collection and rearing of hosts and examination of cadavers under a microscope. When the cause of death is an unknown pathogen, microscopy is usually the only option. Even when more sophisticated molecular techniques are available, microscopy may remain the most efficient technique for processing large numbers of field-collected samples. In recent years, several molecular techniques have been used to detect the presence of known pathogens in field-collected insects.

Immunoassays have been used to detect insect pathogens for many years. Antibodies to insect pathogen-derived proteins are produced in laboratory mammals and then used to detect the presence of the pathogen in extracts from test insects. Modern enhancements of this approach include enzyme-linked immunosorbent assay (ELISA) in which the antigen is bonded with an enzyme that gives a color reaction when extracts containing the virus are tested. For forest insects ELISA has been developed for two pathogens of gypsy moth: *Ld*MNPV (Ma *et al*., 1984) and *E. maimaiga* (Hajek *et al*., 1991a).

DNA hybridization and related techniques provide a reliable and inexpensive way of detecting the presence of particular pathogens within a host. These methods have, to a large extent, supplanted immunoassays due to their specificity and ease of preparation. The methods are easily applied to projects which involve screening large numbers of field-collected individuals. The process involves developing a DNA probe, all or a portion of the pathogen DNA to which a radioactive or colorimetric label is attached. Extracts of test insects are blotted onto nitrocellulose or nylon filters and exposed to the DNA probe which binds to specific sequences of the pathogen DNA on the filter. For example, Keating *et al*. (1991) used this technique to detect gypsy moth *Ld*MNPV in field populations, and the results were compared with traditional microscopic methods (Figure 1). Similar assays have been developed by Kukan and Myers (1995) for *Mp*NPV in western tent caterpillar (*M. californicum pluviale)* and Kaupp and Ebling (1993) for baculoviruses in gypsy moth and spruce budworm.

Restriction enzyme analysis is another molecular tool that has potential applications to field studies. With this technique, pathogen DNA is digested by one or more restriction enzymes that cut the DNA into fragments at specific nucleotide sequences. The resulting fragments are then electrophoretically separated in an agarose gel and the DNA in the gel is either stained directly or blotted onto nitrocellulose filters and probed, as described above. The restriction enzymes cut the DNA from different clones or strains of the same species into fragments of different lengths, which show up as different bands on the filter. For example, in field trials involving application of a baculovirus to a target population, this technique could be used to prove that larval cadavers succumbed to the virus that was applied rather than naturally occurring virus, assuming the applied virus had a unique restriction enzyme banding pattern. Ilyinykh and Chuikova (1989) used this technique to identify different isolates of the nucleopolyhedrovirus of the nun moth, *L. monacha*, associated with different phases of the outbreak cycle. Hajek *et al.* (1990) used this technique to prove that *E. maimaiga* discovered on gypsy moth in North America was identical to the Japanese form of this disease and different from morphologically identical isolates of *Entomophaga aulicae* known to attack other Lepidoptera in North America. She and her colleagues used the same technique to prove that other species of forest Lepidoptera in North America were infected by strains of *E. aulicae* and not *E. maimaiga* (Hajek *et al.*, 1991b).

Polymerase chain reaction (PCR) is a technique for amplification of minute quantities of DNA and, as such, offers unprecedented sensitivity for detecting pathogen DNA within hosts or outside of hosts on environmental substrates. The technique requires a sequenced DNA template that is unique to the target pathogen.

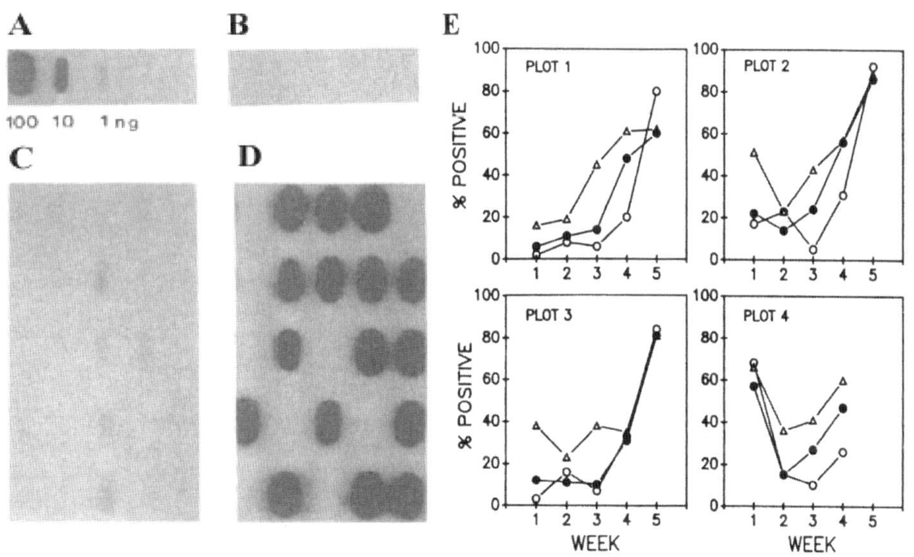

Figure 1. Autoradiographs of slot-blot hybridization of larval gypsy moth DNA extracts on nitrocellulose filters probed with *Ld*MNPV viral DNA and developed after 16 h of exposure (Keating *et al.*, 1989). A) Viral DNA standards: 1, 10, and 100ng of DNA per slot. B) Uninfected control larval extracts. C). Extracts from larvae fed droplets containing 10^4 polyhedral occlusion bodies (OBs)/ml and frozen 2 days post-inoculation. D) Extracts from larvae fed droplets containing 10^4 OBs/ml and frozen 4 days post-inoculation E) Percentage of gypsy moth larvae that contained *Ld*MNPV DNA (O) upon collection each week from 3 control plots (1,2,3) and one plot (4) that had received an aerial application of *Ld*MNPV (Gypchek) compared with the percentage of larvae from the same collection that died from *Ld*MNPV (determined by microscopic examination) after rearing on artificial diet for 6 (•) or 13 (♦) days (Keating *et al.*, 1991)

Although common in laboratory studies, and widely used in field studies of plant pathogens (Henson and French, 1993), there have been relatively few applications of this technique to field studies of forest insect pathogens. D'Amico *et al.* (1999) used PCR to detect the presence of a baculovirus engineered with a *Lac-Z* gene marker in larvae collected from gypsy moth populations in which the engineered virus had been released. The original plans to use a colormetric X-gal assay for the *Lac-Z* gene was compromised by the occurrence of false positives among uninfected larvae caused by the presence of a braconid parasitoid that responded to the *Lac-Z* assay. To solve this problem PCR primers were constructed that would bind to the *Lac* gene and other primers for baculovirus DNA. Both primers together (Figure 2) were used to confirm the presence of the engineered virus in the population.

The extreme sensitivity of PCR is both a strength and a shortcoming. We can use it to detect more minute quantities of pathogen than with any other method. However, it is often difficult to prevent false positives, that is to prove that a positive reading with PCR is not caused by inadvertent contamination of field-collected samples with the target DNA.

Analysis of other substrates for the presence of pathogens is usually done by way of bioassay. Test insects are exposed to the substrate or to substrate extracts and the proportion dying from the pathogen are assessed by one of the detection techniques described above. Concentrations of pathogen outside the host are frequently too low to detect with immunoassay or DNA hybridization directly, but PCR opens the door for detection of minute quantities that might greatly enhance our abilities to detect and quantify particular pathogens in the environment. These techniques, however, will not indicate whether the pathogen is alive or infective; only bioassay can do that. The problem of low concentrations can be partially circumvented if the environmental samples can be processed and the pathogen concentrated prior to bioassay. For example, Podgwaite *et al.* (1979) used bioassay to measure the concentration of gypsy moth *Ld*MNPV on extracts of bark and soil by rinsing the substrates in water and concentrating the pathogen via centrifugation. Carruthers *et al.* (1988) developed a procedure for bioassaying pine beauty moth (*P. flammea*) NPV on pine foliage. Such procedures would only work for pathogen that can survive the extraction process. These bioassay techniques have recently been combined with PCR to evaluate the presence of viable engineered *Ld*MNPV in soil and tree bark collected from a site where the virus had been released in a population of gypsy moths (D'Amico *et al.*, 1999).

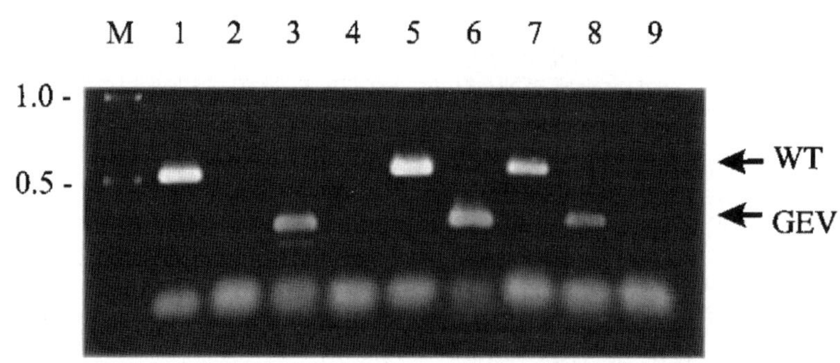

Figure 2. PCR amplification products obtained using primers for wild-type (WT) *Ld*MNPV and the genetically engineered virus (GEV) containing *Lac-Z* gene among gypsy moth larvae collected from a field test of the recombinant virus (D'Amico *et al.* 1999). Lanes 1, 2, 5, and 7, products obtained with wild-type virus primers. Lanes 3, 4, 6, and 8, products obtained with *Lac-Z* gene specific primers. Lanes 1 and 4, wild-type virus; lanes 2 and 3, *Lac-Z* gene recombinant virus; lane 9, DNA-minus control. Lane labeled M contains a molecular weight marker (kb ladder). The 525 bp fragment obtained from wild-type virus and the 302 bp fragment generated from the *Lac-Z* gene containing recombinant virus are indicated by arrows. Only field-collected samples positive for both markers (lanes 5 and 6, sample L2; lanes 7 and 8, sample L4) were considered positive for GEV

3 Measuring impact on forest insect populations

A *Quantifying prevalence and stage specific mortality*

Studies of insect pathogens in the field usually involve collecting hosts from the populations at regular intervals and then assaying or rearing them to determine pathogen prevalence; *i.e.*, the proportion of hosts infected in the sample. Reports of disease prevalence must take account of a variety of factors that might cause biases in the estimated proportion infected. The most obvious of these is that the sample of insects collected from the field may not be representative of the entire population (Fuxa and Tanada, 1987). For example, it is well known that many pathogens cause their hosts to congregate at specific habitat locations such as the tops of trees (Murray and Elkinton, 1992) or the base of trees (Hajek and Soper, 1991). Such behaviors can produce strongly biased estimates of prevalence unless the sampling regime is specifically designed to account for it. Murray and Elkinton (1992), for example, reported that infected larvae climbed to the tops of trees only just before death, so that biased estimates of prevalence could be avoided if larvae that died within one day of collection were excluded from the sample.

Estimates of prevalence obtained from samples collected at one moment in time rarely if ever generate adequate estimates of the total impact or mortality caused by a pathogen over a host generation (Van Driesche, 1983). Prevalence of most diseases changes dramatically as an insect matures (*e.g.*, Woods and Elkinton, 1987). Differences in disease prevalence between populations or treatments may be caused by differences in rates of host development. To estimate impact, one must collect hosts repeatedly as they mature and use some method to compute total disease mortality or prevalence over a stage (*e.g.*, the larval stage). As emphasized by Van Driesche (1983), this is not accomplished by dividing the total number of infected larvae by the total number of larvae collected over a series of sample occasions. Several methods exist to calculate total mortality over a stage and their use depends on the biological details of the system. For many univoltine forest insects, which are nearly all present in a single stage at a given moment, it is valid to collect and rear developing hosts on successive intervals (*e.g.*, weeks) and to compute cumulative mortality C over the entire period consisting of n intervals $C = 1\text{-}S_1S_2S_3 \ldots S_n$ where S_i is the proportion surviving (1- proportion dying) during each interval. Hajek *et al.* (1990), and Woods *et al.* (1991) provide examples of such calculations. Complications with these calculations arise when the pathogen is one of several contemporaneous causes of mortality; for a discussion of this issue and solutions see Elkinton *et al.* (1992).

Additional complications occur when egg hatch or host development is spread out temporally such that several life stages occur simultaneously in the same population (Van Driesche, 1983). Similar complications occur if there are disease-induced effects on larval development. For example, some baculoviruses contain genes that cause delays in host development (O'Reilly and Miller, 1989; Burand and Park, 1992). As a result, it is possible that samples of particular stages collected at a particular time in the season might contain only infected individuals, because uninfected individuals have already graduated to the next stage or instar. The prevalence of the pathogen in the sample would be close to 100%, although the proportion of the life stage that was infected might be quite small. For a thorough discussion of this issue and its possible solutions see Bellows *et al.* (1992).

The calculations described above apply only to diseases that are invariably fatal, so that measures of prevalence can be equated with mortality. Pathogens such as microsporidia produce chronic, non-fatal infections that can occur with high prevalence in many forest insects (Maddox, 1987). They may affect fecundity more than they cause mortality and their primary effect on mortality may be to predispose larvae to other causes of death. The impact of such agents is extremely difficult to quantify. Even baculoviruses that normally kill their hosts may also affect fecundity of adults that survive larval infections (Rothman and Myers, 1994).

In some studies, hosts are deployed in the field for specific intervals, and the proportion of new infections (disease incidence) is measured. These measures are especially useful when the data are to be used to validate host pathogen models as discussed below. For example, Weseloh and Andreadis (1992) deployed gypsy moth larvae in wire-mesh cages to measure incidence of *E. maimaiga* infections. Van Driesche and Bellows (1988) advocated a method for calculating stage specific mortality based on successive measurements of disease incidence (recruitment) in a population. Their method circumvented some of the biases mentioned above that arise when host stages overlap.

B Analysis of mortality data

Once adequate measures of stage-specific mortality are obtained, further analyses are possible. Ecologists frequently summarize such measures in life tables wherein the proportion dying and causes of death are reported for successive instars or life stages. With such data the magnitude of the mortality caused by the disease can be compared to other sources of mortality affecting the host. The effects of factors including pathogens that influence fecundity may also be included in life tables. If life-table data are collected from the same populations for successive generations, several analytic techniques are available to answer two specific questions, namely: are populations stabilized by negative feedback mechanisms and what are the causes of density change? Stabilization of population density or negative feedbacks on population growth entail the action of density-dependent factors whose impact changes proportionately with population density. Mortality from pathogens is inherently density dependent because pathogen transmission is more likely when host densities are high. However, stability and density dependence may also involve factors such as microsporidia that affect fecundity. Detection of density dependence in population systems has been remarkably elusive, mainly because appropriate statistical analyses have been found wanting and data collected for a sufficient number of host generations are rare (Turchin, 1995).

The factors responsible for stabilizing population densities may be different from those that are responsible for fluctuations in density or the onset of outbreaks. For example, the classic study of winter moth dynamics in Great Britain by Varley *et al.* (1973) concluded that soil-dwelling predators of the pupal stage were responsible for stabilizing the population density, whereas factors influencing over-wintering survival of eggs and larvae are responsible for most of the observed fluctuations in density. These authors developed a graphical technique known as key factor analysis to identify which factors were most responsible for changes in density. Subsequent work has extended this approach (Podoler and Rogers, 1975) and offered statistical methods to detect key factors (Manly, 1977), but others have identified the statistical pitfalls associated with these techniques (Kuno, 1971; Royama, 1996).

These general techniques only go so far in elucidating the impact of a pathogen on its host. To go further some sort of model of a population system must be constructed. Many such models have been developed in the past few decades and in the following section, we give a brief overview.

4 Modeling host-pathogen dynamics

A Simple theoretical models

Anderson and May (1979, 1980, 1981) introduced a family of simple models of insect host/pathogen dynamics that became the basis of a large subsequent literature. Their models were derived from earlier models of human diseases (Kermack and McKendrick, 1927). These earlier models simulated the course of an epidemic in a human population which was typically assumed to have constant density. Anderson and May added terms to the models to represent host population growth. In this way the models could be used to explore the effects of the pathogen on host population density. They also added terms to represent free-living pathogen stages outside of the host, a condition that applies to many pathogens, including nucleopolyhedroviruses. They explored a number of alternate forms of their model, but the one most widely used in a forest insect/baculovirus context (Dwyer, 1991) is shown below (Figure 3).

Anderson and May applied their models to the European larch budmoth (*Zeiraphera diniana*), an insect that periodically defoliates larch forests

$$\frac{dS}{dt} = r(S + I) - bS - \nu SP$$

rate of change of susceptibles = reproduction - non-disease deaths - transmission

$$\frac{dI}{dt} = \nu SP - (\alpha + b)I$$

rate of change of infecteds = transmission - deaths of infecteds

$$\frac{dP}{dt} = \lambda I - \mu P - \nu(S + I)P$$

rate of change of pathogens in environment = release from infecteds - pathogen decay - consumption of pathogens by hosts

Figure 3. Host pathogen model G of Anderson and May (1981) following the notation used by Dwyer (1991). Here S is the density (or number) of susceptible hosts, I is the density of infected hosts, P is the density of free-living pathogens outside the host. The model expresses the instantaneous rates of change of these three variables. The per capita rate parameters are as follows: ν is the transmission constant (essentially the encounter rate of host and pathogen), r is the reproductive rate of the host, b is the non-pathogen induced death rate, α is the pathogen induced death rate, λ is the number of pathogen particles (progenies) produced by a cadaver of an infected larva, and μ is the decay rate of the pathogen. Terms in the original model representing host recovery from infection are omitted because recovery is considered negligible in most forest insect/baculovirus associations

in the European Alps. They estimated model parameters from the literature and found that their model predicted oscillations in host density (Figure 4) that closely matched those recorded in the field by Auer (1968). Regular oscillations of density have long been reported for a number of forest insects. Anderson and May's pioneering work suggested that pathogens might be responsible for these oscillations. Furthermore, they showed that there exists a host density threshold below which infections can no longer be sustained in the host population and the persistence of the pathogen depends upon its ability to survive in the environment outside of the host. Thus, even though the prevalence of the pathogen declines to zero in the low density phase of the host, nevertheless the pathogen alone was responsible for the dynamic behavior involving cycles of outbreaks (at least in the model). It is important to note, however, that various studies on larch budmoth have suggested that other factors, including effects of defoliation on host plant quality (Benz, 1974), as well as genetic variation in the budmoth itself (Baltensweiler, 1993) may be the primary causes of density oscillations, and that some outbreaks of budmoth have not been accompanied by virus epizootics (Baltensweiler and Fischlin, 1988).

Bowers *et al.* (1993) extended the Anderson/May model by adding a host carrying capacity K, thus converting host population growth from exponential to logistic in the absence of the pathogen. This model exhibited a stable host-density equilibrium instead of regular cycles for many parameter values. The model enabled definition of the transmission coefficient ν in terms of K and the threshold density of pathogen transmission. This change to the model allowed Bowers *et al.* (1993) estimate values for both these parameters for larch budmoth from the literature, in contrast to Anderson and May (1981), who necessarily chose an arbitrary value of ν for larch budmoth. Bowers *et al.* (1993) concluded that for many reasonable parameter values, the larch budmoth system would remain at equilibrium rather than cycling, and when cycles occurred, they would produce levels of infection that diverged strongly from those observed in nature (Auer, 1968, Figure 4). Dwyer (1994) added a host carrying capacity to the Anderson/May model in a different way. His revised model predicted oscillations of the Douglas fir tussock moth that corresponded well with the 7–10 year cycles that occurred in nature. This contrasted with earlier studies by Vezina and Peterman (1985), who concluded that the virus was too short-lived and the host density growth rate to high to account for the observed oscillation cycle with an Anderson/May model.

Many other studies have extended the original Anderson/May models to include additional features and complications (see Briggs *et al.* (1995) for an excellent review). For example, Hochberg (1989) explored the importance of pathogen reservoirs on Anderson/May dynamics. Many pathogens such as baculoviruses move from a short-lived infective stage on the foliage to a reservoir (typically the soil) where they remain infective but largely inaccessible to their insect hosts indefinitely (Thompson *et al.*, 1981). Hochberg showed that the dynamics of the resulting model were largely governed by the

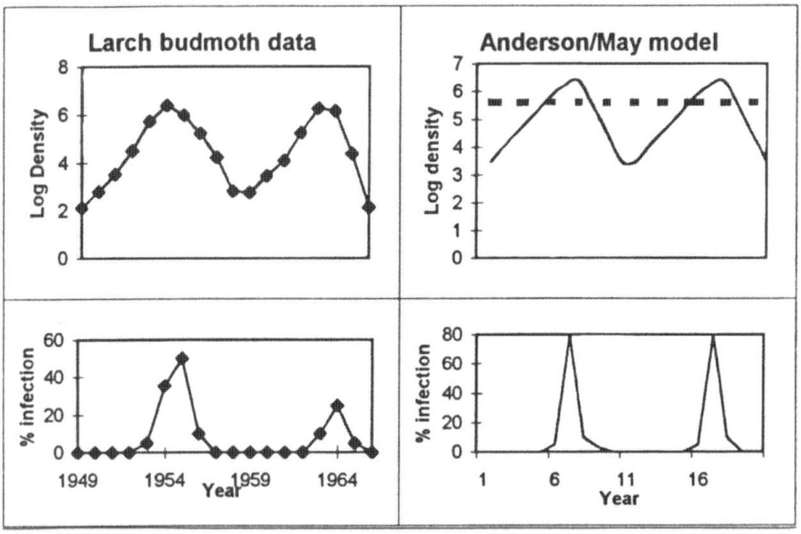

Figure 4. A) Log abundance of the larch budmoth, *Zeiraphera diniana* in the European Alps, and B) the percentage prevalence of infection with a granulovirus in this population (data from Auer (1968)), C) Log host abundance as predicted by the model of Anderson and May (1981; see Fig. 3); the dashed horizontal line indicates the threshold host density for maintenance of the parasite within the host population. D) The prevalence of infection as predicted by the model. Redrawn from Anderson and May (1981)

rates of flow of the pathogen between the free-living stage and the reservoir and the life span of the pathogen in the reservoir. At high and low rates of flow, the host population density oscillated, whereas at intermediate rates of flow the system remained at a relatively low and stable density, a behavior that has been observed with some host/pathogen systems. Hochberg and Holt (1990) extended the Anderson/May model to cases where two pathogens attack the same host and explored the conditions under which they might coexist. Other studies have explored the situation in which the two host species share the same host. (Holt and Pickering, 1985).

Dwyer and Elkinton (1993) explored the within-generation dynamics of an Anderson/May model incorporating an incubation time of the pathogen within the host, and thus a delay between infection and release of infective pathogens into the environment, which for many pathogens occurs when the infected host dies. An important contribution of this paper was to show how to estimate model parameters, particularly the transmission coefficient ν with small-scale field experiments that involved rearing infected and uninfected larvae inside mesh bags on tree branches. Their model provided a reasonably good fit to field data on the progression of an epizootic of *Ld*MNPV during the larval stage of gypsy moths (Woods and Elkinton, 1987). Dwyer *et al.* (1997) extended this work by incorporating variation in host susceptibility to virus in the model and achieved an even better fit to this field data.

Briggs and Godfray (1995) added stage structure (*e.g.*, larvae, pupae, adults) to Anderson/May models and investigated the behavior of several alternative versions, including those in which only the larval stage is susceptible and transmission may or may not occur until after the death of the infected hosts. Their models exhibited complex dynamics, including the occurrence of cycles with durations equal to or less than the developmental time of the host. Briggs and Godfray (1996) explored the behavior of models where the host is regulated at a low density equilibrium by some other factor but occasionally escapes into an outbreak phase which is regulated by the pathogen. In the low host density phase the pathogen resides in a reservoir where it gradually goes extinct. Their model predicted that the density of the outbreak populations would increase with the length of time since the previous outbreak because of the process of pathogen extinction.

Other studies have extended Anderson/May models to explore the rate of spread of pathogens in the environment (Dwyer, 1992; Dwyer and Elkinton, 1995). Such studies are particularly relevant to the recent efforts involving the release of genetically engineered pathogens (*e.g.*, D'Amico *et al.*, 1999). Rate of spread was included in these models by adding terms corresponding to diffusion, an approach that has been widely used to model the movement of organisms from a point source.

The Anderson/May models discussed so far all exhibit fairly simple dynamics. The host population either remains at equilibrium or oscillates with a regular period. Other possibilities exist. For example, May (1985) explored the properties of a very simple host/pathogen model related to his well known discrete logistic model (May, 1974, 1976), with which he introduced the concept of deterministic chaos to population ecology. In model systems with deterministic chaos, the densities of the population fluctuate erratically, often without explicit periods between outbreaks. Such erratic fluctuations in density are common in nature, but, prior to May's pioneering work, ecologists assumed they were caused by stochastic influences (especially weather) rather than by mathematical properties of the host population system. Unlike the discrete logistic which proceeded from a stable equilibrium to regular oscillations to chaos as the reproductive rate increased, the host-pathogen model of May (1985) was chaotic at all values of the reproductive rate. At low values the density fluctuations were very close to a cycle that repeated every two generations. With higher reproductive rates, the predicted density fluctuations became more erratic. This example illustrates the fact that chaotic systems may appear to exhibit regular cycles but with minor variation in cycle amplitude or frequency. Whether or not the erratic fluctuations in density characteristic of real populations, are caused by deterministic chaos or by stochastic factors remains difficult to discern (Turchin and Taylor, 1992).

Dwyer et al. (2000) presented a discrete-time host-pathogen model to represent transmission and host reproduction between host generations coupled with a continuous time model, as discussed above, representing progression of the virus epizootic within a host generation. The combined model exhibited complex dynamics including deterministic chaos when host fecundity was high, as in the models of May (1974, 1985) but either regular oscillations or stable equilibriums at lower and more realistic values of host fecundity. The occurrence of stability versus oscillations in this model was governed in large part by the amount of variation in host susceptibility and the transmission rate of virus between host generations. Dwyer *et al.* (2004) added a generalist predator to a host-pathogen model of this type. The resulting model exhibited very complex dynamics including deterministic chaos under some parameter values. Adding a small amount of random variation to the model produced high variability in time between host outbreaks that qualitatively mimicked some historical time series of gypsy moth fluctuations in density.

This body of work illustrates that simple models with a very limited number of parameters can capture most of the complicated and erratic dynamics exhibited by real host-pathogen systems.

B Complex simulations

An entirely different approach to modeling host-pathogen systems involves construction of complex simulations that contain many variables and much biological detail. The approach arose in the late 1960s with the advent of high speed computers and was applied to many population systems. These models attempted to incorporate much of the existing knowledge about key insect pests and the interactive effects of the host plants, natural enemies and exogenous variables such as weather. Onstad and Carruthers (1990) review this approach to modeling insect pathogen systems. The advantage of these models compared to simple models is that they allowed incorporation of any desired complexity and, hence, biological realism. Indeed there was virtually no limit in the amount of complexity and detail that could be incorporated. A disadvantage of such models was that as the number of functional relationships and estimated parameters grew, so did the overall uncertainty associated with model predictions and the difficulty in understanding model behavior. A number of such models were constructed for major insect pests of forests in North America, such as the western spruce budworm (Sheehan 1989) and the gypsy moth (Sharov and Colbert, 1994). These models included submodels for forest insect pathogens. Most such models were only published in preliminary form and were never tested or validated against field data or used by anyone subsequently. Such models have fallen out of favor with most ecologists. There have been relatively few efforts to determine what lessons have been learned from these models and what the future holds for complex models in ecology (Liebhold, 1994, Logan, 1994). Logan (1994) argued that the future lay with models of intermediate complexity: *i.e.*, models with a limited number of parameters but with a complexity that requires numerical simulation rather than the stability analyses and other mathematical approaches that have been used by theoreticians to study simple theoretical models.

The potential for useful models of intermediate complexity is higher for most insect host-pathogen interactions than it is for entire insect population systems, as discussed above, which include host-plant interactions and other types of natural enemies, such as predators and parasitoids. For one thing, many insect pathogens are quite host specific. Their dynamics involve only the pathogen and a single host species and a small number of exogenous variables such as temperature and rainfall. In contrast, many predators and parasitoids of insects have many hosts and the factors affecting their abundance are largely unknown.

Such models have been used to explore the importance of weather variables such as temperature and rainfall that are critical to the development of epizootics of fungal pathogens of insects. Detailed predictions of epizootic development as a function of these variables are probably too complex a task for simple theoretical models of the type described above. For example, Weseloh *et al.* (1993, Figure 5) and Hajek *et al.* (1993) have independently modeled epizootics of *E. maimaiga*, the fungal pathogen of gypsy moth.

Both models were simulations based upon the temperature-dependent maturation of gypsy moth larvae (Casagrande *et al.*, 1987) and incubation rate of the fungal pathogen in its host. Transmission on each simulated day depended on the

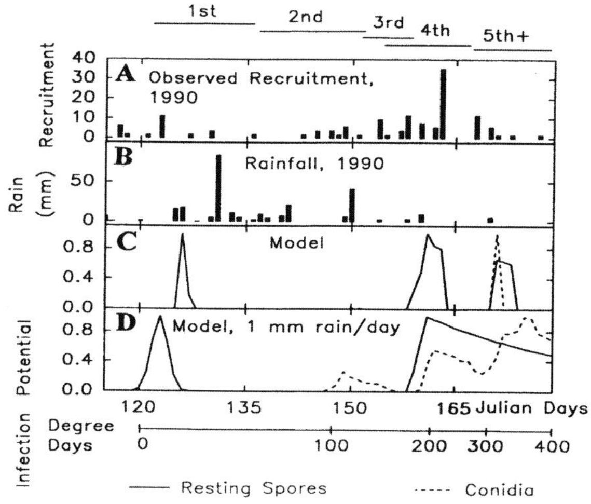

Figure 5. Field data and model predictions of infections of gypsy moth larvae by the fungal pathogen *E. maimaiga* from Weseloh *et al.* (1993). A) Estimated recruitment of healthy larvae to the diseased condition as determined in 1990 at the Mt. Carmel forest site, B) rainfall amounts using weather records of 1990; C) simulated infection potentials from the model using actual climatic data and D) simulated infections assuming 1 mm rain/day. Horizontal lines above the graphs represent the times that each larval instar was common in the forest. The line labeled "Degree Days" represents the accumulated degree-days determined from the climatic data

presence of moisture as determined from rainfall records. These models were used to predict the time course and magnitude of disease prevalence at several sites that differed in temperature and levels of rainfall. The predictions of the Weseloh model were compared with data collected on recruitment of infections among caged larvae placed in the field (Figure 5A). The importance of rainfall was determined by comparing the infections (Figure 5C) predicted from the actual rainfall with that predicted by a model that assumed constant rainfall of 1 mm per day (Figure 5D).

The models of Weseloh *et al.* (1993, Figure 5) and Hajek *et al.* (1993) were also used to explore the relative importance of primary versus secondary infection in explaining the observed time course and prevalence of infection of this pathogen in field populations. Primary infection in this system occurs when the larvae contact germinating azygospores that overwinter in the soil. This occurs when the larvae first hatch in the spring and later in the summer when late instars seek daytime resting locations in the forest litter. Secondary infection occurs when larvae contact windborne conidia released from the fungal-killed insects. Both studies concluded that primary infection occurring late in the season was important in explaining the occurrence of high levels of infection that occur among late instars. This prediction can be seen by comparing the solid vs. the dotted lines in Figures 5C,D.

5 Conclusion

Recent advances in molecular techniques have facilitated the detection of insect pathogens both inside and outside of the host. Many studies over the last few decades some of which have used these techniques have elucidated the basic biology and transmission dynamics of many insect-pathogen systems. Understanding of disease ecology, nevertheless, remains rudimentary. The general analytic tools developed by population ecologists to reveal key factors or detect density dependence in population systems have been largely compromised by statistical problems, and even when they work, provide only preliminary understanding of the effects of a pathogen or any other factor on an insect population. Detailed simulations which incorporate much knowledge about host/pathogen dynamics have often failed to live up to their original promise and have been abandoned by most ecologists. A great deal has been learned from the exploration of simple theoretical models, and general conclusions have been drawn regarding the factors that lead to stable oscillations, persistence of pathogens and host threshold densities required to maintain infections in host populations. As emphasized by Dwyer (1995), some of these conclusions were counterintuitive but were subsequently supported by experimental observations. However, the dynamical predictions of these models vary markedly from one to another depending on which particular biological features are included. It is thus difficult to have much confidence in the actual predictions of any of these models for specific population systems, even if one or more of them appear to fit a set of data, as illustrated above with the larch budmoth. In the budmoth system, as with others including gypsy moth (Elkinton and Liebhold, 1990; Elkinton *et al.*, 1996) field studies suggest that other factors not included in these models are critical to the dynamics of the host insect. Indeed, one is struck by how poorly we understand the role of pathogens and the overall dynamics of even well-studied forest insects such as gypsy moth or spruce budworm. Nevertheless, despite the limitations and short-comings of various approaches to model construction, there exists no alternative but to continue to develop models of host-pathogen dynamics and to couple these with careful field studies. Studies that combine both features remain rare. At least we can say that there is plenty for the next generation of insect pathologists and ecologists interested in insect pathogens to accomplish.

6 Acknowldegments

We thank G. Dwyer for many years of useful discussions; A. Gwynn C. Piazza and G. Witkus for editorial assistance, and T. Tattar for reviewing the manuscript.

7 References

Anderson, R. M. and May, R. M. 1979. Population biology of infectious diseases: Part I. *Nature* **280**, 361–367.

Anderson, R. M. and May, R. M. 1980. Infectious diseases and population cycles of forest insects. *Science* **210**, 658–661.

Anderson, R. M. and May, R. M. 1981. The population dynamics of microparasites and their invertebrate hosts. *Philos. Trans. R. Soc. Lond. Ser. B.* **291**, 451–524.

Andreadis, T. G. and Weseloh, R. M. 1990. Discovery of *Entomophaga maimaiga* in North American gypsy moth, *Lymantria dispar. Proc. Natl. Acad. Sci. U.S.A.* **87**, 2461–2465.

Auer, C. 1968. Erst Ergebnisse einfacher stochastischer modelluntersuchungen uber die ursachen der populationsbewegung des grauen larchenwicklers *Zeiraphera diniana*, Gn. (= *Z. griseana* Hb.) im Oberengadin, 1949/66. *Z. Angew. Entomol.*, **62**, 202–235.

Baltensweiler, W. 1993. A contribution to the explanation of the larch bud moth cycle, the polymorphic fitness hypothesis. *Oecologia* **93**, 251–255.

Baltensweiler, W. and Fischlin, A. 1988. The larch budmoth in the Alps. *In* "Dynamics of Forest Insect Populations: Patterns, Causes, Implications".(A. A. Berryman Ed.), pp 331–351. Plenum Press, New York and London.

Bellows, Jr. T. S., Van Driesche, R. G., and Elkinton, J. S. 1992. Life-table construction and analysis in the evaluation of natural enemies. *Annu. Rev. Entomol.* **37**, 587–614.

Benz, G. 1974. Negative Ruckkoppelung durch Raum- und Nahrungskonkurrenz sowie zyklische Veranderung der Nahrungsgrundlage als Regelprinzip in der Populationsdynamik des Grauen Larchenwicklers, Zeiraphera diniana (Guenee) (Lep. Torticidae). *Z. Angew. Entomol.* **76**, 196–228.

Bird, F. T. and Elgee, D. E. 1957. A virus disease and introduced parasites as factors controlling the European spruce sawfly, *Diprion hercyniae* (Htg) in central new Brunswick. *Can. Entomol.* **89**, 371–378.

Bowers, R. G., Begon, M., and Hodgkinson, D. E. 1993. Host-pathogen population cycles in forest insects? Lessons from simple models reconsidered. *OIKOS* **67**, 529–538.

Briggs, C. J. and Godfray, H. C. J. 1995. The dynamics of insect-pathogen interactions in stage structured populations. *Am. Nat.* **145**, 855–887.

Briggs, C. J., Haites, R. S., Barlow, N. D., and Godfray, H. C. J. 1995. The dynamics of insect-pathogen interactions in stage structured populations. *In* "Ecology of Infectious Diseases in Natural Populations" (B. T. Grenfell and A. P. Dobson, Eds.), pp. 295–326. University of Cambridge Press, Cambridge, UK.

Briggs, C. J. and Godfray, H. C. J. 1996. The dynamics of insect-pathogen interactions in seasonal environments. *Theor. Pop. Biol.* **50**, 149–177.

Burand, J. P. and Park, E. J. 1992. Effect of nuclear polyhedrosis virus on the development and pupation of gypsy moth larvae. *J. Invertebr. Pathol.* **60**, 171–175.

Carruthers, W. R., Corey, J. S., and Entwistle, P. E. 1988. Recovery of pine beauty moth nuclear polyhedrosis virus from pine foliage. *J. Invertebr. Pathol.* **52**, 27–32.

Casagrande, R. A., Logan, P. A., and Wallner, W. E. 1987. Phenological model for gypsy moth, *Lymantria dispar* (Lepidoptera: Lymantriidae), larvae and pupae. *Environ. Entomol.* **16**, 556 562.

Cooper, D., Corey, J. S., Theilmann, D. A., and Myers, J. H. 2003. Nucleopolyhedroviruses of forest and western tent caterpillars: cross-infectivity and evidence for activation of latent virus in high-density field populations. *Ecol. Entomol.* **28**, 41–50.

D'Amico, V., Elkinton, J. S., Podgwaite, J. D., Slavicek J., McManus, M. L., and Burand, J. P. 1999. A field release of genetically engineered gypsy moth (*Lymantria dispar L.*) nuclear polyhedrosis virus (LdNPV). *J. Invertebr. Pathol.* **73**, 260–268.

Doane, C. C. 1970. Primary pathogens and their role in the development of an epizootic in the gypsy moth. *J. Invertebr. Pathol.* **15**, 21–33.

Dwyer, G. 1991. The roles of density, stage, and patchiness in the transmission of an insect virus. *Ecology* **72**, 559–574.

Dwyer, G. 1992. On the spatial spread of insect pathogens: theory and experiment. *Ecology* **73**, 479–484.

Dwyer, G. 1994. Density dependence and spatial structure in the dynamics of insect pathogens. *Am. Nat.* **143**, 533–562.

Dwyer, G. 1995. Simple models and complex interactions. *In* "Population Dynamics: New Approaches and Synthesis." (N. Cappuccino and P. W. Price, Eds.), pp. 209–227. Academic Press, San Diego, CA.

Dwyer, G. and Elkinton, J. S. 1993. Using simple models to predict virus epizootics in gypsy moth populations. *J. Anim. Ecol.* **61**, 1–11

Dwyer, G. A. and Elkinton, J. S. 1995. Host dispersal and the spatial spread of insect pathogens. *Ecology* **76**, 1262–1275.

Dwyer, G., Elkinton, J. S., and Buonaccorsi, J. 1997. Host heterogeneity in susceptibility and disease dynamics: tests of a mathematical model. *Am. Nat.* **150**, 685–707.

Dwyer, G., Dushoff, J., and Elkinton., J. S. 2000. Pathogen-driven outbreaks in forest defoliators revisited: building models from experimental data. *Am. Nat.* **156**, 105–120.

Dwyer, G., Dushoff, J., and Yee, S. H. 2004. The combined effects of pathogens and predators on insect outbreaks. *Nature*. **430**, 341–345.

Elkinton, J. S. and Liebhold, A. M. 1990. Population dynamics of gypsy moth in N. America. *Annu. Rev. Entomol.* **35**, 571–596.

Elkinton, J. S., Buonaccorsi, J. P., Bellows, T. S., and Van Driesche, R. G. 1992. Marginal attack rate, k-values and density dependence in the analysis of contemporaneous mortality factors. *Res. Pop. Ecol.* **34**, 29–44.

Elkinton, J. S., Healy, W. M., Buonaccorsi, J. P., Boettner, G. H., Hazzard, A. M., Smith, H. R., and Liebhold, A. M. 1996. Interactions among gypsy moths, white-footed mice and acorns. *Ecology* **77**, 2332–2342.

Fuxa, J. R. and Tanada, Y. 1987. Epidemiological concepts applied to insect epizootiology. *In* "Epizootiology of Insect Diseases" (J. Fuxa and Y. Tanada, Eds.), pp. 3–21. John Wiley and Sons, New York.

Glaser, R. W. 1915. Wilt of gypsy moth caterpillars. *J. Agric. Res.* **4**, 101–128.

Hajek, A. E., Humber, R. A., Elkinton, J. S., May, B., Walsh, S. R. A., and Silver, J. S. 1990. Allozyme and restriction fragment length polymorphism analyses confirm *Entomophaga maimaiga* responsible for 1989 epizootics in North American gypsy moth populations. *Proc. Natl. Acad. Sci., USA*. **87**, 6979–6982.

Hajek, A. E. and Soper, R. S. 1991. Within-tree location of gypsy moth, *Lymantria dispar*, larvae killed by *Entomophaga maimaiga* (Zygomycetes: Entomophthorales). *J. Invertebr. Pathol.* **58**, 468–469.

Hajek, A. E., Butt, T. M., Strelow, L. I., and Gray, S. M. 1991a Detection of *Entomophaga maimaiga* (Zygomycetes: Entomophthorales) using enzyme linked immunosorbent assay (ELISA). *J. Invertebr. Pathol.* **58**, 1–9.

Hajek, A. E., Humber, R. A., Walsh, S. R. A., and Silver, J. C. 1991b. Sympatric occurrence of two *Entomophaga aulicae* (Zygomycetes: Entomophthorales) complex species attacking forest Lepidoptera. *J. Invertebr. Pathol.* **58**, 373–380.

Hajek, A. E., Carruthers, R. I., Larkin, T. M., and Soper, R. S. 1993. Modeling the dynamics of *Entomophaga maimaiga* (Zygomycetes: Entomophthorales) epizootics in gypsy moth (Lepidoptera: Lymantriidae) populations. *Environ. Entomol.* **22**, 1172–1187.

Henson, J. M. and French, R. 1993. The polymerase chain reaction and plant disease diagnosis. *Annu. Rev. Phytopathol.* **31**, 81–109.

Hochberg, M. E. 1989. The potential role of pathogens in biological control. *Nature* **337**, 262–265.

Hochberg, M. E. and Holt R. D. 1990. The coexistence of competing parasites: the role of cross-species infection. *Am. Nat.* **136**, 517–541.

Holt, R. D. and Pickering, J. 1985. Infectious disease and species coexistence: a model of Lotka Volterra form. *Am. Nat.* **126**, 196–211.

Ilyinykh, K V. and Chuikova, G. V. 1989. Identification of natural isolates of black arches moth (*Lymantria monacha*) nuclear polyhedrosis virus. *Voprosy Virusologii* **34**, 84–89.

Kaupp, W. J. and Ebling, P. M. 1993. Horseradish peroxidase labelled probes and enhanced chemiluminescence to detect baculoviruses in gypsy moth and eastern spruce budworm larvae. *Virol. Methods* **44**, 89–98.

Keating, S. T., Burand, J. P., and Elkinton, J. S.. 1989. DNA hybridization assay for detection of gypsy moth nuclear polyhedrosis virus in infected gypsy moth (*Lymantria dispar L*) larvae. *Appl. Environ. Microbiol.* **55**, 2749–2754.

Keating, S. T., Elkinton, J. S., Burand, J. P., Podgwaite, J. D., and Ferguson, C. S. 1991. Field evaluation of a DNA hybridization assay for nuclear polyhedrosis virus in gypsy moth (Lepidoptera: Lymantriidae) larvae. *J. Econ. Entomol.* **84**, 1329–1333.

Kermack, W. O. and McKendrick, A. G. 1927. A contribution to the mathematical theory of epidemics. *Proc. Royal Soc., Ser. A.* **115**, 700–721.

Kukan, B. and Myers, J. H. 1995. DNA Hybridization assay for detection of nuclear polyhedrosis virus in tent caterpillars. *J. Invertebr. Pathol.* **66**, 231–236.

Kuno, E. 1971. Sampling error as a misleading artifact in "key factor analysis". *Res. Popul. Ecol.* **13**, 28–45.

Liebhold, A. M. 1994. Use and abuse of insect and disease models in forest pest management: past, present and future. *In* "Sustainable ecological systems: implementing an ecological approach to land management" (M. W. Covington and L. F. DeBano, Eds.), pp. 204–210. U.S. Dep. Agric. For. Serv. Tech. Rep. RM-247.

Logan, J. A. 1994. In defense of big ugly models. *Am. Entomol.* **40**, 202–206.

Ma, M., Burkholder, J. K., Webb, R. E., and Hsu, H. T. 1984. Plastic-bead ELISA: an inexpensive epidemiological tool for detecting gypsy moth (Lepidoptera: Lymantriidae) nuclear polyhedrosis virus. *J. Econ. Entomol.* **77**, 537–540.

Maddox, J. V. 1987. Protozoan diseases. *In* "Epizootiology of Insect Diseases" (J. Fuxa and Y. Tanada, Eds.), pp. 417–452. John Wiley and Sons, New York.

Manly, B. F. J. 1977. The determination of key factors from life table data. *Oecologia* **31**, 111–117.

May, R. M. 1974. Biological populations with nonoverlapping generations: stable points, stable cycles, and chaos. *Science* **186**, 645–647.

May, R. M. 1976. Simple mathematical models with very complicated dynamics. *Nature* **261**, 459–467.

May, R. M. 1985. Regulation of populations with nonoverlapping generations by micro-parasites: a purely chaotic system. *Am. Nat.* **125**, 573–583.

Myers, J. H. 2000. Population fluctuations of the western tent caterpillar in southwest British Columbia. *Pop. Ecol.* **42**, 231–241.

Myers, J. H. and Kukan, B. 1995. Changes in fecundity of tent caterpillars: a correlated character of disease resistance or sublethal effects of disease? *Oecologia* **103**, 475–480.

Morris, R. F. 1963. The dynamics of epidemic spruce budworm populations. *Mem. Entomol. Soc. Can.*, 311–332.

Murray, K. D. and Elkinton, J. S. 1992. Vertical distribution of nuclear polyhedrosis virus-infected gypsy moth (Lepidoptera: Lymantriidae) larvae and effects on sampling for estimation of disease prevalence. *J. Econ. Entomol.* **85**, 1865–1871

Onstad, D. W. and Carruthers, R. I. 1990. Epizootiological models of insect diseases. *Annu. Rev. Entomol.* **35**, 399–419.

O'Reilly, D R. and Miller, L. K. 1989. A baculovirus blocks insect molting by producing ecydysteroid UDP-glycosyl transferase. *Science* **245**, 1110–1112.

Podgwaite, J. D., Shields, K. S., Zerillo, R. T., and Bruen, R.B. 1979. Environmental persistence of the nucleopolyhedrosis virus of the gypsy moth, *Lymantria dispar. Environ. Entomol.* **8**, 528–536.

Podoler, H. and Rogers, D. 1975. A new method for the identification of key factors from life-table data. *J. Anim. Ecol.* **44**, 85–114.

Rothman, L. D. and Myers, J. H. 1994. Nuclear polyhedrosis virus treatment effect on reproductive potential of western tent caterpillar (Lepidoptera: Lasiocampidae). *Environ. Entomol.* **23**, 864–869.

Royama, T. 1984. Population dynamics of the spruce budworm *Choristoneura fumiferana. Ecol. Monogr.* **54**, 429–462.

Royama, T. 1992. "Analytical Population Dynamics." Chapman and Hall, New York.

Royama, T. 1996. A fundamental problem in key factor analysis. *Ecology* **77**, 87–93.

Sheehan, K. A. 1989. The western spruce budworm model: structure and content. USDA Forest Service Gen. Tech. Rep. PNW GTR-241. 70 pp.

Sharov, A. A. and Colbert, J. J. 1994. "Gypsy moth life system model, integration of knowledge and a user's guide." Final report on a cooperative research project developed by Department of Entomology, Virginia Polytechnic Institute and State University and USDA Forest Service, Northeastern Experiment Station.

Thompson, C. G., Scott, D. W., and Wickman, B. E. 1981. Long-term persistence of the nuclear polyhedrosis virus of the Douglas-fir tussock moth, *Orgyia pseudotsugata* (Lepidoptera: Lymantriidae), in forest soil. *Environ. Entomol* **10**, 254–255.

Turchin, P. 1995. Population regulation: old arguments and a new synthesis. *In* "Population Dynamics: New Approaches & Synthesis" (N. Cappuccino, and P. W. Price, Eds.), pp. 19–41. Academic Press, San Diego, CA.

Turchin, P. and Taylor, A. D. 1992. Complex dynamics in ecological time series. *Ecology* **73**, 289–305.

Van Driesche, R. G. 1983. Meaning of "percent parasitism" in studies of insect parasitoids. *Environ. Entomol.* **12**, 1611–1622.

Van Driesche, R. G. and Bellows, T. S. 1988. Host and parasitoid recruitment for quantifying losses from parasitism, with reference to *Pieris rapae* and *Cotesia glomerata. Ecol. Entomol.* **13**, 215–222.

Varley, G. C., Gradwell, G. R., and Hassell, M. P. 1973. "Insect Population Ecology an analytical approach." University of California Press, Berkeley.

Vezina, A. and Peterman, R. M. 1985. Tests of the role of a nuclear polyhedrosis virus in the population dynamics of its host, Douglas-fir tussock moth, *Orgyia pseudotsugata* (Lepidoptera: Lymantriidae). *Oecologia* **67**, 260–266.

Weseloh, R. M. and Andreadis, T. G. 1992. Epizootiology of the fungus *Entomophaga maimaiga*, and its impact on gypsy moth populations. *J. Invertebr. Pathol.* **59**, 133–141.

Weseloh, R. M., Andreadis, T. G., and Onstad, D. W. 1993. Modeling the influence of rainfall and temperature on the phenology of infection of gypsy moth, *Lymantria dispar*, larvae by the fungus *Entomophaga maimaiga. Biol. Control* **3**, 311–318.

Woods, S. A. and Elkinton, J. S. 1987. Bimodal patterns of mortality from nuclear polyhedrosis virus in gypsy moth (*Lymantria dispar*) populations. *J. Invertebr. Pathol.***50**, 151–157.

Woods, S. A., Elkinton, J. S., Murray, K. D., Liebhold, A. M., Gould, J. R., and Podgwaite, J. D. 1991. Transmission dynamics of a nuclear polyhedrosis virus and predicting mortality in gypsy moth (Lepidoptera: Lymantriidae) populations. *J. Econ. Entomol.* **84**, 423–430.

SECTION VI

EXOTIC PATHOGENS

Chapter VI-1

Introduction of exotic pathogens and documentation of their establishment and impact

Ann E. Hajek
Department of Entomology
Cornell University
Ithaca, NY 14853, USA

Italo Delalibera Júnior
Departmento de Entomologia
Fitopatologia e Zoologia Agricola
ESALQ-University of São Paulo
Piracicaba
13418-900, Brazil

Michael L. McManus
USDA Forest Service
Hamden, CT 06514, USA

1 Introduction

Importation of exotic natural enemies to establish self-sustained control of pests is one of several major types of biological control. This type of program involves a continuous, interactive process of assimilation of the information that is required to find, import and colonize effective natural enemies (Gonzalez and Gilstrap, 1992). The process itself is generally similar for all natural enemies of arthropods whether they are parasitoids, predators, or pathogens. However, exploring for pathogens, eventually introducing them, and subsequently evaluating their impact on pests present challenges differing from those encountered when working with parasitoids and predators.

During early years in the use of biological control, the majority of programs consisted of introducing exotic natural enemies to areas where they did not already occur in order to control introduced pests that had become established. Thus, the term 'classical biological control' has been used to describe this practice. In fewer instances, exotic natural enemies were introduced to control native instead of exotic pests and this practice has been referred to as use of a 'new association' for classical biological control (Hokkanen and Pimentel, 1984). Although these strategies differ in the origin of the pest, the final goals are the same: to establish a self-sustaining biological control agent in a pest population, resulting in a reduction in the density of hosts and a permanent level of regulation by the natural enemy. As described in this chapter, classical biological control can be extremely successful and cost effective. In addition, many insect pathogens that have no potential for development as microbial products may have great potential as classical biological control agents (Maddox *et al.*, 1992).

Classical biological control has primarily been used for introducing parasitoids and predators against arthropod pests as well as for introducing arthropods to control weeds. Insect pathogens have not been used as commonly for the classical

L.A. Lacey and H.K. Kaya (eds.), Field Manual of Techniques in Invertebrate Pathology, 299–325.

biological control approach. While 131 programs have been conducted to introduce exotic pathogens for establishment (Hajek *et al.*, 2005, 2007), over 5500 programs to introduce parasitoids and predators were conducted between 1888 and 2005 (D. Greathead, personal communication). The majority of programs to introduce exotic pathogens have concentrated on obtaining pathogens from the area where an introduced pest is indigenous. Among the comparatively few programs conducted to introduce pathogens, some have been exceptionally successful (*e.g.*, the non-occluded *Oryctes* virus released against *Oryctes rhinoceros* in the Pacific Islands, the nematode *Deladenus* (= *Beddingia*) *siricidicola* against *Sirex noctilio* in Australia, and the fungus *Entomophaga maimaiga* against the gypsy moth, *Lymantria dispar*, in northeastern North America). Maddox *et al.* (1992) listed several explanations to explain why the use of exotic entomopathogens for biological control has not been enthusiastically embraced by the biological control community. (1) Diagnosis, isolation, and culture of entomopathogens require more specialized procedures and equipment than normally required for collecting and rearing parasitoids and predators. (2) Taxonomy of many groups of entomopathogens is ambiguous, which complicates the process of selecting the appropriate organism and, more importantly, may prohibit satisfying the regulatory requirements for their eventual introduction. (3) Important biological characteristics of entomopathogens such as their host specificity, methods of transmission, and interaction with other organisms in the environment have often not been determined. (4) Little scientific information exists on which characteristics should be used to select the most appropriate pathogens for introduction. And (5) regulations regarding the importation of non-indigenous pathogens have complicated and precluded this process. In the past, regulatory requirements for introducing exotic entomopathogens in many countries have been unclear, although international guidelines are now in place which hopefully will improve this situation (see below).

Although pathogens have been used little for classical biological control, they display great potential for such use. Baculoviruses and fungi may cause massive and dramatic epizootics, killing a high percentage of the host population, and producing many infectious propagules that are redistributed into the host's environment. In contrast, other entomopathogens, such as the microsporidia, produce many subtle and sublethal effects throughout the life cycles of their hosts although they often do not cause recognizable epizootics. However, all groups of pathogens contribute to regulating host populations. Additionally, there are reports that mixed infections by different pathogens may lead to enhancement of infection (Tanada and Fuxa, 1987) and thus exacerbate or accelerate the development of epizootics. As with parasitoids and predators, because of the inherent diversity in the biologies of pathogens and hosts and our lack of knowledge of their ecologies, predicting the outcomes of biological control introductions is challenging, but vital to this field (Ehler,1990b).

The capacity of entomopathogens to spread on their own is usually poorly understood. Past examples have shown that unassisted spread by many pathogens can be slow. Accordingly, introduction programs frequently are composed of two phases: (1) an initial introduction and (2) subsequent redistribution to additional locations within the general area of the introduction. The secondary redistribution phase of introduction can be long-term [*e.g.*, introductions of the nematode *D. siricidicola* continued for at least 20 years in Australia (Bedding, 1993)]. In a few instances, pathogens have been accidentally introduced along with the host or other natural enemies, and such programs have consisted solely of secondary redistribution to new areas.

Pathogen introductions have been attempted most frequently in situations where introduced pest populations are at outbreak levels. Although high host densities are not a requirement, such conditions can aid in pathogen establishment. The most successful pathogen introduction programs have resulted in long-term control and there is rarely need for reapplication at individual locations. This type of control is especially well-suited to stable ecosystems, *e.g.*, pests associated with woody plants, where long-lived stages of the pathogen can easily persist as compared with annual crops in agriculture (Hajek *et al.*, 2005). In addition, classical biological control can be appropriate for systems that can sustain a low density host population so that the pathogen remains present in the host population. However, pathogens with long-lived environmentally resistant stages would not require a low density host population [*e.g.*, the 17-year cicada pathogen *Massospora cicadina* (Lloyd and Dybas, 1966)]. This long-term type of control is also appropriate for areas where applications of synthetic chemicals are undesirable (*e.g.*, urban areas) or are not economically feasible (*e.g.*, rangelands). For situations where farmers cannot afford to use high-priced control tactics or where crops do not yield enough profits to justify the costs of control, introduced pathogens can provide an alternative control requiring negligible inputs by the farmer or land manager.

Throughout this chapter, when we use the term pathogen, we will be referring to microscopic organisms causing disease, so we will be including nematodes when using this term. Methods associated with exploration, importation, and release of parasitoids and predators have been reviewed previously (Zwölfer *et al.*, 1976; Van Driesche and Bellows, 1993). In this chapter, we present methods designed specifically for introducing pathogens and entomopathogenic nematodes attacking arthropods (insect and mite pests) to new areas, drawing predominantly on examples from successful programs. We will also suggest additional methodologies that we feel have potential but which have not yet been used for introductions of exotic pathogens. Finally, we will discuss the importance of post-release evaluation to document the impacts of pathogen introductions on host populations.

2 Surveying for pathogens in potential release areas

A classical biological control program often starts when an exotic pest is detected and conditions indicate that introduction of a pathogen is desirable. Before planning to introduce pathogen(s), surveys should be conducted to document the pathogens already associated with the pest in the potential release area. Such surveys are necessary because we often do not know if pathogens are already associated with the pest population because the taxonomy, distribution, and ecology of both indigenous and non-indigenous entomopathogens, especially associated with a newly introduced pest, are poorly known. Many pathogen strains can differ in virulence so, although the pathogen might already occur in the release area, it may be ineffectual in controlling the pest, and there may clearly be justification for introduction of more virulent, exotic strains. In such cases, pathogen strains from the release areas must be isolated and maintained for future comparison, so that establishment of the exotics to be released can be evaluated. For example, *Zoophthora radicans* isolates retained under liquid nitrogen were vital for developing molecular techniques to evaluate whether strains introduced from Serbia had become established in central New York where endemic strains of this fungus already occurred, but were not providing control of *Empoasca fabae* (Hodge *et al.*, 1995). Pre-release surveys at release sites may also be necessary when applying for introduction permits. While the recommendation for pre-release surveys may seem unnecessary, this procedure should be considered a requirement to document that effective pathogens are not already present.

3 Foreign exploration

Foreign exploration for pathogens is not always necessary for well known hosts or pathogens for which numerous isolates from around the world are available in culture collections. However, even for hosts that have been studied extensively, isolates of known pathogens that are virulent and specific might not be available. Based on our general lack of knowledge of the epizootiology of insect diseases, even for well-studied hosts, it is always possible that pathogens exist that have not been discovered, thereby justifying exploration for pathogens.

Bellows and Legner (1993) provide a detailed overview of the literature on planning for exploration and suggest practical methods that should increase the effectiveness of the foreign exploration process. They separate this process into three phases: (1) selecting favorable search locations, (2) planning the exploration trip, and (3) conducting the exploration. In addition to discussing those specific techniques required for exploration for pathogens, we will summarize the highlights of their excellent synthesis, as appropriate to pathogens, and we suggest that anyone who is contemplating a foreign exploration should read their article in its entirety.

A Selecting favorable search locations

Two approaches have been proposed and utilized for searching for natural enemies. The first approach involves searching within the known native range for natural enemies that attack the target pest and/or hosts taxonomically related to the pest (Bellows and Legner, 1993). As a second approach, Van Driesche and Bellows (1996) suggest that searches in some cases may include natural enemies of taxonomically unrelated organisms which occupy similar ecological niches. Most of the literature relative to this discussion involves searching for parasitoids and predators and not entomopathogens. Considering the concern about non-target effects and ecological host specificity of introduced exotic entomopathogens, attention should be paid to releasing more specific pathogens, thus requiring caution in using this second approach. In addition, pathogens tend to have a narrower host range than predators and some parasitoids and may not reproduce optimally and persist in other hosts (Solter *et al.*, 1997). Furthermore, Waage (1990) suggested that natural enemies obtained from the area of origin of a target pest are more likely to provide control than agents that have had no previous contact with a host (but see Hokkanen and Pimentel, 1984, 1989). Some authors have suggested that searches should be focused on a pest's center of origin if, in fact, that can be determined. Importantly, relevant information can be gleaned from published and unpublished information on the host's geographic distribution, and its abundance throughout its range. Invaluable information can also be obtained from colleagues and researchers from abroad who have had first-hand experience with the target pest.

As an example of the status of our knowledge of pathogens associated with an insect pest, the broad native distribution of the pestiferous gypsy moth, *L. dispar*, extends from northern Africa and Western Europe to the Russian Far East, northeastern China, Japan and Korea. Within this expansive region, gypsy moth populations persist and occasionally reach outbreak levels in diverse forest types (*e.g.*, cork oak in Portugal and Sardinia, oaks in central Europe, locust in Slovakia, willow in the Danube Delta and larch in Japan). Yet, very little is known about the pathogen complex attacking gypsy moth populations in many of these forest habitats or about the prevalence of infection during outbreak versus non-outbreak periods.

Foreign exploration for pathogens should be a continuous process designed to assimilate information on the pathogen complex throughout the range of the host and to determine which organisms demonstrate the greatest potential as candidates for introduction. Several strategies have been proposed for searching for natural enemies within a host's native range: (1) determining the center of distribution of a host species to initiate a search, (2) searching in an area where preferred host plants are located, which is perhaps more suitable when phytophagous host species are monophagous or oligophagous; (3) searching areas where the largest complex of natural enemies is known to occur; and (4) concentrating searches where the host reaches outbreak levels infrequently (Bellows and Legner, 1993). One possible weakness in this latter approach is that in some areas where outbreaks are infrequent, the target pest may be at the ecological limits of its range and therefore may be limited more by climate than by natural enemies.

Foreign exploration should be conducted during various seasons and at different elevations and climates because the natural enemy fauna in each case may vary significantly. In particular, trips must be planned to coincide with the season(s) that pathogen activity has been reported or might be predicted.

Much has been written about the importance of climatic suitability in selecting candidate parasitoids and predators for introduction. Selection of species or races of natural enemies from areas where the climate is similar to that in the intended release area is a prudent approach that has been used frequently for entomophagous species (Van Driesche, 1993). Stiling (1993) conducted an analysis of the causes for failure of releases of introduced natural enemies to reduce pest populations. He listed 14 different reasons why releases failed and determined that, based on 148 releases, 34.5% of failures were related to climate. This conclusion concurs with Messenger (1971) who stated that while there could be several possible causes of failure in biological control, the one most important factor was climate. The role of climate in the search for and release of entomopathogens is not well understood, and microorganisms have not been included in analyses of climatic adaptations versus classical biological control success such as those discussed above. However, abiotic conditions in the microenvironments that pathogens inhabit can have profound influences on propagule survival and subsequent disease transmission. These conditions are not easy to measure or predict in the area of introduction. It remains to be determined to what extent climate matching is an important consideration in introductions of entomopathogens for classical biological control. Certainly no one could have predicted the extent of the epizootics caused by *E. maimaiga* among gypsy moth populations in the eastern USA (Hajek *et al.*, 1995). In Japan, this pathogen usually occurs at low prevalence in low density gypsy moth populations. The impact of this pathogen on gypsy moth populations in the USA has persisted in central New York State since 1989 in both wet and dry years (Hajek, unpublished data). Furthermore, this organism has spread naturally and has been established rather easily in gypsy moth populations ranging from Maine south to Virginia and west to Wisconsin and Illinois. The impact of *E. maimaiga* in several ecological regions in the USA probably exceeds the impact that has been recorded in its native Japan.

As with entomophagous natural enemies, exploration for entomopathogens can be made from throughout a host's geographical range and from regions characterized by different host biotypes. In the quest to isolate and identify microsporidia in European gypsy moth populations, several phylogenetically related isolates differing in morphologies, infectivity and modes of transmission have been recovered from seven different countries, and distribution of microsporidian species varies among host populations (Solter *et al.*, 2000; McManus, unpublished data). Initial observations and recent contacts with colleagues suggest that the prevalence of microsporidia in gypsy moth populations appears to be higher in the Balkan countries that are characterized by a more Mediterranean climate. Based on the early literature, high levels of mortality caused by microsporidia among gypsy moth populations have also been reported from other countries such as the Ukraine (Zelinskaya, 1980) and Yugoslavia (Sidor, 1979). However, because there are no extant culture collections storing such isolates, the characteristics of Ukrainian isolates cannot be compared to isolates collected in recent years in the Balkans.

B Planning the exploration trip

Initial steps in planning and organizing an exploration trip involve obtaining the permits necessary to import the collected pathogens and to coordinate their transport to an authorized quarantine facility. An increasing number of countries may require permits to export biological material from their jurisdiction so it is always judicious to consult with colleagues or the proper authorities in the host country. In Brazil, for example, exploration by foreigners requires a special permit that is only granted if the program is being conducted in cooperation with a Brazilian scientific or technical institution. Current Brazilian legislation states that authorization from the Ministry of Science and Technology is required for shipment of collections by foreigners while exportation of collections by Brazilians requires a specific permit (Moraes and De Nardo, 1996). In general, many exploration trips have failed or have been aborted because prior arrangements and authorizations were not obtained in advance by the explorer/collector.

C Conducting foreign exploration

Bellows and Legner (1993) provide a cookbook approach for planning and conducting a foreign exploration trip and discuss the intricate details associated with foreign travel, obtaining passports and visas, and soliciting close cooperation with scientists and institutes abroad whose assistance is invaluable. Traveling in a foreign country can, at times, be stressful and requires close collaboration and careful planning. Always inquire in depth about the availability of instrumentation and materials that are needed to collect, process, and dissect biological material in order to detect entomopathogens. Whereas foreign exploration for parasitoids and predators usually involves collection of host material from which natural enemies may emerge, or direct collection of the natural enemies themselves, exploration for entomopathogens can be much more complex, depending on the microorganisms being sought. Whereas cadavers can be collected to obtain fungi and some viruses, dissection of individual host life stages may be required to inspect tissues such as salivary glands, fat bodies, Malpighian tubules, and midgut cells for infection by viruses, protists, or bacteria. This can be pain-staking work, which is why it is preferable that an insect pathologist trained in diagnostics should participate in exploration trips for entomopathogens.

Although it is preferable to conduct such a trip when the host population is either increasing or in the outbreak stage, in reality this is a luxury that seldom exists. In many cases, populations of the host may be very low in the area of endemism during an exploration trip. When host populations are low, the prevalence of infection of many microorganisms is usually also very low, which means that the probability of finding an infected individual is slim. One approach that has been successfully used for locating microsporidia in gypsy moth larvae when host populations are low is to dissect individually 30–100 larvae to ascertain the prevalence of infection and then homogenize whatever remaining larvae can be collected. After a period of refrigeration, the homogenate is examined and, if microsporidian spores are detected under light microscopy, the mixture can be filtered or centrifuged to isolate the spores. The spores can be fed back to the host larvae and the microsporidium can be isolated in pure culture from specific infected tissues.

It must be emphasized that exploration in some countries may require that the explorer should be prepared for the unexpected. Common materials and instruments that one might take for granted such as microscopes, slides, cover slips, vials, pipettes, forceps, fixatives, antibiotics and even paper products may not be available. Therefore, these items must be included in the travelers' inventory. Preferably, any entomopathogens that are isolated should be kept refrigerated whenever possible to reduce contamination by bacteria and other organisms.

4 Pathogen selection

The complex of natural enemies associated with a host in its area of endemism often includes parasitoids and predators and several different types of pathogens. For example, in North America, in addition to insect natural enemies, several species of viruses, microsporidia and fungi and one bacterium have been associated with the fall webworm, *Hyphantria cunea* (Kim, 1967; Shu and Yu, 1985; Fuxa *et al.*, 1998). Because relatively few introductions of pathogens have been undertaken, guidelines for selecting which species should be introduced have not been developed. For introductions of parasitoids and predators, while practitioners do not always agree on those traits associated with the most successful introductions, this issue has been discussed extensively (see van Lenteren, 1980; Ehler, 1990a; Waage, 1990; Hopper,1996).

In selecting a natural enemy for introduction, one suggestion by researchers working with parasitoids has been that the first species for introduction should be the one most highly adapted and best synchronized with the host (Pschorn-Walcher, 1977). Relative to pathogens, the first species for release would therefore be one having superior reproductive capacity, persistence (either within the environment or persisting within hosts), and mechanisms for transmission (including, in some cases, dispersal). We frequently know little about the associations between pathogens and hosts in the field, or the effect of the environment on pathogens, hosts or their interactions, making selection of pathogens somewhat based on chance. Unfortunately, highly adapted pathogens are not always the most abundant in nature or the easiest to isolate and culture so that our ability to work with a pathogen can play a large part in determining whether it is considered first for introduction.

Many pathogens have specialized life histories that overcome area-specific unfavorable periods such as winter, dry weather, or periods of very low host densities. Studies have shown that pathogen species or even isolates within a species can vary in environmental tolerances (*e.g.*, Kung *et al.*, 1991; Vidal *et al.*, 1997). Therefore, whether choosing an isolate from foreign exploration trips or from culture collections, the isolate should represent areas with climates most closely matching the area of release to increase the chance of success (see above). Releases of the fungal pathogen *Z. radicans* in Australia against the spotted alfalfa aphid, *Therioaphis trifolii*, utilized a previously isolated strain from Israel (Milner *et al.*, 1982) whose subsequent success could be influenced by pre-adaptation to the climate of the area of release.

Studies should always be conducted to ascertain that the pathotype or species to be introduced is virulent against the intended host. For example, to choose an entomopathogenic nematode for release against the sweetpotato weevil, *Cylas formicarius*, ten strains or species were compared in the laboratory (Mannion and Jansson, 1992). The variables used to select which nematode to introduce included the LC_{50}, pupal and adult mortality, larval mortality rate and progeny production. To choose a nematode for release against the woodwasp *S. noctilio*, seven species were considered. First, these nematodes were screened for species that utilized the symbiotic fungus of *S. noctilio* as an alternate food source as well as parasitizing and sterilizing the host without harming the insect parasitoids (Bedding, 1979). Because these nematodes rely on *S. noctilio* for their dispersal, studies were also conducted to select nematodes that had little to no effect on flight behavior of nematode-parasitized female woodwasps.

5 Evaluating non-target effects

The pathogen to be released should have negligible impact on non-target organisms. This subject is covered in depth in Chapter X-1. Here, we mention that for pathogen introductions, when considering whether a pathogen is a good candidate for release, host range should be investigated. Researchers should scour the literature to find any references to other hosts of this pathogen in order to better understand the breadth of the pathogen's host range. Laboratory bioassays challenging beneficial insects, *e.g.*, honey bees, are usually the first step in host range studies but should be followed by tests of the susceptibility of a diversity of insects, especially including those that might occur in the target host's habitat. Bioassays should use the pathogen strain to be released because strains within a species of microorganism can differ in virulence to host species. Unfortunately, hosts can be infected in the laboratory that are never found infected in the field. Therefore, if possible, field studies should be conducted to evaluate whether hosts other than the target species might become infected. In many cases, this is problematic because the pathogen to be released does not already occur in the release area and cannot be released prior to acquisition and evaluation of this host range information. In such cases, investigating the diversity of host species in the area of endemism that are naturally infected can give an indication of the breadth of pathogen host range prior to release. As an example, Solter *et al.* (2000; 2005; unpublished data) have investigated the host ranges of microsporidia in their areas of origin. Pathogens were recovered from non-target hosts and were then evaluated for morphological and behavioral (e.g. tissue specificity) characteristics, to determine whether they were the same as the pathogens infecting the target host. The target host was then inoculated with any pathogens that were observed to be similar to the released/naturally occurring isolate. Those pathogens that caused typical infections in the target host were subjected to genetic testing to determine the relationship with the pathogen of interest (*i.e.*, to confirm whether the pathogen of interest was the same as the pathogen isolated from non-targets). The conclusions of these authors based on several systems has been that the host ranges of microsporidia in the field are much narrower than would be predicted from laboratory studies.

6 Pathogen importation

There are several concerns regarding shipment of a pathogen from one country to another. First, it is important that the necessary permits accompany the material being imported. These permits are most commonly obtained from the appropriate authorities in the country into which the pathogen is being introduced. However, there are also some countries that require permits for endemic organisms to be relocated for biological control introductions. Thus, the country from which the pathogen is being introduced should be contacted to see whether such regulations are in place (Moraes and De Nardo (1996) (see above).

In order to exclude unwanted microorganisms, it is critical to obtain a purified isolate of a pathogen before its use in introduction programs. Unfortunately, in the case of fastidious organisms, isolation may be impossible to achieve. Pathogen isolation frequently results in segregation of a limited amount of genetic material from the bank of genotypic diversity present in the area of endemism. In this case, the populations being introduced may, by chance, not be the most robust. For example, the 1910–1911 releases of *E. maimaiga* in Massachusetts were based on resting spores from only two cadavers and the resulting strain of this fungus that was introduced was only weakly virulent (Speare and Colley, 1912; see Hajek *et al.*, 1995). Therefore, after isolation, the virulence of the strain to be introduced should be ascertained through laboratory bioassays. Many pathogens are known to lose virulence through prolonged *in vitro* culture in the laboratory or from storage. For example, prolonged storage of *Heterorhabditis* sp. can lead to decreased motility and infectivity (Westerman, 1992). Care should be taken so that such adaptation to laboratory conditions or decreased functioning due to long-term storage, with resulting decreases in activity in the field, do not occur. Accordingly, bioassays should be conducted as quickly as possible using a subsample of the strain to be introduced while maintaining the original pathogen strain in a quiescent state.

It has been argued that introduction of a greater diversity of genetic material could allow for enhanced ability of an organism to adapt to a new environment (Messenger *et al.*, 1976). With introduction of more genetic diversity, there is a greater chance that alleles vital for survival and virulence in the new environment would be included in the subsample of the pathogen collected and released (Roush *et al.*, 1990). There is insufficient evidence from parasitoids and predators to link increased size of the founder population with successful establishment (Mackauer, 1976). While the question of how much genetic variability should be introduced remains unanswered, especially regarding pathogens, it seems prudent to suggest that more rather than less biological material should be collected during foreign exploration and that diversity should be maintained and represented during releases. Unfortunately, this runs counter to the standard methodology of working with pathogen strains that are limited in genetic diversity, *e.g.*, single spore isolates of entomopathogenic fungi.

Agents for biological control introduction generally must be sent to quarantine facilities to ensure that there is no contamination before the material is released (Nickle *et al.*, 1988; Coulson and Soper, 1989). Appropriate quarantine facilities should be contacted so that they are prepared for the shipment. The operational process of shipping pathogens for release has been facilitated by the fast courier services available today. Care must still be taken to package the pathogen securely and to include coolant materials in the packing to minimize overheating that could potentially kill the pathogen or encourage multiplication of contaminants. If the pathogen being sent is fastidious, it will be critical for hosts to be present at the destination. For pathogens that are more difficult to culture *in vitro*, the specific media and instructions on necessary conditions must be prepared in advance. Especially in these latter cases, the scientists who will receive the shipment once it is sent should be notified. Despite these safeguards, extended delays can still occur, especially when the packages must clear customs when pathogens are being shipped. During these delays, temporary refrigerants can lose their effectiveness, thus compromising the shipment. It is always preferable that pathogens should be hand-carried with permits to avoid such delays and assure that fragile pathogens arrive in good condition.

7 Introduction and dissemination of pathogens

The methods used for releasing pathogens and the conditions for their release that will enhance establishment can differ by pathogen group. Most entomopathogens are not actively mobile and do not search for hosts, so releasing inoculum in the correct location for transmission to hosts is critical. In general, entomopathogenic bacteria, viruses, and protists must be eaten to infect because they usually invade the host's body through the gut wall. Therefore, it is critical for inoculum to be deposited in locations where hosts will ingest it, *e.g.*, inoculum should be deposited on leaves for pathogens of phytophagous insects. In contrast, entomopathogenic fungi infect by penetration through the host cuticle and therefore the ultimate objective for fungal release is deposition of propagules on the host itself. Entomopathogenic nematodes differ because many species activity search for hosts (*i.e.*, cruisers) while others have a sit-and-wait strategy (*i.e.*, ambushers). Many species of entomopathogenic nematodes gain access to the host's body through natural openings (*i.e.*, the mouth or anus) and then invade through the gut, although some species can actively penetrate through the cuticle. Therefore, the optimal location for release can differ according to the biology of the nematode species, emphasizing the need to understand the basic biology and ecology of the pathogen or nematode being released.

Although the major pathogen groups differ in their basic biologies, there are some commonalities useful for optimizing releases. For all pathogens, successful introduction and colonization require adequate numbers of hosts and survival of the pathogen in the host habitat, resulting in levels of infection sufficient to allow the pathogen to infect and persist in the environment. In this section, we will discuss the conditions required for successful pathogen establishment as well as different methods that have been or could be used for successful introduction.

A Pathogen identification

The taxonomy of many groups of pathogens is in a state of flux. In order to gain permission to conduct a release, we must be able to identify the pathogen to be released. Therefore, taxonomists should always be involved in research activities related to the pathogen. In addition, voucher specimens of any pathogen to be released should always be maintained for future use in identification or verification.

Before a species or strain is released it is critical that it can be identified after release. Pathogens can be detected and identified using morphological, physiological, behavioral and genetic characteristics. It is rare that only one of these categories of characters is used to identify pathogens; however, morphological characteristics are traditionally used first. Unfortunately, these characters are virtually never adequate for differentiating isolates, pathotypes, and varieties within a species. Currently, sub-cellular methods must be used for adequate discrimination both within and between species of certain pathogens. Biochemical, serological or molecular techniques are powerful tools for distinguishing pathogen strains. Molecular techniques are often used to solve taxonomic problems with pathogens and their applicability in classical biological programs was demonstrated in the introduction of the entomopathogenic fungus *Entomophaga praxibuli* from Australia to North Dakota, USA (Bidochka *et al.*, 1996). Bidochka *et al.* (1995) described a method by which cloned DNA fragments were used to construct pathotype-specific probes to allow positive identification of *E. praxibuli* versus morphologically identical endemic North American *Entomophaga grylli* pathotypes. To evaluate establishment, thick-walled resting spores within field-collected grasshoppers were fractured to extract DNA. Probes were used in dot blot analysis that made screening for pathotype prevalence possible in large numbers of infected grasshoppers. Although some pathogen species with potential for biological control are cosmopolitan in distribution, *e.g.*, *Metarhizium anisopliae* (St. Leger *et al.*, 1992), it may be desirable to introduce more efficient strains than those that already occur in an area. The inability to distinguish among strains within a species has been an impediment to facilitating the importation of pathogens. One recent example is the program for introduction of fungal pathogens into Africa to control the introduced cassava green mite,

Mononychellus tanajoa. Brazilian isolates of the entomophthoralean fungus *Neozygites tanajoae* are highly specific and virulent to *M. tanajoa* (Moraes and Delalibera, 1992; Delalibera and Hajek 2004; Delalibera *et al.*, 2006). Low prevalence of infection of cassava green mite by *N. tanajoae* in Benin, Africa (Yaninek *et al.*, 1996) suggested that, although this pathogen already occurs in Africa, the endemic pathotype is inefficient. Precise discrimination among the closely related *Neozygites tanajoae* isolates from Brazil and Africa was not possible using classical taxonomic criteria but discrimination is critical to evaluation of the establishment and efficacy of released strains. For the development of molecular probes to differentiate among *N. tanajoae* strains, RAPD (Random Amplification of Polymorphic DNA) markers were converted into SCARs (Sequence Characterized Amplified Regions) and specific primers were designed for discrimination between indigenous and exotic isolates. The molecular probes now allow evaluation of the outcome of releases of Brazilian isolates in Africa.

B *Obtaining necessary approval for pathogen release*

Approval by the appropriate regulatory agencies is always required prior to pathogen release. Attention must be paid to the regulatory environment because regulations can always change and it is possible that approval for releases might be needed from other countries on the same continent. In 1996, the Food and Agriculture Organization (FAO) published a Code of Conduct for the import and release of exotic biological control agents (FAO, 1996); this is discussed further in the following section. There is a definite need for the international community to follow such a protocol, especially those agencies in the USA and abroad that are responsible for regulating classical biological control activities. We have not attempted to describe the permitting process required in various countries but rather will focus our discussion on the current situation in the USA as an example.

In the USA, biological control organisms cannot be lumped together for the sake of regulatory simplicity, and, unfortunately, there has not previously been a clear and comprehensive regulatory road map for importing and releasing entomopathogens, although such a system is now being developed. The importation of potentially damaging organisms, including biological control agents such as exotic pathogens, is a significant federal action that requires compliance with the National Environmental Policy Act (NEPA) and the permitting processes of the responsible federal and state agencies (Charudattan and Browning, 1992).

The principal federal agencies responsible for implementing regulations, policies, and procedures for all biological control agents in the USA are the Environmental Protection Agency (EPA), the Animal and Plant Health Inspection Service (APHIS) within the U.S. Department of Agriculture (USDA), and the Fish and Wildlife Service (FWS) within the U.S. Department of the Interior (DOI). A review of the roles of these agencies in regulating the importation and release of biological control organisms is provided by Leppla (1996) and is also discussed in the Office of Technology Assessment report on harmful non-indigenous species (U.S. Congress, 1993).

The statutory authority in APHIS for regulating biological control organisms resides primarily in the Federal Plant Pest Act (FPPA). This act states that APHIS has the legal authority to regulate only those organisms that meet the definition of a "plant pest," *i.e.*, phytophagous organisms intended for the biological control of weeds. Entomophagous organisms, because they are antagonists of plant pests and not of plants, do not meet the definition of "plant pest" and, therefore, their inclusion within the scope of the APHIS FPPA-based regulations was not wholly appropriate.

The case for importation and release of entomopathogens is even more complex because the EPA currently regulates non-indigenous microorganisms intended for use as classical biological control agents as if they are microbial pesticides. The EPA has the authority through the Federal Insecticide, Fungicide, and Rodenticide Act (FIFRA) to regulate pesticides, which they define as "any substance or mixture of substances intended for preventing, destroying, repelling, or mitigating any pest." However, the regulations state that such a pesticide will be exempt from FIFRA if EPA determines that the pesticide is adequately regulated by another Federal Agency. Based on APHIS' historical involvement in biological control regulations, paragraph (a) of that section exempts from FIFRA all biological control agents except (1) eukaryotic organisms such as protists, algae, and fungi, (2) prokaryotic microorganisms such as bacteria, and (3) viruses. Since entomopathogenic nematodes are exempt from regulations under FIFRA, their release as biological control agents continue to be under the responsibility of APHIS. All other entomopathogens are not excluded and therefore these organisms must undergo much more stringent evaluation prior to release.

Since 1984, EPA has had policies in place that require that the EPA must be notified prior to small-scale testing (< 10 acres = 4 ha), for all genetically altered and non-indigenous microbial pesticides. Consequently, scientists who intended to import entomopathogens for classical biological control were urged to consult in advance with state and APHIS authorities to ensure that the necessary permits were secured to import and transfer microorganisms to a certified quarantine facility. If an organism demonstrated promise as a classical biological control agent and a small-scale release was proposed, then it was advised that consultations must be scheduled with APHIS or EPA officials to make sure that the statutes were satisfied; for both agencies, an environmental assessment was required but the required contents of the assessment varied. Obviously these requirements were cumbersome and somewhat limited the potential for using entomopathogens for biological control of plant pests. This is precisely why Miller and Aplet (1993) proposed that new biological control legislation was needed to develop guidelines and provide a road map for the safe use of biological control agents.

Fortunately, progress has been realized in recent years to provide guidance to scientists who are interested in assessing or utilizing non-native entomopathogens for classical biological control. In 2000, the North American Plant Protection Organization (NAPPO), a plant protection organization created in 1976 to coordinate the efforts of Canada, the U.S. and Mexico under the authority of FAO, endorsed a standard similar to that endorsed by the European Plant Protection Organization (EPPO) earlier in the same year (EPPO, 2000). Both of these regional organizations used the FAO Code of Conduct (FAO, 1996) as their baseline document. APHIS, the U.S. representative to NAPPO, then provided a directive for requests to release non-native entomopathogens and entomopathogenic nematodes for biological control of pest insects and mites. The guidelines in the directive are intended to assist researchers in drafting a petition for release of exotic entomophagous agents for biological control and to assist reviewers and regulators in assessing the risks of introductions.

Each permit application (PPQ Form 526) to request the release of a non-native entomopathogen should include the following broad categories of information, each in the form of a separate report: Proposed action; Biological control agent information; Target pest information; Environmental and economic impacts of the proposed release. There are 5–10 specific requests for information within each of these broad categories. The details for entomopathogen permit applicants can be accessed at an APHIS website: http://www.aphis.usda.gov/ppq/permits/biological/entomopathogens.html. A formal review of proposals requesting the release of entomopathogens is facilitated through NAPPO by APHIS-PPQ, Pest Permits Evaluation Branch. A successful test case in 2005–2006 (Solter and McManus) for release of microsporidia from European gypsy moth required approval from EPA before APHIS acted on the NAPPO proposal. At this time (September 2006), it remains unclear what the final mechanism for approval will entail. However, the procedure now in place provides some guidance for scientists to follow and this approval bodes well for future releases.

C Site selection and timing of pathogen introductions

To optimize establishment, releases should be conducted in diverse climatic areas. This enhances the probability of finding areas with appropriate climates and other requirements that are conducive for establishment. Van Driesche and Bellows (1996) described other criteria that might enhance selection and maintenance of release sites for field colonization of new biological control agents. Release areas should be physically secure, stable, and well located so that the introduced agent can persist and spread into new areas. Sites should be protected from destruction or from applications of pesticides. Harvest of crops serving as food for hosts or other agricultural practices should not be carried out if this will compromise the survival of the introduced pathogens. Soil should not be plowed if this practice will destroy resistant stages of the pathogen or hinder their transmission within the appropriate habitat. For example, alfalfa should be left uncut for at least one cutting to allow the establishment and spread of pathogenic organisms (Hall and Dunn, 1958). If the pathogen infects forest or orchard pests as well as pests of annual crops in less stable habitats, inoculations should be conducted in the more stable habitats or in sites with mixed crops. Past efforts to colonize pathogens in new areas have been more successful in habitats that are more stable over time (Hajek *et al.*, 2005).

Site characteristics can be critical to pathogen establishment, activity and persistence. For example, choice of a site with appropriate physical soil characteristics is important for some entomopathogenic nematodes (Jansson, 1993). For pathogens in aquatic systems, water purity can influence pathogen activity. In systems where the host plant of polyphagous phytophagous insects has influenced pathogen activity (*e.g.*, Hajek, 1997a), releases should be made against hosts feeding on the host plant most favorable to the pathogen. Biotic agents in the environment could either compete with pathogens or enhance pathogen activity but detailed knowledge of disease epizootiology is required to use such interactions to enhance establishment.

For many pathogens, high relative humidity and moderate temperatures in the microenvironment are required during the infection process. For example, releases of fungi are frequently conducted most successfully during late afternoon or early morning and a suitable release site would be a moist or irrigated area. Relative humidity can be increased by spraying water within cages used as primary inoculation sites or by using plastic bags to cover plants or stems containing both the inoculum and healthy hosts. In addition, most pathogens are very sensitive to ultraviolet radiation and application to areas protected from direct sunlight should facilitate greater survival of the inoculum.

D Introducing more than one pathogen species?

This subject has been much-debated regarding releases of parasitoids and predators, especially when a sizeable guild of parasitoids attacks the target pests (Ehler, 1990a). Additional discussions have centered around whether releases of multiple species should be simultaneous or sequential. For pathogens, we have much less of an understanding of interactions within pathogen guilds and, in fact, introductions of more than one pathogen species against one pest have rarely been undertaken (see Hajek *et al.*, 2005). If there is an opportunity for introductions of multiple pathogens against one host, we suggest that laboratory bioassays of the pathogens alone and together would lend information about pathogen interactions at the organismal level.

Laboratory bioassays have demonstrated that steinernematid and heterorhabditid nematodes compete for lepidopteran hosts, although total mortality was greater when both were applied (Alatorre-Rosas and Kaya, 1990; 1991). Likewise, laboratory studies showed that the closely related microsporidian species *Nosema lymantriae* and *Vairimorpha disparis* compete within individual gypsy moth hosts (Solter *et al.*, 2002; Pilarska *et al.*, 2006) and have not been observed occurring together in field populations, Laboratory interaction studies of *E. maimaiga* and a nucleopolyhedrovirus (NPV) against *L. dispar* have demonstrated little negative impact of these pathogens on each other during coinfections. However, due to differential speed of kill, order of infection can impact whether both pathogens reproduce during coinfections (Malakar *et al.*, 1999). However, laboratory bioassays do not always reflect activity in the field, *e.g.*, ever since *E. maimaiga* became abundant in North America, epizootics caused by NPV in gypsy moth populations have been less frequent and restricted to drier springs (Hajek, unpublished data). This change in disease dynamics could be due, in part, to the fact that high host densities are required for viral epizootics and *L. dispar* populations have predominantly remained at lower densities since *E. maimaiga* became abundant. Laboratory bioassays did not predict this outcome which is hypothesized to be driven by host density and environmental conditions, demonstrating the importance of understanding interactions at both the organismal and population levels. Questions about pathogen interactions at the population level could be addressed through field studies in the area of endemism of the pest, if both (all) pathogen species of interest are native to the same area.

E Pathogen inoculum for release

Once the choice has been made about which pathogen or pathogens will be introduced and when and where pathogen(s) are to be released, the next step is to decide which stage of the pathogen to release, coupled with how much to release and how to conduct the release. Based on the pathogen, either infective or environmentally resistant stages of pathogens are generally released [*e.g.*, while conidia of anamorphic hypocrealean (= Hyphomycete) fungi are released, for Entomophthorales, resting spores or hyphal bodies are released instead of the delicate conidial stage (Pell *et al.*, 2001)]. Under descriptions of individual release methods, we discuss appropriate pathogen stages for release (see below).

The amount of pathogen to release in each location and the number of releases to make are difficult decisions. The amount released and number of release sites are frequently determined by the ease of pathogen production and the potential for natural dispersal by the pathogen, as well as regulatory requirements. In some cases production of large quantities of inoculum is very difficult based on current methodology. Mass production is not as great a concern with pathogens that readily grow on or in inexpensive media (*e.g.*, some bacteria and fungi).

The quantity of pathogen to release and/or the number of releases for successful establishment and control must be specific to the host/pathogen system. Insufficient quantities of pathogen released can cause introductions to fail, especially with pathogens that need to form a reservoir of resistant or overwintering stages to survive during unfavorable seasons or during low host densities. Some biological control programs using nematodes have been successful with single releases although multiple releases may be needed for other systems (Jansson, 1993). Laboratory bioassays and pre-release studies could help to provide indications about ideal concentrations for release.

F Methods for introduction

In this section, specific methods for introducing pathogens that have been utilized in classical biological programs are described. Among the examples of classical biological control, many are inadequately documented and, in some cases, methods changed as the programs progressed. We have emphasized examples where methods have been described in detail and, as might be expected, these are generally from very successful programs. However, because classical biological control introductions have been conducted relatively rarely with pathogens, poor success in methods used in the past does not prove that these methods cannot succeed. We also discuss methods for release and dissemination of pathogens that we feel have promise but which have not been used previously or have been used to release pathogens in areas where they already occurred, (*i.e.*, augmentation).

Biological control releases can be categorized as inoculative or inundative based subjectively on the amount of material released. For inoculative introductions, it is assumed that if some minimal amount of inoculum is introduced, the organism will then be transmitted in the host population on its own to have an impact. An inundative introduction consists of releasing large amounts of inoculum for immediate control without waiting for the pathogen to increase on its own in the environment. The reason not all programs utilize inundative introductions is that for many pathogens it is not possible to obtain or to produce the massive amounts of inoculum needed for inundative introductions. Specific attributes of host/pathogen systems determine the correct amount of inoculum to release. System specific attributes also play a large part in which introduction strategies are optimal.

1 Inoculative introductions

Inoculative releases are especially appropriate for fastidious organisms or those that are difficult to produce *in vitro*. In these cases, little material is required for release and the inoculum is often produced in field-collected or laboratory-reared hosts. This type of release is also appropriate for introducing pathogens that attack hosts living predominantly in cryptic habitats or in locations difficult to reach using standard spray application technology (*e.g.*, rhinoceros beetles, *O. rhinoceros*, in the crowns of palm trees).

a Release of living-infected hosts and pathogen-killed hosts

This is the simplest method and consists of transferring living-infected or pathogen-killed hosts to a location where the pathogen is not present. The pathogen contained in the transported hosts is usually intended to be transmitted to native hosts during the same season as the release. In the past, many attempts to use pathogens in classical biological programs were made by transporting leaves with living and dead infected hosts, generally collected during an epizootic (Dustan, 1927; Hall and Dunn, 1958). To conduct releases, leaves with diseased hosts of smaller sizes could be attached to the undersides of leaves with potential hosts. This approach was recommended only after checking for high infection levels in the host populations on the leaves to be introduced. While this method was previously considered an excellent and easy way for distributing many kinds of pathogens of phytophagous pests, it is no longer used due to the risk of accidentally introducing plant diseases, other arthropod pests, hyperparasites, etc. along with the intended entomopathogen. However, this technique remains a useful method to speed dispersal to different localities within the same region after initial releases. Conducting large-scale field releases by application of field-collected infected insects is often not feasible because of the difficulties in finding epizootic sites where enough inoculum can be collected. For example, to collect cadavers of *E. maimaiga*-killed *L. dispar* larvae that contain the environmentally resistant resting spores most appropriate for release, epizootics must be detected and visited within ca. 3 weeks of massive mortality because after this time cadavers fall to the ground and decompose, releasing the resting spores into the soil, and the chance to collect spore-bearing cadavers is gone until the next year (Hajek *et al.*, 1998b).

In some cases, healthy hosts can be field-collected or reared in the laboratory, infected, and then released. This method has been used for some entomopathogenic species of Entomophthorales because conidia are short-lived and difficult to produce on artificial media. For these fungal species, an infection procedure utilized to infect larger insects is microinjection with a suspension of protoplasts while smaller hosts can be exposed to infective conidia (Papierok and Hajek, 1997). After a few days, diseased insects die and mummify. When the cadavers are subsequently placed in environments having high humidities, the fungus initiates conidiogenesis. This technique was used for introducing the grasshopper pathogen *E. praxibuli* against *Melanoplus differentialis* populations in North Dakota, USA (Bidochka *et al.*, 1996). From 500 to 2500 grasshoppers were each injected with about 10^4 *E. praxibuli* protoplasts and released at each site. Releasing infected insects was also considered efficient for introducing the entomophthoralean fungus *Neozygites fresenii* against cotton aphids, *Aphis gossypii*, in California (Steinkraus *et al.*, 2002). Despite heavy competition by predators and hot, dry weather, *N. fresenii* was successfully transmitted when 1125 fungal-killed aphids that had not yet produced conidia were released on cotton leaves in the field. In this project, researchers compared releases of lab-produced cadavers with releases of living-infected insects but found that releasing cadavers was less labor intensive and resulted in higher infection levels.

An alternate method for releasing Entomophthorales was used to inoculatively release *Entomophaga aulicae* against gregarious larvae of the browntail moth *(Euproctis chryssorhoea)* in the northeastern USA (Speare and Colley, 1912). To produce *E. aulicae*-infected insects, so-called ‘infection boxes’ were seeded with cadavers containing resting spores and healthy larvae were placed into the boxes for infection. Bags containing these living-infected browntail moth larvae were then suspended in as close proximity to webs or masses of feeding larvae as possible. One side of the bag was opened to allow infected larvae to escape and join the wild population with the hope that the pathogen would thus be transmitted to the healthy population.

A very successful program based on releasing virus-infected insects was mounted against the scarab *O. rhinoceros*, a pest of palms on islands in the South Pacific. As the program developed, researchers tried several different methods for releasing the virus (see below), but the best technique was releasing virus-infected adults (Bedford, 1981). Detailed studies showed that this non-occluded virus multiplies in adult midgut cells and is dispersed during defecation. Based on this information, researchers hypothesized that adults are the principal vectors in nature. Adults were obtained by field-collecting, attractant trapping, rearing from field-collected larvae, and rearing from eggs. To ensure infection of adults for release, they were immersed for 2–3 minutes in a suspension of 2 macerated virus-killed insects/liter of water and then were placed for 24 hours in 1 kg sawdust mixed with one half of a virus-killed grub in 500 ml of water. For release, beetles were placed into the natural habitat and allowed to crawl under logs or into vegetation, and they subsequently dispersed at night. In the Fiji Islands, 30–50 adults inoculated in this way were released per site (Bedford, 1977), and they dispersed widely before dying. Researchers hypothesized that in this way virions were distributed to both breeding sites and palm crowns both before and after adults died.

In a very recent program, Solter and McManus (2006 NAPPO Proposal) plan to release living gypsy moth larvae fatally infected with three species of microsporidia in three sites, one pathogen species per site. This method was chosen to avoid wholesale contamination of the oak foliage environment that would occur with sprays and to allow a slow constant release of viable inoculum via feces and cadavers throughout the spring-time larval development period.

b Release of egg masses treated with pathogens

Several species of microsporidia infecting *L. dispar* have been introduced experimentally in the USA by applying a suspension of spores directly to egg masses (from a laboratory colony) that were then attached to tree trunks (Jeffords *et al.*, 1988; Bauer *et al.*, 1993, 1994). This method was possible because *L. dispar* neonates consume ca. 50% of the chorion when they emerge from eggs. Microsporidian spores were produced in insects from a laboratory colony and were applied to egg masses by soaking pathogen-free egg masses for 5 minutes in suspensions of microsporidian spores containing a few drops of Tween 80. Suspensions varied from 1.0×10^5 to 3.3×10^6 spores/μl, based on the known LD_{50} of microsporidian isolates, and concentrations were chosen that would cause infection but would also minimize larval mortality. These ready-to-hatch egg masses were dried overnight, then encased in 2×2 cm envelopes made from netting and stapled to tree trunks at 60–400 contaminated egg masses per site. This technique resulted in establishment of the pathogen during trials in Maryland (Jeffords *et al.*, 1989) but was not as successful in later trials in Michigan. In Michigan, cold weather delayed host egg hatch for 7–10 days so that the microsporidian spores were exposed to environmental conditions much longer; it is hypothesized that lack of establishment occurred because ultraviolet radiation exposure caused extensive spore mortality (L. S. Bauer, personal communication).

c Releasing pathogens into the host habitat

Spores of bacteria and resistant stages of fungi and microsporidia can be applied as sprays or dusts directly onto the substrates that hosts eat or in which they live, *e.g.*, seeds for seed-feeding hosts or onto the soil for soil-dwelling pests. For entomophthoralean species that produce resting spores, release of such spores can be highly successful (*e.g.*, Hajek *et al.*, 1996). Entomophthoralean resting spores can be stored for long periods (Hajek *et al.*, 2001) and, after inoculation in the field, their germination can occur over numerous years (Weseloh and Andreadis, 1997). For many species of Entomophthorales, these spores are constitutively dormant after production and dormancy requirements must be satisfied before germination is possible (Hajek, 1997a). Small scale studies demonstrated that introduction of *E. maimaiga* was very efficient when overwintered resting spores were released (Hajek and Roberts, 1991). Because at the time it was not possible to produce large quantities of resting spores for release, soil containing resting spores of this pathogen was used for releases. To collect soil containing resting spores, the distribution of resting spores had to be determined; resting spore densities were highest at a soil depth of 0–3 cm from a radius of < 10 cm around the bases of oak trees that had been covered with cadavers of *E. maimaiga*-killed gypsy moth larvae the previous year (Hajek *et al.*, 1998a). In a more extensive project utilizing releases of resting spore-bearing soil, *E. maimaiga* became established in 39 of 41 release sites in areas more recently colonized by the spreading gypsy moth populations in the eastern USA (Hajek *et al.*, 1996). Importantly, movement of soil containing *E. maimaiga* resting spores was problematic; this procedure had to be approved by USDA-APHIS and states to avoid inadvertent movement of insects, nematodes, or plant pathogens and methods to confirm densities of resting spores in soil were time-consuming (Hajek, 1997b). Due to these difficulties, efforts shifted from moving soil containing resting spores to collecting resting spore-filled cadavers after epizootics for release in new areas.

In field studies in the gypsy moth area of origin (Europe), researchers tested methods for releasing microsporidia and evaluating infection in gypsy moth as well as non-target lepidopteran larvae. Microsporidian spores were sprayed on oak tree foliage and infection rates in gypsy moth larvae and in non-target Lepidoptera feeding on the same trees were evaluated (Solter *et al.*, unpublished data). Infection rates varied from 12 to 75% in the gypsy moth, with the more virulent *V. disparis* producing higher rates of infection. Several non-target species, primarily those related to the gypsy moth, were infected by *V. disparis* but no infections were subsequently found in 2 years of follow-up monitoring. Only gypsy moth hosts were infected by sprays of *N. lymantriae*.

In the most successful introduction of an entomogenous nematode for insect control, the nematode *D. siricidicola* was released against the introduced woodwasp *S. noctilio* in Australia (Bedding and Iede, 2005). *D. siricidicola* has two phases in its life cycle: one is mycetophagous, normally feeding on the wasp's symbiotic fungus, while the second is entomophagous, normally infecting woodwasp larvae and living in the host's hemocoel. The program to release *D. siricidicola* involved conducting extensive surveys in Europe, Japan, and New Zealand to compare attributes of strains of this nematode (Bedding, 1979). This nematode can be mass cultured in the mycetophagous phase on wheat and water, yielding 3–10 million nematodes/500 ml flask, a quantity sufficient to inoculate 100 m of timber (Bedding and Akhurst, 1974). Nematodes can be stored for several weeks at 5–10 °C in water within an atmosphere of pure oxygen and they were transported to release sites in this manner. Nematodes were inoculated directly into infested wood using a hypodermic syringe. In order for the nematodes to move into the wood where the host larvae are developing without first desiccating, the precise method of inoculation is critical; a wad punch mounted to form a hammer was used to make inoculation holes and aerated 12% gelatin containing 4000 nematodes/ml was introduced into holes. This medium was used because nematodes will readily emigrate from it (Bedding and Akhurst, 1974). One inoculation hole/m of tree trunk has been shown to produce maximal parasitism. Either trees containing host brood are felled and then infested with the nematodes, or logs containing infested brood are transported to release sites. Once wasps emerge from these logs, the infected females are sterile but will still nemaposit into trees containing healthy hosts, thereby spreading the nematodes throughout the area. Because these nematodes can spread on their own by way of infected female *S. noctilio*, releases of as few as 50 parasitized *S. noctilio* have resulted in establishment throughout an area. As a final note in this long-term program, culture of *D. siricidicola* in the mycetophagous phase in the laboratory for 20 years led to selection of a nematode strain that rarely formed the parasitic phase (Bedding, 1993). Use of this so-called "defective" strain led to increased expense because four times more trees needed to be inoculated than previously. Now, to prevent loss of virulence, fresh nematode cultures are started yearly from the original virulent strain that has been stored in liquid nitrogen. Decline in virulence with *in vitro* culture is common for many diverse groups of pathogens and care must be taken in long-term projects to ensure the maintenance of a strong source of inoculum that retains virulence.

d Attracting hosts to pathogen sources

To release the non-occluded virus that attacks the rhinoceros beetle, *O. rhinoceros*, researchers initially took advantage of the behavior of these beetles in order to expose them to the virus (Bedford, 1977). *O. rhinoceros* breed in a diversity of types of decaying vegetation. Therefore, mounds of decaying leaves and compost or sawdust (3–4 m^2 in area) were mixed with 10–50 macerated virus-killed larvae in water. Adults were attracted to these mounds, became infected and externally contaminated while crawling through them, and then spread the virus to other locations. Unfortunately, a large labor force was necessary to create and maintain these decaying mounds of vegetation. As an alternative, contents of a half sack of sawdust were mixed with 10–50 cadavers and 6–10 1.2 m long pieces of split coconut logs were placed on top of the sawdust. Once again, adults were attracted, became inoculated and spread the virus but, unfortunately, naturally occurring breeding sites were often more attractive and insufficient numbers of adults visited the artificial sites before this non-occluded virus became inactive.

The microsporidium *Paranosema locustae* (= *Nosema locustae*; = *Antonospora locustae*) was released several times from 1978 to 1982 in pastures in Argentina for control of grasshoppers (Lange and Wysiecki, 1996). Releases followed standard protocols developed in the USA for releasing this microsporidium for grasshopper control (Henry and Oma, 1981). *P. locustae* spores were produced in grasshoppers, harvested and sprayed onto wheat bran. Wheat bran functions as bait, attracting grasshoppers that ingest *P. locustae* spores as they feed on the bran. The standard quantity for application, 2.47×10^9 spores on 1.68 kg bran/ha, has typically been dispersed using ground application equipment (Henry and Oma, 1981).

e Pathogen dispersal from in vitro *cultures in the field*

The placement of *in vitro* cultures of pathogens in the field has been a useful method for releasing fragile or short-lived pathogen stages that are not available in abundance for spraying. This procedure was used to establish *Entomophthora* spp. and *Z. radicans* in areas infested with aphids in California (Hall and Dunn, 1957) and Australia (Milner, 1982). Cultures of these fungi were grown on artificial media and then inverted over target pest-infested plants in the field so that forcibly discharged conidia landed on susceptible hosts. Due to the necessity for inverting the cultures over aphid-infested alfalfa plants in the field, a firmer medium with 3% agar was used (Hall and Dunn, 1958). In California, researchers used 118 ml waxed food containers with tight-fitting lids as disposable units to propagate fungi. Over 1700 cultures were shipped throughout the release, accompanied by instructions suggesting that releases be made in irrigated alfalfa having a sufficient aphid population or uncut alfalfa not treated with insecticides. For releases in Australia, 45 liter plastic garbage cans with a hole cut in the bottom were inverted over aphid-infested alfalfa plants in the release field (Milner *et al.*, 1982). Culture dishes with *Z. radicans* that was ready to discharge conidia were inverted over the hole in the base of the garbage can. Water was sprayed within the garbage can which was then covered with a plastic bag to create a very humid environment. Garbage cans with sporulating fungal cultures were left in place overnight and then removed and ca. 50% of aphids that had received conidial showers died of disease 2–5 days later (Milner, 1985). In both of these studies, conidia that were showered infected and killed hosts on plants and both airborne conidia and mobile infected insects are thought to have caused the observed subsequent spread of these pathogens from the initial foci of infection.

2 Inundative introductions

For inundative releases, large amounts of inoculum are broadcast over larger areas. These types of releases are usually not conducted during initial establishment of a pathogen but instead during the redistribution phase. Inundative release is only possible with pathogens that can be mass produced and for hosts that live in habitats where they would be contacted by broadcast applications. For example, many bacteria and some fungi can readily be produced *in vitro*. At present, NPVs can only be efficiently mass produced *in vivo* and suspensions of virions for application are predominantly created from disease-killed hosts. However, methodology for mass production of NPVs in cell culture is being developed.

Application of inoculum into the pest habitat has been widely used for viral introductions (Magnoler, 1974; Zethner, 1976; Gomez and Moscardi, 1991). A very successful example of classical biological control used inundative introduction of NPV against the European pine sawfly, *Neodiprion sertifer*, after the introduction of this pest to North America. Inundative introductions were used to spread this virus after its initial establishment because large amounts of inoculum could be generated and the occlusion bodies were very stable. To produce the virus for application, either large numbers of healthy insects were collected from the field and infected in the laboratory or heavy populations in the field were sprayed with virus and cadavers were collected (Cunningham and Entwistle, 1981). In early programs, viral suspensions were prepared by macerating cadavers to release occlusion bodies and then storing occlusion bodies in water at room temperature for up to several months (Bird, 1953). Occlusion bodies were semi-purified by repeated centrifugation and washing. In the 1970's, methods changed and whole cadavers of virus-killed larvae were freeze-dried for storage (Cunningham and Entwistle, 1981). Before field release, freeze-dried cadavers were finely ground, suspended in water, and sodium omadine was added to kill bacterial contaminants. In some cases, virus for application was prepared by adding dilute sodium dodecyl sulphate (SDS) to slurries of cadavers, filtering through muslin, and purifying using sucrose gradients and zonal rotors (Cunningham and Entwistle, 1981). As a caveat, in the USA, while individual landowners can currently collect cadavers and spray a suspension of occlusion bodies on their own property, for such material to be applied elsewhere, it would have to be approved by the Environmental Protection Agency and then registered (McManus, unpublished data).

During the many years of work with the *N. sertifer* NPV, this virus has been sprayed using different types of equipment both from the ground and air. While studies have shown that application of greater quantities of virus yielded greater sawfly mortality (Bird, 1953), as little as a tenth of one cadaver (1×10^7 occlusion bodies) in 13.6 liters of water sprayed onto trees resulted in some infection (Dowden, 1953). However, for general use, dosages ranged from 8.3×10^6 to 2.4×10^{11} occlusion bodies/ha, wetting and sticking agents were added although not considered essential, and biologically inert chemicals were added to enhance infection (Cunningham and Entwistle, 1981). Total spray coverage was usually attempted, although, during one trial in Sweden, only 300 of 10000 infested ha were treated using a "zebra stripe" application. It was assumed that the virus would spread on its own from the application areas. Spread of the virus may also be facilitated by parasitoids and predators of *N. sertifer*.

The most common method for conducting inundative inoculations of fungi is by applying suspensions of conidia that were produced on artificial media (Shands *et al.*, 1958; Ponomorenko *et al.*, 1975; Gomez, 1987). This method is specifically applicable for fungi that are easily mass-produced. Alternatively, during initial attempts to introduce *Aschersonia aleyrodis* to Florida, conidia were rinsed from citrus leaves and these conidial suspensions were successfully applied for control of the citrus whitefly, *Dialeurodes citri*, and cloudywinged whitefly, *Dialeurodes citrifolii* (Osborne and Landa, 1992). During trials conducted for control of the mosquitoes *Aedes nigromaculis* and *Culex tarsalis*, zoospores of *Lagenidium giganteum*[1] were produced *in vivo*, shipped to application sites and sprayed onto the surfaces of bodies of water at 2.7×10^6 zoospores/m^2 (McCray *et al.*, 1973). Mosquito populations were dramatically

[1]*L. giganteum* is an Oomycete, a group previously considered as belonging in the Kingdom Fungi that is now placed in the Kingdom Chromista (Brasier and Hansen, 1992).

reduced within 3 days of treatment. Another fungal pathogen of mosquitoes, *Culicinomyces clavisporus*, was mass-produced as conidia in inexpensive corn-steep liquor-based media for control of the Australian encephalitis vector *Culex annulirostris*. Applications resulted in 95–100% mortality of larvae when this fungus was sprayed at 5×10^9 to 1×10^{10} conidia/m^2 in unpolluted pools (Sweeney *et al.*, 1983). Efficacy of this fungus was reduced to 80% in a pond polluted by sewage effluent due to warmer conditions that allowed for faster larval development.

Large programs have focused on the introduction of the bacteria *Paenibacillus* (= *Bacillus*) *popilliae* and the less virulent *Paenibacillus* (= *Bacillus*) *lentimorbus* for control of the introduced Japanese beetle (*Popillia japonica*) in the USA; these bacteria target larval stages in the soil. The best method for establishing these bacteria is to apply spores directly to beetle-infested soil. These bacilli cannot be produced efficiently *in vitro* so *P. popilliae* has principally been mass produced *in vivo* in Japanese beetle larvae. Spores were formulated in talc and a total of 103,870 kg of spore dust was distributed in over 160,000 sites throughout at least 14 eastern states and the District of Columbia between 1939 and 1953 (Anonymous, 1955; Fleming, 1968). Once *P. popilliae* became established in an area, control through time has been somewhat erratic, although this bacterium can persist in an area through repeated cycles of host infection for at least 25–30 years (Klein, 1981).

3 *Potential methods*

Methods used as examples above have been employed during pathogen release or redistribution. However, several additional methods that we feel have potential for use are described below.

Most pathogens have a limited capacity to disperse on their own and must rely on physical factors and biotic agents for their dispersal (Tanada and Kaya, 1993). Infected hosts, as well as accidentally contaminated individuals such as parasitoids and predatory insects, other invertebrates, mammals, and birds can transport entomopathogens. In fact, several exotic pathogens are thought to have been accidentally introduced when contaminated or infected parasitoids, or possibly predators, were released for biological control (see Hajek *et al.*, 2005). Although such carriers have not been purposefully used for pathogen releases, we know that vectors can play a significant role in disease transmission. Based on the success of pathogens thought to have arrived through introductions of parasitoids or predators (*e.g.*, the gypsy moth NPV), we suggest that more effort should be dedicated to evaluating the use of contaminated parasitoids and predators to release and/or disperse pathogens.

Congenital pathogen transmission is common among viruses and microsporidia. Research has demonstrated that it is possible to release pathogens via infected or contaminated females because the pathogen can be passed to the progeny within eggs (transovarial transmission) or on the surfaces of eggs (transovum transmission) and thereafter subsequently disperses throughout the host population (Bird, 1961). Cabbage looper, *Trichoplusia ni*, adults, contaminated by feeding or spraying, can mechanically transmit NPV to their progeny (Martignoni and Milstead, 1962; Elmore and Howland, 1964). In the gypsy moth, *L. dispar*, the egg surface and the scales surrounding the egg mass may be contaminated with NPV which can serve as inoculum to infect the newly hatched larvae as they chew their way through the chorion (Doane, 1971, 1975). Adult female gypsy moths infected with the microsporidium *Nosema portugal* produce neonates with high levels of infection (Bauer *et al.*, 1995). Therefore, infected or contaminated host adults that are known to regularly transmit pathogens to progeny could be used for pathogen release and dissemination.

Autodissemination is another method utilizing adult hosts for transmission purposes (see Chapter III-3). Hosts are attracted to a trap containing the pathogen, they become externally contaminated, and are then allowed to escape from the trap and distribute the pathogen throughout the rest of the population. Transmission is from adults that have visited traps and also from production of propagules after death of the hosts. Autodissemination would potentially require low levels of inoculum because the contaminated and/or infected adults selectively go to suitable habitats where healthy insects would be present. Gard and Falcon (1978) demonstrated that the *Helicoverpa zea* NPV could be disseminated by attracting adult moths to light traps where they were mechanically contaminated with virus dusts. The released moths dispersed the virus throughout cotton fields. Traps luring diamond-back moth, *Plutella xylostella*, adults to structures housing cultures of either *Z. radicans* or *Beauveria bassiana* have been designed (Pell *et al.*, 1993) and tested (Furlong *et al.*, 1995). Externally contaminated adults disseminated the fungi in the field both to their larvae and to their mates.

8 Evaluating pathogen establishment and impact

Many of the examples used in this chapter referenced introductions that provided permanent control of the target pest and the decreases in pest population densities were often so convincing that analyses of pest dynamics were not considered necessary. However, it is expected that most releases might not provide such dramatic results. Therefore, careful quantitative measurements of the mortality caused by natural enemies are necessary (Mills, 1997). Pathogens introduced into the environment should hopefully provide long-term control, so evaluations should be implemented over several generations throughout the release areas and adjacent areas, to monitor establishment, efficacy, and spread. To accomplish this, reliable methods are required to detect the disease in the new environment, assess the impact of the pathogen on the target pest population, and measure pathogen persistence and spread. More detailed descriptions of methods for quantifying pathogen activity are provided in Sections V, VII, and VIII.

Pathogens can be detected and identified using morphological, physiological, behavioral and genetic characteristics (see above). It is rare that only one of these categories of characters is used to identify pathogens, but, morphological characteristics are traditionally used first. However, morphological characters are virtually never adequate for differentiating isolates, pathotypes, and varieties within a species. Currently, a variety of molecular methods are used to provide excellent discrimination both within and between species.

After a pathogen is introduced, the first step is to ensure that it has survived and is reproducing in its host in the new environment. A pathogen is considered established if it is continuously recovered for at least a few years after introduction, during which time it would survive unfavorable environmental conditions and low host densities. If a strain is not considered established or impact on the host population is minimal, we suggest that studies should be conducted to help understand why establishment was not successful. For example, an inappropriate strain or species can propagate in the host and establish in a new environment but fail to control the pest population. Wilding *et al.* (1986) reported an example involving a Brazilian strain of *Pandora* (= *Erynia*) *neoaphidis* isolated from pea aphids, *Acyrthosiphon pisum*, and introduced against aphids on winter wheat in Belgium. No changes in infection levels in release areas were observed after introduction of the fungus, suggesting that establishment had not occurred or that this strain was having a minimal impact. The strain released was initially selected because it grew rapidly in liquid medium and had high infectivity for aphids. Further work showed that this strain produced little mycelia and few conidia in infected hosts. These latter attributes certainly would limit the capability of this fungal strain to spread and probably helps explain its failure in field trials.

When pathogen establishment is confirmed, the impact of the pathogen on the host population and the dispersal potential of the pathogen must be evaluated. The impact of an introduced pathogen in a new habitat can be quantified using counts of the target pest population, the pathogen population, or, as more indirect measures, evaluating insect mortality from infection, decreases in numbers of eggs, decreases in fecundity, and reductions in damage due to pests. These approaches aim to estimate directly or indirectly the extent of mortality imposed by the pathogen on the host population. Increased densities of the pathogen in the environment can be measured by sampling and quantifying pathogen propagules (*e.g.*, airborne fungal spores, NPV occlusion bodies in the soil, etc.) or by sampling host populations to determine infection prevalence (see Hajek *et al.*, 2007 for soil analyses). However, quantification of pathogen propagules in the environment is difficult and is rarely done in practice after classical biological control releases. Thus, determination of the percentage of hosts infected or contaminated is virtually always used for quantifying impact. Methods for accurate pathogen sampling and identification are most critical for determining infection levels in host populations. Host individuals should be sampled randomly so that sampling is representative of the entire population. It is known that parasitism can influence the behavior of infected hosts and thereby alter the relative distribution of non-infected and infected hosts (Moore, 1995). A classical example is the tendency of some insects infected by viruses or entomophthoralean fungi to climb to the tops of plants (Roy *et al.*, 2006). In these cases, sampling for pathogen establishment should include collecting individuals from these locations while sampling to determine prevalence should be adjusted to achieve balanced representation of the entire host population.

Care must also be taken to ensure that cross-contamination does not occur between individuals in a sample of hosts after they are collected from the field (Mills, 1997). Unfortunately, it is very easy for pathogens to spread among individuals when they are being reared to determine the prevalence of infection in field populations. To address this problem, specimens can be reared individually while care is taken to prevent contamination and stress. Alternatively, specimens collected can be dissected immediately or immediately preserved for molecular studies to detect and identify pathogens. The method selected for determining infection prevalence should be based on both accuracy desired and time available. For example, determination of infection level by detecting hyphal bodies of the fungus *Neozygites* inside of mites is time consuming and infection is probably underestimated due to the difficulty in diagnosing the initial phase of the disease (Delalibera *et al.*, 2000). Alternatively, for field studies, using the percentage of mites that have conidia adhering to their bodies (contamination) as an indication of infection levels is accurate and less time consuming.

Effectiveness of natural enemies can also be measured using a pesticidal check method. This method was first described by DeBach (1946) and has been used by many authors to measure the efficacy of entomophagous insects. It consists of spraying a dilute concentration of a pesticide that is nontoxic to a given pest but toxic to its natural enemies. The density of the pest population in a treated plot is compared with the pest density in an untreated plot inhabited by the naturally occurring natural enemies. This technique was successfully used to demonstrate the comparative efficacy of the fungal pathogen *N. tanajoae* and predatory mites (Delalibera, unpublished data) against the cassava green mite in its area of endemism to select natural enemies for introduction in Africa. Plots were sprayed weekly with either a fungicide, a diluted concentration of a miticide, or with both pesticides combined. The contribution of each organism to pest mortality was estimated by comparing plots where the fungi and/or the predatory mites were excluded (alone or combined), with plots having untreated pest and natural enemy populations. Such chemical barriers provide a feasible technique for assessing the efficacy of fungi or other pathogens that are selectively susceptible to pesticides, because many of these products (*e.g.*, fungicides) are non-toxic to insect pests. However, some problems can occur in interpreting the results from insecticidal exclusion experiments (Luck *et al.*, 1988). Some insecticides can stimulate reproduction of the prey population, induce sex-ratio bias, and induce physiological changes in the plant. Some fungicides contain microelements (Cu, Zn, Mn) that may have indirect effects on the pest, so preliminary laboratory studies should be conducted prior to interpretation of field data.

A *Non-target effects*

After initial establishment of a pathogen and prior to the secondary phase of dissemination throughout the introduction area, we suggest conducting studies to evaluate the effect of the introduced pathogen on non-target species (see above). At this stage, researchers would be able to collect non-target species in the area of introduction and evaluate whether infection by the introduced pathogen is occurring (see also

Chapter X-1). Results from these studies would help evaluate the validity of planning further dissemination of this pathogen by inoculative or inundative release to increase its distribution.

B Cost-benefit analysis

Classical biological control is frequently cited as being especially economically efficient when compared to other control methods (*e.g.*, Cullen and Whitten, 1995). However, estimated reductions in control costs and costs of introductions have largely been based on parasitoid and predator introductions. Such analyses have seldom been conducted for pathogens. The costs associated with an introduction program can be estimated fairly readily based on the various stages of a program outlined in this chapter. We encourage more researchers to undertake these much-needed analyses.

The longer term benefits of a classical biological program are difficult to determine. For example, numerous control programs are frequently implemented simultaneously after an exotic pest has been introduced and it can be very difficult to distinguish the contribution of each segment of a control program. The benefit of a pathogen introduction can be related to reduction in damage (*e.g.*, increase in yield of a crop) or reduction in the amount or the cost of pesticides used or other control operations undertaken. A preliminary study of the program to control *Oryctes* spp. attacking coconut and oil palms in East Asia and Oceania using the non-occluded *Oryctes* virus estimated a conservative benefit:cost ratio of $> 100:1$ (C. J. Lomer, pers. comm.). While benefits to the environment and human health are perhaps the most important variables to quantify, it can be difficult to assign monetary values to socioeconomic benefits.

9 Acknowledgments

We thank L. Solter and J. Maddox for comments on the regulations section of this manuscript. L. Solter, L. Bauer, C. Lomer and D. Greathead kindly shared unpublished information. J. Becnel, K. Hodge, R. Humber, C. Lange, L. Solter and P. Stock helped with use of correct scientific names. C. Eastburn assisted with manuscript preparation.

10 References

Alatorre-Rosas, R. and Kaya, H. K. 1990. Interspecific competition between entomopathogenic nematodes in the genera *Heterorhabditis* and *Steinernema* for an insect host in sand. *J. Invertebr. Pathol.* **55**, 179–188.

Alatorre-Rosas, R. and Kaya, H. K. 1991. Interaction between two entomopathogenic nematode species in the same host. *J. Invertebr. Pathol.* **57**, 1–6.

Anonymous. 1955. Controlling the Japanese Beetle. USDA, Farmer's Bull. #2004. 13 pp.

Bauer, L. S., Miller, D. L., Onstad, D. W., Maddox, J. V., and McManus, M. L. 1995. Studies on the transmission of an exotic microsporidium that infects the gypsy moth. USDA, Forest Service, Gen. Tech. Rpt. NE-213, 50–51.

Bauer, L. S., Sapio, F. J., McManus, M. L., Maddox, J. V., Jeffords, M. R., and Onstad, D. W. 1993. Interactions of *Microsporidium* and gypsy moth in Michigan field plots. USDA, Forest Service, Gen. Tech. Rpt. NE-179, 22.

Bauer, L. S., Sapio, F. J., McManus, M. L., Maddox, J. V., Jeffords, M. R., and Onstad, D. W. 1994. Interactions of *Microsporidium* and gypsy moth in Michigan field plots. USDA, Forest Service, Gen. Tech. Rpt. NE-188, 5.

Bedding, R. A. 1979. Manipulating the entomophagous-mycetophagous nematode, *Deladenus siricidicola*, for biological control of the woodwasp, *Sirex noctilio*, in Australia. In "Current Topics in Forest Entomology" (W. E. Waters, Ed.), pp. 144–147. USDA, Forest Service, Washington, D. C.

Bedding, R. A. 1993. Biological control of *Sirex noctilio* using the nematode *Deladenus siricidicola*. In "Nematodes and the Biological Control of Insect Pests" (R. Bedding, R. Akhurst and H. Kaya, Eds.), pp. 11–20. CSIRO, East Melbourne, Australia.

Bedding, R. A. and Akhurst, R. J. 1974. Use of the nematode *Deladenus siricidicola* in the biological control of *Sirex noctilio* in Australia. *J. Aust. Entomol. Soc.* **13**, 129–135.

Bedding, R. A. and E. T. Iede. 2005. Application of *Beddingia siricidicola* for Sirex woodwasp control. *In* "Nematodes as Biocontrol Agents." (P. S. Grewal, R.-U. Ehlers, and D. I. Shapiro-Ilan, Eds.), pp. 385–399. CABI Internatl., Wallingford, UK.

Bedford, G. O. 1977. Virus against coconut rhinoceros beetles in Fiji. *So. Pacif. Bull.* **27**, 27–34.

Bedford, G. O. 1981. Control of the rhinoceros beetle by baculovirus. In "Microbial Control of Pests and Plant Diseases 1970–1980" (H. D. Burges, Ed.), pp. 409–426. Academic Press, London.

Bellows, Jr., T. S. and Legner, E. F. 1993. Foreign exploration. In "Steps in Classical Arthropod Biological

Control" (R. G. Van Driesche and T. S. Bellows, Jr., Eds.), pp. 27–41. Thomas Say Publ. Entomol., Entomol. Soc. Am., Lanham, MD.

Bidochka, M. J., Walsh, S. R. A., Ramos, M. E., St. Leger, R. J., Silver, J. C., and Roberts, D. W. 1995. Pathotypes of the *Entomophaga grylli* species complex of grasshopper pathogens differentiated with random amplification of polymorphic DNA and cloned-DNA probes. *Appl. Environ. Microbiol.* **61**, 556–560.

Bidochka, M. J., Walsh, S. R. A., Ramos, M. E., St. Leger, R. J., Silver, J. C., and Roberts, D. W. 1996. Fate of biological control introductions: Monitoring an Australian fungal pathogen of grasshoppers in North America. *Proc. Natl. Acad. Sci. USA* **93**, 918–921.

Bird, F. T. 1953. The use of a virus disease in the biological control of the European pine sawfly, *Neodiprion sertifer* (Geoffr.). *Can. Entomol.* **85**, 437–446.

Bird, F. T. 1961. Transmission of some insect viruses with particular reference to ovarial transmission and its importance in the development of epizootics. *J. Insect Pathol.* **3**, 352–380.

Brasier, C. M. and Hansen, E. M. 1992. Evolutionary biology of *Phytophthora*. Part II. Phylogeny, speciation and population structure. *Annu. Rev. Phytopathol.* **30**, 173–200.

Charudattan, R. and Browning, H. W. (Eds.) 1992. "Regulation and Guidelines: Critical Issues in Biological Control." Proc. USDA/CSRS Natl. Workshop, June 1–12, 1991, Vienna, VA. IFAS, Univ. Florida, Gainesville. 205 pp.

Coulson, J. R. and Soper, R. S. 1989. Protocols for the introduction of biological control agents in the U.S. In "Plant Protection and Quarantine. Vol. III. Special Topics" (R. P. Kane, Ed.), pp. 1–35. CRC Press, Boca Raton, FL.

Cullen, J. M. and Whitten, M. J. 1995. Economics of classical biological control: A research perspective. In "Biological Control: Benefits and Risks" (H. M. T. Hokkanen and J. M. Lynch, Eds.), pp. 270–276. Cambridge Univ. Press, Cambridge, UK.

Cunningham, J. C. and Entwistle, P. F. 1981. Control of sawflies by baculovirus. In "Microbial Control of Pests and Plant Diseases 1970–1980" (H. D. Burges, Ed.), pp. 379–407. Academic Press, London.

DeBach, P. 1946. An insecticidal check method for measuring the efficacy of entomophagous insects. *J. Econ. Entomol.* **39**, 695–697.

Delalibera Jr., I. and Hajek, A. E. 2004. Pathogenicity and specificity of *Neozygites tanajoae* and *Neozygites floridana* (Zygomycetes: Entomophthorales) isolates pathogenic to the cassava green mite. *Biol. Control* **30**, 608–616.

Delalibera Jr., I., Moraes, G. J., Lapointe S. L., Silva, C. A. D. da, and Tamai, M. A. 2000. Temporal variability and progression of *Neozygites* sp. (Zygomycetes: Entomophthorales) in populations of *Mononychellus tanajoa* (Bondar) (Acari: Tetranychidae). *An. Soc. Entomol. Brasil*, **29**, 523–535.

Delalibera Jr., I., Demétrio, C. G. B., Manly, B. F. J., and Hajek, A. E. 2006. Effect of relative humidity and origin of isolates of *Neozygites tanajoae* (Zygomycetes: Entomophthorales) on production of conidia from cassava green mite, *Mononychellus tanajoa* (Acari: Tetranychidae) cadavers. *Biol. Control* **39**, 489–496.

Doane, C. C. 1971. Trans-ovum transmission of nuclear polyhedrosis virus in relation to disease in gypsy moth populations. In Proc. 4th Internatl. Colloq. Insect Pathol., Aug. 1970. Soc. Invertebr. Pathol., College Park, MD, p. 285.

Doane, C. C. 1975. Infectious sources of nuclear polyhedrosis virus persisting in natural habitats of the gypsy moth. *Environ. Entomol.* **4**, 392–394.

Dowden, P. B. 1953. Use of a virus disease to control European pine sawfly. *J. Econ. Entomol.* **46**, 525–526.

Dustan, A. G. 1927. The artificial culture and dissemination of *Entomophthora sphaerosperma* Fres., a fungous parasite for the control of the European apple sucker (*Psyllia mali* Schmidb.). *J. Econ. Entomol.* **20**, 68–75.

Ehler, L. E. 1990a. Introduction strategies in biological control of insects. In "Critical Issues in Biological Control" (M. Mackauer, L. E. Ehler and J. Roland, Eds.), pp. 111–134. Intercept, Andover, Hants., UK.

Ehler, L. E. 1990b. Some contemporary issues in biological control of insects and their relevance to the use of entomopathogenic nematodes. In "Entomopathogenic Nematodes in Biological Control" (R. Gaugler and H. K. Kaya, Eds.), pp. 1–19. CRC Press, Boca Raton, FL.

Elmore, J. C. and Howland, A. F. 1964. Natural versus artificial dissemination of nuclear-polyhedrosis virus by contaminated adult cabbage loopers. *J. Insect Pathol.* **6**, 430–438.

EPPO (European Plant Protection Organization). 2000. Safe use of biological control: import and release of exotic biological control agents. *EPPO Standard* PM 6/2(1).

FAO (Food & Agriculture Organization). 1996. "Code of Conduct for the Import and Release of Exotic Biological Control Agents," ISPM No. 3. FAO, Rome. 12 pp.

Fleming, W. E. 1968. Biological control of the Japanese beetle. U.S.D.A. Tech. Bull. 1383, 78 pp.

Furlong, M. J., Pell, J. K., Ong, P. C., and Syed, A. R. 1995. Field and laboratory evaluation of a sex pheromone trap for the autodissemination of the fungal entomopathogen *Zoophthora radicans* (Entomophthorales) by the diamondback moth, *Plutella xylostella* (Lepidoptera: Yponomeutidae). *Bull. Entomol. Res.* **85**, 331–337.

Fuxa, J. R., Ayyappath, R., and Goyer, R. A. 1998. "Pathogens and Microbial Control of North American Forest Insect Pests." USDA, Forest Service FHTET 97–27.

Gard, I. E. and Falcon, L. A. 1978. Autodissemination of entomopathogens: Virus. In "Microbial Control of Insect

Pests: Future Strategies in Pest Management Systems" (G. E. Allen, C. M. Ignoffo and R. P. Jaques, Eds.), pp. 46–54.

Gomez, D. R. S. 1987. Control microbiano de *Phyllocoptruta oleivora* (Ashm.) y *Eriophyes sheldoni* Ewing mediante pulverizaciones de conidios de tres variedades de *Hirsutella thompsonii* Fisher. In "XI Congresso Brasiliero de Entomologia", pp. 167, Campinas, Brazil.

Gomez, D. R. S. and Moscardi, F. 1991. Microbial control and insect pathology in Argentina. *Ciência e Cultura* **43**, 375–379.

Gonzalez, D. and Gilstrap, F. D. 1992. Foreign exploration: Assessing and prioritizing natural enemies and consequences of preintroduction studies. In "Selection Criteria and Ecological Consequences of Importing Natural Enemies" (W. C. Kauffman and J. E. Nechols, Eds.), pp. 53–70. Thomas Say Publ. Entomol., Entomol. Soc. Am., Lanham, MD.

Hajek, A. E. 1997a. Ecology of terrestrial fungal entomopathogens. *Adv. Microb. Ecol.* **15**, 193–249.

Hajek, A. E. 1997b. *Entomophaga maimaiga*: Where do we go now? USDA, Forest Service, Gen. Tech. Rpt. NE-240, 46–48.

Hajek, A. E., Bauer, L. S., McManus, M. L., and Wheeler, M. M. 1998a. Distribution of resting spores of the *Lymantria dispar* pathogen *Entomophaga maimaiga* in soil and on bark. *BioControl* **43**, 189–200.

Hajek, A. E., Elkinton, J. S., and Witcosky, J. J. 1996. Introduction and spread of the fungal pathogen *Entomophaga maimaiga* (Zygomycetes: Entomophthorales) along the leading edge of gypsy moth (Lepidoptera: Lymantriidae) spread. *Environ. Entomol.* **25**, 1235–1247.

Hajek, A. E., Fuxa, J. R., and Kunimi, Y. 2007. Microorganisms associated with soil arthropods. *In* "Manual of Environmental Microbiology, 3rd ed." (C. J. Hurst, R. L. Crawford, J. L. Garland, D. A. Lipson, A. L. Mills, and L. D. Stetzenbach, Eds.). ASM Press, Washington, D.C. (in press).

Hajek, A. E., Humber, R. A., and Elkinton, J. S. 1995. Mysterious origin of *Entomophaga maimaiga* in North America. *Amer. Entomol.* **41**, 31–42.

Hajek, A. E., McManus, M. L., and Delalibera Jr, I. 2005. "Catalogue of Introductions of Pathogens and Nematodes for Classical Biological Control of Insects and Mites." USDA, Forest Service, FHTET-2005–05. 59 pp.

Hajek, A. E., McManus, M. L., Delalibera Jr, I. 2007. A review of introductions of pathogens and nematodes for classical biological control of insects and mites. *Biol. Control* **41**, 1–13.

Hajek, A. E. and Roberts, D. W. 1991. Pathogen reservoirs as a biological control resource: Introduction of *Entomophaga maimaiga* to North American gypsy moth, *Lymantria dispar*, populations. *Biol. Control* **1**, 29–34.

Hajek, A. E., Tatman, K. M., Wanner, P. H., and Wheeler, M. M. 1998b. Location and persistence of cadavers of gypsy moth, *Lymantria dispar*, containing *Entomophaga maimaiga* azygospores. *Mycologia* **90**, 754–760.

Hajek, A. E., Wheeler, M. M., Eastburn, C. C., and Bauer, L. S. 2001. Storage of resting spores of the gypsy moth fungal pathogen, *Entomophaga maimaiga*. *Biocontrol Sci. Technol.* **11**, 637–647.

Hall, I. M. and Dunn, P. H. 1957. Fungi on spotted alfalfa aphid: Spread of fungi by natural and artificial means is resulting in excellent biological control of aphids in many counties. *Calif. Agric.* **11**, 5 and 14.

Hall, I. M. and Dunn, P. H. 1958. Artificial dissemination of entomophthorous fungi pathogenic to spotted alfalfa aphid in California. *J. Econ. Entomol.* **51**, 341–344.

Henry, J. E. and Oma, E. A. 1981. Pest control by *Nosema locustae*, a pathogen of grasshoppers and crickets. In "Microbial Control of Pests and Plant Diseases 1970–1980" (H. D. Burges, Ed.), pp. 573–586. Academic Press, London.

Hodge, K. T., Sawyer, A. J., and Humber, R. A. 1995. RAPD-PCR for identification of *Zoophthora radicans* isolates in biological control of potato leafhopper. *J. Invertebr. Pathol.* **65**, 1–9.

Hokkanen, H. M. T. and Pimentel, D. 1984. New approach for selecting biological control agents. *Can. Entomol.* **116**, 1109–1122.

Hokkanen, H. M. T. and Pimentel, D. 1989. New associations in biological control: Theory and practice. *Can. Entomol.* **121**, 829–840.

Hopper, K. R. 1996. Making biological control introductions more effective. In "Biological Control Introductions: Opportunities for Improved Crop Protection" (J. K. Waage, Ed.), BCPC Proceedings No. 67, pp. 61–76. British Crop Protection Council, Farnham, Surrey, UK.

Jansson, R. K. 1993. Introduction of exotic entomopathogenic nematodes (Rhabditida: Heterorhabditidae and Steinernematidae) for biological control of insects: Potential and problems. *Fla. Entomol* **76**, 82–96.

Jeffords, M. R., Maddox, J. V., McManus, M. L., Webb, R. E., and Wieber, A. 1988. Egg contamination as a method for the inoculative release of exotic microsporidia of the gypsy moth. *J. Invertebr. Pathol.* **51**, 190–196.

Jeffords, M. R., Maddox, J. V., McManus, M. L., Webb, R. E., and Wieber, A. 1989. Evaluation of the overwintering success of two European microsporidia inoculatively released into gypsy moth populations in Maryland. *J. Invertebr. Pathol.* **53**, 235–240.

Kim, C. W. 1967. Biological control of fall webworm *Hyphantria cunea* Drury in Korea. *Entomol. Res. Bull.* **3**, 5–30.

Klein, M. G. 1981. Advances in the use of *Bacillus popilliae* for pest control. In "Microbial Control of Pests and Plant Diseases 1970–1980" (H. D. Burges, Ed.), pp. 183–192. Academic Press, London.

Kung, S.-P., Gaugler, R., and Kaya, H. K. 1991. Effects of soil temperature, moisture, and relative humidity on entomopathogenic nematode persistence. *J. Invertebr. Pathol.* **57**, 242–249.

Lange, C. E. and Wysiecki, M. L. d. 1996. The fate of *Nosema locustae* (Microsporida: Nosematidae) in Argentine grasshoppers (Orthoptera: Acrididae). *Biol. Control* **7**, 24–29.

Leppla, N. C. 1996. Environmentally friendly methods for reducing damage caused by exotic weeds in natural habitats: Conflicts of interest, safeguards and national policy. *Castanea* **61**, 214–225.

Lloyd, M. and Dybas, H. S. 1966. The periodical cicada problem. I. Population ecology. *Evolution* **20**, 133–149.

Luck, R. F., Shepard, B. M., and Kenmore, P. E. 1988. Experimental methods for evaluating arthropod natural enemies. *Annu. Rev. Entomol.* **33**, 367–391.

Mackauer, M. 1976. Genetic problems in the production of biological control agents. *Annu. Rev. Entomol.* **21**, 369–385.

Maddox, J. V., McManus, M. L., Jeffords, M. R., and Webb, R. E. 1992. Exotic insect pathogens as classical biological control agents with an emphasis on regulatory considerations. In "Selection Criteria and Ecological Consequences of Importing Natural Enemies" (W. C. Kauffman and J. E. Nechols, Eds.), pp. 27–39. Thomas Say Publ. Entomol., Entomol. Soc. Am., Lanham, MD.

Magnoler, A. 1974. Field dissemination of a nucleopolyhedrosis virus against the gypsy moth, *Lymantria dispar* L. *Z. Pflanzenkrank. Pflanzenschutz* **81**, 497–511.

Malakar, R., Elkinton, J. S., Hajek, A. E., and Burand, J. P. 1999. Within-host interactions of *Lymantria dispar* (Lepidoptera: Lymantriidae) nucleopolyhedrosis virus and *Entomophaga maimaiga* (Zygomycetes: Entomophthorales). *J. Invertebr. Pathol.* **73**, 91–100.

Mannion, C. M. and Jansson, R. K. 1992. Comparison of ten entomopathogenic nematodes for control of sweetpotato weevil (Coleoptera: Apionidae). *J. Econ. Entomol.* **85**, 1642–1650.

Martignoni, M. E. and Milstead, J. E. 1962. Trans-ovum transmission of the nuclear polyhedrosis virus of *Colias eurytheme* Boisduval through contamination of the female genitalia. *J. Insect Pathol.* **4**, 113–121.

McCray, E. M., Womeldorf, D. J., Husbands, R. C., and Eliason, D. A. 1973. Laboratory observations and field tests with *Lagenidium* against California mosquitoes. *Proc. Pap. Calif. Mosq. Contr. Assoc.* **41**, 123–128.

Messenger, P. S. 1971. Climatic limitations to biological control. In "Tall Timbers Conf. on Ecol. Anim. Control by Habitat Mgt.", pp. 97–114, Tallahassee, FL.

Messenger, P. S., Wilson, F., and Whitten, M. J. 1976. Variation, fitness, and adaptability of natural enemies. In "Theory and Practice of Biological Control" (C. B. Huffaker and P. S. Messenger, Eds.), pp. 209–231. Academic Press, New York.

Miller, M. and Aplet, G. 1993. Biological control: A little knowledge is a dangerous thing. *Rutgers Law Rev.* **45**, 285–334.

Mills, N. 1997. Techniques to evaluate the efficacy of natural enemies. In "Methods in Ecological and Agricultural Entomology" (D. R. Dent and M. P. Walton, Eds.), pp. 271–291. CAB International, Wallingford, Oxon, UK.

Milner, R. J. 1985. Pathogen importation for biological control: Risks and benefits. In "Pests and Parasites as Migrants" (A. J. Gibbs and H. R. C. Meischke, Eds.), pp. 115–121. Cambridge Univ. Press, Cambridge, UK.

Milner, R. J., Soper, R. S., and Lutton, G. G. 1982. Field release of an Israeli strain of the fungus *Zoophthora radicans* (Brefeld) Batko for biological control of *Therioaphis trifolii* (Monell) f. *maculata*. *J. Aust. Entomol. Soc.* **21**, 113–118.

Moore, J. 1995. The behavior of parasitized animals. *Bioscience* **45**, 89–96.

Moraes, G. J. de and Delalibera, Jr., I. 1992. Specificity of a strain of *Neozygites* sp. (Zygomycetes: Entomophthorales) to *Mononychellus tanajoa* (Acari: Tetranychidae). *Exp. Appl. Acarol.* **14**, 89–94.

Moraes, G. J. de and De Nardo, E. B. 1996. Issues of safety and ownership in biological control introductions: The Brazilian case. In "Biological Control Introductions: Opportunities for Improved Crop Protection" (J. K. Waage, Ed.), BCPC Proceedings No. 67, pp. 135–143. British Crop Protection Council, Farnham, Surrey, UK.

Nickle, W. R., Drea, J. J., and Coulson, J. R. 1988. Guidelines for introducing beneficial insect-parasitic nematodes into the United States. *Ann. Appl. Nematol.* **2**, 50–56.

Osborne, L. S. and Landa, Z. 1992. Biological control of whiteflies with entomopathogenic fungi. *Fla. Entomol.* **75**, 456–471.

Papierok, B. and Hajek, A. E. 1997. Entomophthorales. In "Manual of Techniques in Invertebrate Pathology" (L. A. Lacey, Ed.), pp. 187–212. Academic Press, London.

Pell, J. K., Eilenberg, J., Hajek, A. E., and Steinkraus, D. 2001. Exploring the potential of Entomophthorales in integrated crop management. In "Fungal Biocontrol Agents - Progress, Problems and Potential" (T. Butt, C. Jackson and N. Magan, Eds.), pp. 71–167. CABI Publ., Oxon, UK.

Pell, J. K., Macaulay, E. D. M., and Wilding, N. 1993. A pheromone trap for dispersal of the fungal pathogen *Zoophthora radicans* Brefeld (Zygomycetes: Entomophthorales) amongst populations of the diamondback moth, *Plutella xylostella* L. (Lepidoptera: Yponomeutidae). *Biocontrol Sci. Technol.* **3**, 315–320.

Pilarska, D. K., Solter, L. F., Kereselidze, M., Linde, A., and Hoch, G. 2006. Microsporidian infections in *Lymantria dispar* larvae: interactions and effects of multiple species infections on pathogen horizontal transmission. *J. Invertebr. Pathol.* **93**, 105–113.

Ponomorenko, N. G., Prilepskaja, H. A., Murvanidze, M. Y., and Stoljarova, L. A. 1975. *Aschersonia* against white-flies. *Zashchita Rastenij* **6**, 44.

Pschorn-Walcher, H. 1977. Biological control of forest insects. *Annu. Rev. Entomol.* **22**, 1–22.

Roush, R. T. 1990. Genetic variation in natural enemies: Critical issues for colonization in biological control. In "Critical Issues in Biological Control" (M. Mackauer, L. E. Ehler and J. Roland, Eds.), pp. 263–288. Intercept, Andover, Hants., UK.

Roy, H. E., Steinkraus, D. C., Eilenberg, J., Hajek, A. E., and Pell, J. K. 2006. Bizarre interactions and endgames: Entomopathogenic fungi and their arthropod hosts. *Annu. Rev. Entomol.* **51**, 331–357.

St. Leger, R. J., May, B., Allee, L. L., Frank, D. C., Staples, R. C., and Roberts, D. W. 1992. Genetic differences in allozymes and in formation of infection structures among isolates of the entomopathogenic fungus *Metarhizium anisopliae*. *J. Invertebr. Pathol.* **60**, 89–101.

Shands, W. A., Thompson, C. G., Simpson, G. W., and Wave, H. E. 1958. Preliminary studies of entomopathogenous fungi for the control of potato-infesting aphids in Maine. *J. Econ. Entomol.* **51**, 184–186.

Shu, C. R. and Yu, C. Y. 1985. An investigation on the natural enemies of *Hyphantria cunea*. *Natural Enemies Insects Kunchong Tiandi* **7**, 91–94, 99.

Sidor, C. 1979. The role of insect pathogenic microorganisms in the protection of the environment. *Mikrobiologija* **16**, 173–186.

Solter, L. F., Maddox, J. V., and McManus M. L. 1997. Host specificity of microsporidia (Protista: Microspora) from European populations of *Lymantria dispar* (Lepidoptera: Lymantriidae) to indigenous North American Lepidoptera. *J. Invertebr. Pathol.* **69**, 135–150.

Solter, L. F., Maddox, J. V. and Vossbrinck, C. R. 2005. Physiological host specificity: a model using the European corn borer, *Ostrinia nubilalis* (Hübner) (Lepidoptera: Crambidae) and microsporidia of row crop and other stalk-boring hosts. *J. Invertebr. Pathol.* **90**, 127–130.

Solter, L. F., Pilarska, D. K. and Vossbrinck, C. R. 2000. Host specificity of microsporidia pathogenic to forest Lepidoptera. *Biol. Control* **19**, 48–56.

Solter, L. F., Siegel, J. P., Pilarska, D. K., and Higgs, M. C. 2002. The impact of mixed infection of three species of microsporidia isolated from the gypsy moth, *Lymantria dispar* L. (Lepidoptera: Lymantriidae). *J. Invertebr. Pathol.* **81**, 103–113.

Speare, A. T. and Colley, R. H. 1912. "The Artificial Use of the Brown-tail Fungus in Massachusetts, with Practical Suggestions for Private Experiment, and a Brief Note on a Fungous Disease of the Gypsy Caterpillar," Wright & Potter, Boston.

Steinkraus, D. C., Boys, G. O., and Rosenheim, J. A. 2002. Classical biological control of *Aphis gossypii* (Homoptera: Aphididae) with *Neozygites fresenii* (Entomophthorales: Neozygitaceae) in California cotton. *Biol. Control* **25**, 297–304.

Stiling, P. 1993. Why do natural enemies fail in classical biological control programs? *Amer. Entomol.* **39**, 31–37.

Sweeney, A. W., Cooper, R., Medcraft, B. E., Russell, R. C., O'Donnell, M., and Panter, C. 1983. Field tests of the mosquito fungus *Culinomyces clavisporus* against the Austrialian encephalitis vector *Culex annulirostris*. *Mosq. News* **43**, 290–297.

Tanada, Y. and Fuxa, J. R. 1987. The pathogen population. In "Epizootiology of Insect Diseases" (J. R. Fuxa and Y. Tanada, Eds.), pp. 113–157. John Wiley & Sons, NY.

Tanada, Y. and Kaya, H. K. 1993. "Insect Pathology," Academic Press, London.

U.S. Congress, O. T. A. 1993. "Harmful non-indigenous species in the United States," U.S. Govt. Printing Off., Wash., D.C.

Van Driesche, R. G. 1993. Methods for the field colonization of new biological control agents. In "Steps in Classical Biological Control" (R. G. Van Driesche and T. S. Bellows, Jr., Eds.), pp. 67–86. Thomas Say Publ. Entomol., Entomol. Soc. Am., Lanham, MD.

Van Driesche, R. G. and Bellows, Jr., T. S., Eds. 1993. "Steps in Classical Arthropod Biological Control,". Entomol. Soc. Am., Lanham, MD.

Van Driesche, R. G. and Bellows, Jr., T. S. 1996. "Biological Control," Chapman & Hall, New York.

van Lenteren, J. C. 1980. Evaluation of control capabilities of natural enemies: Does art have to become science? *Netherlands J. Zool.* **30**, 369–381.

Vidal, C., Fargues, J., and Lacey, L. A. 1997. Intraspecific variability of *Paecilomyces fumosoroseus*: Effect of temperature on vegetative growth. *J. Invertebr. Pathol.* **70**, 18–26.

Waage, J. 1990. Ecological theory and the selection of biological control agents. In "Critical Issues in Biological Control" (M. Mackauer, L. E. Ehler and J. Rolands, Eds.), pp. 135–157. Intercept, Andover, UK.

Weseloh, R. M. and Andreadis, T. G. 1997. Persistence of resting spores of *Entomophaga maimaiga*, a fungal pathogen of the gypsy moth, *Lymantria dispar*. *J. Invertebr. Pathol.* **69**, 195–196.

Westerman, P. R. 1992. The influence of time of storage on performance of the insect parasitic nematode *Heterorhabditis* sp. *Fund. Appl. Nematol.* **15**, 407–412.

Wilding, N., Latteur, G., and Dedryver, C. A. 1986. Evaluation of Entomophthorales for aphid control: Laboratory and field data. In "Fundamental and Applied Aspects of Invertebrate Pathology" (R. A. Samson, J. M. Vlak and D. Peters, Eds.), pp. 159–162. Found. 4th Internatl. Colloq. Invertebr. Pathol., Wageningen, NL.

Yaninek, J. S., Saizonou, S., Onzo, A., Zannou, I., and Gnanvossou, D. 1996. Seasonal and habitat variability in the fungal pathogens *Neozygites* cf. *floridana* and *Hirsutella thompsonii*, associated with cassava mites in Benin, West Africa. *Biocontrol Sci. Technol.* **6**, 23–33.

Zelinskaya, L. M. 1980. Role of microsporidia in the abundance dynamics of the gypsy moth, *Porthetria dispar*, in forest plantings along the lower Dnieper River (Ukrinian Republic, USSR). *Vestn. Zool.* **1**, 57–62.

Zethner, O. 1976. Control experiments on the nun moth (*Lymantria monacha* L.) by nuclear-polyhedrosis virus in Danish coniferous forests. *Z. Angew. Entomol.* **81**, 192–207.

Zwölfer, H., Ghani, M. A., and Rao, V. P. 1976. Foreign exploration and importation of natural enemies. In "Theory and Practice of Biological Control" (C. B. Huffaker and P. S. Messenger, Eds.), pp. 189–207. Academic Press, New York.

SECTION VII

EVALUATION OF ENTOMOPATHOGENS IN SPECIFIC SYSTEMS

Chapter VII-1

Application and evaluation of entomopathogens in potato

Stephen P. Wraight
USDA-ARS
Plant, Soil & Nutrition Laboratory
Ithaca, NY 14850, USA

Marc Sporleder
Centro Internacional de la Papa (CIP)
Apartado 1558, Lima, Peru

Tadeusz J. Poprawski
Deceased, formerly with the USDA-ARS
Subtropical Agricultural Research Center
Weslaco, TX 78596, USA

Lawrence A. Lacey
USDA-ARS
Yakima Agricultural Research Laboratory
Wapato, WA 98951, USA

1 Introduction

More than 290 million metric tons of potatoes (*Solanum tuberosum*) are produced throughout the world annually. According to the FAO 2004–2005 statistics, China, Russia, India, Ukraine, and the United States rank 1–5, respectively, in potato production. Potato production ranks fourth among agricultural crops, after wheat, maize, and rice. In some regions, potato is grown as a dryland crop; in others, it is overhead-irrigated with a traveling gun-type irrigation or a central pivot irrigation system. Furrow irrigation is seldom used.

A large number of insect pests attack potatoes. Worldwide, the Colorado potato beetle (CPB), *Leptinotarsa decemlineata*, is the most devastating defoliator, and the potato tuber moth (PTM), *Phthorimaea operculella*, causes greatest damage to the tubers. Other common pests include: green peach aphid, *Myzus persicae* (vector of potato leaf roll virus); buckthorn aphid, *Aphis nasturtii*; potato aphid, *Macrosiphum euphorbiae*; foxglove aphid, *Aulacortum solani*; aster leafhopper, *Macrosteles quadrilineatus* (vector of potato purple-top wilt phytoplasma); potato leafhopper, *Empoasca fabae*; potato psyllid, *Paratrioza cockerelli*; European corn borer, *Ostrinia nubilalis*; potato stalk borer, *Trichobaris trinotata*; potato flea beetle, *Epitrix cucumeris*; wireworms, *Limonius* spp., *Melanotus* spp. and *Agriotes* spp.; leaf mining flies, *Liriomyza huidobrensis*; and Andean potato weevils, *Premnotrypes* spp. Most of these insects are considered secondary pests across their geographic ranges; however, they may be key pests in regions where environmental conditions and cropping systems are optimal for their survival and reproduction or where they vector highly virulent plant pathogens. Microbial control has been investigated for only a few of these insects. Soper (1980) and Duchesne

L.A. Lacey and H.K. Kaya (eds.), Field Manual of Techniques in Invertebrate Pathology, 329–359.

and Boiteau (1987) reviewed the limited work toward development of entomophthoralean fungi for microbial control of aphids affecting potatoes. More extensive research and development efforts have targeted CPB, PTM, potato stalk borer, and Andean potato weevils. This paper will focus on the two most important pests of potato in terms of global impacts, CPB and PTM. Recent field research indicates strong potential for use of pathogenic microbes for sustainable management of these key pests.

2 Colorado potato beetle (CPB)

A Pest status

From its original wild solanaceous hosts (especially weedy nightshades, *Solanum* spp.) in southwestern North America (probably central Mexico), *L. decemlineata* has spread to cultivated crops throughout the continent and to Europe and Asia. Potatoes are preferred, but the beetle may also damage tomato, eggplant, and occasionally pepper; tobacco may be fed upon but is generally a poor host (see Capinera, 2001). CPB may have 1–4 generations per year, depending on the length of the growing season; there is only a partial second generation on potatoes in regions where adults diapause. Adults burrow into the soil at initiation of diapause and emerge at the beginning of the growing season (although some adults remain in diapause for more than a year). Emerging adults that do not find host plants are capable of flying several km in search of new habitats (Voss and Ferro, 1990; Ferro and Lyon, 1991). Eggs are laid in batches of ca. 5–100 eggs, usually on the undersides of leaves. Each adult female deposits multiple batches of eggs, primarily during the first five weeks after emergence; potential fecundity is remarkably high, $> 3,000$ eggs per female under laboratory conditions (Brown *et al.*, 1980). Larvae develop rapidly through four instars and then drop to the soil, burrow in, and pupate. Larval development requires ca. 10 days and the pupation process an additional 12 days at 24 °C (Ferro *et al.*, 1985). Unchecked pest populations can rapidly and completely defoliate favored host plants, ultimately consuming even the plant stems and causing total losses (Hare, 1990).

Historically, management of CPB was a key factor in development of synthetic chemical insecticides and spray-application technologies. This pest is noted for its ability to quickly develop resistance to nearly all insecticides it has encountered, including neonicotinoids, the most effective chemicals relied upon for control in recent years (Zhao *et al.*, 2000; Byrne *et al.*, 2004).

B Biological control agents

1 Overview

CPB has many parasitoids and predators, but most are of little consequence because, without costly augmentation, few have the potential to suppress beetle populations quickly enough to protect young crops from defoliation. Roberts *et al.* (1980), Campbell *et al.* (1985), Duchesne and Boiteau (1987), and Cloutier *et al.* (1995) reviewed the diversity and potential of natural enemies, including entomopathogens, for control of CPB. Microbial control agents may play an important role in potato IPM systems. Known entomopathogens of *L. decemlineata* include the bacterium *Bacillus thuringiensis*, the microsporidium *Vairimorpha necatrix*, steinernematid and heterorhabditid nematodes, and several species of fungi in the genera *Beauveria, Paecilomyces*, and *Metarhizium*.

Although *V. necatrix* has been shown to infect CPB, little research has been conducted toward development of this microsporidium for CPB control. In a study by Jaques and Laing (1989), a semi-purified *V. necatrix* spore suspension was ineffective when applied to potato foliage at a rate of 5×10^{13} spores/ha. Laboratory and greenhouse trials have shown that strains of *Steinernema carpocapsae, S. feltiae*, and *Heterorhabditis bacteriophora* have potential against soil stages of the beetle. However, single applications of *S. carpocapsae* at rates as high as $6.4–7.6 \times 10^5$ nematodes/m^2 had little effect on late larval instars and pupae in several field tests (see Cloutier *et al.*, 1995). Thurston *et al.* (1994) concluded that CPB is not readily infected by *S. carpocapsae*, and Cloutier *et al.* (1995) suggested that massive numbers of steinernematids would be required to control the beetle.

The use of entomopathogenic nematodes against potato beetles is not yet in sight.

2 Bacterial pathogens

Bacillus thuringiensis (*Bt*) is the only bacterial pathogen that has been developed for microbial control of potato beetles. Beetle-control research with this well-known pathogen lagged behind the earlier successes against lepidopteran and other insect pests until the 1982 discovery of a subspecies that was highly pathogenic against chrysomelid beetles (Krieg *et al.*, 1983). This pathogen *B. thuringiensis* subsp. *tenebrionis* (*Btt*) (= *B. thuringiensis* subsp. *san diego*) was rapidly commercialized and used successfully for potato beetle control in the late 1980s and early 1990s (Gelernter and Trumble, 1999). This initial success can be ascribed to the toxin-based activity of this pathogen and its capacity to provide rapid control of larval populations (a distinct advantage over most other biological control agents). On the other hand, efficacy of *Btt* is limited by a number of factors: 1) virulence of the toxin decreases with increasing age of the host, and effective control thus requires targeting of early-instar larvae (Zehnder and Gelernter, 1989); 2) foliar applications have short residual activity (Ferro *et al.*, 1993); 3) *Bt* has virtually no recycling or epizootic potential in potato beetle populations; and 4) potato beetles oviposit over an extended period of time and, because of factors 1–3, effective control requires multiple applications.

As a consequence of these limitations, use of *Btt* for potato beetle control was, from the start, highly vulnerable to competition from synthetic chemical insecticides (Gelernter and Trumble, 1999). Over the past decade, growers have relied primarily on the highly effective, more economical chloronicotinyl chemicals, which have long residual activity due to systemic properties and the added advantage of also being effective against aphid pests (*e.g.*, see Lacey *et al.*, 1999). Much of the recent research effort with *Bt*, in general, has been directed away from sprayable biopesticides to development of transgenic crops that produce *Bt* toxins. This work has included development of potatoes that express the cry3A endotoxin of *Btt* (Perlak *et al.*, 1993; Adang *et al.*, 1993). Although genetically modified potatoes may provide excellent control of CPB, public acceptance of this technology remains a critical issue (Gelernter, 2002).

3 Fungal pathogens

Several insect pathogenic fungi have potential for biological control of CPB. Among these, *Beauveria bassiana* is most promising, being an important natural enemy of the prepupal, pupal and adult stages of the beetle in the soil (Bajan and Kmitowa, 1977; Humber, 1996). Although natural infections are primarily associated with these soil-dwelling stages, larvae feeding in the crop canopy are also susceptible to infection via inundative foliar spray applications. As typical with most fungal pathogens, infection and pathogenesis proceed relatively slowly compared to toxin-based control agents. Under optimal environmental conditions, infected larvae may succumb to infection within a few days, but under high-temperature or high-insolation conditions, fungal development is slowed, and the larvae may not succumb to infection until they enter the soil to pupate. The four larval instars are equally susceptible to infection (Fargues, 1972; M. H. Griggs, personal communication); however, speed of kill is inversely related to host age, and under field conditions most larvae infected after the second instar survive until the prepupal stage. Conidia of *B. bassiana* are rapidly degraded by UV radiation, and residual activity of foliar sprays is limited. Early-instar larvae feeding on the lower surfaces of the leaves are difficult spray targets (Wraight and Ramos, 2002), and larvae nearing molt have reduced susceptibility to fungal infection (Vey and Fargues, 1977). Consequently, multiple applications are needed to provide foliage protection, even when environmental conditions are favorable for the fungus.

Beauveria bassiana has been intensively investigated for microbial control of CPB since the 1950s, with the principal effort aimed at developing foliar application technologies to protect crops from defoliation. Not surprisingly, considering the importance of environmental conditions in *B. bassiana* efficacy, results have been highly variable (Fargues *et al.*, 1980; Sikura and Sikura, 1983; Anderson *et al.*, 1988; Poprawski *et al.*, 1997; Lacey *et al.*, 1999; Martin *et al.*,

2000; Wraight and Ramos, 2002). Generally, under operational or pilot-scale conditions, applications of *B. bassiana* alone have failed to provide adequate early-season crop protection (Lipa, 1985; Hajek *et al.*, 1987). Soil inoculation is a logical alternative to foliar sprays, especially considering the long persistence of *B. bassiana* conidia in this habitat and the natural occurrence of this pathogen in populations of the soil-dwelling stages of the beetle. However, results of soil applications have been as variable as those from foliar applications (Wojciechowska *et al.*, 1977; Watt and LeBrun, 1984; Cantwell *et al.*, 1986; Gaugler *et al.*, 1989). Difficulties in delivering lethal concentrations of inocula to insects that burrow many cm below the soil surface is an important factor. Recent studies suggest that more consistent infection and mortality of prepupae and pupae might be achievable by direct spraying of late-instar or mature larvae before they enter the soil than by applying material to the soil and relying on secondary acquisition of conidia (Wraight, unpublished). Biopesticides based on *B. bassiana* have been available for potato beetle control in Eastern Europe for 50 years, and similar products have been developed more recently in the U.S. Other than *B. bassiana*, little has been published on the field efficacy of *Paecilomyces farinosus, P. fumosoroseus*, and *Metarhizium anisopliae* against CPB.

4 *Integrated microbial biological control*

Despite the limited capacity of microbial control agents to compete directly with the synthetic chemical insecticides currently used for potato beetle control, interest in these agents persists with the demand for sustainable food production systems. The above-described differences in the modes of action of *Bt* and *B. bassiana* indicate a high degree of complementarity and strong potential for integrated use in biologically-based potato pest management. The rapid toxic activity of *Bt* against early-instar larvae can be relied upon to protect plants from defoliation early in the season, and the slower activity of *B. bassiana* against all larval instars and epizootic potential in the soil environment can prevent successful pupation of larvae that survive *Bt* treatments. The combined action can greatly reduce the numbers of beetles surviving to the adult stage and, consequently, the size of the next generation (Groden *et al.*, 2002). Excellent compatibility of these agents is further demonstrated by recent studies showing a low but significant level of synergism between *Bt* and *B. bassiana* applied as a tank mix (Wraight and Ramos, 2005). Intensive use of *Bt* for potato beetle control carries the risk that the beetle will develop resistance to the endotoxin (Whalon *et al.*, 1993). *B. bassiana* represents a unique mode of action suitable for integration with *Bt* in a biologically-based pest management system.

The potential for integrating *Bt* and *B. bassiana* with macrobial biological control agents (predators and parasitoids) is discussed by Cloutier *et al.* (1995). *B. bassiana* has a broader host range than *Btt* and likely poses the greater risk for negative interactions with intraguild biological control agents; however, Lacey *et al.* (1999) noted no significant effects on populations of non-target arthropods in *B. bassiana*-treated potato field plots. Intoxication of beetle larvae (even temporarily) by *Bt* can increase their vulnerability to attack by predators and parasitoids (Lopez and Ferro, 1995; Cloutier and Jean, 1998). A CPB program that compared conventional insecticide treatments to a biologically-based IPM approach that included *Bt, B. bassiana* and a hemipteran predator (*Perillus bioculatus*) was found as effective as a program of conventional insecticide applications, but was more costly (Gallandt *et al.*, 1998; Groden *et al.*, 2002).

C *Protocols for evaluation of microbial control agents (MCAs)*

1 *Aspects of pest biology relevant to field testing*

Because *B. bassiana* acts slowly and *Bt* has low virulence against older larvae and adults, applications of these microbes for immediate foliar protection are targeted against young larvae (1st and 2nd instars). Although *B. bassiana* acts too slowly against late-instar larvae to completely protect the plants from their feeding, infected larvae have been shown to consume less foliage than healthy larvae (Fargues *et al.*, 1994). Also, the fungus is as infectious against late-instar larvae as against the early instars (see above), and it may be targeted against older larvae to reduce adult populations. Healthy adult beetles are much less susceptible to infection than larvae (Blonska, 1957).

Like all insect pests, *L. decemlineata* follows a seasonal life cycle. Depending on latitude and temperature, overwintered adults emerge and lay eggs in May – June. The first larval generation occurs in June – July. After a brief pupation period, adults of the first summer-generation emerge in July. Where a second generation occurs, it starts in July – August; adults emerge, feed, and enter the soil in August – September for winter diapause. Ferro *et al.* (1985) investigated the temperature-dependent growth and feeding rates of CPBs from a laboratory colony that originated in the northeastern U.S. The developmental threshold was estimated at ca. 10 °C, and the developmental optimum was 28 °C. The duration of the egg stage varied from 14.5 days at 12 °C to 4.4 days at 33 °C. The duration of the larval stages at 20/24 °C was 3.7/2.1, 3.8/2.2, 2.5/2.3, and 6.6/3.3 days for the 1st, 2nd, 3rd, and 4th instars, respectively. Each larva consumed a total of 35–45 cm^2 of foliage, 70% of that during the 4th instar. Total consumption remained constant over a broad range of temperatures (15–33 °C). Adult beetles consumed 5–10 cm^2 of foliage per day at temperatures between 20 and 30 °C; consumption was markedly lower at 15 and 33 °C.

2 *Setting up the research plots*

a Experimental designs

Experimental design refers to the rules regulating the assignment of treatments and controls to the experimental plots. It must include replication, randomization, and error control. In potato research, the randomized complete block design is most common. Blocking refers to the assignment of a complete set of treatments and controls to a block of land. Variation within a field can be removed from the experimental error through blocking, thus reducing error and increasing precision of a trial. The larger the differences that exist among blocks, the greater is the reduction in the experimental error; therefore, proper blocking should maximize differences among blocks, and minimize differences between plots within a block. This is accomplish by orienting the long dimension of the blocks perpendicular to any known environmental or other unidirectional gradient (soil fertility, trajectory followed by colonizing overwintered beetles, etc.). Any number of treatments and replicates may be included in a block; however, as block size increases, within-block homogeneity may be lost along with any advantage of blocking (Cochran and Cox, 1957). Testing for interactive effects among treatments (assigned to plots) and blocks in analysis of variance (ANOVA) can be used to determine whether treatment differences, if any, are consistent across the blocks. However, this requires treatment replication within blocks, and testing of block effects in agricultural experiments must be interpreted with caution because of restriction error (Sokal and Rohlf, 1995).

All field operations must be conducted on a per-block basis to control for variations that may arise from differences in the way treatments are applied or data collected. This is especially important if many individuals are applying treatments or collecting and processing samples and there is any possibility that technique or method will vary among individuals. Randomization of treatments is done with a table of random numbers, by drawing lots, or by using computer software. Randomization is important because the estimate of experimental error, which is critical for comparing treatments, assumes that plots within a block are allocated such that each has the same chance of receiving any treatment or control. Multiple samples taken from a single plot over time are subject to repeated measures analysis, which in some forms is equivalent to the analysis of a split-plot design where the plots are split temporally rather than spatially. Steel and Torrie (1980) refer to this as a split-plot in time. If heterogeneity of the field site is suspected across two dimensions, a Latin-square design may be advantageous.

b Size, shape, and orientation of research plots

Experimental plot refers to the unit on which random assignment of treatments is made. The size, shape, and orientation of a plot can greatly affect the magnitude of experimental error in a field trial. Plots that are too small may give unreliable or non-representative results; use of large plots may require more time and resources than necessary. A square plot, with its minimum perimeter, exposes the smallest number of plants to border effects. Orientation of plots can reduce or increase the effects of gradients. Generally, experimental error decreases as plot size increases, but the reduction is not proportional. Plot size not only affects variability but may also be a source of bias. If data from border (edge) rows are considered unreliable, plots must be large enough to provide sufficient samples from the interior rows.

For a given area of land, the number of replications decreases as the plot size is increased. Thus, a gain in precision from increased plot size results in a loss of precision from a reduced number of replications. However, as long as the minimum plot size is reached, a greater increase in precision is expected with an increase in the number of replications. Cultural practices related to the trial can dictate the size and shape of plots for ease of operation. The width of the plot may be governed by equipment used for planting, tilling, spraying, and harvesting. The choice of plot shape is not critical when environmental gradients are as great in one direction as in another. If a gradient is present, plots, as recommended for blocks, should have their longest dimension oriented perpendicular to the direction of the greatest variation. For example, when CPB adult migration pattern is known, a rectangular plot with appropriate orientation will give higher precision. When the pattern is not known, it is safer to use square plots. In potato field experiments, plots for small-scale tests commonly range from 0.005 to 0.05 ha in size. In our field trials using backpack sprayers or small, tractor-mounted research sprayers, plots are usually 0.005–0.01 ha with five or six rows spaced 86–91 cm. We use 0.1–0.2 ha plots for larger tests with operational-scale sprayers and 50 ha plots (with 0.01 ha sampling subplots) for aerial application trials.

c Number of replicated plots

The number of times a treatment or complete set of treatments is repeated in a trial is called the number of replications. Multiple observations from a single treated plot are pseudoreplicates. These observations are subsamples and measures of their variability include sampling error but not experimental error associated with application of the treatment. True replication of the applied treatments is required to account for this error. Increased precision is obtained by increasing the number of replications. The number of replications needed to detect differences among treatment means depends on the magnitude of the variability (including experimental error) likely to be obtained and the degree of precision desired. The degree of precision desired is prescribed as the standard error of the treatment means, or as the magnitude of treatment difference that can be detected. Four or five replications are commonly used in potato field experiments. However, this number of replications may not be sufficient to achieve a desired level of statistical power. In particular, detection of economically significant differences in yield may require substantially greater numbers of replicates (Nault and Kennedy, 1998). Additional information on the problems of pseudoreplication and determination of numbers of replicates needed to detect differences of a specified magnitude is presented in Chapter II-1 of this volume.

3 Pre-treatment assessments

a Sampling the pre-treatment field population

At potato shoot emergence (May – June), each plot is sampled twice weekly for signs of CPB infestation. Fifteen random plants per replicate plot are examined initially. As the crop matures and plants increase in size, samples may be reduced to 10 plants per plot. Sampling is re-randomized for each sampling time, or the plants sampled initially are marked and become ‘permanent’ sampling units. Alternatively, clusters of plants are sampled at each of 10 random sites in each plot, and 20 of the earliest deposited egg masses per plot are marked. Because egg hatch can go from 0 to 30% in 1 day, the plots are sampled every 2 days once egg laying begins. Plants can also be destructively sampled; however, this method may not be possible if plots are too small to allow for removal of large numbers of plants.

Visual, in-field counts of the number of beetles per plant or stem, is a widely used sampling method. Densities of adults, egg masses (hatched and unhatched), small (1st and 2nd instar), and large (3rd and 4th instar) larvae are recorded per plant or stem. Drummond (1996) states that the potato stem (vine) is a changing unit which is not suitable if trial objectives include measuring treatment effects on seasonal densities of insects determined by integration of the insect-count time series, because as stem density changes, insects per stem will increase or decrease without a change in absolute insect abundance (beetles per square meter). Drummond (1996) concludes that the entire plant is the lowest level sampling unit that is stable throughout the growing season. This is essentially true, but it is our experience that, as the season progresses, it becomes increasingly difficult to identify individual plants. Under these circumstances, it may be necessary to make more than one determination of plant density during the season. Because of this problem, some investigators prefer to use the potato hill or a measured length of row as the sample unit (these units are commonly employed when sampling yields). While the whole plant is clearly one of the most efficient sample units for determining absolute beetle densities, this does not mean that lower-level sample units (such as individual stems) are not acceptable. Counts of insects per stem are readily translated into absolute density estimates, provided numbers of stems per square meter are recorded on each sample date. This is comparable to determining plant densities in order to translate counts of beetles per plant to beetles per square meter. Counts on stems (vines) rather than whole plants also may be desirable if efficacy is evaluated in terms of an economic threshold expressed as a specific number of beetles per stem or if the investigator is interested solely in comparing relative efficacy of various treatments. The plant is actually a highly unstable sample unit over time with respect to the amount of foliage produced and thus the number of beetles that can be supported without economic injury. Sampling stems or terminal sections of stems also enables use of destructive sampling methods. Stem samples can be collected rapidly and processed in the laboratory at a later time. All stems from a single plot can be combined in a single bag or other collecting container and stored at 4 °C for several days.

b Action thresholds

Action thresholds for insecticide treatment that are derived from economic threshold levels are available for all life-stages of CPB except the pupa. Unfortunately, the thresholds are as variable as they are numerous and differ between regions and even within regions (reviewed by Holliday, 1995). Moreover, their reliability from year to year is often uncertain. According to action thresholds developed for *Bt* treatment of CPB in New York State (Cornell Cooperative Extension, 2001), application should be made within 1 or 2 days whenever either or both of the following conditions are met: (a) densities of egg masses reach or exceed four per 50 vines and at least 25% of the earliest deposited egg masses have hatched or are in the process of hatching, or (b) densities of small larvae (< 6 mm) reach 76 or more larvae per 50 vines. Zehnder *et al.* (1992) recommended initiation of *Bt* applications within 9 days of peak (33–64%) egg mass hatch and when $\geq 10\%$ of potato vines contain one egg mass.

No action thresholds have yet been developed for *B. bassiana*. Results of field trials at several diverse sites indicate, however, that the recommendations for *Bt* are generally applicable to *B. bassiana*, and spray programs should begin as soon as possible after peak egg hatch. Although Drummond and Groden (1996) indicate that it is best to target *B. bassiana* against the 1st and 2nd instars, a cluster of sprays aimed at the peak of 1st instars is not as effective as a strategy that spaces the sprays over the early 1st to peak 2nd instar occurrence. Control of epigeal populations will be markedly reduced if applications are applied after the larvae have reached the 3rd and 4th instar although high rates of mortality may ultimately be expressed in the prepupal larval population in the soil (Fargues, 1972; Wraight and Ramos, 2002). At times, artificial infestation of research plots will be the only way to conduct a field test. However, this approach should recognize that wild CPB populations are highly aggregated in the field (Harcourt, 1963).

4 Application of MCAs

The objectives of field trials include evaluating the effectiveness of MCAs, comparing their efficacies to recommended chemical insecticides, and comparing application technologies, strategies, formulations, adjuvants, etc. MCAs are applied alone or mixed with synthetic or biorational chemical insecticides; the chemical is often tested at a low or sublethal rate. MCAs have also been tested in combination with other natural enemies of potato beetles, including the pentatomid predator *Perillus bioculatus* and the eulophid egg parasitoid *Edovum puttleri*.

a Application equipment

Technologies for applying microbial biopesticides against field-crop pests are discussed extensively in Bateman (1999) and are also treated in sections I and III of this volume. Here we will present only a brief description of equipment commonly used in applying MCAs to potato crops.

1. Foliar application. Thorough coverage is of primary importance. Bacterial pathogens must be ingested and fungi must come into contact with the pest's cuticle. Materials are applied from the ground or air, depending on the size of the plots or fields. Many different types of ground-based spray equipment may be used. Hydraulic backpack sprayers pressurized by compressed gas or hand-operated piston or diaphragm pumps and motorized backpack airblasters or mistblowers are most commonly used for small-plot applications. Small, tractor mounted hydraulic sprayers designed specifically for research applications (*e.g.*, with multiple spray tanks) are commonly used for larger plots, and conventional hydraulic or airblast spray rigs are used for commercial fields. Electrostatics, air-assist, air-entrainment, rotary atomization, and many other technologies have also been tested, and new sprayer technologies are constantly being developed to address the specific needs of biopesticide delivery.

Many different types of nozzles are available, each designed to produce a particular spray pattern and droplet size range under particular conditions. Attachment of nozzles to drop tubes reduces drift and maximizes deposition on the lower surfaces of potato leaves (where newly hatched larvae aggregate and MCAs are protected from solar radiation). On-target application early in the growing season is achieved by band spraying, where the width of the band corresponds approximately to the width of the potato plant. Banding is an excellent method for increasing the application rates of costly microbial biopesticides or for applying high rates of research materials that are in limited supply. Increases in efficiency can be substantial when treating small plants early in the season. Sprayers should be calibrated periodically throughout the growing season. One of our most effective ground sprayers for small-scale testing is a custom-made backpack hydraulic sprayer equipped with a 2.3–4.5 kg tank of compressed CO_2. The 0.9-m-long horizontal spray boom (the width of a potato bed in North America) is fitted with three nozzles. One nozzle is attached directly to the center of the boom and directed straight downward to spray the dorsal leaf surfaces; two lateral nozzles are attached to swivels on the ends of 25.4-cm-long vertical drop tubes and directed upward at a 45° angle to the ground to achieve coverage of ventral leaf surfaces. The nozzles are D3 hollow cone discs with #45 cores, each with a capacity of 0.87 liters/min at a pressure of 276 kPa. It should be recognized that applications with portable spray equipment (and the resulting efficacy) may be difficult to duplicate with conventional, tractor-mounted sprayers. We have used a Schweizer Ag crop duster (42 hydraulic flat jet nozzles; 207 kPa pressure; 46 liters/ha) for aerial applications of *B. bassiana* against CPB on large commercial fields (110 ha under central pivot irrigation).

2. Soil application. No specialized equipment for soil application of *B. bassiana* has been developed. In most cases, material is applied and then mixed into the soil to a specified depth. Some researchers have hand raked the inoculum into the soil and others have used hand trowels. Using standard equipment for fertilizer application, conidial dusts and wettable powder and granular formulations (both containing conidia) or alginate pellet formulations (containing mycelia) have been spread on the soil surface and then tilled into the soil in spring prior to CPB emergence. Both broadcast and furrow applications have been tested. Specialized drill equipment has been researched in Australasia for applications of *Serratia entomophila* and *M. anisopliae* against scarab beetles (Jackson *et al.*, 1992; Rath, 1992), and similar equipment could be developed for soil applications against CPB.

b Application rates and handling of MCAs

1. Bacteria. Commercial *Bt*-based products are generally easy to handle, and spay mixes can be prepared as follows:

1. Shake or stir the formulated product (water dispersible liquid) thoroughly before use.
2. Fill spray or mixing tank half full of water and begin agitation.
3. Add the recommended amount of product into the water and maintain agitation.
4. Add other spray materials, if any, and add the remainder of water required.
5. Ensure that the tank mix has a pH range of 4 to 8 before the product is added; extreme pH levels may adversely affect product performance.
6. Do not allow diluted spray to remain in the tank for more than 72 hours.

Many local conditions, such as water quality, also may affect the tank-mix, so it is advisable to conduct a small jar-mixing test before field use. These products can be tank-mixed with many commonly used fungicides or insecticides or with *B. bassiana* (without fungicides) to expand control. As for any insecticide, follow label directions. Agricultural spray adjuvants are not required with commercial formulations. It is possible that product performance might be improved by addition of materials such as spreaders or stickers, especially if rain or heavy dews are anticipated. Caution must be advised, however, as untested adjuvants may adversely affect bioinsecticide efficacy. Also, use of any material must be preceded by verification that it is not phytotoxic at the planned rate and volume. Biopesticide manufacturers can often provide information on compatibility of common spray adjuvants.

Rates recommended for the *Bt* products differ with pest population and larval development stage. Small to moderate populations of early-instar larvae (newly hatched to 6 mm in length) require ca. 2.5–7 liters of product/ha (1 liter of the commercial product Novodor® [Valent BioSciences, Libertyville, IL] contains 17.2 million Leptinotarsa Units). Large populations of early-instar larvae are sprayed at 5–7 liters/ha. Mixed populations of younger and older larvae require 7–10 liters/ha. *Bacillus thuringiensis* is not recommended for eggs, late 3rd instars, 4th instars, or adults. Good coverage of the foliage at the feeding site is critical, as the endotoxins must be ingested to be effective. The spray volume depends ultimately on research objectives, but sufficient water must be used to provide thorough coverage. For ground application, apply the recommended amount of product in at least 185 liters water/ha. For aerial application, apply at least 25 liters water/ha. These are only general guidelines, and experimenting with rates and spray volumes continues to be an important research objective. Tests conducted in small plots with hydraulic sprayers typically employ medium spray volumes (defined as 200–600 liters/ha by Matthews (2000)). Medium or higher spray volumes and elevated spray pressures may be tested with the objective of increasing spray penetration of the crop canopy and maximizing spray coverage (especially coverage of leaf undersides to target small larvae and protect MCAs from solar radiation). Alternatively, canopy penetration can be achieved using air-blast or air-assist technologies, which typically involve low spray volumes (50–200 liters/ha). ULV ground-application technologies have not yet been investigated for CPB control with MCAs. Product manufacturers recommend a period of at least 6 hours between *Bt* applications and the beginning of irrigation, and sprays should not be applied if rainfall is imminent. Store containers tightly closed, in a cool, dry place for maximum product stability. *Bt*-based insecticides are not labeled for application through irrigation systems.

2. Fungi. Commercial products incorporating *B. bassiana* are formulated as wettable powders (WP) or emulsifiable oil-based suspensions. Industry has yet to produce guidelines for spraying these mycoinsecticides on solanaceous vegetables to control CPB. Researchers have experimented with application rates ranging from 5×10^{12} to 5×10^{13} viable conidia/ha.

Aerial conidia of *B. bassiana* and many other Hypocreales (formerly placed in the form-class Hyphomycetes) are extremely hydrophobic, and unformulated conidial powders (technical powders) and formulations that do not contain wetting agents are difficult to suspend in water. Many wetting agents are not compatible in tank mixes with *B. bassiana*, being either poor wetting/spreading compounds or lethal to the conidia. Consult the mycoinsecticide label or manufacturer for compatibility with commercially available wetting agents. Organosilicone (polysiloxane) surfactants such as Silwet (Loveland Industries, Greeley, CO) have been found highly effective for creating suspensions of *B. bassiana* conidia (Wraight and Carruthers, 1999). Organosilicones should be dispensed at the time of spray, as hydrolysis will begin to degrade surfactant activity as soon as they are mixed in water (see Stevens, 1993). Mixing hydrophobic conidial powders is easier if the powder is added to the water/surfactant solution (rather than vice versa). For research purposes, large-volume spray suspensions can be prepared by first mixing the conidia in a small amount of water (producing a concentrated slurry) and then pouring this slurry into the spray tank. The following procedure works well for preparing premix concentrates of technical powders.

1. Select a strong container with screw cap and handles; the container should hold 8–20 liters of water with a minimum 25% headspace.
2. Measure the amount of wetting agent required for the final tank mix.
3. Add a small fraction of this to the water in the premix container, seal, and shake until the wetting agent is completely dispersed.
4. Retain the remaining volume of wetting agent for addition to the spray tank.
5. Add the conidial powder slowly to the container, seal tightly, and shake well until a homogeneous suspension is achieved. Shaking for 1–2 min will usually suffice when using organosilicone wetting agents; however, longer periods of agitation may be required when using weaker surfactants or highly aggregated conidial powders.
6. Fill the spray tank to half of the desired volume of water.
7. Add the remaining wetting agent, and operate the tank agitator until it is fully dispersed.
8. Pour the concentrate into the spray tank, maintaining agitation. If the selected wetting agent is toxic to conidia at elevated concentrations, it may be necessary to add this material in carefully measured increments throughout the mixing process.
9. Top off the tank with water to the desired spray volume (this water is used also to rinse the residue from the premix container).
10. An antifoaming agent may be added if foaming is a concern.
11. Maintain some degree of agitation in the spray tank while spraying to prevent settling of the conidia to the bottom of the tank and to ensure even application onto the crop.

Most commercial WP and emulsifiable oil (oil dispersion) formulations of fungal pathogens are readily miscible in water, and measured amounts can be added directly to spray tanks equipped with effective agitators. The above-outlined protocol for *Bt* products can be followed. Special instructions may be found on the product label. A few words of caution are warranted with regard to handling oil-based formulations of fungi. The active ingredient (conidia) in these products settle out of suspension over time, and in materials that have been stored for long periods, the conidia may be densely compacted in the bottom of the container. In such cases, remixing to a homogeneous state can be difficult. The container must be vigorously shaken (often for a prolonged period) to resuspend the active fraction, and it is advised to visually confirm complete removal of all material from the bottom of the container before measuring and dispensing. After thorough mixing of the product, final preparation of the sprayable emulsion should require only gentle agitation. Violent agitation of the final spray mix should be avoided, as this may destabilize the emulsion and produce a sticky, oil/conidial residue. This is most likely to occur when shaking small volumes of material by hand or using a mechanical shaker. We have never experienced this problem with the recirculation agitators found on conventional sprayers. Mycoinsecticides should be applied within a few hours of preparation, because conidia suspended in water for more than 12 hours may lose viability.

The commercial *B. bassiana*-based mycoinsecticides are relatively stable compared with other biological insect control agents. However, these products contain live fungal conidia and some precautions are necessary to maintain their viability. Moisture and high temperatures are detrimental to product shelf life. The commercial WP formulations currently on the market are packaged with low moisture content (< 7%) and tend to absorb water rapidly when exposed to atmospheric conditions. The commercial oil-based formulations are essentially nonhygroscopic. Nevertheless, anhydrous oils are readily contaminated with water (especially from condensation on the walls of containers opened soon after removal from cold storage), and even small amounts of moisture can affect storage stability. All of these products should be stored in tightly sealed, waterproof containers at temperatures below 30 °C. Under these conditions, a shelf life of nearly 1 year can be expected. Refrigeration can extend the life of many products to several years. Avoid exposure of mycoinsecticides to direct sunlight.

The mycoinsecticides are compatible with many chemical insecticides including insect growth regulators, although some insecticides are not compatible with *B. bassiana* in a tank mix or in concurrent treatments (see manufacturers' labels). In a tank mix, most fungicides will kill the fungal conidia. Mycoinsecticides can be integrated with many fungicide treatments, however, by allowing at least a 2-day interval between application of these fungicides and use of mycoinsecticides (*e.g.*, see Jaros-Su *et al.*, 1999). Compatibility of mycoinsecticides tank-mixed with herbicides has not yet been evaluated. Most *B. bassiana*-based biopesticides are not labeled for application through irrigation systems, but there are exceptions (*e.g.*, Mycotrol O); product labels should be consulted for details.

c Number and timing of applications

Multiple foliar applications (usually 3 or 4) are required to achieve effective control with *B. bassiana* or *Bt*. Applications of *B. bassiana* are most effective when applied at 3–4 day intervals (Galaini, 1984; Wraight and Ramos, 2002). As related earlier, control with *B. bassiana* may be delayed under high temperature conditions, and expressed only as a reduction of the population of first-generation adults (providing little early-season foliage protection). Label recommendations for use of *Bt* products generally call for applications at 5–7 day intervals. Zehnder *et al.* (1992) reported effective control from applications made at 5–10 day intervals in the absence of rainfall and at 5-day intervals if heavy rain occurred following the initial spray. The necessary spray interval varies with rate and timing of beetle development; longer spray intervals may be more effective under cool weather (or cool climate) conditions, especially if spring emergence of adults occurs over an extended period (E. Groden, personal communication). High CPB populations, rapid plant growth (which dilutes the pathogens in space), or heavy rainfall (which removes the pathogens from the target site) may necessitate additional applications. Intense insect pressure may also require integrated applications of a chemical insecticide or other control measures, especially to control adults feeding on seedlings. The same general guidelines apply to the second pest generation.

In small-scale field tests, researchers often report making applications of MCAs in late afternoon or evening with the assumption that favorable environmental conditions (moderate temperatures, high humidity, and absence of solar radiation) will enhance efficacy. This may be advantageous, especially in subtropical climates where warm conditions during the night could support high rates of fungal development and larval activity. Larval activity (feeding) is largely determined by temperature (see Chlodny, 1975), and such activity likely results in secondary acquisition of inoculum from the treated foliage. However, in recent studies in New York State comparing evening versus morning applications, we observed no differences in larval mortality, and evening applications resulted in only ca. 10–15% greater control of adult beetle populations (Wraight, unpublished). This small gain might not be sufficient to compensate for the obvious disadvantages associated with restricting applications of MCAs to the end of the day, especially with respect to use on an operational scale. Nevertheless, further research on timing of spray applications is warranted.

d Assessing spray coverage

Fungal conidial coverage may be determined at each occasion of treatment through direct counts taken from plastic coverslips pinned to the adaxial and abaxial leaf surfaces at any desired height in the crop canopy, depending upon experimental objectives (*e.g.*, Wraight and Ramos, 2002). The pin holding the coverslip is simply passed through the leaf and embedded in a small (1 cm) cube of dense foam (such as used in insect pinning boxes).

1. Pin coverslips to upper and lower surfaces of potato leaves on randomly or systematically selected plants in the treated potato rows of each replicate plot. We pin coverslips to the center of the leaves, and it is therefore not possible to place two coverslips (upper and lower surfaces) on the same leaf.
2. Under most conditions, conidial counts on these coverslips will be extremely variable due to imperfect spray coverage and thus a relatively large number of samples must be collected for useful statistical analysis (we use 25–30 coverslips per leaf position divided among the replicate plots of each spray treatment).
3. Record which coverslip is attached to which leaf surface, its location within the row in relation to spray trajectory, and its location within the plant canopy.
4. Coverslips can be placed in the field the day before spraying. Exercise caution as they may become contaminated with dust and dirt, making counting of conidia more difficult.
5. To speed recovery of coverslips, plants may be flagged, especially in large plots, but in small plots, use of bright white foam pinning cubes is usually adequate.
6. Spray plots, allow spray to dry, then collect coverslips. Processing large numbers of conidial samples can become onerous; however, a method for rapid direct enumeration of fungal conidia in surface depositions is described by Wraight *et al.* (1998), and an important advantage to use of coverslips is that the conidial samples may be stored indefinitely before processing.
7. In the laboratory, invert coverslips over a drop of lactofuchsin stain (1 mg acid fuchsin/ml lactic acid) on a microscope slide. Conidia will stain red, unlike debris and dust particles.
8. The slide is viewed at 400x magnification, and counts are made as the microscope stage is slowly moved a measured distance and the conidia cross a measured line on an eyepiece reticle; the number of conidia counted is minimized by reducing the sample area as conidial density increases.
9. In addition to absolute counts, observe the overall distribution of conidia on the coverslip surface. Ideally, the coverslip technique should be validated with actual conidial counts on sprayed leaf surfaces, using optical brighteners and incident UV fluorescence microscopy or by washing conidia from the two substrates and enumerating via hemacytometer or plating of colony-forming units.

Viability of conidia in the spray suspension should be determined immediately following each application. A sample of the residual suspension is collected from one of the spray nozzles, diluted, and a small portion sprayed or spread onto a nutrient substrate (*e.g.*, 1/4 strength Sabouraud dextrose or 0.5% yeast extract agar) containing antibiotics to inhibit growth of bacterial contaminants. Following incubation for 16–18 hours at 25 °C, percentage of germination of conidia is determined from observations of four replicate 100-conidia samples. Conidial counts from the coverslips are then adjusted for conidial viability. Goettel and Inglis (1997) give a thorough description of several methods for determining fungal conidial viability.

Colorimetric and fluorimetric methods for indirect quantification of *Bt* international units (IU)/cm^2 of target surface have been developed (Morris, 1982), but assessment of the applicability of these methods for mycoinsectides is needed. Dye-coated water- or oil-sensitive cards are also useful for assessing spray deposits of bacterial and fungal pathogens. Image processing techniques have been used to translate spray-card information into droplet size distributions and deposition volumes (see Matthews, 2000). Spray cards also enable calculation and immediate visual assessment of spray coverage.

5 *Post-treatment assessments*

a Sampling the post-treatment field population

1. Foliar sampling. Sampling methodology is the same as for the pre-treatment sampling (1 day before the initial application). Sample plots just prior to treatment, and again 2 days after each spray application, and thereafter on 4–7 day intervals until onset of pupation of the first generation of potato beetles (sampling at 4–5 day intervals is recommended under warm conditions favoring rapid larval development). Sampling is resumed if a second generation of beetles must be treated. Because larval populations are aggregated, numbers are transformed to log (n + 1) prior to analysis of variance (Harcourt, 1963). If population levels in the treatment plots within blocks are heterogeneous, it may be useful to control for this variation through analysis of covariance (ANCOVA), with pre-treatment populations (densities of eggs or eggs plus small larvae) as the covariate (*e.g.*, Hajek *et al.*, 1987). However, caution must be exercised in the interpretation of ANCOVA results, and this procedure should only be applied if careful blocking is unable to remove most of the variability between treatment plots. Excellent treatments of the important assumptions and caveats associated with ANCOVA are presented by Little and Hills (1978) and Sokal and Rohlf (1995). An alternative approach is to measure treatment effects in terms of population levels recorded in each plot before and after treatment using the Henderson-Tilton modification of Abbott's formula (Henderson and Tilton, 1955). This method is generally applicable only if the population sampled after treatment is the same as that exposed to the treatment or the post-treatment population comprises the direct offspring of the treated population (thus, it is not recommended for assessing effects of early-season treatments on third- or fourth-generation beetles).

Effects of treatments on populations of small and large larvae can be assessed at any specific point on the population curve, or the seasonal treatment effects may be further examined through analysis on the integrated population curves for these life stages. The differences in the areas under the treatment curves best represent the defoliation pressure. The effectiveness of a treatment in reducing larval populations (% control) is determined by simply calculating the percent reduction in the treated population relative to the control population. This procedure is the same as application of Abbott's formula (Abbott, 1925) to adjust percent mortality, but in this case the actual percentage of treated individuals that died is unknown.

2. *Soil sampling*. Sampling methods vary according to the test protocol and objectives. Infection of mature larvae and pupae in the soil can be determined by removal of soil samples from measured areas beneath the plants and sieving to find all live and dead insects. The depth of soil that must be removed depends upon the soil type and degree of compaction. If an estimate of percentage of infection is desired, the soil samples must be collected before significant numbers of uninfected beetles complete development and emerge.

Cages can be used to expose mature larvae or adults to fungal-treated soils, or to contain mature larvae and adults that were exposed to foliar applications of fungus. In a study by Lashomb *et al.* (1984), over two-thirds of adult beetles that entered a silt loam to overwinter settled at depths between 7.6 and 12.7 cm; few penetrated below the cultivated layer. These researchers introduced adults into bottomless $0.6 \times 0.6 \times 0.6$ m cages with the sides buried to a depth of 30 cm. Soil was removed only from the perimeter, and the cages were lowered over the undisturbed soil columns. Populations of adult beetles emerging in summer or of overwintered adults emerging in early spring can be sampled simply by placing bottomless, coarse-screened cages ($1 \times 1 \times 1$ m) on top of the soil and monitoring daily. It is noteworthy, however, that summer-generation beetles emerging in plots of potatoes that have not been severely defoliated may remain in the plots, and good estimates of adult emergence may be obtained by sampling the uncaged populations (*e.g.*, Poprawski *et al.*, 1997; Wraight and Ramos, 2002).

b Assessing treatment effects from field-collected insects

It may be desirable to verify that observed population reductions are the result of treatment-induced mortality (*e.g.*, *B. bassiana* mycosis), or one may be interested in determining relative contributions to control of multiple pathogens applied concomitantly. This may be accomplished by collecting insects from the field and monitoring in the laboratory. However, many pathogens may be considerably more infectious and pathogenic in the laboratory under optimal conditions of moisture and temperature and in the absence of solar radiation than in the field, and the results must be interpreted with considerable caution (see Chapter IV-4 of this volume).

In such situations, levels of infection and time of death of insects collected after treatment can provide important information relating to the ability of a pathogen to infect and kill CPB under various field conditions. Surface sterilization (described below) and detailed analysis of time of death of field-collected larvae can enable one to determine if larvae were infected under field conditions or were infected only after experiencing favorable conditions in the laboratory.

A good sampling scheme is to collect 50 living, active larvae (25 small and 25 large larvae) from 50 plants in each plot 1 day before the initial treatment (to assess prevalence of endemic disease) and then 2 days after each spray application; final samples are collected 5 and 10 days after the last application.

Collected larvae that have been treated with fungal pathogens are surface-sterilized by submerging in 0.13% Zephiran chloride (benzalkonium chloride; Winthrop Pharmaceuticals, New York, NY), rinsed in clean water, and confined individually in 9-cm plastic Petri dishes lined with moistened filter paper and provisioned with untreated potato foliage (the wet filter paper is needed to maintain the foliage). The dishes are then held at room temperature. Each day, the dishes are wiped clean, and the filter paper and foliage are replaced. Prepupae are transferred to paper cups filled with commercial peat moss.

Mortality in larvae, prepupae, and pupae is recorded daily. Dead insects from fungus-treated plots are incubated in Petri dishes on moistened filter paper to determine the proportion of cadavers with resulting fungal outgrowth and sporulation (overt mycosis) and with secondary infection (septicemia). Mortality data are analyzed by ANOVA following arcsine transformation to minimize correlations between the mean and variance of the data. Mortality of mature larvae and pupae can also be estimated by collecting samples of larvae that drop from the plants, placing them in soil- or peat-filled containers, and incubating in the laboratory. However, as with the evaluations of younger larvae collected from foliage, it must be recognized that field conditions cannot be duplicated in the laboratory. Temperature and moisture, beetle density, and concentrations of *B. bassiana* conidia and of agents antagonistic to *B. bassiana* in the soil (including fungicides that may have been applied for plant disease control) are some important factors that may differ greatly between the field and laboratory.

An alternatively approach for confirming death due to treatment is to collect dead or moribund insects and verify infection by microscopic examination or culture plating in the laboratory. In pathogen interaction studies, infection of individual hosts by more than one pathogen is possible, and determining the actual cause of death may not be possible. With respect to the above protocols, it is important to emphasize that laboratory monitoring of field-collected insects cannot substitute for rigorous field controls. If populations are reduced in treated versus control plots, this should be conclusive evidence that the reduction was due to the treatment and should be considered the bottom line in any field efficacy assessment. It is advisable to include, whenever possible, more than one control plot in each experimental block in randomized block designs (especially if the number of treatments is large). If there is reasonable certainty that the insects will be little affected by the spray carrier blank, a carrier control treatment might serve essentially as a second untreated control check (*e.g.*, Wraight and Ramos, 2002). Use of proper controls is the most efficient method of proving treatment effects. Limited resources may not permit collection (proper sampling) and laboratory monitoring of insects during large-scale field-testing programs.

c Assessing crop damage

1. *Defoliation*. Many different ways of assessing defoliation due to CPB feeding have been proposed (Holliday, 1995; Zehnder *et al.*, 1995). We found visual rating of the plants used in the population counts to be satisfactory. At least two observers rate defoliation independently. Each pair of observers reads all treatments in their assigned block. Defoliation is estimated from zero to 100% in classes of 5 or 10%. Ratings are averaged and expressed as percentage of defoliation of the sampling unit. The values are transformed using the arcsine transformation before ANOVA or ANCOVA.
2. *Yields*. There is little agreement on methodologies for assessing yield. In our small-scale tests, plants

are treated with a chemical defoliant and 3–4 weeks later, the potatoes are dug using a machine that lifts the tubers, separates them from the soil and drops them back onto the soil surface. We then collect and weigh all salable potatoes from the central two meters of each row in each replicate plot (a systematic rather than random sample). The total weight of potatoes from each replicate plot is divided by the total meters of row sampled, and this value represents the sample unit for analysis (our tests usually include 5 or 6 replicate plots). In cases where large gaps occur in the rows (*e.g.*, due to plant loss from disease), it may be necessary to harvest manually on a per-plant or per-hill basis. In cases where small plots are used and a primary objective is comparison of yields to regional standards, it may be desirable to preclude samples from border rows or tractor wheel rows. It may also be desirable to grade the tubers into standard categories (which vary with region and potato variety). Weights are log-transformed prior to ANOVA. Relationships between seasonal densities of CPB, infection rates, and crop yields can be examined using regression analysis.

6 *Supplementary data*

Because agrochemicals, especially insecticides and fungicides, may synergize or antagonize insect pathogens, they should be regarded as epizootiologically relevant factors. Other agronomic practices also may affect insects and their pathogens. In any field trial, anthropogenic factors including planting date, variety, applications of fertilizers, applications of agrochemicals, cultivation, irrigation, etc., must be recorded. Because infectivity and virulence of insect pathogens and susceptibility of their insect hosts may be greatly influenced by weather conditions, precipitation, humidity, temperature, and solar radiation data should be recorded and included in all research reports.

3 Potato tuber moth (PTM)

A *Pest status*

The potato tuber moth (PTM) (also referred to as potato tuberworm), is a cosmopolitan pest of potato and other solanaceous crops. The moth, which likely originated in the tropical mountainous region of South America (Graf, 1917), is now present in almost all tropical and subtropical potato production zones (North Africa, the Middle East, Central America and the inter-Andean valleys of South America) and is considered the most damaging potato pest in the developing world. Population development is limited by the 10 °C annual isotherm in both the southern and northern hemisphere; however, economically significant outbreaks occur occasionally in cooler regions, including New Zealand, southern Europe (Italy and Spain), and in the southwestern USA. Recently, the pest has spread from California into the Colombia Basin of Washington and Oregon and may become a major threat to potato production in that region, especially after mild winters that allow substantial numbers of the pest to over-winter. The host plants of *P. operculella* comprise, almost exclusively, cultivated and wild species of the family Solanaceae (Das and Raman, 1994). Larvae mine the leaves, petioles, stems, and tubers of potatoes. Mining of the tubers causes the greatest damage and occurs both in the field and in storage. During the growing season, the moth larvae mine into foliage, while at plant senescence, adults lay eggs in soil cracks adjacent to potato plants and on exposed tubers. Upon hatching, the larvae mine directly into the tubers. At harvest, tubers do not always show signs of damage but may harbor eggs and early-instar larvae. Pupation takes place in the potato leaves or tubers or in the soil. Transfer of infested tubers to potato stores after harvest leads to further propagation of the pest and increased infestation of tubers. Besides producing direct damage, feeding activities of this pest facilitate entry and damage by secondary pests and plant pathogens, resulting in severe reductions in quality and market value of the tubers. The mining habits of this pest make it a difficult target for control.

PTM has been found to rapidly develop resistance to repeated applications of synthetic chemical insecticides (Richardson and Rose, 1967; Cisneros, 1984). In addition to posing a serious health threat to farmers and consumers, the use of broad-spectrum chemical pesticides has in some cases resulted in outbreaks of secondary pests (*e.g.*, Palacios and Cisneros,

1997). The many problems associated with the extensive use of chemical pesticides have stimulated growing interest in safer alternatives for pest management.

B Biological control agents for pre-harvest control

1 Overview

Numerous IPM components have been identified, including macrobial and microbial control agents, pheromones, and botanical insecticides, and strategies have been developed for targeting *P. operculella* in the field and in storage (Shelton and Wyman, 1979; Kroschel, 1995; Lagnaoui *et al.*, 1996; Palacios and Cisneros, 1997). Establishment of various parasitoids, most notably species of Encyrtidae and Braconidae, have been attempted in at least 20 countries; results have been variable depending on the region (Mitchell, 1978; Horne, 1990; Raman, 1994). Biological control strategies have often called for complementary use of cultural practices such as deeper planting, repeated hilling, use of sprinkler versus furrow irrigation, early harvest, and intercropping to reduce tuber infestation (Foot, 1976; von Arx *et al.*, 1990; Raman, 1994).

Laboratory investigations on two common fungal entomopathogens, *B. bassiana* and *M. anisopliae*, have indicated potential for control of *P. operculella* larvae, especially early instars (Hafez *et al.*, 1997; Sewify *et al.*, 2000; Sun *et al.*, 2004); however, research on use of these agents against this pest is limited. Sabbour and Ismail (2002) reported that toxic extracts from the plants *Atropa belladonna* and *Hyoscyamus niger* synergized the activities of entomopathogenic fungi and bacteria (*B. bassiana* and *Bt*) against *P. operculella*. The fungus *Muscodor albus*, which produces a mixture of antimicrobial volatile organic chemicals could offer promise as a biofumigant. Adult PTM that were exposed to the fungus in a confined chamber responded with highly significant levels of mortality (Lacey and Neven, 2006). Entomopathogenic nematodes (*Steinernema* spp. and *Heterorhabditis* spp.) are also prime candidates for biological control of tuber moth larvae, especially in soil and cull piles. However, their potential is yet to be evaluated. The most extensively studied pathogens for PTM control are the *P. operculella* granulovirus and *Bt*. Use of these agents for both pre- and post-harvest control is dealt with in the remainder of this chapter.

2 Bacterial pathogens

Only one bacterium, *B. thuringiensis* (*Bt*), has been evaluated for PTM control; however, laboratory tests indicate susceptibility to numerous subspecies of this pathogen, including *kurstaki, thuringiensis, tolworthi, galleriae, kenyae, morrisoni*, and *aizawai*. Among these, *Bt* subsp. *kurstaki* (*Btk*) has been identified as one of the most virulent against *P. operculella* (von Arx and Gebhardt, 1990; Salama and Salem, 2000; Salama *et al.*, 1995a; Hernandez *et al.*, 2005), and this subspecies has an extensive history of commercialization for control of lepidopteran larvae.

Investigations assessing foliar applications of *Btk* targeted against larvae mining in the potato leaves have produced mixed results. Kroschel (1995) and Keller (2003) reported minimal to moderate levels of efficacy. Kroschel (1995) attributed this to the cryptic habit of the leaf-mining larvae and the limited persistence of *Bt* on the foliage (due to rapidly degradation by solar radiation and removal from the foliage by rainfall). On the other hand, Awate and Naik (1979) showed that foliar applications of *Btk* at 15-day intervals were almost as effective at controlling tuber infestation as parathion and carbaryl applied to the soil surface. Similarly, Broza and Sneh (1994) reported that three applications at 8-day intervals reduced *P. operculella* larval numbers and fruit damage by nearly 70% in a tomato crop. General methods for field testing of bacterial biopesticides as foliar sprays were presented in the preceding sections on CPB.

3 Viral pathogens

One of the most extensively investigated pathogens of *P. operculella* is a granulovirus (*Po*GV) that attacks the larvae. This pathogen has accompanied the moth in its dispersal from South America and has a nearly worldwide distribution (Reed, 1969; Broodryk and Pretorius, 1974; Hunter *et al.*, 1975; Alcázar *et al.*,

1991; Kroschel and Koch, 1994; Zeddam *et al.*, 1999; Setiawati *et al.*, 1999; Sporleder, 2003). Sporleder (2003) assessed the virulence of 14 geographic isolates of *Po*GV and found a wide range of activity covering several orders of magnitude.

Infected *P. operculella* larvae can be recognized by their opaque, milky-white color and by their behavior (slow reaction when disturbed). Reed (1971) reported on the effect of virus concentration, temperature and larval age on disease progression. Most larvae die within 2–3 weeks of ingesting virus, but high doses can cause death by toxicosis within 48 hours. Generally, infected larvae develop to the final (fourth) instar, but fail to pupate. Like most granuloviruses, *Po*GV has a limited host range; only *P. operculella* and a few other gelechiid species are known to be susceptible to infection. Studies of field populations of *P. operculella* have, in some cases, revealed *Po*GV prevalence levels as high as 35–40% (Kroschel, 1995; Laarif *et al.*, 2003). Such observations, combined with reports of high inherent virulence of the pathogen (exhibited under laboratory conditions) have stimulated investigation of its microbial control potential from the standpoint of inundative augmentation. Results of field tests have been generally promising, though variable in some cases. Tests conducted in Australia (Reed, 1971; Reed and Springett, 1971) indicated that a single, early application of virus at the rate of 6275 larval equivalents/ha could achieve sufficient control and that the virus could spread extensively to untreated areas. It was concluded that virus reached the leaf-mining larvae via stomata and that wind and birds were responsible for spreading the virus. However, subsequent studies by Briese (1982) showed that field populations of *P. operculella* in Australia differed in their susceptibility to the virus. In the Yemeni highlands, Kroschel *et al.* (1996b) observed high rates of infection (82–85%) among larvae collected from fields treated twice at a rate of 5×10^{13} occlusion bodies (OB)/ha. The main factor limiting viral efficacy was identified as rapid inactivation due to solar (UV) radiation, which made at least two applications recommendable. In Tunisia, field applications of *Po*GV to the soil surface before harvest reduced infestation of tubers by 73%, and in-store infestation failed to develop (Ben Salah and Aalbu, 1992). Ben Salah *et al.* (1994) tested a combination of *Po*GV, *Bt*, and extra irrigation for integrated control; in some instances, the integrated controls proved more effective than conventional insecticides. However, a negative interaction has been observed between these two pathogens (Sporleder, 2003; Sporleder *et al.*, 2005b), especially at high application rates, and simultaneous application of *Po*GV and *Bt* is therefore not recommended for *P. operculella* control.

PTM has exhibited a strong potential to develop resistance to *Po*GV (Briese and Mende, 1981, 1983; Briese, 1982; Sporleder, 2003). Briese and Mende (1983) observed a 140-fold increase in LD_{50} following exposure of larvae to *Po*GV over six generations. Similarly, Sporleder (2003) reported high levels of resistance in a population exposed for 12 generations. These findings indicate that resistance management should be incorporated into any control programs that rely extensively on the virus (Cameron *et al.*, 2005).

C Protocols for evaluation of MCAs for pre-harvest control (with emphasis on PoGV)

1 Defining research objectives

The purpose of field experiments using MCAs could be to:

- Assess effects on the pathogen itself, *e.g.*:
  - field stability (kinetics of pathogen inactivation due to UV radiation)
  - enhancement of field stability due to UV-screening formulations
  - levels of horizontal transmission under field conditions
- Assess effects of pathogen applications on the targeted pest population, *e.g.*:
  - lethal effects on different life-stages of the targeted pest
  - sublethal effects on the pest (reduced fecundity, changed feeding habits, etc.)
  - short and long term effects on pest population development (*e.g.*, single application of high rates versus repeated applications of reduced rates)
- Determine effects of pathogen applications on the crop
  - damage, yield, and/or quality

Experimental plans depend on these research questions. When working with MCAs, the Control Window concept presented by Cory and Evans (see Chapter IV-1) should be adapted for good design of experimental series. Detailed knowledge of the target pest biology, derived from field and laboratory studies, should help to plan experiments to test hypotheses.

2 *Mass-rearing of PTM*

In some cases protocols may call for bioassay of leaf samples in the laboratory or release of moths in the field to establish or augment pest populations. Rearing large numbers of *P. operculella* is easily achieved on potato tubers at room temperature under natural photoperiod. Eggs are placed in plastic containers (30 × 20 × 7.5 cm) containing potato tubers as food and sand as the pupation medium. Pupae are easily separated from sand and cocoons by washing in 3% sodium hypochlorite solution. Collected pupae are placed in oviposition cups (0.5 liter) covered with cheesecloth. To provide an oviposition site, a filter paper is placed on the cheesecloth. The filter papers containing the eggs may be collected daily and incubated at ca. 26 °C with relative humidity > 60% to provide a constant supply of insects for bioassays. Alternatively, eggs may be refrigerated (8 °C) for 14 days without significant effects on survival. To reduce prevalence of viral infection in laboratory populations, eggs may be surface sterilized in a 3% sodium hypochlorite solution. Adults are fed 5% sugar solution dropped on the top of the cheesecloth. For additional rearing information, see CIP (1992b), Kroschel (1995), and Sporleder (2003).

3 *Mass-production of the PTM granulovirus*

*Po*GV can be multiplied using the egg-dip method (filter papers containing the eggs are dipped in a *Po*GV suspension containing ca. 5×10^8 OB/ml). Neonate larvae consume the virus with the chorion and can be reared as described above. Infected fourth-instar larvae are collected after 14–20 days of incubation at 25 °C. For further information on *Po*GV purification and quantification see Sporleder *et al.* (2005a)

4 *Aspects of pest biology relevant to field testing*

Modeling has demonstrated that when the age structure of *P. operculella* populations stabilizes (stable-age distributions), adults comprise < 2% of the total population (ca. 50% eggs, 40% larvae, and 5% pupae). Knowledge of the stable-age distribution of *P. operculella* is needed to estimate the effects of biopesticide applications on pest population dynamics. The phenology model developed for *P. operculella* (Sporleder *et al.*, 2004) may be used to define the stable-age distribution of moth populations in different agroecological regions and to estimate the effects of a pathogen application on pest population development. To apply this knowledge, the researcher needs a basic understanding of which life stages (or age-stages) of the pest population are susceptible to the pathogen and which stages are reached by the application. The life stage of *P. operculella* most effectively targeted by the granulovirus is the neonate larva, which ingests virus when entering treated potato leaves or tubers. Reed and Springett (1971) showed that larvae mining inside foliage became infected after application of the virus and suggested that virus could contact mining larvae by penetrating the leaf via the stomata. Susceptibility of PTM larvae to *Po*GV decreases rapidly with larval age (Sporleder, 2003). Virulence of *Bt* is similarly host-age dependent (Salama *et al.*, 1995a).

Development of *P. operculella* is strongly determined by temperature. In subtropical regions this pest may produce over 9 overlapping generations per year, and populations thus comprise individuals of mixed-ages. This is an important factor in predicting efficacy of pathogen applications. In regions with lower temperatures and shorter growing seasons, only 2–3 generations are possible and populations increase most rapidly during the period of maximum vegetative growth of the potato. In these regions, the age structure of the population is more homogeneous, and this may allow for more efficient targeting of young larvae.

5 *Setting up the research plots*

Effective control cannot be achieved unless neighboring fields are free of the pest. Before mating, adult moths exhibit a strong tendency to disperse. Studies by Cameron *et al.* (2002) indicated migration of 6–17% of PTM adults from an infested plot to trap plants located within 40 meters and smaller numbers of moths dispersed > 100 m. Migration of pests from nearby fields or from experimental control plots can easily mask the effects of applied treatment. There are two approaches to overcoming this problem. First, small research plots or areas within larger plots may be enclosed in field cages (*e.g.*, nylon gauze with a 1-mm mesh) (Figure 1). However, this makes the experiments more complicated and expensive, and the cage environment does not genuinely reflect field conditions (caging may affect irradiation, temperature, ventilation, precipitation, irrigation, etc.). Alternatively, the experimental plan may incorporate treated and untreated crop buffer zones (around treatments and controls) of sufficient size to preclude significant pest migration among plots. The principal disadvantages in this case are that the number of experimental treatments or the number of treatment replicates may be limited, depending upon the available area, and treatment of buffer zones requires substantially greater amounts of experimental materials.

6 *Pre-treatment assessments*

a Sampling the pre-treatment field population

Dispersion of *P. operculella* populations in the field is clumped or aggregated with a micro-distribution on individual plants. Kroschel (1995) observed in different potato fields in the Yemeni highlands that the variance of infestation was about 5 times greater than the mean. The distribution can therefore be considered 'contagious' and is the result of oviposition in clusters on single leaves or near individual potato plants. Populations in each replicate plot can be randomly or systematically sampled using approaches similar to those described for CPB. Populations can be assessed as mines per plant or stem or as percentage of plants or stems infested. Kroschel (1995) developed a sequential sampling

Figure 1. Caged potato tuber moth field experiment in the Republic of Yemen (Photo provided courtesy of Jürgen Kroschel)

procedure for Yemeni field conditions. However, the pest intensity-crop loss relationship was highly variable among years. Such an approach will require additional study in greater detail before it will be possible to differentiate among conditions occurring in different agroecological zones.

b Action thresholds

First application to potato foliage should be made when, after shaking potato plants, one or more adults per plant can be observed. Precise control thresholds (level of infestation at which increase in yield gained by control is equal to cost of control) have not been established for PTM control with *Po*GV or *Bt*; however, they would be expected to be lower than thresholds established for chemical insecticides. A threshold of 0.5 mines/plant at formation of leaves and stems to crop extension growth was determined by Kroschel (1995) for fenvalerate (Sumicidin) in the Republic of Yemen. Kroschel (1995) showed that yield losses were significant only when high pest pressures (> 30 larvae per plant) occurred early in the season during crop extension growth, while infestations at the end of crop extension growth and later caused yield losses between 8 and 15%. Because of the great complexity of cropping systems and crop-pest interactions, Keller (2003) emphasizes that threshold values are only one factor among many to be considered in making pest management decisions.

7 Application of MCAs

a Application rates and volumes

Before field-testing, laboratory bioassays employing leaf-disks treated with spray volumes similar to those planned for the field or semi-field tests should be conducted to determine the concentration-mortality relationship. An important consideration for establishment of practical field dosages is the slope of the log concentration-mortality response lines. Probit lines from bioassays of *Po*GV-inoculated leaf disks against neonate *P. operculella* larvae can be expected to have slopes of ca. 0.6 (Sporleder, 2003). This means that augmenting efficacy from 85% mortality to 95% or 99% mortality requires a respective 10-fold and 100-fold increase in concentration. On the other hand, 10- or 100-fold reductions in concentrations would still result in 64% and 38% mortalities, respectively. Thus, control equal to or greater than that provided by a single high-rate application of *Po*GV might be achieved with greater economy by making multiple, low-rate applications at short intervals.

Foliar applications at a rate of 5×10^{13} occlusion bodies (OB)/ha, (corresponding to 10,000 larval equivalents at 5×10^{9} OB/larva) have been shown effective against larvae mining in the foliage (Kroschel *et al.*, 1996), and sprays applied at the same rate and volume onto soil surfaces were also effective in reducing tuber infestation measured at harvest (Ben Salah and Aalbu, 1992). The *Cydia pomonella* granulovirus has been extensively researched at rates of $0.5–5 \times 10^{13}$ OB/ha for codling moth control in orchards (Jaques, 1990; Arthurs *et al.*, 2005), and 5×10^{13} OB/ha seems a reasonable maximum application rate for future research and development of PoGV. Tests of commercial *Bt* products have employed recommended (label) rates, and rates for testing novel isolates can be extrapolated from relative potencies determined in laboratory assays with standard materials.

Increasing spray volume has been hypothesized to enhance penetration of microbial pesticides into the tunnels of leaf-mining *P. operculella* larvae (Broza and Sneh, 1994; Kroschel *et al.*, 1996b). These authors employed air-blast or mist-blower sprayers to deliver MCAs in a volume of 500 liters/ha. Incorporation of surfactant materials into MCA formulations is another important consideration with respect to achieving penetration into the larval mines and leaf folds.

b Number and timing of applications

*Po*GV or *Bt* may be applied to potato foliage during the entire vegetation period or to the soil just before harvest. In temperate regions (under mild temperature conditions), applications should be repeated at intervals of 10 days. In tropical and subtropical regions, due to faster development of pest populations, these intervals should be shorter (5–6 days), and more applications are needed. Differences between treatments are often only visible after 2–3 treatments.

8 *Post-treatment assessments*

a *Sampling the post-treatment field population*

To assess treatment-related changes in pest populations, potato stems or plants can be taken from randomly or systematically selected locations in each replicate plot and examined for tuber moth damage. The number of mines per sample unit provides a measure of population density. Potato tubers can be similarly sampled to compare pest numbers in control versus treatment plots. In low to moderate infestations, it may be possible, by cutting the tubers, to count numbers of mines. Tubers can also be incubated individually in small containers and survivors (emerged adult moths) counted.

b *Assessing treatment effects from field-collected insects*

To assess the initial effects of a foliar application in terms of percent infection, 50 to 100 leaves containing new mines should be collected per replicate plot and returned to the laboratory. Each leaf is then placed on a potato slice (1 cm thick) in a 0.25 liter plastic cup covered with gauze. Following incubation for 2–3 weeks at 25 °C, individual larvae may be classified as infected (failure to pupate with clear symptoms of virus infection) or survived (pupated). Potential for cross-infection between healthy and virus-infected larvae (especially any larvae that might succumb rapidly to infection after ingesting a high dose of virus) must be considered when holding larvae for prolonged incubation periods (Kroschel *et al.*, 1996), and it is best to rear larvae in isolation, if possible (individual larvae with surrounding leaf tissue can be cut from infested leaves). On the other hand, our observations suggest that horizontal transmission of the virus occurs infrequently among *P. operculella* larvae reared in batches. The larvae are not cannibalistic, the virus does not appear to be transmitted via contaminated fecal material, older larvae are highly resistant to viral infection, and bioassays conducted with individually versus batch-reared larvae have produced nearly identical results (Sporleder, unpublished). The above-described sampling method may be repeated at weekly intervals for assessing medium- and long-term effects of the pathogen. It is important to assess infection in larvae from control as well as treatment plots, especially if virus is naturally present in the pest population.

c *Assessing crop damage*

To assess treatment effects on yield, potatoes should be harvested from an area of at least 35 m^2 in each replicate plot. After harvest, not all damage will be visible, but potato tubers may bear young larvae and eggs. At harvest, the number or percentage of damaged tubers should be assessed from 100 tubers selected randomly from the harvest of each plot. Remaining tubers from each plot should be stored as individual batches and the damage assessment repeated at intervals of 30 days.

Damage to potato foliage can be compensated to a high degree by the potato plant, so that significant differences in yields may not be detectable at harvest. Kroschel (1995) showed that yield losses were significant only when high pest pressures (> 30 larvae per plant) occurred early in the season during crop extension growth. PTM populations occasionally reach this level in tropical or subtropical regions during mid- and late-season (*e.g.*, Egypt and Yemen), but such damaging populations are rarely seen in temperate zones. Kroschel (1995) reported that infestations after crop extension growth caused yield losses between 8 and 15%. These reductions were not statistically significant, but could nevertheless represent real losses of considerable economic significance (Nault and Kennedy, 1998).

d *Assessing* virus *persistence*

Following application of virus in the field using the desired application method, residual activity can be assessed at various intervals by collecting leaf samples and conducting laboratory bioassays. This type of experiment is referred to as a semi-field experiment, because field-applied treatments are evaluated in the laboratory. An important advantage of semi-field testing in this case is that field plots are not necessarily required. Testing may be conducted with relatively few plants from small plantings or even potted plants. However, larger-scale testing might ultimately be desirable to assess crop canopy effects or effects of various spray application technologies, and evaluations *in situ* may ultimately be required if field environmental factors are suspected of influencing host susceptibility.

Treatments (completely randomized) should contain virus applied at different concentrations to determine the dosage-mortality relationship. This protocol is more demanding than single-dose studies, but generation of probit regression lines enables calculation of the rate of virus inactivation and estimation of half-life. When a formulated virus is used, it may be desirable to keep the concentrations of formulation ingredients constant while changing the virus concentration. However, this will depend on specific research objectives. In commercial research and development, testing of varying amounts of formulation ingredients over a range of concentrations corresponding to proposed product label rates also may be of interest, especially in cases where formulation ingredients may have phytotoxic, insecticidal or synergistic properties or if the formulation ingredients possess UV-screening properties. Regardless of the objectives, a zero-virus level must be included to allow for correction of virus-caused mortality (formulation without virus). If the formulation itself is expected to contribute directly or indirectly to insect mortality, an additional control, water alone, should be included for comparison.

Probit-log concentration curves for *Po*GV applied to potato plants against *P. operculella* generally exhibit a slope of ca. 0.6 (Sporleder, 2003). This is an exceedingly low regression coefficient, and to obtain clear differences in mortality among treatments, the dilution factor between serial concentrations (dosages) should be between 6 and 9. It is possible to obtain reasonably precise estimations of median lethal concentrations or dosages (LC_{50} or LD_{50}) from assays with as few as four concentrations or dosages (as long as mortalities are within 20 to 80%); however, achieving high levels of precision, especially when estimating extreme values such as an LC_{95}, requires testing at more than five concentrations. Good results can be obtained using a high concentration of 10^9 OB/ml and five successive 6-fold dilutions (this range of concentrations normally produces ca. 20–95% mortality of first-instar *P. operculella* larvae).

Irradiance should be monitored during field exposures. Measuring the irradiance from a specifiable band of wavelengths provides for the most comprehensive analysis of radiation effects on virus activity. If a spectroradiometer is not available, pyranometers in combination with specific UV sensors can be used alternatively (Sporleder, 2003).

The following basic protocol is suggested for semi-field assessment of residual viral activity on potato foliage.

1. Apply treatments during evening hours to avoid UV-inactivation of virus prior to initial sampling. Apply to run-off to ensure as homogeneous a distribution of virus as possible. Replicate sets of control plants should be sprayed with virus and kept shielded from solar radiation.
2. Mark apical leaves to enable identification and exclusion of newly emerged (untreated) leaves during the sampling period. This also allows exclusion of young, growing leaves; leaf expansion post application can markedly reduce the density of viral granules on the leaf surface.
3. Leaf samples for laboratory bioassay should be collected randomly or systematically from each of the replicate plots or plants comprising each treatment. Virus activity is generally monitored at daily intervals (leaf samples collected early each morning) for 6 days to determine residual half-life. An activity baseline and possible effects of exposure to the host plant alone are determined from assaying leaves from the above-described control plants.
4. At least 50 recently emerged first-instar larvae (from a laboratory colony) should be tested per treatment replicate. It is possible to introduce many larvae onto a single leaf, as they will be maintained on the leaf for only a short time; however, minimal contact between larvae, and thus minimal chance for horizontal virus transmission, can be assured if

no more than five larvae are reared per leaf. Leaf condition is maintained by inserting the petiole into a vial of water.

5. After incubation for 2 days at ca. 25 °C, each infested leaf is transferred to a clean container provisioned with pieces of potato tuber to support completion of development (larvae should be provided with 2 g tuber per larva).
6. Mortality is recorded as failure to pupate after an overall incubation period of 14 days.
7. Data can be analyzed using Probit analysis (Finney, 1971), and relative potencies obtained from each sampling day may be plotted against irradiation time or total energy. Negative exponential decay models can be fitted to the data to describe inactivation by differentiating between irradiation time and total energy from specific wavelengths (bands). Experiments reported by Sporleder (2003) showed that virus inactivation follows a bisegmented curve with an initial steep decline followed by a gradual decay (approximately 4 times slower). The two-component model proposed by Hiatt (1964) seems to adequately describe virus inactivation due to solar irradiation in potato fields.

When 99% of the virus has been inactivated, probit analysis becomes problematic because mortalities from the low concentrations may not exceed natural mortality (usually ca. 10%). However, sampling may continue in plots treated with the highest concentration until 99.9% of the virus is inactivated, as this still produces ca. 40% mortality. In general, very rapid rates of viral degradation should be anticipated in the field. Kroschel *et al.* (1996a) observed half-life times of 1.3 days for *Po*GV from field studies conducted in the Yemeni highlands (elevation > 2300 m). Sporleder (2003) observed rapid inactivation of virus (half-life < 0.3 days) during sunny days in Peru (elev. 1814 m; global radiation > 800 W/m^2 at noon), whereas minimal loss of activity was noted during cloudy days (half-life ca. 9 days).

9 *Supplementary data*

During field experiments temperature and precipitation should be recorded in the field. Rainfall is considered to be a significant abiotic factor affecting adult survival. PTM adult populations may collapse due to intense rainfall.

D *MCAs for post-harvest control*

One of the most important factors limiting efficacy of microbial control agents in the field is their high sensitivity to solar radiation. This limiting factor is not present in potato storage, and many researchers consider that the greatest potential of these agents lies in management of *P. operculella* following harvest. Both *Po*GV and *Bt* have been evaluated in many parts of the world for post harvest control of *P. operculella* larvae, particularly in traditional, non-refrigerated potato stores.

1 *Viral pathogens*

A substantial amount of successful testing of *Po*GV has been conducted on stored tubers in the Andean countries of Peru, Ecuador, Bolivia, and Colombia (Alcázar *et al.*, 1992; CIP, 1992a; Zeddam *et al.*, 2003) and in several countries in the Middle East, Northern Africa, and Asia (Amonkar *et al.*, 1979; Islam *et al.*, 1990; Ali, 1991; Das *et al.*, 1992; Setiawati *et al.*, 1999).

The International Potato Center (CIP) in Lima, Peru, initiated research on *Po*GV in the mid-1980s and introduced the virus as an alternative to pesticides in their global IPM program (Winters and Fano, 1997; Gelernter and Trumble, 1999). A simple mass-production and dust-formulation of the virus was developed for protecting stored potatoes, and cottage-type enterprises were launched in Peru, Bolivia, Colombia, Egypt, and Tunisia. Virus-infected larvae are ground and mixed with talc at a rate of 20 larvae per kg talc in one liter of water. The dried product has been applied at a rate of 5 kg per ton of stored potatoes. Such applications have provided high levels of control (usually > 95%) and significantly reduced tuber damage (Alcázar *et al.*, 1992; Alcázar and Raman, 1992; Das *et al.*, 1992; Winters and Fano, 1997). Raman and Alcázar (1990) reported the virus was as effective as the synthetic pyrethroid deltamethrin. Presently, *Po*GV is used in several countries; however, the current propagation and formulation techniques limit its potential. Also, in many potato-growing areas in the Andean region of South America, use of *Po*GV has declined because another gelechiid tuber moth, *Symmetrischema tangolias*, which is not susceptible to *Po*GV, has become a dominant pest. In that region, recent research has focused on *Bt* as a substitute for *Po*GV in the talc-based product.

2 Bacterial pathogens

Bt has been reported by several researchers to provide good protection of treated tubers. Tests in nearly all cases have involved *Btk*. In Yemen, Kroschel and Koch (1996) found that *Btk* (Dipel®; Valent BioSciences, Libertyville, IL) mixed with fine sand and dusted onto tubers was completely effective when applied prior to oviposition. This treatment also controlled 96% of larvae that had already entered the tubers. In Egypt, Farrag (1998) observed that treatment of tubers with a liquid formulation of Dipel eliminated a pest infestation within 60 days. In Tunisia, an integrated control approach comprising *Btk* applied at the onset of storage combined with cultural control (early harvest) effectively replaced parathion sprays when initial levels of tuber infestation were < 20% (von Arx *et al.*, 1987). In tests in Indonesia, tubers treated with *Btk* (Thuricide®; Certis USA, Columbia, MD) or *Po*GV caused 79% and 90% mortality, respectively, after 4 months of storage (Setiawati *et al.*, 1999). In Peru, Raman *et al.* (1987) found a dust formulation of Dipel was effective in reducing tuber damage; however, in other studies, Das *et al.* (1992) observed no protection of tubers treated with a wettable powder formulation of another *Btk* product (Bactospeine®; Abbott Laboratories, Abbott Park, IL).

E Protocols for evaluation of MCAs for post-harvest control of PTM

1 General research techniques

Many fundamental techniques for evaluating MCAs in stored products are presented in Chapter VII-19, and here it is possible to present only a brief overview of methods. A great diversity of research techniques and approaches for testing of both liquid and powder formulations of insecticides are described in the literature, and a number of publications regarding testing of *Po*GV and *Bt* in stored potatoes are cited above.

2 Application of liquid biopesticide formulations

Viral and bacterial biopesticides have been tested as both liquid and powder formulations. Liquids are generally considered easier to work with and present a lower risk of inhalation exposure. They are readily applied onto small stores of potatoes using portable sprayers. Liquids flow over tuber surfaces, and spray applications to near runoff will provide at least some coverage of unexposed surfaces. Alternatively, potatoes can be agitated during spray treatments or simply dipped in biopesticide suspensions. Kroschel and Koch (1996) indicated that 80–100 ml *Bt* suspension was required to spray-coat 10 kg potatoes. Tubers can also be layered into storage, with sprays applied to each layer. von Arx *et al.* (1987) applied *Bt* in a volume of 1 liter/100 kg tubers. In humid environments, it may be necessary to dry the treated potatoes (shaded) before storage. Ultra-low-volume (ULV) spray technology may be useful for large-scale storage applications; however, this technology has not yet been tested.

3 Application of dry biopesticide formulations

Achieving thorough and uniform coverage with dust formulations generally requires more handling than with liquids. A common treatment technique is to shake tubers and biopesticide preparations together in plastic bags or buckets. Tubers are usually treated in 1–10 kg batches at rates of 1–5 g dry-formulated biopesticide per kg (von Arx *et al.*, 1987; Hamilton and Macdonald, 1990; Salama *et al.*, 1995b; Setiawati *et al.*, 1999). In small-scale tests, the potatoes can be held in paper bags, wooden crates, or buckets during storage (each container serving as an experimental replicate). In large-scale tests, treated and untreated (control) potatoes can be stored in replicate piles. If PTM reinfestation (continuous high pest pressure) is desired during a test, treatments and controls can be maintained uncovered in the same storage facility. If reinfestation is not desired, potatoes can be covered with netting (1-mm mesh).

There is a long tradition of using dry, dusty materials such as ash, clay, talc, or fine sand to suppress pest infestations in rustic potato stores. These materials may act as repellents or physical barriers against adult moths or show lethal abrasive/desiccant activity against young larvae. Materials with the latter mode of action may enhance the efficacy of biopesticide formulations (Kroschel and Koch, 1996). However, alkaline materials (*e.g.*, kaolin) should not be uses as inert carriers for PoGV because high pH may adversely affect the viral occlusion body (baculoviruses are tolerant of pH between 4 and 9).

4 Assessment of efficacy

One of the best ways to measure efficacy of tuber treatments is to count numbers of emerging adult moths, but this may be feasible only in small-scale tests. Wooden crates are common storage units in research trials, and these are easily covered with netting to trap emerging adult moths. Crates also serve as convenient test "plots" that can be arranged in randomized complete block or other experimental designs. Crated tubers that have been spray-treated are easily placed in ventilated, shaded areas for drying prior to storage and are readily manipulated for repeated or destructive sampling during storage. Destructive sampling is recommended, but more crates will be necessary depending on the number of evaluations over time. Following treatments, damage is evaluated over the long term by simply sampling the tubers and assessing external and internal damage. Infestation frequency (percentage infestation) and intensity (number of larvae, mines, or holes per tuber) may be determined. Tuber weights may also be recorded. Potatoes in storage generally need to be protected for many months, and damage assessments are usually made at 1–2 month intervals (von Arx *et al.*, 1987; Lal, 1987). To obtain comprehensive data, duration of experiments should allow for monitoring of at least two or three pest generations. Existing phenology models for *P. operculella* (Sporleder, 2004; Roux and Baumgärtner, 1995) may help to define test hypotheses. Roux and Baumgärtner (1998) present a mathematical model for evaluating mortality factors and risk analysis for *P. operculella* in potato stores that can be adopted for further research.

4 References

Abbott, W. S. 1925. A method of computing the effectiveness of an insecticide. *J. Econ. Entomol.* **18**, 265–267.

Adang, M. J., Brody, M. S., Cardineau, G., Eagan, N., Roush, R. T., Shewmaker, C. K., Jones, A., Oakes, J. V., and McBride, K. E. 1993. The reconstruction and expression of a *Bacillus thuringiensis* cryIIIA gene in protoplasts and potato plants. *Plant Mol. Biol.* **21**, 1131–1145.

Alcázar, J., Cervantes, M., and Raman, K. V. 1992. Efectividad de un virus granulosis formulado en polvo para controlar *Phthorimaea* en papa almacenada. *Rev. Peru. Entomol.* **35**, 113–116.

Alcázar, J. and Raman, K. V. 1992. Control de *Phthorimaea operculella* en almacenes rusticos, empleando virus granulosis en polvo. *Rev. Peru. Entomol.* **35**, 117–120.

Alcázar, J., Raman, K. V., and Salas, R. 1991. Un virus como agente de control de la polilla de la papa *Phthorimaea operculella*. *Rev. Peru. Entomol.* **34**, 101–104.

Ali, M. I. 1991. Efficacy of a granulosis virus on the control of potato tuber moth, *Phthorimaea operculella* (Zeller) (Gelechiidae: Lepidoptera) infesting potatoes in Bangladesh. *Bangladesh J. Zool.* **19**, 141–143.

Amonkar, S. V., Pal, A. K., Vijayalakshmi, L., and Rao, A. S. 1979. Microbial control of potato tuber moth (*Phthorimaea operculella* Zell.). *Indian J. Exp. Biol.* **17**, 1127–1133.

Anderson, T. E., Roberts, D. W., and Soper, R. S. 1988. Use of *Beauveria bassiana* for suppression of Colorado potato beetle populations in New York State (Coleoptera: Chrysomelidae). *Environ. Entomol.* **17**, 140–145.

Arthurs, S. P., Lacey, L. A., and Fritts, R. Jr. 2005. Optimizing use of codling moth granulovirus: effects of application rate and spraying frequency on control of codling moth larvae in Pacific Northwest apple orchards. *J. Econ. Entomol.* **98**, 1459–1468.

Awate, B. G. and Naik, L. M. 1979. Efficacies of insecticidal dusts applied to soil surface for controlling potato tuberworm (*Phthorimaea operculella* Zeller) in the field. *J. Maharashtra Agric. Univ.* **4**, 100.

Bajan, C. and Kmitowa, K. 1977. Contribution of entomopathogenic fungi to the natural winter reduction of Colorado potato beetle adults. Pol. Ecol. Stud. 3, 107–114.

Bateman, R. 1999. Delivery systems and protocols for biopesticides. In "Methods in Biotechnology, Vol. 5: Biopesticides: Use and Delivery", (F. R. Hall and J. Menn, Eds.), pp. 509–528. Humana Press Inc., Totowa, NJ.

Ben Salah, H. and Aalbu, R. 1992. Field use of granulosis virus to reduce initial storage infestation of the potato tuber moth, *Phthorimaea operculella* (Zeller), in North Africa. *Agric. Ecosys. Environ.* **38**, 119–126.

Ben Salah, H., Fuglie, K., Temime, A. B., Rahmouni, A., and Cheikh, M. 1994. Utilisation du virus de la granulose de la teigne de la pomme de terra et du *Bacillus thuringiensis* dans la lutte integree contre *Phthorimaea operculella* Zell. (Lepid., Gelechiidae) en Tunisie. *Ann. Inst. Natl. Rech. Agron. Tunis*. **67**, 1–20.

Blonska, A. 1957. Pathogenic fungi of the genus *Beauveria* from the Colorado beetle, *Leptinotarsa decemlineata* Say. *Roczn. Nauk Roln. Ser. A*. **74**, 359–372.

Briese, D. T. 1982. Genetic basis for resistance to a granulosis virus in the potato tuber moth, *Phthorimaea operculella*. *J. Invertebr. Pathol*. **39**, 215–218.

Briese, D. T. and Mende, H. A. 1981 Differences in susceptibility to a granulosis virus between field populations of the potato moth, *Phthorimaea operculella* (Zeller) (Lepidoptera: Gelechiidae). *Bull. Entomol. Res*. **71**, 11–18.

Briese, D. T. and H. A. Mende. 1983. Selection for increased resistance to a granulosis virus in the potato moth, *Phthorimaea operculella* (Zeller) (Lepidoptera: Gelechiidae). *Bull. Entomol. Res*. **73**, 1–10.

Broodryk, S. W. and Pretorius, L. M. 1974. Occurrence in South Africa of a granulosis virus attacking tuber moth, *Phthorimaea operculella* (Zeller) (Lepidoptera: Gelechiidae). *J. Entomol. Soc. South. Afr*. **37**, 125–128.

Brown, J. J., Jermy, T., and Butt, B. A. 1980. The influence of an alternate host plant on the fecundity of the Colorado potato beetle, *Leptinotarsa decemlineata* (Coleoptera: Chrysomelidae). *Ann. Entomol. Soc. Am*. **73**, 197–199.

Broza, M. and Sneh, B. 1994. *Bacillus thuringiensis* spp. *kurstaki* as an effective control agent of lepidopteran pests in tomato fields in Israel. *J. Econ. Entomol*. **87**, 923–928.

Byrne, A. M., Grafius, E. J., Bishop, B. A. and Pett, W. 2004. Susceptibility of Colorado potato beetle populations to imidacloprid and thiamethoxam, 2003. *Arthropod Management Tests* **29**, L12.

Cameron, P. J., Walker, G. P., Penny, G. M., and Wigley, P. J. 2002. Movement of potato tuberworm (Lepidoptera: Gelechiidae) within and between crops, and some comparisons with diamondback moth (Lepidoptera: Plutellidae). *Environ. Entomol*. **31**, 65–75.

Cameron, P. J., Wallace, A. R., Madhusudhan, V. V., Wigley, P. J., Qureshi, M. S., and Walker, G. P. 2005 Mating frequency in dispersing potato tuber moth, *Phthorimaea operculella*, and its influence on the design of refugia to manage resistance in *Bt* transgenic crops. *Entomol. Exp. Appl*. **115**, 323–332.

Campbell, R. K., Anderson, T. E., Semel, M., and Roberts, D. W. 1985. Management of the Colorado potato beetle using the entomogenous fungus, *Beauveria bassiana*. *Am. Potato J*. **62**, 29–37.

Cantwell, G. E., Cantelo, W. W., and Schroder, F. W. 1986. Effect of *Beauveria bassiana* on underground stages of the Colorado potato beetle, *Leptinotarsa decemlineata* (Coleoptera: Chrysomelidae). *Great Lakes Entomol*. **19**, 81–84.

Capinera, J. L. 2001. "Handbook of Vegetable Pests." Academic Press, New York. 729 pp.

Chlodny, J. 1975. Bioenergetics of the larval development of the Colorado potato beetle, *Leptinotarsa decemlineata* (Say), in relation to temperature conditions. *Ann. Zool. (Warsaw)* **33**, 149–187.

CIP. 1992a. "Annual Report 1992." International Potato Center (CIP), Lima, Peru.

CIP. 1992b. Biological control of potato tuber moth using *Phthorimaea* baculovirus. *CIP Training Bulletin* **2**, International Potato Center, Lima, Peru. 27 pp.

Cisneros, F. M. 1984. The need for integrated pest management in developing countries. In "Report of Planning Conference on Integrated Pest Management", pp. 19–30. International Potato Center (CIP), Lima, Peru.

Cloutier, C. and Jean, C. 1998. Synergism between natural enemies and biopesticides: a test case using the stinkbug, *Perillus bioculatus* (Hemiptera: Pentatomidae) and *Bacillus thuringiensis tenebrionis* against Colorado potato beetle (Coleoptera; Chrysomelidae). *J. Econ. Entomol*. **91**, 1096–1108.

Cloutier, C., Jean, C. and Bauduin, F. 1995. More biological control for a sustainable potato pest management strategy. In "Potato Insect Pest Control: Development of a Sustainable Approach", (R.-M. Duchesne and G. Boiteau, Eds.), pp. 15–52. Ministère de l'Agriculture, des Pêcheries et de L'alimentation du Québec, Sainte-Foy.

Cochran, W. G. and Cox, G. M. 1957. "Experimental Designs", 2nd edition. John Wiley & Sons, New York. 611 pp.

Cornell Cooperative Extension. 2001. "2002 Integrated Crop and Pest Management Guidelines for Commercial Vegetable Production." Cornell Cooperative Extension, Cornell University, Ithaca, NY.

Das, G. P., Magallona, E. D., Raman, K. V., and Adalla, C. B. 1992. Effect of different components of IPM in the management of the potato tuber moth in storage. *Agric. Ecosys. Environ*. **41**, 321–325.

Das, G. P. and Raman, K. V. 1994. Alternate hosts of the potato tuber moth, *Phthorimaea operculella* (Zeller). *Crop Prot*. **13**, 83–86.

Drummond, F. A. 1996. Sampling the Colorado potato beetle. In "The Ecology, Economics, and Management of Potato Cropping Systems: A Report of the First Four Years of the Maine Potato Ecosystem Project" (M. C. Marra and B. A. Harrity, Eds.), pp. 175–204. University of Maine, Maine Agricultural and Forest Experiment Station Bulletin 843.

Drummond, F. A. and Groden, E. 1996. Insect pests and natural enemies. In "The Ecology, Economics, and Management of Potato Cropping Systems: A Report of the First Four Years of the Maine Potato Ecosystem Project", (M. C. Marra and B. A. Harrity, Eds.), pp.

80–118. University of Maine, Maine Agricultural and Forest Experiment Station Bulletin 843.

Duchesne, R.-M. and Boiteau, G. 1987. Microbial control of insect pests of potatoes. In "Potato Pest Management in Canada", (G. Boiteau, R. P. Singh and R. H. Parry, Eds.), pp. 112–132. Fredericton, New Brunswick, NJ.

Fargues, J. 1972. Étude des conditions d'infection des larves de doryphore, *Leptinotarsa decemlineata* Say, par *Beauveria bassiana* (Bals.) Vuill. [Fungi Imperfecti]. *Entomophaga* **17**, 319–337.

Fargues, J., Cugier, J. P., and Van de Weghe, P. 1980. Expérimentation en parcelles du champignon *Beauveria bassiana* (Hyphomycete) contre *Leptinotarsa decemlineata* (Col., Chrysomelidae). *Acta Oecol. Appl.* **1**, 49–61.

Fargues, J., Delmas, J. C., and LeBrun, R. A. 1994. Leaf consumption by larvae of the Colorado potato beetle (Coleoptera: Chrysomelidae) infected with the entomopathogen *Beauveria bassiana*. *J. Econ. Entomol.* **87**, 67–71.

Farrag, R. M. 1998. Control of the potato tuber moth, *Phthorimaea operculella* Zeller (Lepidoptera: Gelechiidae) at storage. *Egypt. J. Agric. Res.* **76**, 947–952.

Ferro, D. N., Logan, J. A., Voss, R. H., and Elkinton, J. S. 1985. Colorado potato beetle (Coleoptera: Chrysomelidae) temperature-dependent growth and feeding rates. *Environ. Entomol.* **14**, 343–348.

Ferro, D. N., Tuttle, A. F., and Weber, D. C. 1991. Ovipositional and flight behavior of overwintered Colorado potato beetle (Coleoptera: Chrysomelidae). *Environ. Entomol.* **20**, 1309–1314.

Ferro, D. N. Yuan, Q. C., Slocombe, A., and Tuttle, A. F. 1993. Residual activity of insecticides under field conditions for controlling the Colorado potato beetle (Coleoptera: Chrysomelidae). *J. Econ Entomol.* **86**, 511–516.

Finney, J. R. 1971. "Probit Analysis", 3rd edition. Cambridge University Press. 333 pp.

Foot, M. A. 1976. Cultural practices in relation to infestation of potato crops by the potato tuber moth. 2. Effect of seed depth, re-moulding, pre-harvest defoliation, and delayed harvest. *N. Z. J. Exp. Agric.* **4**, 121–124.

Galaini, S. 1984. The efficacy of foliar applications of *Beauveria bassiana* against *Leptinotarsa decemlineata*. M. S. Thesis, Cornell University, Ithaca, NY. 94 pp.

Gallandt, E. R., Mallory, E. B., Alford, A. R., Drummond, F. A., Groden, E., Liebman, M., Marra, M. C., McBurnie, J. C., and Porter, G. A. 1998. Comparison of alternative pest and soil management strategies for Maine potato production systems. *Am. J. Altern. Agric.* **13**, 146–161.

Gaugler, R., Costa, S. D., and Lashomb, J. 1989. Stability and efficacy of *Beauveria bassiana* soil inoculations. *Environ. Entomol.* **18**, 412–417.

Gelernter, W. D. 2002. The discovery, development and death of *Bacillus thuringiensis* as a microbial control product for the Colorado potato beetle. In "VIII International Colloquium on Invertebrate Pathology and Microbial Control Proceedings", pp. 262–264. Society for Invertebrate Pathology.

Gelernter, W. D. and Trumble, J. T. 1999. Factors in the success and failure of microbial insecticides in vegetable crops. *Integr. Pest Manage. Rev.* **4**, 301–306.

Goettel, M. S. and Inglis, G. D. 1997. Fungi: Hyphomycetes. In "Manual of Techniques in Insect Pathology", (L. A. Lacey, Ed.), pp. 213–249. Academic Press, San Diego.

Graf, J. E. 1917. The potato tuber moth. *USDA Bull.* **427**. 56 pp.

Groden, E., Wraight, S. P., and Drummond, F. A. 2002. Microbial control of Colorado potato beetle in rain-fed potato agroecosystems in the northeastern U. S. In "VIII International Colloquium on Invertebrate Pathology and Microbial Control Proceedings", pp. 265–269. Society for Invertebrate Pathology.

Hafez, M., Zaki, F. N., Moursy, A., and Sabbour, M. 1997. Biological effects of the entomopathogenic fungus, *Beauveria bassiana* on the potato tuber moth *Phthorimaea operculella* (Zeller). *Anz. Schädlingskd. Pflanzenschutz Umweltschutz* **70**, 158–159.

Hajek, A. E., Soper, R. S., Roberts, D. W., Anderson, T. E., Biever, K. D., Ferro, D. N., LeBrun, R. A., and Storch, R. H. 1987. Foliar applications of *Beauveria bassiana* (Balsamo) Vuillemin for control of the Colorado potato beetle, *Leptinotarsa decemlineata* (Say) (Coleoptera: Chrysomelidae): an overview of pilot test results from the northern United States. *Can. Entomol.* **119**, 959–974.

Hamilton, J. T. and Macdonald, J. A. 1990. Control of potato moth, *Phthorimaea operculella* (Zeller) in stored seed potatoes. *Gen. Appl. Entomol.* **22**, 3–6.

Harcourt, D. G. 1963. Population dynamics of *Leptinotarsa decemlineata* (Say) in eastern Ontario. I. Spatial pattern and transformation of field counts. *Can. Entomol.* **95**, 813–820.

Hare, J. D. 1990. Ecology and management of the Colorado potato beetle. *Annu. Rev. Entomol.* **35**, 81–100.

Henderson, C. F. and Tilton, E. 1955. Tests with acaricides against the brown wheat mite. *J. Econ. Entomol.* **48**, 157–161.

Hernández, C. S., Andrew, R., Bel, Y., and Ferré, J. 2005. Isolation and toxicity of *Bacillus thuringiensis* from potato-growing areas in Bolivia. *J. Invertebr. Pathol.* **88**, 8–16.

Hiatt, C. W. 1964. Kinetics of the inactivation of viruses. *Bacteriol. Rev.* **28**, 150–163.

Holliday, N. J. 1995. Economic thresholds and monitoring. In "Potato Insect Pest Control: Development of a Sustainable Approach." (R.-M. Duchesne and G. Boiteau, Eds.), pp. 113–142. Ministère de l'Agriculture, des Pêcheries et de L'alimentation du Québec, Sainte-Foy.

Horne, P. A. 1990. The influence of introduced parasitoids on the potato moth, *Phthorimaea operculella* (Lepidoptera: Gelechiidae) in Victoria, Australia. *Bull. Entomol. Res.* **80**, 159–163.

Humber, R. A. 1996. Fungal pathogens of the Chrysomelidae and prospects for their use in biological control. In "Chrysomelidae Biology, vol. 2: Ecological Studies." (P. H. A. Jolivet and M. L. Cox, Eds.), pp. 93–115. SPB Academic Publishing, Amsterdam.

Hunter, D. K., Hoffmann, D. F., and Collier, S. J. 1975. Observations on a granulosis virus of the potato tuberworm, *Phthorimaea operculella*. *J. Invertebr. Pathol.* **26**, 397–400.

Islam, M. N., Karim, M. A., and Nessa, Z. 1990. Control of the potato tuber moth, *Phthorimaea operculella* (Zeller) (Lepidoptera: Gelechiidae) in the storehouses for seed and ware potatoes in Bangladesh. *Bangladesh J. Zool.* **18**, 41–52.

Jackson, T. A., Pearson, J. F., O'Callaghan, M., Mahanty, H. K., and Willocks, M. J. 1992. Pathogen to product - development of *Serratia entomophila* (Enterobacteriaceae) as a commercial biological control agent for the New Zealand grass grub (*Costelytra zealandica*). In "Use of Pathogens in Scarab Pest Management", (T. R. Glare and T. A. Jackson, Eds.), pp. 191–198. Intercept Ltd., Andover, Hampshire. England.

Jaques, R. P. 1990. Effectiveness of the granulovirus of the codling moth in orchard trials in Canada. In "Proceedings of the Vth International Colloquium on Invertebrate Pathology and Microbial Control", pp. 428–430. Society for Invertebrate Pathology.

Jaques, R. P. and Laing, D. R. 1989. Effectiveness of microbial and chemical insecticides in control of Colorado potato beetle (Coleoptera: Chrysomelidae) on potatoes and tomatoes. *Can. Entomol.* **120**, 1123–1131.

Jaros-Su, J., Groden, E., and Zhang, J. 1999. Effects of selected fungicides and the timing of fungicide application of *Beauveria bassiana*-induced mortality of the Colorado potato beetle (Coleoptera: Chrysomelidae). *Biol. Control* **15**, 259–269.

Keller, S. 2003. "Integrated pest management of the potato tuber moth in cropping systems of different agro-ecological zones." Advances in Crop Research |1|, Margraf Verlag, Weikersheim, Germany. 153 pp.

Krieg, A., Huger, A. M., Langenbruch, G. A., and Schnetter, W. 1983. *Bacillus thuringiensis* var. *tenebrionis*: Ein neuer, gegenüber larven von coleopteren wirksamer pathotyp. *Z. Angew. Entomol.* **96**, 500–508.

Kroschel, J. 1995. Integrated pest management in potato production in the Republic of Yemen with special reference to the integrated biological control of the potato tuber moth (*Phthorimaea operculella* Zeller). Tropical Agriculture |8|, Margraf Verlag, Weikersheim, Germany.

Kroschel, J., Fritsch, E., and Huber, J. 1996a. Biological control of the potato tuber moth (*Phthorimaea operculella* Zeller) in the Republic of Yemen using granulosis virus: biochemical characterization, pathogenicity and stability of the virus. *Biocontrol Sci. Technol.* **6**, 207–216.

Kroschel, J., Kaack, H. J., Fritsch, E., and Huber, J. 1996b. Biological control of the potato tuber moth (*Phthorimaea operculella* Zeller) in the Republic of Yemen using granulosis virus: propagation and effectiveness of the virus in field trials. *Biocontrol Sci. Technol.* **6**, 217–226.

Kroschel, J. and Koch, W. 1994. Studies on the population dynamics of the potato tuber moth (*Phthorimaea operculella* Zell. (Lep., Gelechiidae)) in the Republic of Yemen. *J. Appl. Entomol.* **118**, 327–341.

Kroschel, J. and Koch, W. 1996. Studies on the use of chemicals, botanicals and *Bacillus thuringiensis* in the management of the potato tuber moth in potato stores. *Crop Prot.* **15**, 197–203.

Laarif, A., Fattouch, S., Essid, W., Marzouki, N., Salah, H. B., and Hammouda, M. H. B. 2003. Epidemiological survey of *Phthorimaea operculella* granulosis virus in Tunisia. *Bull. OEPP/EPPO Bull.* **33**, 335–338.

Lacey, L. A., Horton, D. R., Chauvin, R. L., and Stocker, J. M. 1999. Comparative efficacy of *Beauveria bassiana, Bacillus thuringiensis*, and aldicarb for control of Colorado potato beetle in an irrigated desert agroecosystem and their effects on biodiversity. *Entomol. Exp. Appl.* **93**, 189–200.

Lacey, L. A. and Neven, L. G. 2006. The potential of the fungus, *Muscodor albus* as a microbial control agent of potato tuber moth (Lepidoptera: Gelechiidae) in stored potatoes. *J. Invertebr. Pathol.* **91**, 195–198.

Lagnaoui, A., Salah, H. B., and El-Bedewy, R. 1996. Integrated management to control potato tuber moth in North Africa and the Middle East. *CIP Circular* **22**, 10–15.

Lal, L. 1987. Studies on natural repellents against potato tuber moth (*Phthorimaea operculella* Zeller) in country stores. *Potato Res.* **30**, 329–334.

Lashomb, J. H., Ng, Y. S., Ghidiu, G., and Green, E. 1984. Description of spring emergence by the Colorado potato beetle, *Leptinotarsa decemlineata* (Say) (Coleoptera: Chrysomelidae), in New Jersey. *Environ. Entomol.* **13**, 907–910.

Lipa, J. J. 1985. Progress in biological control of the Colorado beetle (*Leptinotarsa decemlineata*) in eastern Europe. *Bull. OEPP/EPPO Bull.* **15**, 207–211.

Little, T. M. and Hills, F. J. 1978. "Agricultural Experimentation: Design and Analysis." John Wiley and Sons, Inc., New York.

Lopez, R. and Ferro, D. N. 1995. Larviposition response of *Myiopharus doryphorae* (Diptera: Tachinidae) to Colorado potato beetle (Coleoptera: Chrysomelidae) larvae treated with lethal and sublethal doses of *Bacillus thuringiensis* Berliner subsp. *tenebrionis*. *J. Econ. Entomol.* **88**, 870–874.

Martin, P. A. W., Schroder, R. F. W., Poprawski, T. J., Lipa, J. Z., Hausvater, E., and Rasocha, V. 2000. Temperature effects on the susceptibility of the Colorado potato beetle to *Beauveria bassiana* (Balsamo) Vuillemin in Poland, the Czech Republic and the United States. *J. Entomol. Sci.* **35**, 251–258.

Matthews, G. W. 2000. "Pesticide Application Methods", 3rd edition. Blackwell Science, Oxford, UK. 432 pp.

Mitchell, B. L. 1978. The biological control of potato tuber moth *Phthorimaea operculella* (Zeller) in Rhodesia. *Rhodesia Agric. J.* **75**, 55–58.

Morris, O. N. 1982. Bacteria as pesticides: forest applications. In "Microbial and Viral Pesticides" (E. Kurstak, Ed.), pp. 239–287. Marcel Dekker, Inc., New York.

Nault, B. A. and Kennedy, G. G. 1998. Limitations of using regression and mean separation analyses for describing the response of crop yield to defoliation: a case of the Colorado potato beetle. *J. Econ. Entomol.* **91**, 7–20.

Palacios, M. and Cisneros, F. 1997. Integrated pest management for the potato tuber moth in pilot units in the Andean Region and the Dominican Republic. In "Program Report: 1995–1996." International Potato Center (CIP), Lima, Peru.

Perlak, F. J., Stone, T. B., Muskopf, Y. M., Petersen, L. J., Parker, G. B., McPherson, S. A., Wyman, J., Love, S., Reed, G., Biever, D., and Fischhoff, D. A. 1993. Genetically improved potatoes: protection from damage by Colorado potato beetles. *Plant Mol. Biol.* **22**, 313–321.

Poprawski, T. J., Carruthers, R. I., Speese, J. III, Vacck, D. C., and Wendel, L. E. 1997. Early-season applications of the fungus *Beauveria bassiana* and introduction of the hemipteran predator *Perillus bioculatus* for control of Colorado potato beetle. *Biol. Control* **10**, 48–57.

Raman, K. V. 1994. Pest management in developing countries. In: "Advances in Potato Pest Biology and Management" (G. W. Zehnder, M. L. Powelson, R. K. Jansson, and K. V. Raman, Eds.), pp 583–596. The American Phytopathological Society Press, St. Paul, MN, USA.

Raman, K. V. and Alcázar, J. 1990. Biological control of potato tuber moth (*Phthorimaea operculella*) using granulosis virus. *Am. Potato J.* **67**, 574 (abstract).

Raman, K. V., Booth, R. H., and Palacios, M. 1987. Control of potato tuber moth *Phthorimaea operculella* (Zeller) in rustic potato stores. *Trop. Sci.* **27**, 175–194.

Rath, A. C. 1992. *Metarhizium anisopliae* for control of the Tasmanian pasture scarab *Adoryphorus couloni*. In "Use of Pathogens in Scarab Pest Management", (T. R. Glare and T. A. Jackson, Eds.), pp. 217–227. Intercept Ltd., Andover, Hampshire, England.

Reed, E. M. 1969. A granulosis virus of potato moth. *Aust. J. Sci.* **31**, 300–301.

Reed, E. M. 1971. Factors affecting the status of a virus as a control agent of the potato moth (*Phthorimaea operculella* (Zell.) (Lep., Gelechiidae)). *Bull. Entomol. Res.* **61**, 207–222.

Reed, E. M. and Springett, B. P. 1971. Large-scale field testing of a granulosis virus for the control of the potato moth (*Phthorimaea operculella* (Zell.) (Lep., Gelechiidae)). *Bull. Entomol. Res.* **61**, 223–233.

Richardson, M. E. and Rose, D. J. W. 1967. Chemical control of potato tuber moth, *Phthorimaea operculella* (Zell.) in Rhodesia. *Bull. Entomol. Res.* **57**, 271–278.

Roberts, D. W., LeBrun, R. A., and Semel, M. 1980. Control of the Colorado potato beetle with fungi. In "Advances in Potato Pest Management" (J. H. Lashomb and R. Casagrande, Eds.), pp. 119–137. Hutchinson Ross, Stroudsburg.

Roux, O. and Baumgärtner, J. 1995. Life table statistics and cohort parameters for potato tuber moth *Phthorimaea operculella* (Zeller). *Boll. Zool. Agrar. Bachic.* **27**, 123–142.

Roux, O. and Baumgärtner, J. 1998. Evaluation of mortality factors and risk analysis for the design of an integrated pest management system. *Ecol. Model.* **109**, 61–75.

Sabbour, M. and Ismail, I. A. 2002. The combined effect of microbial control agents and plant extracts against potato tuber moth *Phthorimaea operculella* Zeller. *Bull. Natl. Res. Cent. (Cairo)* **27**, 459–467.

Salama, H. S., Ragaei, M., and Sabbour, M. 1995a. Larvae of *Phthorimaea operculella* (Zell.) as affected by various strains of *Bacillus thuringiensis*. *J. Appl. Entomol.* **119**, 241–243.

Salama, H. S. and Salem, S. A. 2000. *Bacillus thuringiensis* and neem seed oil (*Azadirachta indica*) effects on the potato tuber moth *Phthorimaea operculella* Zeller in the field and stores. *Arch. Phytopathol. Plant Prot.* **33**, 73–80.

Salama, H. S., Zaki, F. N., Ragaei, M., and Sabbour, M. 1995b. Persistence and potency of *Bacillus thuringiensis* against *Phthorimaea operculella* (Zell.) (Lep., Gelechiidae) in potato stores. *J. Appl. Entomol.* **119**, 493–494.

Setiawati, W., Soeriaatmadja, R. E., Rubiati, T., and Chujoy, E. 1999. Control of potato tubermoth (*Phthorimaea operculella*) using an indigenous granulosis virus in Indonesia. *Indones. J. Crop Sci.* **14**, 10–16.

Sewify, G. H., Abol-Ela, S., and Eldin, M. S. 2000. Effects of the entomopathogenic fungus *Metarhizium anisopliae* (Metsch.) and granulosis virus (GV) combinations on the potato tuber moth *Phthorimaea operculella* (Zeller) (Lepidoptera: Gelechiidae). *Bull. Fac. Agric. Cairo Univ.* **51**, 95–106.

Shelton, A. M. and Wyman, J. A. 1979. Potato tuberworm damage to potatoes under different irrigation and cultural practices. *J. Econ. Entomol.* **72**, 261–264.

Sikura, A. I. and Sikura, L. V. 1983. Use of biopreparations. *Zasch. Rast.* **5**, 38–39.

Sokal, R. R. and Rohlf, F. J. 1995. "Biometry", 3rd", edition. W. H. Freeman and Company, New York. 887 pp.

Soper, R. S. 1980. Role of entomophthoran fungi in aphid control for potato integrated pest management. In "Advances in Potato Pest Management" (J. H. Lashomb and R. Casagrande, Eds.), pp. 153–177. Hutchinson Ross, Stroudsburg.

Sporleder, M. 2003. "The Granulosis of the Potato Tuber Moth *Phthorimaea operculella* (Zeller): Characterisation and Prospects for Effective Mass Production and Pest Control." Advances in Crop Research |3|, Margraf Verlag, Weikersheim, Germany.

Sporleder, M., Kroschel, J., Gutierrez Quispe, M. R., and Lagnaoui, A. 2004. A temperature-based simulation model for the potato tuberworm, *Phthorimaea operculella* Zeller (Lepidoptera; Gelechiidae). *Environ. Entomol.* **33**, 477–486.

Sporleder, M., Kroschel, J., Huber, J., and Lagnaoui, A. 2005a. An improved method to determine the biological activity (LC_{50}) of the granulovirus *Po*GV in its host *Phthorimaea operculella. Entomol. Exp. Appl.* **116**, 191–197.

Sporleder, M., Mamani, D., Huber, J. and Kroschel, J. 2005b. Interaction between the granulovirus *Po*GV and *Bacillus thuringiensis* (Berliner) against the potato tuber moth, *Phthorimaea operculella* (Zeller). In "Program and Abstracts for the 38th Annual Meeting of the Society for Invertebrate Pathology" p. 84 (abstract). Society for Invertebrate Pathology.

Steel, R. G. D. and Torrie, J. H. 1980. "Principles and Procedures of Statistics - A Biometrical Approach", 2nd edition. McGraw-Hill, New York. 633 pp.

Stevens, P. J. G. 1993. Organosilicone surfactants as adjuvants for agrochemicals. *Pestic. Sci.* **38**, 103–122.

Sun, Y. X., Li, Z. Y., Gui, F. R., Yan, N. S., Chen, B., Xu, R. Q., and He, S. Q. 2004. Susceptibility of *Phthorimaea operculella* (Zeller) larvae to eight isolates of *Beauveria. Southwest. Chin. J. Agric. Sci.* **17**, 627–629.

Thurston, G. S., Yule, W. N., and Dunphy, G. B. 1994. Explanations for the low susceptibility of *Leptinotarsa decemlineata* to *Steinernema carpocapsae. Biol. Control* **4**, 53–58.

Vey, A., Fargues, J., 1977. Histological and ultrastructural studies of *Beauveria bassiana* infection in *Leptinotarsa decemlineata* larvae during ecdysis. *J. Invertebr. Pathol.* **30**, 207–215.

von Arx, R. and Gebhardt, F. 1990. Effects of a granulosis virus, and *Bacillus thuringiensis* on life-table parameters of the potato tubermoth, *Phthorimaea operculella. Entomophaga* **35**, 151–159.

von Arx, R., Goueder, J., Cheikh, M., and Temime, A. B. 1987. Integrated control of potato tubermoth *Phthorimaea operculella* (Zeller) in Tunisia. *Insect Sci. Appl.* **8**, 989–994.

von Arx, R., Roux, O., and Baumgärtner, J. 1990. Tuber infestation by potato tubermoth, *Phthorimaea operculella* (Zeller), at potato harvest in relation to farmers' practices. *Agric. Ecosys. Environ.* **31**, 277–292.

Voss, R. H. and Ferro, D. N. 1990. Phenology of flight and walking by Colorado potato beetle (Coleoptera: Chrysomelidae) adults in western Massachusetts. *Environ. Entomol.* **19**, 117–122.

Watt, B. A. and LeBrun, R. A. 1984. Soil effects of *Beauveria bassiana* on pupal populations of the Colorado potato beetle. *Environ. Entomol.* **13**, 15–18.

Whalon, M. E., Miller, D. L., Hollingworth, R. M., Grafius, E. J., and Miller, J. R. 1993. Selection of a Colorado potato beetle (Coleoptera: Chrysomelidae) strain resistant to *Bacillus thuringiensis. J. Econ. Entomol.* **86**, 226–233.

Winters, P. and Fano, H. 1997. The economics of biological control in Peruvian potato production. Working Paper No. 1997–7, International Potato Center (CIP), Lima, Peru.

Wojciechowska, M., Kmitowa, K., Fedorko, A, Bajan, C., 1977. Duration of activity of entomopathogenic organisms introduced into the soil. *Pol. Ecol. Stud.* **3**, 141–148.

Wraight, S. P. and Carruthers, R. I. 1999. Production, delivery, and use of mycoinsecticides for control of insect pests of field crops. In "Methods in Biotechnology, Vol. 5: Biopesticides: Use and Delivery" (F. R. Hall and J. Menn, Eds.), pp. 233–269. Humana Press Inc., Totowa, NJ.

Wraight, S. P., Carruthers, R. I., Bradley, C. A., Jaronski, S. T., Lacey, L. A., Wood, P. and Galaini-Wraight, S. 1998. Pathogenicity of the entomopathogenic fungi *Paecilomyces* spp. and *Beauveria bassiana* against the silverleaf whitefly, *Bemisia argentifolii. J. Invertebr. Pathol.* **71**, 217–226.

Wraight, S. P. and Ramos, M. E. 2002. Application parameters affecting field efficacy of *Beauveria bassiana* foliar treatments against Colorado potato beetle *Leptinotarsa decemlineata. Biol. Control* **23**, 164–178.

Wraight, S. P. and Ramos, M. E. 2005. Synergistic interaction between *Beauveria bassiana*- and *Bacillus thuringiensis tenebrionis*-based biopesticides applied against field populations of Colorado potato beetle larvae. *J. Invertebr. Pathol.* **90**, 139–150.

Zeddam, J. L., Pollet, A., Mangoendiharjo, S., Ramadhan, T. H., and López-Ferber, M. 1999. Occurrence and virulence of a granulosis virus in *Phthorimaea operculella* (Lep., Gelechiidae) populations in Indonesia. *J. Invertebr. Pathol.* **74**, 48–54.

Zeddam, J.-L., Vasquez Soberon, R. M., Vargas Ramos, Z., and Lagnaoui, A. 2003. Producción viral y tasas de aplicación del granulovirus usado para el control biológico de las polillas de la papa *Phthorimaea operculella* y *Tecia solanivora*

(Lepidoptera: Gelechiidae). *Bol. Sanidad Vegetal, Plagas* **29**, 659–667.

Zehnder, G. W., Gelernter, W. D. 1989. Activity of the M-ONE formulation of a new strain of *Bacillus thuringiensis* against the Colorado potato beetle (Coleoptera: Chrysomelidae): Relationship between susceptibility and insect life stage. *J. Econ. Entomol.* **82**, 756–761.

Zehnder, G. W., Ghidiu, G. M., and Speese, J. III. 1992. Use of occurrence of peak Colorado potato beetle (Coleoptera: Chysomelidae) egg hatch for timing of *Bacillus thuringiensis* spray application in potatoes. *J. Econ. Entomol.* **85**, 281–288.

Zehnder, G., Vencill, A. M., and Speese, J. III. 1995. Action thresholds based on plant defoliation for management of Colorado potato beetle (Coleoptera: Chrysomelidae) in potato. *J. Econ. Entomol.* **88**, 155–161.

Zhao, J.-Z., Bishop, B. A., and Graphius, E. J. 2000. Inheritance and synergism of resistance to imidacloprid in the Colorado potato beetle (Coleoptera: Chrysomelidae). *J. Econ. Entomol.* **93**, 1508–1514.

Chapter VII-2

Application and evaluation of entomopathogens in crucifers and cucurbits

John D. Vandenberg
and Stephen P. Wraight
*USDA-ARS, Plant
Soil & Nutrition Laboratory
Ithaca, NY 14853, USA*

Anthony M. Shelton
*Department of Entomology
Cornell University
New York State Agricultural
Experiment Station
Geneva, NY 14456, USA*

1 Introduction

Cruciferous and cucurbitaceous vegetable row crops are comprised of many species and varieties and are cultivated worldwide in many climates. In this chapter, we present a brief overview of selected pests and microbial control agents that are effective against some of them. We confine our remarks primarily to crops grown in open field settings. Management of pests of crops grown in protected environments (*e.g.*, under glass) is covered in Chapter VII-8. Using lepidopterous pests of crucifers and whitefly pests of cucurbits as examples, we will present protocols for the application and evaluation of microbial control agents, with emphasis on the bacterium *Bacillus thuringiensis* and the fungus *Beauveria bassiana*.

Cultivated cruciferous vegetables include many closely related varieties, including cabbage, broccoli, collard, kale, cauliflower, and others. The most destructive insect pest of crucifers worldwide is the diamondback moth (DBM), *Plutella xylostella*. It is estimated that the cost of annual control of this insect is US$1 billion (Talekar and Shelton, 1993). Many other lepidopterous pests & attack crucifers, and in North America include the beet armyworm, *Spodoptera exigua*, cabbage looper, *Trichoplusia ni*, and imported cabbageworm, *Pieris rapae*. Other major pests of crucifers include the silverleaf whitefly (SLW), *Bemesia argentifolii*, sweet potato whitefly, *Bemisia tabaci*, onion thrips, *Thrips tabaci*, flea beetles, *Phyllotreta* spp., the cabbage root maggot, *Delia radicum*, and the cabbage aphid, *Brevicoryne brassicae* (Finch and Thompson, 1992; Capinera, 2001).

Cultivated cucurbits include varied species such as cucumbers, melons, squash and pumpkin. Major insect pests in the USA include SLW, the melon aphid, *Aphis gossypii*, striped cucumber beetles, *Acalymma* spp., spotted cucumber beetle, *Diabrotica undecimpunctata*, the squash bug, *Anasa tristis*, and several Lepidoptera species including pickleworms and melonworms, *Diaphania* spp., and squash vine borers, *Melittia* spp. (York, 1992; Capinera, 2001).

Many Lepidoptera species infesting crucifers and cucurbits can be managed using pesticides containing spores and crystals of certain varieties of the bacterium *Bacillus thuringiensis* (*Bt*). Several pesticide companies have sprayable

L.A. Lacey and H.K. Kaya (eds.), Field Manual of Techniques in Invertebrate Pathology, 361–374.

Bt-based products labeled for use against a variety of lepidopterous pests at a wide range of application rates (Anonymous, 2005). *Bt* formulations are produced and packaged much like chemical insecticides and have shelf-lives equivalent to them. General procedures for handling, preparing and applying *Bt* are much the same as for chemical insecticides, but proper application is more critical since the material has to be ingested. Coverage of the undersides of the leaves is important since DBM and other insects typically inhabit leaf undersurfaces (Harcourt, 1957), and this habitat is shielded from UV radiation, which rapidly degrades *Bt* toxin. *Bt* is compatible with many other products for tank mixing prior to application. In general, only moderate control of lepidopterans attacking cucurbits has been achieved using *Bt*, perhaps because of the cryptic feeding habits of these insects (mining or boring) (York, 1992). Repeated applications are usually recommended.

Genetic resistance of DBM to *Bt* has emerged in several parts of the world (Tabashnik *et al.*, 1990; Shelton *et al.*, 1993; Mohan and Gujar, 2002; Sayyed *et al.*, 2004; Baxter *et al.*, 2005) and must be carefully monitored to avoid control failures. This issue must be considered in the design and testing of any novel control strategies involving this pathogen. The occurrence of *Bt* resistance and strategies for managing it are discussed in Chapter IX-1. Crucifers engineered to express one or more toxins produced by *Bt* are under development (Metz *et al.*, 1995; Cao *et al.*, 1999; Cao *et al.*, 2002; Wei *et al.*, 2005; Anderson *et al.*, 2005) (see also Chapter VIII-1).

Beauveria bassiana has been commercialized in numerous countries (Feng *et al.*, 1994; Shah and Goettel, 1998; Wraight *et al.*, 2001; Khachatourians *et al.*, 2002; Alves *et al.*, 2003). In the U.S., conidia of *B. bassiana* comprise the active ingredient in Mycotrol® and BotaniGard® (Laverlam International Corporation, Butte, MT) and in Naturalis® (Troy BioSciences, Phoenix, AZ). Product labels list many pests and cropping systems, including greenhouse and vegetable field crops. Use of *B. bassiana* for control of whiteflies on cucurbits and crucifers was extensively investigated in the 1990s; numerous studies demonstrated potential against the SLW at rates ranging from $1–5 \times 10^{13}$ conidia/ha (Jaronski and Lord, 1996; Poprawski, 1999; Liu *et al.*, 1999; Wraight *et al.*, 2000; Orozco-Santos *et al.*, 2000; Liu and Meister, 2001). However, *B. bassiana* treatments were generally found to be only moderately effective under field conditions, and interest in this and other fungi as whitefly-control agents has declined over the past decade due to competition from new highly effective chemical insecticides (especially the systemic neonicotinoids). In the U.S., current use of *B. bassiana*-based pesticides is primarily for control of insects infesting greenhouse crops. Formulations of *B. bassiana* have been developed also for use in organic crop production (*e.g.*, Mycotrol-O and Naturalis-O). Recommendations for field use of fungus-based biopesticides against *Bemisia* spp. are outlined by Faria and Wraight (2002).

Numerous other entomopathogenic fungi have been reported from SLW and other economically important whiteflies (see Osborne and Landa, 1992; Faria and Wraight, 2002; Lacey *et al.*, 2007). *Paecilomyces fumosoroseus* can cause spectacular epizootics in whitefly populations under humid conditions (Carruthers *et al.*, 1993; Lacey *et al.*, 1996). Preparations of this pathogen have been evaluated against *Bemisia* spp. infesting cucurbits, but field testing has been limited (Hernandez *et al.*, 1995; Wraight *et al.*, 2000). Development has been pursued more vigorously for greenhouse vegetables, especially cucumber (Bolkmans *et al.*, 1995; Fang *et al.*, 1996); strain Apopka-97 of *P. fumosoroseus* has been commercialized in several European countries under the trade name PreFeRal® (Biobest N. V., Westerlo, Belgium) and has been registered, but not yet commercialized, in the U.S. as PFR-97® (Certis, Inc., Columbia, MD). *Lecanicillium* spp. (formerly identified as *Verticillium* spp.) also have been commercially developed for *Bemisia* and *Trialeurodes* whitefly control in protected crops. Most notable of these is *L. muscarium* which is commercialized as Mycotal®) (Koppert Biological Systems, Berkel en Rodenrijs, The Netherlands) (see Hajek *et al.*, 2001; Lacey *et al.*, 2007). *Metarhizium anisopliae* and *Aschersonia aleyrodis* also have potential as whitefly biological control agents (Malsam *et al.*, 2002; Lacey *et al.*, 2007).

DBM has been found highly susceptible to an equally diverse array of fungal pathogens. Using $1.25–5 \times 10^{13}$ conidia/ha, Vandenberg *et al.* (1998) and Shelton *et al.* (1998) showed the potential of *B. bassiana* (Mycotrol) to control DBM on cabbage. Both *Bt*-resistant and non-resistant DBM populations tested by Shelton *et al.* (1998) were susceptible to fungal infection. Spray applications of this pathogen at a concentration of 10^8 conidia/ml resulted in 66% mortality of DBM in a screenhouse test with Chinese cabbage in Korea (Yoon *et al.*, 1999), and applications at 5×10^7 conidia/ml produced 65–88% reductions in DBM populations in cabbage fields in Costa Rica (Acuña and Carballo, 2000). Ibrahim and Low (1993) achieved significant reductions in DBM populations on cabbage in Malaysia when they applied 3.75×10^{13} conidia/ha of laboratory-produced strains of *B. bassiana* and *P. fumosoroseus*. In a laboratory screening study, Silva *et al.* (2003) found *M. anisopliae* substantially more virulent than *B. bassiana* against DBM larvae; however, Sabbour and Sahab (2005) reported equal efficacy of these two pathogens against DBM infesting cabbage in the field (mean reductions in infestation of 58–62%). *Zoophthora radicans*, an entomophthoralean fungus, frequently causes natural epizootics in DBM larval populations (see Furlong and Pell, 2001), and the potential to exploit this capacity, especially via autodissemination, is under investigation (Pell *et al.*, 1993; Furlong *et al.*, 1995, Furlong and Pell, 2001; Vickers *et al.*, 2004). Parallel studies with *B. bassiana* also suggest potential for autodissemination and horizontal transmission of this fungus in DBM populations (Furlong and Pell, 2001; Vickers *et al.*, 2004; see also Chapter III-3); however, epizootics of *B. bassiana* occur only rarely among insects inhabiting crop foliage, and induction of disease on a large scale will likely prove more difficult than with *Z. radicans*. Levels of DBM control achievable with fungal pathogens, though quite high in some cases, are generally not sufficient to provide adequate crop protection, and these agents are usually recommended for use in well-developed IPM systems.

A number of baculoviruses have been tested for control of a variety of lepidopteran pests of crucifers (see reviews by Cunningham, 1998; Huber, 1998; Federici, 1999; Moscardi, 1999). Extensive work by R. P. Jaques (cited in Cunningham, 1998) demonstrated the use of the nucleopolyhedrovirus of the alfalfa looper, *Autographa californica*, (*Ac*MNPV) for control of *T. ni* and a granulovirus (*Pr*GV) for control of *Pieris (=Artogeia) rapae* (also see Webb and Shelton, 1990, 1991). The *Ac*MNPV and the *Pr*GV were both pathogenic to *P. xylostella*, and *Pr*GV provided good control of a synchronous host population in a glasshouse trial (Kadir, 1992). The *S. exigua* baculovirus (*Se*NPV) is registered and formulated as SPOD-X® (Certis, Inc., Columbia, MD) and is labeled for use on both crucifers and cucurbits. The nucleopolyhedrovirus of *Mamestra brassicae* (*Mb*MNPV) is registered in Europe as Mamestrin® (Calliope SAS, Nagueres, France). This product has shown potential for control of several lepidopteran pests, including *P. xylostella* in synergy with a pyrethroid insecticide (Huber, 1998). Recent research and development have been focused on the use of recombinant baculoviruses engineered to improve speed of kill for control of a variety of Lepidoptera, including the cabbage looper (Wood *et al.*, 1994; Treacy *et al.*, 1997; Treacy, 1999; Bonning *et al.*, 2002; Kamita *et al.*, 2005).

Nematodes have potential for control of soil-dwelling stages of several insects. The entomopathogenic nematode, *Steinernema feltiae*, can control *D. radicum* in the field (Schroeder *et al.*, 1996), but this agent has not made it into commercial cabbage production systems. *Steinernema carpocapsae* is labeled for use under numerous product names in several cropping systems against a variety of insects, including lepidopteran and coleopteran pests of crucifers and cucurbits (Copping, 2004). Kaya *et al.* (1995) and Choo *et al.* (1996) showed that *S. carpocapsae* could contribute to control of *Diabrotica undecimpunctata* larvae. Baur *et al.* (1997, 1998) found limited efficacy of *S. carpocapsae* against DBM in the field and stressed the need for improvements in formulation and efficacy.

2 Protocols for application and evaluation of selected microbial control agents in crucifers

We will confine our suggested protocols primarily to evaluation of microbials against lepidopterous pests, particularly DBM, of cabbage. Simultaneous evaluation of microbials against lepidopterous pests of crucifers is essential because it can be impossible to attribute signs of plant damage or yield loss to only one species. Similar procedures may, of course, be adopted for other crops. For example, procedures for evaluation of sprays of products containing *Bt* spores and crystals used against lepidopterans are similar among many cropping systems. Likewise, procedures for evaluation of *B. bassiana* conidia applied as a mycoinsecticide are similar for other pests. The relative advantages or disadvantages of options for experimental design and analysis are beyond the scope of this chapter (see Chapters II and VII-1). There are many useful references covering these areas (Little and Hills, 1978; Sokal and Rohlf, 1995; Zar, 1999).

A *Setting up the plots*

Choose a cabbage variety recommended or suitable for growing conditions in the area of the study. A variety that is tolerant to thrips or other pests not being evaluated will limit confusion when evaluating plant productivity and damage (Shelton *et al.*, 1998). Start crucifer seedlings in seedling trays or procure them commercially. Transplant seedlings at the 2- to 4-leaf stage. In our trials, we typically space the plants 45 cm apart in rows spaced 90 cm apart, but plant spacing varies considerably throughout the world for cruciferous vegetables so local standards should be followed. Provide irrigation and fertilizer as recommended, throughout the season, for your area. Management of non-target pests is essential for a controlled study (see Section 3, below).

There are two distinct types of field testing strategies, but many variations on each. One strategy which we often employ is to allow natural populations of the target insect to build up to high levels, then spray once and evaluate one or more times. Treatment differences can be more readily apparent using a single application. Growers commonly switch materials during the season, so this approach can provide a more realistic idea of the level of control growers may obtain with a single product. The disadvantage of this approach is that it is not compatible with marketability studies (see below), since one has already allowed the insects to build up on the plant and some damage would have occurred. However, we often use this approach to evaluate rates, formulations, applications, etc., since it allows us to gain considerable information in an efficient manner. This method is also compatible with artificial infestations of plants (see below).

The other approach is to use multiple applications through the season. Each strategy will influence the size of the plots needed. For the season-long evaluation, plots must include a sufficient number of plants for periodic destructive sampling as well as yield and damage evaluation at season's end. For example, at the plant spacing we use (see above), a plot size of 4.6 m × 4.6 m will include 50 plants. This will allow up to 10 destructive samples of 3 plants/plot during the course of the season, leave 10 plants/plot for marketability evaluation, and allow extra plants in case of accidental loss. Leave an uncultivated buffer zone of approximately 3 m between plots to minimize insect movement and deposition of spray drift.

In cropping areas where an infestation of DBM can be assured, no artificial infestation may be needed. However, infestations in more temperate latitudes may be seasonally variable. Artificial infestations ensure the presence and abundance of the target pest, and we have used this method frequently for evaluation of materials for control of DBM. Artificial infestation can be done by supplementing any field populations of DBM with additional laboratory-reared insects. This can help ensure sufficient pest populations are present to obtain a valid comparison among treatments. Early in the season, pin egg masses containing approximately 100 eggs deposited on foil or wax paper to the underside of one leaf of each plant you have chosen to sample throughout the season (Vandenberg *et al.*, 1998).

Replication is the key to obtaining sound conclusions from numerical data collected from field studies because of tremendous variability due to many causes (artificial infestations can help reduce this variability). Provided adequate plot sizes are obtained, use at least 4 replicate plots/treatment. Block replicates according to any known or anticipated productivity or pest population gradients (*e.g.* soil fertility or prevailing winds). Align the blocks across the gradient. Assign each treatment randomly to the plots within each block.

Microbial control agents (MCAs) generally kill their hosts more slowly than chemical insecticides; thus, crop plants may continue to sustain damage for one or more days after application. Because of the slow speed-of-kill and because pests in early stages of development are, in many cases, more susceptible to MCAs than later-stage pests, apply the MCAs early in the season. This will help limit damage. Since published treatment thresholds (Shelton *et al.*, 1983; Reiners *et al.*, 1999) are based on quick-kill chemical insecticides, adjust such thresholds downward for more slow-acting MCAs. In DBM-infested cabbage, this may be as early as the seedling stage (Reiners *et al.*, 1999). Shelton *et al.* (1996) showed that applying Mycotrol to DBM-infested seedlings significantly reduced foliar damage.

B Pre-treatment sampling of target insect and environmental conditions

1 Sampling methods

Pre-treatment sampling allows a baseline estimate of pest populations in all plots. Sample results may be used in experimental design; you may use initial infestation level as a blocking variable for assignment of treatments. That is, if pre-treatment populations vary across a field, align treatment blocks perpendicular to the pest population gradient.

Cabbage plants are easily enumerated within plots. This allows randomization of plants for sampling purposes. Sample 3 randomly selected plants/plot. Place a bag over each plant and pull it up by the roots. Take the bags to a work area and, while cutting the leaves from each plant, count and identify all lepidopterans. For analysis of treatment differences, use the number of each pest species/plot, *i.e.*, the number of insects on three plants. Record life stages and any signs of microbial infection. Life-stage distributions may also be used to evaluate possible effects of MCAs on insect development. Also note the presence and numbers of other pests (*e.g.*, aphids) or signs of their damage (*e.g.*, flea beetles). This information provides the basis for treatment decisions regarding control of other pests (see Section 3, below).

2 Monitoring environmental conditions

Meteorological and agronomic data should be collected for the field sites under study. These data may be essential to fully evaluate success or failure of pest control using MCAs. This is especially true for fungal pathogens, as the persistence, infectivity, and virulence of these agents may be substantially influenced by temperature, solar radiation, and moisture conditions (*e.g.*, relative humidity, rainfall, irrigation).

C Application of the microbial control agent (MCA)

Although MCAs are reduced-risk materials, handling and application of any MCA should still be done with all the normal precautions that apply to handling any pesticide. Furthermore, applying pesticides of any type is regulated in most states in the USA and other countries and applicators must be licensed. These policies and procedures are beyond the scope of this chapter; refer to pesticide applicator information available for your state or country.

1 Handling of MCA

Handling instructions for registered microbial pesticides, including those containing *Bt*, *B. bassiana* or viruses, are provided by manufacturers on their pesticide label. By definition, MCAs consist of living microorganisms. Any handling must ensure optimal survival conditions. Environmental extremes should be avoided, *i.e.*, limit exposure to moisture, high or low temperatures, and direct sunlight. Pesticide formulations should be mixed and applied soon after mixing according to label instructions. Commercial products based on state-of-the-art formulations generally do not require additional adjuvants to enhance such characteristics as miscibility, coverage or rain-fastness. This may not be the case for first-generation formulations or unformulated pathogens applied during research and development. For example, the conidia of many common entomopathogenic fungi, including *B. bassiana*, are extremely hydrophobic and strong wetting agents are required for preparing large volumes of spray suspensions. A number of surfactants are compatible with *B. bassiana*; particularly effective is the organosilicone wetting agent Silwet L77® (Loveland Industries, Greeley, CO) at concentrations of 0.01–0.03% (Wraight and Carruthers, 1999). The MCA manufacturer should be consulted before using any adjuvant not specifically recommended on the product label. The pH of the water is commonly thought to influence the effect of *Bt*. However, in one trial, when a *Bt kurstaki* product was used within a few hours of mixing, no effect of pH was observed on the level of control (Perez *et al.*, 1996).

Pesticide labels may also contain information on known compatibilities or incompatibilities with other pesticides. Of particular relevance to *B. bassiana* applications is the possible incompatibility with fungicides applied for crop disease control (Clark *et al.*, 1982; Loria *et al.*, 1983; Todorova *et al.*, 1998). However, field tests have demonstrated potential compatibility of *B. bassiana* with certain fungicides applied for potato late blight prevention (Jaros-Su *et al.*, 1999). Further discussion of the compatibility of fungi with other agrochemicals is presented in Chapter IV-4. For detailed information on particular compatibility issues, consult the manufacturer.

2 *Brief description of the equipment used*

Application to small plots is most easily done using backpack mounted spray equipment. For larger plots, a number of tractor-mounted sprayers are suitable. Electrostatic sprayers have proven effective for application of *Bt* in cruciferous crops (Law and Mills, 1980; Perez *et al.*, 1995), but this technology has not been widely adopted by growers (Matthews, 2000; and see Chapter IV-4). Wraight and Carruthers (1999) described several application alternatives using Mycotrol against SLW in crucifers and other crops. A more detailed review of spray-application equipment is provided in Chapter III-1. We cannot overemphasize that proper coverage is essential for materials such as *Bt* and fungal products, perhaps more so than for conventional insecticides.

3 *How to apply the MCA*

With few exceptions, fungal entomopathogens invade their hosts after germinating on external cuticle. Lepidopterous pests of crucifers may be found anywhere on the plant. Because of these considerations, thorough coverage of plant surfaces is essential to ensure adequate contact with target insects. Apply *B. bassiana* at a rate between 1 and 5×10^{13} conidia/ha, depending on experimental objectives. Using a hydraulic sprayer, apply the fungus in a volume of 280 to 470 liters/ha at a pressure of 2.8 to $4.2\,\text{kg/cm}^2$ (Vandenberg *et al.*, 1998). Better coverage may be obtained using a portable air-blast sprayer and may result in improved efficacy (Wraight and Carruthers, 1999). For application to crucifers, spray nozzles should be directed downward toward the top of the plant as well as laterally or upward, from drop-tubes, from the sides of the plant (Vandenberg *et al.*, 1998; Wraight and Carruthers, 1999). This arrangement should allow coverage of all exposed leaf surfaces.

Conidia of *B. bassiana* require nearly saturated moisture conditions for germination and host infection and are rapidly inactivated upon exposure to the germicidal UV rays in sunlight. Consequently, recommendations for use of this agent often include making applications in the evening or late afternoon to take advantage of favorable environmental conditions. However, we are not aware of any published studies directly comparing efficacy of daytime versus evening applications of fungal entomopathogens. Evening application have been found to provide only small improvements in efficacy of *B. bassiana* against Colorado potato beetle (S. P. Wraight and M. E. Ramos, unpublished).

4 *Number of treatments and intervals between treatments*

Adult moths arrive or emerge throughout the season and many lepidopterans, DBM in particular, infest crops as multiple, overlapping generations. For season-long trials, as compared with trials involving only one spray and one evaluation, weekly treatments will ensure that individual larvae are exposed at least once. The number of treatments will depend of course upon experimental objectives (see above for two strategies for evaluating products in the field). However, weekly treatments of *B. bassiana* have proven effective against DBM in crucifers (Vandenberg *et al.*, unpublished).

5 Assessing coverage

Evaluation of deposition is easily accommodated within a replicated field plot design. Plastic cover slips may be pinned to leaves prior to application and collected afterwards. An estimate of spore deposition on plants in the field may then be obtained by counting spores deposited on cover slips using a microscope. For studies of cabbage, pin plastic cover slips to three leaves on 25 plants/treatment prior to application. Pin one cover slip facing downward on a lower leaf and pin one each on the upper and lower surfaces of a frame leaf near the developing cabbage head. Data may be used to obtain estimates of deposition on various parts of the plant and modify application tactics accordingly. Refer to Chapter VII-1 for a full description of the method and for techniques for cover slip processing and data collection. Other methods for assessing persistence of conidia are presented in Chapter IV-4.

D Post-treatment sampling of the target population

1 Sampling methods

Sampling pest populations after initial MCA application is done using the same protocol as for pre-treatment sampling (Section 2 B, above). For season-long trials, periodic sampling is needed throughout the season to fully document the potential effect of the MCA relative to untreated controls or chemical pesticide-treated checks. Sampling cabbage for Lepidoptera should be done until harvest, since caterpillar damage just prior to harvest can greatly affect cabbage head marketability. Weekly sampling of 3 plants/plot will allow close tracking of populations and may help indicate the need for control of non-target pests (Section 4, below)

2 Processing of target insects for subsequent microbiological studies

Fungal isolation from diseased insects observed in the field will help confirm diagnosis. Techniques for fungal isolation and identification have been described (Goettel and Inglis, 1997; Humber, 1997). Collection of living insects from treated plots for the purpose of assessing infection following laboratory or cage incubation is problematic and should be done only if the pest/crop/pathogen/environment interactions are thoroughly understood or if these interactions are the subject of the research. Cage or laboratory retention studies by definition provide an altered environment for insect maintenance and may stress insects to the point of enabling infection that might not have occurred under more natural conditions. See Chapter IV-4 for additional detail.

3 Assessment of persistence

Persistence of *B. bassiana* is probably limited primarily by conidial exposure to solar radiation. However, persistence of viable conidia or germlings for even a few days may contribute to efficacy (see Chapter IV-4). Vandenberg *et al.* (1998) used both leaf washes and bioassay to evaluate persistence and observed insect mortality and the presence of *B. bassiana* colony-forming units up to 6 days after spray. Persistence of conidia may also be measured using the coverslip and other methods at specified intervals following application. To evaluate *B. bassiana* spray persistence by bioassay:

1. Collect leaf samples from both treated and untreated plots prior to spray and at 0-, 1-, 3-, 5-, and 7-days post-spray.
2. Place 4 replicate 15 cm diameter leaf discs/treatment in 15 cm Petri dishes.
3. Add 15 laboratory-reared second-instar DBM to each dish.
4. Incubate at 25 °C with a 15:9 h (L:D) photoperiod.
5. Monitor survival daily for 7 days.
6. Remove dead larvae to separate dishes with moistened filter paper and incubate for 1–2 days. Death due to mycosis may be diagnosed when *B. bassiana* hyphae and conidia are evident. The change in percent mycosis over time post-spray gives a direct estimate of persistence. However,

as noted above, infection levels achieved under laboratory conditions may not accurately reflect the efficacy of residual inoculum in the field.

E Assessing crop damage

Except for conditions of extremely high caterpillar infestations, combined with poor growing conditions or the occurrence of other pest species not be affected by the MCA (*e.g.* flea beetles or some aphid species), cabbage head weight is not likely to be affected by larval feeding (Shelton *et al.*, 1982). Thus, yield data based on head weight may not always be useful. Application of a cabbage head marketability index allows better discrimination among treatments and is more directly related to crop value. A common scale for damage ratings is the one used by Greene *et al.* (1969) which gives a scale of 1–6. Ratings of 1–3 have no damage to the head but increasing damage to the frame leaves whereas ratings above 3 have some damage to the head and increasing damage to the frame leaves. An alternative method is to evaluate only damage to the head (Sears *et al.*, 1985). Using this method, randomly select and score 5 heads from each plot on a scale from 1 (no damage) to 4 (severe damage). Analyze the variance among treatments for average scores or for percentage of heads that are marketable (scores of 1 or 2).

3 Protocols for application and evaluation of selected microbial control agents (MCAs) in cucurbits

Many protocols recommended in studies of MCAs of crucifer pests are also useful in cucurbit cropping systems. Different approaches are noted below.

A Setting up the plots

Planting schemes for cucurbits are highly variable, especially with respect to row spacing. Bush varieties are planted in rows spaced 0.6–1.0 m apart, while vining varieties are planted in beds that may be more than 2 m wide. A reasonable minimum size for research plots arranged in randomized complete block designs is 10 m by 4 rows. As with crucifers, selection of plot size may depend on destructive sampling needs. An uncultivated buffer zone of at least 3 m should be used between plots.

Many cucurbits require long growing seasons and are produced in arid regions under irrigation. Methods used for irrigation of research fields should take standard production practices into consideration. In particular, MCAs are susceptible to wash off, and use of overhead irrigation may be inappropriate in research trials of technologies under development for commercial production systems that rely on drip or ditch irrigation (and vice versa).

B Pre-treatment sampling of target insect and environmental conditions

1 Insect sampling

In arid climates, the currently most devastating pest of cucurbit crops is SLW. During outbreak conditions, adults of this insect enter new plantings in great numbers, creating sharp population gradients across research fields. It is imperative that developing populations of this and other targeted pests be monitored closely prior to treatment, in order to properly orient experimental blocks. Failure to do so will seriously jeopardize collection of useful data from randomized complete block experiments. Use of Latin-square designs may be effective alternatives in cases where pest infestations are unknown or unpredictable. Even with optimal experimental design, detection of significant differences among pre-treatment populations infesting plots assigned to different treatments is common. In these situations, statistical adjustments may be possible through analysis of covariance and application of the Henderson-Tilton formula (Henderson and Tilton, 1955).

Development and use of slow acting biological control agents such as entomopathogenic fungi may call for modifications to standard insect sampling methods. For example, because whitefly adults are attracted to the young, terminal growth of plants and the nymphs are sedentary and located on the undersides of leaves, SLW populations are highly stratified (adults, eggs and early instars are concentrated on terminal foliage, while later instars are found on older leaves). Many recommended sampling protocols call for enumeration of SLW on leaves at specific nodes determined to harbor the highest levels of infestation, and these nodes are typically identified by counting down from the terminal growing point (Lynch and Simmons, 1993; Naranjo and Flint, 1994). Following this protocol early in the season, when plants are growing rapidly, results in the sampling of a different group of nymphs on each sample date. This is an efficient and effective procedure for assessing efficacy of fast-acting chemical insecticides; however, in tests of *B. bassiana* and *P. fumosoroseus* against SLW nymphs, this method does not enable accurate estimation of nymphal mortality due solely to mycosis (S. P. Wraight, unpublished). Better estimates of efficacy of fungi and other slow-acting control agents against whitefly nymphs are achieved by tracking a targeted group (or cohort) and sampling one or more times at predetermined appropriate times post-treatment. Monitoring of a specific group of nymphs is achievable by tagging the plant, but is more easily accomplished by identifying leaf nodes on the basis of age. Counting nodes from the base of the plant (or from the crown of vining plants), is the simplest way to track leaves of a specific age and monitor mortality among individuals of an associated group of whitefly nymphs. The most important caveats with respect to adoption of this sampling protocol are that estimation of season-long control requires sampling of potentially many consecutive cohorts of nymphs developing progressively farther from the base of the plant, and comprehensive evaluation of efficacy also requires sampling of adult whiteflies, which are attracted to new foliage produced at the growing tips of the vines.

2 *Monitoring environmental conditions*

As discussed previously for crucifers, monitoring and reporting of environmental conditions is necessary for thorough evaluation of field trials with MCAs. The large, horizontally oriented leaves of cucurbits provide unique opportunities for study of phyllosphere (or near-phyllosphere) temperature and moisture conditions relative to ambient field conditions. Miniaturization of temperature and relative humidity probes has reached the point that they can be readily affixed to leaf surfaces. Portable electronic data loggers can be placed directly in the field with probes regularly adjusted to maintain readings at a specified level in the canopy or positioned to record conditions at various levels simultaneously.

C *Application of MCA*

Development of delivery systems to maximize efficacy of MCA applications is an active area of research. In outlining protocols here, it is our intent to stimulate innovation rather than to recommend specific methods.

For MCAs to act, they must survive exposure to damaging environmental conditions and contact the host cuticle or be ingested. A principal focus of application research has thus involved development of methods to deliver pathogens to lower surfaces of foliage and other cryptic habitats where SLW and many other insect pests reside (especially during early stages of development) and where environmental conditions are greatly moderated (reduced irradiation, temperature, and desiccation). With erect plants, this is best achieved by directing sprays from below-canopy level; however, this is difficult for many cucurbits because their vining, recumbent growth habit gives them an extremely low profile.

This dense, spreading growth habit of cucurbits creates another significant complication. Seedlings on wide beds are obviously most efficiently treated by banding of applications (limiting sprays to the green row). But, while an efficacious dosage may be applied economically in this way, efficacy may be lost when inoculum is diluted over plants that have grown to cover a wide bed. Many cucurbits (*e.g.*, hybrid melons) grow profusely and may cover the entire field with a dense mat of vegetation. The horizontally oriented leaves comprising this mat are extremely difficult to penetrate with conventional overhead sprayers.

With any spray application, investigators must carefully note and report all planting and spray parameters. Banding of applications represents a useful strategy for improving the efficacy and economics of microbial control, and it is actually a method more widely utilized than generally recognized. Field researchers using backpack hydraulic sprayers with lateral drop tubes, for example, often report rates only as active ingredient/hectare regardless of the width of the spray band. Row spacing is clearly another critical factor in dose determination. It is important that research reports provide sufficient information to allow calculation of dose in terms of both treated hectares and the size of the entire field.

For research applications, these difficulties are most effectively overcome by using portable, motorized mist blowers (air-blast sprayers) with extension tubes permitting sprays to be directed laterally from near ground level (Wraight and Carruthers, 1999; Wraight *et al.*, 2000). In tests of *B. bassiana* against SLW in hybrid melons, best coverage with tractor mounted equipment was achieved using a high-pressure hydraulic sprayer with hollow-cone nozzles configured to spray downward at a 45° angle from near canopy level (lateral nozzles directed perpendicular to the row, center nozzles directed forward or rearward, parallel to the row) (Wraight and Bradley, 1996). Using this sprayer, effective control of whitefly nymphs was achieved in pickle cucumbers planted in rows spaced 0.6 m and in melons during the early part of the season; however control could not be maintained in melons after the plants matured and the sprayer was extended to cover the 2-m beds. (Wraight and Carruthers, 1999). Air-assist and air-blast sprayers have provided mixed results, but have considerable potential (Jaronski and Lord, 1996; Wraight *et al.*, 1996; Wraight and Carruthers, 1999; Wraight *et al.*, 2000). Research is needed on the efficacy of fungal biocontrol agents applied with tractor-mounted air-blast sprayers, varying spray volume and configuration (especially boom height).

D Post-treatment sampling of the target population

Because of the growth habit of vining cucurbits, sampling of individual plants is not practical. Sampling schemes most commonly involve collection of individual leaves, flowers or fruit (Brewer and Story, 1987; Tonhasca *et al.*, 1994) or visual observation or suction sampling of measured quadrants (Edelson, 1986). In the case of SLW nymphs, high population densities often necessitate subsampling of individual leaves. The distribution of immature SLW on individual melon leaves is quite uniform (Tonhasca *et al.*, 1994), and in most recent studies where populations were too high for whole-leaf counts, single areas, usually 2–4 cm^2 marked in a standardized location on each leaf, were selected as the sample unit (see Pre-treatment sampling, Section 3.B.1, above). Wraight *et al.* (2000) counted nymphs in $2-cm^2$ areas on 10 melon or cucumber leaves per replicate plot.

E Assessing crop damage

With respect to cucurbit yields, counts and weights of salable fruit will generally suffice for testing differences between research treatments. However, the value of cucurbit fruit is greatly affected by a number of factors such as size, sugar content and appearance. For example, SLW can damage maturing cucurbit fruit by production of honeydew that supports the growth of sooty mold. Grading of fruit according to specific market demands is thus required for any economic analysis of pest control methods. Because of these grading factors, and because cucurbit growth is indeterminate, harvests must be carefully timed and are usually repeated over a prolonged period. In small plot trials, total yield can be measured from each replicate plot. For large plots, yield may be determined from randomly located subplots (row sections).

4 Supplementary information

Non-target, secondary pests may infest the crop and reach population levels that require some management or control. For example, aphid or flea beetle damage to cabbage may be severe enough to hinder plant productivity, regardless of the success of an MCA against target Lepidoptera. Consequently, treatment options must be available for the non-targets that will not adversely affect the target populations. A detailed review of these options is beyond the scope of this chapter. Consult a pest control manual or pesticide guide (*e.g.*, Cornell Cooperative Extension, 2006; Anonymous, 2005) for options.

5 References

Acuña, A. and Carballo, M. 2000. Comparación de una cepa de *Beauveria bassiana* con insecticidas utilizados para el control de *Plutella xylostella*. *Manej. Integr. Plagas (Costa Rica)* **56**, 52–57.

Alves, S. B., Pereira, R. M., Lopes, R. B., and Tamai, M. A. 2003. Use of entomopathogenic fungi in Latin America. *In* "Advances in Microbial Control of Insect Pests" (Upadhyay, R. K., Ed.), pp. 193–211. Kluwer Academic/Plenum Publishers, New York.

Anderson, P., Sharma, S. R., Sresty, T. V. S., Devi, S., Bala, S., Kumar, P. S., Saradhi, P. P., Frutos, R., Altosaar, I., and Kumar, P. A. 2005. Transgenic cabbage (*Brassica oleracea* var. *capitata*) resistant to diamondback moth (*Plutella xylostella*). *Indian J. Biotechnol.* **4**, 72–77.

Anonymous 2005. "Crop Protection Reference." 21st ed. C & P Press, New York. 2431 pp.

Baur, M. E., Kaya, H. K., Gaugler, R., and Tabashnik, B. 1997. Effects of adjuvants on entomopathogenic nematode persistence and efficacy against *Plutella xylostella. Biocontrol Sci. Technol.* **7**, 513–525.

Baur, M. E., Kaya, H. K., Tabashnik, B. E., and Chilcutt, C. F. 1998. Suppression of diamondback moth (Lepidoptera: Plutellidae) with an entomopathogenic nematode (Rhabditida: Steinernematidae) and *Bacillus thuringiensis* Berliner. *J. Econ. Entomol.* **91**, 1089–1095.

Baxter, S. W., Zhao, J. Z., Gahan, L. J., Shelton, A. M., Tabashnik, B. E. and Heckel, D. G. 2005. Novel genetic basis of field-evolved resistance to *Bacillus thuringiensis* toxins in *Plutella xylostella. Insect Mol. Biol.* **14**, 327–334.

Bolckmans, G., Sterk, J., Eyal, J., Sels, B., and Stepman, W. 1995. PreFeRal (*Paecilomyces fumosoroseus* strain Apopka 97) a new microbial insecticide for the biological control of whiteflies in greenhouses. *Med. Fac. Landbouww. Univ. Gent.* **60**, 707–711.

Bonning, B. C., Boughton, A. J., Jin, H., and Harrison, R. L. 2003. Genetic enhancement of baculovirus insecticides. *In* "Advances in Microbial Control of Insect Pests" (Upadhyay, R. K., Ed.), pp. 193–211. Kluwer Academic/Plenum Publishers, New York.

Brewer, M. J. and Story, R. N. 1987. Larval spatial patterns and sequential sampling plan for pickleworm, *Diaphania nitidalis* (Stoll) (Lepidoptera: Pyralidae), on summer squash. *Environ. Entomol.* **16**, 539–544.

Cao, J., Tang, J. D., Strizhov, N., Shelton, A. M., and Earle, E. D. 1999. Transgenic broccoli with high levels of *Bacillus thuringiensis* Cry1C protein control diamondback moth larvae resistant to Cry1A or Cry1C. *Mol. Breeding* **5**, 131–141.

Cao, J., Zhao, J. Z., Tang, J. D., Shelton, A. M., and Earle, E. D. 2002. Broccoli plants with pyramided cry1Ac and cry1C *Bacillus thuringiensis* genes control diamondback moth resistant to cry1A and cry1C proteins. *Theor. Appl. Genet.* **105**, 258–264.

Capinera, J. L. 2001. Handbook of Vegetable Pests. Academic Press, New York.

Carruthers, R. I., Wraight, S. P., and Jones, W. A. 1993. An overview of biological control of the sweet potato whitefly, *Bemisia tabaci*. *In* "Proceedings: Beltwide Cotton Conferences, Vol. 2" (Herber, D. J. and Richter, D. A., Eds.), pp. 680–685. National Cotton Council of America, Memphis, TN, USA.

Choo, H. Y., Koppenh¬fer, A. M., and Kaya, H. K. 1996. Combination of two entomopathogenic nematode species for suppression of an insect pest. *J. Econ. Entomol.* **89**, 97–103.

Clark, R. A., Casagrande, R. A., and Wallace, D. B. 1982. Influence of pesticides on *Beauveria bassiana*, a pathogen of the Colorado potato beetle. *Environ. Entomol.* **11**, 67–70.

Copping, L. G. (Ed.) 2004. "The Manual of Biocontrol Agents." British Crop Protection Council, Alton, UK.

Cornell Cooperative Extension. 2006. "2006 Integrated Crop and Pest Management Guidelines for Vegetables." Cornell University, Ithaca, NY, USA.

Cunningham, J. T. 1998. North America. *In* "Insect Viruses and Pest Management" (Hunter-Fujita, F. R., Entwistle, P. F., Evans, H. F. and Crook, N. E., Eds.), pp. 313–331. Wiley, New York.

Edelson, J. V. 1986. Comparison of sampling methods for insect pests of canteloupe. *J. Econ. Entomol.* **79**, 266–270.

Fang, Q. X., Yang, S. F., Hu, Y. M., Zhou, Y. Y., Gong, Y. X., and Kang, Z. J. 1986. Control of whitefly, *Trialeurodes vaporariorum* by *Paecilomyces fumosoroseus* var. *beijingensis* Fang et Q. T. Chen, in greenhouses. *Chin. J. Biol. Control* **2**, 129–131.

Faria, M. and Wraight, S. P. 2002. Biological control of *Bemisia tabaci* with fungi. *Crop Prot.* **20**, 767–778.

Federici, B. A. 1999. Naturally occurring baculoviruses for insect pest control. *In* "Methods in Biotechnology, vol. 5: Biopesticides: Use and Delivery" (Hall, F. R. and Menn, J. J., Eds.), pp. 301–320. Humana Press, Totawa, NJ.

Feng, M. G., Poprawski, T. J., and Khachatourians, G. G. 1994. Production, formulation and application of the entomopathogenic fungus *Beauveria bassiana* for insect control: Current status. *Biocontrol Sci. Technol.* **4**, 3–34.

Finch, S. and Thompson, A. R. 1992. Pests of cruciferous crops. *In* "Vegetable Crop Pests" (McKinlay, R. G., Ed.) pp. 87–138. CRC Press, Boca Raton, FL.

Furlong, M. J. and Pell, J. K. 2001. Horizontal transmission of entomopathogenic fungi by the diamondback moth. *Biol. Control* **22**, 288–299.

Furlong, M. J., Pell, J. K., Choo, O. P., and Rahman, S. A. 1995. Field and laboratory evaluation of a sex pheromone trap for the autodissemination of the fungal entomopathogen *Zoophthora radicans* by the diamondback moth, *Plutella xylostella. Bull. Entomol. Res.* **85**, 331–337.

Goettel, M. S. and Inglis, G. D. 1997. Fungi: Hyphomycetes. *In* "Manual of Techniques in Insect Pathology" (Lacey, L. A., Ed.), pp. 213–249. Academic Press, London.

Greene, G. L., Genung, W. C., Workman, R. B., and Kelsheimer, E. G. 1969. Cabbage looper control in Florida – a cooperative program. *J. Econ. Entomol.* **62**, 798–800.

Hajek, A. E., Wraight, S. P., and Vandenberg J. D. 2001. Control of arthropods using pathogenic fungi. *In* "Bio-Exploitation of Filamentous Fungi" (Pointing, S. B. and Hyde, K. D., Eds.), pp. 309–347. Fungal Diversity Press, Hong Kong.

Harcourt, D. G. 1957. Biology of the diamondback moth *Plutella maculipennis* (Curt.) (Lepidoptera: Plutellidae) in eastern Ontario. II. Life-history, behaviour, and host relationships. *Can. Entomol.* **89**, 554–564.

Henderson, C. F. and Tilton, E. 1955. Test with acaricides against the brown wheat mite. *J. Econ. Entomol.* **18**, 265–267.

Hernandez Velazquez, V. M., Garza Gonzalez, E., and Berlanga Padilla, A. M. 1995. Control microbial de mosquitas blancas con *Paecilomyces* spp. en Mexico. *In* "Memoria, Simposio sobre Control Biologico de Mosquita Blanca," pp. 29–36. Sociedad Mexicana de Control Biologico.

Huber, J. 1998. Western Europe. *In* "Insect Viruses and Pest Management" (Hunter-Fujita, F. R., Entwistle, P. F., Evans, H. F. and Crook, N. E., Eds.), pp. 201–215. Wiley, New York.

Humber, R. A. 1997. Fungi: Identification. *In* "Manual of Techniques in Insect Pathology" (Lacey, L. A., Ed.), pp. 153–185. Academic Press, London.

Ibrahim, Y. B. and Low, W. 1993. Potential of mass-production and field efficacy of isolates of the entomopathogenic fungi *Beauveria bassiana* and *Paecilomyces fumosoroseus* against *Plutella xylostella*. *Int. J. Pest Manag.* **39**, 288–292.

Jaros-Su, J., Groden, E., and Zhang, J. 1999. Effects of selected fungicides and the timing of fungicide application in *Beauveria bassiana*-induced mortality of the Colorado potato beetle (Coleoptera: Chrysomelidae). *Biol. Control* **15**, 259–269.

Jaronski, S. T. and Lord, J. C. 1996. Evaluation of *Beauveria bassiana* (Mycotrol® WP) for control of whitefly in spring cantaloupes. *Arthropod Manag. Tests* **21**, 103.

Kadir, H. A. 1992. Potential of several baculoviruses for the control of diamondback moth and *Crocidolomia binotalis* on cabbages. *In* "Diamondback moth and other crucifer pests: Second international workshop, 1990" (Talekar, N. S., Ed.) pp. 185–192. Asian Vegetable Research and Development Center, Tainan, Taiwan.

Kamita, S. G., Kang, K.-D., Hammock, B. D., and Inceoglu, A. B. 2005 Genetically modified baculoviruses for pest control. *In* "Comprehensive Molecular Insect Science, Vol. 6" (Gilbert, L. I., Iatrou, K., and Gill, S., Eds.), pp. 271–322. Elsevier Pergamon, Oxford, UK.

Khachatourians, G. G., Valencia, E. P., and Miranpuri, G. S. 2002. *Beauveria bassiana* and other entomopathogenic fungi in the management of insect pests. In "Microbial Biopesticides" (Koul, O. and Dhaliwal, G. S., Eds.), pp. 239–275. Taylor & Francis, London.

Kaya, H. K., Burlando, T. M., Choo, H. Y., and Thurston, G. S. 1995. Integration of entomopathogenic nematodes with *Bacillus thuringiensis* or pesticidal soap for control of insect pests. *Biol. Control* **5**, 432–441.

Lacey, L. A., Fransen, J. J., and Carruthers, R. 1996. Global distribution of naturally occurring fungi of *Bemisia*, their biologies and use as biological control agents. *In* "*Bemisia* 1995: Taxonomy, Biology, Damage, and Management" (Gerling, D. and Mayer, R., Eds.), pp. 401–433. Intercept, Andover, UK

Lacey, L. A., Wraight, S. P., and Kirk, A. A. 2007. Entomopathogenic fungi for control of *Bemisia* spp.: foreign exploration, research and implementation. *In* "Classical Biological Control of the Sweetpotato Whitefly (Biotype B): Research and Implementation of a National Five-Year Plan" (Gould, J., Goolsby, J., and Hoelmer, K., Eds.). *USDA, Agricultural Research Service Publication.* (in press).

Law, S. E. and Mills, H. A. 1980. Electrostatic application of low-volume microbial insecticide spray on broccoli plants. *J. Am. Soc. Hortic. Sci.* **105**, 774–777.

Little, T. M. and Hills, F. J. 1978. "Agricultural Experimentation: Design and Analysis." Wiley, New York.

Liu, T.-X. and Meister, C. W. 2001. Managing *Bemisia argentifolii* on spring melons with insect growth regulators, entomopathogens and imidacloprid in south Texas. *Subtrop. Plant Sci.* **53**, 44–48.

Liu, T.-X., Stansly, P. A., Sparks, A. N. Jr., Knowles, T. C., and Chu, C. C. 1999. Application of Mycotrol and Naturalis-L (*Beauveria bassiana*) for management of *Bemisia argentifolii* (Homoptera: Aleyrodidae) on vegetables, cotton, and ornamentals in southern United States. *Subtrop. Plant Sci.* **51**, 41–48.

Loria, R., Galaini, S., and Roberts, D. W. 1983. Survival of inoculum of the entomopathogenic fungus *Beauveria bassiana* as influenced by fungicides. *Environ. Entomol.* **12**, 1724–1726.

Lynch, R. E. and Simmons, A. M. 1993. Distribution of immatures and monitoring of adult sweetpotato whitefly, *Bemisia tabaci* (Gennadius) (Homoptera: Aleyrodidae) in peanut, *Arachis hypogaea*. *Environ. Entomol.* **22**, 375 –380.

Malsam, O., Kilian, M., Oerke, E.-C., and Dehne,H.-W. 2002. Oils for increased efficacy of *Metarhizium aniso pliae* to control whiteflies. *Biocontr. Sci. Technol.* **12**, 337–348.

Matthews, G. W. 2000. "Pesticide Application Methods", 3rd edition. Blackwell Science, Oxford.

Metz, R. D., Roush, R. T., Tang, J. D., Shelton, A. M., and Earle, E. D. 1995. Transgenic broccoli expressing a *Bacillus thuringiensis* insecticidal crystal protein: implications for pest resistance management strategies. *Mol. Breeding* **1**, 309–317.

Mohan, M. and Gujar, G. T. 2002. Geographic variation in larval susceptibility of the diamondback moth, *Plutella xylostella* (Lepidoptera: Plutellidae) to *Bacillus thuringiensis* spore-crystal mixtures and purified crystal proteins and associated resistance development in India. *Bull. Entomol. Res.* **92**, 489–498.

Moscardi, F. 1999. Assessment of the application of baculoviruses for control of Lepidoptera. *Annu. Rev. Entomol.* **44**, 257–289.

Naranjo, S. E. and Flint, H. M. 1994. Spatial distribution of preimaginal *Bemisia tabaci* (Homoptera:Aleyrodidae) in cotton and development of fixed-precision sequential sampling plans. *Environ. Entomol.* **23**, 254–266.

Orozco-Santos, M., Farias-Larios, J., López-Pérez, J., and Ramírez-Vázquez, N. R. 2000. Uso de *Beauveria bassiana* para el control de *Bemisia argentifolii* en melón. *Manej. Integr. Plagas (Costa Rica)* **56**, 45–51.

Osborne, L. S. and Landa, Z. 1992. Biological control of whiteflies with entomopathogenic fungi. *Fla. Entomol.* **75**, 456–471.

Pell, J. K., Macaulay, E. D. M., and Wilding, N. 1993. A pheromone trap for dispersal of the pathogen *zoophthora radicans* Brefeld. (Zygomycetes: Entomophthorales) amongst populations of the diamondback moth, *Plutella xylostella* L. (Lepidoptera: Yponomeutidae). *Biocontrol Sci. Technol.* **3**, 315–320.

Perez, C. J., Shelton, A. M., and Derksen, R. C. 1995. Effect of application technology and *Bacillus thuringiensis* Subspecies on management of B. *thuringiensis* subsp. *kurstaki*-resistant diamondback moth (Lepidoptera: Plutellidae). *J. Econ. Entomol.* **88**, 1113–1119.

Perez, C. J., Shelton, A. M., and Erickson, N. 1996. Testing for water pH effect on *Bacillus thuringiensis* toxicity to diamondback moth. *Arthropod Manag. Tests* **21**,399–400.

Poprawski, T. J. 1999. Control of *Bemisia argentifolii* on collards using different formulations and rates of *Beauveria bassiana*. *Arthropod Manag. Tests* **24**, 122–123.

Reiners, S., Petzoldt, C. H., Hoffmann, M. P., and Schoenfeld, C. C., Eds. 1999. "Integrated crop and pest management recommendations for commercial vegetable production." Cornell University Cooperative Extension, Ithaca, New York. 305 pp.

Sabbour, M. M. and Sahab, A. F. 2005. Efficacy of some microbial control agents against cabbage pests in Egypt. *Pak. J. Biol. Sci.* **8**, 1351–1356.

Sayyed, A. H., Raymond, B., Ibiza-Palacios, M. S., Escriche, B. and Wright, D. J. 2004. Genetic and biochemical characterization of field-evolved resistance to *Bacillus thuringiensis* toxin cry1Ac in the diamondback moth, *Plutella xylostella. Appl. Environ. Microbiol.* **70**, 7010–1017.

Schroeder, P. C., Ferguson, C. S., Shelton, A. M., Wilsey, W. T., Hoffman, M. P., and Petzoldt, C. 1996. Greenhouse and field evaluations of entomopathogenic nematodes (Nematoda: Hetrorhabditidae and Steinernematidae) for Control of cabbage maggot (Diptera: Anthomyiidae) on cabbage. *J. Econ. Entomol.* **89**, 1109–1115.

Sears, M. K., Shelton, A. M., Quick, T. C., Wyman, J. A., and Webb, S. E. 1985. Evaluation of partial plant sampling procedures and corresponding action thresholds for management of Lepidoptera on cabbage. *J. Econ. Entomol.* **78**, 913–916.

Shah, P. A. and Goettel, M. S., Eds. 1998. "Directory of microbial control products and services." Society for Invertebrate Pathology, Division of Microbial Control, Gainesville, Florida. 26 pp.

Shelton, A. M., Andaloro, J. T., and Barnard, J. 1982. Effects of cabbage looper, imported cabbageworm, and diamondback moth on fresh market and processing cabbage. *J. Econ. Entomol.* **75**, 742–745.

Shelton, A. M., Kroening, M. K., Eigenbrode, S. D., Petzold, C., Hoffmann, M. P., Wyman, J. A., Wilsey, W. T., Cooley, R. J., and Pedersen, L. H. 1996. Diamondback moth (Lepidoptera: Plutellidae) contamination of cabbage transplants and the potential for insecticide resistance problems. *J. Entomol. Sci.* **31**, 347–354.

Shelton, A. M., Robertson, J. L., Tank, J. D., Perez, C., Eigenbrode, S. D., Preisler, H. K., Wilsey, W. T., and Cooley, R. J. 1993. Resistance of diamonback moth (Lepidoptera: Plutellidae) to *Bacillus thuringiensis* subspecies in the field. *J. Econ. Entomol.* **86**, 697–705.

Shelton, A. M., Sears, M. K., Wyman, J. A., and Quick, T. C. 1983. Comparison of action thresholds for lepidopterous larvae on fresh market cabbage. *J. Econ. Entomol.* **76**, 196–199.

Shelton, A. M., Vandenberg, J. D., Ramos, M., and Wilsey, W. T. 1998. Efficacy and persistence of *Beauveria bassiana* and other fungi for control of diamondback moth (Lepidoptera: Plutellidae) on cabbage seedlings. *J. Entomol. Sci.* **33**, 142–151.

Shelton, A. M., Wilsey, W. T., and Schmaedick, M. A. 1998. Management of onion thrips on cabbage using plant resistance and insecticides. *J. Econ. Entomol.* **91**, 329–333.

Silva, V. C. A., Barros, R., Marques, E. J., and Torres, J. B. 2003. Susceptibility of *Plutella xylostella* (L.) (Lepidoptera: Plutellidae) to the fungi *Beauveria bassiana* (Bals.) Vuill. and *Metarhizium anisopliae* (Metsch.) Sorok. *Neoptrop. Entomol.* **32**, 653–658.

Sokal, R. R. and Rohlf, F. J. 1995. "Biometry." 3rd ed. Freeman, New York. 887 pp.

Tabashnik, B. E., Cushing, N. L., Finson, N., and Johnson, M. W. 1990. Field development of resistance to *Bacillus thuringiensis* in diamondback moth (Lepidoptera: Plutellidae). *J. Econ. Entomol.* **83**, 1671–1676.

Talekar, N. S. and Shelton, A. M. 1993. Biology, ecology and management of the diamondback moth. *Annu. Rev. Entomol.* **38**, 275–302.

Todorova, S. I., Coderre, D., Duchesne, R. M., and Cote, J. C. 1998. Compatibility of *Beauveria bassiana* with selected fungicides and herbicides. *Environ. Entomol.* **27**, 427–433.

Tonhasca, A., Jr., Palumbo, J. C., and Byrne, D. N. 1994. Distribution patterns of *Bemisia tabaci* (Homoptera: Aleyrodidae) in cantaloupe fields in Arizona. *Environ. Entomol.* **23**, 949–954.

Treacy, M. F. 1999. Recombinant baculoviruses. *In* "Methods in Biotechnology, vol. 5: Biopesticides: Use and Delivery" (Hall, F. R. and Menn, J. J., Eds.), pp. 321–340. Humana Press, Totawa, NJ.

Treacy, M. F., All, J. N., and Ghidiu, G. M. 1997. Effect of ecdysteroid UDP-glucosyltransferase gene deletion on efficacy of a baculovirus against *Heliothis virescens* and *Trichoplusia ni* (Lepidoptera: Noctuidae). *J. Econ. Entomol.* **90**, 1207–1214.

Vandenberg, J. D., Shelton, A. M., Wilsey, W. T., and Ramos, M. 1998. Assessment of *Beauveria bassiana* sprays for control of diamondback moth (Lepidoptera: Plutellidae) on crucifers. *J. Econ. Entomol.* **91**, 624–630.

Vickers, R. A., Furlong, M. J., White, A., and Pell, J. K. 2004. Initiation of fungal epizootics in diamondback moth populations within a large field cage: proof of concept for auto-dissemination. *Entomol. Exp. Appl.* **111**, 7–17.

Webb, S. E. and Shelton, A. M. 1990. Effect of age structure on the outcome of viral epizootics in field populations of imported cabbageworm. *Environ. Entomol.* **19**, 111–116.

Webb, S. E. and Shelton, A. M. 1991. A simple action threshold for timing applications of a granulosis virus to control *Pieris rapae*. *Entomophaga* **36**, 379–389.

Wei, W., Schuler, T. H., Clark, S. J., Stewart, C. N. and Poppy, G. M. 2005. Age-related increase in levels of insecticidal protein in progenies of transgenic oilseed rape and its efficacy against a susceptible strain of diamondback moth. *Ann. Appl. Biol.* **147**, 227–234.

Wood, H. A., Hughes, P. R., and Shelton, A. 1994. Field studies of the co-occlusion strategy with a genetically altered isolate of the *Autographa californica* nuclear polyhedrosis virus. *Environ. Entomol.* **23**, 211–219.

Wraight, S. P. and Bradley, C. A. 1996. Use of Mycotrol (*Beauveria bassiana*) for biological control of *Bemesia argentifolii* (Homoptera: Aleyrodidae) in field crops. *In* "Technology transfer in biological control: from research to practice" (Silvy, C., Ed.) pp. 151. IOBC WPRS Bulletin 19(8), Montpellier, France.

Wraight, S. P. and Carruthers, R. I. 1999. Production, delivery and use of mycoinsecticides for control of insect pests of field crops. *In* "Biopesticides: Use and Delivery" (Hall, F. and Menn, J., Eds.), pp. 233–269. Humana Press, Totowa, NJ.

Wraight, S. P., Carruthers, R. I., and Bradley, C. A. 1996. Development of entomopathogenic fungi for microbial control of whiteflies of the *Bemisia tabaci* complex. *In* "Simpósio de Controle Biológico, Anais: Conferências e Palestras." pp. 28–34. V SICONBIOL, Foz do Iguaçu, Brazil.

Wraight, S. P., Carruthers, R. I., Jaronski, S. T., Bradley, C.A., Garza, C. J., and Galaini-Wraight, S. 2000. Evaluation of the entomopathogenic fungi *Beauveria bassiana* and *Paecilomyces fumosoroseus* for microbial control of the silverleaf whitefly, *Bemisia argentifolii*. *Biol. Control* **17**, 203–217.

Wraight, S. P., Jackson, M. A., and De Kock, S. L. 2001. Production, stabilization, and formulation of fungal biocontrol agents. *In* "Fungi as Biocontrol Agents: Progress, Problems and Potential" (Butt, T. M., Jackson, C. W., and Magan, N., Eds.), pp. 253–287. CAB International, Wallingford, UK.

Yoon, C.-S., Sung, G.-H., Park, H.-S., Lee, S.-G., and Lee, J.-O. 1999. Potential of the entomopathogenic fungus *Beauveria bassiana* strain CS-1 as a biological control agent of *Plutella xylostella* (Lep.: Yponomeutidae). *J. Appl. Entomol.* **123**, 423–425.

York, A. 1992. Pests of cucurbit crops: marrow, pumpkin, squash, melon and cucumber. *In* "Vegetable Crop Pests" (McKinlay, R. G., Ed.), pp. 139–161. CRC Press, Boca Raton, Florida.

Zar, J. H. 1999. "Biostatistical Analysis," fourth edition. Prentice Hall, Upper Saddle River, NJ.

Chapter VII-3

Microbial control of insect pests of corn

Leslie C. Lewis
USDA-ARS, Corn Insects & Crop Genetics Research Unit, Iowa State University Ames, IA 50011, USA

Denny J. Bruck
USDA-ARS, Horticultural Crops Research Laboratory Corvallis, OR 97330, USA

Jan J. Jackson
Pioneer Hybrid International, South Dakota Research Center Brookings, SD 57006, USA

1 Introduction

Hybrid field corn, open-pollinated flint, popcorn, and sweet corn are grown on every continent, except Antarctica. Six nations (USA, China, Brazil, Mexico, France, and Argentina) produce 75% of the worlds corn supply. The USA alone accounts for 39% of the total with the majority of production occurring in the Corn Belt stretching from Nebraska to Ohio. Most corn is grown for grain and used for animal feed and human food. In 2004, 27.6 million bushels of grain was produced worldwide.

Corn is grown over a wide geographical range and is vulnerable to insect attack during all phenological stages. Several species of insects are pests of corn. These are commonly categorized as soil insects (root and seedling feeders) or foliar pests (leaf, stem, tassel, and ear feeders).

A Soil insect pests

Most soil insect pests feed on corn roots as larvae. Others, like those in the cutworm complex, live in the soil but feed on young corn seedlings at the soil-surface interface. *Diabrotica virgifera virgifera* typifies an insect that feeds on corn roots during the larval stages but as an adult, feeds on above ground plant tissues including leaves, pollen, and silks. Females oviposit in soil in late summer, the eggs pass the winter in diapause and hatch in late spring, and larvae feed primarily on the roots of corn. *Agrotis ipsilon* is a representative of the cutworm complex. Eggs are laid on weeds and plant residue in spring; early instars feed on these weeds until seedling corn is available. Larvae develop through six instars, pupate in the soil, and the adults emerge 7–10 days after pupation to begin another generation, or depending on location and time of the year, migrate.

B Foliar pests

Larvae of Lepidoptera are the primary foliar feeders on corn. While there are several pest species, plant damage is often similar because it is strongly influenced by similarities in plant

L.A. Lacey and H.K. Kaya (eds.), *Field Manual of Techniques in Invertebrate Pathology*, 375–392.

Table 1. Major insect pests of corn

Pathogen	Pest insect	Plant damage	Key references
Bacteria			
Bacillus thuringiensis	African maize stalk borer	Stalk borer, foliage feeder	Brownbridge, 1990
B. thuringiensis	Corn earworm	Ear	Bartels and Hutchison, 1995
B. thuringiensis	European corn borer	Stalk borer	Bartels and Hutchison, 1995; Hazzard *et al.*, 2003; Langenbruch, 1979; Lynch *et al.*, 1980; Lynch *et al.*, 1977a; Lynch *et al.*, 1977b; McGuire *et al.*, 1990; Nolting and Poston, 1982
B. thuringiensis	Fall armyworm	Stalk borer, foliage feeder	All *et al.*, 1994; Hazzard *et al.*, 2003
B. thuringiensis	South western corn borer	Stalk borer, foliage feeder	Nolting and Poston, 1982
Fungi			
Beauveria bassiana	Asiatic corn borer	Stalk borer	Cooperative Group of Corn borer control by *Beauveria bassiana* of Kirin Province, 1997; Zhang *et al.*, 1990
B. bassiana	Corn earworm	Ear	Cheung and Grula, 1982
B. bassiana	European corn borer	Stalk borer	Bing and Lewis, 1991; Feng *et al.*, 1988; Lewis *et al.*, 1996; Lewis and Bing, 1990; Lewis *et al.*, 2002a; Lewis and Cossentine, 1986
Nomuraea rileyi	Corn earworm	Ear	Mohamed *et al.*, 1978
N. rileyi	Fall Armyworm	Stalk borer, foliage feeder	Hamm and Hare, 1982
Microsporidia			
Nosema heliothidis	Corn earworm	Ear	Brooks *et al.*, 1978
N. pyrausta	European corn borer	Stalk borer	Lewis *et al.*, 1983; Lewis and Johnson, 1982
Vairimorpha sp.	Fall armyworm	Stalk borer, foliage feeder	Hamm and Hare, 1982
V. heterosporum	Fall armyworm	Stalk borer, foliage feeder	Hamm and Hare, 1982
V. necatrix	Corn earworm	Ear	Mitchell and Cali, 1994; Fuxa, 1979
V. necatrix	European corn borer	Stalk borer	Lewis *et al.*, 1983; Lewis and Johnson, 1982
Nematodes			
Heterorhabditis bacteriophora	Western corn rootworm	Root	Jackson, 1996
Steinernema carpocapsae	Asiatic corn borer	Stalk borer	He *et al.*, 1991
S. carpocapsae	Black cutworm	Stem cutter	Levine and Oloumi-Sadeghi, 1992; Shapiro *et al.*, 1999
S. carpocapsae	Corn earworm	Ear	Cabanillas and Raulston, 1996b.
S. carpocapsae	European corn borer	Stalk borer	Ben-Yakir *et al.*, 1998; Lewis and Raun, 1978
S. carpocapsae	Northern corn rootworm	Root	Ellsbury *et al.*, 1996; Wright *et al.*, 1993
S. carpocapsae	Western corn rootworm	Root	Jackson, 1996; Jackson and Hesler, 1995; Journey and Ostlie, 2000; Wright *et al.*, 1993
S. feltiae	Corn earworm	Ear	Richter and Fuxa, 1990
S. feltiae	Asiatic corn borer	Stalk borer	He *et al.*, 1991
S. feltiae	Fall armyworm	Stalk borer, foliage feeder	Richter and Fuxa, 1990
S. riobrave	Corn earworm	Ear	Cabanillas and Raulston, 1996a.
Viruses			
Agrotis ipsilon MNPV	Black cutworm	Stem cutter	Boughton *et al.*, 2001
Anagrapha falcifera MNPV	Corn earworm	Ear	Pingel and Lewis, 1997a; Pingel and Lewis, 1997b

Autographa californica MNPV	Black cutworm	Stem cutter	Johnson and Lewis, 1982
A. californica MNPV	European corn borer	Stalk borer	Lewis and Johnson, 1982
Heliothis NPV	Corn earworm	Ear	Ignoffo *et al.*, 1965; Ignoffo *et al.*, 1980
Rachiplusia ou MNPV	Black cutworm	Stem cutter	Johnson and Lewis, 1982; Lewis and Johnson, 1982
R. ou MNPV	European corn borer	Stalk borer	Lewis and Johnson, 1982
Spodoptera frugiperda MNPV	Fall armyworm	Stalk borer, foliage feeder	Hamm and Hare, 1982

phylogeny and insect life cycles. For example, multi-voltine Lepidoptera will damage leaf and stalk tissues in one generation, but also ear tissue in the next generation. The most serious insect pests of corn, their distribution, plant phenology at time of insect attack, and microbial agents used against the pest are presented in Table 1.

A generalized life cycle of a foliar pest is represented by *Ostrinia nubilalis*. *O. nubilalis* overwinters as a mature 5th instar in cornstalk debris. The time of pupation and adult emergence depends heavily on temperature and moisture in late winter and early spring and whether the population is uni- or multivoltine. Adults congregate in the moist habitat produced by tall, dense vegetation that is ideally suited for mating and resting (Showers *et al.*, 1976). Females leave these grassy areas at night to lay eggs, with the majority of oviposition occurring between dusk and midnight. Eggs are generally laid on the underside of corn leaves. The number of generations per year ranges from one to four depending on geographic location (Showers, 1979).

When infestation occurs during the vegetative stages (V4–V12) of plant development (Ritchie *et al.*, 1997), larval establishment is primarily near the moisture-leaf interface within the spirally rolled leaves of the whorl (Dicke and Guthrie, 1988). Early instars feed primarily on leaves, sheath collar, and midrib tissue. Later instars bore into the stalk and usually pupate and eclose before the plant is in the reproductive stages (R1-R3). When infestations occur during these reproductive stages of plant development, larvae feed on tissues associated with the inflorescence, in florets, on pollen accumulations at the axils of leaves, or on ear structures. Later instars bore into tissues of the stalk, shank, and ear to diapause.

2 Entomopathogens of corn insect pests

A Nematodes

Nematodes in the families Steinernematidae and Heterorhabditidae have been used in biological control programs for corn insect pest management. Infective juveniles (IJs) actively penetrate their hosts cuticle directly or enter the host through a natural opening (*i.e.*, mouth, anus, or spiracle) and begin development in the hemocoel. IJs release their associated bacterium which causes a septicemia in the host and serves as a food source for the nematode. Nematodes are effective against both soil and foliar insects (see Chapter IV-5).

Applied to V7–V12 corn at 50,000 IJs/plant, the Mexican strain of *Steinernema carpocapsae* effectively reduced the number of *O. nubilalis* larvae. A similar application to ears of sweet corn reduced the percentage of damaged ears from 20 to 5% (Ben-Yakir *et al.*, 1998). When *S. carpocapsae* were applied to corn residue at 9×10^4 IJs/m^2, few diapausing 5th instar *O. nubilalis* were killed, even though nematodes were isolated from the soil two months after application (Lewis and Raun, 1978). Spraying of *S. carpocapsae* on vegetative corn caused infection in 33–43% of *Spodoptera frugiperda* larvae, occasionally causing a significant reduction in numbers of larvae per plant. Also, spray application of a *S. carpocapsae* suspension on ears of sweet corn caused an infection in 71% of larvae and up to 53% reduction in larval numbers of a mixed population of *Helicoverpa zea* and *S. frugiperda* (Richter and Fuxa, 1990). An aqueous application of 200 IJs/plant of *S. carpocapsae* (Agrotis strain) into the whorl of the plant killed

80 and 91% of *Ostrinia furnicalis* larvae in each year of a 2-year study (He *et al.*, 1991).

In experiments with soil insects, *S. carpocapsae* applied at 1.3 and 2.5×10^9 IJs/ha to the soil surface was effective as or better than several chemical insecticides in protecting seedling corn from *A. ipsilon* (Levine and Oloumi-Sadeghi, 1992). *Steinernema riobrave* applied to the soil surface at 200,000 IJs/m^2 infected 95% of the prepupae and pupae of *H. zea*. Applications were most effective when applied through furrow irrigation and when applied to soil prior to irrigation. Application of *S. carpocapsae* was also effective against *H. zea* pupae (Cabanillas and Raulston, 1996a,b). Application of the All strain of *S. carpocapsae* through an irrigation system and by a hand-held sprayer reduced feeding on corn roots by *D. virgifera virgifera* and *Diabrotica barberi*. Root damage, determined by a root-feeding index (Oleson, 2005), was lower in the nematode treatment than in the untreated control but higher than a chemical insecticide treatment (Ellsbury *et al.*, 1996). Wright *et al.* (1993) showed that *S. carpocapsae* applied through a center-pivot irrigation system at 1.2 and 2.5×10^9 IJs/ha reduced root feeding by *D. virgifera virgifera* and *D. barberi* better than a chemical insecticide. *Steinernema carpocapsae* and *Heterorhabditis bacteriophora* were efficacious in reducing root feeding by *D. virgifera virgifera* when applied at 200,000 IJs/plant as a single soil-surface application after planting (Jackson, 1996). Efficacy was also demonstrated by applying 10^6 and 10^7 *S. carpocapsae*/30.5 m of row against 2nd and 3rd instar *D. virgifera virgifera*, respectively (Journey and Ostlie, 2000).

Plants under attack from herbivores often emit volatile compounds that attract natural enemies of the herbivore. An enhanced understanding of the tritrophic interaction between corn, *D. virgifera virgifera* and entomopathogenic nematodes has shed light on the insect-induced below ground plant signals which strongly attract nematodes (Rasmann *et al.*, 2005). Corn roots release the sesquiterpene, (*E*)-β-caryophyllene, in response to larval *D. virgifera virgifera* feeding which is strongly attractive to entomopathogenic nematodes. Most commercial corn hybrids from North America do not release (*E*)-β-caryophyllene, whereas European lines and the wild maize ancestor, teosinte, do in response to herbivory (Rasmann *et al.*, 2005). Field experiments showed a fivefold increase in nematode infection rate of *D. virgifera virgifera* larvae on corn producing (*E*)-β-caryophyllene, compared to varieties that do not. These data suggest that varietal differences in (*E*)-β-caryophyllene, production may explain some of the variability in nematode efficacy from trial to trial. As these interactions between trophic levels become well understood, they can be utilized to enhance the efficacy and consistency of biological control agents.

B Bacteria

Bacillus thuringiensis (Bt) is the only bacterium used to manage insect pests of corn. It is not only widely used in crop protection but also toxinregulating genes from *Bt* have been incorporated into corn, primarily to control *O. nubilalis* and *Diabrotica* spp. However, development of new corn lines is underway for other members of the corn pest complex as well. The discussion of *Bt* in this chapter will be limited to its use as a microbial control agent (MCA). For information on transgenic corn, see Chapter VIII-1.

Early studies using *Bt* to control corn insects were conducted before microbiologists had techniques to key *Bt* to subspecies. Chorine (1930) observed that *Bt* reduced the number of *O. nubilalis* larvae per plant by 92%. Raun and Jackson (1966) successfully reduced damage to corn by *O. nubilalis* using application of *Bt* subsp. *thuringiensis*. Dulmage (1970) reported a new isolate, *Bt* subsp. *alesti* designated HD-1, which was later designated as *kurstaki* (DeBarjac and Le Mille, 1970). In field tests, *Bt* subsp. *kurstaki* and *thuringiensis* reduced damage by *O. nubilalis* larvae by 67 and 28%, respectively (McWhorter *et al.*, 1972). Majority of *Bt* research to manage *O. nubilalis* has since been conducted with *Bt* subsp. *kurstaki*.

Several isolates representing *Bt* subsp. *aizawai, galleriae, kurstaki*, and *kumamotoensis*, were effective in reducing larval populations of *Heliothis virescens* and *Spodoptera exigua* when formulated as granules (Tamez-Guerra *et al.*, 1998). The combination of *Bt* subsp. *kurstaki* and corn oil nearly doubled the percentage of marketable ears of sweet corn (Hazzard *et al.*, 2003). A novel strain of *Bt*, EG4961 subsp. *tenebrionis* (Cry IIIA), was toxic to larvae and adults of *Diabrotica undecimpunctata howardii* in the laboratory; however, field data are not available (Johnson *et al.*, 1993). Genetically modified *Bt* (recombinant DNA) successfully reduced populations of *S. frugiperda* in vegetative stage corn with three spray applications on a 7-day schedule (All *et al.*, 1994).

Bt has been applied to corn using aqueous, granular, dust and foam formulations. An aqueous application of *Bt* (500 g of formulated material/ha) reduced larval populations of *Sesamia cretica* (Al-Adil *et al.*, 1986). *Bt* applied through a center-pivot sprinkler irrigation system was efficacious in reducing populations of both *O. nubilalis* and *Diatraea grandiosella* (Nolting and Poston, 1982). Aerial applications of spray formulations of *Bt* for control of *O. nubilalis* and *H. zea* in sweet corn were effective, but timing was critical (Bartels and Hutchison, 1995). *Bt* was effective against *Busseola fusca* in both aqueous and granular formulations (Brownbridge, 1990). In experiments with *O. nubilalis* to evaluate efficacy of several formulations, granules and foams were more effective than a spray in reducing larval populations (Lynch *et al.*, 1980). Granular formulations were even more efficacious when ultraviolet (UV) protectants were included that extend the residual activity (McGuire *et al.*, 1990, 1994). The Dipel 2x formulation (Abbott Laboratories, N. Chicago, IL) was efficacious against *S. exigua* when applied at 400g/ha (Salama *et al.*, 1993). In all studies, *Bt* was superior when formulated as a granule. This delivery system places the bacterium in the whorl of the plant in vegetative corn or behind the leaf-sheath-collar in R1-stage corn where neonatal larvae begin their feeding (Lynch *et al.*, 1977a, b). Sprays, however, place the bacterium on the surface of leaves where it is vulnerable to the environment. However, the inclusion of oils in aqueous applications can enhance efficacy (Hazzard *et al.*, 2003). As the corn plant grows during the vegetative stages prior to tassel-bud formation, the unfurling leaves form a funnel. This funnel captures the granules and channels them to the interface where the leaves are tightly furled and extending upward. The first flight of *O. nubilalis* in the US Corn Belt coincides with vegetative corn. The second flight of *O. nubilalis* usually occurs when the plant is in the R1-R3 stages. A granular application of *Bt* at this time collects in the area of the leaf-sheath collar where neonatal larvae are feeding on the collected pollen prior to establishing behind the collar.

Timing and placement were also very important. *Ostrinia nubilalis* were effectively reduced with an application of *Bt* when applied to corn 10–14 days after larval hatch (Tancik and Cagan, 1998). Application of a spray to the underside of the leaf was more effective than application to the top of the leaf (Langenbruch, 1979).

C Fungi

Entomopathogenic fungi are distributed widely and play a major role in natural insect control (Chenwonogrodzky, 1980; Bruck and Lewis, 2001). *Beauveria bassiana* has been evaluated extensively as a component for the management of insect pests of corn. Up to 50% mortality can occur when *B. bassiana* is applied when adult *D. virgifera virgifera* are present in the field (Mulock and Chandler, 2000; Bruck and Lewis, 2002). *Beauveria bassiana* significantly reduced larval populations of *O. nubilalis* when applied to V7–V9 corn (Lewis and Bing, 1990; Bing and Lewis, 1991; Lewis *et al.*, 2002a) and killed 60% of *O. nubilalis* larvae that were placed on plants the same day (Feng *et al.*, 1988). Granular applications of *B. bassiana* significantly reduced larval populations of *O. nubilalis* when applied to both vegetative and reproductive stage corn (Lewis *et al.*, 1996). Field application of *B. bassiana* to whorl stage corn as spray, dust, and granules reduced larval populations of *O. furnicalis* by 87, 81, and 79%, respectively (Zhang *et al.*, 1990). The spray was assessed as cheaper, less labor-intensive, and easier to store than the other formulations. When applied to corn stubble (crop residue) as a dust, *B. bassiana* killed 70–80% of overwintering *O. furnicalis* and reduced corn borer damage the following year by > 70% (Cooperative, 1977). *Beauveria bassiana* placed in whorls of corn plants reduced larval populations of the first and the second generations of *O. nubilalis*, indicating maintenance on or in the plant for up to 60 days (Lewis and Cossentine, 1986).

Research (Bing and Lewis, 1991, 1992a, b, 1993; Wagner and Lewis, 2000; Lewis *et al.*, 2002a) confirmed that *B. bassiana* forms an endophyte with corn. Endophytic *B. bassiana* does not result in pathology in the corn plant (Lewis *et al.*, 2001).

Spray formulation of *Nomuraea rileyi* applied to ears of sweet corn killed 95 and 88% of 4th and 5th instar *H. zea*, respectively (Mohamed *et al.*, 1978). Applications of *N. rileyi* through an overhead irrigation system were efficacious against *S. frugiperda* feeding on hybrid corn (Hamm and Hare, 1982). *Nomuraea rileyi* was also efficacious against *Heliothis armigera* when applied as an aqueous suspension of conidia (Tang and Horn, 1998).

Biology is also beginning to play a more prominent role in the selection and utilization of fungal entomopathogens. The traditional approach to the biological control of soil-borne insects with entomopathogenic fungi historically has not been very successful, and when success has been demonstrated, it has not been consistent. A new approach which shifts our focus away from laboratory bioassay data and onto fungal biology may very well lead to enhanced success. Recently, two isolates of *Metarhizium anisopliae* (ARSEF 1080 and F52, Novozymes Biologicals Inc., Salem, VA) were demonstrated to be rhizosphere competent (Hu and St. Leger, 2002; Bruck, 2005). While the notion of fungal populations increasing in the rhizosphere is not new, these were the first reports of entomopathogenic fungi doing so. In field studies, the population of *M. anisopliae* in the bulk soil decreased from 10^5 to 10^3 propagules/g soil after several months while populations in the inner rhizosphere of cabbage plants remained at 10^5 propagules/g soil (Hu and St. Leger, 2002). In soil less potting media, the population of *M. anisopliae* declined slowly over a one year period in the bulk soil while the fungal population in the rhizosphere soil was maintained at nearly a log increase (Bruck, 2005). Bruck (2005) was also the first to demonstrate that fungal inoculated roots were effective at managing soil insects. Seventy six percent of the *Otiorhynchus sulcatus* (Coleoptera: Curculionidae) feeding on fungal inoculated *Picea abies* roots became infected (Bruck, 2005). Inoculation of roots with *M. anisopliae*, or potentially other entomopathogenic fungi, represents a novel method for delivering entomopathogenic fungi and would greatly reduce application costs.

D Viruses

The only viruses with demonstrated activity against corn pest insects are baculoviruses. The nucleopolyhedrovirus (NPV) of *H. zea* is efficacious on sweet corn as an aqueous suspension (Ignoffo *et al.*, 1965, 1980). The multicapsid NPV (*Agip*MNPV) of the black cutworm, *Agrotis ipsilon*, is high infectious in laboratory bioassays (Boughton *et al.*, 1999) and also significantly reduced feeding damage to corn seedlings in the greenhouse and field (Boughton *et al.*, 2001). The NPV of the celery looper, *Anagrapha falcifera*, is efficacious against *H. zea* on sweet corn ears as an aqueous suspension or as an aqueous suspension with a starch-based additive (Pingel and Lewis, 1997a, b). Aqueous suspensions of NPVs from *Autographa californica* or *Rachiplusia ou* reduced damage to corn caused by *O. nubilalis* larvae (Lewis and Johnson, 1982). These same viruses formulated on wheat bran, and applied manually at 7×10^{11} polyhedral occlusion bodies (OB)/ha were efficacious in reducing plant cutting by *A. ipsilon* larvae (Johnson and Lewis, 1982). The addition of fluorescent brighteners to aqueous suspensions has also been shown to enhance the activity of NPV on the fall armyworm, *Spodoptera frugiperda*. The use of viruses has been limited because of lack of persistence in the field. The most detrimental environmental factor to virus efficacy in the field is UV (200–400 nm) (Young and Yearian, 1974) and UV-B radiation (Shapiro and Domek, 2002).

E Microsporidia

Several microsporidia have been isolated from corn insects and evaluated as a component of an integrated pest management program. Most microsporidia are variable in efficacy when applied as population suppressants. *Nosema pyrausta*, a microsporidium infecting *O. nubilalis*, was efficacious in reducing larval populations. Spores applied to whorls of corn plants in an aqueous suspension increased the percentage of *O. nubilalis* population infected and the severity of infections (Lewis *et al.*, 1983).

Nosema pyrausta and *Vairimorpha necatrix* were applied to corn but did not cause a reduction in damage by *O. nubilalis*; however, there was an increase in infection in the larvae (Laing and Jaques, 1984). In other work (Lewis *et al.*, 1983), aqueous suspensions of *N. pyrausta* and *V. necatrix* spores were applied separately to corn foliage in the vegetative and reproductive stages. Both species of microsporidia caused infection of *O. nubilalis* larvae 12 days after application (Lewis, 1982). Evaluation of microsporidian efficacy is typically measured by the percentage of a population infected and not by reduced damage to host plants. Per os infection of *O. nubilalis* larvae with *N. pyrausta* is important in maintaining the infection in the population as well as reducing the vitality of the population (Sajap and Lewis, 1992).

3 Generalized protocol for microbial application

A Field plots

A typical small-field plot consists of two treatment rows 10 m in length, 15 cm plant spacing with two guard rows between treatments replicated six to eight times. Data obtained are statistically more powerful from an increased number of replications versus an increased number of samples from a treatment. Also, MCAs should be evaluated in multiple locations to be sure that treatment differences are not related to the environment or other factors unique to a single location. Choice of plant size for application of MCA is dependent on the insect and time in the growing season. Corn in V2–V4 stage is used for research with *A. ipsilon*, V7–V10 for first flight of *O. nubilalis*; R1 for second flight of *O. nubilalis* and R1-R3 for *H. zea*. Infesting plants with laboratory-reared insects may not be necessary if a consistent perennial infestation occurs, but may be used to assure that an adequate and uniform population of the pest is present.

Generalized techniques are used for rearing large quantities of insects to augment field populations. *Ostrinia nubilalis* are produced by techniques of Guthrie *et al.* (1971). Adults are field collected, allowed to oviposit in the laboratory on waxed paper and the resulting larvae are grown on a wheat-germ based diet (Lewis and Lynch, 1969). For augmentation, eggs are held for 4 days at 27 °C, 70% RH. Once the eggs hatch larvae are mixed with corn cob grits (40 mesh), placed in 500 ml bottles and metered onto corn plants with a hand-held applicator (Davis and Oswalt, 1979). Approximately 50 neonatal larvae are placed in the whorl or leaf-sheath-collar area of the plant.

When using laboratory-reared insects for field evaluations, one must be cognizant of insect quality to assure that feeding on the corn plant mimics feeding by wild insects. Also, be sure insects are free of pathogens and are vigorous, and the laboratory colony has been infused with individuals from a wild population within the past year.

B Pretreatment sampling

Timing of pretreatment sampling is again dependent on insect species and plant phenology. Sampling for *Diabrotica* spp. is done one of two ways, *i.e.*, count the number of beetles per plant during the previous growing season or count the number of eggs per unit of soil, following the previous growing season. Plant counts are taken in U-shaped stratified random design with two plants visually inspected for *Diabrotica* spp. adults at 27 sites within a field during peak beetle emergence. An average of one beetle per plant indicates that there will be an economic population of corn rootworm larvae the following year (Steffey *et al.*, 1982). Soil samples are taken using a golf-cup cutter (10.2 cm diameter). Four cores are taken to a depth of 20 cm on either side of the corn row and two cores taken near the plant base. Samples are mixed and a subsample of 0.47 liter is taken (Hein and Tollefson, 1985). Samples are washed and eggs separated and counted using a binocular microscope (Shaw *et al.*, 1976).

If the insect feeds in the whorl, the whorl is physically removed by grasping the unfurled leaves and pulling upward. Carefully unroll the leaves, counting the number of insect larvae. A sample size of 5 plants/row/replication will provide reliable population estimation. If the insect feeds behind the leaf-sheath collar, pull the collar away from the corn stalk and visually count the insects. Also, count insects feeding on any pollen accumulated at the interface of leaf axil and leaf-sheath collar. If the insect feeds on the ear, pull back the husks at the ear tip and count the larvae. Cutworm populations are determined by counting the number of cut plants/3 m of row length. Five such samples should be taken at random per research plot.

C Application of MCA

1 Methods of application

Application for soil insects, *i.e.*, *Diabrotica* spp., has been made with a knife applicator; however, over the row applications of a liquid suspension were found to be most efficacious for parasitic nematodes (Jackson, 1996). MCAs for the cutworm complex are applied when plants are in an early V stage using a hand-broadcast method. This is usually accomplished by preweighing the formulation to give a known amount of MCA per unit of area.

Application of granules for managing whorl-feeding insects and those feeding behind the sheath-collar is accomplished by using a metered device that drops granules into the whorl or in the leaf-axils. General requirements for an applicator of granulated insecticides to control *O. nubilalis* in field corn are as follows:

1. A positive metering mechanism that will handle materials having weight per volume of 150 to 1200 kg/m^3.
2. For control of whorl-feeding insects, ground clearance of 1 to 1.5 m is needed; for insects on pollen-shed corn; clearance of 2.5 to 3.5 m is needed.
3. If ground driven, machines should have a discharge rate that is directly proportional to the speed of the metering mechanism or an adjustable discharge opening and a means for maintaining constant field speeds.
4. If the metering device is not ground driven, the discharge rate should be constant for the normal operating speed at a given opening.
5. Spreaders should be used over each row to obtain a width of spread of 30 to 45 cm.
6. A positive shut-off device should be used to prevent excessive waste.
7. Physical breakdown of particles should be held at a minimum.
8. Provisions should be made to make loading, unloading, cleaning, and calibration of the equipment as easy as possible.

2 Formulations

Some pathogens are formulated commercially. If not, they can be formulated in the laboratory for small plot tests. Prepare a stock of the pathogen and quantify the pathogen to units of microbe per unit of weight or volume.

a. *Aqueous suspensions.* A quantity of the candidate microbe is suspended in a volume of distilled water plus wetting agent and adjusted to microbes/ml or/mg.
b. *Granules.* Apply microbe to blank granules which are usually clay or organics such as corn cob or corn kernel with a plant mister within a large rolling jar on a roller-mixer.
c. *Bait.* Apply the microbe to wheat bran or other organic material such as apple or grape pomace with a plant mister.

D Equipment

Examples of equipment for the application of MCAs include those used for chemical pesticides to special equipment based on the ingenuity of the researcher. If a substantial amount of a MCA is available, applications against *O. nubilalis* are applied with a high clearance machine operated at a ground speed of 6.4 km/h. The machine is fitted with a hydraulic motor to drive calibrated metering devices for applying granular formulations. To insure uniform distribution, granular formulations are applied through a "fish tail" distributor positioned directly over the corn plant. Liquid formulations are applied through one flat fan agricultural nozzle positioned over each row to be treated. Each nozzle is calibrated to deliver 93.5 liters/ha when operated at 40 psi (276 kPa). If the amount of MCA is limited, then application is made with backpack sprayers, pipettes, graduated cylinders, and other suitable containers. Nematodes to manage *A. ipsilon* are applied to soil with the use of a 5-cc disposable syringe and a 21 gauge needle, 1000 nematodes suspended in 4 ml of water are injected into the soil to a depth of approximately 3 cm. The syringes are then washed with 5 ml of water, which is expelled onto the soil surface. To manage corn rootworm, nematodes are applied with a Wheaton self-filling syringe and a 14 gauge needle (Jackson and Hesler, 1995). When seeking efficacy and practicality with an MCA on a large scale, however, equipment used for application of conventional insecticides should be considered.

E Timing of MCA application

When working with a stalk-boring insect, it is imperative that the MCA, regardless of whether it acts as a contact or oral pathogen, be applied before the insect enters the plant. Once in the plant, it is nearly impossible to deliver the pathogen to the insect. The same is true for insects that feed on the corn ear, they must come in contact with the pathogen before they move within the emerging corn silks. An exception to this may be the use of *B. bassiana* as an endophyte within the corn plant. However, even when using *B. bassiana* as an endophyte, the fungus must be applied prior to the presence of the pest to allow for formation of the endophytic relationship (Bing and Lewis, 1992a, b). An application of *B. bassiana* to control adult *Diabrotica* spp. should be applied when the adults are present on the plant to maximize the level of infection (Mulock and Chandler, 2000; Bruck and Lewis, 2002).

Application of MCAs to manage soil insects i.e., *Diabrotica* spp. should be made shortly before the occurrence of the target or most susceptible stage (*i.e.*, second instars when using *S. carpocapsae*). This is done by applying MCAs when plants are V6–V9 (Journey and Ostlie, 2000).

F Number of treatments

Number of treatments is associated with timing, *i.e., O. nubilalis* is an external feeder through the third instar and bores into the plant during the late fourth or early fifth instar. Because *O. nubilalis* becomes cryptic within 7–10 days, multiple applications of MCAs are not warranted. Furthermore, an increase of 3–5x of inoculum is necessary to kill later instars. With foliar feeding insects, additional treatments are warranted if the ovipositional period is such that the plant is continually reinfested. *Helicoverpa zea* for example, may lay eggs for several days requiring multiple applications of a MCA.

Single applications are used against *Diabrotica* spp. because all eggs hatch within a very short time, thus any additional application would be to kill older instars which, like *O. nubilalis*, are very ineffective. *Agrotis ipsilon* is a migratory insect with a very short ovipositional period; therefore, only one application is warranted.

G Assessing coverage

Plant coverage with a MCA is rarely quantified when conducting research to manage insect pests of corn. For information on coverage, see Chapters III-1 and III-2 that deal with ground and aerial application, respectively.

H Post-treatment sampling of target populations

1 Sampling methods

Sampling of the target pest after application uses the same techniques as the pre-treatment sampling. The appropriate sampling method is dependent on the life stage of the pest at the time of sampling and the structures of the plant to be searched. The length of time between MCA application and subsequent sampling of the target insect will depend on the mode of action of thc MCA and its speed of kill. Pathogens applied to corn that kill quickly, such as viruses and *Bt*, may be sampled for their efficacy as soon as a few days after application, whereas those that cause a slow, debilitating effect such as microsporidia may be sampled 6–8 weeks after application. Sampling also depends on plant phenology. In V-stage corn, sampling of *O. nubilalis* is conducted 7 days after application of *Bt* or a virus by selecting 5 plants/treatment/replication, pulling the whorls and counting alive and dead larvae.

If larvae are feeding behind the leaf-sheath collar, sampling is conducted 40–60 days after application by selecting 5 plants/treatment/replication, splitting the plant from tassel to base and counting alive and dead insects and the length of the tunnels. When a microsporidium is the MCA, the target insects are sampled several days after application, but prior to pupation. If *O. nubilalis* is the target, 5 plants/treatment/replication are dissected and the larvae are retrieved, frozen, and stored for later evaluation (Lewis, 1982). Sampling of target insects that are soil inhabitants, *i.e., Diabrotica* spp. and *A. ipsilon*, is usually not done because cadavers are extremely difficult to find in the soil.

The number of generations of the pest and the number of MCA applications also influence the sampling regime. Pests with multiple generations that require an MCA application for each generation should be sampled prior to the insect completing its life cycle. MCAs which are applied early in the growing season with the intent of season-long control of a single or multigeneration insect pest require sampling of insects during each generation.

2 Processing of insects

Processing of insects for confirmation of an infection is not common practice when MCAs are used against corn pests, except for microsporidia. In such instances insects are examined in the laboratory by homogenizing the larvae and evaluating the homogenate for spores (Raun *et al.*, 1960).

The larva is weighed and placed in a homogenizing tube, water is added at a rate of 0.1 ml/3 mg, and the larva is homogenized. The homogenate is shaken to assure mixture, a bacteriological loopful is placed on one side of a double chambered Improved Neubauer Hemacytometer and covered with a cover slip. Concentrations of spores are determined by counting the number of spores in 20 squares. The four corner squares are counted from each of the four corner quadrants and from the center quadrant of the hemacytometer (each square contains $1/4000\,mm^3$). Calculate spores per mg of larva by multiplying the total number of spores by 6666.

3 Assessment of persistence and spread

Persistence of the pathogen can be determined by either assaying the soil and or plant tissue for the microbe or by using the target insect for a bioassay. To assess persistence of *Bt*, plants in the whorl stage are sampled by cutting a 7.6 cm section of the whorl beginning at the point where the leaves are tightly furled and extending upward (Lewis *et al.*, 2002b). During the reproductive stage of corn two, 3.8 cm samples are taken. One directly below the primary ear, and one directly above the primary ear. Samples taken from five plants per treatment per replication are immediately sealed in a labeled plastic bag and stored in a freezer until analyses.

The analysis is done on a per plant basis by homogenizing each sample in 100 ml sterile distilled water in a sterile blender for 2 min. The homogenate is serially diluted, and each dilution plated in triplicate on nutrient agar plates by spreading 0.1 ml of the diluent across the surface with a sterile glass spreader. The nutrient agar plates are amended with filter-sterilized polymyxin B sulfate at 8 ppm (Saleh *et al.*, 1969) and neomycin sulfate at 1 ppm. Colony counts are made after the plates have incubated for 24–30 h at 30 °C.

Persistence can also be determined by placing laboratory-reared insects on plants at intervals following application of a MCA. This procedure requires an additional complement of experimental plants for each interval tested. For example, *Bt* is applied to five sets of 10 plants each. On day one, insects are placed on one set of plants and mortality is determined 7 days later; on day 3, insects are placed on an additional set of plants and mortality is assessed 7 days later, etc. (Lewis *et al.*, 2002b).

A pathogen can spread between plants in a corn ecosystem. This occurs in the case of a microsporidium when an infected insect moves to an adjacent plant, defecates and leaves infective spores to be consumed by a noninfected insect (Lewis, 1978). This spread is measured by placing noninfected insects on the plants where the contaminated frass has been deposited, retrieving them after 7–10 days and quantifying spores using the method of Raun *et al.* (1960).

I Assessing insect control and crop damage

Assessment of the amount of control obtained by application of an MCA may be done during or at the end of the growing season. Entomological data are the most informative for assessing the level of control in corn. Data are obtained by completely dissecting the corn plant to determine the amount of leaf or stalk feeding, and the number of healthy and diseased insects. With the exception of marketable ears of sweet corn, grain yield is generally not a good determinate of efficacy because many factors other than insect damage impact yield. However, if very large plot sizes are used (4–6 rows wide by 122 m in length) and grain harvested from the entire plot using a conventional machine harvester, yield benefits to MCA application can be observed (Lewis *et al.*, 2002a). Efficacy of MCAs on sweet corn is determined by measuring centimeters of insect feeding from tip of ear towards the base of the ear; see 4 D3 for details (Pingel and Lewis, 1997a, b).

4 Application of pathogens against specific pest insect groups

A *Root feeders – corn rootworm complex*

1 *Plot design*

A typical field experiment for testing the efficacy of an insect parasitic nematode against *Diabrotica* spp. is set up in a randomized complete block with four to ten replications. Individual treatments are placed in an 11 m row of corn separated within a row by a 3 m unplanted area and from an adjoining treatment row by an untreated guard row. Row spacing is on 102 cm centers and plant spacing is 30 cm within the treatment areas. Treatments consist of different nematode species (potentially at several rates), an untreated check, and a soil insecticide. A hybrid seed corn adapted to the test area is used.

2 *Application and evaluation of nematodes*

a. To ensure a uniform insect population, treatment rows are infested with 800 viable *Diabrotica* spp. eggs/30.5 cm of row length. Eggs are applied in 0.15% agar at a depth of 10 cm in two furrows 15 cm on each side of the row or sometimes in a single furrow in the row (Sutter and Branson, 1980).
b. Nematodes are applied as a knife application (at or after planting) at 200,000 IJs/30.5 cm of row or 200,000 IJs/plant in 20 ml of water as a surface application. Applications are made when the target insects are 2nd instars. Nematodes can be either purchased from commercial sources or produced in the laboratory. Late instar wax moth, *Galleria mellonella*, serve as a laboratory host for small scale production (Kaya and Stock, 1997). IJs are collected and stored at 10 °C in distilled water 1–4 weeks before application.

Nematode efficacy can be compared by visually ranking the root damage on 10 plants per treatment replication (Oleson *et al.*, 2005). Efficacy can also be compared by quantifying the number of *Diabrotica* adults which are collected from an emergence cage (Hein *et al.*, 1985) placed in each treatment replication.

B *Vegetative emergence – cutworm complex*

1 *Plot design*

Field studies to evaluate efficacy of nematodes for control of *A. ipsilon* are usually conducted in small plots. Experimental units consist of individual plots (60 × 80 cm) arranged in the field according to the experimental design used such as randomized complete block. Two rows of corn are planted by hand within each plot on 0.75 m centers with 23 cm separating each plant within a row. To prevent larval movement outside the plot, a 10-cm wide barrier of aluminum is placed around each plot at a depth of 3–7 cm, and petroleum jelly is applied in a ring around the top 2 cm of the barrier. Each plot may be covered with fiberglass netting to prevent predation of larvae by birds (Capinera *et al.*, 1988).

2 *Application and evaluation of nematodes*

a. Place 10 fourth instar or larger *A. ipsilon* in the center of each plot at the V-1 stage of corn development.
b. Apply nematodes to the plots at a rate of at least 3.0×10^5 IJs/m^2. Nematodes are suspended in distilled water and poured onto or hand mixed into the soil in the corn rows. Rinse nematode containers with a small volume (2 or 3 ml) of water and place onto the site of nematode application.
c. Evaluation of treatments is determined by the amount of plant damage. Classification of damage may include leaf feeding (holes), severed leaves, and plants cut at ground level. Observe plots daily to 10 days post-treatment. The plants within each treatment are numbered consecutively and any

insect activity recorded. In addition to observing damage to corn plants, the level of nematode activity can be verified through use of a susceptible host bioassay. An assay using last instar *G. mellonella* can be easily performed. Combine subsamples from each plot to form a 200 g sample per plot (Capinera *et al.*, 1988). Samples are placed in petri dishes, kept moist, inoculated with last instar *G. mellonella*, and incubated for 48 h at 24 °C. After incubation, larvae are removed from the soil and placed on moist blotter paper for another 48 h at which time larval mortality is recorded (Capinera *et al.*, 1988; Shapiro *et al.*, 1999). Persistence of the nematodes can be evaluated once or throughout the study depending on the experimental objectives and expected treatment effects.

Efficacy studies with other pathogens (*e.g.*, viruses) against soil pests (*A. ipsilon*) are conducted with techniques similar to those used to evaluate nematodes (Johnson and Lewis, 1982; Boughton *et al.*, 2001). The main difference is that viruses are formulated on wheat bran as bait and applied to the soil surface manually (see section 3 C1). Evaluations are made by daily categorizing the damage to individual plants as outlined above.

C Vegetative stage of corn

Bt is a MCA that is effective in controlling many lepidopteran pests, including *O. nubilalis*. Commercial formulations of *Bt* for corn pest insects are available as liquids and granules. Again, one of the most limiting aspects of *Bt* is the short half-life due to UV inactivation.

Bt formulations are applied at different rates depending on instar and stage of plant development. In the major corn-growing areas of the USA, there are generally two generations of *O. nubilalis* per year (Mason *et al.*, 1996). Therefore, studies are designed to evaluate control of this insect infesting vegetative and reproductive stages of corn.

1 Plot design

A randomized complete block design with six to eight replications is most often used for field evaluation of *Bt*. The randomized complete block design can then be easily modified to include split-plots if a number of variables are to be evaluated simultaneously. Formulations are sometimes made in the laboratory, limiting the amount of test material. Thus, evaluations may also be conducted in small plots, *i.e.*, two rows 15–20 m in length separated by two guard (border or buffer) rows.

2 Application and evaluation of Bt

Because *Bt* must be ingested by *O. nubilalis* to be effective, it is critical that the formulations be applied to the area of the plant where the young larvae are feeding. Applications are made by over-the-row equipment, airplane, irrigation systems, or hand-held applicators (see section on 3 D for detail).

a. *Post-treatment sample.* Five plants per treatment per replication are sampled for larvae by removing the whorl or by removing the leaf-sheath-collar tissue and counting the number of live larvae.
b. *Evaluation.* Efficacy determinations during the vegetative stages can be done in two ways: 3 to 5 days after application, *i.e.*, by pulling the whorl of the plant (5/plot), unrolling the leaves, and counting the number of live and dead larvae, or 30–40 days after application by splitting stalks from tassel to base and recording dead and live larvae and length of stalk tunneling. This latter method is the most effective means of evaluating efficacy when MCAs are applied at R1 stage of plant phenology.

D Reproductive stage of corn

Helicoverpa zea is one of the most damaging pests of sweet and field corn (Tanada and Reiner, 1962; Dicke and Guthrie, 1988). It damages emerging tassels but causes most of the damage on ears. A baculovirus, *Af*MNPV, will be used for illustrative purposes to manage an insect feeding on corn ears. This virus has a wide host range including *H. zea* (Hostetter and Puttler, 1991).

1 Production of inoculum

Many viruses are not available commercially, and must be propagated in the laboratory. This can be done in 30-ml plastic cups containing *H. zea* larval diet (Young *et al.*, 1976) treated with an aliquot of stock viral suspension, providing 10^3 OBs/mm^2 of diet surface. A 4th instar *H. zea* is placed in each cup and incubated at 27 °C in total darkness. Dead larvae are collected, allowed to putrificate, and homogenized in distilled water; the homogenate is poured through four layers of cheese cloth to separate insect parts from OBs. The filtrate is centrifuged at 500 g for 30 min and the pellet resuspended in distilled water. After thoroughly mixing the suspension, the number of OBs per ml is counted in an Improved Neubauer Hemocytometer (Pingel and Lewis, 1997b).

2 *Plot designs*

Field plots are a single row 10 m in length, arranged in a randomized complete block design. Two guard rows separating treatments and four guard rows at the end of the plots prevent confounding between treatments. Uniform plants are essential for evaluation. Ten to 15 plants within each treatment row are selected based on their development, R1 to early R2, with any plants not in the appropriate stage removed.

3 *Application and evaluation of baculoviruses*

a. Application of the virus is directed at the primary ear of each plant with a back-pack sprayer or other appropriate delivery device. The sprayer is calibrated to deliver a volume of 20 ml/sec.
b. Immediately after applying the virus, place 3 to 5 neonate *H. zea* on each ear with a camel's hair brush (Pingel and Lewis, 1997b).
c. When the plants reach the R3 stage, harvest and evaluate for damage in the laboratory. Grade for damage and index by depth of larval penetration from the ear tip on a scale of 0 to 5.0 = no damage; 1 = damage from tip to 0.5 cm penetration; 2 = damage from 0.5 to 1.9 cm; 3 = damage from 1.9 to 3.8 cm; 4 = damage from 3.8 to 6.3 cm; and 5 = damage > 6.3 cm of penetration (Douglas and Eckhart, 1957).

4 *Assessment of persistence of virus in the field*

Two h after viral application, collect silks from the field on the day of application and subsequent days after application. Place silks from six ears of each treatment in plastic bags and freeze at −15 °C, thaw three of the silks from each treatment, cut into 4-mm pieces and add to blender jars containing 100 ml of a solution of 0.25 g of FDC blue dye no. 1 and 12.5 ml of a 1% solution of Tween 80/500 ml of phosphate buffered saline (Pingel and Lewis, 1997b). Blend for 3 min and filter through cheese cloth. Use the filtrates for bioassay.

Before being bioassayed, allow filtrates to equilibrate to room temperature for 1 h, agitate and feed to neonate *H. zea* with a droplet assay (Hughes *et al.*, 1986). Thirty larvae are assayed for each treatment and collection date for each replication. Larvae are placed individually in diet cups and maintained at 27 °C and 60% RH in total darkness. Controls consist of larvae fed the filtrate from silks with a dye solution only. Larval mortality is determined 8 days later (Pingel and Lewis, 1997b). Other MCAs can be evaluated with the same procedure. Appropriate modifications must be made to accommodate differences in the pathogen and the target insect.

E *Testing* Beauveria bassiana *for endophytism*

Isolates of *B. bassiana* will colonize corn plants killing *O. nubilalis* larvae that attempt to invade the plant (Lewis and Bing, 1990; Bing and Lewis, 1991, Wagner and Lewis, 2000). The following protocol describes techniques for application of *B. bassiana* to corn plants and to assessment of endophytism occurrence.

1 *Plot design*

Small field plots are used to evaluate *B. bassiana* for proclivity to colonize corn. Plant corn using accepted agronomic practices *i.e.*, tillage, fertilizer, and herbicides. Use one row plots separated by two guard rows, with length sufficient to grow 25 plants. Arrange treatments (isolates of *B. bassiana* and an untreated control) in a completely randomized block design replicated four times.

2 *Application of inoculum*

Place an aqueous or a granular formulation of *B. bassiana* into the whorl of a V7–V9 stage plant at the rate of 10^7 conidia/plant or 10^{12} conidia/ha.

Apply with a hand-held inoculator or an over-the-row high clearance machine. Assessment of endophytism is made 30 days after application and again at plant maturity.

Ten plants are randomly removed per treatment row per replication and the leaves are removed. The resulting corn stalk is sanitized by wiping the exterior with 95% ethyl alcohol. The plant is split from tip to base, and pith tissue is excised from two nodal plates below and three nodal plates above the primary ear. In addition, samples are excised from internodal areas. Plating of all nodal tissue is performed using standard microbiological techniques. All tissue is placed on agar plates that favor the growth of *B. bassiana* and *Metarhizuim anisopliae* (Doberski and Tribe, 1980). Agar plates with tissue are placed in a microbial growth chamber maintained at 26 °C, 70% relative humidity, and total darkness. Ten days later tissues are observed for growth of *B. bassiana*. Fungal colonies are viewed with a binocular scope (50–60X) to confirm the species. The following materials will be needed:

Corn knife or any sharp blade to slice stalk from tip to base such as a xacto knife or scalpel; forceps; cutting board; 95% ethyl alcohol; alcohol lamp; petri dishes; and ingredients to prepare medium described by Doberski and Tribe (1980). See Appendix I.

5 Appendix I

Medium used in testing for *Beauveria bassiana* endophytism (Doberski and Tribe, 1980).

Glucose 40 g; Neopeptone (Difco) 10 g; agar 15 g; crystal violet 0.01 g; cycloheximide 0.25 g; chloramphenicol 0.5 g; distilled water 1 liter. The cycloheximide is autoclaved separately at 10 mg/cm^3 concentration and added to the rest of the medium when cool after sterilization.

Beauveria bassiana colonies show good growth of white aerial mycelium on this medium which is very distinct against the violet background. The reverse of colonies is dark violet, apparently due to selective uptake of dye from the medium. This is a rather characteristic feature which assists rapid identification of *B. bassiana* colonies.

6 References

Al-Adil, K. M., Al-Jassany, R. F. and Al-Samariee, A. I. 1986. Integration dimensions in controlling corn stem borer *Sesamia cretica* Led. (Lepidoptera: Noctuidae). 1. Chemical and biological control. *J. Agric. Water Resource Res. Plant Prod.* **5**, 127–139.

All, J. N., Stancil, J. D., Johnson, T. B. and Gouger, R. 1994. A genetically modified *Bacillus thuringiensis* product effective for control of the fall armyworm (Lepidoptera: Noctuidae) on corn. *Fla. Entomol.* **77**, 437–440.

Bartels, D. W. and Hutchison, W. D. 1995. On-farm efficacy of aerially applied *Bacillus thuringiensis* for European corn borer (Lepidoptera: Pyralidae) and corn earworm (Lepidoptera: Noctuidae) control in sweet corn. *J. Econ. Entomol.* **88**, 380–386.

Ben-Yakir, D., Efron, D., Chen, M. and Glazer, I. 1998. Evaluation of entomopathogenic nematodes for biocontrol of the European corn borer, *Ostrinia nubilalis*, on sweet corn in Israel. *Phytoparasitica* **26**, 101–108.

Bing, L. A. and Lewis, L. C. 1991. Suppression of *Ostrinia nubilalis* (Hübner) (Lepidoptera: Pyralidae) by endophytic *Beauveria bassiana* (Balsamo) Vuillemin. *Environ. Entomol.* **20**, 1207–1211.

Bing, L. A. and Lewis, L. C. 1992a. Endophytic *Beauveria bassiana* (Balsamo) Vuillemin in corn: the influence of the plant growth stage and *Ostrinia nubilalis* (Hübner). *Biocontrol Sci. Technol.* **2**, 39–47.

Bing, L. A. and Lewis, L. C. 1992b. Temporal relationships between *Zea mays, O. nubilalis* (Lepidoptera: Pyralidae) and endophytic *Beauveria bassiana. Entomophaga* **37**, 525–536.

Bing, L. A. and Lewis, L. C. 1993. Occurrence of the entomopathogen *Beauveria bassiana* (Balsamo) Vuillemin in different tillage regimens and in *Zea mays L.* and virulence towards *Ostrinia nubilalis* (Hübner). *Agric. Ecosyst. Environ.* **45**, 147–156.

Boughton, A. J., Harrison, R. L., Lewis, L. C. and Bonning, B. C. 1999. Characterization of a nucleopolyhedrovirus from the black cutworm, *Agrotis ipsilon* (Lepidoptera: Noctuidae). *J. Invertebr. Pathol.* **74**, 289–294.

Boughton, A. J., Lewis, L. C. and Bonning, B. C. 2001. Potential of *Agrotis ipsilon* nucleopolyhedrovirus for suppression of the black cutworm (Lepidoptera: Noctuidae) and effect of an optical brightener on virus efficacy. *J. Econ. Entomol.* **94**, 1045–1052.

Brooks, W. M., Cranford, J. D. and Pearce, L. W. 1978. Benomyl: effectiveness against the microsporidian *Nosema heliothidis* in the corn earworm, *Heliothis zea. J. Invertebr. Pathol.* **31**, 239–245.

Brownbridge, M. 1990. Evaluation of *Bacillus thuringiensis* for the control of cereal stem borers. *In* "Proceedings 5th International Colloquium on Invertebrate and Microbial Control." Adelaide, Australia. p. 495.

Bruck, D. J. and Lewis, L. C. 2001. Adult *Diabrotica* spp. (Coleoptera: Curculionidae) infection at emergence with indigenous *Beauveria bassiana* (Deuteromycotina: Hyphomycetes). *J. Invertebr. Pathol.* **77**, 288–289.

Bruck, D. J. and Lewis, L. C. 2002. Whorl and pollen-shed stage application of *Beauveria bassiana* for suppression of adult western corn rootworm. *Entomol. Exp. Appl.* **103**, 161–169.

Bruck, D. J. 2005. Ecology of *Metarhizium anisopliae* in soilless potting media and the rhizosphere: implications for pest management. *Biol. Control* **32**, 155–163.

Cabanillas, H. E. and Raulston, J. R. 1996a. Effects of furrow irrigation on the distribution and infectivity of *Steinernema riobravis* against corn earworm in corn. *Fund. Appl. Nematol.* **3**, 273–281.

Cabanillas, H. E. and Raulston, J. R. 1996b. Evaluation of *Steinernema riobravis, S. carpocapsae*, and irrigation timing for the control of corn earworm, *Helicoverpa zea. J. Nematol.* **28**, 75–82.

Capinera, J. L., Pelissier, D., Menout, G. S. and Epsky, N. D. 1988. Control of black cutworm, *Agrotis ipsilon* (Lepidoptera: Noctuidae), with entomogenous nematodes (Nematoda: Steinernematidae, Heterorhabditidae). *J. Invertebr. Pathol.* **52**, 427–435.

Chenwonogrodzky, J. W. 1980. Microbial agents as pesticides. *Residue Rev.* **76**, 73–96.

Cheung, P. Y. K. and Grula, E. A. 1982. In vivo events associated with entomopathology of *Beauveria bassiana* for the corn earworm (*Heliothis zea*) fungal pathogen. *J. Invertebr. Pathol.* **39**, 303–313.

Chorine, V. 1930. On the use of bacteria in the fight against the corn borer. *Intern. Corn Borer Invest. Sci. Rep.* **3**, 94–98.

Cooperative Group of corn borer control by *Beauveria bassiana* of Kirin Province. 1977. Application of *Beauveria bassiana* in corn stalk-stubble piles for controlling corn borers. *Acta Entomol. Sinica* **3**, 269–275.

Davis, F. M. and Oswalt, T. G. 1979. Hand inoculator for dispensing lepidopterous larvae. United States Department of Agriculture. Science and Education Administration. *Adv. Agric. Tech.* **9**.

DeBarjac, H. and Le Mille, F. 1970. Presence of flageller antigenic subfactors in serotype 3 of *Bacillus thuringiensis. J. Invertebr. Pathol.* **15**, 139–140.

Dicke, F. F. and Guthrie, W. D. 1988. The most important corn insects. *In* "Corn and Corn Improvement." pp. 767–867. Agron. Monogr. 18, 3rd ed.

Doberski, J. W. and Tribe, H. T. 1980. Isolation of entomogenous fungi from elm bark and soil with reference to ecology of *Beauveria bassiana* and *Metarhizium anisopliae. Trans. Br. Mycol. Soc.* **74**, 95–100.

Douglas, W. A. and Eckhart, R. C. 1957. Dent corn hybrids inbreds and hybrids resistant to the corn earworm in the south. *U.S. Dept. Agric. Tech. Bull.* **1160**.

Dulmage, H. T. 1970. Insecticidal activity of HD-1, a new isolate of *Bacillus thuringiensis* variety *alesti. J. Invertebr. Pathol.* **15**, 232–239.

Ellsbury, M. M., Jackson, J. J., Woodson, W. D., Beck, D. L. and Stange, K. A. 1996. Efficacy, application distribution, and concentration by stemflow of *Steinernema carpocapsae* (Rhabditida: Steinernematidae) suspensions applied with a lateral-move irrigation system for corn rootworm (Coleoptera: Chrysomelidae) control in maize. *J. Econ. Entomol.* **89**, 71–81.

Feng, Z., Carruthers, R. I., Larkin, T. S. and Roberts, D. W. 1988. A phenology model and field evaluation of *Beauveria bassiana* (Bals.) Vuillemin (Deuteromycotina: Hyphomycetes) mycosis of the European corn borer, *Ostrinia nubilalis* (Hbn.) (Lepidoptera: Pyralidae). *Can. Entomol.* **120**, 133–144.

Fuxa, J. R. 1979. Interactions of the microsporidium *Vairimorpha necatrix* with a bacterium, virus, and fungus in *Heliothis zea* control experiments with the corn earworm in field crops. *J. Invertebr. Pathol.* **33**, 316–323.

Guthrie, W. D., Russell, W. A. and Jennings, C. W. 1971. Resistance in maize to second-brood European corn borers. *In* "Proceedings of the 26th Annual Corn and Sorghum Research Conference." (J. I. Sutherland and R. J. Falasca, Eds.), pp. 165–179. American Seed Trade Association, Washington, D.C.

Hamm, J. J. and Hare, W. W. 1982. Application of entomopathogens in irrigation water for control of fall armyworms and corn earworms (Lepidoptera: Noctuidae) on corn. *J. Econ. Entomol.* **75**, 1074–1079.

Hazzard, R. V., Schultz, B. B., Groden, E., Ngollo, E. D. and Seidlecki, E. 2003. Evaluation of oils and microbial pathogens for control of lepidopteran pests of sweet corn in New England. *J. Econ. Entomol.* **96**, 1653–1661.

He, K. L., Zhou, D. R. and Yang, H. W. 1991. Biological control of Asian corn borer with the entomopathogenic nematode, *Steinernema feltiae agriotos. Chinese J. Biol. Control* **7**, 1–6.

Hein, G. L., Bergman, M. K., Bruss, R. G. and Tollefson, J. J. 1985. Absolute sampling technique for corn rootworm (Coleoptera: Chrysomelidae) adult emergence that adjusts to fit common-row spacing. *J. Econ. Entomol.* **78**, 1503–1506.

Hein, G. L. and Tollefson, J. J. 1985. Use of the Pherocon AM trap as a scouting tool for predicting damage by corn rootworm (Coleoptera: Chrysomelidae) larvae. *J. Econ. Entomol.* **78**, 200–203.

Hostetter, E. L. and Puttler, B. 1991. A new broad spectrum nuclear polyhedrosis virus isolated from the celery looper, *Anagrapha falcifera* (Lepidoptera: Noctuidae). *Environ. Entomol.* **20**, 1480–1488.

Hu, G. and St. Leger, R. 2002. Field studies using a recombinant mycoinsecticide (*Metarhizium anisopliae*) reveal that it is rhizosphere competent. *Appl. Environ. Microbiol.* **68**, 6383–6387.

Hughes, P. R., van Beek, N. A. M. and Wood, H. A. 1986. A modified droplet feeding method for rapid assay of *Bacillus thuringiensis* and baculoviruses in noctuid larvae. *J. Invertebr. Pathol.* **48**, 187–192.

Ignoffo, C. M., Chapman, A. J. and Martin, D. F. 1965. The nuclear-polyhedrosis virus of *Heliothis zea* (Boddie) and *Heliothis virescens* (Fabricus). III. Effectiveness of the virus against field populations of *Heliothis* on cotton, corn, and grain sorghum. *J. Invertebr. Pathol.* **7**, 227–235.

Ignoffo, C. M., Zuber, M. S., Garcia, C., Darrah, L. L. and Deutscher, D. 1980. Evaluation of *Baculovirus heliothis, Bacillus thuringiensis, Nomuraea rileyi* and carbaryl against *Heliothis zea* on hand-pollinated sweet corn ears. *J. Kans. Entomol. Soc.* **3**, 485–489.

Jackson, J. J. 1996. Field performance of entomopathogenic nematodes for suppression of western corn rootworm (Coleoptera: Chrysomelidae). *J. Econ. Entomol.* **89**, 366–372.

Jackson, J. J. and Hesler, L. S. 1995. Placement and application rate of the nematode *Steinernema carpocapsae* (Rhabditida: Steinernematidae) for suppression of the western corn rootworm (Coleoptera: Chrysomelidae). *J. Kans. Entomol.* **68**, 461–467.

Johnson, T. B. and Lewis, L. C. 1982. Evaluation of *Rachiplusia ou* and *Autographa californica* nuclear polyhedrosis viruses in suppressing black cutworm damage to seedling corn in greenhouse and field. *J. Econ. Entomol.* **75**, 401–404.

Johnson, T. B., Slaney, A. C., Donovan, W. P. and Rupar, M. J. 1993. Insecticidal activity of EG4961, a novel strain of *Bacillus thuringiensis* toxic to larvae and adults of southern corn rootworm (Coleoptera: Chrysomelidae). *J. Econ. Entomol.* **86**, 330–333.

Journey, A. M. and Ostlie, K. R. 2000. Biological control of the western corn rootworm (Coleoptera: Chrysomelidae) using the entomopathogenic nematode *Steinernema carpocapsae. Environ. Entomol.* **29**, 822–831.

Kaya, H. K. and Stock, S. P. 1997. Techniques in insect nematology. *In* "Manual of Techniques in Insect Pathology." (L. A. Lacey, Ed.), pp. 281–324. Academic Press, London.

Laing, D. R. and Jaques, R. P. 1984. Microsporidia of the European corn borer (Lepidoptera: Pyralidae) in southwestern Ontario: natural occurrence and effectiveness as microbial insecticides. *Proc. Entomol. Soc. Ontario* **115**, 13–17.

Langenbruch, G. 1979. Vergleich zweier Spritzgestange zur biologischen Maiszundebekampfung. *Nachrichtenbl. Dtsch. Pflanzenschutzdienst. (Berl)* **31**, 185–189.

Levine, E. and Oloumi-Sadeghi, H. 1992. Field evaluation of *Steinernema carpocapsae* (Rhabditida: Steinernematidae) against black cutworm (Lepidoptera: Noctuidae) larvae in field corn. *J. Entomol. Sci.* **27**, 427.

Lewis, L. C. 1978. Migration of larvae of *Ostrinia nubilalis* (Lepidoptera: Pyralidae) infected *Nosema pyrausta* (Microsporida: Nosematidae) and subsequent dissemination of this microsporidium. *Can. Entomol.* **110**, 897–900.

Lewis, L. C. 1982. Persistence of *Nosema pyrausta* and *Vairimorpha necatrix* measured by microsporidiosis in the European corn borer. *J. Econ. Entomol.* **75**, 670–674.

Lewis, L. C., Berry, E. C., Obrycki, J. J. and Bing, L. A. 1996. Aptness of insecticides (*Bacillus thuringiensis* and carbofuran) with endophytic *Beauveria bassiana*, in suppressing larval populations of the European corn borer. *Agric. Ecosyst. Environ.* **57**, 27–34.

Lewis, L. C. and Bing, L. A. 1990. *Bacillus thuringiensis* Berliner and *Beauveria bassiana* (Balsamo) Vuillemin for European corn borer control: program for immediate and season-long suppression. *Can. Entomol.* **123**, 387–393.

Lewis, L. C., Bruck, D. J. and Gunnarson, R. D. 2002a. On-farm evaluation of *Beauveria bassiana* for control of *Ostrinia nubilalis* in Iowa, USA. *BioControl* **47**, 167–176.

Lewis, L. C., Bruck, D. J. and Gunnarson, R. D. 2002b. Measures of *Bacillus thuringiensis* persistence in the corn whorl. *J. Invertebr. Pathol.* **80**, 69–71.

Lewis, L. C., Bruck, D. J., Gunnarson, R. D. and Bidne, K. G. 2001. Assessment of plant pathogenicity of endophytic *Beauveria bassiana* in Bt transgenic and non-transgenic corn. *Crop Sci.* **41**, 1395–1400.

Lewis, L. C. and Cossentine, J. E. 1986. Season long intraplant epizootics of entomopathogens, *Beauveria bassiana* and *Nosema pyrausta*, in a corn agroecosystem. *Entomophaga* **31**, 363–369.

Lewis, L. C., Cossentine, J. E. and Gunnarson, R. D. 1983. Impact of two microsporidia, *Nosema pyrausta* and *Vairimorpha necatrix*, *Vairimorpha* in *Nosema pyrausta*-infected European corn borer (*Ostrinia nubilalis*) larvae. *Can. J. Zool.* **61**, 915–921.

Lewis, L. C. and Johnson, T. B. 1982. Efficacy of two nuclear polyhedrosis viruses against *Ostrinia nubilalis* (Lep.: Pyralidae) in the laboratory and field. *Entomophaga* **27**, 33–38.

Lewis, L. C. and Lynch, R. E. 1969. Rearing the European corn borer, *Ostrinia nubilalis* (Hübner), on diets containing corn leaf and wheat germ. *Iowa State J. Sci.* **44**, 9–14.

Lewis, L. C. and Raun, E. S. 1978. Laboratory and field evaluation of the DDB136 strain of *Neoaplectana carpocapsae* (nematodes) for control of the European corn borer, *Ostrinia nubilalis. Iowa State J. Res.* **52**, 391–396.

Lynch, R. E., Lewis, L. C. and Berry, E. C. 1980. Application efficacy and field persistence of *Bacillus thuringiensis* when applied to corn for European corn borer control. *J. Econ. Entomol.* **73**, 4–7.

Lynch, R. E., Lewis, L. C., Berry, E. C. and Robinson, J. F. 1977a. European corn borer control with *Bacillus thuringiensis* standardized as corn borer international units. *J. Invertebr. Pathol.* **30**, 169–174.

Lynch, R. E., Lewis, L. C., Berry, E. C. and Robinson, J. F. 1977b. European corn borer: granular formulations of *Bacillus thuringiensis* for control. *J. Econ. Entomol.* **70**, 389–391.

Mason, C. E., Rice, M. E., Calvin, D. E., Van Duyn, J. W., Showers, W. B., Hutchison, W. D., Witkowski, J. F., Higgins, R. A., Onstad, D. W. and Dively, G. P. 1996. European corn borer, ecology and management. *North Central Region. Exten. Publ.* **327**.

McGuire, M. R., Shasha, B. S., Lewis, L. C., Bartelt, R. J. and Kinney, K. 1990. Field evaluation of granular starch formulations of *Bacillus thuringiensis* against *Ostrinia nubilalis* (Lepidoptera: Pyralidae). *J. Econ. Entomol.* **83**, 2207–2210.

McGuire, M. R., Shasha, B. S., Lewis, L. C. and Nelsen, T. C. 1994. Residual activity of granular starch-encapsulated *Bacillus thuringiensis. J. Econ. Entomol.* **87**, 631–637.

McWhorter, G. M., Berry, E. C. and Lewis, L. C. 1972. Control of the European corn borer with two varieties of *Bacillus thuringiensis. J. Econ. Entomol.* **65**, 1414–1417.

Mitchell, M. J. and Cali, A. 1994. *Vairimorpha necatrix* (Microsporida: Burenellidae) affects growth and development of *Heliothis zea* (Lepidoptera: Noctuidae) raised at various temperatures. *J. Econ. Entomol.* **87**, 933–940.

Mohamed, A. K. A., Bell, J. V. and Sikorowski, P. P. 1978. Field cage tests with *Nomuraea rileyi* against corn earworm larvae on sweet corn. *J. Econ. Entomol.* **71**, 102–104.

Mulock, B. and L. Chandler. 2000. Field-cage studies of *Beauveria bassiana* (Hyphomycetes: Moniliaceae) for the suppression of adult western corn rootworm, *Diabrotica virgifera virgifera* (Coleoptera: Chrysomelidae). *Biocontrol Sci. Technol.* **10**, 51–60.

Nolting, S. P. and Poston, F. L. 1982. Application of *Bacillus thuringiensis* through center-pivot irrigation systems for control of the western corn borer and European corn borer (Lepidoptera: Pyralidae). *J. Econ. Entomol.* **75**, 1069–1073.

Oleson, J. D., Tollefson, J. J., Nowatzki, T. M. and Park, Y. L. 2005. Node-injury scale to evaluate root injury by corn rootworms (Coleoptera: Chrysomelidae). *J. Econ. Entomol.* **98**, 1–8.

Pingel, R. L. and Lewis, L. C. 1997a. Field evaluation of *Bacillus thuringiensis* and *Anagrapha falcifera* multiple nulceopolyhedrovirus against the corn earworm (Lepidoptera: Noctuidae). *J. Econ. Entomol.* **90**, 1195–1199.

Pingel, R. L. and Lewis, L. C. 1997b. Field persistence of a multiple nucleopolyhedrovirus of the celery looper, *Anagrapha falcifera*, on sweet corn silks. *J. Invertebr. Pathol.* **69**, 282–284.

Rasmann, S., Köllner, T. G., Degenhardt, J., Hiltpold, I., Toepfer, S., Kuhlmann, U., Gershenzon, J. and Turlings, T. C. J. 2005. Recruitment of entomopathogenic nematodes by insect-damaged maize roots. *Nature* **434**, 732–737.

Raun, E. S. and Jackson, R. D. 1966. Encapsulation as a technique for formulating microbial and chemical insecticides. *J. Econ. Entomol.* **59**, 620–623.

Raun, E. S., York, G. T. and Brooks, D. L. 1960. Determination of *Perezia pyraustae* infection rates in larvae of the European corn borer. *J. Insect Pathol.* **2**, 254–258.

Richter, A. R. and Fuxa, J. R. 1990. Effect of *Steinernema feltiae* on *Spodoptera frugiperda* and *Heliothis zea* (Lepidoptera: Noctuidae) in corn. *J. Econ. Entomol.* **83**, 1286–1291.

Ritchie, S. W., Hanway, J. J. and Benson, G. O. 1997. How a corn plant develops. Special Report 48. Iowa State University Extension, Ames, IA. 21 pp.

Sajap, A. S. and Lewis, L. C. 1992. Chronology of infection of European corn borer (Lepidoptera: Pyralidae) with the microsporidium *Nosema pyrausta*: effect on development and vertical transmission. *Environ. Entomol.* **21**, 178–182.

Salama, H. S., Zaki, F. N., Salem, S. A., Hamdy, M. K., and Megahed, M. 1993. Activity of *Bacillus thuringiensis* Berl. on the lesser cotton leafworm *Spodoptera exigua* Hbn. (Lep., Noctuidae) in corn and sunflower cultivations in Egypt. *Anzeiger-fuer-Schaedlingskunde,-Pflanzenschutz,-Umweltschultz (Germany)* **66**, 29–31.

Saleh, S. M., Harris, R. F. and Allen, O. N. 1969. Method for determining *Bacillus thuringiensis* var. *thuringiensis* Berliner in soil. *Can. J. Microbiol.* **15**, 1101–1104.

Shapiro, M. and Domek, J. 2002. Relative effects of ultraviolet and visible light on the activities of corn earworm and beet armyworm (Lepidoptera: Noctuidae) nucleopolyhedroviruses. *J. Econ. Entomol.* **95**, 261–268.

Shapiro, D. I., Obrycki, J. J., Lewis, L. C. and Jackson, J. J. 1999. Effects of crop residue on the persistence of *Steinernema carpocapsae. J. Nematol.* **31**, 517–519.

Shaw, J. T., Ellis, R. O. and Luckmann, W. H. 1976. Apparatus and procedure for extracting corn rootworm eggs from soil. Illinois Natural History Survey. Biological Notes, **96**.

Showers, W. B. 1979. Effect of diapause on the migration of the European corn borer in the southeastern United States, Chapter 37 *In* "Movement of Highly Mobile Insects," (R. L. Rabb and G. G. Kennedy, Eds), pp. 456. North Carolina State University Graphics, Raleigh, NC.

Showers, W. B., Reed, G. L., Robinson, J. F. and Derozari, M. B. 1976. Flight and sexual activity of the European corn borer. *Environ. Entomol.* **5**, 1099–1104.

Steffey, K. L., Rice, M. E., All, J., Andow, D. A., Gray, M. E. and Van Duyn, J. W. 1999. Handbook of corn insects. Entomol. Soc. Amer., Lanham, MD.

Steffey, K. L., Tollefson, J. J. and Hinz, P. N. 1982. Sampling plan for population estimation of northern and western corn rootworm adults in Iowa cornfields. *Environ. Entomol.* **11**, 287–291.

Sutter, G. R. and Branson, T. F. 1980. A procedure for artificially infesting field plots with corn rootworm eggs. *J. Econ. Entomol.* **73**, 135–137.

Tamez-Guerra, P., Castro-Franco, R., Medrano-Roldan, H., McGuire, M. R., Galan-Wong, L. J. and Luna-Olvera, H. A. 1998. Laboratory and field comparisons of strains of *Bacillus thuringiensis* for activity against noctuid larvae using granular formulations (Lepidoptera). *J. Econ. Entomol.* **91**, 86–93.

Tanada, Y. and Reiner, C. 1962. The use of pathogens in the control of corn earworm, *Heliothis zea* (Boddie). *J. Invertebr. Pathol.* **4**, 139–154.

Tancik, J. and Cagan, L. 1998. Control of the European corn borer (*Ostrinia nubilalis* Hbn.) with chemical and biological insecticides. *J. Agric. Sci.* **44**, 100–109.

Tang, L. C. and Horn, R. F. 1998. Potential application of the entomopathogenic fungus, *Nomuraea rileyi*, for control of the corn earworm, *Helicoverpa armigera. Entomol. Exp. Appl.* **88**, 25–30.

Wagner, B. L. and Lewis, L. C. 2000. Colonization of corn, *Zea mays*, by the entomopathogenic fungus *Beauveria bassiana. Appl. Environ. Microbiol.* **66**, 3468–3473.

Wright, J. R., Witkowski, J. F., Echtenkamp, G. and Georgis, R. 1993. Efficacy and persistence of *Steinernema carpocapsae* (Rhabditia: Steinernematidae) applied through a center-pivot irrigation system against larval corn rootworms (Coleoptera: Chrysomelidae). *J. Econ. Entomol.* **86**, 1348–1354.

Young, J. R., Hamm, J. J., Jones, R. L., Perkins, W. D. and Burton, R. L. 1976. Development and maintenance of an improved laboratory colony of corn earworms. U.S. Dept. Agric., *Agric. Res. Serv. Rep.* **110**.

Young, S. Y. and Yearian, W. C. 1974. Persistence of *Heliothis* NPV on foliage of cotton, soybean, and tomato. *Environ. Entomol.* **3**, 253–254.

Zhang, A. W., Liu, W. Z., Deng, C. S., Nong, K. Q., Wu, Z. K., Guo, W. L., Jiang, B., Wang, S. F., Song, L. W. and Chen, F. 1990. Field control of Asian corn borer, *Ostrinia furnacalis* (Lep.: Pyralidae) with different preparation forms of *Beauveria bassiana. Chinese J. Biol. Control* **6**, 118–120.

Chapter VII-4

Evaluation of microbial agents against rice pests

H. Y. Choo
Department of Agricultural Biology
Gyeongsang National University
Chinju, Gyeongnam
660–701, Republic of Korea

W. C. Rice
USDA-ARS
Renewable Energy and Manure Management
Research Unit
Bushland, TX 79012, USA

1 Introduction

Rice is unique among the world's major food crops by virtue of the extent and variety of its uses and adaptability to a broad range of climatic, edaphic, and cultural conditions (Mikkelsen and Datta, 1991). It provides 20% of the per capita energy and 15% of the per capita protein for humans worldwide. Annual world rice production is approximately 460 million tons grown on more than 145 million ha. Over 90% of this area lies in Asia, while the remainder is divided among Latin America, Africa, Australia, Europe, and the USA (see Way and Bowling, 1991). Rice production should be increased to supply a rapidly expanding population; however, it has been hindered by a number of diseases and insect pests. Moreover, rapid changes in rice production technologies have created greater frequencies of pest epidemics (Reissig *et al.*, 1986).

More than 70 species of insect pests are known to feed on rice, and at least 20 of them can seriously affect rice production. They attack all parts of the rice plant at all growth stages and some serve as vectors of viruses that adversely affect the plant (Mikkelsen and Datta, 1991). Control of insect pests has primarily depended on the application of chemical insecticides, but because of the many disadvantages of chemicals, alternative biological control approaches are needed. In the rice agroecosystems, many types of entomopathogens may suppress insect pests. In this chapter, we detail techniques on the use and evaluation of entomopathogens for the potential control of a few key insect pests (*i.e.*, the brown planthopper, the white-backed planthopper and the rice water weevil) of rice.

2 Major insect pests of rice

A number of insect pests that attack rice plants account for yield losses of 24% worldwide. Some of the major ones are listed in Table 1. Below, we provide information on a few key pests.

A *Homoptera (brown planthopper and white-backed planthopper)*

The brown planthopper (BPH), *Nilaparvata lugens*, and white-backed planthopper (WBPH), *Sogatella furcifera*, are serious pests in many rice-growing regions of the Far East

L.A. Lacey and H.K. Kaya (eds.), Field Manual of Techniques in Invertebrate Pathology, 393–409.

(Choo *et al.*, 1989). They occur in tropical to temperate areas and can cause extensive damage by feeding on rice plants or serving as vectors for grassy or rugged stunt virus of rice. Both sexes of BPH adults have short-winged (brachypterous) and long-winged (macropterous) types, whereas WBPH adult females have both short-winged and long-winged types, but all males are the long-winged type. The short-winged types cannot fly and remain in the field to feed and reproduce, whereas the long-winged types can disperse, mate, and reproduce (*i.e.*, BPH). A macropterous BPH female lays about 100 individual eggs and a brachypterous female lays about 300 eggs in the tillers using its saw-like ovipositor (Reissig *et al.*, 1986). A WBPH female lays from 300–500 eggs. Nymphs and adults suck sap from the base of plants, just above the waterline. In heavy infestations, these planthoppers can cause hopperburn resulting in browning and wilting of some or all tillers in a hill.

In the tropics, the planthoppers have multiple generations per year. In temperate regions, they have 2 or 3 generations per year. They cannot overwinter in the temperate regions, and reinfestion of rice occurs each summer when tropical storms assist their dispersal northward. Currently, chemical insecticides and host plant resistance are the major control tactics for BPH and WBPH.

B Lepidoptera (rice stem borer)

Although many lepidopterous species are associated with rice (Table 1), the rice stem borer, *Chilo suppressalis*, has been one of the most important pests, especially in Asia. However, the release of the short-stature, high-yielding rice varieties, planting of early maturing varieties, improved cultural practices including destruction of post-harvest stubble, and the use of synthetic insecticides have reduced the stem borers from a key to a secondary pest (Way and

Table 1. Some major insect pests of rice occurring in the USA and Asia

Insect species	Plant part attacked	Region	References
Homoptera			
*Nephotettix cincticeps**	Leaf and stem	Asia	Reissig *et al.* (1986)
*Nephotettix virescens**	Leaf and stem	Asia	Reissig *et al.* (1986)
*Laodelphax striatelius**	Stem	Asia	Reissig *et al.* (1986)
*Macrosteles fascifrons**	Stem	USA	Elliott *et al.* (1994)
*Nilaparvata lugens**	Stem	Asia	Reissig *et al.* (1986)
Sogatella furcifera	Stem	Asia	Reissig *et al.* (1986)
Coleoptera			
Oulema oryzae	Leaf	Asia	Reissig *et al.* (1986)
Lissorhoptrus oryzophilus	Root	Asia, USA	Reissig *et al.* (1986)
Diptera			
Chironomus spp.	Grain	USA	Elliott *et al.* (1994)
Cricotopus spp.	Grain	USA	Elliott *et al.* (1994)
Tanytarsus spp.	Grain	USA	Elliott *et al.* (1994)
Hydrellia griseola	Leaf	Asia	Reissig *et al.* (1986)
		USA	Elliott *et al.* (1994)
Pseudonapomyza asiatica	Leaf	Asia	Reissig *et al.* (1986)
Atherigona exigua	Tiller	Asia	Reissig *et al.* (1986)
Atherigona oryzae	Tiller	Asia	Reissig *et al.* (1986)
Lepidoptera			
Mythimna separata	Leaf and stem	Asia	Reissig *et al.* (1986)
Pseudaletia unipuncta	Leaf and stem	USA	Elliott *et al.* (1994)
Sesamia inferens	Stem	Asia	Reissig *et al.* (1986)
Spodoptera frugiperda	Stem	USA	Elliott *et al.* (1994)
Chilo suppressalis	Stem	Asia	Reissig *et al.* (1986)
Cnaphalocrocis medinalis	Leaf	Asia	Reissig *et al.* (1986)
Scirpophaga incertulas	Stem	Asia	Reissig *et al.* (1986)

* Virus vector

Bowling, 1991). Yet, it can still cause significant damage by reducing tiller numbers even on resistant varieties and, in spite of the successes with chemical insecticides, remain difficult to control because of its plant-boring habit (Reissig *et al.*, 1986).

The adults are nocturnal with a female capable of laying 200–300 eggs in masses near the base of rice leaves or leaf sheaths. The larvae penetrate tillers and feed on the inner surface of the stem walls, interrupting the movement of water and nutrients. The central leaves of damaged tillers of young plants turn brown (called dead hearts). If the damage occurs after spikelets form, panicles turn white (called whiteheads) and no grain filling occurs (Reissig *et al.*, 1986). The plant often dies and the larva moves to another stem. Therefore, one larva is capable of damaging several stems. There are 1 to 6 generations a year depending on the cultivation region. For example, *C. suppressalis* generally has two generations a year in temperate regions with the second generation causing more serious damage than the first generation.

C Coleoptera (rice water weevil)

The rice water weevil, *Lissorhoptrus oryzophilus*, is the most important and ubiquitous insect pest of rice in the USA (Bunyarat *et al.*, 1977; Bowling, 1980). Although Kuschel (1951) lists 6 species of *Lissorhoptrus* in the USA, *L. oryzophilus* is the major pest of rice in the southern states expanding to other rice growing regions of the world. It was found in California in 1959 (Lange and Grigarick, 1959) and was accidentally introduced to Japan in 1976 and has spread to other Asian rice-producing countries including Korea and China. The species is bisexual in the southern states but only parthenogenetic females occur in California and Asia.

The adults overwinter in sheltered areas in debris or leaf litter of forests, at the bases of perennial types of grasses, or on elevated areas such as irrigation levees. The overwintering adults may feed on aquatic grasses and sedges following emergence. They fly to the rice fields between April and June. The number of accumulated day degrees needed for muscle development in relation to peak departure flights from overwintering sites has been determined in Arkansas and Louisiana (Morgan *et al.*, 1984). Muscle degeneration occurs after landing in flooded fields of the southern states and California. Oviposition soon follows and may continue for several weeks. Adults feed on the rice leaves during this period, which produces characteristic linear slits on the dorsal surface. The earlier infestation in flooded rice paddies is generally more serious.

Most eggs are inserted individually in the submerged sheath tissue near to and above the crown (Grigarick and Beards, 1965). After hatching in 6–9 days, the legless white larvae mine the stem for a short time (about 1 day) and then move to and complete 4 instars of development in or on the roots. The larva obtains oxygen by tapping roots with its dorsal abdominal tracheal hooks (Way and Bowling, 1991). Pupation occurs in a waterproof, round cell that has an outside coating of mud and is attached to the root. Adults begin emerging from these cells in early to mid-July in California and continue to do so through the rice growing season depending on such factors as the oviposition period, temperature and possibly interrupted feeding on the rice roots. A small proportion of the emerging generation produce eggs in California (Grigarick and Beards, 1965) and Arkansas (Muda *et al.*, 1981), but the economic significance of this "partial" second generation (Tucker, 1912; Webb, 1914) is not known. Under optimal conditions, 4 generations of the rice water weevil can develop in southern Louisiana; however a 2nd, and perhaps a "partial" 3rd generation occurs more frequently. The "last" generation of emerging adults feed, build up fat reserves and fly to overwintering sites in late July and August (Knabke, 1970; Muda *et al.*, 1981).

Extensive adult feeding on the leaves can injure very small seedlings, but root pruning by the larvae causes the major problems. Plants with damaged roots may become stunted and produce less grain because of reduction of the number of tillers and panicles or cause grain maturity to be uneven at the time of harvest. The earlier the root pruning is to the plant growth stages, the greater is the level of injury for a given number of larvae. Distribution of adult activity and resultant injury in Arkansas and California are greater along the levees and field margins (Socksai and

Tugwell, 1978; Lange and Grigarick, 1959) and, in general, are greater in the sections of fields associated with open water. The use of chemical insecticides is the main control tactic for this insect.

3 Entomopathogens of major rice pests

The insect pests are attacked naturally by a number of microbial control agents such as nematodes, fungi, bacteria, and viruses (Table 2). These entomopathogens can reduce pest populations but often they do not occur at a high enough frequency to suppress the population. By understanding the biology and ecology of these entomopathogens, we may be able to use them effectively in the integrated pest management of rice through augmentation or inundative releases.

A Mermithid parasite of planthoppers

More than 200 natural enemies (parasitoids, predators, and entomopathogens) have been recorded from BPH (see Benrey and Lamp, 1994), and most of them have an overlapping host range with WBPH. Although BPH and WBPH have an impressive list of natural enemies, a mermithid nematode parasite, *Agamermis unka*, is the most important and common natural enemy in temperate regions. This unique parasite is widely distributed and commonly found in rice paddies at high densities in the southern part of Korean (Cho *et al.*, 2002). Parasitism is generally high. Over 50% of BPHs (Choo *et al.*, 1989; Choo and Kaya, 1990 1993, 1994; Choo *et al.*, 1995) and WBPH (Choo and Kaya, 1990) were parasitized by *A. unka* in Korea. Esaki and Hashimoto (1931) found that > 40% of BPH and > 70% of WBPH populations were parasitized by *Agamermis* in southern Japan. In Hunan Province, China, no control measures are needed when the population density of BPH is under 2000 insects on 100 rice plants, provided that natural infection rate is above 75% (Wang and Li, 1987). *Agamermis* parasitism castrates the reproductive organs of BPH and WBPH (Choo and Kaya 1990). Parasitism of the host usually occurs at the lower part of the rice stem where most planthoppers are found (Choo and Kaya, 1993).

The newly-hatched second stage mermithid is the infective stage (pre-parasite). Once the mermithid contacts the planthopper nymph, it uses its stylet to penetrate through the cuticle into the host hemocoel to initiate the parasitic phase. The 3rd and 4th stage juveniles occur in the hemocoel. Two to three weeks after parasitization, the 4th stage juvenile (postparasite) exits its adult host by boring through the thin intersegmental area of the abdominal segment, always causing death of the host. After emergence, the postparasite burrows into the soil, molts, and overwinters as an adult (Choo and Kaya, 1993).

Agamermis can be redistributed by artificial releases. When the pre-parasites of *A. unka* were released at a mermithid to BPH ratio of 10:1, parasitism of BPHs ranged from 33 to 63% (Choo *et al.*, 1995). Because egg production and hatchability of *Agamermis* are high, inoculative releases into areas where the mermithid population is low or nonexistent appear feasible.

B Entomopathogenic fungi from planthoppers

A number of entomopathogenic fungi have been isolated from BPH and WBPH (Table 2). In addition, other fungi such as *Beauveria brongniartii* and *Hirsutella saussurei* have been isolated from other planthopper and leafhopper species. Interestingly, several hoppers in the rice ecosystem are infected with *Nomuraea rileyi*, a common fungus of lepidopterous larvae (Li, 1985). Some of these pathogens have been evaluated for the control of planthoppers (Aguda and Litsinger, 1984; Aguda *et al.*, 1987). BPH and WBPH are susceptible to *B. bassiana*, but some isolates are more efficacious than others (Aguda and Litsinger, 1984). *B. bassiana* and *Metarhizium anisopliae* appear to be the most useful against planthoppers because of their ease of mass production, storage, virulence, and ease of application.

C Entomopathogens from lepidopterous species

Nematodes and entomopathogenic fungi, bacteria, and viruses have been isolated from lepidopterous pests (Table 2). *Bacillus thuringiensis* can be used against various lepidopterous pests. In addition, several transgenic rice plants have been developed that

Table 2. Some entomopathogens isolated from or tested against insect pests of rice

Insect species	Entomopathogens	Region	References
Homoptera			
Nephotettix cincticeps	Fungi		
	Entomophthora spp.	Asia	Reissig *et al.* (1986); Rombach *et al.* (1987b)
	Beauveria bassiana	Asia	Reissig *et al.* (1986)
	Nematode		
	Agamermis unka	China	Zhao *et al.* (1987)
Nephotettix virescens	Fungi		
	Entomophthora spp.	Asia	Reissig *et al.* (1986); Rombach *et al.* (1987b)
	B. bassiana	Taiwan	Chu and Hirashima (1981)
	Conidiobolus coronatus	Philippines	Rombach *et al.* (1987b)
	Metarhizium album	Philippines	Rombach *et al.* (1987a)
	Paecilomyces farinosus	Thailand	Rombach *et al.* (1987b)
	Nematode		
	Hexamermis sp.	Asia	Reissig *et al.* (1986)
Nilaparvata lugens	Fungi		
	Erynia delphacis	Japan	Rombach *et al.* (1987b)
	B. bassiana	Asia	Chu and Hirashima (1981) Rombach *et al.* (1987b) Li (1985) Rombach *et al.* (1987a)
	B. brongniartii	Asia	Reissig *et al.* (1986)
	Cephalosporium sp.	India	Rombach *et al.* (1987b)
	C. coronatus	Japan	Rombach *et al.* (1987b)
		Philippines	Rombach *et al.* (1987b)
	Entomophthora spp.	Asia	Rombach *et al.* (1987b)
	Metarhizium anisopliae	Asia	Reissig *et al.* (1986)
	M. flavoviride	Philippines	Rombach *et al.* (1986b)
	Paecilomyces farinosus	Japan	Rombach *et al.* (1987b)
	Nematode		
	A. unka	Japan	Kaburaki and Imamura (1932); Esaki and Hashimoto (1931)
		Korea	Choo *et al.* (1989, 1995); Choo (1991)
		China	Yan *et al.* (1986); Zhao *et al.* (1987)
	Hexamermis sp.	India	Manjunath (1978)
	Mermis sp.	India	Manjunath (1978)
Sogatella furcifera	Fungi		
	B. bassiana	China	Li (1985)
	B. brongniartii	Asia	Reissig *et al.* (1986)
	Entomophthora sp.	Fiji island	Rombach *et al.* (1987b)
	Hirsutella citriformis	Rep. Solomon Islands	Rombach *et al.* (1987b)
	Nematode		
	A. unka	Korea	Choo and Kaya (1990)
		Japan	Esaki and Hashimoto (1931) Kaburaki and Imamura (1932)
		China	Zhao *et al.* (1987)

(Continued)

Table 2. (Continued)

Insect species	Entomopathogens	Region	References
Coleoptera			
Lissorhoptrus oryzophilus	Fungi		
	B. bassiana	USA	Rice *et al.* (1994) Urtz and Rice (1997)
Lepidoptera			
Scirpophaga incertulas	Fungus		
	B. bassiana	Asia	Reissig *et al.* (1986)
	Nematode		
	Hexamermis cathetospicula	Malaysia	Poinar and Chang (1985)
Chilo suppressalis	Fungi		
	B. bassiana	Asia	Reissig *et al.* (1986)
	M. anisopliae	Philippines	Rombach *et al.* (1987b)
	Paecilomyces farinosus	Japan	Rombach *et al.* (1987b)
	Nematode		
	Amphimermis zuimushi	Japan	Kaburaki and Imamura (1932)
	Virus		
	Chilo iridovirus	Japan	Rombach *et al.* (1987b)
Cnaphalocrocis medinalis	Fungus		
	B. bassiana	India China	Rombach *et al.* (1987b) Li (1985)
	Virus		
	Granulovirus	India	Rombach *et al.* (1987b)

express *B. thuringiensis* toxin genes for insect control. These include *B. thuringiensis cryIA(b)* and *cryIA(c)* genes conferring resistance to the yellow stem borer (Nayak *et al.*, 1997, Cheng *et al.*, 1998).

D Entomopathogens of the rice water weevil

The role of natural entomopathogenic enemies in the suppression of the rice water weevil has not been determined (Reissig *et al.*, 1986). However, an unidentified mermithid nematode was isolated from the larvae (Bunyarat *et al.*, 1977). Weevils collected in mid-June had the highest parasitization rate (26%), with peak nematode emergence occurring during the last week in June. The mermithid decreased egg production of weevils. In a laboratory study, the entomopathogenic nematode, *Steinernema carpocapsae* Mexican strain, was applied against the weevil, but this nematode did not reduce weevil population on the roots even at 120,000 infective juveniles/liter/pot (Nagata, 1987).

Recently, the potential role that entomopathogenic bacteria and fungi may play in controlling the rice water weevil has been investigated. Both *B. thuringiensis* and *B. bassiana* isolates found in the rice ecosystem or obtained from other sources were pathogenic to this weevil.

A systematic survey of suspected overwintering rice water weevil sites produced 17 potential *B. bassiana* isolates from diapausing and nondiapausing weevils (Rice *et al.*, 1994, Urtz and Rice, 1997). The methodologies developed during this survey are useful for isolating various fungal pathogens. This involves breaking the diapause of overwintering adult rice water weevil by submerging leaf litter in 36 °C water for a minimum of 4 h to a maximum of 18 h.

Protocol for isolation of fungi pathogenic for rice water weevil.

1. Collect 4 (1 m^2) samples of leaf litter by raking until the soil is exposed.
2. Transfer the leaf litter and debris to individual 151 liter rubber garbage cans,
3. Add tap water at 36 °C until the litter is submerged. Place on the top of the leaf litter an aluminum trash can lid fitted with a wire screen (40 cm diameter with 1 cm space between wires) into a hole cut in the center of the lid to keep the litter submerged. Place an inverted funnel on the trash can lid over the screen so that the dormant rice water weevil adults float on the surface of the water.
4. Wait 4 h and then collect the viable adult rice water weevil that climb to the top of the funnel. Adult rice water weevil may be collected for up to 18 h.
5. Incubate adult weevils in glass jars (30–50/liter jar) at 25 °C containing paper tissue saturated with sterile H_2O, and periodically (2–3 day intervals) examine the jar for cadavers.
6. Remove the cadavers, suspend them individually in a test tube in 0.1% Tween 80, vortex vigorously, and plate 0.1 ml on an oatmeal-dodine-chlorotetracycline agar (ODTA), a medium selective for *Beauveria* species (Chase *et al.*, 1986).
7. Incubate ODTA plates at 25 °C until distinct colonies appear that are 2–5 mm in diameter.
8. Transfer subsamples of individual colonies to Sabouraud dextrose agar (SDA) plates containing 0.2% yeast extract.
9. Identify *B. bassiana* isolates based on colony appearance, the ability to grow on ODTA plates, and conidial morphology (Mugnai *et al.*, 1989). Koch's postulate needs to be conducted to verify the pathogenicity of the fungal isolates to the rice water weevil.

Other cadavers can be dissected and examined for other potential pathogens such as microsporidia or bacteria. An individual experienced with microsporidian spores can identify them with a compound microscope at 400 X. The spores can be partially purified by centrifugation and used to confirm their pathogenicity using Koch's postulates. The bacterial populations are more difficult to assess and will require isolation of individual colonies on bacteriological media selective for *B. thuringiensis*. Several types of media used in selective isolation strategies for *B. thuringiensis* are acetate (inhibiting the germination of *B. thuringiensis* spores) in conjunction with heat (Travers *et al.*, 1987), and antibiotic selection (Delucca *et al.*, 1981) in conjunction with media selective for *B. cereus* (Sekar *et al.*, 1997). Unless the individual is qualified, this isolation may be best done in conjunction with a bacteriologist.

In addition to isolating pathogens from rice water weevils, pathogens that are being developed for microbial control of other insects may be used for rice water weevil control. For example, a novel *B. thuringiensis* vegetative insecticidal protein, vip3A(a), with a wide spectrum of activities has been described (Estruch *et al.*, 1996). Vip3A(a) and its homologue vip3A(b) show no homology to the Cry class of insecticidal toxin but appear to have a similar mode of action as observed on the midgut of lepidopterous insects.

B. thuringiensis spore crystal-toxin mixture(s) of selected strains containing potentially coleopteran active Cry toxins was produced using Schaeffer sporulation media (Schaeffer *et al.*, 1965) and evaluated in greenhouse and field trials. Results of a greenhouse trial involving spore crystal-toxin mixtures produced from various *B. thuringiensis* strains are shown in Table 3. This study compared several *B. thuringiensis* strains that contained different cry toxin genes (based on PCR analysis). The experimental protocol was conducted as described (see greenhouse assay) with rice water weevil feeding scars (no. of feeding scars on newest leaf) and larval densities being recorded to measure effectiveness of treatments. With respect to feeding scars, a significant reduction was observed with strains 4AC1 and 4H2 on the 1st sample date, while on the 2nd date (5 and 11 days following infestation with two adult rice water weevil per plant) a reduction was observed with 4H2 and HD201. However, rice water weevil feeding activity is not always correlated with rice water weevil larval densities. Application of strain 4G1 resulted in the greatest reduction of rice water weevil larvae on both sample dates (21 and 28 days post-permanent flood). These results suggest that there may be differential activity of *B. thuringiensis* toxins against adult versus larval rice water weevil.

Table 3. The effect of *Bacillus thuringiensis* (*Bt*) strains after treatment on rice plants on leaf scaring by adult rice water weevil (RWW) and number of larvae/plant on two different sample dates

Bt strains	Leaf scars		RWW larvae/plant	
	1st date	2nd date	1st date	2nd date
4AC1	9.1	5.4	22.5	35.5
4G1	15.7	4.3	0	0
4H2	10.1	0.5	8.5	11.5
HD201	18.6	2.1	4	10.5
HD537	20.6	4.5	9.5	5
Control	19.4	8.2	7	13

4 Protocols for studying, applying and evaluating selected entomopathogens

A Agamermis unka

1 Naturally occurring Agamermis *in BPH and WBPH populations*

BPH and WBPH generally feed on the lower part of the rice stem close to the water surface. Thus, sweeping using a general-purpose net is not efficient for these insects. Instead, beating with hand or sucking with an aspirator or D-Vac machine is recommended for BPH and WBPH sampling.

To evaluate parasitism of *A. unka*, one approach is to transplant or seed BPH and WBPH susceptible rice cultivars at intervals of 20 × 20 cm between hills in a 662 m field. The paddy is divided into three main sampling plots of the same size. The plot is bordered by 5 rows of rice hills between plots. Three to five subsampling sites within each plot are randomly chosen to collect insects. The two or three rows of rice plants along the paddy banks should be excluded from the study to avoid edge effects. The size of the subsampling sites is 2 × 2 m with 3 rows of rice hill border.

1. Collect insects randomly from within a plot with a mouth aspirator for 20 min and place them into a plastic container (*e.g.*, 10 cm diameter × 6 cm height) containing two or three rice seedlings. The top of the container should be covered with a gauze or screen and provisions made to place the insects into the container through an aperture in the gauze or screen. The aperture can be plugged with cotton.
2. Label each container and transport to the laboratory in a cooler.
3. Segregate the insects by planthopper species, sex and developmental stage in the laboratory and dissect them individually in Ringer's solution to determine *Agamermis* parasitism.

A second approach is to collect BPH and WBPH with a mouth aspirator from 50 rice plants from a farmer's field. The sampling site can be adjusted according to shape or size of the paddy field. A third method is to use an insect-collecting plate (*e.g.*, 23 cm × 30 cm plastic or wooden plate) that is painted black or white to obtain BPH and WBPH. Place the plate at a right angle to the bottom of the rice plant and shake the plant three to four times. Insects that drop onto the plate are knocked into a container (*e.g.*, 20 × 15 × 10 cm) that includes rice seedlings. Thirty to 50 rice plants are shaken to obtain a representative sample of insects. This method will collect all stages except the macropterous adults that may disperse before they can be placed into the cage. A fourth method described by Pea and Shepard (1986) sampled BPH using a D-Vac machine to determine parasitoid parasitism in farmers' rice fields. They obtained planthoppers and leafhoppers by sampling for 1 min at each of five to seven sites at each location.

2 Evaluation of Agamermis *against planthoppers in the laboratory*

1. Rear BPH or WBPH continuously on susceptible rice seedlings (*e.g.*, cv. Hwaseong or Chucheong or other suitable rice cultivars) in screened cages (*e.g.*, 30 × 25 × 30 cm) in an enclosed room or other suitable sites at a 15 h light: 9 h dark cycle. Add new rice seedlings into the cages as needed. Augment laboratory colony with field-collected adults every year.
2. Select 100 BPH or WBPH in the second and third instars from the colony and release them into a plastic box (40 × 30 × 20 cm) containing 45-day-old rice seedlings susceptible to the planthoppers. BPH-susceptible rice cultivars are recommended because they reduce insect mortality. Maintain the water level in the plastic box at about 3 cm in depth.

3. Place screen over the plastic box by inserting a 30 cm high wooden stake at each corner and the center of the box. Secure the bottom of the screen to the box with string or wire.
4. Add *A. unka* to the water at a ratio of 10 pre-parasites:1 BPH or WBPH.
5. Segregate the leafhoppers 10 to 13 days later to males and females and macropterous and brachypterous adults and dissect to determine the presence of the mermithid. Record the number of mermithids per host as more than one may occur in a host. Three or four replicates are sufficient to evaluate mermithid parasitism (Choo *et al.*, 1995).

3 *Evaluation of* Agamermis *against planthoppers in the field*

There are two approaches to assess *Agamermis* in the field. One approach is to select rice paddies that are free of this mermithid and add pre-parasites into the plots. The second method is to use plots that have an established population of the mermithid. The protocol below is to assess the impact of the mermithid in areas where it does not occur.

1. Transplant 30-day-old rice seedlings, susceptible to planthoppers, in the rice paddy free of *Agamermis* (240 m^2). Susceptible rice varieties should be used because they are better for the reproduction of BPH and WBPH.
2. Divide the paddy into three plots and place 1.2 × 1.2 × 1.2 m screened cages in each plot with three or four replicates. Prepare a temporary levee with paddy soil or a plastic panel. This will keep the applied mermithids (see next step) in the plots.
3. Apply a ratio of 100 to 200 *Agamermis* pre-parasite: 1 BPH or WBPH. (This rate is recommended over the 10:1 laboratory ratio because a higher ratio is needed for field parasitism.)
4. Place 30 pairs of BPH or WBPH adults into each cage for 45 days.
5. Sample the planthopper progeny 25 days later using an aspirator. Segregate the planthoppers into males and females and brachypterous and macropterous adults and categorize the nymphs into their respective developmental stage.
6. Dissect the planthoppers to determine mermithid parasitism (Choo, 1991)

4 *Sampling for adult* Agamermis *in soil*

The presence and the population density of *Agamermis* adults can be determined when the rice fields are fallow or standing water is absent. If the impact of a natural population of this mermithid is to be determined, the above protocol (*i.e.*, *Evaluation of* Agamermis *against planthoppers in the field*) can be followed with slight modifications. The area selected will have the mermithid present and the temporary levees are not needed. The following protocol can be used to determine the density of the mermithid.

1. Dig a soil sample, 20 × 20 × 20 cm, with a shovel to confirm the presence of the mermithid (Choo *et al.*, 1995).
2. Break the soil sample with your hands into smaller pieces and pick up *Agamermis* adults with your hands or an applicator stick because they are large (28.3 × 18.8 mm in female and 16.3 × 9.1 mm in male) and can be seen with the naked eye.
2. Count the number of adults per sample
3. Take several (15 to 20) soil samples from various locations in the rice paddy to obtain the number of nematodes per cm^3.
4. Return mermithids to the soil or collect them for laboratory studies.

Collected nematodes are placed in a plastic container (*e.g.*, 10 cm diameter × 6 cm height or other suitable containers) with field soil to avoid desiccation and transported in an ice chest to the laboratory. In the laboratory, they may be placed individually or in a group in a Petri dish containing distilled water (Choo *et al.*, 1995). They may be stored at 4 °C until they are needed. Nematodes stored for 4 months will produce viable eggs (Choo, unpublished data); when stored for 12 months, they are alive and vigorous but egg production cannot be guaranteed. Eggs can be collected daily or every other day depending upon the purpose of the study.

5 *Sampling for pre-parasite* Agamermis *in the field*

a. *Sampling from rice stems*

1. Transplant a planthopper-susceptible rice cultivar in the field as above and delineate

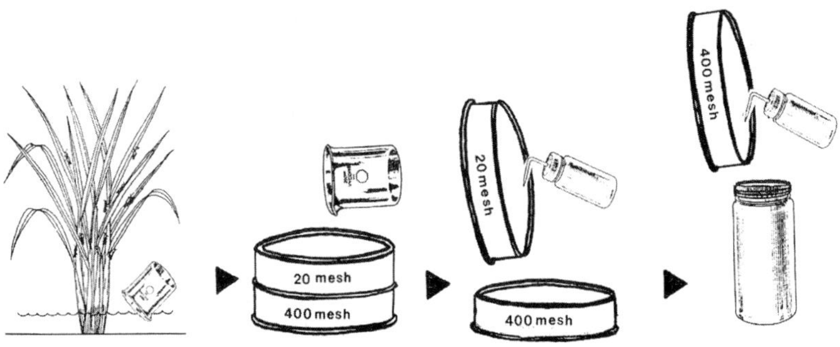

Figure 1. Method for sampling *Agamermis unka* pre-parasites from a flooded rice field.

2.5 × 2.5 m plots with a 0.5 m border between plots to determine pre-parasite activity and behavior in the rice paddy.

2. Follow normal rice cultural techniques except that no insecticides or nematicides are used.
3. Examine the rice stems for presence of the pre-parasites by cutting 10 stems just above the water level (> 1 cm above), and rinse in 60 ml of water in an 80 ml (or similar size) container for 1 min. This can be done at various time during the rice growing season.
4. Place the containers in an ice chest for transport to the laboratory.
5. In the laboratory, pour 10 ml of water into a Petri dish (100 × 15 mm) and count the number of pre-parasite under the microscope. This procedure is repeated until the entire water sample is examined for pre-parasites.

b. Sampling from the water

1. Obtain the density of pre-parasites in a flooded rice field by scooping rice field water from a 2.5 × 2.5 m plot into a 300 ml beaker and pouring the water sample through a 20 mesh (850 μm) and a 400 mesh (38 μm) sieve.
2. Rinse the 400 mesh sieve with distilled water several times; then rinse the sieve material into a 100 ml bottle (Figure 1).
3. Take 10 samples from each 2.5 × 2.5 m plot.
4. Place the 100 ml bottles into an ice chest and transport to the laboratory.
5. Pour the water into a Petri dish and count the number of pre-parasites.

In an unflooded field, a 30 × 25 cm pool is made between the rice plants, and water is collected and scooped into a beaker. In addition, 300 ml beaker is inserted at the level of soil surface until sufficient water is collected. The beaker is removed and the water is processed as above from Section IVA5b, step 1.

6 *Sampling time for* Agamermis

The sampling time for each developmental stage of *Agamermis* is different. In temperate regions, adults are sampled after rice harvest in September or October until planting in May the following year. Pre-parasite sampling is done during the rice growing season from July to August. In tropical regions, adult samples should be taken during periods when no water is present in the paddy and pre-parasite samples taken when water is present as well as planthopper hosts.

B Beauveria *and* Metarhizium *spp.*

Conidial suspensions of both entomopathogenic fungi have been used against a limited number of rice insect pests. The target pest species and its distribution in the field will affect how the plot size is determined. Other considerations for determining the size of the plots include fungal species, the number of replicates, paddy size, and availability of labor. Plots 50 m^2 in size is recommended for evaluation of low mobility pests, but plots 200 or more m^2 is required for highly mobile insects such as BPH or leafhoppers. In addition, the spray equipment and application time are dependent upon the target pest species. The fungal rate required to effect a desired mortality is a function of the susceptibility of the host and the virulence of the microorganisms (Reichelderfer, 1985).

1 *Protocol and evaluation of entomopathogenic fungi against planthoppers*

a. Pot assay in small field cages

1. Plant three rice seedlings susceptible to planthoppers in a suitable pot and maintain in the greenhouse in water. (The number of rice seedlings may vary according to the size of the pot.)
2. Cover each pot with a screen cage.
3. Release 100 third instar BPH nymphs on the plants and allow them to acclimatize for 1 day.
4. Place the pots into the rice paddy plots with a 0.3 m border between pots with plants of similar growth age. The pots can be placed into a random design or a random block design depending upon the experiment and size of the rice paddy.
5. Apply control and fungal treatments with a hand sprayer until runoff.
6. Take live insect counts every 3 to 4 days after treatment up to 21 days; if possible, collect dead insects and examine for fungal infection.
7. For multiple treatments, apply second or third treatment at 7 to 10 day intervals and take counts as in step 6 above.

b. Field treatment

A randomized block design with three or four replicates are usually employed for the field evaluation of entomopathogenic fungi for BPH or WBPH control.

1. Establish a plot size of 2×2 m (18–20 hills) and plant susceptible rice cultivars with a 20×20 cm spacing between rice hills. There should be a minimum of four replicates for each treatment.
2. One day before treatment, enclose each plot with a screen cage to prevent fungus-infected planthoppers from moving to neighboring plots (Aguda *et al.*, 1987).
3. Apply the conidial suspension at a rate of $4.0 \sim 5.0 \times 10^{12}$ conidia/ha (Rombach *et al.*, 1986b) to 7.5×10^{12} conidia/ha (Aguda *et al.*, 1987). The concentrations can be varied depending on the fungus or field situations. Dry mycelia of *B. bassiana* can be also applied at rates equivalent to 200g/ha in a formulation with 5% Liqua Gel®.
4. Three hundred ml of suspension is applied per plot by ultra low volume applicator. The nozzle of the applicator is adapted for the mycelium size and application is made at panicle stage of the rice.
5. Live BPHs from 12 randomly selected hills in each cage are counted because infected insects fall into the water and are difficult to find (Aguda *et al.*, 1987).

2 *Protocol and evaluation of entomopathogenic fungi against rice water weevil*

Adult rice water weevil can be bioassayed using one of two methods; (1) a test tube or submersion bioassay and (2) a droplet bioassay. Although this is a laboratory procedure, the rice water weevil is difficult to assess in the field, and more controlled tests against laboratory and/or greenhouse procedure (see protocol outlined for greenhouse assay below in Section 4C2) may be needed before field tests are done.

a. Submersion bioassay

1. Collect either over-wintering adults or adults from rice fields.
2. Place 10 weevils in a 15 ml polypropylene test tube containing a 1 ml suspension of conidia (10^4 to 10^8 conidia/ml).
3. Gently vortex for 5 sec and incubate for 30 to 60 sec.
4. Transfer weevils to a 100 mm Petri dish containing a moist filter paper (sterile H_2O) and rice leaves for 1–20 days. Replace with fresh rice leaves every 2–3 days.
5. Transfer Petri dishes containing inoculated weevils to a partially closed plastic container containing H_2O saturated paper towels and incubate at 28 °C. Re-wet filters as necessary to maintain a humid environment. Alternatively, transfer individual adult weevils to a test tube (1.8×10.5 cm) that has a rice seedling and assess mortality for the next 20 days. Replace with fresh rice seedling as needed.
6. Record adult mortality at 3, 5, 7, 10, 14 and 20 days following treatment.

Special precautions are required to record mortality accurately. Mortality is determined by positioning adult weevils on their backs. After a 10 min interval, viable adult weevils will right themselves. At this time an accurate viability determination can be made. Observation of fungus sporulating on or dissection of cadavers can confirm mycosis.

b. Droplet bioassay

1. Transfer 10 adult rice water weevils/sterile 100 mm Petri dish containing a pre-wetted filter paper (1 ml sterile H_2O).
2. Dilute conidia serially to the appropriate suspension in 0.1% Tween 80 to prevent aggregation.
3. Place a 10 μl drop of the desired dose to the dorsal side of the adult rice water weevil using a pipetman.
4. Follow steps 4–6 described above in the "Submersion bioassay."

C *Bacillus thuringiensis assay against rice water weevil*

Two types of assays (leaf dip and greenhouse) can be used to evaluate *Bacillus thuringiensis* suspensions against rice water weevil adults.

1 *Leaf dip bioassay*

1. Streak a fresh nutrient agar plate from an appropriate culture of *B. thuringiensis* strain and incubate at 30 °C overnight.
2. Transfer a single colony to 1 ml of LB Medium (Difco) and incubate at 30 °C overnight.
3. Inoculate 20 ml of Schaeffer sporulation media in a 250 ml Erlenmeyer flask with 0.02 ml of an overnight *B. thuringiensis* culture and incubate at 30 °C for 5 days.
4. Harvest the spore crystal-toxin suspension by centrifugation at 5,000 rpm for 10 min.
5. Resuspend the spore crystal-toxin pellet in 10 ml of TNT (20mM Tris, 10mM NaCl, 0.05% Triton X-100); wash by centrifugation 2 times with TNT and resuspend in 5 ml of TNT.
6. Determine the spore crystal-toxin concentration using a hemocytometer as described by Klein (1997). Spore counts may also be made using the dilution and plating method described by Thiery and Frachon (1997).
7. Dip the rice leaves from young plants (at the 4–6 leaf stage of growth) into *B. thuringiensis* spore crystal-toxin suspension containing 0.04% Sil-Wet 77 or other compatible wetting agent (aids in dispersing and attaching the spore crystal-toxin suspension on the rice leaves).
8. Air-dry the rice leaves and transfer to bioassay dishes as previously described except that the filter paper is not saturated with water.
9. Remove old treated leaves and replace with freshly-treated leaves every 3–4 days.
10. Record mortality at 3, 5, 7, 10 and 14 days after initial treatment.

2 *Greenhouse assay*

This protocol is used for *B. thuringiensis* but may also be used for testing conidial suspensions of fungal pathogens such as *B. bassiana* or *M. anisopliae*.

1. Seed pots (10 cm × 15cm height) with 6–10 rice seeds per pot that contain an equal mixture of potting soil and topsoil. Rice is grown in small pots in greenhouse conditions at 14 h light (30 °C):10 h dark (22 °C)
2. At the 2–3 leaf stage, thin to 3 plants/pot to achieve uniform plant size and maturity.
3. Transfer 10 pots each into aquariums (45 cm wide × 60 cm long × 45 cm height).
4. At this time, apply a *B. thuringiensis* spore crystal-toxin mixture to the rice foliage and soil surface using a CO_2 powered sprayer fitted with a single 8015VS nozzle at a rate of 5×10^{12} spore crystal-toxin particles/ha.
5. Add rice water weevil to the aquariums at a density of 50 adult weevils/aquarium. Adult weevil may be collected from either overwintering sites or from field plots
6. Immediately cover the aquarium with a cage (made with a wood frame and gauze mesh) of equal size to that of the aquarium to prevent the adult weevils from escaping.
7. Make a second treatment application 3–4 days following the initial application.
8. Establish a permanent flood in the aquariums 1–2 days following the 2nd pre-flood application and follow with 1–2 more post-flood treatment(s) of the biological agent.
9. Check for rice water weevil mortality at 14 days after the second treatment.

5 Field experimentation for rice water weevil

A randomized complete block plot design is generally used for field trials evaluating the effectiveness of fungi (*B. bassiana* or *M. anisopliae*) or bacteria (*B. thuringiensis*) for their potential for rice water weevil control. Generally, there are 4–6 replications per treatment with 3 to 5 subsamples obtained per treatment. Treatments (1–2) are made prior to and after the establishment of the permanent flood. Sampling may involve the counting of adults (number of rice water weevil adults within a 0.093 m^2 ring), adult weevil leaf feeding scars (on newly-emerged leaf), and rice water weevil egg and/or larval samples. Rice plant responses to treatments are evaluated by measuring root mass (volume and dry weight) and root length, plant height, plant maturity, and grain yield.

A *Preparation of soil bed – water seeded or drill seeded rice*

1. Till soil using conventional tillage practices to prepare a fine seed bed. Rice plant health is assured by a pre-plant incorporation of nitrogen-phosphorus-potash at a rate of 112-67-67 kg/ha.
2. Water seed the rice crop for experiments to be planted early in the growing season and drill seed the rice crop for planting later in the growing season. Draw the levees around the bays (paddies containing the experimental plots) to allow for flooding of the rice plots.
3. Soak the rice seed for 24 h in 2–3 volumes of water (non-chlorinated), drained, and allowed to set for another 24 h at ca 30 °C. Examine the rice seed for the presence of pips (sprouting).
4. Broadcast by hand the pre-germinated seed into the previously flooded designated rice plots. The rice seed is allowed to develop for 3–5 days (dependent on temperature) under flood or until radicle and coleoptile have developed sufficiently. Drain the water and allow the rice plant to peg (for the root to firmly attach to the soil) for 2–3 days after which a shallow permanent flood is applied.
5. Prior to application of the permanent flood, apply a top dress of N (34–67 kg/ha).
6. Follow the protocol outlined below (*i.e.*, section 5B) for treating rice water weevil adults.

For drill-seeded rice, use conventional drills to deliver the seed to a depth of 1–2 cm. If the soil contains enough moisture the seed will germinate. Alternatively a flush (temporary flood) can be applied to cause uniform plant germination. The desired outcome in either case is to achieve uniformity in emergence of the rice plant.

B *Transplanted rice cultivation*

1. Transplant rice with a 20 × 20 cm spacing between rice hills. The plot size is 2 × 2 m (18 ~ 20 hills) and is blocked to avoid contamination of the entomopathogenic fungus to control and/or other plots.
2. Enclose the plot with screen or gauze to prevent the overwintering adults from moving to neighboring plots. The transplanting time of rice, adult activity, oviposition, adult longevity, persistence of pathogens, and fungus species determine application time.
3. Apply the fungus once during the peak time of oviposition at a given concentration of *B. bassiana* or *M. anisopliae* evenly during the evening on a windless day.
4. Collect weevil adults 7 and 14 days after treatment and examine for fungus infection. A randomized block design with three replicates can be employed for the field evaluation of entomopathogenic fungi for rice water weevil. For the first generation adult of rice water weevil, entomopathogenic fungi mixed with white carbon are applied with a duster at the beginning of adult occurrence. The adults are collected with a sweep net for 3 weeks at 2- to 5-day intervals and fungus infection determined.

C *Pre- and post-treatment sampling*

Before treatment, the rice fields should be sampled for rice water weevil adults directly or indirectly. In the indirect method, the presence of adult weevils can be assumed by the presence of feeding scars on the leaves. For the direct method, sampling rings (0.093 m^2) can be placed in the rice plots after the establishment of the permanent flood. A treatment threshold has been reached if there are 1 or more adult rice water weevils per sample ring (2–3 subsamples) per plot. Weevil densities should be checked 2 days before treatment and from 7 days after treatment and at 3- or 4-day intervals after that. Dead insects can be collected with forceps along the rows by zigzagging and making 10 rounds. Thus, insect density and pathogen infection can be determined.

D Inoculum

Inoculum may consist of fungal conidial powders or formulated *B. thuringiensis*. Dry, powdered conidia from various fungal isolates containing approximately 4×10^{10} conidia/g have been formulated in either water or oil-based suspensions by Mycotech Corporation (Butte, MT) using a proprietary novel semi-dry fermentation system. Selected strains of *B. thuringiensis* containing potentially active coleopteran Cry toxins (based on PCR analysis, *e.g.*, cry3Aa1, cry3Aa2, cry1Ia1, etc.) can be produced using Schaeffer sporulation media (Schaeffer *et al.*, 1965) as previously described. A wetting agent such as Sil-Wet 77 is needed to suspend either the conidia or the bacteria for wettable suspensions. Alternatively a 5% crop oil-water suspension can be applied for the conidia or *B. thuringiensis*.

E Application methodology

Application of various conidial or *B. thuringiensis* suspensions can be made using CO_2 powered backpack sprayers at a rate of 140 liters/ha using VS110015 nozzles (R & D Sprayers, Opelousas, LA) or similar spray equipment. Conidia or *B. thuringiensis* can be applied at a rate of 6.2×10^{12} particles/ha (particles are either the number of conidia or number of spore crystal-toxin particles). At the time applications are made, the leaf area index of the rice crop needs to be determined. The leaf area index will rapidly increase in value at this stage of rice plant growth and thus indicates the relative amount of rice leaf foliage to which applications are being made as opposed to soil or water if the field is flooded. Generally, 1–2 pre-flood and 1–2 postflood applications of the biological control agent should be made due to the life cycle of the rice water weevil.

F Treatment assessment – efficacy

Several methods (direct and indirect) can be used to assess the effectiveness of conidial formulations in controlling adult rice water weevil populations. Immediately following application of treatments, adult rice water weevils are collected from treated and control plots and transferred to Petri dishes with food. They are held for 14 days, checked for mortality every other day, and food added as needed. In this way a determination of initial contact of the pathogen and insect can be made. At various times following application (generally at 12 h intervals), adult rice water weevils are collected and bioassayed as described above. To assess the potential of leaf surface contact for infecting rice water weevil; leaves from control and treated plots can be collected at random from experimental plots (5 subsamples per plot) and placed in Petri dishes. Adult rice water weevils are collected from field plots several hundred meters away from the experimental plots and transferred to the bioassay plates. Rice water weevil mortality is determined as previously described. Rice water weevil egg densities are obtained in a time period of 5–9 days following establishment of the permanent flood, while rice water weevil larvae are sampled over a 1-month period starting approximately 3 weeks following establishment of the permanent flood

G Treatment assessment – persistence

Leaf samples can be collected at random in experimental field plots and transferred to the lab at the time of collection. Subsequently, the center of the leaves can be cut into $^1/_2$ cm length and placed into 5 ml of an extraction buffer (phosphate buffered saline pH 6.7) in 20 ml screw-capped vials. The sample vials are incubated at 27 °C for 2 h on an orbital shaker rotating at 250 RPM. An aliquot is removed from each vial, diluted appropriately and plated on ODTA plates for fungal conidia or Schaeffer's medium for *B. thuringiensis*. After incubation for 7 to 9 days for conidia and 2 to 3 days *for B. thuringiensis* at 27 °C, the colony forming counts (CFU) are recorded. The total numbers of CFU/ml are determined and from the leaf surface area of the samples, a determination of the number of CFU/cm^2 of leaf surface can be made.

H Environmental data

1. Obtain environmental information (abiotic factors) to aid in assessing the effectiveness of entomopathogens in field trials against rice insect

pests. The relevant environmental data that are obtained are temperature, relative humidity, solar radiation, and rainfall.

2. Obtain the information from a weather station located from a nearby weather station or make provisions for weather data to be collected on site. From this information, it is possible to determine the influence of abiotic factors on the persistence and stability of entomopathogens applied to rice plants.

6 References

Aguda, R. M. and Litsinger, J. A. 1984. Pathogenicity of *Beauveria bassiana* on brown planthopper (BPH), whitebacked planthopper (WBPH), and green leafhopper (GLH). *Intern. Rice Res. Newsl.* **9**, 20.

Aguda, R. M., Rombach, M. C., In, D. J., and Shepard, B. M. 1987. Suppession of populations of the brown planthopper, *Nilaparvata lugens* (Stål) (Hom.; Delphacidae) in field cages by entomogenous fungi (Deuteromycotina) on rice in Korea. *J. Appl. Entomol.* **104**, 167–172.

Benrey, B. and Lamp, W. O. 1994. Biological control in the management of planthopper populations. *In* "Planthoppers Their Ecology and Management, Part Five Ecological Approaches to Planthopper Management." (R. F. Denno and T. J. Perfect, Eds.), pp. 519–550. Chapman & Hall, New York.

Bowling, C. C. 1980. Insect pests of the rice plant. *In* "Rice: Production and Utilization" (B. S. Luh, Ed.), pp. 260–288. AVI Publishing Co., Westport, CT.

Chase, A. R., Osborne, L. S., and Ferguson, V. M. 1986. Selective isolation of the entomopathogenic fungi *Beauveria bassiana* and *Metarhizium anisopliae* from an artificial potting medium. *Fla. Entomol.* **69**, 285–292.

Bunyarat, M., Tugwell, P., and Riggs, R. D. 1977. Seasonal incidence and effect of a mermithid parasite on the mortality and egg production of the rice water weevil, *Lissorhoptrus oryzophilus. Environ. Entomol.* **6**, 712–714.

Cheng, X., Sardana, R., Kaplan, H., and Altosaar, I. 1998. Agrobacterium-transformed rice plants expresssing synthetic cryIA(b) and cryIA(c) genes are highly toxic to striped stem borer and yellow stem borer. *Proc. Natl. Acad. Sci. USA* **95**, 2767–2772.

Cho, S. R., Lee, D. W., Choo, H. Y., and Kim, H. H. 2002. Distribution of *Agamermis unka* (Nematoda: Mermithidae), a mermithid parasite of brown planthopper (*Nilaparvata lugens*) in Korean rice paddies. Korean *J. Appl. Entomol.* **41**, 129–136. (In Korean with English summary)

Choo, H. Y. 1991. Ecological study of *Agamermis unka*, a nematode parasite of brown planthopper *Nilaparvata lugens* (Homoptera: Delphacidae). *Res. Rept. Rural Development* Agency **33**, 185–191 (In Korean with English summary)

Choo, H. Y. and Kaya, H. K. 1990. Parasitism of brown planthopper and whitebacked planthopper by *Agamermis unka* in Korea. *J. Nematol.* **22**, 513–517.

Choo, H. Y. and Kaya, H. K. 1993. Parasitism of the brown planthopper by the mermithid nematode *Agamermis* in rice *In* "Nematodes and the Biological Control of Insect Pests." (R. Bedding, R. Akhurst, and H. Kaya, Eds.), pp. 21–26. CSIRO, East Melbourne, Australia.

Choo, H. Y. and Kaya. H. K. 1994. Biological control of the brown planthopper by a mermithid nematode. Korean J. *Appl. Entomol* **33**, 207–215.

Choo, H. Y., Kaya, H. K., and Kim, H. H. 1995. Biological studies on *Agamermis unka* (Nematoda: Mermithidae), a parasite of the brown planthopper *Nilaparvata lugens. Biocontrol Sci. Technol.* **5**, 209–223.

Choo, H. Y., Kaya, H. K., and Kim, J. B. 1989. *Agamermis unka* (Mermithidae) parasitism of *Nilaparvata lugens* in rice fields in Korea. *J. Nematol.* **21**, 254–259.

Choo, H. Y., Kaya, H. K., Kim, J. B., and Park, Y. D. 1991. Evaluation of entomopathogenic nematodes, *Steinernema carpocapsae* (Steinernematidae) and *Heterorhabditis bacteriophora* (Heterorhabditidae) against rice stem borer *Chilo suppressalis* (Walker) (Lepidoptera: Pyralidae). *Korean J. Appl. Entomol* **30**, 50–53. (In Korean with English summary)

Chu, Y. I. and Hirashima, Y. 1981. Survey of Taiwanese literature on the natural enemies of rice leafhoppers and planthoppers. *Esakia Occas. Pap. Hikosan Biol. Lab. Entomol.* **16**, 33–37.

Delucca, A. J., Simonson, J. G., and Larson, A. D. 1981. *Bacillus thuringiensis* distribution in soils of the United States. *Can. J. Microbiol.* **27**, 865–870.

Elliott, N. C., Hein, G. L., and Shepard, B. M. 1994. Sampling arthropod pests of wheat and rice. *In* "Handbook of Sampling Methods for Arthropods in Agriculture" (L. P. Pedigo and G. D. Buntin, Eds.), pp. 627–666. CRC Press, Boca Raton, FL.

Esaki, T. and Hashimoto, S. 1931. Report on the leafhoppers injurious to the rice plant and their natural enemies. Entomological Laboratory, Department of Agriculture, Kyushu Imperial University Publication **2**, 54–59. (In Japanese).

Estruch, J. J., Warren, B. W., Mullins, M. A., Nye, G. J., Craig, J. A., and Koziel, M. G. 1996. Vip3A, a novel *Bacillus thuringiensis* vegetative insecticidal protein with a wide spectrum of activities against lepidopterous insects. *Proc. Natl. Acad. Sci. USA* **93**, 5389–5394.

Grigarick, A. A. and Beards, G. W. 1965. Ovipositional habits of rice water weevil in California as related to a greenhouse evaluation of seed treatments. *J. Econ. Entomol.* **58**,1053–1056.

Kaburaki, T. and Imamura, S. 1932. Mermithid-worm parasitic in leafhoppers, with notes on its life history and habits. *Proc. Imperial Acad.* **8**, 139–141.

Klein, M. G. 1997. Bacteria of soil-inhabiting insects. *In* "Manual of Techniques in Insect Pathology." (L. Lacey, Ed.), pp. 101–116. Academic Press, London.

Knabke, J. J. 1970. Diapause in the rice water weevil in California. *In* Proceedings of the 13th Rice Technical Working Group, pp. 52–53 (Abst.).

Kuschel, G. S. U. D. 1951. Revision de *Lissorhoptrus* Lecorte y generos vecores de America. *Revista Chilena Entomol.* **1**, 23–74.

Lange, W. H. and Grigarick, A. A. 1959. Rice water weevil beetle pest in rice growing areas of southern states discovered in California. *Calif. Agric.* **13**, 10–11.

Li, H. 1985. Entomopathogenic microorganisms of rice planthoppers and leafhoppers in China. *Intern. Rice Res. Newsl.* **10**, 13–14.

Manjunath, T. M. 1978. Two nematode parasites of rice brown planthopper in India. *Intern. Rice Res. Newsl.* **3**, 11–12.

Mikkelsen, D. S. and Datta, S. K. D. 1991. Rice culture. *In* "Volume Rice Production." (B. S. Luh, Ed.), pp. 103–186. Van Nostrand Reinhold, New York.

Morgan, D. R., Slaymaker, P. H., Robinson, J. F., and Tugwell, N. P. 1984. Rice water weevil (Coleoptera:Curculionidae) indirect flight muscle development and spring emergence in response to temperature. *Environ. Entomol.* **13**,26–28.

Muda, A. R. B., Tugwell, N. P., and Haizlip, M. B. 1981. Seasonal history and indirect flight muscle degeneration and regeneration in the rice water weevil. *Environ. Entomol.* **10**,*685–690.*

Mugnai, L., Bridge, P. D., and Evans, H. C. 1989. A chemotaxonomic evaluation of the genus *Beauveria. Myco. Res.* **92**, 199–209.

Nagata, T. 1987. Infectivity of *Steinernema feltiae* on the rice water weevil, *Lissorhoptrus oryzophilu* Kuschel. *In* "Recent Advances in Biological Control of Insect Pests by Entomogenous Nematodes in Japan," pp. 65–70. Grant-in Aid for Developmental Scientific Research Ministry of Education, Culture and Science, Japan.

Nayak, P., Basu, B., Das, S., Basu, A., Ghosh, D., Ramakrishnan, N. A., Ghosh, M., and Sen, S. K. Transgenic elite indica rice plants expressing CryIAc delta-endotoxin of *Bacillus thuringiensis* are resistant against yellow stem borer (*Scirpophaga incertulas*). *Proc. Natl. Acad. Sci. USA* **94**, 2111–2116.

Pea, N. and Shepard, M. 1986. Seasonal incidence of parasitism of brown planthoppers, *Nilaparvata legens* (Homoptera: Delphacidae), green leafhoppers, *Nephotettix* spp., and whitebacked planthoppers, *Sogatella furcifera* (Homoptera: Cicadellidae) in laguna province, Philippines. *Environ. Entomol.* **15**, 263–267.

Poinar, G. O., Jr. and Chang, P. M. 1985. *Hexamermis cathetospiculae* n. sp. (Mermithidae: Nematoda), a parasite of the rice stemborer, *Tryporyza incertulas* (Wik.) (Pyralidae: Lepidoptera) in Malaysia. *J. Nematol.* **17**, 360–363.

Reichelderfer, C. F. 1985. Biological assays with insect pathogens. *In* "Handbook of Natural Pesticides: Methods, Volume Theory, Practice, and Detection." (N. B. Mandava, Ed.), pp. 489–516. CRC Press, Boca Raton, FL.

Reissig, W. H., Heinrichs, E. A., Litsinger, J. A., Moody, K., Fiedler, L., Mew, T. W., and Barrion, A. T. 1986. "Illustrated Guide to Integrated Pest Management in Rice in Tropical Asia." Intern. Rice Res. Inst. Los Banos, Laguna, Philippines.

Rice, W. C., Trahan, G. B., and Bollich, P. A. 1994. Effect of formulation on the efficacy and environmental persistence of fungal pathogens to the rice water weevil. Annu. Res. Rpt., Rice Res. Stn., La. Agric. Exp. Stn., LSU Agricultural Center. 5 pp.

Rombach, M. C., Aguda, R. M., Shepard, B. M., and Roberts, D. W. 1986a. Infection of rice brown planthopper, *Nilaparvata lugens* (Homoptera: Delphacidae), by field application of entomopathogenic Hyphomycetes (Deuteromycotina). *Environ. Entomol.* **15**, 1070–1073.

Rombach, M. C., Humber, R. A., and Roberts, D. W. 1986b. *Metarhizium flavoviride* var. *minus*, var. nov., a pathogen of plant-and leafhoppers on rice in the Philippines and Solomon islands. *Mycotaxon* **27**, 87–92.

Rombach, M. C., Humber, R. A., and Evans, H. C. 1987a. *Metarhizium album*, a fungal pathogen of leaf- and planthoppers of rice. *Trans. Br. Mycol. Soc.* **88**, 451–459.

Rombach, M. C., Rombach, G. M., and Roberts, D. W. 1987b. Pathogens of insect pests of rice: a bibliography. *Insect Sci. Appl.* **8**, 197–210.

Schaeffer, P., Millet, J., and Aubert, J. P. 1965. Catabolic repression of bacterial sporulation. *Proc. Natl. Acad. Sci. USA*, **54**, 704–711.

Sekar, V., Held, B., Tippett, J., Amirhusin, B., Robert, P., Wang, K., and Wilson, H. M. 1997. Biochemical and molecular characterization of the insecticidal fragment of CryV. *Appl. Environ. Microbiol.* **63**, 2798–2801.

Socksai, S. and Tugwell, P. 1978. Adult rice water weevil feeding symptoms: number samples required, spatial and seasonal distribution in a rice field. *J. Econ. Entomol.* **71**, 145–148.

Thiery, I. and Frachon, E. 1997. Identification, isolation, culture and preservation of entomopathogenic bacteria. *In* "Manual of Techniques in Insect Pathology" (L. Lacey, Ed.), pp.55–77. Academic Press, London.

Travers, R. S., Martin, P. A. W., and Reichelderfer C. F. 1987. Selective process for efficient isolation of soil *Bacillus* sp. *Appl. Environ. Microbiol.* **53**, 1263–1266.

Tucker, E. S. 1912. The rice water weevil and methods for its control. U.S. Bur. Entomol. Circ. No. 152.

Urtz, B. E. and Rice, W. C. 1997. RAPD-PCR characterization of *Beauveria bassiana* isolates from the rice water weevil *Lissorhoptrus oryzophilus*. *Letters Appl. Microbiol.* **25**, 405–409.

Wang, J. X. and Li, L. Y. 1987. Article bibliographic: Entomogenous nematode research in China. *Rev. Nematol.* **10**, 483–489.

Way, M. O. and Bowling, C. C. 1991. Insect pests of rice. *In* "Volume Rice Production." (B. S. Luh, Ed.), pp. 237–268. Van Nostrand Reinhold, New York.

Webb, J. L. 1914. Notes on the rice water weevil, *Lissorhoptrus simplex* Say. *J. Econ. Entomol.* **7**, 432–328.

Yan, S., Qian, H., Xia, K., and Ling, L. 1986. Life cycle of *Amphimermis* sp., a nematode parasite of the brown planthopper in China. *Acta Phytophylacica Sinica* **13**, 79–84. (In Chinese with English summary).

Zhao, H., Hu, Z., and Zhang, S. 1987. Field occurrence of *Agamermis* sp., a nematode parasite of rice hoppers. *Chinese J. Biol. Control* **3**, 73–77. (In Chinese with English summary).

Chapter VII-5

Microbial control of insect pests of soybean

Flávio Moscardi
and Daniel R. Sosa-Gómez
Embrapa National Soybean Research Center
Londrina, PR, 86001-970, Brazil

1 Introduction

Soybean is the major leguminous crop grown worldwide and the most important producers are the United States, Brazil, Argentina, China and India. Insects attacking soybean germinating seeds, roots, stem, leaves, pods and seeds may limit production, depending on their abundance and damage potential in each soybean-growing region (Kogan and Turnipseed, 1987; Moscardi, 1993; Gazzoni *et al.*, 1994; Hoffmann-Campo *et al.*, 2003). Although entomopathogens can be used in different ways to control insect pests (Lacey *et al.*, 2001), this chapter will deal mostly with the "microbial insecticide approach," when entomopathogens are applied as many times as needed to maintain pests below economic injury levels. In this sense, the pathogen should be highly virulent to the host insect and should reach the pest in its most susceptible stage. Our objective is to provide field techniques and protocols for application and evaluation of entomopathogens being used or with potential to be employed in soybean systems as microbial agents. Emphasis will be given to Lepidoptera (especially defoliators) and stink bugs (Hemiptera: Pentatomidae) due to their importance in most soybean growing regions and the greater availability of information on application and evaluation of entomopathogens for these groups.

2 Insect pests of soybean and their entomopathogens

A *Lepidoptera*

The most important defoliating species in soybean are noctuids. The velvetbean caterpillar (VBC), *Anticarsia gemmatalis*, is by far the most important defoliator in the Americas, usually occurring in high populations from northern Argentina to the southeastern USA (Herzog and Todd, 1980). Different species of the Plusiinae complex (semi-loopers) may be important in distinct soybean growing regions. *Pseudoplusia includens* is the most common soybean Plusiinae in the Americas (Herzog, 1980). *Rachiplusia nu* in southern Brazil, Uruguay and Argentina, and *Chrysodeixis* spp. and *Thysanoplusia orichalcea* are other species of Plusiinae attacking soybeans in some African and Asian countries. *Plathypena scabra* is a widespread defoliating noctuid in North America (Pedigo, 1980). Other important soybean-defoliating species of noctuids include *Spodoptera littoralis* (Africa), *S. litura* (Asia), *Spilosoma virginica* (Argentina), and *Helicoverpa/Heliothis* spp. (worldwide) (Gazzoni *et al.*, 1994). All of these species are susceptible to the fungus *Nomuraea rileyi*, which

L.A. Lacey and H.K. Kaya (eds.), *Field Manual of Techniques in Invertebrate Pathology*, 411–426.

is often observed causing natural epizootics to the point of maintaining hosts below damaging levels, especially in VBC and Plusiinae populations (Moscardi and Sosa-Gómez, 1992). In Brazil, VBC, *P. includens* and *R. nu* are infected by other fungi in the Hypocreales such as *Paecilomyces tenuipes* (Sosa-Gómez, unpublished data), and Entomophthoraceae such as *Pandora gammae* (R. Humber, pers. com.) and *Furia crustosa* (Moraes *et al.*, 1991).

Lepidopteran pests that have more cryptic habits and act as stem or pod borers (*e.g., Epinotia aporema, Maruca testulalis, Etiella zinckenella*, and *Elasmopalpus lignosellus*) and leafrollers (*e.g., Omiodes indicatus* and *Laprosema indicata*) are also important in specific regions of soybean production (Gazzoni *et al.*, 1994). However, because of their cryptic habits and their intermittent nature as pests in most regions, studies related to entomopathogens as microbial control agents (MCAs) of these pests have been very scanty.

Although an impressive volume of research has been generated in different countries regarding the natural occurrence of entomopathogens associated with soybean insects and their potential as MCAs, use of chemical insecticides is still the predominant control tactic employed in soybean cultivation systems. Only a very small proportion of the pathogens that attack soybean insects have been used as MCAs at the farmer level.

Among viruses, the family Baculoviridae (nucleopolyhedrovirus and granulovirus) has been the most studied for control of different species of Lepidoptera (Moscardi, 1999). The most notable example of virus use for insect control in soybean is the multiple-embedded nucleopolyhedrovirus of VBC (*Ag*MNPV) that has been used annually in more than 2,000,000 ha in Brazil (Moscardi,1999; Hoffmann-Campo *et al.*, 2003).

The most studied group of fungi used for control of soybean pest insects is the Hypocreales (formerly placed in the form-class Hyphomycetes), with *N. rileyi* presenting the highest potential for development as a MCA. Despite their importance, fungi have not been practically used by soybean growers. One of the exceptions is *Beauveria bassiana* developed as a commercial formulation to control defoliating caterpillars (mainly *Chrysodeixis acuta* and *S. litura*) in India (Sharma, 1995). To date no microbial control program has been carried out with the Entomophthorales, although this group of fungi is prevalent in several species of soybean insects. Commercial formulations of *Bacillus thuringiensis (Bt)* have been used against soybean-defoliating species of Lepidoptera.

B Hemiptera

The most common species of Hemiptera encountered worldwide on soybeans are mainly the pentatomid stink bugs *Nezara viridula* (cosmopolitan species), *Piezodorus* spp., *Euschistus* spp. and *Acrosternum* spp. (Panizzi and Slansky, 1985). Other species in the genus *Riptortus* have been reported from Africa and Asia (Jackai *et al.*, 1990). Stink bugs are attacked by several fungi in the Hypocreales including *B. bassiana, Metarhizium anisopliae*, and *Paecilomyces fumosoroseus*. Although *B. bassiana* is the most common fungus isolated from these insects, *M. anisopliae* offers better potential for their control. Field studies carried out with *B. bassiana* and *M. anisopliae* showed that application of the latter resulted in higher mortality of soybean stink bugs than *B. bassiana* (Sosa-Gómez and Moscardi, 1998).

Other pathogens that have been found consistently in stink bug populations are protozoans from the trypanosomatid group. They can reach prevalence ranging from 0% to 95% depending on the stink bug species, with the highest prevalence in *N. viridula* and *P. guildinii* (Fuxa *et al.*, 2000; Sosa Gómez *et al.*, 2005). Despite their prevalence, they cause chronic diseases, limiting their potential as biological control agents in short term conditions.

3 Application and evaluation of entomopathogens for control of Lepidoptera

A Baculovirus for control of Lepidoptera

Although baculoviruses of important insect pests of soybean have been developed commercially or been field tested, such as NPVs against the *Helicoverpa/Heliothis* complex, *S. littoralis*, and *S. litura*, their application has been mostly directed to control these pests in other crops. Use of baculovirus in soybean is exclusively performed with *Ag*MNPV to control *A. gemmatalis* in extensive areas in Brazil, and on a small scale with the NPV of *T. orichalcea* in some Asian and African countries (for review see Moscardi, 1999).

1 *Preparation of inoculum*

a *Baculovirus field production*

Baculoviruses can be produced in the field in countries where cost of labor is low and the insect habit (*i.e.*, defoliator) allows field collection of virus-infected caterpillars, as exemplified with *Ag*MNPV production in Brazil (Moscardi, 1999). Fields with moderate to high caterpillar populations [20 to 25/m of row, estimated by the ground-cloth method (Kogan and Pitre, 1980]) can be treated with double the rate recommended for routine control purposes [1.5×10^{11} occlusion bodies (OBs)/ha]. *Ag*MNPV treated larvae should be collected from the 7th to 10th day after application (Moscardi, 1989). Baculovirus-infected larvae can be collected manually in large numbers, as diseased caterpillars climb to the top of the canopy, becoming sluggish and pale green. Through applications of *Ag*MNPV in different fields each day in a certain region, experienced technicians can select the fields that will result in highest *Ag*MNPV yields by inspecting larvae showing advanced symptoms of the disease around the 5th to 6th day after treatment.

The quality control of the material being collected should be initiated in each collection site by trained field technicians through removal of host larvae not typically infected with *Ag*MNPV and other extraneous materials such as larvae of other lepidopteran species, leaves, etc. If the weather is not rainy and moisture does not favor mycoses (*N. rileyi* and Entomophthorales) in caterpillar populations at collection sites, field production can average 1.8 kg of NPV-killed larvae/collector/day (Moscardi, 1999). At each collection site the material passing through initial quality control approximately every 2 h should be rapidly stored in plastic bags in cooler boxes on ice or in freezers to avoid bacterial growth and darkening of larvae by phenoloxidase activity (melanization). At the laboratory storage site, the collected material may be subjected to another quality control (elimination of extraneous materials and larvae not typically killed by the virus or in decomposition) and washing with water, if necessary, before storage in freezers for further processing and formulation.

Other methods for baculovirus production include release of laboratory-reared insects (larvae or adults) in field screen cages for further treatment of established larval populations and collection of NPV-killed larvae, as well as continuous production in laboratory facilities (Moscardi, 1989; Sosa-Gómez and Moscardi, 1996). The latter method is very dependable, as it allows planning virus production outputs on a weekly or monthly basis, while any type of field production may vary enormously from season to season, as host abundance is affected by abiotic and biotic factors. Therefore, despite its higher cost, laboratory production of baculoviruses may be important to complement field production. More recently, *in vivo* laboratory commercial production of the virus has been implemented in Brazil by one a private company (Coodetec). The company inoculates about 800,000 larvae per day in order to produce the *Ag*MNPV needed for treatment of about 2,000,000 ha per year.

b *Processing and formulation*

*Ag*MNPV can be formulated using a very simple process:

1. Homogenize larvae in large blenders (1 kg of larvae/100 ml of water).*
2. Filter through layers of gauze or fine cloth to obtain insect hemolymph free of large insect debris.*
3. Mix the suspension with kaolin or other fine clay material in a proportion of 1:1 (v/w), preparing a homogeneous "slurry."
4. Pour a thin layer of slurry on plastic trays previously covered with a resistant plastic film and let dry at 30–35 °C overnight.

* Steps 1 and 2 above can be replaced by an adapted juice extractor.

5. Use fans and dehumidifiers to accelerate the drying process.
6. Mill dried flakes of *Ag*MNPV and kaolin into a fine powder. A grain mill equipped with sieves with pores ≤ 0.8 mm diameter should be used to obtain appropriate particle sizes in the formulation and to avoid nozzle clogging duringbreak application.
7. Store virus formulation in a freezer (−5 to −10 °C).

c Quality control of the formulated product

This process should be conducted to ensure that the produced virus has appropriate biological activity, and for standardization of the different production batches. One sample (10–20 g) of each 30–60 kg formulated batch should be evaluated in the laboratory for determination of the number of OBs/unit weight and potency through bioassays with the host larvae (see Evans and Shapiro, 1997). In Brazil, samples of formulated batches of *Ag*MNPV produced by five private companies are sent every year to the Insect Pathology Laboratory of the National Soybean Research Center (Embrapa) for quality control assessment in relation to a known standard of the virus before packaging and selling to farmers.

The bioassays are performed by incorporating each formulated sample into the artificial diet of VBC (Hoffmann-Campo *et al.*, 1985), after it cools to 50 °C, at 50, 150, 450, 1350 and 4050 OBs/ml of diet. The different dilutions are poured into 50 ml plastic cups, and after the diet cools and solidifies, three 3rd instar VBC larvae are placed into each cup. A minimum of 30 larvae should be used for each concentration. The bioassays should be kept at constant temperature and humidity. They can be performed at temperatures ranging from 26 to 29 °C; at 29 °C the time to mortality is reduced by about one day, saving time and operational costs (Sosa-Gómez and Moscardi, 1996). Mortality data should be recorded every day until pupation. The number of adults emerging from pupae should also be considered. Mortality data can be processed using different models, such as probit, logistic analysis or log-log complement, according to the data fitness (McCullagh and Nelder, 1989).

2 Plot design

To assess control of larval populations of VBC by *Ag*MNPV, as well as baculoviruses of other soybean-defoliating Lepidoptera, plots of at least 10 rows of soybean ×10 m length should be used, and applications directed to a mean larval population of at least 20 larvae/ground cloth (or 10/row m). When possible, borders should be left between plots (at least one m) and blocks (at least five rows) to minimize drift effects and contamination of plots by *Ag*MNPV through its dissemination by insect predators. Also, field experiments with the virus should be conducted only in soybean areas where no caterpillars are found infected by the fungus *N. rileyi* or where the prevalence of this fungus is still very low. If mean prevalence is one or more infected larvae/ground cloth (2 m of row) in pre-samplings, the experimental area should be avoided for trials of *Ag*MNPV or other microbial insecticides (*e.g., Bt*), especially when evaluations are extended beyond 10 days after treatments.

The experimental design should be defined according to the objectives of the experiment. Usually, the most common designs are randomized blocks and completely randomized. The latter is very useful when the experiment site is relatively uniform or when no recognizable basis for grouping exists (Petersen, 1994). This is the case for VBC populations in relatively large areas (50 × 50 m or greater). In larger plots, it is necessary to conduct a survey prior to treatment to obtain the population distribution. Criteria for grouping plots in randomized block design or more complex designs can be found in Petersen (1994). When the objective of the trials is to test a package of recommendations at the farmer level that have been generated by small plot experiments, "pilot experiments" are first recommended. These should be conducted in large paired fields (virus treated, insecticide treated, and check) of 1.0 ha or more in different locations of a region for at least two consecutive seasons (Moscardi and Corrêa-Ferreira, 1985) before the technology is transferred to farmers.

3 Equipment and supplies for field applications

The machinery used for the application of MCAs in soybean is usually the same as those employed for conventional pesticides (see Chapters III-1 and III-2). For small-plot experiments, the pathogen can be applied with hand–held or knapsack sprayers operated with CO_2 or air under pressure, equipped with nozzles that deliver 80 to 150 liters/ha, or ground application equipped with a boom sprayer configured with cone or flat fan nozzles with a flow rate of 100 to 150 liters/ha. These alternatives enabled effective control of VBC larval populations using *Ag*MNPV (Silva, 1987). This variation in spray volume is related to the plant phenological stage at the time of application, insect species and habits, and type of pathogen applied to attain proper plant coverage to reach the target insect in its most susceptible stage. However, the current trend is towards using lower spray volumes.

Aerial application can be made with small planes or helicopters, but the latter type of aircraft is rarely used in soybeans. The volume of water usually used is 18–40 liters/ha (NPVs and *Bt*), but not less than 15 liters of water/ha should used (Silva, 1986). Oil suspensions (*e.g.*, crude soybean oil) can be used at volumes as low as 5 liters/ha in aircraft equipped with rotary atomizers (*e.g.*, Micronair). In this case, the pressure in the hydraulic system should be regulated according to the flow rate. The width of the application swath varies from 15 to 18 m according to the aircraft used. The height of the flight should be 3 to 5 m above the soybean canopy and the application is not recommended if wind speed exceeds 7 km/h. The airplane should fly at 169–193 km/h and guided by flaggers, as commonly used for application of chemical insecticides.

Spray deposits can be sampled with microscope slides or kromekote cards placed in the treated plots. Amount of spray deposit can be determined by fluorescent methods (Yates and Akesson, 1963) or through counting the number of droplets per unit area. Diameter of individual droplets can be measured with an ocular micrometer. Trial runs are recommended to check quality of application (*i.e.*, plant coverage) and adjust application parameters as needed for use of the *Ag*MNPV and *Bt* commercial formulations in soybean, to ensure ingestion of the pathogens.

4 *Baculovirus application*

Application of *Ag*MNPV in the field can be made by using 50 larval equivalent (LE)/ha (one LE $\cong 3 \times 10^9$ OBs). This number of LE corresponds to approximately 20–25 g of *Ag*MNPV-killed larvae/ha or 1.5×10^{11} OBs/ha in formulated products. Viral suspension can be prepared by macerating diseased larvae in water and straining the material through layers of cheesecloth or fine cloth. This type of inoculum suspends well in water. When wettable powder formulation is used, it is best to use a small volume of water (1–2 liters) and agitate the suspension thoroughly before pouring it into the spray tank.

Application of the virus early in the growing season can also be made to favor the early spreading of the disease in VBC populations, with the purpose of keeping the density of the insect larvae below damaging levels. In this case, virus application can be made either alone or simultaneously with post emergence herbicides in tank mixes. This strategy has been adopted to some extent by soybean growers in Brazil and with success in certain regions under no-till soybean cultivation (Moscardi, unpublished observations). Farmers may save money by mixing the *Ag*MNPV with post emergence herbicides, when the use of these herbicides is mandatory in no-till systems. At this time, VBC larval populations are still low, but usually sufficient to allow multiplication and dissemination of the virus to hold the populations below damaging levels. A similar approach has been considered by Young and Yearian (1979).

5 *Timing*

In field assays, VBC populations can be controlled when the majority of the population is composed of small larvae (*i.e.*, from the first to third instar or < 1.5 cm). The virus should be applied against small larvae at a maximum population of 40 larvae/ground cloth (20 larvae/m of row), or when 30 small larvae and no more than 10 fourth instar larvae or bigger are detected. This recommendation is based on the differential susceptibility to the virus according to larval age or size, and response of field caged soybean (defoliation, yield) to different VBC larval populations treated with the pathogen (Moscardi and Sosa-Gómez, 1992). Fifth instar larvae were 40–48, 7–16 and 4–6 times less susceptible to *Ag*MNPV than 2nd, 3rd and 4th instar larvae, respectively (Boucias *et al.*, 1980; Moscardi, 1983). Soybean plants can stand defoliation (30% or more) in the vegetative stages without yield losses (Kogan and Turnipseed, 1980). For this reason, virus applications against early VBC instars are important to avoid larvae from causing more than 30% defoliation.

6 *Evaluation*

Usually the parameters evaluated are number of live (healthy) and diseased larvae/ground cloth (2 row m) and % defoliation. Methods involving others stages (eggs, pupae and adults) are time consuming and/or do not represent the immediate impact of treatments on larval population. Sampling by the ground-cloth method is most adequate for defoliators, such as VBC. For cryptic insects, such as *E. aporema*, evaluation of control with baculoviruses, *Bt* or fungi should be performed by plant examination in different sites of the experimental plots and recording the number of attacked plants. In addition, attacked plants can be collected, and the larvae reared on the host plant or on artificial diet to assess the impact of the treatments.

Ground cloth sampling should be performed by two persons, in at least two sites per plot (small plot experiments). For review on sampling methods for soybean insects, see Kogan and Herzog (1980). The first sampling should be made before virus application to record the density and homogeneity of the population. Additional samplings can be made at approximately 4, 7, 10, 13 and 16 days after application. If the plots are larger (> 0.1 ha), at least 10 samples should be taken per plot/date. To better understand virus activity, effect on number of small (< 1.5 cm) and large (> 1.5 cm) larvae should be recorded separately.

Defoliation estimates are important and they can be made visually in the field (less precise) or by using leaf area measurements (more precise) of foliar samples taken to the laboratory. The latter is laborious and time consuming, but leads to more precise estimates of defoliation between treatments, provided there is a good standard for comparison (*i.e.*, plants treated with an efficient insecticide that does not allow progression of defoliation after treatment). However, visual % defoliation ratings can be very useful, as these can be made at each plot during each sampling, provided there are substantial differences between treatments and the sampling personnel are well trained for this type of evaluation (see Kogan and Turnipseed, 1980). Other useful records are phenological stage of plants (see Fehr *et al.*, 1971) at each sampling date, prevalence of natural enemies, temperature (amplitude and means), relative humidity and rain. This information will be useful to understand unexpected population density changes.

Records of the occurrence of *N. rileyi* are important because this natural control agent interferes with experiments being conducted with *Ag*MNPV and other entomopathogens (*e.g., Bt*) because of its high prevalence in VBC populations. Fungicide application using benomyl or difenoconazole based compounds can help to delay *N. rileyi* infections in experimental plots but does not prevent its outbreak in lepidopteran populations (Sosa-Gómez and Moscardi, unpublished).

Collection of yield data may be also important, provided data on % defoliation differ greatly between treated plots and controls. Small differences in defoliation usually do no reflect in yield reductions, as soybean compensates for high defoliation (Kogan and Turnipseed, 1980). Efficacy can be calculated using Abbott (1925) formula or Henderson and Tilton (1955) formula to compensate for mortality in control plots. The latter is more appropriate if significant differences are found between plots in samplings before application of treatments (Gazzoni *et al.*, 1984).

B *Fungi for control of Lepidoptera*

There are few studies considering microbial control of VBC populations with entomopathogenic fungi, and most of the attempts have been made with *N. rileyi*. In one of these attempts, a Brazilian isolate of *N. rileyi* was produced in VBC larvae in the laboratory, and applied at 5×10^{11} viable conidia/ha. This treatment resulted in over 90% mortality of VBC larvae compared to less than 10% mortality caused by the fungus in the untreated plots (Moscardi and Corrêa-Ferreira, 1985). However, this fungus is not used in microbial control, probably due to its high production cost, unstable conidia production, mainly after repeated subcultures, and loss of virulence to the host (Morrow *et al.*, 1989). It never reached a stage of commercialization, although studies have been made on its production and cost under laboratory conditions (Bell *et al.*, 1982; Holdom and Klashorst, 1986). Therefore, formulation studies with this species still remain unconsidered. Advances in the last years with packaging process of *M. anisopliae* formulation (Krueger *et al.*, 1995) will be valuable to improve shelf life of *N. rileyi* and other entomopathogenic fungi.

1 *Preparation of inoculum*

In order to provide uniform distribution of inoculum over a large area, the fungal material can be suspended in liquid (*i.e.*, water, kerosene, mineral or vegetable oil) or mixed with a powder (*i.e.*, kaolin, talc, lactose). Compatibility with the diluent or carrier should be evaluated before bioassay or field trials are conducted. Devi and Prasad (1996) showed that different vegetable oils were compatible with *N. rileyi*, but they did not increase mortality on *S. litura*. Prior to any fungal application, the viability of infective units should always be determined using appropriate culture medium and following the appropriate protocols for a given fungus (Goettel and Inglis, 1997). For most entomopathogenic fungi, media such as SDAY (Sabouraud dextrose agar + yeast extract) or PDA (potato dextrose agar) plus yeast extract can be routinely used; for some fungi, such as *N. rileyi*, SMAY (Sabouraud maltose agar + yeast extract) is adequate.

2 *Plot design*

Experimental plots should be far enough from one another to avoid contamination between treatments, especially when evaluations are made over an extended period of time (more than 12 days after treatments). The distance between plots also should not be so far that the experimental area becomes a source of heterogeneity. Powder application will require large isolated areas or areas isolated with untreated buffer zones. The plots can be isolated with barriers, by planting rows of corn, pigeon pea (*Cajanus cajan*) or other plants on plot boundaries. Since VBC populations are usually homogeneously distributed in the field, it is possible to set up experiments in plots with a minimum size of 10 rows $\times$ 10 m length (as referred in section 3A2). There should be a minimum of four replicates/treatment, including untreated control plots and other check treatments such as plots sprayed only with the carriers (water, oil, clay, etc.) used in different fungal preparations.

3 *Fungal application*

There are no studies related to application technology with entomopathogenic fungi in soybean. The appropriate concentration should be selected depending on the susceptibility of the pest, and its efficacy to control the pest under variable weather conditions. The application should lead to an amount of inoculum to cause infection and control of a large proportion of the host insect population. The density of the inoculum can be expressed as conidia/ha as well as conidia or colony forming units (CFU) that remain on the surface of the leaves. Quality of application (plant coverage and number of conidia/unit leaf area) is important to increase the probabilities of fungal infection on host insects. Methods for detecting conidial coverage are presented in Chapter IV-4 and by Goettel and Inglis (1997).

Insects that attack early stages of the crop require a lower fungal concentration considering that better plant coverage occurs with liquid or powder formulations. Theoretically, *N. rileyi* can cause mortality close to 50% in first instar larvae when 2.4 conidia/mm^2 are present on soybean leaves (Ignoffo *et al.*, 1976) or 90% in third instar VBC with 97 conidia/mm^2 (Sosa-Gómez, unpub. data). Based on these findings, *N. rileyi* use on soybean at stages from V5 to R1 (Fehr *et al.*, 1971) will require from 2.7×10^{10} to 4.7×10^{10} conidia/ha, to provide at least 97 conidia/mm^2 to cause mortalities of ca. 90% in VBC populations, when most of the larvae are in the third instar. However, these concentrations were based on laboratory bioassay data, as mean lethal concentrations on soybean leaves, without the benefit of field data. On the other hand, application of *N. rileyi* at 5.0×10^{11} conidia/ha, when most of the larvae were small (< 1.5 cm), resulted in over 90% larval mortality in the field, compared to ca. 10% in the untreated plots (Moscardi and Corrêa-Ferreira, 1985). Application of very high concentrations (2.7×10^{13} to 5.4×10^{13} conidia/ha) of *N. rileyi* initiated earlier epizootics on *P. scabra* and *Trichoplusia ni* by two weeks (Ignoffo *et al.*, 1976). Also, application of 2.7×10^{13} conidia/ha against *H. zea* reduced larval population by ca. 77%, but the lower concentration (2.75×10^{11}/ha) failed to provide efficient control of this pest (Ignoffo *et al.*, 1978).

The fungal material (conidia or mycelia) can be applied using several procedures. Some of them that have been used in experimental plots are laborious and time consuming. For example Sprenkel and Brooks (1975) distributed mummified cadavers with viable *N. rileyi* to induce epizootics in the field. The introduction of mummified cadavers also can be used for small arthropods such as aphids, thrips, whiteflies and mites, infected with fastidious fungi (*i.e., Neozygites*). This technique has been used extensively for aphid fungi (Wilding *et al.*, 1986; Wilding *et al.*, 1990).

Ground applications can be selected according to the size of the plot and the objective of the application. For small experimental plots, dry mycelia can be applied by hand or can be used as water suspensions and applied with hand–held or knapsack sprayers. It is desirable to maintain constant pressure using compressed-air or CO_2 tanks, equipped with a pressure gauge. Dry mycelia can also be delivered with a hand duster or knapsack sprayer with a fan-powered engine. Plots of 0.5 ha or larger should be treated with tractor mounted, row crop boom sprayer that delivers from 50 to 150 liters/ha. Ground or aerial applications can also be used for ULV application of oil-conidia formulations.

Several authors add wetting agents, such as Triton and Citowett, to improve the wettability of hydrophobic conidia, but their effect on conidial adhesion to the host is unknown. It should be kept in mind that dry fungal conidia (*e.g., Nomuraea, Beauveria, Metarhizium, Paecilomyces*) mediate their attachment to hosts by hydrophobic interactions (Boucias and Pendland, 1991); therefore, powder formulations could improve the infection process of these fungi.

4 *Timing*

To our knowledge there are no studies about timing applications of entomopathogenic fungi in soybean systems. One of the reasons is the inconsistency of the results obtained with fungal applications against their hosts. Since the foliar consumption by VBC larvae infected with *N. rileyi* is not much reduced (Carvalho *et al.*, 1991), the possibilities of economic defoliation could be diminished by applications of this fungus against low VBC larval densities and early larval instars (1st to 3rd).

5 *Evaluation of efficacy*

Non-destructive pre-treatment samples should be taken to verify the density and homogeneity of the population before setting up a field test. The ground cloth method is rapid and precise to sample defoliating caterpillars and stink bugs. Post application samples (at least two/small plot) should be taken at intervals of 3–5 days until target inset populations die out or the fungus is no longer effective. Insects counted by the ground-cloth method should be returned to the respective sampling sites. More detailed studies to evaluate the prevalence of different species or strains of the fungal agents, individual samples of alive larvae should be taken to the laboratory and kept until death or adult emergence to determine the causes of diseases (for review see Carner, 1980; Fuxa and Tanada, 1987). When insects brought from the field do not feed readily on artificial diet, soybean leaves treated with sodium hypochlorite (2.5 %) and three washes of sterile water can be used. Excess water on the leaves should be removed by blotting with filter paper to avoid bacterial contamination. Hemolymph from live larvae collected after field applications of fungi can be smeared on glass slides and examined microscopically for the presence of hyphal bodies. Rapid diagnosis of fresh specimens can be made by using optical brighteners and an UV equipped microscope

(Magalhães, 1994; see also Chapter IV-4). Evaluation of defoliation and yield can be performed as discussed earlier for Baculovirus applications (section 3A6).

For evaluation of fungal deposit and persistence, a selective medium can be used, depending on the fungus species or group being evaluated (see Chapter IV-4). Indirect methods can also be used, such as evaluating the presence of the pathogen using a susceptible insect as an indicator, taking samples of leaves or soil to the laboratory and carrying out bioassays with the susceptible host (Goettel and Inglis, 1997). The level of insect mortality by the pathogen may give an estimate of the concentration in the sample, provided previous trials have been conducted to relate pathogen activity to different concentrations of the pathogen on the substrate to be sampled (leaves, soil, etc.). Insecticidal activity can be translated to % mortality and plotted against time, to determine residual activity. This method has been used to evaluate *N. rileyi, Nosema necatrix* (Gardner *et al.*, 1977), *B. thuringiensis*, and the NPV of *Helicoverpa/Heliothis* (Ignoffo *et al.*, 1974) in soybean.

C Bacteria for control of Lepidoptera

There are different *Bt* formulations available for control of soybean Lepidoptera. The *Bt* formulations can be evaluated under field conditions by comparing their efficacy with other recommended insecticides, studying the effect of application methods on efficacy using different concentrations, coverage or droplet size, determining the impact of environmental conditions and climatic factors, determining the interaction of *Bt* with biological and chemical insecticides, etc.

1 Plot design

Plots, replicates, and design for *Bt* evaluation are similar to those presented for *Ag*MNPV.

2 Equipment calibration

After the desired nozzles are installed, the output of the system should be checked to verify if the correct flow rate is taking place. To test the spray pattern, a dye or a fluorescent marker can be added to the water in the tank, and cards can be laid out at regular intervals in the field. After application is made at a desired speed, pressure and height, cards are collected to determine droplet distribution pattern and size. Based on this information, orientation and localization of nozzles, as well as other parameters, can be modified to attain good plant coverage.

The coverage of *Bt* can also be determined using bacteriological techniques. Prior to the spray applications, cover slips with semi-selective medium (Thiery and Frachon, 1997) can be placed at different locations in the experimental area and heights of the canopy or on the ground. On foliar surfaces, evaluation of coverage can be made by taking aliquots of suspensions from washed leaves and assaying them for viable spores using the pour plate technique (Thiery and Frachon, 1997). Counts based on four or more replicates can be used to calculate the number of spores per leaflet or leaflet area.

3 Bacterial application

The activity of *Bt* against different target pests should be taken into account. Insecticidal activity of *Bt* depends on very specific host-pathogen interactions; therefore the selection of the bacterial strain could be an important determinant of the control efficacy. Laboratory studies have compared the relative activity of several strains against *Spodoptera frugiperda* (Polanczyk *et al.*, 2003), *P. includens* and *A. gemmatalis* (Barreto, 2005). Applications against the soybean looper, *P. includens*, require higher rates than those used against VBC, since *P. includens* larvae are 4 to 5 times less susceptible to *Bt* present in commercial formulations with *Bt kurstaki* (Morales *et al.*, 1995). Also, soybean-defoliating Lepidoptera larvae in India (*C. acuta* and *S. litura*) require application of 1 kg of commercial formulations of *Bt*/ha for effective control (Sharma, 1995), while VBC in Brazil can be controlled with 300 g of *Bt*/ha or even lower rates (Moscardi, 1984).

Application of *Bt* can be made with knapsack hand operated sprayers or, more precisely, with

CO_2 pressurized knapsack sprayers that assure uniform pressure during application. Ground application equipment commonly employed for synthetic insecticides can also be used. The nozzles used are similar to those (XR2-3 Teejet, Twinjet or Conejet) for application of conventional insecticides. Electrostatic charged droplets of *Bt* suspensions may improve deposition on abaxial leaf surfaces, but this type of application technology has not been widespread in most soybean growing regions.

Usually the volume of *Bt* water suspensions applied at the R2 soybean developmental stage (Fehr *et al.*, 1971) range from 100 to 250 liters/ha, and the application pressure varies from 30 to 60 psi (207 to 413 kPa). At earlier stages of soybean development, lower volumes can be applied, ranging from 50 to 100 liters/ha. Alkaline water (pH > 8) should be avoided, although the natural buffering properties of *Bt*-based products often result in a pH that does not affect its insecticide activity. Chlorine concentrations of 1–3 ppm in water used in application tank mixtures do not affect *Bt* activity (Neisess, 1980).

Aerial application can be made with aircraft equipped with rotary atomization systems (Micronair, Mini-spin, Airfoil) that deliver 80–100 μm droplets or with conventional boom and nozzles systems. *Bt*-based products are usually suspensions. Depending on the formulation and on the size of the nozzle screens, between 50 and 100 mesh can be used. Filters should always be cleaned after spray operations. The working pressure can be 25 to 45 psi (172 to 310 kPa).

ULV equipment can deliver volumes from a few hundred ml to 2 liters/ha. Concentrated suspensions can be used with the advantage that lower volumes can be applied/unit area, increasing efficiency and saving time and labor costs, diminishing evaporation and rainfastness. The use of these concentrated formulations allows for a fast rate of particle deposition, reducing drift hazards. The drift will be proportional to the wind velocity and aircraft height and inversely proportional to the droplet size. Droplets of 100 μm can drift easily to a distance of 130 m with wind velocity of 5 km/h (Overhults, D.G., http://www.uky.edu/Agriculture/PAT/cat11/cat11.htm).

4 *Timing*

Bt-based products should be used during the vegetative stage of soybean and when VBC larvae are 40/ground cloth (or 20 larvae/m of row) and the defoliation is a maximum of 30%. If the application is made too early (initial stages of the plant and against very low caterpillar populations), another application may be needed on later stages of the plant, as the residual effect of *Bt* products is short (7–10 days). After flowering, a defoliation of 15% can be tolerated, but during pod set and seed development stages, caterpillar populations are usually declining because of high prevalence of fungi such as *N. rileyi* and Entomophthorales.

5 *Evaluation of efficacy*

Non-destructive pre-sampling is recommended to know the density and homogeneity of the population in the experimental plots. Post-treatment samples of live larvae should be taken at regular intervals, using the ground cloth method at least in two sites per small plot. For a more detailed interpretation of the data, larvae may be divided into arbitrary groups as: 1) small (1st, 2nd and 3rd instar); and 2) large (4th, 5th and 6th instar). However, in practice, it has been found that no significant information is lost by counting larvae as small (< 1.5 cm) and large (> 1.5 cm), as control decisions are based on number of large larvae for *Bt* and chemical insecticide applications (Gazzoni *et al.*, 1994). The short residual activity of *Bt* after the application favor the increased survival of newly hatched larvae, demanding records of small larvae during all samplings. When *Bt* is applied against VBC, perceptible differences between *Bt* treatments and check plots can be observed 2–3 days after treatment. When the data of the pre-treatment samples are homogeneous, Abbott's formula (Abbott, 1925) can be applied to estimate % control of the target insect. When pre-treatment populations are not homogeneous, the Henderson and Tilton (1955) formula should be used.

In *Bt* trials it is very important to evaluate defoliation, as this pathogen usually leads to a sharp decrease in foliage consumption, even at VBC control levels below 80%. Applications of rates as low as 200 g of formulated product (fp)/ha usually inhibit leaf consumption, avoiding economic and yield losses, although they may not cause the same level of larval reduction as the recommended rate of *Bt* (500 g fp/ha) (Moscardi, 1984). Yield loss assessments may be valuable only if differences of at least 30% in defoliation are observed between treated and check plots.

4 Application and evaluation of fungi for control of hemipteran pests (Stink Bugs)

1 Preparation of inoculum

M. anisopliae and *B. bassiana* products are available commercially in some countries (Brazil, South Africa and USA, among others). Usually, the production has been made using semi-solid media (Alves and Pereira, 1989) or the di-phasic process (Soper and Ward, 1981). The production on semi-solid media results in bioinsecticides with a concentration ranging from 10^{10} to 2×10^{11} conidia/g for *M. anisopliae* and *B. bassiana*, respectively. Conidial viability should be checked always before field tests.

2 Experimental design

Field cage experiments can be used for insects that have high mobility stages. For instance, Sosa-Gómez and Moscardi (1998) investigated the dynamics of the mycoses caused by *M. anisopliae* and *B. bassiana* in stink bug populations using caged adults. With less mobile stages, like nymphs, field plots are appropriate. Caged experiments have the disadvantages that stink bug populations are favored inside the cages and microclimatic conditions are different of the soybean open-air environment. In experiments designed to the control of nymphs, plots of 20 rows × 15 m of length can be used, with a minimum density of 4 stink bugs per sampling unit (ground cloth).

Contamination between treatments should be checked, by taking samples from the control plots and verifying the causes of mortality, using semi-selective media (Goettel and Inglis, 1997). For *Beauveria* and *Metarhizium* treatments, dodine based medium can be used at a concentration of 10 to 50 μg/ml to allow good recovery of CFU from both fungi. However, their interactions with the type of soil and fungal strains involved should be checked previously to allow the best recovery rates for each condition. For example, differential germination responses among *M. anisopliae* strains have been observed on dodine media (Dettenmaier *et al.*, 2005).

3 Fungal application

The control of stink bugs with fungi in soybean systems requires high concentrations because at the time of stink bugs attack, the crop canopy the leaf area index is at its maximum. In soybean the application methods that can be used against stink bugs are pulverization of conidial suspensions in oil emulsion (peanut, soybean or mineral oil) and powdering of dry conidia (mixtures with different compatible clays, talc, powdery milk). Due to the low susceptibility of stink bugs to fungal infections (Sosa-Gómez *et al.*, 1997; Sosa-Gómez and Moscardi, 1998), the mixture of conidia with stress agents (*e.g.*, insecticides at low concentrations) was tested in the laboratory. Some commercial formulations of insecticides, such as metamidophos and carbaryl, did not interfere with conidial germination, while methyl parathion and endosulfam caused a strong inhibition of fungal germ tube growth. These types of laboratory studies are important to determine the compatibility of low concentrations of insecticides with different species or strains of fungi. However, the adverse effects observed in the laboratory should be confirmed in the field.

In soybean fields, Sosa-Gómez and Moscardi (1998) applied 1.5×10^{13} conidia/ha of a powder formulation of *M. anisopliae* conidia in kaolin (4%) against stink bugs, through a knapsack motor duster (Hatsuta). Powdering applications are easily homogenized and quickly applied, but their major disadvantage is that they can be drifted away easily. On the other hand, dry conidial distribution diminishes the possibility of some interference of wetting agents on attachment interactions between the surface of the conidium and the integument of the target insect. Applications of conidia formulated in soybean oil resulted in better distribution than powdering, and apparently also showed a protective effect or favored germination (Sosa-Gómez *et al.*, unpublished). Protective additives as skim milk, molasses or carbon black are referred into the literature, but according to Soper and Ward (1981) their success has been limited. The formulation should be fine enough to avoid clogging spray nozzles, usually equipped with sieves of 50 mesh. This is especially critical with wettable powder formulations.

4 *Timing*

Considering the different species of stink bugs, the mean time to mortality observed in the field varies from 13.9 to 25.6 days (Sosa-Gómez and Moscardi, 1998). The application should be made early in the colonization phase, when the first stink bugs reach the field, and when the nymphs, the most susceptible stage, can be controlled more easily. Possibly, the timing should be different for each stink bug species, since *P. guildinii* and *N. viridula* are potentially more harmful to soybean than *E. heros*. *P. guildinii* can cause foliar retention ("green beans" syndrome) at lower densities than the other referred species (Sosa-Gómez and Moscardi, 1995).

5 *Evaluation*

When population density is too low, mass reared insects can be introduced into the plots. A methodology for *N. viridula* mass rearing is available (Harris and Todd, 1981; Corrêa-Ferreira, 1985). For other stink bugs (*E. heros* and *P. guildinii*) for which there is no available methodology for mass rearing, field collected adults can be used to lead to subsequent generations of undisturbed and unstressed nymphs in the plots. Sampling should be taken, when possible, at regular intervals depending on the speed of action of the pathogen or combinations of stress agents plus the pathogen. Sosa-Gómez and Moscardi (1998) sampled once a day to determine mean time to mortality of fungi in field caged conditions, an important parameter to estimate the lethal potential of the diseases. Some unpredictable factors can interfere with the evaluation, as stink bug cadavers are readily carried away by fire ants (*Solenopsis* spp.). Usually, in epizootiological studies the parameter most commonly evaluated is the number of hosts dying due to a disease per unit of time divided by the total number in the host population during that time (Fuxa and Tanada, 1987).

In open field evaluations, each plot should be sampled four times/sampling date, and some insects should be taken to the laboratory to determine the causes of mortality and percentage of infection. Mortality and infection should be evaluated discriminating the nymphal size, as the dynamics of the diseases can be different through time. Hemolymph samples can be taken and observed for the presence of hyphal bodies. Dead or infected insects can be examined microscopically and the percentage of diseases or fungal caused mortality determined. The use of optical brigtheners can help to detect fungal structures and facilitate the diagnostic process of fungal infections (Magalhães, 1994).

Indirect evaluations can be made to assess stink bug damage on pods or the seeds, by using different methods. Number of feeding punctures (stylet sheaths) can be determined by soaking pods or seeds in an acid fuchsin solution (1%) for 3 h, and after this period the substrate should be washed in running tap water to remove the excess of fuchsin and the number of punctures can be recorded (Panizzi *et al.*, 1995). Another technique is the tetrazolium test. The principle of this test relies on the synthesis of a red compound indicative of viable tissues in the seeds, and consists of:

1. Preparing a solution with tetrazolium salt (2,3,5-triphenyl tetrazolium chloride) at 0.075% (v/v).
2. Seed preconditioning, keeping the seeds under evaluation in a moist germination paper towel at 25 °C for 16 h.
3. Placing the seeds in a beaker with the tretazolium salt solution. The beaker should be kept in darkness at 35–40 °C for 165 ± 15 min.
4. Rinsing the stained seeds in tap water several times to stop the staining reaction, and keeping the seeds in water in a refrigerator for 12 h before they are evaluated.
5. Evaluating through a microscope or magnifying lenses under fluorescent light. Each seed should be

sectioned in half and the coat should be removed to expose the outer surface of the cotyledons, because internal tissues can be colonized by the fungus *Nematospora coryli* (causal agent of the yeast spot diseases) that is usually not revealed on the external seed coat.

Affected tissues are flaccid, white, greenish, yellowish, or grayish white. Isolated lesions show a distinct dark-red boundary. In confluent lesions, this region can be undistinguishable. For more details about this method, França Neto *et al.* (1998) should be consulted. The viability and vigor of the seeds can also be evaluated using germination tests. Yield evaluations, although it is a valuable information, may be masked by many factors (soil fertility, diseases, weather, genetic components and their interactions).

The evaluation of fungal deposit on the phylloplane could be made by using direct methods, as conidial enumeration on cover slips or leaves, or by using indirect methods as selective medium to evaluate fungal CFU (Goettel and Inglis, 1997). The latter method have some disadvantages, as colonies can be derived from mycelia fragments rather than conidia, and the presence of different fungal species could inhibit the growth amongst them. However, direct methods (conidial traps) also have disadvantages, as in the case of fungi with similar conidia that can often lead to misidentification.

Randomly selected leaflets from treated plots should be collected in pre- and post-application dates, at regular intervals or on a daily basis for more accurate results. After collection, leaflets can also be stored at $-15\,^{\circ}\mathrm{C}$ or lower temperatures for further analysis. Leaflets should be washed separately, by strong agitation, in a sterile solution of Tween 80 (0.01%) and plating the resulting suspension on semi-selective media for colony quantification. The half life of entomopathogenic fungi can be determined plotting the numbers of infection units or colony forming units against time. Equations can be adjusted to the observed values and the best model can be selected according to statistical parameters (Draper and Smith, 1981). Apparently, for the majority of cases, exponential cubic equations are the model that better explain fungal persistence (see Sosa-Gómez and Moscardi, 1998).

5 Conclusions

Considerable achievements have been attained in soybean with regard to application and evaluation of entomopathogens, especially for NPV and *Bt* against defoliating Lepidoptera. Fungi have not been practically used at the farmer level against Lepidoptera or Hemiptera in spite of progress attained with some fungi. In general, this entomopathogen group still deserves much research efforts towards improving techniques for production, field application, and evaluation. Among the major insect pests in soybean, only a small proportion of those are amenable to control by entomopathogens and for which necessary information for use as a microbial insecticide is available. There is a general lack of basic and applied information on entomopathogens associated with important insect groups in soybean. These include, soil insects such as scarabaeids, burrowing bugs (*Scaptocoris* spp.), stem-attacking girdle beetles, stem borers (*M. testulalis, E. aporema*), the stem fly complex, leaf rollers and pod borers. Despite the difficulties in controlling these insects because of their feeding habits, increased research effort involving entomopathogens and these insect pests is needed. The same is true for other invertebrates, such as slugs, snails and millipedes, which have become increasingly important in no-till soybean systems in Brazil.

6 References

Abbott, W. S. 1925. A method of computing the effectiveness of an insecticide. *J. Econ. Entomol.* **18**, 265–267.

Alves, S. B. and Pereira, R. M. 1989. Produção do *Metarhizium anisopliae* (Metsch.) Sook e *Beauveria bassiana* (Bals.)Vuill em Bandejas. *Ecossistema* **14**, 188–192.

Barreto, M. 2005. Prospecção e caracterização de genes de *Bacillus thuringiensis* com potencial para o controle de insetos-praga da cultura da soja. Universidade Federal do Paraná. PhD Thesis. 93 pp.

Bell, J. C., Hamalle, R. J. and Ignoffo, C. M. 1982. Methods and costs of producing *Nomuraea rileyi* conidiospores. New Orleans, USDA Southern Series 24, 7p.

Boucias, D. G., Johnson, D. W. and Allen, G. E. 1980. Effects of host age, virus dosage, and temperature on the infectivity o nucleopolyhedrosis virus against velvetbean caterpillar, *Anticarsia gemmatalis*, larvae. *Environ. Entomol.* **9**, 59–61.

Boucias, D. G. and Pendland, J. C. 1991. Attachment of mycopathogens to cuticle. The initial event of mycosis in arthropod hosts. *In* "The fungal spore and diseases initiation in plants and animals" (G. T. Cole and H. C. Coch, Eds.), pp 101–127. Plenum Publishing Corporation.

Carner, G. R. 1980. Sampling pathogens of soybean onsect pests. *In* "Sampling Methods in Soybean Entomology" (M. Kogan and D. C. Herzog, Eds.), pp. 559–574. Springer-Verlag, New York.

Carvalho, R. C. Z. de, Moscardi, F. and Foerster, L. A. 1991. Efeito do fungo *Nomuraea rileyi* (Farlow) Samson e de um vírus de poliedrose nuclear no consumo e na utilização de alimento por *Anticarsia gemmatalis* Hübner (Lepidoptera, Noctuidae). *ANn. Soc. Entomol. Brasil* **20**, 379–393.

Corrêa-Ferreira, B. S. 1985. Criação massal do percevejo verde *Nezara viridula* (L.). Londrina, Embrapa. CNPSo. 16 p. Documentos 11.

Dettenmaier, S. J., Rangel, D. E. N., Evans, E. W., and Roberts, D. W. 2005. Comparative susceptibility of *Metarhizium anisopliae* varieties *anisopliae* and *acridum* to the selective fungicide dodine. Program and abstracts for the 38th Annual Meeting of the Society for Invertebrate Pathology. Anchorage, Alaska. p. 48.

Devi, P. S. V. and Prasad, Y. G. 1996. Compatibility of oils and antifeedants of plant origin with the entomopathogenic fungus *Nomuraea rileyi*. *J. Invertebr. Pathol.* **68**, 91–93.

Draper, N. R. and Smith, H. 1981. "Applied Regression Analysis". 2nd ed. John Wiley and Sons, New York.

Evans, H. and Shapiro, M. 1997. Viruses. *In* "Manual of Techniques in insect Pathology". (L. A. Lacey, Ed.), pp. 17–53. Academic Press, London.

Fehr, W. R., Caviness, C. E., Burmood, D. T., and Pennington, J. S. 1971. Stage of development descriptions for soybeans, *Glycine max* (L.) Merrill. *Crop Sci.* **11**, 929–930.

Fuxa, J R. and Tanada, Y. 1987. Epidemiological concepts applied to insect epizootiology. In "Epizootiology of Insect Diseases" (J. R.. Fuxa and Tanada, Y., Eds.), pp. 3–21. John Wiley and Sons, New York.

Fuxa, J. E., Fuxa, J. R., Richter, A. R., and Weidner, E. H.. 2000. Prevalence of a trypanosomatid in the Southern green stink bug, *Nezara viridula*. *J. Eukaryot. Microbiol.* **47**, 388–394.

França Neto, J. B., Krzyzanowski, F. C., and Costa, N. P. da. 1998. The tetrazolium test for soybean seeds. Londrina. Embrapa CNPSo. 71 p. Documentos 115.

Gardner, W. A., Sutton, R. M., and Noblet, R. 1977. Persistence of *Beauveria bassiana, Nomuraea rileyi*, and *Nosema necatrix* on soybean foliage. *Environ. Entomol.* **6**, 616–618.

Gazzoni, D. L., Oliveira, E. B., and Moscardi, F. 1984. Estudos comparativos entre as fórmulas de Abbott e Henderson e Tilton. III. Validação do modelo matemático. In "Seminário Nacional de Pesquisa de Soja 3", Campinas 1984. Anais Embrapa CNPSo, Londrina. p. 201–217.

Gazzoni, D. L., Sosa-Gómez, D. R., Moscardi, F., Hoffmann-Campo, C. B.; Corso, I., Corrêa-Ferreira, B. S., and Oliveira, L. 1994. Insects. *In* "Tropical Soybean Improvement and Production" Plant Production and Protection Series, 27 (FAO, Ed.). pp. 82–112. Londrina, Brazil.

Goettel, M. S. and Inglis, G. D. 1997. Fungi: Hyphomycetes. *In* "Manual of Techniques in Insect Pathology" (L. A. Lacey, Ed.), pp. 213–249. Academic Press, London.

Harris, V. E. and Todd, J. W. 1981. Rearing the southern green stink bug, *Nezara viridula*, with relevant aspects of its biology. *J. Ga. Entomol. Soc.* **16**, 203–210.

Herzog, D. C. 1980. Sampling soybean looper in soybean. *In* "Sampling Methods in Soybean Entomology" (M. Kogan and D. C. Herzog, Eds.), pp. 141–168. Springer-Verlag, New York.

Herzog, D. C. and Todd, J. W. 1980. Sampling velvetbean caterpillar in soybean. *In* "Sampling Methods in Soybean Entomology". (M. Kogan and D. C. Herzog, Eds.), pp. 107–139. Springer-Verlag. New York.

Henderson, C. F. and Tilton, E. W. 1955. Tests with acaricides against the brown wheat mite. *J. Econ. Entomol.* **48**, 157–161.

Hoffmann-Campo, C. B., de Oliveira, E. B., and Moscardi, F. 1985. Criação massal da lagarta da soja (*Anticarsia gemmatalis*). Londrina. Embrapa CNPSo. Documentos 10, 23 p.

Hoffmann-Campo, C. B., Oliveira, L. J., Moscardi, F., Gazzoni, D. L., Corrêa-Ferreira, B. S., Lorini, I. A., Borges, M., Sosa-Gómez, D. R., Corso, I. C., and Panizzi, A. R.. 2003. Integrated Pest Management in Brazil,. In: "Integrated Pest Management in the Global Arena" (K. M Maredia, D. Dakouo, and D. Mota-Sanches, Eds.), pp. 285–299. CABI Publishing. Wallingford, UK and Cambridge, USA.

Holdom, D. G., and Klashorst, G. V. de. 1986. Inexpensive culture media and methods for *Nomuraea rileyi*. *J. Invertebr. Pathol.*, **48**, 246–248.

Ignoffo, C. M., Hostetter, D. L., Biever, K. D., Garcia, C., Thomas, G. D., Dickerson, W. A., and Pinnell, R. 1978. Evaluation of an entomopathogenic bacterium, fungus and virus for control of *Heliothis zea* on soybeans. *J. Econ. Entomol.* **71**, 165–168.

Ignoffo, C. M., Hostetter, D. L., and Pinell, R. E. 1974. Stability of *Bacillus thuringiensis* and *Baculovirus heliothis* on soybean foliage. *Ev. Entomol.* **3**, 117–119.

Ignoffo, C. M., Marston, N. L., Hostetter, D. L., Puttler, B. and Bell, J. V. 1976. Natural and induced epizootics of *Nomuraea rileyi* in soybean caterpillars. *J. Invertebr. Pathol.* **27**, 191–198.

Jackai, L. E. N., Panizzi, A. R., Kundu, G. G. and Srivastava, K. P. 1990. Insect pests of soybean in the tropics. *In* "Insect Pest of Food Legumes". (S. R. Singh, Ed.), pp. 91–156. John Wiley and Sons Ltd. New York.

Kogan, M. and Herzog, D. C. 1980. "Sampling Methods in Soybean Entomology". Springer-Verlag, New York.

Kogan, M. and Pitre, H. N., Jr. 1980. General sampling methods for above-ground populations of soybean arthropods. *In* "Sampling Methods in Soybean Entomology". (M. Kogan and D. C. Herzog, Eds.), pp. 30–60. Springer-Verlag, New York.

Kogan, M. and Turnipseed, S. G. 1980. Soybean growth and assessment of damage by arthropods. In "Sampling Methods in Soybean Entomology". (M. Kogan and D. C. Herzog, Eds.), pp. 3–29. Springer-Verlag, New York.

Kogan, M. and Turnipseed, S. G. 1987. Ecology and management of soybean arthropods. *Annu. Rev. Entomol.* **32**, 507–538.

Krueger, S. R., Duan, H., Rath, A. C., and Rotramel, G. L. 1995. Development of Bio-Blast™ biological termiticide. In "SIP 28th Annual Meeting. Society for Invertebrate Pathology,". p.35. Cornell University, Ithaca, New York.

Lacey, L. A., Frutos, R., Kaya, H. K., and Vail, P. 2001. Insect pathogens as biological control agents: Do they have a future? *Biol. Control* **21**, 230–248.

Magalhães, B. P. 1994. The use of fluorochromes in insect and plant mycopathology. *Rev. Microbiol.* **25**, 90–96.

McCullagh, P. and Nelder, J. A. 1989. Generalized linear models. Monographs on statistics and applied probability 37. Chapman & Hall, London. 511 pp.

Moraes, R. R., Loeck, A. E., and Belarmino, L. C. 1991. Inimigos naturais de *Rachiplusia nu* (Guenée, 1852) e de *Pseudoplusia includens* (Walker, 1857) (Lepidoptera: Noctuidae) em soja no Rio Grande do Sul. *Pesq. Agropec. Bras.* **26**, 57–64.

Morales, L., Moscardi, F., Kastelic, J. G., Sosa-Gómez, D. R., Paro, F. E., and Soldorio, I. L. 1995. Suscetibilidade de *Anticarsia gemmatalis* Hübner e *Chrysodeixis includens* (Walker) (Lepidoptera: Noctuidae) a *Bacillus thuringiensis* (Berliner). *Ann. Soc. Entomol. Brasil.* **24**, 593–598.

Morrow, B. J., Boucias, D. G., and Heath, M. A. 1989. Loss of virulence in an isolate of an entomopathogenic fungus, *Nomuraea rileyi*, after serial *in vitro* passage. *J. Econ. Entomol.* **82**, 404–497.

Moscardi, F. 1983. Utilização de *Baculovirus anticarsia* para o controle da lagarta da soja, *Anticarsia gemmatalis*. Comunicado Técnico No. 23. Centro Nacional de Pesquisa de Soja. 13 pp.

Moscardi, F. 1984. Efeito de aplicações de *Bacillus thuringiensis* sobre populações de *Anticarsia gemmatalis* em soja. *In* Seminário Nacional de Pesquisa de Soja 3, Campinas 1984. Anais Embrapa CNPSo. Londrina. p. 158–165.

Moscardi, F. 1989. Use of viruses for pest control in Brazil: the case of the nuclear polyhedrosis virus of the soybean caterpillar, *Anticarsia gemmatalis. Mem. Inst. Oswaldo Cruz* **84**, 51–56.

Moscardi, F. 1993. Soybean integrated pest management in Brazil. *Plant Prot. Bull.* **41**, 91–100.

Moscardi, F. 1999. Assessment of the application of Baculoviruses for control of Lepidoptera. *Annu. Rev. Entomol.* **44**, 257–289.

Moscardi, F. and Corrêa-Ferreira, B. S. 1985. Biological control of soybean caterpillars. *In* Proceedings of World Soybean Research Conference III (R. Shibles, Ed.), pp. 703–711. Westview Press, Boulder., CO.

Moscardi, F. and Sosa-Gómez, D. R. 1992. Use of viruses against soybean caterpillars in Brazil. *In*: "Pest Management in Soybean" (L. G. Copping, M. B. Green, and R. T. Rees, Eds.), pp. 98–109. Elsevier Applied Science, London.

Neisess, J. 1980. Effect of pH and chlorine concentrations on activity of *Bacillus thuringiensis* tank mixtures. *J. Econ. Entomol.* **73**, 186–188.

Panizzi, A. R. and Slansky, F. 1985. Review of phytophagous pentatomids (Hemiptera: Pentatomidae) associated with soybean in the Americas. *Fla. Entomol.* **68**, 184–214.

Panizzi, A. R., Niva, C. C. and Hirose, E. 1995. Feeding preference by stink bugs (Heteroptera: Pentatomidae) for seeds within soybean pods. *J. Entomol. Sci.* **30**, 333–341.

Pedigo, L. P. 1980. Sampling green cloverworm on soybean. *In* "Sampling Methods in Soybean Entomology". (M. Kogan and D. C. Herzog, Eds.), pp. 169–186. Springer-Verlag. New York.

Petersen, R. G. 1994. Agricultural Field Experiments. Design and Analysis. Marcel Dekker, Inc. New York.

Polanczyk, R. A., Silva, R. P. da., Fiuza, L. M. 2000. Effectiveness of *Bacillus thuringiensis* strains against *Spodoptera frugiperda* (Lepidoptera: Noctuidae). *Brazil. J. Microbiol.* **31**, 165–167.

Sharma, A. N. 1995. Insect-pests of soybean and their management. *In* "Technology for Increasing Soybean Production in India. (P. S. Bhatnagar and S. P. Tiwari, Eds.), pp. 26–41. Natl. Res. Center for Soybean. Technical Bull. 1.(revised), Indore, India.

Silva, M. T. B. 1986. Estudo comparativo sobre o uso do Baculovirus anticarsia: por diversos equipamentos de aplicação. *Trigo e Soja.* **87**, 22–29.

Silva, M. T. B. 1987. Baculovirus anticarsia: o vírus da lagarta da soja. *Trigo e Soja.* **94**, 24–31.

Soper, R. S. and Ward, M. G. 1981. Production, formulation and application of fungi for insect control. *In* "Biological Control in Crop Production" (G. C. Papavizas, Ed.), pp. 161–180. BARC Symposium number 5.

Sosa-Gómez, D. R. Borges, E., Licursi Viera I. H. T., Costa F. and Oliveira C. N. 2005. Trypanosomatid Prevalence of *Nezara viridula* (L.), *Euschistus heros* (Fabricius) and *Piezodorus guildinii* (Westwood) (Heteroptera: Pentatomidae) populations in Northern Parana, Brazil. *Neotrop. Entomol.* **34**, 341–347.

Sosa-Gómez, D. R., Boucias, D. G., and Nation, J. L. 1997. Attachment of *Metarhizium anisopliae* to the Southern Green Stink Bug *Nezara viridula* cuticle and fungistatic effect of cuticular lipids and aldehydes. *J. Invertebr. Pathol.* **69**, 31–39.

Sosa-Gómez, D. R. and Moscardi, F. 1995. Retenção foliar diferencial em soja provocada por percevejos (Heteroptera: Pentatomidae). *Ann. Soc. Entomol. Brasil* **24**, 401–404.

Sosa-Gómez, D. R. and Moscardi, F. 1996. Producción de vírus patógenos de ácaros e insectos. *In* "Microorganismos Patógenos Empleados en el Control Microbiano de Insectos Plagas". (R. E. Lecuona, Ed.), p. 223–236. Taller Mariano Mas. Buenos Aires. Argentina.

Sosa-Gómez, D. R. and Moscardi, F. 1998. Laboratory and field studies on the infection of stink bugs, *Nezara viridula, Piezodorus guildinii*, and *Euschistus heros* (Hemiptera: Pentatomidae) with *Metarhizium anisopliae* and *Beauveria bassiana* in Brazil. *J. Invertebr. Pathol.* **71**, 115–120.

Sprenkel, R. K. and Brooks, W. M. 1975. Artificial dissemination and epizootic initiation of *Nomuraea rileyi*, and entomogenous fungus of lepidopterous pests of soybeans. *J. Econ. Entomol.* **68**, 847–851.

Thiery, I. and Frachon, E. 1997. Identification, isolation, culture and preservation of entomopathogenic bacteria. *In* "Manual of Techniques in Insect Pathology" (L. A. Lacey, Ed.), pp. 55–77. Academic Press, London.

Wilding, N., Mardell, S. K. and Brobyn, P. J. 1986. Introducing *Erynia neoaphidis* into a field population of *Aphis fabae*: form of the inoculum and effect of irrigation. *Ann. appl. Biol.* **108**, 373–385.

Wilding, N., Mardell, S. K., Brobyn, P. J., Wratten, S. D. and Lomas, J. 1990. The effect of introducing the aphid-pathogenic fungus *Erynia neoaphidis* into populations of cereal aphids. *Ann. appl. Biol.* **117**, 683–691.

Yates, W. E. and Akesson, N. B. 1963. Fluorescent tracers for quantitative microresidue analysis. *Trans ASAE* **6**, 104–107.

Young, S. Y. and Yearian, W. C. 1979. Soil application of Pseudoplusia NPV: persistence and incidence of infection in soybean looper caged on soybean. *Environ. Entomol.* **8**, 860–864.

Chapter VII-6

Microbial insecticide application and evaluation: Cotton

D. C. Steinkraus and S. Y. Young
Department of Entomology
University of Arkansas
Fayetteville, AR 72701, USA

D. H. Gouge
Department of Entomology
University of Arizona, MAC
Maricopa, AZ 85239, USA

J. E. Leland
Novozymes Biologicals Inc.
Salem, VA 24153, USA

1 Introduction

Microbial insecticides have been widely tested on cotton pests in the USA and other countries. Application of microbial insecticides to cotton has consisted primarily of evaluating *Bacillus thuringiensis* (*Bt*) and *Helicoverpa zea* nucleopolyhedrovirus (*Hz*NPV) formulations. However, progress has also been made with entomopathogenic fungi, microsporidia, nematodes, and protozoa and these entomopathogens will also be considered in this chapter.

Microbial insecticides on cotton were studied and developed, in part, in response to development of chemical insecticide resistance by heliothines. Introduction in the early 1980s of the highly efficacious and cost-effective pyrethroid insecticides curtailed development of microbial insecticides. However, resistance of *Heliothis virescens* to pyrethroids renewed interest in microbials (Plapp and Campanhola, 1986; Leonard *et al.*, 1987; Luttrell *et al.*, 1987), and widespread use of *Bt* products in tank mixtures with chemical insecticides against heliothines occurred in the 1990s in the southern USA prior to the introduction of transgenic cotton varieties containing a *Bt* toxin gene.

Testing of microbial insecticides for cotton pests began in the 1960s (Ignoffo *et al.*, 1965; Allen *et al.*, 1966). Promising results from these and other similar tests were the impetus for widespread testing by industry in an effort to develop microbial pesticides against *H. virescens* and *H. zea*. By 1972, at least 28 tests had been conducted with *Hz*NPV alone (Yearian and Young, 1978). Many field tests were also carried out with *Bt*, although these were usually industry sponsored and results were often not reported in the literature. A major effort on application of microbial insecticides on cotton continued into the early 1980s (Roome, 1975; Bell and Kanavel, 1977; Durant, 1977; Pieters *et al.*, 1978; Luttrell *et al.*, 1981, 1983; Johnson, 1982).

In more recent years (late 1990s and 2000s) there has been a shift in emphasis from control of heliothines to control of sucking insect pests of cotton (Williams, 2004, 2005). This was in large part due to the introduction of *Bt*-cotton varieties reducing the need for chemical (or microbial) insecticides, the intro-

L.A. Lacey and H.K. Kaya (eds.), *Field Manual of Techniques in Invertebrate Pathology*, 427–455.

duction of Lepidoptera-selective insecticides, and the reduced need for applications targeting the boll weevil (*Anthonomus grandis*) after eradication programs in the USA. Thus, broad-spectrum chemical insecticides targeting heliothines and boll weevils that had previously provided coincidental control of sucking insect pests were greatly reduced in cotton systems. As a result, sucking insect pests emerged as the new challenge for entomologists to develop specific and sustainable insect control. Most research on microbial control of sucking insect pests of cotton has been done with entomopathogenic fungi because of their contact mode of action. Recent research has focused on management of sucking insect pests in a regional context, targeting populations on crops and wild host plants in which pest populations develop before moving to cotton. Management of populations in these alternate hosts will be considered later in this chapter.

Microbial insecticides have typically been less effective than the most efficacious chemical insecticides against pests on cotton. Successful use of microbial insecticides is dependent on satisfactory plant coverage and proper timing of applications. Failure to deliver the required concentration to feeding sites of pests may well be the single factor most limiting pathogen efficacy. This is especially true because many cotton pests, such as heliothines, the pink bollworm, and the boll weevil, are principally secluded feeders. This need has led to the testing of microbial pesticides with an array of applicators, nozzle arrangements and adjuvants (Young and Yearian, 1986). Although application of microbial insecticides on cotton has used many formulations, these have been restricted by the need for their commercial application to conform to application equipment used by producers. It would rarely be economically feasible for growers to purchase separate applicators to apply only biological insecticides. The similarity of producers' spray equipment and the need to limit plot size has restricted application equipment of microbial insecticide tests to boom-type sprayers applied from ground level. Although most testing of biological insecticides on cotton has been *Hz*NPV and *Bt* formulations, other groups of microbial insecticides and other viruses have been tested against several lepidopterous larval pests of cotton (Falcon *et al.*, 1974; Bell and Kanavel, 1975; Lindegren *et al.*, 1992; Young *et al.*, 1999). Only in a few instances have they been tested on cotton against insect pests other than lepidopterans (Akey and Henneberry, 1994; Brown *et al.*, 1997c; Hinz and Wright, 1997).

There have also been considerable efforts to manage cotton insect pests on a regional basis, targeting populations developing on alternate host plants before they move into cotton. Much of this work has been done in intensively-farmed agricultural areas where weed management within fields throughout the year restricts wild host plants to field margins representing a small fraction of the total land area. Area-wide management programs have targeted pest populations by destruction of wild host plants (Snodgrass *et al.*, 2000, 2003, 2005) and by application of pathogens such as viruses for heliothines (Bell and Hayes, 1994; Hayes and Bell, 1994; Bell and Hardee, 1994, 1995; Hardee and Bell, 1995; Streett *et al.*, 1998; Hardee *et al.*, 1999). More recently application of fungi for plant bugs (McGuire *et al.*, 2006; Leland *et al.*, 2005b) shows promise for selective targeting of these populations. There are several advantages to the use of pathogens in managing populations on alternate hosts in area-wide management programs: 1) in intensively-farmed agricultural areas alternate hosts are geographically restricted reducing the area that needs to be treated and making the relatively high cost of microbial pesticides more affordable; 2) chemical insecticides may not be appropriate for targeting populations wild host plant areas because of non-target effects; 3) reducing insect pressures in cotton without using chemical insecticides may slow the development of insecticide resistance that occurs within seasons and over the long term; and 4) alternate host plant habitat often serve as refuges for beneficial insects and selective control of pests with pathogens rather than by weed control or chemical insecticides may be a more sustainable option.

In previous years microbial biocontrol in cotton was only given serious consideration when chemical insecticides failed due to resistance development. When microbial biocontrol options were being considered, they were primarily evaluated following the chemical paradigm for treating populations in cotton. Microbial insecticides were discarded as control options whenever inexpensive effective chemical insecticide and/or effective transgenic cotton options arose. However, by considering special situations such as regional suppression of pest populations, we may be able to make use of the advantages pathogens have over chemical insecticides. Microbial biocontrol may provide options that are sustainable both biologically and commercially.

2 Major cotton pests in North America

A *Heliothines*

Heliothines (Noctuidae) are major pests of cotton worldwide. In the USA, Central, and South America, *H. zea* and *H. virescens* must be controlled annually on much of the cotton acreage. *Heliothis virescens* is resistant to conventional chemical insecticides such as organophosphates and pyrethroids. The high costs of insecticide applications and/or technology fees for transgenic cottons, along with crop losses, greatly reduce cotton profitability in areas where *H. virescens* is a pest. Throughout the remainder of the world, *Helicoverpa armigera* is a major pest, with insecticide resistance levels and crop loss comparable to *H. virescens*. *Helicoverpa armigera* is a susceptible host for entomopathogenic nematodes (EPN) and significant mortalities have been reported and enhanced by mixing antidesiccants with nematodes applied to cotton foliage (Glazer *et al*., 1992). In the future, genetically-modified baculoviruses may play a role in heliothine control (Sun *et al*., 2002). One disadvantage of baculoviruses is their slow speed of kill. Recombinant baculoviruses with genes for various toxins could remove this limitation.

The heliothine species complex's pest status and susceptibility to a variety of virulent entomopathogens made it the pest of choice in most studies involving microbial insecticides prior to the introduction of *Bt*-cotton. Since much of Section 3 "Application of Viral and *Bt*-based Insecticides" will be dedicated to work with heliothines, a description of these studies will be reserved for that section.

B *Pink bollworm*

The pink bollworm, *Pectinophora gossypiella* (PBW), is a major pest of cotton in the subtropics and tropics and is found throughout much of the cotton producing areas of the world. It is an important pest from Texas westward in the USA. The fruiting structures, primarily the seeds in bolls, are attacked, and it is most damaging late in the season. Heavy infestations can cause 100% loss of the crop. Management programs include the use of *Bt* cotton, pheromones for mating disruption and trapping, post season stalk destruction, and insecticide applications.

Several entomopathogens are known to affect the larval stage of PBW including: *Bt* (Ignoffo and Graham, 1967), a cypovirus (CPV) (Ignoffo and Adams, 1966), a nucleopolyhedrovirus (NPV) (Bell and Henneberry, 1980), and various entomopathogenic nematode species in the families Steinernematidae and Heterorhabditidae (Gouge *et al*., 1997). These entomopathogens are effective in the laboratory, but only entomopathogenic nematodes had a significant impact on PBW in the field (Graves and Watson, 1970; Bell and Henneberry, 1980; Gouge *et al*., 1996). Control using baculovirus has been improved with gustatory baits but not to the level desired for commercial use (Bell and Kanavel, 1978). *Bt*, when ingested by PBW larvae, is extremely toxic (Bartlett, 1993).

A number of entomopathogens have been tested against PBW. For example, first-instar PBW larvae were susceptible to a lactose dust formulation of *Bt* (Ignoffo and Graham, 1967). The commercial development of the HD1 isolate of *Bt* was an important step in making *Bt* products viable. Commercial production of such isolates would not be difficult as industrial methods for production were already known. Development of a larval feeding stimulant for PBW also increased the potential of *Bt* as a viable PBW control agent. Ignoffo and Adams (1966) described a CPV, *Smithiavirus pectinophorae*, found in PBW from a laboratory culture. Feeding larvae on virus-treated diet resulted in reduced pupal weight and pupation rate as well as larval mortality. Vail *et al.* (1972) demonstrated that the *Ac*MNPV (a multiple-embedded virus) isolated from the alfalfa looper, *Autographa californica*, also infected PBW larvae. Field tests found a low level of efficacy of this virus in cotton that was improved somewhat by its application in a bait formulation (Vail *et al.*, 1978). Ignoffo and Garcia (1965) found that PBW larvae were susceptible to the protozoan, *Mattesia grandis*, isolated from the boll weevil (McLaughlin, 1965). Like the bacteria and viruses, protozoa must be ingested to infect a host.

Entomopathogenic nematodes have a potential application in PBW management (Lindegren *et al.*, 1993a; Henneberry *et al.*, 1995a,b, 1996a). Larvae are highly susceptible to *Steinernema carpocapsae* and *S. riobrave* (Lindegren *et al.*, 1992, 1993b, 1994; Henneberry *et al.*, 1995a, b, 1996a). *Steinernema riobrave* has better host searching efficiency than *S. carpocapsae* (Lindegren *et al.*, 1993a) and is more tolerant of high temperatures (Henneberry *et al.*, 1996b), an advantage under desert growing conditions. Treatment of commercial cotton fields in Arizona (2.5 billion nematodes/ha) showed that *S. riobrave* persisted in large numbers for 19 days and were recovered up to 75 days after treatment (Gouge *et al.*, 1996). The numbers of cotton bolls infested by the PBW during the season were reduced and the cotton yield increased 19% compared with untreated cotton fields. Similar results were obtained with *S. riobrave* and *S. carpocapsae* in cotton fields in Texas (Gouge *et al.*, 1997). The timing of application, application method, and application rates have been further refined in recent research, and the implementation of entomopathogenic nematodes in area-wide management programs appears promising.

C Other Lepidoptera

The beet armyworm, *Spodoptera exigua*, is an occasional pest of cotton in the southern USA. Severe outbreaks of this pest can destroy the entire crop. It mainly feeds on the foliage of cotton but will also damage fruiting structures. Hot dry summers are favored by this pest. It is difficult to control on cotton with conventional chemical insecticides. *S. exigua* outbreaks often occur following insecticide applications directed against more susceptible pests on cotton due to reduction of natural enemies. *Bt* products have limited effectiveness against *S. exigua* larvae on cotton. Some control of this pest occurs on transgenic cotton varieties expressing a *Bt* derived Cry IA_2 insecticidal protein (*Bt*-cotton), but the survival rate is much higher than with other lepidopteran cotton pests (Stapel *et al.*, 1997).

The fall armyworm, *Spodoptera frugiperda*, is also an occasional pest of cotton, and is primarily a fruit feeder on cotton causing damage similar to heliothines. This pest is often found in mixed populations with heliothines, and chemical insecticides directed against these pests are also used to control *S. frugiperda*. *Bt* sprays and *Bt*-cotton do not appear to provide adequate protection of cotton from these pests (Adamczyk *et al.*, 1998; S. Y. Young, unpubl. data).

Gouge *et al.* (1999) demonstrated the relative susceptibility of several common US cotton pests to entomopathogenic nematodes. In laboratory assays, infection levels varied by host and temperature. In general, *P. gossypiella* was the most susceptible followed by *H. virescens, Trichoplusia ni*, and *S. exigua*. Two lepidopterous pests not occurring in the USA, the Egyptian cotton leafworm, *Spodoptera littoralis*, a polyphagous insect attacking a number of plant species, and *Earias insulana*, have both been shown to be susceptible to EPNs combined with antidesiccants (Glazer *et al.*, 1992).

The soybean looper, *Pseudoplusia includens*, is a foliage feeding pest of cotton, especially along the coastal areas of southeastern USA. It, like *S. exigua*, is an occasional pest that seldom occurs unless chemical insecticides have been applied against other pests and have eliminated its natural enemies. Eggs of *P. includens* are laid low on the plant and small larvae feed on the lower side of leaves low in the canopy. Larvae move up the plant as they consume foliage. They are not well exposed to insecticides until they move to the upper canopy which usually means they are large and difficult to kill with biological insecticides. This includes *Bt* products on cotton, where the control levels obtained have seldom been acceptable (Adamczyk *et al.*, 1997). Results of tests with the *P. includens* NPV on cotton have not been reported, but it is efficacious against *P. includens* on soybean (McLeod *et al.*, 1982).

D Boll weevil

The boll weevil was a major pest of cotton in the USA for many years. It is considered a key pest because early season chemical insecticide applications are usually required to control it, thereby setting the stage for outbreaks of other pests. This pest has several generations annually and overwinters as diapausing adults. Both larvae and adults attack all fruiting stages, with eggs usually laid in fruiting structures where larvae develop. Recent successful eradication programs in the southeastern USA, midsouthern USA, and Texas have greatly reduced the impact of this pest. Boll weevils can be managed with insecticides and pheromone traps. Natural enemies provide little control.

Boll weevils are not susceptible to baculoviruses or *Bt*. Immature stages are protected within bolls, and therefore, immature stages are unlikely to be affected by pathogens. Adults feeding and ovipositing on cotton are potentially vulnerable to pathogens as are overwintering adults in leaf litter. These adults have been the target for *Mattesia grandis* (a neogregarine) and *Glugea gasti* (a microsporidium) (McLaughlin *et al.*, 1968; McLaughlin and Bell, 1970) and the mycoinsecticide *Beauveria bassiana* (McLaughlin, 1962). Third-instar boll weevils were susceptible to *Steinernema riobrave, S. glaseri, Heterorhabditis indica* and *H. bacteriophora* in petri dish assays and in buried cotton bolls (Cabanillas, 2003).

E Plant bugs

Plant bugs (Miridae) are serious pests of cotton throughout the USA. In the southern USA, the tarnished plant bug, *Lygus lineolaris*, is the primary hemipteran pest. In some regions the clouded plant bug, *Neurocolpus nubilus*, is a significant pest. In western cotton-growing regions, *Lygus* species also include the western tarnished plant bug, *Lygus hesperus*, and the pale legume bug, *Lygus elisus*. Damage is caused by both the nymphal and adult stages. Fruiting structures, primarily squares, are fed upon causing them to die and fall to the ground. In addition, loss of apical meristematic tissue may cause secondary growth and excess branching of the plants.

Lygus spp. overwinter as diapausing adults. Plant bugs have extremely broad host ranges and the adults are highly mobile. Therefore, management of *Lygus* populations in cotton requires understanding, and potentially targeting, population development on alternate hosts. In southern cotton-growing regions of the USA, *L. lineolaris* populations often build up in flowering weeds surrounding cotton (Snodgrass *et al.*, 1984a, b) and weed control may help reduce plant bug infestations as has been evaluated in area-wide management programs (Snodgrass *et al.*, 2000, 2003, 2005). In the drier western cotton-growing regions, flowering weeds support populations during the wet months of winter and early spring. Populations then develop in alternate crops, such as alfalfa, before moving to summer crops such as cotton (Goodell *et al.*, 2000; Carriere *et al.*, 2006).

Typically plant bugs in cotton are managed with chemical insecticides. Historically, plant bug control was often coincidental with control of boll weevils and heliothines. Recent reductions in insecticide applications for boll weevil following eradication, and for heliothines following the introduction of *Bt*-cotton, have resulted in plant bugs increasing as a problem. In spite of these reductions, *Lygus* resistance to insecticides has become widespread (Hollingsworth *et al.*, 1997; Snodgrass, 1996; Snodgrass and Scott, 2002; Zhu and Snodgrass, 2003; Zhu *et al.*, 2004). This has led to an increase in crop damage and insecticide use against this pest. Trap cropping may be a useful tool in plant bug management (Stewart and Layton, 2000).

Pathogens, other than fungi, are unlikely to provide satisfactory control of sucking pests such as plant bugs. There has been some research on screening of *Bt*-isolates and toxins against plant bug species in laboratory bioassays but to date only isolates containing the broad spectrum ß-exotoxin have exhibited toxicity against plant bugs (Brandt *et al.*, 2004; Wellman-Desbiens and Côté, 2004, 2005).

There has been considerably more work on the use of entomopathogenic fungi for plant bug control. *B. bassiana* isolates from *Lygus* spp. have been discovered (Steinkraus and Tugwell,

1997; McGuire, 2002; Leland and Snodgrass, 2004), characterized in the laboratory (Noma and Strickler, 2000; Liu *et al.*, 2002; 2003; Leland, 2005; McGuire *et al.*, 2005; Leland *et al.*, 2005b) and evaluated in field trials against *Lygus* spp. in cotton and alternate hosts (Snodgrass and Elzen, 1994; Brown *et al.*, 1997a, b, c; Steinkraus and Tugwell, 1997; Noma and Strickler, 1999; McGuire and Leland, 2006; Lund *et al.*, 2006; McGuire *et al.*, 2006). Fungi may provide an alternative to chemicals for control of *Lygus* spp. developing on non-cotton hosts in a regional context, with reductions in non-target impacts and slowing development of resistance compared with current chemical insecticide-based management strategies in cotton.

F *Stink bugs*

Three species of stink bugs are common cotton pests in the southeastern USA: the green stink bug, *Acrosternum hilare*, the southern green stink bug, *Nezara virudula*, and the brown stink bug, *Euschistus servus*. Their life histories in the southeastern USA are similar as those described above for plant bugs. Adults overwinter in diapause and wild host plants play a major role in spring population development. However, stink bugs have two major differences from plant bugs. First, soybeans also play a major role in population development and second, stink bugs are damaging to cotton later in the season than plant bugs. Because stink bugs have piercing-sucking mouthparts, entomopathogenic fungi may be the best microbial biocontrol option for controlling stink bugs. However, very little research has been done to date on microbial biocontrol of stink bugs and some research indicates stink bugs may have natural defense mechanisms that are fungistatic to fungi (Sosa-Gomez *et al.*, 1996; Sosa-Gomez and Moscardi, 1998). Therefore, microbial control of stink bugs may not be feasible. Efforts should be made to identify new species of natural pathogens of stink bugs.

G *Whiteflies*

The silverleaf whitefly (SLWF), *Bemisia argentifolii*, is a serious pest of cotton. It is especially difficult to control because it is found on hundreds of different host plants. The short distances between weed hosts and cultivated crops lead to rapid colonization. Population numbers can increase exponentially on cotton. In addition, conventional chemical control technologies have been unsatisfactory because SLWF infests the undersides of leaves and has also developed resistance to some chemicals (Butler and Henneberry, 1994). The SLWF feeds by sucking phloem sap from the leaves and can excrete more than 300 kg of honeydew/ha in heavily infested cotton fields. When honeydew falls on open bolls, it causes the cotton fibers to become sticky and eventually moldy. Sticky cotton causes severe problems for cotton gins and textile mills (Hendrix *et al.*, 1996). Like most sucking insects, it is difficult to use entomopathogens that must be ingested for control of this pest. Therefore, the most likely candidates for microbial control of sucking pests are entomopathogenic fungi. One fungal species, *Paecilomyces fumosoroseus*, is a common pathogen of *Bemisia* spp. on cotton and other crops (Lacey *et al.*, 1996). It has been responsible for epizootics in *Bemisia* spp. populations and has been evaluated against SLWF with promising results (Carruthers *et al.*, 1993; Lacey *et al.*, 1996; Akey and Henneberry, 1998). Commercial formulations of *B. bassiana* also show promise for control of this pest (Akey and Henneberry, 1994; Hinz and Wright, 1997). Recently a new species of fungal pathogen, *Orthomyces aleyrodis*, was described that caused natural epizootics in bandedwinged whitefly, *Trialeurodes abutilonea*, populations in cotton in Alabama, USA (Steinkraus *et al.*, 1998). Further study of *O. aleyrodis*, and discovery of other new fungal pathogens of whiteflies, could yield new candidates for microbial pesticides.

H *Aphids and other pests*

The cotton aphid, *Aphis gossypii*, is a pest of cotton throughout the USA. It sucks phloem sap from cotton leaves and excretes honey dew. Honey dew contaminates open bolls, causing sticky cotton, a serious problem for cotton gins and mills. High populations that damage the crop often occur after chemical insecticides used against other pests have eliminated beneficial insects. Chemical insecticides may be somewhat effective early in the season, but resistance in the population develops rapidly and insecticides will often not provide effective control later in the season. The fungus, *Neozygites fresenii*, attacks the aphid and in the southern USA and epizootics will often control the aphid populations (Hollingsworth *et al.*, 1995). More information on the effectiveness of *N. fresenii* and a program designed to maximize its effectiveness is presented in Chapter V-1.

Many other pests damage cotton, such as spider mites (Tetranychidae) and thrips (Thysanoptera). The entomopathogenic fungus, *Neozygites floridana*, has been reported from spider mites on cotton (Carner and Canerday, 1968); however, few attempts to control these pests with microbial insecticides in cotton have been made and these pests will not be considered here further. Additional information on the many arthropod pests of cotton can be found in Matthews and Tunstall (1994).

3 Application of viral and *BT*-based insecticides

A *Background*

Microbial insecticide tests on cotton, with few exceptions, have been conducted using ground equipment against heliothines. This pest complex will be emphasized in the description of the application methods used on cotton. Control of heliothines with NPVs and *Bt*-based insecticides has been erratic. Since heliothines are primarily secluded feeders, feeding on plant terminals and fruiting structures, it is a challenge to apply the pathogen to feeding sites. Feeding sites differ with stage of larval development, stage of crop and location of egg deposition on the plant. In addition, cotton plants grow rapidly and within days the new terminals grow away from microbial insecticide deposits. Consequently, many trials using viruses or *Bt* products have not resulted in significant increases in yield. While viral and *Bt* products often provide a level of control, applications usually have not been carefully timed for optimum efficacy or were applied against high population levels that cannot be managed with microbial insecticides alone. Efficacious treatment of cotton with pathogens for lepidopteran control must be timed against neonates to second instar.

B *Application equipment*

Various types of equipment have been used to apply microbial insecticides. However, with few exceptions, ground applications of microbial insecticides in field trials have attempted to mimic coverage by tractor mounted boom-type sprayers, but usually on a much smaller scale. Much of the variability in equipment used in tests to evaluate viral pathogens, especially in large tests, is due to the equipment available to the researcher. Also, the equipment type has to fit the size of the test area. In most tests with microbial insecticides, the treatments will be less efficacious than recommended chemical insecticides; thus, crop damage, crop value and potential financial loss will often dictate that ground equipment be used that is designed for small areas. Plot size is often 0.04 ha or less. Some equipment choices are determined by the need to fulfill application parameters in tests designed to improve efficiency through improved formulations and application methodology. These changes have ranged from changing sprayer type (Stacey *et al.*, 1977a) to altering spray volume (Ali and Young, 1993).

Many tests have involved adding adjuvants to the formulation, or more often, to the tank mixture, just prior to application. These adjuvants may require changes in application equipment, especially nozzle size, spray pressure and volume, etc., to apply the desired rate and obtain uniform coverage. In commercial cotton production in developed countries, application equipment consists almost entirely of fixed-boom, high clearance ground sprayers and airplanes. Most late season commercial cotton is sprayed by airplane, but airplanes are not used in replicated experimental plots on cotton because plot sizes are too small. Experimental plots have been sprayed with boom-type ground equipment such as a high-clearance tractor mounted sprayer (Young *et al.*, 1997), a bicycle type sprayer (Stacey *et al.*, 1977b), or a CO_2-powered back-pack sprayer (Bell and Kanavel, 1977). Few replicated efficiency evaluations have had plots large enough to use a tractor mounted sprayer.

However, area-wide management programs with viruses for heliothines in cotton-growing regions of the Midsouth have been conducted with aerial applications allowing coverage of areas over 80,000 ha (Bell and Hayes, 1994; Hayes and Bell, 1994; Bell and Hardee, 1994, 1995; Hardee and Bell, 1995; Streett *et al.*, 1998; Hardee *et al.*, 1999). Aerial applications were calibrated to deliver 100 larval equivalents (LE) (one LE equals 6 billion polyhedral occlusion bodies) of *Hz*NPV (ELCAR, Sandoz Crop Protection, Inc., Des Plaines, IL) in 9.4 liters of water containing 4% (v/v) crop oil spray additive (SuperPost, Walker Formulating Co., Olive Branch, MS) per ha by fixed wing aircraft. Aircraft were configured with 30 tee-jet, flat-fan, 8006 tips with 25 mesh slotted screens (Spray Systems Co., Wheaton, IL) at a boom pressure of 2.8×10^3 g/cm^2 (40 psi = 275 kPa) while traveling at 192 km/h. The virus (2,271 liters) was mixed thoroughly in 3,785-liter tanks allowing space for mixing. During some wet springs of this area-wide management program, truck-mounted mist blowers were used to treat excessively weedy areas along major roadways. Truck mounted mist blowers were calibrated to deliver 100 LE in 20 liters/ha along the side of the truck in an 18-m swath at 8 km/h.

In addition to airplanes and tractor mounted or manual boom-type sprayers, viral insecticides have occasionally been applied with dusters (Montoya and Ignoffo, 1966), aerosol generators, granulators (Stacey *et al.*, 1977b), and knapsack sprayers (Roome, 1975). However, these types of applicators are rarely used in cotton production systems.

C Field application

1 Plot setup

The necessity of replicating treatments for statistical analysis of data and the potential for significant economic yield loss with products of less than desired efficacy usually limit test size in cotton. In order to maximize the chances of having a heliothine population of sufficient number to result in economic damage and effectively test the products, it is necessary to place the test in a locality that has or can be expected to have a damaging population of the pest. Most cotton fields are much larger than the test area, and conventional insect pest management in the area of the field that is not used in the test will usually be carried out, often by airplane. It is desirable to locate the experiment in a corner of the field and clearly flag the field boundaries to minimize over spraying of plots from pilot error or indifference or spray drift. Leave an unsprayed border of at least 4 m around your test if the remainder of the field is to be sprayed with ground equipment and 20 m if it is to be sprayed by airplane.

A typical experimental design is a randomized, complete block with a minimum of three replications, with four replications preferred. Plot size is usually 0.04 ha and seldom smaller than 0.02 ha, with two rows for a buffer between plots and 3.0–4.5 m of buffer at ends of plots. Plots are 4 to 16 1.0 m rows and 30 m in length. Larger plots, up to 0.4 ha, may be obtained using a tractor mounted sprayer, but often result in sufficient yield loss that financial contributions to the grower will be necessary. The monetary value of such losses cannot be predicted but can easily be hundreds of US dollars per ha. If the cotton plots are routinely scouted for insect pests and damage, crop loss may be minimized by over spraying the test with the most effective chemical insecticide but will usually require that the experiment be terminated prematurely. Large plots such as 0.4 ha are necessary if commercial tractor-mounted sprayers, which cover 16 or more rows 1.0 m in width, are to be used.

After plots are setup remove plants from the alleyways between blocks and the ends of the test area so that the spray equipment may be easily moved about and the plots can be easily identified. To identify plots, each one must be clearly marked at both ends. This may be done using a small plastic flag on a wire stake or a narrow wooden stake. The stakes must extend above the height of the plants when the test is completed. This may be as high as 1–2 m. Place poles that extend approximately 2 m above the plants around the borders of the test area and mark the top of these with colorful ribbon or some other means so that the applicator spraying the remainder of the field can easily identify the test area.

2 Environmental data

Environmental data are not usually recorded in field efficacy evaluations involving viruses or *Bt*. The necessary equipment to obtain data may not be available, or the responsible scientist may believe that microbial insecticide tests must be identical to those of commercial chemical insecticides. The amount and date of rainfall should be noted since spray deposits may be washed from the plant. Collection of additional data, such as wind speed and direction, temperature, humidity, ultraviolet light from sunlight, and spray droplet size and number could be useful and may be required for field tests designed to test the effects of selected environmental factors on efficacy of experimental formulations (Burges, 1998).

3 *Stability and initial virulence of microbial insecticides*

The concentration of active ingredient in the product is determined by bioassay for either viral or *Bt* products if produced in the laboratory or unknown when obtained from another source (Evans and Shapiro, 1997; McGuire *et al*., 1997). Commercial products will have the product concentration on the label and usually recommended rates to use are attached or in their protocol. However, one cannot be certain that products have the desired insecticidal activity without confirmation by bioassay. Viruses should have been kept cold, preferably refrigerated. Microbial products do not always remain stable long in farm storage facilities if exposed to the high temperatures experienced in the cotton producing areas of the world, or in some instances at room temperature. Aliquots of the inoculum need to be taken at the beginning and end of the application period and stored refrigerated until assayed to determine if any activity was lost before or during the test period. Many field evaluations, some involving many localities, have been conducted on microbial insecticides that had reduced or even negligible activity. However, these data are seldom published and the scientist may never know the reason for failure of the treatment.

4 *Pre-treatment data*

Take pre-treatment data to determine heliothine numbers and crop damage. The sampling method used may differ but is usually that used commercially in the state or cotton growing area. Take data from the centermost rows of the plot. The test data taken will differ somewhat between cotton growing areas of a country or within an area. Data will be derived from a visual search for heliothines on terminals, fruiting structures, or the whole plant, at several locations in the field. It must be such that it can be analyzed statistically. This often includes counting eggs, larvae, and damaged and total fruiting structures in the terminals or on whole plants within a certain length of row (typically 2–5 m) within each plot. Eggs are usually found on terminal leaves, but are also found on older leaves and bracts of fruiting structures. Larvae are usually in terminals and fruiting structures. Pre-application data should include all that will be taken in post-application counts. The sampling method we use is a point sample technique in which whole plant searches are conducted on 4.3 m of row at four locations within a field (Lincoln *et al*., 1970). Data taken include total large and small larvae, eggs, and total healthy and damaged squares, blooms, and bolls. The treatment level will be based on the number of eggs or small and large larvae per area. This level will differ with stage of crop development.

5 *Sprayer setup*

The sprayer will include a boom with nozzles, evenly spaced, usually two nozzles per row, such as two nozzles 51 cm apart per 102 cm wide row. The boom may cover any number of rows depending on the spray equipment used. A hand-held back-pack sprayer could have a one- or two-row boom, whereas the boom on a tractor mounted sprayer often covers up to 24 rows. The nozzle size and speed of travel will vary with the volume of spray. The volume of choice for a back-pack or small automated sprayer is often 96 liters/ha, with TX-6 nozzles (Spraying Systems Co., Wheaton, IL) and a speed of 3.25 km/h. A change in nozzle size or type will necessitate a change in sprayer pressure and(or) speed of travel. A tractor sprayer used in commercial production would likely apply $\leq$48 liters/ha and travel $\geq$8 km/h. The sprayer must be calibrated before use to determine that the desired volume per area will be applied. This is usually done using water or water plus any tank mix material that might alter the flow rate of the product to be tested (see Chapter III-1). Rates of *Bt* will usually vary between 1×10^{10} international toxic units (ITU)/ha and 4×10^{10} ITU/ha and NPV rates will vary from 5×10^{12} to 5×10^{13} polyhedral occlusion bodies (OB)/ha. Lower rates of *Bt* or viruses are sometimes applied in mixtures with other insecticides.

The microbial insecticides to be tested may be mixed with water before or after they are placed in the sprayer. Some method of agitation in the sprayer is desired and sometimes necessary to assure that the product stays in suspension. Since microbial products often contain large particulate materials from insect or diet formulation, a screen in the spray line is needed to assure that nozzle tips and screens are not clogged. Ideally, apply treatments when wind is low (8 km/hour or less) to assure good coverage and minimize drift. This will often be soon after dawn or near dusk. However, optimal conditions are not always available and it may be necessary to spray when the wind is higher than desired.

The number of products to be tested and the rate(s) of each to be used in the test will be influenced by the plot space available, labor needed to apply the materials and make evaluations, the quantity of product available, and information desired by those supplying the products. The number of treatments chosen will often be less than desired due to labor, space, time, and cost constraints. Usually, the application rates recommended by industry are those that are considered financially necessary in order to market the product. If mixtures are to be used, each component of the mixture should be tested separately, although this may not be in the protocol if the test is being done for industry. The treatments may be applied one or more times according to the protocol. When applied more than once, there are usually 5 days between applications on cotton. A control and insecticide standard treatment should be included in the test. The insecticide standard is one commonly used by growers in the area. This allows comparison of the test products with those currently in commercial use.

6 *Post-treatment data*

Post-treatment samples are taken in the same manner as the pre-treatment data for both *Bt* and viral products. Take data from the centermost rows of the plots. This typically means that data are taken at 3–4 and 7 days, and sometimes in addition, at weekly intervals, thereafter. The interval between sampling dates may be altered one or more days by weather or the availability of labor. Data are often taken throughout the remainder of the growing season and crop yields are taken. In addition to taking counts of insect numbers, it may be desirable to make one or more live larval collections from plots, particularly at 3–4 and 7 days. The larvae are reared on semisynthetic diet until death or pupation, and growth rate and mortality determined. It is often desirable to hold the collected larvae until adult emergence for species determination. Larvae from virus-treated plots are examined for viral infection (see procedures of Evans and Shapiro, 1997). This will verify mortality due to the virus and indicate if epizootics are initiated in the field. Even though epizootics have seldom been observed in cotton fields, they have been induced when larval densities were high (S. Y. Young, unpubl. data).

D *Unconventional application methods*

Unconventional applicators have also been used to test microbial insecticides on cotton. One approach is to use small diameter droplets as a way to provide more thorough coverage than conventional boom-type sprayers (Falcon *et al.*, 1974; Smith *et al.*, 1978). Small diameter droplets often result in increased coverage and increased efficacy of chemical insecticides. Airbase (air-blast) sprayers were used to test this approach with NPV and *Bt* products against *H. zea* and *H. virescens* on cotton (Falcon *et al.*, 1974; Stacey *et al.*, 1980), but only sketchy procedures were given for either test. A plot size larger than that for boom-type sprayers is typically required for an airbase sprayer. The high pressure and smaller droplet size result in greater drift than obtained with other spray systems. A self-propelled, boom-type hydraulic nozzle, high clearance sprayer can be modified to permit use with an airbase system, such as a model 141 "Span Spray" system to make an air-blast sprayer (Stacey *et al.*, 1980). Stacey *et al.* (1980) mounted two AB fans on the back of a sprayer boom. Coverage width with the fans was 10 rows. Plots were 15 rows, 100 cm in width by 30.5 m in length, with 5 row buffers between plots. Spray volume was regulated by changing orifice plates, with speed and pressure constant at

3.6 km/h and 2.1 kg/cm^2, respectively, and low spray volumes, such as 96 liters/ha. Since this system can be used in replicated plots, products can be mixed for application and data taken as with the conventional boom-type sprayer. Limitations of the airbase sprayer in evaluating microbial insecticides are that much larger plot size is required and application must be made with little wind to minimize drift.

Two cold fogger generators for aerosol application of *Hz*NPV and *Bt* were used by Falcon *et al.* (1974) in unreplicated tests against *H. zea, Trichoplusia ni* and *S. exigua* in California. The generators were tractor mounted. Two machines were tested. A Microgen fogger equipped with 4 BelAir nozzles operated at 100 psi (690 kPa) and 10–15 psi (69–103 kPa) fluid resulted in droplets of 10–40 μm. A second machine that produced a coarse aerosol was equipped with a Calblower and two twin fluid nozzles, and used pressures up to 120 psi (825 kPa) and 70 psi (480 kPa) fluid. Plot size was not given but swath width was approximately 77 m. Coverage was improved at the higher pressures. Larval mortality in the test was as high as 100% near to the spray source but decreased with the distance from the source.

1 Granular application

Granular application is another method used in one study to test dry viral and *Bt* formulations in small-plot efficacy trials (Stacey *et al.*, 1977b). This may offer an advantage when using a large concentration of adjuvants, such as gustatory stimulants, with the insecticide. To prepare the microbial insecticide, the virus is partially purified, freeze-dried, and diluted with powdered cellulose to the desired concentration (6×10^9 OB/g) before mixing with dry bait. A variety of dry baits that are gustatory stimulants for heliothines may be used. The bait materials are mixed with a small amount of the viral preparation to obtain the concentration to be applied. The viral granules may be applied with a hand-held Casoran granular applicator at a rate of several kg/ha. Although the investigators were able to apply the treatments using this applicator, the granular baits provided less control than the virus sprayed alone. In addition, it must be noted that large scale applicators are unlikely to be available to use this method commercially.

2 Dust application

Dusts have rarely been used to apply chemical insecticides during the past 50 years and have received little attention for the application of microbial insecticides. One test reported was the application of NPV against *H. zea* on cotton (Montoya and Ignoffo, 1966). The virus was applied alone and with a corn-based gustatory stimulant.

To prepare the viral dust, 40 ml of a viral suspension plus corn extract in water was mixed with 25 g attapulgite clay, the mixture freeze-dried under vacuum, triturated, sifted through a 200-mesh screen, and blended into a homogeneous dust. The viral dust was applied at 12.9 g/ha using a hand-dust gun. Viral rates in the dust ranged from 9.7×10^5 to 4.9×10^6 OB/g. A viral spray at 6×10^{11} OB/ml was included for comparison. The treatments were applied to 0.04 ha plots replicated four times, with 10 applications made over the length of the season. Results from damaged fruiting structures and yield data showed that the viral dust at all rates was less effective than the viral spray at 6×10^{11} OB/ml.

E Adjuvants

The lack of desired efficacy in most tests with microbial insecticides on cotton, thought to be due in part to their rapid inactivation under field conditions and inadequate coverage of the plant, has encouraged the development of adjuvants for tank mixtures at application that will enhance their performance. Efforts to develop adjuvants have been particularly emphasized for *Bt* and viral insecticides against cotton pests. Rapid sunlight inactivation under field conditions has been thought to play an important role in limiting efficacy of microbial insecticides on cotton. Tests have also been conducted with emulsifiers and spreader-stickers with a wide variety of products tested by industry and governmental agencies. The other group of adjuvants that has received attention has been gustatory stimulants. The most promising of these adjuvants have been those that include properties of sunlight screen, gustatory stimulant and thickener. Coax (see 3. Gustatory stimulants below) is one such adjuvant (Potter and Watson, 1983a, b). Subsequently, stilbene fluorescent brighteners, such as Tinopal B, have been tested as adjuvants for NPVs (Hamm *et al.*, 1994). However, there are no reports of stilbenes being used on cotton.

Many of the field tests conducted with microbial insecticides on cotton have been designed to test the benefits obtained from including adjuvants in the tank mixes and are usually designed to increase pathogen persistence or act as gustatory stimulants for the larval pest. These adjuvants are prepared from a variety of sources and by different means and can require adjustments in the application equipment. Regardless of their designed use, they vary widely in the amount of particles and their size, density and viscosity of particulate matter. The following describes the types of adjuvants that have been used with microbial insecticides against heliothines on cotton.

1 Sunlight protectants

Types of sunlight protectants often tested are sugars (molasses), charcoal, starches, optical brighteners, and lignin sulfate (Roome, 1975; Pieters *et al.*, 1978; Yearian *et al.*, 1980; Ignoffo *et al.*, 1990; Shapiro and Vaughn, 1995). Sunlight protectants are usually added at concentrations of 1% or more. At these high concentrations, care must be taken to have adequate agitation in the spray tank to place them in solution or keep them in suspension. Failure to accomplish this can result in clogged spray lines or nozzles. These additions often alter spray properties; for example, sugars increase the viscosity of the liquid and the droplet size of the spray, but decrease drift. When this happens, the tank mixture often does not spray well at lower pressures. A novel approach has been to add oxidative enzymes and antioxidants as spray tank adjuvants to protect *Hz*NPV against inactivation by sunlight (Ignoffo and Garcia, 1994).

As mentioned above, attention has been focused on stilbene fluorescent brighteners as adjuvants which are sunlight protectants as well as synergists for NPV of lepidopterous larval pests. These materials have UV-absorbing properties and are usually solids. These usually have adequate spray properties at low concentrations, 0.1% or less, but under field conditions have only been effective at concentrations of 0.33–1.0%. At these high concentrations, the optical brighteners will have a high percentage of solids and the aggregates will rapidly settle out of suspension, plugging the sprayer hoses, screen and nozzle tips.

2 Surfactants

Surfactants are liquid surface-active agents, such as emulsifiers, wetting agents, and spreader-stickers, that are designed to improve spray characteristics and increase plant coverage. They are added to the spray tank at low concentrations, such as 0.1% of the spray mixture. A commercial spreader-sticker is usually added to the tank mixture when microbial insecticides are used on cotton to increase spread of droplets over the plant surface and/or to reduce wash off of the active ingredient by rain or dew. Commercial microbial insecticides, however, usually have surfactants in the formulated product.

3 Gustatory stimulants

Gustatory stimulants tested against lepidopteran pests of cotton have often been food products. Some of these have been in powdered form such as wheat or powdered milk, or sugars, such as sucrose, or in liquid form, such as molasses, and invert sugars, or cotton seed oil and used without further preparation (Montoya *et al.*, 1966). Extracts, often freeze-dried, prepared from host plant seeds such as garbanzo beans, cotton, corn and clover have also been tested (Stacey *et al.*, 1977a). Furthermore, industry has tested a variety of adjuvants, some of which had a high concentration of large particulate material. As with sunlight protectants, many gustatory stimulants need agitation to place them in solution or to stay suspended. It is recommended that a surfactant be included in the tank mix. Plant seed extracts are prepared by soaking seeds for several hours in a minimum of water, homogenizing, then filtering to remove larger particulate materials. Extracts must be stored at a low temperature to prevent spoilage. If adequately filtered, extract adjuvants should be easily sprayed. However, as with most other types of adjuvants, the higher the concentration used, the greater is the chance of improving microbial insecticide performance, but at the cost of increased difficulty in spraying.

Very complex crop seed-based adjuvants have been tested for use with viral and *Bt* insecticides against heliothines on cotton. These adjuvants often improve crop protection more than the individual components used separately. As an example, Andrews *et al.* (1975) developed a cotton seed oil-based bait by mixing cotton seed oil, 21.9%; invert sugars, 10.8%; Dacagin™, 0.5%; Thixin R™, 0.5%; hydroxyethylcellulose, 1.1%; and water, 65.2%. This bait was costly and not easily sprayed using commercial spray equipment, but at 5.8 kg/ha, it significantly improved performance of *Hz*NPV against heliothines on cotton. This bait was later simplified to consist of cotton seed flour, 5%; cotton seed oil, 1%; and sugar, 2% (Bell and Kanavel, 1978). Other similar baits have since been developed and sold by industry. Coax AE (Wilbur-Ellis Co., Ft. Worth, TX) is cotton seed flour and oil based, and Gustol AE (Sandoz, Inc. San Diego, CA) has a high concentration of soybean meal and oil. As can be seen, adjuvants with similar properties are readily made. An emulsifier should be included at 0.1% to improve spray properties. Rates of Coax or Gustol must be high to be effective on cotton making the products prohibitively expensive.

4 Application of nematode, protozoan, micropsoridian, and fungal-based insecticides

A Introduction

While the majority of tests of microbial insecticides on cotton have been with baculoviruses and *Bt* for control of heliothines, a number of tests have been made using other pathogens, such as entomopathogenic fungi, protozoa, microsporidia, and nematodes, against various cotton pests. These other insects, include some of the most serious pests of cotton including PBW, *A. grandis, L. lineolaris*, and SLWF. Selected entomopathogens that have been tested and the methods used will be discussed for each pest.

The following data should be collected and recorded during most tests of entomopathogens for cotton pests: cotton variety and planting date, field location and size, soil type, temperature and rainfall, growth stage(s) of cotton, and dates of treatments and sampling. The equipment used for application should be described in detail, especially the boom pressure, nozzle types and arrangements, liters of material/ha, and applicator speed. Additives used, such as chemicals or adjuvants, should be described. If water is used in preparing the pathogen, it is good practice to use non-chlorinated water because chlorine might have a negative effect on the pathogen. Apply entomopathogens at dusk or at night in order to prevent deactivation by sunlight. The source of insect pests used in the test should be recorded, *i.e.* laboratory culture, wild collected, natural populations, etc.

In almost all cases, field experiments in cotton involve a randomized complete block design, with 4 replications of each treatment including a control. Plot sizes vary depending on the budget of the researchers, the intent of the experiment, the quantities of microbial agent available, and labor available. Generally, plot sizes are less than 1 ha. Optimally, data will be collected on population densities of the target insect, mortality induced by the pathogen, insect damage to the crop, and yield. In most cases, means are derived from the data, and ANOVA and various tests, such as LSD t-tests, are used to separate the means (for further statistical information, see Chapter II-1). Whenever possible, microbial insecticides should be simultaneously compared with standard recommended chemical insecticides.

B Nematodes for pink bollworm control

PBW larvae enter the soil to pupate, and remain in the top 1.3 cm of soil, making this stage vulnerable to nematodes. Most pathogen research for PBW control, besides that with *Bt*, has been with entomopathogenic nematodes, particularly *S. riobrave* and *S. carpocapsae* (Forlow Jech and Henneberry, 1997; Forlow Jech *et al.*, 1998). Nematodes have an advantage over other microbial insecticides because they can move into PBW infested plant material or soil and locate a host. Nematodes can be produced *in vivo* in *Galleria mellonella* (or other hosts) or *in vitro* in large fermentation systems. Methods for initial production and handling of nematodes can be found in Kaya and Stock (1997). More conveniently, nematodes can be obtained from commercial sources. In either case, it is essential to quantify viability of the nematodes prior to use. The use of *G. mellonella* larvae as "bait" for nematodes is also an excellent method to test for nematodes in the soil (Bedding and Akhurst, 1975).

Two stages of PBW are susceptible to entomopathogenic nematodes in the soil: prepupae (Henneberry *et al.*, 1995b) and diapausing larvae (Gouge *et al.*, 1997). Pupae are not susceptible. Nematodes have been applied during the pre-sowing irrigation cycle for control of diapausing larvae. Mid-season applications of steinernematid nematodes have focused on controlling the prepupal stage during the severe heat during the growing season (Lindegren *et al.*, 1992, 1994; Gouge *et al.*, 1996, Henneberry *et al.*, 1996a, b). Excellent PBW mortality has been achieved using application rates of 2.5 billion infective juveniles (IJs)/ha, during early season applications (Henneberry *et al.*, 1996a) and 3.25 billion IJs/ha during mid-season (Gouge *et al.*, 1996).

Application of nematodes during pre-plant irrigation to control diapausing larvae is undoubtedly the most convenient strategy, utilizing the advantages of cooler soil temperatures and uniform distribution within fields unoccupied by plants. However, PBW moths emerging from diapause are highly mobile (Flint and Merkle, 1981) and unless this strategy is adopted on an area-wide basis, protection of localized areas will be short lived. After cotton plants flower, moth movements become increasingly restricted (Flint and Merkle, 1981) and applications of nematodes to smaller areas is more successful. Small-scale plots can be artificially infested with laboratory-reared PBW if insect emergence is to be quantified as an indication of efficacy.

As Gaugler (1988) pointed out, behavioral and ecological barriers have restricted the effective field-host range of nematodes primarily to the larval stages of certain insects in cryptic and soil inhabit. Attempts to control foliar insects with steinernematids have been discouraging (Kaya, 1985). Poor control has been attributed to the failure of the nematodes to withstand UV light and rapid desiccation.

S. carpocapsae can survive desiccation if dehydration occurs slowly (Simons and Poinar, 1973; Womersley, 1990). Kung *et al.* (1990) found that steinernematids became inactive and survived best in soils with 2–4% moisture. The nematodes were able to enter into a state similar to anhydrobiosis because, when the soil was rehydrated to 16% moisture, the nematodes remained pathogenic. Because steinernematids can survive for long periods in soil with little moisture, moisture levels that occur normally in natural soils are not necessarily limiting.

Temperature is an important factor that limits the pathogenicity of steinernematids either by its influence on nematode activity, their bacterial symbionts, or both (Kaya, 1990; Griffin and Downes, 1991; Kung *et al.*, 1991). Temperatures above 12 °C are required by most commercially-available nematodes (Georgis, 1992). With the exception of *S. riobrave*, temperatures above 28 °C reduce nematode pathogenicity (Gouge *et al.*, 1999).

Soil texture affects the ability of nematodes to infect hosts (Molyneux and Bedding, 1984; Geden *et al.*, 1985). Nematode movement is impeded in soils with high clay content (Georgis and Poinar, 1983a, b, c; Kaya, 1990). Acidic soils with pH 4 or below and above pH 10 may limit the nematode's host-finding ability and pathogenicity (Fischer and Führer, 1990; Kung *et al.*, 1990).

1 Plot setup

The sizes of field study plots and number of replicates used to evaluate entomopathogenic nematode control of PBW have varied immensely from replicated plots of equal size [24.7 m^2 (Henneberry *et al.*, 1996a) to 1.1 ha (Gouge *et al.*, 1996)] arranged in a completely randomized block design or as paired strip plots. Larger scale studies using commercial fields as replicate plots have varied in size from 1.9–19.3 ha (Gouge *et al.*, 1997). The most useful and reliable efficacy data have been collected from trials that utilized smaller field plots where the population of PBW was uniform throughout the experimental area. Plots within the same field should be separated by buffer zones to avoid contamination due to the movement of nematodes, particularly in irrigated crops because of the transport of nematodes in water.

2 Environmental and agricultural data

Soil samples are usually taken prior to nematode application for soil structure analysis and pH determination. Soil temperature may be monitored using waterproof data loggers. Temperatures at 0-, 3-, and 15-cm depths as well as the ambient temperature 1 m above ground are useful. Rainfall should be monitored using rainfall data loggers. Data loggers have an advantage over the usual variety of rain gauges, as accuracy is not compromised due to evaporation of collected rain. Wind speed and direction should be recorded at the time of nematode application if aerial spray equipment is used.

When using commercial cotton growing land it is important to note basic cultural practices that may have significant effects upon collected data sets. Record the species and cultivar of cotton grown, row spacing, planting and harvesting dates, and the frequency and amount of irrigation water applied during the season. Irrigation of every row is advisable as opposed to every second row, and laser-leveled fields are an advantage if relying upon water movement to distribute nematodes. A history of application of other chemical pesticides, herbicides, fertilizers, nematicides, and defoliants should always be recorded.

3 Pre-treatment sampling

PBW populations are often monitored using pheromone (gossyplure) baited sticky traps (Foster *et al.*, 1977; Toscano *et al.*, 1979). Trap catches are monitored daily to weekly depending upon population density. The rubber septas containing pheromone should be changed at weekly intervals. Traps are best placed near the center of fields at near-canopy height with approximately one trap/3 ha (Chu and Henneberry, 1990). A more direct method of confirming adult PBW population levels emerging at specific times can be determined using pyramid emergence cages (base area 1 m^2) placed directly over cotton plants (Watson *et al.*, 1975). Emerging PBW fly up into the cages and through metal screen funnels into collection jars attached to the apex of the cages. The insects can then be counted on a daily basis. Spiders, ants, and other predatory arthropods must be removed regularly from the collection jars for accurate counts of PBWs.

Initially, soil samples from experimental fields are baited with live *G. mellonella* larvae to locate native entomopathogenic nematodes. From each replicate field plot, ten 100-ml soil samples are collected randomly from row tops and ten from furrow bases. The samples are baited with 8 late-instar *G. mellonella* larvae in large Petri dishes. Dry soil is moistened with distilled water until the sample is moist, but not wet (approximately 10–15% water, weight/weight). After 4 days incubation at 27 °C, larvae are collected from the soil samples, washed in distilled water, and then dissected in quarter-strength Ringer's solution using a stereo dissecting microscope (Woodring and Kaya, 1988) and the presence of infected insects recorded. The method described here is only useful for determining the presence or absence of entomopathogenic nematodes. Identification and quantification techniques used to establish described species are detailed in Lacey (1997). A very useful method for estimating entomopathogenic nematode densities in soil samples is described in Koppenhöfer *et al.* (1998) and Chapter IV-5.

4 Application of nematodes

There are physiological differences between steinernematid and heterorhabditid species (Poinar, 1990), and as a result several different formulation media are available. The shelf life of nematode products has been extended by reducing nematode metabolism (either chemically or by using low temperature storage), by immobilizing, or by partially desiccating the nematodes (Georgis, 1992). Nematodes formulated on a moist substrates such as sponge, vermiculite, or peat require continuous refrigeration to maintain nematode viability (Georgis, 1990). Clays, alginate, and polyacrylamide gels immobilize nematodes, partially desiccate them, or reduce their metabolism, improving their storage stability (Georgis, 1990).

Most formulations simply require mixing with water prior to application. Clay-based water-dispersible granule formulations, require no prior extraction procedure and the nematode formulation is added directly into a spray tank. Nematodes may take several hours to regain activity. Nematodes formulated on foam or sponge must be washed from the inert media before adding to the spray tank. Usually nematodes extracted from sponge media are instantly active and able to move into the soil media. Methods used for quantifying viable nematodes are presented by Kaya and Stock (1997).

Irrigation during nematode application and continued moderate soil moisture levels are essential for nematode movement, persistence and pathogenicity (Georgis, 1990; Georgis and Gaugler, 1991). Henneberry *et al.* (1996a, b) noted an increase in PBW mortality in plots treated with *S. riobrave* after successive field irrigations.

The potential of combining nematodes with other control strategies has been reviewed by Kaya (1985) and Georgis (1990). Nematodes can be mixed safely with commercial preparations of *Bt* (Poinar *et al.*, 1990), pyrethroids (Rovesti *et al.*, 1988), and some other pesticides and fertilizers (Georgis, 1990). Hara and Kaya (1983), Rovesti *et al.* (1988, 1990), Kaya and Burlando, (1989), and Forschler *et al.* (1990) discuss pesticides which adversely affect nematodes when tank mixed. The use of nematicides with entomopathogenic nematodes is not recommended.

5 *Application equipment and strategies*

Calibration of spray equipment varies widely but is usually determined by calculating the following parameters: number of nematodes/unit volume, flow rate (volume dispensed/unit time) and application speed (time for delivery system to cover a unit of area).

Nematodes should be collected from the formulation media and examined under the microscope before being incorporated into an application system to determine viability. Nematodes should also be collected after they have been applied through whatever delivery system is in use to determine if the application equipment has harmed them. Viability assessments are made visually with the aid of a dissecting microscope at 40x.

a. *Tractor-mounted sprayers*. Gouge *et al.* (1997) applied IJs using a standard tractor mounted spray rig, delivering 374 liters/ha at 275 kPa, through Tee Jet VS 80-O8 nozzles. The tractor speed was 10 km/h, delivering nematodes over 8 rows using 18 nozzles spaced at 50.8 cm. Fields were furrow irrigated immediately after application. Gouge *et al.* (1996) applied nematodes mid-season using a similar spray boom with drop nozzles attached to deliver the nematode treatment below the cotton canopy.
b. *Aerial application*. Gouge *et al.* (1997) used preplant, aerial application utilizing an Ag Husky fixed wing aircraft. Nematodes were applied in 94 liters/ha. A total of 31 nozzles were used, fitted with D-12 cores and sprayed at 221 kPa. For nematode protection, whirl plates and strainers were removed from the nozzles, which were orientated directly backwards producing a 10.7-m swath. A Satellite Locking GPS guidance system and a prototype flow control unit aided accurate field application. Plane ground speed was 193 km/h. Fields were furrow irrigated prior to aerial spraying.
c. *Irrigation application*. Application of nematodes at the time of irrigation offers an extremely inexpensive and easy application option. Forlow Jech and Henneberry (1997) used a PVC pipe manifold system described by Forlow Jech and Henneberry (1996) which maintained the nematodes in suspension by agitation and pumped nematodes directly into the irrigation furrows. Gouge *et al.* (1997) mixed nematodes in a 2,590-liter agitation nurse tank, which fed into a constant flow battery box situated over an irrigation ditch at the water inlet. Fields received water through base gates.
d. *Subterranean drip application*. Lindegren *et al.* (1992) applied *S. carpocapsae* through a subterranean drip system in a commercial cotton field, delivering the nematodes to the root zone with minimal exposure to UV light and desiccation. IJs of *S. carpocapsae* are negatively geotropic and move naturally towards the soil surface (Moyle and Kaya, 1981; Georgis and Poinar, 1983a; Ferguson *et al.*, 1995) where PBW pupate (University of California, 1996). However, other nematode species may have different dispersal behaviors.

Only single applications of entomopathogenic nematodes per cotton growing season have been evaluated. Multiple treatments have not been tested due to the expense. Multiple treatments using lower nematode rates offer no obvious advantage for the control of diapausing populations; however, maintenance of pink bollworm populations at low levels during mid-season may be economically viable in the future.

6 *Assessment of nematode coverage*

The uniformity of nematode coverage should be determined. The most direct way of monitoring ground coverage is by burying collection cups at ground level. After nematode application, cups are collected, aliquots withdrawn from the nematode suspension and the number of IJs/ml counted under a dissecting scope. This method is useful for assessing ground spray application methods. For systems delivering nematodes directly into the soil immediately after or during irrigation, soil sampling and nematode extraction (as described previously) may be used. Aerial and ground spray applications can also be assessed using water sensitive dye cards, although this method assumes uniform distribution of the nematodes in the spray media.

7 *Post-treatment sampling*

Prior to nematode application, laboratory-reared or field-collected PBW larvae may be caged within plastic biopsy cassettes (Tissue Path IV, $4 \times 2.8 \times 0.7$ cm) and buried in field plots 2.5 cm below the soil surface. Cassettes are buried along five rows within each replicate plot. Cassettes should be buried at five points along each row with one cassette buried in the row top and one in the furrow base. Cassettes are recovered 48 h post-application. Larvae or cadavers are removed from cassettes, dissected in quarter strength Ringer's solution, and the presence of nematodes is recorded.

Ten 100-ml soil samples from each replicate plot are collected randomly from row tops and furrow bases. Samples should be baited as previously described with late instar *G. mellonella* larvae in large Petri dishes (15×2.5 cm). Nematode persistence over time can be monitored weekly until entomopathogenic nematodes are no longer detected.

Delta traps (sticky, cardboard, enclosures) use gossyplure-impregnated rubber septa to attract male moths. Septa are placed on a platform of sticky glue that traps moths as they arrive. Traps should be examined daily to weekly depending upon population density, and the rubber baits changed at weekly intervals. Trap placement in a field is as described above. Funnel traps can be used if sticky delta traps cannot be used due to high levels of dust and wind. Any funnel trap design may be used when baited with gossyplure. Mineral oil, cotton seed oil, or anti-freeze (ethylene glycol, 50% mixture with water) can be used to catch and preserve the moths until collection.

8 *Assessing crop damage*

The number of rosette blooms can be used early in the season to compare plant damage between treatments. Thirty-meter sections of central rows are measured in each plot, and the numbers of rosette blooms recorded at sunrise. When sampling bolls, select bolls 14–21 days old, the stage when they are most likely to be infested. Bolls at this stage are fully-grown but have only partially-developed lint. The bolls feel slightly spongy but firm when squeezed between the thumb and forefinger. Crack the bolls open and check each lock for larvae or entrance mines where neonate larvae have entered. Young larvae remain close to the entrance mine, have a brown head and white body which renders them almost invisible against a lint background. Older larvae are usually found near the seeds and stain the lint around the feeding site.

For standard sampling, pick 25 bolls at random in each quarter of the field. If the field is larger than 3.2 ha, pick one sample of 25 bolls in each 8 ha (University of California, 1996). Sequential sampling at weekly intervals will give the most accurate assessment of actual boll damage.

Fye (1976) described a method for more accurate estimation of boll infestations. Fye suggested that bolls should be collected and incubated at ambient temperature for a 2-week period in well ventilated but secure clear plastic boxes. During incubation smaller larvae and eggs, which are difficult to detect, have a chance to develop to last instar larvae, where they can be counted more easily in the box. Henneberry *et al.* (1977) compared the relative efficiency of counting larvae from freshly collected bolls and bolls which had been held for 2 weeks. Almost twice as many PBW larvae were detected by holding the bolls and allowing the insects to cut out naturally.

Cotton is usually harvested between October and December using two, four, or five-row commercial cotton pickers. Cotton may be weighed by dumping from pickers into a boll buggy situated on digital trailer scales next to a module maker. Samples of mechanically picked cotton (0.2–0.5 kg) may be collected directly from the buggy for analysis. These samples can be ginned using a 20-saw laboratory cotton gin (Porter Morrison and Son, Dennis Manufacturing Co., Woodward, OK) or similar machine and the seed x-rayed for damage evaluation (Wilson and Wilson, 1975). Lint samples should be graded by an official cotton analysis laboratory using United States HVI standards of quality.

From each of the experimental field plots, 0.2–0.5-kg samples of cotton may be hand picked from each or a selection of specific plant nodes. These samples should be ginned as before and subjected to seed x-ray evaluation and lint grading. Several fiber characteristics should be evaluated using HVI analysis including fiber strength, reflectance, spottiness, color grade, length, length uniformity, elongation, and micronaire. Once the seed cotton samples are ginned, the lint and seed can be weighed separately to establish the percent lint from the total seed cotton weight.

C A protozoan and microsporidium for boll weevil control

During the late 1960s, research was conducted on two organisms that attack the boll weevil. A neogregarine, *Mattesia grandis*, and a microsporidium, *Glugea gasti*, were tested extensively with promising results against the boll weevil (McLaughlin *et al.*, 1968, 1969). Because these organisms are obligate pathogens, material for testing must be produced *in vivo*. McLaughlin and Bell (1970) reported on mass production of spores of these two pathogens by spraying spores onto plates of artificial diet on which boll weevil larvae were being fed. Spores were harvested from infected adults by blending them in a household-type blender, then filtering and purifying the spores before storage in distilled water at 4 °C.

In order to infect boll weevils, the neogregarine or microsporidian spores must be ingested. Therefore, field experiments tested mixtures of either type of spores with baits and feeding stimulants applied to boll weevil feeding sites on cotton, primarily terminals and fruiting structures (McLaughlin *et al.*, 1969). Water suspensions of neogregarine spores had almost no effect on the boll weevil populations, but when spores were formulated into viscous liquids and granules with sugar, agar, honey, and other materials, disease prevalence ranged from 18–93% (McLaughlin, 1966). McLaughlin *et al.* (1969) described field tests in which the spores and bait were mixed with red dye (Daum *et al.*, 1967). The red dye permitted field identification of boll weevils that had fed on the bait. Spore concentrations were 3.0×10^6 spores/ml and 22.7 liters or 45.4 liters/0.4 ha were applied on young and fully mature plants, respectively, using motor-driven pumps on a high clearance self-propelled spray machine. The bait was viscous and they attempted to apply as much bait as possible on cotton squares and terminals.

When plants were fruiting, all the fruit on 15.4 row-m were searched and boll weevils collected. Later, plants were shaken into canvas bags to collect boll weevils. The weevils were examined in the field for red coloration in the abdomen which indicated they had fed on the bait. Samples of weevils were also held in the laboratory for at least 7 days and then tissues from each weevil were examined with a microscope for signs of infection. Therefore, the data collected were based on whether weevils had ingested the bait or not, and infection. They found that in August 50–70% of the boll weevils were marked with dye and diseased. They concluded that 96% winter mortality occurred in diapausing weevils from treated fields compared with 85% in untreated fields. Further work by Bell and McLaughlin (1970) showed that boll weevils infected with *M. grandis* were more susceptible to low concentrations of insecticides than uninfected weevils.

D B. bassiana *for boll weevil and plant bug control*

1 General considerations for working with B. bassiana

There are four methods of obtaining *B. bassiana* for use in experiments: obtaining your own isolate from nature, obtaining an isolate from the USDA-ARS Collection of Entomopathogenic Fungal Cultures (Ithaca, NY) or American Type Culture Collection (Rockville, MD), obtaining conidia from USDA or other research lab, or obtaining formulated material from a company. Isolation and production of fungi for research are covered by Goettel and Inglis (1997). The most practical way for most researchers to obtain *B. bassiana* for trials is to request quantities of commercial formulations, such as Naturalis-L (Troy Biosciences Inc., Phoenix, AZ) or BotaniGard ES (Emerald BioAgriculture Corporation, Lansing, MI). While these products are formulated and the conidial concentration should be accurately quantified, conidial numbers and viability should still be confirmed by the researcher using methods in Goettel and Inglis (1997).

In all tests with *B. bassiana* there are certain considerations to keep in mind. The fungal conidia are living organisms and may be harmed by excessive heat, sunlight, and by chemicals in the spray mix, such as fungicides. Therefore, it is good practice to keep the material in a cool dark place until used, and best to apply treatments late in the day, to minimize the effect of heat and sunlight on the conidia. If the conidia were produced in the laboratory, then the conidial numbers and viability must be quantified.

It is essential that the viability of conidia used in field trials be assessed prior to treatment. This is because heat, sunlight, and other environmental factors during transport, formulation and handling, may reduce their viability. Serial dilutions of conidial suspensions are prepared in deionized water plus 0.01% Tween 80 or 0.05% Silwet L77. Conidial suspensions can be checked for viability by plating 0.1 ml of a 1×10^7 suspension of conidia onto a 9-cm Petri dish containing Sabouraud dextrose agar plus yeast extract agar (SDAY). After incubation for 24 h at 25 °C in the dark, the percentage germination may be determined by counts using a phase microscope at 400x (Steinkraus and Tugwell, 1997).

Steinkraus and Tugwell (1997) cultured *B. bassiana* on solid media (SDAY) for 2 weeks in the dark at 25 °C, then harvested the conidia and stored them at 6 °C until used. Production on SDAY will produce sufficient conidia for small scale field testing. Leland *et al.* (2005a) produced conidia for field trials in biphasic culture on liquid media in shake flasks followed by production on barley in spawn bags and recent advances in production have improved conidial yield (Leland, unpublished data). An excellent manual on production of *B. bassiana in vitro* is "Solid Substrate Fermentation of *B. bassiana* and *Metarhizium anisopliae*" by Grace and Jaronski (2005).

2 B. bassiana *for boll weevil control*

McLaughlin (1962) reported that *A. grandis* was susceptible to *B. bassiana*. Since then there have been reports of various methods of applying *B. bassiana* for boll weevils with mixed success (Frank and Slosser, 1990; Wright and Chandler, 1992; Wright, 1993).

Wright and Chandler (1992) reported that the *B. bassiana* formulation, Naturalis-L, combined with a pheromone attractant, grandlure, was effective against overwintered boll weevil populations and in the field (Wright, 1993). The isolate originated from an infected boll weevil from Texas. The fungal material was formulated in an emulsifiable oil. A feeding substrate, Konsume (Fermone, Phoenix, AZ) and a wetting agent, Nufilm-17 (Miller Chemical and Fertilizer, Hanover, PA) were added. Rates of 1.0×10^{10} and 1.7×10^{10} conidia/ha were used in 1991 and 1992, respectively. Treatments were applied with a high-clearance tractor mounted sprayer equipped with a 14-row boom and 5003 nozzles in 1991 and TX8 conejet nozzles (Wylie Manufacturing, Petersburg, TX) in 1992. The boll weevil pheromone, grandlure (1,000 mg/ha), was added to some treatments in 1991. The results suggested that Naturalis-L was effective and could be used in an integrated approach with chemicals for control of boll weevils.

3 B. bassiana *for plant bug control*

a B. bassiana *isolates used in plant bug tests*

Entomopathogenic fungi have the best potential of all the pathogen groups for control of *Lygus* species. Steinkraus and Tugwell (1997) reported finding natural *B. bassiana* infections in *L. lineolaris* in Arkansas. Leland and Snodgrass (2004) reported finding naturally infected *L. lineolaris* populations in the Mississippi Delta region at approximately a 0.3% natural infection rate. Natural *B. bassiana* infection rates were found to be much higher in *Lygus* spp. of the San Joaquin Valley, California, with rates averaging approximately 10% across the season and high infection rates even observed in hot summer months (McGuire, 2002).

B. bassiana (ARSEF 3769) isolated from naturally infected *L. lineolaris* was used in field tests on cotton in Arkansas (Steinkraus and Tugwell, 1997; Brown *et al.*, 1997 a, b, c); isolates from naturally infected bugs from California and Mississippi were used in field trials on alfalfa in California (McGuire *et al.*, 2006), on cotton, pigweed, and radish in Mississippi (Leland *et al.*, 2005a; McGuire and Leland, 2006; unpublished data), and on cotton and trap crops in Arkansas (Lund *et al.*, 2006).

b B. bassiana *tests on trap crops and alternate hosts*

Field trials in California and Mississippi as well as recent trap crop studies in Arkansas have focused on controlling populations on alternate hosts with *B. bassiana* with the long term objective of incorporating *B. bassiana* into area-wide management and trap cropping approaches. Trials in California on alfalfa and Mississippi on pigweed both compared three isolates of *B. bassiana*; one used in the commercial product Mycotrol® (GHA), one isolated from *L. hesperus* in California, and one isolated from *L. lineolaris* in Mississippi. These isolates were obtained from surveys of natural infection levels of *B. bassiana* in plant bug populations of the San Joaquin Valley of California (McGuire, 2002) and the Mississippi Delta (Leland and Snodgrass, 2004) and were selected for field trials based on a number of characteristics (McGuire *et al.*, 2005; Leland, 2005; Leland *et al.*, 2005b).

c *Handling of plant bugs and plot setup*

Plant bugs for field studies can be collected from wild flowering plants or from flowering canola or mustard using sweep nets or beat sheets. In some cases laboratory colonies of plants bugs are available. In all cases, plant bugs collected for and from tests must be kept out of direct sunlight in coolers in order to ensure their quality.

Because of the mobility of adult plant bugs, it is often desirable to cage some or all of the bugs used in field tests. Sleeve cage studies were used in some Arkansas and Mississippi field trials. In an Arkansas study, a randomized complete block design was used with 15 replicates per treatment. Each replicate consisted of a 1-mm mesh nylon tulle sleeve cage 0.6 by 0.3 m covering a cotton plant. Cages were separated by 5 m within rows and 3 m across rows. Test insects were collected by sweep net from flowering mustard planted nearby. Ten adult tarnished plant bugs were placed in each cage, either before or after application of *B. bassiana*. A *B. bassiana* conidial suspension that contained 5.8×10^7 viable conidia/ml was applied to each plant. The plants were sprayed either before or after adding bugs using a trigger-pump hand sprayer containing either water (control) or the conidial suspension (Steinkraus and Tugwell, 1997).

Field plots for trials in Mississippi (Leland *et al.*, 2005a) and California (McGuire *et al.*, 2006) were divided into four replicate blocks of approximately 0.1 ha that were each divided into four, 0.03 ha plots for the three fungal treatments and carrier control. In trials against *L. lineolaris* on pigweed, field plots were surrounded with six rows of corn and 4×9 m cages were placed in each plot to restrict the migration of insects between plots. Also in trials against *L. lineolaris* on pigweed, sleeve cages were used hold plant bugs to estimate the contribution of direct spray, indirect uptake from plant surfaces and 24-h conidial residues to infection of *L. lineolaris* adults.

Field plots were treated with 1×10^{13} viable conidia/ha to 2.5×10^{13} conidia/ha using conidia suspended in 0.05% Silwet L77. Making an initial suspension at a higher Silwet concentration (*e.g.*, 0.2% Silwet L77) then diluting to 0.05% Silwet improved suspension of the conidia and pouring the conidial suspensions through a nylon mesh stocking prior to application removed any residual mycelial debris that could clog spray nozzles. In Mississippi trials, field plots were treated using a CO_2 back-pack sprayer equipped with four 8004 flat fan T-jet nozzles (Spraying Systems, Wheaton, IL) spaced at 51 cm at 234 kPa (34 psi) and 187 liters/ha at approximately 45.7 cm above the plant canopy and directed straight down.

d Determination of infection and mycosis

Collections of treated and control plant bugs from cages or plants in plots after treatment are extremely important in accurately assessing the effect of *B. bassiana* treatments. Percentage infection and mortality rates can be determined for adults and nymphs collected prior to application and at intervals after application. Plant bugs are manually collected from cages using insect aspirators or forceps, or captured from plants by sweep nets or beat sheets. In the laboratory, plant bugs are placed individually in labeled small plastic cups and monitored for mortality over time and mycoses in dead plant bugs. Plant bugs may be kept alive by adding a piece of fresh green bean, sweet corn, or broccoli. Alternatively, holding *L. lineolaris* on florist foam moistened with 10% honey water in cups rather than green beans, sweet corn, or broccoli appears to reduce control mortality, handling mortality, and contamination of cadavers by saprophytes when determining mycoses.

Five days after treatment, Steinkraus and Tugwell (1997) removed all live and dead *L. lineolaris* from field sleeve cages. Remaining live *L. lineolaris* from the cages were placed in individual plastic cups in the laboratory with a piece of fresh green bean for food. Mortality was recorded daily starting with the day the insects were removed from the cages. Dead *L. lineolaris* were held on moist filter paper in Petri dishes to permit those infected with *B. bassiana* to form conidia providing positive confirmation of infection. This test and others (Brown *et al.*, 1997a, b, c) suggest that *B. bassiana* could play a role in *L. lineolaris* management in cotton if *B. bassiana* could be produced at a low enough cost to treat large areas of cotton.

In some tests it is worthwhile to collect non-target insects to determine the effect of *B. bassiana* on beneficial arthropods. Non-target insects found in field plots can be collected from plots after treatment and returned to the laboratory to determine mortality rates and percentage infection in a similar manner as described for plant bugs. Fungi from dead non-target insects can be isolated and cultured then identified using molecular markers to determine the impact of the applications on non-target arthropods. Hagler *et al.* (1992) did this and in most cases found that *B. bassiana* infecting non-targets were not the isolates applied during the treatments. In *L. hesperus* field trials molecular markers were used to determine the identity of isolates from subsamples of *L. hesperus* cadavers exhibiting mycoses to distinguish infection from applied isolates from background infection and dispersion of infected insects.

Plant bugs and other insects can also be marked with inexpensive proteins (*e.g.*, egg whites) followed by detection by enzyme-linked immunosorbent assay (ELISA). This may be a valuable technique for evaluating dispersion of treated insects or percentages of insect populations contacted by *B. bassiana* applications if these proteins markers can be mixed with *B. bassiana* suspensions prior to application (Hagler *et al.*, 1992).

4 Analysis of data

In California field trials (McGuire *et al.*, 2006), populations at each sample date and results of overall infection were analyzed by analysis of variance and means were compared among treatments each sample day using the protected least significant difference test. Comparisons among percentage mycosis or 7 day mortality for insects treated with different isolates at a single sample date were made using analysis of variance (Proc GLM, Tukey's HSD, SAS, 1999). Percentage mycosis was defined as the number of insects exhibiting frank mycosis divided by the total number of insects that were held for mortality $\times$ 100% rather than the total number of cadavers. The interactions of treatment and insect caging with populations over time was estimated using a mixed model treating time as a simple split unit but modeled with a repeated statement for Log_{10} $(x + 1)$ transformed data

(Proc Mixed, SAS, 1999). Mortality of *L. lineolaris* and beneficial insect species over time were analyzed by survivorship analysis (Proc Lifereg, SAS) (Leland, 2005; unpublished data).

In 2 years of field trials against *L. lineolaris* on pigweed, rates of infection ranged from 30 to 80% across all isolates. Infection was generally higher in adults than nymphs. Mortality after 7 day incubation generally corresponded well with infection rates. In 2 years of field trials against *L. hesperus* on alfalfa, infection in adults was approximately 80% across all isolates. Despite high infection rates of insects collected from the field and held under laboratory conditions, population reduction was inconsistent in field trials. Further research is needed to determine mortality rates of infected insects under field conditions and multiple applications may be needed to effectively manage populations of this multivoltine insect. Advances in formulations for UV protection (Leland and Behle, 2005) may improve the field persistence of *B. bassiana*. Recent advances in production techniques and facilities have resulted semi-commercial production that will facilitate more extensive field evaluation of *B. bassiana* isolates and their potential incorporation into area-wide management programs (Leland, unpublished data).

E Entomopathogenic fungi for whitefly control

Entomopathogenic fungi are among the few promising microbial control agents for homopteran pests. The *B. bassiana* products, Naturalis-L and Mycotrol WP, and the fungus *Paecilomyces fumosoroseus* were used by Akey and Henneberry (1994, 1998) in tests against the SLWF.

Plots were 6 rows wide by 58.7 m long and buffers of 2 rows by 2.4 m separated the plots. The entomopathogenic fungal treatments were part of a 16 treatment randomized block design with 4 replications/treatment. Treatments were applied weekly and included an unsprayed control. Applications were made by ground with 3 nozzles per row: one overhead, and 2 with swivel nozzles angled upward on drops. Because whiteflies live on the underside of leaves, getting good coverage is important. *Paecilomyces fumosoroseus* PFR-97® was obtained from Thermo Trilogy Corp. and contained 1×10^9 conidia/g. The rate/ha of PFR-97® was not clearly stated. *B. bassiana* (Naturalis-L® and Mycotrol®WP) were obtained from Troy Biosciences Inc. and Mycotech Corp., respectively. The Naturalis-L® product contained 2.3×10^7 conidia/ml and was applied at 297 g/0.4 ha, while the Mycotrol® WP contained 2×10^{13} conidia/373 g and was applied at 186 g per 0.4 ha. Sprays were applied at 113.5 liters/0.4 ha.

Eggs, small nymphs, and large nymphs were sampled from leaves removed from 5 plants/plot, taking the 5th main-stem leaf below the first expanded terminal leaf. All the SLWF forms present on a 2.2 cm diameter leaf disk, taken from between the main leaf vein and the next lateral vein, were counted. The *P. fumosoroseus* and the two *B. bassiana* materials gave effective control of the SLWF in this experiment.

Rektorik and Wright (1992) tested various spray application machinery and techniques for applying Naturalis-L for whitefly control, including a mist blower, conventional nozzles, boom sprayer with drop nozzles, and mechanical air downdraft and updraft sprayers. Plots were 0.4 ha in size. They concluded that mechanical air updraft sprayers were most effective. An essential aspect of using a microbial insecticide on cotton for whiteflies, aphids, and other pests, is getting good coverage on the underside of the leaves.

5 References

Adamczyk, J. J., Fife, J. H., and Leonard, B. R. 1997. Efficacy of selected insecticides against Lepidoptera in cotton. *Arthropod Management Tests* **22**, 236–237.

Adamczyk, J. J., Holloway, J. W., Church, G. E., Leonard, B. R., and Graves, J. B. 1998. Larval survival and development of fall armyworm (Lepidoptera: Noctuidae) on normal and transgenic cotton expressing the *Bacillus thuringiensis* CryIA (c) δ-endotoxin. *J. Econ. Entomol.* **91**, 539–545.

Akey, D. H. and Henneberry, T. J. 1994. Sweetpotato whitefly control by Naturalis-L, the fungus *B. bassiana*, in furrow and sub-drip irrigated upland cotton. *Proc. Beltwide Cotton Conf.*, pp. 817–819.

Akey, D. H. and Henneberry, T. J. 1998. Control of silverleaf whitefly with the entomopathogenic fungi,

Paecilomyces fumosoroseus and *B. bassiana* in upland cotton in Arizona. *Proc. Beltwide Cotton Conf.*, pp. 1073–1077.

Ali, A. and Young, S. Y. 1993. Effects of rate and spray volume of *Bacillus thuringiensis* (Lepidoptera: Noctuidae) and persistence in cotton terminals. *J. Econ. Entomol.* **86**, 735–738.

Allen, G. E., Gregory, B. G., and Brazzel, J. R. 1966. Integration of the *Heliothis* nuclear polyhedrosis virus into a biological control program on cotton. *J. Econ. Entomol.* **59**, 1333–1336.

Andrews, G. L., Harris, F. A., Sikorowski, P. P., and McLaughlin, R. E. 1975. Evaluation of *Heliothis* nuclear polyhedrosis virus in a cotton seed oil bait for control of *Heliothis virescens* and *H. zea* on cotton. *J. Econ. Entomol.* **68**, 87–90.

Bartlett, A. C. 1993. Response of the pink bollworm to transgenic cotton leaf material. *Proc. Beltwide Cotton Conf.*, pp. 1038–1040.

Bedding, R. A. and Akhurst, R. J. 1975. A simple technique for the detection of insect parasitic rhabditid nematodes in soil. *Nematologica* **21**, 109–110.

Bell, M. R. and Hardee, D. D. 1994. Early season application of a baculovirus for area-wide management of *Heliothis/Helicoverpa* (Lepidoptera: Noctuidae): 1992 field trial. *J. Entomol. Sci.* **29**, 192–200.

Bell, M. R. and Hardee, D. D. 1995. Tobacco budworm and cotton bollworm: methodology for virus production and application in large-area management trials. *Proc. Beltwide Cotton Conf.*, 857–858.

Bell, M. R. and Hayes, J. L. 1994. Areawide management of cotton bollworm and tobacco budworm (Lepidoptera: Noctuidae) through application of a nuclear polyhedrosis virus on early-season alternate hosts. *J. Econ. Entomol.* **87**, 53–57.

Bell, M. R. and Henneberry, T. J. 1980. Entomopathogens for pink bollworm control. *In* "Pink Bollworm Control in the Western United States" (H. M. Graham, Ed.), pp. 76–81. Science and Education Administration, ARM, USDA-ARS-W-16, Washington, D.C.

Bell, M. R. and Kanavel, R. F. 1975. Potential of bait formulation to increase effectiveness of nuclear polyhedrosis virus against the pink bollworm. *J. Econ. Entomol.* **68**, 389–391.

Bell, M. R. and Kanavel, R. F. 1977. Field test of a nuclear polyhedrosis virus in a bait formulation for control of pink bollworms and *Heliothis* spp. in cotton in Arizona. *J. Econ. Entomol.* **70**, 625–629.

Bell, M. R. and Kanavel, R. F. 1978. Tobacco budworm; Development of a spray adjuvant to increase effectiveness of a nuclear polyhedrosis virus. *J. Econ. Entomol.* **71**, 350–352.

Brandt, S. L., Coudron, T. A., Habibi, J., Brown, G. R., Ilagan, O. M., Wagner, R. M., Wright, M. K., Backus, E. A., and Huesing, J. E. 2004. Interaction of two *Bacillus thuringiensis* O-endotoxins with the digestive system of *Lygus hesperus*. *Current Microbiol.* **48**, 1–9.

Brown, J. Z., Steinkraus, D. C., and Tugwell, N. P. 1997a. Evaluation of Mycotrol WP9504 (*B. bassiana*) and Provado (imidacloprid) to control tarnished plant bug in cotton, 1996. *Arthropod Management Tests* **22**, 241–242.

Brown, J. Z., Steinkraus, D. C., and Tugwell, N. P. 1997b. Persistence of Mycotrol WP9504 (*B. bassiana*) and Provado (imidacloprid) to control tarnished plant bug in cotton, 1996. *Arthropod Management Tests* **22**, 242.

Brown, J. Z., Steinkraus, D. C., Tugwell, N. P., and Teague, T. G. 1997c. The effects and persistence of the fungus *B. bassiana* (Mycotrol) and imidacloprid (Provado) on tarnished plant bug mortality and feeding. *Proc. Beltwide Cotton Conf.*, pp. 1302–1305.

Burges, H. D. (Ed.). 1998. "Formulation of Microbial Biopesticides: Beneficial Microorganisms, Nematodes, and Seed Treatments." Kluwer Academic, Dordrecht, The Netherlands.

Butler, G. D., Jr. and Henneberry, T. J. 1994. *Bemisia* and *Trialeurodes* (Hemiptera: Aleyrodidae). *In* "Insect Pests of Cotton" (G. A. Matthews, and J. P. Tunstall, Eds.), pp. 325–352. CAB International, Wallingford, UK.

Cabanillas, H. E. 2003. Susceptibility of the boll weevil to *Steinernema riobrave* and other entomopathogenic nematodes. *J. Invertebr. Pathol.* **83**, 188–197.

Carner, G. R. and Canerday, T. D. 1968. Field and laboratory investigations with *Entomophthora fresenii*, a pathogen of *Tetranychus* spp. *J. Econ. Entomol.* **61**, 956–959.

Carriere, Y. Ellsworth, P. C., Dutilleul, P. E-K., Christa Barkley, V., and Antilla, L. 2006. A GIS based approach for areawide pest management: the scales of *Lygus hesperus* movements to cotton from alfalfa, weeds, and cotton. *Entomol. Exp. Appl.* **118**, 203–210.

Carruthers, R. I., Wraight, S. P., and Jones, W. A. 1993. An overview of biological control of the sweetpotato whitefly, *Bemisia tabaci*. *Proc. Beltwide Cotton Conf.*, pp. 680–685.

Chu, C. C. and Henneberry, T. J. 1990. Pink bollworm seasonal distribution, yearly variation, and male moth trap catch relationships to population increases in cotton. *Southwest. Entomol.* **15**, 273–280.

Daum, R. J., McLaughlin, R. E., and Hardee, D. D. 1967. Development of the bait principle for boll weevil control: cottonseed oil, a source of attractants and feeding stimulants for the boll weevil. *J. Econ. Entomol.* **60**, 321–325.

Durant, J. A. 1977. Field efficacy of insecticides against the *Heliothis* spp. complex on cotton in South Carolina. *J. Ga. Entomol. Soc.* **12**, 163–166.

Evans, H. and Shapiro, M. 1997. Viruses. *In* "Manual of Techniques in Insect Pathology" (L. A. Lacey, Ed.), pp. 17–54. Academic Press, London.

Falcon, L. A., Sorensen, A., and Abesson, N. B. 1974. Aerosol application of insect pathogens. *Calif. Agric.* **28**, 11–15.

Ferguson, C. S., Schroeder, P. C., and Shields, E. J. 1995. Vertical distribution, persistence, and activity of entomopathogenic nematodes (Nematoda: Heterorhabditidae and Steinernematidae) in alfalfa snout beetle (Coleoptera: Curculionidae) infested fields. *Environ. Entomol.* **24**, 149–158.

Fischer, P. and Führer, E. 1990. Effect of soil acidity on the entomophilic nematode *Steinernema kraussei* Steiner. *Biol. Fertility Soil* **9**, 174–177.

Flint, H. M. and Merkle, J. R. 1981. Early-season movements of pink bollworm male moths between selected habitats. *J. Econ. Entomol.* **74**, 366–371.

Forlow Jech, L. and Henneberry, T. J. 1996. A method of distributing *Steinernema riobravis* in cotton furrow irrigation and pink bollworm larval mortality response. *Proc. Beltwide Cotton Conf.*, pp. 707–710.

Forlow Jech, L., and Henneberry, T. J. 1997. Pink bollworm mortality following application of *Steinernema riobravis* entomopathogenic nematodes in cotton furrow irrigation. *Proc. Beltwide Cotton Conf.*, pp. 1192–1194.

Forlow Jech, L., Henneberry, T. J., Panter, M. J., and Faulconer, S. 1998. Horizontal and vertical movement of *Steinernema riobravis* and *S. carpocapsae* (Rhabditida: Steinernematidae) in soil in the laboratory. *Proc. Beltwide Cotton Conf.*, pp. 1296–1299.

Forschler, B. T., All, J. N., and Gardner, W. A. 1990. *Steinernema feltiae* activity and infectivity in response to herbicide exposure in aqueous and soil environments. J. *Invertebr. Pathol.* **55**, 375–379.

Foster, R. N., Staten, R. T., and Miller, E. 1977. Evaluation of sticky traps for pink bollworm. *J. Econ. Entomol.* **70**, 289–291.

Frank, W. A. and Slosser, J. E. 1990. Efficacy of *B. bassiana* (Balsamo) Vuillemin to control overwintering boll weevils, *Anthonomus grandis* Boheman. *Southwest. Entomol.* **15**, 77–78.

Fye, R. E. 1976. Improved method for holding cotton bolls for detecting pink bollworms. USDA-ARS W-37, 3 pp.

Gaugler, R. 1988. Ecological considerations in the biological control of soil-inhabiting insects with entomopathogenic nematodes. *Agr. Ecosys. Environ.* **24**, 351–360.

Geden, C. J., Axtell, R. C., and Brooks, W. M. 1985. Susceptibility of the lesser mealworm, *Alphitobius diaperinus* (Coleoptera: Tenebrionidae) to the entomogenous nematodes *Steinernema feltiae, S. glaseri* (Steinernematidae) and *Heterorhabditis heliothidis* (Heterorhabditidae). *J. Entomol. Sci.* **20**, 331–339.

Georgis, R. 1990. Formulation and application technology. *In* "Entomopathogenic Nematodes in Biological Control" (R. Gaugler and H. K. Kaya, Eds), pp. 173–191. CRC Press, Boca Raton, FL.

Georgis, R. 1992. Present and future prospects for entomopathogenic nematode products. *Biocontrol Sci. Technol.* **2**, 83–99.

Georgis, R. and Gaugler, R. 1991. Predictability in biological control using entomopathogenic nematodes. *J. Econ. Entomol.* **84**, 713–720.

Georgis, R. and Poinar, G. O., Jr. 1983a. Effect of soil texture on the distribution and infectivity of *Neoaplectana carpocapsae* (Nematoda: Steinernematidae). *J. Nematol.* **15**, 308–311.

Georgis, R. and Poinar, G. O., Jr. 1983b. Effect of soil texture on the distribution and infectivity of *Neoaplectana glaseri* (Nematoda: Steinernematidae). *J. Nematol.* **15**, 329–332.

Georgis, R. and Poinar, G. O., Jr. 1983c. Vertical migration of *Heterorhabditis bacteriophora* and *H. heliothidis* (Nematoda: Heterorhabditidae) in sandy loam soil. *J. Nematol.* **15**, 652–654.

Glazer, I, Klein, M., Navon, A., and Nakache, Y. 1992. Comparison of efficacy of entomopathogenic nematodes combined with antidesiccants applied by canopy sprays against three cotton pests (Lepidoptera: Noctuidae). *J. Econ. Entomol.* **85**, 1636–1641.

Goettel, M. S. and Inglis, G. D. 1997. Fungi: Hyphomycetes. *In* "Manual of Techniques in Insect Pathology" (L. A. Lacey, Ed.), pp. 213–250. Academic Press, London.

Goodell, P. B., Wright, S. D., and Carter, M. W. F. 2000. Managing Western tarnished plant bug (*Lygus hesperus*) in a regional context. *Proc. Beltwide Cotton Conf.*, pp. 1123–1125.

Gouge, D. H., Lee, L. L., and Henneberry, T. J. 1999. Effect of temperature and lepidopteran host species on entomopathogenic nematode (Nematoda: Steinernematidae, Heterorhabditidae) infection. *Environ. Entomol.* **28**, 876–883.

Gouge, D. H., Reaves, L. L., Stoltman, M. M., Van Berkum, J. R., Burke, R. A., Forlow-Jech, L. J., and Henneberry, T. J. 1996. Control of pink bollworm *Pectinophora gossypiella* (Saunders) larvae in Arizona and Texas cotton fields using the entomopathogenic nematode *Steinernema riobravis* (Cabanillas, Poinar, & Raulston) (Rhabditida: Steinernematidae). *Proc. Beltwide Cotton Conf.*, pp. 1078–1082.

Gouge, D. H., Smith, K. A., Payne, C., Lee, L. L., Van Berkum, J. R., Ortega, D., and Henneberry, T. J. 1997. Control of pink bollworm, *Pectinophora gossypiella*, (Saunders) (Lepidoptera: Gelechiidae) with biocontrol and biorational agents. *Proc. Beltwide Cotton Conf.*, pp. 1066–1072.

Grace, J. and Jaronski, S. 2005. Solid substrate fermentation of *B. bassiana* and *Metarhizium anisopliae*. USDA-ARS Northern Plains Workshop, Feb. 16–18, 2005 (available online at www.sidney.ars.usda.gov/personnel/sjaronski.html).

Graves, G. N. and Watson, T. F. 1970. Effect of *Bacillus thuringiensis* on the pink bollworm. *J. Econ. Entomol.* **63**, 1828–1830.

Griffin, C. T. and Downes, M. J. 1991. Low temperature activity in *Heterorhabditis* sp. (Nematoda: Heterorhabditidae). *Nematologica* **37**, 83–91.

Hagler, J. R., Cohen, A. C., Bradley-Dunlop, D., and Enriquez, F. J. 1992. New approach to mark insects for feeding and dispersal studies. *Environ. Entomol.* **21**, 20–25.

Hamm, J. J., Chandler, L. D., and Sumner, H. R. 1994. Field test with a fluorescent brightener to enhance infectivity of fall armyworm (Lepidoptera: Noctuidae) nuclear polyhedrosis virus. *Fla. Entomol.* **77**, 425–437.

Hara, A. H. and Kaya, H. K. 1983. Toxicity of selected organophosphate and carbamate pesticides to infective juveniles of the entomogenous nematode *Neoaplectana carpocapsae* (Rhabditida: Steinernematidae). *Environ. Entomol.* **12**, 496–501.

Hardee, D. D. and Bell, M. R. 1995. Area-wide management of *Heliothis/Helicoverpa* in the delta of Mississippi. *In* Challenging the Future: Proceedings of the World Cotton Research Conference-1, Brisbane Australia, February 14–17, 1994, (G. A. Constable and N. W. Forrester, Eds), pp. 434–436. CSIRO, Melbourne, Australia.

Hardee, D. D., Bell, M. R., and Street, D. A. 1999. A review of area-wide management of *Helicoverpa* and *Heliothis* (Lepidoptera : Noctuidae) with pathogens (1987–1997). *Southwest. Entomol.* **24**, 62–75.

Hayes, J. L. and Bell, M. 1994. Evaluation of early-season baculovirus treatment for suppression of *Heliothis virescens* and *Helicoverpa zea* (Lepidoptera: Noctuidae) over a wide area. *J. Econ. Entomol.* **87**, 58–66.

Hendrix, D. L., Steele, T. L. and Perkins, H. H. Jr. 1996. *Bemisia* honeydew and sticky cotton. *In* "Bemisia 1995: Taxonomy, Biology, Damage, Control and Management" (D. Gerling and R. T. Mayer Eds.), pp. 189–199. Intercept, Andover, UK.

Henneberry, T. J., Bariola, L. A., Fry, K. E., and Kittock, D. L. 1977. Pink bollworm infestations and relationships to cotton yield in Arizona. USDA-ARS, W-49, 9 pp.

Henneberry, T. J., Forlow Jech, L., and Burke, R. A. 1996a. Pink bollworm adult and larval susceptibility to steinernematid nematodes and nematode persistence in the soil in laboratory and field tests in Arizona. *Southwest. Entomol.* **21**, 357–367.

Henneberry, T. J., Forlow Jech, L., Burke, R. A., and Lindegren, J. E. 1996b. Temperature effects on infection and mortality of *Pectinophora gossypiella* (Lepidoptera: Gelechiidae) larvae by two entomopathogenic nematode species. *Environ. Entomol.* **25**, 179–183.

Henneberry, T. J., Lindegren, J. E., Forlow Jech, L. J., and Burke, R. A. 1995a. Pink bollworm (Lepidoptera: Gelechiidae): effect of steinernematid nematodes on larval mortality. *Southwest. Entomol.* **20**, 25–32.

Henneberry, T. J., Lindegren, J. E., Forlow Jech, L., and Burke, R. A. 1995b. Pink bollworm (Lepidoptera: Gelechiidae), cabbage looper, and beet armyworm (Lepidoptera: Noctuidae) pupal susceptibility to steinernematid nematodes (Rhabditida: Steinernematidae). *J. Econ. Entomol.* **88**, 835–839.

Hinz, S. E. and Wright, J. E. 1997. Naturalis-L: a biological product (*B. bassiana* JW-1) for the control of cotton pests. *Proc. Beltwide Cotton Conf.*, pp. 1300–1302.

Hollingsworth, R. G., Steinkraus, D. C., and McNew, R. W. 1995. Sampling to predict fungal epizootics in cotton aphids (Homoptera: Aphididae). *Environ. Entomol.* **24**, 1414–1421.

Hollingsworth, R. G., Steinkraus, D. C., and Tugwell, N. P. 1997. Responses of Arkansas populations of tarnished plant bugs (Heteroptera: Miridae) to insecticides, and tolerance differences between nymphs and adults. *J. Econ. Entomol.* **90**, 21–26.

Ignoffo, C. M. and Adams, J. R. 1966. A cytoplasmic-polyhedrosis virus, *Smithiavirus pectinophorae* sp. n. of the pink bollworm, *Pectinophora gossypiella* (Saunders). *J. Invertebr. Pathol.* **8**, 59–66.

Ignoffo, C. M., Chapman, A. J., and Martin, D. F. 1965. The nuclear polyhedrosis virus of *Heliothis zea* (Boddie) and *Heliothis virescens* (Fabricius) III. effectiveness of the virus against field populations of *Heliothis* on cotton, corn, and grain sorghum. *J. Invertebr. Pathol.* **7**, 227–235.

Ignoffo, C. M. and Garcia, C. 1965. Infection of the cabbage looper, bollworm, tobacco budworm, and pink bollworm with spores of *Mattesia grandis* McLaughlin collected from boll weevils. *J. Invertebr. Pathol.* **7**, 260–262.

Ignoffo, C. M. and Garcia, C. C. 1994. Antioxidant and oxidative enzyme effects on the inactivation of inclusion bodies of the *Heliothis* baculovirus by simulated sunlight-UV. *J. Econ. Entomol.* **23**, 1025–1029.

Ignoffo, C. M. and Graham, H. M. 1967. Laboratory and field cage tests with *Bacillus thuringiensis* against pink bollworm larvae. *J. Invertebr. Pathol.* **9**, 390–394.

Ignoffo, C. M., Shasha, B. S., and Shapiro, M. 1990. Sunlight ultraviolet protection of the *Heliothis* nuclear polyhedrosis virus through starch-encapsulation technology. *J. Invertebr. Pathol.* **57**, 134–136.

Ignoffo, C. M., Yearian, W. C., Young, S. Y., Hostetter, D. L., and Bull, D. L. 1976. Laboratory and field persistence of new commercial formulations of the *Heliothis* nuclear polyhedrosis virus, *Baculovirus heliothis*. *J. Econ. Entomol.* **69**, 233–236.

Johnson, D. R. 1982. Suppression of *Heliothis* spp. on cotton by using *Bacillus thuringiensis, Baculovirus heliothis*, and two feeding adjuvants. *J. Econ. Entomol.* **75**, 207–210.

Kaya, H. K. 1985. Entomogenous nematodes for insect control in IPM systems. *In* "Biological Control in Agricultural IPM Systems" (M. A. Hoy and D. C. Herzog, Eds.), pp. 283–302. Academic Press, London.

Kaya, H. K. 1990. Soil ecology. *In* "Entomopathogenic Nematodes in Biological Control" (R. Gaugler and H. K. Kaya, Eds.), pp. 93–115. CRC Press, Boca Raton, FL.

Kaya, H. K. and Burlando, T. M. 1989. Infectivity of *Steinernema feltiae* in fenamiphos-treated sand. *J. Nematol.* **21**, 434–436.

Kaya, H. K. and Stock, S. P. 1997. Techniques in insect nematology. *In* "Manual of techniques in Insect Pathology" (L. A. Lacey, Ed.), pp. 281–324. Academic Press, London.

Koppenhöfer, A. M., Campbell, J. F., Kaya, H. K., and Gaugler, R. 1998. Estimation of entomopathogenic nematode population density in soil by correlation between bait insect mortality and nematode penetration. *Fund. Applied Nematol.* **21**, 95–102.

Kung, S. P., Gaugler, R., and Kaya, H. K. 1990. Influence of soil pH and oxygen on persistence of *Steinernema* spp. *J. Nematol.* **22**, 440–445.

Kung, S. P., Gaugler, R., and Kaya, H. K. 1991. Effects of soil temperature, moisture, and relative humidity on entomopathogenic nematode persistence. *J. Invertebr. Pathol.* **57**, 242–249.

Lacey, L. 1997. "Manual of Techniques in Insect Pathology." Academic Press, London.

Lacey, L. A., Fransen, J. J., and Carruthers, R. 1996. Global distribution of naturally occurring fungi of *Bemisia*, their biologies and use as biological control agents. *In* "Bemisia 1995: Taxonomy, Biology, Damage, and Management" (D. Gerling and R. Mayer, Eds.), pp. 401–433. Intercept, Andover, UK.

Leland, J. E. 2005. Characteristics of *B. bassiana* isolates from *Lygus lineolaris* populations of Mississippi. *J. Agri. Urban Entomol.* **22**, 57–71,

Leland, J. E., and Behle, R. W. 2005. Coating *B. bassiana* with lignin for protection from solar radiation and effects on pathogenicity to *Lygus lineolaris. Biocontrol Sci. Technol.* **15**, 309–320.

Leland, J. E., McGuire, M. R., Gore, J., and Jaronski, S. T. 2005a. Field evaluation of *B. bassiana* isolates for control of *Lygus* plant bug. *Proc. Beltwide Cotton Conf.*, pp. 1290–1301.

Leland, J. E., McGuire, M. R., Grace, J. A., Jaronski, S. T., Ulloa, M., Park, Y.-H., and Plattner, R. D. 2005b. Strain selection of a fungal entomopathogen, *B. bassiana*, for control of plant bugs (*Lygus* spp.) (Heteroptera : Miridae). *Biol. Control* **35**, 104–114

Leland, J. E. and Snodgrass, G. L., 2004. Prevalence of naturally occurring *B. bassiana* in *Lygus lineolaris* (Heteroptera: Miridae) populations from wild host plants of Mississippi. *J. Agri. Urban Entomol.* **21**, 157–163

Leonard, B. R., Graves, J. B., Sparks, T. C., and Pavloff, A. M. 1987. Susceptibility of bollworm and tobacco budworm larvae to pyrethroid and organophosphate insecticides. *Proc. Beltwide Cotton Conf.*, pp. 320–324.

Lincoln, C., Boyer, W. P., Dowell, G. C., Barnes, G., and Dean, G. 1970. Six years experience with point-sample cotton insect scouting. *Univ. Ark. Agr. Exp. Sta. Bull.* 754.

Lindegren, J. E., Henneberry, T. J., and Jech, L. F. 1992. Mortality response of pink bollworm to the entomopathogenic nematode *Steinernema carpocapsae. Proc. Beltwide Cotton Conf.*, pp. 930–931.

Lindegren, J. E., Henneberry, T. J., Raulston, J. R., Forlow Jech, L. J., and Valero, K. A. 1994. Current status of pink bollworm control with entomopathogenic nematodes. *Proc. Beltwide Cotton Conf.*, pp. 1242–1243.

Lindegren, J. E., Meyer, K. F., Henneberry, T. J., Vail, P. V., Forlow Jech, L. J., and Valero, K. A. 1993a. Susceptibility of pink bollworm (Lepidoptera: Gelechiidae) soil associated stages to the entomopathogenic nematode *Steinernema carpocapsae* (Rhabditida: Steinernematidae). *Southwest. Entomol.* **18**, 113–120.

Lindegren, J. E., Valero, K. A., and Jech, L. J. F. 1993b. Host searching response of various entomopathogenic nematodes; sand barrier bioassay. *Proc. Beltwide Cotton Conf.*, pp. 1044–1045.

Lund, J., Teague, T. G., Steinkraus, D. C., and Leland, J. E. 2006. Control of tarnished plant bug (*Lygus lineolaris*) in midsouth cotton using the entomopathogenic fungus, *B. bassiana* and insect growth regulator Diamond®. *Proc. 2006 Beltwide Cotton Conf.*, San Antonio, TX. National Cotton Council, Memphis, TN. (in press).

Luttrell, R. G., Roush, T., Ali, A., Mink, J. S., Reid, M. R., and Snodgrass, G. L. 1987. Pyrethroid resistance in field populations of *Heliothis virescens* (Lepidoptera: Noctuidae) in Mississippi in 1986. *J. Econ. Entomol.* **80**, 985–989.

Luttrell, R. G., Young, S. Y., Yearian, W. C., and Horton, D. L. 1981. Evaluation of *Bacillus thuringiensis* - spray - adjuvant - viral insecticide combinations against *Heliothis* spp. on cotton. *Environ. Entomol.* **11**, 783–787.

Luttrell, R. G., Yearian, W. C., and Young, S. Y. 1983. Effect of spray adjuvants on *Heliothis zea* (Lepidoptera: Noctuidae) nuclear polyhedrosis virus efficacy. *J. Econ. Entomol.* **79**, 162–167.

Liu, H., Skinner, M., Brownbridge, M. and B. L. Parker. 2003. Characterization of *B. bassiana* and *Metarhizium anisopliae* isolates for management of tarnished plant bug, *Lygus lineolaris* (Hemiptera: Miridae). J. Invert. Pathol. **82**, 139–147.

Liu, H., Skinner, M., Parker, B. L., and Brownbridge, M. 2002. Pathogenicity of *B. bassiana, Metarhizium anisopliae* (Deuteromycotina: Hyphomycetes), and other entomopathogenic fungi against *Lygus lineolaris* (Hemiptera: Miridae). J. Econ. Entomol. **95**, 675–681.

Matthews, G. A. and Tunstall, J. P. 1994. "Insect Pests of Cotton." CAB International, Wallingford, UK.

McGuire, M. R. 2002. Prevalence and distribution of naturally occurring *B. bassiana* in San Joaquin Valley populations of *Lygus hesperus* (Heteroptera: Miridae). *J. Agric. Urban Entomol.* 19, 237–246.

McGuire, M. R., Galan-Wang, L. J., and Tamez-Guerra, P. 1997. Bacteria: bioassay of *Bacillus thuringiensis* against lepidopteran larvae. *In* "Manual of Techniques in Insect Pathology" (L. A. Lacey, Ed.), pp. 91–100. Academic Press, London.

McGuire, M. R. and Leland, J. E. 2006. Field trials with *B. bassiana* against *Lygus* spp. in California and Mississippi. *Proc. Beltwide Cotton Conf.*, pp. 1389–1392.

McGuire, M. R., Ulloa, M., and Park Y.-H. 2005. Biological and molecular characteristics of *B. bassiana* isolates from California *Lygus hesperus* (Hemiptera: Miridae) populations. *Biol. Control* 33, 307–314.

McGuire, M. R., Leland, J. E., Dara, S., Park, Y.-H., and Ulloa, M. 2006. Effect of different isolates of *B. bassiana* on field populations of *Lygus hesperus. Biol. Control*, **38**, 390–396.

McLaughlin, R. E. 1962. Infectivity tests with *B. bassiana* (Balsamo) Vuillemin on *Anthonomus grandis. J. Insect Pathol.* **4**, 386–388.

McLaughlin, R. E. 1965. *Mattesia grandis* n. sp., a pathogen of the boll weevil, *Anthonomus grandis* Boheman (Coleoptera: Curculionidae). *J. Protozool.* **12**, 405–413.

McLaughlin, R. E. 1966. Infection of the boll weevil with *Mattesia grandis* induced by a feeding stimulant. *J. Econ. Entomol.* **59**, 909–911.

McLaughlin, R. E. and Bell, M. R. 1970. Mass production in vivo of two protozoan pathogens, *Mattesia grandis* and *Glugea gasti*, of the boll weevil, *Anthonomus grandis. J. Invertebr. Pathol.* **16**, 84–88.

McLaughlin, R. E., Cleveland, T. C., Daum, R. J., and Bell, M. R. 1969. Development of the bait principle for boll weevil control IV. Field tests with a bait containing a feeding stimulant and the sporozoans *Glugea gasti* and *Mattesia grandis. J. Invertebr. Pathol.* **13**, 429–441.

McLaughlin, R. E., Daum, R. J., and Bell, M. R. 1968. Development of the bait principle for boll-weevil control, III. Field-cage tests with a feeding stimulant and the protozoans *Mattesia grandis* (Neogregarinida) and a microsporidian. *J. Invertebr. Pathol.* **12**, 168–174.

McLeod, P. J., Young, S. Y., and Yearian, W. C.. 1982. Application of a baculovirus of *Pseudoplusia includens* to soybean: efficacy and seasonal persistence. *Environ. Entomol.* **11**, 412–416.

Meredith, J. A., Smart, G. C., and Overman, A. J. 1987. Evaluation of drip irrigation systems for the delivery of *Neoaplectana carpocapsae. Univ. Florida Ann. Rep.* 120, Mole Cricket Research.

Molyneux, A. S. and Bedding, R. A. 1984. Influence of soil texture and moisture on the infectivity of *Heterorhabditis* sp. D1 and *Steinernema glaseri* for larvae of the sheep blowfly, *Lucilia cuprina. Nematologica* **30**, 358–365.

Montoya, E. L. and Ignoffo, C. M. 1966. Laboratory technique and apparatus for testing viruses applied as dusts. *J. Invertebr. Pathol.* **8**, 251–254.

Montoya, E. L., Ignoffo, C. M., and McGraw, R. L. 1966. A feeding stimulant to increase effectiveness of, and a field test with a nuclear polyhedrosis virus of *Heliothis. J. Invertebr. Pathol.* **8**, 320–324.

Moyle, P. L. and Kaya, H. K. 1981. Dispersal and infectivity of the entomogenous nematode, *Neoaplectana carpocapsae* Weiser (Rhabditida: Steinernematidae), in sand. *J. Nematol.* **13**, 295–300.

Noma, T. and Strickler, K., 1999. Factors affecting *B. bassiana* for control of *Lygus* bug (Hemiptera: Miridae) in alfalfa seed fields. *J. Agric. Urban. Entomol.* **16**, 215–233.

Noma, T. and Strickler, K. 2000. Effects of *B. bassiana* on *Lygus hesperus* (Hemiptera: Miridiae) feeding and oviposition. *Environ. Entomol.* **29**, 394–402.

Pieters, E. P., Young, S. Y., Yearian, W. C., Sterling, W. L., Clower, D. F., Melville, D. R., and Gilliland, F. R. 1978. Efficacy of microbial pesticide and chlordimeform mixture for control of *Heliothis* spp. on cotton. *Southwest. Entomol.* **3**, 237–240.

Plapp, F. W. and Campanhola, C. 1986. Synergism of pyrethroid by chlordimeform against susceptible and resistant *Heliothis. Proc. Beltwide Cotton Conf.*, pp. 167–169.

Poinar, G. O., Jr. 1990. Taxonomy and biology of Steinernematidae and Heterorhabditidae. In "Entomopathogenic Nematodes in Biological Control" (R. Gaugler and H. K. Kaya, Eds.), pp. 23–61. CRC Press, Boca Raton, FL.

Potter, M. F. and Watson, T. F. 1983a. Garbanzo bean as a potential feeding stimulate for use with a nuclear polyhedrosis virus of the tobacco budworm (Lepidoptera: Noctuidae). *J. Econ. Entomol.* **76**, 449–451.

Potter, M. F. and Watson, T. F. 1983b. Timing of nuclear polyhedrosis virus - bait spray combination for control of egg and larval stages of tobacco budworm (Lepidoptera: Noctuidae). *J. Econ. Entomol.* **76**, 446–448.

Rektorik, R. J. and Wright, J. E. 1992. Application technology for whitefly control with Naturalis™, a biorational insecticide. *Proc. Beltwide Cotton Conf.*, pp. 858–859.

Roome, R. E. 1975. Field trials with a nuclear polyhedrosis virus and *Bacillus thuringiensis* against larvae of *Heliothis armigera* (Hb.) (Lepidoptera, Noctuidae) on sorghum and cotton in Botswana. *Bull. Entomol. Res.* **65**, 507–514.

Rovesti, L., Fiorini, T., Bettini, G., Heinzpeter, E. W., and Tagliente, F. 1990. Compatibility of *Steinernema* spp. and *Heterorhabditis* spp. with pesticides. *Informatore Fitopatologico* **40**, 55–61.

Rovesti, L., Tagliente, F., Deseö, K. V., and Heinzpeter, E. W. 1988. Compatibility of pesticides with the entomopathogenic nematode *Heterorhabditis bacteriophora* Poinar (Nematoda: Heterorhabditidae). *Nematologica* **34**, 462–476.

SAS. 1999. SAS Windows Version 8. Copyright 1999–2001. SAS Institute Inc., Cary, NC, USA.

Shapiro, M. and J. L. Vaughn. 1995. Enhancement in activity of homologous and heterologous baculoviruses infectious to cotton bollworm (Lepidoptera: Noctuidae) by an optical brightener. *J. Econ. Entomol.* **88**, 265–269.

Simons, W. R. and Poinar, G. O. Jr. 1973. The ability of *Neoaplectana carpocapsae* (Steinernematidae: Nematoda) to survive extended periods of desiccation. *J. Invertebr. Pathol.* **22**, 228–230.

Smith, D. B., Hostetter, D. L., and Ignoffo, C. M. 1978. Formulation and equipment effects on application of a viral (*Baculovirus heliothis*) insecticide. *J. Econ. Entomol.* **71**, 814–817.

Snodgrass, G. L. 1996. Pyrethroid resistance in field populations of the tarnished plant bug (Heteroptera: Miridae) in cotton in the Mississippi Delta. *J. Econ. Entomol.* **89**, 783–790.

Snodgrass, G. L. and Elzen, G. W. 1994. Efficacy of Naturalis-L® for adults and nymphs of the tarnished plant bug in cotton. *Proc. Beltwide Cotton Conf.*, pp. 1103–1104.

Snodgrass, G. L. and Scott, W. P. 2002. Tolerance to acephate in tarnished plant bug (Heteroptera: Miridae) populations in the Mississippi river delta. *Southwest. Entomol. Suppl.* **27**, 191–199.

Snodgrass, G. L., Scott, W. P., Abel, C. A., Robbins, J. T., Gore, J., and Hardee, D. D. 2005. Tarnished plant bug (Heteroptera; Miridae) populations near fields after early season herbicide treatment. *Environ. Entomol.* **34**, 705–711.

Snodgrass, G. L., Scott, W. P., Robbins, J. T., and Hardee, D. D. 2000. Area wide management of the tarnished plant bug by reduction of early-season wild host plant density. *Southwest. Entomol. Suppl.* **23**, 59–67.

Snodgrass, G. L., Scott, W. P., Robbins, J. T., and Hardee, D. D. 2003. Suppression of tarnished plant bugs in cotton by treatment of early season wild host plants with herbicides in nine-square-mile areas of the Mississippi Delta. *Proc. Beltwide Cotton Conf.* (no page numbers available).

Snodgrass, G. L., Scott, W. P., and Smith, J. W. 1984a. An annotated list of the host plants of *Lygus lineolaris* (Hemiptera: Miridae) in the Arkansas, Louisiana, and Mississippi delta. *J. Georgia. Entomol. Soc.* **19**, 93–101.

Snodgrass, G. L., Scott, W. P., and Smith, J. W. 1984b. Host plants and seasonal distribution of the tarnished plant bug (Hemiptera: Miridae) in the Delta of Arkansas, Louisiana, and Mississippi. *Environ. Entomol.* **13**, 110–116.

Sosa-Gómez, D. R. and Moscardi, F. 1998. Laboratory and field studies on the infection of stink bugs, *Nezara viridula, Peizodorus guildinii*, and *Euschistus heros* (Hemiptera: Pentatomidae) with *Metarhizium anisopliae* and *B. bassiana* in Brazil. *J. Invertebr. Pathol.* **71**, 15–120.

Sosa-Gómez, D. R., Boucias, D. G., and Nation, J. L. 1996. Attachment of *Metarhizium anisopliae* to southern green stink bug *Nezara viridula* cuticle and fungistatic effect of cuticular lipids and aldehydes. *J. Invertebr. Pathol.* **69**, 31–39.

Stacey, A. L., Yearian, W. C., and Young, S. Y. 1977a. Evaluation of *Baculovirus heliothis* with feeding stimulants for control of *Heliothis* larvae on cotton. *J. Econ. Entomol.* **70**, 779–784.

Stacey, A. L., Yearian, W. C., and Young, S. Y. 1977b. Effectiveness of *Baculovirus heliothis* and *Bacillus thuringiensis* applied in dry baits. *Ark. Farm Res.* **26**, 3.

Stacey, A. L., Yearian, W. C., and Young, S. Y. 1980. Field evaluation of *Baculovirus heliothis* on cotton by using selected application methods. *J. Georgia Entomol. Soc.* **15**, 365–372.

Stapel, J. O., Waters, D. J., Ruberson, J. R., and Lewis, W. J. 1997. Development and behavior of *Spodoptera exigua* (Lepidoptera: Noctuidae) larvae in choice tests with food substrates containing toxins of *Bacillus thuringiensis. Biol. Control* **11**, 29–37.

Steinkraus, D. C., Oliver, J. B., Humber, R. A., and Gaylor, M. J. 1998. Mycosis of bandedwinged whitefly (*Trialeurodes abutilonea*) (Homoptera: Aleyrodidae) caused by *Orthomyces aleyrodis* gen. & sp. nov. (Entomophthorales: Entomophthoraceae). *J. Invertebr. Pathol.* **72**, 1–8.

Steinkraus, D. C. and Tugwell, N. P. 1997. *B. bassiana* (Deuteromycotina: Moniliales) effects on *Lygus lineolaris* (Hemiptera: Miridae). *J. Entomol. Sci.* **32**, 79–90.

Stewart, S. D. and Layton, M. B. 2000. Cultural controls for the management of *Lygus* populations in cotton. *Southwest. Entomol. Suppl.* **23**, 83–95.

Streett, D. A., Bell., M. R., and Hardee, D. D. 1998. Evaluation of the area-wide budworm/bollworm management program with virus in the Mississippi delta. *Proc. Beltwide Cotton Conf.*, pp. 1232–1235.

Sun, X., Chen, X., Zhang, Z., Wang, H., Bianchi, F. J. J. A., Peng, H., Vlak, J. M., and Hu, Z. 2002. Bollworm response to release of genetically modified *Helicoverpa armigera* nucleopolyhedroviruses in cotton. *J. Invertebr. Pathol.* **81**, 63–69.

Toscano, N. C., Sevacherian, V., and Van Steenwyk, R. A. 1979. "Pest Management Guide for Insects and Nematodes of Cotton in California." Univ. California, Agr. Sci. Pub. Berkeley, California. 63 pp.

University of California. 1996. "Integrated Pest Management for Cotton in the Western Region of the United States." Div. Agr. Natural Resources Pub. 3305.

Vail, P. V., Henneberry, T. J., and Bariola, L. A. 1978. Evaluation of several control techniques as potential components of an integrated system for pink bollworm suppression in the Southwest. USDA Production Res. Rep. **172**, 18 p.

Vail, P. V., Jay, D. L., Hunter, D. K., and Staten, R. T. 1972. A nuclear polyhedrosis virus infective to the pink bollworm, *Pectinophora gossypiella. J. Invertebr. Pathol.* **20**, 124–128.

Watson, T. F., Langson, D. T., and Crowder, L. A. 1975. Comparison of cage size in evaluating spring moth emergence of the pink bollworm. *J. Econ. Entomol.* **68**, 128–129.

Wellman-Desbiens, E. and Côté, J. C. 2004. Screening of the insecticidal activity of *Bacillus thuringiensis* strains against *Lygus hesperus* (Hemiptera: Miridae) nymphal population. *J. Econ. Entomol.* **97**, 251–258.

Wellman-Desbiens, E. and Côté, J. C. 2005. Development of a *Bacillus thuringiensis*-based assay on *Lygus hesperus. J. Econ. Entomol.* **95**, 1469–1479

Williams, M. R. 2004. Cotton Insect Loss Estimates, 2003. *Proc. Beltwide Cotton Conf.*, pp. 1258–1312.

Williams, M. R. 2005. Cotton Insect Loss Estimates, 2004. *Proc. Beltwide Cotton Conf.*, pp. 1105–1160.

Wilson, R. L. and Wilson, F. D. 1975. Comparison of an X-ray and green-boll technique for screening cotton for resistance to pink bollworm. *J. Econ. Entomol.* **68**, 636–638.

Womersley, C. Z. 1990. Dehydration survival and anhydrobiotic potential. *In* "Entomopathogenic Nematodes in Biological Control" (R. Gaugler, and H. K. Kaya, Eds.), pp. 117–137. CRC Press, Boca Raton, FL.

Woodring, J. L. and Kaya, H. K. 1988. "Steinernematid and Heterorhabditid Nematodes: A Handbook of Techniques." *Ark. Agr. Exp. Sta. So. Coop. Ser. Bull.* 331.

Wright, J. E. 1993. Control of the boll weevil (Coleoptera: Curculionidae) with Naturalis-L: a mycoinsecticide. *J. Econ. Entomol.* **86**, 1355–1358.

Wright, J. E. and Chandler, L. D. 1992. Development of a biorational mycoinsecticide: *B. bassiana* conidial formulation and its application against boll weevil populations (Coleoptera: Curculionidae). *J. Econ. Entomol.* **85**, 1130–1135.

Yearian, W. C., Luttrell, R. G., Stacy, A. L., and Young, S. Y. 1980. Efficacy of *Bacillus thuringiensis* and *Baculovirus heliothis* - chlordimefom spray mixtures against *Heliothis* spp. on cotton. *J. Ga. Entomol. Soc.* **15**, 260–271.

Yearian, W. C. and Young, S. Y. 1978. Application of microbial insecticides on field crops. *Entomol. Soc. Am. Misc. Publ.* **10**, 21–26.

Young, S. Y., Johnson, D. R., Lorenz, G. M., and Fisher, A. M. 1999. Evaluation of biological insecticides for beet armyworm on cotton. *Arthropod Management Tests*, **24**, 258.

Young, S. Y., Kring, T. J., Johnson, D. R., and Klein, C. D. 1997. *Bacillus thuringiensis* alone and in mixtures with chemical insecticides against *Heliothis* and effects on predator densities in cotton. *J. Entomol. Sci.* **32**, 183–191.

Young, S. Y. and Yearian, W. C. 1986. Formulation and application of baculoviruses. *In* "The Biology of Baculoviruses" (R. R. Granados, and B. A. Federici, Eds.), pp. 157–180. CRC Press, Boca Raton, FL.

Zhu, Y. C. and Snodgrass, G. L. 2003. Cytochrome P450 CYP6X1 cDNAs and mRNA expression levels in three strains of the tarnished plant bug *Lygus lineolaris* (Heteroptera: Miridae) having different susceptibilities to pyrethroid insecticide. *Insect Mol. Biol.* **12**, 39–43.

Zhu, Y. C., Snodgrass, G. L., and Chen, M. S. 2004. Enhanced esterase gene expression and activity in a malathion-resistant strain of the tarnished plant bug, *Lygus lineolaris. Insect Biochem. Mol. Biol.* **34**, 1175–1186.

Chapter VII-7

Mushroom pests

Parwinder S. Grewal
Department of Entomology
Ohio State University, OARDC
Wooster, OH 44691, USA

1 Introduction to mushroom culture

More than a dozen species of edible fungi are cultivated worldwide. The white button mushroom, *Agaricus bisporus*, is cultivated in over 100 countries and accounts for approximately 70% of all mushroom production (Chang and Miles 1989). *A. bisporus* mushrooms are cultivated on pasteurized compost prepared either from horse manure or agricultural wastes such as wheat or rice straw. The raw materials are mixed with nitrogen-rich materials such as chicken manure or chemical fertilizers and wetted in long stacks on concrete floors 'out-of-door.' The stacks are shaken up and re-stacked ('turning') at 2–3 day intervals. After 7–14 days, the compost is moved into rooms fully equipped with environmental controls and is pasteurized at 55–60 °C for periods up to 8 h. This process kills most pathogens, nematodes, and insects. Following an appropriate cooling period, the pasteurized compost is used to fill trays, built-in shelves, or tunnels, and inoculated with mushroom mycelium ('spawning'). A recent trend is to spawn compost in tunnels where it is pasteurized and moved to shelves or trays only after the mushroom mycelial colonization ('spawn-run') is complete. Mushroom mycelium colonizes the compost within 10–14 days at 24 °C. To induce fructification, compost is covered to a depth of 3–5 cm with a layer of soil, or more often a mixture of moss-peat and chalk. This is called the casing layer. Mushrooms start to appear from the surface of the casing within 12–15 days after its application. At first to ensure good growth of mycelia into the casing layer, the temperature is kept at about 20 °C, but after a few days, when fruiting body initiation has begun, the growing rooms are maintained between 16 and 20 °C. Ventilation with fresh air is important but air humidity has to be kept above 90%. Following the production of the first mushrooms, the crop is produced in a succession of fruiting bodies known as 'flushes.' Each flush of mushroom is harvested over a period of about 4 days, and flushes occur at intervals of 7–14 days. A single crop may occupy a growing room for 7–10 weeks.

2 Major mushroom pests and their distribution

Sciarid flies (*Lycoriella* spp., Diptera: Sciaridae) are major pests of *A. bisporus* mushrooms. In Europe, the dominant species is *Lycoriella auripilla* and in the USA, Canada, Asia, and Australia, it is *L. mali* (Fletcher *et al.*, 1989; Nickle and Cantelo 1991). In Poland and some parts of Asia, *L. solani* also appears to be prevalent. All three species pose similar problems for growers. Adult flies invade mushroom buildings containing freshly-pasteurized or spawned compost and lay eggs in the compost. Emerging larvae feed on compost, destroying structure and water retention capacity. Mycelial colonization of the compost is inhibited, reducing mushroom yields. Larvae also feed at the growing mycelial front and on the developing fruiting primordia and make mushrooms unsaleable by tunneling through stipe tissue.

L.A. Lacey and H.K. Kaya (eds.), *Field Manual of Techniques in Invertebrate Pathology*, 457–461.

Adult flies are a constant nuisance to mushroom picking staff and some workers develop allergies. Adult flies can also serve as vectors for fungal pathogens, nematodes, and mites. Reliance on organophosphate insecticides, such as diazinon and permethrin, has resulted in widespread resistance and also cause yield losses due to mycotoxic side effects. Two insect growth regulators, diflubenzuron and methroprene, give acceptable control of only the first generation *L. auripilla* and *L. mali* when incorporated into the casing layer.

Phorid flies (*Megaselia* spp., Diptera: Phoridae) are usually the pests of secondary importance in mushroom cultivation. *Megaselia halterata* is perhaps the most prevalent species in the mushroom growing areas worldwide. Direct crop damage by the phorid larvae is less severe than the sciarid larvae (White, 1986). In the USA, the action thresholds of adults on black light traps used by growers are about 100 times higher for *M. halterata* compared to *L. mali* (Keil, 1991). However, the phorid adults mechanically serve as vectors for the mushroom pathogen *Verticillium fungicola* var. *fungicola* (White, 1981), and therefore must be managed.

3 Biological control options for mushroom pests

A Nematodes

Entomopathogenic nematodes of the genera *Steinernema* (Rhabditida: Steinernematidae) and *Heterorhabditis* (Rhabditida: Heterorhabditidae) are already used commercially for sciarid fly control in the USA and Europe (Grewal and Georgis, 1998). *Steinernema feltiae* has proven to be most effective against all three *L. auripilla, L. mali*, and *L. solani*. *S. feltiae* provides better control of sciarid flies than the most effective insecticides, diflubenzuron and methro prene, under most circumstances (Grewal and Smith, 1995). The third and fourth-stage larvae are more susceptible to nematodes than the smaller and younger larvae. Nematode efficacy depends mainly on three factors: the stage of the mushroom crop at the time of nematode application, dosage rate, and compatibility with chemicals used in mushroom production. Nematode applications to the casing layer are usually more effective than to the compost applications (Richardson and Grewal, 1991). This is mainly due to the higher moisture in the casing layer. Rapid depletion of moisture by the growing mycelium in the compost restricts nematode motility and survival. Under some circumstances, split applications of *S. feltiae* made 2–3 days prior to casing and 2–7 days after casing have proven more effective than single applications at casing (Grewal and Smith, 1995; Scheepmaker *et al.*, 1997). *S. feltiae* can persist in the casing layer for the entire duration of the crop and can therefore also be applied preventatively (Grewal *et al.*, 1992). *S. feltiae* also recycles in the fly larvae and maintains its effectiveness longer than commonly used insecticides (Grewal and Richardson, 1993). Usually application rates of $1.0–1.5 \times 10^6$ infective juvenile *S. feltiae*/m^2 provide both economic and effective control of sciarid fly larvae (Grewal and Smith, 1995).

The mushroom crop is intensively managed, and nematode compatibility with chemicals used in mushroom growing must be considered. Although *S. feltiae* is compatible with several fungicides and insecticides, nematode efficacy is significantly reduced by some of the chemicals routinely used in mushroom growing. Formalin and chlorine are two such compounds that are used routinely to control bacterial and fungal diseases and for general sanitation. Both compounds are highly toxic to *S. feltiae* (Grewal *et al.*, 1998). Therefore, it is recommended to apply the nematodes either 3 days prior to or 3 days after applications of formalin or chlorine.

Efficacy of entomopathogenic nematodes against phorid flies has been variable. Scheepmaker *et al.* (1997) obtained up to 75% reduction in *M. halterata* emergence following a single application of *S. feltiae* to the casing layer. Later, Scheepmaker *et al.* (1998) reported that *S. carpocapsae* was more effective than *S. feltiae, Heterorhabditis bacteriophora* or *H. megidis* to control *M. halterata*. They obtained 65 and 73% reduction in *M. halterata* emergence at 6 and 15×10^6 infective juvenile *S. carpocapsae*/m^2. This concentration, however, is too expensive to have any commercial appeal.

B Bacteria

The bacterium, *Bacillus thuringiensis*, has the potential for the control of both sciarid and phorid flies. The toxicity of *B. thuringiensis* subsp. *israelensis (Bti)* to *L. mali* larvae was first demonstrated by Cantwell and Cantelo (1984). Keil (1991) reported that the LC_{50} of *B. thuringiensis* subsp. *israelensis* against *L. mali* was four times higher in compost than in agar. He demonstrated that in six of the seven field tests, a formulation of *B. thuringiensis* subsp. *israelensis* (ABG 6193) reduced emergence of *L. mali* and *M. halterata* as well as or better than methroprene and diflubenzuron. Field efficacy of *Bti* against *L. auripilla* in England has also been demonstrated (White, 1990). In addition to the reduction in the activity of the insecticidal proteins in compost, the bacterial toxins can be totally inactivated by the excessive use of chlorine in mushroom culture (Keil, 1991). Therefore, the compatibility of chlorine and other routinely used compounds such as formalin with insecticidal bacterial protein formulations should be evaluated prior to the use of *Bti* for pest control in mushrooms.

C Fungi

The use of *Pandora gloeospora* (Entomophthorales: Zygomycetes) has been attempted for the control of *L. mali* (Keil, 1994). Fungal conidia may infect late instar larvae and pupae but mycosis is not evident until after adult emergence. Adult flies, whether they are infected as larvae or adults, succumb to mycosis within 24 h. Female flies surviving the initial infection have substantially reduced fecundity. The fungal mycelium has been encapsulated into alginate beads for small scale field trials. High density application of the mycelium in alginate beads provided economic fly control. Application must be timed to coincide with adult emergence.

4 Field experiment techniques

The following protocol is suggested for evaluation of microbial agents against mushroom fly pests.

1. Set up plots in a randomized block design with a minimum of three treatments and 6–12 replications. A nematode or bacteria treatment, a standard insecticide treatment, and an untreated control treatment should be included at a minimum. Commercial growing trays or boxes can be used as individual plots. Plots can also be established on mushroom beds when continuous shelves are used for mushroom growing (Figure 1A). A 30–60 cm alleyway should be kept between plots if using the continuous shelf system.

2a. Drench apply nematode, bacterial, or fungal product to the surface of compost or casing using a suitable container such as a watering can or a bottle with holes in the lid. Extreme care should be taken to avoid settling of nematodes in the application containers. Amount of water required to treat each plot should be determined prior to making the applications.

2b. Alternatively, mix the entomopathogen in casing material at the time of its preparation (Figure 1B).

3a. Remove three cores (3.3 cm deep and 4 cm dia each) from each plot and place them in a plastic pot. Place a sticky trap (6×2.5 cm) on the surface of casing material. Cover the pots with a tightly secured polyethylene sheet or muslin cloth and incubate at 22 °C. Count the numbers of flies emerged and stuck to each trap 2 weeks later. Repeat this procedure each week for up to 6 weeks after treatment.

3b. Alternatively, place emergence traps on the surface of plots to estimate fly populations. The traps (30×25 cm) are easily constructed from a wire mesh with two edges bent to lift the trap off the surface (Figure 1C). White freezer paper (28×21.5 cm) is clipped securely to the inside of the wire mesh, and sprayed with a suitable insect adhesive, such as Tanglefoot insect adhesive (Tanglefoot Company, Grand Rapids, MI). Tether the traps with the shelves to allow their removal during watering. The numbers of flies stuck to each trap can be counted weekly or at the end of each flush.

4. Assess nematode persistence in the plots using the *Galleria mellonella* bait technique (Fan and Hominick, 1991).
5. Apply water and fungicides as scheduled.
6. Record total yield of mushrooms (both the numbers and weights of sporophores) from each plot to determine the effects of treatments.

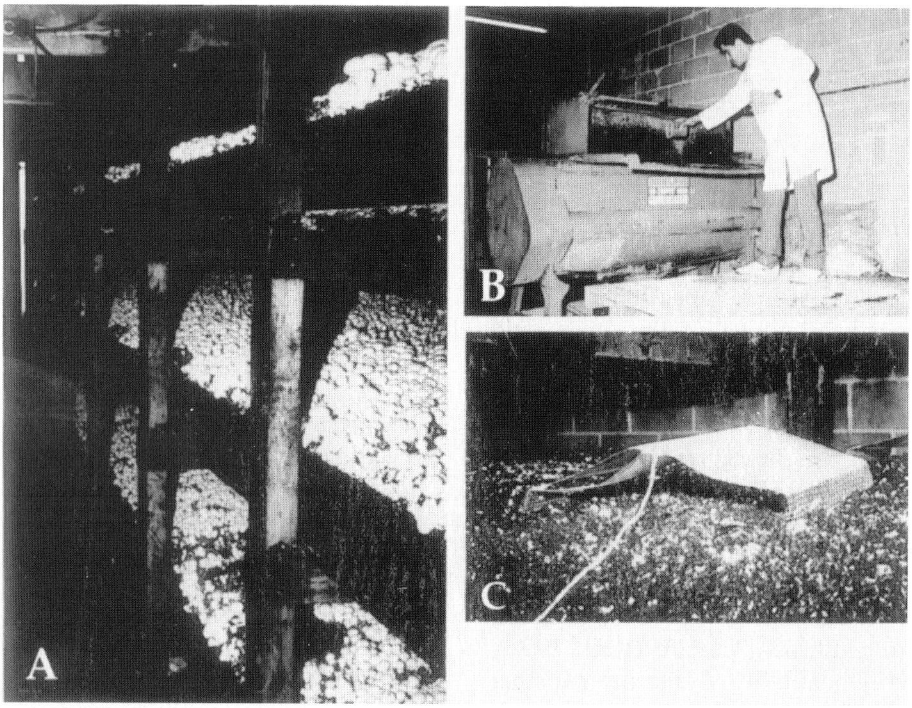

Figure 1. Mushroom cultivation system and application of entomopathogenic nematodes. **A**. A multi-layered continuos-shelve-system for mushroom cultivation; **B**. Application of entomopathogenic nematodes to the casing material during its preparation from peat and chalk; **C**. A fly emergence trap

The 'core-sampling' technique of assessing fly emergence is more convenient and reliable than the 'emergence traps' method. When mushrooms are present, the cores should be removed immediately after the harvest of mushrooms to minimize influence on mushroom primordia. Emergence traps should be placed on the beds after about 1 week of casing to allow egg-laying by the invading flies. Another limitation of the traps is that they have to be removed for watering and other operations and can only be effectively used for the first generation of flies. As the mushroom crop is sensitive to small disturbances, the influence of microbial treatments on mushroom yields and flushing patterns should be evaluated in the presence and absence of the target pests. The impact of treatments on flushing patterns of mushrooms can be assessed by plotting the daily average of mushroom yields (Richardson and Grewal, 1991). In order to obtain best results, the timing of application of microbial control agents should be evaluated in comparison to other chemical inputs during mushroom production.

5 References

Cantwell, G. E. and Cantelo, W. W. 1984. Effectiveness of *Bacillus thuringiensis* var. *israelensis* in controlling a sciarid fly, *Lycoriella mali*, in mushroom compost. *J. Econ. Entomol.* **77**, 473–475.

Chang, S. T. and Miles, P. G. 1989. *Edible mushrooms and their cultivation*, CRC Press, Boca Raton, FL.

Fan, X. and Hominick, W. M. 1991. Efficiency of the *Galleria* bait technique for recovering infective stages of entomopathogenic rhabditids (Steinernematidae and Heterorhabditidae) from sand and soil. *Rev. Nematol.* **14**, 381–387.

Fletcher, J. T., White, P. F., and Gaze, R. H. 1989. *Mushrooms: Pests and disease control*, 2nd Edition, Intercept Ltd, Andover.

Grewal, P. S. and Georgis, R. 1998. Entomopathogenic Nematodes, in F. R. Hall and J. J. Menn (eds), *Methods in Biotechnology, Biopesticides: Use and Delivery*, Humana Press, Totowa, N, pp. 271–297.

Grewal, P. S. and Richardson, P. N. 1993. Effects of application rates of *Steinernema feltiae* (Nematoda: Steinernematidae) on control of the mushroom sciarid fly *Lycoriella auripilla*. *Biocontrol Sci. Technol.* **3**, 29–40.

Grewal, P. S., Richardson, P. N., Collins, G., and Edmondson, R. N. 1992. Comparative effects of

Steinernema feltiae (Nematoda: Steinernematidae) and insecticides on yield and cropping of the mushroom, *Agaricus bisporus*. *Ann. appl. Biol.* **121**, 511–520.

Grewal, P. S. and Smith, C. 1995. Insect-parasitic nematodes for mushroom pest control, *Mushroom News* **14**, 15–25.

Grewal, P. S., Weber, T. A., and Batterley, D. A. 1998. Compatibility of *Steinernema feltiae* with chemicals used in mushroom production, *Mushroom News* **46**, 6–10.

Keil, C. B. 1991. Field and laboratory evaluation of a *Bacillus thuringiensis* var. *israelensis* formulation for the control of fly pests of mushrooms. *J. Econ. Entomol.* **84**, 1180–1188.

Keil, C. B. 1994. *Pandora gloeospora*: A fungal pathogen of mushroom pest, *Lycoriella mali, Proc.6th Intern. Coll. Invertebr. Pathol Microbial Control*, Montpellier, France, p. 88.

Nickle, W. R. and Cantelo, W. W. 1991 Control of mushroom-infesting fly, *Lycoriella mali*, with *Steinernema feltiae*. *J. Nematol.* **23**, 145–147.

Richardson, P. N. and Grewal, P. S. 1991. Comparative assessment of biological (Nematoda: *Steinernema feltiae*) and chemical methods of control for the mushroom fly *Lycoriella auripilla* (Diptera: Sciaridae), *Biocontrol Sci. Technol.* **1**, 217–228.

Scheepmaker, J. W. A., Geels, F. P., Smits, P. H., and Van Griensven, L. J. L. D. 1997. Control of the mushroom pests *Lycoriella auripilla* (Diptera: Sciaridae) and *Megaselia halterata* (Diptera: Phoridae) by *Steinernema feltiae* (Nematoda: Steinernematidae) in field experiments. *Ann. appl. Biol.* **131**, 359–368.

Scheepmaker, J. W. A., Geels, F. P., Rutjens, A. J., Smits, P. H., and Van Griensven, L. J. L. D. 1998. Comparison of efficacy of entomopathogenic nematodes for the biological control of the mushroom pests *Lycoriella auripilla* (Sciaridae) and *Megaselia halterata* (Phoridae). *Biocontrol Sci. Technol.* **8**, 277–288.

White, P. F. 1981. Spread of the mushroom disease *Verticillium fungicola* by *Megaselia halterata* (Diptera: Phoridae). *Protection Ecol.* **3**, 17–24.

White, P. F. 1986. The effect of sciarid larvae (*Lycoriella auripilla*) on the yield of the cultivated mushroom (*Agaricus bisporus*), *Ann. appl. Biol.* **109**, 11–17.

White, P. F. 1990. Laboratory and field tests with *Bacillus thuringiensis* for the control of the mushroom sciarid *Lycoriella auripilla, Brighton Crop Protection Confer.-Pests Dis.*, **1**, 373–378.

Chapter VII-8

Techniques for testing microbials for control of arthropod pests in greenhouses

H. Denis Burges
21 Withdean Ave, Goring-by-Sea
Worthing, West Sussex BN12 4XD
UK. Retired from Horticulture Research
International, Wellesbourne CV35 9EF, UK

1 Introduction

A Greenhouse environment

Glass or plastic sheathed structures are used to protect growing plants from environmental extremes and to increase the number of hours favorable for good plant growth and development. They range from small (2 × 3 m) glasshouses and large, high or low, multi-span glasshouses covering many hectares to long (3–15 m), wide (6–60 m or more) plastic tunnels. The only environmental modification equipment may be automatic or hand-operated ventilation, or the environment may be partially or almost completely controlled by combinations of heaters, coolers and humidifiers. Light intensity and day length may be increased by artificial light. Blackout sheets of various materials may be drawn over the plants diurnally to modify day length during part of the plant's development; transparent sheets may be suspended over the plants as extra insulation to economize on heating costs. The sheets may also increase humidity around the plants. The most sophisticated greenhouses may have complex controls, programmed and regulated to optimize growing conditions and limit energy costs. Paradoxically greenhouse crops may suffer greater extremes of high temperature and low humidity than occur concurrently outdoors. Inevitably weather influences internal greenhouse environment and hence the performance of insect pathogens (Curtis, 1998). Figure 1 illustrates some greenhouse conditions.

B Greenhouse arthropods

Characteristically, greenhouse crops grow very rapidly and have high value with low economic damage tolerance; their pests also increase rapidly and can cause costly damage. Pest control problems depend largely on the range of organisms encountered, illustrated in this chapter mostly by UK species (Table 1). Pest impact depends mainly on indoor climate (Figure 1), crop (Table 2) and mode of attack on the crop (Table 1). Greenhouse climate influences primarily the rate of pest increase, but it must be remembered that this climate is controlled entirely with plant welfare and cost of energy and labor in mind – not pest control! Pest attack tends to be either by sucking or chewing (Table 1). Economic impact depends primarily on which parts of the plants are attacked (Table 2); cosmetic damage to saleable parts is the most costly, while high levels of damage may impair plant development and hence crop timing and yield. Pest control strategy is determined by the source and spread of pests, also by whether they are permanent greenhouse residents or invaders. Control may be aimed at keeping pest numbers below an economic threshold or preferably at elimination (section 3D). The technical difficulty of applying control agents depends on pest position on the crop (Table 1), the most difficult

L.A. Lacey and H.K. Kaya (eds.), Field Manual of Techniques in Invertebrate Pathology, 463–479.

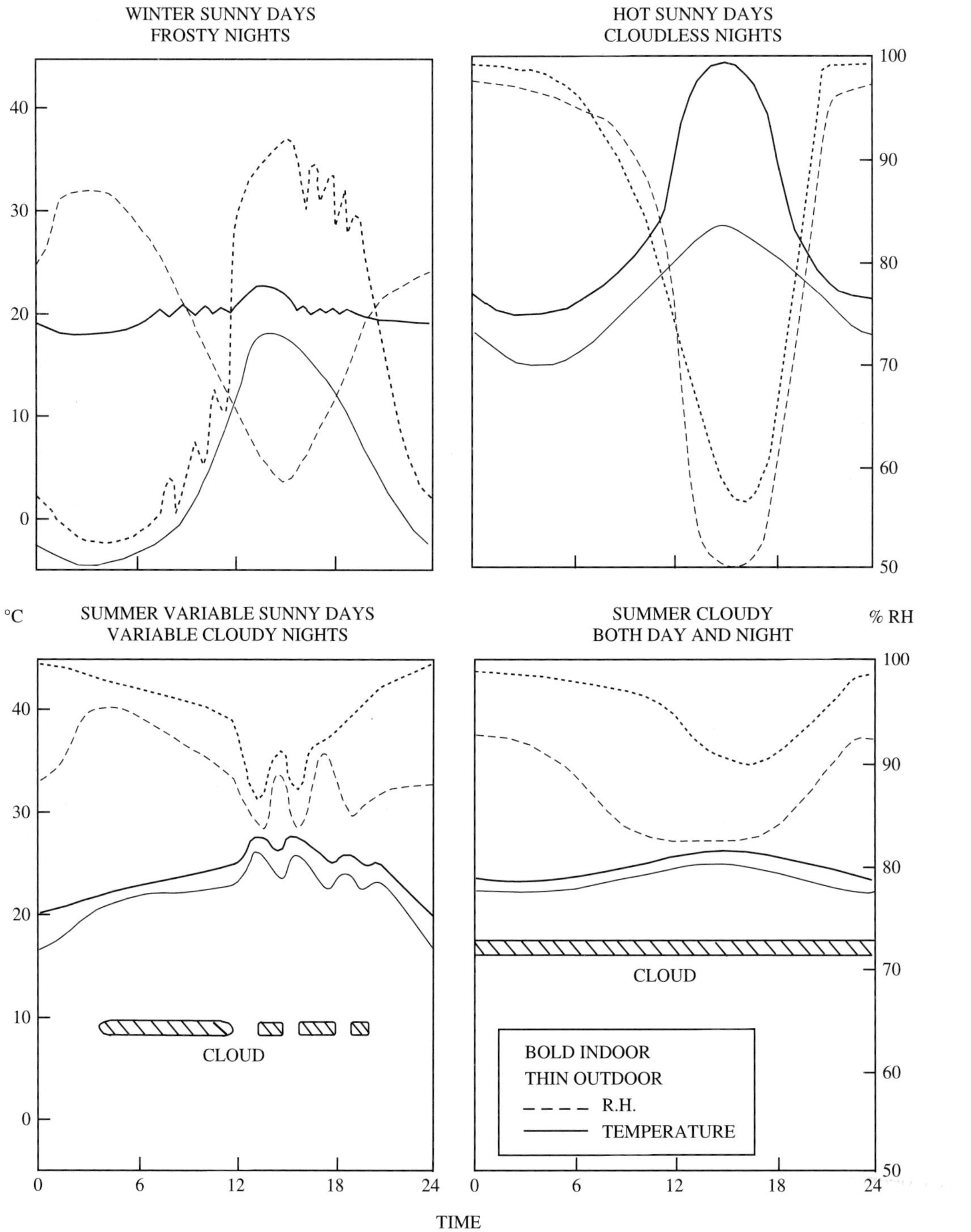

Figure 1. Schematic graphs of temperature and humidity outdoors on four contrasting typical days and the corresponding indoor conditions in a glasshouse equipped with heating, switching at 20°C (differential 1°C) and able to maintain a temperature 22°C above minimum UK outdoor ambient, with vents opening at 22°C and closing at 18°C, in a temperate climate

Table 1. Main UK greenhouse pests (Jarrett, 1985) and some common integrated pest management (IPM; Copping, 1998) options for their control

Group	Feeding habit: damage	Integrated pest management
Aphid	Sucking. Disfigure foliage. Sooty mold. Transmit plant virus	Pathogens: *Lecanicillium (Verticillium) longisporum.*[1]; *Beauveria bassiana.* Parasitoids: *Aphelinus abdominalis* (wasp) likes large aphids; *Aphidius* spp. (wasps) and *Aphidoletes aphidimyza* (gall midge) good searchers. Predators: *Chrysoperla carnea* (lacewing) good for dense spots of infestation; *Harmonia axiridis* (ladybird) rapid control
Leaf miner	Burrow into leaves. Disfigure	Parasite: *Dacnusa sibirica* (wasp) good in cool conditions. Predators: *Diglyphus isaea* (wasp) best > 22 °C, good searcher
Lepidoptera	Chewing. Distort flowers	Pathogens: Nucleopolyhedrovirus of *Spodoptera exigua; Bacillus thuringiensis.* Parasitoids: *Cotesia* spp. (wasps) and *Trichogramma* spp., good searchers
Mealybug	Sucking. Disfigure leaves	Pathogen: *Lecanicillium lecanii s.l.*[1] Predator: *Cryptolaemus montrouzieri* (ladybird) not on tomatoes nor < 21 °C nor dry conditions; *Leptomastix* spp. (wasps) good searchers. Insecticide: buprofezin
Plant hopper	Browse, sucking	Parasitoid: *Anagrus atomus* (wasp) attacks eggs
Sciarid fly	Eat roots	Nematodes: *Steinernema carpocapsae, S. feltiae.* Insect growth regulators
Spider mite	Surface browser, sucking. Disfigure leaves	Predators: *Amblyseius californicus* (mite) prefers high temperature, good searcher; *Feltiella acarisuga* cool conditions, good searcher; *Galendromus occidentalis* (mite) high temperature, exists without mites; *Phytoseiulus persimilis* well proven; *Typhlodromus occidentalis* better in hot dry conditions. Acaricide: fenbutatin oxide
Thrips	Browse. Disfigure leaves and flowers. Virus vector	Pathogen: *B. bassiana; L. lecanii s.l*[1] Predators: *Amblyseius* spp. survive on mites and pollen *Hypoaspis* spp. (mites); *Orius* spp. (bugs) survive on pollen, good searchers
Vine weevil	Eat roots. Disfigure leaves	Pathogen: *B. bassiana; Metarhizium anisopliae.* Nematodes: *S. carpocapsae* and *Heterorhabditis bacteriophora* and *H. megidis*
Whitefly	Sucking. Sooty mold	Pathogens: *Paecilomyces fumosoroseus; L. muscarium*[1]; *B. bassiana.* Parasitoids: *Encarsia formosa* (wasp) well proven, better on *T. vaporariorum* than *B. tabaci.* Predator: *Delphastus pusillus* (ladybird) good searcher

[1] The revision of Zare and Gams (2001) assigns different species names for some of the isolates in the species complex previously and collectively referred to as *Lecanicillium* (=*Verticillium*) *lecanii*. For example, the whitefly active species used in the Koppert product Mycotal is *Lecanicillium muscarium* while the aphid active isolate used in the Koppert product Vertalec is *Lecanicillium longisporum*.

pests to reach being those on the undersides of leaves deep in a foliage canopy, or those that burrow into plants or soil.

C Greenhouse crops

Greenhouse crops comprise vegetables, fruits and ornamental plants (Table 2). They are relatively delicate, labor intensive, much handled and confined to limited space. Chemical pest control involves poisons, possibly phytotoxic and frequently unpopular with greenhouse staff. In contrast, microbial control is worker friendly, not phytotoxic and desirable for many other reasons. Pest control measures are intensive and individual components must be mutually compatible. Accessibility for spraying varies with plant growth habit from low, sessile rosettes of lettuce to upstanding row crops *e.g.* cucumbers and peppers; tomatoes may be layered or arched forming dense,

Table 2. Main crops grown under glass in UK, damage and damage tolerance (Malais and Ravensberg, 1992)

Crop	Part sold	Damage tolerated at sale	Type and cause of damage
Tomato Cucumber	Partly ripe fruit Table-ready fruit	None	Holes (caterpillars). Skin blemishes (browsers). Sooty mold (aphids, whitefly scales). Damaged fruit removed at packing. Flower failure is earliest sign of phytotoxicity (chemical pesticides)
Foliage plants, lettuce	All	Little	Holes (caterpillars) and blemishes (browsers and suckers) lower quality grade. Damage to lower discarded foliage not important
Cut and potted flowers	Flowers, buds, upper foliage	Very little	Buds and flowers distorted (caterpillars). Foliage blemishes (browsers and suckers). Flower infestation (all). Trimming at packing, time consuming
Many plants	Various	Limited	Hidden root damage (vine weevil), lettuce (root aphid) stunt and kill plants, lowering yield

high canopies (see section 3B); bedded crops are often dense such as year-round chrysanthemums (section 4B) or the almost impenetrable, 6-foot (1.8 m) high carnation beds. Arthropod pests occur on the upper and/or lower surfaces of foliage, but less often in soil, except vine weevils and lettuce root aphids.

D Arthropod pathogens used in greenhouses

A limited number of arthropod pathogens are used on protected crops (Table 3). When used correctly and in the right conditions they are extremely effective. Because crop quality is vital and a rapid impact on the pest is necessary, so strategy with pathogens is inundative. The high value of crops permits relatively frequent applications and high dosages. In the most favorable conditions, pathogens may give crop-long control after one or a few applications but not extending reliably to neighboring plants or subsequent crops. Integrated pest management (IPM) involving predators and parasites is common on protected crops and bee pollinators may be used, so the compatibility typical of insect pathogens is a very valuable asset.

2 Experimental design specific to greenhouses

In a literature search, techniques specific to greenhouse design and described in post 1985 papers tended to be relatively simple and briefly

Table 3. Arthropod pathogens used or potentially useful in greenhouses

Pathogen	Pests mainly from UK	Main crops
Bacillus thuringiensis (Bt) subsp. *kurstaki*	Tomato moth, *Autographa gamma*, tortricids	Tomato, pepper, rose, chrysanthemum and other ornamentals
Bt subsp. *aizawai*	*Mamestra* spp., *Spodoptera* spp.	
Lecanicillium (Verticillium) species[1]	Aphid, whitefly, thrips, mealybug	Chrysanthemum, cucumber, tomato, other ornamentals
Paecilomyces fumosoroseus	Whitefly	Cucumber
Beauveria bassiana	Aphid, thrips, whitefly, vine weevil	Most horticultural crops
Metarhizium anisopliae	Vine weevil	Cyclamen, other ornamentals
Granulovirus of tomato moth	Tomato moth (potential)	Tomato
Nucleopolyhedroviruses	Species specific in many Lepidoptera (potential)	Crops infested by only one moth species (potential)
Entomopathogenic nematodes	Vine weevil	Cyclamen, other ornamentals

[1] See footnote in Table 1 and revision of Zare and Gams (2001).

described, so many of these references have been omitted in favor of more comprehensive earlier work. Details of technology described in other chapters, *e.g.* spraying and spore enumeration, have also been omitted. A number of features common to all greenhouse protocols are included in this section and will not be repeated. Initial experiments on a new pathogen should be designed to favor the pathogen and later experiments should progressively involve conditions approaching commercial practice. A series of protocols will be described to grow, infest, treat and record in detail experimental crops. As far as possible, IPM systems (Table 1) are used to maintain crops in a state that is healthy apart from the effects of the target insects. All the protocols assume that an IPM system is already in place. The effects of the pests on crops are fully detailed, including events after sale. Protocols should include records of operator times for different activities and material costs as data for practical cost analysis. At the end of an experiment, residues of crops should be removed or sterilized and greenhouses efficiently cleaned to prevent pests and their diseases, as well as IPM components, causing complications in subsequent crops.

A *Experimental units*

By design or necessity, an experimental unit ranges from a potted plant to a whole large greenhouse full of plants. Pots are easy to handle and have the advantage of flexibility. They can be sprayed with insect pathogens in the laboratory, if necessary in a containment cabinet to avoid contaminating the laboratory or other plants. Uniform spray deposit can be facilitated by a turntable, or by a gantry in a special building with moving spray jets, or by conveying the potted plants between fixed jets (McGuire *et al.*, 1997). Treated potted plants should be arranged in a fixed array during incubation in a greenhouse using random distribution to compensate for position effects in the array. Edge effects can be avoided by using outer guard rows of plants.

Beds of plants, or partitioned sections of beds, in a single greenhouse may be used as experimental units. These have similar position considerations as arrays of potted plants described above. Contamination by spray drift can be prevented by covering with plastic sheet all rows not being sprayed; alternatively it can be minimized by using directional high volume (HV) sprays; it can also be minimized by starting with the most dilute concentration and working up the series.

Use of individual small houses or partitioned compartments of a larger house as units avoids the spray drift problem and partly avoids the problem of target insect migration. Even so, position effects between greenhouses must be allowed for, particularly differences in internal greenhouse climate.

The unit may be a large cropped greenhouse in the ultimate trials on commercial crops. Untreated controls are not feasible with commercial crops so comparisons have to be made with previous crops or positive controls using other pest limitation methods. Experimentation on this scale is the only means of testing individual ultra low volume (ULV) or very low volume (VLV) treatments using fan-assisted machines (Jones and Burges, 1998) often directed inwards sequentially from each doorway around the greenhouse.

B *Infestation with pests*

Initial insect infestation for experimental crops may be natural or applied. In commercial crops, natural infestation is usually the only option and account must be taken of patchy insect distribution (section 3D). Untreated controls cannot be used and a grower will allow infestation to develop only to a limited extent to avoid costly crop damage.

On experimental crops target infestations usually have to be initiated, or natural infestations supplemented, taking due account of pest biology (Malais and Ravensberg, 1992). Adults may be released (section 4B). If highly mobile they may cause very asymmetrical infestations, *e.g.* whiteflies lay eggs mainly on new foliage and moths lay most eggs on the part of the crop that faces the setting sun (Burges and Jarrett, 1976). Release of Lepidoptera adults may require the greenhouse to be insect proofed to avoid excessive escapes.

More uniform infestations can be obtained by distributing egg masses of Lepidoptera (section 3B) or pieces of leaf bearing dense spider mite infestations. Infestations are most uniform if young larvae or insects like aphids are distributed systematically (sections 3E and 4A). Alternatively, the use of greenhouse infestations can be avoided by exposing samples of treated foliage to insects in the laboratory (section 3C).

It is usually desirable to start with a crop free from incidental pests and plant diseases. If necessary, effective treatment(s) with *transient* broad-spectrum chemical insecticides and acaricides will kill resident pests (and insect predators and parasites). Also *specific* plant disease treatments and prophylactics, compatible with the microbial, will ensure plant health (Table 1). There must be a sufficient time lapse before reintroducing an IPM system that will maintain incidental pest control during an experiment and also before introducing the target pest. The sufficiency of this lapse should be experimentally confirmed. Copping (1998) and Table 1 list and describe the use of IPM agents that are currently available.

With some pests, especially Lepidoptera, greenhouses must be bird proofed (Burges and Jarrett, 1976). A single pair of wrens, nesting outside and entering via a small hole in one pane of glass, soon stripped caterpillars off an experimental chrysanthemum crop (H. D. Burges, unpublished).

C Spraying in greenhouses

Traditionally spraying under glass is HV (> 500 liters/ha) and medium volume (51–500 liters/ha) at high pressure (*e.g.*, 7 bars = 700 kPa or 100 psi) using knapsack sprayers, or hand-held lances from powered sprayers (Jones and Burges, 1998). Frequently nowadays semi-automatic booms of nozzles are mounted horizontally above crops or vertically between tall row crops. To cover the undersides of leaves, nozzles commonly are angled *e.g.* Allman hollow cone nozzles with a cone angle of 60°. Most formulations for HV are emulsions, water dispersible granules or wettable powders sprayed in water (Jones and Burges, 1998; Burges and Jones, 1998; Burges, 1998). For pathogens, chlorinated water should be avoided or aired overnight in containers, *e.g.* buckets, to remove chlorine; acidic or alkaline waters should be neutralized or avoided. All pathogens are particulate so, to prevent settling, spray tanks should be frequently shaken, stirred, or fitted with a mechanical agitator.

Increasingly, very low volume (VLV, 5–50 liters/ha) and ultra low volume (ULV, < 5 liters/ha) sprays are being used to reduce labor input, usually with fan-assisted spinning-disc machines, hand-held or mounted on wheels (Sopp *et al.*, 1990). Leaf cover in some canopies is equal to or better than with HV (Jones, 1998). With hand-held machines, treatment is rapid, the operator walking along every two or three rows, or along every row with high canopies *e.g.* tomato. Hand-held equipment is tiring to use for long periods and may cause health problems like 'tennis elbow', so spraying should be spaced in time or shared between several operators. Mist can be blown into four doors spaced around a hectare greenhouse in a total operational time of ca 1 h compared with > 4 h with HV. However, windy weather must be avoided in all but plastic sheathed houses, because wind penetrates glass structures and blows the mist to one side or out and thus giving asymmetric deposit on a crop. Frequently, spray contractors invest in this specialized equipment. Suspensions in oil may be used, or the same formulations as above for HV in water, but the sprays are more concentrated (Burges and Jones, 1998; Burges, 1998; Copping, 1998).

To cover all aspects of experimental techniques two organisms will be used, *Bacillus thuringiensis* and *Lecanicillium (Verticillium) lecanii senso lato*, in separate sections because of their diverse technical requirements.

3 Use of *Bacillus thuringiensis* to control Lepidoptera

Larvae of most Lepidoptera (Table 1; Jarrett, 1985) can be controlled specifically by *B. thuringiensis (Bt)* (Table 3; Burges and Jarrett, 1978, 1979; Jarrett and Burges, 1982, 1986; Copping, 1998). Eggs, pupae and adults are not affected. The youngest larvae are almost universally the most susceptible and in some species only instars I-III are susceptible enough to be effectively controlled. Anyway older larvae may tunnel into fruit (*e.g.* tomatoes and peppers) and stems, where they no longer eat surface deposits of bacteria. *Bt* acts perorally and has no contact action, so must be eaten to take effect. Consequently good even cover is required over the larva's foliage food, particularly on the undersides of leaves, the feeding site of the highly susceptible, newly hatched larvae (Burges and Jarrett, 1978). Phagostimulants (Burges and Jones, 1998) can be added to formulations to improve control. Light or pheromone traps can be used to detect adult moths and improve spray timing, but caution is needed if setting traps inside houses to ensure that extra adults are not attracted into the house and that extra light does not alter photoperiods to influence crop timing, which can be affected even by intermittent light.

The range of physical conditions experienced by glasshouse crops (section 1; Figure 1) has virtually no effect on *Bt*. Equable temperature accelerates larval feeding and development. Temperatures low enough to curb feeding, which would impair control, are rare (Figure 1; below 10–15 °C depending on species). The UV component of sunlight, which kills spores and more slowly inactivates the crystals of toxic protein, is largely filtered out by glass but not by plastic (Jones and Burges, 1998). Pathogenicity was retained under glass for > 28 days (Jarrett, 1981). Sunlight protectants (Burges and Jones, 1998), added to formulations to improve control, are relevant only under plastic.

Development of larval resistance to *Bt* has not been a problem in greenhouse crops. Precautions to prevent its development are often unnecessary because most species are immigrants from outside and populations are eliminated every time a greenhouse is cleared out and recropped. In countries where resistance has developed outdoors, Lepidoptera moving into greenhouses may, of course, already be resistant.

The behavior pattern of *Lacanobia oleracea*, the tomato moth (Table 1), makes it particularly suitable for control with *Bt*. Eggs are laid on undersides of leaves but 2 days after hatching, they eat through to the upper surface, which normally is the more heavily treated. In daytime larvae assume inconspicuous defensive positions along stems, leaf veins and edges and move to a new feeding site every day, which increases the likelihood of ingesting bacteria. Larvae move down crops away from new foliage (where the bacterial cover is in effect diluted by new growth), particularly by parachuting on silken threads, many reaching the ground to be taken by predators, *e.g.* spiders. On untreated control crops, populations halved every 3–4 days. Good control on mature tomato and pepper crops can be obtained with 5% aqueous VLV spray at 5 liters/ha or 0.04% spray with HV at 5600 liters/ha, using products with 32,000 international activity units/mg (Burges, 1974; Burges and Jarrett, 1978, 1979; Jarrett and Burges, 1982).

A *Measuring potency of* Bacillus thuringiensis *products*

The potency of modern *Bt* products is defined in various ways on the label. In publications, it is important to include all label information about potency. For experimental work, particularly if substantial, it is desirable to confirm potency levels because product batches may vary considerably and may deteriorate. The main active component is the toxin crystal, but many leaf-feeding pests are also sensitive to live spores. A complete measure of potency can be obtained only by bioassay. Different pest species react differently to different bacterial strains and batches of test insect larvae vary. The most reliable potency measures can be obtained as ratios against a standard product included with every batch of assays. The traditional use of *Trichoplusia ni* and the internationally available HD1 standard is useful for products based on the *kurstaki* subspecies (McGuire *et al.*, 1997). For other subspecies and for products containing strains with modified toxin spectra, the pest insect species itself should be used in assays against a personal standard. This standard should be a dry product for the best storability, kept at 2–10 °C, in absolutely moisture-proof containers. Every time the standard powder is used, it should be allowed to equilibrate to room temperature before opening the container. This procedure avoids moisture from the air condensing on the powder, which would seriously impair storability. After every use the container should be resealed and the standard returned to low temperature.

Because a basic measurement of potency is required, not influenced by factors such as crop plant species, an artificial-diet assay is best, *e.g.* that described in detail by McGuire *et al.* (1997). Jarrett and Burges (1986) and McGuire *et al.* (1997) describe assays on leaf. Some modification of diet components may be necessary for different larval species.

B Low volume spraying of tomatoes

The distribution of a thermal fog is shown in Figure 2 as an example of application by ULV and VLV (Burges and Jarrett, 1979). Typically, thermal fogs move to the end of the greenhouse, then up the end wall and back again, reaching the wall at the opposite end if the house is relatively small. The largest drops are deposited near the sprayer, smaller droplets further on, the distance traveled before settling increasing with reduction in droplet size. A protocol is described below for investigating deposit distribution, also infestation and damage by the tomato moth (Burges and Jarrett, 1978, 1979):

1. Bird proof the greenhouse (section 2B).
2. Infestation of mature crop. Estimate numbers of eggs in laboratory-reared egg masses laid on paper (Jarrett and Burges, 1982) by visually assessing area covered by eggs and multiplying by the number of eggs per unit area. Cut squares of paper ca 2×2 cm, each bearing an egg mass, and pin squares individually to undersides of leaves on selected plants spaced equidistant along rows. These form foci of infestations. A minimum of 500 eggs should be placed at each focus, preferably many more, because of the rapid natural loss of larvae from the population (section 3 above). On neighboring plants after spraying place 10 II-III instar larvae on each of seven leaves and confine them with fine white cotton cloth bags (20×30 cm) firmly tied at the necks around petioles to prevent natural loss of larvae. Every 2 days count the larvae in bags and transfer to fresh leaves.
3. Initial count of free larvae and damage assessment. When most larvae have reached instar II (approximately 15 days after oviposition), count the free larvae on selected labeled leaves spaced up the focus plants. Visually estimate area of holes eaten through these leaves. Count fruit of ≥ 2 cm diameter, recording ripeness and damage. Disturb foliage minimally to avoid shaking larvae off, or stimulating them to fall or parachute down.
4. Deposit measurement. Light a thin strip of magnesium ribbon and hold a 7.5×2.5 cm glass microscope slide in the smoke until one side is evenly coated with a thick layer of magnesium oxide (May, 1950). Fix slides horizontally with cellotape at positions sited at regular intervals along one row of plants in the glasshouse and along three traverses at 90° angles, one traverse near the middle of the house and one near each end (Figure 2a). At each position fix slides at three or more heights (bottom, middle and top leaves of the plants and – if possible – some higher in the structure of the glasshouse and on the soil. At representative positions, fix one extra slide with the oxide facing vertically towards and another vertically away from the direction of the spray, as well as at least two others at angles mimicking the angles of the upper and lower surfaces of typical leaves. Fog droplets make pits in the very soft oxide coating, easily measured under low power microscope to reveal size spectra (May, 1950; Burges and Jarrett, 1978). The diameter of droplets still in their spherical shape is measured, which is more accurate than measurements made after deposit on special cards.
5. Treatment (see section 2C). Mix the *Bt* (maximum final concentration 5%) into ≤ 500 ml of water. If the formulation is a wettable powder, ensure that all powder balls are dispersed. Make up to the required volume. Fill spray tank through a funnel equipped with a mesh screen to minimize the chance of powder balls or debris blocking the sprayer. Do not allow to stand before use. Apply with a hand-held, pulse-jet type thermal fogger (*e.g.*, Pulsfog K10 with jet size 12, or Swingfog SN11P with no. 1.2 jet) at 40 liters/ha, in water with or without 5% w/v Nevolin (=VK2 carrier), containing water, methanol and 2-ethoxyethanol (Burges and Jarrett, 1979). Alternatively use a fan assisted mister or cold fogger. Starting at one end of the house, walk backwards along a path between rows to minimize operator exposure to the spray (Burges and Jarrett, 1978, 1979).
6. Observation of deposit. When all spray droplets and particles have settled (no sooner than 2 h after spraying) collect slides and store in holders, oxide upwards. Droplet diameter is measured by the diameter of its pit (May, 1950). Initially measure the diameter of selected pits using an eye piece reticule marked in calibrated squares. Visually classify pits in seven size ranges. Count pits in

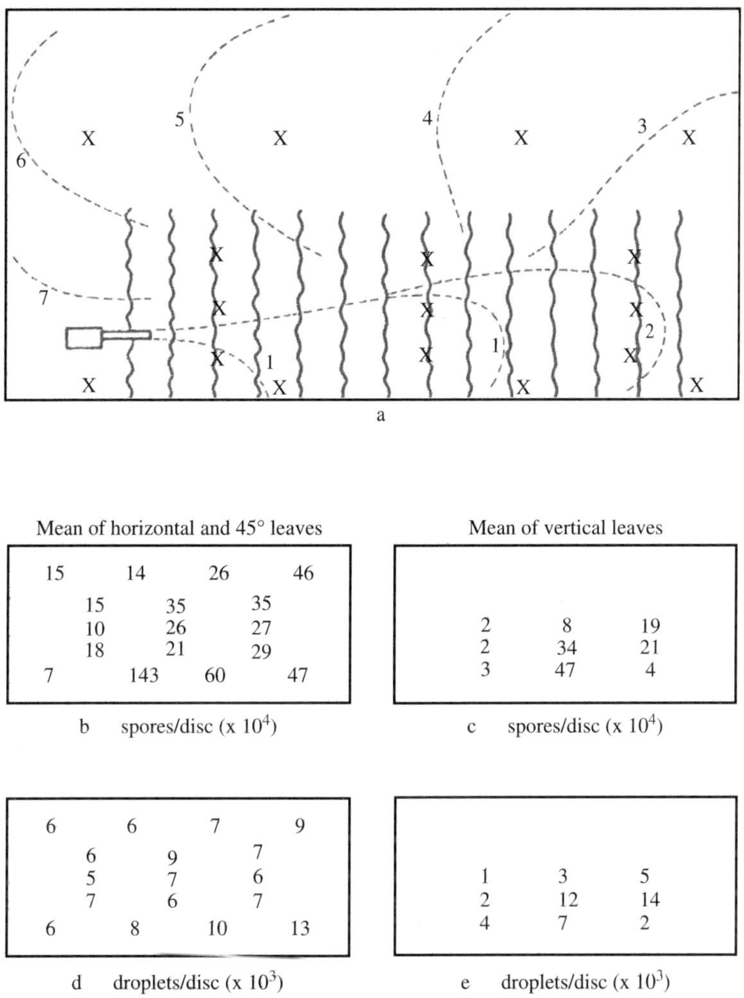

Figure 2. Distribution of *Bacillus thuringiensis* obtained with a thermal fogger (bottom left) along a single span glasshouse 10 x 9 m. a. Elevation through a row of tomato plants trained by the S-hook layering method (wavy lines) to show sampling positions (x) and fog path (broken lines 1 to 7); b-e. Spores and spray droplets per four sample leaf discs, 2 cm diameter, at each position from leaves at three angles to the fog path (Burges and Jarrett, 1979).

these sizes over a known slide area, aiming at a minimum of 30 pits if the deposit rate is high enough. Deposit per cm^2 is given by droplet number/area counted. For alternative assessment of coverage using dyes and spore counts, see section 3C.

7. Observation of larvae. Every 2 days count live and dead larvae on the labeled leaves (see step 3) and on fruit trusses, visually estimating the instar of live larvae. At each fruit harvest count, and weigh sound and damaged fruit.
8. Observation of foliage damage. Periodically the oldest, lowest leaves are cut off to expose fruit trusses to sunlight. Estimate damage to any cut leaves labeled at step 3. End the experiment when all larvae have pupated, or sooner, and estimate damage on all labeled leaves. If damage is noted on new leaves growing during the experiment, label and record extra leaves.

The larvae in bagged leaves (step 2) help to solve the problem of having enough free insects to give meaningful mortality and damage results. A higher ratio of infestation to foliage area can be used with potted plants.

C High volume Bacillus thuringiensis *spray*

Since *Bt* needs good, even coverage on undersides of leaves, HV is often the best spray method, but time consuming, hence labor intensive (Burges and Jarrett, 1978). Spraying a hectare with 4000 liters takes 1 working day. Early instars are far more susceptible than late instars so the objective is to prevent their survival, because older larvae (instar IV and later) of many species – unlike the tomato moth – move up the plants at night and attack growing points, buds and flowers. On beds of chrysanthemums, thorough spraying becomes progressively more difficult as plants grow taller, requiring lances with twin nozzles angled upwards to be laboriously moved between rows in the beds; at ≥ 60 cm high, beds of plants present a solid phalanx of foliage. In commercial practice it is too time consuming to move lances between individual rows in the beds, so several sprays are needed as the plants grow. A partly eaten flower bud grows to produce a grotesque flower, so if a few large larvae survive deep in the beds, buds and upper parts of stems may have to be sprayed with a fast-acting contact insecticide after they have grown above the mass of foliage. It is important to assess spray coverage (Burges and Jarrett, 1978), *e.g.* using a fluorescent dye (Courshee and Ireson, 1961) as in the following protocol:

1. Mix 0.1% of a fluorescent dye with 0.1–0.3% aqueous *Bt* spray plus a non-ionic wetting agent, *e.g.* 0.1% Agral. A suitable dye is Saturn Yellow Wettable Powder MF-7 with particles of 1–2 μm (H. Haeffrom and Co. Ltd, Chepstow, UK).
2. Apply a fine spray almost to run off across the top and along the sides of a bed of chrysanthemums 60 cm high (*e.g.* 10 plants × 50 + rows at 10 cm spacings each way) with a high pressure sprayer through a hollow cone nozzle, *e.g.* No. 3 Allman hollow cone nozzle, cone angle 60°, delivering 1.6 liters/min at 7 bars (700 kPa).
3. After the spray has thoroughly dried, place into Petri plates leaves selected systematically at different heights on plants positioned representatively in the rows.
4. View leaves under low power microscope with UV light. Erect seven arbitrary categories of intensity of fluorescent cover, including apparent zero cover, and score leaves in these categories.
5. Estimate the larvicidal efficacy of the categories by adding groups of 20 instar II larvae to sample leaves of each category on moist filter paper in Petri dishes and scoring larval mortality daily as long as the leaves remain in reasonable condition. Leaf damage provides a more sensitive efficacy assessment than the mortality assay. It can be measured by putting the leaves on squared paper or by using a photoelectric area meter (*e.g.* Evans Electro-selenium Ltd Halstead, Essex, UK).
6. Bacterial coverage in terms of viable spores can be measured by cutting 1.4 cm circular leaf discs with a punch or cork borer. Coverage on only one leaf surface can be measured by masking before the spraying the opposite surface with tape bearing a central disc of paper that stops the tape sticking to the leaf disc to be sampled. Place each disc in 10 ml of phosphate buffer + detergent, pH 7.2, in a screw capped vial with fine glass balls. Shake on a vibrating shaker for 10 min. The spores and crystals are enumerated microscopically by bacterial counting chamber, or viable spores by plating (Jarrett *et al.*, 1978; Burges and Jarrett, 1979; Goettel and Inglis, 1997).

D Elimination of tortrix moth on roses

Cacoecimorpha pronubana forms resident year round populations on perennial rose bushes under glass, while immigrants from outside are infrequent. Larvae tunnel into buds – often unnoticed until flowers open after marketing, appearing shot by a machine gun. Frequent sprays with chemical insecticides harden bushes, reducing the number of side shoots, thus lowering crop yield; *Bt* has not had this effect. The glasshouse moth population can be eliminated as follows:

1. Commercially, unopened buds from the whole nursery are harvested and boxed for market twice weekly. Destructively sample 100 buds at random to establish initial infestation rate. Count adults resting on the glass, where they show up well, all round a house up to 2 m above ground. This is rapid and avoids a risk with attractive traps of enticing more adults into the greenhouses from outside (section 3).
2. Treatment. Load the spray tank (as in section 3B step 5) of a high pressure HV sprayer with a 1 m lance. Apply a 0.1% w/v, *kurstaki* or *aizawai* based formulation (Table 3), use powerful, fine spray nearly to run off (as in section 3C, step 2), directed at the undersides of leaves with extra spray on hibernaculae of spun-together leaves. To improve spray penetration of these feeding tunnels

use a high concentration (0.2%) of a wetting agent compatible with *Bt* (many listed with data by Bernhard *et al*., 1998; Burges and Jones, 1998).

3. At day 6 after treatment, start a routine of sampling another 100 buds, counting adults and respraying (steps 1 and 2). If bud infestation is < 5%, sample 100 more buds.
4. Repeat this step 3 routine at 6–7 day intervals until infestation is < 1%, then extend interval to 14 days and increase sample size to 300 buds.
5. When no larvae or adults are found, increase the routine interval to 1 month. After none are found on four consecutive counts the resident moth population can be deemed eliminated.
6. After elimination, continue searches for adults on the glass at 2-week intervals as an early warning of possible immigrants from outside. If two or more moths are found in a search, spray three times at 2-week intervals as a prophylactic to prevent re- establishment of another resident population.

E Potted plant experiments with caterpillars

Potted plants are useful for testing pathogen effectiveness and observing behavior of treated larvae on a small experimental scale in greenhouses. Each plant is stood on a square metal tray that extends beyond the span of the foliage and larvae are established by placing them on the leaves. The tray sides should be vertical, 2–3 cm high, rough and easily climbed to give an indication of the health and co-ordination of larvae moving or falling from plants. Each tray is housed in a similar tray at least 20 cm larger and containing light machine oil to trap escapers. Plants can be sprayed (section 2A) and leaf damage estimated (section 3C, step 5). Live and dead larvae can be counted on the plant, pot soil, tray and in the oil. Larvae moving or falling onto the inner tray are unlikely to climb back up the pot on to the plant because - on an arena such as a tray – larvae walk at random until they reach a barrier, *i.e.* the tray walls, which they tend to follow or climb rather than walk away from; furthermore, the distance around the tray walls is much greater than the circumference of a pot which is less likely to be followed anyway. Some of the healthiest larvae scale the tray walls and die in the oil.

4 Use of fungi to control plant sucking pests

Entomopathogenic fungi (Table 3) are an important option in IPM programs (Table 1). Because fungi usually invade insect hosts by penetrating the cuticle, they are the only pathogens that have been developed for control of insects with piercing and sucking mouthparts such as aphids (Latgé and Papierok, 1988) and whiteflies (Lacey *et al.* 1996). Protocols will be described to investigate this option in greenhouses, incorporating techniques to examine the factors involved. Ancillary laboratory techniques examine key factors in more detail.

Fungi infect when conidia attach to insect cuticle, *i.e.* a contact action not involving feeding, so target pests include sucking insects (Tables 1 and 3). Some target insects are hit directly by spray droplets and others pick up conidia from leaf surfaces, although with some difficulty (Hall, 1976b). A hiatus period of 2–7 days follows when nothing appears to happen and pests continue to feed and reproduce. Then fungus breaks through the intersegmental membranes of infected insects and sporulates. Some fungi break through while the insect is still mobile – a walking conidia disperser until the day before death (Hall, 1976a). Dead insects commonly stick to the leaf, shrouded in sporulating mycelia. In humid conditions infections spread, even wiping out pest populations, and sometimes spread extends irregularly to neighboring crops.

For experiments, good strain handling and maintenance techniques are vital (Hall, 1980a). In my opinion, each strain should be isolated from a single conidium to minimize within-strain variability. Also, to avoid possible change during serial laboratory culture, the isolate should be grown up on agar, harvested as an aqueous suspension, then many aliquots stored in liquid nitrogen, deep frozen or stored in an equivalent inert condition depending on survivability of the fungus species (Humber, 1997). A fresh aliquot should be activated and cultured for each batch of experiments.

The activity of fungal products is measured by conidia enumeration and by bioassay. Conidia are enumerated by total microscopic counts and by viable counts on agar plates or on slides coated with agar, which allow direct microscopic observation of germination. Innate conidia potency can be measured only by bioassay at a virtual 100% R.H. in conditions favoring the mode of action of the fungus species in use. If stored conidia are to be assayed shortly before spraying, comparison with a newly-produced batch of conidia is desirable. The exact conditions of production and harvest of the new conidia are often critical so these conditions should be defined and described in detail (Burges, 1998). Conidia enumeration and bioassay are described by Hall (1976a, 1979, 1982b, 1984), Goettel and Inglis (1997), and Kerwin and Petersen (1997).

Physical conditions, particularly the fluctuation in greenhouse temperature and humidity (Figure 1), dominate fungal pest control. Techniques and apparatus for the laboratory simulation of fluctuating greenhouse conditions are described by Milner and Lutton (1986) and Curtis (1998).

The first fungus pathogen studied and applied in greenhouses, *Lecanicillium (Verticillium) lecanii s.l.* (see Zare and Gams, 2001 for revision of the species complex), will be used here as an example. At constant temperatures its range for good growth is 15–27 °C, with no growth above 31.5 °C. At diurnally fluctuating temperatures the range is wider (Kanagaratnam *et al.*, 1982; Hall and Papierok, 1982; Curtis, 1998). These ranges are narrower than both the ranges encountered in glasshouses (Figure 1), even within foliage canopies, and the ranges of target pests (Malais and Ravensberg, 1992). Thus both low and high greenhouse temperatures curb the fungus. Other fungus species may be less dependent on low humidity; *Metarhizium anisopliae* var. *acridum* (=*M. flavoviride*) formulated in oil is able to infect a range of Orthoptera at very low humidity in desert conditions (Burges, 1998).

For strains of *L. lecanii* low humidity is limiting. Unformulated conidia, on various surfaces *in vitro* at constant humidities, survive well only at 80–100% R.H. (Hall, 1980a). Germination of conidia requires higher humidity than survival, being optimal at ≥ 97% R.H., poor at 95% R.H. and ceasing at 92–93% R.H. (Milner and Lutton, 1986). *In vivo*, low humidity reduces pest mortality resulting from direct hits by conidia and curbs fungal sporulation on dying or dead insects, all of which reduce spread of *L. lecanii* in aphid populations (Ekbom, 1979; Hall, 1982a; Hall and Papierok, 1982; Kanagaratnam *et al.*, 1982; Milner and Lutton, 1986; Curtis, 1998).

In humid conditions, formulations containing nutrients allow mini fungal colonies of *L. lecanii*, usually invisible to the naked eye, to grow and sporulate, mainly on the undersides of leaves. The extra conidia can increase foliage conidia cover by ×40 (Hall and Papierok, 1982) and accelerate fungal spread in a pest population. This is prevented at low humidities. At > 95% R.H. alternating diurnally with 70% R.H., germination and growth on the leaf were poor (Curtis, 1998).

In a moisture retaining and nutrient formulation, at constant 70% R.H., germination on leaf was good but aphid infection poor. At > 95% R.H. for 15 to 18 h, alternating with 40 or 70% R.H., fungal growth, sporulation and aphid infection were good (Curtis, 1998). Humidity in the crop canopy (Ravensberg *et al.*, 1990) and presumably microhumidities intimately associated with the live insect body are above ambient greenhouse humidity. Unfortunately, humidity at the microsites on the leaf where the insect lives cannot be measured accurately. At best, instruments/sensors should be sited in critical regions of a greenhouse near the insects and their pathogens. If compact sensors are available, with a low-growing crop for example, records are desirable at soil level, half way up plants, at the top of plants, inside and outside any covers, all for comparison with the ambient air record of the greenhouse and meteorological records from a nearby weather station.

Humidity is thus of equal or greater importance than temperature, and the ability of the fungus to sense humidity is probably superior to our ability to measure high humidities consistently!

A Lecanicillium (Verticillium) lecanii s. l. *to control aphids on chrysanthemums*

Chrysanthemums are grown in both pots and in dense beds in soil. Crops are timed to flower in succession by controlling day length with black covers. Polyethylene can create high humidity ideal for fungi and insect control, but water vapor-porous covers do not have this effect.

As well as humidity and temperature, it is important to investigate other factors that influence pest control by fungi *e.g.* density of insect population (Hall and Burges, 1979; Quinlan, 1988; Curtis, 1998) and factors that affect conidial germination, including formulation (Curtis 1998; Burges, 1998).

An important factor is the characteristic behavior patterns of different aphid species (Hall, 1976b; Hall and Burges, 1979). For example, *Myzus persicae* breeds throughout a crop, particularly on foliage deep in the center of a bed where insect control is good, probably because humidity is highest there and this aphid is restless, which probably aids fungal spread. *Macrosiphoniella sanborni* clusters on stems, especially below buds, a less humid position where control may be poor. Before flowering, *Brachycaudus helichrysi* gathers within the humid space around growing points where control is good; then later it aggregates below buds where control is often poor; finally it moves into flowers as buds open, where control may be good in humidity sites, but poor in dry sites. These differences in degree of control are not related to innate susceptibilities of the three species, which are similar (Hall, 1976a).

Aphids almost invariably appear on UK chrysanthemums, so treatment is best made prophylactically 2 weeks after planting, particularly as control is ineffective in dense populations (Quinlan, 1988; Curtis, 1998). Varying with species, adult aphids fly from bed to bed, often spreading the fungus, so individual houses are ideal – one for an untreated control and one for each conidia concentration, although short term experiments using individual beds and even sections of beds can be satisfactory. Treatments should be replicated, either over time or at one time.

The following protocol describes a comprehensive investigation on bedded chrysanthemums (Hall and Burges, 1979):

1. Infest (1 aphid/plant) rooted cuttings of an aphid-susceptible cultivar 2 weeks after planting 10 cm apart in beds in six greenhouses. Black polyethylene sheets are drawn on supporting wires over the beds for 16 h daily.
2. After 3 days, count aphids on 10 plants/bed, spaced systematically to represent key positions in the bed. Apply a fine high pressure (7bars = 700 kPa) HV spray of conidia nearly to run off, *e.g.* 0, 10^6, 10^7 and 10^8 conidia/ml, in that order (Hall, 1980b), using an angled nozzle on a lance, preferably late in the day. If a dry formulation is used, *e.g.* Vertalec, presoak for 2 hours unless specifically instructed otherwise on the label (Quinlan, 1988; Burges, 1998).
3. After the spray has dried, calibrate and site temperature and humidity sensors (Appendix 1). Sensors are temporarily removed while watering plants.
4. At intervals count aphids (classifying into healthy adults and nymphs, various stages of disease, and dead) on representative plants. If plants grow too large to see lower leaves, pick ca 10% of the leaves on a plant for examination (*e.g.* top, middle and lower leaves or otherwise depending on the behavior/feeding site of the aphid species in use) moving to a neighboring plant at each examination date. Assess fungus germination, growth and sporulation on leaves by applying the sticky side of cellotape firmly to the leaf surface, pulling off, staining in 0.1% acid fuchsin in lactophenol and examining microscopically, sticky side up (Curtis, 1998).
5. Respray with conidia at the various concentrations 1 week before infestation is expected to approach disfiguring levels at – for example – the second lowest conidia concentration. Alternatively, if the treated infestations are eliminated by the fungus, introduce more aphids to simulate reinfestation and respray 3 days later. Continue regular aphid counts.
6. Just before the first buds show some color, count aphids on buds and on the elongated stems. Spray with a selective chemical aphicide, *e.g.* pirimicarb (poor against some species), across half the crop and thereafter do comparative aphid counts in the two zones. This is designed to test for loss of fungal pest control after blooms are cut and taken into dry environments in homes.
7. Harvest the crop when the first 20% of flowers have opened and destructively examine sample plants and flowers.
8. Place bunches of flowers in vases in representative sites to mimic various typical after-sale conditions and monitor for aphid infestation. Cast skins falling on to table reveal aphids.

B Lecanicillium (Verticillium) lecanii s.l.
to control whiteflies on cucumbers

Cucumber plants grow rapidly up wires in rows or arches (now uncommon), often reaching the roof of low greenhouses, forming dense canopies. Leaves past maturity are removed to keep the crop open; they contribute little to cucumber yield. The crop is planted directly in soil, in plastic 'grow bags' or in slabs of rock wool (mineral rock fibers). Bags and slabs rest on sheet polyethylene covering all the soil, so the climate is drier than in greenhouses with bare soil.

Adult whiteflies fly towards light and oviposit mainly on undersides of young foliage. Instar I is brief and mobile, eventually settling abaxially on leaves. Later instars (scales) and pupae are static, waxy and firmly pressed down on the leaf surface. Larvae secrete honeydew, which falls on foliage and cucumbers, where it supports disfiguring black fungal growth – often the most important form of crop damage (Malais and Ravensberg, 1992).

Since adults are free-flying, whole greenhouses are the best experimental units to avoid cross infestation. To provide infestations of defined age range, adults can be killed after release without harming other stages by low dosages of dichlorvos strips overnight (caution, use only for experiments, because dichlorvos is banned in the Netherlands for all commercial use). The following experimental protocol extends over the duration of the crop to investigate the effect of different conditions on fungal control of whiteflies (Kanagaratnam *et al.*, 1982; Hall, 1982a):

1. Collect newly emerged adults with a suction collector, using minimal suction because adults are delicate. Release 15/plant across the lower leaves in separate greenhouses, each house with a different experimental test condition.
2. When small scales are well established after about 14 days, label leaves and fruit chosen representatively at all heights (including some young leaves on which larvae have not yet appeared) and at all relative positions in the greenhouse. Record condition of leaves and numbers of all stages of all insect species (scales are plump when live, dry and flat when dead). If obviously > 100 insects occur on a leaf, count numbers on a representative sector and multiply up to give an estimate for the whole leaf. Record condition of representative fruit.
3. Spray (section 2C) with a single concentration of fungus (*e.g.* 10^7 conidia/ml) in water plus 0.02–0.05% Tween 80, almost to run off, late in the day when humidity is rising (*e.g.* 1800 h), with a high pressure sprayer, ensuring good coverage abaxially.
4. When the spray is dry, site climatic sensors representatively and reposition them as the crop grows.
5. Weekly after the spray repeat step 2 observations on all labeled leaves and fruit.
6. At each cucumber harvest, sample fruit, assess and classify.
7. Periodically, if control is good, reinfest to simulate natural reinfestation in commerce; if not, repeat the sprays.
8. To investigate fungal spread, fold and pin areas of leaf to mask them during spraying. Also after a spray, strategically site unsprayed potted plants pre-infested with scales 1–3 day old and remove just before adults emerge from them.

5 Future requirements of research into techniques

Scientifically the techniques used to investigate microbial control of pests in greenhouses are relatively simple, *e.g.* counting insects and conidia. It is in the future orientation of these techniques where most research is needed, particularly towards pest, quality control, labor and cost problems faced by the grower. In addition, some aspects can be related particularly to each of the two major pathogen groups used presently under glass, bacteria and fungi, and also to insect pathogenic nematodes.

A major grower concern is economics. Although crops are valuable, costs are high, markets competitive and profit margins typically low. Techniques need better orientation towards cosmetic appearance and quality crops. Greenhouse staff reaction to pest control safety measures and working conditions is becoming more important, hence the greater use of IPM systems with insect pathogens as an important component.

Prediction of expected quality of pest control with insect pathogens is another major grower problem. Especially at the hiatus period of apparent inaction after spraying, technical help is needed in assessing potential control, *e.g.* definition of early signs given by insect appearance, change in activity and in behavior (*e.g.* section 3E). Strategy for bacterial control is mainly to kill one larval generation produced by moths entering a UK greenhouse seasonally, but strategy with fungi involves ongoing breeding of pests inside the greenhouse over many generations throughout a season. Thus, observations of insect populations treated with fungi should be designed to enable modeling in relation to physical conditions. A great opportunity for predictive modeling for long term pest control is provided by the sophisticated control, measurement and recording of temperature, light intensity and sometimes of humidity available in some greenhouses. Some studies are available on the effect of diurnal environmental fluctuation, particularly Curtis (1998). Modeling could be based on cumulative day-degrees and day-percentages of relative humidity, which should include the impact of extremes.

To facilitate cost analysis, experiments should include records of labor input and material costs. For example, Helyer and Wardlow (1987) and Sopp *et al.* (1990) found that several treatments with rapidly applied ULV mist of *L. lecanii* at low dosage were better than one high dosage HV treatment; also the cost of wetting agent needed at HV was avoided by ULV. Techniques for the better incorporation of lipophobic *L. lecanii* conidia into ULV oil sprays (already available with *Beauveria bassiana*) need development in an attempt to emulate the success of *M. flavoviride* in dry conditions. *L. lecanii* may best be used strategically in an IPM system with other natural enemies to regain short term control if predators or parasitoids are about to be swamped.

With *Bt*, better measurement is needed of evenness of spray cover. This is particularly important in relation to the topography of foliage in canopies and of individual leaves, *e.g.* the arching back of leaves along the midrib in some tomato varieties. The reactions of larvae immediately post spray, their falling off foliage and subsequent fate are important. The virtually universal greater susceptibility of young than older larvae is now established fact and needs no more research, but spray timing in relation to population age structure, pitted against cost of crop monitoring, needs more study.

6 Appendix 1

A *Calibration and use of temperature and humidity recorders*

Good greenhouse climatic records are vital to proper analysis and understanding of experimental results. Sensors controlling greenhouse systems are typically protected, often above crop level, in a box to avoid direct sunlight and the box is fan-ventilated to ensure rapid response to changing ambient air conditions. The recorder is often sited in this box, frequently a standard meteorograph with bimetallic strip temperature sensor and hair humidity sensor, alternatively a wet and dry bulb hygrometer. For experiments, records are needed close to the insects and their pathogens, most conveniently obtained with small distant-reading sensors and an electronic recorder. Sensor settings tend to drift and humidity sensors are notoriously variable at very high humidities, one supplier of data loggers quoting accuracy as ±4% R.H. and, above 95% R.H., ±10% R.H., although another claims ±2% from 0–90% R.H. and ±3% from 90–100% R.H. Thus literature about humidity effects with fungi in greenhouse experiments is only as good as the accuracy of humidity sensors and controlling mechanisms; their probable shortcomings make many published results incomparable. Recorders should be calibrated at three points: 1. maximum likely to be encountered, 2. median and 3. minimum, giving priority to critical levels, *e.g.* high humidity. Readings should be made in sequence, *i.e.* down the scale, then back again to detect any hysteresis, enabling ranges to be reset.

Calibrating temperature is easy against mercury in glass thermometers placed very near sensors in rooms, cabinets or cellars in which temperature change is slow. In controlled facilities, it is useful to turn off controls ca 5 min before readings to avoid possible rapid fluctuations when heaters and/or humidifiers come on. Avoid interference by operator body heat and breath on sensors.

Humidity is far more difficult to calibrate than temperature. Minimal temperature variation is essential because small temperature changes cause large relative humidity changes. The same facilities as for temperature can often be used, but any cabinets must be large enough to allow humidity measurement without the measuring activities affecting cabinet temperature. Spot readings can be made with a chilled mirror dew point meter or a wet and dry bulb hygrometer, but care must be taken to operate these instruments to the limit of their accuracy, *e.g.* with a humidity recorder:

1. Starting at the highest humidity, read the recorder being calibrated.
2. Ensure that the hygrometer wick is clean/new and loaded with distilled water until it glistens. Do not handle the wick to avoid finger grease. Whirl the hygrometer at arms length at least 40 times. Rapidly read the thermometers.
3. Repeat step two until thermometer readings are constant, checking wick wetness each time.
4. Read the recorder under calibration again. It should not have changed. If it has, use a mean value and reset the calibration.
5. Move the recorder in turn to two other facilities, allowing full equilibration before reading.
6. Repeat readings in these three facilities in reverse order, then make final adjustment to calibration.
7. When finished, empty the water reservoir (if present) on the hygrometer. Allow wick to dry rapidly before storing the instrument in its box to avoid mould growth.

Small humidity sensors can be calibrated in sealed containers at controlled temperature, with solutions of sulphuric acid to control humidity (Solomon, 1951). The ratio of liquid surface area to air volume must be high to ensure rapid equilibration. Recorders should be calibrated regularly. The minimum is once before and after each greenhouse experiment.

7 References

Bernhard, K., Holloway, P. J. and Burges, H. D. 1998. Appendix I: a catalogue of formulation additives: function, nomenclature, properties and suppliers. In "Formulation of Microbial Biopesticides, Beneficial Microorganisms, Nematodes and Seed Treatments" (H. D. Burges, Ed.), pp. 333–365. Kluwer Academic Publishers, Dordrecht, The Netherlands.

Burges, H. D. 1974. Modern pest control in glasshouses. *Span* **17**, 32–34.

Burges, H. D. 1998. Formulation of mycoinsecticides. *In* "Formulation of Microbial Biopesticides, Beneficial Microorganisms, Nematodes and Seed Treatments" (H. D. Burges, Ed.), pp. 131–185. Kluwer Academic Publishers, Dordrecht, The Netherlands.

Burges, H.D and Jarrett, P. 1976. Adult behaviour and oviposition of five noctuid and tortricid moth pests and their control in glasshouses. *Bull. Entomol. Res.* **66**, 501–510.

Burges, H.D and Jarrett, P. 1978. Caterpillar control with *Bacillus thuringiensis. Grower* **90**, 589–590, 593–595.

Burges, H.D and Jarrett, P. 1979. Application and distribution of *Bacillus thuringiensis* for control of tomato moth in glasshouses. *Proc. 1979 British Crop Protection Conf. – Pests and Diseases* **2**, 433–439.

Burges, H. D. and Jones, K. A. 1998. Formulation of bacteria, viruses and Protozoa to control insects. *In* "Formulation of Microbial Biopesticides, Beneficial Microorganisms, Nematodes and Seed Treatments" (H. D. Burges, Ed.), pp. 33–127. Kluwer Academic Publishers, Dordrecht, The Netherlands.

Copping, L. G. 1998. The BioPesticide Manual. British Crop Protection Council, Farnham, UK, 333pp.

Courshee, R. J. and Ireson, M. J. 1961. Experiments on the subjective assessment of spray deposits. *J. Agric. Engin. Res.* **6**, 175–182.

Curtis, J. E. 1998. Enhancing the effectiveness of *Verticillium lecanii* against the green peach aphid *Myzus persicae*. Ph.D. Thesis. La Trobe University, Bundoora, 226 pp.

Ekbom, B. S. 1979. Investigations on the potential of a parasitic fungus (*Verticillium lecanii*) for biological control of the greenhouse whitefly (*Trialeurodes vaporariorum*) *Swedish J. Agric. Res.* **9**, 129–138.

Goettel, M. S. and Inglis, G. D. 1997. Fungi: Hyphomycetes. *In* "Manual of Techniques in Insect Pathology" (L. A. Lacey, Ed.) pp. 213–249. Academic Press, London.

Hall, R. A. 1976a. A bioassay of the pathogenicity of *Verticillium lecanii* conidiospores on the aphid *Macrosiphoniella sanborni. J. Invertebr. Pathol.* **27**, 41–48.

Hall, R. A. 1976b. *Verticillium lecanii* on the aphid *Macrosiphoniella sanborni. J. Invertebr. Pathol.* **28**, 389–391.

Hall, R. A. 1979. Comparison of the pathogenicity of*Verticillium lecanii* conidia and blastospores against aphids. *Entomophaga* **24**, 191–198.

Hall, R. A. 1980a. Effect of relative humidity on survival of washed and unwashed conidiospores of *Verticillium lecanii. Acta Oecol.: Oecol. Appl.* **1**, 265–274.

Hall, R. A. 1980b. Effect of repeated subculturing on agar and passaging through an insect host on pathogenicity, morphology and growth rate of *Verticillium lecanii. J. Inverbr. Pathol.* **36**, 216–222.

Hall, R. A. 1981. Laboratory studies on the effects of fungicides, acaricides and insecticides on the entomopathogenic fungus, *Verticillium lecanii*. *Entomol. Exp. Appl.* **29**, 39–48.

Hall, R. A. 1982a. Control of whitefly, *Trialeurodes vaporariorum* and the cotton aphid, *Aphis gossypii*, in glasshouses by *Verticillium lecanii*. *Ann. Appl.Biol.* **101**, 1–11.

Hall, R. A. 1982b. Deuteromycetes: virulence and bioassay design. *In* "Invertebrate Pathology and Microbial Control" pp. 191–196. Proc. IIIrd International Colloquium on Invertebrate Pathology, 6–10 Sept. 1982. University of Sussex, Brighton.

Hall, R. A. 1984. Epizootic potential for aphids of different isolates of the fungus *Verticillium lecanii*. *Entomophaga* **29**, 311–321.

Hall, R. A. and Burges H. D. 1979. Control of aphids in glasshouses with the fungus *Verticillium lecanii*. *Ann. Appl. Biol.* **93**, 235–246.

Hall, R. A. and Papierok, B. 1982. Fungi as biological control agents of arthropods of agricultural and medical importance. *Parasitology*. **84**, 205–240.

Helyer, N. L. and Wardlow, L. R. 1987. Aphid control on chrysanthemums using frequent, low dose applications of *Verticillium lecanii*. *Bull. SROP* **10**, 62–65.

Humber, R. A. 1997. Fungi: preservation of cultures. *In* "Manual of Techniques in Insect Pathology" (L. A. Lacey, Ed.) pp. 269–279. Academic Press, London.

Jarrett. P. 1981. Persistence of *Bacillus thuringiensis* on tomatoes. Rep. Glasshouse Crops Res. Inst., 1980, 117.

Jarrett, P. 1985. Biology of greenhouse caterpillar pests. *In* "Biological Pest Control: the Glasshouse Experience" (N. W. Hussey and N. Scopes, Eds.) pp. 99–103. Blandford Press, Dorset.

Jarrett, P. and Burges, H. D. 1982. Control of the tomato moth, *Lacanobia oleracea*, by *Bacillus thuringiensis* on glasshouse tomatoes and the influence of larval behaviour. *Entomol. Exp. Appl.* **31**, 239–244.

Jarrett, P. and Burges, H. D. 1986. Isolates of *Bacillus thuringiensis* active against *Mamestra brassicae* and some other species: alternatives to the present commercial isolate HD1. *Biol. Agric. Hort.* **4**, 39–45.

Jarrett, P., Burges, H. D. and Matthews, G. A. 1978. Penetration of controlled drop spray of *Bacillus thuringiensis* into chrysanthemum beds compared with high volume spray and thermal fog. *In* Symposium on Controlled Drop Application. British Crop Protection Council Monograph, No. 22, 75–81.

Jones, K. A. 1998. Spray application criteria. *In* "Formulation of Microbial Biopesticides, Beneficial Microorganisms, Nematodes and Seed Treatments" (H. D. Burges, Ed.), pp. 367–375. Kluwer Academic Publishers, Dordrecht, The Netherlands.

Jones, K. A. and Burges, H. D. 1998. Technology of formulation and application. *In* "Formulation of Microbial Biopesticides, Beneficial Microorganisms, Nematodes and Seed Treatments" (H. D. Burges, Ed.), pp. 7–30. Kluwer Academic Publishers, Dordrecht, The Netherlands.

Kanagaratnam, P., Hall, R. A. and Burges, H. D. 1982. Control of glasshouse whitefly, *Trialeurodes vaporariorum* by an 'aphid' strain of the fungus *Verticillium lecanii*. *Ann. Appl. Biol.* **100**, 213–219.

Kerwin, J. L. and Petersen, E. E. 1997. Fungi: Oomycetes and Chytridiomycetes. *In* "Manual of Techniques in Insect Pathology" (L. A. Lacey, Ed.) pp. 251–268. Academic Press, London.

Lacey, L. A., Fransen, J. J. and Carruthers, R. 1996. Global distribution of naturally occurring fungi of *Bemisia*, their biologies and use as biological control agents. *In* "*Bemisia* 1995: Taxonomy, Biology, Damage, and Management" (D. Gerling and R. Mayer, Eds.),, pp.401–433. Intercept, Andover, UK.

Latgé, J. P. and Papierok, B. 1988. Aphid pathogens. *In* A. K. Minks and P. Harrewijn, eds., Aphids Their Biology, Natural Enemies and Control, Vol. B, pp.323–335. Elsevier Science Publishers B. V., Amsterdam.

Malais, M. and Ravensberg, W. I. 1992. "Knowing and Recognizing. The Biology of Glasshouse Pests and their Natural Enemies". Koppert, Berkel en Rodenrijs. 109 pp.

May, K. R. 1950. The measurement of airborne droplets by the magnesium oxide method. *J. Sci. Instr.* **27**, 128–130.

McGuire, M. R., Galan-Wong, L. J. and Tamez-Guerra, P. 1997. Bacteria: bioassay of *Bacillus thuringiensis* against lepidopteran larvae. In "Manual of Techniques in Insect Pathology" (L. A. Lacey, Ed.), pp. 91–99. Academic Press, London.

Milner, R. J. and Lutton, G. G. 1986. Dependence of *Verticillium lecanii* (Fungi: Hyphomycetes) on high humidities for infection and sporulation using *Myzus persicae* (Homoptera: Aphididae) as host. *Environ. Entomol.* **15**, 380–382.

Quinlan, R. J. 1988. Use of fungi to control insects in glasshouses. In "Fungi in Biological Control Systems" (M. N. Burge, Ed.), pp. 19–36. Manchester University Press, Manchester, UK.

Ravensberg, W. J., Malais, M. and Van Der Schaaf, D. A. 1990. *Verticillium lecanii* as a microbial insecticide against glasshouse whitefly. In "Brighton Crop Protection Conference – Pests and Diseases", pp. 265–268. The British Crop Protection Council. Lavenham Press, Lavenham, UK.

Solomon, M. E. 1951. Control of humidity with potassium hydroxide, sulphuric acid, or other solutions. *Bull. Entomol. Res.* **42**, 543–554.

Sopp, P. I., Gillespie, A. T. and Palmer, A. 1990. Application of *Verticillium lecanii* for aphid control on chrysanthemums. *Crop Prot.* **9**, 177–184.

Zare, R. and Gams, W. 2001. A revision of *Verticillium* section *Prostrata*. IV. The genera *Lecanicillium* and *Simplicillium*. *Nova Hedwigia* **73**, 1–50.

Chapter VII-9

Forest defoliators

K. van Frankenhuyzen
Canadian Forest Service
Great Lakes Forestry Centre
Sault Ste. Marie, ON P6A 5M7, Canada

R. C. Reardon
USDA Forest Service, Forest Health
Technology Enterprise Team
Morgantown, WV 26505, USA

N. R. Dubois
We dedicate this chapter to the memory of Normand Dubois, formerly with the USDA Forest Service, Northeast Research Station in Hamden, CT; Norm had accepted the challenge to write this chapter, we accepted the honor to complete the task

1 Introduction

Large-scale outbreaks of forest defoliators have provided a unique impetus to the development of entomopathogens as microbial control agents. In fact, forestry is one of the few markets in which a microbial control agent has replaced broad-spectrum chemical insecticides. This applies in particular to *Bacillus thuringiensis* subsp. *kurstaki* (*Btk*). The development of *Btk* during the 1970s and 1980s for control of lepidopteran forest defoliators in North America paved the way for broader commercialization during the late 1980s and 1990s. Today, *Bt* is the most successful microbial insecticide on the market with worldwide application for protection of forests, crops, and human and animal health (Entwistle *et al.*, 1993). Forestry has also played a prominent role in demonstrating the potential of insect viruses as effective control agents. Because practical use of entomopathogens in forestry is limited to *Btk* and some viruses, in particular nucleopolyhedroviruses (NPVs), this chapter provides a user-oriented description of how to apply and evaluate those entomopathogens in forest protection. Three prominent pest-pathogen associations are used as leading examples throughout this chapter: the use of *Btk* against spruce budworm, *Choristoneura fumiferana*, and the use of *Btk* and NPV against gypsy moth, *Lymantria dispar*.

2 Development and current use

A Bacillus thuringiensis

1 Development

The discovery of *Bt* and its early history as an insect control agent were reviewed previously (Beegle and Yamamoto, 1992). The focus of commercialization shifted to North America with

L.A. Lacey and H.K. Kaya (eds.), *Field Manual of Techniques in Invertebrate Pathology*, 481–504.

the production and registration of Thuricide® in 1957. Field tests during the 1960s produced inconsistent results and *Bt* did not measure up against synthetic insecticides. The introduction of formulations based on the more active *kurstaki* variety of HD-1 improved field efficacy during the 1970s. Most of the field development during that decade took place in North America, where large-scale aerial spraying against defoliating forest Lepidoptera, such as spruce budworm and gypsy moth, presented an opportunity for field-testing at a scale that was commercially attractive (van Frankenhuyzen, 1990; Reardon *et al*., 1994). Effectiveness improved steadily as appropriate application and formulation prescriptions were developed, but results remained inconsistent. Cost-effectiveness started to improve in the late 1970s when manufacturers made significant advances in both production and formulation technologies. By the end of the decade, *Btk* was considered an operational alternative. During the next decade, new formulations with potencies of 12.7–16.9 billion international units (BIU)/liter permitted undiluted application. The use of ultra-low application volumes improved spray plane production rates while the higher product potency increased reliability of control operations. A shift in political climate in favor of biological products finally pushed *Btk* into full operational use in the mid 1980s.

2 *Current use*

In Canada, the use of *Btk* for control of the spruce budworm increased from < 5% of the total area sprayed in the early 1980s to 100% 15 years later. Between 1980 and 1998, a total of almost 6 million ha of budworm-infested forest were sprayed (van Frankenhuyzen , 2000). Another 2.3 million ha were treated during that same period to control other lepidopteran defoliators. Current use involves aerial application of undiluted high-potency products ranging from 12.7 to 25.4 BIU/liter at application volumes of 1.2–2.5 liters/ha, using rotary atomizers to produce small droplets.

Operational use in the USA followed a similar pattern. In Maine, *Btk* was used in 2% of the spray programs in 1979 compared to 81% in 1985 when the spruce budworm outbreak collapsed, for treatment of 460,000 ha in total. In the Pacific Northwest, aerial spray programs to control western spruce budworm, *C. occidentalis*, switched from primarily carbaryl in 1983 to 85–100% *Btk* after 1985, for a total use on 547,000 ha (Sheehan, 1996). In the eastern USA, about 2.4 million ha of forests were treated with *Btk* between 1980 and 1992 for suppression of gypsy moth, with the insect growth regulator diflubenzuron (Dimilin®) being the second most widely used insecticide on 1.7 million ha (Reardon *et al*., 1994). Operational use of *Btk* for population suppression and foliage protection usually involves one or two applications of undiluted high-potency products at 50–90 BIU/ha, using a wide range of aircraft types and spray equipment.

Improvements in the formulation and application of *Btk*, resulting from extensive operational experience in North America, prepared the stage for its use in Europe for control of forest insects. High-potency products and undiluted application in ultra-low volumes were introduced by manufacturers in the early 1990s and quickly became standard practice (Glowacka, 1996). Between 1990 and 1998 *Btk* was used on almost 1.8 million ha of deciduous and coniferous forests for control of various defoliators.

B *Baculoviruses*

1 *Development*

The first recorded attempt to use a virus for insect control involved the "Wipfelkrankheit" of the nun moth in 1892 in Germany. In the early 1900s, a viral disease of gypsy moth was reported in the USA, and its potential for biological control was demonstrated in limited field tests (Lewis, 1981). The establishment of insect pathology as a scientific discipline, together with budding environmental concerns about insecticide use in the mid 1900s, renewed interest in the use of insect viruses as insecticides. Aerial application of the European sawfly NPV in Canada in 1950 heralded the development of viral insecticides for forest protection (Cunningham, 1995).

a *Gypsy moth NPV*

Worldwide research on gypsy moth NPV intensified in the 1960s and 1970s with field tests in Europe, the USSR and the USA (Lewis, 1981). Ground spray trials in the early 1960s were followed by aerial spray trials in the 1970s and culminated in the USDA Forest Service obtaining registration of a product called Gypchek® in 1978. The USDA Forest Service and USDA APHIS produce annually sufficient virus to treat up to 2,000 ha. Since its registration, Gypchek has been the subject of intense research to maximize its production and field efficacy. Various application rates and volumes were tested using various types of aircraft and nozzles (Reardon and Podgwaite, 1994; Reardon *et al.*, 1996). Efficacy improved with the introduction of a new tank mix in 1987. Aqueous flowable carriers were evaluated in the early 1990s, and permitted application of the recommended dose rate in 9.7 liters/ha as compared to 19.4 liters/ha for the conventional tank mix. The Canadian Forest Service obtained acceptable results at even lower application volumes (2.5–5.0 liters/ha) and received registration for Disparvirus® in 1997 (Cunningham, 1998). The same virus that is present in Gypchek and Disparvirus is now being produced under the name Biolavirus-LD® in the Czech Republic. In the former Soviet Union, the development of a different NPV of gypsy moth with a broader host range was pioneered in the early 1960s. It is now registered under the product name Virin-ENSh®, and is the most important viral product in terms of production volume (Lipa, 1998). Commercialization of gypsy moth NPV was also attempted in Germany in response to large-scale outbreaks (Huber, 1998).

b Sawfly NPVs

The NPV of the European pine sawfly, *Neodiprion sertifer*, is endemic in Eurasia where natural epizootics can cause severe larval mortality (Cunningham and Entwistle, 1981; Cunningham, 1998). It is highly effective because the virus replicates in midgut cells and is rapidly transferred to healthy larvae through defensive regurgitation and contaminated frass, aided by their gregarious feeding habits. Because transmission of the virus is so efficient, a single application is generally sufficient to obtain extended population control. It was first tested in an aerial spray in Ontario in 1950, which was followed by field tests in many countries throughout the Holarctic. The USDA Forest Service obtained registration of Neochek-S® in 1983, and the same data were used for registration of Virox® in the United Kingdom.

Another example of a technical success is the NPV of the redheaded pine sawfly, *N. lecontei*, in Canada (Cunningham and Entwistle, 1981; Cunningham, 1998). This sawfly is found only in North America and shares *N. sertifer*'s gregarious feeding habits. Trials conducted by the Canadian Forest Service in the 1970s resulted in the registration of Lecontvirus® in 1983.

A more recent success is the development of the NPV of the balsam fir sawfly, *Neodiprion abietis*. Outbreaks of this insect in coniferous forests of North America are usually of short duration, due to the actions of natural enemies including a nucleopolyhedrovirus (Thurston, 2002). An outbreak in pre-commercially thinned balsam fir stands on the island of Newfoundland that started in 1991 became a particular concern when it persisted much longer than expected and expanded over the years to encompass 40,000 ha by the end of the decade (Thurston, 2002). The potential loss to the forest industry precipitated an intense collaboration between the Canadian Forest Service, the forest industry and provincial agencies to develop its NPV as a biological control alternative for aerial application (Moreau *et al.*, 2005). The multi-stakeholder collaboration and a $6.5 million investment over 9 years resulted in the registration of Abietiv® in 2006.

c Douglas-fir tussock moth NPV

The NPV of the Douglas-fir tussock moth, *Orgyia pseudotsugata*, is an example of how viruses can be used to prevent outbreaks by inducing an epizootic early in the outbreak cycle. Outbreaks of this insect occur periodically in southern British Columbia and the western USA and are generally terminated by epizootics of a naturally-occurring NPV, but not before severe damage occurs (Otvos *et al.*, 1995). Field tests conducted by the USDA Forest Service and the Canadian Forest Service resulted in the registration of a multicapsid strain produced in *O. pseudotsugata* by the USDA Forest Service under the name TM BioControl-1® in 1976 (Martignoni, 1999). The Canadian Forest Service received registration in 1987 for the same virus produced in *O. leucostigma* under the name Virtuss® (Cunningham, 1995). Further tests in the early 1980s demonstrated that early introduction of the virus can bring about population collapse before tree mortality occurs (Otvos *et al.*, 1995). Tussock moth populations are being monitored with pheromone traps and treated with NPV when populations reach a threshold density (Shepherd, 1994).

2 Current use

Despite decades of field tests and several technical successes, the practical use of baculoviruses in forestry is insignificant compared to *Btk*, despite the fact that viral insecticides have distinct advantages. Their high degree of host specificity is desirable from an ecological point of view, as they provide a means of controlling defoliator populations without impacts on non-target Lepidoptera. Such impacts can pose concerns for *Btk* sprays (*e.g.*, Whaley *et al.*, 1998; Strazanac and Butler, 2005). Another advantage is their efficient horizontal (within generations) and vertical (between generations) transmission. In some cases, this enables limited introduction into the pest population to initiate an epizootic, rather than repeated inundative introduction as is needed for *Btk*. Ironically, both features are liabilities from a commercial point of view and have, together with the need for production in living cells, discouraged commercialization. Registrations of most viral forestry products are held by government agencies in anticipation of commercialization.

Operational use of insect viruses in forestry is limited to NPVs for control of gypsy moths, sawflies and tussock moths. The USDA Forest Service has treated about 16,500 ha of gypsy moth infestations with Gypchek between 1980 and 1998. Sufficient Gypchek to protect 2,000 ha of infested forests is produced annually for use in government-sponsored area-wide gypsy moth suppression programs. Virin-ENSh was used on *ca.* 150,000 ha in the former Soviet Union between 1972 and 1978. The Forestry Commission in the United Kingdom used European sawfly NPV (Virox) on 7,500 ha of lodgepole pine between 1984 and 1988. In Finland, 8,700 ha were treated with a commercial product between 1974 and 1983 (Huber, 1986). Neither product is registered at present. In the USA, Neochek-S was used on 258 ha between 1976 and 1987, but its registration has lapsed since then. In Canada, a similar product (Sertifervirus) was used on 152 ha between 1975 and 1993 but was never registered as a pest control product. Lecontvirus has an active registration and is used routinely, but on a small scale; between 1976 and 1994 about 6,000 ha of plantations were treated in Ontario and Quebec. Balsam fir sawfly virus was applied to 22,500 ha of balsam fir sawfly infested forests in Newfoundland between 2000 and 2005 in research and demonstration trials (Moreau *et al.*, 2005) that resulted in its registration in early 2006. Operational use of Abietiv is anticipated in 2006. The USDA Forest Service produced TM BioControl-1 between 1985 and 1995 and stockpiled enough product to treat 400,000 ha (Martignoni, 1999). Bioassays of samples showed only a 30% loss of biological activity after 5 to 15 years of storage at −10 °C (Otvos *et al.*, 2006). The virus was used operationally to suppress outbreaks of the Douglas-fir tussock moth in British Columbia (1,850 ha were treated between 1991 and 1993) and recently in the Pacific Northwest, with 16,200 ha treated in 2,000 and 6,100 ha in 2001.

3 Application of entomopathogens

A General considerations

1 Use strategies

There are three possible strategies for the use of entomopathogens in forestry. Firstly, the pathogen can be introduced into the pest population to spread and permanently regulate the population in concert with other mortality factors. An example is the fortuitous introduction of a NPV of the European sawfly, *Gilpinia hercyniae*, concomitant with parasitoid introductions into Canada, which are credited with maintaining populations of this insect at endemic levels since the early 1940s (Cunningham, 1998). Secondly, applications can be made to initiate epizootics. Resulting population suppression is long-term but not permanent so that treatment is repeated each time a new outbreak occurs. Examples are the NPVs for control of sawflies and Douglas-fir tussock moth. Thirdly, inundative releases involve repeated broad scale application, such as the use of *Btk* for control of budworm and other defoliators. Which strategy to use depends on ecological characteristics of the particular pathogen-pest association. Operational use in forestry includes inundative release of *Btk* and inundative or limited release of NPVs to initiate epizootics. Both involve spray application of the pathogen formulated into an insecticide to the pest infestation, either from the air or from the ground.

2 *Spray application*

Spray application of microbials is based on many of the same principles and technologies as for the application of synthetic insecticides (see Chapters I-2 and III-1, III-2). Even though microbial control products have unique characteristics that require special consideration (see Chapter I-2), little work has been done to optimize their application. True optimization requires detailed knowledge of pest susceptibility and dose acquisition processes (see Chapter IV-1), knowledge which is often not available. The pressure of finding immediate solutions to acute infestation problems traditionally favors a trial and error approach, resulting in considerable variation in operational protocols. This presents an important caveat: current protocols are operationally acceptable (*i.e.*, they work), but are not necessarily optimal.

Because most spray applications in forestry are done from the air, minimal effort is spent on refining ground-based applications. Ground application of microbials is practical only when dealing with small-scale problems. Examples are the application of *Btk* by homeowners or arborists to control gypsy moth, the use of sawfly baculoviruses in pine plantations, or the use of *Btk* to augment aerial pest eradication efforts. Ground applications are commonly used in experimental trials as a cost-effective way of comparing pathogen strains (Thorpe *et al.*, 1998), testing formulation additives (Webb *et al.*, 1994), etc. Ground-spray equipment used in forestry includes conventional hydraulic sprayers, mist blowers, and air blast sprayers (see Chapter III-1).

In aerial spray programs, the size and remoteness of spray blocks determine the type of aircraft used. Four-engine aircraft (DC-4 and DC-6) are used for application of *Btk* to large contiguous spray blocks; examples include control of spruce budworm in Quebec in the late 1980s and eradication of Asian gypsy moth in Vancouver (1992) and white spotted tussock moth in Auckland (1996). Single-engine aircraft are the more common choice for treatment of smaller spray blocks, and range in size from small agricultural aircraft to much larger military aircraft (Prebble, 1975). Rotary-wing aircraft are often preferred in mountainous locations or in areas with a lack of adequate airstrips. Spray aircraft can be equipped with rotary atomizers, such as Micronair® (spinning cage) and Beecomist® (spinning sleeve) atomizers, or hydraulic nozzles (see Chapter III-2).

B *Specific considerations*

When designing a microbial control spray trial, the user must define the dose rate to be applied, the application volume, the desirable droplet size spectrum, and the timing and frequency of spray application. Once those parameters are selected, the equipment must be calibrated to ensure delivery of the correct application rate per ha, application volume, and droplet size spectrum (see Chapters III-1, III-2). Selection of appropriate spray parameters is often done empirically, but should be based on spray technology as well as biological factors. Spray technological factors, such as topography, canopy structure, weather conditions, and the type of application equipment, were discussed in Chapter III-2. Biological factors, such as innate susceptibility of the target pest, its feeding habits, its spatial distribution in the host canopy, phenology of pest and host, and pest population quality should be considered in the context of the control objectives.

1 *Target insect susceptibility*

The target insect's relative susceptibility to the pathogen is often determined prior to field application to facilitate selection of the most effective isolate or product. Bioassay techniques that permit comparison of pathogen concentrations required to achieve 50% and 95% mortality (LC_{50} and LC_{95}) include diet-based assays (diet incorporation or surface-contamination assays) or natural substrate assays (leaf-dip or leaf-disk assays). Protocols can be found in Lacey (1997). Determining lethal median doses (LD_{50} and LD_{95}) requires the use of bioassay techniques that permit ingestion of a known dose within a short and fixed time period. Suitable techniques include the leaf-disk, diet-plug or droplet-feeding methods (Lacey, 1997), or the droplet-feeding method described by van Frankenhuyzen *et al.*, (1997). In the case of baculoviruses, lethal dose estimates can be combined with information on larval feeding rates to obtain a first approximation of required application parameters, as illustrated in Chapter IV-1 (Control Window model). In the case of *Btk*, such calculations are confounded by feeding inhibition at sublethal dose levels, but can still provide insights into desired product potency and droplet size spectrum (*e.g.*, van Frankenhuyzen and Payne (1993)).

2 Target insect vulnerability and phenology

Even a highly susceptible target species will not die if exposure to the pathogen is insufficient. Since *Bt* and NPVs need to be ingested, the degree of exposure is affected by feeding habits of the insect, its distribution in time and space, and its synchrony with host tree phenology. Target-insect vulnerability affects selection of application parameters, such as application volume. For example, spruce budworm larvae feed on new shoots, which are distributed along the periphery of the tree crown. This presents a relatively exposed target for spray droplet impaction, and adequate coverage can be achieved with application volumes as low as 0.5–2.4 liters/ha. In contrast, distribution of gypsy moth larvae throughout the hardwood canopy presents a much greater challenge for canopy penetration by spray droplets, and higher application volumes are often deemed necessary. Larval phenology affects the timing and frequency of sprays. For instance, early spruce budworm instars are highly susceptible to *Btk* (Masse *et al.*, 2000) but vulnerability is reduced by their cryptic feeding habits inside needles and swelling buds. Effective delivery requires delaying spray application until the larvae are feeding on flaring buds, which usually coincides with peak fourth instar. In the case of gypsy moth, all larval stages are highly vulnerable. However, sprays are typically targeted against early instars because later instars are less susceptible and require a higher application rate for effective control (Dubois *et al.*, 1994).

3 Target insect population quality

Control objectives are generally easier to achieve in older, declining populations where natural mortality factors have greater impact than in vigorous and building populations. Previous outbreak history and the density and health of the pest population are important considerations, particularly when selecting the dose to be applied. For instance, adequate foliage protection is achievable with one or two applications of *Btk* at 15 BIU/ha in declining spruce budworm populations, whereas two applications of 30 BIU/ha or one application at 50 BIU/ha may be required in healthy, high-density populations (Bauce *et al.*, 2004; Carter, 1991; Régnière and Cooke, 1998).

4 Control objectives

Objectives of the spray program or trial need to be clearly defined before appropriate application parameters can be selected. Control objectives range along a sliding scale from 'easy' to 'difficult' to achieve, with nuisance control and eradication at extreme ends of this continuum, and foliage protection and population suppression taking up intermediate positions. Spray objectives also influence selection of appropriate application parameters, such as timing of application. For example, in eastern Canada, the objective of spruce budworm control programs is to limit current defoliation to less than 40–50%. This is generally achieved through one application of *Btk* at 30 BIU/ha targeted at peak fourth instar or as soon as buds flare (Bauce *et al.*, 2004; Carter, 1991). In northern Alberta, foliage protection is more difficult to achieve, because at northern latitudes larvae develop more rapidly so that they spend more time in unflared buds (Volney and Cerezke, 1992). This necessitates targeting spray application against fifth and sixth instars, with the goal of reducing larval populations to a level that will result in acceptable defoliation the following year. Control objectives also affect application frequency and dose rate. For example, in most of the eastern USA where gypsy moth is well established, aerial control programs are conducted to limit annual defoliation to less than 60%. This can be achieved by one or two applications of *Btk* at 50–90 BIU/ha/application targeted at early instars. In the western USA, however, efforts to prevent establishment of gypsy moth necessitates repeated (3–4) applications at higher rates (*e.g.*, Whaley *et al.*, 1998).

C Application parameters

This section provides parameters for the application of entomopathogens in forestry. However, the reader should keep in mind the lack of conclusive studies on parameter optimization. Problems associated with quantitative assessment of treatment effects in forestry (see section 4D) often make it difficult to demonstrate significant differences in efficacy between various application prescriptions. A notable exception is the recent work by Bauce *et al.*, (2004) on optimization of *Btk* spray application prescriptions for control of spruce budworm. Generally, there is usually insufficient data from the scientific literature to conclusively recommend one set of application parameters over another. The parameters included here should, therefore, be seen as an illustration of how application of microbials in forestry could be done rather than as a prescription of how it should be done, with few exceptions.

1 General aspects of Btk use

The number of commercial formulations available for use in forestry applications has shrunk in recent years as a result of market consolidation. The available products were concentrated in fewer corporate hands when Valent BioSciences Corporation, a joint venture between Valent USA and Sumitomo Chemical, purchased Abbott's forestry, public health and agricultural *Bt* products in 2000, and Certis's *Bt* products (the original Sandoz products) in 2003. Only a small subset of the original slate of products is currently available for use in forestry (Canada and USA) (Table 1). All formulations are suspensions of crystals, spores and fermentation solids, with other inert ingredients and additives, in either water (aqueous flowables) or oil (non-aqueous emulsions). Exact physical properties vary with each formulation, so the user should consult manufacturer's technical bulletins (*e.g.*, Anonymous, 1998) and product labels for specific handling instructions. Sometimes it is desirable to add a tracer dye to permit quantification of spray deposits (see 4C). Powder dyes for aqueous formulations should be dissolved in the minimum amount of water and then added to the airplane hopper or spray tank, followed by vigorous recirculation. In recent years, so-called Dry Flowable (DF) products have been developed and are being tested for efficacy to circumvent the costs of hauling high volumes of liquids over large distances, and to permit longer-term storage of products.

2 Use of Btk against spruce budworm

This is probably the best characterized and optimized system known to date, benefiting from decades of research on spruce budworm ecology and budworm-*Btk* interactions honed by more than two decades of operational, large-scale use. Detailed knowledge of dose acquisition and population processes has permitted the construction of a model that adequately describes *Btk* efficacy resulting from aerial applications (Régnière and Cooke, 1998). The model can

Table 1. Formulations of *Bacillus thuringiensis* available in 2005 for use in forestry

Type of formulation	Potency[1]	Product name
Aqueous flowable suspension	12.7	Thuricide 48LV®
	20.0	Thuricide 76LV®
	12.7	Foray 48B®
	20.0	Foray 76B®
	25.0	Foray 96B®
Emulsifiable suspension	16.9	Dipel 176®

[1] Expressed as BIU/liter

be used by spray operators to optimize application prescriptions for local population and stand conditions (*e.g.*, optimum volume and dose application rate, timing of application and application frequency based on larval density and insect-host phenology). Recent work in Quebec has, by incorporating the best available knowledge on budworm population ecology into assessment methods (see section 4D), for the first time demonstrated conclusive differences in treatment efficacy, permitting a more explicit definition of optimum application prescriptions (Bauce *et al.*, 2004).

- *Mixing* – Undiluted application of high-potency products containing 12.7 BIU/liter or more has become the accepted standard for control of spruce budworm, so that tank mixing is no longer required.
- *Dose application rate* – Recommended dose rate is 30 BIU/ha in one or two applications. However, dose rate may be adjusted down to 15 BIU/ha for populations of less than 15–20 larvae/45-cm branch (Bauce *et al.*, 2004; Carter, 1991). In Canada, the maximum registered rate is 30 BIU/ha/application, but higher application rates of up to 50 BIU may be needed to afford foliage protection against high larval populations (30–40 larvae/branch) (Bauce *et al.*, 2004; Régnière and Cooke, 1998).
- *Volume application rate* – Volume application rates are determined by the choice of product potency (BIU/liter) and the dose rate (BIU/ha). Thus, the recommended rate of 30 BIU/ha is applied in 2.4 liters/ha when using products containing 12.7 BIU/liter, and in 1.5 liters/ha for 20 BIU/liter (Bauce *et al.*, 2004).
- *Droplet size spectrum* – Undiluted application in ultra-low volumes necessitates fine atomization to generate enough spray droplets for adequate coverage of the target foliage. This is usually achieved by using Micron air rotary atomizers, in particular AU4000 and AU5000, set to spin at 8,000–10,000 rpm, while limiting the flow rate per atomizer to less than 4 liters/min (Picot and Kristmanson, 1997). The droplet size spectrum on coniferous foliage resulting from such application is dominated by droplets in the 25–125 μm diameter range. Typically, 50% of the droplets are less than 50–60 μm (NMD or number median diameter) while 50% of the deposited volume is in droplets of less than about 100 μm (VMD or volume median diameter) (van Frankenhuyzen and Payne, 1993; Picot and Kristmanson, 1997; Régnière and Cooke, 1998). However, several jurisdictions in Canada and the USA prefer coarser atomization by using an atomizer rpm of about 5,000–6,000. Equivalent efficacy resulting from coarser atomization (van Frankenhuyzen *et al.*, 1996) suggests that there is considerable leeway in the selection of the most appropriate droplet size distribution. The optimum tradeoff among droplet size, product potency and application volume has yet to be determined. Although droplets of $< 125\,\mu$m may be optimum for spray impaction on coniferous foliage (Picot and Kristmanson, 1997), a larger droplet size may be needed to deliver a biologically optimum dose to feeding spruce budworm larvae (van Frankenhuyzen and Payne, 1993).
- *Timing of application* – Spray application for foliage protection is timed according to development of the insect and the host tree. Usually, a Larval Development Index (LDI) and Bud Development Index (BDI) are obtained from branches collected at regular intervals from areas that are slated for treatment. A protocol for determining these indices is provided in Figure 1. On balsam fir, sprays are applied as soon as the buds are starting to flare (BDI = 3.5–4.0), which normally coincides with peak fourth instar (LDI = 4.0–4.5). However, during unusually warm spells budworm larvae develop more rapidly than the buds. The insects can inflict a lot of damage before treatment, because relatively large larvae are feeding on relatively small buds. Cool, wet weather, on the other hand, slows larval development so that larvae are still small when buds start to flare, making it easier to obtain foliage protection. Recent field data suggest that the standard window of spray application for obtaining foliage protection can be extended by spraying early: sprays against third-instar larvae were as efficacious as sprays using conventional timing (Bauce *et al.*, 2004). For sprays aimed at population suppression, it is better to target late larval stages (fifth and sixth instars), which are important in determining overall generation survival. Large larvae are also more vulnerable due to their higher innate susceptibility (van Frankenhuyzen *et al.*, 1997) and more exposed feeding habits.
- *Frequency of application* – One application is standard practice. However, a second application within one week of the first is recommended

Field collection

1. For every 50 ha to be treated, select 5-10 samples trees of target host species
2. Collect one mid-crown branch from each tree every 2-3 days, starting at peak larval emergence
3. Ship branches to field laboratory for examination

Larval Development Index (LDI)

4. Remove all larvae from branches up to a total of at least 100 larvae
5. Record instar for each larva by measuring head capsule widths
6. Calculate the LDI as per example:

Instar	2	3	4	5	6	total
#	18	38	14	0	0	70

LDI=[(18x2)+(38x3)+(14x4)]/70=2.9

Instar	Head capsule width (mm)
1	< 0.25
2	0.25-0.32
3	0.33-0.50
4	0.51-0.90
5 male	0.91-1.30
5 female	1.00-1.50
6 male	1.30-1.80
6 female	1.55-2.10

Bud Development Index (BDI)

7. Randomly remove 20 buds from each of 5 branches
8. Record the development stage of each bud, using Auger's bud development scale:

Class 1: bud resinous, enclosed in membrane
Class 2: buds swollen, 10-35% of needles visible
Class 3: all needles visible but buds not yet flaring
Class 4: buds flaring
Class 5: buds flared, shoots elongating

9. Calculate the BDI as per example:

Class	1	2	3	4	5	total
# of buds	0	9	10	80	1	100

BDI=[(0x1)+(9x2)+(10x3)+(80x4)+(1x5)]/100=3.7

Figure 1. Calculation of phenological indices for timing of eastern spruce budworm sprays on balsam fir

when dealing with high population densities (> 30 larvae/branch) (Bauce *et al.*, 2004), or when host or insect phenology is not uniform throughout the spray block.

- Weather conditions – Small droplets have negligible sedimentation velocity and, therefore, require instantaneous airflow (turbulence) to reach the target foliage. When using small droplets in ultra-low-volume sprays, it is important to avoid spraying when there is no wind (optimum range: 5–15 km/h wind speeds) or when atmospheric conditions are highly stable.

3 *Use of* Btk *against other coniferous defoliators*

a Jack pine budworm (C. pinus pinus)

A close relative of the spruce budworm in Central Canada and the Great Lakes states, the jack pine budworm erupts periodically at 8–15 year intervals in outbreaks that typically last 1–5 years. Near complete defoliation sustained for 2–3 consecutive years can result in extensive top-kill, reduced radial increment and jack pine mortality. This can be prevented by aerial application of *Btk*. For example, between 1985 and 1996 *ca.* 900,000 ha of infested jack pine stands in Ontario were protected by one application of 20 BIU/ha in 1.57 liters, using undiluted products containing 12.7 BIU/liter (van Frankenhuyzen, 2002). Application parameters are the same as for control of spruce budworm with the exception of timing. Sprays should be applied when the jack pine needles are beginning to escape their fascicle sheets, which usually coincides with the fourth instar.

b Western spruce budworm (C. occidentalis)

A western relative of the spruce budworm, this species is found from South Central British Columbia down to Arizona and New Mexico. Although its principal host is Douglas-fir, *Pseudotsuga menziesii*, it also feeds on *Abies* and *Picea* species. Repeated defoliation of developing buds and new foliage reduces tree growth and increases top kill and general mortality. *Btk* has been used extensively for operational control on hundreds of thousands of budworm-infested hectares in Washington and Oregon between 1984 and 1996 and at a smaller scale in British Columbia. In the western USA, operational use involved predominantly diluted application of products containing 8.4–12.7 BIU/liter in volumes ranging from 3.5 to 7.0 liters/ha to deliver 30–60 BIU/ha (Sheehan, 1996). In western Canada, high-potency products are applied undiluted at 30 BIU in 2.4 liters/ha (Otvos *et al.*, 2002) Timing of spray application is the same as for eastern spruce budworm

c Hemlock looper (Lambdina fiscellaria fiscellaria)

A major pest of balsam fir in eastern coniferous forests, this insect has cyclic outbreaks occurring about once every 10 years. Early instars feed predominantly on new shoots but later instars feed wastefully on both old and new foliage. Severe defoliation may kill trees within 1 or 2 years. *Btk* is used extensively for operational suppression of eastern hemlock looper populations. In eastern North America, *ca.* 300,000 ha were treated between 1985 and 1999 (van Frankenhuyzen *et al.*, 2002). Operational use involves undiluted application of high-potency products at 30 BIU/ha for each of two applications. The first application is made when egg hatch is nearly complete and the second about one week later when at least 50% of the larvae is second instars (West *et al.*, 1989).

4 Use of Btk *against gypsy moth*

Work is currently in progress to develop a process-oriented model to describe *Btk* efficacy in gypsy moth, based on the spruce budworm model of Régnière and Cooke (1998). This model is expected to help eliminate some of the uncertainties regarding optimization of *Btk* aerial applications, in particular the selection of application volumes and targeted droplet size spectra.

- *Mixing* – Undiluted application of high-potency products containing 12.7 BIU/liter is the accepted standard.
- *Dose application rate* – The current recommended rate in the USA is 60–90 BIU/ha in one application. The maximum registered rate is 100–125 BIU/ha/application, depending on the formulation. The maximum dose rate is recommended when sprays are applied to later instars, or for population eradication. In Canada, the recommended procedure for foliage protection is two applications of 30–50 BIU/ha.
- *Volume application rate* – Until the mid-1980s volume rates of 9.5–19 liters/ha were used. The current recommendation in the USA is 4.7 liters/ha/application or higher, since tests with lower application rates produced inconsistent results. Higher application volumes can be used to increase the droplet density and coverage throughout the canopy (Dubois *et al.*, 1993), but should not be done at the expense of tank mix concentration through excessive dilution. The lower the concentration in the tank mix, the more droplets are required per unit leaf area to obtain efficacious results (van Frankenhuyzen 1991; Dubois *et al.*, 1993). In Canada, the standard volume application rate is 2.4–3.9 liters/ha.
- *Droplet size spectrum* – Research data on the optimum droplet size and density for gypsy moth are minimal and often inconsistent. The current recommendation in the United States is a density of 10–20 drops/cm^2 with a volume median

diameter (VMD) of 100–200 μm. For example, Bryant and Yendol (1988) showed that a given dose per cm^2 oak leaf was more effective when delivered at a higher density of small droplets (50–150 μm) than at a lower density of larger droplets (> 150 μm). The use of different nozzles (Micronair, Flat fan, Twin Jet) did not affect efficacy or spray deposition (Smitley and Davis, 1993; Dubois *et al.*, 1994). In Canada, most sprays are applied with rotary atomizers set to produce a VMD of 70–150 μm.

- *Timing of application* – Application should be timed according to the development of larvae and foliage. Apply when the majority of the larvae are second instars (*i.e.*, 2-week period in the spring from late-first instar to early-third instar) and, of less importance, when oak leaves are approximately 40–50% expanded. If necessary, a second application is made 7–10 days later. Application should not be delayed beyond the early-third instar, because efficacy decreases significantly when sprays are applied against later instars (Dubois *et al.*, 1994).
- *Frequency of application* – Standard practice is one (USA) or two (Canada) applications for both foliage protection and population reduction. Proper timing is critical when using a single application. Multiple applications may be needed in mountainous terrain where hatch of gypsy moth eggs and foliage expansion can be extended over weeks, or if the objective is population eradication.
- *Weather conditions* – Many different dispersion regimes can exist above forest canopies and conditions can change rapidly. Spraying can be done under many of these conditions but they require different approaches to ensure effective deposition on targeted leaf surfaces, as outlined in Chapter III-2.

5 *Use of NPV against gypsy moth*

a *Aerial application*

- *Mixing* – Two tank mixes are available for aerial application (Table 2). For either tank mix, a slurry of Gypchek and water needs to be prepared first. Mix the required amount of Gypchek powder with non-chlorinated water in a clean pail until it is evenly dispersed. Whenever practical, shield the powder from direct sunlight while preparing the slurry. Always wear a surgical mask and safety glasses for eye protection.

  Carrier 038A tank mix: Fill the mixing tank approximately half-full with Carrier 038 and agitate. Pour the Gypchek-water slurry into the tank. Rinse the pail with a small amount of water and add the rinsate to the tank. Agitate for a few minutes and slowly add the rest of the Carrier 038A. Add the Bond sticker (Loveland Industries, Greeley, CO). Continue to mix thoroughly for 15–20 min before loading into the aircraft hopper.

  Lignosulfonate-molasses tank mix: Fill the mixing tank with the needed amount of non-chlorinated water (pH < 7.5). While circulating, slowly add the necessary Lignosite AN powder. When the Lignosite AN is in solution, add the molasses and mix thoroughly for about 5 min. The Lignosite AN-molasses mix can stand overnight or two

Table 2. Recommended tank mixtures for ground and aerial application of Gypchek

Application	Tank mix	Ingredient	Amount (per 3.8 liters)
Ground	Lignosulfonate	Gypchek	10^{10} OBs
		Lignosite AN	225 g
		Bond	0.07 liter
		Water	3.7 liters
Air	Lignosulfonate-molasses	Gypchek	1 to 2.5×10^{11} OBs
		Lignosite AN	225 g
		Feed-grade molasses	0.5 liter
		Bond	0.07 liter
		Water	3.2 liters
	Carrier 038A	Gypchek	2 to 5×10^{11} OBs
		Carrier 038A	3.6 liters
Bond	0.07 liter		
		Water	0.18 liter

nights in cool weather. Add the Bond sticker and then the Gypchek slurry slowly to avoid clumping. Circulate for 15–20 min prior to loading.

- *Application rates and frequency* – Recommendations depend on program objectives. For population suppression, apply twice at 3-day interval, using 5×10^{11} polyhedral occlusion bodies (OBs)/ha/application in 18.7 liters (Lignosulfonate-molasses tank mix) or 4.7 or 9.4 liters (Carrier 038A tank mix). Alternatively, a single application of 10^{12} OBs/ha in 4.7 or 9.4 liters of the Carrier 038A tank mix can be used. For population eradication, use at least two applications of 1.25×10^{12} OBs/ha/application in 18.7 liters (Lignosulfonate-molasses tank mix) or 9.4 liters (Carrier 038A tank mix).
- *Droplet size spectrum* – There are no specific recommendations. Both hydraulic nozzles and rotary atomizers (*e.g.*, Micronair AU-5000, 4000, 7000) are acceptable and should be set to generate droplets with a VMD within the 100–450 μm range.
- it Timing of application – When using two applications, the first should be made as soon as hatch is complete, all larvae are actively feeding, the majority of the larvae are in the late-first instar, and foliage is at least 20% expanded.
- *Weather conditions* – Preferred conditions for spray application include winds of 5–15 km/h, a temperature below 20 °C and a relative humidity above 40%. Do not apply if rain is predicted within 24 h.

b Ground application

- *Mixing* – To prepare the recommended tank mix (Table 2), slowly add Lignosite AN powder to non-chlorinated water (pH < 7.5) while circulating, and add the Gypchek as a slurry once the Lignosite AN is in solution.
- *Application rates and frequency* – The recommended application rate is 10^{12} OBs/ha/application in 380 liters/ha for treatment of wood lots and small areas, and in 95 liters/tree for treatment of individual trees. Sprays are applied until run off so that droplet size considerations are immaterial.
- *Timing of application* – The same as for aerial application.
- *Weather conditions* – Do not apply if rain is predicted within 24 h.

6 Use of NPV against pine sawflies

- *Mixing* – Sawfly NPV preparations are available as wettable powders, consisting of virus-infected larvae that have been freeze-dried and ground to a fine powder. Tank mix formulations have varied widely. Viral preparations can be mixed with water alone for either aerial or ground application. Sometimes molasses is added at 10–25% of total volume, together with a commercial spreader sticker and sunlight protectant (Sandoz Shade at 60 g/liter). However, additives are not essential. Excellent control has been achieved with highly purified OBs suspended in water (Cunningham and Entwistle, 1981), or in a 20%molassess solution (Moreau *et al.*, 2005) but there are no firm scientific data to relate spray performance to tank mix formulation (Young and Yearian, 1986). For mixing, weigh the quantity of powder needed for application and suspend in water using 100 ml water/5 g powder. Mix the suspension thoroughly in a blender and then dilute the concentrate with the appropriate volume of water in a holding tank. Mix thoroughly by repeated recirculation. Avoid highly alkaline or chlorinated water. Final tank mix pH should be between 5.0 and 8.0. If only chlorinated water is available, let stand for 24 h before use. Because bacteria will grow once the powder has been mixed with water, made-up tank mix should not stored for any amount of time. When using aqueous tank mixes, care should be taken to spray under high relative humidity conditions to avoid excessive droplet evaporation.
- *Dose application rate* – Sawfly NPVs are highly infectious and relatively low application rates can provide excellent control. Application rates have ranged from 8×10^6 to 2.4×10^{11} OBs/ha but the current recommended rate is 5×10^9 OBs or about 50 virus-killed larvae per ha in one application for either air or ground sprays. Moreau *et al.* (2005) reported efficacious results against the balsam fir sawfly at rates as low as 1×10^9 OBs/ha.
- *Volume application rate* – Wide variation exists in the volumes sprayed. Cunningham (1995) recommended 5–10 liters/ha for aerial application and 20 liters/ha for ground sprays. Moreau *et al.* (2005)

used an application volume of 2.5–3.0 liters/ha. With pressurized hand-held hydraulic sprayers, individual insect colonies can be treated on small trees over a limited area like plantations. Vehicle-mounted mist blowers, foggers or high-pressure, high-volume hydraulic sprayers can be used for treatment of larger trees or sizable acreage of small trees, using volumes of 50–100 liters/ha. Since the virus spreads rapidly through the population, it is often sufficient to spot or strip applications. With backpack mist blowers, every third or fourth row in a plantation can be treated while utilizing any wind to carry the spray across adjacent rows of trees. The emphasis should be placed on treating colonies near the tops of trees, as the virus will readily spread to colonies lower on the trees. With aerial application, it can be sufficient to space swaths at 100-m intervals (rather than the usual 30- or 40-m) to obtain strip coverage from where the virus can spread throughout the plantation (Cunningham and Entwistle, 1981).

- *Droplet size spectrum* – In Canada, aerial spray systems are calibrated to deliver droplets of 100–150 μm, but optimum droplet size requirements are not known. The recent work by Moreau *et al.* (2005) demonstrates that sawfly NPVs can be applied successfully in ultra-low spray volumes of a few liters per ha and using small droplets (target diameter of 80 μm).
- *Timing of application* – Sprays should be applied as soon after egg hatch as possible to prevent defoliation and to provide sufficient time for the virus to spread and initiate secondary infection. Applications can be made up the fourth instar, but a higher dose rate may be needed when treating third instar or larger. With sprays aimed at second and third instars, larvae die as fourth instars. One application is often sufficient to free plantations of sawfly populations for several years.

4 Evaluation of entomopathogens

To evaluate the effectiveness of microbial sprays in forestry, the user has to: (1) ensure a proper experimental design that permits assessment of the agent's effects on the target defoliator population, (2) monitor appropriate environmental conditions during and after spray application, (3) monitor the entomopathogen in the forest environment to aid the interpretation of observed treatment effects, and (4) monitor the effect of spray treatment on the pest population. Protocols used for evaluating effectiveness in experimental trials are often much more rigorous than in large-scale operational use programs. The emphasis in this section is on experimental trials.

A Experimental design

Efficacy trials in forestry are designed to control as many sources of variation as is feasible. Stands of similar age, species composition, topography, and pest outbreak history are selected within the general infestation. An extensive survey should be conducted to determine population levels so that initial density can be used as a blocking criterion in a randomized block design. However, forestry trials are often compromised by sources of variation that are beyond the control of the experimenter. Variation in stand type, age structure, species composition, tree vigor, soil conditions and topography, to name a few, often invalidates the concept of homogeneity. Add to that variation among 'similar' stands in pest population characteristics (*e.g.*, age of outbreak, population density), and the inevitable variation in spray deposition among similarly treated blocks (see 4C), and it becomes evident that true replication does not really exist in the forestry environment. Furthermore, cost and safety considerations often have an overriding influence on experimental design in forestry. What is feasible and affordable often takes precedence over what is desirable.

A critical factor in designing a field efficacy trial is selection of untreated check plots to account for normal seasonal changes in the pest population. In practice, efficacy is assessed by calculating the percent of foliage protected or percent difference in mortality between sprayed and unsprayed plots, using Abbott's formula. Implicit in the use of Abbott's correction is the assumption that ecological influences on treated and untreated populations are the same at the time of treatment. That assumption is often invalidated by considerable and unpredictable spatial heterogeneity in local population processes. Inappropriate matching of treated and untreated plots in terms of population density or age (outbreak history) can introduce major biases in efficacy estimates (Fleming and Retnakaran, 1985; Cooke and Régnière, 1999). The best way to safeguard against this is to select untreated plots in the same general outbreak area (close proximity) and to use as many untreated checks as can be accommodated.

Size of treated plots is generally determined by the method of application, population sampling requirements, and availability of the forest resource (*e.g.*, size of infested stand with desired species composition or size of the woodlot), but generally range from 10 to 100 ha for aerial applications. The plot should be long enough to allow for the atomizers to stabilize. When using a small fixed-wing aircraft with Micronairs flying at 160 km/h, a minimum length of 400–500 m is needed to permit a stabilization time of 10 sec. A 40-m lane separation, and at least 5 flight lines to allow for optimum deposit by overlapping swaths, pegs the minimum width at 200 m, thus producing a minimum plot size of about 10 ha. Allowing twice the time for atomizer stabilization and doubling the number of passes will increase the plot size to 35–40 ha. That is often used as a minimum plot size in spruce budworm spray trials, in which the applicator relies on cross winds to drift the finely atomized spray cloud into the canopy. In gypsy moth trials, employing a larger droplet size spectrum with less swath displacement, plots down to about 10 ha are more frequently used.

B *Monitoring environmental conditions*

Key meteorological conditions are usually monitored on site to assess suitability of spray conditions (see Chapter III-2). Having the equipment on site permits continued monitoring of meteorological conditions that could influence effectiveness of the treatment. For example, a local downpour following application could explain poor efficacy in specific plots. We recommend continuous recording of at least temperature (hourly averages, daily minimum and maximum) and rainfall from spray application until the end of the larval feeding (exposure) period. If spray plots are widely separated, monitoring should be site-specific. If clustered, it is often sufficient to record in one central location.

C *Monitoring the entomopathogen*

1 *In the forest canopy*

Although spray dispersal systems are carefully calibrated to ensure emission of the correct application rates and droplet size spectrum, deposition of the spray cloud into the forest canopy is a complex process that is governed by many factors (see Chapter III-2). Variation in spray aircraft flying height, wind speed, wind direction, above-canopy turbulence, atmospheric stability, and flight lane separation, to name a few, can result in huge variation in how much of the emitted spray cloud actually reaches the target foliage (Picot and Kristmanson, 1997). Spray deposition is, therefore, a key variable that needs to be taken into account when evaluating the effectiveness of entomopathogens, as poor treatment success can be due to poor spray deposition rather than inadequacy of the control agent. The need is less critical in ground-based sprays where the applicator has more control over ensuring uniform coverage.

Most techniques for spray deposit assessment were developed for chemical insecticides. They rely on the use of artificial collectors, such as cards (Kromekote cards, water-sensitive spray cards) on or near the ground, or foliage simulators (aluminum combs, Kromekote card strips, foil brushes, etc.) that are deployed in the forest canopy (see Picot and Kristmanson, 1997). However, artificial collectors do not necessarily have the same droplet collection efficiencies as natural foliage and require extensive calibration. When spraying microbial control agents that require ingestion, the most relevant way to measure on target spray deposition is to collect foliage from larval feeding habitats. The method used for quantifying the spray deposit depends on the information that is required, being either an estimate of the total amount of active ingredient deposited, or a quantification of droplet densities and size distribution.

a Estimation of active ingredient

Active ingredients of microbials are not amenable to conventional analytical techniques. As a result, practical methods to obtain direct estimates for microbials are not readily available. An exception is the recent commercialization of an immunoassay-based method for quantification of *Btk* toxin proteins, the so-called Abbott Deposit Assessment Method (ADAM). The ADAM kit is produced by Agdia Inc. (Elkhart, IN) on behalf of Abbott Laboratories. Two versions of the kit are available: a simplified and qualitative version for field assessment of *Btk* spray deposits on the target foliage, and a version for more quantitative laboratory use. The kits include detailed protocols.

Various techniques are available to obtain indirect estimates of the total amount of active ingredient deposited. Spray deposits are washed off the foliage and the concentration of a tracer or dye is measured by either fluorometry or colorimetry. The dye or tracer concentration is assumed to be proportional to the concentration of active ingredient. This may not be an appropriate assumption for microbials, due to the highly particulate and often clumped nature of the active ingredient (OBs, fermentation solids). Nevertheless, dyes or other marker substances can provide an approximation of how much of the emitted spray volume reached the target foliage (volumetric assessment).

b Estimation of droplet density and size distribution

Active ingredient or volumetric estimates do not provide information on how the deposit was distributed on the foliar surface. Spray efficacy is determined not only by the dose per unit area, but also by the distribution, density and size of the spray droplets (Bryant and Yendol, 1988; Falchieri *et al.*, 1995; Payne and van Frankenhuyzen, 1995). Such spray deposit characteristics can be ascertained by visual examination of foliage. A simple but laborious technique for visual assessment of *Btk* spray deposits on coniferous foliage is outlined in Figure 2. Direct microscopic examination under a 10-to 30-X magnification is the only option for sprays that employ predominantly small droplets ($< 125\,\mu m$). Optical image analyzers can be used for deposit assessment of coarsely atomized, diluted sprays in broadleaf canopies (see Bryant and Yendol, 1991; Duboi *et al.*, 1993).

c Compatibility of dyes and tracers

Visual and volumetric assessment techniques require the use of a dye or tracer substance. A wide range of substances is available, but choices are limited by several key considerations, including compatibility with the microorganism and the formulation; no impact on physical properties such as surface tension and viscosity; and safety for non-targets and the environment. An overview of tracers and markers that have been used in spray trials with microorganisms is provided in Hunter-Fujita *et al.* (1998). Fluorescent particulate tracers are the most common choice for use with non-aqueous formulations, such as Brilliant Sulfaflavine (Bryant and Yendol, 1991), which is compatible with pathogens (Hunter-Fujita *et al.*, 1998). For aqueous formulations we recommend the use of FD&C Blue No. 1 (BF Goodrich Hilton Davis, Cincinnati, OH) at 0.3%, a dye which is safe for human consumption, provides good contrast with foliage, and does not interfere with virulence of NPVs or *Btk*.

2 In the target insect

Disease prevalence in the target pest population after application is often used as an additional parameter to assess treatment success, in particular for baculoviruses. With some insect species, infection estimates can be made reliably in the field. For instance, in the case of pine sawflies, the prevalence of NPV-infected colonies is easily determined. For most Lepidoptera, accurate assessment of infection depends on collection of larvae followed by individual diagnosis.

a Direct microscopic examination

The presence of OBs in larval tissues can be detected by direct microscopic examination of larval tissue smears, using phase contrast or after staining. Specific procedures can be found in Lacey (1997) and Hunter-Fujita *et al.*, (1998). With a sensitivity of generally 10^6 OBs/larva (Kaupp and Ebling, 1993), optical detection is ambiguous at low infection levels and only advanced infections can be detected reliably. Thus, optical detection of infection immediately upon collection will underestimate the real level of infection, especially shortly after spraying.

b Individual rearing

Individual rearing of larvae until pupation or death, followed by microscopic examination of cadavers to confirm presence of the pathogen, can circumvent this problem. However, this method tends to overestimate the real impact of the pathogen on the insect population, unless rearing is done under site-specific conditions in

---

Field collection

1. Collect one mid-crown branch from each sample tree about 1 h after spray application
2. Randomly remove 5 buds (early sprays) or shoots (late sprays) from each branch
3. Place shoots individually in 12-dram plastic vial or equivalent container (*e.g.*, blood serum vials)
4. Transport in cooler to field laboratory

Droplet density estimates

5. Determine fresh weight or length of each bud or shoot, respectively
6. Examine the entire bud under a dissecting microscope at 12-X magnification; at this magnification, droplets down to 15–20 μm are easily observed
7. While counting, remove needles with droplets that have been counted to prevent double counts
8. If shoots are elongated, use only a 2-to 3-cm section
9. Express droplet density as number of droplets per gram fresh weight (buds), or number of droplets per needle length (shoots)

Droplet size spectrum

10. Select buds with high droplet density to facilitate diameter measurements
11. Examine buds under dissecting microscope at 25- to 30-X magnification and measure diameter of each droplet, using an ocular micrometer or equivalent and properly calibrated, automated device (*e.g.*, Wild-Leitz Digital length Measuring Unit)
12. Measure at least 300 droplets per treatment for construction of droplet size frequency distribution

Important notes

1. Droplet size measurements and accurate density estimates are possible only if the foliage is dry during spray application; moisture on the needles (dew) dilutes the dye and disintegrates the droplets into ill-defined and faint dye smears
2. By using airtight containers, buds can be stored under ambient conditions for at least 48 h
3. This method is suitable only for aqueous, high-potency Btk products after undiluted application
4. Water-based products typically produce spherical to hemispherical droplets, with clearly defined diameters (SEM -1); larger (> 100 μm) droplets tend to form pancake-like stains (SEM-2)
5. Oil-based products typically produce ill-defined and irregularly shaped droplets (SEM-3); the use of particulate (fluorescent) tracers permits detection and counting of stains (density estimates) but size determinations cannot be done
6. In our experience, a sample of 5 buds per branch is sufficient to obtain an accurate estimate of the mean branch deposit, while one mid-crown branch per sample tree provides an accurate estimate of the mean plot deposit

Figure 2 cont'd

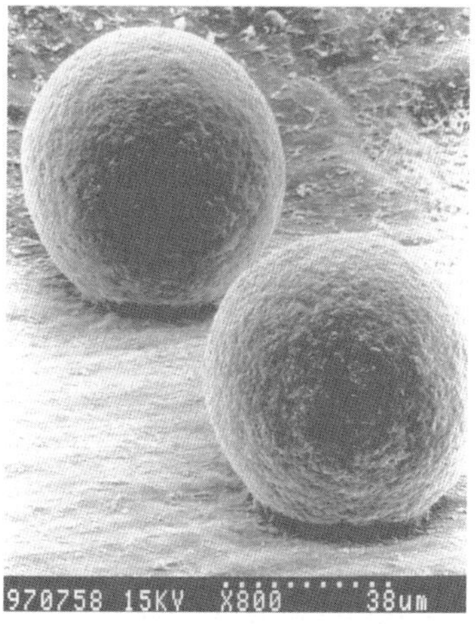

SEM-1: Two spherical droplets (ca. 60-µm in diameter) of Foray 48B on a fir needle

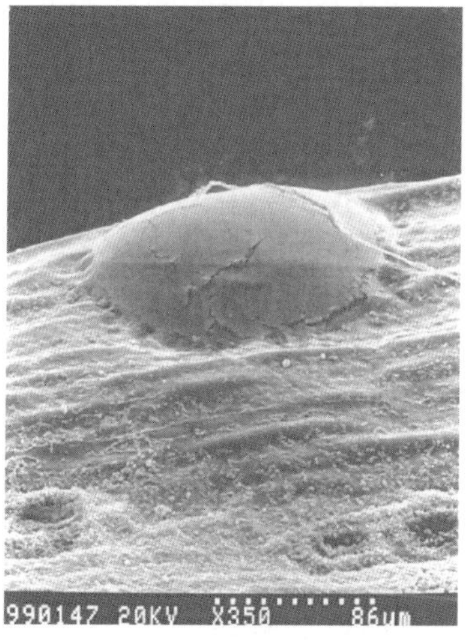

SEM-2: One pancake-like droplet (ca. 160-µm in diameter) of Foray 48B on a fir needle

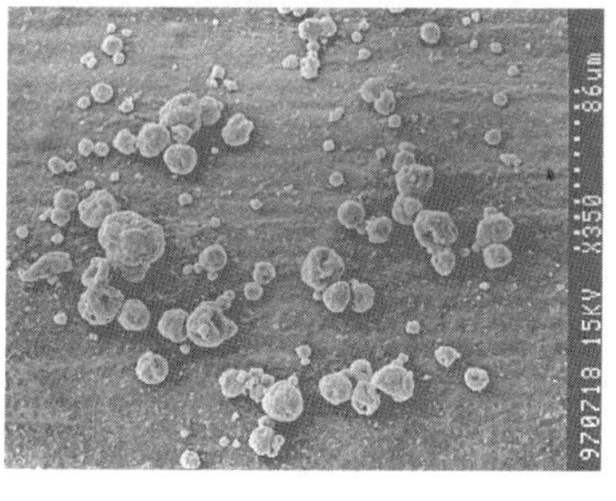

SEM-3: Dispersion of formulation solids over a 260-µm area after impaction of a a120-µm droplet of Dipel 12L on a fir needle

(photos by C.N. Davis, courtesy Canadian Forest Service)

Figure 2. A protocol for visual assessment of *Btk* spray deposits on coniferous foliage

outdoor insectaries, or for an appropriately short duration to prevent overestimation of infection due to contamination and secondary acquired infections (Woods and Elkinton, 1987).

c DNA dot-blots

Viral infection can be detected by using viral DNA probes that are allowed to hybridize with viral DNA in larval homogenates. The bound probe can be detected by enhanced chemiluminescence, using the Amersham ECL kit, or by using ^{32}P. The DNA dot-blot hybridization assay was found to be more sensitive and less ambiguous than optical detection (Kaupp and Ebling, 1993), and provided the same estimates of viral prevalence as individual rearing of larvae followed by microscopic diagnosis (Kukan and Myers, 1995). The dot-blot method is suitable for processing large numbers of larvae at a reasonable cost, and has been used for detection of baculovirus in tent caterpillars, gypsy moth, spruce budworm and balsam fir sawfly (Keating *et al.*, 1989; Kaupp and Ebling, 1993; Kukan and Myers, 1995; Moreau *et al.*, 2005). Detailed protocols can be found in those papers.

D Monitoring efficacy

1 General considerations

Effects of pathogen treatment on the target pest population are determined by comparing changes in treated populations with those occurring simultaneously in untreated populations. This requires sampling before the treatment is applied, but as close to the date of treatment as possible, followed by one or more samples after treatment. Sampling methods and frequency of sampling are different for each insect species and for each life stage within species, and also depend on the objectives of the control program (Table 3). In this section we provide sampling protocols for spruce budworm and gypsy moth, as examples of a coniferous and deciduous defoliator, respectively. However, the reader should keep in mind that a limited ability to apply rigorous statistical control to tests in forestry makes it difficult to standardize assessment protocols. Different jurisdictions use different protocols (Dorais and Kettela, 1982), as determined by factors that place practical constraints on experimental design (see 4A) as well as on design of optimal sampling regimes.

One key issue is that the most frequently used estimators of spray efficacy are inherently biased (Fleming and van Frankenhuyzen, 1992; Cooke and Régnière, 1999). Spray efficacy is usually assessed as the % population reduction or % foliage protection due to treatment. Both are estimated with Abbott's formula from pre-and post-spray estimates of insect density and % defoliation (see 4A). Each of those estimates is subject to considerable sampling error due to spatial heterogeneity in population processes. Combine this with the problem of matching treated and untreated plots (see section 4A), and one can see why efficacy is difficult to measure in a forestry context. Care should be taken to select efficacy estimators with the least amount of bias. It is for this reason that Cooke and Régnière (1999) recommend avoiding the use of corrected population reduction and foliage protection, and instead to directly compare mean larval survival and mean defoliation between spray and control treatments. An example is the recent study by Bauce *et al.*, (2004), who used analysis of variance methods (using initial larva density as

Table 3. General approaches for evaluating the effectiveness of microbial control agents against various target insect groups in relation to the continuum of possible control objectives

Pest insect	Control objective: Nuisance control	Foliage protection	Suppression	Eradication
Lymantria spp.	public survey visual observation	aerial survey defoliation	large larvae (burlap) pupae (burlap) egg mass density	moth survey larval survey egg mass survey
Orgyia spp.	public survey visual observation	aerial survey defoliation	small larvae (twigs) large larvae (twigs) egg mass density	moth survey larval survey egg mass survey
Choristoneura spp.	not applicable	aerial survey defoliation	larval density defoliation	not applicable
Lambdina spp.	not applicable	aerial survey defoliation	larvae (twigs) pupae (burlap)	not applicable
Malacosoma spp.	public survey visual observation	not applicable	not applicable	not applicable
Neodiprion spp.	not applicable	not applicable	colony counts	not applicable

a covariate when needed) to analyze treatment effects.

a Spruce budworm

- *Experimental design* – (1) Treatment plots vary from 10–100 ha, depending on the objectives of the trial and aircraft used for application. (2) Three replicate plots per treatment are the absolute minimum. Cooke and Régnière (1999) recommend the use of 6 replicates. With only 2 treatments (treated and control), such replication provides 10 degrees of freedom to compute the experimental error term. The same authors provide an alternative method for efficacy testing when extensive replication is not possible, which is based on the use of their thoroughly validated simulation model (Cooke and Régnière, 1996; 1998). (3) Treatment plots should be matched with unsprayed check plots of similar outbreak history and initial larval density. (4) Suitable sample trees (dominant and co-dominant spruce or fir) are selected at 20- to 40-m intervals along one or more transects perpendicular to the anticipated aircraft flight direction. The same trees are resampled for pre- and post-spray samples. See Bauce *et al.*, (2004) for an example of an effective experimental design with sufficient replication.
- *Sampling method* – A practical summary of sampling techniques for each budworm life stage can be found in Schmitt *et al.*, (1984). The most common unit for sampling large larvae (third-sixth instar), the target of aerial sprays, is the 45-cm branch tip collected from the mid-crown of each sample tree. A sectional pole pruner is used and the cutting head is adapted to incorporate a clasping device so that the branches can be lowered to the ground with a minimum effort. Basket attachments should be used for sampling fifth and sixth instars, because they readily drop when disturbed. Branch samples are placed individually in labeled bags and forwarded to a field lab for manual removal of the larvae and assessment of defoliation.
- *Sampling intensity* –The number of samples required for a specified level of precision can be found in Régnière and Sanders (1983). Efficacy assessments are often based on 20–60 sample trees/plot, depending on plot size and population level, with limited replication. However, because sampling error is estimated from all samples taken in a replicated trial, a more powerful approach is to reduce the sample size and increase replication (Cooke and Régnière, 1999). In other words, six small (10–30 ha) replicated plots with 15 sample trees each (as used by Bauce *et al.*, 2004) is better than three larger (30–100 ha) plots with 30 trees each. To maximize the amount of information obtained with limited manpower, the use of a single branch from the upper mid-crown is recommended (*e.g.*, Walton and Lewis, 1982).
- *Population reduction* – To evaluate spray efficacy in terms of mortality, one pre-spray sample just before peak fourth instar (the recommended time of application) and at least one post-spray sample at the end of the larval feeding period (peak pupation or later) are required. Two or three additional post-spray samples between application and the final sample are highly recommended. Because the distribution of budworm larvae is aggregated at higher densities (Régnière and Sanders, 1983), a logarithmic transformation is needed to stabilize variances (Cooke and Régnière, 1999).
- *Foliage protection* – Defoliation of current year's growth can be assessed concurrently with the last larval/pupal sample, using the Fettes method (Fettes, 1950), which is provided in Figure 3. If spray application is delayed beyond peak third instar, another defoliation estimate should be made just before application to obtain a pre-spray baseline. Since defoliation is highly density-dependent, pre-spray larval density should be used as a covariate in comparisons of post-spray defoliation (Cooke and Régnière, 1999; see Bauce *et al.*, 2004 for an applied example). Because the relationship between larval density and defoliation is nonlinear, a logarithmic transformation should be applied to pre-spray density. The significance of varying levels of annual budworm defoliation depends on such things as extent of damage in previous years, stand conditions, tree species, and management objectives. As a rule of thumb, a tree will survive if it retains 50–60% of the current year's foliage.

b Gypsy moth

- *Experimental design* – (1) Treatment plots vary most commonly between 20 and 120 ha. (2) Five replicates per treatment are highly recommended. (3) Treated plots must be matched with unsprayed check plots with a similar range in initial egg mass

1. Collect one mid-crown branch from each sample tree in treated and untreated plots at the end of the larval feeding period
2. Assign 20 shoots from each branch to one of the 12 defoliation categories according to Fettes:

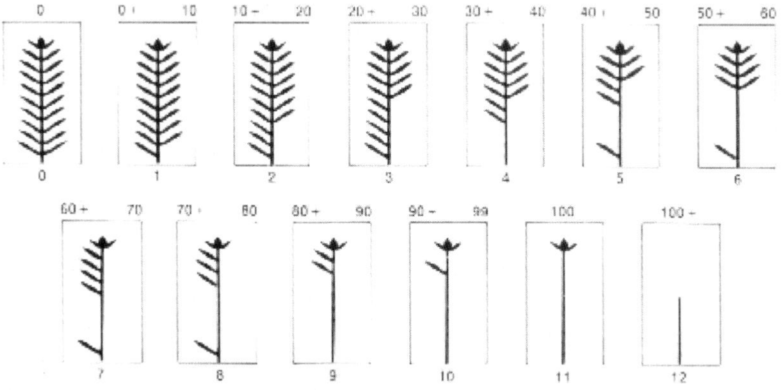

3. Calculate mean % defoliation for each branch as per example:

Defoliation category	% defoliation/category	Number of shoots
0	0	0
1	5	3
2	15	5
3	25	22
4	35	13
5	45	22
6	55	15
7	65	8
8	75	5
9	85	6
10	95	1
11	100	0
12	100	0

$$\% \text{ Defoliation} = [(3 \times 5) + (5 \times 15) + (22 \times 25) + (13 \times 35) + (22 \times 45) + (15 \times 55) + (8 \times 65) + (5 \times 75) + (6 \times 85) + (1 \times 95)]/100 = 44.1$$

Figure 3. The Fettes method to assess defoliation of current year's growth by spruce budworm

densities. (4) Plots for estimating egg mass density, larval monitoring and defoliation should be evenly distributed throughout the sprayed area, or laid out along transects.

- *Sampling methods* – The most appropriate sampling technique depends on the objectives of the control program (Table 3). Evaluation of foliage protection efforts requires estimates of defoliation. Population suppression efforts are evaluated on the basis of reduction in egg mass density, and are usually supplemented by relative estimates of larval and pupal abundance at regular intervals between pre- and post-treatment egg mass counts. Quantitative sampling of gypsy moth larvae is difficult because of their distribution and movement. Assessment of eradication efforts relies on extensive egg mass surveys combined with monitoring of adult populations. Sampling methods are different for each of these life stages.

  Egg masses: This is the most convenient stage to sample, as egg masses are available from late summer to early spring and are easily visible from the ground. The objective is to estimate egg mass density, expressed as number of egg masses/ha. The simplest and recommended method is the use of "fixed-radius" plots (Fleischer *et al.*, 1992). All egg masses within a circle (*e.g.*, 0.01 ha) are counted with plots replicated throughout the stand. Do not use timed- or fixed-distance walk methods due to lack of precision and observer effects. Be sure to distinguish between old (previously hatched) and new egg masses. New egg masses are usually darker in color and appear less ragged, but need to be touched, as new egg masses feel firm and full while old egg masses feel soft and spongy. Egg mass viability can be assessed by collecting a random sample of new egg masses in early spring (above and below the snow line). Egg masses are placed individually in a Petri dish and allowed to hatch in the laboratory, so that non-viable and viable (hatched) eggs can be counted.

Larvae: There are no recommended methods for quantifying abundance of early instars (first through third). Upon reaching fourth instar, larvae generally seek refuge during the day and return to the foliage to feed at night. Relative abundance of older instars (fourth through sixth) can be monitored by using barrier bands, such as 30-cm wide flaps of burlap wrapped around tree boles at breast height (Elkinton and Liebhold, 1990). The bands need to be checked every other day at approximately the same time of day to minimize the influence of time, weather and other variables on larval behavior and distribution. Frass collections can also be used as an indirect measure of larval abundance or activity (Liebhold and Elkinton, 1988a, 1988b).

Pupae: As larvae often pupate in protected sites, the same barrier bands used for monitoring larval abundance can be used to obtain relative pupal counts. Count only live pupae (move when touched).

Adults: Adult activity can be monitored with the gypsy moth pheromone, Disparlure (Hercon Environmental, Emigsville, PA). Male moths can be trapped by using delta pheromone traps, which are coated on the inside with adhesive. Because trap efficiency declines if more than 10 moths are caught, this trap is not recommended for monitoring population trends. Milk carton pheromone traps contain an insecticidal strip and do not get saturated as easily. The primary use of pheromone traps is to detect the presence of adult populations and to detect infestation centers in areas with low and dispersed populations.

- *Sampling intensity* – Estimation of egg mass densities is imprecise at all densities but more so at low population densities (*i.e.*, < 250 egg masses/ha). The number of survey plots required to obtain a specified level of precision depends on egg mass density. Sequential sampling plans can be found in Fleischer *et al.* (1992).
- *Population reduction* – Pre-spray egg mass counts need to be made either in the previous fall or spring prior to egg hatch (preferred). Post-treatment counts can be made anytime after moth flight has ceased, but are easier to do in late fall after the leaves have dropped. Burlap bands should be checked at least 6 times to count larvae, once to count pre-pupae, and once for a final pupal count.
- *Foliage protection* – Defoliation is usually assessed in the 0.01-ha fixed-radius plots by eyeballing the degree of defoliation on all preferred host trees within the plot. Defoliation surveys should be conducted just before spray application and again as soon as all larvae have pupated. Because deciduous trees will refoliate after heavy defoliation, the post-treatment survey should not be delayed beyond full pupation.

5 References

Anonymous, 1998. Protecting Forests with Foray and Dipel, a handbook for program managers and applicators. Abbott Laboratories, Chicago, IL.

Bauce, E., Carissey, N., Dupont, A., and van Frankenhuyzen, K. 2004. *Bacillus thuringiensis* subsp. *kurstaki* aerial spray prescriptions for balsam fir stand protection against spruce budworm (Lepidoptera: Tortricidae). *J. Econ. Entomol.* **97**, 1624–1634.

Beegle, C. C. and Yamamoto, T. 1992. History of *Bacillus thuringiensis* Berliner research and development. *Can. Entomol.* **124**, 587–616.

Bryant, J. E. and Yendol, W. G. 1988. Evaluation of the influence of droplet size and density of *Bacillus thuringiensis* against gypsy moth larvae. *J. Econ. Entomol.* **81**, 130–134.

Bryant, J. E. and Yendol, W. G. 1991. Deposition characteristics of aerially applied *Bacillus thuringiensis* to a hardwood forest with hydraulic nozzles. *J. Econ. Entomol.* **84**, 542–551.

Carter, N. E. 1991. Efficacy of *Bacillus thuringiensis* in New Brunswick, 1988–1990. *In* Proceedings of the 72nd Annual Meeting, Woodlands Section, Canadian Pulp and Paper Association, pp. 113–116, Montreal.

Cooke, B. J. and Régnière, J. 1996. An object-oriented process-based stochastic simulation model of *Bacillus thuringiensis* efficacy against spruce budworm, *Choristoneura fumiferana* (Lepidoptera: Tortricidae). Int. J. Pest Management **42**, 291–306.

Cooke, B. J. and Régnière, J. 1999. Predictability and measurability of *Bacillus thuringiensis* efficacy against spruce budworm, *Choristoneura fumiferana* (Lepidoptera: Tortricidae). *Environ. Entomol.* **28**, 711–721.

Cunningham, J. C. 1995. Baculoviruses as microbial insecticides. *In* "Novel Approaches to Integrated Pest Management" (R. Reuveni, Ed.), pp. 261–292. CRC Press, Boca Raton, FL.

Cunningham, J. C. 1998. North America. *In* "Insect Viruses and Pest Management" (F. R. Hunter-Fujita, P. E. Entwistle, H. F. Evans and N. E. Crook, Eds.), pp 313–331. J. Wiley & Sons, Chichester, UK.

Cunningham, J. C. and Entwistle, P. F. 1981. Control of sawflies by baculovirus. *In* "Microbial control of pests and plant diseases, 1970–1980" (H. D. Burges, Ed.), pp. 379–407. Academic Press, London.

Dorais, L. and Kettela, E. G. 1982. A review of entomological survey and assessment techniques used in the regional spruce budworm surveys and in the assessment of operational spray programs. A report of the committee for standardization of survey and assessment techniques, Eastern Spruce Budworm Council, Quebec Department of Energy and Resources, Quebec.

Dubois, N. R., Reardon, R. C., and Mierzejewski, K. 1993. Field efficacy and deposit analysis of *Bacillus thuringiensis*, Foray 48B, against gypsy moth. *J. Econ. Entomol.* **86**, 26–33.

Dubois, N. R., Mierzejewski, K., Reardon, R. C., McLane, W., and Witcosky, J. J. 1994. *Bacillus thuringiensis* field applications: effect of nozzle type, drop size, and application timing on efficacy against gypsy moth. *J. Environ. Sci. Health* **B29**, 679–695.

Elkinton, J. and Liebhold, A. 1990. Population dynamcis of gypsy moth in North America. *Annu. Rev. Entomol.* **35**, 571–596.

Entwistle, P. F., Cory, J. S., Bailey, M. J., and Higgs, S. 1993. "*Bacillus thuringiensis*, an Environmental Biopesticide: Theory and Practice". J. Wiley & Sons, New York.

Falchieri, D., Mierzejewski, K., and Maczuga, S. 1995. Effects of droplet density and concentration on the efficacy of *Bacillus thuringiensis* and carbaryl against gypsy moth larvae. *J. Environ. Sci. Health* **B30**, 535–548.

Fettes, J. 1950. Investigations of sampling techniques for population studies of the spruce budworm on balsam fir in Ontario. Technical Report, Forest Insect Laboratory, Canadian Forest Service, Sault Ste. Marie, Ontario.

Fleischer, S., Carter, J., Reardon, R., and Ravlin, F. W. 1992. Sequential sampling plans for estimating gypsy moth egg mass density. Appalachian IPM Technology Transfer NA-TP-07-92, USDA Forest Service, Morgantown, WV.

Fleming, R. A. and Retnakaran, A. 1985. Evaluating single treatment data using Abbott's formula with reference to insecticides. *J. Econ. Entomol.* **78**, 1179–1181.

Fleming, R. A. and van Frankenhuyzen, K. 1992. Forecasting the efficacy of operational *Bacillus thuringiensis* Berliner applications against spruce budworm, *Choristoneura fumiferana* Clemens. *Can. Entomol.* **124**, 1101–1113.

Glowacka, B. 1996. The control of the nun moth, *Lymantria monacha* L., with the use of *Bacillus thuringiensis* in Poland. *IOBC (wprs) Bull.* **19**, 57–60.

Huber, J. 1986. Use of baculoviruses in pest management programs. *In* "The Biology of Baculoviruses. II. Practical Application for Insect Control" (R. R. Granados and B. A. Federici, Eds.), pp 181–202. CRC Press, Boca Raton, FL.

Huber, J. 1998. Western Europe. *In* "Insect Viruses and Pest Management" (F. R. Hunter-Fujita, P. E. Entwistle, H. F. Evans and N. E. Crook, Eds.), pp 201–215. J. Wiley & Sons, Chichester, UK.

Hunter-Fujita, F. R., Entwistle, P. F., Evans, H. F., and Crook, N. E. 1998. "Insect Viruses and Pest Management". J. Wiley & Sons, Chichester, UK.

Kaupp, W. J. and Ebling, P. M. 1993. Horseradish peroxidase-labelled probes and enhanced chemiluminescence to detect baculoviruses in gypsy moth and eastern spruce budworm larvae. *J. Virol. Methods* **44**, 89–98.

Keating, S. T., Burand, J. P., and Elkinton, J. S. 1989. DNA hybridization assay for detection of gypsy moth nuclear polyhedrosis virus in infected gypsy moth larvae. *Appl. Environ. Microbiol.* **55**, 2749–2754.

Kukan, B. and Myers, J. H. 1995. DNA hybridization for detection of nuclear polyhedrosis virus in tent caterpillars. *J. Invertebr. Pathol.* **66**, 231–236.

Lacey, L. 1997. "Manual of Techniques in Insect Pathology." Academic Press, San Diego, CA.

Lewis, F. B. 1981. Control of the gypsy moth by a baculovirus. *In* "Microbial Control of Pests and Plant Diseases, 1970–1980" (H. D. Burges, Ed.), pp. 363–377. Academic Press, London.

Liebhold, A. and Elkinton, J. 1988a. Techniques for estimating the density of late-instar gypsy moth populations using frass drop and frass production measurements. *Environ. Entomol.* **17**, 381–384.

Liebhold, A. and Elkinton, J. 1988b. Estimating the density of larval gypsy moth using frass drop and frass production measurements: sources of variation and sample size. *Environ. Entomol.* **17**, 385–390.

Lipa, J. J. 1998. Eastern Europe and the former Soviet Union. *In* "Insect Viruses and Pest Management" (F. R. Hunter-Fujita, P. E. Entwistle, H. F. Evans and N. E. Crook, Eds.), pp 216–231. J. Wiley & Sons, Chichester, UK.

Masse, A., van Frankenhuyzen, K., and Dedes, J. 2000. Susceptibility and vulnerability of third-instar larvae of the spruce budworm (Lepidoptera: Tortricidae) to *Bacillus thuringiensis* subsp. *kurstaki*. *Can. Entomol.* **132**, 573–580.

Martignoni, M. 1999. History of TM BioControl-1: the first registered virus-based product for control of a forest insect. *Am. Entomol.* **45**, 30–37.

Moreau, G., Lucarotti, C. J., Kettela, E. G., Thurston, G. S., Holmes, S., Weaver, C., Levin, D. B., and Morin, D. B. 2005. Aerial application of nucleopolyhedrovirus induces decline in increasing and peaking populations of *Neodiprion abietis*. *Biol. Control* **33**, 65–73.

Otvos, I. S., Cunningham, J. C. and Shepherd, R. F. 1995. Douglas-fir tussock moth, *Orgyia pseudotsugata*. *In* "Forest Insect Pests in Canada" (J. A. Armstrong and W. G. H. Ives, Eds.), pp127–132. Natural Resources Canada, Canadian Forest Service, Ottawa.

Otvos, I. S., Condor, N., and van Frankenhuyzen. 2002. Choristoneura occidentalis, western spruce budworm (Lepidoptera: Tortricidae). *In* "Biological control programmes in Canada, 1981–2000" (P. G. Mason and J. T. Huber, Eds.), Chapter 13, pp. 69–75. CABI publishing, Oxon, UK.

Otvos, I. S., Kukan, B., Reardon, R., and Ragenovich, I. 2006. Effects of long-term storage on potency of TM Biocontrol-1, the registered viral insecticide of *Orgyia pseudotsugata* (Lepidoptera: Lymantriidae). *J. Econ. Entomol.* **99**, 14–22.

Payne, N. J. and van Frankenhuyzen, K. 1995. Effect of spray droplet size and density on efficacy of *Bacillus thuringiensis* against the spruce budworm, *Choristoneura fumiferana*. *Can. Entomol.* **127**, 15–23.

Picot, J. J. C. and Kristmanson, D. D. 1997. "Forestry Pesticide Aerial Spraying. Spray Droplet Generation, Dispersion and Deposition." Environmental Science and Technology Library volume 12, Kluwer Academic Publishers, Dordrecht, the Netherlands.

Prebble, M. L. 1975. "Aerial Control of Forest Insects in Canada." Department of the Environment, Ottawa.

Reardon, R., Dubois, N. R., and McLane, W. 1994. *Bacillus thuringiensis* for managing gypsy moth: a review. FHM-NC-01-94, National Center of Forest Health Management, USDA Forest Service, Morgantown, WV.

Reardon, R. and Podgwaite, J. 1994. Summary of efficacy evaluations using aerially applied Gypchek against gypsy moth in the USA. *J. Environ. Sci. Health* **B29**, 739–756.

Reardon, R., Podgwaite, J., and Zerillo, R. 1996. Gypchek-the gypsy moth nucleopolyhedrosis virus product. FHTET-96–16, Forest Health Technology Enterprise Team, USDA Forest Service, Morgantown, WV.

Régnière, J. and Sanders, C. J. 1983. Optimal sample size for the estimation of spruce budworm (Lepidoptera: Tortricidae) populations on balsam fir and white spruce. *Can. Entomol.* **115**, 1621–1626.

Régnière, J. and Cooke, B. J. 1998. Validation of a process-oriented model of *Bacillus thuringiensis* variety *kurstaki* efficacy against spruce budworm (Lepidoptera: Tortricidae). *Environ. Entomol.* **27**, 801–811.

Schmitt, D. M., Grimble, D. G., and Searcy, J. L. 1984. Spruce budworms handbook. Managing the spruce budworm in eastern North America. Agricultural Handbook No. 620, USDA Forest Service, Cooperative State Research Service, Washington, DC.

Sheehan, K. A. 1996. Effects of insecticide treatments on subsequent defoliation by western spruce budworm in Oregon and Washington: 1982–1992. General Technical report PNW-GTR- 367, USDA Forest Service, Pacific Northwest Research Station, Portland, OR.

Shepherd, R. F. 1994. Management strategies for forest insect defoliators in British Columbia. *For. Ecol. Man.* **68**, 303–324.

Smitley, D. R. and Davis, T. W. 1993. Aerial application of *Bacillus thuringiensis* for suppression of gypsy moth in *Populus-Quercus* forests. *J. Econ. Entomol.* **86**, 1178–1184.

Strazanac, J. S. and Butler, L. 2005. Long-term evaluation of the effects of *Bacillus thuringiensis kurstaki*, gypsy moth nucleopolyhedrosis virus product Gypchek, and *Entomophaga maimaiga* on nontarget organisms in mixed broadleaf-pine forests in the central Appalachians. FHTET-2004-14, 81 pp. Forest Health Technology Enterprise Team, USDA Forest Service, Morgantown, WV.

Thorpe, K. W., Podgwaite, J. D., Slavicek, J. M., and Webb, R. E. 1998. Gypsy moth (Lepidoptera: Lymantriidae) control with ground-based hydraulic applications of Gypchek, in vitro-produced virus, and *Bacillus thuringiensis*. *J. Econ. Entomol.* **91**, 875–880.

Thurston, G. S. 2002. *Neodiprion abietis*, balsam fir sawfly (Hymenoptera: Diprionidae). *In* "Biological control programmes in Canada, 1981–2000" (P. G. Mason and J. T. Huber, Eds.), Chapter 38, pp. 196–198. CABI publishing, Oxon, UK.

van Frankenhuyzen, K. 1990. Development and current status of *Bacillus thuringiensis* for control of defoliating forest insects. *For. Chron.* **66**, 498–507.

van Frankenhuyzen, K. 2000. Application of *Bacillus thuringiensis* in forestry. *In* "Entomopathogenic Bacteria: from Laboratory to Field Application"

(J. C. Charles, A. Delecluse and C. Nielsen LeRoux. Eds.), Chapter 6.2, pp. 371–382. Kluwer Academic Publishers, Dordrecht, the Netherlands.

van Frankenhuyzen, K. 2002. *Choristoneura pinus pinus*, jack pine budworm (Lepidoptera: Tortricidae). *In* "Biological control programmes in Canada, 1981–2000" (P. G. Mason and J. T. Huber, Eds.), Chapter 14, pp. 75–68. CABI publishing, Oxon, UK.

van Frankenhuyzen, K., Wiesner, C. J., Riley, C. M., Nystrom, C., Howard, C. A., and Howse, G. M. 1991. Distribution and activity of spray deposits in an oak canopy following aerial application of diluted and undiluted formulations of *Bacillus thuringiensis* Berliner against the gypsy moth, *Lymantria dispar*. *Pest. Sci.* **33**, 159–168.

van Frankenhuyzen, K. and Payne, N. J. 1993. Theoretical optimization of *Bacillus thuringiensis* Berliner for the control of eastern spruce budworm, *Choristoneura fumiferana* Clem. (Lepidoptera: Tortricidae): estimates of lethal and sublethal dose requirements, product potency, and effective droplet sizes. *Can. Entomol.* **125**, 473–478.

van Frankenhuyzen, K., Payne, N. J., Cadogan, L, Mickle, B., and Robinson, A. 1996. Effect of droplet size spectrum and application rate on field efficacy of *Bacillus thuringiensis*. Spray Efficacy Research Group report 1995/02/final, 38 pp.

van Frankenhuyzen, K., Gringorten, L., Dedes, J., and Gauthier, D. G. 1997. Susceptibility of different instars of the spruce budworm (Lepidoptera: Tortricidae) to *Bacillus thuringiensis* var. *kurstaki* estimated with a droplet-feeding method. *J. Econ. Entomol.* **90**, 560–565.

van Frankenhuyzen, K., West, R. J., and Kennis, M. 2002. *Lambdina fiscellaria fiscellaria*, hemlock looper. *In* "Biological control programmes in Canada, 1981–2000" (P. G. Mason and J. T. Huber, Eds.), Chapter 30, pp. 141–144. CABI publishing, Oxon, UK.

Volney, W. J. A., and Cerezke, H. F. 1992. The phenology of white spruce and the spruce budworm in northern Alberta. *Can. J. For. Res.* **22**, 198–205.

Walton, G. S. and Lewis, F. B. 1982. Spruce budworm core Bt test-1980. Research Paper NE-506, USDA Forest Service, Northeastern Forest Experiment Station, Broomall, PA.

Webb, R. E., Dill, N. H., Podgwaite, J. D., Shapiro, M., and Ridgway, R. L. 1994. Control of third and fourth instar gypsy moth with Gypchek combined with stilbene disulfonic acid additive on individual shade trees. *J. Entomol. Sci.* **29**, 82–91.

West, R. J., Raske, A. G., and Sundaram, A. 1989. Efficacy of oil-based formulations of *Bacillus thuringiensis* against the hemlock looper, *Lambdina fiscellaria fiscellaria* (Lepidoptera: Geometridae). *Can. Entomol.* **121**, 5–63.

Whaley, W. H., Anhold, J., and Schaalje, G. B. 1998. Canyon drift and dispersion of *Bacillus thuringiensis* and its effects on select nontarget lepidopterans in Utah. *Environ. Entomol.* **27**, 539–548.

Woods, S. A. and Elkinton, J. S. 1987. Bimodal patterns of mortality from nuclear polyhedrosis virus in gypsy moth (*Lymantria dispar*) populations. *J. Invertebr. Pathol.* **50**, 151–157.

Young, S. Y. and Yearian, W. C. 1986. Formulation and application of baculoviruses. *In* "The Biology of Baculoviruses. II. Practical Application for Insect Control" (R. R. Granados and B. A. Federici, Eds.), pp. 157–179. CRC Press, Boca Raton, FL.

Chapter VII-10

Microbial control of wood-boring insects attacking forest and shade trees

Ann E. Hajek
Department of Entomology
Cornell University
Ithaca, NY 14853, USA

Leah S. Bauer
USDA Forest Service
North Central Research Station
and Department of Entomology
Michigan State University
East Lansing, MI 48823, USA

1 Introduction

Wood-boring insect pests that feed on the bark, phloem, or xylem (wood) of living trees pose unique management challenges because their immature stages live in cryptic, often inaccessible, habitats within host trees. The eggs of wood borers are laid in or on tree trunks, branches, terminal shoots, or roots. After the eggs hatch, neonates tunnel in and feed on internal target tissues, making infestation both difficult and expensive to detect and control. Adult wood borers typically emerge from the tree to feed, disperse, mate, and oviposit, thus occupying very different habitats and exhibiting different behaviors than immature stages. Both immature and adult wood borers can be relatively long-lived.

The majority of wood-boring insect species that attack live trees are in the order Coleoptera (beetles), with fewer in Lepidoptera (moths) and Hymenoptera (horntails and sawflies), and still fewer in Diptera (flies) (Solomon, 1995). In this chapter, we will review those species for which microbial control has been attempted to control insects that feed in tree bark, phloem, or xylem (wood) of living trees. We will predominantly cover wood borers attacking above-ground woody parts of trees growing in forests or as shade trees or windbreaks, etc. Wood-borers that principally attack orchard trees or woody plants in nurseries are covered elsewhere in this volume (Chapters VII-11, VII-12, VII-13 and VII-16). Insect pests feeding on bark or wood of roots are covered in chapters on citrus trees (VII-13) and woody plants in nurseries (VII-16).

The wood-boring insects covered in this chapter include native species that outbreak periodically in numerous types of habitats: 1) in forested areas due to natural abiotic and biotic conditions such as drought or forest maturation; 2) in monocultures such as fiber farms, tree plantations, nurseries and windbreaks; and 3) in trees under stress after transplantation along streets, in parks, in other urban or suburban environments. We will also review the management of invasive wood-boring beetles, which often become pests after their inadvertent movement between countries in solid-wood packing materials used to transport goods and/or in commodities such as bonsai trees, nursery stock, wooden products, logs, and lumber. The increased movement of non-native wood-boring insects among countries is one consequence of our global economy and threatens

L.A. Lacey and H.K. Kaya (eds.), Field Manual of Techniques in Invertebrate Pathology, 505–525.

the sustainability of forests throughout the world (Chornesky *et al.*, 2005). Between 1985 and 2005, at least 25 exotic species of wood borers were found to have become established in the continental United States, including two species of Buprestidae (metallic wood-boring beetles), five Cerambycidae (long-horned beetles), and 18 Scolytinae (formerly Scolytidae) (bark beetles) (Haack, 2006). Due to the lack of both natural enemies and coevolved tree-resistance mechanisms to control these borers, these exotic species may become destructive and uncontrollable invasive tree pests. One example is the European woodwasp, *Sirex noctilio* (Hymenoptera: Siricidae), which has become a serious pest in each country to which it has been introduced. For example, in the Australian 'Green Triangle' area, *S. noctilio* killed ca. 4.8 million pine trees in 1987–1989 alone (Bedding and Iede, 2005). Another example is the emerald ash borer, *Agrilus planipennis* (Coleoptera: Buprestidae), discovered in North America in 2002 (Haack *et al.*, 2002). In Michigan, land managers estimate that larvae of this phloem-feeding beetle have killed more than 20 million ash (*Fraxinus* spp.) trees to date.

Control of wood borers by spraying synthetic chemical insecticides was made more difficult and less effective when chlorinated hydrocarbons, with extended residual persistence, were taken off the market in many countries. Methods to control or eradicate wood-boring insects using conventional materials are still being developed, but the newest methods usually involve application of systemic insecticides, such as imidacloprid through direct trunk injection, soil injection, or soil drench. This chemical moves within trees to reach cryptic larvae as well as affecting adults of some species that feed externally. The desired level of control, however, is seldom achieved by this approach due to unpredictable translocation of insecticides within trees (Poland *et al.*, 2006) and asynchronous larval development, both of which allow insects to evade treatment. Moreover, chemical insecticides are usually broadly toxic, resulting in risks to non-target organisms and potential groundwater contamination. Many chemical insecticides and formulations require handling by licensed applicators resulting in prohibitive costs. The use of conventional insecticides to control insect borers, therefore, is restricted to a relatively small number of high-value urban trees or periodic borer outbreaks in tree plantations or nurseries and, unfortunately, few management options are available for borers attacking trees in forested and riparian areas. Management of borers in such environmentally-sensitive areas is thus more amenable to microbial control agents (MCAs), which are generally accepted by the public, biodegradable, compatible with other management strategies, and which can provide adequate control in these systems with relatively high damage thresholds.

Wood-boring insect pests for which microbial control methods have been developed predominantly occur in the Orders Coleoptera and Lepidoptera (Table 1). This chapter will cover augmentative strategies for control of these pests. Major classical biological control programs have been undertaken to introduce the parasitic nematode *Deladenus* (= *Beddingia*) *siricidicola* for permanent control of *S. noctilio*. This nematode has been introduced and become established in 8 geographic areas (Hajek *et al.*, 2005) in many cases providing excellent long-term control. Classical biological control using *D. siricidicola* is described briefly in Chapter VI-1 and in more detail by Bedding and Iede (2005).

Many types of pathogens and nematodes (see Sections IV and V) are known to naturally infect and parasitize insect pests that bore in wood (Fuxa *et al.*, 1998). However, only some of these species demonstrate a large enough impact to justify development as MCAs. In other cases, MCAs are developed when preliminary laboratory, greenhouse and field studies demonstrate high levels of mortality. The majority of microbes investigated and utilized as MCAs for control of wood borers are insect-pathogenic fungi, nematodes, and the bacterium, *Bacillus thuringiensis (Bt)*.

Pestiferous borers in their native range seldom kill large healthy trees after initial attacks. However, after repeated attacks, either during the same season, as for scolytines, or during subsequent seasons, as for cerambycids and buprestids, even healthy trees may succumb. Some species are better known for attacking and killing trees that are already stressed, such as over-mature trees that might already be declining or urban trees exposed to air pollution and with roots stressed by soil compaction and inadequate

Table 1. Major wood-boring pests attacking shade and forest trees for which control using entomopathogens has been evaluated in the field

Order Family	Pest species, Common name	Country where microbial control investigated	MCA[1]
Order Coleoptera			
Cerambycidae	*Anoplophora glabripennis*, Asian longhorned beetle	China, U.S.	F, N
	Monochamus alternatus, Japanese pine sawyer	Japan, China	F, N
	Plectrodera scalator, Cottonwood borer	U.S.	F
Buprestidae	*Agrilus planipennis*, Emerald ash borer	U.S.	F, B
	Melanophila decastigma, Ten-blotched poplar flatheaded borer	China	N
Curculionidae	*Cryptorhynchus lapathi*, Poplar-and-willow borer	Italy	N
	Hylobius abietis, Large pine weevil	Europe	N
	Ips typographus, European spruce bark beetle	Germany	F
	Scolytus scolytus, European elm bark beetle	United Kingdom	N
Order Lepidoptera			
Cossidae	*Squamura discipuncta*, Acacia pseudo carpenter moth	China	F
	Cossus cossus, Willow goat moth	China	N
	Holcocerus insularis, Carpenterworm	China	N
	Prionoxystus robiniae, Carpenterworm	U.S.	B, N
	Zeuzera multistrigata, Leopard moth	China	F, N
Sesiidae	*Paranthrene robiniae*, Western poplar clearwing	U.S.	N
	Paranthrene simulans, Oak clearwing borer	U.S.	B
	Paranthrene tabaniformis, Dusky clearwing	Italy, Poland	F, N
	Podosesia aureocincta, Banded ash clearwing	U.S.	N
	Podosesia syringae, Lilac borer	U.S.	N
	Synanthedon culiciformis, Large red-belted clearwing	U.S.	N
	Synanthedon resplendens, Sycamore borer	U.S.	N
	Synanthedon scitula, Dogwood borer	U.S.	N
Pyralidae	*Euzophera batangensis*, Persimmon bark borer	China	F, N
	Hypsipyla spp. Mahogany shoot borers'	China, Cuba, Honduras, India, Mexico	B, F

[1] F = Fungus; N = Nematode; B = Bacterium.

water and nutrients. The behavior and location of the destructive stages of wood borers, in part, determine which MCAs are most appropriate. Often, the immature stage of a wood borer causes the most damage to live trees due to their feeding on the bark, phloem-cambial region, and/or xylem. In most species of scolytines and buprestids, this stage remains just beneath the tree bark within the phloem-cambial region. In many cerambycids, the immature stages begin feeding just under the bark, and later instars move into the xylem where they are less accessible by MCAs. The larvae of some wood-boring pests, such as cerambycids and buprestids, are solitary, with one immature per gallery, thus limiting horizontal transmission of MCAs. Transmission between individuals is also limited by the oviposition behavior of many bark beetle species that oviposit in a central maternal gallery, and after hatching, neonate larvae tunnel away from the main gallery as they feed on the phloem and cambium. A greater chance of contact among individuals occurs for some species of Cossidae (carpenterworms), where several larvae share the same gallery. Another factor affecting MCA development for wood borers is the prolonged larval stage, especially for larger wood borers or those predominantly living in the xylem. These larval stages can be long-lived, with some species of wood borers taking several years to develop from egg to adult. In addition, some wood-borer species are quite host specific, while others can have broad host ranges.

Investigations toward development of MCAs begin with laboratory bioassays to determine microbe pathogenicity, followed by comparisons of virulence among different microbial isolates or strains. Thus, many species of wood borers have been challenged with different pathogens and parasitic nematodes in the laboratory. Due to the complexity of larval and adult wood borer habitat and biology, however, laboratory bioassays only demonstrate pathogenicity and virulence under idealized circumstances and cannot reflect effectiveness of MCAs in the field. Therefore, in this chapter we will focus on describing methods and systems for which use of pathogens and nematodes has progressed past laboratory bioassays to field trials evaluating efficacy and developing application methods. Microbial control methods for Coleoptera and Lepidoptera will be surveyed below, followed by in-depth descriptions of methods for microbial control agents to control the Asian longhorned beetle (*Anoplophora glabripennis*) and the emerald ash borer (*A. planipennis*) as case studies.

2 Bark- and Wood-Boring insect pests targeted with microbial control

To date, the major groups of pathogens developed for control of wood borers are the fungi and nematodes. Both of these groups of microbes can infect without being eaten by hosts, which is advantageous because applying microbes that require ingestion to infect (*e.g.*, viruses and bacteria) is not appropriate for many wood-borer larvae. For example, late-instar Asian longhorned beetle larvae feed in galleries deep within tree trunks, while the long-lived adults emerge and feed on petioles throughout tree canopies for several months. These characteristics make it difficult for many MCAs to survive long enough or be delivered at an effective rate for ingestion by either larvae or adults. However, *Bt* has been most successfully evaluated in systems where larvae of wood-boring Lepidoptera can feed on it.

Different MCAs can be more appropriate for life stages in different habitats. Species of entomopathogenic nematodes that search for hosts and can remain alive while partially desiccated for some time are more often used to target borer stages living within wood. For applications against wood-boring larvae, cruiser nematodes that disperse to find cryptic hosts within galleries may prove efficacious, although researchers suggest that ambusher nematodes may also prove effective in larval galleries of host species that move in and out of the gallery to feed and expel frass (*e.g.*, some cerambycids). In contrast, conidia of fungal pathogens have also often been applied against wood-borer adults or larval stages. Throughout studies conducted in different host/pathogen systems, emphasis has been placed on development of application technologies that are system specific as well as both effective and economically feasible for land managers.

Fungi developed for control of beetles are all conidial fungi (anamorphs = asexual states) now classified in the Family Clavicipitaceae (Order

Hypocreales) but previously listed in the Class Hyphomycetes of the Division Deuteromycota (a taxonomic group presently being abandoned as the phylogenetic affinities of so many conidial fungi become known). The predominant species that have been developed into MCAs are *Beauveria bassiana, Beauveria brongniartii* and *Metarhizium anisopliae.*

Nematodes that have been investigated for control of wood-boring Coleoptera and Lepidoptera belong to the Order Rhabditida. Most work has been done with species within the genus *Steinernema* (Family Steinernematidae) although in a few instances species of *Heterorhabditis* (Family Heterorhabditidae) have also been investigated, predominantly in the laboratory.

A Coleoptera (Cerambycidae, Buprestidae, Curculionidae)

1 General biologies

Beetles are the most economically important order of insects attacking trees. Within the Coleoptera, species in the families Buprestidae, Cerambycidae, Curculionidae and its Subfamily Scolytinae, are by far the most destructive wood borers. MCAs have targeted species in all of these beetle groups. Larvae of bark beetles and buprestids remain in the phloem-cambial region just below tree bark and kill trees by mass-attacking and girdling them. Larvae of cerambycids often feed just under the bark for some time after which some species bore extensively into the wood; these feeding zones within trees vary by species. Biologies of these families are described in more detail in Solomon (1995).

2 Control with entomopathogenic fungi

a Japanese pine sawyer

Usually the activity of cerambycids alone kills or weakens trees, but in the case of the Japanese pine sawyer, *Monochamus alternatus*, adult beetles are the problem because they serve as vectors for the pinewood nematode, *Bursaphelenchus xylophilus*, the causal agent of pine wilt disease, which kills pines. Interestingly, this cerambycid was not a problem until *B. xylophilus* was inadvertently introduced from North America to Japan in the early 1900s. *M. alternatus* is now considered a destructive pest of pines throughout Japan and other Asian counties, and the importation of coniferous chips, unseasoned lumber and logs is regulated worldwide to limit the spread of both the insect and vectored nematode. Despite regulation, the Japanese pine sawyer is frequently intercepted at ports of entry throughout the world, *e.g.*, live *M. alternatus* adults emerged from infested wood crating in a plumbing warehouse in western New York in June 1998. It is critical to develop management methods for this and related species.

M. alternatus is susceptible to the insect-pathogenic fungus *B. bassiana* (Figure 1), which has been developed in Japan as an MCA for control of larvae and young adults to prevent transmission of *B. xylophilus* from infected to healthy trees. Spraying *B. bassiana* directly on larval-infested pine logs yielded a maximum of 75% larval mortality (Shimazu and Kushida, 1980), but results were variable during practical use. Infection was improved through release of pine bark beetles, *Cryphalus fulvus*, contaminated with *B. bassiana* conidia; these small bark beetles carried fungal conidia under the bark, resulting in *M. alternatus* larval infections

Figure 1. Monochamus alternatus adult killed by *Beauveria bassiana* (photo taken by Mitsuaki Shimazu)

(Shimazu *et al.*, 1999). Another strategy for delivering *B. bassiana* conidia to *M. alternatus* larvae in nematode-infested pines was implantation under host-tree bark of wheat-bran pellets on which *B. bassiana* had been cultured. This strategy yielded 13–81% beetle mortality in logs and 43–45% mortality in standing trees, but was difficult to use in infested standing trees and smaller branches (Shimazu *et al.*, 1992). *B. bassiana* conidia must come into contact with *M. alternatus* larvae living under the bark to infect its host, and the most effective strategy for infecting larvae has been placement of non-woven fabric strips, impregnated with cultures of *B. bassiana* (hereafter called 'fungal bands'), on top of infested logs and branches in piles on the ground (Shimazu *et al.*, 1995) (Figure 2). These fungal bands were more effective when applied to wood with younger *M. alternatus* larvae because first through third instars feed within the cambial region while later instars bore into the sapwood (Shimazu and Sato, 2003). It is actually not known how conidia reach the larval microhabitat by this method, but it has been hypothesized that water flow from rain or small animals inhabiting this microhabitat move the conidia.

Young *M. alternatus* adults are also a target for MCA development because *B. xylophilus* is not vectored during a period of maturation feeding after adult emergence. Initially, methods for applying *B. bassiana* for control of *M. alternatus* adults involved spraying conidia onto infested pine trees just before adult emergence (Shimazu *et al.*, 1982) and during adult maturation feeding (Shimazu *et al.*, 1983). More recently, fungal bands have been tested against *M. alternatus* adults, and this approach yielded more promising results. A field trial resulting in high levels of infection, involved covering infested trees with a plastic sheet, forcing emerging adults to walk across fungal bands to escape (Okitsu *et al.*, 2000). Adults died from infection within 14 days of emergence, prior to transmission of *B. xylophilus* (M. Shimazu, personal communication). At present, a product based on *B. bassiana* bands for control of *M. alternatus* is being registered for commercialization in Japan. *B. xylophilus* was first discovered in China in 1982 and field trials with fungal band plus host attractants for control of *M. alternatus* to prevent transport of nematodes to healthy pines are being conducted in Anhui Province (Z. Li, personal communication).

b Asian longhorned beetle

The Asian longhorned beetle, *A. glabripennis*, was accidentally introduced from China to several urban areas in northeastern North America and was first detected in 1996 in the New York City area (Hajek, 2007). *A. glabripennis*

Figure 2. Applications of *Beauveria bassiana* bands to logs for control of *Monochamus alternatus* larvae within the wood (photo taken by Mitsuaki Shimazu)

was subsequently found in Chicago, Toronto and the state of New Jersey and also in Europe. This species is presently the target of eradication programs in the U.S., Canada, and several European countries. *A. glabripennis* kills numerous species of trees and has caused catastrophic levels of damage in China. Although this species is known for killing living trees, this effect results from successive generations repeatedly attacking the same tree. At first upper branches are infested and killed, and in subsequent years, eggs are laid and larval galleries are present on the trunk (Haack *et al.*, 1997). Eventually, with repeated attacks, the entire tree will die.

The principal microbial approach for control of this species has targeted adults using fungal bands placed around tree trunks and branches. After emergence from trees, adults undergo maturation feeding for > 1 week, and during this time, these reluctant fliers walk on tree trunks and contact the fungal bands. In addition, adults contaminated with conidia after contacting a band can transmit conidia to another adult during mating (Hajek, unpublished data). Band production was developed by amending methods developed by Nitto Denko (Osaka, Japan) for commercial production of fungal bands made with *B. brongniartii*, which are used to control cerambycid pests (*e.g.*, *Psacothea hilaris, Anoplophora malasiaca*, and *Apriona japonica*) in Japanese orchards (Higuchi *et al.*, 1997); the fungal band product used in Japanese orchards is named Biolisa Kamikiri. For control of *A. malasiaca*, bands are placed at the bases of citrus trees where females oviposit. However, for control of other species such as *P. hilaris* that oviposit higher in trees, bands are placed around limbs in tree canopies (Tsutsumi, 1998).

Field studies on the use of fungi for control of *A. glabripennis* in the U.S. have been conducted in China because field populations of this pest in North America are low and it is being eradicated. Trials began with a comparison of adults caged with tree trunks that had been sprayed with either *B. bassiana* or *B. brongniartii* or caged with fungal bands of these fungal species (Dubois *et al.*, 2004a) (Figure 3). Longevity and oviposition of adults receiving the two treatments were similar. However, conidial viability was high on bands 10 days

Figure 3. Cage containing a *Beauveria brongniartii* band and 5 *Anoplophora glabripennis* adults to evaluate effects of fungal bands on adult longevity and oviposition (photo taken by Thomas Dubois and reprinted from Dubois *et al.* 2004a by permission of the Entomological Society of America)

after the experiment began but was drastically reduced from sprays. Thus, subsequent field trials were based on use of fungal bands. Bands containing cultures of *B. brongniartii* and *M. anisopliae* demonstrated faster mortality of adults from treatment plots compared with adults from control plots, as well as greatly reduced oviposition in treatment plots (Dubois *et al.*, 2004b; Hajek *et al.*, 2006). At present, *M. anisopliae* F-52 is being used for fungal bands in the U.S. because this virulent strain was already registered with the U.S. Environmental Protection Agency. Because adults emerge asynchronously for several months, it is difficult to apply materials for control that impact all adults. However, fungal bands are excellent in this regard because activity has been shown to persist for well over one month in the field (Higuchi *et al.*, 1997; Hajek, unpublished data).

c Emerald ash borer

The emerald ash borer, *A. planipennis*, was identified as the cause of extensive ash mortality (*Fraxinus* spp.) in areas of Michigan and Ontario in 2002 (Haack *et al.*, 2002). Although control programs initially focused on eradication, *A. planipennis* spread into 5 states due to transport of infested nursery stock, firewood, timber, and natural spread (USDA FS NCRS, 2006), resulting in abandonment of eradication programs in parts of four Midwestern states, which are now considered generally infested. *A. planipennis* typically has a one-year life cycle and overwinters as mature larvae under the bark. After eclosion in spring to early summer, the long-lived adults mate and maturation-feed on ash foliage for almost three weeks before females begin to oviposit in bark crevices and between bark layers. Newly hatched larvae bore directly into the bark until reaching the phloem, where they feed in the cambial region, forming characteristic serpentine galleries. High larval densities, as is typically observed in ash trees in North America, result in overlapping galleries, contact between individuals, and cannibalism, providing opportunities for improved efficacy of MCAs through horizontal transmission. Research on conventional insecticides for management and containment or control of *A. planipennis* in high value landscape trees is ongoing (Poland and McCullough, 2006). Suitable methods, however, including microbial control, are also needed for management of this pest in environmentally-sensitive forested and riparian areas.

Comparative laboratory bioassays demonstrated *B. bassiana* strain GHA, registered as BotaniGard® for control of a variety of pests, was highly virulent against *A. planipennis* immatures and adults compared to other fungal isolates (Liu and Bauer, 2006b). Median lethal concentrations of petroleum- or vegetable-based formulations ranged from 17 to 800 conidia/cm^2 and lethal times ranging from 4 to 10 days. In the greenhouse, BotaniGard application methods for control of *A. planipennis* were compared at a single rate (1×10^{14} conidia/ha) applied in a spray tower to 1) potted ash trees or 2) uninfested ash logs which were then caged with *A. planipennis* adults, or 3) infested ash logs prior to beetle emergence. When adults were caged with treated ash trees or logs, only 10% or 18% of adults became infected, whereas 61% of adults became infected when emerging from treated logs. Pre-emergent BotaniGard sprays were then field-tested by spraying and caging infested ash trunks at two rates (1×10^{14} and 1×10^{15} conidia/ha), resulting in 43% and 76% fungal infection of emerging adults, respectively (Bauer *et al.*, 2004). In a separate experiment, fungal bands grown with *B. bassiana* GHA were evaluated on infested, caged ash trunks, resulting in 32% infection of the emergent adults (Liu and Bauer, unpublished data). Fall applications of BotaniGard (3×10^{14} conidia/ha) applied to infested ash trunks resulted in ca. 20% larval infection, presumably through infiltration of sprays into cracks and splits that form in ash tree trunks and branches after *A. planipennis* attack. These findings led to a larger field trial in a 20-year old infested ash plantation in which infested ash trees (canopy and trunk) were sprayed with BotaniGard (3×10^{14} conidia/ha) to drip four times during the *A. planipennis* emergence period. These sprays resulted in 36% fewer larvae infesting the trunks in the fall and 68% fewer adults emerging in the spring compared to control trees (Liu and Bauer, 2006a), and ca. 26% less decline in condition of the ash canopy in the treated trees when compared to untreated trees (Liu and Bauer, unpublished data). Additional field trials are ongoing to increase the efficacy of BotaniGard for *A. planipennis* management with fewer applications.

d Cottonwood borer

Taking a different approach to cerambycid control using fungi, *B. bassiana* was applied at different rates as a soil drench for control of the cottonwood borer, *Plectrodera scalator*, in a nursery where adults laid eggs into root collars and larvae tunneled in roots (Forschler and Nordin, 1989). After application, adults were collected and reared throughout the flight period (June-August) and > 60% of adults were infected with *B. bassiana*. The authors hypothesized that adults became infected both when emerging from the rootstocks but also from contact with *B. bassiana*-contaminated soil in the treated plots. For the highest rate applied (1×10^{11} colony forming units/m^2), neonate larval establishment was lower in the treatment than controls. However, there were no differences among treatments for second-year larvae.

e European spruce bark beetle

A *B. bassiana* product from Fytovita (Praha, Czech Republic) named Boverol has been investigated for control of the European spruce bark beetle (*Ips typographus*) (Kreutz *et al.*, 2004). Field-cage studies in a spruce stand demonstrated horizontal transmission among adults after *B. bassiana*-treated beetles were released into a healthy population. As a result of the treatment, the length of maternal galleries and numbers of *I. typographus* larvae and pupae were reduced. In a second study, beetles were lured into cages using commercial pheromone traps and some of those adults entering cages were autoinoculated with *B. bassiana* conidia. Numbers of holes constructed by females to form egg galleries and lengths of maternal galleries were reduced, and no larvae or pupae were found under the bark. The authors state that the next step is to determine whether *B. bassiana* plus pheromones is more effective for population control than mass trapping using pheromones alone.

f Poplar-and-willow borer

Larvae of the weevil *Cryptorhynchus lapathi* bore into wood of poplars and willows and can kill, weaken or deform trees. Larvae tunnel in xylem and the openings of their galleries are plugged with aggregations of wood particles. In one study in Italy, wood plugs were treated with suspensions of *B. bassiana* conidia in growing poplars, resulting in about 50% larval mortality (Cavalcaselle, 1975).

3 Use of entomopathogenic nematodes for control of wood-boring beetles

a Japanese pine sawyer

Entomopathogenic nematodes have been investigated for control of cerambycid larvae within wood. *Steinernema carpocapsae* was sprayed onto horizontally oriented logs to control *M. alternatus*, the vector of *B. xylophilus*, in Japan; spray volume, timing and rate were investigated (Yamanaka, 1993). Larval mortality was greater (69.2–72.2%) after exposure to *S. carpocapsae* than when fenitrothion was applied. However, the most efficacious nematode concentration was much higher than concentrations used commercially for control of other coleopteran pests. Although nematodes were applied only to the tops of logs, *M. alternatus* larvae under bark and within tunnels became infected, both those on tops of logs and those beneath logs; the nematodes clearly dispersed to find hosts. The author stated that use of nematodes to control larvae of *M. alternatus* needed to be part of an integrated control approach because nematode applications alone would not prevent spread of *B. xylophilus*.

b Asian longhorned beetle

Species of *Steinernema* were also tested against *A. glabripennis* larvae in China. Using *S. feltiae*, 2000 infective juveniles (IJs)/ml were injected into galleries of later instars until galleries were full, resulting in an average 62% larval mortality (Qin *et al.*, 1988). Galleries of *A. glabripennis* larvae do not intersect and are constructed upwards, so injection of nematodes into larval tunnels is not practical. It was likely that *A. glabripennis* larvae became infected with entomopathogenic nematodes when they traveled to the bark surface to discard frass from their galleries (Qin *et al.*, 1988). In another study, IJs were injected directly into the gallery openings from which larvae expel frass or sponges soaked in a suspension of IJs were placed in the openings; different concentrations and application methods for two strains of *S. carpocapsae* were compared in the field. Injection of 7,500 IJs/ml into each gallery opening resulted in an 86.4% reduction in active frass removal, suggesting larval mortality (Liu *et al.*, 1998). Laboratory studies have shown that *S. carpocapsae* infectivity is inhibited by exposure to aqueous extracts from *A. glabripennis* although pathogenicity was not affected (Fallon *et al.*, 2004). *S. feltiae* juveniles were positively attracted to *A. glabripennis* frass extracts, in agreement with laboratory bioassays during which *S. feltiae* IJs were more infectious than *S. carpocapsae*.

c European elm bark beetle

For control of the bark beetle *Scolytus scolytus*, a suspension of *S. carpocapsae* sprayed onto heavily infested logs in the field was considered ineffective (Finney, 1977; Finney and Walker, 1979). Although logs treated in the summer yielded some infected larvae and adults beneath the bark, no overall difference in numbers of emergence holes were found in comparisons with controls.

d Ten-blotched poplar flatheaded borer

The only wood-boring buprestid tested with entomopathogenic nematodes has been *Melanophila decastigma* in poplars in China (Liu *et al*., 1998). Two strains of *S. carpocapsae* were injected into gallery openings, which active larvae had formed to eject frass. Equivalent results were obtained for different strains, with a maximum reduction of 89.5% active larval gallery openings after application of 10,000 nematodes in 1 ml. However, the authors suggested that for economic reasons, 5,000 nematodes in 1 ml should be applied, although in their study this only yielded a 66.7% reduction in active galleries.

e Large pine weevil

Among the weevils, a large program using entomopathogenic nematodes has been developed in Europe for control of the large pine weevil, *Hylobius abietis* (Torr *et al*., 2005). Larvae of this species develop within stumps and roots of dying and dead conifers and can reach high population densities after clear cutting a stand. Most damage is inflicted after young trees are transplanted at a site and adults feed on bark and phloem of transplants, at times girdling them. Transplants can be weakened or killed by the long-lived *H. abietis* adults, resulting in up to 100% loss of plantation restocks in some sites. To reduce populations of this weevil, larvae are targeted by nematodes that can search within stumps for hosts. However, it is critically important that the phenology of *H. abietis* is monitored because nematodes should be applied close to the time when larvae are pupating in stumps, because this is a particularly vulnerable stage. At present, researchers recommend applying a single application of *S. carpocapsae* at 3.5×10^6 nematodes in 500 ml water around the base of each stump. Because it can be difficult to reach all stumps after extensive tree felling has occurred, Torr *et al.* (2005) describe use of a forwarder-mounted spray rig to deliver nematodes to target stumps using hand-held lances. This equipment can treat 5 ha per day and was used for treatment of 200 ha in the United Kingdom in 2003. Using this methodology, Torr *et al.* (2005) estimate adult emergence is reduced by 60–75%.

f Poplar-and-willow borer

Nematodes have also been applied against another weevil, *C. lapathi*, whose larvae bore in stems of young willows and poplars. When cotton swabs soaked with suspensions of 20,000 IJs of *S. carpocapsae* were applied to entrances of larval galleries, 100% mortality was obtained (Cavalcaselle and Deseö, 1984). These trials were followed with application of three species of nematodes at different concentrations, applied to bark before larvae entered the wood. However, the highest mortality resulting from these applications (75% from *S. feltiae*) was not considered sufficient for practical control.

4 Use of bacteria for control of wood-boring beetles

Among MCAs targeting all insect pests, the most widely used is the insect-pathogenic bacterium *Bacillus thuringiensis (Bt)*. Although most products target larval insect pests in the orders of Lepidoptera and Diptera, many *Bt* strains are active against species of Coleoptera. *Bt* isolates showing potential for control of wood-boring beetles in laboratory bioassays include isolates of *Bt* subsp. *tenebrionis* for species of Bostrichidae, Cerambycidae, Curculionidae, and Scolytidae (Cane *et al*., 1995; Beegle, 1996; Weathersbee *et al*., 2002; Chen *et al*., 2005); *Bt* subsp. *thuringiensis* and *Bt* subsp. *entomocidus* for species of Scolytinae (Jassim *et al*., 1990); and *Bt* subsp. *israelensis* for species of Cerambycidae and Scolytinae (Alfazairy, 1986; Méndez-López *et al*.,

2003). The use of *Bt* for controlling larvae of wood borers, however, may only be achieved through the expression of *Bt* Cry toxin genes in genetically modified trees.

a. Asian longhorned beetle. Use of *B. thuringiensis* was proposed for management of *A. glabripennis* and mulberry longicorn beetle, *Apriona germari*, in China with a Cry3Aa toxin gene cloned from *Bt* strain 866 (Chen *et al.*, 2005).

b. Emerald ash borer. After the discovery of *A. planipennis* in North America, four registered *Bt*-based MCAs were evaluated for activity against adults, which feed on ash leaves throughout their adult life (Bauer *et al.*, 2004). Although some activity was observed at high rates of formulated *Bt*s in laboratory and small field trials, further laboratory bioassays demonstrated that the major Cry toxins from *Bt* subsp. *kurstaki, Bt* subsp. *tenebrionis*, and *Bt* subsp. a*izawai* were not toxic to adult *A. planipennis*. Therefore, different *Bt* isolates, with known coleopteran activity, were acquired from public and private culture collections and are now being screened for activity against this buprestid (Bauer *et al.*, 2006).

B *Lepidoptera (Cossidae, Sesiidae, Pyralidae)*

1 *General biologies*

Larvae of numerous lepidopteran families bore in trees (Solomon, 1995). Larvae of species in 3 families have been targets for MCAs. The Cossidae includes carpenterworms, with caterpillars of some species living gregariously in tunnels within wood. Also within the Cossidae are the leopard moths, named for the spotted patterns on wings and bodies of adults. Larvae of species in the family Sesiidae, the clearwing moths, bore in wood, living one per tunnel. Larvae of wood-boring pyralids feeding in shoots and in cambium are often most damaging to young trees. General biologies of cossids and sesiids are described briefly in Chapter VII-16 and in Solomon (1995).

2 *Control with entomopathogenic nematodes*

a Cossidae in China

Entomopathogenic nematodes have been investigated for control of cossids and sesiids in both China and the U.S., based on the ability of nematodes to survive in the moist larval galleries within wood and travel to reach larvae within the wood. In northern China, the carpenterworm *Holcocerus insularis* is a major pest of ash (*Fraxinus pennsylvanica*), the Chinese scholar tree (*Sophora japonica*) and willows (*Salix* spp.) which are grown as shade trees, as well as Chinese hawthorn, *Crataegus pinnatifida* var. *major*, which is grown for fruit and medicine. This wood borer often occurs with several immatures per gallery within tree trunks. In southern China, the leopard moth *Zeuzera multistrigata* is a pest of Australian pine, *Casuarina equisetifolia*, a tree species planted extensively for windbreaks. *Z. multistrigata* attacks younger trees and has only one larva per tree but larvae will migrate from tree to tree several times, causing serious damage before completing development (Kaya *et al.*, 2006). In Gansu Province in northwestern China, *Cossus cossus* is a major pest of willows. Four species of entomopathogenic nematodes have been tested against *H. insularis*, demonstrating that *S. carpocapsae* was the most effective species (Yang *et al.*, 1993). Field trials with *Z. multistrigata* demonstrated that *S. carpocapsae* is also effective against this species. Studies included comparisons of application rates, demonstrating highest susceptibility of *Z. multistrigata* (50 IJs in 2 ml for 93% mortality), intermediate susceptibility of *H. insularis* (100,000 IJs in 100 ml for > 90% mortality), while *C. cossus* was quite resistant (800,000 IJs in 100 ml killed only 80%). The authors suggested that while innate susceptibility could explain these results, factors such as depth and structure of galleries, densities and ages of larvae and moisture levels within galleries would also affect the ability of nematodes to disperse to borer larvae and infect. Methods for the application of nematodes were compared using *Z. multistrigata*, demonstrating equivalent results from injection of nematodes into borer tunnels and blocking the borer holes with a sponge laced with a suspension of *S. carpocapsae*. The sponge plug method was easier to use and was applied to 25.3 ha in 1990, resulting in 90% larval mortality (Yang *et al.*, 1990). Comparison of control by *S. carpocapsae* versus dichlorvos, an organophosphate fumigant, against *H. insularis* demonstrated higher levels of mortality due to *S. carpocapsae* than the insecticide after 40 days (96% versus 76%, respectively). Interestingly, one Chinese study challenging *H. insularis* with *S. carpocapsae* in the field demonstrated two peaks in infection, showing that this MCA completes two life cycles, with numbers of dead larvae peaking 2–4 days after application and then again on the 14th day after application (Qin *et al.*, 1988). Following demonstrated success in controlling *H. insularis* using *S. carpocapsae*, this nematode was used extensively in five areas in China (Yang *et al.*, 1993). Control has been so successful that this nematode has been used to protect street trees in Tianjin since 1987, reducing infestations from 12.6 to 4% of trees (Kaya *et al.*, 2006).

b Carpenterworm in the U.S.

In the U.S., two methods for applying *Steinernema carpocapsae* and *S. feltiae* were compared for control of the carpenterworm, *Prionoxystus robiniae* (Cossidae) (Forschler and Nordin, 1988). Gallery injection yielded mortality of 70–100% of 5th and 6th instar larvae, while results from bark surface application were more variable. Because gallery injection is more labor intensive, only bark surface applications were tested further although this method requires that nematodes survive on the bark surface to reach and enter galleries. Concentrations of *S. carpocapsae* ranging from $0.5–2.9 \times 10^4$ nematodes/gallery applied to the bark surface resulted in 50–85% larval mortality. Percent mortality from applications made during moist weather conditions at ca. 21–23 °C was higher than when applications were made during drier and cooler weather; researchers hypothesized that more nematodes survived to disperse on bark and reach galleries under the moister and warmer weather conditions.

c Sesiidae

Entomopathogenic nematodes have also been successfully used for control of sesiids in diverse hardwood tree species (see Table 1). Among the hosts that have been investigated, 88.9% of western poplar clearwing moths (*Paranthrene robiniae*) were controlled by *S. carpocapsae* on heavily infested birch and poplar (Kaya and Lindegren, 1983). *S. carpocapsae* was also used against the dusky clearwing, *Paranthrene tabaniformis*, in poplars, applying nematodes with cotton swabs soaked with suspensions of 2×10^4 IJs to entrances of larval galleries to yield 97.5% larval mortality (Cavalcaselle and Deseö, 1984). *S. carpocapsae* was sprayed on the bark of lightly infested dogwood trees (*Cornus* spp.) to control dogwood borer (*Synanthedon scitula*), resulting in an 84.6% reduction in borer abundance (Davidson *et al.*, 1992). Percent control of the large red-belted clearwing, *Synanthedon culiciformis*, in alder reached 77–84% when 6.5 or 11.5×10^6 *S. carpocapsae* IJs were applied directly to bark. However, control increased to 86–93% when 1.8 or 3.6×10^4 IJs were applied directly to each gallery opening (Kaya and Brown, 1986). These researchers found that *S. carpocapsae* was more effective at controlling sycamore borer, *Synanthedon resplendens*, than *S. feltiae* and hypothesized it was more difficult for the larger *S. feltiae* to enter the smaller gallery openings of this host species. Results suggested that entomopathogenic nematodes were more effective against sesiid hosts with larvae living in moist heartwood habitats (*e.g., S. culiciformis*) compared with species with larvae living in drier bark galleries (*e.g., S. resplendens*) (Kaya and Brown, 1986; Kaya, 1988).

Several nematode species have been assayed against the banded ash clearwing moth, *Podosesia aureocincta*, and the lilac borer, *Podosesia syringae*, attacking ash (*Fraxinus* spp.) trees (Gill *et al.*, 1994; Smith-Fiola *et al.*, 1996). Nematode applications reduced numbers of living larvae when tree bark was thoroughly sprayed with water prior to application whereas control was meager when bark was not sprayed with water prior to nematode applications (Smith-Fiola *et al.*, 1996). The authors hypothesized the water spray improved nematode survival, allowing more dispersal into larval galleries.

d Persimmon bark borer

Eight strains of *Steinernema* were tested against the pyralid, *Euzophera batangensis*, in a coastal casuarina windbreak in Fujian, China. *S. carpocapsae* provided the best control at 72% larval mortality (Huang, 1995).

3 *Control with entomopathogenic fungi*

a Cossids and sesiids

Relatively few trials have been conducted using entomopathogenic fungi against wood-boring Lepidoptera. In one study in China, cossid larvae were targeted using a novel method to apply conidia so that these would reach larvae within the wood (Huang *et al.*, 1990). *B. bassiana* conidia were mixed with waste molasses and sweet potato starch to form a paste which was smeared into excretion holds of larvae of the cossids *Z. multistrigata* and *Squamura discipuncta (=Arbela discipuncta)* in casuarina trees. Resulting larval mortality was 93.6–96.8%. This paste retained conidial infectivity for 90 days at room temperature. In 1988–89, the paste was applied to 583.7 ha of forests at a cost of 0.128 yuan (U.S. $0.02) per ha, resulting in 88.5–98.4% larval mortality.

Beauveria bassiana was used against larvae of the clearwing *P. tabaniformis* in 3-year-old infested poplars in Poland. Conidial suspensions injected into larval galleries in July, yielded 94–96% control (Schnaiderowa and Swiezynska, 1977).

b Pyralids

Beauveria bassiana has been tested against mahogany shoot borers in diverse countries by spraying conidia on trees. When applied to a plantation of 10-month old red cedars (*Cederela odorata*) in Mexico once a month and once every 3 months, 71% mortality of *Hypsipyla. grandella* larvae was reported (Sanchez-Monsalvo and Velasquez-Estrada, 1998). In Uttar Pradesh, India, 80% mortality of *Hypsipyla robusta* larvae in shoots of young red cedars (*Toonia ciliata*) was recorded (Misra, 1993). In Cuba, *B. bassiana* and *M. anisopliae* have been tested in nurseries and plantations at 4 kg of suspended spore powder/ha. In the plantation, 40.7% and 39.6% infection was achieved for *B. bassiana* and *M. anisopliae*, respectively. Also in Cuba, a mixture of *B. bassiana*, chemical insecticides during the first year after planting and silvicultural treatments are recommended to control *H. grandella* in mahogany plantations (Casanova *et al.*, 2001).

When *B. bassiana* conidia were applied in water and diesel oil against persimmon bark borer, *E. batangensis*, in coastal windbreaks of casuarina in Fujian, China, 86.1–100% control was reported (Huang, 1995).

4 *Control with entomopathogenic bacteria*

a. Cossids and sesiids. As many isolates of *Bt* provide effective microbial control of lepidopteran larvae, *Bt* has been investigated for control of cossid and sesiid larvae. *Bt* was injected into phloem-cambial mines and galleries of the cossid *P. robiniae* and the sesiid *Paranthrene simulans* but was ineffective even at high rates (Solomon, 1985).

b. Mahogany shoot borer. Bacillus thuringiensis was evaluated for control of *H. grandella* in one- and two-year old mahogany plantations in Honduras. In these trials, *Bt* that was sprayed on entire trees weekly provided better control than untreated trees (Goulet *et al.*, 2005). *Bt* was also tested against *H. grandella* in Mexico at different frequencies. When applied every month, 91% of larvae died, while mortality dropped to 67% when Bt was applied every 3 months (Sanchez-Monsalvo and Velasquez-Estrada, 1998).

3 Application and evaluation of entomopathogens for control of Coleoptera

A *Case Study: Fungal control of* Anoplophora glabripennis: *use of fungal bands*

1 *Preparation of inoculum*

This application method targets adult *A. glabripennis* that contaminate themselves when walking on tree trunks and branches. Fungi are applied as non-woven fiber bands impregnated with fungal cultures that are wrapped around trees. For these studies, the methodology for growing bands of *B. brongniartii* for control of *A. glabripennis* was amended from the method developed by Nitto Denko (Osaka, Japan) for control of orchard pests (Higuchi *et al.*, 1997). We could not confirm that *B. brongniartii* is native to North America (Hajek, unpublished data). Therefore, registration of this fungal species by the U.S. Environmental Protection Agency (EPA) would probably be lengthy, if this was even eventually possible. Bioassays demonstrated that *M. anisopliae* F-52, a strain already registered with the EPA for control of other pest species and marketed by Earth Biosciences (New Haven, Connecticut), was virulent against *A. glabripennis* adults. Therefore, this strain was chosen for further development.

To produce fungal bands, fungal cells are grown in liquid media on shakers (Dubois *et al.*, 2004a). Once cell number is maximal, additional media and molten agar are added. Quickly, pieces of non-woven fiber material are soaked in media and then laid on racks at 100% RH until surfaces of the bands are covered with conidia (approximately 7–10 days depending on the media composition and temperature). It is important that after sporulation, bands are allowed to dry slowly under high humidity, usually for several days. Bands are 5 cm wide and usually 50 cm long. The material for bands that is used in Japan is created from wood pulp and this is advantageous because bands are then biodegradable and do not have to be removed from trees. We have been unable to find comparable biodegradable material in the U.S. and therefore have substituted non-woven polyester-based quilt batting (*e.g.*, Soft and Bright, The Warm Co., Seattle, WA).

2 *Experimental design*

Evaluating the efficacy of fungal band applications is difficult because *A. glabripennis* is a long-lived beetle that is cryptic as a larva, living within wood. Adults are not easily seen in tree canopies. Adults preferentially feed on the bark of small twigs, often high in tree canopies, and they are not attracted to lights. Long distance pheromones or host attractants for quantifying or collecting adults have not been identified (Hajek, 2007). Adults lay eggs under the bark in branches as small as 3 cm in diameter but also in tree trunks. During field studies, many adults were found at > 3.5 m high in trees that were ca. 8 m tall, with fewer adults found at 2 m or lower on trunks (Hajek *et al.*, 2006). Thus, infested wood is often high in trees, also making sampling more difficult.

Ovipositing females chew shallow niches or pits in the bark, which they use to lay eggs slightly under the bark. Early instars feed in the phloem-cambium directly beneath the bark while later instars tunnel into the xylem. Until larvae reach later instars that created holes in the bark for discarding frass from galleries, there is little evidence that larvae are present in trees (oviposition niches are very small). Thus, sampling larvae requires cutting down trees and carefully dissecting the wood to collect living or dead larvae. It is not always possible to fell trees but even if this is possible, it can be difficult to split large diameter wood without damaging larvae under the bark and within the sapwood. Thus, to evaluate the effects of fungal bands, we have used the methods described below (Hajek *et al.*, 2006; Dubois *et al.*, 2004a, b).

Because *A. glabripennis* is being eradicated in the U.S. and populations are very low, field trials have all been conducted in China. Most importantly, stands of trees with abundant *A. glabripennis* populations must be located for conducting studies. Experienced professionals locate infested sites during winter; experts are needed to ensure that the borers in trees are indeed the species of interest. To do this, a few sample trees are cut down, dissected, cerambycid larvae are collected from within trees, and identified to species. It is critical to locate sites with adequate densities of *A. glabripennis*, which can be estimated from numbers of larvae within trees that were felled.

At appropriate sites where we have conducted studies, groups of 40 trees adjacent to each other were chosen as plots within larger plantings of trees. For our most basic studies, 3–5 replicate plots where fungal bands were hung on tree trunks were established along with 3–5 control plots (*e.g.*, Hajek *et al.*, 2006). We have tried to separate replicate plots from each other by at least 50–100 m to try to eliminate, or at least minimize, movement of adults and fungal conidia among plots; although *A. glabripennis* adults can fly, they are reluctant fliers and it is not thought that they regularly disperse very far in the presence of abundant host trees (Huang, 1991; Huang and Zhou, 1992).

3 Fungal application

Fungal bands are hung around tree trunks or branches, usually attaching them with nails or staples. In our studies using 15–19 cm diameter poplars and willows in China, we attached bands at a height of 2.0–2.5 m. It is critical that fungal bands not be hung in the sun or conidia will quickly die. Therefore, band height should be adapted to make sure bands are shaded. All studies in China have been conducted with one band/tree, as is recommended for use of Biolisa Kamikiri in Japanese orchards (Higuchi *et al.*, 1997). These guidelines from Japan were created for small citrus and fig trees and perhaps more bands/tree should be applied for larger shade trees in the U.S. However, we hypothesize that addition of attractants to bands, as is our eventual goal, would reduce the numbers of fungal bands that are needed per tree.

4 Timing

Adults of *A. glabripennis* emerge from trees asynchronously over several months although the peak numbers of adults at field sites in Anhui Province in 2002 occurred during late June throughout July to early August (Hajek *et al.*, 2006). For our studies, bands were hung early in July when many adults had only recently emerged. Our collections of adults continued over 37–42 days, ending 19–20 August.

5 Evaluation

Following, we will outline the different procedures required to evaluate fungal band treatments:

1. Collect adults (5 per plot) before bands are hung as a pre-application sample of natural levels of fungal infection. We have collected these large adults by climbing trees or by using long bamboo poles to knock beetles out of trees. Adults are reared individually in plastic cups between 20 and 25 °C, providing fresh twigs from host trees every 5 days as food. After collection, beetles are monitored daily for 40 days. It is important to open these pre-treatment rearing cups as seldom as possible to prevent potential cross-contamination.
2. After bands are hung, adult beetles are collected every 5 days and reared to detect fungal infections. During 2000 and 2002, we established a maximum of 5 beetles for each collection per plot, but often were unable to find this maximum number. This limit was established to ensure that we were collecting a small enough proportion of the population on each sampling date so that productive sampling could continue through that season (populations of these beetles are often not sufficiently abundant that large numbers of beetles could be removed from the population at regular intervals throughout the season) without depleting the population before the end of the season. For the first 24 hours after collection, a wet cotton ball is placed in each adult rearing cup to allow high humidity so that any *M. anisopliae* conidia potentially on the beetle cuticle will be able to germinate. Rearing procedures are as described above. Any beetles dying are maintained under moist conditions and fungal outgrowth from cadavers is used to identify the species of entomopathogenic fungus killing the beetle.
3. The background population densities of adult beetles can be monitored by regularly walking transects through the replicate plots, using binoculars to record locations and numbers of adult beetles. During studies conducted in 2006, approximately 43% of the trees in each plot were scanned by experienced researchers every 5 days to count adults, which were on tree trunks and in tree canopies.
4. To evaluate the effects of fungal band treatments on oviposition, we used ladders or climbed trees to count oviposition niches and emergence holes. To make sure that adult beetles in plots are disturbed as little as possible, we created a subset of trees where adults were collected and a subset of trees where oviposition and emergence data were taken. In this way, we were not counting oviposition and emergence on the same trees where we were hoping to collect adults to determine whether they were infected. It is important that this sampling is done before bands are hung to count oviposition niches and emergence holes from previous years as baseline data. After bands are hung, trees are ascended to quantify oviposition and emergence every 5 days. To estimate the numbers of adult females that are active in a plot, we divided the number of emergence holes in all trees in that plot

by two (because the sex ratio is approximately 1:1). When using ladders and climbing trees, for each replicate plot, the number of new oviposition niches is divided by the number of females in that plot to derive the number of eggs laid per female.

5. During 2006, we wanted to evaluate entire trees (when climbing trees or using ladders, we could not evaluate smaller diameter wood) so we arranged with tree owners and cut down a subsample of trees from each treatment plot approximately 20 days after bands were hung. Trees were then carefully dissected to count the numbers of oviposition niches and emergence holes as well as numbers and locations of larvae in the wood. Eggs were maintained under moist conditions on moist filter paper until hatch or for 10 days. Larvae are placed in cups with artificial diet (Dubois *et al.*, 2002) and monitored daily for 30 days. Necropsies were performed on all unhatched eggs and larval or pupal cadavers to determine the cause of death. This type of sampling would be enhanced if trees were felled for quantification before bands were hung as well as several times after bands are hung. However, permission to cut trees and cost of trees can limit this type of sampling.

6 *Persistence studies*

To evaluate persistence of activity of conidia on bands, bands were placed at approximately 3 m in the shade, around trunks of *A. glabripennis* host trees (predominantly *Acer* spp.) in Queens, New York. We chose this location because it is a location where *A. glabripennis* has been present in the field. At 1–4 week intervals for up to 3 months after band placement on trees, three randomly chosen fungal bands are removed from trees and brought to the laboratory. If bands were wet from rain they could not be processed immediately, and were dried briefly at high humidity before transport. Densities of conidia on bands were then quantified by blending several 5 cm^2 pieces of bands, filtering through cheesecloth and counting conidia with a hemocytometer. Germination of conidia from bands is quantified by spreading conidia on water agar for 24 hours. The density of germinated conidia per cm^2 of band is plotted by time to evaluate the persistence of living conidia on fungal bands.

B *Case study: fungal control of* Agrilus planipennis: *pre-emergent trunk sprays*

1 *Preparation of inoculum*

BotaniGard ES (BioWorks, Inc., Fairport, NY), a petroleum-based conidial suspension of *B. bassiana* GHA was suspended in water within label rates prior to field application.

2 *Experimental design*

Initially, BotaniGard trunk sprays were evaluated for efficacy against emerging *A. planipennis* adults in a 20-year old plantation of green ash trees (*Fraxinus pennsylvanica*) moderately infested with 50 to 100 larvae/m^2 of bark. The trees, spaced 2 m apart, ranged in diameter from 9 to 14 cm and in height from 8 to 10 m. A completely randomized block design was used to compare the efficacy of the three treatments (two fungal concentrations and the control) with five trees/treatment. In early June, prior to emergence of *A. planipennis* adults, BotaniGard was applied to 180-cm trunk sections, which were then caged with aluminum screening (Brite-Kote™, 18 × 16 mesh, Phifer Wire Products, Inc, Tuscaloosa, AL) to contain emerging *A. planipennis* adults. Untreated, caged trees were used as controls (Figure 4).

In the same ash plantation the following year, a field trial targeting both adult and immature *A. planipennis* was conducted by spraying "to drip" the canopies and trunks of ca. 180 infested green ash trees (now heavily infested with > 100 larvae/m^2) with BotaniGard. Trees were sprayed individually every two weeks from late June to early August within the treatment plot; trees in the control plot were not sprayed. The condition of each tree crown was estimated in early June the year of treatment as well as one year after treatment. Crown condition was based on standardized observations of ash crown die-back observed after *A. planipennis* attack, using a scale of 0 to 100%. Trees with low, medium, and high infestation levels exhibit crown die-back of 0–24%, 25–50%, and > 50%, respectively. Due to the high variability in *A. planipennis* attack rates between individual trees, 50 sample trees (25 treatment and 25 control trees) were selected by crown die-back category

Figure 4. Screen cages trapped *Agrilus planipennis* emerging from infested green ash trees, *Fraxinus pennsylvanica*, following a pre-emergent spray of *Beauveria bassiana* strain GHA in late June. After emergence was complete, cadavers of *A. planipennis* adults were evaluated for fungal infection in the laboratory (photo taken by Houping Liu)

to include 30 with low infestation, 10 with medium, and 10 with high infestation. The 50 samples trees were felled at the end of the study to quantify larval infestation levels and adult emergence. Measures of treatment efficacy included a) crown condition before and after treatment; b) *A. planipennis* larval density and fungal infection prevalence; and 3) numbers of *A. planipennis* adults emerging the following year.

3 Fungal application

For the initial study, the trunk treatment targeting *A. planipennis* adults was sprayed with a professional 11.4-L sprayer equipped with a flat fan nozzle and pressure gauge for delivery of a known quantity of *B. bassiana* GHA conidial suspension to the bark surface of each tree. Fungal suspensions were applied at 35 psi (300 kPa) to the north and south sides of each trunk at two application rates (1×10^{14} and 1×10^{15} conidia/ha); control trees were untreated. An average of 166 ml (5×10^{7} conidia/ml) and 169 ml (5×10^{8} conidia/ml) were sprayed on each tree to achieve the two treatment application rates. Before treatment, the surface area was calculated for each trunk section and used to calibrate the amount of fungal suspension needed for each tree at the designated application rate.

The following year, tree canopy and trunk applications targeting all *A. planipennis* life stages required a truck-mounted hydraulic sprayer to accommodate the volume needed to spray 8–11 liters of BotaniGard suspension on each tree to achieve the rate of approximately 3×10^{14} conidia/ha.

4 Timing

For the initial study, trunk treatments targeting adults, sprays were conducted several days before *A. planipennis* adults were expected to emerge (26 June, 2003); adult emergence was estimated by periodic dissection of trees in the plantation and removal of pupae to determine their age. Adults were removed from the screen cages 45 days after spraying on 11 August. At this time adult emergence was complete. Adults were evaluated for mortality and fungal infection prevalence.

The following year, tree canopy and trunk applications targeting all life stages were sprayed every two weeks starting 23 June and ending 3 August 2004, for a total of 4 treatments. Crown conditions were evaluated 31 May and 14 June 2005. Sample trees were dissected for immature stages between December 2004 and March 2005, while adult emergence from these trees was monitored during summer 2005 from logs cut and held in the laboratory. In addition, canopy condition was monitored in summer 2005.

5 Evaluation

For the initial study, trunk sprays targeting emerging adults were evaluated following adult

Figure 5. Adult *Agrilus planipennis* killed by *Beauveria bassiana* strain GHA while emerging from an ash tree trunk after pre-emergent trunk spray with BotaniGard ES. It is hypothesized that adults became infected after eclosion, while chewing out of trees. Adults are approximately 1.3 cm long (photo taken by Houping Liu)

emergence in the field and death within cages, ca. 6 weeks later. When screen cages were removed all *A. planipennis* adults were collected, including those on the bark, inside bark crevices, and in exit holes on treated trunk section (Figure 5). Tools used for handling insects were surface sterilized with 70% alcohol between uses. *A. planipennis* adults were placed in individual wells of sterile 24-well plastic plates under saturated humidity conditions to assess cause of death. Mycosis was confirmed after two weeks if a white conidial bloom appeared on the cadaver.

To evaluate the canopy and trunk applications, 25 treatment and 25 control trees from each plot were felled with a chainsaw. The main trunk was cut into 100-cm log sections from the base up to wood that was 2 cm in diameter. Each 100-cm section was cut again so that 30-cm log-sections could be transported to the laboratory for dissection and removal of insects. Live and dead *A. planipennis* larvae, prepupae, pupae, and adults were collected from each trunk-section and assessed for fungal infection as described above. The remaining 70-cm log sections from 15 treatment and 15 control trees were incubated for adult emergence in individual cardboard rearing tubes (20–30 cm in diameter, 80 cm in length) for 8 weeks at room temperature. Data on each *A. planipennis* life stage were collected and recorded from each log. Crown conditions of the remaining trees in both the treated and control plot were reassessed one year after the initial assessment.

4 Acknowledgments

We thank Mitsuaki Shimazu and Houping Liu for sharing information, providing images, and translations, and Zengzhi Li for sharing unpublished information. Ted Edwards, Kathie Hodge, Richard Humber, Ryan Shanley, Patricia Stock and Paul Weston all assisted with improving accuracy of this chapter. Rodney Poland assisted with images. This chapter was written, in part, with support from the Alphawood and Milstein-Litwin Foundations.

5 References

Alfazairy, A. A. 1986. The pathogenicity of *Bacillus thuringiensis*, isolated from the *Casuarina* stem borer *Stromatium fulvum* (Coleoptera: Cerambycidae), for larvae of 12 species of mosquitoes. *Insect Sci. Appl.* **7**, 633–36.

Bauer, L. S., Liu, H-P, and Miller, D. L. 2004. Microbial control of the emerald ash borer. *In* "Proceedings of the 2003 Emerald Ash Borer Research and Technology Development Meeting" (Mastro, V. and

Reardon, R., Eds), pp. 31–32. USDA Forest Service, FHTET-2004-02.

Bauer, L. S., Dean, D. H., and Handelsman, J. 2006. *Bacillus thuringiensis*: potential for management of emerald ash borer. *In* "Proceedings of the 2005 Emerald Ash Borer Research and Technology Development Meeting" (Mastro, V. and Reardon, R., Eds.), pp. 38–39. USDA Forest Service,.FHTET-2005-16.

Bedding, R. A. and Iede, E. T. 2005. Application of *Beddingia siricidicola* for *Sirex* woodwasp control. *In* "Nematodes as Biological Control Agents" (Grewal, P. S., Ehlers, R.-U. and Shapiro-Ilan, D. I., Eds.), pp. 385–399. CABI Publ., Wallingford, UK.

Beegle, C. C. 1996. Efficacy of *Bacillus thuringiensis* against lesser grain borer, *Rhyzopertha dominica* (Coleoptera: Bostrichidae). *Biocontrol Sci. Tech.* **6**,15–21.

Cane, J. H., Cox, H. E., and Moar, W. J. 1995. Susceptibility of *Ips calligraphus* (Germar) and *Dendroctonus frontalis* Zimmerman (Coleoptera: Scolytidae) to coleopteran active *Bacillus thuringiensis* and Avermectin B I. *Can. Entomol.* **127**, 831–837.

Casanova, A. D., Torres, J. M. M., Smith, M. De. C. B., Barroso, J. R. M., and Rito, A. A. 2001. Integrated management of *Hypsipyla grandella* in nurseries and plantations of Meliaceae in Cuba. *In* "*Hypsipyla* Shoot Borers in Meliaceae: Proceedings of an International Workshop" (Floyd, R. B., and Hauxwell, C., Eds.), pp. 175–178. Austral. Centre Internatl. Agric. Res., Canberra, Australia.

Cavalcaselle, B. 1975. Possibility of using products based on *Beauveria bassiana* against the larvae of some wood-eating insects. *Meded. Facult. Landbouw. Rijksuniv. Gent* **40**, 437–442.

Cavalcaselle, B. and Deseö, K. V. 1984. Prove di lotta contro le larve di due insetti xilofagi del pioppo con nematodi entomopatogeni. *Att. Gior. Fitopatol.* **2**, 393–402 [in Italian; BIOSIS English abstract]

Chen, J., Dai, L-Y., Wang, X-P., Tian, Y-C., and Lu, M-Z. 2005. The *cry3Aa* gene of *Bacillus thuringiensis* encodes a toxin against long-horned beetles. *Appl. Microbiol. Biotechnol.* **76**, 351–356.

Chornesky, E. A., Bartuska, A. M., Aplet, G. H., Britton, K. O., Cummings-Carlson, J., Davis, F. W., Eskow, J., Gordon, D. R., Gottschalk, K. W., Haack, R. A., Hansen, A. J., Mack, R. N., Rahel, F. J., Shannon, M. A., Wainger, L. A., and Wigley, T. B. 2005. Science priorities for reducing the threat of invasive species to sustainable forestry. *BioScience* **335**, 335–348.

Davidson, J. A., Gill, S. A., and Raupp, M. J. 1992. Controlling clearwing moths with entomopathogenic nematodes: The dogwood borer case study. *J. Arboricult.* **18**, 81–84.

Dubois, T., Hajek, A. E., and Smith, S. 2002. Methods for rearing the Asian longhorned beetle, *Anoplophora glabripennis* (Coleoptera: Cerambycidae) on artificial diet. *Ann. Entomol. Soc. Am.* **95**, 223–230.

Dubois, T., Hajek, A. E., Hu, J-F., and Li, Z-Z. 2004a. Evaluating the efficacy of entomopathogenic fungi against the Asian longhorned beetle, *Anoplophora glabripennis* (Coleoptera: Cerambycidae), by using cages in the field. *Environ. Entomol.* **33**, 62–74.

Dubois, T., Li, Z-Z., Hu, J-F., and Hajek, A. E. 2004b. Efficacy of fiber bands impregnated with *Beauveria brongniartii* cultures against Asian longhorned beetle, *Anoplophora glabripennis* (Coleoptera: Cerambycidae). *Biol. Control* **31**, 320–328.

Fallon, D. J., Solter, L. F., Keena, M., McManus, M., Cate, J. R., and Hanks, L. M. 2004. Susceptibility of Asian longhorned beetle, *Anoplophora glabripennis* (Motschulsky) (Coleoptera: Cerambycidae), to entomopathogenic nematodes. *Biol. Control* **30**, 430–438.

Finney, J. R. 1977. The DD-136 strain of *Neoaplectana* sp. as a potential biological control agent for the European elm bark beetle, *Scolytus scolytus*. *J. Invertebr. Pathol.* **29**, 7–9.

Finney, J. R. and Walker, C. 1979. Assessment of a field trial using the DD-136 strain of *Neoaplectana* sp. for the control of *Scolytus scolytus*. *J. Invertebr. Pathol.* **33**, 239–241.

Forschler, B. T. and Nordin, G. L. 1988. Suppression of carpenterworm, *Prionoxystus robiniae* (Lepidoptera: Cossidae), with the entomopathogenic nematodes, *Steinernema feltiae* and *S. bibionis*. *J. Kans. Entomol. Soc.* **61**, 396–400.

Forschler, B. T. and Nordin, G. L. 1989. Impact of the cottonwood borer, *Plectrodera scalator* (Coleoptera: Cerambycidae), in a commercial cottonwood nursery. *J. Entomol. Sci.* **24**, 186–190.

Fuxa, J. R., Ayyappath, R., and Goyer, R. A. 1998. "Pathogens and Microbial Control of North American Forest Insect Pests." USDA, For. Serv. FHTET-97-27.

Gill, S., Davidson, J., MacLachlan, W., and Potts, W. 1994. Controlling banded ash clearwing moth borer using entomopathogenic nematodes. *J. Arboricult.* **20**, 146–149.

Goulet, E., Rueda, A., and Shelton, A. 2005. Management of mahogany shoot borer, *Hypsipyla grandella* (Zeller) (Lepidoptera: Pyralidae), through weed management and insecticidal sprays in one- and two-year-old *Swietenia humilis* Zucc. plantations. *Crop Protection* **24**, 821–828.

Haack, R. A. 2006. Exotic bark- and wood-boring Coleoptera in the United States: Recent establishments and interceptions. *Can. J. For. Res.* **36**, 269–288.

Haack, R. A., Law, K. R., Mastro, V. C., Ossenbruggen, H. S., and Raimo, B. J. 1997. New York's battle with the Asian longhorned beetle. *J. For.* **95**, 11–15.

Haack, R. A., Jendek, E., Liu, H-P., Marchant, K. R., Petrice, T. R., Poland, T. M., and Ye, H. 2002. The emerald ash borer: a new exotic pest in North America. *Newsl. Michigan Entomol. Soc.* **47**, 1–5.

Hajek, A. E. 2007. Asian longhorned beetle: Ecology and control. *In* (D. Pimentel, Ed.) Encyclopedia of Pest Management, Marcel Dekker, NY. (in press).

Hajek, A. E., McManus, M. L., and I. Delalibera, Jr., I. 2005. "Catalogue of Introductions of Pathogens and Nematodes for Classical Biological Control of Insects and Mites." USDA, For. Serv. FHTET-2005-05. 59 pp.

Hajek, A. E., Huang, B., Dubois, T., Smith, M. T., and Li, Z. 2006. Field studies of control of *Anoplophora glabripennis* (Coleoptera: Cerambycidae) using fiber bands containing cultures of the entomopathogenic fungi *Metarhizium anisopliae* and *Beauveria brongniartii*. *Biocontrol Sci. Technol.* **16**, 329–343.

Higuchi, T., Saika, T., Senda, S., Mizobata, T., Kawata, Y., and Nagai, J. 1997. Development of biorational pest control formation against longicorn beetles using a fungus, *Beauveria brongniartii* (Sacc.) Petch. *J. Ferment. Bioengin.* **84**, 236–243.

Huang, J. 1991. Current status and problems in the control of poplar wood-boring insect pests. *J. For. Dis. Insect Pests* **1**, 52–56.

Huang, J. S. 1995. Study on the occurrence and integrated management of *Euzophera batangensis*. *Sci. Silv. Sinicae* **31**, 421–427.

Huang, J. L. and Zhou, Z. Y. 1992. New progress in *Anoplophora glabripennis* Motch. study in China. *Shaanxi For. Dis. Insect Pests* **1**, 52–56.

Huang, J. S., He, Y. L., and Lin, Q. Y. 1990. Use of a new paste preparation of *Beauveria bassiana* in the forest to control *Zeuzera multistrigata* (Lep.: Cossidae). *Chin. J. Biol. Control* **6**, 121–123.

Jassim, H. K., Foster, H. A., and Fairhurst, C. P. 1990. Biological control of Dutch elm disease: *Bacillus thuringiensis* as a potential control agent for *Scolytus scolytus* and *S. multistriatus*. *J. Appl. Bacteriol.* **69**, 563–568.

Kaya, H. K. 1988. Princes from todes. *Am. Nurseryman* **167**, 63–69.

Kaya, H. K. and Brown, L. R. 1986. Field application of entomogenous nematodes for biological control of clear-wing moth borers in alder and sycamore trees. *J. Arboricult.* **12**, 150–154.

Kaya, H. K. and Lindegren, J. E. 1983. Parasitic nematode controls western poplar clearwing moth. *Calif. Agric.* **37 (3 & 4)**, 31–32.

Kaya, H. K., Aguillera, M. M., Alumai, A., Choo, H. Y., de la Torre, M., Fodor, A., Ganguly, S., Hazir, S., Lakatos, T., Pye, A., Wilson, M., Yamanaka, S., Yang, H., and Ehlers, R. U. 2006. Status of entomopathogenic nematodes and their symbiotic bacteria from selected countries or regions of the world. *Biol. Control* **38**, 134–155.

Kreutz, J., Zimmermann, G., and Vaupel, O. 2004. Horizontal transmission of the entomopathogenic fungus *Beauveria bassiana* among the spruce bark beetle, *Ips typographus* (Col., Scolytidae), in the laboratory and under field conditions. *Biocontrol Sci. Technol.* **14**, 837–848.

Liu, H-P. and Bauer, L. S. 2006a. Potential of *Beaveria bassiana* strain GHA for management of the emerald ash borer. *In* "Proceedings of the 2005 Emerald Ash Borer Research and Technology Development Meeting" (Mastro, V. and Reardon, R.), pp. 42–43. USDA Forest Service, FHTET-2005-16.

Liu, H-P. and Bauer, L. S. 2006b. Susceptibility of *Agrilus planipennis* (Coleoptera: Buprestidae) to *Beauveria bassiana* and *Metarhizium anisopliae*. *J. Econ. Entomol.* **99**, 1096–1103.

Liu, H. X., Chang, Q., and Ma, F. 1998. Field experiment of *Anoplophora glabripennis* and *Melanophila decastigma* control with pathogenic nematode *Steinernema carpocapsae*. *Ningxia J. Agric. For. Sci. Technol.* **5**, 15–17.

Méndez-López, I., Basurto-Ríos, R., and Ibarra, J. 2003. *Bacillus thuringiensis* serovar *israelensis* is highly toxic to the coffee berry borer, *Hypothenemus hampei* Ferr. (Coleoptera: Scolytidae). *FEMS Microbiol. Lett.* **11131**, 1–5.

Misra, R. M. 1993. *Beauveria bassiana* (Balsamo) Vuillemin, a fungal pathogen of *Hypsipyla robusta* Moore (Lepidoptera: Pyralidae). *Indian J. For.* **16**, 236–238.

Okitsu, M., Kisii, Y., and Takagi, Y. 2000. Control of adults of *Monochamus alternatus* Hope (Coleoptera: Cerambycidae) by application of non-woven fabric strips containing *Beauveria bassiana* (Deuteromycotina: Hyphomycetes) on infested tree trunks. *J. Jpn. For. Soc.* **32**, 276–280.

Poland, T. M. and McCullough, D. G. 2006. Emerald ash borer: Invasion of the urban forest and the threat to North America's ash resource. *J. Forestry* **104**, 118–124.

Poland, T. M., Haack, R. A., Petrice, T. B., Miller, D. L., Bauer, L. S., and Gao, R. 2006. Field evaluations of systemic insecticides for control of *Anoplophora glabripennis* (Coleoptera: Cerambycidae) in China. *J. Econ. Entomol.* **99**, 383–392.

Qin, X., Kao, R., Yang, H., and Zhang, G. 1988. Study on application of entomopathogenic nematodes *Steinernema bibionis* and *S. feltiae* to control *Anoplophora glabripennis* and *Holcocerus insularis*. *Forest Res.* **1**, 179–185.

Sanchez-Monsalvo, V. and Velasquez-Estrada, C. 1998. Evaluacion de dos insecticidas biologicos en el control de *Hypsipyla grandella* (Zeller), barrenador de brotes de las meliaceas. *Ciencia For. Mex.* **23**, 33–39.

Schnaiderowa, J. and Swiezynska, H. 1977. Proby zwalczania przeziernika topolowca (*Paranthrene tabaniformis* Rott. - Aegeriidae, Lep.) owadobojczym grzybem *Beauveria bassiana* (Bals.) Vuill. *Sylwan* **121**, 23–29. [in Polish: BIOSIS English abstract]

Shimazu, M. and Kushida, T. 1980. Microbial control of *Monochamus alternatus*—Treatment of pathogens on the infested pine trees. *Trans. 32nd Mtg. Kanto Branch Jpn. For. Soc.*, pp. 93–94 [in Japanese].

Shimazu, M. and Sato, H. 2003. Effects of larval age on mortality of *Monochamus alternatus* Hope (Coleoptera: Cerambycidae) after application of nonwoven fabric strips with *Beauveria bassiana. Appl. Entomol. Zool.* **38**, 1–5.

Shimazu, M., Kushida, T., and Katagiri, K. 1982. Microbial control of *Monochamus alternatus*—Spray of pathogens onto the infested pine trees just before adult emergence. *Trans. 93rd Mtg. Jpn. For. Soc.*, pp. 399–400 [in Japanese]

Shimazu, M., Kushida, T., and Katagiri, K. 1983. Microbial control of *Monochamus alternatus*—Spraying of pathogens during maturation feeding. *Trans. 94th Mtg. Jpn. For. Soc.*, pp. 485–486 [in Japanese]

Shimazu, M., Kushida, T., Tsuchiya, D., and Mitsuhashi, W. 1992. Microbial control of *Monochamus alternatus* Hope (Coleoptera: Cerambycidae) by implanting wheat-bran pellets with *Beauveria bassiana* in infested tree trunks. *J. Jpn. For. Soc.* **74**, 325–330.

Shimazu, M., Tsuchiya, D., Sato, H., and Kushida, T. 1995. Microbial control of *Monochamus alternatus* Hope (Coleoptera: Cerambycidae) by application of nonwoven fabric strips with *Beauveria bassiana* (Deuteromycotina: Hyphomycetes) on infested tree trunks. *Appl. Entomol. Zool.* **30**, 207–213.

Shimazu, M., Sato, H., and Maehara, N. 1999. Microbial control of *Monochamus alternatus* using *Beauveria bassiana. Proc. Internatl. Symp. Sustainability of Pine Forests in Relation to Pine Wilt and Decline, Tokyo, 27–28 October, 1998*, pp. 200–204.

Smith-Fiola, D. C., Gill, S. A., and Way, R. G. 1996. Evaluation of entomopathogenic nematodes as biological control against the banded ash clearwing borer. *J. Environ. Hort.* **14**, 67–71.

Solomon, J. D. 1985. Comparative effectiveness of gallery-injected insecticides and fumigants to control carpenterworms (Lepidoptera: Cossidae) and oak clearwing borers (Lepidoptera: Sesiidae). *J. Econ. Entomol.* **78**, 485–488.

Solomon, J. D. 1995. Guide to insect borers of North American broadleaf trees and shrubs. Agric. Handbk. 706. Washington, DC: U.S. Dept. Agric., Forest Service. 735 p.

Torr, P., Wilson, M. J., and Heritage, S. 2005. Forestry applications. *In* "Nematodes as Biological Control Agents" (Grewal, P. S., Ehlers, R.-U., and Shapiro-Ilan, D. I., Eds.), pp. 281–293. CABI Publ., Wallingford, UK.

Tsutsumi, T. 1998. Microbial control of longicorn beetles using non-woven fabric sheet containing *Beauveria brongniartii* in the orchard. *Proc. Soc. Invertebr. Pathol., Sapporo, Japan*, pp. 203–209.

USDA Forest Service NCRS. 2006. Emerald ash borer maps. http://ncrs.fs.fed.us/4501/eab/maps/

Weathersbee III, A. A., Tang, Y. Q., Doostdar, H., and Mayer, R. T. 2002. Susceptibility of *Diaprepes abbreviatus* (Coleoptera: Curculionidae) to a commercial preparation of *Bacillus thuringiensis* subsp. *tenebrionis. Fla. Entomol.* **85**, 330–335

Yamanaka, S. 1993. Field control of the Japanese pine sawyer, *Monochamus alternatus* (Coleoptera: Cerambycidae), larvae by *Steinernema carpocapsae* (Nematoda: Rhabditida). *Jpn. J. Nematol.* **23**, 71–78.

Yang, H., Zhang, S., Zhang, G., Huang, J., Zheng, H. 1990. Control of the leopard moth, *Zeuzera multistrigata leuconota* (Lep.: Cossidae) on Casuarina trees by plugging sponge carrying *Steinernema feltiae. Chin. J. Biol. Control* **6**, 157–160.

Yang, H., Zhang, G., Zhang, S., and Jian, H. 1993. Biological control of tree borers (Lepidoptera: Cossidae) in China with the nematode *Steinernema carpocapsae. In* "Nematodes and the Biological Control of Insect Pests" (R. Bedding, R. Akhurst, and H. Kaya, Eds.), pp. 33–40. CSIRO, East Melbourne, Australia.

Chapter VII-11

Microbial control of lepidopteran pests of apple orchards

Lawrence A. Lacey, Steven P. Arthurs
and Alan L. Knight

USDA-ARS
Yakima Agricultural Research Laboratory
Wapato, WA 98951, USA

Jürg Huber

BBA, Institut für Biologischen Pflanzenschutz
64287 Darmstadt, Germany

1 Introduction

A broad diversity of lepidopteran pest species inhabit apple orchards throughout the world, feeding on leaves, fruit, and vascular systems of trees. The major families, important genera or species and common names of tree fruit insect pests in North America are listed in Table 1. The economic damage inflicted on apple by this group is significant and affects production both by direct feeding damage on the fruit and reduction in fruit quality from indirect feeding on the tree. Species in the family Tortricidae comprise the most serious pests of apple. In most major tree fruit growing regions, there is a suite of 2 to 10 tortricid species that feed on both the foliage and fruit of apple (Croft and Hull, 1991). Many of these species are minor pests and rarely feed on the fruit or occur only in non-commercial orchards. Other species are key pests and are targeted by pest managers with one or more insecticidal sprays (Croft and Bode, 1983; Beers *et al.*, 1993).

The most devastating pest of apple on a global scale is the codling moth, *Cydia pomonella* (Barnes, 1991). Neonate larvae bore into the fruit and remain there throughout their feeding stages. When larvae are full grown, they leave the fruit in search of cryptic habitats, such as rough bark, in which to spin their cocoons and pupate. There are 1–4 generations per year depending on climatic conditions (Audemard, 1991). In the fall of the year, diapausing 5th instar larvae overwinter within the cocoon (hibernaculum). The two stages most amenable to microbial control are neonates and cocooned 5th instars. Neonates provide a very narrow window for treatment with microbials since they bore into fruit within hours of eclosion. Mature larvae, particularly overwintering larvae, provide a much broader window of opportunity for control, but do not feed making pathogens with *per os* entry unusable.

After codling moth, leafrollers and other species of Lepidoptera are pests of apple production to varying degrees. More than 20 leafroller species (Tortricidae) are considered as regular apple pests in Europe alone (Blommers, 1994). As the name implies, leafrollers are usually found feeding within leaves they have rolled together with silk. Although they are predominantly defoliators, they may also feed on the surface of fruit.

Control of insect pests of orchards for the most part has been accomplished or attempted using broad spectrum conventional insecticides, such as azinphos methyl (Guthion®). A number of health and environmental concerns have encouraged more environmentally friendly integrated pest management (IPM) approaches

L.A. Lacey and H.K. Kaya (eds.), Field Manual of Techniques in Invertebrate Pathology, 527–546.

Table 1. Major lepidopteran pests of apple in North America

Family	Common name and major genera or species
Sesiidae	Tree borers, *Synanthedon* spp.
Arctiidae	Fall webworm, *Hyphantria cunea*
Geometridae	Winter moth, *Operophtera brumata*
Gracillaridae	Tentiform leafminers, *Phyllonorycter* spp.
Lasiocampidae	Tent caterpillars, *Malacosoma americanum*
Noctuidae	Various species fruitworms, including *Lacanobia subjuncta* and *Orthosia hibisci*
Tortricidae	Codling moth, *Cydia pomonella*
	Oriental fruit moth, *Grapholita* (= *Cydia*) *molesta*
	Lesser appleworm, *Grapholita prunivora*
	Eyespotted budmoth, *Spilonota ocellana*
	Tufted apple budmoth, *Platynota idaeusalis*
	Leafrollers, major genera: *Acleris, Argyrotaenia, Archips, Clepsis, Choristoneura, Pandemis, Sparganothis*
Yponomeutidae	Apple ermine moth, *Yponomeuta malinellus*

in which microbial control and other soft technologies are central (Lacey and Shapiro-Ilan, 2003; Lacey and Unruh, 2005). The adoption by commercial orchardists of the codling moth female sex pheromone to disrupt mating has enabled effective control of the moth with reduced use of conventional insecticides (Vickers and Rothschild, 1991; Calkins and Faust, 2003). This strategy works best when codling moth population density is low. Supplemental codling moth control with selective interventions, such as microbial control agents (MCAs), that have little or no effect on nontargets, would complement mating disruption and help conserve natural control factors. One consequence of mating disruption has been the increase of some leafroller species that are normally controlled by insecticides used for codling moth (Gut and Brunner, 1998). Soft approaches for their control, such as the use of MCAs, would conserve insect natural enemies with no deleterious effects on the environment.

With the impending restriction on the use of conventional chemical insecticides for control of orchard pests (Calkins and Faust, 2003), microbials and other selective interventions will be needed to fill the void. The use of microbial insecticides in apple will depend on their efficacy, as well as the availability and economics of alternative tactics. At present, few microbial insecticides are used in apple. The major exceptions are the bacterial pathogen *Bacillus thuringiensis* (*Bt*) for tortricid leafrollers, and granuloviruses (GV) for control of codling moth (*Cp*GV) and in Europe, the summer fruit tortrix moth, *Adoxophyes orana* (*Ao*GV).

2 Entomopathogens of tortricid pests of apple

Representatives of each major pathogen group have been reported from lepidopteran orchard pests. An overview of viruses, bacteria, nematodes, and fungi (including Microsporidia) found in or used against tortricid pests of apple orchards is presented below. A review of the development of several entomopathogens for use against insect pests of apple and pear in Europe and North America is presented by Cross *et al.* (1999), Lacey and Shapiro-Ilan (2003) and Lacey and Unruh (2005).

A Viruses

Several types of virus are found in Lepidoptera. Those with the most potential as MCAs are in the family Baculoviridae [granuloviruses (GV) and nucleopolyhedroviruses (NPV)]. Transmission of Baculoviruses is usually *per os* (*i.e.*, the virus must be ingested).

The only viral MCA used for control of neonate codling moth larvae is the granulovirus of *C. pomonella* (*Cp*GV). Its specificity for codling moth and some closely related species and safety to nontarget organisms are well documented and its use contributes to the conservation of other natural enemies in the orchard agroecosystem. It is one of the most virulent baculoviruses; the LD_{50} for neonate larvae was determined at 1.2 virus granules/larva (Huber, 1986). Falcon *et al.* (1968) first successfully demonstrated its potential for codling moth control. Over the past 30 years, numerous field trials have demonstrated good activity against codling moth in a variety of settings across Europe, South Africa, Australia, New Zealand, South America, and North America (Huber, 1986; Jaques, 1990; Falcon and Huber, 1991; Guillon and Biache, 1995; Cross *et al.*, 1999; Lacey *et al.*, 2004; Arthurs *et al.*, 2005).

Despite its effectiveness in controlling codling moth, some of the concerns expressed by growers regarding *Cp*GV are: the increased number of shallow stings to the fruit, short residual activity of the virus due to ultraviolet (UV) radiation, the need for multiple applications, slow speed of kill, lower efficacy against high density codling moth populations and expense of the virus (Glen and Clark, 1985; Jaques *et al.*, 1987). A range of spray adjuvants has been tested with *Cp*GV with the goal of improving virus uptake and/or increasing persistence on the surface of foliage or fruit. Substances such as molasses, sucrose, skimmed milk powder and oxybenzone have been reported to improve *Cp*GV persistence slightly, although the rates used are considered high for routine field use (Keller, 1973; Charmillot *et al.*, 1998; Ballard *et al.*, 2000). More recently a non-commercial microencapsulated formulation of *Cp*GV has been shown to provide extended UV protection (Arthurs *et al.*, 2006). *Cp*GV is commercially available in Europe marketed as Madex, Granupom, Carpovirusine and Virgo and more recently in North America, marketed as Carpovirusine, Cyd-X and Virosoft.

Although *Cp*GV is efficacious against the vast majority of codling moth populations, Fritsch *et al.* (2005) and Sauphanor *et al.* (2006) reported development of resistance to the virus in Germany and France in certain populations that have received regular virus applications for several years. Sauphanor *et al.* (2006) noted that the resistance appears to be highly dominant and not related to resistance to chemical insecticides. However, management of resistance along the same lines as for chemical insecticides has been recommended to prevent its spread. Eberle and Jehle (2006) concluded that CpGV resistance is incompletely dominant and autosomally inherited.

There are several GV and NPV reported from leafrollers and other tortricid orchard pests (Zimmermann and Weiser, 1991). Viruses of apple pests include the NPV of *Epiphyas postvittana, Platynota idaeusalis* and *Choristoneura rosaceana*, NPV and GV of *Adoxophyes orana*, and GV of *Archips argyrospila*. The viruses of *A. orana* have been extensively tested in Europe and Japan. Using the NPV of *A. orana*, Flückiger (1982) and Blommers *et al.* (1987) obtained effective control in European apple orchards. The granulovirus of *A. orana* is produced in Europe and marketed under the name Capex. It has been credited with persistent control; however, two major problems associated with it are low tolerance to UV radiation and slow rate of kill; mortality is usually observed in mature larvae. Improvement of *Ao*GV activity through formulation is reported by Luisier and Benz (1994). A variety of *Ao*GV that is effective in apple orchards in Japan has been reported by Shiga *et al.* (1973), Yamada and Oho (1973), Sekita *et al.* (1984), and others. The virus is safe for non-targets; however, virus-infected larvae are often not suitable for parasitoid development (Nakai and Kunimi, 1997). Most recently a GV of *Pandemis pyrusana* (*Pp*GV) was discovered in Washington State (Pfannenstiel *et al.*, 2004). Larval infection rates monitored for three years varied from 2.6 to 67%, suggesting *Pp*GV has the potential to regulate natural populations in a density dependent manner.

Most of the viruses of leafroller and other tortricids are very specific and compatible with integrated pest management relying on preserving natural enemies for other orchard pests. A disadvantage is that multiple virus types would be required to control a complex of pests. Given the development of virus within living cells increases the cost and complexity of production, this requirement may dissuade commercial production of viruses for non-major pests.

B Bacteria

Bacillus thuringiensis (*Bt*) is the most widely used biopesticide (Lacey *et al.*, 2001) and the only bacterium used for control of insect pests in orchards. The Cry 1 and Cry 2 endotoxins of *Bt* are active to varying degrees against nearly all lepidopteran species (Navon, 1993). The toxins must be ingested in order to be effective and the specific feeding habit of the various lepidopteran pests of apple is one major factor that affects which pests can be targeted with *Bt*. For example, species that feed inside of fruits such as codling moth, or within leaves such as tentiform leafminers, or under bark such as tree borers are not good candidates for control with *Bt* (Andermatt *et al.*, 1988). In contrast, insects that feed on the surface of foliage such as leafrollers, budmoths, and fruitworms can be readily controlled with *Bt* formulations (Jaques, 1965; De Reede *et al.*, 1985; Westigard *et al.*, 1986; Hardman and Gaul, 1990; Huber and Hassan, 1991; Knight, 1994). In 2003, 13% of the apple acreage in the USA was treated with at least one application of *Bt* (USDA, 2004). Use of *Bt* for apple tortricids has been most frequently adopted in certified-organic orchards (Knight, 1994), in conjunction with mating disruption for codling moth (Knight, 1995), or in orchards where resistance to organophosphate insecticides is a concern (Vakenti *et al.*, 1984).

A number of factors can affect the performance of *Bt* formulations against tortricid pests of apple. Differences in the susceptibility of leafrollers to a range of *Bt* Cry 1 endotoxins can affect the performance of the various commercial formulations (Knight *et al.*, 1998). Spray coverage and spray concentration are also two important factors affecting the efficacy of *Bt* (Sorenson and Falcon, 1980; Li and Fitzpatrick, 1996). The short residual life of *Bt* (Nyouki and Fuxa, 1994) in combination with rapid foliage growth in the spring and early summer limits efficacy and increases the need for multiple applications to successfully control pests (Brunner, 1994). Environmental factors such as rainfall and sunlight are major determinants of the residual life of *Bt* (Frye *et al.*, 1973). *Bt* activity is temperature dependent and susceptibility among larval instars often varies due to differences in feeding rates (Li *et al.*, 1995a, b). Avoidance of treated foliage and feeding inhibition following ingestion of *Bt* residues can reduce the effectiveness of control and cause a range of sublethal effects such as delay in larval growth (Harris *et al.*, 1997; Knight *et al.*, 1998). Optimal use of *Bt* products may require concentrated spray applications to ensure mortality of the target pest following consumption of one or a few droplets of formulation (Payne and Van Frankenhuyzen, 1995). The addition of feeding stimulants to enhance the activity of *Bt* has also been shown to be effective for leafrollers (Brunner, 1994; Li and Fitzpatrick, 1997; Pszczolkowski *et al.*, 2004). Finally, optimal use of *Bt* in orchards depends on having adequate knowledge of the phenology of both pest and natural enemy species present. Cossentine *et al.* (2003) reported how spring *Bt* treatments on apple may be timed to maximize the survival of parasitoids of *C. rosaceana* found in the southern interior of British Columbia, Canada. The availability of predictive phenology models (Knight and Croft, 1991) and reliable, low-cost sampling methods (Beers *et al.*, 1994) are additional prerequisites to using *Bt* effectively in orchards.

Spinosad, derived from the soil actinomycete, *Saccharopolysporea spinosa*, has also become widely used in apple orchards in recent years for control of Lepidoptera; used on 31% of apple acreage on the USA in 2003 (USDA, 2004). The rotation of *Bt* with spinosad was recommended as an insecticide resistance management strategy for leafrollers (Smirle *et al.*, 2003).

C Entomopathogenic nematodes

Cryptic habitats, such as those used by codling moth and other tortricids for their cocoons in overwintering sites (under loose bark, in litter at the base of trees, in nearby woodpiles, fruit bins and the like), are favorable environmental sites for entomopathogenic nematodes. Natural infections of cocooned codling moth larvae with nematodes in the families Steinernematidae and Heterorhabditidae have been reported in Europe and North

America when larvae are located near the base of trees and close to the soil (Weiser, 1955; Vega *et al.*, 2000; Lacey *et al.*, 2006a). Control of the overwintering stage targets the entire population, at a time when other interventions cannot be used, and may reduce or eliminate damage to fruit early in the following growing season. Previous studies have evaluated post harvest application of nematodes against cocooned overwintering codling moth in orchards (Kaya *et al.*, 1984; Nachtigall and Dickler, 1992; Unruh and Lacey, 2001; Lacey *et al.*, 2006a, 2006b) and fruit bins (Lacey and Chauvin, 1999; Cossentine *et al.*, 2002; Lacey *et al.*, 2005). In general, good rates of control have been achieved with *Steinernema carpocapsae* and *S. feltiae* under optimal conditions of temperature and moisture. Low moisture, resulting in rapid drying and death of infective juvenile (IJ) stages, and extremes in temperature are the major limiting factors for nematodes in orchards. Habitat modification, particularly the use of irrigation before and after treatment, application timing, formulation with antidesiccants and the use of strip mulches around tree bases are strategies that may enhance or extend the activity of IJs in orchards and other cryptic habitats. Selection and testing of cold active nematodes would also be beneficial for control of overwintering larvae in the late fall and early spring.

Control of other lepidopteran pests of apple and other tree fruit using entomopathogenic nematodes has also been reported. In China, for example, the peach fruit moth, *Carposina niponensis*, a major apple pest that pupates in the soil, was controlled by *S. carpocapsae* (Wang, 1993). Population reduction of the overwintering larvae in the soil resulted in the subsequent protection of the fruit. More recently, however, *S. carpocapsae* has not been used against this insect because better quality apples can be obtained by wrapping the fruit (Kaya *et al.*, 2006). A number of other examples include wood boring insects and others with soil stages that are most effectively controlled by *Steinernema* and *Heterorhabditis* spp. (Jaques *et al.*, 1968; Deseö *et al.*, 1984a, b; Deseö and Miller, 1985; Cossentine *et al.*, 1990; Kahounova and Mráček, 1991; Jie *et al.*, 1994; Parvizi, 2003). Application of *S. carpocapsae* to foliage, however, results in low efficacy due to rapid desiccation of IJs (Jaques, 1967).

D Fungi

A number of naturally occurring fungi are reported from lepidopteran pests of apple (Zimmermann and Weiser, 1991). The most prevalent fungi attacking tortricids are in the genera: *Paecilomyces, Beauveria, Metarhizium, Hirsutella, Lecanicillium* (= *Verticillium*) and *Entomophthora*. The fungus most frequently isolated is *Beauveria bassiana* (Jaques and MacLellan, 1965; Subinprasert, 1987; Falcon and Huber, 1991). Falcon and Huber (1991) and Lacey and Unruh (2005) summarized published research on *B. bassiana* as an MCA of codling moth. Application of the fungus on tree trunks for control of overwintering larvae has resulted in negligible mortality in some cases and in those studies with the highest level of control, up to 50% mortality was observed (Ferron and Vincent, 1978). However, Garcia-Gutierrez *et al.* (2004) reported effective control of neonate codling moth larvae using a strain native to Mexico and two commercial formulations of *B. bassiana* applied at 1.2×10^{12} conidia/ha. However, fruit damage was significantly higher for the *B. bassiana* treatments than for azinphos methyl treatments.

The Microsporidia (formerly included with the Protozoa) have recently been placed in the Kingdom Fungi. Microsporidian diseases of insects are ubiquitous and comprise an important regulatory role in insect populations (Chapter IV-3). They require a living host in which to develop and are generally host specific and slow acting, most often producing chronic infections. Microsporidia are commonly reported in tortricids (Zimmermann and Weiser, 1991). Although decimating epizootics caused by Microsporidia are rarely observed, their function in regulation of insect populations may be quite significant (Maddox, 1987; Brooks, 1988). However, Falcon and Huber (1991) state that the Microsporidia have not been observed to be an important factor in natural mortality of codling moth. The microsporidium, *Nosema carpocapsae*, can be found in some laboratory colonies and field collections of codling moth (Malone and Wigley, 1981; Siegel *et al.*, 2001). No attempt to manipulate this or other Microsporidia for control of lepidopteran orchard pests has been reported.

3 Efficacy assessment of entomopathogens against codling moth and leafroller larvae

Protocols for the assessment of efficacy of the wide variety of entomopathogens of tortricid pests of fruit trees could fill a small volume. Space limitations necessitate a more selective presentation. We will provide protocols for the evaluation of a virus and nematodes against codling moth and for *Bt* against leafrollers. Comparable feeding strategies of several pests of apple will help to facilitate use of similar efficacy assessment protocols. A number of variables will determine the exact protocol that will be needed for evaluation of entomopathogens of tortricids in the orchard agroecosystem. These include application rate, formulation effects, equipment and application methods, treatment interval and number and timing of applications, climatic conditions and biotic factors. The instructions and information presented below will provide a flexible framework for the application and evaluation of MCAs against codling moth and other lepidopteran pests of apple.

A Efficacy assessment of granulovirus against codling moth larvae

1 Preparation of inoculum

Insecticidal *Cp*GV formulations are available through several commercial producers in Europe: Carpovirusine (Natural Plant Protection, France, distributed in the USA by Arvesta Corp.), Granupom (PROBIS GmbH, Germany), Madex (Andermatt Biocontrol, Switzerland), Virgo (Sipcam, Italy), and in North America, Virosoft CP4 (BioTepp Inc., Quebec, Canada) and Cyd-X (Advan, USA). Protocols for *in vivo* production methods in *C. pomonella* larvae are presented by Glen and Payne (1984). An average production of 9×10^9 virus capsules/larva was reported by Glen and Payne (1984). Purification and laboratory quality assessment are also presented by Glen and Payne (1984) and Laing and Jaques (1980). In order to optimize coverage and persistence of virus produced in the laboratory, spreaders, stickers and UV screens can be added to virus suspensions. Examples of these are presented by Keller (1973), Glen and Payne (1984), Wearing (1993) and Ballard *et al.* (2000). Care should be given to the quality of the water used for spraying. Water from the public water supply system is usually chlorinated. This can have an adverse effect on the activity of the virus. The pH of the water should be in the range of 5 to 9 and free of silt or other particles that may block nozzles.

2 Experimental design

Plot sizes can range from individual trees, to orchards of one or more hectares. Plot size will depend on the purpose of the study. Small plots are best used to compare different experimental treatments (such as dosage or use of formulation additives) where localized conditions and pest population pressure will be most easily standardized. Single tree plots are normally replicated 10 times. Larger experimental plots are better able to quantify the seasonal population reduction in a localized area by reducing contamination of moths migrating in from adjacent untreated areas, but often at the expense of number of experimental treatments that can be compared. Depending on the experimental treatments, plots can be arranged in completely randomized blocks or factorial design to account for variation found within the orchard (see Chapter II-1). Experimental design also depends on the spraying equipment used in the trials (see below). With regard to untreated controls, it should be noted that the dose-response curve for *Cp*GV is extremely flat when compared to chemical insecticides. Therefore, care has to be taken to avoid spray drift into untreated blocks since even very low virus contamination result in a noticeable effect (a 100-fold dilution of the normal concentration of *Cp*GV, for instance, can still provide 30–50% reduction of codling moth). When choosing the location of untreated plots it is, therefore, recommended to also take the direction of prevailing winds into consideration. Moveable screens can help protect the untreated plots during spraying (Figure 1).

3 Spraying equipment

*Cp*GV can be applied with a variety of sprayers ranging from small compressed air or CO_2 pressurized models to handgun orchard

Figure 1. Application of codling moth granulovirus using a backpack airblast sprayer. Photo by Heather Headrick

sprayers, to air-blast sprayers (Figure 2) (see Chapter III-1). Tractor-mounted air-blast sprayers, also called speed sprayers, are the equipment of choice for orchardists. Our experience shows commercial air-blast sprayers are suited to treat larger blocks of 0.2 ha or more while backpack airblast (Figure 1) or hydraulic handgun sprayers can treat single tree plots with less associated spray drift (Lacey *et al.*, 2004; Arthurs *et al.*, 2005). Screens and buffer rows can also be employed to minimize drift onto nearby trees. We have also used a small diesel-powdered air-blast sprayer towed with an all terrain vehicle (ATV) to treat 16 tree blocks.

4 Rate and timing of applications

a Rate

Rates of *Cp*GV ranging from 10^9 to 10^{14} granules/ha have been utilized for testing. In Canadian trials summarized by Jaques (1990), 7.5×10^{12} to 5×10^{13} capsules/ha were evaluated. Operational rates of 2.1×10^{12} to 1.3×10^{13} granules/ha are recommended for commercial products in North America. The dilution, *i.e.* volume application rate of water in which the virus is applied, is another important consideration. Amounts of virus from 7×10^9 to 7×10^{10} capsules/liter were applied by Glen and Payne (1984) in 2.25 liters of water/tree (ca. 900 liters/ha). Huber and Dickler (1977) applied 1.5 liters of water/tree. Best results are achieved by spraying the trees until run off. Even when the same amount of virus/ha was used, low, or ultra low volume sprays did not result in the same level of control as spraying until run off (Payne *et al.*, 1984). The addition of kaolin clay or similar inert compounds which are commercially available and form a physical barrier on drying, can be quickly and easily used to assess spray coverage in the canopy.

b Timing of applications

Insect viruses have to be ingested by larvae to be effective. Because codling moth larvae may feed little on the exposed surface of apples or leaves, virus sprays should be timed to coincide with hatching of neonate larvae, which ingest virus granules before or during entry into fruit. In the Pacific Northwest, a temperature-based phenology model is used to predict egg hatch and timing of sprays (Beers *et al.*, 1993). Wing-type (delta) monitoring traps baited with pheromone (codlemone) lures that

Figure 2. Tractor-pulled airblast sprayer configured for application of codling moth granulovirus. Photo by Steve Arthurs

are commercially available should be placed in the orchard prior to moth flight (approx. full bloom) to determine 'biofix' (first consistent moth catch) and to monitor seasonal flight patterns. Traps should be placed in the upper third of the canopy in areas subject to infestations, such as adjacent to abandoned orchards or bin piles, or where codling moth damage was observed the previous season. Lures and sticky trap liners should be changed every 2–3 weeks. Shade temperature measured within the orchard is used to calculate degree days throughout the season. Initial virus applications are made at 250 degree days post male biofix (approx. 5% hatch) and continued until approximately 90% of hatch. Subsequent larval generations can be treated in a similar manner, starting approx 1200 degree days in the Pacific Northwest. Timing of treatments can be confirmed through visual inspections of damaged fruit or egg surveys in the orchard. In some cases, the phenology model described by Beers *et al.*, (1993) may be less accurate in orchards with mating disruption or with pesticide resistant populations, due to delayed mating or development respectively. Recently, a chemical derived from pear fruit, pear ester, ethyl (E, Z)-2,4-decadienoate, has been shown to be attractive to both male and female codling moths. Monitoring female moths directly with pear ester may allow growers to more accurately time female emergence in the season and the occurrence of egg laying by females and subsequent egg hatch (Knight and Light, 2005).

c *Frequency of applications*

In commercial operations, virus applications are made at 7–21 day intervals during egg hatch periods. The total number of virus applications required in orchards will depend on the localized pest pressure. For example, while 6–8 treatments per generation may be required in high pressure areas, reduced input strategies may still be acceptable if they prevent outbreaks in low pressure areas such as orchard interiors (Jaques *et al.*, 1994; Arthurs *et al.*, 2005). Additional virus applications may be required in more southerly latitudes where a significant third codling moth generation is found, while fewer may be required in short growing seasons of northern Europe or Canada. For example, in Nova Scotia where there is only one codling moth generation per year, Jaques *et al.* (1994) reported that only two applications of *Cp*GV were usually needed. In many organic programs, acceptable control will be a level at which a mating disruption (MD) program continues to be effective. In Switzerland, Charmillot and

Pasquier (2002) reported that a combination of MD and *Cp*GV was successful, but took several years to bring codling moth populations back to a very low level, when the virus was applied at 10-day intervals. Field trials of *Cp*GV have shown that weekly spraying using only 1/10 of the normal concentration provides the same, or even better efficacy as spraying at two week intervals with the normal concentration (Dickler and Huber, 1988). The expense of frequent application of *Cp*GV can be reduced by tank mixing with other materials, such as fungicides. Tank mixing the virus with *Bt* enables simultaneous treatment of codling moth and leafroller larvae.

5 *Assessment of efficacy*

a *Fruit damage*

Most growers applying fast acting chemical pesticides determine their effectiveness by assessing damage to fruit. Because *Cp*GV is slower acting, some damage can be expected following treatment because infected larvae may survive long enough to cause shallow entries or 'stings' in sprayed fruit. This may be a problem in cases of high pest pressure, but economic damage from low populations may be controlled through routine manual thinning practices (Arthurs and Lacey, 2004; Arthurs *et al.*, 2005). Fruit with shallow stings may also be suitable for processing (juice, canning, etc.). Fruit damage is normally assessed mid season (following 1st flight) and again prior to harvest (after the second flight). Fruit damage may also be assessed 7–14 days following a single application. Sample sizes for fruit range from 30–100 fruits/tree in single tree plots up to hundreds of fruits from large plots. The number and condition of infested fruit that has fallen prior to harvest also provide an indication of codling moth activity (Huber and Dickler, 1977; Glen and Payne, 1984). Because virus is more effective at population suppression than preventing damage, accurate assessments require that fruit be dissected to determine mortality of larvae within. Alternatively, tree bands can be used to monitor the surviving larvae emerging from fruit (see below).

b *Numbers of larvae*

Monitoring of the larval population outside of fruit is normally accomplished using corrugated cardboard bands (B flute, 2–9 cm wide) at the end of the growing season, starting at approx. 2000 degree days. Bands placed around the trunks of sprayed trees are utilized by full grown larvae as sites in which to spin cocoons and are later removed and examined for cocooning codling moth. This approach can also be used to monitor parasitism by some larval parasitoids in the fall or spring. There may be 1–4 generations of codling moth during the year depending on latitude and weather (Audemard, 1991). In mid-summer in areas with two or more generations, non-diapausing larvae will spin cocoons and pupate. The pupal period for non-diapausing generations ranges from 10 to 21 days (30–20 °C, respectively). Larvae that spin cocoons in the fall will overwinter as prepupae and do not pupate until the following spring. The density of the diapausing population is the best indication of population suppression in orchards that have received applications of *Cp*GV or other microbials during the growing season. See Section 3B5 for more detail on banding.

c *Persistence of virus*

The main reason for lack of residual activity of the virus is inactivation by UV radiation. A 50% reduction in activity has been observed between 2 and 8 days following application, depending on the locations and time of season (Huber, 1980; Glen and Payne, 1984; Jaques *et al.*, 1987; Arthurs and Lacey, 2004; Lacey *et al.*, 2004). To assess the persistence of virus in the orchard, vegetation or fruit are collected at various intervals following application of *Cp*GV and exposed to neonate larvae in laboratory bioassays. This approach may be used to evaluate adjuvants providing possible UV protection. Leaf or fruit samples should be removed immediately after spraying and at 1–3 day intervals for the week following application. Subsequent samples can be taken less frequently until activity declines below 25% mortality in neonate larvae. Sample sizes ranged from 30 to 50 leaves/fruit per sample date. Using fruit for bioassay comes closest to the conditions in the field and is preferred for testing because larvae placed on the fruit do not need to be removed until the end of the test. Fruits are placed in plastic cups at 25 °C and larval mortality assessed after 7–10 days. Larvae on leaf discs samples should be removed to artificial diet after a defined feeding period, but additional handling involved may increase mortality. An alternative method for assessing virus persistence in the laboratory using a solar simulator has also been developed (Lacey and Arthurs, 2005).

Because of the shallow dose response curve, the decline in larvicidal activity does not proportionally reflect virus inactivation. Glen and Payne (1984) describe a technique to measure inactivation of *Cp*GV. Virus washed off the leaves of sprayed trees (at various intervals post spraying) is incorporated into an artificial diet and used for bioassay. The amount of infectious virus remaining is calculated by comparing the sample mortality against a standard dilution series of known concentration. Using this approach in one trial, Glen and Payne (1984) showed that although *Cp*GV infectivity was reduced by half in 3 days, some activity persisted 4–8 weeks after spraying.

6 Protocol for operational evaluation of CpGV *in small plots*

1. Select treatments and variables to be included in the study. Randomly allocate experimental blocks according to orchard conditions and allow buffer zones to prevent contamination between treatments and controls; see 3A2.
2. Monitor environmental conditions such as shade temperature, rainfall, irrigation, solar radiation for the duration of the test.
3. If possible, time treatments based on orchard observations and a phenology model. The number of applications and time interval between treatments will depend on the number of codling moth generations, population density and the desired level of control, see 3A*4*.
4. Thoroughly suspend *Cp*GV formulation in sufficient water to apply 1–2 liters/tree. If application volume is one of the treatment variables being evaluated, adjust the volume of water accordingly; see 3A*1*.
5. Use a compressed air or CO_2 powered sprayer and evenly apply the suspension to individual trees to the point of run-off. If an airblast sprayer is used, plots must large enough to minimize drift into control plots or those receiving different treatments; see 3A*3*.
6. Fruit damage can be assessed at regular intervals following treatment and shortly before harvest. Dissections of fruit should be performed to confirm the extent of damage and that damage was caused by codling moth larvae and to see if larvae are dead; see 3A*5*.
7. Assess the surviving larval population by banding trees in early fall when the overwintering generation start exiting fruit. This is most effective in larger plots. Overwintering larvae found on treated trees in single tree plots may not necessarily have originated from that tree. Larvae in bands may be attacked by birds necessitating their protection by covering the bands with aluminum flashing or the like; see 3A*5*.

B Use of entomopathogenic nematodes for codling moth control

1 Preparation of inoculum

IJs of entomopathogenic nematodes are available from several commercial producers (see Chapter IV-5). For small-scale trials, small to moderate amounts of IJs may be produced *in vivo*; laboratory protocols for this are presented by Kaya and Stock (1997). *In vivo* production is usually accomplished by infecting wax moth larvae (*Galleria mellonella*) and collecting emerging IJs. Artificial media are also available, but require careful management of the nematode's symbiotic bacteria in order to ensure good larvicidal activity (Kaya and Stock, 1997).

2 Experimental design

A randomized block design is well suited for experiments with entomopathogenic nematodes in the orchard environment. In our tests, a minimum of 6 replicated blocks are utilized. The individual replicate may involve one or more trees depending on application method. Hand applied nematode suspensions can be accurately applied to individual trees with minimal contamination of adjacent trees.

3 Equipment

With modifications, the same range of equipment used for application of virus can be used for application of entomopathogenic nematodes. The object will be to apply the IJs to the trunk and scaffold branches of the tree, leaf litter or mulch strip (depending on location of larvae), not to foliage. Application equipment and tractor speed will be configured for increased droplet size, increased volume of water and reduced friction. For example, if an airblast sprayer is used for application, screens and swirl plates should be removed, pressure reduced to 690 kPa (100 psi) or lower, and nozzles changed to ones that permit coarser droplets (*e.g.*, D10 cone nozzles). Tractor speed should permit a delivery rate of 3740 liters/ha (400 gal./acre) with the top 2–3 nozzles per side turned off. Compressed air or CO_2 powered backpack sprayers equipped with flat fan nozzles can be used for treatment of individual trees.

4 Rate and timing of applications

An application rate of 10^6 IJs of *S. carpocapsae*/tree can provide up to 100% mortality if temperatures are 20–30 °C and IJs are kept moist (Kaya *et al.*, 1984; Lacey and Unruh, 1998; Lacey *et al.*, 2006a). Some variables that will affect dosage are species of nematode, temperature, size and variety of tree and type of surface to be sprayed (tree trunk, artificial substrate, leaf litter, etc.). In addition to dosage, the volume of water in which the IJs are applied can be important. One or more liters of water plus wetting agent/tree may be necessary to bring IJs into contact with codling moth larvae and prevent them from drying before penetration of the host. Pre- and post-treatment wetting of treated trees is necessary to optimize the survival and efficacy of applied IJs.

The target population of codling moth for entomopathogenic nematodes is the overwintering cocooned larvae. In practice, post harvest applications will ensure that all larvae have excited the fruit and are ensconced in their hibernacula, thus constituting a captive audience. The efficacy of *S. carpocapsae* declines below 20 °C, while *S. feltiae* is active at temperatures as low as 10 °C (Lacey *et al.*, 2006a). Heterorhabditids may also provide control under cooler conditions, but require further evaluation. Therefore, application should be made when temperatures are favorable for the IJs. Nematode applications can also be made during warm spring days before emergence of the first generation moths. However, because codling moth pupae are significantly less susceptible than larvae (Lacey *et al.*, 2005), spring applications are best made before larvae pupate.

5 Pre- and post-treatment sampling

The codling moth stages that are targeted for control using entomopathogenic nematodes are cocooned prepupae. Because mature larvae seek cryptic habitats for spinning their cocoons, natural populations are very difficult to monitor, especially when population density is low. To overcome this limitation in our tests, we use diapausing sentinel larvae from laboratory colonies that can be attached to trees (using infested substrates) or other habitats immediately before application of nematodes. Sentinels can be removed following treatment, incubated in the laboratory, and then assessed for larval mortality (see timing and conditions below). To prepare sentinels, perforated cardboard strips (double faced, B flute) may be placed directly over infested rearing media or infested apples to intercept larvae in search of sites in which to spin cocoons. Alternatively, strips may be individually infested by placing them in plastic bags or other containers and adding the desired number of mature larvae. Leaving the containers in well-lighted conditions will encourage the larvae to enter the flutes or crevices. The use of diapausing larvae allows cold storage of the strips for long periods of time and prevents rapid pupation. Singh and Ashby (1986) present instructions for inducing diapause in lab-reared codling moth larvae.

Banding trees with cardboard strips is often employed to monitor overwintering larvae. Bands are placed in the orchard before larvae start exiting fruit. Bands of cardboard and some other materials that are placed around tree trunks are readily utilized as hibernacula by larvae crawling down from the fruit bearing branches of the tree or up from the ground. The cardboard type and width used may vary. In our studies we use two single faced, B flute, cardboard (Weyerhauser, Tacoma, WA) 2-cm wide strips secured to trees with staples. The strips are placed back to back such that the corrugations interlock and the smooth sides face outward and against the trunk. When the bands will be in place for several weeks before treatments are made and/or where predation by birds and other natural enemies is high, they should be protected by covering them with screen, aluminum flashing or another protective material. Because the surface of the cardboard is an impenetrable barrier to nematodes, perforations along the strips (3–5 rows of 3–5 perforations [0.5 mm diam.]/cm) before infestation with larvae will facilitate passage of the IJs. A tailor's pattern marker or sewing machine are ideal for making perforations 0.5 mm in diameter.

In addition to placement on trees, cardboard strips can also be placed in other habitats likely to be used by larvae for hibernacula. These may include leaf litter, cracks in soil, prop piles or fruit bins. Protocols for treatment of fruit bins with nematodes for control of cocooned codling moth larvae are presented by Lacey and Chauvin (1999), Cossentine *et al.* (2002) and Lacey *et al.* (2005).

Another use of sentinels under more natural conditions is by allowing access of lab-reared mature larvae to apple or pear logs. Sawing grooves (10 cm long, 2–3 mm wide, 5–10 mm deep) in apple logs with a table saw creates sites that are readily utilized by larvae in which to spin cocoons. The logs can be attached to the trees with wire, elastic bands, twine, etc. just before application of nematodes. Older pear logs with scaly bark can be used without the necessity of sawing grooves. Larvae will readily utilize the natural sites under the bark for spinning cocoons.

In our studies we remove sentinel logs and cardboard strips from the orchard 24–48 hours after treatment and incubate them under conditions that will allow field-infected individuals to show patent signs of infection (25°C for 4–6 days). This prevents excessive mortality in the field due to predators and allows rapid differentiation of nematode-killed larvae versus larvae that have died of other causes. Nematodes die when the substrates become dry, so exposure of sentinel prepupae to IJs ends before logs are removed from the orchard.

6 *Treatment protocol using sentinel prepupae*

The exact protocol used will be determined by the specific objectives of the experiment. Treatment variables could include: comparison of application methods (comparison of application equipment, volume of water/ha, etc.), nematode species (or strains), environmental conditions, formulations, and methods for keeping IJs moist until penetration of hosts takes place. The following generic protocol can be modified to accommodate evaluation of any or all of the above variables.

1. Select trees to be treated and assign treatment variables. Lay out plots in randomized block design. A minimum of 6 replicated blocks should be utilized. The individual replicate may involve one or more trees depending on application method. For example, in applications made with an airblast sprayer, at least three adjacent trees should be treated with sampling from the middle tree. Depending on the type of application method, a buffer zone between treatments and control trees may be required to prevent contamination of adjacent treated/control trees.
2. Prior to application of nematodes, place sentinel prepupae in cardboard strips and/or logs in the orchard. Exact placement will depend on the objectives of the test. Under optimal conditions the sentinel substrates should receive good coverage with nematode suspensions.
3. Nematode viability and quality control should be assessed shortly prior to spraying. One method involves applying nematode suspensions on cocooned larvae in cardboard strips in Petri dishes as described by Lacey and Unruh (1998). Counting of viable nematodes required for application and determination of infectivity should be done just prior to spraying. See Kaya and Stock (1997) for details of dilution counting. Methods for determination of infectivity are presented by Kaya and Stock (1997) using wax moth larvae and by Lacey and Unruh (1998) using codling moth larvae.
4. Thoroughly mix the concentrated suspension of nematodes with non-chlorinated water to provide the desired concentration of IJs just prior to application. Calibration of spray equipment to

determine the time, pressure, nozzle sizes, etc. required to apply the desired volume/concentration of IJs should be done in advance to avoid delays once nematodes are placed in the tank of the sprayer.

5. For optimal effect, trees should be wetted just prior to application of IJs. Over-tree irrigation that provides thorough wetting of the tree, or water applied with an air blast sprayer (as configured in Section 3B*3*.), or misting nozzles on hoses may be used.

6a. Using a backpack sprayer (flat fan nozzles, 275 kPa [40 psi]) apply nematodes evenly from scaffold branches to the trunk of the tree at the soil level using at least one liter of water/tree.

6b. Using an air blast sprayer (large orifice cone nozzles, $\leq$ 690 kPa), apply nematodes only to scaffold branches and trunk by using the bottom 3–4 nozzles on each side of the sprayer (*i.e.* closing the top 2–3 nozzles). Remove sieves and swirl plates from the sprayer prior to application. An application rate of 3740 liters/ha and the addition of a wetting agent (for example, Silwet L77®) to the tank will enhance penetration of larval habitats and cocoons.

6c. Protocols for evaluation of entomopathogenic nematodes for control of cocooned codling moth in mulch strips is presented by Lacey *et al.* (2006b).

7. Take samples of nematodes directly from the nozzles at one or more times during the application process. The viability and infectivity of samples should be assessed using procedures prescribed by Kaya and Stock (1997) and bioassay of Lacey and Unruh (1998). At the very least, IJ samples should be taken toward the end of applications.

8. Ideally, trees should be kept moist for 6 hour or longer following application of nematodes. In some cases this will be possible with existing irrigation. A mist rather than stream of water is preferred to limit wash-off of IJs from treated surfaces. In applications of IJs made to mulch strips, the amount of post treatment wetting can be as little as an hour.

9. Monitor environmental conditions such as temperature, rainfall, irrigation, solar radiation for the duration of the test. If possible, use devices that enable constant recording of environmental conditions.

10. Remove sentinels 24 hours after application of nematodes and incubate in the laboratory at 25 °C.

11. Assess mortality 4 or more days after application. Nematode-killed larvae will appear tan to creamy brown (steinernematids) or rose to red (heterorhabditids). Those suspected of dying from other causes should be dissected to confirm the absence of nematodes.

C *Evaluation of* Bt *for tortricid control in apple*

Various types of field studies can be conducted with *Bt* formulations to evaluate their performance against tortricid species. Field efficacy trials can be used to compare the following factors: different formulations, comparison with other insecticides, the effect of spreaders, stickers, protectants, or feeding stimulants, the residual life of the products, environmental or climatic factors, spraying methods (coverage, concentration or droplet size) or sublethal effects on larvae, pupae, and adults. Various experimental designs can be employed to complete these studies. The following section briefly covers some of these.

1 Plot design

Studies can be conducted in small replicated plots consisting of one or more trees/plot. The size of the plot should ensure that an adequate number of larvae are present. Untreated plots or trees sprayed with only water plus inert ingredients should be included in the study as a control. Plots treated with standard insecticides for comparison with *Bt* may also be included. Treatments should be laid out using a randomized complete block design. The study should include as many replicates of each treatment as feasible; however, studies with 5–10 replicates are usually sufficient to statistically compare treatments. Plots should be established in orchard(s) with a uniform, high-level of larval or egg infestation. It is also important to standardize tree size, pruning method, cultivar, etc. among plots. Make a map of the blocks showing locations and numbers of trees; and delimit the edges of plots with color flagging to avoid mistakes in spraying or sampling. Plots should be adequately separated to avoid contamination due to drift. In general, one can use a single untreated row or two trees within a row (> 10 m) to separate plots if applications are being made with a handgun sprayer. With air blast applications, plots should be separated by at least two untreated rows or five trees within a row (> 15 m). Spray applications should not be applied to wet foliage and irrigation that will wet the foliage during the test should be avoided.

2 Spray application

Plots can be sprayed with hydraulic or electrostatic air blast, backpack, or handgun sprayers. The type of sprayer and its operational characteristics will affect the number of droplets deposited/area, the spectrum of droplet sizes, and the coverage attained in the orchard. Thus, it is important to use the same sprayer for all treatments. Do not spray in windy conditions. All applications should be applied under similar environmental conditions. With air blast applications, turn off the outside nozzles when spraying the edge rows of each plot to minimize drift onto the other plots. Small plots being sprayed with a backpack or a handgun sprayer can be protected with the use of a tarp to minimize drift onto the adjacent trees (Figure 1). This method will allow the experimenter to minimize the distance between plots and thus more fully utilize the trees in the experimental orchard. Sprays with a handgun are applied until runoff. Nozzles can be selected to simulate applications of 150–600 liters/ha. Sprays applied in the spring are timed with the phenology of the plant, *i.e.*, pink, full bloom, or petal fall. Summer sprays are timed against young larvae based on a phenology model in conjunction with moth captures in a sex pheromone-baited trap ("biofix"). Record environmental conditions at the time of spray application (temperature and wind conditions, sunlight levels). Record temperature and sunlight levels daily during for the duration of the test.

3 Sampling

Tortricid leafrollers and budmoths overwinter either as eggs deposited in masses on the bark, second-third instars under bark scales, or as variable larval stages in leaf litter and annual weeds on the orchard floor. Tortricid larvae feeding on the tree tend to concentrate in the terminal leaves of new shoot growth. There are three time periods to use *Bt* to control leafrollers in apple: between pink and petal fall for the overwintering generation; July - August for the summer generation of multiple-voltine species; and in the fall prior to harvest for the generation destined to overwinter as larvae. Sampling of larvae is conducted just prior to and 7–10 days after the treatments are applied.

Sampling of larvae early in the season (green tip to early pink) requires careful dissections of the developing buds. Detection of larval feeding sites later in the season from pink through petal fall is easier. Sampling methods can include either timed visual counts or inspection of a specific number of terminals/tree or plot. In general, an adequate visual inspection for leafroller feeding sites is 1–2 minutes/tree. In small plots (1–5 trees), every tree can be inspected. In larger plots, a subsample of trees are sampled (*i.e.*, 25 trees/0.5 ha sampled for 1 minute each). Visual sampling on trees taller than 2 m requires that both the top and bottom of the canopy be sampled using a ladder. The alternative sampling method is to inspect 10–40 terminals/shoots/tree on each tree or up to 20 trees/plot for larger plot sizes. Sample terminals randomly. The terminals are collected and later inspected for leafrollers in the laboratory. This method is acceptable if the percentage of terminals sampled is small relative to the total number of terminals in each plot. In some cases, such as if the plot size is small or if the trees are small, terminals may be inspected for larvae without removal and disturbance of the larvae. However, because larvae may or may not be present in leaf shelters, an effort should be made to confirm their presence. When sampling tall trees, collect terminals from both low and high in the canopy. Place samples in paper not plastic bags. Roll the end of the bag over at least twice and staple to prevent larvae from crawling out during transport. Label each bag with the treatment and replicate number and date and keep samples in a cooled container (ice chest) during transport to the laboratory.

During the summer months, larvae are often concentrated in the terminal ends of vegetative shoot growth. In larger trees, sampling of these vertical shoots is made easier with a telescoping pruner tool. Either visual counts or a fixed number of terminals/tree can be sampled similar to spring sampling. Plots with smaller trees can be sampled by complete census of each tree. With large plots, the sampling can be subdivided into quadrants or border versus interior subplots.

4 Infesting orchards

Studies of the effect of *Bt* on leafrollers can be facilitated in some instances by seeding orchard plots with either field-collected or lab-reared insects. First, it is important to reduce the density of biological control agents such as spiders in the plots. The plots can be sprayed 1–2 weeks before they will be seeded with a short-lived, broad spectrum insecticide. Plots can then either be infested by releasing adults, or by placing egg masses, neonates, or small larvae on the foliage. Our experience has been that releasing virgin adults into the orchard is the easiest method, but also the least likely to be successful. Moths may emigrate out of the plots, the distribution of egg masses and subsequent larvae may be highly clumped, and failures can occur due to the lack of mating and egg laying by the released females. Sleeve cages with adults can be used to ensure that mating and egg laying will occur in designated areas of the plots. Seeding orchards with egg masses allows one to also control the distribution of larvae more carefully. Egg masses produced in a laboratory colony or collected from an infested field site can be clipped to the foliage just prior to hatch. Seeding plots with neonates is another useful method. Larvae are placed in terminals with a fine brush. However, with these methods the number of established larval shelters will be affected by the high level of mortality that occurs during the neonate stage. Instead, we have found that the best method to evaluate the effectiveness of *Bt* against larvae within established feeding sites has been to seed orchards with second and third instars. Larvae are reared on artificial diet in the laboratory and are transferred by brush individually to terminal shoots. This method is time consuming, but for small plots, it has allowed acceptable densities of larvae to be tested. Inherent differences in response to treatments between lab-reared and field populations, however, may limit these approaches.

5 Studies of Bt *persistence*

Field/lab studies can be conducted with field-treated foliage to compare the performance of different *Bt* formulations, spray rates, the addition of surfactants or feeding stimulants, etc. The potency of these field-aged residues can be determined by spraying trees and then collecting leaves at specific periods after application. Conduct the bioassay initially with leaves prior to spraying as a check and at intervals up to 14 days post-treatment. Record environmental conditions throughout the test. It is important to standardize the cultivar and age of the leaves sampled. It is also important to avoid having active residues of any other spray material on the leaves and to avoid using dusty leaves from the edges of orchard plots. The standard method is to collect foliage from the top and bottom of a specified number of trees in each replicate. Leaves are mixed together and a sufficient number of leaf disks (20–23 mm) are punched (1–2/leaf) for the bioassay. Handle leaves and leaf disks carefully to avoid removing any spray deposit. Two to four leaf disks are placed in each Petri plate depending on the size of the Petri plate or larval age. For example, one can either use the 50 × 9 mm Petri plates or the standard 100 × 15 mm Petri plates. These larger Petri plates typically need to be lined with filter paper moistened with distilled water. A third bioassay method is to use the entire leaf instead of punching leaf disks. The leaf petiole can be placed into a small closed vial containing either distilled water or a floret solution. The entire leaf is then placed inside of a large plastic or paper cup. It is important to place Petri plates or cups inside of larger containers that maintain the RH above 50%. Bioassays are typically conducted at 20–25 °C. These types of bioassays can be conducted with either neonates or older larvae. Tests with older larvae should standardize their age and weight. Leaf disk assays are run for only 4–7 days due to the deterioration of leaf tissue. Assays using entire leaves can usually be run for longer periods (7–14 days) or until the leaves in the untreated checks are entirely consumed. After the exposure to field-treated leaves, larvae can be transferred to clean leaves or artificial diet to measure any delayed mortality or sublethal effects.

6 Studies of sublethal effects

Prior to the test, determine the age of a subsample of larvae in the orchard plots by either measuring the width of their head capsules or by weight. Seven to 10 days after the spray application, sample larvae again from each plot/treatment. In the laboratory, larvae can be reared until pupation at a constant temperature (20–25 °C) either on foliage or artificial diet. Record the number of days until pupation, pupal weight, and sex of pupae. Reduction in the number of viable eggs produced by surviving females is another indicator of sublethal effects. At least 20–40 larvae from each treatment should be collected, measured, and reared to allow meaningful comparisons of delays in larval development, pupal weights, changes in the sex ratio, fecundity, and parasitism.

4 References

Andermatt, M., Mani, E. Wildbolz, T. H. and Lüthy, P. 1988. Susceptibility of *Cydia pomonella* to *Bacillus thuringiensis* under laboratory and field conditions. *Entomol. Exp. Appl.* **49**, 291–295.

Arthurs, S. and Lacey, L. A. 2004. Field evaluation of commercial formulations of the codling moth granulovirus (CpGV): persistence of activity and success of seasonal applications against natural infestations in the Pacific Northwest. *Biol. Control* **31**, 388–397.

Arthurs, S., Lacey, L. A. and Behle, R. W. 2006. Evaluation of spray-dried lignin-based formulations and adjuvants as ultraviolet light protectants for the granulovirus of the codling moth, *Cydia pomonella* (L). *J. Invertebr. Pathol.* **93**, 88–95.

Arthurs, S., Lacey, L. A. and Fritts, R. Jr. 2005. Optimizing the use of the codling moth granulovirus: effects of application rate and spraying frequency on control of codling moth larvae in Pacific Northwest apple orchards. *J. Econ. Entomol.* **98**, 1459–1468.

Audemard, H. 1991. Population dynamics of the codling moth. *In* "Tortricid Pests, Their Biology, Natural Enemies and Control" (L. P. S. van der Geest and H. H. Evenhuis, Eds.), pp. 329–338. Elsevier Science Publishers B. V., Amsterdam.

Ballard, J., Ellis, D. J. and Payne, C. C. 2000. The role of formulation additives in increasing the potency of *Cydia pomonella* granulovirus for codling moth larvae, in laboratory and field experiments. *Biocontrol Sci. Technol.* **10**, 627–640.

Barnes, M. M. 1991. Tortricids in pome and stone fruits, codling moth occurrence, host race formation, and damage. *In* "Tortricid Pests, Their Biology, Natural Enemies and Control" (L. P. S. van der Geest and H. H. Evenhuis, Eds.), pp. 313–327. Elsevier Science Publishers B. V., Amsterdam.

Beers, E. H., Hull, L. A. and Jones, V. P. 1994. Sampling pest and beneficial arthropods of apple. *In* "Handbook of Sampling Methods for Arthropods in Agriculture" (L. P. Pedigo and G. D. Buntin, Eds.), pp. 383–416. CRC Press, Boca Raton, FL, USA.

Beers, E. H., Brunner, J. F., Willett, M. J. and Warner, G. M. 1993. Orchard Pest Management: A Resource Book for the Pacific Northwest. Good Fruit Grower, Yakima, WA. 276 pp.

Blommers, L. H. M. 1994. Integrated pest management in European apple orchards. *Annu. Rev. Entomol.* **39**, 213–241.

Blommers, L., Vaal, F., Freriks, J. and Helsen, H. 1987. Three years of specific control of summer fruit tortrix and codling moth on apple in the Netherlands. *J. Appl. Entomol.* **104**, 353–371.

Brooks, W. M. 1988. Entomogenous Protozoa. *In* "Handbook of Natural Pesticides, Vol. V: Microbial Insecticides, Part A: Entomogenous Protozoa and Fungi" (C. M. Ignoffo and N. B. Mandava, Eds.), pp. 1–149. CRC Press, Boca Raton, FL, USA.

Brunner, J. F. 1994. Using Bt products as tools in pest control. *Good Fruit Grower* **45**, 34–38.

Calkins, C. O. and Faust, R. J. 2003. Suppression of codling moth in western USA. *Pest Manage. Sci.* **59**, 601–604.

Charmillot, P. J. and Pasquier, D. 2002. Combination of mating disruption (MD) technique and granulosis virus to control resistant strains of codling moth *Cydia pomonella. Rev. Suisse Viticul. Arboricul. Horticul.* **34**, 103–108.

Charmillot, P. J., Pasquier, D. and Scalco, A. 1998. Granulosis virus of codling moth *Cydia pomonella*: 2. Efficacy in microplots, persistence and effect of adjuvants. *Rev. Suisse Viticul. Arboricul. Horticul.* **30**, 61–64.

Cossentine, J. E., Banham, F. L. and Jensen, L. B. 1990. Efficacy of the nematode, *Heterorhabditis heliothidis* (Rhabditida: Heterorhabditidae) against the peachtree borer, *Synanthedon exitiosa* (Lepidoptera: Sesiidae) in peach trees. *J. Entomol. Soc. Brit. Columbia* **87**, 82–84.

Cossentine, J. E., Jensen, L. B. and Deglow, E. K. 2003. Strategy for orchard use of *Bacillus thuringiensis* while minimizing impact on *Choristoneura rosaceana* parasitoids. *Entomol. Exp. Appl.* **109**, 205–210.

Cossentine, J. E., Jensen, L. B. and Moyls, L. 2002. Fruit bins washed with *Steinernema carpocapsae* (Rhabditida: Steinernematidae) to control *Cydia pomonella* (Lepidoptera: Tortricidae). *Biocontrol Sci. Technol.* **12**, 251–258.

Croft, B. A. and Bode, W. M. 1983. Tactics for deciduous fruit IPM. *In* "Integrated Management of Insect Pests of Pome and Stone Fruits" (B. A. Croft and S. C. Hoyt, Eds.), pp. 219–270. John Wiley & Sons, New York.

Croft, B. A. and Hull, L. A. 1991. Tortricid pests of pome and stone fruits, chemical control and resistance to pesticides. *In* "Tortricid Pests their Biology, Natural Enemies and Control" (L. P. S. Van Der Geest and H. H. Evenhuis, Eds.), pp. 473–486. Elsevier Science Publishers B. V., Amsterdam.

Cross, J. V., Solomon, M. G., Chandler, D., Jarrett, P., Richardson, P. N., Winstanley, D., Bathon, H., Huber, J., Keller, B., Langenbruch, G. A. and Zimmermann, G. 1999. Biocontrol of pests of apples and pears in Northern and Central Europe: 1. Microbial agents and nematodes. *Biocontrol Sci. Technol.* **9**, 125–149.

De Reede, R. H., Gruys, P. and Vaal, F. 1985. Leafrollers in apple integrated pest management under regimes based on *Bacillus thuringiensis*, diflubenzuron or epofenonane. *Entomol. Exp. Appl.* **37**, 263–274.

Deseö, K. V. and Miller, L. A. 1985. Efficacy of entomogenous nematodes, *Steinernema* spp., against clearwing moths, *Synanthedon* spp., in North Italian apple orchards. *Nematologica* **31**, 100–108.

Deseö, K. V., Govoni, I. and Miller, L. A. 1984a. Impiego di nematodi entomopatogeni (Rhabditida, Steinernematidae) contro diverse specie di *Synanthedon* (Lepidoptera, Aegeriidae) in meleti e diospireti. *Difesa Piante*. **7**, 315–330.

Deseö, K. V., Grassi, S., Foschi, F. and Rovesti, L. 1984b. Un sistema di lotta biologica contro il rodilegno giallo (*Zeuzera pyrina* L.; Lepidoptera, Cossidae). *Atti Giornate Fitopatol.* **2**, 403–414.

Dickler, E. and Huber, J. 1988. Modified strategy for the use of codling moth granulosis virus (CpGV). *In* "Production and Application of Viral Bio-Pesticides in Orchards and Vegetables" (H. Audemard and R. Cavalloro, Eds.), pp. 37–42. Office for Official Publications of the European Communities, Luxembourg.

Eberle, K. E. and Jehle, J. A. 2006. Field resistance of codling moth against *Cydia pomonella* granulovirus (CpGV) is autosomal and incompletely dominant inherited. *J. Invertebr. Pathol.* **93**, 201–206.

Falcon, L. A. and Huber, J. 1991. Biological control of the codling moth. *In* "Tortricid Pests their Biology, Natural Enemies and Control" (L. P. S. Van Der Geest and H. H. Evenhuis, Eds.), pp. 355–369. Elsevier Science Publishers B. V., Amsterdam.

Falcon, L. A., Kane, W. R. and Bethell, R. S. 1968. Preliminary evaluation of a granulosis virus for control of the codling moth. *J. Econ. Entomol.* **61**, 1208–1213.

Ferron, P. and Vincent, J. J. 1978. Preliminary experiments on the use of *Beauveria bassiana* against *Carpocapsa pomonella*. *Mittl. Biol. Bund. Land Fort.* **1978**, 84–87.

Flückiger, C. R. 1982. Untersuchungen über drei Baculovirus-Isolate des Schalenwicklers, *Adoxophyes orana* F. v. R. (Lep., Tortricidae), dessen Phänologies und erste Feldversuche, als Grundlagen zur mikrobiologischen Bekämpfung dieses Obstschädlings. Mitt. Schweiz. Entomol. Gesell. **55**, 241–288.

Fritsch, E., Undorf-Spahn, K., Kienzle, J., Zebitz, C. P. W. and Huber, J. 2005. Apfelwickler-granulovirus: erste hinmeise auf unterschiede in der empfindlichkeit lokaler apfelwickler-populationen. *Nachrichtenbl. Dtsch. Pflanzenschutzd.* **57**, 29–34.

Frye, R. D., Scholl, C. G., Scholz, E. W. and Funke, B. R. 1973. Effect of weather on a microbial insecticide. *J. Invertebr. Pathol.* **22**, 50–54.

Garcia-Gutierrez, C., Gonzales-Maldonado, M. B., Medrano-Roldan, H. and Chairez-Hernandez, I. 2004. Evaluación de la cepa BbP1 de *Beauveria bassiana*, Mycotrol®, Meta-Sin® y azinfos-metilico contra *Cydia pomonella* L. (Lepidoptera: Tortricidae) en laboratorio y campo. *Folia Entomol. Mex.* **43**, 1–7.

Glen, D. M. and Clark, J. 1985. Death of *Cydia pomonella* larvae and damage to apple fruit, after field application of codling moth granulosis virus. *Entomol. Exp. Appl.* **38**, 93–96.

Glen, D. M. and Payne, C. C. 1984. Production and field evaluation of codling moth granulosis virus for control of *Cydia pomonella* in the United Kingdom. *Ann. Appl. Biol.* **104**, 87–98.

Guillon, M. and Biache, G. 1995. IPM strategies for control of codling moth (*Cydia pomonella* L.) (Lepidoptera: Olethreutidae) interest of CmGV for long term biological control of pests complex in orchards. *Med. Fac. Landbouww. Univ. Gent.* **60**, 3a, 695–705.

Gut, L. J. and Brunner, J. F. 1998. Pheromone-based management of codling moth (Lepidoptera: Tortricidae) in Washington apple orchards. *J. Agri. Entomol.* **15**, 387–405.

Hardman, J. M. and Gaul, S. O. 1990. Mixtures of *Bacillus thuringiensis* and pyrethroids control winter moth (Lepidoptera: Geometridae) in orchards without causing outbreaks of mites. *J. Econ. Entomol.* **83**, 920–936.

Harris, M. O., Mafile, F. and Dhana, S. 1997. Behavioral responses of lightbrown apple moth neonate larvae on diets containing *Bacillus thuringiensis* formulations or endotoxins. *Entomol. Exp. Appl.* **84**, 207–219.

Huber, J. 1986. Use of Baculoviruses in pest management programs. *In* "The Biology of Baculoviruses. Vol. II Practical Application for Insect Control (R. R. Granados and B. A. Federici, Eds.), pp. 181–202. CRC Press, Boca Raton.

Huber, J. 1980. Field persistence of the codling moth granulosis virus *Laspeyresia pomonella*. *IOBC/WPRS Bull.* **3**, 58–59.

Huber, J. and Dickler, E. 1977. Codling moth granulosis virus: Its efficiency in the field in comparison with organophosphorus insecticides. *J. Econ. Entomol.* **70**, 557–561.

Huber, J. and Hassan, S. A. 1991. Biological control of totricid pests in pome and stone fruits. *In* "Tortricid Pests their Biology, Natural Enemies and Control"

(L. P. S. Van Der Geest and H. H. Evenhuis, Eds.), pp. 497–505. Elsevier Science Publishers B. V., Amsterdam.

Jaques, R. P. 1965. The effect of *Bacillus thuringiensis* Berliner on the fauna of an apple orchard. *Can. Entomol.* **97**, 795–802.

Jaques, R. P. 1967. Mortality of five apple insects induced by the nematode DD136. *J. Econ. Entomol.* **60**, 741–743.

Jaques, R. P. 1990. Effectiveness of the granulosis virus of the codling moth in orchard trials in Canada. *Proc. Fifth Int. Colloq. Invertebr. Pathol.*, pp. 428–430.

Jaques, R. P. and MacLellan, C. R. 1965. Fungal mortality of overwintering larvae of the codling moth in apple orchards in Nova Scotia. *J. Invertebr. Pathol.* **7**, 291–296.

Jaques, R. P., Laing, J. E., Laing, D. R. and Yu, D. S. K. 1987. Effectiveness and persistence of the granulosis virus of the codling moth *Cydia pomonella* (L.) (Lepidoptera: Olethreutidae) on apple. *Can. Entomol.* **119**, 1063–1067.

Jaques, R. P., Stultz, H. T. and Huston, F. 1968. The mortality of the pale apple leafroller and winter moth by fungi and nematodes applied to soil. *Can. Entomol.* **100**, 813–818.

Jaques, R., Hardman, J., Laing, J. and Smith, R. 1994. Orchard trials in Canada on control of *Cydia pomonella* (LEP: Tortricidae) by granulosis virus. *Entomophaga*, **39**, 281–292.

Jie, L., N. Liu, R. Xie, Y. Zhang and C. Yin. 1994. Applications of a *Steinernema* nematode against peach fruit-borer in apple orchards. *Act. Phytophyl. Sinica* **21**, 221–224.

Kahounova, L. and Mráček, Z. 1991. Larval mortality of *Synanthedon myopaeformis* (Lepidoptera, Sesiidae) in apple trees sprayed with *Steinernema* sp. strain *Hylobius* (*Steinernema*, Steinernematidae). *Acta Entomol. Bohemoslov.* **88**, 205–210.

Kaya, H. K. and Stock, S. P. 1997. Techniques in insect nematology. *In* "Manual of Techniques in Insect Pathology" (L. A. Lacey, Ed.), pp. 281–324. Academic Press, London.

Kaya, H. K., Aguillera, M. M., Alumai, A., Choo, H. Y., de la Torre, M., Fodor, A., Ganguly, S., Hazir, H., Lakatos, T., Pye, A., Wilson, M., Yamanaka, S., Yang, H. and Ehlers, R-U. 2006. Status of entomopathogenic nematodes and their symbiotic bacteria from selected countries or regions of the world. *Biol. Control* **38**, 134–155.

Kaya, H. K., Joos, J. L., Falcon, L. A. and Berlowitz, A. 1984. Suppression of the codling moth (Lepidoptera: Olethreutidae) with the entomogenous nematode, *Steinernema feltiae* (Rhabditida: Steinernematidae). *J. Econ. Entomol.* **77**, 1240–1244.

Keller, S. 1973. Mikrobiologische Bekämpfung des Apfelwicklers (*Laspeyresia pomonella* [L.]) (= *Carpocapsa pomonella*) mit spezifischem Granulosisvirus. *Z. Angew. Entomol.* **73**, 137–181.

Knight, A. L. 1994. Insect pest and natural enemy populations in paired organic and conventional apple orchards in the Yakima Valley, Washington. *J. Entomol. Soc. Brit. Columbia* **91**, 27–31.

Knight, A. L. 1995. The impact of codling moth (Lepidoptera: Tortricidae) mating disruption on apple pest management in Yakima Valley, Washington. *J. Entomol. Soc. Brit. Columbia* **92**, 29–38.

Knight, A. L. and Croft, B. A. 1991. Modeling and prediction technology. *In* "Tortricid Pests their Biology, Natural Enemies and Control" (L. P. S. Van Der Geest and H. H. Evenhuis, Eds.), pp. 301–312. Elsevier Science Publishers B. V., Amsterdam.

Knight, A. L. and Light, D. M. 2005. Timing of egg hatch by early-season codling moth (Lepidoptera: Tortricidae) predicted by moth catch in pear ester and codlemone-baited traps. *Can. Entomol.* **137**, 728–738.

Knight, A. L., Lacey, L. A., Stockhoff, B. A. and Warner, R. L. 1998. Activity of Cry 1 endotoxins of *Bacillus thuringiensis* for four tree fruit leafroller pest species (Lepidoptera: Tortricidae). *J. Agri. Entomol.* **15**, 92–103.

Lacey, L. A. and Arthurs, S. P. 2005. New method for testing solar sensitivity of commercial formulations of the granulovirus of codling moth (*Cydia pomonella*, Tortricidae: Lepidoptera). *J. Invertebr. Pathol.* **90**, 85–90.

Lacey, L. A. and Chauvin, R. L. 1999. Entomopathogenic nematodes for control of codling moth in fruit bins. *J. Econ. Entomol.* **92**, 104–109.

Lacey, L. A. and Shapiro-Ilan, D. L. 2003. The potential role for microbial control of orchard insect pests in sustainable agriculture. *J. Food Agric. Environ.* **1**, 326–331.

Lacey, L. A. and Unruh, T. R. 1998. Entomopathogenic nematodes for control of codling moth: effect of nematode species, dosage, temperature and humidity under laboratory and simulated field conditions. *Biol. Control* **13**, 190–197.

Lacey, L. A. and Unruh, T. R. 2005. Biological control of codling moth (*Cydia pomonella*, Tortricidae: Lepidoptera) and its role in integrated pest management, with emphasis on entomopathogens. *Vedalia* **12**, 33-60.

Lacey, L. A., Arthurs, S., Knight, A., Becker, K. and Headrick, H. 2004. Efficacy of codling moth granulovirus: effect of adjuvants on persistence of activity and comparison with other larvicides in a Pacific Northwest apple orchard. *J. Entomol. Sci.* **39**, 500–513.

Lacey, L. A., Arthurs, S. P., Unruh, T. R., Headrick, H. L. and Fritts, R. Jr. 2006a. Entomopathogenic nematodes for control of codling moth (Lepidoptera: Tortricidae) in apple and pear orchards: effect of nematode species and seasonal temperatures, adjuvants, application equipment and post-application irrigation. *Biol. Control* **37**, 214–223.

Lacey, L. A., Frutos, R., Kaya, H. K. and Vail, P. 2001. Insect pathogens as biological control agents: Do they have a future? *Biol. Control* **21**, 230–248.

Lacey, L. A., Granatstein, D., Arthurs, S. P., Headrick, H. and Fritts, R. J. 2006b. Use of mulches in conjunction with entomopathogenic nematodes (Steinernematidae) for control of overwintering codling moth (Lepidoptera: Tortricidae). *J. Entomol. Sci.* **41**, 107–119.

Lacey, L. A., Neven, L. G., Headrick, H. L. and Fritts, R. Jr. 2005. Factors affecting entomopathogenic nematodes (Steinernematidae) for the control of overwintering codling moth (Lepidoptera: Tortricidae) in fruit bins. *J. Econ. Entomol.* **98**, 1863–1869.

Laing, D. R. and Jaques, R. P. 1980. Codling moth: Techniques for rearing larvae and bioassaying granulosis virus. *J. Econ. Entomol.* **73**, 851–853.

Li, S. Y. and Fitzpatrick, S. M. 1997. Responses of larval *Choristoneura rosaceana* (Harris) (Lepidoptera: Tortricidae) to a feeding stimulant. *Can. Entomol.* **129**, 363–369.

Li, S. Y., Fitzpatrick, S. M. and Isman, M. B. 1995a. Effect of temperature on toxicity of *Bacillus thuringiensis* to the obliquebanded leafroller (Lepidoptera: Tortricidae). *Can. Entomol.* **127**, 271–273.

Li, S. Y., Fitzpatrick, S. M. and Isman, M. B. 1995b. Susceptibility of different instars of the obliquebanded leafroller (Lepidoptera: Tortricidae) to *Bacillus thuringiensis* var. *kurstaki. J. Econ. Entomol.* **88**, 610–614.

Li, S. Y. and Fitzpatrick, S. M. 1996. The effects of application rate and spray volume on efficacy of two formulations of *Bacillus thuringiensis* Berliner var. *kurstaki* against *Choristoneura rosaceana* (Harris) (Lepidoptera: Tortricidae) on raspberries. *Can. Entomol.* **128**, 605–612

Luisier, N. and Benz, G. 1994. Improving the efficiency of the granulosis virus of *Adoxophyes orana* F. v. R. for microbiological control. *Bull. OILB SROP* **17**, (3) 274.

Maddox, J. V. 1987. Protozoan diseases. *In* "Epizootiology of Insect Diseases" (J. R. Fuxa and Y. Tanada, Eds.), pp. 417–452. Wiley and Sons, New York.

Malone, L. A. and Wigley, P. J. 1981. The morphology and development of *Nosema carpocapsae*, a microsporidian pathogen of the codling moth, *Cydia pomonella* (Lepidoptera: Tortricidae) in New Zealand. *J. Invertebr. Pathol.* **38**, 315–329.

Nachtigall, G. and Dickler, E. 1992. Experiences with field applications of entomoparasitic nematodes for biological control of cryptic living insects in orchards. *Act. Phytopathol. Entomol. Hung.* **27**, 485–490.

Nakai, M. and Kunimi, Y. 1997. Granulosis virus infection of the smaller tea tortrix (Lepidoptera: Tortricidae): effect on the development of the endoparasitoid, *Ascogaster reticulatus* (*Hymenoptera: Braconidae*). *Biol. Control* **8**, 74–80.

Navon, A. 1993. Control of lepidopteran pests with *Bacillus thuringiensis. In "Bacillus thuringiensis*, an Environmental Biopesticide: Theory and Practice" (P. F. Entwistle, J. S. Cory, M. J. Bailey and S. Higgs, Eds.), pp. 127–146. John Wiley and Sons, New York.

Nyouki, F. R. and Fuxa, J. R. 1994. Persistence of natural and genetically engineered insecticides based on *Bacillus thuringiensis. J. Entomol. Sci.* **29**, 347–356.

Parvizi, R. 2003. An evaluation of the efficacy of the entomopathogenic nematodes *Steinernema* sp. and *Heterorhabditis bacteriophora* in controlling immature stages of the apple clearwing *Synanthedon myopaeformis. Iranian J. Agric. Sci.* **34**, 303–311.

Payne, C. C., Richards, M. G., Crook, N. E., Glen, D. M. and Cranham, J. E. 1984. Research on the production and application of the granulosis virus of codling moth, *Cydia pomonella*, in the United Kingdom (1979–1983). *In* "Programme on Integrated and Biological Control: Final Report 1979/1983" (R. Caralloro and A. Piavaux, Eds.), pp. 53–57. Office for Official Publications of the European Communities, Luxembourg.

Payne, N. J. and Van Frankenhuyzen, K. 1995. Effect of spray droplet size and density on efficacy of *Bacillus thuringiensis* Berliner against the spruce budworm, *Choristoneura fumiferana* (Lepidoptera: Tortricidae). *Can. Entomol.* **127**, 15–23.

Pfannenstiel, R. S., Szymanski, M., Lacey, L. A., Brunner, J. F. and Spence, K. 2004. Discovery of a granulovirus of *Pandemis pyrusana* (Lepidoptera: Tortricidae), a leafroller pest of apples in Washington. *J. Invertebr. Pathol.* **86**, 124–127.

Pszczolkowski, M. A., Brunner, J. F., Doerr, M. D. and Brown, J. J. 2004. Enhancement of *Bacillus thuringiensis* with monosodium glutamate against larvae of obliquebanded leafroller (Lep.:Tortricidae). *J. Appl. Entomol.* **128**, 474–477.

Sauphanor, B., Berling, M., Toubon, J.-F., Reyes, M. and Delnatte, J. 2006. Carpocapse des pommes: cas de résistance aux virus de la granulose dans le Sud-Est. *Phytoma* **590**, 24–27.

Sekita, N., Kawashima, K., Aizu, H., Shirisaki, S. and Yamada, M. 1984. A short term control of *Adoxophyes orana fasciata* Walsingham (Lepidoptera: Tortricidae) by a granulosis virus in apple orchards. *Appl. Entomol. Zool.* **19**, 498–508.

Shiga, M., Yamada, H., Oho, H., Nakazawa, H. and Ito, Y. 1973. A granulosis virus, possible biological agent for control of *Adoxophyes orana* (Lepidoptera: Tortricidae) in apple orchards. Semipersistent effect of artificial dissemination into an apple orchard. *J. Invertebr. Pathol.* **21**, 149–157.

Siegel, J. P., Lacey, L. A. and Vossbrinck, C. R. 2001. Impact of a North American isolate of the microsporidium *Nosema carpocapsae* on a laboratory population of the codling moth, *Cydia pomonella. J. Invertebr. Pathol.* **78**, 244–250.

Singh, P. and Ashby, M. D. 1986. Production and storage of diapausing codling moth larvae. *Entomol. Exp. Appl.* **41**, 75–86.

Smirle, M. J., Lowery, D. T. and Zurowski, C. L. 2003. Susceptibility of leafrollers (Lepidoptera: Tortricidae) from organic and conventional orchards to azinphos-methyl, spinosad, and *Bacillus thuringiensis. J. Econ. Entomol.* **96**, 879–884.

Sorensen, A. A. and Falcon, L. A. 1980. Comparison of microdroplet and high volume application of *Bacillus thuringiensis* on pear: suppression of fruit tree leafroller (*Archips argyrospilus)* and coverage on foliage and fruit. *Environ. Entomol.* **9**, 350–358.

Subinprasert, S. 1987. Natural enemies and their impact on overwintering codling moth populations (*Laspeyresia pomonella* L.) (Lep., Tortricidae) in South Sweden. *J. Appl. Entomol.* **103**, 46–55.

United States Department of Agriculture. 2004. Agricultural chemical usage 2003 fruit summary: apple. Http://jan.mannlib.cornell.edu/

Unruh, T. R. and Lacey, L. A. 2001. Control of codling moth, *Cydia pomonella* (Lepidoptera: Tortricidae) with *Steinernema carpocapsae*: effects of supplemental wetting and pupation site on infection rate. *Biol. Control* **20**, 48–56.

Vakenti, J. M., Campbell, C. J. and Madsen, H. F. 1984. A strain of fruittree leafroller, *Archips argyrospilus* (Lepidoptera: Tortricidae), tolerant of azinphos-methyl in an apple orchard region of the Okanagan Valley of British Columbia. *Can. Entomol.* **116**, 69–73.

Vega, F. E., Lacey, L. A., Reid, A. P., Herard, F., Pilarska, D., Danova, E., Tomov, R. and Kaya, H. K. 2000. Infectivity of a Bulgarian and an American strain of *Steinernema carpocapsae* against codling moth. *BioControl* **45**, 337–343.

Vickers, R. A. and Rothschild, G. H. L. 1991. Use of sex pheromones for control of codling moth. *In* "Tortricid Pests, Their Biology, Natural Enemies and Control" (L. P. S. van der Geest and H. H. Evenhuis, Eds.), pp. 339–354. Elsevier Science Publishers B. V., Amsterdam.

Wang, J. 1993. Control of the peach fruit moth, *Carposina niponensis,* using entomopathogenic nematodes. In "Nematodes and the Biological Control of Insect Pests" (R. Bedding, R. Akhurst and H. Kaya, Eds.), pp. 59–65. CSIRO Publications, East Melbourne, Australia.

Wearing, C. H. 1993. Control of codling moth with a commercial preparation of granulosis virus. *Proc. 46th N. Z. Plant Protection Conf. 1993*, 146–151.

Weiser, J. 1955. *Neoaplectana carpocapsae* n. sp. (Anguillulata, Steinernematinae), a new nematode parasite of the codling moth, *Carpocapsa pomonella* L. *Ves.. Ceskoslovenske Spolecnosti Zool.* **19**, 44–52.

Westigard, P. H., Gut, L. J. and Liss, W. J. 1986. Selective control program for the pear pest complex in Southern Oregon. *J. Econ. Entomol.* **79**, 250–257.

Yamada, H. and Oho, N. 1973. A granulosis virus, possible biological agent for control of *Adoxophyes orana (Lepidoptera: Tortricidae)* in apple orchards. *J. Invertebr.* Pathol. **21**, 144–148.

Zimmermann, G. and Weiser, J. 1991. Pathogens and diseases. *In* "Tortricid Pests, Their Biology, Natural Enemies and Control" (L. P. S. van der Geest and H. H. Evenhuis, Eds.), pp. 253–271. Elsevier Science Publishers B. V., Amsterdam.

Chapter VII-12

Microbial control of insect pests of stone fruit and nut crops

David I. Shapiro-Ilan
USDA-ARS
Southeastern Fruit and Tree Nut Research Unit
Byron, GA 31008, USA

Lawrence A. Lacey
USDA-ARS
Yakima Agricultural Research Laboratory
Wapato, WA 98951, USA

Joel P. Siegel
USDA-ARS
San Joaquin Valley Agricultural Sciences
Center, Parlier, CA 93648, USA

1 Introduction

The focus of this chapter is on application and evaluation of microbial control agents in stone fruit and nut crops. The key crops of interest, *i.e.*, those of significance where entomopathogens have been or are being evaluated, include peach, plum, nectarine, cherry, almond, and pecan and pistachio. Reviews of the biology, taxonomy, origin and distribution of these crops may be found elsewhere (Rehder, 1940; Westwood, 1978; Janick and Moore, 1996; Wood, 2003). A wide variety of insect pests, particularly among the orders Coleoptera, Diptera, Hemiptera, Homoptera, and Lepidoptera attack stone fruit and nut crops and cause severe economic damage. Further information on the diversity of insects that attack peach, plum, nectarine, cherry, almond, and pecan and pistachio may be found in the following references: Boethel and Eikenbary (1979), Strand (1999), Dutcher *et al.* (2003), Horton and Johnson (2005).

Stone fruit and nut crops are grown in perennial orchard plantings. Orchards tend to contain attributes that are amenable to insect pest suppression using certain entomopathogens such as hosts being available through much of the year, favorable soil conditions (moist, sandy), and partial or full protection from UV radiation due to shading (Shapiro-Ilan *et al.*, 2002a; Lacey and Shapiro-Ilan, 2003). Additionally, crops produced in orchards are often relatively high in value, which facilitates economic feasibility of microbial control applications. As a result, a number of orchard pests have been extensively studied for their potential to be controlled with entomopathogens, and moderate to very high levels of efficacy have been achieved. A list of some of the most notable examples is shown in Table 1, and it is these pest and pathogen combinations that will be our focus in the remainder of the chapter. More extensive reviews and analyses of efficacy and potential for use of microbial control agents in stone fruit and nut crops may be found in Lacey and Shapiro-Ilan (2003) and Shapiro-Ilan *et al.* (2005).

L.A. Lacey and H.K. Kaya (eds.), Field Manual of Techniques in Invertebrate Pathology, 547–565.

Table 1. Field efficacy of microbial control agents against selected insect pests of stone fruit and nut crops

Insect (stage)[1]	Crop	Pathogen	Applied to	Rate[3]	% Suppression	Reference
Amyelois transitella -L	Pistachio	Sc	Tree base/soil	1.0×10^9 IJs	43–75%	Siegel *et al.* (2006)
Anastrepha ludens -L	Mango	Hb	Soil	2.5×10^{10}	74%	Toledo *et al.* (2006)
Conotrachelus nenuphar -L	Peach	Sf	Soil	1.0×10^{10} IJs	NS[4]	Shapiro-Ilan *et al.* (2004b)
		Sf	Soil	2.0×10^{10} IJs	40%	Alston *et al.* (2005)
		Sr	Soil	1.0×10^{10} IJs	78–100%	Shapiro-Ilan *et al.* (2004b)
Curculio caryae -A	Pecan	Bb	Soil	Unknown	72%	Gottwald and Tedders (1983)
	Pecan	Bb	Soil	3×10^{14} con	69%[5]	Shapiro-Ilan *et al.* (2004a)
	Pecan	Ma	Soil	Unknown	50%	Gottwald and Tedders (1983)
	Pecan	Sc	Soil	1.0×10^{10} IJs	50–80%	Shapiro-Ilan *et al.* (unpub)
C. caryae -L	Pecan	Bb	Soil	1.2×10^8 con	34%	Harrison *et al.* (1993)
Rhagolites indifferens -L	Cherry	Sc	Soil	$5–10 \times 10^9$ IJs	69–72%	Yee and Lacey (2003)
		Sf	Soil	$5–10 \times 10^9$ IJs	78–83%	
Synanthedon exitiosa -L	Peach	Hb	Tree base/soil	4×10^4/tree	80%	Cossentine *et al.* (1990)
	Peach	Sc	Tree base/soil	3×10^5/tree	88%	Cottrell and Shapiro-Ilan (2006)
	Peach	Sr	Trcc basc/soil	3×10^5/trcc	NS	Cottrell and Shapiro-Ilan (2006)

[1] Stage: A = adult, L = larva, P = pupa
[2] Pathogen: Bb = *Beauveria bassiana*, Ma = *Metarhizium anisopliae*, Sc = *Steinernema carpocapsae*, Sf = *S. feltiae*, Sr = *S. riobrave*,
[3] Rate (of application) is per ha unless otherwise noted, con = condia, IJs = infective juveniles
[4] NS = not significant
[5] Average mycosis during first week post-application

2 Site selection and plot design

Essentially all commercial stone fruit and nut species should be suitable for field tests (as well as most wild plums). It is preferable, however, if the test site includes only one variety. Consideration should be given to cultural practices such as pesticide applications and fertilization as some inputs can interact with pathogen applications (*e.g.*, see Wraight *et al.* Chapter IV-4, and Koppenhöfer Chapter IV-5 this volume). Adequate moisture/relative humidity is required for survival and infection of some pathogens (nematodes and fungi). Thus, irrigation is essential for soil application with nematodes, and in some cases, fungi as well. If irrigation is required, an adequate irrigation system such as micro-jet or solid set should be calibrated and in place prior to application, or, in small plot studies, a water tank with hose or watering can may be used. Aspects of the test site that should be recorded include soil parameters (*e.g.*, soil type, texture, pH, and nutrients), orchard size, ground cover, and the age, size, density and variety of trees.

It is preferable to select plots that have not previously received microbial control applications and that have low or negligible natural entomopathogen populations. To test for high densities or uneven distributions of endemic entomopathogen populations, pretreatment sampling is recommended. For example (for soil applications), 2–4 weeks prior to treatment application, a minimum of four soil cores (ca. 50 ml each) should be taken from each plot and placed in a 150 mm Petri dish with five target insects; if the target insect is not available then another susceptible host might be used, *e.g.*, *Galleria mellonella* can be used for baiting entomopathogenic nematodes and fungi. Insect mortality can then be assessed after 7–14 days of incubation at 25 °C (or whatever temperature and time period is deemed optimum for the suspected pathogen present). If a high level or uneven distribution of endemic entomopathogen population is detected, appropriate statistical measures or selection of another site should be considered.

Rectangular or circular plots within stone fruit and nut orchards are conducive to entomopathogen application and efficacy assessment. To avoid biasing trends in field conditions, it is recommended to use a randomized block design (blocking by row), but completely randomized designs may also be suitable. In all experiments, a minimum of 5 replicates for each treatment and untreated control is recommended. In some cases it may be useful to use a standard recommended treatment (*e.g.*, chemical insecticide) as a positive control. Furthermore, all field experiments should be repeated in time (*e.g.*, two consecutive seasons or years) at the same location, and if possible, additional locations with differing conditions (such as soil or crop type or climactic conditions) should tested before broad conclusions on efficacy are made. Plot size can vary. Thus, “micro-plots” in the range of 0.02 to 1 m^2 may be useful in early stages of efficacy assessment, when multiple treatments are being compared, or if availability of entomopathogen inoculum is limited. Larger scale tests, which (depending on the system) may be more conducive to assessing efficacy on a commercial scale, can be conducted with plots in the range of several trees to several hectares of orchard space. It is important that the plots are spaced at a sufficient distance to avoid contamination among treatments and controls. The required distance between plots will vary based on the size of plots (the larger the plots the more distance required), and manner of application, *e.g.*, aerial/foliar sprays will surely result in more drift than pouring inoculum onto soil in microplots. Perhaps as a rule of thumb for soil applications, we can recommend an absolute minimum of > 5 m between plots.

3 Preparation and application of inoculum

A Fungi

Inoculum for orchard applications of entomopathogenic fungi, *Beauveria bassiana* and *Metarhizium anisopliae*, can be obtained through laboratory production or commercial sources. The fungi grow readily on Petri dishes containing suitable solid media such as Sabouraud dextrose agar or potato dextrose agar (Goettel and Inglis, 1997). Petri dish level production is appropriate for routine lab culture and experiments and could be suitable to very small field applications (such as in buried pots or PVC pipes, see below), but when inoculum must be grown for larger scale applications, other approaches must be sought. *Beauveria bassiana* and *M. anisopliae* conidia can be grown in quantity in large containers, trays, or sterile bags containing a nutritive source such as grains (Bartlett and Jaronski, 1988; Wraight and Carruthers, 1999; Bradley *et al.*, 2002). Growing conidia under such conditions is suitable for small plot trials particularly when the goal is to directly compare the efficacy of different strains and species when cultured in parallel. However, for large plot studies (*e.g.*, 0.4 ha per plot or larger), it will likely be necessary to obtain inoculum from commercial sources. Although it is best to use the inoculum as soon as possible, conidia can generally be stored under refrigeration at 4 °C for short periods (*e.g.*, a few weeks or months) (Goettel and Inglis, 1997). In addition to application of conidia, fungal blastospores or mycelia may also be assessed for efficacy although their stability and infectivity or efficacy may be reduced (Roberts and Yendol, 1971; Wraight and Carruthers, 1999).

To achieve efficacy, a minimum application rate in the range of 10^{13} to 10^{14} *B. bassiana* or *M. anisopliae* conidia per ha is required (Bartlett and Jaronski, 1988; Wraight and Carruthers, 1999). The concentration of conidia in the inoculum can be verified prior to application through dilution counts on a hemacytometer (Goettel and Inglis, 1997). Additionally, the viability of inocula should be determined by checking for germination after streaking a diluted quantity onto an appropriate solid medium, or by adding a conidial mixture to liquid media and shaking for 12 h (Goettel and Inglis, 1997; Wraight *et al.*, Chapter IV-4; McCoy *et al.*, Chapter VII-13).

During transport and application, the inoculum must be protected against harmful environmental conditions (Wraight *et al.*, Chapter IV-4). Sunlight should be avoided. Although *B. bassiana* and *M. anisopliae* can survive and germinate at temperatures of 37 °C or higher (Rangel *et al.*, 2005; Quesada-Moraga *et al.*, 2006), optimum temperatures for infection are more moderate (*e.g.*, 20–30 °C, Wraight *et al.* (Chapter IV-4), and inoculum should be kept cool (*e.g.*, 12–25 °C) for as long as possible prior to application. High levels of relative humidity are generally required for germination of most entomopathogenic fungi (*e.g.*, > 90%, Roberts and Yendol, 1971). However, high levels of ambient relative humidity are not always necessary because the requirements for germination can be met within the microclimate of the target site (*e.g.*, on the insect-cuticle); in fact, in under certain conditions, a lower relative humidity may be more suitable for efficacy (Lord, 2005), and irrigation or precipitation can be detrimental (Shapiro-Ilan *et al.*, 2004a). For more details on effects of environmental conditions on entomopathogenic fungi see Wraight *et al.* (Chapter IV-4).

The fungal inoculum can be applied using a variety of equipment. For small plot applications the conidial suspension can be applied using watering cans or a backpack sprayer. For larger plot applications, equipment with large capacities is needed such as herbicide rigs, hydraulic or air-blast sprayers, or microjet irrigation systems. It is important to ensure adequate agitation during application.

B Nematodes

Nematode inoculum can be grown in the laboratory or obtained commercially. It may be convenient for researchers to culture their own inoculum particularly if the number of nematodes required is not too large (*e.g.*, less than several billion), and if novel strains are being tested that are not easily available commercially. A number of sources may be consulted for information on culturing entomopathogenic nematodes, *e.g.*, Kaya and Stock (1997), Shapiro-Ilan and Gaugler (2002), Shapiro-Ilan *et al.* (2002b). Inoculum stored under refrigeration (4–15 °C depending on species or strain) may be kept in culture flasks or for larger quantities bubbled in buckets or jugs (Kaya and Stock 1997). It is best not to keep inoculum for more than one to two weeks prior to application. If it is not feasible to produce the inoculum "in house" (due to the quantity needed or a lack of facilities or required expertise), the inoculum should be obtained from commercial sources. If one of the objectives is to compare innate virulence or efficacy of different nematode strains or species, all inoculum should be produced together under parallel conditions. If, however, the goal is to compare one product versus another, it may be appropriate to incorporate several variables (such as differing nematode sources, strain or species, formulation, production technique) into each treatment. The inoculum should always be the same age (or as close as possible) and optimum storage and handling procedures should be followed for each product.

As with all nematode applications, consideration should be given to the environmental conditions during and post-treatment (see Shapiro-Ilan *et al.*, 2006; Koppenhöfer, Chapter IV-5). To avoid desiccation in soil applications, the ground should be wet prior to application, watered after application to assist nematode movement below the soil-surface, and irrigated thereafter. Optimum soil moisture and corresponding irrigation levels for nematode survival and efficacy will vary with soil type (Koppenhöfer *et al.*, 1995). The potential for desiccation to reduce nematode survival and efficacy is exacerbated in aboveground treatments, and therefore, addition of wetting agents or other adjuvants that may enhance nematode efficacy aboveground should be considered (Schroer and Ehlers, 2005; Wright *et al.*, 2005). During transport and application of inoculum, exposure to extremes in temperature or UV should be avoided. Applications made late in the afternoon or in the evening will facilitate avoiding UV radiation. Following application, the nematodes can be shallowly cultivated into the soil surface to avoid exposure to UV, or simulation thereof may be achieved by covering with approximately

1 cm field soil. Applications should be made during periods when the targeted insect stage occurs naturally in the regional environment. Temperature data (soil and or ambient as appropriate), *e.g.*, daily highs and lows, should be recorded during the course of the experiment. Viability of the nematodes being applied should be confirmed (preferably prior to and during the treatment) (see Koppenhöfer, Chapter IV-5, and McCoy *et al.*, Chapter VII-13).

Entomopathogenic nematode applications may be made using various spray equipment or hand held equipment (Wright *et al.*, 2005; Shapiro-Ilan *et al.*, 2006). Application using simple watering cans may be optimum for small plot application (*e.g.*, Shapiro-Ilan *et al.*, 2004b). In these small plot applications, a single volume of water in a watering can (*e.g.*, totaling up to approximately 11 liters) containing the approximate exact number of nematodes designated for each plot can be applied (and variation per plot minimized). Similarly, in micro-plot applications (*e.g.*, Alston *et al.*, 2005), an exact quantity of nematodes can simply be poured onto the soil surface in a standardized quantity of water. If nematodes must be applied to larger plots (*e.g.*, $> 100\,m^2$), a suitable spraying apparatus such as a back-pack sprayer or herbicide rig should be considered. Conceivably, applications could also use other methods such as through mircrojet irrigation systems (see McCoy *et al.* Chapter VII-13), in infected host insects (Shapiro-Ilan *et al.*, 2003), or subsurface injection or baits (Wright *et al.*, 2005); although these methods have not been reported in stone fruit and nut crops, in some cases they could be efficacious and thus should be investigated.

To achieve efficacy, rates of application should generally be 25 infective juveniles (IJs) per cm^2 ($= 2.5 \times 10^9$/ha) (Georgis and Hague, 1991; Shapiro-Ilan *et al.*, 2006). However, lower rates may be appropriate for highly susceptible pests. On the other hand, higher rates are often necessary due to the pest's susceptibility or to soil or other environmental conditions. In some cases high rates, *e.g.*, > 200 IJs/cm^2, may be justified economically if the crop value is especially high or if the total area of application is only a small proportion of the orchard (such as applications for *Synanthedon exitiosa* control).

4 Protocols for assessment of entomopathogen efficacy

A *Stone fruits*

1 *Plum curculio,* Conotrachelus nenuphar

The plum curculio, *Conotrachelus nenuphar*, is a major pest of stone and pome fruit in North America (Racette *et al.*, 1992). Adult weevils enter orchards from overwintering sites in the spring, and feed, and oviposit in fruit. Attacked fruit aborts or is disfigured rendering it non-saleable. Larvae continue to develop in fallen fruit, exit as fourth instars, and burrow into the soil (1–8 cm) to pupate (Racette *et al.*, 1992). Following emergence, adults feed on fruit and migrate to litter surrounding the orchard to overwinter (Racette *et al.*, 1992; Olthof and Hagley, 1993). In the southern United States, an additional generation may occur on many peach cultivars prior to overwintering (Horton *et al.*, 2003).

Microbial control agents that have exhibited pathogenicity to *C. nenuphar* larvae and or adults include entomopathogenic fungi (*B. bassiana* and *M. anisopliae*) and nematodes (*Heterorhabditis* spp. and *Steinernema* spp.) (Tedders *et al.*, 1982; Olthof and Hagley, 1993, Shapiro-Ilan *et al.*, 2002c, Alston *et al.*, 2005). Suppression in stone fruits under field conditions, however, has thus far only been reported using steinernematid nematodes against *C. nenuphar* larvae; trials were conducted in peach and wild plum (Shapiro-Ilan *et al.*, 2004b; Alston *et al.*, 2005).

Plot size and application rates used in *C. nenuphar* studies can vary. Plot size can be relatively small such as those previously reported in the literature - ca. 175 cm^2 (Alston *et al.*, 2005) or 4000 to 9000 cm^2 (Shapiro-Ilan *et al.*, 2004b), or, in large scale tests, can encompass the drip line of several trees. Rates of application for *C. nenuphar* suppression that have caused significant mortality vary from 1.0 to 2.0×10^{10} IJs/ha (100 to 200 per cm^2 (Shapiro-Ilan *et al.*, 2004b; Alston *et al.*, 2005). Higher or low rates may be appropriate to the test as well depending on various factors such as soil type and nematode species and strain.

One approach to measuring entomopathogenic nematode efficacy in suppressing *C. nenuphar* is to apply nematodes to soil containing the larval stage and subsequently determine the number of adults that emerge (Shapiro-Ilan *et al.*, 2004b). Larvae can be collected by placing infested fruit (indicated by the distinctive crescent shaped scars on the fruit [Racette *et al.*, 1992]) on a screen with mesh size just large enough to hold the fruit. The screen should be in a container with a layer of sterile soil or potting mix below; the larvae will drop into the soil; the soil or potting mix containing the larvae can be stored at 15 °C for a short period until used (Shapiro-Ilan *et al.*, 2002c). Two or three days prior to treatment applications, *C. nenuphar* larvae, ca. at least one per every 65 cm^2 within the plot (Shapiro-Ilan *et al.*, 2004b), are placed approximately 5 cm below the soil surface within the plots' perimeters. To mimic natural conditions, larvae are then covered with soil and a layer of peaches (from the same location within the orchard). After nematodes are applied, cone traps made of aluminum screening (*e.g.*, hole size 0.03 cm diameter, Figure 1) fitted with boll weevil traps on top (Boethel *et al.*, 1976; Duncan *et al.*, 2001) are placed over each plot and secured around the edges with potting soil. Once adult weevils begin emerging, the number of weevils in each trap is recorded daily until zero emergence is recorded on three consecutive sample dates.

Alternatively, rather than burying larvae in the soil, infested fruit can be placed directly on the soil allowing larvae to emerge and burrow down naturally. Approximately one infested fruit per 40 cm^2 of plot area is placed on the soil surface within the plot (Shapiro-Ilan *et al.*, 2004b). In order to estimate the number of larvae emerging in each plot, 100 infested peaches can be placed in each of five Berlese funnels and larval exit recorded approximately every 2–3 days until no larvae are recorded on two consecutive sample dates. The Berlese funnels should be placed in the field under the peach trees adjacent to the cone cage plots about 25 cm above the ground. After nematode application, adult emergence is monitored with cone cages as described above.

Regardless of whether the *C. nenuphar* larvae are added directly to the soil or allowed to emerge from fruit, analysis of treatment effects is the same (Shapiro-Ilan *et al.*, 2004b). The average total number of weevils emerged per plot is analyzed for treatment effects through analysis of variance, and mean separation may be elucidated through Tukey's multiple range test or other similar tests (SAS, 2001). Percentage control (relative to the number of weevils emerged in

Figure 1. A cone trap in a peach orchard. The trap can be used for capturing emerging plum curculio (*Conotrachelus nenuphar*)

the non-treated check) can be calculated using Abbott's formula (Abbott, 1925).

Another approach, instead of monitoring adult emergence, is to directly assess larval mortality (Alston *et al.*, 2005). This approach may be particularly conducive to small ("micro") plots that enable the researcher to relocate larvae that have been released. For example, micro-plots can be made with plastic pots (such as 15 cm diam., 13.5 cm deep) with the bottoms removed and replaced with wire mesh (1.5 mm openings), which allow water to flow through but prevent larval escape (Alston *et al.*, 2005). The pots should be filled with soil from the experimental site. Approximately one larva should be added for every 15–20 cm^2 of plot area; larvae that do not burrow down within 5 minutes should be replaced. Treatments can be applied prior to addition of larvae (if the goal is a "barrier" approach), or after (if a more curative approach is sought). Mortality of *C. nenuphar* larvae is assessed 7 days after treatment. Treatment differences are detected through ANOVA, and percentage control (relative to the number of weevils emerged in the non-treated check) can be calculated using Abbott's formula (Abbott, 1925). An advantage to direct assessment of larval mortality relative to trapping the emerging adults is that control mortality may be lower in the former because the insects are not in the soil as long and thus natural mortality factors have less time to act. On the other hand, one can argue that a more realistic or "natural" measure of overall mortality and treatment effects may be obtained by measuring adult emergence.

2 *Fruit flies*

Fruit flies comprise some of the most serious pests of fruit, including stone fruit, throughout the world. Fungi and entomopathogenic nematodes have been evaluated as alternatives to conventional insecticides for control of several important pest species (Table 1). Two of the more recently studied fruit flies in terms of microbial control using entomopathogenic agents are the western cherry fruit fly (WCFF), *Rhagoletis indifferens* and the Mexican fruit fly, *Anastrepha ludens*.

WCFF is a key pest of cherry in all cherry-growing regions of the western U. S. (Beers *et al.*, 1993). Larvae develop within cherries and when full grown, exit the fruit and drop onto the ground and enter the soil where they pupate within 4 to 8 hours (Frick *et al.*, 1954). The majority of the univoltine life cycle of WCFF (about 10 months, beginning in mid-June) is spent as pupae 0–10 cm below the soil surface (Frick *et al.*, 1954; AliNiazee, 1974). Adults emerge in the late spring and oviposit on fruit.

Based on laboratory studies by Patterson Stark and Lacey (1999) and field and laboratory studies by Yee and Lacey (2003), WCFF larvae are the most vulnerable stage to infection by entomopathogenic nematodes. Of several steinernematid and heterorhabditid species tested, *S. carpocapsae* and *S. feltiae* were the most effective against WCFF larvae. Mortalities of larvae at 0, 2, 4, and 6 days following their introduction into soil previously treated with *S. carpocapsae* and *S. feltiae* at 100 IJs/cm^2 were 90.0, 92.0, 100.0, and 84.0% and 90.0, 50.0, 42.0, and 40.0%, respectively. Adult WCFF remain in the soil briefly (< 1 day) after they emerge from puparia. Some adults (11–53%) that emerged through soil treated with 100 IJs/cm^2 of *S. carpocapsae* or *S. feltiae* IJs in the laboratory became infected. Although some WCFF larvae that were exposed to *S. carpocapsae* or *S. feltiae* IJs died as pupae, fully formed pupae were not infected (Yee and Lacey, 2003). Based on field trials (see Table 1), Yee and Lacey (2003) proposed the use of entomopathogenic nematodes for control of WCFF in isolated and abandoned lots or in yards of homeowners as a means to reduce invasion of commercial orchards.

Unlike WCFF, the Mexican fruit fly develops in a variety of species of fruit crops and is especially damaging in mango and citrus. It is widely distributed in Mexico, most of Central America and in certain states in the southern United States. Females oviposit in ripening fruit and larvae burrow into the pulp. When full grown, larvae exit the fruit, usually after it has fallen to the ground, and pupate in the soil. Peak abundance of this and other *Anastrepha* species occurs with maximum abundance of host fruit (Aluja, 1994). Although *A. ludens* is susceptible to a variety of enomopathogenic nematode species under laboratory conditions (Lezama-Gutiérrez *et al.*, 1996; Toledo *et al.*, 2001, 2005), extremely high application rates are required for control in the field (Table 1).

Similarly, laboratory and field research conducted on the effectiveness of EPNs as control agents of the Mediterranean fruit fly, *Ceratitis capitata*, revealed susceptibility of larvae to several nematode species (Poinar and Hislop, 1981, Lindegren and Vail, 1986; Gazit *et al*., 2000; Laborda *et al*., 2003), but elevated application rates are required for control in the field (Lindegren *et al*., 1989).

Testing of entomopathogenic fungi on fruit flies has been predominantly on *M. anisopliae* and *B. bassiana*. Laboratory and field research reported by Ekesi *et al*. (2002, 2003, 2005) and Dimbi *et al*. (2003, 2004) on *M. anisopliae* against *Ceratitis* spp. elucidated the effect of a variety of factors on the activity of the fungus including temperature, moisture, gender, life stage and fly species. Laboratory and field studies on *M. anisopliae* activity against *A. ludens* were reported by Lezama-Gutiérrez *et al*. (2000). In field cages when *M. ansiopliae* was applied at the rate of 2×10^5 colony forming units/cm^2, adult emergence was reduced by up to 43% in loam soil. Yee and Lacey (2005) reported on the activity of *M. anisopliae* against WCFF larvae and adults. They found that soil treated with *M. anisopliae* (7.5×10^5 conidia/g of soil) did not reduce emergence, but up to 68% of teneral adults became infected as they emerged. Third instar WCFF larvae were not infected when exposed to soil treated with up to 4.8×10^6 conidia/cm^2.

Konstantopoulou and Mazomenos (2005) reported on the laboratory evaluation of *B. bassiana* and *B. brongniartii* against adults of *C. capitata* and the olive fly, *Bactrocera oleae*. Rosa *et al*. (2002) studied the effects of *B. bassiana* on *A. ludens*. In both studies, adult flies were very susceptible to infection by conidia, but Rosa *et al*. (2002) reported negligible effects on larvae and pupae.

a Evaluation protocol for entomopathogenic nematodes – use of WCFF sentinel larvae in contained field plots

Due to its univoltine life cycle, rearing of WCFF has only been met with limited success. However, larvae can be readily obtained from infested sweet and sour cherries, *Prunus avium* and *P. cerasus*, respectively, from trees that have not been sprayed with insecticides. Infested fruit is placed on wire screens over water-filled tubs to collect larvae. Emerging larvae will drop into the water, where they remain viable for at least 24 hours. The water prevents them from pupating too quickly. This method allows collection of sufficient larvae for experiments and apparently has little or no effect on control mortality (Yee and Lacey, 2003). Larvae that have been allowed to enter soil and pupate may be kept in cold storage until diapause is broken (3–4 °C for 6 months). Adults will then emerge under warmer conditions (25–27 °C for ≈ 1 month with adequate moisture).

Because there is zero tolerance for WCFF in commercial cherries, evaluations conducted in orchards should eliminate the potential for larvae to escape. To ensure containment of larvae in commercial orchards, containers can be filled with soil from the orchard and used in the orchards as removable mini-plots (microcosms). In the studies conducted by Yee and Lacey (2003), small soil-filled pails (15.5 cm diameter, 18.5 cm high) were inserted into holes in the ground (such that the tops of the pails were level with the orchard soil surface) in a randomized block design that blocked on individual trees. Each pail had a hole on the bottom covered with organdy that enabled drainage, but prevented escape of larvae. The tops of the pails were covered with lids and organdy to prevent larval escape. The following protocol should be used:

1. Each treatment and control should be replicated in 4 or more clearly labeled containers spaced 25–30 cm apart underneath the north (shady) side of trees. Starting soil moisture should be ≈ 20–25%. Depending on availability of technical assistance and larvae, tests can also be replicated on separate dates.
2. Pre-moisten soil immediately before inoculation to ensure initial survival of IJs.
3. Apply IJs of the selected species and desired concentration to the surface of the soil in the mini-plots using sufficient water to ensure penetration of the top few mm of soil. In studies of Yee and Lacey (2003), 50 or 100 IJs/cm^2 of *S. carpocapsae* or *S. feltiae* were applied with a pipette. For larger scale trials using conventional equipment; an application rate of 3800 liters/ha (400 gal/ac)

will ensure that adequate water assists penetration of the soil by IJs.

4. Soon after inoculation of the mini-plots with IJs, infest each pail with 50 third-instar larvae directly from the holding container. Alternative treatments could involve addition of larvae to separate mini-plots at various intervals following IJ treatments.
5. Remoisten mini-plots. This can be one of the treatment variables (*i.e.*, how much water and how often is required for effective control).
6. Remove mini-plots from the field and store at 10 °C until mortality is assessed by sieving and/or careful search.

Soil temperatures should be measured, using a continuously recording data logger if possible, for the time the containers are left in the field. It will be important to determine the type of soil (proportion of loam, clay, sand) as this has a strong influence on IJ movement.

b Use of natural infestations

The area around isolated heavily infested trees can be monitored for adult emergence with the use of ammonia-baited yellow sticky traps (Frick *et al.*, 1954) in the season following nematode application. Caging individual trees has been used for other studies on WCFF without a great deal of success (W. L. Yee, personal communication), but deliberate infestation of soil with sentinel larvae beneath caged trees may provide a realistic setting for single tree plots. Monitoring adult emergence the following spring using the methods of Frick *et al.* (1954) could be used to determine the effectiveness of entomopathogenic nematode applications.

3 Clear winged moths (Sesiidae)

The lesser peachtree borer, *Synanthedon pictipes*, and the peachtree borer, *Synanthedon exitiosa*, are important pests of peach and other *Prunus* spp. in North America (Johnson *et al.*, 2005a, b). *Synanthedon pictipes* attacks the tree's trunk and limbs often laying eggs in the crotch or near injured areas (Bobb, 1959, Johnson *et al.*, 2005b). Because *S. pictipes* does not spend any part of its life cycle in the soil, microbial control applications need to be directed to the insect-bored tunnels aboveground, which is a tactic that has proved successful for other *Synanthedon* spp. in apples, (Deseö and Miller, 1985, Nachtigall and Dickler, 1992), alder (Kaya and Brown, 1986), and black currants, (Bedding and Miller, 1981; Miller and Bedding, 1982). Laboratory studies indicated relatively high virulence of *S. carpocapsae, S. feltiae*, and *S. riobrave* to *S. pictipes* (Shapiro-Ilan and Cottrell, 2006). Field testing of nematodes (or other pathogens) for suppression of *S. pictipes*, however, has yet to be conducted. In contrast, field studies have been conducted on microbial control of *S. exitiosa* using nematodes (Cossentine *et al.*, 1990; Cottrell and Shapiro-Ilan, 2006), and therefore, the remainder of this section will focus on this pest.

Synanthedon exitiosa larvae burrow into the tree and feed on the trunk's cambium and inner bark and on the roots. The larvae form galleries that are found at the base of the trunk to a depth of about 25 cm (Figure 2) (Johnson *et al.*, 2005a). Young trees are highly susceptible to severe damage even from a single larva. Larvae overwinter in their burrows or in silken coverings or hibernaculae on the bark outside their burrows. Mature larvae exit their burrows and form a cocoon to pupate. The pupal case usually remains protruding after the moth emerges. Adult emergence can begin in early spring with moths being present in the orchard until November; in the Southeastern US, the majority of *S. exitiosa* moths emerge during late summer and early fall (Johnson *et al.*, 2005a).

Nematode applications for *S. exitiosa* control should be made to or in the vicinity of the base of the tree. In general, with young trees, *e.g.*, trunk diameter of 5–10 cm, applications can remain with 30 cm of the trunk and should be in the range of 1.0×10^5 to 5×10^5 IJs/tree. Single or multiple applications (2–4 weeks apart) can be tested. For experimental purposes, applications can be poured onto the trunk base from a flask or may be applied via watering can or sprayer. If applications are being made as a curative treatment to trees that are already infested, then trees with observable signs of active *S. exitiosa* damage must be identified, *e.g.*, by frass at the burrow's entrance. Alternatively, prophylactic applications can be made during egg-laying (preferably in orchards where the infestation is known to be substantial). Assessment of season long damage may be made in late fall when adult flight has

Figure 2. Damage on a peach tree from *Synanthedon exitiosa*, the peachtree borer

ended. Active infestations may be discerned by removing soil from around the base of the trunk and looking for signs of active infestation and also by opening feeding galleries when obvious signs of infestation were not visible (Johnson *et al.*, 2005a; Cottrell and Shapiro-Ilan, 2006). Alternatively, pupal cases or cocoons can be sought on the trunk or in the surrounding soil (Johnson *et al.*, 2005a; T. Cottrell, personal communication). If nematodes were applied to known infestations, efficacy can be determined 2 to 8 weeks post application by examination of galleries for frass activity. Percentage infestation among treatments and controls can be compared using standard ANOVA and appropriate multiple range tests.

B Nuts

1 Navel orangeworm, Amyelois transitella

The navel orangeworm (NOW), *Amyelois transitella*, is a multivoline pyralid that is the major pest of pistachios and almonds in California. Three or more generations occur in the Central Valley of California depending on degree-day accumulation. The navel orangeworm overwinters in unharvested nuts (mummies/sticktights) as a larva or pupa and continues to develop throughout the winter as temperature permits (Bentley *et al.*, 2000). Adults begin emergence in late winter (first flight) and emergence continues until mid July and under some circumstances even into August. Adults oviposit on mummy nuts and new crop nuts when they become available. Larvae can develop more than twice as fast in new crop nuts than in mummies, and as a consequence, several generations overlap throughout the summer and two generations overwinter in the Central Valley. The density of NOW decreases through the winter and by February less than 0.2% of the nuts on the ground may be infested. Overwintering NOW is currently controlled by orchard sanitation. Unharvested almonds are shaken or knocked from the trees, piled in windrows and then shredded. Pistachios are usually allowed to fall naturally (the stems become weakened by winter rain and fog) and the fallen pistachios are tilled into the soil or left to rot (Zalom *et al.*, 2002). Cultural control is more efficacious in almond than pistachio orchards because almonds are readily shredded compared to pistachios and the smooth orchard floor in almonds enables most to be swept up and gathered into piles.

Entomopathogenic nematodes may be an ideal agent to target overwintering larvae in fallen pistachios because they can seek out infested nuts. Insecticides are used to protect almonds from the first and second flight and pistachios from the third and fourth flight of NOW. The techniques listed below, although developed for NOW in pistachios, are applicable for almonds as well as other insect pests infesting fallen nuts.

The experimental design is dependent on the choice of outcome variable when assessing field efficacy of nematode applications. Although the ultimate goal is population reduction, the easiest outcome variable to assess is mortality within a sentinel nut. The principal advantage of using sentinels is that the researcher can control the prevalence of NOW, even when the actual prevalence in the field is low. Sentinels are appropriate for initial assessment of candidate nematodes and/or application rates in small plot studies, and are also appropriate to use in large plot studies to assess application quality and coverage.

Sentinel nuts are infested by adding second or third instar larvae to split nuts and then incubating the infested nuts at 27 °C for at least 10 days before using them in the field. These later instars are more likely to successfully infest the nuts than neonates. Commercially hulled and dehydrated nuts ($\leq$ 5% moisture) or freshly picked nuts that are hulled in the laboratory can be used (Siegel *et al.*, 2006). The key factor is the availability of nuts and the amount of time the researcher can spend infesting them; it is easier to use commercially hulled nuts. Although it is important to avoid overcrowding the larvae, the rate of successful establishment is low and therefore adding two larvae for every nut is recommended. If the larvae are overcrowded, infection with *Bacillus thuringiensis* often occurs (Siegel *et al.*, 2000). Pistachios are placed in rectangular plastic containers filled with nuts no greater than 4 cm deep. Once the larvae are established, either inside the nuts or webbed on the surface, they can be held at 10 °C for as long as 2 weeks until use. The containers should then be removed and held at room temperature for at least 1 day and the nuts from several containers blended in case some containers have a higher infestation level. The nuts should then be dispensed into containers (100–150 nuts/container) and one container used per plot. Larval establishment at this point ranges from 20–50%.

In small plot studies, we recommend 1 m^2 plots containing sentinel nuts at a density of 150 nuts/plot. This design was successfully used to evaluate the efficacy of *S. carpocapsae* and *S. feltiae* (Siegel *et al.*, 2004, 2006). Arrange the plots so that there are at least two replicates of each treatment and control within a row and maximize the number of rows used to eliminate localized effects. Two replicates per treatment row are recommended to insure that a row is not dropped from analysis in case animals remove the nuts in a replicate or ants remove larvae. All plots should have the same orientation and be placed within range of the irrigation system to ensure wetting. Orchards irrigated by flood irrigation or drip irrigation are unsuitable. The nuts should be placed on soil covered netting (625 cm^2) to facilitate their removal and lightly sprinkled with soil and possibly lightly covered with leaves in order to replicate field conditions. Once the application has been applied to the plots the nuts should be covered with a fine screen as soon as possible to minimize removal by birds or small mammals (Figure 3). Nuts can be collected 10 days after application; there was no appreciable difference in mortality between nuts collected 10 days and 21 days after application (Siegel *et al.*, 2004). However, if the experiment is conducted when temperature facilitates rapid insect development, the nuts may need to be collected sooner so that adult emergence is not confounded with mortality. The sentinels are collected after treatment, dissected, and the number of living and dead individuals counted to determine percent mortality. If there is reason to suspect that nematode-killed larvae may be preferentially removed from the study site, reduction in potential emergence or actual adult emergence should be used as the outcome variable (Siegel *et al.*, 2006; Zhou *et al.*, 2002; Baur *et al.*, 1998). This is calculated by comparing the average number of living larvae in treated and control plots, or the sentinel nuts can be placed in marked cages and the number of emerged adults compared. The data generated in small plot trials can be analyzed using ANOVA or multiple regression techniques (Siegel *et al.*, 2006).

Large-scale studies (> 16 ha) pose a different challenge because the target is a naturally infested mummy rather than a sentinel nut. Since the prevalence of NOW in mummies can be quite low, tens of thousands of mummy nuts must be collected in order to obtain a sufficient number

Figure 3. Screen covering sentinel nuts (*e.g.*, pistachio or almond) that are infested with navel orangeworm, *Amyelois transitella*

of insects for meaningful comparisons. Sentinel nuts can still play a role in the large-scale study if they are used as a check on the quality and extent of coverage of the application. Unfortunately, birds and mammals will remove many of the sentinels and the starting density should be at least 300 sentinels per hectare. If sufficient resources exist, the following experimental design employing treated and control plots within the orchard should be used.

Use flags or tape to mark areas containing mummies. One day before application, rake some of the nuts in the marked areas into piles. Ideally, at least 10,000 mummies should be collected Clean the nuts by removing leaves and twigs, and put them into buckets, filling the buckets up to a depth of eight centimeters. Screen the buckets, incubate them at 27 °C for 40–50 days and record adult emergence. When the experiment is terminated count the number of nuts in a subset of buckets to determine the average number of nuts per bucket. Use this average to standardize adult emergence per 1,000 nuts in each plot. This is the pre-treatment density. Determine the post-treatment density by returning to the marked areas within the rows 7–10 days after treatment and filling the buckets with the remaining nuts. Monitor adult emergence as described above. The treatment effect is determined by subtracting the difference between the pre-treatment and post-treatment in nematode-treated plots and comparing that to the pre-treatment minus post-treatment densities in the control plots. The drawback to this technique is the space required to store tens of thousands of nuts at constant temperature as well as the time and labor to check buckets. If NOW is present at higher density, such as in almonds, fewer nuts need to be used.

The grower cooperator may pose an additional challenge if he/she desires to treat an entire block rather than leave part of the orchard untreated as a control. If the design described above cannot be followed then the following alternative design can be used to address the grower's concern as well as problems of inadequate labor. The alternate approach is similar to the design described above except that the same plots are used as control and treatment in order to maximize the number of nuts collected. Sentinel nuts are included in this design to assess the impact of confounding factors such as weather-associated mortality. Pre-treatment insect density is determined in plots as described above. Determine the post-treatment density by returning to the marked areas 7–10 days after treatment and collecting a similar quantity of mummies. Monitor adult emergence as described above. The treatment effect is determined by subtracting the difference in adult emergence between the pre-treatment and post-treatment samples. Use sentinel nuts in this design to assess the possible impact of weather-associated mortality. The control sentinels should be placed in adjacent orchards and along the outside of treated block. Compare mortality or adult reduction between control and treated sentinels, and then compare the magnitude of the effect to that observed in the collected mummies.

2 *Pecan weevil,* Curculio caryae

The pecan weevil, *Curculio caryae*, is a key pest of pecans throughout the Southeastern US as well as Oklahoma, Kansas, and parts of Texas. The insects have a 2- or 3-year life cycle (Harris, 1985). Adults emerge from soil in late July-August, feed on developing nuts, and oviposit into the nuts after dough stage begins (Harris, 1985). Larvae develop within the nut and fourth instars drop to the soil (generally from October through December) where they burrow to a depth of 8–25 cm, form a pupal cell, and overwinter. The following year, approximately 90% of the larvae pupate and spend the next 9 months in the soil as adults (Harris, 1985). The remaining 10% of the population spend 2 years in the soil as larvae, emerging as adults in the third year (Harris, 1985).

Microbial control can be considered at certain points in *C. caryae*'s life cycle. It is unlikely that any control strategy, chemical or biological, could be successful against *C. caryae* when larvae are protected in the nut; thus the insects must be targeted during periods when they are exposed. Microbial control of *C. caryae* currently emphasizes the use of entomopathogenic fungi and nematodes targeting emerging adults or larvae after they drop from the nut. The literature pertaining to pathogens of *C. caryae* has been reviewed previously by Sikorowski (1985), Fuxa *et al.* (1998), and Shapiro-Ilan (2003).

a Fungal applications

B. bassiana and *M. anisopliae* have been applied for *C. caryae* control. Assessing efficacy of adult *C. caryae* control relies on determining percentage infection after insects move through the inoculum; the intention is that the adults will either emerge directly through the inoculum as they exit their overwintering site or crawl over the inoculum as they move up toward the tree canopy. The fungal conidia can be applied to the soil surrounding the tree trunk using watering cans or a sprayer (*e.g.*, back-pack sprayer). The fungal conidia can also be applied to the tree trunk (*e.g.*, up to 2 m) using a back-pack sprayer or similar. Irrigation is not recommended for fungal control of adult *C. caryae* because it may cause leaching of conidia below the soil surface where the insects do not occur. They are active at the soil surface (*e.g.*, crawling to the tree) or irrigation may also decrease fungal persistence through increased microbial degradation (Shapiro-Ilan *et al.*, 2004a). When applying to soil, the area of application should generally consist of the tree's dripline (the weevil will not occur beyond the dripline). However, it is also possible to assess the impact of an application to a reduced area (*e.g.*, 1 to 4 m radius from the trunk) within the dripline because a significant proportion of *C. caryae* may become infected as they crawl to the trunk (Shapiro-Ilan *et al.*, 2004a). The dripline (or a smaller area within) within a single tree should constitute the plot for application purposes; thus, in most pecan orchards sufficient distance between plots (*e.g.*, $>$ 10 m) will be inherent in the design and potential for cross-contamination among treatments will be minimal. If necessary for statistical purposes (such as when weevil populations are low), multiple trees can constitute a replicate. The rate of application per unit area should be equivalent to a minimum of 2×10^{13} conidia/ha but can be as much as 5×10^{14}/ha; effective rates will depend on fungal strain, environmental conditions, formulation, etc. It should be noted that even relatively high application rates could conceivably be cost effective because a large proportion of the orchard (minimum of between the driplines) do not need to be treated.

Adult *C. caraye* infections are assessed by capturing the insects in Circle traps (Figure 4), which are passive trap that capture weevils crawling up the trunk (Mulder *et al.*, 2003). The traps can be made of wire mesh (1.5 mm pore size) with an open area (*e.g.*, approximately

Figure 4. A Circle trap attached to a pecan tree. The trap is used for capturing adult pecan weevils (*Curculio caryae*)

44 cm wide) facing toward the soil and tapering up to a removable screw-on top. Prior to application Circle traps should be attached to the tree trunk using small nails or a staple gun. The traps should be placed approximately 1.5 to 2 m above the ground so that weevils that are crawling or flying to the trunk will be captured. Alternatively, the traps may be placed low to the ground, *e.g.*, 30 cm above the soil surface, so that (almost) exclusively crawling insects are captured. Ideally, enough traps should be placed on the tree so that they cover the entire circumference. However, if weevil populations are high enough to provide sufficient numbers for statistical analysis, and as long as the number of traps per tree is consistent throughout the experiment, fewer traps may be used.

The tops to the traps should be screwed on just prior to application. Trapped weevils should be collected 24 hours post-application. Assessment of the treatment effect over time can be made thereafter by replacing trap tops and again collecting captured insects 24 hours later. Weevils should not be left in the traps longer than 24 hours due to the potential of predation. Collections can be made daily or several times per week. Pretreatment collections can be made in a similar manner in order to assess endemic pathogen populations in the plot areas.

On each day that *C. caryae* are trapped in the field, the insects captured in each trap top should be placed in separate plastic bags and brought to the laboratory to determine levels of fungal infection. The weevils are placed individually in 30 ml plastic cups (3–4 cm i.d., 3.5 cm deep) with a 3 cm cotton wick moistened with approximately 2.1 ml of tap water. Cups are then incubated at 25 °C. After 7 and 14 days of incubation (*e.g.*, at 25 °C), the percentage *C. caryae* mycosis is estimated by examining the cadavers for signs of fungal infection (Goettel and Inglis, 1997). The percentage of total *C. caryae* mortality (mycosis plus other causes) should also be recorded. Treatment effects compounded throughout the experimental periods (averaged over all days post-treatment) can be analyzed using repeated measures analysis and LSMEANS (*e.g.*, Proc Mixed, SAS, 2001). Additionally, treatment effects can be analyzed by day (for each sampling date separately) using ANOVA and a multiple range test (SAS, 2001).

It must be noted that incubating the insects at high humidity after collection may offer more ideal conditions for infection to develop than in the field. Thus, the method described above to assess adult *C. caryae* infection estimates the potential efficacy level rather than the actual. To estimate infection when the weevils remain in the field, conceivably, the captured insects could be placed in sleeve cages that are placed over nut clusters and mycosis/mortality could be assessed 7 and 14 days later. However, confining the weevils to a cage on a terminal branch (on a nut cluster) deprives the weevils of their natural movement and the ability to find refugia if environmental conditions are adverse; thus, such an approach is also not reflective of natural conditions and is not ideal.

Fungal control of *C. caryae* larvae has been assessed in small plots. Larvae, collected as they emerge from infested nuts (Gottwald and Tedders, 1983; Shapiro-Ilan, 2001a), are allowed to burrow into the plot area. To maximize the chances of recovery, it is best to confine the larvae, by using buried pots (with sufficient drainage capacity), PVC pipe, or drain tiles (Tedders *et al.*, 1973; Harrison *et al.*, 1993). For example, Harrison *et al.* (1993) hammered three PVC pipes (16 dia., 30 cm length) into soil in 1 m^2 plots; the pipes were equidistant from each other within the plot. Fungal conidia are then applied to the plot area with a back-pack sprayer, watering can, or similar. The rate of application should be similar to that suggested for adult weevil control. The idea behind fungal applications for *C. caryae* larval control is that the larvae will contact the fungus as they fall from the tree and burrow into the soil. Thus, the larvae should be added to plots after the fungus inoculum is applied (*e.g.*, 4 hours to 4 days post-treatment [Tedders *et al.*, 1973; Harrison *et al.*, 1993]). Harrison *et al.* (1993) added 20 larvae to each 16 cm diameter PVC pipe (48 larvae/1 m^2 plot). Approximately 3–4 weeks post-treatment the pipes, pots, or tiles, are removed and soil within them is sifted to recover larvae, which are then examined for mortality/mycosis. Treatment effects are determined through ANOVA and multiple range tests. A potential drawback to the above described approach to assessing fungal control of *C. caryae* larvae is that the experiment must be conducted in the fall during the period of larval drop, and thus if cool temperatures occur in the soil efficacy can be reduced.

b Nematode applications

Entomopathogenic nematodes have been tested for control of *C. caryae* adults and larvae in the field and in greenhouse studies. Laboratory screening indicates that the nematodes are generally more effective against adult *C. caryae* than to the larvae, and *S. carpocapsae* appears to be the most virulent to adults of those tested thus far (Shapiro-Ilan, 2001a,b, 2003). Methods to determine adult *C. caryae* suppression using entomopathogenic nematodes are essentially the same as those described above for assessing fungal control except that nematode applications should only be applied to the soil within the dripline (trunk applications are not recommended for nematodes due to susceptibility to desiccation), and irrigation is required. Nematodes should be applied at a rate ranging from approximately 25 to 200 IJs/cm^2 using a watering can or sprayer. The nematodes should be applied to moist soil, watered in after treatment, and plots should be irrigated thereafter. The amount of irrigation required will vary by soil type and depend on the amount of precipitation during the experimental period. However, it is important to keep the soil moist without over-irrigating and causing standing water or run off. Generally irrigating 2–3 times per week at 0.5 to 2.0 liters per m^2 should be sufficient (again, will depend on soil type and other environmental conditions). Plot size considerations, capture of adult weevils in Circle traps, and assessment of treatment effects are handled the same as with fungal experiments except that, for nematode applications the insects only need to be incubated (*e.g.*, at 25 °C) for 4 to 7 days after capture before determining percentage mortality (Shapiro-Ilan, 2001b).

Also similar to fungal efficacy experiments, the potential for nematodes to suppress *C. caryae* larvae can be determined in pots, PVC pipes, or similar mini-plots. Nyczepir *et al.* (1992) and Smith *et al.* (1993) placed containers in the greenhouse to assess larval control with entomopathogenic nematodes. Smith *et al.* (1993) used square plant containers (57.8 cm^2, 22.9 cm deep) that was packed with a 10.2 cm clay soil base (to facilitate the weevils in making soil cells and mimic the natural soil profile) and covered with 7.6 cm topsoil. Each container received 5 larvae 5 hours prior to nematode application (to simulate nematodes seeking hosts already in the ground) or 2 hours after nematode application (to simulate a barrier treatment). Nematodes were applied at varying rates from 31 to $62\,\mathrm{IJs/cm^2}$ (Smith *et al.*, 1993), but higher rates could be tested, particularly due to the relatively low susceptibility of the *C. caryae* larval stage (Shapiro-Ilan, 2001a, b, 2003). Treatment effects can be determined 2–3 weeks post-application by screening the soil for remaining live weevils (Smith *et al.*, 1993; Shapiro-Ilan, unpublished data), and through ANOVA and multiple range tests. Similar pot/container experiments can be conducted in the field by burying the containers to the level of the soil surface (as described in the fungal experiments), but seasonal temperatures during larval drop may affect efficacy.

It is also conceivable to assess multi-season applications. Larvae, obtained from infested nuts, can be allowed to burrow in small plots (*e.g.*, 4000 to $10,000\,\mathrm{cm^2}$). To ensure sufficient recovery a high density of larvae is recommended *e.g.*, at least 1 weevil/$10\,\mathrm{cm^2}$. The nematodes should be applied prior to releasing the weevils (to mimic a barrier treatment) at rates ranging from 25 to $200\,\mathrm{IJs/cm^2}$, and irrigation should be provided for at least 2 weeks post-treatment. The plots can be marked with a flag. Nematodes can be re-applied one or several times the following late spring and summer, as well as in the mid and late summer in the second year prior to weevil emergence. Cone traps should be placed over the plots before adult emergence, and treatment effects assessed based on number of adults emerged in treatment and control plots (In an identical manner as described for *C. nenuphar* experiments, Shapiro-Ilan *et al.*, 2004b).

5 References

Abbott, W. S. 1925. A method of computing the effectiveness of an insecticide. *J. Econ. Entomol.* **18**, 265–267.

AliNiazee, M. T. 1974. The western cherry fruit fly, *Rhagoletis indifferens* (Diptera: Tephritidae) I. Distribution of the diapausing pupae in the soil. *Can. Entomol.* **106**, 909–912.

Alston, D. G., Rangel, D. E. N., Lacey, L. A., Golez, H. G., Kim, J. J., and Roberts, D. W. 2005. Evaluation of novel fungal and nematode isolates for control of *Conotrachelus nenuphar* (Coleoptera: Curculionidae) larvae. *Biol. Control* **35**, 163–171.

Aluja, M. 1994. Bionomics and management of *Anastrepha. Annu. Rev. Entomol.* **39**, 155–178.

Bartlett, M. C. and Jaronski, S. T. 1988. Mass production of entomogenous fungi for biological control of insects. *In* "Fungi in Biological Control Systems" (M. N. Burge, Ed.) pp. 61–85. Manchester University Press, Manchester UK.

Baur, M. E., Kaya, H. K., and Strong, D. R. 1998. Foraging ants as scavengers on entomopathogenic nematode-killed insects. *Biol. Control* **12**, 231–236.

Bedding, R. A. and Miller, L. A. 1981. Disinfecting blackcurrant cuttings of *Synanthedon tipuliformis*, using the insect parasitic nematode *Neoaplectana bibionis. Environ. Entomol.* **10**, 449–453.

Beers, E. H., Brunner, J. F., Willett, M. J., and Warner, G. M. 1993. Orchard Pest Management: A Resource Book for the Pacific Northwest: *Good Fruit Grower*, Yakima, WA. 276 pp.

Bentley, W. J., Rice, R. E., Beede, R. H., and Daane, K. 2000. UC IPM Pest Management Guidelines: Pistachio Insects and Mites. University of California ANR Publication 3461.

Bobb, M. L. 1959. The biology of the lesser peach tree borer in Virginia. *J. Econ. Entomol.* **52**, 34–636.

Boethel, D. J. and Eikenbary, R. D. 1979. Pecan weevil: Seasonal emergence of larvae from three pecan cultivars in Oklahoma. *J. Ga. Entomol. Soc.* **14**, 75–83.

Boethel, D., Eikenbary, R., Morrison, R., and Criswell, J. 1976. Pecan weevil, *Curculio caryae* (Coleoptera: Curculionidae): comparison of adult sampling techniques. *Can. Entomol.* **108**, 11–18.

Bradley, C. A., Wood, P. P., Black, W. E., Kearns, R. D., and Britton, J. 2002. Solid culture substrate including barley. US Patent Application Number 20020006650.

Cottrell, T. E. and Shapiro-Ilan, D. I. 2006. Susceptibility of the Peachtree borer, *Synanthedon exitiosa*, to *Steinernema carpocapsae* and *Steinernema riobrave* in laboratory and field trials. *J. Invertebr. Pathol.* **92**, 85–88.

Cossentine, J. E., Banham, F. L., and Jensen, L. B. 1990. Efficacy of the nematode *Heterorhabditis heliothidis* (Rhabditida: Heterorhabditidae) against peachtree borer, *Synanthedon exitiosa* (Lepidoptera: Sesiidae) in peach trees. *J. Entomol. Soc. Brit. Columbia* **87**, 82–84.

Deseö, K. V. and Miller, L. A. 1985. Efficacy of entomogenous nematodes, *Steinernema* spp., against clearwing moths, *Synanthedon* spp., in north Italian apple orchards. *Nematologica* **31**, 100–108.

Dimbi, S., Maniania, N. K., Lux, S. A., and Mueke, J. M. 2004. Effect of constant temperatures on germination, radial growth and virulence of *Metarhizium anisopliae* to three species of African tephritid fruit flies. *BioControl* **49**, 83–94.

Dimbi, S., Maniania, N. K., Lux, S. A., and Mueke, J. M. 2003. Host species, age and sex as factors affecting the susceptibility of the African tephritid fruit fly species, *Ceratitis capitata, C. cosyra* and *C. fasciventris* to infection by *Metarhizium anisopliae. Anzeiger Schadlingskunde* **76**, 113–117.

Duncan, L. W., Genta, J. G., and Zellers, J. 2001. Soil texture and the efficacy of *Steinernema riobrave* against *Diaprepes abbreviatus* L. *J. Nematol.* **33**, 256 (abstract).

Dutcher, J. D. Harris, M. K., and Dean, D. A. (Eds.). 2003. "Integration of Chemical and Biological Insect Control in Native, Seedling, and Improved Pecan Production". *Southwest. Entomol. Supplement* 27.

Ekesi, S., Maniania, N. K., and Lux, S. A. 2002. Mortality in three African tephritid fruit fly puparia and adults caused by the entomopathogenic fungi, *Metarhizium anisopliae* and *Beauveria bassiana. Biocontrol Sci. Technol.* **12**, 7–17.

Ekesi, S., Maniania, N. K., and Lux, S. A. 2003. Effect of soil temperature and moisture on survival and infectivity of *Metarhizium anisopliae* to four tephritid fruit fly puparia. *J. Invertebr. Pathol.* **83**, 157–167.

Ekesi, S., Maniania, N. K., Mohamed, S. A., and Lux, S. A. 2005. Effect of soil application of different formulations of *Metarhizium anisopliae* on African tephritid fruit flies and their associated endoparasitoids. *Biol. Control* **35**, 83–91.

Frick, K. E., Simkover, H. G., and Telford, H. S. 1954. Bionomics of the cherry fruit flies in eastern Washington. *Wash. Agric. Expt. Sta. Tech. Bull.* **13**, 66 pp.

Fuxa, J. R., Ayyappath, R., and Goyer, R. A. 1998. "Pathogens and Microbial Control of North American Forest Insect Pests". Forest Health Technology Enterprise Team, USDA Forest Service, Morgantown, WV.

Gazit, Y., Rössler, Y., and Glazer, I. 2000. Evaluation of entomopathogenic nematodes for the control of Mediterranean fruit fly (Diptera: Tephritidae). *Biocontrol Sci. Technol.* **10**, 157–164.

Georgis, R., Hague, N. G. M. 1991. Nematodes as biological insecticides. *Pesticide Outlook* **3**, 29–32.

Goettel, M. S. and Inglis, G. D. 1997. Fungi: Hyphomycetes. *In* "Manual of Techniques in Insect Pathology" (L. A. Lacey, Ed.) pp. 213–249. Academic Press, San Diego, CA.

Gottwald, T. R. and Tedders, W. L. 1983. Suppression of pecan weevil (Coleoptera: Curculionidae) populations with entomopathogenic fungi. *Environ. Entomol.* **12**, 471–474.

Harris, M. K. 1985. Pecan phenology and pecan weevil biology and management. *In* "Pecan Weevil: Research Perspective" (W. W. Neel, Ed.) pp. 51–58. Quail Ridge Press, Brandon, MS.

Harrison, R. D., Gardner, W. A., and Kinard, D. J. 1993. Relative susceptibility of pecan weevil fourth instars and adults to selected isolates of *Beauveria bassiana. Biol. Control* **3**, 34–38.

Horton, D. and Johnson, D. (Eds.). 2005. Southeastern Peach Growers Handbook. The University of Georgia Cooperative Extension Service, Athens, GA.

Horton D., Bellinger, B., and Ritchie, D. 2003. 2003 Southeastern peach, nectarine, and plum pest management and culture guide, Bulletin 1171. University of Georgia, Athens, GA.

Janick, J. and Moore, J. N. (Eds.). 1996. "Fruit Breeding: Volume I Tree and Tropical Fruits". John Wiley & Sons, New York, NY.

Johnson, D., Cottrell, T., and Horton, D. 2005a. Peachtree borer. *In* "Southeastern Peach Growers Handbook (D. Horton and D. Johnson, Eds.) pp. 266–269. The University of Georgia Cooperative Extension Service, Athens, GA.

Johnson, D., Cottrell, T., and Horton, D. 2005b. Lesser peachtree borer. *In* "Southeastern Peach Growers Handbook (D. Horton and D. Johnson, Eds.) pp. 270–272. The University of Georgia Cooperative Extension Service, Athens, GA.

Kaya, H. K. and Brown, L. R. 1986. Field application of entomogenous nematodes for biological control of clearwing moth borers in alder and sycamore trees. *J. Arboric.* **12**, 150–154.

Kaya, H. K., and Stock, S. P. 1997. Techniques in insect nematology. *In* "Manual of Techniques in Insect Pathology" (L. A. Lacey, Ed.) pp. 281–324. Academic Press, San Diego, CA.

Konstantopoulou, M. A. and Mazomenos, B. E. 2005. Evaluation of *Beauveria bassiana* and *B. brongniartii* strains and four wild-type fungal species against adults of *Bactrocera oleae* and *Ceratitis capitata. BioControl* **50**, 293–305.

Koppenhöfer, A. M., Kaya, H. K., and Taormino, S. P. 1995. Infectivity of entomopathogenic nematodes (Rhabditida: Steinernematidae) at different soil depths and moistures. *J. Invertebr. Pathol.* **65**, 193–199.

Laborda, R., Bargues, L., Navarro, C., Barajas, O., Arroyo, M., Garcia, E. M., Montoro, E., Llopis, E., Martinez, A., and Sayagues, J. M. 2003. Susceptibility of the Mediterranean fruit fly (*Ceratitis capitata*) to entomopathogenic nematode *Steinernema* spp. ("Biorend C"). Bull. *OILB/SROP*. **26(6)**, 95–97.

Lacey, L. A. and Shapiro-Ilan, D. I. 2003. The potential role for microbial control of orchard pests in sustainable agriculture. *Food, Agric. Environ.* **1**, 326–331.

Lezama-Gutiérrez, R., Molina, O. J. Contreras-Ochoa, O. L., Gonzáles-Ramírez, M., Trujillo-de la Luz, A., and Rebolledo-Domínguez, O. 1996. Susceptibilidad de larvas de *Anastrepha ludens* (Diptera: Tephritidae) a diversos nemátotodos entomopatógenos (Steinernematidae y Heterorhabditidae), *Vedalia* **3**, 31–33.

Lezama-Gutiérrez, R., Trujillo de la Luz, A., Molina-Ochoa, J., Rebolledo-Dominguez, O., Pescador, A. R., Lopez-Edwards, M. and Aluja, M. 2000. Virulence of *Metarhizium anisopliae* (Deuteromycotina:

Hyphomycetes) on *Anastrepha ludens* (Diptera: Tephritidae) : laboratory and field trials. *J. Econ. Entomol.* **93**, 1080–1084.

Lindegren, J. E. and Vail, P. V. 1986. Susceptibility of Mediterranean fruit fly, melon fly, and oriental fruit fly (Diptera: Tephritidae) to the entomogenous nematode *Steinernema feltiae* in laboratory tests. *Environ. Entomol.* **15**, 465–468.

Lindegren, J. E., Wong, T. T., and McInnis, D. O. 1989. Response of Mediterranean fruit fly (Diptera: Tephritidae) to the entomogenous nematode *Steinernema feltiae* in field tests in Hawaii. *Environ. Entomol.* **19**, 383–386.

Lord, J. C. 2005. Low humidity, moderate temperature, and desiccant dust favor efficacy of *Beauveria bassiana* (Hyphomycetes: Moniliales) for the lesser grain borer, *Rhyzopertha dominica* (Coleoptera: Bruchidae). *Biol. Cont*rol **34**, 180–186

Miller, L. A. and Bedding, R. A. 1982. Field testing of the insect parasitic nematode *Neoaplectana bibionis* [Nematoda: Steinernematidae] against currant borer moth, *Synanthedon tipuliformis* [Lep: Sesiidae] in blackcurrants. *Entomophaga* **27**, 109–114.

Nachtigall, G. and Dickler, E. 1992. Experiences with field applications of entomogenous nematodes for biological control of cryptic living insects in orchards. *Acta Phytopath. Entomol. Hung.* **27**, 485–490.

Mulder P. G., Reid, W., Grantham, R. A., Landgraf, S., Taliaferro, L., Payton, M. E., and Knutson, A. 2003. Evaluations of trap designs and a pheromone formulation used for monitoring pecan weevil, *Curculio caryae*. *In*: "Integration of Chemical and Biological Insect Control in Native, Seedling, and Improved Pecan Production" (J. D. Dutcher, M. K. Harris, and D. A. Dean, Eds.) pp. 85–99. *Southwest. Entomol. Supplement* 27.

Nyczepir, A. P., Payne J. A., and Hamm, J. J. 1992. *Heterorhabditis bacteriophora*: a new parasite of pecan weevil *Curculio caryae*. *J. Invertebr. Pathol.* **60**, 104–106.

Olthof, T. H. and Hagley, E. C. 1993. Laboratory studies of the efficacy of steinernematid nematodes against the plum curculio (Coleoptera: Curculionidae). *J. Econ. Entomol.* **86**, 1078–1082.

Patterson Stark, J. E. and Lacey, L. A. 1999. Susceptibility of western cherry fruit fly (Diptera: Tephritidae) to five species of entomopathogenic nematodes in laboratory studies. *J. Invertebr. Pathol.* **74**, 206–208.

Poinar Jr., G. O. and Hislop, R. G. 1981. Mortality of Mediterranean fruit fly adults (*Ceratitis capitata*) from parasitic nematodes (*Neoaplectana* and *Heterorhabditis* spp.). *IRCS Med.* **9**, 641.

Quesada-Moraga E., Maranhao, E. A. A., Valverde-Garcıa, P., Santiago-Alvarez, C. 2006. Selection of *Beauveria bassiana* isolates for control of the whiteflies *Bemisia tabaci* and *Trialeurodes vaporariorum* on the basis of their virulence, thermal requirements, and toxicogenic activity. *Biol. Control* **36**, 274–287.

Racette, G., Chouinard, G., Vincent, C., and Hill, S. B. 1992. Ecology and management of plum curculio, *Conotrachelus nenuphar* [Coleoptera: Curculionidae], in apple orchards. *Phytoprotection* **73**, 85–100.

Rangel, D. E. N., Braga, G. U. L., Anderson A. J., and Roberts, D. W. 2005. Variability in conidial thermotolerance of *Metarhizium anisopliae* isolates from diverent geographic origins. *J. Invertebr. Pathol.* **88**, 116–125.

Rehder, A. 1940. "Manual of Cultivated Trees and Shrubs: Hardy in North America". The Macmillan Company, New York, NY.

Roberts, D. W. and Yendol, W. G. 1971. Use of fungi for microbial control of insects. *In* "Microbial Control of Insects and Mites" (H. D. Burges and N. W. Hussey, Eds.) pp. 125–149. Academic Press, New York, NY.

Rosa, W. de la, Lopez, F. L., and Liedo, P. 2002. *Beauveria bassiana* as a pathogen of the Mexican fruit fly (Diptera: Tephritidae) under laboratory conditions. *J. Econ. Entomol.* **95**, 36–43.

Schroer, S. and Ehlers, R-U. 2005. Foliar application of the entomopathogenic nematode *Steinernema carpocapsae* for biological control of diamondback moth larvae (*Plutella xylostella*). *Biol. Control* **33**, 81–86.

SAS, 2001. SAS Software: Version 8.2. SAS Institute, Cary, NC.

Shapiro-Ilan, D. I. 2001a. Virulence of entomopathogenic nematodes to pecan weevil larvae *Curculio caryae* (Coleoptera: Curculionidae) in the laboratory. *J. Econ. Entomol.* **94**, 7–13.

Shapiro-Ilan, D. I. 2001b. Virulence of entomopathogenic nematodes to pecan weevil adults (Coleoptera: Curculionidae). *J. Entomol. Sci.*, 325–328.

Shapiro-Ilan, D. I. 2003. Microbial control of the pecan weevil, *Curculio caryae*. *In*: "Integration of Chemical and Biological Insect Control in Native, Seedling, and Improved Pecan Production" (J. D. Dutcher, M. K. Harris, and D. A. Dean, Eds.) pp. 100–114. *Southwest. Entomol. Supplement* 27.

Shapiro-Ilan, D. I. and Cottrell, T. E. 2006. Susceptibility of the lesser peachtree borer (Lepidoptera: Sesiidae) to entomopathogenic nematodes. *Environ. Entomol.* **35**, 358–365.

Shapiro-Ilan D. I. and Gaugler, R. 2002. Production technology for entomopathogenic nematodes and their bacterial symbionts. *J. Ind. Microbiol. Biotech.* **28**, 137–146.

Shapiro-Ilan, D. I., Gouge, D. H., and Koppenhöfer, A. M. 2002a. Factors affecting commercial success: case studies in cotton, turf and citrus. *In* "Entomopathogenic Nematology" (R. Gaugler, Ed.) pp. 333–355. CABI, New York, NY.

Shapiro-Ilan D. I., Gaugler, R. Tedders, W. L., Brown, I., and Lewis, E. E.. 2002b. Optimization of inoculation for in vivo production of entomopathogenic nematodes. *J. Nematol.* **34**, 343–350.

Shapiro-Ilan, D. I., Mizell, R. F. III, and Campbell, J. F. 2002c. Susceptibility of the plum curculio, *Conotrachelus nenuphar*, to entomopathogenic nematodes. *J. Nematol.* **34**, 246–249.

Shapiro-Ilan, D. I., Lewis, E. E., Tedders, W. L., and Son, Y. 2003. Superior efficacy observed in entomopathogenic nematodes applied in infected-host cadavers compared with application in aqueous suspension. *J. Invertebr. Pathol.* **83**, 270–272.

Shapiro-Ilan, D. I., Cottrell, T., and Gardner W. A. 2004a. Trunk perimeter applications of *Beauveria bassiana* to suppress adult *Curculio caryae* (Coleoptera: Curculionidae). *J. Entomol. Sci.* **39**, 337–349.

Shapiro-Ilan, D. I, Mizell, R. F., Cottrell, T. E., and Horton, D. L. 2004b. Measuring field efficacy of *Steinernema feltiae* and *Steinernema riobrave* for suppression of plum curculio, *Conotrachelus nenuphar*, larvae. *Biol. Control* **30**, 496–503.

Shapiro-Ilan, D. I., Duncan, L. W., Lacey, L. A., and Han, R. 2005. Orchard crops. *In* "Nematodes as Biological Control Agents" (P. Grewal, R-U Ehlers, and D. Shapiro-Ilan Eds.) pp. 215–230. CABI Publishing, New York, NY.

Shapiro-Ilan, D. I., Gouge; D. H., Piggott, S. J., and Patterson Fife, J. 2006. Application technology and environmental considerations for use of entomopathogenic nematodes in biological control. *Biol. Control* **38**, 124–133.

Siegel, J. P., Tebbets J. S., and Vail, P. V. 2000. Prevalence and transovum transmission of *Bacillus thuringiensis* Berliner in a navel orangeworm colony. *J. Invertebr. Pathol.* **75**, 171–173.

Siegel, J. P., Lacey, L. A., Fritts Jr., R., Higbee, B. S., and Noble, P. 2004. Use of steinernematid nematodes for post harvest control of navel orangeworm (Lepidoptera: Pyralidae, *Amyelois transitella*) in fallen pistachios. *Biol. Control* **30**, 410–417.

Siegel, J. P., Lacey, L. A., Higbee, B. S., Noble, P., and Fritts Jr, R. 2006. Effect of application rates and abiotic factors on *Steinernema carpocapsae* for control of overwintering navel orangeworm (Lepidoptera: Pyralidae, *Amyelois transitella*) in pistachios. *Biol. Control* **36**, 324–330.

Sikorowski, P. P. 1985. Pecan weevil pathology. *In* "Pecan Weevil: Research Perspective" (W. W. Neel, Ed.) pp. 87–101. Quail Ridge Press, Brandon, MS.

Smith, M. T., Georgis, R., Nyczepir, A. P., and Miller, R. W. 1993. Biological control of the pecan weevil, *Curculio caryae* (Coleoptera: Curculionidae), with entomopathogenic nematodes. *J. Nematol.* **25**, 78–82.

Strand, L. L. (Ed.). 1999. "Integrated Pest Management for Stone Fruits". University of California, Oakland, CA.

Tedders, W. L, Weaver D. J., and Wehunt, E. J. 1973. Pecan weevil: suppression of larvae with the fungi *Metarhizium anisopliae* and *Beauveria bassiana* and the nematode *Neoaplectana dutkyi*. *J. Econ. Entomol.* **66**, 723–725.

Tedders, W. L., Weaver, D. J., Wehunt, E. J., and Gentry, C. R. 1982. Bioassay of *Metarhizium anisopliae, Beauveria bassiana,* and *Neoaplectana carpocapsae* against larvae of the plum curculio, *Conotrachelus nenuphar* (Herbst) (Coleoptera: Curculionidae). *Environ. Entomol.* 11, 901–904.

Toledo, J., Gurgúa, J. L., Liedo, P., Ibarra, J. E., and Oropeza, A. 2001. Parasitismo de larvas y pupas de la mosca mexicana de la fruta, *Anastrepha ludens* (Loew) (Diptera: Tephritidae) por el nemátodo *Steinernema feltiae* (Filipjev) (Rhabditida: Steinernematotidae). *Vedalia* **8**, 27–36.

Toledo, J., Ibarra, J. E., Liedo, P., Gomez, A., Rasgado, M. A., and Williams, T. 2005. Infection of *Anastrepha ludens* (Diptera: Tephritidae) larvae by *Heterorhabditis bacteriophora* (Rhabditida: Heterorhabditidae) under laboratory and field conditions. *Biocontrol Sci. Technol.* 15, 627–634.

Toledo, J., Rasgado, M. A., Ibarra, J. E., Gomez, A., Liedo, P., and Williams, T. 2006. Infection of *Anastrepha ludens* following soil applications of *Heterorhabditis bacteriophora* in a mango orchard. *Entomol. Exp. Appl.* **119**, 155–162.

Westwood, M. N. 1978. "Temperate Zone Pomology". W. H. Freeman and Company, San Francisco, CA.

Wood, B. W. 2003. Pecan production in North America. *Southwest. Entomol. Suppl.* **27**, 1–19.

Wraight, S. P. and Carruthers, R. I. 1999. Production, delivery, and use of mycoinsecticides for control of insect pests on field crops. *In* "Methods in Biotechnology. Vol 5 Biopesticides Use and Delivery" (F. R. Hall and J. J. Menn, Eds.) pp. 233–269. Humana Press, Totowa, NJ.

Wright, D. J., Peters, A., Schroer, S., and Fife, J. P. 2005. Application technology. *In* "Nematodes as Biological Control Agents" (P. Grewal, R-U Ehlers, and D. Shapiro-Ilan Eds.) pp. 91–106. CABI Publishing, New York, NY.

Yee, W. L. and Lacey, L. A. 2003. Stage-specific mortality of *Rhagoletis indifferens* (Diptera: Tephritidae) exposed to three species of *Steinernema* nematodes. *Biol. Contr.* **27**, 349–356.

Yee, W. L. and Lacey, L. A. 2005. Mortality of different life stages of *Rhagoletis indifferens* (Diptera: Tephritidae) exposed to the entomopathogenic fungus *Metarhizium anisopliae*. *J. Entomol. Sci.* **40**, 167–177.

Zalom, F. G., Van Steenwyk, R. A., Bentley, W. J., Coviello, R. L., Rice, R. E., Hendricks, L. C., Pickle, C. L., and Freeman, M. W. 2002. UC IPM Pest Management Guidelines: Almond Insects and Mites. University of California ANR Publication 3431.

Zhou, X. S., Kaya, H. K., Heunges, K., and Goodrich-Blair, H. 2002. Response of ants to a deterrent factor(s) produced by the symbiotic bacteria of entomopathogenic nematodes. *Appl. Environ. Microbiol.* **68**, 6202–6209.

Chapter VII-13

Application and evaluation of entomopathogens for citrus pest control

Clayton W. McCoy, Robin J. Stuart and Larry W. Duncan
Citrus Research and Education Center
University of Florida
Lake Alfred, FL 33850, USA

David I. Shapiro-Ilan
USDA-ARS
Southeast Fruit and Tree Nut Research Unit
Byron, GA 31008, USA

1 Introduction

Originally from Southeast Asia, citrus has been introduced into many arid or humid, subtropical to tropical regions around the world where it is cultivated in a diverse array of quasi-permanent tree-crop plantings that range from small inter-planted hillside patches to vast monocultural enterprises (Webber, 1948). Numerous arthropods infest citrus and some are considered serious pests locally or more globally either because of the direct damage they cause or because they transmit or otherwise facilitate important plant pathogens (*e.g.*, the brown citrus aphid, *Toxoptera citricida*, vectors the citrus tristeza virus (CTV), Poprawski *et al.*, 1999). Among the most widespread and chronic pests are several mite species, and their importance is often dictated by climate and horticultural practices. Homopterous pests (*e.g.*, aphids, scales, whiteflies, psyllids, etc.) are also common and exhibit considerable diversity and specialization. By comparison, relatively few hemipteran and lepidopteran insects are found feeding on mature fruit and foliage, respectively. Coleopterans (mainly root weevils) and ants are the predominant soil pests whereas fruit flies are important direct pests of fruit (Talhouk, 1975; Browning *et al.*, 1995).

As with other perennial tree crops grown in warm climates, numerous parasites, predators and pathogens inflict a high level of natural control on citrus arthropods (McCoy, 1985; Browning *et al.*, 1995). Overall, fungi appear to be the most prevalent entomopathogens attacking citrus arthropods, especially in humid regions such as shaded rainforest where citrus is planted as an under story crop. However, in monocultural plantings where tree canopies are hedged and topped regularly and cover cropping is not practical, the incidence of fungal pathogens is much lower. To date, natural infections of at least twenty-four mycopathogens, one virus, one gregarine, one microsporidium, and several species of entomopathogenic nematodes have been reported from citrus arthropods (McCoy, 1998; Duncan *et al.*, 2003).

Entomopathogens are widespread on citrus and some, such as *Aschersonia aleyrodes*, a fungal pathogen of citrus whitefly, *Dialeurodes citri*, have been isolated and used as microbial control agents of greenhouse pests (Landa and Jiranova, 1989). However, relatively

L.A. Lacey and H.K. Kaya (eds.), Field Manual of Techniques in Invertebrate Pathology, 567–581.

Table 1. Some arthropod pests of citrus and the entomopathogens known to be effective for their control

Target host	Entomopathogen	Rate of application	Selected references
Citrus rust mite, *Phyllocoptruta oleivora*	Fungus: *Hirsutella thompsonii*	2.2–4.4 kg/ha (2.8×10^8 CFUs/kg)	McCoy, 1985; 1996; 1998; McCoy and Couch, 1982
Citrus red mite, *Panonychus citri*	Non-occluded virus	0.1% aqueous suspension	Shaw *et al.*, 1968; van der Geest, 1985
Brown citrus aphid *Toxoptera citricida*	Hypocreales fungi	2.5–5.0 $\times 10^{13}$ conidia/ha	Poprawski *et al.*, 1999
Citrus root weevils, *Diaprepes, Pachnaeus, Asynonychus*	Fungus: *Beauveria bassiana*	53.8 ml/m^2 (2.1×10^{10} CFUs/ml)	Quintela and McCoy, 1998
	Fungus: *Metarhizium anisopliae*	5 g/m^2 (mycelia)	Quintela and McCoy, 1998
	Entomopathogenic nematodes: *Steinernema* spp., *Heterorhabditis* spp.	2.0–4.0 $\times 10^9$/ha; high rate for heavy soils[a]	McCoy, 1985; Duncan *et al.*, 1996, 2003; Morse and Lindegren, 1996; Shapiro-Ilan *et al.*, 2002

[a] Recommendation of Becker Underwood for control of *Diaprepes abbreviatus* with *Steinernema riobrave.*

few entomopathogens have been used either commercially or experimentally for citrus pest control (McCoy, 1998). Some major citrus pests and the entomopathogens known to be efficacious for their control are listed in Table 1.

The fungal pathogen, *Hirsutella thompsonii*, is infectious to mites at the conidial stage. Following conidial germination and cuticular entry into the host, hyphae ramify within the hemocoele and subsequently kill the host. This pathogen has been mass-produced, formulated and applied in the field for control of the citrus rust mite, *Phyllocoptruta oleivora*, and other acarine pests of turf, coconut, and greenhouse crops around the world (McCoy, 1996). Mite control can be achieved within 1–2 weeks when environmental conditions are optimal for fungal survival and infection. The first commercial formulation of *H. thompsonii* available in the USA was produced in 1976 by Abbott Laboratories (North Chicago, IL) as ABG 6065 and subsequently under the trade name Mycar®. An Experimental Use Permit was issued by the EPA in 1978 and registration granted in 1981, but this fungus is not available commercially today.

The citrus red mite, *Panonychus citri*, is a serious pest in California and other arid citrus-growing regions and causes premature defoliation and yield loss. A viral disease of this mite was reported around 1955 from Florida and California (van der Geest, 1985). This rod-shaped, non-occluded virus forms inside the nuclei of midgut epithelial cells. The virus causes epizootics in California and Arizona, and growers are advised to delay chemical sprays to allow natural epizootics to occur. Aqueous suspensions of titrated diseased mites sprayed on citrus trees gave satisfactory control in California (Shaw *et al.*, 1968). This virus has potential as a microbial control agent but instability and the necessity of growing the virus in living organisms are impediments to further development.

The brown citrus aphid, *Toxoptera citricida*, is considered the most efficient vector of citrus tristeza virus (CTV) and occurs in various regions around the world (Poprawski *et al.*, 1999). Citrus varieties grafted to a susceptible rootstock (*e.g.*, sour orange, *Citrus aurantium*) are subject to viral epidemics and subsequent destruction. Numerous hypocrealean (formerly classified within the class Hyphomycetes) fungi are pathogenic to the brown citrus aphid and commercial preparations of *Beauveria bassiana* (Mycotrol GHA®, Mycotrol ES9601, Mycotech Corp., Butte, MT), *Metarhizium anisopliae* and *Paecilomyces fumosoroseus* have been field-tested for microbial control of this pest (Poprawski *et al.*, 1999).

A complex of tropical root weevils including the genera *Diaprepes* and *Exophthalmus* are

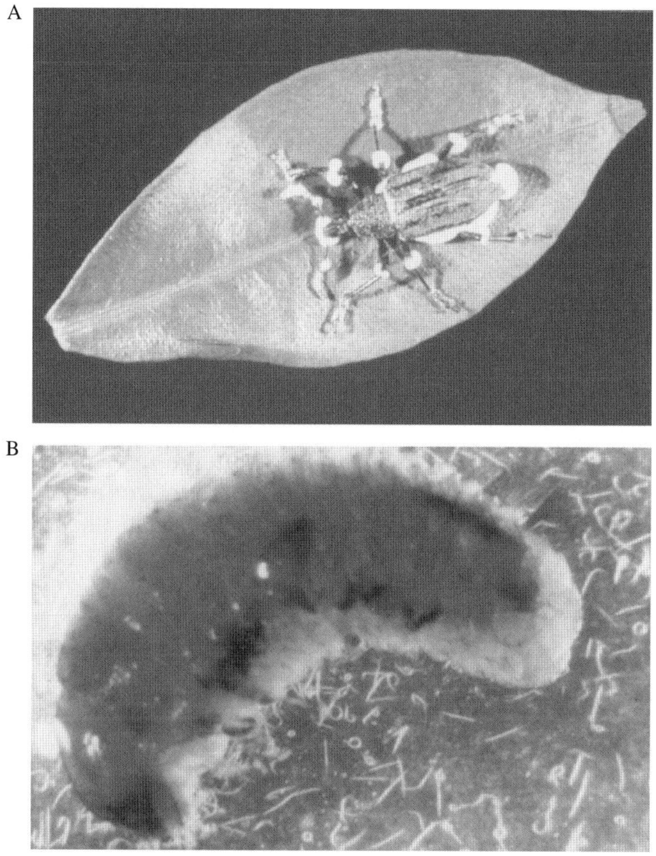

Figure 1. Diagnostic characteristics of cadavers infected with entomopathogens of selected citrus arthropods: A. Adult *Diaprepes abbreviatus* root weevil with fungal hyphae of *Beauveria bassiana* protruding from intersegmental sites of the exoskeleton; B. Infective juvenile nematodes of *Heterorhabditis bacteriophora* dispersing from the body of a late instar larva of *D. abbreviatus*

serious pests of citrus and other agricultural crops worldwide (McCoy, 1998; McCoy *et al.*, 2003). In citrus, the arboreal adults are of minor economic importance but larvae feeding on the roots reduce productivity and can cause tree death either directly or by facilitating infection by soil-borne pathogens (*e.g., Phytophthora* spp., Graham *et al.*, 2003). Both indigenous entomopathogenic fungi and nematodes infect adults and larvae in the soil (Figure 1). Experimental studies show that the fungus, *B. bassiana*, as a commercial product formulated in oil (Mycotrol ES9601) or a combination of fungus and a sublethal concentration of imidacloprid (a chloronicotinyl insecticide) are efficacious against neonate larvae and teneral adults when applied as a soil barrier (Quintela and McCoy, 1998). However, the efficacy of Mycotrol as a weevil control agent is hindered by its poor persistence in soil.

Numerous laboratory and field studies have been conducted in Florida citrus on the use of entomopathogenic nematodes (Steinernematidae and Heterorhabditidae) for biological control of citrus root weevils, particularly *Diaprepes abbreviatus* (McCoy *et al.*, 2002; Shapiro-Ilan *et al.*, 2002; Duncan *et al.*, 2003; Stuart *et al.*, 2004). In Florida, entomopathogenic nematodes have been marketed to growers for weevil control for over 15 years with two species, *Steinernema riobrave* (BioVector®, Becker Underwood, Ames, IA) and *Heterorhabditis indica* (Symbion-South®, BioControl Systems, Aurora, IN), currently available. These nematodes appear to be most effective at high temperatures ($27 \pm 2\,°C$) in sandy loam soils. Larval mortality rates of over 90% have been reported for field trials with

S. riobrave when applied at a rate of 1.2×10^{10} infective juveniles/ha (Duncan *et al.*, 1996). In California citrus, larvae of the Fuller rose weevil, *Asynonychus godmanni*, were suppressed by 80% using *S. carpocapsae* at 50–500 infective juveniles/cm^2 (Morse and Lindegren, 1996).

This chapter provides a description of field protocols developed to evaluate the performance of some of the above entomopathogens when applied to the fruit and foliage of citrus or to the soil beneath the tree canopy to suppress specific pests.

2 Protocols for application and evaluation of selected microbial control agents

A Selection of test site

In the selection of a citrus planting for the application of entomopathogens, numerous factors must be considered including the plant system (arrangement of trees and tree density, age, and health), target and non-target pests (abundance, distribution, natural enemies) and the microbe being applied (natural prevalence, abiotic factors influencing survival). One or more of these factors can influence the outcome of an experiment. For example, the selection of immature or atypical trees with small or weak canopies is inappropriate because less canopy shading will result in greater solar radiation that can influence the survival of entomopathogens applied to either the foliage or the soil beneath the tree (Duncan *et al.*, 1996).

Entomopathogens such as fungi and nematodes require irrigation to maximize survival and efficacy in the soil. Microsprinkler irrigation under the tree canopy supplies necessary moisture and offers a good way to apply microbes. Other horticultural practices such as fertilizer, pesticide, hedging, and weed control schedules can be of importance in selecting a test site. Information regarding these factors should be obtained from the cooperator. A close relationship between the scientist and grower is imperative to a successful experiment because an unexpected fungicide application can doom a mycopathogen field test.

The final step in the selection of a test site is the preparation of a detailed plot diagram with each tree within the experimental site designated on a map. A copy of the diagram should be retained by the grower cooperator.

B Plot design

Different plot designs have been used for comparative testing of entomopathogens as foliar or soil treatments on citrus. Both completely randomized and randomized block designs have been used for both small and large-scale tests. Plot size can vary widely from single-tree to multiple-tree plots centrally located with surrounding untreated or treated buffer rows. Replication can differ widely and is influenced by plot size or by quantity of entomopathogen available for testing. Buffer zones and plot sizes can also be influenced by motility of target insects and rates of infestation (continual, intermittent).

Plot size can be an especially important factor. For example, tests have shown that both healthy and diseased mites are readily dispersed among trees and even among groves by the wind (Bergh and McCoy, 1997). Thus, when a fungus or virus is applied as a foliar spray to control phytophagous mites in a comparative study on citrus, small plots are ineffective because untreated controls are readily contaminated by other treatments (Shaw *et al.*, 1968; McCoy and Couch, 1982). For these studies, plot size should be no less than 1 ha and only a small central area of the plot (*e.g.*, 4 trees) utilized for sampling. Larger plots (> 0.5 ha) also tend to be better for assessing treatment effects on certain natural enemies or various other non-target organisms.

C Plot modifications

Larvae and teneral adults of root weevils appear to aggregate within the rhizosphere of the citrus tree at relatively low population densities and their distribution from tree to tree can be highly variable and unpredictable (*e.g.*, McCoy *et al.*, 2003). Therefore, artificial seeding of laboratory-reared larvae has been used to increase larval numbers in plots for experimental purposes. The procedure for artificial infestation is as follows:

1. Pre-irrigate the grove for about 2 h (5-cm depth).
2. Flag designated trees for infestation and record them in the plot diagram.
3. Place 400 neonate larvae (~ 48 h old) obtained from a greenhouse weevil colony into snap cap micro-centrifuge tubes (volumetric estimate or visual count). Micro-centrifuge tubes should contain about 1 g of moistened soil to prevent larval injury.
4. Place larvae in a cooler (10–12 °C) for transport to the field.
5. Scatter contents of each vial on the soil beneath the tree within 10–20 cm of the trunk.
6. Perform larval infestation late in the day or under cloudy skies.
7. Assess experimental results by adult trapping at emergence or destructive tree sampling and counting larvae recovered from the rhizosphere via soil sieving.

D Preparation of the field inoculum

The foliage of citrus trees is tolerant of a wide range of spray adjuvants (spreader-stickers, surfactants and anti-transparents) used to enhance the efficacy and persistence of sprayable mixtures. Spray mixtures have been tested in the field in combination with fungal inocula in free form as adjuvants or with conidial preparations commercially formulated as wettable powder (*e.g.*, Mycar®, Abbott Laboratories) or as oil mixtures (*e.g.*, Mycotrol ES, Mycotech Corp.) to improve fungal deposition on the leaf (McCoy and Couch, 1982). Petroleum-based spray oils are widely used as combination spreader-sticker/pesticides. When the adjuvant is not included in the formulated product, it can be added to the tank mix after the fungus has been thoroughly mixed with water under vigorous agitation. Generally, adjuvants are applied at 0.05–5.0% (v/v) without affecting the inoculum or causing phytotoxicity to the plant.

Prior to application of unformulated or formulated fungal preparations with and without adjuvants, the viability of the conidial suspension should be determined. Plating techniques are the preferred method to measure viable propagules per unit volume, but at least 3 days must elapse before a reading can be taken. This delay is generally unacceptable. Therefore, quantification methods using a hemacytometer have been used with some success (Goettel and Inglis, 1997). Conidial mixtures (20–30 ml) should be placed in nutrient broth in a test tube and agitated for 30 sec. The suspension is then incubated on a shaker at 27 °C at a moderate speed (100–120 rpm) for 12 h to allow for germination. Ten random samples/treatment are taken via pipette and transferred to a hemacytometer for counting with phase microscopy (420X). Conidia are considered viable if germ tube lengths are twice the diameter of the propagule in question and a count of 50 to 100 conidia/sample is considered appropriate. With some adjuvants, visibility of the conidia in the hemacytometer may be limited by its properties. In such cases, the suspension should be diluted with water at the time of the count.

Prior to the application of nematodes in the field, viability should be determined by examining a sample of 50–100 nematodes via a dissecting microscope and counting the live and dead nematodes (Kaya and Stock, 1997). When probed with a needle, live nematodes that are merely immobile will respond to the stimulus. Ideally, viability should be checked before transport to the field, at the spray tank, and at the nozzle to assure viability during the entire application process. The spray tank should be triple rinsed if chemicals have been used previously because nematodes and fungi are not compatible with some agri-chemicals (Table 2). However, in some cases, pesticides act synergistically with entomopathogens and could improve the efficacy of applications (Quintela and McCoy, 1998; Grewal, 2002).

Fungal and nematode preparations should always be stored out of direct sunlight and kept cool (12–25 °C) prior to use in the field. When tank mixing nematode preparations in the field, an estimate of the number of nematodes per container can be obtained by counting the number volumetrically. With the sprayer agitator running, the nematodes are added to the holding tank containing a known amount of water (50% of capacity). Then, the remaining water is added to bring the tank mix to a designated concentration. The pH of the water should always fall between 4 and 8. Water dispersible granule formulations of nematodes can be premixed in water before being added

Table 2. Nematodes can be used simultaneously with most citrus-registered pesticides and fertilizers at standard rates with the following exceptions

Schedule	Chemical name	Trade name(s)
Do not use 1 week before / after nematode application	**HERBICIDES**	
	2, 4-D	Various names
	Triclopyr	Turflon®, Confront®
	INSECTICIDES	
	azinphosmethyl	Guthion®
	bendiocarb	Rotate®
	chlorphyrifos	Dursban® (non-encapsulated)
	methomyl	Lannate®
Do not use 2 weeks before/after nematode application	**INSECTICIDES**	
	carbofuran	Furadan®
	ethoprop	Mocap®
	NEMATICIDES	
	aldicarb	Temik®
	fenaminphos	Nemacur®
	oxamyl	Vydate®

Recommendation of Becker Underwood for *Steinernema riobrave.*

to the spray tank. When this is not possible, the agitator should be run fast enough to thoroughly dissolve the dry preparation almost immediately as it is slowly poured into the tank. Fungi require similar procedures when tank mixing. Many different formulations exist (see Grewal, 2002), and the manufacturer's instructions should be followed closely for mixing and application.

E Application of inoculum

1 Foliar

Fungal and viral spray mixtures applied as foliar sprays to citrus trees have been delivered successfully using hydraulic and air blast systems designed for chemical pesticide application. Two basic sprayers have been used. For small plot work, a hydraulic hand sprayer with a single nozzle is normally used. The amount of liquid applied per tree can be estimated by simply measuring the discharge in the holding tank. From this information and a knowledge of the concentration of the spray mix, an estimate of foliar coverage can be calculated.

The general procedures for applying a microbial using a hydraulic system are as follows:

1. Apply water to 3 or more trees to spray runoff after setting spray pattern for hand applicator. Spray applicator should always apply the finished product at a fixed distance from the tree (approximately 3 m) and travel clockwise around the tree during application.
2. Calculate the average volume of spray required to treat a whole tree to runoff.
3. Based on the number of trees/treatment, tank mix the total finished spray needed to complete the spray operation.
4. For mature canopy trees, set pump pressure at 3100–3450 kPa (450–500 psi) to assure penetration of the tree canopy.
5. Always apply the lowest spray concentration first and continue to the highest.
6. Monitor weather conditions, and never spray when the wind speed exceeds 16 km/h.

The conventional sprayer for large plot studies is the air blast type. Tractor-drawn air blast sprayers function on the principle of air displacement within the tree canopy. Coverage is more precise and application is less labor intensive than for a hydraulic hand sprayer. Nozzles are attached to the sprayer at fixed locations and travel past the trees at a constant speed (3–5 km/h). Normally one-half the spray volume is directed to the upper half of the tree and one-half to the lower by arranging the nozzles and air deflectors. Both high and low volumes (preferred for microbials) can be applied with an air blast sprayer. Smaller trees can be sprayed with the same concentration of microbial control agent by using fewer nozzles.

Generally, microbials have been applied in the range of 2,366–4,733 liters/ha. The pH of the tank mixture should be maintained in a range of 4.0–8.5 and application made late in the day to reduce solar effect. The general procedures for application using an air blast system are as follows:

1. Select intended application rate by:
   a. Determining nozzle arrangement (based on tree height).
   b. Measuring spray discharge for nozzle arrangement.
   c. Calibrating sprayer ground speed (distance traveled in time).
2. Prepare tank mix based on plot size and ensure proper agitation and spray mixture compatibility.
3. Monitor weather conditions (wind speed, rainfall, temperature) before and during application.
4. Maintain a two-row buffer between plots.

2 *Soil*

Both fungal and nematode mixtures have been applied beneath the tree canopy for control of weevil larvae by using herbicide applicators and chemigation systems (injection via low volume irrigation systems). Using a tractor mounted herbicide spray unit with a 1.5–2.5 m boom, a treated band can be uniformly applied to the soil surface beneath the tree between the trunk and the canopy edge. Fungal conidial preparations are applied in 285–475 liters water/ha at 2.4 km/h. For uniform coverage #10 flood jets at 30 cm spacing with a #80C outside nozzle assures coverage of both sides of the tree. Nematodes should be mixed in a minimum of 300 liters of water allowing for the extra volume remaining in the tank after spraying. Apply under moderate agitation to assure thorough distribution of the tank mix and adequate aeration during application. For best results, nematodes should be applied at 900–950 liters/ha and operating pressures not exceeding 2000 kPa (290 psi). However, for at least one nematode species (*H. megidis*), operating pressures should not exceed 1380 kPa (200 psi) (Fife *et al.*, 2003). The pH of the water should be maintained in the range of 4–8. Pre-irrigate the grove to assure soil moisture to a depth of 5 cm.

The application of fungi and nematodes via under tree microsprinkler irrigation is a feasible strategy, which offers the advantage of the treated area being limited to the soil surface beneath the tree. This can result in cost savings in material used in application, especially for young trees. Sprinkler emitters should be positioned about 30 cm from the tree trunk and be of a type to assure about 80% coverage of the root zone. Ideally, they are equipped with 360° nozzles (Fan-Jet, Bowsmith Inc., Avon Park, FL) capable of covering a 1.8–2.4 m radius with a volume output of 38–76 liters/h (10–20 gal/h). Volume output at the injection pump should be adjusted to a rate based on tree number and type of irrigation system. By knowing the volume of the holding tank and rate of output, the inoculum concentration can be determined. Add sufficient nematodes to the tank to achieve the desired inoculum concentration and inject all during a 30–60 min period. Whether applying nematodes or fungi, the content of the holding tank must be agitated throughout the application. Applications should be carried out in late afternoon or evening to reduce exposure to UV radiation and increase the number of hours of high humidity following application. In line variability at the emitters and variation in coverage beneath the tree can be determined during application by collecting spray mix in collecting pans placed on the ground beneath the tree. Viability of the inoculum also should be checked at the holding tank and at the nozzle to ensure effective application. Inoculum concentration is often reduced near the ends of irrigation lines (Duncan, unpublished).

The following stepwise procedure is advised:

1. Pre-irrigate the grove if necessary to achieve uniform soil moisture to at least 5 cm (deeper for sandy soils).
2. Prepare the irrigation system by removing screens and filters if possible. If screens must remain, a U.S. mesh screen size of 50 or coarser should be used.
3. Do not subject nematodes to excessive pump pressures (usually < 2000 kPa or 290 psi).
4. Calibrate the injection pump to deliver the concentrate over a 30–60 min period.
5. Irrigate for 15–30 min immediately prior to starting injection to cool the lines and establish a wetting pattern under the trees.
6. Water dispersible granules can be poured directly into the tank containing water or a slurry can be

pre-mixed and then added to the tank water. The water temperature should be between 10 and 30 °C.

7. The nematode suspension can be held in the agitated tank for up to 24 h, but the tank must be kept cool (under 27 °C).
8. Irrigate the grove after application to a depth of 1.5–2.5 cm of water.

F Pre- and post-treatment sampling of target insects

Knowledge of the seasonal population dynamics of an arthropod pest is basic to designing an appropriate sampling and application plan for microbial control. Each year, citrus produces new leaves that have a life expectancy on the tree of about 1.5 years. For foliar and soil pests, seasonal population dynamics are often governed by abiotic and biotic factors, and diseases are often involved. Insect diapause is virtually non-existent in citrus growing areas, and seasonal sampling for pests should begin when the rate of increase in the abundance of the pest on feeding sites such as new citrus flush or fruit begins to rise. Many entomopathogens such as the fungus, *H. thompsonii*, cause predictable natural epizootics at high host densities during the summer (McCoy, 1996), and knowledge of when epizootics occur is an important consideration for the sampling plan and the timing of applications of microbial control agents. If the microbial control agent is being applied as a "prophylactic" treatment prior to the onset of a natural epizootic then it is likely that cadavers resulting from the treatment will be mixed with those dying from the natural inoculum.

Methods for estimating mite and aphid populations involve counting the numbers of one or more developmental stages of the arthropod on fruit or leaves in the field or removing the arthropod stage(s) from the fruit or leaves and counting a known fraction in the laboratory. Because of their small size, mite counts are frequently performed with the aid of a hand lens in the field or a stereoscopic microscope in the laboratory. Counts can involve the whole leaf or fruit or most often just a fixed amount of the surface area.

Sampling of the diminutive citrus rust mite and its fungal pathogen, *H. thompsonii*, involves different methods in the field and laboratory. For large field trials, an estimate of mite population density and disease prevalence on fruit or leaves before and after a spray application is made using a gridded hand lens (20X). Live, dead and dead diseased adult mites observed within a $1-\text{cm}^2$ area are easily distinguished: live mites are clear and mobile, dead mites are deformed, and dead diseased mites are deformed with filamentous hyphae protruding from their collapsed bodies. However, live mites with disease cannot be distinguished from those without disease using this method.

1 Sampling procedures for citrus rust mites and its fungal pathogen

a Field (direct method)

1. Sample four adjacent trees from the center of each plot at any time of the day.
2. Select at random 5 fruit/tree or 20 fruit/4 tree plot.
3. Using a hand lens (20X) with a 1 cm^2 grid attachment, count the number of live, dead and diseased mites.
4. Count the mite forms at 2 random sites/fruit. A minimum sample/treatment should be in the range of 100–120 fruit.
5. Perform 2 pre-treatment samplings one week apart and post-treatment sampling at weekly intervals.

Using this sampling frequency, the immediate effect of the microbial application and its persistence can be determined with a declining mite population and increasing prevalence of infection in the field. The fact that this sampling procedure can be performed quickly in the field is an important advantage and makes this the preferred method.

An indirect sampling procedure also has been used with good results in estimating more accurately the prevalence of diseased mites in a population. This procedure is as follows:

b Laboratory (indirect method)

1. Perform 2 pre-treatment samplings one week apart and post-treatment sampling at weekly intervals.
2. Sample four adjacent trees from the center of each plot at anytime of the day.
3. Collect at random 10 leaves/tree from the inner and outer canopy at 1.5 m above the ground.
4. Place leaves for each plot in separate paper bags. Keep leaves cool during transport to the laboratory.

5. In the laboratory, count the number of live, dead and visibly diseased mites in 2 microscope fields selected from the upper and lower leaf surface at 100X magnification using a stereoscopic microscope.
6. Select randomly no more than 50 live and dead mites from a given treatment, clear separately in 1.0% lactic acid or Nesbitt's clearing reagent by placing the mite in a droplet of clearing agent and heating for 30 sec.
7. After mounting cleared mites in Hoyer's solution, examine with a phase contrast microscope to confirm infection.

The accumulated percentage of diseased mites for each treatment is obtained by dividing the accumulated total of diseased mites by the total number sampled. Although more time consuming, this method is more precise than the field method.

2 *Sampling citrus red mite and its virus disease*

To estimate the effect of a foliar spray of an aqueous suspension of titrated diseased mites with a non-occluded virus on citrus red mite populations, an indirect sampling procedure has been used in the field. The procedure is as follows:

1. Count adult female citrus red mites twice per week for 2 weeks pre-treatment and weekly thereafter.
2. Collect 20 leaves randomly per treatment in paper bags. Keep leaves cool during transport to the laboratory.
3. Mites are removed from leaves manually or with the aid of a mite brushing machine (Morgan *et al.*, 1955).
4. If available, 20 live and 20 dead mites (or any combination equaling 40 individuals) are mounted in Hoyer's medium and examined microscopically.
5. Virus infection is confirmed by the presence of birefringent crystals.

3 *Sampling brown citrus aphid and its fungal agents*

The brown citrus aphid presents different circumstances in terms of sampling. Aphids develop on the newly-formed leaf flush of growing terminals forming on the tree particularly in the spring. Therefore, the sampling unit is the terminal. Pre-treatment and post-treatment population estimates are made by counting the live and dead aphids on each terminal with a hand lens (10X). According to Poprawski *et al.* (1999), fine mesh sleeve cages are needed to exclude predatory insects. Cages are placed carefully over the treated and untreated terminals immediately after treatment.

4 *Sampling entomopathogens of citrus root weevils*

Numerous sampling methods have been employed in citrus groves to measure directly or indirectly the effect of either fungi or nematodes on adult root weevils on the tree and larvae in the soil. For adult weevil control on the tree with a fungal agent, trees are generally sprayed with an air blast sprayer. The following procedures are used to assess efficacy:

1. Immediately after spray application, 3–5 adult weevils are placed in a standard screened limb bag (30 × 46 cm).
2. For treated and untreated foliage, infested bags are placed over clumps of shoots with new leaves (preferred food source) and attached to the limb with cord to prevent weevil escape.
3. After one week, the bags are removed and weevil mortality recorded.
4. Samples from treatments are replicated a minimum of 5 times (15–25 weevils/treatment).
5. Dead weevils are returned to the laboratory and placed in a humidity chamber at 95–100% RH for 3–4 days to confirm fungal mycosis.
6. The procedure is repeated weekly on selected terminals to measure fungal contact effect and persistence.

Citrus root weevil females deposit their eggs in masses sealed between leaves in the tree canopy. Hatching neonate larvae drop from the canopy and enter the soil to feed on roots and will emerge from the soil as adults mainly in the spring and summer months of the next year (McCoy *et al.*, 2003). Two types of ground traps have been used to monitor adults that emerge from soil in the year following treatments to control larvae (Stansly *et al.*, 1997; Duncan *et al.*, 2001; McCoy *et al.*, 2003). These traps are referred to as cone traps and Tedders traps (Figure 2). For both types of traps, the trap top or "collector" generally consists of the top portion of a cotton

A

B

Figure 2. Ground traps used to catch adult root weevils beneath the tree canopy: A. cone trap; B. Tedders trap (Photos by Ian W. Jackson)

boll weevil trap (Great Lakes IPM, Vestaburgh, MI). The collector consists of a small screen cone (10 cm high × 11.5 cm dia) inside a plastic frame that has a removable ventilated plastic cup fitted on top. The collector is secured to the trap base with screws. The cup holds the weevils after they enter the collector and pass through a 1-cm dia. hole at the top of the screen cone.

The bottom portion of the cone trap is constructed of standard galvanized hardware cloth (0.3-cm mesh size) with a base of 0.9 m in diameter, and covers about 0.6 m^2 of soil surface. The cone trap is placed beneath the tree as close to the trunk as possible with the bottom rim of the cone buried in the soil (Figure 2A). Adults emerging from the soil beneath the cone trap crawl up the wire mesh, pass through a 1-cm dia hole in the top of the cone, and are captured in the collector.

A modified Tedders trap (or citrus Tedders trap) has a pyramid-shaped base 61 cm high. The base is formed from two roughly triangular-shaped pieces of black masonite or corrugated polyethylene (PBE Graphics, West Palm Beach, FL) that are slotted in the middle to half their lengths in opposite directions so that they fit together and thereby form a pyramid with four flanges extending in different directions (Figure 2B). The Tedders trap is placed just inside the margin of the tree canopy and is anchored tightly to the soil with wire loops or hooks running over or through the flanges (Wireco, Winter Haven, FL). Newly emerged weevils tend to crawl up the trunks of trees and the Tedders trap catches adults exhibiting this behavior. It has been estimated that 100 traps/ha are generally necessary to collect meaningful data on weevil emergence (Bullock *et al.*, 1999). A statistical comparison of Tedders trap catches from various sites indicates that the optimal number of traps depends on weevil abundance and the degree of precision desired in the results (Duncan *et al.*, 2001).

A Tedders trap usually catches more adults than a cone trap because, unlike the cone trap, it is not restricted to catching weevils emerging from the soil directly beneath the trap. Moreover, weevil catches in cone traps and Tedders traps are not always correlated, and Tedders trap catches might be a better indication of general weevil activity than weevil emergence *per se* (McCoy *et al.*, 2003). Because Tedders traps capture weevils from a broad area of unknown size, they should be used to monitor weevils only in relatively large experimental plots (*e.g.*, 0.5–1.0 ha), whereas cone traps are appropriate for either small or large-plot research. Because of lengthy larval developmental times, both types of traps should be monitored weekly throughout the year of a larval treatment and through the following spring (late June) to properly assess the efficacy of a larval treatment. Adult catches should be examined in the laboratory for microbial infection. For fungi, adults should be incubated in a humidity chamber at 95–100% RH for a week after being captured.

The most common method for quantifying fungal inoculum on or in the soil is based on dilution plating (Goettel and Inglis, 1997). The abundance and persistence hypocrealean fungi at the soil surface beneath the tree can be obtained by collecting a minimum of 10 g of soil from randomly selected sites with a spatula. If soil depth is a consideration, a #9 cork borer is used to cut soil columns to a depth of 15 cm. Usually 5 cm increments are separated and then pooled. A pooled soil sample of about 30 g is adequate to estimate propagule density using the soil dilution plate method. The procedure for estimating fungal colony-forming units (CFUs) is as follows:

1. In the laboratory, the soil sample contained in a plastic bag is thoroughly mixed by hand shaking.
2. A 10 g sample is placed in a 50 ml conical centrifuge tube containing 30 ml of sterile distilled water.
3. After vigorously shaking the tube for 30 sec, serial dilutions are performed using a 100 μ*l* inoculum. Additional dilutions (10^{-2}, 10^{-3}, and 10^{-4}) can be made based on an estimate of propagule density in the soil.
4. Two-hundred microliters of diluted suspension is spread on divided Petri plates (two inoculations/plate) containing oatmeal-dodine selective media.
5. After 7–10 days incubation at 27–28 °C, the number of fungal propagules per 200 μ*l* is determined per plate.
6. Plate values are then converted to CFUs/g of soil.

In addition to quantifying fungal inoculum in treated and untreated field soil, soil column bioassays can be performed simultaneously to confirm pathogenicity using laboratory-reared neonate larvae. The procedures are as follows:

1. Sample all treatments at 0, 7, 14, 21, and 30 days post-treatment.
2. Collect 20 soil cores using a #9 cork borer (area 1.27 cm^2) to a depth of 2.5 cm randomly from an area halfway between the trunk and tree skirt beneath the tree.
3. Upon collection, each intact soil core is carefully placed in a plastic tube with a screen base (20 mm mesh) to allow for larval escape into the well of a plastic tissue culture plate (Figure 3).
4. In the laboratory, 10 vigorous neonate larvae (48-h-old) are placed on the soil surface of each column.
5. After 72 h at 25 °C, the number of larvae capable of moving through the soil column to the catch well of the bioassay unit are counted as live, dead or missing.
6. Dead larvae are held at 27 °C at 95–100% RH for 10 days to assess fungal mycosis.

Generally, nematode density in the soil after an inoculative application is correlated with host infection in time (Duncan *et al.*, 1996).

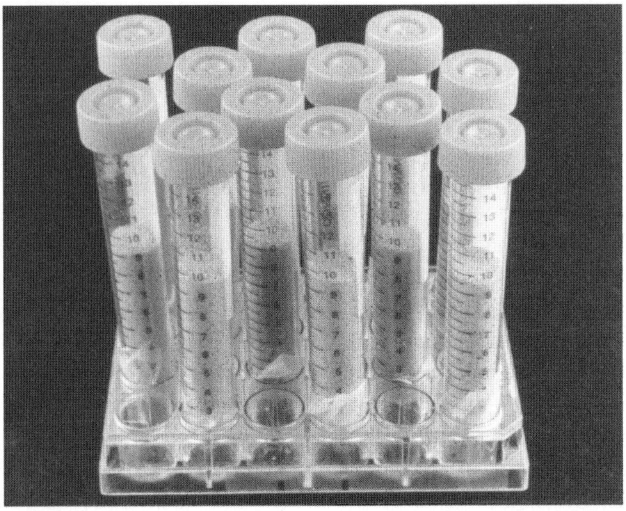

Figure 3. The bioassay unit used to test the efficacy and persistence of fungal conidial inocula applied to the soil surface

Nematode density and persistence can be readily measured in time using the following procedures:

1. At predetermined intervals after application, 10–15 soil probe samples to a depth of 15 cm are taken from beneath the tree canopies in each plot and pooled to form a single sample.
2. Samples are obtained from the same location relative to the irrigation emitter about halfway between the trunk and the dripline of the tree.
3. The soil samples are collected in a plastic bag for transport to the laboratory in a cooler.
4. In the laboratory, soil is thoroughly mixed, and then a 60 ml aliquot is placed in a Baermann funnel (Kaya and Stock, 1997).
5. After 72 h, nematodes trapped in the water are poured into test tubes (100 ml) and allowed to settle overnight.
6. After aspirating excess water, the infective juvenile (dauer stage) nematodes collected from the bottoms of the tubes are counted microscopically using a gridded Petri dish.
7. Taxonomic keys and known samples of live *Heterorhabditis* or *Steinernema* are used as a reference in counting.

The biological activity of nematodes applied to the soil surface for suppressing weevil larvae in the citrus rhizosphere can be estimated using a tree removal/soil sampling procedure (Duncan *et al.*, 1996). Briefly, trees are uprooted using a tractor-mounted front-end loader with a fork head attachment. By vigorously shaking the tree *in situ*, most of the soil adhering to the roots is removed along with any larvae. The tree is then placed on a plastic tarpaulin in the row middle and the soil lodged in the tree crown area is removed by flushing the root zone with high-pressure water delivered by a sprayer applicator. Larvae from the crown area are then collected from the tarp. To estimate the number of larvae remaining in the upper soil dislodged from roots (30 cm depth) and lower rhizosphere (30–60 cm depth), a soil sample of about 0.25 m^3 is collected using a shovel and then sieved through a coarse screen to separate the larvae from the soil.

Two field methods using caged weevil larvae, either with or without a host plant, have been devised to estimate nematode efficacy. A buried cage without a plant was designed for use in the soil at various depths. This cylindrical cage (7.6 cm length and 1.9 cm dia) was constructed from a 6.5 ml plastic vial and galvanized steel screen (2-mm mesh). The vial was cut into two parts to form the ends of the cage and the screen formed a cylindrical portion in the middle that was glued to both ends. Unfortunately, this cage leaves the larva open to predation by ants, which is problematic for interpreting experimental results. Consequently, a predator proof cage was designed using a 225 mesh stainless steel in-line sprayer filter (7 cm length × 3 cm dia.) (Chemical Container, Lake Wales, FL). The cylindrical filter is capped on both ends using polyethylene snap caps (3.0 cm outside diameter). For both types of buried cages, it is useful to secure a 40 cm wire through one or both ends of the cage. When the cage is buried vertically, the wire protrudes through the soil surface and can be flagged so that the cage is more easily located and excavated after field exposure.

Larval instars of *D. abbreviatus* vary considerably in their susceptibility to *S. riobrave* with younger instars being much more susceptible than older instars (Stuart and McCoy, unpublished). Thus, the use of a particular instar (*e.g.*, 6th instar) as indicated by head width measurements (see Quintela *et al.*, 1998) is essential for larval assays of nematode efficacy to produce consistent results. The procedure for use of the cages is as follows:

1. Place one 6th instar larva in each cage and fill the cage with sterilized sand at 8% moisture by weight.
2. Place the cage in the soil beneath the tree using a probe to cut a hole to a given depth (15–20 cm). Fill the open hole with soil.
3. Bury at least 4 cages per plot about 1 m inside the margin of the tree canopy.
4. Remove the cage from the soil after 7 days exposure in treated and untreated plots.
5. Repeat this process at 0, 1, 2, 3, and 4 weeks post-treatment.
6. Record the number of healthy and dead larvae in the laboratory; place each dead larva in a Petri dish containing moist filter paper and a wax moth (*Galleria mellonella*) larva to detect nematode parasitism. *G. mellonella* larvae are extremely susceptible to entomopathogenic nematodes (Kaya and Stock, 1997).

A different field cage was designed for burial to a depth of 15 cm with the top (16.5 cm dia) open to the surface at the soil line. This cage was made of polyvinyl chloride pipe (16.5 cm diameter) with a fine mesh screen base (225 mm mesh) on the bottom to allow for water leaching and predator exclusion. This pot-like cage can be used to support a citrus plant as food for weevil larvae added to the trap. The procedure for use of the cage is as follows:

1. Inoculate each soil-filled cage with 15, 6th instar larvae a few days before placing the cages in the field.
2. Bury larval infested cages, 2/plot, in the field in a row, one week prior to nematode application. At this time, cut the top off the seedling to expose the soil surface to the treatment. A bead of Tanglefoot (The Tanglefoot Company, Grand Rapids, MI) is then placed on the upper edge of the cage to prevent invasion by predators.
3. Remove the cages from the soil at one week post-treatment.
4. Separate the plant from the soil and sieve all soil to recover live and dead larvae.
5. In the laboratory, diagnose nematode parasitism using the procedure described above.

G *Assessing tree health*

The following methods have been developed for assessing tree health (or tree decline) as influenced by citrus pests and their biological control agents.

1 *Fibrous root density*

Root mass is directly related to shoot growth of a citrus tree. Procedures for measurement are as follows (*e.g.*, Duncan *et al.*, 1994):

1. Perform sampling in June and November annually.
2. Using a standard soil probe, take 1 sample/tree to a depth of 25 cm halfway between trunk and canopy dripline from 16–20 trees/plot or increase the number of cores/tree to result in 20–30 samples.
3. Pool soil from each plot into one plastic bag.
4. Store soil samples in cold room at 4 °C until processed.
5. Separate citrus fibrous roots from soil and debris via wet sieving (100 mm mesh) and hand picking with forceps.
6. Determine dry root weight/sample.

2 *Tree canopy size and density*

The foliage of a citrus tree canopy generates the photosynthate that in turn determines fruit yield. Procedures for measurement of photosynthetically active radiation (PAR) transmittance and citrus canopy volume are as follows (Albrigo *et al.*, 1975; McCoy *et al.*, 2004):

1. Perform sampling in June and November annually.
2. Measure photosynthetically active radiation (PAR) for a given number of trees using a Accupar (Decagan Devices, Inc., Seattle, WA). Perform 2 readings/tree beneath the tree canopy at a fixed

location at mid-day with full sunlight. Calculate percent PAR transmittance.
3. Determine canopy volume for a given number of trees by taking the measurements necessary for use of the following formula. The canopy of a citrus tree can approximated as being a half prolate spheroid above a cylinder, and the volume of the tree canopy (vc) can be calculated using the formula: $vc = \partial R^2(2X/3+Y)$ where $X = HT–HD$ and $Y = HD–HS$ with R=half the tree diameter at the widest point, HT=the overall tree height, HS=the skirt height, and HD=the height from the ground to the widest point of the tree.

3 Tree trunk growth

Citrus trees in decline will show little or no increase in trunk growth from year to year. Procedures for measurements are as follows (McCoy *et al.*, 2004):
1. Perform sampling in June and November annually.
2. Select at random a given number of control and treatment trees and flag for identification.
3. Measure the diameter of the tree trunk at approximately 5 cm above the budunion with calipers.
4. Compare the relative growth rate of the cross sectional area of the trunk between control and treatment trees over time.

4 Fruit yield

Yield is the most difficult measurement of tree health to be collected mainly because of labor requirements, yet it is the best measure of treatment profitability. Procedures for measurements are as follows:
1. Harvest a minimum 100 trees/ha or 40% of a given plot.
2. Weigh the total fruit/tree.

5 Fruit quality

The citrus rust mite inflicts a peel injury to the fruit that is visible to the naked eye. Fruit quality in experimental plots can be determined for various treatments using the following procedure:
1. Flag a given number of trees selected randomly per plot (4–8).
2. Insert a frame representative of $0.37\,m^2$ at 4 cardinal directional sites within the tree canopy (Stout, 1962).
3. Examine each fruit within an imaginary tunnel extending from the frame to the center of the tree for the presence or absence of mite injury.
4. Calculate percent fruit injury.

3 References

Albrigo, L. G., Anderson, C. A., and Edwards, G. J. 1975. Yield estimation of 'Valencia' orange research plots and groves. *Proc. Fla. State Hort. Soc.* **88**, 44–49.

Bergh, J. C., and McCoy, C. W. 1997. Aerial dispersal of citrus rust mite from Florida citrus groves. *Environ. Entomol.* **26**, 256–264.

Browning, H. W., McGovern, R. J., Jackson, L. K., Calvert, D. V., and Wardowski, W. F. 1995. *Florida Citrus Diagnostic Guide*. Florida Science Source, Inc., Lake Alfred, FL. pp. 1–244.

Bullock, R. C., Pelosi, R. R., and Killer, E. E. 1999. Management of citrus root weevils on Florida citrus with soil-applied entomopathogenic nematodes. *Fla. Entomol.* **82**, 1–7.

Duncan, L. W., Graham, J. H., Dunn, D. C., Zellers, J., McCoy, C. W., and Nguyen, K. 2003. Incidence of endemic entomopathogenic nematodes following application of *Steinernema riobrave* for control of *Diaprepes abbreviatus. J. Nematol.* **35**, 178–186.

Duncan, L. W., McCoy, C. W., and Terranova, A. C. 1996. Estimating sample size and persistence of entomogenous nematodes in sandy soils and their efficacy against the larvae of *Diaprepes abbreviatus* in Florida. *J. Nematol.* **28**, 56–67.

Duncan, L. W., McCoy, C. W., Stansly, P. A., Graham, J. H., and Mizell, R. F. 2001. Estimating the relative abundance of adult citrus root weevils (Coleoptera: Curculionidae) with modified Tedders traps. *Environ. Entomol.* **30**, 939–946.

Duncan, L. W., El-Morshedy, M. M., and McSorley, R. 1994. Sampling citrus fibrous roots and *Tylenchulus semipenetrans. J. Nematol.* **26**, 442–451.

Fife, J. P., Derksen, R. C., Ozkan, H. E., and Grewal, P. S. 2003. Effects of pressure differentials on the viability and infectivity of entomopathogenic nematodes. *Biol. Control* **27**, 65–72.

Goettel, M. S., and Inglis, G. D. 1997. Fungi: Hyphomycetes. *In* "Manual of Techniques in Insect Pathology" (L. A. Lacey, Ed.), pp. 213–249, Academic Press, London.

Graham, J. H., Bright, D. B., and McCoy, C. W. 2003. *Phytophthora-Diaprepes* weevil complex: *Phytophthora* spp. relationship to citrus rootstocks. *Plant Dis.***87**, 85–90.

Grewal, P. S. 2002. Formulation and application technology. *In* "Entomopathogenic Nematology" (R. Gaugler, Ed.), pp. 265–287, CABI, NY.

Kaya, H. K., and Stock, S. P. 1997. Techniques in insect nematology. *In* "Manual of Techniques in Insect Pathology" (L. A. Lacey, Ed.), pp. 281–324, Academic Press, London.

Landa, A., and Jiranova, R. 1989. Entomopathogenic fungi as additional selective pest suppressing agents of greenhouse whitefly populations on greenhouse cucumbers. *Proc. Conf. Biopesticides-Theory Practice*. Ceske Budejovice, pp. 120–130.

McCoy, C. W. 1985. Citrus: current status of biological control in Florida. *In* "Biological Control in Agricultural IPM Systems" (M. Hoy and D. Herzog, Eds.), pp. 481–499, Academic Press, London.

McCoy, C. W. 1996. Pathogens of eriophyoid mites. *In* "Eriophyoid Mites. Their Biology, Natural Enemies and Control", Vol. 6 (E. E. Lindguist, M. W. Sabelis, and J. Bruim, Eds.), pp. 481–490, Elsevier Press, NY.

McCoy, C. W. 1998. Microbial control of arthropod pests of citrus. *Proc. VII*th *Intern. Colloq. Invertebr. Pathol. Microbial Control*. Sapporo, Japan, pp. 160–165.

McCoy, C. W., Castle, W. S., Graham, J. H., Syvertsen, J. P., Schumann, A. W., and Stuart, R. J. 2004 Pesticide suppression of *Diaprepes abbreviatus* (L.) (Coleoptera: Curculionidae) promoted differential growth and survival of "Hamlin" orange trees budded to five rootstocks in a *Phytophthora* infested grove. *Proc. Fla. State Hort. Soc.* **117**, 167–173.

McCoy, C. W. and Couch, T. L. 1982. Microbial control of the citrus rust mite with the mycoacaricide, Mycar. *Fla. Entomol.* **65**, 116–126.

McCoy, C. W., Stuart, R. J., Duncan, L. W., and Nguyen, K. 2002. Field efficacy of two commercial preparations of entomopathogenic nematodes against larvae of *Diaprepes abbreviatus* (Coleoptera: Curculionidae) in alfisol type soil. *Fla. Entomol.* **85**, 537–544.

McCoy, C. W., Stuart, R. J., and Nigg, H. N. 2003. Seasonal life stage abundance of *Diaprepes abbreviatus* in irrigated and non-irrigated citrus plantings in central Florida. *Fla. Entomol.* **86**, 34–42.

Morgan, C. V. G., Chant, D. A., Anderson, N. H., and Ayre, G. L. 1955. Methods for estimating orchard mite populations, especially with the mite brushing machine. *Can. Entomol.* **87**, 189–199.

Morse, J. G., and Lindegren, J. E. 1996. Suppression of Fuller rose beetle on citrus with *Steinernema carpocapsae*. *Fla. Entomol.* **79**, 373–384.

Poprawski, T. J., Parker, P. E., and Tsai, J. H. 1999. Laboratory and field evaluation of hyphomycete insect pathogenic fungi for control of brown citrus aphid. *Environ. Entomol.* **28**, 315–321.

Quintela, E. D., and McCoy, C. W. 1998. Synergistic effect of imidacloprid and two entomopathogenic fungi on the behavior and survival of larvae of *Diaprepes abbreviatus* in soil. *J. Econ. Entomol.* **91**, 110–122.

Quintela, E. D., Fan, J., and McCoy, C. W. 1998. Development of *Diaprepes abbreviatus* (Coleoptera: Curculionidae) on artificial and citrus root substrates. *J. Econ. Entomol.* **91**, 1173–1179.

Shapiro-Ilan, D. I., Gouge, D. H., Koppenhöfer, A. M. 2002. Factors affecting commercial success: case studies in cotton, turf and citrus. *In* "Entomopathogenic Nematology" (R. Gaugler, Ed.), pp. 333–356, CABI, NY.

Shaw, J. G., Chambers, D. L., and Tashiro, H. 1968. Introducing and establishing the non-inclusion virus of the citrus red mite in citrus groves. *J. Econ. Entomol.* **61**, 1352–1355.

Stansly, P. A., Mizell, R. R., and McCoy, C. W. 1997. Monitoring *Diaprepes abbreviatus* with Tedders traps in southwest Florida citrus. *Proc. Fla. State Hort. Soc.* **110**, 22–26.

Stout, R. G. 1962. Estimating citrus production by use of frame count survey. *J. Farm Econ.* **44**, 2–3.

Stuart, R. J., Shapiro-Ilan, D. I., James, R. R., Nguyen, K. B., and McCoy, C. W. 2004. Virulence of new and mixed strains of the entomopathogenic nematode *Steinernema riobrave* to larvae of the citrus root weevil *Diaprepes abbreviatus*. *Biol. Control* **30**, 439–445.

Talhouk, A. J. 1975. Citrus pests throughout the world. *Citrus Tech. Mon.* **4**, 21–24.

van der Geest, L. P. S. 1985. Pathogens of spider mites. *In* "Spider Mites. Their Biology, Natural Enemies and Control, Vol. 1B" (W. Helle and M. W. Sabelis, Eds.), pp. 247–258, Elsevier, Amsterdam, The Netherlands.

Webber, J. J. 1948. History and development of the citrus industry. *In* "The Citrus Industry, Vol. I. History, Botany and Breeding" (H. J. Webber and L. D. Batchelor, Eds.), pp. 1–40, Univ. California Press, Berkeley, CA.

Chapter VII-14

Small fruits

Steven R. Booth
2523 B 44th Ave. N. W.
Olympia, WA 98502, USA

Frank A. Drummond and Eleanor Groden
Department of Biological Sciences
University of Maine
Orono, ME 04469, USA

1 Introduction

The small fruit crops covered in this chapter share many aspects of pest management that justify the research and development of microbial control agents. As perennial crops grown in relatively undisturbed soils, they offer subterranean entomopathogens a better chance to persist. In the past, consumer demands for high quality small fruit with little cosmetic damage and for products with no insect contamination have precluded the use of microbial controls. Recently, consumers have become more apprehensive about pesticides in their food and legislation has curtailed the use of many pesticides in small fruits. The agricultural community has responded by developing tactics to determine the relative efficacy of both commercially produced and naturally occurring entomopathogens. This chapter reflects that response, and we review the potential of entomopathogens for small fruit pest management and present research guidelines for cranberry, lowbush blueberry, caneberry (red raspberry, blackberry, boysenberry, loganberry, marionberry, etc.), grape, and strawberry production.

2 Insect pests of small fruit

Biologies and distributions of insect pests of small fruits have been comprehensively reviewed for each crop elsewhere (*e.g.*, Bournier, 1976; Shanks, 1981; Jennings, 1988; Crandall, 1995; Collins and Drummond, 1998). They are numerous, diverse, and distributed among all common terrestrial taxonomic orders. Often, key pests are the same despite the crop. In addition to the crop type and its structure, the functional group or feeding guild of the pest figures prominently in the ways microbials are applied and assessed. Below is a condensed summary of the more important insect pests of small fruit, especially those with high potential to be suppressed with microbial control agents, presented according to feeding guild.

A Fruit feeders

Most insect pests that feed directly on small fruits are lepidopteran larvae. In cranberry, these include the blackheaded fireworm (*Rhopobota naevana*), yellowheaded fireworm (*Acleris minuta*), *Sparganothis* fruitworm (*Sparganothis sulfureana*), and cranberry fruitworm (*Acrobasis vaccinii*). The grape berry moths comprise two

L.A. Lacey and H.K. Kaya (eds.), Field Manual of Techniques in Invertebrate Pathology, 583–598.

genera (*Endopiza* spp. and *Lobesia* spp.) and are common in all major growing regions.

Species in other insect orders are also important direct pests of small fruit. The blueberry maggot, *Rhagoletis mendax*, causes crop rejection if more than 4 larvae/liter are present in berries at processing (Brown and Ismail, 1981). Adult clay colored weevils (*Otiorhynchus singularis*) feed on developing caneberry buds in early spring in the Pacific Northwest (Crandall, 1995) and Scotland (Gordon *et al.*, 1990). Raspberry beetles (*e.g.*, raspberry fruitworm, *Byturus tomentosus*) are a problem mostly in Great Britain. Tarnished plant bug nymphs (*Lygus* spp.) feed on individual achenes of developing strawberries and destroy auxin production which results in deformed fruit ('catfacing'). Strawberry bud weevil females (*Anthonomus signatus*) damage strawberries on the east coast of the USA by ovipositing in the flower receptacles (Vincent *et al.*, 1990).

B Defoliators

Insects that feed on and reside within leaves are perhaps more diverse taxonomically than those of other guilds or functional groups. Some foliage feeders are host specific and serious defoliators, (grapeleaf skeletonizer, *Harrisina brillians*) (Bournier, 1976), whereas others are common to multiple crops and pose a greater threat as a contaminant at harvest (orange tortrix, *Argyrotaenia citrana*) and the obliquebanded leafroller, *Choristoneura rosaceana*) (Crandall, 1995). Blueberry spanworm (*Itame argillacearia*), blueberry leaf beetle (*Pyrrhalta vaccinii*), blueberry flea beetle (*Altica sylvia*), blueberry case beetle (*Neochlamisus cribripennis*), and blueberry sawflies (*Neopareophora litura* and *Pristiphora cincta*) are all key leaf feeding pests of blueberry (Phipps, 1930), whereas the red-striped fireworm (*Aroga trialbamaculella*) is mostly a nuisance pest interfering with raking of the berries.

C Sucking insects

In grape, over 20 species of leafhoppers and sharpshooters (Homoptera: Cicadellidae) serve as vectors for the bacterium (*Xylella fastidiosa*) that causes Pierce's disease. In Washington, a complex of three leafhoppers do not serve as disease vectors, but their feeding can cause defoliation of vines and affect fruit maturation. The eastern grape leafhopper (*Erythroneura comes*) is a problem in New York (Martinson *et al.*, 1991). In California, untreated sharpshooter populations can become dense enough to disrupt grape harvest (Bentley *et al.*, 1990). Mealybugs (*Pseudococcus maritimus*, *P. ficus*, *P. citri*, and *Maconellicoccus hirsutus*) damage grapes worldwide (California, France, Southern Europe, and India) by secreting honeydew which promotes the development of sooty molds. Aphids both cause sooty molds and serve as vectors for several viruses, especially in strawberry. Midges are also vectors for several viruses and fungi in grape (Bournier, 1976).

D Stem borers

Stem-boring insects of cranberry, strawberry, and lowbush blueberry are not as common or economically significant as those of other small fruit crops. The grape cane borer (*Melalqus confertus*) infests the stems of grapevines (Bentley *et al.*, 1990). The raspberry crown borer (*Pennisetia marginata*, Sesiidae: Lepidoptera) infests crowns and canes and causes significant economic losses if not controlled. Ambrosia beetle (Scolytidae) attacks the trunks of grape vines in Virginia (Pfeiffer *et al.*, 1990), but caterpillars (*Cossus cossus* and *Parobta paradoxus*) are more destructive in Europe and the Middle East, respectively (Bournier, 1976).

E Root feeders

The black vine weevil (*Otiorhynchus sulcatus*), as well as several other root weevils, (*O. ovatus*, *Sciopithes obscurus*, and *Nemocestes incomptus*) are key subterranean pests of all small fruit crops in most northern and some southern temperate growing regions. Scarab grubs are also fairly ubiquitous. In cranberry, both root weevils and the cranberry girdler (*Chrysoteuchia topiaria*), also known as the sod webworm, can destroy several years of perennial growth and create patches of bare soil within a single season.

In addition to root weevils and scarabs, the grape root borer (*Vitacea polistiformis*)

(Sesiidae) ranges throughout the American midwest (Alm *et al.*, 1989; Townsend, 1990) and southeast (Pfeiffer *et al.*, 1990). The grape phylloxera (*Daktulosphaera vitifoliae*) is the most important insect pest of grapes in nearly all growing regions, and threatens areas where it is yet to be introduced (Bournier, 1976; Mullins *et al.*, 1992). Its status varies depending on regional differences in climate, soil type, and especially rootstock. Grape phylloxera is best suppressed by grafts of resistant North American rootstocks.

3 Entomopathogens for small fruit pests

A Nematodes

Although extensively studied, entomopathogenic nematodes have yet to be fully utilized in small fruit pest management programs. In the Pacific Northwest, the use of the most commonly commercialized nematode, *Steinernema carpocapsae*, has been modest because soil temperatures are often below its threshold for activity (15 °C) during optimum application times in most small fruit crops. *S. carpocapsae* has also been applied with little success against the grape root borer (All *et al.*, 1981), but application rates were low and non-uniform. Difficulties in formulation and marketing may have also slowed the implementation of entomopathogenic nematodes. Smaller scale and less expensive production techniques, and the research and development of strains adapted to a particular host or region may encourage greater use of entomopathogenic nematodes in small fruits.

B Fungi

The potential entomopathogenic fungi as microbial control agents of insect pests may be especially high in cranberry, given the crop's relatively moist soils and moderate temperatures. Over a dozen indigenous strains of *Metarhizium anisopliae*, *Beauveria bassiana*, and *Paecilomyces farinosus* have been isolated from soil and black vine weevil cadavers in Washington and Oregon cranberry bogs (Humber, 1992). Three of these strains were more virulent than a strain of *M. anisopliae* from Florida at 14 °C (Booth, unpublished data). Despite this potential, only a few preliminary trials with entomopathogenic fungi have been conducted against soil-dwelling cranberry pests in the field (Booth and Shanks, 1998). Preliminary laboratory bioassays using *B. bassiana* against sod webworm (Hall, 1954) and white grub (Kreuger *et al.*, 1992) were promising but have not been pursued.

Beauveria bassiana, produced as Mycotrol™ES, was evaluated against several blueberry pests (Drummond and Collins, 1997; Collins and Drummond, 1998). Blueberry flea beetle was susceptible as both larvae and adults, as were blueberry leaf beetle adults. Red-striped fireworm in the laboratory and blueberry thrips in the field were not susceptible.

Naturally occurring entomopathogenic fungi appear to be quite diverse in blueberry soils. Strongman *et al.* (1997) found 9 species in eastern Canadian blueberry fields, including *B. bassiana, Entomophaga aulicae*, *Hirsutella nodulosa*, *Hirsutella* sp., *Paecilomyces farinosus*, *P. fumosoroseus*, *Tolypocladium niveum*, *Lecanicillium* (=*Verticillium*) *lecanii*, and an undetermined species. Of these, *B. bassiana* was responsible for about 10% of the total insects killed by fungi.

Several strains of entomopathogenic fungi have been isolated from grape phylloxera (Humber, 1992), but trials have not been conducted. The potential of entomopathogenic fungi to suppress mealybugs in grapes seems to have been similarly overlooked, as *B. bassiana* reportedly caused natural epizootics in three mealybug species in Romanian vineyards (Mirica *et al.*, 1969). *M. anisopliae* and *B. bassiana* contributed to the natural mortality of the grape root borer in Georgia vineyards (Dutcher and Hall, 1978).

C Bacteria

Currently, several subspecies of the entomopathogenic bacterium, Bacillus thuringiensis (*Bt*) are produced commercially. *Bt*-based products have potential against blackheaded fireworm and Sparganothis fruitworm in cranberry (Polavarapu and Peng, 1997), but 2–3 applications are required to blanket the egg hatch period. In

Maine, blueberry spanworm was suppressed to below action thresholds with Javelin™ WG and Able™ 50WP, but not Agree™ 50WP (Collins and Drummond, 1998). *Bt* also has potential against leafrollers in red raspberry (Li and Fitzpatrick, 1996). *Bt* subspecies *kurstaki* (*Btk*) has been applied against the grape berry moth in Israel (Ben-Yehuda *et al.*, 1993) and Crete (Roditakis, 1986). In California, the grape berry moth, the grape leaffolder, and the western grapeleaf skeletonizer are suppressed with *Btk* (Bentley *et al.*, 1990). Multiple applications of *Btk* suppressed a complex of caterpillars in Missouri, but damage levels remained above economic thresholds (Townsend, 1990).

D Viruses

Because of the host-specific nature of many insect pathogenic viruses and because small fruits are a minor crop, little research effort has focused on virus control of small fruit insect pests. A granulovirus provides substantial biological control against the western grapeleaf skeletonizer in California (Stern and Federici, 1990). The virus was unintentionally introduced when the parasitoids *Apanteles harrisinae* and *Amedoria misella* (*Sturmia harrisinae*) were collected in other parts of the Southwest and mass released in San Diego County. The virus was introduced to other areas, where it reduces skeletonizer egg production by as much as 70% and larval survival from contaminated eggs by 90%.

4 Specific crop protocols for application and evaluation of entomopathogens

Despite the immense abundance, diversity, and economic importance of insect pests of small fruit, the number of field studies featuring microbial control agents to suppress them has been small (Table 1). The remainder of this chapter briefly describes the factors which determine specific protocols for the application and evaluation of microbials against insect pests of small fruit crops, followed by examples of step-by-step procedures for selected pathogen/pest/crop problems.

A Crop and plot characteristics

Field methods of microbial control in small fruits depend to a greatly on particular characteristics of the pest and its pathogen, especially the ability of each to disperse and persist, but certain characteristics of the crop are also important. Crop age, structure, value, and sensitivity to sampling are important considerations of plot size, placement, and replication. One way to maintain crop value and still do destructively sampling is to minimize plot size. Small plots are compatible with the biology of many soil-dwelling insect pests. Root weevil larvae feed on the roots of a single plant and lateral movement is negligible, so small plot trials are suited to those pests. Small plots are also suited to applications of subterranean entomopathogens, but soil structure, temperature, and moisture may affect their dispersal, survival, and efficacy. More mobile pests (terrestrial adults) require plots to be either larger or contained. Although flightless, black vine weevil adults can still walk several meters in a single night (Maier, 1978) and can climb almost any barrier except those coated with Fluon® (polytetrafluorethylene, Northeast Chemical Co., Woonsocket RI). However, very large plots are sometimes difficult to treat and barriers may affect more than the dispersal of pest and entomopathogen. These and other considerations vary more specifically among crops and are discussed in context below.

In a study in western Washington, small field plots comprising three partial rows (3 m long) of weevil-infested second year strawberries were enclosed by a 20 cm tall aluminum fence coated with Fluon® and treated with liquid formulations of *M. anisopliae* conidia.

1 Cranberry

Field trial methodology in cranberry is strongly determined by two distinctive crop characteristics. First, pesticides are commonly applied through the set sprinkler system (Bicki, 1997), especially on the west coast of North America, where farms are smaller and less contiguous than on the east coast. Second, destructive sampling is often discouraged in cranberry, due to the crop's

Table 1. Field studies of microbial control agents against key insect pests of small fruit crops

Crop	Guild	Key target pests	Family: order	Microbial agent	Application rate (ha)	References
cranberry						
	fruit	*Rhopobota naevana*	Tortricidae: Lep.	*Bacillus thuringiensis (Bt)*	26×10^9 BIU	Polavarapu and Peng (1997)
	root	*Otiorhynchus sulcatus*	Curculionidae: Col.	*Steinernema carpocapsae*	7.4×10^9 IJs	Shanks and Agudelo-Silva (1990)
				Heterorhabditis marelatus	2.5×10^9 IJs	Berry and Liu (unpublished)
		Chrysoteuchia topiaria	Pyralidae: Lep.	*S. carpocapsae*	7.4×10^9 IJs	Henderson (unpublished)
				H. bacteriophora	7.4×10^9 IJs	Shanks and Agudelo-Silva (1990)
				Beauveria bassiana	5.4×10^{10} conidia	Broaddus (unpublished)
bluberry						
	leaf	*Itame argillacearia*	Chrysomelidae: Col.	*Bt kurstaki* (*Btk*)	14.5–29.0 BIU	Yarborough and Collins (1997)
				B. bassiana	4.8×10^{13} conidia	Collins and Drummond (1998)
	root	*Altica Sylvia*	Geometridae: Lep.	*B. bassiana*	4.8×10^{13} conidia	Collins and Drummond (1998)
caneberry						
	leaf	*Choristoneura rosaceana*	Tortricidae: Lep.	*Btk*	18.0–53.0 BIU	Li and Fitzpatrick (1996)
	root	*O. sulcatus*	Curculionidae: Col.	*S. carpocapsae*	4.4×10^9 IJs	Booth *et al.* (2002)
				B. bassiana	20–60 kg mycelia	Booth *et al.* (2002)
grape						
	leaf	*Desmia funeralis*	Pyralidae: Lep.	*Btk*	2.8–18.8 BIU	AliNiazee and Jensen (1973)
	leaf	*Harrisina* spp.	Zygaenidae: Lep.	*Hb* granulovirus	not available	Stern and Federici (1990)
	root	*Amathes c-nigrum*	Noctuidae: Lep.	*S. feltiae*	$2.5–5 \times 10^9$ IJs	Wennemann (1997)
				S. carpocapsae	$2.5–10 \times 10^9$ IJs	Wennemann (1997)
				S. riobrave	10×10^9 IJs	Wennemann (1997)
				H. bacteriophora	10×10^9 IJs	Wennemann (1997)
		Vitacea polistiformis	Sesiidae: Lep.	*S. carpocapsae*	5.0×10^9 IJs	All *et al.* (1981)
		O. sulcatus	Curculionidae: Col.	*S. feltiae*	$2.5–5 \times 10^9$ IJs	Wennemann (1997)
				S. carpocapsae	$2.5–5 \times 10^9$ IJs	Wennemann (1997)
				S. riobrave	$2.5–5 \times 10^9$ IJs	Wennemann (1997)
				H. bacteriophora	$2.5–5 \times 10^9$ IJs	Wennemann (1997)
strawberry						
	root	*O. sulcatus*	Curculionidae: Col.	*H. marelatus*	$2–14 \times 10^9$ IJs	Berry *et al.* (1997)
		O. sulcatus		*H. bacteriophora*	$2–14 \times 10^9$ IJs	Berry *et al.* (1997)
		Otiorhynchus spp.		*S. carpocapsae*	$11–13 \times 10^9$ IJs	Simser and Roberts (1994)
				H. bacteriophora	$11–13 \times 10^9$ IJs	Simser and Roberts (1994)
				M. anisopliae	20–60 kg mycelia	Booth and Shanks (1998)
				B. bassiana	13.2×10^{13} conidia	Booth *et al.* (2002)
				Heterorhabditis sp.	2.5×10^5 IJ/plant	Curran (1992)
				Steinernema sp.	2.5×10^5 IJ/plant	Curran (1992)
				S. carpocapsae	$2.2–7.6 \times 10^9$ IJs	Kakouli-Duarte *et al.* (1997)
				H. megidis	0.5×10^9 IJs	Kakouli-Duarte *et al.* (1997)
				S. carpocapsae	2.2×10^9 IJs	Booth *et al.* (2002)
				H. bacteriophora	2.2×10^9 IJs	Booth *et al.* (2002)

[1] Unless otherwise noted; BIU = billion international units; IJs = infective juveniles

high value and extreme sensitivity to disturbance. Severing a single vine near the root eliminates several years of vine growth and opens the canopy for weeds to establish, many of which are better competitors than the cranberry. Grower cooperators are more likely to allow larger plots and untreated areas to be destructively sampled in severely damaged or abandoned bogs scheduled for renovation or removal, but such bogs may be managed very differently than bogs under commercial production.

Small field plots ($\sim 1\,m^2$) are well suited to studies of entomopathogenic nematodes and fungi against root weevil larvae in cranberry. Both S.carpocapsae and Heterorhabditis bacteriophora moved only 4 cm in 14 days in a cranberry bog on the east coast (Simser and Roberts, 1994), whereas a small percentage of the latter species (*H. bacteriophora* strain HP88) moved 30 cm in 60 days in a west coast cranberry bog (Booth unpublished data). A similarly low percentage of inoculated *H. bacteriophora* moved more than a meter during a 10 month period in a very sandy bog (Shanks and Agudelo-Silva, 1990). Both conidia and mycelia of entomopathogenic fungi can spread short distances from inoculation sites in cranberry, probably due to high volumes of irrigation water (Booth and Shanks, 1998). Small plots are also easier to protect from pesticides chemigated to the rest of the cranberry bog. Small plots or blocks of plots can be temporarily covered with tarpaulins that must be removed soon after chemigation. Alternatively, the nozzles of sprinklers can be plugged or turned off during application of unwanted chemigated materials, provided the system's water pressure and the delivery of other sprinklers are not disrupted. Cranberry girdler larvae are more mobile than root weevil larvae, so larger plots may be required, especially if the study is scheduled to last several weeks. Alternatively, cranberry girdler larvae can be contained in small buckets inserted into cranberry bogs (R. Berry, personal communication).

2 *Lowbush blueberry*

The clonal nature of lowbush blueberry affects the size and placement of plots in trials of entomopathogens. Lowbush blueberry plants are established from seed, but a substantial portion of total reproduction is due to clones, which grow from underground rhizomes. An average clone covers 7 to $22\,m^2$ and older clones are usually larger (Yarborough, 1998). A hectare of wild blueberries may contain 500 to 1000 clones.

Small plots ($0.5–5.0\,m^2$) are easier to replicate either within a single clone or among separate clones (Collins and Drummond, 1998). Bottomless buckets or cut stove pipe sections ($452\,cm^2$) pushed 10–15 cm into the soil are useful to evaluate *B. bassiana.* against blueberry thrips. Small field cages ($1\,m^2$) have been used in field trials against blueberry spanworm or blueberry flea beetle. Small plots are also well suited for highly aggregated pests such as the red-striped fireworm.

Most blueberry foliar pests require large plots of $50–100\,m^2$ with 2 m buffers between plots (Collins and Drummond, 1998). Because blueberries are clonal, a minimum of 5 to 6 replicates are required for such large plots. Highly mobile blueberry pests like grasshoppers or blueberry maggot require aerial applications to exceptionally large plots up to 2 ha in size, with at least 100 m buffers between them. Blueberry maggot flies initially colonize the borders of fields before moving to the center, and because of this, statistical blocks should be placed parallel to the border.

3 *Caneberry and grapes*

In caneberry and grapes, farms and vineyards are often large and crop rows (alleys) are separated by trellis and wires, so infestations of patchily distributed insect pests are difficult to locate and incorporate into field plots. Trellised rows are conventionally sprayed with mist or air blast sprayers that may be harmful to entomopathogenic nematodes, but studies in other cropping systems may demonstrate otherwise (see Chapter IV-5). Chemigation practices through both above ground and below ground irrigation systems are becoming more common in both caneberry and grape.

In caneberry, small plots (5 plants) were appropriate for application of *Bt*-based insecticides against sedentary obliquebanded leafroller larvae on the foliage (Li and Fitzpatrick, 1996). Plastic sheets were placed between plots to prevent cross contamination by spray drift (Li and Fitzpatrick, 1996). More mobile pests such as adult root weevils or raspberry beetles require larger plots and application by conventional mist or air blast sprayers. Applications through drip systems usually require long (entire row) plots, but shorter rows and smaller plots were created at a red raspberry farm in western Washington by hanging a temporary overhead dripline adjacent to the permanent line (Booth *et al.*, 2002).

Perhaps more than other crops, root weevil infestations in caneberry are generally located at field edges and usually extend along trellised rows rather than across them. The perimeter row should be discarded or treated as an unsampled buffer row, because weather and irrigation patterns, as well as pest densities, may be atypical there. Plots may be placed adjacent to one another within the same row, provided the infestation is uniform among plots and that the buffer is large enough.

Larger plots (12 to 15 rows) were used in a series of studies designed to evaluate the efficacy of entomopathogenic nematodes applied through dripline systems in Washington vineyards (Wennemann, 1997). Evaluation was simplified by randomly sampling only 5 rows of each treatment plot, but such a protocol assumes uniform distributions of pest and pathogen; that assumption was not tested. If microbial control agents are applied through a dripline, then treatments should be blocked among areas not directly connected to the same water source. If not, treatments may receive different amounts of post-application irrigation water, as the simplest way to apply nematodes to a particular treatment block is by restricting the flow to subsequent rows with plugs (Wennemann, 1997).

4 Strawberry

Strawberry is a perennial plant but is often cultured as an annual crop in warm locales, and bi- or tri-annuals in cooler regions. Most varieties bloom briefly in the spring, but ever-bearing varieties have an extended bloom. Dormant crowns are usually planted in fumigated fields during spring in cool regions or during fall at warmer sites. During spring, foliage and inflorescence tissues develop quickly and plants are quite large by early summer. Simultaneously, axillary buds on the crown develop stolons from vegetative tendrils ('runners') that sprout new 'daughter' plants. Such a simple method of vegetative reproduction is advantageous for fast and efficient plant propagation, but can also result in a redirection of nutrients away from developing fruit. Accordingly, multi-year plantings are usually pruned post-harvest to promote crown growth before winter causing a thick understory of foliage and duff to develop by late season. Overwintering plants are often further protected by mulching strawberry hills with straw or black plastic. These ground layer habitats are conducive to the survival and development of microbials, deleterious or otherwise. Fields are usually fumigated between plantings to suppress harmful pathogens, but that practice limits persistence of entomopathogens.

Field plots are often easier to establish in strawberry than in other small fruit crops, as the size and location of most infestations can be more easily discerned. Plots may be partial rows, but buffer rows should be included for most pests and pathogens. Strawberries grow in a variety of soil types, which can strongly affect efficacy of entomopathogenic nematodes (Georgis and Poinar, 1983), so this should be considered when planning nematode trials.

B Application timings and methods

For most temperate small fruit crops, entomopathogenic nematodes have the greatest potential against black vine weevil during mid-spring, when most of the population is in susceptible, larger larval lifestages. However, the window for application and evaluation is usually short, especially at northern latitudes, where soil temperatures routinely remain below the thermal activity threshold for most commercial strains of entomopathogenic nematodes (15 °C) until late spring (Booth *et al.*, 2002). Weevils begin to pupate and emerge as adults when soils are at similar temperatures.

The time frame for application of entomopathogenic nematodes against late instar root weevil larvae may be extended in strawberry beds by using mulches. Black plastic can raise soil temperature several degrees during early spring (Kakouli-Duarte *et al.*, 1997), but such a practice could also accelerate larval development. Alternatively, entomopathogenic nematodes can be applied in early fall when soil temperatures are still above 15 °C, but most weevil larvae are still very small, less susceptible, and more difficult to detect.

Entomopathogenic nematodes should be applied against cranberry girdler during mid-summer, when soil temperatures are warm and larvae are at least medium sized. Because moth emergence is often extended, the life stage distribution of girdler larvae is correspondingly mixed. Thus, an initial application at 2 weeks after peak adult emergence, as indicated by pheromone trap catches, should be followed by a second application 10 to 14 days later.

Soil-dwelling pests of caneberry might be better suppressed with post-harvest applications when soil temperatures are warm. Unfortunately, irrigation is often reduced during late summer in some crops, (caneberry and strawberry) limiting the potential of both entomopathogenic nematodes and fungi because they require moderate soil moisture.

Date of application is less critical for trials of entomopathogenic fungi against root weevil larvae. Although cool temperatures can slow the infection process, both mycelial and conidial stages of entomopathogenic fungi can persist in soils for months (Studdert and Kaya, 1990), until soil temperatures rise above threshold levels of infection. Conidial formulations should not be applied during wet winter months, as conidia may be less effective in water saturated soils (Stenzel, 1992).

Phenology of both host plant (time and duration of bloom) and pest (time and duration of egg hatch) strongly affect time of pesticide application. Because flowers are pollinated by bees, broad-spectrum organophosphate or carbamate compounds cannot be applied during bloom. Some microbial control agents, mostly *Bt*-based products, are not harmful to bees, so are a good alternative for lepidopteran pests. However, most commercial *Bt*-based products are sensitive to ultra-violet radiation and do not persist well, so several applications are required to bracket the period of larval vulnerability. This interval may be especially brief for leafrollers (Tortricidae) that envelope themselves between leaves shortly after egg hatch. In Washington cranberry bogs, blackheaded fireworm hatch from eggs during spring and move to developing flower buds within 2 or 3 days. They remain in the buds until they drop to the soil and pupate. At least 2 applications of Bt-based products at 4 or 5 day intervals are required to bracket the period of egg hatch. A similar spray schedule is required in Maine blueberry fields against early instar blueberry spanworm caterpillars (Drummond and Collins, 1998). In the Pacific Northwest caneberry fields, both the obliquebanded leafroller and the orange tortrix overwinter as early instar larvae, but early-season densities are usually too low to warrant treatment. Pheromone traps and field observations are used to monitor adult emergence and predict hatch of the first summer generation larvae. Two applications at 7 to 14 day intervals should bracket egg eclosion. Six applications of *Btk* were made every 3 to 4 days during the 2 week period of egg eclosion against the grape berry moth in Missouri (Biever and Hostetter, 1975). The complex of lepidopteran pests on grapes often develop simultaneously, so applications targeted against one pest may suppress another.

In field trials featuring small plots, entomopathogenic nematodes are most easily applied against soil-dwelling insect pests using sprinkling cans (Booth *et al.*, 2002). A standard drench depth of 1.25 cm of water (2000 liters/ha) has been established for application of entomopathogenic nematodes to cranberry (Shanks and Agudelo-Silva, 1990), but additional water volumes should be applied before and after application to stimulate nematode movement into the soil and wash them from foliage. Because nematodes are much denser than water, prepared suspensions must be continually agitated and distribution at application should be carefully monitored.

A fumigant injector was used to apply entomopathogenic nematodes to deep soil layers (> 10 cm) in grapes, but 50 separate injections of a 10 ml nematode suspension were required for each 1 × 0.5 m plot and treatments were not effective (All *et al.*, 1981). Another type of subsurface injector was more efficient, but treatments still failed to effectively suppress grape root borer (All and Dutcher, 1977).

In western Washington red raspberry fields, entomopathogenic nematodes were applied through temporary overhead driplines to simulate potential commercial applications (Booth *et al.*, 2002). In that study, nematodes were suspended in 1 liter of water and measured into the line with a Mini-Dos® water-powered injector. Application time was rather lengthy (30 min/line), so the nematode suspension had to be frequently stirred by hand. Larger-scale injectors may be more efficient. If microbial control agents are to be applied through permanent commercial driplines then they should be thoroughly checked for leaks or plugs prior to application. Studies of dripline application to grapes featured injection of entomopathogenic nematodes by syringe (Wennemann, 1997), but until that method is more thoroughly tested, we recommend nematodes be introduced into driplines with an injection pump. The injection pump will more evenly distribute the pathogen and will also utilize a larger volume of nematodes in suspension which is easier to prepare and assess.

The width of the irrigated band, including subsurface flow, should be measured and included in computations of application rates. Pump pressure, emitter type, and even the slope of the field (especially vineyards) may create different widths of irrigated soil. For example, if row width is 50 cm, row length is 100 m, and the desired application rate is 5×10^9 infective juveniles (IJs)/ha, then 2.5×10^4 IJs are needed per row. This amount is equivalent to 5×10^3 IJs/ml if suspended in 5 liters water and at least a 100-fold dilution is required to accurately assess concentration and vigor. If injected into a fully pressurized dripline (2 liters/ha) for 30 min, the suspension will be further diluted to 25 IJs/ml at the site of the emitter, a concentration which can be quickly and accurately assessed in the field.

Nematode distribution beneath an overhead dripline can be checked for uniformity in both space and time by collecting 10 to 15 ml of the irrigated suspension in small cups held beneath emitters at several distances along the row during application. If the application rate, suspension concentration, and dripline traits are similar to those described above, the samples need not necessarily be diluted before examination. An additional volume of water should be run through the dripline immediately after application to wash the entomopathogen off the foliage and enhance movement into the soil.

Kakouli-Duarte *et al.* (1997) demonstrated that infective juvenile entomopathogenic nematodes can successively pass through below ground T-Tape® drip irrigation systems in raised strawberry beds. As in all field plot trials, but especially those featuring novel application techniques, the concentration and viability of nematodes should be assessed at regular spatial and temporal intervals.

C Sampling and monitoring

1 Pests

Pre-treatment samples are useful to help plan plot size and placement and to more accurately assess microbial efficacy but can be expensive in terms of both space and time, as well as crop value. Efficacy can often be sufficiently demonstrated by comparing pest survival or mortality between treated and untreated plots.

A large diameter (10 cm) soil corer is a preferred tool to sample subterranean pests (Southwood, 1978). A golf cup cutter is especially useful in cranberry, where stems and roots are otherwise difficult to penetrate. Vines will also reestablish if the holes are small and filled with sand. Cranberry is a relatively shallow rooted plant and most weevil larvae are found at depths of less than 20 cm. Cranberry girdler larvae reside in the shallower duff layer. For both pests, the optimum number of cores per plot can be determined by examining a few cores from untreated plots, but 8–10 cores/1×1 m plot are usually required. In other crops, sandy soils can be sampled with shovels and sieved through large mesh screen to detect large larvae or cadavers more quickly. Caneberry plantings are less dense than matted crops such as cranberry, so peripheral roots can be sampled without damage to the main crown. Root weevil larvae also infest nonfruit-bearing primocanes which can be sampled instead of more valuable floricanes. In both caneberry and grape, each side of the row receives sunlight at different times of the day, so soil should either be taken equally from both sides of the row or consistently from the same side.

Unfortunately, root weevil larvae do not float in water and some soil samples contain a lot of vegetative debris (cranberry), making it difficult to wash larval specimens from soil samples (Montgomery *et al.*, 1979). Most larvae must be painstakingly sorted from soil by hand. In strawberry, root weevil larvae feed on the primary root (stele) of strawberry, so the entire plant and the surrounding soil should be carefully examined. More mobile subterranean pests such as cranberry girdler larvae can be extracted from soil samples with Berlese funnels, provided only a small volume of soil is put in each funnel.

Small cadavers, especially those with entomopathogenic nematodes, degrade quickly, so samples should be stored at cool temperatures and processed within a week of collection. Larvae or cadavers recovered from soil sampled at short post-application intervals (< 7 days) may not have been exposed to the microbial agent long enough to display disease symptoms, especially if soil temperatures are cool, so incubating field collected larvae at 22–25 °C for a few more days may more precisely indicate infestaton levels. Larvae from trials of entomopathogenic fungi should be surface sterilized according to techniques described elsewhere (Chapter IV-4; Goettel and Inglis, 1997). Because efficacy of many microbial agents depends partially on size of the target host, recovered larvae and cadavers should be graded into size classes.

For cranberry girdler, destructive sampling can be avoided altogether by using cages to sample emerging adults. In two separate trials, emergence cages were used to sample adults in plots treated the previous summer (Booth and Shanks, 1998). Small (30 × 25 cm at base) cages were fitted with an adhesive covered Petri dish furnished with part of a pheromone cap and inspected regularly for live moths inside the cages and dead moths on the adhesive. Adults of both cranberry girdler and blackheaded fireworm can be counted as they flutter from resting spots as the observer walks a transect across the bog.

Adult root weevils are nocturnal but usually fully active within an hour after sunset. In cranberry, weevils should be sampled at night with a sweep net (at least 25 sweeps) but sampling may need to be curtailed during bloom. In caneberry, weevils and raspberry beetles can be easily counted with a flashlight on a large (2–3 m long × 1 m wide) white tarp placed beneath the row. Vigorously shake the trellis a specified number of times (3–6) to dislodge foliar insects to the tarp. In grape, Cone *et al.* (1990) suggested sampling weevils in traps hung on the trellis wires. In strawberry, weevils can be directly observed on the foliage with the aid of a flashlight. Note that both traps and visual observations may bias samples toward healthy insects.

Sweep nets are used to sample abundance and development of blackheaded fireworm larvae in cranberry and most foliar pests of blueberry. In Maine blueberry fields, a 30 cm diameter net is used in a minimum of 20 samples (10 sweeps/sample) in sampling path resembling a “Z” pattern (Yarborough *et al.*, 1993). To avoid cross contamination, dip the sweep net and wipe the investigator’s boots in a 10% bleach solution (0.5 % sodium hypochlorite) between plots. Allow the net to dry between plots. Insects should be returned to some plots after counting, but the procedure should be limited to trials featuring small plots or low insect densities. If possible, additional plots should be sampled less frequently and to account for sampling artifacts.

Many other foliar pests, especially lepidopteran larvae, are best sampled by collecting and examining plant parts. Dead, diseased, or live leafroller larvae can be collected in the field and assessed for pathogenicity according to surface sterilization and incubation protocols specified elsewhere (Chapter IV-4; Goettel and Inglis, 1997). To prevent cross-contamination, larvae from each replicate should be handled and stored separately. To detect fireworm infestation of cranberry, dissect 15 to 20 flower buds per plot. To sample red-striped fireworm in blueberry, count the number of leaf curls along a transect across the plot, or randomly sample 100 stems and examine them for infestation.

Although not always an accurate indicator of abundance, pheromone traps provide the easiest method to monitor adult emergence and predict dates of egg hatch of many lepidopteran pests (grape leaffolder, western grapeleaf skeletonizer, and grape berry moth) (Bentley *et al.*, 1990; Townsend, 1990). Sample blueberry maggot flies with baited (ammonium) yellow Pherocon® apple maggot trap suspended from stakes at 15 to 25 cm above the plants. Bend the trap into an inverted 'V' and place sticky-side down (Collins and Forsythe, 1987).

Tarnished plant bug can be sampled in strawberry by a variety of methods (Zalom *et al.*, 1993) including those which utilize home-made beating trays or leaf-blowers modified to vacuum samplers (Strand, 1994). Tarnished plant bug nymphs are highly mobile and are best sampled in early morning when cool temperatures decrease their activity.

2 *Entomopathogens*

Because indigenous strains of entomopathogenic fungi are not uncommon in soils of some small fruit crops and because entomopathogenic nematodes and fungi have the potential to persist in soils for several months (Shanks and Agudelo-Silva, 1990; Booth and Shanks, 1998), and also to reproduce or recycle among infected hosts for much longer time periods, levels of these pathogens should be assessed both before and after treatment. Techniques to isolate and quantitatively assess entomopathogens directly from the soil are described elsewhere (Goettel and Inglis, 1997). In most cases, the presence/absence of background entomopathogens can be determined using a wax moth (*Galleria mellonella*) baiting technique (Zimmermann, 1986) that can be standardized to specific conditions and features an experimental host which is inexpensive and available through the mail in the USA and some other countries. Alternatively, infection of entomopathogens can also be assayed directly from target hosts collected from field plots according to standard protocols (Goettel and Inglis, 1997; Kaya and Stock, 1997).

In some crops, microbial persistence can be assessed by burying baits in small screen cages for later recovery (Berry *et al.*, 1997). Soils of cranberry bogs, however, are highly structured and strongly layered, so the burial of even very small cages would disrupt that structure and allow greater movements of entomopathogens through the soil. Strawberry fields are more frequently tilled, so buried baits may better approximate the density and distribution of entomopathogenic nematodes and fungi.

Microbial persistence can be monitored by continued periodic sampling and assay of inoculated foliage after application. A stepwise protocol to assay for *B. bassiana* on blueberry is detailed below.

1. Randomly sample 30 stems from treated and control plots and cut into 2–3 cm lengths.
2. Select a 1 g sample of the pooled stems from each plot and vortex in 30 ml of 0.1% Tween 20 solution for 2 min to wash conidia from the foliage.
3. Serially dilute the supernatant 2–3 times and plate four 0.5 ml aliquots of each suspension onto a selective medium (Chase *et al.*, 1986).
4. At the same time, create a standard for comparison by plating a standard quantity of *B. bassiana* conidia in 10 ml of Tween 20 solution on the selective medium.
5. Count the number of *B. bassiana* colony forming units on each plate after a 10 day incubation period at 21–23 °C.
6. Mark a subsample of the observed *B. bassiana* colonies on each plate and incubate the dishes for an additional 10 days until the fungal colonies have fully sporulated. Collect a small sample from the marked colonies, place them in a drop of distilled water on a microscope slide to verify species identity.

3 *Crop*

In most crops, yield is the most meaningful assessment of long term horticultural practices, including pest management tactics. Yield may not accurately measure immediate pest levels, especially for slow growing crops like cranberry or grape. Damage from fruit feeders, such as fireworm larvae, is easier to detect. An individual larva infests two or three cranberries during development, so fruit should be dissected and examined for both previous and current infestation. Because infested fruits usually drop from the vines, several random samples should be taken over time. In recent trials, 40–80 fruit/9 m^2 plot gave adequate estimates of damage (Booth, unpublished data).

In blueberry, yield estimates are best if whole plots are sampled by rake. Six plots of 100 m^2 generally result in a precision of 20% (percentage of the standard error to mean), a level usually adequate for field trials. If the entire plot cannot be raked, 20 subplots, each $1\,m^2$, will give an acceptable level of precision. Infestation by blueberry maggots can be assessed by floating berries in a 75% sugar solution or by boiling them for 3–5 min in a liter jar (Neilson, 1987). Once the boiled berries have split, pour them into a black painted tray where the white fly larvae are clearly visible. Sample 10–15 liters of berries per plot by rake from different locations within a plot when 80 to 100% of the fruit is ripe but before fruit drop (mid-late August) (Collins and Drummond, 1998).

Severe root weevil infestations in strawberry are often visible by the end of the growing season, so number of dead plants may accurately assess efficacy of entomopathogens for that pest. Damage by tarnished plant bugs are not as easily distinguished, as catfacing can also be caused by cool weather during fruit development, so the number of deformed fruit does not necessarily accurately indicate infestation levels. Achenes injured by tarnished plant bugs are of normal size but hollow.

5 Field plot case studies

A Entomopathogenic nematodes and fungi against cranberry girdler in cranberry

1. In mid to late August
   Establish small plots (1 × 1 m) in areas symptomatic of current girdler damage (bronzed foliage). Assess soil for background levels of indigenous entomopathogenic fungi or nematodes according to wax moth bait or other techniques of pathogen isolation (see above). Using sprinkling can and 4 liters of water for each step: pre-moisten soil, apply treatments in suspension, wash entomopathogen from foliage.
2. At 14–28 days post-application
   Sample for mortality due to infection among 6–8 10-cm diameter soil cores per plot. Incubate all larvae, live or dead, individually in a Petri dish at 25 °C dark conditions.
3. At 14 day intervals post-application
   Assess for microbial persistence or recycling according to methods described elsewhere (Section C.4 above; Zimmerman, 1986).

B Entomopathogenic fungi against blueberry spanworm in lowbush blueberry

1. In mid to late April
   Use a 30-cm diameter sweep net to sample blueberry stems in a bearing field for first instar spanworm larvae. Take 10 sweeps/sample and place plots in areas with at least 10 larvae/sample. If densities are too low, wait 3–4 days for additional eggs to hatch. However, if densities remain low and larvae develop beyond the first instar, the field is not suitable. Layout plots in a randomized block design using first instar spanworm density as a blocking variable. If multiple fields are necessary, establish complete blocks in each field. At pest densities of 10–15 larvae/sample, optimum plot size is 7 × 7m but can be adjusted to fit the width of the spray boom. A buffer zone of at least 2 m should exist between each plot. Replicate each treatment 4–5 times. If possible, add two more sets of replicated plots of the unsprayed control and entomopathogen treatments to serve as sources of larvae for laboratory assays of infection rate (see step 4 below).
2. At 1 day before treatment
   Measure pre-application density by taking 10 sweeps/plot, but return spanworm larvae to their respective plots after counting. Derive sampling order among plots according to a random number table. Calibrate sprayer to insure delivery of the proper rate.
3. On spray day
   Apply the treatments just after dawn or when conditions for spray drift are minimal. To further reduce risk, spray height should be 4–8 cm above the canopy. Record crop phenology, crop management practices, and pertinent weather conditions (wind speed and direction, temperature, and presence of leaf wetness) at the time of application.
4. At 1, 3, 7, and 14 days after application
   Sample plots with a sweep net (10 sweeps/plot) with sampling order again determined according to a random number table. To minimize contamination among treatments, sample entomopathogen-treated plots with sweep nets designated to those plots only. To reduce sampling affects on smaller plots, return larvae to each plot after counting. Also at each sample date: collect 20–100 larvae from the extra entomopathogen-treated and control plots (see step 1), and incubate them in the laboratory

to determine infection rates. Surface sterilize field-collected larvae with a 0.5% sodium hypochlorite solution and place individually in a Petri dish provided with moistened filter paper and fresh foliage and incubate at 25 °C. Change foliage for live larvae and monitor for mortality and sporulation daily for 7 days.

5. During late July to August, after infested but before healthy berries drop
   Harvest by hand raking entire plots, measuring yield in both liters and kg per plot. To assess treatment effects on pollination, weigh 30–50 berries individually and dissect them to count number of seeds per berry.
6. During analysis
   Assess treatment effects on abundance of blueberry spanworm larvae and blueberry yield using analysis of variance and analysis of covariance if a density effect exists. Trapezoidal integration of spanworm larval abundance per plot over time can be used as an estimate of seasonal density, a dependent variable for the analysis of variance. Laboratory rearing results can be used to confirm results found in the field and to provide estimates of the level of infection over time.

C *Entomopathogenic nematodes against black vine weevil larvae in caneberry*

1. In mid-spring
   Establish partial row plots (at least 8 plants) in areas symptomatic of root weevil damage. Presample the soil surrounding young floricanes to determine infestation levels and stage of larval development. Begin monitoring soil temperature at 10 cm depth.
2. When soil temperatures exceed 15 °C or ~ 50% of the population are pupae.
   Assess background levels of entomopathogenic nematodes or fungi according to wax moth bait technique.
3. Immediately before application
   Prepare suspension of entomopathogen and assay concentration and viability by examining several samples. Entomopathogenic nematodes can be easily counted in 0.3 ml aliquots using a stereomicroscope. At lower application rates ($1–2 \times 10^9$ IJs/ha), the suspension need not be diluted, but higher rates may require dilution prior to counting. Continually stir the suspension until application.
4. At application
   Using sprinkling can and 4 liters water per step, conduct the following three application steps: pre-moisten soil, apply treatments, wash entomopathogens from foliage.
5. At 7–10 days post-application, but before adult emergence
   Assess larval mortality and mortality due to infection among 3–4 plants per plot by sampling soil surrounding young floricanes 20 cm from main primocane on both sides of the row. Incubate all larvae, live or dead, at 22–25 °C for a few days, then assess for entomopathogen development.
6. At several weekly intervals post-application
   Assess persistence or recycling of entomopathogens in soil according to wax moth bait technique.

D *Entomopathogenic fungi against black vine weevil adults in strawberry*

1. In late spring to early summer
   Monitor black vine weevil life stage development in soil samples and night collections. Newly emerged adults have very soft integument. After adults have fully emerged, but before egg laying begins (~ 30 days), establish plots of multiple partial rows (at least 4 adjacent 10 m sections) with at least 2 buffer rows between each plot. Plot size and placement may be best determined by counting weevils on plants at 1 h after sunset with a flashlight.
2. At application time
   Prepare suspensions of entomopathogens, assess viability and concentration, and apply as described above.
3. Beginning at 5 days after application and at weekly intervals for at least 3 weeks
   Count number of surviving weevils in middle 2 rows only of each plot at night with flashlight and collect at least 6 representative weevils per middle of each plot to assess entomopathogenic infection. Adult black vine weevils are mobile, but a 3 m buffer at both ends of each plot should suffice, especially if only short-term treatment effects are assessed. Place collected weevils individually inside a Petri dish and inspect daily for survival. Adults will survive for several days without food, but will die within 3 days without water. Cadavers should be surface sterilized and placed individually inside an incubation chamber (small vial or Petri dish containing moistened tissue paper),

held at 25 °C dark conditions for at least 2 weeks, and inspected at least every 2–3 days for entomopathogenic development.

6 References

AliNiazee, M. T. and Jensen, F .L. 1973. Microbial control of grape leaffolder with different formulations of *Bacillus thuringiensis*. *J. Econ. Entomol.* **66**, 157–158.

All, J. N. and Dutcher, J. D. 1977. Subsurface and surface insecticide applications to control subterranean larvae of the grape root borer. *J. Econ. Entomol.* **70**, 649–651.

All, J. N., Saunders, M. C., Dutcher, J. D., and Javid, A. M. 1981. Susceptibility of grape root borer larvae, *Vitacea polistiformis* (Lepidoptera: Sesiidae) to *Neoaplectana carpocapsae* (Nematoda: Rhabditidae) and potential of host kairomones for enhancement of nematode activity in grape vineyards. *Misc. Publ. Entomol. Soc. Am.* **12**, 9–14.

Alm, S. R., Williams, R. N., Pavuk, D. M., Snow, J. W., and Heinlein, M. A. 1989. Distribution and seasonal flight activity of male grape root borers (Lepidoptera: Sesiidae) in Ohio. *J. Econ. Entomol.* **82**, 1604–1608.

Ben-Yehuda, S., Izhar, Y., Wysoki, M., and Argaman, Q. 1993. The grape berry moth, *Lobesia botrana* Denis & Schiffermueller (Lepidoptera: Tortricidae), in pear orchards in Israel. *Int. J. Pest Manag.* **39**, 149–151.

Bentley, W. J., Zalom, F., Granett, J., Smith, R. J., and Varela, L. 1990. UC IPM Pest Management Guidelines: Grapes. Univ. Calif. Dept. Agric. and Nat. Res. Pub. 3339.

Berry, R. E., Liu, J., and Groth, E. 1997. Efficacy and persistence of *Heterorhabditis marelatus* (Rhabditida: Heterorhabditidae) against root weevils (Coleoptera: Curculionidae) in strawberry. *Environ. Entomol.* **26**, 465–470.

Bicki, T. J. 1997. Best management practices to optimize chemigation applications. Ocean Spray Environ. Bull. Series #5, 21 pp.

Biever, K. D. and Hostetter, D. L. 1975. *Bacillus thuringiensis*: against lepidopterous pests of wine grapes in Missouri. *J. Econ. Entomol.* **68**, 66–68.

Booth, S. R. and Shanks, C. H. 1998. Potential of a dried rice/myceliam formulation of entomopathogenic fungi to suppress subterranean pests in small fruits. *Biocontrol Sci. Technol.* **8**, 197–206.

Booth, S. R., Tanigoshi, L. K., and Shanks, C., Jr. 2002. Evaluation of entomopathogenic nematodes to manage root weevil larvae in Washington State cranberry, strawberry, and red raspberry. *Environ. Entomol.* **31**, 895–902.

Bournier, A. 1976. Grape insects. *Annu. Rev. Entomol.* **22**, 355–376.

Brown, H. L. and Ismail, A. A. 1981. 1980 blueberry Integrated pest management program. *Univ. Maine Coop. Ext. Ser. Bull.* **639**, 19 pp.

Chase, A. R., Osborne, L. S., and Ferguson, V. M. 1986. Selective isolation of the entomopathogenic fungi *Beauveria bassiana* and *Metarhizium anisopliae* from an artificial potting medium. *Fla. Entomol.* **69**, 285–291.

Collins, J. A. and Drummond, F. A. 1998. Blueberry insect control. 1997. *Arthr. Manag. Tests: 1998.* **23**, 47–50.

Collins, J. A. and Forsythe, H. Y. 1987. Monitoring for blueberry maggot. Univ. Maine Coop. Ext. Ser. Wild Blueberry Fact Sheet No. **201**, 3 pp.

Cone, W. W., Wright, L. C., and Conant, M. M. 1990. Management of insect pest populations in a developing, cool-climate grape industry. *In* 'Monitoring and Integrated Management of Arthropod Pests of Small Fruit Crops' (N. J. Bostanian, L. T. Wilson, and T. Dennehy, Eds.) pp. 27–43. Intercept Ltd., Andover, UK.

Crandall, P. T. 1995. 'Bramble production: the management and marketing of raspberries and blackberries.' Food Products Press, Binghamton, NY.

Curran, J. 1992. Influence of application method and pest population size on the field efficacy of entomopathogenic nematodes. *J. Nematol.* **24**, 631–636.

Drummond, F. A. and Collins, J. A. 1997. Laboratory tests for blueberry insect control. *Arthr. Manag. Tests: 1996.* **22**, 404–409.

Drummond, F. A. and Collins, J. A. 1998. Laboratory tests for blueberry insect control. *Arthr. Manag. Tests: 1997.* **23**, 365–368.

Dutcher, J. D., and Hall, J. N. 1978. Survivorship of the grape root borer in commercial grape vineyards with contrasting cultural practices. *J. Econ. Entomol.* **71**, 751–754

Georgis, R. and Poinar, G. O. 1983. Effect of soil texture on the distribution and infectivity of *Neoaplectana carpocapsae* (Nematoda: Steinermatidae). *J. Nematol.* **15**, 308–311.

Goettel, M. S. and Inglis, G. D. 1997. Fungi: Hyphomycetes. *In* 'Manual of Techniques in Insect Pathology' (L. A. Lacey, Ed.), pp. 213–250. Academic Press, London.

Gordon, S. C., Woodford, J. A. T., and Barrie, I. A. 1990. Monitoring pests of red raspberry in the United Kingdom, and the possible implementation of an integrated pest management system. *In* 'Monitoring and Integrated Management of Arthropod Pests of Small Fruit Crops' (N. J. Bostanian, L. T. Wilson, and T. Dennehy, Eds.) pp. 1–26. Intercept Ltd., Andover, UK.

Hall, D. F. 1954. Studies of microorganisms pathogenic to the sod webworm. *Hilgardia* **22**, 535–561.

Humber, R. A. 1992. Collection of entomopathogenic fungal cultures: Catalog of strains, 1992. USDA, ARS/ARS-110, 117 pp.

Jennings, D. L. 1988. 'Raspberries and Blackberries: Their Breeding, Diseases, and Growth.' Academic Press, London.

Kaya, H. K. and Stock, S. P. 1997. Techniques in insect nematology. *In*. 'Manual of Techniques in Insect Pathology' (L. A. Lacey, Ed.), pp. 281–324. Academic Press, London.

Kakouli-Duarte, T., Labuschagne, L., and Hague, N. G. M. 1997. Biological control of the black vine weevil, *Otiorhynchus sulcatus*, (Coleoptera: Curculionidae) with entomopathogenic nematodes (Nematoda: Rhabditidae). *Ann. Appl. Biol.* **131**, 11–27.

Kreuger, S. R., Villani, M. G., Martins, A. S., and Roberts, D. W. 1992. Efficacy of soil applications of *Metarhizium anisopliae* (Metsch.) Sorokin conidia, and standard and lyophilized mycelial particles against scarab grubs. *J. Invertebr. Pathol.* **59**:54–60.

Li, S. Y. and Fitzpatrick, S. M. 1996. The effects of application rate and spray volume on efficacy of two formulations of *Bacillus thuringiensis* Berliner var. *kurstaki* against *Choristoneura rosaceana* (Harris) (Lepidoptera: Tortricidae) on raspberries. *Can. Entomol.* **128**, 605–612.

Maier, C. T. 1978. Dispersal of adults of the black vine weevil, *Otiorhynchus sulcatus* (Coleoptera: Curculionidae), in an urban area. *Environ. Entomol.* **7**, 854–857.

Martinson, T. E., Hoffman, C. J., Dennehy, T., Kamas, J. S., and Weigle, T. 1991. Risk assessment of grape berry moth and guidelines for management of the eastern grape leafhopper. New Yorks' Food and Life Sci. Bull. **138**, pp 1–10.

Mirica, A., Savescu, A., and Mirica, I. 1969. Factors in forecasting for the control of the vine mealybug (*Pulvinaria betulae* L.). *Ann. Inst. Cerc. Pentru Protect. Plant.* **5**, 387–393.

Montgomery, M. E., Musick, G. J., Polivka, J. B., and Nielsen, D. G. 1979. Modifiable washing-flotation method for separation of insect eggs and larvae from soil. *J. Econ. Entomol.* **72**, 67–69.

Mullins, M. G., Bouquet, A., and Williams, L. E. 1992. 'Biology of the Grapevine.' Cambridge University Press, Cambridge, UK.

Neilson, W. 1987. Recovery of maggots (larvae) from lowbush blueberries. Kentville Research Station Annual Report: 33–34. Agriculture Canada Research Station, Kentville, Nova Scotia.

Pfeiffer, D. G., Boucher, T. J., LaChance, M. W., and Killian, J. C. 1990. Entomological research in Virginia (United States) vineyards. *In* 'Monitoring and Integrated Management of Arthropod Pests of Small Fruit Crops' (N. J. Bostanian, L. T. Wilson, and T. Dennehy, Eds.) pp. 45–61. Intercept Ltd., Andover, UK.

Phipps, C. R. 1930. Blueberry and Huckleberry Insects. Maine Agric. Exp. Stn. Bull. **356**, pp. 1–232.

Polavarapu, S. and Peng, H. 1997. Evaluation of insecticides against blackheaded fireworm in cranberries, 1997. *Arthr. Manag. Tests: 1997.* **22**, 53.

Roditakis, N. E. 1986. Effectiveness of *Bacillus thuringiensis* Berliner var. *kurstaki* on the grape berry moth *Lobesia botrana* Den. and Shiff. (Lepidoptera, Tortricidae) under field and laboratory conditions in Crete. *Entomol. Hellenica* **4**, 31–35.

Shanks, C. H., Jr. 1981. Insects and mites. *In* 'The Strawberry' (N. F. Childers, Ed.) pp 293–308. Horticultural Publications. Gainesville, FL.

Shanks, C. H., Jr. and Agudelo-Silva, F. 1990. Field pathogenicity and persistence of heterorhabditid and sternernematid nematodes (Nematoda) infecting black vine weevil larvae (Coleoptera: Curculionidae) in cranberry bogs. *J. Econ. Entomol.* **83**, 107–110.

Simser, D. and Roberts, S. 1994. Suppression of strawberry root weevil, *Otiorhynchus ovatus*, in cranberries by entomopathogenic nematodes (Nematoda: Steinernematidae and Heterorhabditidae). *Nematologica* **40**, 456–462.

Southwood, T. R. E. 1978. 'Ecological Methods.' Chapman and Hall, London.

Stenzel, K. 1992. Mode of action and spectrum of activity of BIO 1020 (*Metarhizium anisopliae*). *Pflanzenschutz-Nachrichten Bayer* **45**, 143–158.

Stern, V. M. and Federici, B. A. 1990. Biological control of western grapeleaf skeletonizer, *Harrisina brillians* Barnes and McDunnugh (Lepidoptera: Zygaenidae), with a granulosis virus in California. *In*. 'Monitoring and Integrated Management of Arthropod Pests of Small Fruit Crops' (N. J. Bostanian, L. T. Wilson, and T. Dennehy, Eds.) pp. 117–129. Intercept Ltd., Andover, UK.

Strand, L. L. 1994. Integrated pest management for strawberries. Univ. Calif. Statewide IPM Project. Div. of Agric. and Nat. Res. Pub. 3351.

Strongman, D., MacKenzie, K., and P. Dixon. 1997. Entomopathogenic fungi in lowbush blueberry fields. *Acta Hort.* **446**, 465–476.

Studdert, J. P. and Kaya, H. K. 1990. Effect of water potential, temperature, and clay-coating on survival of *Beauveria bassiana* conidia in a loam and peat soil. *J. Invertebr. Pathol.* **55**, 417–427.

Townsend, H. G. 1990. Grape insect pest monitoring and IPM in the Missouri Ozarks. *In* 'Monitoring and Integrated Management of Arthropod Pests of Small Fruit Crops' (N. J. Bostanian, L. T. Wilson, and T. Dennehy, Eds.) pp. 117–129. Intercept Ltd., Andover, UK.

Vincent, C., de Olivera, D. D., and Belanger, A. 1990. The management of insect pollinators and pests in Quebec strawberry plantations. *In*. 'Monitoring and Integrated Management of Arthropod Pests of Small Fruit Crops' (N. J. Bostanian, L. T. Wilson, and T. Dennehy, Eds.) pp. 1–26. Intercept Ltd., Andover, UK.

Wennemann, L. G. J. 1997. Early season application of entomopathogenic nematodes (Nematoda: Steinernematidae, Heterorhabditidae) against two pests of wine and juice grapes (*Vitis vinifera* L. and *V. labrusca* L.). Ph.D. dissertation. Washington State University. Pullman, Washington.

Yarborough, D. E. 1998. Wild blueberry culture in Maine. Wild Blueberry Fact Sheet. Fact Sheet #**220**, Bull. #2088. University of Maine Cooperative Extension, Orono, ME. 4 pp.

Yarborough, D. E. and Collins, J. A. 1997. 1997 Insect control guide for wild blueberries. Wild Blueberry Fact Sheet #**209**, Bull. #2001. University of Maine Cooperative Extension, Orono, ME. 4 pp.

Yarborough, D. E., Hundhammer, M., and Collins, J. A. 1993. Integrated crop management field scouting guide for lowbush blueberries. Wild Blueberry Fact Sheet. Fact Sheet #**204**, Bull. #2275. University of Maine Cooperative Extension, Orono, ME. 9 pp.

Zalom, F. G., Pickel, C., Walsh, D. B., and Welch, N. C. 1993. Sampling for *Lygus hesperus* (Hemiptera: Miridae) in strawberries. *J. Econ. Entomol.* **86**, 1191–1195.

Zimmermann, G. 1986. The "*Galleria* bait method" for detection of entomopathogenic fungi in soil. *Z. angew. Entomol.* **102**, 213–215.

Chapter VII-15

Application and evaluation of entomopathogens for control of pest insects in mint

Ralph E. Berry

Emeritus Professor of Entomology
Oregon State University
18160 Cottonwood Road, PMB #798
Sunriver, OR 97707, USA

1 Introduction

Peppermint (*Mentha piperita*), native spearmint (*M. spicata*), and Scotch spearmint (*M. cardiaca*) are grown commercially for their essential oils. Essential oils, which are produced in leaf glands primarily on the lower leaf surface, contain volatile terpenes and sesquiterpenes that produce the unique flavoring for candy, chewing gum, toothpaste, medicines and other products. The oil, initially recovered by on-farm steam distillation, is further processed and blended to produce oils to meet specifications requested by product manufacturers.

Commercial peppermint production in the USA started in Ashfield, Massachusetts in 1812, and subsequently its cultivation moved rapidly to New York and the Midwest. Cultivation of peppermint reached the northwestern United States in 1909 and was grown in western Oregon and central Washington and then in Idaho and central Oregon (Sievers, 1948; Green, 1963; Landing, 1969; Matzat, 1981). Presently, peppermint and spearmint is produced commercially in the Midwest in Indiana, Michigan and Wisconsin (Lacy *et al.*, 1981) and in the West in Oregon, Washington, Idaho and Montana. Smaller production areas include California, Nevada, and North and South Dakota. The principal peppermint cultivars are black Mitcham, Todd's Mitcham, Murray's Mitcham, redefined Murray's Mitcham, and Robert's Mitcham. Over 49,776 ha of peppermint were in production in the USA in 1998. In 1997, 4,004,310 kg of oil, valued at about $124 million, were produced. Mint is cultivated commercially in fields ranging from 8 to 50 ha producing a gross value of approximately $2,818/ha.

Peppermint is a perennial crop that spreads by underground rhizomes or stolons. Propagation is through digging and replanting rhizomes or propagating tip cuttings in greenhouses. New fields are planted in rows either in the fall or spring and the rhizomes spread in all directions sending up new growth at the nodes or joints. Peppermint grows best on rich, well-drained, mineral sandy loam or clay loam soils or soils high in organic matter, such as the muck soils found in some areas of the Midwest (Martin *et al.*, 1976; Jackson *et al.*, 1983). Optimum production occurs in areas with a midsummer day length of at least 15 h and cool nighttime temperatures, such as those found just north and south of the 45th parallel. In most production areas, irrigation is required for maximum mint production because of its shallow root system. Irrigation creates a cool, humid environment that enhances the habitat for survival of entomopathogens.

In many production areas, mint is not tilled or cultivated after it is established because of the likelihood of spreading verticillium wilt (*Verticillium dahliae*). This practice allows populations of soil arthropods to increase and reduces the opportunity to incorporate soil pesticides. Therefore, the increase in soil arthropods and

L.A. Lacey and H.K. Kaya (eds.), Field Manual of Techniques in Invertebrate Pathology, 599–607.

other pests such as weeds, nematodes, and verticillium wilt may limit the production life of mint stands to 5 or 6 years.

2 Insect pests

The major insect pests in mint are all subterranean and either feed on the roots and crowns or bore into the underground rhizomes. In the West, larvae of mint root borer, *Fumibotys fumalis*, strawberry root weevil, *Otiorhynchus ovatus*, and mint flea beetle, *Longitarsus ferrugineus*, are the most damaging pests. The mint stem borer, *Pseudobaris nigrina*, only occurs in Idaho. Mint root borer and strawberry root weevil are not pests in the Midwest, but the mint flea beetle frequently causes serious damage. Subterranean cutworms, *Euxoa, Agrotis* and *Mamestra* spp., occur in the West and cause sporadic injury to mint in the spring (Berry, 1975; Berry and Fisher, 1993). Other species of cutworms, such as variegated cutworm, *Peridroma saucia*, and loopers (*Autographa californica* and *Trichoplusia ni*) are important pests in most production areas.

3 Entomopathogens

Table 1 summarizes the entompathogens that have been used against key insect pests of mint. The primary entomopathogens have been entomopathogenic nematodes and *Bacillus thuringiensis* subsp. *kurstaki* (*Btk*).

A Entomopathogenic nematodes

1 Mint root borer

Steinernema carpocapsae was evaluated in large field plots against mint root borer (Takeyasu, 1994) and was used commercially by growers on approximately 1,000 ha in 1993 and 1994. Post-harvest applications in mid-August through sprinkler irrigation at a rate of 2.5 or 4.9×10^9 infective juveniles (IJs)/ha controlled 75.2% and 85.5% of the mint root borer larvae, respectively (Takeyasu, 1994). The low rate was recommended for fields with a root borer density less than 1.0 larva/929 cm^2, and the high rate was recommended for fields with densities greater than 1.0 larva/929cm^2. Split applications of nematodes (applied pre-harvest in late July and post-harvest in mid-August) also have been evaluated at rates of 1.2 and 1.8×10^9 IJs/ha. Pre-harvest applications of 1.2 and 1.8×10^9 IJs/ha controlled 28.9% and 60.9% of the mint root borer larvae, respectively. A second post-harvest application of 1.2 and 1.8×10^9 IJs/ha increased control of mint root borer larvae to 88.9% and 94.3%, respectively (Takeyasu, 1994).

S. riobrave injected into the top 2.5 cm of soil in furrow irrigated mint reduced the mint root borer population by 54% (Gillespie *et al.*, 1996; Research Reports, Mint Industry Research Council, unpublished). Application of nematodes through sprinkler irrigation allows for a more even distribution than injection into the soil on

Table 1. Insect pests in mint and their control with entomopathogenic nematodes in the field

Key target pests	Nematode species	Application rates (10^9 IJs/ha)	References
Mint root borer	*Steinernema carpocapsae*	3.7–4.9	Takeyasu, 1994
	S. oregonense	4.9	
	S. riobrave	4.9	
	Heterorhabditis marelatus	4.9	
	H. bacteriophora	4.9	
Strawberry root weevil	*S. carpocapsae*	7.4	
	H. marelatus	4.9	
	H. bacteriophora	7.4	
Mint flea beetle	*S. carpocapsae*	4.9–9.8	Morris, 1990
	S. feltiae	4.9–9.8	
	S. riobrave	3.7	
	H. bacteriophora	7.4	Morris, 1990
Variegated cutworm	*S. carpocapsae*	4.0	

the furrow ridge and may help explain the reason for the low level of control.

Large quantities of *S. carpocapsae* were not available after 1995 and growers discontinued using entomopathogenic nematodes to control this pest. However, growers are likely to use entomopathogenic nematodes if large quantities become available at costs competitive with insecticides. Additional nematode species, such as *Heterorhabditis marelatus, H. bacteriophora, S. riobrave* and *S. oregonense*, have been evaluated in the laboratory for control of mint root borer. In these experiments, all nematode species were applied at rates equivalent to 4.9×10^9 IJs/ha and control ranged from 90 to 100%.

2 *Strawberry root weevil*

Entomopathogenic nematodes, *S. carpocapsae, H. marelatus* and *H. bacteriophora*, provide control of strawberry root weevil larvae and pupae in mint in late April and early May. Higher rates of *S. carpocapsae* are required to control strawberry root weevil and pupae than *H. marelatus* and *H. bacteriophora*. For example, at a rate of 7.4×10^9 IJs/ha, *S. carpocapsae* only reduced the weevil population 68.1% when applied through sprinkler irrigation after harvest (Takeyasu, 1994). However, 4.9×10^9 IJs of *H. marelatus*/ha reduced the weevil population by 96%. Applications of entomopathogenic nematodes in late April or early May will control weevil larvae, pupae, and teneral adults. Post-harvest applications in August will control both strawberry root weevil larvae and mint root borer larvae, but application rates may need to be adjusted depending on the insect pest and the species of nematodes being applied. Applications to control strawberry root weevil in the spring have been less successful largely because soil temperatures are often below 16 °C and irrigation may not be available.

3 *Mint flea beetle*

Control of mint flea beetle larvae has been achieved with *S. carpocapsae* and *H. bacteriophora* (Morris, 1990). Spring applications of *S. carpocapsae* and *H. bacteriophora* at a rate of 7.4×10^9 IJs/ha in late May or early June may provide control of mint flea beetle larvae, prepupae, pupae, and teneral adults. For example, Morris (1990) showed that *S. carpocapsae* reduced the mint flea beetle population by 68.6% and *H. bacteriophora* reduced the population 93.9%. Mahr and Wyman (1996 Research Reports, Mint Industry Research Council, unpublished) demonstrated that *S. feltiae, S. carpocapsae* and *S. riobrave* could disperse in muck soils and infect *Galleria mellonella mellonella* larvae buried in the soil. They suggested that these nematode species might be good candidates for mint flea beetle control in the Midwest. Morris (1990) reported that applications should be made in the evening through sprinkler irrigation after soil temperatures exceed 16 °C in the spring and after about 400 day-degrees have been accumulated. Early spring applications may not be effective because of cold soil temperature, lack of adequate irrigation water, and small size of larvae and their concealment in rhizomes.

4 *Variegated cutworm*

S. carpocapsae applied at 7.4×10^9 IJs/ha reduced populations of variegated cutworm by 94.5% in experiments conducted in 532 cm^2 plots. However, when applied to 1 ha plots through sprinkler irrigation at 4.9×10^9 IJs/ha, the population was only reduced 45.8%. Early instars of variegated cutworm feed on the foliage and may avoid infection by *S. carpocapsae*. Later instars remain on the soil surface or under debris and may be more susceptible to infection, but additional research is needed to determine whether *S. carpocapsae* will provide adequate control of this pest.

B Bacillus thuringiensis *subsp. kurstaki (Btk)*

Btk controls populations of small cutworms and loopers feeding on mint leaves. However, *Btk* failed to control subterranean cutworms, such as the redbacked cutworm, in mint because this species feeds below ground and does not consume above ground plant parts treated with *Btk* until nearly mature. The major disadvantage of *Btk* is that the dense mint foliage prevents penetration and adequate coverage of the lower leaves. Further studies are needed to determine if new formulations of *Btk* may be effective against foliage feeding cutworms and loopers.

4 Protocols for application of entomopathogens

A Crop characteristics

Destructive sampling of mint rhizomes is the most effective method to evaluate pest densities and determine efficacy of entomopathogens. In most situations, a 464 cm^2 or 929 cm^2 soil sample taken 10 cm deep is sufficient to estimate the population. Soil samples may be either sorted or sieved directly in the field or transported to the laboratory and extracted using Berlese funnels. A core sampler also may be used for root weevils and mint flea beetles, but the core size should be no less than about 10 to 15 cm diameter and samples should be taken 10 cm deep. When the samples are processed in the field, the rhizomes may be replaced, which reduces the impact on the plant stand.

B Plot characteristics

Evaluation of entomopathogens in mint may be conducted either in small plots ranging from 1 m^2 to 15 m^2 or in larger plots ranging from 0.5 to 3.0 ha sized plots, depending on the insect pest and its distribution and the methods of application. Buckets with the bottoms removed (ca. 532 cm^2) or PVC pipe cylinders 16 cm inside diameter and 32 cm long pushed 10 cm into the soil also may be used to evaluate entompathogens in mint. A known number of larvae collected from the field site may be placed in these micro-plots or the micro-plots may be placed within a known area of high insect infestation. Insect larvae are the most damaging stages on mint rhizomes and have limited powers of dispersal. Because of the host-seeking behavior of entomopathogenic nematodes, a 1 to 2 m buffer should be left between adjacent plots. Except for cutworms, adults of the major insect pests on mint have limited powers of dispersal, but if left unchecked their populations can reach economic levels in a few years. Also, since mint rhizomes are dug and used to establish new mint plantings, several of the serious insect pests may be transported with the rhizomes and soil, resulting in infestation of new plantings. Mint rhizomes or tip cuttings, certified free of insect pests, should be used to plant new mint fields.

C Application methods

Entomopathogenic nematodes are packaged differently depending on the supplier. For example, some are packaged in inert carrying material such as a clay-based formulation or in finely ground vermiculate. Both formulations are suspensible in water, but since the particle size may vary, all filters should be removed from sprayers. Spray nozzle opening should be at least 500 μm. Others supply nematodes in a water soluble gel formulation or on small pieces of sponge. The nematodes are released from the gel or sponge when agitated in water. Entomopathogenic nematodes also are shipped as "nematode wool" that can be added directly to water. It is important to keep nematodes cool and protected from sunlight until application. Nematodes may be stored up to 3 months at 2 to 6 °C in the dark, but the length of storage and nematode viability after storage will depend on the formulation. To check for viability, follow directions given by Koppenhöfer (see Chapter IV-5).

Entomopathogenic nematodes may be applied to small plots with a backpack pressure sprayer or a small garden watering can, but for larger plots, tractor drawn pressure tank sprayers with 0.38 to 1.0 m^3 capacity may be used.

1. If needed, pre-moisten the soil with a minimum of 6 mm of water prior to application of nematodes,
2. Estimate the actual nematode concentration applied in the plots, place five or more small cups (3.6 cm inside diameter, 4.0 cm deep) in each treatment plot before applying the nematodes.
3. Collect the cups immediately after application, and measure the quantity of water and count the number of nematodes in each cup to estimate the actual concentration reaching the soil surface. This technique should be used in small and large plots.
4. For large plots, directly inject the nematodes into the irrigation system over a 30–60 min interval to ensure an even distribution (Takeyasu, 1994).
5. After application, irrigate the plots weekly with approximately 25 mm of water.

Btk applied before harvest in July at a rate of 5.6 to 11.7 billion international units (BIUs)/ha provides control of variegated cutworm and loopers feeding on mint leaves. *Btk* is commonly applied with a ground sprayer or airplane in a minimum of 93.5 liters of water/ha. The addition of a spreader sticker or wetting agent improves efficacy. *Btk* is more effective if applied at night, when larvae are actively feeding on the foliage. *Btk* incorporated into baits also has been effective against cutworms and loopers, but they are not widely used because bait formulations are generally unavailable and are difficult to apply uniformly with ground equipment during the growing season.

D Pattern and type of infestation

The major soil insect pests on mint (except for cutworms) have an aggregated or clumped distribution and five or more replications are required for field efficacy testing. The number of replications should be increased for small plot experiments. Pretreatment soil samples may provide some indication of the relative infestation level in a particular area. To do this, small replicated plots can be randomly established within the field. Artificially infesting small plots with a known number of insect larvae collected from the field is frequently the best method to evaluate entompathogens in mint. This method is particularly useful if bottom less buckets or plastic cylinders are used as the plots. Burying small screen cages such as screen mesh tea strainers (4 cm diameter) artificially infested with a known number of larvae (except large cutworms) also is a good method to evaluate efficacy of entomopathogenic nematodes. Burying screen cages containing insect or *G. mellonella* larvae at different depths in the soil may be used to determine nematode vertical movement.

To determine persistence of nematodes in the soil:

1. Bury *G. mellonella* larvae in screen tea strainers in the soil or collect treated soil and bait it with 5 to 10 *G. mellonella*.
2. After 2 days, remove tea strainers containing *G. mellonella* larvae or *G. mellonella* that were placed in treated soil and count the number of infected larvae.
3. Rinse all *G. mellonella* larvae in distilled water, and store them at room temperature in petri dishes with moistened filter paper for 3 days.
4. Dissect *G. mellonella* cadavers before the nematodes reproduce (within 3 days in studies with *Steinernema* spp. or within 4 days for studies with *Heterorhabditis* spp.) to validate that they were killed by nematodes and count the number of entomopathogenic nematodes (Koppenhöfer *et al.*, 1998). The percentage of nematodes recovered from the cadavers in the soil samples can be calculated relative to the nematode concentration applied.

E Application dates

1 Mint root borer

Application of entomopathogens against insect pests in mint depends on the pest phenology. For mint root borer, *S. carpocapsae* provides control of larvae when applied pre-harvest in late July or early August or post-harvest from mid-August to mid-September. Studies in the laboratory have shown that *H. marelatus* also controls root borer larvae. Since pre-harvest applications target earlier instars, crop damage is minimized. However, the nematodes may be applied too early. The mint root borer has a prolonged adult emergence over a 2-month period (Berry, 1974). This, in combination with the short persistence of *S. carpocapsae* in the soil, makes timing an important factor in achieving good control. To benefit from a pre-harvest application, nematodes should be applied as early as possible to minimize crop damage, yet late enough to ensure control of larvae resulting from adults that emerge later in the summer.

Just as pre-harvest applications may be applied too early, a post-harvest application may be applied too late. Prior to entering the prepupal stage, late instar mint root borers construct a silk-lined earthen cell, a hibernaculum, in which to overwinter (Berry, 1974; Pike *et al.*, 1988; Berry and Fisher, 1993). Once it enters the prepupal stage, it is no longer susceptible to *S. carpocapsae* (Takeyasu, 1994). Therefore, fields must be treated before hibernacula are formed. Since nematode applications must be accompanied by irrigation to move nematodes into the soil and to ensure nematode survival, proper timing of a post-harvest application may be hampered by the lack of irrigation immediately after harvest. Also interfering with post-harvest timing is the time-consuming nature of diagnosing and treating fields. Processing soil samples either by hand or in Berlese funnels is a slow, labor-intensive process. If an infestation is found, growers typically need up to a week to treat a field. In most situations, applications of entomopathogenic nematodes at 3.7 to 4.9×10^9 IJs/ha should be timed to coincide with early instar mint root borers. Preharvest treatments have generally been less effective than a single post-harvest application. Split applications, such as one application before harvest at 1.8 to 2.4×10^9 IJs/ha and a second application at 1.8 to 2.4×10^9 IJs/ha after harvest, are effective, but the labor required may outweigh the advantages of a single application after harvest. Perhaps, a higher pre-harvest rate would provide control and minimize crop damage earlier in the season, followed by a lower post-harvest rate, if necessary. This combination may give better control than a split application using the same rate. In mint, entomopathogenic nematodes persist in the soil for about 2 to 3 weeks so a single application, if properly timed, will effectively reduce the root borer population.

2 *Strawberry root weevil*

To control strawberry root weevil larvae and pupae, entomopathogenic nematodes may be applied in mint in late April or early May. Applications after harvest in mid-August or early September will control root weevil larvae. In the spring, cool soil temperature and lack of irrigation may limit efficacy of *S. carpocapsae*. *H. marelatus* is a cool-temperature adapted nematode that has been effective against root weevil larvae and pupae at temperatures as low as 10 °C (Berry *et al.*, 1997). If properly timed, a single application of 4.9×10^9 IJs/ha after harvest may provide control of both root weevils and mint root borer larvae.

3 *Mint flea beetle*

Mint flea beetle eggs hatch in April and early May and larvae begin feeding on small mint roots and later tunnel into rhizomes. Larvae complete development during late May and June and the pupal stage is completed in about 3 to 4 weeks. Spring applications of entomopathogenic nematodes at a rate of 7.4×10^9 IJs/ha should be made in late May or early June to control larvae, prepupae, pupae, and teneral adults. Applications of *S. carpocapsae* should be made in the evening through sprinkler irrigation after soil temperatures exceed 16 °C. In areas East of the Cascade Mountains, cool spring temperatures and the lack of irrigation water may limit the effectiveness of early spring applications.

4 *Variegated cutworm*

Variegated cutworm larvae are susceptible to *S. carpocapsae*, but only when applied preharvest in late June and July. Takeyasu (1994) demonstrated that there was a 94.5% reduction of variegated cutworms in small plots, but only about a 46% reduction in the population in large plots. Early instars feed on foliage and later instars feed near the soil surface or beneath crop debris. However, because of the overlap in larval stages, a single application may not provide adequate control because larvae feeding on above ground foliage may escape infection by *S. carpocapsae*.

F *Sampling and monitoring*

1 *Mint root borer*

Mint root borer larvae cause damage by feeding inside peppermint and spearmint rhizomes from late July through mid-September (Berry, 1974). Damage resulting from feeding injury weakens mint stands, which overwinter poorly and regrow slowly in the spring.

Inspection of fields during July and August should be performed when adults are active and can be detected by observing moth flight as you walk through the field. When disturbed, adults fly a few meters and land on the underside of a leaf. Adults may be collected by sweeping foliage, but in most instances the wing scales have been lost and adults cannot be easily distinguished from other small moths. A sex pheromone is available commercially to sample for mint root borer males (Davis *et al.*, 1984, 1991). The successful development of this sex pheromone in a trapping system could provide valuable early-season information to growers concerning the need to control mint root borer larvae. However, Takeyasu (1994) was unable to correlate adult density in pheromone trap catches with larval infestation. Therefore, the use of pheromone traps for mint root borer may be limited to revealing the presence or absence of mint root borer adults in a field or concentration of adult activity in a certain area.

Presence of adults in late June and July serves as a signal to take postharvest soil samples for larvae in August or early September. Late September and early October soil samples will usually only detect mature larvae and hibernacula after damage is done and control with nematodes will be ineffective.

1. Soil samples ($929\,cm^2$) should be taken from several locations in the field. It is recommended that one sample be taken/ha with a minimum of 25 samples per field. Samples should include soil and rhizomes to a depth of 5 to 10 cm. These samples also may be used to estimate the population of root weevil larvae in the fall, although the depth of samples may be too shallow to collect a representative sample of weevils.
2. Samples may be sorted by hand in the field or with the use of a Berlese funnel. If using Berlese funnels, separate the soil and rhizomes to enhance recovery of larvae. If samples are processed in the field, examine rhizomes for damage and larvae that may be within the rhizome.
3. Depending on soil texture and moisture, the use of screens to sieve soil may speed up the process of locating larvae in the samples that have left rhizomes. Treatment is justified if an average of 2 to 3 larvae is found/$929\,cm^2$.

2 *Strawberry root weevil*

Strawberry root weevil larvae are best estimated by taking $929\,cm^2$ soil samples after harvest in late August and September or the following spring in April or early May (Emmenegger, 1976; Emmenegger and Berry, 1978). Samples for root weevils may be combined with samples for mint root borer in the fall or combined with samples for cutworms and flea beetles in April and May. Take at least 25 soil samples from different areas of the field (approximately one sample/ha), screen the soil, count the number of larvae and calculate the average number per sample. Treatment is warranted if the average number of root weevil larvae exceeds $2.0/929\,cm^2$.

A sequential sampling program has been developed for strawberry root weevil in central Oregon (Cacka, 1982).

1. This sampling plan is based on taking a minimum of 25, $929\,cm^2$ soil samples and accumulating the number of adults, pupae, and larvae found in each sample.
2. Using this method, treatment is recommended if the total accumulated number of adults, pupae, and larvae exceed $5.5/929\,cm^2$. Treatment is not recommended if the total weevils/sample is less than $2.7/929\,cm^2$.
3. These sequential sampling plans are based on treatment thresholds, which were calculated using an oil value of $22.00/kg and a cost of treatment of $47.35/ha. However, as the value of the oil and the cost of treatment change, the recommended treatment thresholds also change (Berry and Fisher, 1993).
4. In practice, mint fields suspected of being infested with root weevils should be sampled for adults with a sweep net in the evening, a couple of hours after sunset. Still, warm and dry evenings are best. Windy, cool and/or rainy periods produce fewer weevils/sweep, giving the impression of a smaller field population than is actually present.
5. Take 10 straight-line sweep samples in at least five different sites in fields up to 12 ha. Add one additional sample site for each additional 4 ha.

3 *Mint flea beetle*

Root damage caused by mint flea beetle larvae can be seen in the spring. Regrowth of mint injured by this pest is slow and characterized by spotty stands and reddish plants that are stunted and appear stressed. However, these symptoms also may be typical of other factors, such as water stress, inadequate fertility or other root feeding pests, such as plant-parasitic nematodes and root weevils.

Recognition of adult flea beetle feeding damage on mint leaves is the easiest clue to determine if this pest is present. Feeding is interveinal and appears as many small "shot holes" through both leaf surfaces.

1. Use a sweep net in late June and July to monitor for adult mint flea beetles. The small, yellowish-brown beetles are most efficiently sampled early in the morning on dry foliage.
2. Sample at least five sites for every 8 ha taking 10 to 20 straight-line sweep net samples through the foliage at each site. No treatment threshold has been developed for adult flea beetles, but yield reductions have been observed for an average of 5 to 10 adults/sweep sample.
3. To sample for larvae, take soil samples in late May and early June or after 300 to 400 day-degrees have been accumulated (Morris, 1990).
4. Take a 12.8 by 18 cm sample of soil and roots to a depth of 10 cm (350 cm^3 core) every 0.4 to 0.8 ha. Take a minimum of 25 samples/field.
5. Screen the soil samples in the field or use Berlese funnels to collect larvae or pupae (Morris, 1990; Shields *et al.*, 1981). If Berlese funnels are used, it is recommended that the soil and rhizomes be separated to enhance recovery. Carefully inspect the soil and roots for small cream-colored larvae (1 to 2 mm in length with brown heads). They also can be found in black or brown "tracks" within tissue just below the surface of the root. Occasionally they will be in the soil near the root or protruding from the root itself.
6. At an oil price of \$33.07/kg, the economic injury level ranges from 0.61 to 1.12 larvae/350 cm^3 depending on cost of control (Morris, 1990).

4 *Variegated cutworm*

The decision to control variegated cutworm is usually based on the average number of larvae found in 929 cm^2 samples taken on the soil surface. To estimate populations of fourth, fifth, and sixth instars:

1. Inspect the soil surface by first vigorously shaking mint foliage and closely observing and recording the number of larvae/929 cm^2 randomly through the field.
2. Take a ground search sample every 2 ha for fields up to 12 ha. Add an additional site for every 4 ha in fields that exceed 12 ha. Look very closely for small and curled-up larvae under and in folded leaves on the ground. When leaf chewing is quite evident and cutworm counts from ground searches are low, return after dark and sample with a sweep net to collect larvae actively feeding on the foliage.
   a. Sequential sampling plans have been developed for variegated cutworm using sweep net samples to estimate larvae (instars 2 to 4) and for ground search samples (929 cm^2) to estimate instars 4 to 6 (Coop, 1987; Coop *et al.*, 1995). These sequential sampling plans are based on treatment thresholds which were calculated using an oil value of \$26.46/kg and a cost of treatment of \$47.35/ha; as the value of the oil and cost of treatment change, the recommended treatment thresholds also change (Berry and Shields, 1980; Berry and Fisher, 1993).
   b. Using these plans, treatment of instars 2 to 4, sampled with a sweep net, is recommended if 60 larvae are collected from a minimum of 11 different field sites (a minimum of 10 sweep net samples should be taken at each site). Treatment is not recommended if fewer than 44 larvae are collected in sweep net samples.
   c. For ground search sampling, treatment of instars 4 to 6 is recommended if 24 larvae are collected in 929 cm^2 samples taken from a minimum of 18 different sites. Treatment is not recommended if fewer than 17 larvae are collected in the samples.

A sex pheromone is commercially available and can be used to detect and monitor adult males of the variegated cutworm in the spring. Trapping males could provide valuable early season information to growers concerning the potential need to control cutworm larvae during June and July (Coop, 1987). Sticky traps baited with this pheromone can be set in fields in late April and monitored weekly or biweekly through June. Although action levels for treatment have not been developed relative to moth catches, it is likely that large and continuous catches greater than 25 adults/week will result in similarly large populations of larvae being observed approximately 2 weeks following peak trap counts. The real value of pheromone traps lies in the fact that they signal when to begin inspecting fields for larvae, thereby greatly improving timing of control. However, large trap catches will not always result in large larval populations. This is particularly true when the mint field does not have a resident population of variegated cutworm and the trap catches are a result of males being lured into traps from other crops. Conversely, a small trap catch does not necessarily mean that the larvae will not exceed an action level.

5 References

Berry, R. E. 1974. Biology of *Pyrausta fumalis* on peppermint in Oregon. *Ann. Entomol. Soc. Am.* **67**,580–582.

Berry, R. E. 1975. Redbacked cutworm: flight period and egg development under field conditions. *Environ. Entomol.* **4**, 603–605.

Berry, R. E. and Fisher, G. 1993. A guide to peppermint insect and mite identification and management. PNW 182. Oregon State Univ. Ext. Ser. Corvallis, OR.

Berry, R. E., Liu, J., and Groth, E. 1997. Efficacy and persistence of *Heterorhabditis marelatus* (Rhabditida: Heterorhabditidae) against root weevils (Coleoptera: Curculionidae) in strawberry. *Environ. Entomol.* **26**, 465–470.

Berry, R. E. and Shields, E. J. 1980. Variegated cutworm: leaf consumption and economic loss in peppermint. *J. Econ. Entomol.* **73**, 607–608.

Cacka, J. F. 1982. "Biology, distribution and economic threshold of the strawberry root weevil, *Otiorhynchus ovatus* (L.) in peppermint." M. S. Thesis, Oregon State Univ., Corvallis, OR.

Coop, L. B. 1987. "Phenology modeling to predict occurrence and treatment thresholds for variegated cutworm, *Peridroma saucia* Hübner, on peppermint." Ph.D. Thesis. Oregon State Univ., Corvallis, OR.

Coop, L., Berry, R. E., Fisher, G., and Kogan, M. 1995. Integrated Pest Management on Peppermint (IPMP v. 2.0). Computer Software Publication CS 195. Integrated Plant Protection Center. Oregon State Univ., Corvallis, OR.

Davis, H. G., McDonough, L. M., and Pike, K. S. 1984. Attraction of male *Fumibotys fumalis* to females of the species. *J. Entomol. Soc. British Columbia.* **81**, 25–28.

Davis, H. G., McDonough, L. M., Smithhisler, C. L., Brown, D. F., Leonhardt, B. A., Voerman, S., and Chapman P. S. 1991. Mint root borer, *Fumibotys fumalis* (Guneeé): identification and field tests of female sex pheromone gland components. *J. Chem. Ecol.* **17**, 249–258.

Emmenegger, D. B. 1976. "Biology and population regulation of the strawberry root weevil, *Otiorhynchus ovatus* (Coleoptera: Curculionidae)." M. S. Thesis, Oregon State Univ., Corvallis, OR.

Emmenegger, D. B. and Berry, R. E. 1978. Biology of strawberry root weevil on peppermint in western Oregon. *J. Econ. Entomol.* **7**, 495–498.

Green, R. J. 1963. "Mint Farming". Agric. Information Bull. No. 212. USDA, Washington, DC.

Jackson, T. L., Gardner, E. H., and Doerge, T. A. 1983. Fertilizer Guide: peppermint (western Oregon-West of Cascades). Oregon State Univ. Ext. Ser, Corvallis, OR.

Koppenhöfer, A. M., Campbell, J. F., Kaya, H. K., and Gaugler, R. 1998. Estimation of entomopathogenic nematode population density in soil by correlation between bait insect mortality and nematode penetration. *Fundam. Appl. Nematol.* **21**, 95–102.

Lacy, M. L., Stephens, C. T., Green, R. J., and York, A. C. 1981. Mint production in the midwestern United States. North Central Regional Ext. Pub. 155. Coop. Ext. Ser., Michigan State Univ., East Lansing, MI.

Landing, J. E. 1969. "American Essence - A history of the Peppermint and Spearmint Industry in the United States". A. M. Todd Foundation, Kalamozoo, MI.

Martin, J. H., Leonard, W. H., and Stamp, D. L. 1976. "Principles of Field Crop Production." Macmillan Publishing Co., Inc., New York.

Matzat, J. H. 1981. "The peppermint industry: situation and outlook with emphasis on Oregon." M. S. Thesis, Oregon State Univ., Corvallis, OR.

Morris, M. A. 1990. "Sampling methods, biology and management of the mint flea beetle in central Oregon peppermint". M. S. Thesis. Oregon State Univ., Corvallis, OR.

Pike, K. S., Baird, C. R., and Berry, R. E. 1988. Mint root borer in the Pacific Northwest, PNW 322. Washington State Univ. Ext. Ser., Pullman, WA.

Shields, E., Berry, R. E., and Fisher, G. 1981. Sampling for soil insect pests: an inexpensive way to build a Berlese funnel. Oregon State Univ. Ext. Circ. 1079.

Sievers, A. F. 1948. Mint Farming. USDA Bull. No. 1988.

Takeyasu, J. 1994. "Control of mint root borer, *Fumibotys fumalis*, with the entomopathogenic nematode, *Steinernema carpocapsae*". M. S. Thesis. Oregon State Univ., Corvallis, OR.

Chapter VII-16

Insect and Mite Control on Nursery and Landscape Plants with Entomopathogens

Denny J. Bruck
USDA-ARS
Horticultural Crops Research Laboratory
Corvallis, OR 97330, USA

Ralph E. Berry
Emeritus Professor of Entomology
Oregon State University
18160 Cottonwood Road, PMB #798
Sunriver, OR 97707, USA

Jack D. DeAngelis
College of Agricultural Science
Oregon State University
Corvallis, OR 97331, USA

1 Introduction

Over 2,500 species of insects and mites are destructive to nursery crops in the United States (Johnson and Lyon, 1991). Likewise, there is a large and diverse number of nursery plant species used in different landscape settings and geographic regions. Nursery and landscape plants represent a complex challenge for insect control products because of the large number of plant species, insect pests, growing conditions, potting media, and production techniques (Martin, 1997). In addition, feeding injury on many ornamental plants directly affects plant marketability or aesthetic value, and the threshold for damage is relatively low. The use of entomopathogens on nursery plants is an effective alternative to chemical control because of the problems associated with the use of pesticides, such as homeowner concerns, worker safety, toxicity, and potential for resistance.

Because of the large diversity of host plants and the insects and mites that attack them, we have chosen to separate the major arthropod pests into guilds and discuss case histories that illustrate the use of entomopathogens for control of selected pests in each guild (Table 1). Individuals working in pest management of ornamental plant pests should find the information useful as examples for their particular pest situation (also see Chapters VII-9, VII-10, VII-11, VII-12, VII-14 and VII-18).

2 Sampling considerations

Nursery and landscape plants represent a diverse assemblage of problems for researchers evaluating pest control methods and materials. The principal problem is determining how to sample for the pest to obtain a reliable estimate of pest density (or damage) before, and at intervals after, treatments are applied. The methods used

L.A. Lacey and H.K. Kaya (eds.), *Field Manual of Techniques in Invertebrate Pathology*, 609–626.

Table 1. Pest guilds in nursery and landscape plants and entomopathogens used for their control

Guild	Bacteria	Nematodes	Fungi	Key references
Wood-Borers Peach Tree Borer		*Steinernema* spp. *Heterorhabditis* spp.		Kaya and Brown, 1986; Cossentine *et al*., 1990; Curran, 1992; Gill *et al*., 1992
Root Feeders Black Vine Weevil		*Steinernema* spp. *Heterorhabditis* spp.	*Metarhizium anisopliae* *Beauveria bassiana*	Easterbook *et al*., 1992; Moorhouse *et al*., 1993a, 1993b; Schwarz, 1995; Kakouli-Duarte *et al*., 1997; Bruck, 2004; Bruck, 2005; Bruck *et al*., 2005
Defoliators Elm Bark Beetle	*Bacillus thuringiensis* subsp. *tenebrionis*	*Steinernema* spp.		Kaya *et al*., 1981; Cranshaw *et al*., 1989; Cranshaw and Klein, 1994; Thurston, 1998
Sap Feeders Woolly Apple Aphid Green Peach Aphid Whiteflies			*Lecanicillium* (= *Verticillium*) *lecanii* [1] *Beauveria bassiana* *Paecilomyces fumosoroseus*	Hall, 1981; Khalil *et al*., 1985; Xu and Feng, 2000; Faria and Wraight, 2001
Phytophagous Mites Tetranychidac and Eriophyidae			*Hirsutella thompsonii* [2] *Beauveria bassiana*	McCoy, 1981; Poinar and Poinar, 1998; Shi and Feng, 2004; Chandler *et al*., 2000, 2005

[1]Available in Europe.
[2]No longer commercially available.

are largely dictated by the location of the pest (above or below ground), and the density and distribution of the host plants (single landscape plants versus homogeneous, field-planted or container-grown perennials). Because plant type and growing location can vary so widely, only suggestions regarding pre- and post-treatment sampling are presented here.

1. For soil-inhabiting pests in potted plants, such as root weevil larvae infesting potted rhododendron, pre-treatment pest densities can be estimated by destructive sampling or by artificially infesting pots with a known pest density, such as a known number of root weevil larvae or eggs/pot. The presence or absence of larval infestations can also be determined with acoustic detection (Mankin and Fisher, 2002). Post-treatment counts usually require destructive sampling of treated pots, which severely limits the number of post-treatment samples and sampling intervals. Larvae may be recovered by sieving the soil or handpicking larvae from the roots.
2. For perennial landscape plants, it may be necessary to sub-sample individual plants, thus treating the whole plant as a small "field". For example, divide the plant into quadrants based on cardinal direction (N, S, E, W) and height above ground. These quadrants can then be randomized and sub-sampled through time starting at time zero (pre-treatment). Samples of soil may be taken around the base of plants. Larvae may be recovered by sieving the soil or handpicking.
3. Above ground pests, such as aphids, can be sampled by dividing fields into plots and sub-plots. Samples can be taken pre- and post-treatment (1, 3, 7, 14 and 21 days) using beating trays or counting insects on leaf samples.

3 Wood-borers

Insect pests in the wood-borer guild in ornamental and nursery stock are represented by species in the orders Lepidoptera and Coleoptera.

For example, wood-boring Lepidoptera are either clear winged moths in the family Sesiidae or carpenter moths in the family Cossidae. The majority of the wood-boring beetles are in the families Buprestidae (flatheaded wood-borers), Cerambycidae (roundheaded wood-borers) and Scolytidae (bark beetles).

A *Life history of wood-boring Lepidoptera*

Wood-boring Lepidoptera, such as the peach tree borer (PTB), lesser peach tree borer (LPTB), and other *Synanthedon* spp., have strong mandibles for chewing wood and are concealed within galleries in the wood. Young shoots or saplings are particularly attractive to these insects and females usually lay their eggs singly or in masses in thin bark or small cracks or depressions in the bark, often near the soil line. PTB adults are active in August and September in the Northern Hemisphere, whereas adults of the LPTB are active throughout the season from March through October, with peak flight activity in August. Eggs of most species hatch in about 10 days and young larvae bore into the trunk or tunnel into the root crown just beneath the soil surface, where they girdle the plant. The tender bark and sapwood of nursery seedlings and young transplants are particularly vulnerable to attack. Young trees may be killed by the girdling injury while older trees become more vulnerable to secondary pests (Madsen and Procter, 1982). An accumulation of frass (sawdust-like material) or plant sap on the bark near the bases of plants or just beneath the mulch layer indicates the presence of borers. Leaves on infested branches may wilt and the entire branch may die. Feeding injury also may cause unsightly knots on infested wood or injury may stimulate adventitious growth. Larvae feed in the wood for one to three years, depending on the species, and then pupate. Adults exit the wood leaving an exit hole, sometimes with a portion of the old pupal case remaining in the opening. Pheromone traps may be used to monitor adult emergence and flight activity (Tumlinson *et al.*, 1974; Gentry *et al.*, 1978) and to disrupt mating (Yonce and Gentry, 1982; Gentry and Snow, 1984; Pfeiffer *et al.*, 1991). Control of larvae is difficult once they have become established in a tree, but the potential for the successful use of entomopathogens to control larvae is high.

B *Entomopathogenic nematodes for control of wood-boring Lepidoptera*

Entomopathogenic nematodes have been used successfully to control PTB on cherry, laurel and peach (Cossentine *et al.*, 1990; Gill *et al.*, 1992;), *Synanthedon culiciformis* on alder (Kaya and Brown, 1986), *S. resplendens* on sycamore (Kaya and Brown, 1986), *Vitacea polistiformis* on grape (Williams *et al.*, 2002), and *S. tipuliformis* on black currant (Bedding and Miller, 1981a).

1 *Control of wood-borers on high-value trees*

a. *Tree selection*
   Locate active galleries on trees by looking for an accumulation of frass or sap on the trunk or just under the mulch layer around the base. To determine if galleries are active, remove the frass, mark the galleries with paint, and wait a few days to see if fresh frass appears. The number of trees/treatment, number of replicates, and plot size depends on the number of infested trees available at specific sites.
b. *Application*
   Spray active galleries with nematodes on the trunk or around the base of trees with a backpack sprayer or inject the nematodes directly into the active galleries using a 500 ml plastic wash bottle.
   1. Determine the viability of the nematodes before preparing the concentrations. Take a small amount of the formulation, mix in water, and count the number of live nematodes under a microscope (see Kaya and Stock, 1997). Some formulations may require several hours of rehydration before infective juveniles (IJs) start moving.
   2. Spray *Steinernema carpocapsae* at 500 infective juveniles (IJs)/2.5 cm^2 of bark or ground area (Gill *et al.*, 1992). Spray tree trunks with 4,000 to 8,000 IJs/ml to runoff (approximately 6.5 to 11.5×10^6 nematodes/tree) (Kaya and Brown, 1986). Alternatively, spray around the base of trees with 200 ml of water containing 200 *H. bacteriophora*/ml (Cossentine *et al.*, 1990).
   3. Inject 4.5 ml of water containing 4,000 to 8,000 *S. carpocapsae*/ml into each active tunnel (approximately 18,000 to 36,000

nematodes/tunnel) (Kaya and Brown, 1986). Alternatively, inject 2 ml of water containing 500 *H. bacteriophora*/ml into each active tunnel (Cossentine *et al.*, 1990).

c. *Evaluation*
Determine efficacy after 1 to 2 weeks by examining tunnels to see if fresh frass has been produced. Removing larvae from their galleries with a knife is not recommended because due to the likelihood of killing the larvae and damaging the plant. Trees may be wrapped with screen cages to capture adults, which can be counted to determine nematode efficacy. Gill *et al.*, (1992) dug a root ball around the plants and counted the number of larvae in the soil near the roots and in the bark below ground level. This method, useful for some research, is not usually recommended because of the destructive nature of this sampling technique. Correct for natural background mortality using Abbott's formula (Abbott, 1925).

2 *Control of wood-borers on nursery cuttings*

a. *Selection of cuttings*
If nursery cuttings are suspected of being infested with wood-borer, bundles of 50 cuttings may be sprayed with entomopathogenic nematodes or dipped in a suspension containing entomopathogenic nematodes. A minimum of four replicates should be used/treatment.

b. *Application*
1. Determine the viability of the nematodes before preparing the concentrations.
2. Spray bundles of cuttings to run-off with *S. feltiae, S. carpocapsae* or *H. bacteriophora* at a rate of 2,000 IJs/ml (Bedding and Miller, 1981a). Immediately after spraying, wrap the bundles in polyethylene bags and place in the dark at 22 °C for 4 days. Bedding and Miller (1981a) also sprayed commercial-sized quantities of cuttings (16,000 to 30,000 cuttings/bundle) with 2 to 3 ml/cutting of a suspension containing 50,000 IJs/ml of *S. feltiae* when wood-borers were in diapause or with 2 to 3 ml/cutting of a suspension containing 30,000 IJs/ml when larvae were active.
3. Dip bundles of 50 cuttings in 10 liters water containing a suspension of 80,000 IJs/ml (Bedding and Miller, 1981a).

c. *Evaluation*
Dissect cuttings after 14 days to determine mortality of wood-borer larvae. Place living larvae in individual petri dishes, store at 22 °C and re-examine after 3 days. When treating larger quantities of cuttings, remove a subsample of cuttings and dissect cuttings to determine larval mortality. Correct for natural background mortality using Abbott's formula (Abbott, 1925).

3 *General considerations*

Successful control of wood-boring insects with entomopathogenic nematodes is perhaps due to the moist habitat in which borers live, allowing the nematodes to remain viable and locate their host (Kaya and Brown, 1986). Kaya and Brown (1986) concluded that it was important to consider which nematode species to use against wood-boring Lepidoptera. They suggested that the size of the IJs might be an important factor in determining efficacy because large IJs may not be able to enter the gallery openings. Also, the strain of nematode used and the size of the wood-borer larvae may influence the nematode's ability to find and infect larvae. Mortality of PTB on cherry laurel sprayed with *S. carpocapsae* (All strain) averaged 66% (Gill *et al.*, 1992). *S. carpocapsae* controlled 84% of *S. culiciformis* on alder and 61% of the sycamore borer on sycamore trees (Kaya and Brown, 1986). When injected into active galleries, mortality of *S. culiciformis* reached 93% (Kaya and Brown, 1986). Efficacy of *H. bacteriophora* on PTB sprayed on the bark of peach trees reduced the number of emerging adults by 80%, but did not significantly reduce the adults when injected or sprayed on the outside of borer galleries (Cossentine *et al.*, 1990).

Mortality of currant borers in cuttings sprayed with *S. feltiae* (T335 strain) ranged from 86.8 to 99.1% (Bedding and Miller, 1981a) and this treatment controlled diapausing and actively feeding larvae. Mortality of wood-borers in cuttings sprayed with *S. carpocapsae* (Agroitos strain) or *H. bacteriophora* (T310 strain) at rates of 2,000 IJs/ml or higher exceeded 70%. Mortality of larvae in cuttings dipped in a nematode suspension exceeded 90% (Bedding and Miller, 1981a). Successful control of wood-borer larvae depends on maintaining a post-treatment environment of high humidity long enough to allow nematodes to infect larvae. Bedding and Miller (1981a) suggested spraying at night, applying during rainy periods, tenting or bagging, or other measures to maintain high humidity of treated plant material.

4 Root feeders

A *Life history of black vine weevil*

Black vine weevil (*Otiorhynchus sulcatus*) and other root weevils are serious pests in nurseries and landscape plantings. Black vine weevil is more serious on ornamental shrubs than the strawberry root weevil, *O. ovatus*, which is more frequently associated with small fruit crops. The host range of black vine weevil includes over 140 plant species (Warner and Negley, 1976). Black vine weevil, which are all parthenogenetic females and incapable of flight, lay 500 to 800 eggs beginning 3 to 4 weeks after eclosion in May and continuing through September. Eggs are laid singly or in small groups in the soil or beneath debris on the soil surface. Eggs hatch in 1 to 2 weeks and larvae tunnel through the soil in search of roots. Larvae feed from mid-summer into the fall, and again the following spring, when the most serious injury occurs. In mild climates, larvae may feed during the winter (Smith, 1932). Larger larvae consume feeder roots and bark on larger roots in the spring, which results in reduced vigor of the plant and, in severe infestations, death of the plant (Moorhouse *et al.*, 1992a; LaLone and Clarke, 1981; Garth and Shanks, 1978). Pupation occurs in soil 5 to 10 cm deep in the spring and adults begin emerging as early as March or as late as June, depending on the region. There is usually one generation each year, but some adults overwinter in protected areas. See Moorhouse *et al.*, (1992b) for a more detailed description of the biology of *O. sulcatus*.

B *Entomopathogenic nematodes for control of black vine weevil*

Entomopathogenic nematodes have high potential for the control of black vine weevil larvae in several different crop systems (Poinar, 1986; Rutherford *et al.*, 1987; Klein, 1990; Miller *et al.*, 1992; Gill *et al.*, 2001; Willmott *et al.*, 2002; Shapiro-Ilan *et al.*, 2003; Lola-Luz *et al.*, 2005; Bruck *et al.*, 2005). The entomopathogenic nematodes, *S. carpocapsae, S. feltiae, Heterorhabditis marelatus* (OH10 strain), *H. megidis, H. bacteriophora* and *H. downesi*, provide control of root weevil larvae and pupae on container-grown nursery plants and landscape ornamentals. However, the use of nematodes on certain nursery plants with very low damage thresholds, such as cyclamen (1 larva/pot) (Moorhouse, 1990) and rhododendron (3 larvae/plant) (LaLone and Clarke, 1981), may require more frequent applications at higher rates or their use may need to be integrated with chemical treatments (Kakouli-Duarte *et al.*, 1997).

Selecting areas within a landscape that are infested with root weevils is difficult because root weevils have an aggregated or clumped distribution. Infestations in nurseries and landscapes usually occur through the movement of infested plant material from areas where suitable host plants are concentrated (Stimmann *et al.*, 1985; Bruck, 2003). Therefore, regularly scheduled monitoring of host plants for the presence of adults or leaf notching is perhaps the easiest and least costly method of detecting black vine weevil infestations (Stimmann *et al.*, 1985). Acoustic detection can also be utilized to detect the presence of weevil larvae (Mankin and Fisher, 2002).

1 *Plot selection*

a. *Landscape plants*

1. To locate plants in the landscape that may be infested with black vine weevil, examine plants at night for the presence of adults or leaf notching. Take a minimum of four soil samples 9 cm in diameter by 30 cm long, equidistantly 20 cm from the base of the plant to determine the presence of larvae (Brandt *et al.*, 1996). Sort or sieve samples in the field or extract samples in Berlese or modified Tulgren funnels in the laboratory.
2. Arrange plots in a randomized block design with at least four replicates of each treatment and control. Leave a buffer area of at least

1 m between plots. The number of plants/plot depends on the specific nursery or landscape situation.

b. Container-grown plants

1. Infest containers with 5 to 10 larvae confined in screen tea strainers (4 cm diameter) buried 8 cm deep. Alternatively, infest containers with a minimum of 10 to 15 late instars. Containers also may be infested with 50–100 darkened weevil eggs, but application and evaluation will need to be delayed (8–10 weeks) until larvae are large enough to recover. Black vine weevil can be mass reared following the procedure outlined by Fisher and Bruck (2004). Containers with visible adult feeding injury may be used, but there is no practical method to accurately estimate the larval population without destructively sampling the container. Acoustic detection can be utilized to determine the presence or absence of larvae, but can not be used to accurately estimate the larval population (Mankin and Fisher, 2002).
2. Arrange infested containers in plots and replicate each treatment and untreated control a minimum of five times in a randomized block design. The number of infested containers/plot depends on the size of the plots, *e.g.*, 1 m^2 plot will hold 16 containers, 40 cm in diameter.

2 *Application*

1. Rinse or flush application equipment prior to use to ensure that pesticide or fertilizer residues have been removed. Cool irrigation lines that may have been lying in the sun.
2. Determine the viability of the nematodes before preparing the concentrations.
3. Pre-wet the soil surface with water prior to application (about 10 liters of water/m^2 or 1 ml/cm^2 of the surface area of containers).
4. Prepare the nematode suspension according to the directions on the product label. Constantly agitate the suspension to prevent the nematodes from settling on the bottom of the container. Apply the nematodes immediately after preparation.
5. Apply nematodes at 40 to 100 IJs/cm^2 in 10 liters of water/m^2 or in 1 ml/cm^2 as a drench (Simons, 1981, Stimmann *et al.*, 1985; Bruck *et al.*, 2005) or apply 20 ml of a suspension containing 750 to 2,000 IJs/ml to the surface of containers (Bedding and Miller, 1981b; Georgis and Poinar, 1984). Irrigate immediately after application with 6 to 12 mm water and once or twice each week for 14 days after treatment.
6. At the time of application, collect samples of the IJ suspension in small cups and count the nematodes in the samples to assess viability of nematodes being delivered to the target.

3 *Alternate application methods*

1. Inject 40 to 60 ml of a nematode suspension containing 750 to 1,000 IJs/ml into containers (inject 20 ml at 2 to 3 different locations in the containers) (Bedding and Miller, 1981b).
2. Dip containers in a suspension containing 3.3×10^6 IJs so each container receives approximately 25 ml of the suspension (Pye and Pye, 1985) or in a suspension containing 1,000 IJs/ml so that each container receives an average of 56,000 to 62,000 nematodes/liter of soil (Klingler, 1988). To determine the actual nematode dose, weigh the containers before and after dipping. After immersion, keep the containers at 20 °C for 5 to 7 days before evaluating efficacy.
3. Apply nematodes in overhead sprinkler or drip irrigation system (Kakouli-Duarte *et al.*, 1997; Curran and Patel, 1988). Before adding nematodes to the header tank, clean the filters and run the system for 30 min. Add 1 to 2×10^8 IJs in the header tank in 40 liters of water. Irrigate for 30 min to deliver the nematode suspension. Collect samples of the suspension in small cups by placing five or more cups in each treatment before applying the nematodes. Collect the cups immediately after application, measure the quantity of water and count the number of nematodes in each cup to estimate the actual number of nematodes delivered.
4. Nematodes have also been shown to be effective when applied as infected-host cadavers (Jansson *et al.*, 1993; Jansson and Lecrone, 1994; Shapiro-Ilan *et al.*, 2003; Bruck *et al.*, 2005). *Tenebrio molitor* cadavers seven days post infection are applied, three at a time, evenly spaced around the surface of a 15 cm pot and buried to a depth of 1–2 cm (Bruck, 2005). Water all containers with approximately 1 cm depth of water before and after cadaver application. Depending on soil temperatures, efficacy can be evaluated 14 to 28 days after application.

4 *Evaluation*

a. Landscape plants

1. Determining efficacy of nematodes on nursery plants in established landscapes is very difficult because of the destructive nature of the sampling methods and the variability and diversity of host plants in the landscape setting.
2. Determine larval mortality by comparing larval density from pre-treatment core soil samples (see above) with samples taken 3 to 4 days after treatment.
3. Count the number of live weevils in the core samples and calculate the percentage infested with weevils.
4. Correct for natural background mortality using Abbott's formula (Abbott, 1925).
5. Take a small amount of treated soil or potting mix and bait it with 5 to 10 *Galleria mellonella* larvae to determine nematode persistence. Count the number of infected *Galleria* after 7 days (see Kaya and Stock, 1997 and Chapter IV-5 of this volume for additional methods to determine persistence).

b. Container-grown plants

1. Remove screen tea strainers from treated and control containers after 3 to 4 days and count dead and alive larvae.
2. Place larvae in individual petri dishes with moist filter paper.
3. After 7 days, dissect dead larvae to determine whether or not mortality was caused by nematodes. If containers were artificially infested with weevil larvae, dump out the contents of the container and recover dead and live larvae. Set aside all larvae as before to determine cause of mortality.
4. Take a small amount of treated soil or potting mix and bait it with 5 to 10 *G. mellonella* larvae to determine nematode persistence. Count the number of infected *Galleria* after 7 days (see Kaya and Stock, 1997 and Chapter IV-5 of this volume for additional methods to determine persistence).

5 *General considerations*

Soil temperature is critical for activity and survival of entomopathogenic nematodes and should be measured at the time of treatment and during subsequent sampling intervals. Continuous sampling is feasible with relatively inexpensive datalogger type devices. Acceptable temperature ranges for survival, infection, and development vary with the nematode species. In general, steinernematids have a lower temperature range (4–14 °C) for storage than heterorhabditids (10–16 °C) (Woodring and Kaya, 1988). Soil moisture is critical for the survival and dispersal of nematodes, and it is important to measure the soil moisture before and after applying the nematodes and during subsequent sampling periods. Methods to determine soil water potential (pF) have been described by McQueen and Miller (1968) and Al-Khafaf and Hanks (1974).

Entomopathogenic nematodes should be targeted against early instars and before serious root injury occurs. Fall applications should be made after adults have completed egg laying, usually by September. However, newly hatched larvae may be less susceptible to nematodes than later instars (Klingler, 1986; Kakouli-Duarte *et al.*, 1997) so treatment in the fall must be carefully timed (Klingler, 1988). Since larvae are very small and difficult to recover in the fall, assessment of efficacy is not possible unless the larvae are extracted in the laboratory or assessment is delayed until the following spring when larvae are large enough to count.

Applying nematodes in March or April in the Northern Hemisphere provides control of black vine weevil larvae and delaying applications until April has the added advantage of controlling pupae and teneral adults, but is disadvantageous because much of the plant damage has already occurred. Also, since soil temperature is one of the most important factors affecting efficacy of nematodes against black vine weevil, applications in the spring before the soil temperatures reach about 14 °C are not very effective (Schirocki and Hague, 1994; Bruck *et al.*, 2005). Utilization of cool-temperature active nematodes, such as *H. marelatus* (OH10 strain), *H. megidis* and *S. kraussei* may permit applications earlier in the spring when soil temperatures are lower than 14 °C (Berry *et al.*, 1997; Willmott *et al.*, 2002).

Since it is not possible to sample containers for the intensity of larval infestation before treatment, the use of entomopathogenic nematodes is usually a prophylactic rather than a curative treatment. If a source of root weevils is available, artificially infesting containers with a known number of eggs or larvae is a preferred technique. Infest containers with 10 larvae/pot placed 5 cm deep in the soil (Georgis and Poinar, 1984) or with 20 to 40 eggs/pot placed under the soil (Simons, 1981; Kakouli-Duarte *et al.*, 1997). The number of eggs added to containers depends on egg viability, the plant species, and the size of the container.

Higher rates of *Steinernema* spp. are generally required to control root weevil larvae and pupae than *Heterorhabditis* spp. There have been several reports that heterorhabditids provide better control of black vine weevil than steinernematids. The difference in efficacy between heterorhabditids and steinernematids is likely due to their different host penetration and foraging strategies (Georgis and Hague, 1981; Bedding and Molyneux, 1982; Kaya and Gaugler, 1993). Stimmann *et al.* (1985) reported that *H. heliothidis* (= *bacteriophora*) killed up to 90% of black vine weevil larvae, whereas *S. carpocapsae* killed 78%. Shanks and Agudelo-Silva (1990) reported that *S. carpocapsae* (All strain) controlled 76% of *O. sulcatus* larvae and pupae in cranberry, whereas *H. bacteriophora* (HP88 strain) controlled 100% applied at the rate of 75 IJs/cm^3. Results with *H. marelatus* (OH10 strain) applied to container-grown azalea at rates of 12.5, 25.0 and 50.0 IJs/cm^2 controlled 64.5%, 80.6%, and 83.9% weevil larvae, respectively (Berry, Liu and Poinar, unpublished data). Zimmermann and Simons (1986) and Ferguson *et al.* (1990) suggested that *H. bacteriophora* applied at 50 to 100 IJs/cm^2 would be sufficient to control black vine weevil larvae, if conditions for nematode survival are suitable.

C Entomopathogenic fungi for control of black vine weevil

The fungus, *Metarhizium anisopliae*, formulated as dry granules (F52, Novozymes Biologicals Inc., Salem VA), infects larvae of black vine weevil and has been shown to be an effective biological control agent for black vine weevil on several different container-grown nursery crops (Soares *et al.*, 1983; Zimmermann, 1986; Easterbrook *et al.*, 1992; Moorhouse *et al.*, 1992a, 1993a; Bruck, 2005) (see Chapter VII-12). *Beauveria bassiana* is also effective against black vine weevil (McCoy, 1990; Bruck, 2004), but will not be discussed in this chapter.

The experimental design used to evaluate efficacy of entomopathogenic fungi against root weevils is similar to the design described above for evaluation of nematodes against black vine weevil in containers.

1 Application

1. Mix 1 g F52 (or its equivalent) granules/liter of potting media (Stenzel *et al.*, 1992; Bruck, 2005) or 0.5 to 1 g conidia/liter of potting media (Reinecke *et al.*, 1990). Premixing conidia in the potting media and incubating at a minimum temperature of 15 °C for 4 to 7 days before potting increases the titer to approximately 10^6 spores/g dried media (Stenzel *et al.*, 1992; Bruck, 2005).
2. Alternatively, drench or top-dress containers with 0.7 and 1.3 g conidia/liter of media (Schwarz, 1995).
3. For biological activity of *M. anisopliae* against weevils in the field, soil temperatures should not be lower than 15 °C.

2 Evaluation

Efficacy of *M. anisopliae* is best evaluated by artificially infesting container-grown nursery plants with either eggs or larvae of black vine weevil.

1. Infest pots containing conidia incorporated in the media at potting with 100 darkened eggs or 10 to 15 larvae and evaluate experiments 90 and 28 days later, respectively.
2. Alternatively, infest individual containers with 100 to 200 weevil eggs 4 to 6 weeks after treatment with conidia.
3. Alternatively, infest individual containers with 100 darkened eggs or 10 to 15 larvae and wait 30 and 10 days, respectively before applying *M. anisopliae* as a drench. Evaluate the pots 3 to 6 weeks after treatment.
4. Dump the contents of the containers and pick through the roots and media or sieve the soil to recover the larvae. Larvae recovered from soil 7 to 14 days after treatment with *M. anisopliae* may not have been exposed long enough for symptoms to occur. Accordingly, larvae recovered

from treated media can be incubated a few more days to determine if signs of infection can be induced. Place individual surface sterilized larvae on moist filter paper in petri dishes inside plastic bags. Place bags in incubation chambers at 20 to 25 °C and assess mortality daily by examining for *M. anisopliae* conidia on the cadavers (Moorhouse *et al.*, 1994).

5. Correct for natural background mortality using Abbott's formula (Abbott, 1925).
6. Since all stages of black vine weevil are susceptible to infection by *M. anisopliae* and the conidia are viable for a long time, one prophylactic application can provide control of black vine weevil for 3 to 12 months (Bruck, 2005).

3 *General considerations*

Adding 1 g of *M. anisopliae* conidia/liter of potting media controlled 70 to 80% of black vine weevils after 28 days (Reinecke *et al.*, 1990). Mixing 1 g F52 (or its equivalent) granules/liter of potting media controlled 93.5 to 97.5% of the larvae (Bruck, 2005), but controlled only 0 to 57% when applied as a top dressing (Stenzel *et al.*, 1992). However, in studies reported by Schwarz (1995), top dressing *M. anisopliae* at rates of 0.7 and 1.3 g/liter of soil reduced black vine weevils 73 and 86%, respectively. The advantage of using a top dressing is that it is possible to treat second year potted perennials without repotting each plant (Schwarz, 1995). However, since *M. anisopliae* has limited dispersal capability in soil, the pest must be present in the treated area (Stenzel *et al.*, 1992).

The optimum temperatures for germination and growth of *M. anisopliae* are strain dependent and range from 20 to over 30 °C (Moorhouse, 1990; Moorhouse *et al.*, 1994). Therefore, it is important to select strains on the basis of their efficacy over the temperature ranges commonly found in the cropping system (Soares *et al.*, 1983; Moorhouse *et al.*, 1994). Stenzel *et al.*, (1992) found that when temperatures were 10°C or lower, it took longer to produce a conidia titer sufficient to control black vine weevil. Schwarz (1995) and Andersch (1992) found that *M. anisopliae* remained viable for 70 weeks at 4 °C, whereas at 20 °C viability was maintained for 20 weeks. Bruck (2005) found that F52 persisted well at ambient outdoor temperatures for nearly one year.

5 Defoliators

Numerous insects feed on and defoliate nursery and landscape plants and many can be controlled with entomopathogens, particularly with products containing the bacterium *Bacillus thuringiensis (Bt)*. *B. thuringiensis* is probably the most widely used entomopathogen for control of insects affecting nursery and landscape plants. It is readily mass-produced and commercially available. *B. thuringiensis* subsp. *kurstaki* (*Btk*) is widely used to control insects on forest and shade trees. *Btk* is labeled for control of important lepidopteran pests such as tent caterpillars, fall webworm, spruce budworm, and gypsy moth (addressed in Chapters IV-2 and VII-9). *Bacillus thuringiensis* subsp. *tenebrionsis* (*Btt*) has been used successfully to control elm leaf beetle. *Btt* selectivity is a distinct advantage in pest management programs that target pest species while trying to minimize impacts on non target organisms (Cranshaw and Klein, 1994).

Although not discussed in this chapter, the use of baculoviruses against leaf-feeding Lepidoptera and sawflies has received considerable attention, particularly for control of gypsy moth. Like other entomopathogens, viruses are sensitive to environmental conditions and their use must be carefully timed to coincide with the most susceptible stage(s) of the target insect. The greatest advantage of using baculoviruses is their host specificity, but this also restricts the market potential for these products (see Chapter IV-1).

A *Life history of elm leaf beetle*

Elm leaf beetle, (*Xanthogaleruca luteola*), was accidentally introduced into the USA from Europe and now occurs essentially everywhere its host trees are grown. All species of elm (*Ulmus*) and Japanese zelkova (*Zelkova serrata*) serve as hosts. Both trees are widely planted as ornamental and shade trees. Adults overwinter in protected places and often become a nuisance in homes, sheds and barns during the fall and winter. During spring, when host trees are producing the first flush of leaf growth, adults fly back to trees where they feed and lay eggs. Larvae skeletonize lower leaf surfaces during their development. Damaged leaves may be lost. Repeated, severe defoliation can result in tree death. Mature larvae pupate at the base of the host tree where they migrated at the completion of development. Two generations/year are typical for this insect in North America (see Johnson and Lyon, 1991).

B Bacillus thuringiensis *subsp.* tenebrionis *for control of elm leaf beetle*

Mortality of elm leaf beetle larvae treated with *Btt* has been shown to exceed 90% (Cranshaw *et al.*, 1989) and with significant reduction of defoliation (Thurston, 1998).

1 Site selection

1. Depending on the number of infested trees, use a randomized complete block design with two or more trees/replicate. A minimum of four replications is recommended for each treatment and control.
2. Mark 6 to 10 branch terminals on each tree and count the number of larvae or visually estimate the percentage of defoliation (Dahlsten *et al.*, 1993).

2 Application

1. Apply when larvae are small (first or second instar) or at peak egg hatch.
2. Rate of application depends primarily on instar. Infestations composed mostly of early instars (less than 0.6 cm) should be treated at the rate of 0.95 – 2.84 liters of *Btt* containing $16.3–48.9 \times 10^6$ Leptinotarsa units (LTU) in 75 liters of water. Later instars should be treated at a higher rate of 2.84 – 3.78 liters of *Btt* containing $48.9–65.2 \times 10^6$ LTU in 75 liters of water. Treat foliage to run-off. Since *Btt* must be consumed to be effective, thorough coverage of leaf surfaces is essential. Spray control trees with an equal amount of water.

3 Evaluation

1. Count the number of live larvae on the marked terminals 2 to 3 days after treatment.
2. Compare the number of larvae in pre-treatment samples with post-treatment samples and with controls.
3. Correct for natural background mortality using Abbott's formula (Abbott, 1925).
4. Treatment efficacy can also be assessed using a visual standard on branch terminals by comparing the change in percent defoliation of treated versus untreated trees 2 to 3 weeks after treatment (Dahlsten *et al.*, 1993).

4 General considerations

Excellent coverage is key to good control of elm leaf beetle with *Btt*. For large trees, roto-mist sprayers have proven to be superior (Thurston, 1998). Because ingestion of *Btt* toxin may only paralyze the gut, but not cause rapid mortality, care should be exercised in assessing true mortality. However, cessation of feeding may be a better criterion than mortality, especially at short post-application intervals (less than 72 h). Because *Btt* has a residual activity of only a few days in the field (Cranshaw *et al.*, 1989), two applications at 14 day intervals may be necessary to protect foliage. In areas where elm leaf beetle has more than one generation, treatments should be repeated after the second generation larvae begin feeding.

C Entomopathogenic nematodes for control of elm leaf beetle

Steinernema carpocapsae controlled over 90% of migrating elm leaf beetle larvae when added to a cellulose mulch at a rate of 200,000 IJs in 200 ml water/liter of mulch placed in tree bands (Thurston, 1998), but was not effective when applied as a foliar spray (Kaya *et al.*, 1981). Kaya *et al.*, (1981) demonstrated that elm leaf beetle adults, larvae, and pupae were highly susceptible to *S. carpocapsae* in the laboratory at all rates tested from 50 to 400 IJs applied in 0.5 ml of water to a 5.5 cm filter paper. In field tests, nematodes were less effective and may not significantly reduce leaf damage. Kaya *et al.*, (1981) suggested that pupae are the most vulnerable stage and application of nematodes in litter around the base of trees may provide control of larvae entering the soil to pupate.

The use of tree bands containing *S. carpocapsae* to control elm leaf beetle larvae migrating down the tree trunk to locate pupation site depends on the material used in the band. Thurston (1998) found that cellulose wood fiber mulch (hydromulch) placed in a tree band provided a suitable habitat for nematode survival for at least one month.

1 Application

1. Add 200,000 IJs in 200 ml of water/liter of mulch.
2. Place the nematode-treated mulch in bands 45 cm high constructed of polyethylene film lined with burlap, open at the top and securely attach to the bottom of the tree trunk.
3. Place the bands around the tree trunk 2 to 3 m above the ground at the beginning of larval migration down the trunk.
4. Band 5 to 10 trees with nematode-treated bands or untreated bands, to serve as control trees.

2 Evaluation

1. Determine efficacy by removing subsamples of the mulch containing elm leaf beetle larvae and pupae after 2 to 3 weeks.
2. Count the number of dead larvae and pupae in treatment and control samples and determine the cause of mortality.

3 General considerations

Even though Thurston (1998) found high mortality of elm leaf beetle larvae in bands treated with *S. carpocapase*, it is unlikely that nematodes would provide season long control in areas where elm leaf beetle has more than one generation. Also, since some larvae drop from the tree or pupate in crevices above the bands, the tree-banding technique may help reduce populations locally, but would not provide area wide control.

6 Sap feeders

Insects that suck plant sap, such as aphids, adelgids and scales represent important insect pests on nursery and landscape plants. Most have highly variable life histories and attack a large number of nursery and landscape plants in different parts of the world (Johnson and Lyon, 1991).

A Life history of aphids

Aphids vary greatly in color and size and the symptoms of feeding damage are directly related to the population size and the susceptibility of the host plants. In some instances, a small number may cause severe symptoms on one host plant, but not on another, and even large numbers may not seriously injure some host plants. Aphid behavior is dependent largely on the species of host plant, the feeding site, and the time of year. Some species feed underground on roots, others on bark, twigs, leaves, flowers and fruit. The piercing-sucking feeding activities may produce inconspicuous signs of injury, but leaves may be depleted, which gradually weakens or kills the plant. Heavy infestations result in an unthrifty appearance of host plants and distorted or wilted leaves. A black sooty fungus often grows on the honeydew excreted by aphids, giving the foliage a black or dirty appearance. Some aphid species also transmit viruses, which may cause stunting, discoloration, distorted leaves, or galls.

Aphids are distributed worldwide and few ornamental plants escape injury from these pests. In general, their life history is as follows: the winter is spent in the egg stage or as adults and nymphs in protected areas such as in the soil, under bark or soil litter, in stems or twigs, on leaves or within galls. In the spring, eggs hatch into nymphs that feed for a short time before maturing. Mature females give birth to living young, females. There are several overlapping generations of aphids during spring and summer on many different plants. Some of these aphids produce wings and disperse to other suitable hosts. In late summer and fall, winged forms may be produced, which migrate back to overwintering hosts. The progeny of these fall migrants produce male and female aphids that mate and lay overwintering eggs or seek protected areas in which to overwinter. In regions where winter temperatures are mild, aphids may be abundant on hosts throughout the year and apparently never enter the egg stage to overwinter.

B Entomopathogenic fungi for control of aphids

Fungi are the only entomopathogens used to control aphids on plant surfaces. Epizootics caused by entomopathogenic fungi are often responsible for decimation of aphid and whitefly populations (Latgé and Papierok, 1988; Lacey *et al.*, 1996; McLeod *et al.*, 1998; Faria and Wraight, 2001). However, because many naturally-occurring fungal pathogens of insects are density dependent, epizootics often occur after economic thresholds have been surpassed. Fungi that can be produced on artificial media, stored until use and inundatively applied to pest populations; enable predictable control under appropriate environmental conditions. Fungi in the Hypocreales, such as *Lecanicillium* (= *Verticillium*) *lecanii, M. anisopliae*, and *B. bassiana*, provide these advantages. *L. lecanii* has been successfully used to control of aphids, scale insects and whiteflies (Hall, 1981; Khalil *et al.*, 1983; Lacey *et al.*, 1996; Faria and Wraight, 2001). Khalil *et al.* (1985) suggested that *L. lecanii* could be recommended for biological control of some aphid species in glasshouses, but not under field conditions. In glasshouses, *L. lecanii* controlled 100% of *Myzus persicae* after 25 days at 25 °C and 100% RH (see Chapter VII-8). More studies are needed on nursery and landscape plants to determine the potential of entomopathogenic fungi for biological control of aphids.

1 Application

1. Spray aphid colonies on plants with a blastospore suspension of *L. lecanii* containing 10^8 spores/ml in 0.003% phosphate buffer containing a wetting agent. Alternatively, prepare a spore suspension of *L. lecanii* 10^{10} spores/g by adding 100 g/100 liters of water. Thus, the final spray liquid contains 10^7 spores/ml.
2. On a large scale, for potted plants and cuttings, apply 1.5 kg in 1,500 liters of water/hectare.
3. For tall plants, apply 3 kg in 3,000 liters of water/hectare.

2 Evaluation

1. Determine the mean number of aphids/plant or leaf before application. The number of pre-treatment samples depends on the number of plants available and on the species of plants used in experiments. Treat a minimum of 25 plants, replicated four times for each treatment. In general, count the aphids on a minimum of 10 leaves/plant.
2. Determine the mean number of aphids/plant (as above) weekly for two months after treatment. Aphids infected with *L. lecanii* die before the fungus is visible, usually 7 to 10 days after treatment; thus it is difficult to attribute aphid mortality to *L. lecanii* in samples taken too soon after treatment. Dead aphids can be surface sterilized and placed on water agar to induce mycosis to verify if death was caused by the fungus (Goettel and Inglis, 1997).

3 General considerations

The timing and number of applications are important because 7 to 10 days are required for *L. lecanii* to kill the target insects. In general, 2 to 4 applications, 7 days apart, will provide control. Also, *L. lecanii* does not spread quickly because the conidia do not become airborne and epizootics are slow to develop. However, because of its hydrophilic nature, it can be spread in water.

L. lecanii requires a relative humidity of 80% for 10 to 12 h/day and temperatures ranging from 18 to 28 °C for successful control of aphids. Spraying in the late afternoon or early evening when the relative humidity is high will stimulate the development of *L. lecanii*. The mode of action of *L. lecanii* is based on direct contact between the conidia and the target insects. Sprays should be directed to the undersides of leaves and growing tips.

C Entomopathogenic nematodes for control of root-feeding aphids

Brown *et al.* (1992) demonstrated that *S. carpocapsae* controlled woolly apple aphid, *Eriosoma lanigerum*, on roots of small apple trees. For spray application, the authors reported significantly fewer aphid colonies in the treated trees compared to untreated controls (a reduction of about 55%). However, there were no differences in the number of aphid colonies when nematodes were mixed with peat moss and applied to the soil surface around the base of plants.

1 Application

1. The experimental design and number of replications/treatment depend on the number and species of plants being evaluated. In general, five plants should be evaluated in five replications/treatment as well as a control.
2. Pre-moisten the soil around the base of trees with 10 liters of water.
3. Spray 375,000 IJs/m^2 mixed in 10 liters of water at the base of each tree trunk, using a hydraulic sprayer, covering the entire area under the tree branches. Spray 10 liters of water around the base of control trees.
4. Alternatively, mix 10^7 IJs/4,700 cm^3 of peat moss. Spread approximately 940 cm^3 of the peat moss on the soil surface beneath the branches. Spread an equal amount of untreated peat moss around the base of control trees.
5. After application, apply an additional 10 liters of water around the base of each tree.

2 Evaluation

1. Evaluate spray treatments 10 to 12 weeks after spraying by uprooting treated and untreated plants and counting the number of woolly apple aphid colonies on the roots. Record the aphid colonies in size classes of 1–5, 6–20, 21–50, or more than 50 aphids/colony.
2. Evaluate top-dress treatments 30 days after treatment and count the colonies of woolly apple aphid as described above.

3 General considerations

Even though small host size prevented successful reproduction of *S. carpocapsae* in woolly apple aphid, Brown *et al.* (1992) observed live nematodes in droplets of honeydew and found live nematodes in the body of aphids. They found that *S. carpocapsae* caused mortality of woolly apple aphid, and with mortality likely having been caused by the symbiotic bacteria or physical damage caused by IJs puncturing the aphid's body.

7 Phytophagous mites

Phytophagous mites (Acari: Tetranychidae and Eriophyidae) are important worldwide and can be significant pests in nurseries and landscapes. Damage is especially high late in the season as plant growth slows and water is often in short supply. Mites, especially tetranychids (spider mites), are frequently treated with conventional miticides to ameliorate foliar damage.

A Life History of the two-spotted spider mite

The two-spotted spider mite is distributed throughout the world and is a common pest of nearly all crops, including ornamentals (Walter and Proctor, 1999). They can be particularly problematic in greenhouse production. Adult mites (0.3–0.5 mm long) are greenish yellow with a pair of distinctive red spots on the back. Stages include a six-legged larva, and eight-legged protonymph and an eight-legged deutonymph. The pale yellow eggs, laid in the spring are attached to undersides of leaves. Adults overwinter as bright orange females under ground litter or in tree bark. A typical life cycle takes 7 to 21 days, with adult females living up to 9 weeks. Numerous generations occur per year.

B Entomopathogenic fungi for control of two-spotted spider mites

While research into mite entomopathogens is generally lacking, all major pathogen groups (bacteria, rickettsiae, fungi, protozoans, nematodes and viruses) have been associated with mite diseases (Poinar and Poinar, 1998). This fact alone suggests that future work in this area is warranted. However, the use of entomopathogenic fungi is the most promising agent for microbial control of these pests (Chandler *et al.*, (2000). Fungal pathogens are able to invade and infect their host directly and are not required to be taken *per os* to be pathogenic. Chandler *et al.*, (2005) demonstrated that *B. bassiana* (Naturalis-L, Troy Biosciences Inc., Phoenix, AZ, USA) controlled two-spotted spider mite, *Tetranychus urticae*, on tomato plants in the greenhouse. For spray application, the authors reported significantly fewer mobile *T. urticae* and eggs per leaf in the treated plants compared with the untreated controls. In addition, the use of the predatory mite, *Phytoseiulus perimilis*, in combination with *B. bassiana* reduced the numbers of *T. urticae* adults, nymphs and eggs (98% reduction in all stages).

1 *Application*

1. Arrange the experiment in a randomized complete block design. In general, perform a minimum of 4–6 replications/treatment including a control. Separate plants to be treated with two untreated guard plants.
2. Apply treatments (one treatment/plant) to three marked leaves per plant, at the top, middle and bottom of the plant.
3. Release 40 *T. urticae* onto one leaflet of each marked leaf, followed by 20 more mites 7 days later.
4. Prepare fungal suspensions in 0.01% Triton X-100 at concentrations between 10^5–10^7 conidia/ml. Apply the fungus 7 days following the last mite release. Spray fungal suspensions to run-off, with a hand held sprayer, onto the leaves on which the mites were released. Spray 0.01% Triton X-100 alone to run-off onto the leaves of control plants. Apply a second treatment 7 days later.
5. After application, lightly wet the floor of the greenhouse with water to increase the humidity and simulate a full canopy.

2 *Evaluation*

1. Seven days following the final fungal application, remove the marked leaves and determine the number of motile *T. urticae* (nymphs + adults) and eggs per leaf.

3 *General considerations*

There may be a trend for increasing numbers of motile *T. urticae* in the upper canopy of the tomato plants. Chandler *et al*., (2005) found that a commercial *B. bassiana* product had the greatest impact on the *T. urticae* population. Fungal isolates often provided much better control in the greenhouse than anticipated from laboratory bioassays. Further research is needed to design bioassays that make the selection of virulent fungal isolates more reliable.

8 Summary

Entomopathogens are a rich source of control options for management of insects on nursery and landscape plants, but for many, much remains to be done before they will be widely adopted for practical use in nursery and landscape systems. For example, most are species specific and must be carefully timed with the host's life cycle to coincide with the susceptible stage(s) of the target insect. Some require special application methods and, for some, practical and economic application techniques have not yet been developed. Most are very sensitive to environmental conditions and may not be effective if temperature and moisture requirements are not met or maintained. The commercial production, distribution and storage of viable products are perhaps the most important considerations for their successful use.

9 References

Abbott, W. S. 1925. A method of computing the effectiveness of an insecticide. *J. Econ. Entomol.* **18**, 265–267.

Al-Khafaf, S. and Hanks, R. J. 1974. Evaluation of the filter paper method for estimating water potential. *Soil Sci.* **117**, 194–199.

Andersch, W. 1992. Production of fungi as crop protection agents. *In* "Biological Crop Protection, Symposium, 24 May 1991" (M. Esters, Ed.). pp 129–141. *Pflanzenschutz-Nachr. Bayer* **45**, 128–141.

Bedding, R. A. and Miller, L. A. 1981a. Disinfesting blackcurrant cuttings of *Synanthedon tipuliformis*, using the insect parasitic nematode, *Meoplectana bibionis*. *Environ. Entomol.* **10**, 449–453,

Bedding, R. A. and Miller, L. A. 1981b. Use of a nematode, *Heterorhabditis heliothidis*, to control black vine weevil, *Otiorhynchus sulcatus*, in potted plants. *Ann. Appl. Biol.* **99**, 211–216.

Bedding, R. A. and Molyneux, A. S. 1982. Penetration of insect cuticle by infective juveniles of *Heterorhabditis* spp. (Heterorhabditidae: Nematoda). *Nematologica* **28**, 354–359.

Berry, R. E., Liu, J., and Groth, E. 1997. Efficacy and persistence of *Heterorhabditis marelatus* (Rhabditida: Heterorhabditidae) against root weevil (Coleoptera: Curculionidae) in strawberry. *J. Econ. Entomol.* **26**, 466–470.

Brandt, J. P., Smith, S. M., and Hubbes, M. 1996. Distribution and sampling of root weevil larvae in young ornamental conifer plantations. *Can. Entomol.* **128**, 1125–1133.

Brown, M. W., Jaeger, J. J., Pye, A. E., and Schmitt, J. J. 1992. Control of edaphic populations of woolly apple aphid using entomopathogenic nematodes and a systemic aphicide. *J. Entomol. Sci.* **27**, 224–232.

Bruck, D. J. 2003. Stopping Weevil Deeds. *American Nurseryman* **199**, 43–50.

Bruck, D. J. 2004. Natural occurrence of entomopathogens in pacific northwest nursery soils and their virulence to the black vine weevil, *Otiorhynchus sulcatus* (F.) (Coleoptera: Curculionidae). *Environ. Entomol.* **33**, 1335–1343.

Bruck, D. J. 2005. Ecology of *Metarhizium anisopliae* in soilless potting media and the rhizosphere: implications for pest management. *Biol. Control* **32**, 155–163.

Bruck, D. J, Shapiro-Ilan, D. I., and Lewis, E. E. 2005. Evaluation of application technologies of entomopathogenic nematodes for control of the black vine weevil. *J. Econ. Entomol.* **98**, 1884–1889.

Chandler, D., Davidson, G., Pell, J. K., Ball, B. V., Shaw, K. and Sunderland, K. D. 2000. Fungal biocontrol of Acari. *Biocontrol Sci. Technol.* **10**, 357–384.

Chandler, D., Davidson, G., and Jacobson, R. J. 2005. Laboratory and glasshouse evaluation of entomopathogenic fungi against the two-spotted spider mite, *Tetranychus urticae* (Acari: Tetranychidae), on tomato, *Lycopersicon esculentum. Biocontrol Sci. Technol.* **15**, 34–54.

Cossentine, J. E., Banham, F. L., and Jenson, L. B. 1990. Efficacy of the nematode, *Heterorhabditis heliothidis* (Rhabditida: Heterorhabditidae) against the peachtree borer, *Synanthedon exitiosa* (Lepidoptera: Sesiidae) in peach trees. *J. Entomol. Soc. Brit. Columbia.* **87**, 82–84.

Cranshaw, W. S., Day, S. J., Gritzmacher, T. J., and Zimmermann, R. J. 1989. Field and laboratory evaluations of *Bacillus thuringiensis* strains for control of elm leaf beetle. *J. Arboric.* **15**, 31–43.

Cranshaw, W. S. and Klein, M. G. 1994. Microbial control of insect pests of landscape plants. *In* "Handbook of Integrated Pest Management for Turf and Ornamentals" (A. R. Leslie, Ed.). pp. 503–520. Lewis Publishers, Boca Raton, FL.

Curran, J. and Patel, V. 1988. Use of a trickle irrigation system to distribute entomopathogenic nematodes (Nematoda: Heterorhabditidae) for the control of weevil pests (Coleoptera: Curculionidae) of strawberries. *Aust. J. Exp. Agric.* **28**, 639–645.

Curran, J. 1992. Influence of application method and pest population size on the field efficacy of entomopathogenic nematodes. *J. Nematology. Suppl.* **24**, 631–636.

Dahlsten, D. L., Tait, S. M., Rowney, D. L., and Gingg, B. G. 1993. A monitoring system and development of ecologically sound treatments for elm leaf beetle. *J. Arboric.* **19**, 181–186.

Easterbrook, M. A., Cantwell, M. P., and Chandler, D. 1992. Control of the black vine weevil, *Otiorhynchus sulcatus*, with the fungus *Metarhizium anisopliae. Phytoparasitica.* **20**, Suppl. 17S–19S.

Faria, M. and Wraight, S. P. 2001. Biological control of *Bemisia tabaci* with fungi. *Crop Prot.* **20**, 767–778.

Ferguson, C. M., Barratt, B. I. P., Jones, P. A., and Garnham, M. L. 1990. "Control of black vine weevil (*Otiorhynchus sulcatus*) larvae with different rates of a parasitic nematode (*Heterorhabditis bacteriophora*)", pp. 67–69. Proc. 43rd New Zealand Weed and Pest Control Conf. Lincoln, Canterbury.

Fisher, J. R. and Bruck, D. J. 2004. A technique for continuous mass rearing of the black vine weevil, *Otiorhynchus sulcatus. Entomol. Exp. Appl.* **113**, 71–75.

Garth, G. S. and Shanks, C. H. 1978. Some factors affecting infestation of strawberry fields by black vine weevil in western Washington. *J. Econ. Entomol.* **71**, 443–448.

Gentry, C. R., Holloway, R. L., and Pollet, D. K. 1978. Pheromone monitoring of peachtree borer and lesser peachtree borer in South Carolina. *J. Econ. Entomol.* **72**, 247–253.

Gentry, C. R. and Snow, J. W. 1984. Disruption of mating by lesser peachtree borers and peachtree borers in a pheromone permeated peach orchard. *J. Ga. Entomol. Soc.* **19**, 350–356.

Georgis, R. and Hague, N. G. M. 1981. A neoaplectanid nematode in the web-spinning larch sawfly *Cephalacia lariciphila* (Hymenoptera: Pamphiliidae). *Ann. Appl. Biol.* **99**, 171–177.

Georgis, R. and Poinar, G. O. Jr. 1984. Greenhouse control of the black vine weevil, *Otiorhynchus sulcatus* (Coleoptera: Curculionidae) by heterorhabditid and steinernematid nematodes. *Environ. Entomol.* **13**, 1138–1140.

Gill, S., Davidson, J. A., and Raupp, M. J. 1992. Control of peachtree borer using entomopathogenic nematodes. *J. Arboric.* **18**, 184–187.

Gill, S., Lutz, J., Shrewberry, and Raupp, M. 2001. Evaluation of biological and chemical control methods for black vine weevil, *Otiorhynchus sulcatus* (Fabricius) (Coleoptera: Curculionidae), in container grown perennials. *J. Environ. Hort.* **19**, 166–170.

Goettel, M. S. and Inglis, G. D. 1997. Fungi: Hyphomycetes. In "Manual of Techniques in Insect Pathology" (L. A. Lacey, Ed), pp. 211–249. Academic Press, London.

Hall, R. A. 1981. The fungus *Verticillium lecanii* as a microbial insecticide against aphids and scales. *In* "Microbial Control of Pests and Diseases 1970–1980" (H. D. Burges, Ed.), pp. 438–498. Academic Press. London.

Jansson, R. K., Lecrone, S. H., and Gaugler, R. 1993. Field efficacy and persistence of entomopathogenic nematodes (Rhabditida: Steinernematidae, Heterorhabditidae) for control of sweet potato weevil (Coleoptera: Apionidae) in southern Florida. *J. Econ. Entomol.* **86**, 1055–1063.

Jansson, R. K. and Lecrone, S. H. 1994. Application methods for entomopathogenic nematodes

(Rhabditida: Heterorhabditidae): aqueous suspension versus infected cadavers. *Fla. Entomol.* **77**, 281–284.

Johnson, W. T. and Lyon, H. H. 1991. "Insects that feed on trees and shrubs: An illustrated practical guide". Second Edition. Cornell University Press, Ithaca. 560 pp.

Kakouli-Duarte, T., Labuschagne, T. L., and Hague, N. G. M. 1997. Biological control of black vine weevil, *Otiorhynchus sulcatus* (Coleoptera: Curculionidae) with entomopathogenic nematodes (Nematoda: Rhabditida). *Ann Appl. Biol.* **131**, 11–27.

Kaya, H. K. and Brown, L. B. 1986. Field application of entomogenous nematodes for biological control of clearwing moth borers in alder and sycamore trees. *J. Arboric.* **12**, 150–154.

Kaya, H. K. and R. Gaugler. 1993. Entomopathogenic nematodes. *Annu. Rev. Entomol.* **38**, 181–206.

Kaya, H. K., Hara, A. H., and Reardon, R. C. 1981. Laboratory and field evaluation of *Neoaplectana carpocapsae* (Rhabditida: Steinernematidae) against the elm leaf beetle (Coleoptera: Chrysomelidae) and the western spruce budworm (Lepidoptera: Tortricidae). *Can. Entomol.* **113**, 787–793.

Kaya, H. K. and Stock, S. P. 1997. Techniques in insect nematology. In "Manual of Techniques in Insect Pathology" (L. A. Lacey, Ed), pp. 281–324. Academic Press, London.

Khalil, S. K., Bartos, J., and Landa, Z. 1985. Effectiveness of *Verticillium lecanii* to reduce populations of aphids under glasshouse and field conditions. *Agric. Ecosyst. Environ.* **12**, 151–156.

Khalil, S. K., Taborsky, V., and Bartos, J. 1983. "Studies on *Verticillium lecanii* for the biological control of aphids". *In* Proc. 10th Int. Congr. Plant Protection, 20–26 November 1983, p. 788. Brighton.

Klein, M. G. 1990. Efficacy against soil-inhabiting insect pests. In "Entomopathogenic nematodes in biological control" (R. Gaugler and H. Kaya, Eds.), pp. 195–214. CRC Press. Boca Raton, FL.

Klingler, J. 1986. Einsatz und wirksamkeit insektinparasitischer nematoden gegen den Gefurghten Dickmaulrüssler. *Gartnermeister.* **89**, 277–279.

Klingler, J. 1988. Investigation on the parasitism of *Otiorhynchus salicicola* and *O. sulcatus* (Coleoptera: Curculionidae) by *Heterorhabditis* sp. (Nematoda). *Entomophaga* **33**, 325–331.

Lacey, L. A., Fransen, J. J., and Carruthers, R. 1996. Global distribution of naturally occurring fungi of *Bemisia*, their biologies and use as biological control agents. *In "Bemisia* 1995: Taxonomy, Biology, Damage, and Management: (D. Gerling and R. Mayer, Eds.) pp 401–433. Intercept. Andover.

LaLone, R. S. and R. G. Clarke. 1981. Larval development of *Otiorhynchus sulcatus* (Coleoptera: Curculionidae) and effects of larval density on larval mortality and injury to rhododendron. *Environ. Entomol.* **10**, 190–191.

Latgé, J. P. and Papierok, B. 1988. Aphid pathogens. *In* "Aphids, Their Biology, Natural Enemies and Control. Vol. B" (A. K. Minks and P. Harrewijn, Eds.) pp 323–433. Elsevier Science Publishers B. V., Amsterdam.

Lola-lutz, T., Downes, M. and Dunne, R. 2005. Control of the black vine weevil larvae *Otiorhynchus sulcatus* (Fabricius) (Coleoptera: Curculionidae) in grow bags outdoors with nematodes. *Agric. For. Entomol.* **7**, 121–126.

Madsen, H. F. and Procter, P. J. 1982. "Insects and mites of tree fruits in British Columbia". Ministry of Agric. and Food. Victoria, B. C. 79 pp.

Mankin, R. W. and Fisher, J. R. 2002. Acoustic detection of black vine weevil, *Otiorhynchus sulcatus* (Fabricius) (Coleoptera: Curculionidae) larval infestations in nursery containers. *J. Environ. Hort.* **20**, 166–170.

Martin, W. R., Jr. 1997. Using entomopathogenic nematodes to control insects during stand establishment. *Hort. Sci.* **32**, 196–200.

McCoy, C. W. 1981. Pest control by the fungus *Hirsutella thompsonii. In* "Microbial Control of Pests and Plant Diseases 1970–1980" (H. D. Burgess, Ed.), pp 499–512. Academic Press, London.

McCoy, C. W. 1990. Entomogenous fungi as microbial pesticides. *In* "New Directions in Biological Control: Alternatives for Suppressing Agricultural Pests and Diseases" (R. R. Baker, and P. E. Dunn, Eds.), pp. 139–159. Alan R. Liss, New York.

McLeod, P. J., Steinkraus, D. C., Correll, J. C., and Morelock, T. E. 1998. Prevalence of *Erynia neoaphidis* (Entomophthorales: Entomophthoraceae) infections of green peach aphid (Homoptera: Aphididae) on spinach in the Arkansas river valley. *Environ. Entomol.* **27**, 796–800.

McQueen, I. S. and Miller, R. A. 1968. Calibration and evaluation of a wide-range gravimetric method for measuring moisture stress. *Soil Sci.* **106**, 225–231.

Miller, L. A., Bedding, R. A., and Akhurst, R. J. 1992. "Control of black vine weevil, *Otiorhynchus sulcatus*, with insect parasitic nematodes". Abstract Collection, Abstr. No. 1019. Proc. 21st International Horticultural Congress, Hamburg.

Moorhouse, E. R. 1990. "The potential of the entomogenous fungus *Metarhizium anisopliae* as a microbial control agent of the black vine weevil, *Otiorhynchus sulcatus*". Ph.D. Thesis, University of Bath.

Moorhouse, E. R., Charnley, A. K., and Gillespie, A. T. 1992a. Effect of potting media on the control of *Otiorhynchus sulcatus* larvae on outdoor strawberry plants using the entomogenous fungus *Metarhizium anisopliae. Biol. Control* **2**, 238–243.

Moorhouse, E. R., Charnley, A. K., and Gillespie, A. T. 1992b. A review of the biology and control of the vine weevil, *Otiorhynchus sulcatus* (Coleoptera: Curculionidae). *Ann. Appl. Biol.* **121**, 431–454.

Moorhouse, E. R., Gillespie, A. T., and Charnley, A. K. 1993a. Selection of virulent and persistent *Metrrhizium anisopliae* isolates to control black vine weevil (*Otiorhynchus sulcatus*) larvae on glasshouse Begonia. *J. Inverteb. Pathol.* **62**, 47–52.

Moorhouse, E. R., Easterbrook, M. A., Gillespie, A. T., and Charnley, A. K. 1993b. Control of *Otiorhynchus sulcatus* (Fabricius) (Coleoptera: Curculionidae) larvae on a range of hardy ornamental nursery stock using the entomogenous fungus *Metarhizium anisopliae. BioControl Sci. Technol.* **3**, 63–72.

Moorhouse, E. R., Gillespie, A. T., and Charnley, A. K. 1994. The influence of temperature on the susceptibility of vine weevil, *Otiorhynchus sulcatus* (Fabricius)(Coleoptera: Curculionidae), larvae to *Metarhizium anisopliae* (Deuteromycotina: Hyphomycetes). *Ann. Appl. Biol.* **124**, 185–193.

Pfeiffer, D. G., Killian, J. C., Rajotte, E. G., Hull, L. A., and Snow, J. W. 1991. Mating disruption for reduction of damage by lesser peachtree borer (Lepidoptera: Sesiidae) in Virginia and Pennsylvania peach orchards. *J. Econ. Entomol.* **84**, 218–223.

Poinar, G. O., Jr. 1986. Entomophagous nematodes. *In* “Biological Plant and Health Protection: Biological Control of Plant Pests and of Vectors of Human and Animal Diseases” (J. M. Franz, Ed.), pp. 95–121. G. Fischer Verlag, New York.

Poinar, G, Jr. and Poinar, R. 1998. Parasites and pathogens of mites. *Annu. Rev. Entomol.* **43**, 449–469.

Pye, A. E. and Pye, N. L.. 1985. Different applications of the insect parasitic nematode *Neoaplectana carpocapsae* to control the large pine weevil, *Hylobius abietis. Nematologica* **31**, 109–116.

Reinecke, P., Andersch, W., Stenzel, K., and Hartwig, J. 1990. BIO 1020, a new microbial insecticides for use in horticultural crops. Proc. Brighton Crop Protection Conference - Pests and Diseases. **1**, 49–54.

Rutherford, T. A., Trotter, D., and Webster, J. M. 1987. The potential of heterorhabiditid nematodes as control agents of root weevils. *Can. Entomol.* **119**, 67–73.

Schirocki, A. G. and Hague, N. G. M. 1994. The effect of temperature on the susceptibility of the black vine weevil, *Otiorhynchus sulctus*, to different isolates of *Steinernema* and *Heterorhabditis. IOBC/WPRS Bull.* **17**, 61–64.

Schwarz, M. R. 1995. *Metarhizium anisopliae* for soil pest control. *In* “Biorational Pest Control Agents: Formulation and Delivery” (Hall, F. R. and J. W. Barry, Eds.), pp. 183–196. ACS Symposium Series 595.

Shanks, C. H., Jr. and Agudelo-Silva, F. 1990. Field pathogenicity and persistence of heterorhabditid and steinernematid nematodes (Nematoda) infecting black vine weevil larvae (Coleoptera: Curculionidae) in cranberry bogs. *J. Econ. Entomol.* **83**, 107–110.

Shapiro-Ilan, D. I., Lewis, E. E., and Tedders, W. L. 2003. Superior efficacy observed in entomopathogenic nematodes applied in infected-host cadavers compared with application in aqueous suspensions. *J. Invertebr. Pathol.* **83**, 270–272.

Shi, W.-B. and Feng, M.-G. 2004. Lethal effect of *Beauveria bassiana, Metarhizium anisopliae*, and *Paecilomyces fumosoroseus* on the eggs of *Tetranychus cinnabarinus* (Acari: Tetranychidae) with a description of a mite egg bioassay system. *Biol. Control* **30**, 165–173.

Simons, W. R. 1981. Biological control of *Otiorhynchus sulcatus* with heterorhabditid nematodes in the glasshouse. *Neth. J. Plant Pathol.* **87**, 149–158.

Smith, F. F. 1932. Biology and control of the black vine weevil. USDA Technical Bulletin 325.

Soares, G. G., Jr., Marchal, M., and Ferron, P. 1983. Susceptibility of *Otiorhynchus sulcatus* (Coleoptera: Curculionidae) larvae to *Metarhizium anisopliae* and *Metarhizium flavoviride* (Dueteromycotina: Hyphomycetes) at two different temperatures. *Environ. Entomol.* **12**, 1886–1890.

Stenzel, K. Holters, J., and Andersch, W. 1992. BIO 1020: granular *Metarhizium* - a new product for biocontrol of soil pests. Proc. Brighton Crop Protection Conference - Pests and Diseases. **4**, 363–368.

Stimmann, M. W., Kaya, H. K., Burlando, T. M., and Studdert, J. P. 1985. Black vine weevil management in nursery plants. *Calif. Agric.* **39**, 25–26.

Tumlinson, J. H., Yonce, C. E., Doolittle, R. E., Heath, R. R., Gentry, C. R., and Mitchell, E. R. 1974. Sex pheromones and reproductive isolation of the lesser peachtree borer and the peachtree borer. *Science.* **185**, 614–616.

Thurston, G. S. 1998. Biological control of elm leaf beetle. *J. Arboric.* **24**, 154–159.

Walter, D. E. and Proctor, H. C. 1999. Mites: Ecology, Evolution and Behaviour. Wallingford, UK: CABI Publishing.

Warner, R. E. and Nagley, R. B. 1976. The genus *Otiorhynchus* in America north of Mexico (Coleoptera: Curculionidae). *Proc. Entomol. Soc. Wash.* **78**, 240–262.

Williams, R. N., Fickle, D. S., Grewal, P. S., and Meyer, J. R. 2002. Assessing the potential of entomopathogenic nematodes to control the grape root borer *Vitacea polistiformis* (Lepidoptera: Sessiidae) through laboratory and greenhouse bioassays. *Biocontrol Sci. Technol.* **12**, 35–42.

Willmott, D. M., Hart, A. J., Long, S. J., Edmondson, R. N., and Edmondson, P. N. 2002. Use of a cold-active entomopathogenic nematode *Steinernema kraussei* to control overwintering larvae of the black vine weevil *Otiorhynchus sulcatus* (Coleoptera: Curculionidae) in outdoor strawberry plants. *Nematology* **4**, 925–932.

Woodring J. L. and Kaya, H. K. 1988. “Steinernematid and Heterorhabditid Nematodes: A Handbook of Techniques". Southern Cooperative Series Bull. 331. 30 pp.

Xu, J.-H. and Feng, M.-G. 2000. The time-dose-mortality modeling and virulence indices for two entomophthoralean species, *Pandora delphacis* and *P. neoaphidis*, against the green peach aphid, *Myzus persicae*. *Biol. Control* **17**, 29–34.

Yonce, C. E. and Gentry, C. R. 1982. Disruption of mating of peachtree borer. *In* "Insect Suppression with Controlled Release Pheromone Systems". (A. F. Kydonieus and M. Beroza, Eds.). **II**: 99–106. CRC, Boca Raton, FL.

Zimmermann, G. 1986. Insect pathogenic fungi as pest control agents. *In* "Biological Plant and Health Protection: Biological Control of Plant Pests and of Vectors of Human and Animal Diseases" (J. M. Franz, Ed.). pp. 217–231. G. Fischer Verlag, New York.

Zimmermann, G. and W. R. Simons. 1986. Experiences with biological control of the black vine weevil, *Otiorhynchus ovatus* (F.). pp. 529–533. *In* "Fundamental and Applied Aspects of Invertebrate Pathology" (R. A. Samson, J. M. Flak, and D. Peters, Eds.). Proc. IV Int. Colloq. Invertebr. Pathol. Wageningen, Netherlands.

Chapter VII-17

Grasshoppers and locusts

G. Douglas Inglis and Mark S. Goettel
Agriculture & Agri-Food
Canada Research Centre
Lethbridge, AB T1J 4B1, Canada

Martin A. Erlandson
Agriculture & Agri-Food
Canada Research Centre
Saskatoon, SK S7N 0X2, Canada

David K. Weaver
Department of Land Resources
and Environmental Sciences
Montana State University
Bozeman, MT 59717, USA

1 Introduction

Grasshoppers and locusts belong to the class Orthoptera, superfamily Acridoidae. While the majority of pest species are found within the family Acrididae, species that cause economic losses in some circumstances are found in other families within the superfamily. In this chapter we refer to this group of insects collectively by the common name, "acridoids." Acridoids have a tremendous impact on agricultural production throughout the world. They are capable of decimating cultivated crops, pastures and rangeland, and historically have been indirectly responsible for death by starvation or death by diseases associated with starvation, of untold thousands of people. While the impact of acridoids in North America has abated in recent years, "plagues" of locusts and "outbreaks" of grasshoppers continue to be responsible for food shortages in many parts of the world, where they destroy substantial quantities of all crops. In these areas, locusts and grasshoppers must be managed or controlled to supply the human species' increasing need for food and fiber. Although there have been some advances in the implementation of integrated methods for managing acridoids, existing control strategies still rely almost exclusively on the application of chemical insecticides. The use of chemical insecticides imposes great expenditures annually and despite considerable research into the development of new generations of chemicals and formulations that lessen their adverse effects on the environment by narrowing the range of target organisms, the need for less obtrusive management strategies within an integrated pest management (IPM) framework is clear.

A complex of parasites, predators, and microorganisms (Tables 1 and 2) affect acridoid populations under natural conditions, but entomopathogenic microorganisms may be the only biological agents with the reproductive potential to control grasshoppers during outbreak periods. To date, the vast majority of work with entomopathogens against grasshoppers and locusts has been developed under a chemical paradigm, and unfortunately, this approach to

L.A. Lacey and H.K. Kaya (eds.), *Field Manual of Techniques in Invertebrate Pathology*, 627–654.

Table 1. Pathogens of grasshoppers and locusts

Pathogen	Group	Examples
Virus	Poxviridae	Entomopox viruses
	Baculoviridae	Nucleopolyhedrovirus[a]
	Iridoviridae	Iridovirus[b]
	Picornaviridae	Crystalline array virus[ab]
	Reoviridae	Cypovirus[a]
Bacteria	Bacillaceae	*Bacillus*
	Enterobacteriaceae	*Serratia*[b], *Xenorhabdus*[b], *Photorhabdus*[b], *Enterobacter*[b]
	Pseudomonadaceae	*Pseudomonas*
	Rickettsiaceae	*Rickettsiella*[b]
Fungi	Zygomycota	*Entomophaga*
	Ascomycota[c]	*Aspergillus, Beauveria, Lecanicillium, Metarhizium, Paecilomyces, Sungliocladium*
	Microsporidia[d]	*Johenrea, Paranosema, Perezia, Vairimorpha*
Protozoa	Rhizopoda	*Malamoeba*[b]
	Apicomplexa	*Gregarinia*[b], *Leidyana*
Gordian worm	Nematomorpha	*Gordius* (nematomorphs)
Nematodes	Adenophorea	*Mermis, Amphimermis, Imamura, Hexamermis, Agamermis* (mermithids)
	Secernentea	*Steinernema, Heterorhabditis, Diplotriaena*

[a] Rare.
[b] Weakly virulent (often found only in laboratory or rearing settings).
[c] All the listed genera are anamophic taxa (*i.e.* mitosporic fungi) with ascomycetous affinities. The telomorph for *Lecanicillium* is *Torrubiella*.
[d] Members of the phylum Microsporidia have traditionally been considered primitive protozoa. However, recent evidence indicates that they are actually highly evolved intracellular fungi (Hirt *et al.*, 1999).

their deployment has more than often yielded unsatisfactory results. The factors that are responsible for the initiation and development of epizootics in acridoid populations are extremely complex, involving interactions among the pathogen(s), insect host, environment and time. The main objective of this chapter is to provide the reader with the background information on current field experimentation techniques that will advance our knowledge of the factors responsible for epizootic initiation, with the ultimate goal of using entomopathogenic microorganisms for the efficacious suppression of acridoid populations. In the six years since the initial version of this chapter was published, testing of microbial control agents against field populations of acridoids has continued. The results of these trials in suppressing acridoid populations have been mixed. If microbial control is to become an important component of IPM programs against acridoid pests, it is clear that the continued development and field testing of these agents is of paramount importance. Every attempt should be made to design field tests in such a manner that addresses potential constraints on the efficacy of microbial control agents with the goal of designing strategies to overcome them.

2 Propagation, formulation, and application

The type of microorganism (*e.g.*, biotrophic versus saprotrophic) has a direct bearing on the type of propagation, formulation, and application strategy employed. The three application strategies utilized against acridoids include: (1) classical; (2) inoculative; and (3) inundative approaches. The site of infection is also of paramount importance in the selection of formulation and application strategy. Bacteria, microsporidia, and viruses primarily infect *per os*, whereas fungi and nematodes generally penetrate the body through natural openings or directly through the external integument.

Table 2. Key to the pathogen groups affecting acridoids based on macroscopic and microscopic (light microscope level) signs of disease

1.	White or green cottony or powder-like external growth, sometimes limited to intersegmental regions of grasshopper cadavers	Fungi
1.	No external growth observed on cadavers	2
2.	Worm-like (non-segmented) organisms observed on inside or exterior of cadaver.	Nematode/Gordian worm
2.	No worm-like organisms observed	3
3.	Insects usually normal in appearance but may be stunted or malformed; upon dissection body may contain one or more multicellular organisms with many body segments and a distinct to indistinct head region with mandibles	Insect parasites
3.	Parasitic insects not observed	4
4.	Upon dissection of cadaver, hyphae observed in the hemocoel	Fungi[a]
4.	Tissues not containing hyphae	5
5.	Unicellular or relatively undifferentiated particles present	6
5.	Particles absent	Viruses or abiotic
6.	Particles in hemocoel spherical, iridescent, and stain red with Sudan III	Fat globules
6.	Particles absent or if present, variable to uniform in size, do not stain red with Sudan III	7
7.	Particles relatively large (> 5 μm) and melanized with prominent walls ("chlamydospores" or cells are variable in size and yeast-like in appearance ("hyphal bodies" or "protoplasts")	Fungi
7.	Particles relatively uniform in shape, not yeast-like in appearance and are commonly < 5 μm in diameter, if> 5 μm lack a conspicuous cell wall and are hyaline (*i.e.*, colorless)	8
8.	Particles uniform in shape and size, rod-shaped, usually motile exhibiting Brownian movement, prevalent in the hemocoel, and cadaver tissues frequently liquefied	Bacteria
8.	Particles are generally non-motile and frequently develop within specific host cells or tissues	9
9.	Infectious particles rod-shaped or pleomorphic in shape, highly refringent, often occurring in pairs or chain-like structures, small, just visible under the light microscope, occur in fat bodies, and at late stages of infection also present in the hemocoel, hemocytes, oenocytes, and salivary glands, usually exhibit Brownian movement	Rickettsia
9.	Infectious particles or structures containing infectious particles rod-shaped, ellipsoidal, or spherical and generally conspicuous under the light microscope	10
10.	Particles variable in size, ovoid to pyriform (*i.e.*, pear-shaped), 2–5 μm in length, refringent, non-motile or motile, may possess a polar filament, and may be present in Malpighian tubules	11
10.	Particles spheroid in shape, variable in size ranging from 1–25 μm in diameter ("occlusion bodies"), primarily observed in the cytoplasm of fat body cells	Entomopoxvirus
11.	Spores and/or sporoblasts present within fat bodies which may be grossly hypertrophied, spores may extrude a polar filament (*e.g.*, when pressure is applied)	Microsporidia
11.	Cells (*e.g.*, cysts, trophozoites) typically associated with Malpighian tubules and/or with gut tissues	Protozoa

[a] In this key, "fungi" refers to eufungi excluding the microsporidia.

A Inoculum

It is often necessary for researchers to produce their own inoculum for experimental testing of entomopathogens. In general, there are two major strategies employed to propagate entomopathogens for use against acridoids: (1) *in vitro* propagation for nematodes, bacteria and most mitosporic fungi; and (2) *in vivo* propagation for viruses, microsporidia, *Entomophaga grylli*, and nematodes. For *in vivo* production, propagules can be produced in living hosts or in cell cultures (primarily viruses and microsporidia). We summarize two methods which have been used to produce entomopathogens in sufficient quantities to conduct large-scale field experimentation: production of *Paranosema locustae* (synonyms, *Antonospora locustae* and *Nosema locustae*) and *Melanoplus sanguinipies* entomopoxvirus 'O' (MSEV) within living hosts; and diphasic fermentation of *Metarhizium anisopliae* var. *acridum* (synonym, *M. flavoviride*).

1 Paranosema locustae and MSEV

The *in vivo* production of *P. locustae* spores as described by Henry (1985) is by oral infection of 5th instar nymphs of *Melanoplus bivittatus*, and recovery of spores from infected individuals between 20 and 35 days post-inoculation. Cadavers can be stored from −20 to −80 °C until they are homogenized in aqueous buffers, filtered, and centrifuged to concentrate the spores. This method has been developed for commercial-scale production of *P. locustae* spores (Menely and Sluss, 1988). For the production of MSEV occlusion bodies (OB), grasshoppers are typically infected as 3rd or 4th instar nymphs with 5.0×10^6 OB bodies/g of wheat bran (Oma and Streett, 1983). Higher yields can be achieved with individual oral inoculation of nymphs with 1×10^4 OB/nymph on lettuce discs or by injection of virions released from OB directly into the hemocoel of grasshopper nymphs, but these methods are too labor intensive if a large field trial is anticipated. MSEV OB can be harvested from cadavers and moribund nymphs in a fashion similar to *P. locustae* spores (McGuire and Henry, 1989).

2 Metarhizium anisopliae var. *acridum*

The *in vitro* production of *M. anisopliae* var. *acridum* conidia is described by Jenkins *et al.* (1998). Initially, biomass is produced in an appropriate liquid medium in shake culture (*e.g.*, a medium containing a source of carbon such as 2% sucrose or glucose, nitrogen provided by 1 to 2% yeast, yeast extract or peptone, and vitamins provided in materials such as yeast extract). For aerial conidia production, the biomass produced in liquid culture is transferred to a solid substrate. Although a number of substrates can be utilized, those high in starch (*e.g.*, rice) are commonly used; porous substrates facilitate air exchange and provide a good surface area. The use of autoclavable mushroom spawn bags equipped with gas-exchange filters are frequently used for the solid-substrate fermentation stage of production. Once conidiogenesis is complete, conidia are mechanically extracted from the substrate by sieving and dried for storage.

3 Propagule storage

The conditions that ensure optimum storage of inoculum will differ among entomopathogens. While lyophilization and storage at ultralow temperatures (*e.g.*, over liquid nitrogen) often provide excellent results, it is not logistically possible to use these methods to store large quantities of inoculum. For many microorganisms, drying of inoculum followed by storage at low (*ca.* 5 °C) or moderate temperatures (room temperature) maintains propagule viability adequately to allow field experimentation. Storage of microsporidian spores and MSEV OB present special problems. While aqueous suspensions of *P. locustae* spores retain some viability when stored at −20 °C, freshly prepared spores from cadavers give consistently higher viability (Henry and Oma, 1974). MSEV OB stored in frozen cadavers also retain infectivity and virulence, but McGuire *et al.* (1991) found that OB formulated in starch granules retained activity when stored for up to 9 months at 4 °C. Additional information on the storage of the different pathogen groups is presented in Lacey (1997).

4 Propagule quality assessments

Measurements of inoculum quality prior to or during application is essential but is often ignored in field tests, particularly if the entomopathogen being tested is a commercial product. Assessments should be made on the active ingredient prior to formulation, after formulation, and at the field site, preferably by collecting inoculum as it is being applied. Two factors that should be measured are viability and virulence.

a Propagule viabilities

Viabilities can be determined using vital stains or germination assessments (Lacey and Brooks,

1997). The primary advantage of using vital staining methods is that they provide rapid results and can be used with entomopathogens that do not grow saprotrophically, except for MSEV OB for which the only current test of viability is via bioassays. A common method used to quantify the viability of many entomopathogens is propagule germination. The viability of fungal propagules (*e.g.*, conidia or hyphal bodies) is usually assessed by measuring germination in or on an appropriate medium after a certain period of time under optimal conditions (Goettel and Inglis, 1997). Since the fungi that are pathogenic to acridoids possess ascomycetous affinities (*e.g.*, *Beauveria, Metarhizium, Paecilomyces, Aspergillus*), cell replication is affected by the fungicide, Benlate (active ingredient, benomyl). Using low concentrations of Benlate (0.001–0.005% Benlate) in a medium, propagules form germ tubes but cell division is halted and hyphal growth is prevented thus eliminating problems with overgrowth of slow-germinating propagules (Goettel and Inglis, 1997). Conidia and resting spores of *E. grylli* can be germinated on water agar since entomophthoralean fungal propagules do not require exogenous nutrients to germinate (Papierok and Hajek, 1997). However, the suspension medium can influence germinability and glycine-NaOH adjusted to a pH of 7 to 9 supported the highest overall germination. Resting spores of *E. grylii* can remain dormant for several years before germination regardless of their environment, and the lack of germ tube formation does not necessarily indicate lack of viability.

Spores of *P. locustae* can be induced to extrude their polar filament *in vitro* by placing them in an alkaline (pH 9.0 to 11.0) solution (Undeen and Vávra, 1997), but this method does not always provide a reliable predictor of spore viability. Assessments *in vivo* or the use of fluorescent vital stains can be a more accurate means of assessing viability (Undeen and Epsky, 1990). For *in vivo* assessment, *P. locustae* spores are fed to adult grasshoppers on disks of lettuce. Following ingestion, the grasshoppers are maintained at 30 °C for 1 to 2 h at which point the adults are killed by freezing, cut transversely with a razor blade (Albrect, 1953), the midgut region removed, its contents thawed in the presence of pH 9.5 glycine-NaOH and the spores examined with a phase contrast microscope. Because polar filaments are rapidly digested in the grasshopper gut, the researcher must rely on the presence of "black spores" as an indication of viability. In general, 2 to 5% of spores in each batch will be black prior to ingestion by grasshoppers.

b Virulence assessments

At present, the only established way to assess the viability of MSEV virions and the relative virulence of propagules of all entomopathogenic microorganisms of acridoids is via bioassay. Bioassays should be conducted under defined conditions and extreme care should be taken in extrapolating bioassay results to field efficacy (Butt and Goettel, 2000). Here we present commonly used bioassay designs for grasshoppers and locusts. The researcher should always keep in mind that a bioassay should be repeatable and strict control of dosage and treatment method must be exercised. It is important that a "standard" be included in experiments (*i.e.*, a properly preserved, and preferably a well-characterized genotype) when comparisons are being made among numerous isolates. It is also critical that the health of the grasshopper or locust colony from which test animals are drawn be of the highest quality possible (*e.g.*, pathogen-free), particularly for bioassays with microsporidian and virus pathogens that take much longer to kill the host. The stage at which the acridoid is targeted in field experiments is often used in bioassays. For example, nymphs are primarily targeted in control efforts against grasshoppers in North America, whereas locust adults are frequently targeted in Africa. Eggs are usually obtained from laboratory colonies or from field-collected adults or late-instar nymphs that are allowed to mate and oviposit in captivity. Whether a laboratory colony (with or without genetic infusion) or individuals from field-collected grasshoppers are used will depend on the requirements of the bioassay.

There are five primary inoculation strategies used in bioassays with acridoid entomopathogens: (a) topical application of specific volumes usually with an oil carrier;

(b) topical application of non-specific volumes usually accomplished by dipping; (c) *per os* inoculation using bait carriers; (d) surface infestation using bait carriers; and (e) injection (Lacey, 1997). If possible, the bioassay should include a number of concentrations of the pathogen so that a dose-mortality response curve can be obtained (Goettel and Inglis, 1997). The time to death of a group of insects receiving a single concentration of a pathogen can also be used as a measure of virulence, although temporal data does not provide as much information as a dose-mortality experiment. It is very important that the researcher control a number of potentially confounding factors such as host species, age and physiological condition, size of bioassay chamber and incubation conditions (Goettel and Inglis, 1997). Furthermore, some pathogens such as MSEV exhibit a biphasic temporal pattern of mortality as a function of dosage (Olfert and Erlandson, 1991; Woods *et al.*, 1992).

Behavioral thermoregulation by acridoids via habitat selection is a well documented phenomenon, and it is common to observe acridoids basking in field environments (Figure 1A). The ability of acridoids to elevate their body temperatures higher than ambient has been shown to provide significant survival benefits by delaying or reducing the impact of infectious pathogens (*e.g.*, Carruthers *et al.*, 1992). In some instances, infected individuals have been shown to behaviorally elevate their body temperatures higher than non-infected individuals (*i.e.*, via a behavioral fever) (Boorstein and Ewald, 1987). The impacts of behavioral thermoregulation on microbial control of acridoids by mitosporic fungi (*e.g., B. bassiana* and *M. anisopliae* var. *acridum*) have since received considerable attention, and this phenomenon is one example of a variable that researchers should consider in designing bioassays. For example, Inglis *et al.* (1996a) designed a simple bioassay environment that addressed the thermoregulatory behavior of acridoids on the efficacy of entomopathogens by fitting cages with an incandescent bulb (*i.e.*, a heat source) placed adjacent to a vertical climbing mesh that allowed grasshoppers to voluntarily bask (Figure 2). It should be considered that not all species of acridoids may be active behavioral thermoregulators. For example, variegated grasshoppers (*Zonocerus variegatus*) in humid tropical ecosystems do not exhibit any overt behavioral postures or microclimate selection, either in healthy individuals or in individuals infected with *M. anisopliae* var. *acridum* (Blanford *et al.*, 2000). Acridoids reared under crowded conditions may be more resistant than solitary animals to entomopathogens (Wilson *et al.*, 2002), and this also should be considered in designing a bioassay.

B Formulation

The properties of the microorganism to be tested (*e.g.*, surface hydrophobicity and the mode of infection), storage conditions, equipment

Figure 1. Gregarized Madagascar migratory locusts (*Locust migratoria capito*). (A) Nymphs basking on the sunny side of a termite mount in the morning, and (B) a nymphal band in a savannah ecosystem

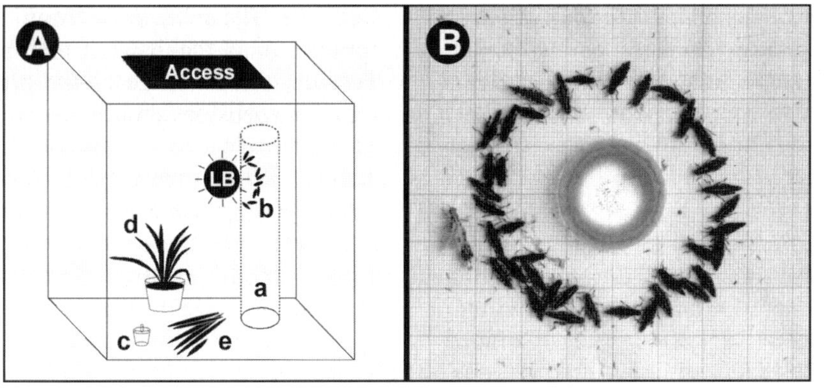

Figure 2. A bioassay cage designed to allow acridoids to thermoregulate. (A) Drawing of a cage containing a climbing mesh (a) on which acridoids can climb the mesh and orient themselves (b) relative to a heat source, in this case an incandescent light bulb (LB). Sources of water, in this case water presented in a cup fitted with a lid with a central hole containing a dental pad designed to wick water (c), and food, in this case growing grass in a pot (d) or grass that was harvested and subsequently placed on the floor of the cage (e). For omnivorous species, other sources of food such as bran can be presented in cups. Temperatures at the bottom of the cage were ambient, and acridoids had to climb the mesh to raise their body temperature higher than ambient. (B) *Melanoplus sanguinipes* nymphs orienting themselves relative to a 40W incandescent light bulb (note: wire thermistors within grasshoppers basking relative to the incandescent light bulb determined that insects were optimizing their internal their body temperature) (Image provided courtesy of Dr. Derek Lactin)

available for application, environmental conditions (both macro and microclimate) and the characteristics of the terrain in which acridoids occur will ultimately influence decisions on which formulation to use. The primary carriers used to apply entomopathogens against acridoids include: particulate formulations (*e.g.*, baits and encapsulates); oil formulations; oil emulsion formulations; and water formulations. Another aspect of formulation is the inclusion of adjuvants for enhanced storage or efficacy (*e.g.*, synergists).

1 Oil emulsion formulation

Although conidia of entomopathogenic fungi (*e.g.*, *Beauveria bassiana* and *M. anisopliae* var. *acridum*) can be uniformly suspended in water using mechanical wetting methods, it is a laborious procedure and water formulations of fungi possessing hydrophobic cell walls has been limited to laboratory and small plot work. For medium and large-scale field trials, conidia can be applied in an oil-in-water emulsion with or without a suspender (*e.g.*, suspension or flowable concentrate). A suspension concentrate is defined here as a suspension of particles (propagules plus a suspender) in a liquid. Suspension concentrates of entomopathogens may contain 10 to 40% suspender, 1 to 5% dispersant (to reduce the rate of sedimentation by flocculation), 3 to 8% surfactant (acts as emulsifier, a wetter and a spreader), and 35 to 65% carrier liquid (Burges and Jones, 1998). A number of materials can be used as suspenders such as colloidal clays (*e.g.*, attapulgite), polysaccharide gums, cellulose, and synthetic polymers. To prepare a suspension concentrate of fungal propagules to be applied at a rate of 2×10^{13} viable conidia in 100 liters/ha:

1. Estimate the number of viable conidia per g of powder using a hemocytometer and germination medium (*e.g.*, 5×10^{10} viable conidia/g) (Goettel and Inglis, 1997).
2. Calculate the amount of conidial powder required (*e.g.*, 2×10^{13} conidia/ha $\div$ 5×10^{10} conidia/g $=$ 400 g/ha).
3. Mix the dry conidia into an appropriate oil (if the concentration of oil is 2% by volume, the volume of oil required is 0.02×100 liters/ha $=$ 2 liters/ha). Confirm the density of conidia using a hemocytometer.
4. Combine the conidia in oil, suspender (*e.g.*, attapulgite clay), surfactant, and dispersant with the water carrier. This is often done in the application tank under agitation. If the concentration of the suspender, surfactant and dispersant are each 5% (0.05×100 kg $=$ 5 kg/ha), the total volume of water/ha required would be

[100 - (0.4 kg of conidia +2 kg of oil +5 kg of suspender +5 kg of surfactant +5 kg of dispersant)] = 82.6 liters. For simplicity, the specific gravity of oil and other materials are often rounded to 1.

2 Oil formulation

A number of advantages to the formulation and application of entomopathogenic fungi in oil have been suggested (*i.e.*, reduced volumes, more efficient droplet distributions, enhanced spreading, increased efficacy under conditions of low humidity). However, more detailed experimentation is warranted on the advantages/disadvantages of oil formulation of entomopathogens. A number of different types of oils that can be used include mineral oils and plant-derived oils. Viscosity, as influenced by temperature, is an important consideration. Vegetable oils possess relatively high viscosities whereas mineral oils possess relatively low viscosities. In addition, the adverse effect of rancidity on propagule viability is an important consideration in the use of plant-derived oils. Oil formulations (mineral oils alone and in combination with plant-derived oils) have been used for the application of *M. anisopliae* var. *acridum* and *B. bassiana* against acridoids (Bateman, 1997).

3 Bait formulation

The primary method for application of pathogens which are active *per os* and therefore must be ingested (*e.g.*, *P. locustae* and MSEV) is through the use of bait formulations. However, grasshoppers and locusts can become surface-infested with propagules during ingestion of the bait substrate. Therefore, baits that promote tactile contact may also be an efficacious method of targeting acridoids with pathogens which infect externally, such as fungi (Inglis *et al.*, 1996b; Moore and Caudwell, 1997). For preparations of bait formulations of *P. locustae*, spores with or without sticking agents (*e.g.*, methylcellulose or molasses) are suspended in water. If the final rate is to be 1.5×10^9 spores on 1 kg of bran, suspend the spores in the water carrier, enumerate the spores with a hemocytometer, adjust the concentration to 1.5×10^8 spores/ml, and apply 10 ml of the suspension/kg of bait substrate. The spore suspension should be evenly sprayed onto the bait carrier during agitation. To facilitate coverage of the bran particles, the spore suspension should be applied onto bran during mixing (*e.g.*, tumble using a cement mixer). Wheat bran is a bait that is very palatable to grasshoppers (Onsager *et al.*, 1980), but other bait materials such as oats can be used. For the control treatment, water with the sticking agent without spores should be applied to the bran alone; care must be taken to ensure that the bran is not contaminated with spores from equipment. A major problem with wheat bran baits is that the shelf life of *P. locustae* spores is limited. Cadavers killed by entomopathogens often provide an important source of secondary inoculum in field environments, particularly for entomopathogens transmitted *per os* (*i.e.*, either through scavenging of cadavers or by indirect ingestion of propagules liberated from the cadavers). Although not a bait *sensu stricto*, evidence now suggests that sporulation of mitosporic fungi on and/or in cadavers can be an important source of inoculum in field settings, even in arid ecosystems.

4 Encapsulation

Despite the potential advantages of encapsulation (*i.e.*, enhanced persistence), the encapsulation of entomopathogens for managing acridoids is still in its infancy. Encapsulation is defined here as the incorporation of the microorganisms into a packet or capsule, and in this chapter we include pelletization as a type of encapsulation. McGuire *et al.* (1991) described the following method for encapsulation MSEV OB:

1. While mixing, add corn or wheat germ oil (3 ml) to 2.5 g of molasses, 130 mg of Congo red (an ultraviolet radiation-absorbing compound) or carbon (a sunlight blocker), 8 g of corn starch, and 18 ml of water containing 1.3×10^8 MSEV OB.
2. When a gelatinous consistency is obtained, place the mixture at 5 °C overnight.
3. Break the resultant mass into pieces in a blender while slowly adding 5 g of pearl starch or finely milled wheat germ.
4. Air dry the particles and pass them through mesh sieves. Separate the two particle sizes; those that pass through a 16 but not a 20 mesh sieve

(0.9–1.2 mm), and those that pass through a 20 but not a 40 mesh screen (0.4–0.9 mm). Encapsulation of MSEV in starch improved storage quality (good viability was retained for up to 9 months at 4 °C).

Capinera and Hibbard (1987) reported a method for the incorporation of *Steinernema carpocapsae* into calcium alginate capsules for use against acridoids using the following method:

1. Culture nematodes in *Galleria mellonella* and store them at 10 °C at a concentration of 3.4×10^4 nematodes/ml (Kaya and Stock, 1997).
2. Dissolve sodium alginate (LF-60; 2 g) in 100 ml of water in a blender for 4 to 5 min, and add 5 g of wheat germ and infective juvenile (IJs) nematodes to the solution.
3. Drop the alginate into a complexing solution of $CaCl_2 \cdot 2H_2O$ (continuous stirring) to form the calcium alginate capsules.
4. Allow the capsules to complex for 20 to 30 min and then separate them from the complexing solution by sieving while rinsing in deionized water.
5. Store pellets in Petri dishes or plastic bags at 4 °C. The concentration of nematodes in the calcium alginate pellets is calculated from the concentration in the sodium alginate solution and the weight of the average capsule after formulation (approximately 20 to 60 mg). Target concentrations are approximately 400 IJs/capsule.

5 *Adjuvants*

A number of adjuvants may be used to increase persistence and to enhance attachment, penetration and disease development. Some adjuvants that have been investigated with entomopathogens of acridoids include ultraviolet (UV) protectants, physiological stressors, behavioral modifiers, stickers, and phagostimulants. For example, the ability of grasshoppers and locusts to elevate their body temperature higher than ambient by basking can be detrimental to entomopathogens and the use of strategies to overcome this constraint (*e.g.*, behavioral modifiers, such as sublethal dosages of compounds that prevent or delay basking) may be very beneficial. Care must be taken so that the appropriate control treatments are included in field experiments (*e.g.*, untreated and formulation control treatments). A number of UV protectants have been field tested with MSEV, *B. bassiana* and *M. anisopliae*. For example, a water-compatible fluorescent brightener, Tinopal (5%), and a clay emulsion (12%) enhanced survival of *B. bassiana* conidia (Inglis *et al.*, 1995). However, oil-compatible UV protectants (*i.e.*, oxybenzone, octyl-salicylate and ethyl-cinnamate) did not protect conidia of either *B. bassiana* or *M. anisopliae* (Inglis *et al.*, 1995; Shah *et al.*, 1998).

C *Experimental design and application of entomopathogens*

The formulation, crop, area to be treated, target acridoid (*e.g.*, species and stage of development), frequency and urgency of treatments, and the capital investment available will determine the experimental design, plot setup, and the best equipment for the application of entomopathogenic propagules.

1 *Experimental design and plot setup*

a *Design and replication*

Coverage of the basic factors in experimental design with microbial agents against grasshoppers and locusts are given in Johnson (1992) and further practical discussion of field tests is given in Lomer *et al.* (1997). Microbial agents are different from chemical insecticides and often the hypotheses tested are related to persistence of the agent, whether damage or loss is reduced before mortality occurs, as well as the expected measure of the rate of kill and percentage of the population impacted. For any given situation, the interaction between infective agents, microclimate, plant community, and prevailing weather will be unique. All trials need to be conducted as a randomized block design, with an adequate number of blocks, plots, and subplots. If the area of interest includes different substrates (*e.g.*, cultivated crops versus feral vegetation), each main factor should then have its own set of blocks, plots, and subplots. A basic field experiment involves a minimum of three treatments (entomopathogen, a carrier control, and an untreated control) with a minimum of three replicates (blocks).

Measurements are generally conducted over time in all trials, and sampling methods sensitive to immobilization and to movement of infected individuals should be used concomitantly (see section 3D). The goal is to have repeatability of all efficacy data for each treatment across the blocks. A temporal replicate of a field experiment is an admirable goal, but given the nature of the environmental inputs, it is best to look for relative similarity from one test to the next. In some large-scale field trials, treatments are applied to large areas and it is not possible to replicate the experiment (*i.e.* each treatment is applied to one large block). This design is referred to as a pseudoreplicated block design (Hurlbert, 1984). It is not statistically possible to separate treatments and block effects with this design, and comparisons of acridoid populations among treatments must be made with extreme caution.

b Plot size

The size of test plots should be carefully considered to maximize the information given the limitations on resources for each situation. With a well-defined and low mobility population or when the experiment does not deal directly with acridoids (*e.g.*, when the objective is to study application and deposition rates, persistence of propagules, or environmental constraints, etc.), a small field plot (*e.g.*, 50 m^2) can be used. When population size and location are not easily pinpointed, the pest is highly mobile and/or the entomopathogen is relatively slow-acting, it will be necessary to increase the area for each plot to reduce variability. Random fixed transects can be limiting in large scale trials and a uniform grid of subplots can be much more informative. For the scenario in which the target is a community of species, the block size has to be large, and plots carefully chosen so that variability in experimental design does not bias community data. In this situation in particular, careful pre-sampling of density (Onsager, 1991) and community composition (Quinn *et al.*, 1991) is necessary to ensure covariance can be addressed, but in all cases samples collected before treatment are necessary to properly define the host population.

c Age structure and species composition

A pre-application knowledge of the species composition of the field site is very important. Many microbial agents have differential efficacy against different species and across development stages. Thus, it is necessary to define the age and species structure by some form of repeated sampling. Thompson (1987) discusses methodology to improve accuracy in assessing population age structure as a function of sample unit size and sample number. Caution must be taken for species which become gregarious and highly mobile as they mature, to avoid a situation in which application is made concomitantly to two physiologically distinct groups as they enter phase transition. To determine age structure and the species composition, sweepnet samples are usually taken at arbitrarily-selected locations or along predetermined transects or grids within each plot (see section 3A). There are often cryptic species present in acridoid communities, and these must be sampled using fixed quadrats or another form of an absolute census. Samples within each plot are usually bulked, examined (frequently with the aid of a stereomicroscope), and frequencies of individual species and stages of development determined using an appropriate taxonomic key and reference collections if available. The superfamily Acridoidea, contains six families with the vast majority of pest species falling within the family, Acrididae. There are more than 7000 species within Acrididae, and for a review on the taxonomy and identification of this group of insects, the reader is referred to Vickery (1997) and references therein.

d Migration considerations

Migration of individuals presents major challenges for the design of field experiments, particularly with entomopathogens which are slow acting (*e.g.*, *P. locustae* and MSEV can take several weeks or longer to fully express mortality in susceptible hosts). This has obvious implications for the selection of plot sizes and on border regions between plots. The problem with migration is further confounded by the fact that uninfected grasshoppers or locusts from control plots are likely to be more mobile than infected insects leading to the dilution of treatment effect. For slow acting entomopathogens such as *P. locustae*, a plot size of 16 ha is considered to be the minimum size required (Johnson, 1992). Within the treated plots, large buffer zones ($\geq$ 50 m) that are treated, but in which no population assessments are conducted, should also be utilized in an attempt to minimize the effects of migration (*i.e.*, an "edge effect." Barriers can be erected to reduce migration if the insect stage is not particularly mobile, but often it is necessary to contain the treated individuals (*i.e.*, in cages). While a number of markers (*e.g.*, dyes) have been used to monitor droplet deposition, these markers are generally too short-lived to provide a measure of migration. However, the utilization of elemental markers, such as RbCl in bait formulations, are stable in grasshopper populations for at least 7 days following ingestion of baits and have been used to estimate both the level of bait uptake as well as population dispersal from control plots (Woods and Streett, 1996a, b).

2 *Equipment and application rate*

A variety of types of equipment have been used to apply entomopathogens against acridoids, and here we summarize some of the most common application methods used. For more information, the reader is referred to the chapters in Section III.

a Hydraulic spray application

Water-based formulations (including oil-emulsion and wettable powders) are primarily applied using conventional boom hydraulic systems (Figure 3A–B). The application volume depends on pressure, nozzle type, and rate of travel. In general, this type of application results in a wide range of droplet sizes (in most situations droplets will be between 100 to 500 μm in diameter but the distribution of droplet sizes is a function of nozzle type and operating pressure). In arid agroecosystems, water in smaller droplets may evaporate before deposition on the target but the influence that evaporation has on targeting and efficacy of entomopathogens has not been adequately examined. The choice of nozzle type depends on a number of factors. In rangeland settings where the crop canopy is generally low, the 110° angle is generally selected because it allows the boom height to be lowered which reduces the influence of wind on droplet dispersal. In most instances,

Figure 3. Commonly used application methods. (A) A boom hydraulic spray system mounted on a tractor three-point hitch used to apply large volumes of water or oil-emulsion in relatively uniform-topography grassland, (B) a "bicycle sprayer" used for applying entomopathogens in water or oil emulsion in small plot, (C) hand-held spinning disk applicator used to apply entomopathogens at very low or ultra low volumes in oil, and (D) a truck-mounted positive displacement rotor system for applying granular formulations such as *P. locustae* formulated on bran

the pressure selected is in the 100 to 200 kPa (15 to 29 psi) range. With the application of water-based formulations, the spray volume per unit area can range from 50 to 250 liters/ha. Since water can be limited in arid agroecosystems, volumes on the lower end of the range are generally selected.

Tractor-mounted or trailed equipment are commonly available and can be used for applying entomopathogens in relatively large-scale field trials (Figure 3A). One of the major disadvantages of using tractor-mounted application equipment is that rough terrain can cause tremendous boom movement which adversely affects the spray pattern and can result in damage to nozzles and the boom itself if it comes in contact with the ground. Although expensive, specialized spray equipment for the application of hydraulics such as self-propelled applicators and sprayer equipped fixed-wing aircraft and helicopters can increase the efficacy of entomopathogen application in large-scale field trials. Aerial application is particularly important when the terrain or accessibility to the field site precludes the use of conventional equipment. Other types of hydraulic sprayers are very useful for small-scale field trials or small plot work, and include knapsack-style and CO_2 pressurized sprayers. *Beauveria bassiana* has been applied against acridoids using a variety of hydraulic spray application equipment. Total volumes have ranged from 100 to 250 liters/ha, and the concentration of propagules has ranged from 1 to 3×10^{13} conidia/ha.

b Spinning disk application

Entomopathogenic microorganisms can be formulated in oils and applied at very low (VLV; 5 to 20 liters/ha) or ultra-low (ULV; < 5 liters/ha) volume application rates. In most instances, the application of entomopathogens at VLV or ULV has been conducted using oil formulations due to the susceptibility of water droplets to evaporation. The target droplet size is usually 50 to 150 μm diam. Using application volumes less than 1 liter/ha, droplets become excessively small and become very susceptible to drift. Spinning disk sprayers are commonly used, and they consist of a spinning disk rotated by an electric motor with the speed of the disk dependent on the voltage of the power supply. Hand-carried spinning disk sprayers (Figure 3C) are frequently used, but they can be mounted on a short boom, on a truck or aircraft (Matthews and Thornhill, 1995; Thornhill and Matthews, 1995). Both *B. bassiana* and *M. anisopliae* var. *acridum* have been applied in oil at ULV against acridoids (*ca.* 1 to 3×10^{13} conidia/ha).

c Bait and granule application

The application of dry formulations of entomopathogens is generally more difficult than spraying a liquid because the individual particles are irregular in shape and they do not flow as easily through the equipment. The simplest and least accurate metering system consists of an orifice through which granules flow by gravity; the flow is controlled by the size of the orifice which is adjustable. A much more accurate metering system consists of a positive displacement rotor (Figure 3D). A popular design is the toothed rotor in which granules fall by gravity into the gaps between the rotor teeth and are then carried by the rotation to the low side of the rotor where they fall into the delivery system. In most instances, granules are broadcast onto the soil surface using systems such as cyclone seeders, truck-mounted bran applicators equipped with a gasoline engine-powered fan, fixed wing aircraft and helicopter. Application rates of *P. locustae* granules vary from 1.5×10^7 to 1.5×10^{10} spores on 1 to 2.5 kg of bran/ha but a typical field application rate is 2.5×10^9 spores on 1.7 kg bran/ha. Very few trials have been conducted with non-microsporidian pathogens with either bait or granular formulations. In one field trial, MSEV formulated in starch granules was applied at a rate of 1×10^9 and 1×10^{10} OB/ha (Streett *et al.*, 1997). In another trial, *B. bassiana* was applied in a bait formulation, but evidence indicated that conidia were liberated from a substrate during application (Johnson and Goettel, 1993). This underscores the need for adequate coverage and targeting assessments when applying bait formulations.

D Coverage and targeting

An extremely important aspect of field experimentation is the quantification of the number of propagules that come in contact with the insect target or a variety of other substrates such as vegetation. This is usually accomplished using microbiological or microscopic techniques. However, chemical and molecular methods, and bioassays can be used as well, and in some instances, will be the method of choice (Lacey, 1997).

1 Deposition of droplets and granules

The deposition of droplets is frequently measured using water- or oil-sensitive papers or strips (*e.g.*, Spraying Systems Inc.) (Figure 4A–B). For qualitative assessments of size and density, cards can be visually compared to standards. For quantitative estimation of droplet density, the number of droplets/area (usually $1\,cm^2$) can be counted with the aid of a magnifying lens or stereomicroscope. In addition to droplet densities, image analysis systems can provide information on droplet area, size, and percent coverage. Another method which has been used to chemically assess targeting involves the use of tracer dyes (Cilgi and Jepson, 1992).

To enumerate bait or granule particles/area, containers of defined area are placed on the soil surface and bait particles in the container are counted following application. Sticky surfaces can also be used to trap the bait particles to prevent potential loss or redistribution of particles by wind or rain. Elemental markers have been used to estimate bait uptake by grasshoppers (Woods and Streett, 1996a,b), but elemental analysis requires access to relatively sophisticated equipment and it is quite laborious.

2 Microbiological and microscopic methods

Bacteria and fungi can be recovered directly from field-collected acridoids or from non- target substrates such as leaves, soils, or objects placed in the field for that purpose (*e.g.*, coverslips) using microbiological techniques (Figure 5). Samples should be kept cool if possible to avoid the possible adverse effects of high temperatures and sunlight on the microorganisms. Prior to processing, captured acridoids are usually identified to species, stage of development and weighed. Dislodgement of propagules from the surfaces of the acridoid or plant cuticles is usually accomplished by homogenization or vigorous washing in sterile water or buffer; 50 to 100 mM phosphate buffer (pH 7) containing a surfactant such as Tween 80 (0.01%). Surfactants can be toxic to propagules (*e.g.*, many of the organosilicone wetting agents are toxic to *M. anisopliae*

Figure 4. Commercially available oil- and water-sensitive papers used to measure coverage of entomopathogens applied against acridoids in field environments. (A) Oil-sensitive paper. (B) Water-sensitive paper. Papers are situated at defined locations within a field plot before application, collected and the density and size of droplets are estimated by eye or calculated more specifically by image analysis

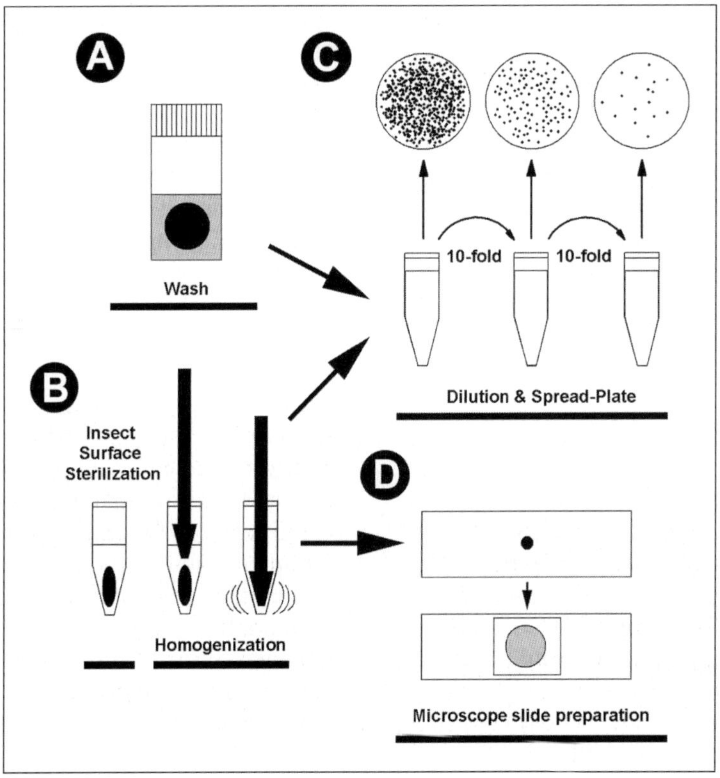

Figure 5. Methods commonly used to assess targeting of and/or infection by entomopathogens. (A) A cadaver or inanimate object (*e.g.*, a round coverslip placed in a field plot just before application) is placed in an appropriate liquid (*e.g.*, phosphate buffer) with or without a surfactant added, and the entomopathogen is physically removed from the surface of the object by vigorous washing typically using a rotary or reciprocal shaker, or a vortex, (B) a cadaver (depending on the microorganism and the objective of the procedure, the insect may be surface-sanitized to remove viable cells on the integument) is homogenized in an appropriate liquid using a sterile pestle (in this case, a polypropylene pestle sized to homogenize an insect within a microcentrifuge tube), (C) for "culturable" entomopathogens, the wash solution or homogenate is serially diluted, and aliquots of each dilution are spread onto an appropriate medium, the medium incubated under appropriate conditions and length of time, and colonies typically counted at the dilution yielding 30 to 300 colonies per dish, and (D) the homogenate or wash solution is examined microscopically

and caution must be exercised in selecting a surfactant. A number of homogenization devices can be used to recover propagules from the surfaces of insects, leaves, and coverslips (see Chapter IV-4) but recovery of propagules using homogenates from intact insects can provide erroneous results since propagules may originate from within the hemocoel or the alimentary canal. The number of propagules is usually recovered as colony-forming units (CFU) on semiselective or general purpose media using the spread-plate method (Figure 6A), but the most probable number (MPN) technique can also be used. For details on the use and limitations of the spread-plate and MPN methods for isolating entomopathogenic fungi, the reader is referred to Goettel and Inglis (1997) and to Chapter IV-4. The density of fungal propagules on glass or plastic coverslips (placed at various positions before application with pins or double-sided sticky tape) can also be measured directly using a light microscope (see Chapter IV-4). Another option is to prepare slides of washed or homogenized substrates (Figure 5) and examine propagule density by microscopy. In some instances, it is desirable to examine spatial distribution and density of entomopathogens directly on a particular substrate, and scanning electron microscopy can be used for this purpose (Figure 6C). To avoid the potential removal of poorly adhered propagules (*i.e.*, in liquid

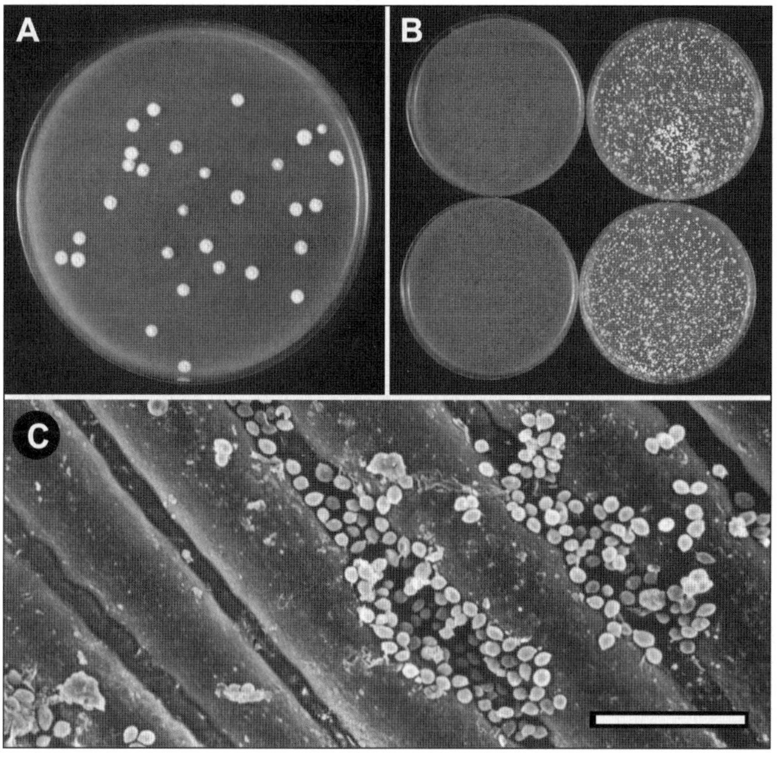

Figure 6. Assessing targeting of entomopathogens. For some taxa of fungi (*e.g.*, *Beauveria bassiana* and *Metarhizium anisopliae* var. *acridum*) targeting can be assessed quantitatively and/or qualitatively using semi-selective microbiological media, but targeting of unculturable entomopathogens must be assessed using microscopic and/or molecular methods. (A) Densities of *B. bassiana* conidia deposited on vegetation quantitatively determined using a dilution spread-plate method, (B) qualitative assessments of *B. bassiana* conidia applied in a grassland ecosystem. For qualitative assessments, plates containing a semi-selective medium (lid removed) were placed at soil level at defined locations within a field plot before application of the fungus (note: the two dishes on the left were placed in control treatment plots, whereas the two dishes on the right were placed in plots inoculated with *B. bassiana* conidia), and (C) scanning electron micrograph of field-applied *B. bassiana* conidia on a grass leaf (bar = 20 μm)

during fixation for electron microscopy), osmium tetroxide vapor as a fixative can be used.

A simple way to qualitatively assess targeting of culturable entomopathogens is to place Petri dishes containing an appropriate semiselective medium at soil level at defined locations within field plots before application. The dishes are recovered, placed in an appropriate environment, and growth of the applied entomopathogen assessed (Figure 6B).

3 Bioassays

The most frequently used method to assess targeting of acridoids involves their capture in sweepnets (see section 3A) and subsequent maintenance in enclosures in either controlled or field environments. In selecting an environment for maintaining individuals, a number of factors such as the type of enclosure and the physiological status of the insect (*e.g.*, nutrition, water, stress) should be considered. When the primary goal is to quantify targeting (in contrast to predicting field efficacy), a cage environment that is conducive for disease development is frequently selected. The cages are either designed for solitary or large numbers of individuals (Figure 7A–D). The primary advantages of housing grasshoppers individually are that the possibility of horizontal transmission is eliminated, mortality is easily monitored, and the cadavers will not be cannibalized. The main advantages to maintaining grasshoppers communally are that the feeding process is expedited and more individuals can be processed. When selecting a diet substrate, it

Figure 7. Four cages used to study entomopathogens infecting acridoids. (A) A plastic rearing cup for solitary confinement, (B) a plexiglass cage fitted with a wire screen bottom (*i.e.*, to reduce contact with frass) for the incubation of moderate numbers of individuals (*e.g.*, 20–30) in a controlled environment chamber, (C) a cage fitted with wire mesh sides and elevated screen bottom (*i.e.*, to allow frass to be eliminated from the cage) used for the incubation of large numbers of insects, typically in a greenhouse or rearing facility, and (D) a cage fitted with wire mesh sides and bottom for maintaining acridoids in a field environment

should be free of entomopathogenic propagules. During the incubation period, cadavers should be removed daily and living individuals examined for evidence of infection such as malaise. Mortality should be compared relative to the control treatment (*i.e.*, individuals captured in an identical manner but from non-pathogen-treated plots), and infection confirmed (see section 3C). To quantify the degree of targeting, linear dose-mortality responses of field-collected individuals from control plots or adjacent non-treated areas can be inoculated with known quantities of propagules and maintained under the same conditions as those individuals collected from treated plots.

Targeting of non-insect substrates can also be assessed using bioassay methods. Samples (*e.g.*, foliage or soil) are collected and presented to a susceptible host (usually but necessarily an acridoid). The substrates can be processed to extract either the propagules or the substrates presented directly to the hosts and mortality monitored.

4 *Molecular methods*

Molecular techniques have not been extensively employed to quantify targeting and infection of acridoids with entomopathogens. The primary techniques used involve immunological and nucleic acid methodologies. Monoclonal antibodies were used to detect *P. locustae* in grasshoppers (Knoblett and Youssef, 1996; Keohane *et al.*, 2001), but Streett and McGuire (1988) found that monoclonal antibodies were not sensitive enough to conclusively identify MSEV-infected individuals in homogenates of field-collected grasshoppers using enzyme linked immunosorbent assay. Diagnostic polymerase chain reaction (PCR) has revolutionized detection and quantification of animal pathogens, although the utilization of this technology in field assessments of entomopathogens against acridoids has not been extensively applied to date. Nested PCR is often used to ensure specificity at high levels of sensitivity, but extreme care must be taken in validating primers to reduce the possibility of non-specific amplification and thus false positive results. The inclusion of an internal amplification control (to ensure adequate removal of PCR inhibitors and/or adequate amplification) is also highly recommended to avoid false negative results. Quantitative PCR using real time technology is another PCR-based method that could be used to study the fate of entomopathogens applied against acridoids in field environments, but has yet to be applied in the study of acridoid pathogens. A variety of nucleic acid genotyping methods also are available. A description of the various methods is beyond the scope of this chapter, but the reader should be aware that each method possesses strengths and weakness. Important factors to consider in choosing a method are reproducibility, sensitivity, through put, and cost (capital and operational costs). Often it is exceptionally valuable, particularly in habitats in which the entomopathogenic taxon that is being applied is endemic, to know whether the genotype that caused infection and mortality is the genotype that was applied. Genotyping technology is also particularly powerful (in combination with bioassay data) in selecting strains for development in field control programs.

3 Efficacy assessments

The nature of acridoids and their pathogens makes assessments on pathogen efficacy difficult. Pathogens are relatively slow-acting and grasshoppers and locusts are very mobile. With a few exceptions, it is usually not possible to quantify cadavers of grasshoppers or locusts killed by entomopathogens in a field environment due to rapid removal of cadavers by scavengers. As a result, it is necessary to assess the efficacy of entomopathogenic microorganisms using indirect techniques such as crop damage assessments, changes in population size and species composition, and using methods to measure infection and disease progression. While a number of entomopathogens have been shown to exhibit sublethal effects in controlled environment studies, very few studies have attempted to address this possibility in a field setting.

A Acridoid collection

The collection of acridoids is usually accomplished using sweepnets. However, very little experimental information is available on the degree of cross-contamination of individuals during capture in sweepnets and a potentially major limitation with this type of collection is that one cannot preclude the possibility of cross-infesting insects from each other or from infested vegetation. To prevent contamination among treatments, it is essential that clean sweepnets be used to collect individuals, particularly from control treatment plots. It is a good practice to use dedicated sweepnets for each treatment and to decontaminate nets after each collection time. While the collection and maintenance of acridoids individually will eliminate secondary inoculation of other individuals or from vegetation and/or soil, this is often logistically not possible in most field experiments.

The goal of sweepnet collection is to arbitrarily-collect individuals so that the sample is representative of acridoid populations within the field plot. Therefore, it is essential that the plot be sampled evenly to avoid bias due to aggregation. Extreme care should be taken while sweeping to minimize injury to acridoids. A relatively rapid but gentle side-to-side movement is desired, and the researcher should avoid any rapid downward or "whipping" movement which will increase the chance of damage due to the collected individuals striking the soil surface. In addition, the sweepnet should be frequently emptied. A good practice is to regularly transfer the collected individuals into a collection cage or directly into a sorting cage. The sorting cage should be fitted with a relatively large access port preferably with an articulating and locking door. A collection cage is usually smaller than the sorting cage, and it is often used as an intermediate between sweepnet collections and the larger sorting cages. The collection cage should be designed so that it is small and robust enough to easily transport within the field and from the field to the laboratory. An all-terrain vehicle facilitates movement throughout the field, and the collecting cage can be attached to the vehicle. The objective of the collector is to minimize heat and other stresses to the collected individuals. Therefore, avoid over crowding in the cages, do not permit prolonged exposures to direct sunlight during the day, and sort the collected individuals as quickly as possible. Not all species of acridoids can be easily maintained in captivity, and in general, species that are omnivorous are preferred.

B Cage assessments

Many assessments of entomopathogen efficacy against acridoids have been based on mortality of caged individuals. Immediately following pathogen application and at regular intervals thereafter, insects are collected and placed into field cages. Mortality is generally assessed daily and compared to insects collected from control treatment plots. Care must be taken to ensure that the caged individuals are appropriately fed uncontaminated food, and in situations where multiple species of acridoids are present, as in the great plains of North America, it is important that individuals be identified prior to confinement. This is an essential step so that appropriate nutritional requirements can be met.

Results obtained from cage assessments must be used with extreme caution with respect to extrapolation to field efficacy. Cages provide a microclimate that is very different from that of the natural environment and numerous examples have shown that mortality in cages does not necessarily reflect field efficacy (Inglis *et al.*, 1997a, b). Despite the inherent problems with cage studies to predict field efficacy, they have an essential role to play in assessing the efficacy of entomopathogens. In some instances, it is not possible to use traditional population density measurement methods (*e.g.*, gregarized locusts including adults and marching nymphal bands) and cage assessments are used in such circumstances. When the goal of the cage assessments is to predict field efficacy, the cage design is of paramount importance. Ideally the cage design should mimic field conditions as closely as possible and environmental conditions within and outside of the cages should be monitored. The cages should be large enough to minimize cage wall effects, since cage screening provides substantial shading and protection from wind.

The thermoregulatory behavior of acridoids can adversely affect entomopathogens, and cage designs should take this into consideration (see previous discussion on virulence assessments). For example, basking perches (*e.g.*, vegetation, sticks, or simulated termite mounds) could be provided within field cages. The behavior of individuals in entomopathogen-treated and control cages should be monitored on a daily basis and if possible, body temperatures of individuals exhibiting typical behavior should be recorded. The cage ceiling can have a tremendous influence on light transmission which in turn can influence the ability of caged acridoids to elevate their body temperatures. In some instances, it may be possible to design cages without a cage ceiling. For example, *Locusta migratoria* nymphs are unable to jump more than a few cm and by incorporating a climbing barrier on the mesh walls of a cage such as a plexiglass or metal strip sprayed with a teflon spray, it is possible to prevent escapes. However, in some situations it may be necessary to utilize a coarse mesh ceiling to prevent predation by birds. The utilization of cage designs that mimic field conditions has not been widely adopted, and the cage design will to a large extent be dictated by the habitat and parameters being tested. Cage assessments can also be an extremely effective tool for manipulating environmental conditions with the goal of elucidating constraints on entomopathogens (Figure 8).

Sublethal effects of entomopathogens on grasshoppers and locusts have primarily been assessed in cage environments (*e.g.*, Johnson and Pavlikova, 1986). To accomplish this, bottomless field cages are arranged in a block design. An appropriate number of individuals are placed in each cage, mortality is monitored daily, and surviving individuals are counted at the end of the experimental period. Remaining vegetation is cut and oven dried at the end of the experiment. Estimates of the quantity of vegetation removed by feeding in each cage are calculated by subtracting the dry weights of the grass remaining in the cages with

Figure 8. Cage designs used to empirically test the impact of various environmental parameters on field efficacy. (A) cages exposed to direct sunlight, shaded, and protected from the ultraviolet-B radiation, (B) a cage containing field-collected grasshoppers that was protected from UV-B radiation using a UV-B absorbing plastic film (< 355 nm), and (C) a cage containing field-collected grasshoppers that was shaded for most of day (0900 to 1700). Visible light, temperature and relative humidity were measured within and next to representative cages in each environment with a photocell, and thermistor/relative humidity monitor. In addition, precipitation, and wind speed and direction were also monitored (note the anemometer adjacent to the field cage in figure C)

acridoids from those not containing insects. To account for differences in growth of vegetation among plots, adjustments are made on the basis of growth potential determinations made the previous year. Since mortality will differ among treatments, the confounding affects of mortality must be removed. To accomplish this, per capita consumption is calculated by dividing the quantity of dry matter removed from each caged plot by the corresponding number of acridoid-days (Onsager and Hewitt, 1982). Instead of measuring food consumption directly, Thomas *et al.* (1997) quantified frass production in cages fitted with bottoms. Measurements of sublethal effects of entomopathogens on non-caged acridoids will require long-term monitoring of populations, feeding, fecundity, and crop damage assessments in large-scale field trials.

C *Infection assessments*

Although frequently neglected, qualitative and quantitative assessments of infection can provide valuable information, particularly when there are no detectable changes in the populations or measurable effects on other parameters, despite excellent targeting. Accurate diagnosis of diseases requires considerable experience and expertise, particularly when the diagnostician is dealing with infectious agents that have not been applied in field trials. Furthermore, it should not be overlooked that acridoids may be co-infected with pathogens which may have implications on field efficacy (Thomas *et al.*, 2003). For detailed descriptions of methods used in disease diagnoses, the reader should consult general references dedicated to this topic (*e.g.*, Steinhaus, 1963; Poinar and Thomas, 1978).

1 *Gross pathology*

The reader is referred to Albrecht (1953) for a reference on the dissection and anatomy of acridoids. During dissection, abnormalities in tissue appearance should be noted. For example, the Malpighian tubules of grasshoppers infected with some protozoa (*e.g.*, *Malamoeba locustae* and *Perezia dichroplusae*) exhibit conspicuous hypertrophy, and the fat bodies in grasshoppers infected with MSEV and *P. locustae* can also be hypertrophied. The salivary glands and gonads of individuals infected with microsporidia may also appear hypertrophied, darker and more granular than normal glands. In some instances, it may be possible to observe the presence of large numbers of resting structures such as chlamydospores of the fungus, *Synglioeladium*, or spores of *P. locustae* and *E. grylli*. The limitation of using gross pathology examinations is that in the early stages of disease, symptoms or signs are quite often poorly manifested.

For entomopathogenic fungi that grow saprotrophically (*e.g.*, *B. bassiana* and *M. anisopliae* var. *acridum*), it is a common practice to place cadavers in a high humidity environment soon after death. The fungus is then allowed to completely colonize the cadaver and produce external hyphae and conidiogenesis cells, and this is often used as an indicator of death by mycosis (Figure 9). High humidity environments can be achieved using very simple (*e.g.*, moistened filter paper or in Petri dishes with saturated cotton) or more sophisticated (*e.g.*, saturated salt solutions) methods.

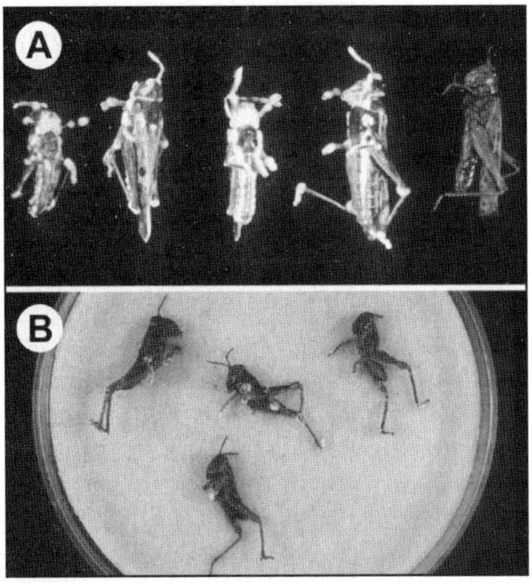

Figure 9. To assess infection by mitosporic fungi, cadavers are placed in a moist environment at a moderate temperature typically for 2 to 7 days, and characteristic hyphal growth erupting through intersegmental folds in the integument are observed microscopically (*i.e.*, for conidiogenesis) and/or macroscopically. (A) Four grasshoppers (on the left) colonized by *Beauveria bassiana* as evidenced by the characteristic emergence of white mycelial mats (note: the individual on the right did not produce the characteristic growth of *B. bassiana*, was colonized by bacteria and was classified not to have died due to mycosis), and (B) field-collected grasshopper nymphs placed on moistened filter paper within a Petri-dish (*i.e.*, a method commonly used to obtain a high-humidity environment) for 2 days (note the white hyphae erupting from the legs)

2 *Microscopic assessments*

In many instances, whole tissues are excised and, if small enough, mounted directly on a microscope in an appropriate liquid. If the tissues are large, they can either be teased apart, thick sectioned with a hand held-blade, or homogenized using a small tissue grinder such as sterile Kontes micropestle (Figure 5). The tissue pieces or homogenate are then mounted for light microscopy. Homogenates are examined at an appropriate power using phase-contrast, Nomarski, or bright field (usually in combination with tissue stains) for hyphal bodies of fungi, OB of MSEV, nematodes, spores of microsporidia or *E. grylli*, chlamydospores of *Synglioeladium*, and/or specific types of bacteria such as spore-forming rods. The reader is referred to Becnel (1997) for additional information on embedding and staining of insect tissues for detection of entomopathogens. For smaller microorganisms such as occluded and non-occluded viruses and rickettsiae, it may be necessary to embed tissues for subsequent observation with a transmission electron microscope (Becnel, 1997).

3 *Microbiological assessments*

Prior to isolation of microorganisms, acridoids are frequently surface-sterilized using chemicals such as formaldehyde (10 to 40%), sodium hypochlorite (1 to 5%), ethanol (70%), used alone or in combination (Lacey, 1997) (Figure 5). A commonly used method for surface sterilization involves the submersion of the substrate in 70% ethanol for 1 min, then in a 1% solution of sodium hypochlorite for 1 min, followed by two to three rinses in sterile water. The use of surface sterilization methods does not preclude the isolation of entomopathogens from the surface of the integument (methods are rarely 100% efficacious) or from the alimentary canal. The only way to ensure that propagules are within the hemocoel is to aseptically remove hemolymph or to remove the intact alimentary canal prior to homogenization. Extreme caution must be taken to avoid rupture of the gut during dissection and removal.

Aseptically excised tissues may be placed directly on an appropriate medium, and bacterial and fungal growth from the tissues is facilitated. The microorganism can then be transferred to another medium and standard purification protocols conducted. In some instances, bacteria or fungi may be isolated directly from hemolymph, although hemolymph can usually only be collected from individuals that have recently died due to rapid drying of the cadavers in the field. Once the hemolymph is removed, it can be placed (*e.g.*, drops) onto an appropriate medium and bacteria or fungi isolated. To quantify entomopathogenic bacteria and fungi, collected individuals are surface-sterilized, homogenized, the homogenate diluted and plated on a general purpose or semi-selective medium, and the number of CFU/unit weight or individual are quantified (Goettel and Inglis, 1997).

It is also possible to isolate propagules of a number of entomopathogens using filtration and/or centrifugation relying on either differential size or density of the entomopathogens. While filtration removes materials by size, centrifugation separates particles on the basis of density. Centrifugation is the primary technique used to isolate many entomopathogenic microorganisms of grasshoppers including *E. grylli*, *P. locustae*, MSEV and *Syngliocladium*.

D *Quantification of field populations*

1 *Sampling frames*

The most commonly used method to measure population densities of non-gregarized acridoids relies on counts of individuals in permanent quadrats. The "quadrats" are usually constructed of metal or plastic, are square or circular in shape, are generally 0.25 to 1 m^2 in area, and they are usually positioned along predefined transects within the plot (Figure 10). For most grasshopper species and locusts in a solitary phase (*i.e.*, non-aggregated), the sampling frame method provides a relatively unbiased estimate of population density of both active and lethargic or moribund insects. Counts are made before and at various times following the application. In addition to aggregation considerations, the accuracy of frame counts is influenced by the observer, time of day and climatic conditions (*e.g.*, temperature and cloud cover), vegetation, and development stage of the acridoids. In instances where it is necessary to use multiple data collectors (*i.e.*, with large-scale field trials), individual researchers should be assigned to do counts in one block for each treatment within a specific time period in an attempt to reduce bias. To obtain counts, the observer should move along each transect and identify each sampling frame as counts are made. The observer should move towards the sampling frame at a moderate pace, and as they approach the frame, the insects that jump from the quadrat are recorded. The utilization of a hand counter is recommended. When the observer reaches the frame, they should stand over the frame and count any remaining individuals; some researchers use a sweeping motion of their foot over the sampling frame to stimulate individuals to jump. When insects are moribund as is typically the case at advanced stages of infection, more detailed observations will be necessary to detect cryptic individuals (*e.g.*, parting of vegetation and observations of individuals within the vegetation and on the soil surface). In instances where one species or a few species of acridoids are predominant, numbers of individuals for that species should be recorded.

Although the fixed sampling frame method is preferred by many researchers, a variety of related methods have been used. Examples of other methods include: transect counts; cage counts; mean distance between individual counts; and imaginary quadrat counts. For transect counts, the observer travels (by foot or by vehicle) along a transect of defined length and counts the number of acridoids that jump or fly from the transect. In some instances, the vegetation is beat with a stick to stimulate insect movement when traveling by foot. When traveling by vehicle, a constant speed is maintained and acridoids jumping or flying in front of the vehicle are recorded. While this method allows the researcher to rapidly evaluate relatively large areas, it is much more subject to bias from the observer, vegetation, and climatic conditions, than the fixed quadrat method. The cage method is most commonly used when the population consists primarily of

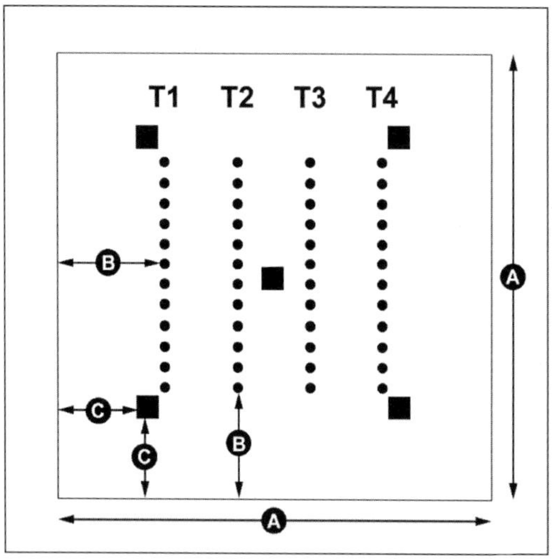

Figure 10. A typical field plot design used to test the efficacy of entomopathogens against acridoids. The plot is 200 m (A) by 200 m (A) in size (4 ha). Sampling frames (circles) are arranged in four transects (T1-T4) of 12 frames per transect, and are situated 50 m (border area) in from the edge of the plot boundary (B). One central and four peripheral (squares) subplots are used for coverage and propagule persistence evaluations. The peripheral subplots are located 40 m from the edge of the plot boundary (C)

nymphs, and is often used in conjunction with fixed quadrats. As the observer approaches the sampling frame, they quickly place a bottomless cage over the sampling site (Smith and Stewart, 1946). The number of individuals in the cage is then counted. Quantification of the mean distance between individuals is a very rapid method but is not generally used to assess the efficacy of entomopathogens. The imaginary quadrat method is very similar to the fixed quadrat method with the exception that the sampling area is estimated by the observer.

2 *Sweepnet sampling*

Quantitative sweepnet sampling is almost always accompanied by other methods of population density assessments such as the sampling frame method, and is usually conducted with the same sampling schedule. Consistency in sweepnet sampling is of paramount importance. A defined number of sweeps at a defined sweep angle (*e.g.*, 180°) and height (*e.g.*, 5–15 cm through vegetation), at a specific time (to control environmental conditions as best as possible) are used. Ideally the sweeps are made along a predefined transect or grid to reduce bias from factors such as differences in vegetation, but often sweeps are conducted at arbitrarily selected locations throughout the plot. The number of sweeps required per plot will depend on factors such as the size of the plot, the diversity in habitat, the diversity and population density of acridoids, but most researchers complete a minimum of 50 sweeps per ha. Following collection, individuals are usually bulked in transfer cages, transported to the laboratory, frozen, and the numbers of individuals recorded.

Sweepnet counts have been used relative to sampling frame counts as a measure of grasshopper health (Johnson *et al.*, 1986). Grasshoppers at various stages of disease development are less active and often become cryptic relative to healthy individuals and therefore they are less prone to be captured. To determine whether an entomopathogen influences sweepnet sampling efficiency, total sweep catches are adjusted against acridoid density by plot, date, and treatment. Concomitant sweepnet data and sampling frame information can be used to create contingency tables that can illuminate changes in activity or behavior of a species post-treatment.

3 *Species composition and age structure*

For all field trials, it is extremely important to accurately identify temporal shifts in the immature and adult members of the species present. If possible, this should be accomplished using more than one sampling method to avoid sampling bias against immature individuals with sweepnets. Following collection, captured individuals are often transported to the laboratory and killed by freezing. Proper taxonomic determination is the first step, and the experimenter should ensure that descriptions of the immature stages are available. A common approach is to use a method to view dispersion about the mean such as Taylor's power law (Taylor, 1961, 1971) to evaluate the benefit of further sampling. When increasing sample size has no further benefit, then the accuracy of the number reported will be maximal. In many cases, early and late treatments are compared using orthogonal contrasts and the measure of the percent of individuals in any species in a particular stage must be repeatable for the species of interest. This allows for good assessment of the field activity of the agent against different developmental stages.

4 Quantification of gregarized nymphs

Many economically-important species of acridoids undergo gregarization induced by crowding (Figure 1B). Accurate quantification of gregarized acridoids, such as the migratory locust, is a major obstacle facing researchers assessing efficacy of entomopathogens. This is particularly difficult because bands are dynamic (*i.e.*, bands in proximity to each other typically coalesce and divide). Therefore, extreme care must be exercised in treating bands as independent units of measure. A number of researchers have described methods of measuring population densities of acridoids in bands. All methods are based on estimating the area of the band. The following is the method used by Hunter *et al.* (1999) to estimate population densities of locust bands in Australia. Once located, each band is marked and followed as often as possible (*ca.* at 1-day intervals). Band area is estimated from the length and width of the band. To measure band length, flags are placed every 30–40 m along its length, and the length of the band is recorded. To measure band width, width is estimated by walking at right angles to the main band front at each of the flags. The area of acridoids at the following densities is calculated: area at dense band (1000–2500 nymphs/m^2); area at band (30–100 nymphs/m^2); area at sub-band (30–100 nymphs/m^2); and area at numerous transects (5–30 nymphs/m^2). From the area of locusts in each density category and the means of each category, total numbers of locusts in each band are estimated. The accuracy of this method in estimating population densities requires experience in assessing densities by eye and also depends on the behavior of the acridoid species being evaluated. For example, the migratory locust forms discrete swarms amenable to the above method. It is often necessary to determine densities of individuals in satellite bands that are in proximity to main bands, and add these counts to that of the main band. Data are then typically adjusted to percentage changes in densities (adjusted for control treatments) over time.

5 Statistical methods

A number of statistical designs have been used to analyze acridoid population density data. A split-plot design is the most common. With this design, treatment effects are tested with experimental error (block by treatment interaction), and experimental and sampling error are evaluated with the split-plot error term. Since sampling frame positions are fixed, analysis of covariance is often applied with the pre-treatment observation data used as the concomitant variable. Data for each post-treatment sampling date is then analyzed separately. To test appropriate treatment comparisons (*e.g.*, treated versus untreated), orthogonal contrasts are often used.

In many instances the researcher is most interested in measuring the impact of an entomopathogen on an acridoid population over time (*i.e.*, the treatment by time interaction). However, acridoid populations are almost always recorded from the same experimental units over time (*i.e.*, repeated measures) due to their high mobility and the requisite for relatively large field plots. Some researchers have included time in the statistical model using split-split-plot in time (block, treatment, subplot and time are included in the model) or split-plot in time (the mean density per plot is calculated across subplots, and therefore, subplot is not included in the model) designs. To avoid committing a Type I error (see Chapter II-1), researchers have applied conservative corrections, such as the Box correction where the degrees of freedom for time, the time by treatment interaction and the split-plot error are reduced by time-1 (Milliken and Johnson, 1984). In other instances, regression or non-linear curve fitting methods have been applied to temporal acridoid population data. For example, the mean number of individuals per unit area in each plot for each sample period is calculated (unadjusted or adjusted for control mortality using Abbott's formula) and fit to linear models using an appropriate transformation (see Chapter II-1). Analysis of covariance (*i.e.*, heterogeneity of slopes) is then used to test for differences in regression relationships (intercepts and slopes) among treatments. Logistic regression or probit analysis can be applied to correlated data (Throne *et al.*, 1995a) and Throne *et al.* (1995b) discuss the assessment of fitted lines for these specialized forms of regression.

Advancements in mixed model analysis such as those utilized in the MIXED procedure of SAS (Littell *et al.*, 2006; SAS Institute Inc, 2005) are facilitating analysis of repeated measure data, and will likely become the preferred method for analyzing temporal changes in acridoid populations. The MIXED procedure handles variance heterogeneity among treatments and correlations among the data, and it assumes both a normal distribution and a linear model. The non-linear counterparts to the MIXED procedure are the NLMIXED procedure and the NLINMIX macro (SAS Institute Inc, 2005). NLINMIX can be used to analyze non-linear data that is correlated but has normally distributed errors, while the NLMIXED procedure cannot handle correlated data but does not require the data to follow a normal distribution. The GLIMMIX procedure can also be used to handle correlated data that can have any distribution in the exponential family (*e.g.*, binary, binomial, Poisson, beta, gamma, normal, and chi-square). This procedure is new and is currently provided as an add-on to SAS 9.1; it can be downloaded from http://support.sas.com/rnd/app/da/glimmix.html. In addition to handling non-normally distributed data, the GLIMMIX procedure possesses all the functionality of the MIXED procedure plus it has many other useful features.

Many new approaches are available in the study of spatial patterns in insect distributions, using geostatistical approaches in applied insect ecology (Liebhold *et al.*, 1993) and pest management (Brenner *et al.*, 1998). An interesting possibility is to use large plots for field evaluations with a smaller central subplot being intensively grid-sampled to provide a spatial representation of infectivity and mortality over time. The same approach could be used on the controls and resulting plots of density collected over time to describe the movement of infected and uninfected individuals. These approaches which address spatial continuity using variography, may very well help to define and measure what had previously been confounding factors, such as migration, dilution, and clumping of treated individuals in the total population. For direct comparisons of population associations with spatially explicit tests for statistical comparison, the freeware program SADIE (Spatial Analysis of Distance Indices) provides a useful template for assessing pre- and post-treatment effects spatially (Perry, 1995; Perry *et al.*, 1996) and can be utilized to spatially characterize species associations (Weaver *et al.*, 2005).

E Crop damage assessments

The ultimate goal of acridoid management is to prevent economic loss, even though microbial control agents are often deployed against a refugial population which may only impact human food supplies at a later time.

1 Defoliation indices

There are many types of defoliation indices that can be used to indirectly measure the impact of slow acting or sublethal infections. These can be devised relatively easily for any crop, and are most commonly used in assessment of antifeedant properties of botanical insecticides (Passerini and Hill, 1993). One approach is to assign numerical ranks to increasing levels of leaf consumption in collected subsamples of leaves from a group of plants in a sample subplot. For example, leaves consumed 0 to 20% are given a rank of one, while leaves consumed 80 to 100% are given a rank of five. Mean separation after analysis of variance can give some idea of immediate treatment effects. An interesting possibility is to combine such indices with intensive grid-samples to view spatial continuity of defoliation changes in the area of interest using geostatistical approaches. An example of acridoid consumption indices for seed heads of pearl millet are presented in Kogo and Krall (1997).

2 Yield

In terms of true economic benefit, it is important to consider actual yield losses. Again, this approach is not often used given the pesticide paradigm under which grasshopper and locust management strategies are evaluated. Yield measurement for large areas can be accomplished by using conventional harvest equipment or manpower by adequate subsampling of both the treated area and the control. When manpower is limited, sample subplots can be used to collect representative samples at harvest. A transect or grid system can be used, with the advantage of the grid being the ability to spatially evaluate yield over the entire treated and control area. An example would be the random collection of 20 seed heads of wheat in a meter of planting row for each of 100 sample points at equally spaced 25-m intervals (10 × 10 grid) in the center of a 10 ha block. After threshing, mean values can be calculated for the treatment(s) and controls, and spatial analysis can be conducted using the individual data entries. Yield measurement should be included as part of the evaluation of any control agent, but needs to be viewed with caution with acridoid pests. The mobility of these insects can cause serious problems in relation to crop losses in the control treatment. The control plots may be missed by a moving band of locusts, or individuals can emigrate thereby resulting in an inaccurate measure of crop damage. Recovery of partially denuded crops is also critical, but difficult to reconcile, because it may make the uncertainty of the value of treatment even more confounding. The long-term population suppression benefits caused by significant mortality may also be difficult to evaluate in a situation where only a small potential yield loss was averted.

3 Remote sensing

The use of remote sensing and forecast modeling has been a significant part of developing strategies for control of locusts. The scale and resolution of early remote sensing strategies was sufficient for large geographical areas, but consistent technological breakthroughs have increased resolution to the order of tens of meters or less. Multispectral satellite images (Voss and Dreiser, 1997) or aerial hyperspectral images can be used to make detailed maps of vegetation for breeding habitats, and can also provide repeated imagery that can be used to visualize defoliation in the path of marching bands. At present, the technology is relatively expensive (requiring a ground truth survey), and the resolution of relative defoliation is still in its infancy, requiring sophisticated correspondence and correlation statistical techniques. However, as these meter scale approaches become more common and technological costs decrease, it will be possible to use these technologies to monitor pathogen treated acridoid bands and view their movement, subsequent decrease in feeding and ultimate decrease in density both spatially and temporally. This will ultimately be the best approach for evaluation of candidate entomopathogens in the field.

F Monitoring environmental parameters

As in any field trial, it is extremely important to monitor as many environmental parameters as possible during pathogen application and the subsequent assessment period. This information is essential for identifying environmental constraints of entomopathogens of acridoids. Parameters for consideration include temperature, humidity, precipitation, solar radiation including the UV-B portion of the solar spectrum (280–320 nm), and wind. In many instances it is desirable to monitor microclimatic as well as macroclimatic conditions (*e.g.*, conditions within crop canopies or on the surface of basking surfaces). A number of devices can be used to measure environmental parameters. Combination probes can be used to record temperature and relative humidity (vapor pressure), but a radiation shield should be used when the probe is exposed to direct sunlight. For microclimatic measurements of temperature such as acridoid body temperatures, wire thermistors (*e.g.*, 0.125-mm-diam copper constant thermocouple) connected to a data logger or digital thermometer are commonly used. A number of devices are available for electronic measurements of precipitation (*e.g.*, tipping bucket rain gages) but simple rain gages may suffice. Measurements of global solar radiation are primarily obtained using a pyranometer (calibrated for daylight spectrum) but quantum sensors are used to measure photosynthetic wavelengths within and outside of plant canopies. For the UV-B range, radiometers are used. Wind speed is measured using anemometers that are ideally positioned at increasing distances above the ground surface. For detailed information on measurements of microclimatic conditions, the reader is referred to Rosenberg (1974) and Unwin (1981).

4 Conclusion

Despite the control potential of entomopathogens, acridoid mobility and in some instances, gregarious behavior, present serious obstacles for evaluating entomopathogens of grasshoppers and locusts. Furthermore, entomopathogens are often relatively slow acting or their effects are sublethal, and the application of a chemical paradigm for their evaluation is inadequate. A major challenge facing researchers is the development and implementation of novel methods to evaluate microbial control agents of acridoids under field conditions. It is also important that field experiments be designed to elucidate environmental constraints on entomopathogens with the goal of developing strategies to overcome them and thereby provide an efficacious alternative to chemical insecticides for the management of grasshopper and locust pests.

5 References

Albrecht, F. O. 1953. "The Anatomy of the Migratory Locust." The Althone Press, London.

Bateman, R. 1997. Methods of application of microbial pesticide formulations for the control of grasshoppers and locusts. *In* "Microbial Control of Grasshoppers and Locusts" (M. S. Goettel and D. L. Johnson, Eds.), pp. 69–81. *Mem. Entomol. Soc. Canada*, Vol. 171.

Becnel, J. J. 1997. Complementary techniques: preparations of entomopathogens and diseased specimens for more detailed study using microscopy. *In* "A Manual of Techniques in Insect Pathology" (L. A. Lacey, Ed.), pp. 337–353. Academic Press, London.

Blanford, S., Thomas, M. B. and Langewald, J. 2000. Thermal ecology of *Zonocerus variegatus* and its effects on biocontrol using pathogens. *Agric. For. Entomol.* **2**, 3–10.

Boorstein, S. M. and Ewald, P. W. 1987. Costs and benefits of behavioral fever in *Melanoplus sanguinipes* infected by *Nosema acridophagus. Physiol. Zool.* **60**, 586–595.

Brenner, R. J., Focks, D. A., Arbogast, R. T., Weaver, D. K., and Shuman, D. 1998. Practical use of spatial analysis in precision targeting for integrated pest management. *Am. Entomol.* **44**, 79–101.

Burges, H. D. and Jones, K. A. 1998. Formulation of bacteria, viruses and protozoa to control insects. *In* "Formulation of Microbial Biopesticides" (H. D. Burges, Ed.), pp 33–127. Kluwer Academic Publishers, Dordrecht, The Netherlands.

Butt, T. M. and Goettel, M. S. 2000. Bioassays of entomogenous fungi. *In* "Bioassays of Entomopathogenic Microbes and Nematodes" (A. Navon and K. R. S. Ascher, Eds.), pp. 141–195. CABI International Press, Wallingford, UK.

Capinera, J. L. and Hibbard, B. E. 1987. Bait formulations of chemical and microbial insecticides for suppression of crop-feeding grasshoppers. *J. Agric. Entomol.* **4**, 337–344.

Carruthers, R. I., Larkin, T. S., and Firstencel, H. 1992. Influence of thermal ecology on the mycosis of a rangeland grasshopper. *Ecology* **73**, 190–204.

Cilgi, T. and Jepson, P. C. 1992. The use of tracers to estimate the exposure of beneficial insects to direct pesticide spraying in cereals. *Ann. Appl. Biol.* **121**, 239–247.

Goettel, M. S. and Inglis, G. D. 1997. Fungi: Hyphomycetes. *In* "A Manual of Techniques in Insect Pathology" (L. A. Lacey, Ed.), pp. 213–249. Academic Press, London.

Henry, J. E. 1985. Effect of grasshopper species, cage density, light intensity, and method of inoculation on mass production of *Nosema locustae* (Microsporidia: Nosematidae). *J. Econ. Entomol.* **78**, 1245–1250.

Henry, J. E. and Oma, E. A. 1974. Effect of prolonged storage of spores on field applications of *Nosema locustae* (Microsporida: Nosematidae) against grasshoppers. *J. Invertebr. Pathol.* **23**, 371–377.

Hirt, R. P., Logsdon, J. M., Healy, B., Dorey, M. W., Doolittle, W. F., and Embley, T. M. 1999. Microsporidia are related to fungi: evidence from the largest subunit of RNA polymerase II and other proteins. *Proc. Natl. Acad. Sci. USA* 96, 580–585.

Hunter, D. M., Milner, R. J., Scanlan, J. C., and Spurgin, P. A. 1999. Aerial treatment of the migratory locust, *Locusta migratoria* (L.) (Orthoptera: Acrididae) with *Metarhizium anisopliae* (Deuteromycotina: Hyphomycetes) in Australia. *Crop Prot.* **18**, 699–704.

Hurlbert, S. H. 1984. Pseudoreplication and the design of ecological field experiments. *Ecol. Monogr.* **54**, 187–211.

Inglis, G. D., Goettel, M. S., and Johnson, D. L. 1995. Influence of ultraviolet light protectants on persistence of the entomopathogenic fungus, *Beauveria bassiana. Biol. Control* **5**, 581–590.

Inglis G. D., Johnson, D. L. and Goettel, M. S. 1996a. Effects of temperature and thermoregulation on mycosis by *Beauveria bassiana* in grasshoppers. *Biol. Control* **7**, 131–139.

Inglis, G. D., Johnson, D. L., and Goettel, M. S. 1996b. Effect of bait substrate and formulation on infection of grasshopper nymphs by *Beauveria bassiana. Biocontrol Sci. Technol.* **6**, 35–50.

Inglis, G. D., Johnson, D. L., and Goettel, M. S. 1997a. Effects of temperature and sunlight on mycosis (*Beauveria bassiana*) (Hyphomycetes: Sympodulosporae) of grasshoppers under field conditions. *Environ. Entomol.* **26**, 400–409.

Inglis, G. D., Johnson, D. L., and Goettel, M. S. 1997b. Field and laboratory evaluation of two conidial batches of *Beauveria bassiana* against grasshoppers. *Can. Entomol.* **129**, 1–16.

Jenkins, N. E., Heviefo, G., Langewald, J., Cherry, A. J., and Lomer, C. 1998. Development of mass production technology for aerial conidia for use as mycopesticides. *Biocontrol News Info.* **19**, 21N–31N.

Johnson, D. L. 1992. Introduction: biology, ecology, field experimentation and environmental impact. *In* "Biological Control of Locusts and Grasshoppers" (C. J. Lomer and C. Prior, Eds.), pp 267–278. CAB International, Wallingford, UK.

Johnson, D. L. and Goettel, M. S. 1993. Reduction of grasshopper populations following field application of the fungus *Beauveria bassiana*. *Biocontrol Sci. Technol.* **1**, 165–175.

Johnson, D. L., Hill, B. D., Hinks, C. F., and Schaalje, G. B. 1986. Aerial application of the pyrethroid deltamethrin for grasshopper (Orthoptera: Acrididae) control. *J. Econ. Entomol.*, **79**, 181–188.

Johnson, D. L. and Pavlikova, E. 1986. Reduction of consumption by grasshoppers (Orthoptera: Acrididae) infected with *Nosema locustae* Canning (Microsporida: Nosematidae). *J. Invertebr. Pathol.* **48**, 232–238.

Kaya, H. K. and Stock, S. P. 1997. Techniques in insect nematology. *In* "Manual of Techniques in Insect Pathology" (L. A. Lacey, Ed.), pp. 281–324. Academic Press, London.

Keohane, E. M., Takvorian, P. M., Cali, A., Tanowitz, H. B., Wittner, M. and Weiss, L. M. 2001. Monoclonal antibodies to cytoplasmic antigens of *Nosema locustae* (Microsporida: Nosematidae). *J. Invertebr. Pathol.* **77**, 81–82.

Knoblett, J. N. and Yousseff, N. N. 1996. Detection of *Nosema locustae* (Microsporidia; Nosematidae) in frozen grasshoppers (Orthoptera: Acrididae) by using monoclonal antibiodies. *J. Econ. Entomol.* **89**, 841–847.

Kogo, S. A. and Krall, S. 1997. Yield losses on pearl millet panicles due to grasshoppers: a new assessment method. *In* "New Strategies in Locust Control" (S. Krall, R. Peveling and D. Ba Diallo, Eds.), pp. 415–423. Birkhäuser, Basel, Switzerland.

Lacey, L. A. 1997. "Manual of Techniques in Insect Pathology." Academic Press, London.

Lacey, L. A. and Brooks, W. M. 1997. Initial handling and diagnosis of diseased insects. *In* "Manual of Techniques in Insect Pathology" (L. A. Lacey, Ed.), pp. 1–16. Academic Press, London.

Liebhold, A. M., Rossi, R. R., and Kemp, W. P. 1993. Geostatistics and geographic information systems in applied insect ecology. *Annu. Rev. Entomol.* **38**, 303–327.

Littell, Ramon C., George A Milliken, Walter W. Stroup, Russell D. Wolfinger, and Oliver Schabenberger. 2006. "SAS for Mixed Models," Second Edition. SAS Institute Inc., Cary, NC.

Lomer, C. J., Prior, C., and Kooyman, C. 1997. Development of *Metarhizium* spp for the control of grasshoppers and locusts. *In* "Microbial Control of Grasshoppers and Locusts" (M. S. Goettel and D. L. Johnson, Eds.), pp. 265–286. *Mem. Entomol. Soc. Canada.*, Volume171.

Matthews, G. A. and Thornhill, E. W. 1995. "Pesticide Application Equipment for Use in Agriculture. Vol. 1. Manually Carried Equipment". FAO, Rome.

McGuire, M. R., and Henry, J. E. 1989. Production and partial characterization of monoclonal antibodies for detection of entomopoxvirus from *Melanoplus sanguinipes. Entomol. Exp. Appl.* **51**, 21–28.

McGuire, M. R., Streett, D. A., and Shasha, B. S. 1991. Evaluation of starch-encapsulation for formulation of grasshopper (Orthoptera: Acrididae) entomopoxviruses. *J. Econ. Entomol.* **84**, 1652–1656.

Menely, J. C. and Sluss, T. P. 1988. Development of "NOLO Bait" (*Nosema locustae*) for the control of grasshoppers and locusts. Proc. Brighton Crop Prot. Conf., Pest Dis., pp. 597–602.

Milliken, G. A. and Johnson, D. E. 1984. "Analysis of Messy Data. Volume 1: Designed Experiments." Van Nostrand Reinhold Company, New York.

Moore, D. and Caudwell, R. W. 1997. Formulation of entomopathogens for the control of grasshoppers and locusts. *In* "Microbial Control of Grasshoppers and Locusts" (M. S. Goettel and D. L. Johnson, Eds.), pp. 49–67. *Mem. Entomol. Soc. Canada*, Vol. 171.

Olfert, O. and Erlandson, M. A. 1991. Wheat foliage consumption by grasshoppers (Orthoptera: Acrididae) infected with *Melanoplus sanguinipes* entomopoxvirus. *Environ. Entomol.* **20**, 1720–1724.

Oma, E. A. and Streett, D. A. 1983. Production of a grasshopper entomopoxvirus (Entomopoxbirinae) in *Melanoplus sanguinipes* (F.) (Orthoptera: Acrididae). *Can. Entomol.* **125**, 1131–1133.

Onsager, J. A. 1991. Toward an optimum strategy for estimating density of grasshoppers (Orthoptera: Acrididae) from a Poisson distribution. *Environ. Entomol.* **20**, 939–944.

Onsager, J. A., Henry, J. E., Foster, R. N., and Staten, R. T. 1980. Acceptance of wheat bran bait containing insecticide, *Nosema locustae* by species of rangeland grasshoppers. *J. Econ. Entomol.* **73**, 548–551.

Onsager, J. A. and Hewitt, G. B. 1982. Rangeland grasshoppers: Average longevity and daily rate of mortality among six species in nature. *Environ. Entomol.* **11**, 127–133.

Papierok, B. and Hajek, A. E. 1997. Fungi: Entomophthorales. *In* "Manual of Techniques in Insect Pathology." (L. A. Lacey, Ed.), pp. 187–212. Academic Press, London.

Passerini, J. and Hill, S. B. 1993. Field and laboratory trials using a locally produced neem insecticide against the Sahelian grasshopper, *Kraussaria angulifera* (Orthoptera: Acrididae), on Millet in Mali. *Bull. Entomol. Res.* **83**, 121–126.

Perry, J. N. 1995. Spatial aspects of animal and plant distribution in patchy farmland habitats. *In* "Ecology and Integrated Arable Farming Systems" (D. Glen, M. Greaves, and H. M. Anderson, Eds.), pp. 221–242. Wiley, Chichester, UK.

Perry, J. N., Bell, E. D., Smith, R. H. and Woiwod, I. P. 1996. SADIE: software to measure and model spatial pattern. *Aspects Appl. Biol.* **46**, 95–102.

Poinar, G. O. and G. M. Thomas. 1978. "Diagnostic Manual for the Identification of Insect Pathogens." Plenum Press, New York.

Quinn, M. A., Kepner, R. L., Walgenbach, D. D., Bohls, R. A., Pooler, P. D., Nelson Foster, R., Reuter, K. C., and Swain, J. L. 1991. Habitat characteristics and grasshopper community dynamics on mixed-grass rangeland. *Can. Entomol.* **123**, 89–105.

Rosenberg, N. J. 1974. "Microclimate: The Biological Environment". J. Wiley and Sons, New York.

SAS Institute Inc. 2005. SAS OnlineDoc® 9.1.3. Cary, NC: SAS Institute Inc.

Shah, P. A., Douro-Kpindou, O. K., Sidibe, A., Daffe, C. O., Van Der Pauuw, H., and Lomer, C. J. 1998. The effects of the sunscreen oxybenzone on field efficacy and persistence of *Metarhizium flavoviride* conidia against *Kraussella amabile* (Orthoptera: Acrididae) in Mali, West Africa. *Biocontrol Sci. Technol.* **8**, 357–364.

Smith, R. W. and Stewart, W. W. A. 1946. A useful cage for sampling field populations of grasshoppers. *Annu. Rep. Entomol. Soc. Ont.* **76**, 32–35.

Steinhaus, E. A. 1963. Background for the diagnosis of insect diseases. *In* "Insect Pathology: An Advanced Treatise." (E. A. Steinhaus, Ed.), Vol. 2, pp. 549–589. Academic Press, NY.

Streett, D. A., Woods, S. A., and Erlandson, M. A. 1997. Entomopoxviruses of grasshoppers and locusts: Biology and biological control potential. *In* "Microbial Control of Grasshoppers and Locusts" (M. S. Goettel and D. L. Johnson, Eds.), pp. 115–130. *Mem. Entomol. Soc. Canada*. Vol. 171.

Taylor, L. R. 1961. Aggregation, variance and the mean. *Nature* **189**, 732–735.

Taylor, L. R. 1971. Aggregation as a species characteristic. *Stat. Ecol.* **1**, 357–377.

Thomas, M. B., Blanford, S., and Lomer, C. J. 1997. Reduction of feeding by the variegated grasshopper, *Zonocerus variegatus*, following infection by the fungal pathogen, *Metarhizium flavovirde. Biocontol. Sci. Technol.* **7**, 327–334.

Thomas, M. B., Watson, E. L., and Valverde-Garcia, P. 2003. Mixed infections and insect pathogen interactions. *Ecol. Lett.* **6**, 183–188.

Thompson, D. C. 1987. Sampling rangeland grasshoppers. *In* "Integrated Pest Management on Rangeland" (J. L. Capinera, Ed.), pp. 219–233. Westview Press, Boulder, CO.

Thornhill, E. W. and Matthews, G. A. 1995. "Pesticide Application Equipment for Use in Agriculture. Vol. 2. Mechanically Powered Equipment". FAO, Rome.

Throne, J. E., Weaver, D. K., Chew, V., and Baker, J. E. 1995a. Probit analysis of correlated data: multiple observations over time at one concentration. *J. Econ. Entomol.* **88**, 1510–1512.

Throne, J. E., Weaver, D. K., and Baker, J. E. 1995b. Probit analysis: assessing goodness-of-fit based on backtransformation and residuals. *J. Econ. Entomol.* **88**, 1513–1516.

Undeen, A. H. and Epsky, N. D. 1990. *In vitro* and *in vivo* germination of *Nosema locustae* (Microspora: Nosematidae) spores. *J. Invertebr. Pathol.* **56**, 372–379.

Undeen, A. H. and Vávra, J. 1997. Research methods for entomopathogenic protozoa. *In* "Manual of Techniques in Insect Pathology" (L. A. Lacey, Ed.), pp. 117–151. Academic Press, London.

Unwin, D. J. 1981. "Microclimate Measurement for Ecologists." Academic Press, New York.

Vickery, V. R. 1997. Classification of thc Orthoptera (*sensu stricto*) or Caelifera. *In* "The Bionomics of Grasshoppers, Katydids and Their Kin" (S. K. Gangwere, M. C. Muralirangan, and M. Muralirangan, Eds.), pp. 5–40. CAB International, Wallingford,UK.

Voss, F. and Dreiser, U. 1997. Mapping of desert locust habitats using remote sensing techniques. *In* "New Strategies in Locust Control" (S. Krall, R. Peveling and D. Ba Diallo, Eds.), pp. 37–45. Birkhäuser, Basel.

Weaver, D. K., Nansen, C., Runyon, J. B. Sing, S. E., and Morrill. W. L. 2005. Spatial distribution of *Cephus cinctus* Norton (Hymenoptera: Cephidae) and its braconid parasitoids in Montana wheat fields. *Biol. Control* **34**, 1–11.

Wilson, K., Thomas, M. B., Blanford, S., Doggett, M., Simpson, S. J., and Moore, S. L. 2002. Coping with crowds: density dependent disease resistance in desert locusts. *Proc. Nat. Acad. Sci. USA* **99**, 5471–5475.

Woods, S. A., Streett, D. A., and Henry, J. E. 1992. Temporal patterns of mortality from an entomopoxvirus and strategies for control of the migratory grasshopper (*Melanoplus sanguinipes* F.). *J. Invertebr. Pathol.* **60**, 33–39.

Woods, S. A. and Streett, D. A. 1996a. Assimilation of rubidium by *Melanoplus* grasshoppers (Orthoptera: Acrididae). *Environ. Entomol.* **25**, 906–911.

Woods, S. A. and Streett, D. A. 1996b. Grasshopper (Orthoptera: Acrididae) bait feeding studies using rubidium chloride. *Environ. Entomol.* **25**, 912–918.

Chapter VII-18

Lawn, turf and grassland pests

Michael G. Klein
and Parwinder S. Grewal
Department of Entomology
Ohio State University, OARDC
Wooster, OH 44691, USA

Trevor A. Jackson
AgResearch, Lincoln, New Zealand

Albrecht M. Koppenhöfer
Department of Entomology
Rutgers University
New Brunswick, NJ 08901, USA

1 Introduction

Grasses are the dominant vegetation in many environments which vary in size and composition from the great prairies to manicured golf courses, bowling greens, and lawns around our homes. Natural grasslands cover millions of hectares throughout the world, providing sustenance for vast numbers of wildlife. Grasslands, improved by sowing and managing desirable species, support livestock industries around the world. Hard-wearing grass species are used to create recreational spaces in the urban environment. Such amenity turfgrass occupies > 12 million ha in the USA alone, comprised of over 60 million lawns, 16,000 golf courses, many parks, athletic fields, cemeteries, and sod farms (Potter, 1998; Emmons, 2000). Besides its recreational uses, grass controls soil erosion, captures and cleans run-off water from urban areas, provides soil improvement and restoration, moderates temperature, reduces glare and noise, reduces pests, pollen and human disease exposure, creates good wildlife habitats, and improves physical and mental health of urban populations (Beard and Green, 1994).

Grasses become the dominate form of vegetation either through evolutionary adaptation to harsh environments or through management to provide forage for animals or spaces for urban aesthetics and recreation. While there are many differences among grassland ecosystems, there are similarities as well. Most grasslands are permanent or semi-permanent and persist through natural seeding or by spread of vegetative structures such as tillers, rhizomes, and stolons. The utility of grass comes through its ability to form a turf mat of intertwined plants forming a solid ground cover with an extensive root mass, and its ability to regenerate from the crown after defoliation. With these properties, grassland turf can provide a mass of nutritive plant material for regular grazing by browsing animals, or alternatively, a hard-wearing amenity ground cover.

2 Major turfgrass pests and their distribution

Permanent turf provides a stable habitat for many species of invertebrates, most of which feed on vegetation and detritus without causing obvious

L.A. Lacey and H.K. Kaya (eds.), Field Manual of Techniques in Invertebrate Pathology, 655–675.

damage or loss of productivity. Spectacular outbreaks of grasshoppers, armyworms, or white grubs can occur over large expanses of natural grassland, but such attacks are uncommon. More intense management of grasslands by sowing palatable species and increasing fertility has provided greater energy resources for some herbivorous species that have become key pests of forage systems. Amenity turf is under constant critical scrutiny from the public, and its high cosmetic value and low damage thresholds have led to a large number of insect species being regarded as pests. Indeed, in the USA, about 36 million people spend 3.6 billion hours on golf courses each year, and about 56 million take part in their lawn care. Between professional lawn care management and homeowner lawn care, turf maintenance has become a $45 billion per year industry. A substantial amount of this budget and time is spent on insect and mite management (Danneberger, 1993).

Insect pests of turfgrass vary in their behavior and position of feeding. White grubs, larvae of the Scarabaeidae, usually feed on the grass roots, while larvae of webworms, hepialids, and cutworms (Lepidoptera) create burrows in the soil from which they emerge at night to feed on the grass shoots. Armyworms (Lepidoptera) live on the surface, feeding on grass foliage and stems, whereas some weevils including billbug and fly larvae (Diptera) feed or bore into stems and/or crowns killing tillers or entire plants. At times, outbreaks of a particular pest will be of concern, but frequently it is the combined feeding rate of the pest complex that exceeds the growth of the grass, resulting in visible damage and the need for control. Pests of grasslands and turf have been reviewed by Delfosse (1993), Potter (1998), Vittum *et al.* (1999), and Niemczyk and Shetlar (2000).

While grasses support a wide variety of living organisms, less than 1% of these organisms acquires pest status requiring control. Major pests and the part of the plants they attack are listed in Table 1. Root feeding white grubs, stem and crown feeding weevils and stem and foliage feeding Lepidoptera are pests worldwide, but other groups have a more limited distribution.

White grubs are an extremely diverse group, with both endemic and exotic species becoming serious turf pests in different parts of the word (Jackson, 1992). Endemic masked chafers,

Table 1. Major turf pests, part of the plants they attack, geographical problem areas, and potential microbial controls

Part attacked-damage	Pests	Pest life stage	Location	Potential microbial controls
Root-chewing	White grubs	Larva	World wide	Bacteria, Nematodes, Fungi
	Mole crickets	Nymph, adult	s.e. USA	Nematodes, Fungi
Stem/crown-chewing	Annual bluegrass weevil	Larva	n.e. USA	Nematodes, Fungi
	Billbugs	Larva	USA, Japan, NZ-Australia	Nematodes, Fungi
	Crane fly	Larva	n.w./n.e. USA, Europe	Bacteria, Nematodes
Leaf/stem-chewing	Armyworms	Larva	Worldwide	Nematodes, Bacteria
	Cutworms	Larva	Worldwide	Nematodes
	Sod webworms	Larva	USA	Nematodes, Bacteria
Leaf/stem-sucking	Chinch bugs	Nymph, adult	c./e. USA, s.e. Canada, Japan,	Fungi
	Greenbug, aphids	Nymph, adult	USA	Fungi
	Mites	Nymph, adult	USA	–
	Spittlebugs	Nymph, adult	e. USA, Brazil	Fungi
	Scales	Nymph, adult	s. USA, Japan	–
	Mealybugs	Nymph, adult	s. USA, NZ	–

c.=central; e.=eastern; n.e.=northeastern; n.w.=northwestern; s.=southern; s.e.=southeastern

Cyclocephala spp., and May/June beetles, *Phyllophaga* spp., are pests in many parts of the Americas. Throughout Asia and Africa *Holotrichia* and *Heteronychus* spp. abound, whereas *Melolontha, Amphimallon*, and *Phyllopertha* spp. predominate in Europe, and *Anomala* spp. are most common in Japan. Exotic species which have invaded new regions include the Japanese beetle, *Popillia japonica*, the oriental beetle, *Anomala (=Exomala) orientalis*, the European chafer, *Rhizotrogus majalis*, and the Asiatic garden beetle, *Maladera castanea*, in North America, and the South African black beetle, *Heteronychus arator*, in New Zealand and Australian pastures. Some pest species have a more limited distribution. Billbugs, *Sphenophorus* spp., are common pests throughout most of the USA and Japan, with different species assuming dominance in different regions. Serious damage by mole crickets, *Scapteriscus* spp., is limited to the southeastern USA.

Turf is also a suitable habitat for large numbers and a great diversity of microbes. A healthy gram of pasture soil will contain about 10^{5-8} bacteria, 10^{6-7} actinomycetes, 10^{5-6} fungal colony forming units (cfu), and 10^{2-3} nematodes (Metting, 1993). Most microbes are actively involved in organic matter breakdown and soil processes but a small proportion are invertebrate pathogens. The close association between soil dwelling pests and microbes seems to have led to a high number of diseases associated with those insects (*e.g.*, Jackson and Glare, 1992). Paradoxically, soil-dwelling pests appear to be resistant to generalist pathogens, but some specific species and strains of microbes have overcome host defenses and proven particularly effective as microbial control agents of soil dwelling pests (Jackson, 1996).

3 Microbial control options for turfgrass pest management

Several microbial organisms are available for control of turf and grassland pests (Table 2). These products can be divided into two categories, pathogenic microbes and microbial derivatives. Pathogenic microbes cause infection in a target host (pest) and have the potential to spread (recycle) in host populations. They include pathogenic viruses, fungi, nematodes, and bacteria including some strains of *Bacillus thuringiensis* (*Bt*). Microbial derivatives are toxins, such as the δ-endotoxin of *Bt* (see below), recovered from cultures of bacteria and fungi. Products based on microbial derivatives often do not contain live microorganisms, and therefore cannot spread in the pest populations. Brief descriptions of major microbial insecticides are given below.

A Bacillus thuringiensis

Several strains of the soil bacterium *Bt* produce insecticidal proteins called δ-endotoxins. For example, *Bt* subsp. *kurstaki* produces spores and the δ-endotoxin during fermentation and these are found in products registered for the control of turf pests such as armyworms and sod webworms (Table 2). While generally lepidopteran active, they are not effective against this group of pests in turf. Products based on *Bt* are not widely used on turf because of their short residual activity due to rapid photo-degradation of the toxins, slow activity of the endotoxin, lack of contact activity, and an inability to kill late instar larvae. Some turf Lepidoptera like black cutworm larvae are also generally resistant to *Bt*. Successful use of these products is dependent on recognition of the presence of early instars of the pest. If egg hatch occurs over an extended period, multiple treatments of *Bt* may be needed. Efficacy of *Bt* products against caterpillars can be improved by treating late in the day so that the night active caterpillars acquire the toxins before they have been degraded by sunlight.

Recently, a new strain of *Bt* subsp. *japonensis* (Buibui strain) was isolated in Japan. Research shows that this strain has good activity against scarab larvae such as *P. japonica, Anomala orientalis*, and green June beetle (*Cotinis nitida*) grubs (Alm *et al.*, 1997). Commercial development of this strain continues primarily in Japan. Another strain, *Bt* subsp. *israelensis*, has activity against the European crane fly (*Tipula paludosa*) in The Netherlands (Smits *et al.*, 1993), and may be developed commercially for the USA turf market.

Table 2. Some microbial products for turfgrass pest management

Pathogenic microbes	Target insect	Product name
Fungi		
Metarhizium anisopliae	*Adoryphorus couloni*	
Beauveria bassiana	*Dermolepida albohirtum*	Biogreen Bio-Cane™
	Chinch bugs	Naturalis T Botanigard
B. brongniartii	*Melolontha melolontha*	Engerlingspilz, Betel™
	Hoplochelus marginatus	Melocont®-Pilzgerste
Nematodes		
Steinernema carpocapsae	Annual bluegrass weevil, Armyworms, Billbugs, Cutworms, Fleas, European crane fly, Sod webworms	Biosafe, Carponem, Millenium, Nemastar, No Flea™, Ecomask, NEMAgräs
S. glaseri	White grubs	Biotopia
S. riobrave	Mole crickets	BioVector
S. scapterisci	Mole crickets	Nematac™ S
Heterorhabditis bacteriophora	Billbugs, White grubs	Nemasys G, Terranem H&G nema-green, Heteromask, Symbion-Temperate
H. marelatus	White grubs	Symbion-North
Bacteria		
Paenibacillus (=Bacillus) popilliae	*Popillia japonica*	Milky Spore
Serratia entomophilia	*Costelytra zealandica*	Invade
Bacillus thuringiensis (Bt)		
Bt subsp. *aizawai*	Sod webworm, armyworms	XenTari
Bt subsp. *kurstaki*	Sod webworm, armyworms	Deliver, Dipel, Javelin, Lepinox, Crymax
Bt subsp. *israelensis*	*Tipula* spp.	Teknar, Bactimos
Bt subsp. *japonensis* Strain buibui	White grubs	MYX

B *Other insect-pathogenic (entomopathogenic) bacteria*

Other entomopathogenic bacteria are common in most soils. Two types of bacterial diseases, "milky disease" and "amber disease," are commercially available for turf insects. The milky disease bacterium, *Paenibacillus (=Bacillus) popilliae*, was the first microbial agent registered as an insecticide in the USA, and has been used in suppressing Japanese beetle populations for over 50 years (Klein, 1995). Although only the strains of *P. popilliae* that infect *P. japonica* grubs are commercially available, other strains infect different grub species such as black turfgrass ataenius, *Ataenius spretulus*, and *Cyclocephala* spp. These strains are highly specific, infecting only one host species. Infection is initiated when the host insect ingests bacterial spores. Spores germinate in the gut and bacteria move into the insect hemolymph where they multiply and sporulate. As the spores with their characteristic parasporal bodies fill up the body cavity, the larvae show a typical milky appearance that gives the disease its name. Each larva may contain up to $2\text{–}5 \times 10^9$ spores. Infected larvae take up to a month to die, and upon host death, the spores are released into the soil leading to a buildup of the disease. The spores can persist in the soil for several years. The milky disease bacterium is an excellent pathogen for permanent establishment in turf. However, the primary value of the disease is its contribution to general suppression of larval populations, not as a primary biological insecticide. Establishment and buildup of *P. popilliae* is dependent on several factors. Soil temperatures should be above 21 °C for the rapid development of the disease. High densities (300 or more/m^2) of Japanese beetle larvae and neutral soil pH enhance buildup of milky disease spores in soil. The major hindrance to the large-scale commercial development of the milky disease bacterium is the lack of an artificial culture medium. Current commercial production relies on the collection of grubs from nature to mass-produce the bacterium. Only one company is currently producing *P. popilliae* products (Table 2).

Amber disease is caused by certain strains of *Serratia* spp., a genus of bacteria commonly found in turf soils. The pathogenic bacteria are ingested by the host, cause a cessation of feeding, and grow in the gut of susceptible insects, eventually leading to starvation and death of the host (Jackson *et al.*, 1993). In New Zealand, *S. entomophilia* is mass-produced *in vitro* and is used commercially as the product Invade™ against the grass grub *C. zealandica* (Jackson *et al.*, 1992). Similar bacteria have been isolated from Japanese beetle and masked chafer grubs in the USA, but no commercial development has taken place. Nevertheless, amber disease may be a significant natural mortality factor in grub populations in turfgrass soils.

C Entomopathogenic fungi

Two genera of entomopathogenic fungi show potential for the biological control of insects in turf. *Beauveria bassiana* has been associated with large-scale natural mortality of chinch bugs, especially under hot and humid conditions, and is commercially available as Naturalis® H&G and Botanigard™. Another fungus, *B. brongniartii*, has shown good suppression of the cockchafer grub, *Melolontha melolontha*, in Europe. This fungus is commercially available there as Engerlingspilz™, Betel™, and Melocont(R)-Pilzgerste. The green muscardine fungus, *Metarhizium anisopliae*, can infect many turf pests including mole crickets in the USA, Japanese beetle larvae in the USA and the Azores (Portugal) and *Adoryphorus couloni* in Australia. The green muscardine fungus is commercially available in Brazil for the control of froghoppers and spittlebugs on pasture turf, and as Biogreen™ for control of *A. couloni* grubs in Australia. Commercial products containing *B. bassiana* are available for the control of chinch bugs, aphids, and leafhoppers in the USA. Testing against white grubs and other insects in turf is being pursued aggressively.

D Entomopathogenic nematodes

Entomopathogenic nematodes (EPN) (*Steinernema* and *Heterorhabditis* spp.) are well suited for pest control in turfgrass because they attack a broad range of pests and can be easily applied with conventional spray equipment (Grewal and Georgis, 1998; Wright *et al.*, 2005; Shapiro-Ilan *et al.*, 2006; Chapter IV-5). EPN occur naturally in almost all soils and reproduce only in insect hosts which they have killed. More than 50 species have been discovered worldwide. Due to the ease of EPN mass-production, several EPN-based products have been developed for the control of turfgrass pests (Table 2). EPN species and strains differ in their activity against different insect pests. These differences are partially due to differences in their searching behaviors, and partially to the type of symbiotic bacteria carried by the infective juveniles.

Two EPN species are effective against mole crickets in turf and pastures in the southern USA. *S. riobrave* can only provide short-term curative control because it does not reproduce in mole crickets. However, the mole cricket-specific *Steinernema scapterisci* can provide long term suppression of mole cricket populations after inoculative releases and may become a major means of mole cricket management in pastures in Florida (Parkman and Smart, 1996). *Steinernema scapterisci* is more effective against *Scapteriscus borellii* than *S. vicinus* and particularly *S. abbreviatus*, whereas it is substantially less effective against nymphal mole crickets.

EPN are the most extensively studied parasites of white grubs (see Klein, 1993; Grewal *et al.*, 2005; Georgis *et al.*, 2006). At least five EPN species, *S. anomali, S. glaseri, S. kushidai, S. scarabaei*, and *Heterorhabditis megidis*, were originally collected and described from naturally infected white grubs and many more species have been documented to use white grubs as natural hosts. Four species, *H. bacteriophora, H. zealandica, H. marelatus*, and *S. glaseri* are currently available commercially for grub control in the world. However, white grubs are among the more difficult insects to control because they developed various barriers to EPN infection including EPN-impenetrable sieve plates on their spiracles, frequent defecation, aggressive grooming behaviors, evasive behaviors, and a thick peritrophic membrane that delays infections through thc midgut epithelium.

The efficacy of EPN can vary considerably with white grub species, nematode species, and white grub larval stage. Among white grub species that are important pest of turfgrass in the USA, *P. japonica* appears to be the most EPN-susceptible species, whereas larvae of other white grub species including *Cyclocephala* spp., *Anomala orientalis, R. majalis, M. castanea*, and *Phyllophaga* spp. appear to be less susceptible to the commonly used EPN, *i.e., H. bacteriophora* and *S. glaseri (e.g.*, Koppenhöfer *et al.*, 2004; Grewal *et al.*, 2005; Koppenhöfer *et al.*, 2006). Larval stages may also differ in their susceptibility but the effect varies with white grub and EPN species (*e.g.*, Koppenhöfer and Fuzy, 2004).

Grewal *et al.* (2005) give an extensive summary of studies on the efficacy of various species and/or strains of EPN against white grubs. Nematodes that have provided good field control of *P. japonica* include *S. scarabaei* (100%), *H. bacteriophora* (strain GPS11) (34–97%), *H. bacteriophora* (strain TF) (65–92%), and *H. zealandica* (strain X1) (73–98%). *S. scarabaei* is the only nematode species that has provided high field control of *Anomala orientalis* (87–100%), *M. castanea* (71–86%), and *R. majalis* (89%). Against *C. borealis*, *H. zealandica* (X1) (72–96%), *S. scarabaei* (84%), and *H. bacteriophora* (strain GPS11) (47–83%) appear to be the most promising nematodes.

For more detailed information on the efficacy of EPN against turfgrass and pasture pests and factors affecting it, see Shapiro-Ilan *et al.* (2002), Georgis *et al.* (2006), and Grewal *et al.* (2005).

4 Field application techniques

Pasture and turf form reasonably homogeneous surfaces for establishment of trial plots. Unfortunately grassland pests, especially subterranean species, usually occur in aggregations which make trial design much more difficult. Insect populations are present in a dynamic state and potential soil interactions will not be obvious to the researcher. The effect of applied entomopathogens on the population may not be immediately evident, and in fact, the objective may be to establish a long term suppressive agent for the pest rather than a quick control. Trials with microbials will also differ from those with conventional chemical pesticides as the applied agent may spread from the treated areas to untreated locations. All these characteristics will influence trial design, as well as the objectives of the microbial application.

A Trial design

The trial design chosen is dependent on the experimental objectives, which in turn is influenced by the target pest stage and microbial agent life cycle. Field testing usually follows isolation, identification, and preliminary laboratory testing. The first stage of field testing will often involve screening of several pathogen species, strains, or isolates to determine which are the most efficacious under field conditions. This evaluation can be carried out in small contained areas in the field – microplots. As the number of potential pathogens is narrowed down, and the production methods are developed so that the quantity of material available for testing increases, the scale of trials can be increased to include replicated small field plots. In a commercial phase, the selected strain will be evaluated for efficacy through the monitoring of large field plots or commercial applications.

1 Microplots or cylinder studies

Microplot or cylinder studies can be extremely useful for assessing the efficacy of microbial control agents against either root-feeding white grubs, or surface-feeding pests such as armyworms, cutworms, sod webworms, and chinch bugs. The microplots are contained arenas in the field that have been artificially infested with known numbers of the target species before application of the microbe. These plots are usually less than 1 m^2 in surface area, and the boundaries can be constructed of various materials including wood, concrete, metal, PVC, hard plastic garden edging material, and gauze bags. A common method for use of microplots is described below.

1. Construction and establishment – Arenas are constructed from PVC drainage pipe 30 cm in diameter cut to 35 cm lengths. The pipe lengths

are placed within a 45 cm deep sand pit on a layer of sand (for testing deep burrowing species the lower end of the pipe can be covered with screen). The containers are then filled with friable sieved soil or can be filled with cores of undisturbed soil carefully extracted from turf or grasslands. In the former case grass seed must be sown in the soil and allowed time to grow to provide a food source for the test insects. Where intact soil cores are used, it is important that these fit tightly within the PVC tube and it may be necessary to pack the edges with soil or fine sand to ensure a tight fit. If 10-cm pipe is used, soil cores can be taken with a standard golf cup changer and fitted directly into the PVC pipe (Figure 1A, D). Fifteen and 20 cm cores can be extracted with an irrigation head cutter (Figure 1C). Once the tubes are filled with soil, the area between them is packed with sand so that conditions within the tubes are similar to the general soil environment. As an alternative, PVC or iron cylinders are inserted into the ground on a suitable turf site (Figure 2) and the host larvae are introduced before or after treatment with the microbial control agent. Cylinders of various diameters can be driven into a moist turfgrass site. Selected sites must be relatively free of rocks, the soil should be moist, and tests should not be established in hardpan clay soils. The cylinders

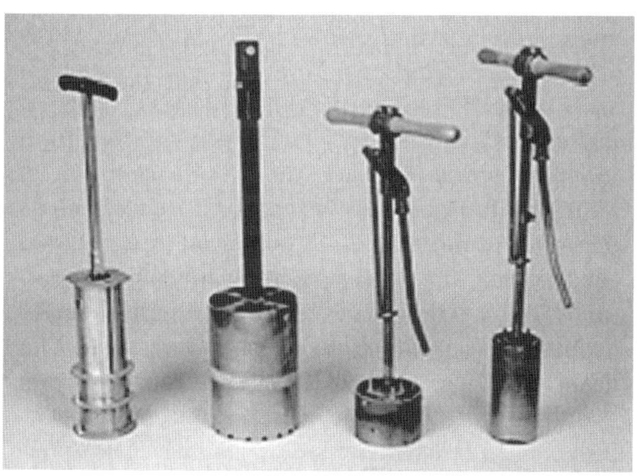

A B C D

Figure 1. Cup cutters. From left to right. A. Standard golf hole cup cutter (10.8 cm). Pounded into ground – sliding semicircles for ejecting plug. B. Cast iron template for cutting 20 cm circular slit with tractor or post-hole digger. C. Irrigation head cup cutter (15 cm). Pushed into ground – lever for ejecting plug. D. Standard golf hole cup cutter (10.8 cm). Pushed into ground – lever for ejecting plug

can be sunk into the ground by pounding directly on the cylinder with a mallet or iron bar, or by pounding on a board placed on top of the cylinder to reduce impact and breakage of the PVC. Care should be taken to prevent injury to fingers or toes, and to protect eyes from potential flying objects. A template was constructed by Mobay Corp. (Bayer, Indianapolis, IN) to cut a circular slit in the ground to enable placement of 20-cm diam. PVC cylinders with less pounding. A 4 mm thick hollow stainless steel cylinder, 20 cm in diam. by 30 cm long, forms the soil cutting device (Figure 1B). The bottom of the cylinder is sharpened and scalloped to facilitate cutting. A 50 cm long, 2.5 × 2.5 cm thick, bar is welded to crossbars at the top of the cylinder. The distal end of the bar has a fitting for attachment to a tractor mounted post-hole digger. When in operation, the circular movement of the cutter makes a thin circular slice in the soil where the PVC cylinder can be inserted.

2. Collection of pest insects – Many soil pests are univoltine or have even longer life cycles and cannot easily be reared in the laboratory. Therefore, it is often necessary to collect test insects from the field. The quality of the test insects is of crucial importance for the successful outcome of the trial. A high level of death among the untreated control insects will make interpretation of results extremely difficult. Survival of field collected insects will vary with the amount of care in handling as well as their condition at the time of sampling. If handled roughly, larvae will suffer internal bruising and breakdown of tissues that leads to septicemia within a few days. Many scarabs and late instar black cutworms are highly aggressive and cannot be maintained in crowded conditions. The effects of combat can be minimized by storage at low temperature immediately after collection. Insects should be collected carefully from the soil, preferably with gloves to prevent skin irritation of grubs, and placed in individual compartments in trays, which can be quickly placed in storage bins at low temperature. Welled tissue culture plates (with 24 or 12 wells) work very well for grub storage for up to several months. Intrinsic factors such as disease can also have an effect on the survival of any cohort, so test insects should be collected only from healthy populations, and some prescreening should be undertaken if possible to ensure that

Figure 2. PVC cylinders in place for field test

this is the case. A microplot study should reflect conditions in the field. Therefore, larvae should be added at densities which approximate those found in natural grasslands. For many grubs this will vary from 100–500/m^2.

3. Placement of larvae in cylinders – After establishment of the microplot, an appropriate number of target insect larvae can be released in the cylinder. Healthy subterranean insects will bury themselves in the soil in 20–30 min. Those insects still on the surface after 30 min should be replaced. Care must be taken when introducing the insects into the microplots; they should be well spaced or provided with a small entry hole so that they do not injure each other during entry into the soil. Predatory ants in the cylinders or on the surface may remove larvae and may require treatment with a short-lived insecticide prior to a field test.
4. Placement of adults in cylinders – Cylinders or microplots are also useful for caging adults to conduct tests against eggs or early instar larvae (Figure 3). Cages are particularly useful where adult females can be readily collected in the field or respond to lures or lights in traps. Cages are usually constructed of metal or fiberglass hardware cloth. Care should be taken to keep the cages low to the ground so that adults do not cling on the screen where they can be eaten by predators or remain in position and do not enter the ground to oviposit. Field-collected adult scarab beetles can assume to be mated, and while only females are needed in the cages, it is often easier to add a mixed population once the sex ratio is known. For the Japanese beetle, Catch-Can™ Japanese Beetle Traps (Trece, Salinas, CA) baited with the standard floral, or food-type, attractant (phenethyl propionate/eugenol/geraniol, 3:7:3) (Klein and Edwards, 1989) can be used to capture large numbers of adults. Bag type traps should be avoided since beetles quickly die in a plastic bag, and the sex pheromone (Japonilure) should not be used since it will shift the captures toward males. About 25 female beetles are usually sufficient for standard cylinder sizes. If the beetles

Figure 3. A 20-cm PVC cylinder with a screen lid to hold in adult beetles

feed as adults, they can be given apple slices, or a variety of leaves, as a food source. However, care should be taken since the food source can attract animals such as birds, raccoons, and squirrels which dislodge the cages in an attempt to reach the food. An alternative to feeding the beetles is to place additional beetles into the cage after 5–7 days.

5. Introduction of pathogens – Pathogens can be introduced in several ways. The microbial agent can be distributed as a suspension into the cylinder with a pipette or a beaker using enough suspension to allow for a uniform distribution. Alternately, the microbial agent can be sprayed into the cylinders or applied in a water drench. Pathogens can also be injected or applied in a manner simulating field application (*e.g.*, into grooves made in the soil). The entire turf plot can also be treated with the test agent prior to, or after the insertion of the cylinders.
6. Evaluation – For testing pathogens against grubs, tubes are left in the ground from 1 to 6 months. On extraction of the cylinders from the ground, surviving insects can be removed from the soil by hand and an assessment made of disease among the survivors. Pathogens can also be isolated directly from the soil (see Klein, 1997). Tests with Lepidoptera in cylinders are often evaluated in 2–3 days. Surviving and dead larvae can be separated from the soil as with grubs, or irritant flushes (see 4B 3, below) can be used to drive the surviving larvae to the surface.

There are advantages and disadvantages to the use of different materials in microplot or cylinder construction. (1) Wood is readily available, but is hard to keep clean, it absorbs materials, and is difficult to drive into the ground. (2) Plastic is easy to cut to size and relatively light to carry. Cylinders can be cleaned with steam or bleach. The plastic tubes should be reasonably large as an edge effect develops where roots meet the plastic. (3) Metal is very durable and relatively easy to drive into the ground. It is more difficult to cut to length, and very heavy to move around. (4) Gauze allows free flow of water and nutrients through the experimental arena. Placement into the soil may leave cracks or channels. Gauze may allow lateral movement and a dilution of pathogens if there are lower numbers of the pathogen in the surrounding soil.

2 *Small scale field plots*

Evaluation of microbial control agents against natural insect populations in turfgrass can be conducted in field plots. Plot size may vary according to the target insect and specific purpose of the study. A standard procedure is described below:

1. Select a suitable turfgrass site with availability of water for irrigation and presence of the target insect.
2. Establish plot boundaries (*e.g.*, 1×1 m) by using wooden pegs or iron rods at the corners, just at or below ground level. Record the distances from corner pegs to two permanent objects. A metal detector can help to find burred iron rods even years later.
3. Arrange plots in a suitable experimental design such as a randomized block design, Latin square design, or completely randomized design. Use at least 4–6 replicates.
4. Alleyways should be kept to separate the plots and the size of the alleyways would vary according the mobility of the target insect and the to-be-applied pathogen. Plastic fences or barriers can be placed on the surface or buried between plots. Paths between plots (1 m) can be created by killing grass with a herbicide to prevent movement of insects like tipulid larvae.
5. Take a pre-treatment count of the target insect in plots using a suitable sampling technique (see below).
6. Record environmental parameters including air and soil (at 2.5-cm depth) temperature, cloudiness, wind direction and speed, etc.
7. Determine plant species composition (grass types and weeds) in 4–6 randomly selected plots.
8. Take soil samples from 4–6 randomly selected plots for soil analyses including soil texture, organic matter content, moisture, pH, cation exchange capacity, etc.
9. Apply a pre-treatment irrigation if the soil is dry.
10. Apply microbial control agents at dusk or on a cloudy day when lethal UV is minimal. If the microbial agent is applied to the surface, UV intensity should be recorded with a UV meter at the time of application. Microbial agents may be applied with hand held sprayers, spreaders, or shaker jars. Liquid formulations can be applied using a CO_2 pressured sprayer with TeeJet 8010 nozzles at 20.7 kPa (3 psi)

calibrated to deliver about 4 liters/93 m^2. Granular formulations can be applied using a fertilizer spreader, modified seed drill, or shaker jar. Treatments could be applied using dyes to avoid mistakes.

11. Post-application irrigation should be applied especially when the microbial control agents are used against soil-dwelling insects. Irrigation serves as a rinse to wash the microbial control agents off the foliage and help in their movement into the soil.
12. The number of insects in each plot should be estimated using an appropriate sampling method depending on the target insect. Assessment time and interval would vary depending on the target insect. For example, the numbers of dead and living armyworms and cutworms may be monitored 2–3 days after treatment, chinch bugs can be counted at 7–14 days, and the number of surviving white grubs is usually counted 20–30 days after treatment. Subsequent sampling will depend upon the purpose of the study.

3 *Large scale field plots*

The value of many biological control agents is sometimes best seen after application to large plots. Where possible, commercial application of microbial agents should be monitored. Often, the evaluation of large scale treatments tends to be anecdotal, but with appropriate design, sound statistical analysis can be applied. Replication of paired treated and untreated areas can provide reliable results on the efficacy of a microbial treatment. Additional information can be obtained if trial conditions are monitored and it may be possible to relate degree of efficacy or even failure to some secondary factor. However, for each different pest and control agent, the criteria for success need to be carefully established before the start of the trial. Large scale field plots can vary considerably in size and shape. Several designs used in the past are listed below.

1. Replicated grass plots 10 × 50 m with buffer strips between plots and at each end.
2. Split golf course fairways, with half treated and half as the control.
3. Front and back lawns or split lawns.
4. Whole golf course greens, with alternating greens having a treatment or acting as a control.
5. Use of matched treated and untreated pastures.

Parkman *et al.* (1993) established large scale field plots using the EPN *S. scapterisci* against mole crickets in Florida.

1. Three bahiagrass pastures to receive treatments and an additional pasture as a control were selected.
2. Two pitfall traps (see section 4B 1. below) were installed at least 100 m apart in each pasture.
3. Fifty m^2 of pasture was treated with about 200,000 infective juveniles/m^2 around each trap.
4. Nematodes were applied as a drench with a watering can, or buried 2–4 cm under the ground in mole cricket cadavers (4/m^2).
5. Electronic mole cricket callers were used at each site to increase the population density of mole crickets.
6. Mole crickets were collected from the pitfall traps and returned to the laboratory for examination.
7. Rates of infection in the mole crickets and population levels at the sites were followed for 5 years.

B *Sampling methods*

Microbial agents often lack a rapid knock-down activity and work best by preventing pest damage to turf. They should, therefore, be applied at a time before damage becomes apparent and while the target insects are in a susceptible phase. Thus, for rational application it is important to determine the presence and number of pest insects in the turf prior to application. Post application sampling is necessary to determine the efficacy of treatment. Techniques developed to sample insects in turf and grassland have been described in detail by Barrett *et al.* (1990), Potter (1998), Vittum *et al.* (1999), and Niemczyk and Shetlar (2000).

1 *Trapping*

Black light traps can be used to determine when night flying insects are present in an area. A variety of trap models are available which can plug directly into an electrical outlet, run off a generator, or be hooked up to a battery for use at a remote location.

Pitfall traps can be used to monitor mobile, crawling insects such as chinch bugs and billbugs near the soil surface (Niemczyk and Shetlar, 2000).

1. The trap consists of a plastic cup, with a funnel and collection container placed inside the cup.
2. A hole the size of the diameter and length of the cup is placed in the turf.
3. The lip of the cup is situated at the thatch soil interface.
4. Water or alcohol is placed in the collection container.
5. Traps should be emptied daily to keep them from drying out.
6. Deflectors may be placed above the traps to keep rain out.

A linear pitfall trap for mole crickets was developed by Lawrence (1982) and modified by Parkman *et al.* (1993). This trap is also useful to monitor billbug adults.

1. A 19-liter plastic bucket is sunk in the ground.
2. Four arms consist of 3-m sections of 7.6-cm diam. PVC pipe made into troughs by cutting lengthwise slits 2.5 cm wide.
3. The arms are placed in the ground so that the open slits are at ground level.
4. One end of the arm is plugged and the other inserted through the wall of the bucket.
5. A pail with soil placed inside the bucket expedites emptying the traps.

2 *Collection from the surface*

Sweep nets can be effective in detecting turf pests such as aphids and chinch bugs.

1. Use a sturdy handle and solid, not mesh, bag.
2. Sweep back and forth over the turf so that the rim just hits the grass blades.
3. Turn bag inside out and examine contents.

Billbugs and other weevil larvae can often be found by looking for adults in gutters or sidewalks along turf areas. The weevils fall into the gutters and can be collected and counted as an index of abundance. Surface inhabiting species can also be collected by a simple devise like an aspirator (Niemczyk and Shetlar, 2000), or more sophisticated suction apparatus (Goldson *et al.*, 1984).

1. Adult weevils can be sampled with a vacuum cleaner connected to a generator for mobile field use.
2. All litter and loose material from a predefined area is sucked into a collecting bag.
3. When the bag contents are spread onto a warm metal plate (about 50 °C), the mobile weevils will become active and seek to escape.
4. While moving, weevils can be easily collected from the sample and counted.

Some insects can be attracted from the turf onto a warm surface. Goldson and Proffit (1991) showed that an electric blanket could be used to attract weevils out of grassland turf in cool winter conditions.

3 *Flushes and flotation*

Pest insects that form borrows in the turf can be sampled by drenching the soil with an irritant solution, usually of a detergent or weak insecticide (see Niemczyk and Shetlar, 2000). Irritant flushes are effective for examination of plots infested by sod webworms, cutworms, some weevils, or mole crickets.

1. A lemon scented household detergent [such as Joy^{R} (Proctor & Gamble, Cincinnati, OH)] can be very effective.
2. Use 32 ml of liquid detergent to 8 liters of water.
3. As an alternative, a few drops of a synthetic pyrethroid insecticide can be used.
4. Apply the solution with a watering can to about 1 m^2 of turf.
5. Pest insects normally surface within 10 min, but small sod webworms and some other insects may take up to 20 min.
6. Irrigate the sampled area to prevent sun scalding of the grass.

Particularly for mole crickets the method generally works better when soil is warm and moist. Pre-irrigation may improve extraction efficiency if soil is dry.

Vittum *et al.* (1999) note that chinch bugs can be sampled by flooding a turfgrass area and then covering the area with a piece of white cloth. The bugs will crawl up and cling to the underside of the cloth where they can be counted. The following flotation technique can also be an effective way of sampling (Niemczyk and Shetlar, 2000):

1. Cut both ends from a one kg coffee can or similar container.
2. Remove one end with a tin snips to provide a sharp edge.

3. Push the sharp end solidly into the turf.
4. Fill the can to the top with water.
5. Stir the container to agitate the thatch and help dislodge insects.
6. Add more water if needed to keep water level near top of can.
7. Chinch bugs will float to the surface in 5 to 10 min where they can be removed with a small brush

4 *Sampling from the soil*

For insects that are relatively immobile and live deep within the soil, there is no alternative but to extract a turf/soil sample. Where soils are stony, a regular digging spade can be used to cut a cube of soil from the turf. A standard method has used a straight edged spade with a 17.5-cm blade. Three squares cut with such a spade are equal to $0.09\,m^2$. The depth of the cut should exceed that of most of the insect population. A more accurate sample of grub populations can be obtained from several smaller samples through a plot than from fewer larger samples. The number of insects per spade square can be converted to numbers per meter square by simple multiplication (*e.g.*, for a 17.5-cm spade, multiply by 32.6).

Where soils are deep and free of stones, a more accurate assessment can be made with a round corer. A defined volume of soil can be obtained by driving a metal cylinder a defined depth into the ground.

1. A standard golf cup cutter can be used to obtain samples with a diameter of 10.8 cm.
2. The cup cutter is pushed into the ground to the required depth and the sample is ejected by a lever (Figure 1D).
3. Cup cutters are available with 35-cm blades which can be pounded into harder soil and then separated to allow access to the soil core (Figure 1A).
4. A robust, hinged, split corer has been developed for sampling pasture soils which allows rapid release of the core (Figure 4). This can be adapted for hydraulic operation with a tractor to increase the speed of sampling.

After taking the cores or samples, the insects must be extracted. Large insects can be extracted by hand sorting through the soil. But for small insects and delicate life stages, other methods may be necessary. Cores can be separated by hand to retrieve pest insects.

Small mobile insects can be removed from soil cores with the aid of a Berlese funnel (Niemczyk and Shetlar, 2000), which uses heat to drive insects from a sample.

1. Samples are taken with a golf cup cutter or other coring device.
2. Samples are placed grass side down on a screen ledge inside the funnel.
3. A heat source such as a 25 watt light bulb is positioned about 5 cm above the funnel.
4. Heat is applied to the sample for 24 to 48 h.
5. Insects are collected in 2.5 cm of 70% alcohol in the collecting jar at the bottom of the funnel.
6. Berlese funnels are most effective at removing insects from thatch or litter since insects are often unable to escape from soil before dying.

Small and soft bodied insects can be isolated from soil by flotation in concentrated salt solutions. Kain and Atkinson (1976) developed a rapid extraction system for isolating all stages of *C. zealandica* and other scarabs from soil with a combination of flotation and mechanical extractions. Vlug and Paul (1986) demonstrated a flotation technique based on a salt solution for sampling tipulid larvae.

1. Take soil samples from field with 10-cm soil corer, or take larger grass sods 50 × 50 cm.
2. Make a stock of salt water (1 kg of salt to 5 liters of water).
3. Put individual samples in small buckets, larger sods in containers.
4. Add salt water until grass is fully submerged
5. Stir mixture and let stand for 15–30 min.
6. Tipulids will float to the surface and can be collected with a sieve and counted.

5 *Damage surveys*

The presence of soil insects can be determined indirectly from the damage caused by them. Cobb and Mack (1989) developed a damage rating system for mole crickets which was shown to have a direct correlation with numbers of mole crickets collected following a soap flush (see section 4B 3 above).

1. A $0.6\,m^2$ frame (76 × 76 cm) was divided into 9 equal, square sections (25.3 × 25.3 cm).
2. The frame is placed on the ground.

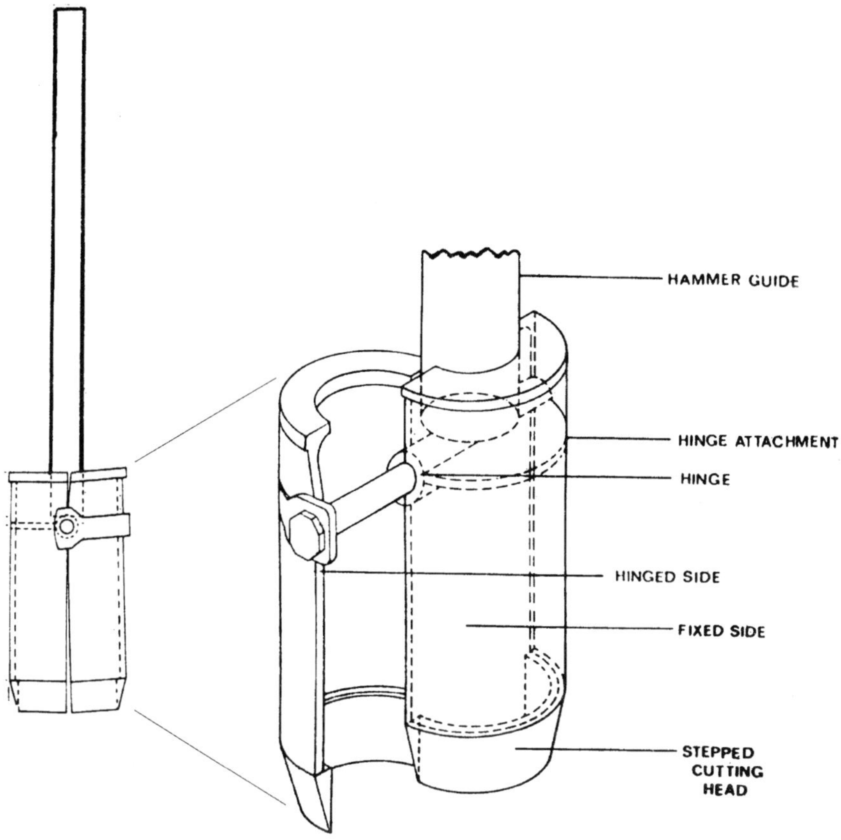

Figure 4. Hinged, split, soil corer. (From Kain and Young, 1975)

3. The number of sections which contain mounds and/or tunnels (visually and by touch) is counted.
4. Damage rating is between 0 (no damage) and 9 (damage in all nine sections).

6 *Sampling plans*

Most insect populations in turf are clumped or aggregated with the population distributed according to a negative binomial distribution. The degree of aggregation will determine the number of samples required to achieve a population estimate with a predetermined level of accuracy. A detailed discussion of this problem can be found earlier in this book in Chapter II-1.

7 *Pre-treatment sampling*

Pre-treatment sampling should be carried out prior to testing organisms for microbial control in the field. Ideally, the test microbe should be applied to a healthy, feeding population at a medium density that is evenly distributed throughout the experimental area. In practice, the ideal test population in the field can be difficult to locate. In addition, the mobility of the pest may cause problems. Melolonthid grubs and tipulid larvae can move several meters in search of suitable forage. When a potential site for testing is identified, a sample of the target insects should be collected from several locations in the test area and examined in the laboratory for pathogens, feeding and general health. Soil samples should also be taken and tested for the background presence of the organism to be applied (nematode, fungi, or bacteria). If the population appears to be healthy, an initial quick sampling should be carried out to define the area occupied by the target insect. The plots for the trial should then be laid out. A presample before pathogen application can aid in stratification of the trial into blocks covering a range of host densities and provide an initial population estimate which can be used as a covariate in analysis of treatment effect.

C Application – examples from the main groups of microbial control agents

1 Bacteria

a. Spore forming

Spore forming bacteria and their toxins are available in stable powder or wettable concentrate formulations.

1. *Bacillus thuringiensis*
   Alm *et al.* (1997) used the following protocol for field testing *Bt* subsp. *japonensis* strain Buibui against Japanese and oriental beetle larvae.
   1. Bacteria were applied to early instar larvae (late August – early September) with densities between 45 and $200/m^2$.
   2. Plot size was $1.5\,m^2$, with 3–4 replicates in a randomized complete block design.
   3. All treatments were applied with a watering can in 10–19 liters/plot.
   4. The addition of a wetting agent or post-treatment irrigation did not improve results.
   5. Treatments were evaluated between 21 and 48 days after treatment by taking samples with a standard golf course cup cutter.

   Leatherjacket larvae (Diptera: Tipulidae) have been shown to be susceptible to *Bt* subsp. *israelensis* (Smits *et al.*, 1993) using the following protocol for field testing.
   1. Bacteria were applied to $1\,m^2$ plots each surrounded by a plastic barrier.
   2. Treatments were applied at equivalents of 5, 15, 45, and 135 liters/ha of product.
   3. Each treatment was replicated 3 times.
   4. The bacteria were applied with a propane driven hand sprayer at 4 bar pressure in a volume of $500\,ml/m^2$.
   5. 25 ml of Citowett (BASF Canada, Toronto, Canada) spreader-sticker added to each 100 ml water.
   6. 200 first instar laboratory-reared *Tipula oleracea* were added to each plot.
   7. Plots evaluated after 4 weeks.
   8. Top 15 cm of the whole plot was removed and placed in a container with salt water (see section 4B 4 above)
   9. Larvae floating on the surface after 30 min were collected with a sieve and counted.
2. *Paenibacillus (Bacillus) popilliae*
   A method for field testing milky disease bacteria against Japanese beetle larvae was first outlined by Dutky (1941) and has been modified only slightly here.
   1. Bacteria should be applied to moderate grub populations (90 to 225 larvae/m^2 is suggested).
   2. Spore powder with 1×10^8 million spores/g should be used. Spore powder can be purchased commercially or obtained by infection of larvae of the same species to be treated, followed by production of the spore powder (Klein, 1997).
   3. Plots of at least $40\,m^2$ should be used, with at least three replicates per treatment. There is no indication that the use of small or mini plots give a satisfactory evaluation of milky disease products.
   4. One gram of spore powder should be applied in spots at intervals between 0.5 and 1 m.
   5. Plots should be examined monthly when larvae are active following treatment.
   6. Although the ideal would be to obtain 100 larvae from each plot, at least 10 samples of $0.9\,m^2$ should be taken in each plot.
   7. Macroscopic evidence of disease should be confirmed by microscopic examination. Microscopic examination of all specimens is needed to obtain an accurate evaluation of milky disease infected larvae (Kaya *et al.*, 1992; Klein, 1997).

b. Non-spore forming bacteria

Serratia entomophila – Non-spore forming bacteria such as *Serratia* spp. are sensitive to UV light, heat and desiccation. Cultures should be handled carefully and stored in cool conditions prior to application.

Cultures of *S. entomophila* are applied to pasture for control of the New Zealand grass grub (*C. zealandica*) as the product Invade™. The application method has been described by Jackson *et al.* (1989, 1992) and is updated below from product literature.

1. The bacterial product Invade™ should be stored under refrigeration prior to use where a cell viability $> 4 \times 10^{10}$/ml can be maintained for > 3 months. Bacterial cultures should be transported under refrigeration where possible and held in chillers in the field prior to application to avoid extreme changes in temperature.
2. *S. entomophila* should be applied to pastures containing late 2nd/early 3rd instar larvae (Southern hemisphere summer – February–April) when the insects are actively feeding. For best results application should be made to moderate grass grub populations (75–150 larvae/m^2).
3. Bacteria are applied at the rate of 1 liter/ha (4×10^{13} bacteria) suspended in 100 liters clean, non-chlorinated water.
4. The bacterial suspension is applied to the soil using a modified seed drill (Figure 5).
5. Establishment of bacteria in the field can be determined from soil samples using the method of O'Callaghan and Jackson (1993).
6. The level of amber disease in the population can be determined by visual examination of larvae 1–2 months after application. Infected larvae have the clear gut characteristic of amber disease (Jackson *et al.*, 1993). Alternatively, the presence of applied bacteria can be determined from larval macerates using an amber disease specific DNA probe (Jackson *et al.*, 1997).
7. The long term effects of application over the next few years can be determined by monitoring grass grub population trends and the persistence of the applied bacteria.

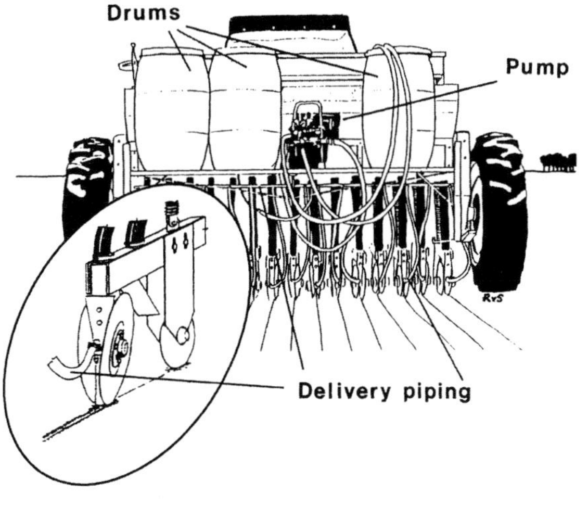

Figure 5. A standard, triple-disk seed drill modified for bacterial application by the addition of water tanks, a pump and pipes and nozzles for delivery. Tanks holding up to 1000 liters are mounted on the drill or towed behind it. The pump is powered by a motor mounted onto the drill. Flow of bacterial suspension is controlled by a cab mounted switch operating a solenoid in a 25 mm feed line to a 30 mm diameter stainless steel manifold mounted across the back of the drill. The suspension is delivered through individual 12 mm flexible feed lines to solid stream spray jets mounted on metal plates between the discs and directly above the drill row (See figure insert). Normal operating pressure is 200 kPa (30 psi), although this can be varied to maintain a constant application rate at different working speeds. Strainers and ball check valves are incorporated into the delivery line above the nozzles to prevent clogging and avoid loss of suspension when delivery is turned off. During operation, the drill is lowered into the soil to cut grooves to a depth of 3–5 cm and a solid stream of suspension is squirted into the "V" shaped groove made by the drill in the turf surface. Where there is good root mass and the soil is moist, the drill row will collapse, recover and be indistinguishable from the surrounding turf within one month. If the turf is dry or friable it may be necessary to follow application with a roller to restore integrity of the turf. To restore the turf after insect damage, grass seed can be applied through the drill at the same time as the bacterial suspension

2 *Fungi*

a Handling

Most fungi used in microbial control of turf pests belong to the Deuteromycete genera of *Metarhizium* or *Beauveria*. These fungi are usually grown on grains of rice, wheat or barley and applied as conidia. An alternative method of production is to produce blastospores in liquid fermentation but these tend to be very sensitive to environmental conditions during storage. Fungal conidia can be stored in cool, dry conditions on the grain or separated prior to reformulation in oils, talc or granules. Concentrates of pure conidia of *M. flavoviride* can be kept in sealed aluminum foil pouches for > 12 months at room temperature (http://www.lubilosa.org/index.htm). *M. anisopliae* granules are more sensitive to temperature. Storage life of vacuum packed granules is extended from 20 weeks at 20 °C to 70 weeks at 4 °C (Schwarz, 1995). While storage requirements will vary between species and strains, the general rule is that fungal preparations must be maintained in cool, dry conditions.

Care must also be taken during production and handling of fungi to avoid inhalation of airborne conidia. Production/formulation facilities should have adequate ventilation and a face mask should be worn during culture preparation. Stored cultures should be tested for viability by dilution plating onto the appropriate selective medium (Goettel and Inglis, 1997).

b Application

Fungal formulations for control of surface active pests can be sprayed directly onto the turf, but when soil-dwelling stages are the target, achieving contact between the fungal conidia and the target insect can be more difficult. As with bacteria, best results have been obtained with fungi by direct placement of fungal conidia into the soil. The following procedure is used for application of BioGreen™ for control of the red headed cockchafer (*A. couloni*) in pastures, and is based on the research of Rath *et al.* (1995).

1. The fungal conidial culture is contained in a grain medium and should be stored in the closed original container in a dry area within a temperature range of 5–10 °C prior to application.
2. The fungus is applied at the rate of 10 kg/ha (2.5×10^{13} conidia/ha) to pasture through a modified seed drill.
3. The fungus should be drilled into the soil to a depth of 20–25 mm.

A similar approach has been used to control the European cockchafer (*M. melolontha*) in Swiss meadows (Keller *et al.*, 1979; Keller, 1978). Difficulties in the use of a seed drill on steep terrain led Bondaz and Valet (1996) to develop a method for application of *B. brongniartii* to fresh snow and allow the snow melt in spring to carry the conidia into the soil. Successful methods for the autodissemination of fungi into an insect population are covered earlier in this book in Chapter III-3.

3 Entomopathogenic nematodes (EPN)

a General guidelines

EPN don't have a true resting stage and therefore require a little extra care to ensure that they remain viable and able to kill their host insects. Check with nematode suppliers and extension agents, or the Insect Parasitic Nematode Website (http://www2.oardc.ohio-state.edu/nematode) to see if your target pest requires modifications which are critical to achieving the best pest control.

b Nematode selection

There are several species and strains of EPN on the market today that are specialized for particular host insects. In addition, temperature tolerances differ among species. It is critical to choose the right species and product for the job. Check the Insect Parasitic Nematode Web Page listed above for precise information on hosts and effective EPN. Briefly, nematodes such as *S. carpocapsae* use an "ambush" strategy as they wait for an insect host to move by. They work well against pests that are moving around on the soil surface. Heterorhabditids can be used against relatively immobile soil pests because these nematodes are "cruisers". That is, they actively search for hosts deeper in the soil profile. It is important to choose a reputable supplier who will likely sell different EPN species and provide information on pests and application.

c Treatment timing

EPN have to be applied when the target pest is present (see section 4B above). In some situations, several applications may be more effective than one. If unsure when to apply for which insect, check with the web site, nematode suppliers, or extension agents.

d Nematode storage and handling

Proper storage and handling are essential to EPN health and efficacy. Nematodes must be stored in a cool and dry location, out of direct sunlight. Some products require refrigerated storage. Always follow the package instructions for the best method of mixing nematodes. Formulations vary depending on the species and target insect. They may be gels, dry granules, clay, vermiculite, or water-filled sponges. No matter what the formulation, EPN should be used as soon as possible after receiving them from the supplier. Most *Steinernema* species can be refrigerated (but not frozen) from a few days up to a month before they are mixed with water, but once diluted in water, they cannot be stored. Also, nematodes can't be left in hot vehicles or remain in spray tanks for more than a few hours.

e Application

EPN work best at moderate soil moisture. Watering the insect-infested area before and after applying nematodes keeps the soil moist and helps move the EPN down and the pest insects up in the soil profile. Care should be taken not to soak the area because EPN perform poorly in water-saturated soil. Nematodes are not very mobile in clay soils. EPN can be applied with most standard application equipment, and are compatible with pressurized, mist, electrostatic, fan and aerial sprayers. Hose-end sprayers, pump sprayers, and watering cans are effective applicators as well. Nematodes are even applied through irrigation systems on some crops. Check the label of the nematode product or the EPN Web Site to determine the best application method.

1. Nematodes should be applied (250–500,000 infective juveniles/m^2) during early morning, late evening, or on a cloudy, rainy day to avoid exposure to UV light or high temperatures which are lethal.
2. If the soil is dry or hot, apply at least 2.5 mm of irrigation to lower the soil temperature before application. Optimal temperatures for most EPN are between 15 and 30 °C. Some cold-tolerant species such as *S. feltiae* are effective between 15 and 27 °C whereas some heat-tolerant species such as *S. riobrave* are effective between 20 and 35 °C.
3. Carrier volume should be 750–1890 liters of water per ha.
4. Agitation must be provided in the spray tank to ensure proper mixing of the nematodes and dispersion of the product during the spraying process.
5. Sprayer screens must be 50 mesh or coarser, or the screens can be removed to prevent clogging.
6. The application should be followed by post-application irrigation within 30–120 min to wash the nematodes off the foliage into the soil. The lower the spray volume and the warmer the air temperature, the quicker the nematodes should be watered in. For applications against surface active insects 2 mm of irrigation should be applied, for applications against soil insect at least 6 mm should be used.

If nematodes are to be applied by fertigation equipment, the following additional precautions should be followed:

1. Lightly water (syringe) the area to be treated to moisten the turf surface and cool irrigation lines.
2. Pre-mix the required amount of the nematode product in sufficient water to uniformly inject the entire irrigation system.
3. Inject EPN during the second watering. Nematodes need to be in a clean feeder tank. The EPN suspension must be continuously agitated during injection to prevent nematodes from settling. After injection, rinse feeder tank with clean water and inject into system to purge nematodes from fertigation system.
4. After EPN application, immediately irrigate treated area with a minimum of 2 mm of water.

When applying EPN, there is no need for utilizing masks or specialized safety equipment. EPN are safe for plants and animals (earthworms, birds, pets, humans). They are either exempt from registration as in the USA or have been registered with minimal requirements in other countries. Because EPN leave no residues, application can be made anytime before a harvest and there is no re-entry time after application.

Fertilizers should be avoided roughly 2 weeks prior to and after nematode application, because EPN may be adversely affected by their high nitrogen content. Some insecticides work well with nematodes when their mutual exposure is limited while others may kill nematodes (Table 3; Koppenhöfer and Grewal, 2005). Check labels, or the EPN web site for additional information. Incompatible with EPN are the fungicides anilazine, dimethyl benzyl, ammonium chloride, fenarimol, and mercurous chloride, the herbicides 2,4-D and trichlopyr, and the nematicide fenamiphos.

Assessing efficacy of EPN may require time. While activity on Lepidoptera hosts may be observed in a few days, results on grubs and weevil larvae may not be noticeable for 2–3 weeks, or considerably longer if it is a soil pest that would not emerge until the next season (spring). EPN-infected insect

Table 3. Some chemicals that should not be tank mixed with entomopathogenic nematodes

Chemical	Trade name	Chemical	Trade name
Anilazine	Dyrene	Fipronil Chipco	Choice
Azadirachtin	Azatin	Insectidal soap	Various names
Bendiocarb	Turcam	Methomyl	Lannate
Carbofuran	Furadan	Oxamyl	Vydate
Carbaryl	Sevin	2-4-D	Various names
Chlorpyriphos	Dursban	Trichlorfon	Dylox
Ethoprop	Mocap	Triclorpyr	Turflon, Confront
Fenamiphos	Nemacur		

turn color depending on the EPN species and insect species, typically various shades of brown, from ochre to almost black, for *Steinernema* spp. and various shades or red, from yellow-orange to brick-red or purple for *Heterorhabditis* spp.

D *Post-treatment evaluation of the success of application*

The level of post-treatment sampling will depend on the objectives of the application. In the early stages of development of a microbial agent for soil pest control, it is important to closely follow the establishment of the microbial control agent and its uptake in the host population. This requires intensive sampling of specific trials. Once the potential of an organism is established, variation in effect becomes more important; selected parameters should be tested over a wide range of different conditions. When the organism is established as a commercial product, a few key indicators should be selected which allow user evaluation of the efficacy of control.

1 *Microbial establishment in the soil*

For a number of organisms the success of the application method can be determined by quantitative sampling of the organism from the soil immediately after application. This is greatly facilitated by the use of selective media. Caprylate-thallous agar (CTA) allows growth of *Serratia* spp. but is inhibitory to most other bacteria. This property allows it to be used for primary isolation of *Serratia* spp. from soil after application. Usually 20 2.5-cm soil cores are taken to a depth of 10 cm directly over the drill rows. Soil is bulked, mixed and bacteria are isolated and characterized as described by O'Callaghan and Jackson (1993) and Klein (1997). Fungi can be enumerated in a similar manner. Rath *et al.* (1995) used Doberski and Tribe's agar for enumeration of *M. anisopliae* DAT F-001 after application and noted that this medium is selective for a narrow selection of strains of the fungus. Nematodes can also be enumerated directly from soil samples [see Van Bezooijen (1999) and Chapter IV-5].

2 *Level of infection in the population*

For pathogens that produce a distinctive pathology in the infected host, the number of diseased insects can provide a good indicator of the success of microbial application. The level of infection in the population may be assessed after sampling from treated and untreated areas and enumerating insects displaying symptoms of infection by the microbial control agent. The number of diseased insects found will usually underestimate the impact of microbial application (Kaya *et al.*, 1992). This is especially true where the insect cadaver breaks down rapidly after infection [*e.g.* nematode- or milky disease-killed insects in warm climates (Cherry and Klein, 1997)] or the agent recycles through the population for an extended period of time. Thus, it must be emphasized that percentage infection is only an indicator of the success of application. For some diseases infection is obvious from the visual appearance of cadavers. For chronic diseases it may be necessary to prepare smears for microscopic examination or use molecular methods for identification of nonapparent infections.

3 Effect on the population

The short term objective of microbial control of turf insects is to reduce pest populations to below the damage or economic threshold. While the relationship between pest numbers and damage in turf is complex and not always well understood, the immediate objective will be to reduce pest numbers. Sampling methods and possible trial designs were described previously and statistical methods for trial analysis are outlined in Chapter II-1.

4 Long term effects

The overall objective of grassland and turf pest management is to minimize the impact of pest species on production and quality of the sward. This can be achieved in the short term by pest suppression but the true benefits of microbial control may come from establishment of the insect pathogenic microbe in the turf which may confer a resistance to further pest outbreaks. Most successful microbial control agents in turf and grasslands have shown the ability to persist for several years after application (Jackson and O'Callaghan, 1997). Klein and Georgis (1992) demonstrated the persistence of heterorhabditid nematodes and the control of Japanese beetle larvae for more that one year in golf course turf. To determine the long term effects of microbes in pest control it is necessary to monitor both pest population and soil pathogen levels for several years after application and establish appropriate comparisons with non-treated areas (See Section 4A 3 above).

5 References

Alm, S. R., Villani, M. G., Yeh, T., and Shutter, R. 1997. *Bacillus thuringiensis japonensis* strain Buibui for control of Japanese and oriental beetle larvae (Coleoptera: Scarabaeidae). *Appl. Entomol. Zool.* **32**, 477–484.

Barrett, B. I. P., van Toor, R. F., Ferguson, C. M., and Stewart, K. M. 1990. "Grass grub and porina in Otago and Southland: a guide to management and control." Tablet, Dunedin, N. Z.

Beard J. B., and Green, R. L. 1994. The role of turfgrass in environmental protection and their benefits to humans. *J. Environ. Quality* **23**, 452–460.

Bondaz, F. and Valet, S. 1996. Essai d'epandage a la vol de graines enrobees de *Beauveria brongniartii* instauration et penetration dans le sol. *IOBC/wprs Bulletin (Bulletin OILB/srop)* **19**, 65–68.

Cherry, R. H. and Klein, M. G. 1997. Mortality induced by *Bacillus popilliae* in *Cyclocephala parallela* (Coleoptera: Scarabaeidae) held under simulated filed conditions. *Fla. Entomol.* **80**, 261–265.

Cobb, P. P. and Mack, T. P. 1989. A rating system for evaluating tawny mole cricket, *Scapteriscus vicinus* Scudder, damage (Orthoptera: Gryllotalpidae). *J. Entomol Sci.* **24**, 142–144.

Danneberger, T. K. 1993. "Turfgrass Ecology and Management," Franzak and Foster, G. I. E., Cleveland, OH.

Delfosse, E. S. 1993. "Pests of Pastures. Weed, Invertebrate and Disease Pests of Australian Sheep Pastures." CSIRO Entomology.

Dutky, S. R. 1941. Testing the possible value of milky disease for control of soil-inhabiting larvae. *J. Econ Entomol.* **34**, 217–218.

Emmons, R. D. 2000. "Turfgrass Science and Management." 3rd ed. Delmar, Albany, NY.

Georgis, R., Koppenhöfer, A. M., Lacey, L. A., Bélair, G., Duncan, L. W., Grewal, P. S., Samish, M., Tan, L., Torr, P., and van Tol, R., W. H. M 2006. Successes and failures in the use of parasitic nematodes for pest control. *Biol. Control* **38**, 103–123.

Goettel, M. S. and Inglis, G. D. 1997. Fungi: Hyphomycetes. *In* "Manual of Techniques in Insect Pathology" (L. A. Lacey, Ed.), pp. 213–249. Academic Press, San Diego, CA.

Goldson, S. L., Frampton, E. R., Barrett, B. I. P., and Ferguson, C. M. 1984. Seasonal biology of *Sitona discoideus* Gyllenhall (Coleoptera: Curculionidae), and introduced pest of New Zealand lucerne. *Bull. Entomol. Res.* **74**, 249–259.

Goldson S. L. and Proffitt J. R. 1991. Use of an electric blanket for winter field collection of Argentine stem weevil, *Listronotus bonariensis* (Kuschel) (Coleoptera: Curculionidae). *N. Z. Entomol.* **14**, 44–47.

Grewal, P. S. and Georgis, R. 1998. Entomopathogenic nematodes. *In* "Methods in Biotechnology, Vol. 5: Biopesticides: Use and Delivery." (F. R. Hall and J. J. Menn, Eds.), pp. 271–299. Humana Press, Totowa, NJ.

Grewal. P. S., Koppenhöfer, A. M., Choo, H. Y. 2005. Lawn, turfgrass, and pasture applications. *In* "Nematodes as Biocontrol Agents." (P. S. Grewal, R.-U. Ehlers, and D. I. Shapiro-Ilan, Eds.), pp 115–146. CABI Publishing, Wallingford, UK.

Jackson, T. A. 1992. Scarabs -pests of the past or future? *In* "Use of Pathogens in Scarab Pest Management." (T. A. Jackson and T. R. Glare, Eds.), pp. 1–10. Intercept Ltd., Andover, UK.

Jackson, T. A. 1996. Soil dwelling pests – is specificity the key to successful microbial control? *In* "Proceedings of the 3rd International Workshop on Microbial Control of Soil Dwelling Pests." (T. A. Jackson and T. R. Glare, Eds.), pp. 1–6. AgResearch. Lincoln, NZ.

Jackson, T. A., Broad, T. M., O'Callaghan, M., and Martin, G. 1989. Development of a subsurface application system for *Serratia entomophila* for control of grass grub in pasture. *Proc. 5th Australasian Conf. Grassland Invertebr. Ecol.* pp. 63–68.

Jackson, T. A. and Glare, T. R. 1992. "Use of Pathogens in Scarab Pest Management." Intercept, Andover, UK.

Jackson, T. A., Pearson, J. F., O'Callaghan, M. O., Mahanty, H. K., and Willocks, M. J. 1992. Pathogen to product-development of *Serratia entomophila* (Enterobacteriaceae) as a commercial biological control agent for the New Zealand Grass grub (*Costelytra zealandica*). *In* "Use of Pathogens in Scarab Pest Management." (T. A. Jackson and T. R. Glare, Eds.), pp. 191–198. Intercept Ltd., Andover, UK.

Jackson, T. A., Huger, A. M., and Glare T. R. 1993. Pathology of amber disease in the New Zealand grass grub *Costelytra zealandica* (Coleoptera: Scarabaeidae). *J. Invertebr. Pathol.* **61**, 123–130.

Jackson, T. A. and O'Callahan, M. 1997. Environmental competence – an essential characteristic of successful microbial control agents for soil-dwelling pests. *In* "Soil Invertebrates in 1997." (P. G. Allsopp, D. J. Rogers, and L. N. Robertson, Eds.), pp. 74–77. BSES, Brisbane, Australia.

Jackson, T. A., Townsend, T. R., Nelson, T. L., Richards, N. K., and Glare, T. R. 1997. Estimating amber disease in grass grub populations by visual assessment and DNA colony blot analysis. *Proc. 50th N. Z. Plant Protection Conf.*, pp. 165–168.

Kain W. M. and Young, J. 1975. Population studies of *Costelytra zealandica* (White). I. A soil sampler suitable for sampling pasture land. *N. Z. J. Exp. Agric.* **3**, 177–180.

Kain W. M. and Atkinson, D. M. 1976. Population studies of *Costelytra zealandica* (White). II. A rapid mechanical extraction method suitable for intensive sampling of *Costelytra zealandica* and other scarabaeids. *N. Z. J. Exp. Agric.* **4**, 391–397.

Kaya, H. K., Klein, M. G., Burlando, T. M., Harrison, R. E., and Lacey, L. A. 1992. Prevalence of *two Bacillus popilliae* Dutky morphotypes and blue disease in *Cyclocephala hirta* LeConte (Coleoptera: Scarabaeidae) populations in California. *Pan-Pacific Entomol.* **68**, 38–45.

Keller, S. 1978. Infektionsversuche mit dem Pliz *Beauveria tenella* an adulten Maikafern (*Melolontha melolontha* L.). *Bull. Soc. Entomol. Suisse.* **51**, 13–19.

Keller, S., Keller, E., and Ramser, E. 1979. Ergebnisse eines Versuches zur mikrobiologischen Bekampfung des Maikafern (*Melolontha melolontha* L.) mit dem Pliz *Beauveria tenella. Bull. Soc. Entomol. Suisse.* **52**, 35–44.

Klein, M. G. 1993. Biological control of scarabs with entomopathogenic nematodes. *In* "Nematodes and the Biological Control of Insects." (R. Bedding, R. Akhurst and H. Kaya, Eds.), pp. 49–57. CSRIO, East Melbourne, Australia.

Klein, M. G. 1995. Microbial control of turfgrass insects. *In* "Handbook of Turfgrass Insect Pests." (R. L. Brandenburg and M. G. Villani, Eds.), pp. 95–100. Entomol. Soc. Am., Lanham, MD.

Klein, M. G. 1997. Bacteria of soil-inhabiting insects. *In* "Manual of Techniques in Insect Pathology" (L. A. Lacey, Ed.), pp. 101–116. Academic Press, San Diego, CA.

Klein, M. G. and Edwards, D. C. 1989. Captures of *Popillia lewisi* (Coleoptera: Scarabaeidae) and other scarabs on Okinawa with Japanese beetle lures. *J. Econ Entomol.* **82**, 101–103.

Klein, M. G. and R. Georgis. 1992. Persistence of control of Japanese beetle (Coleoptera: Scarabaeidae) larvae with steinernematid and heterorhabditid nematodes. *J. Econ. Entomol.* **85**, 727–730.

Koppenhöfer, A. M. and Fuzy, E. M. 2004. Effect of white grub developmental stage on susceptibility to entomopathogenic nematodes. *J. Econ. Entomol.* **97**, 1842–1849.

Koppenhöfer, A. M., Fuzy E. M., Crocker R., Gelernter W., and Polavarapu S. 2004. Pathogenicity of *Steinernema scarabaei, Heterorhabditis bacteriophora* and *S. glaseri* to twelve white grub species. *Biocontrol Sci. Technol.* **14**, 87–92.

Koppenhöfer, A. M., and Grewal, P. S. 2005. Compatibility and interactions with agrochemicals and other biocontrol agents. *In* "Nematodes as Biocontrol Agents." (P. S. Grewal, R.-U. Ehlers, and D. I. Shapiro-Ilan, Eds.), pp 363–381. CABI Publishing, Wallingford, UK,.

Koppenhöfer, A. M., Grewal, P. S., and Fuzy, A. M. 2006. Virulence of the entomopathogenic nematodes *Heterorhabditis bacteriophora, H. zealandica*, and *Steinernema scarabaei* against five white grub species (Coleoptera: Scarabaeidae) of economic importance in turfgrass in North America. *Biol. Control* **38**, 397–404.

Lawrence, K. O. 1982. A linear pitfall trap for mole cricket and other soil arthropods. *Fla. Entomol.* **65**, 376–377.

Metting, F. B. 1993. Structure and physiological ecology of soil microbial communities. *In* "Soil Microbial Ecology." (F. B. Metting, Ed.), pp. 3–25. Marcel Decker Inc., New York. NY.

Niemczyk, H. D., and Shetlar, D. J. 2000. "Destructive Turf Insects," HDN Book, Wooster, OH.

O'Callaghan, M. and Jackson, T. A. 1993. Isolation and enumeration of *Serratia entomophila* – a bacterial pathogen of the New Zealand grass grub, *Costelytra zealandica*. *J. Appl. Bacteriol.* **75**, 307–314.

Parkman, J. P., Hudson, W. G., Frank, J. H., Nguyen, K. B., and Smart, G. C., Jr. 1993. Establishment and persistence of *Steinernema scapterisci* (Rhabditida: Steinernematidae) in field populations *of Scapteriscus* spp. mole crickets (Orthoptera: Gryllotalpidae). *J. Entomol. Sci.* **28**, 182–190.

Parkman, J. P. and Smart, G. C. Jr. 1996. Entomopathogenic nematodes, a case study: introduction of *Steinernema scapterisci* in Florida. *Biocontrol Sci. Technol.* **6**, 413–419.

Potter, D. A. 1998. "Destructive Turfgrass Insects: Biology, Diagnosis, and Control." Ann Arbor Press, Chelsea, MI.

Rath, A. C., Worledge, D., Koen, T. B., and Rowe, B. A. 1995. Long-term field efficacy of the entomogenous fungus *Metarhizium anisopliae* against the subterranean scarab, *Adoryphorus couloni*. *Biocontrol Sci. Technol.* **5**, 439–451.

Schwarz, M. R. 1995. *Metarhizium anisopliae* for soil pest control. *In* "Biorational Pest Control Agents." (F. R. Hall and J. W. Barry, Eds.) pp. 183–196. ACS Symposium Series 595.

Shapiro-Ilan, D. I., Gouge, D. H., Koppenhöfer, A. M. 2002. Factors affecting commercial success: case studies in cotton, turf and citrus.*In* "Entomopathogenic Nematology." (R. Gaugler. Ed.), pp. 333–355. CABI Publishing, Wallingford, UK.

Shapiro-Ilan, D. I., Gouge, D. H., Piggott, S. J., and Fife, J. P. 2006. Application technology and environmental considerations for use of entomopathogenic nematodes in biological control. *Biol. Control* **38**, 124–133.

Smits, P. H., Vlug, H. J., and Wiegers, G. L. 1993. Biological control of leatherjackets with *Bacillus thuringiensis*. *Proc. Exp. Appl Entomol., N. E. V. Amsterdam* **4**, 187–192.

Van Bezooijen, J. 1999. Nematode extraction. *In* "EUR – Cost 819 – Entomopathogenic nematodes Application and persistence of entomopathogenic nematodes" (R. L Gwynn, P. H. Smits, C. Griffin, R.-U. Ehlers, N. Boemare, and J.-P. Masson, Eds.) pp. 61–72. European Commission, Luxembourg.

Vittum, P. J., Villani, M. G., and Tashiro, H. 1999. "Turfgrass Insects of the United States and Canada." Cornell University Press, Ithaca, NY.

Vlug, H. J. And Paul, H. 1986. Sampling leatherjackets. *Med Fac. Landbouww. Rijksuniv. Gent* **51**, 939–942.

Wright, D. J., Peters, A., Schroer, S., and Fife, J. P. 2005. Application technology. *In* "Nematodes as Biocontrol Agents." (P. S. Grewal, R.-U. Ehlers, and D. I. Shapiro-Ilan, Eds.), pp 91–106. CABI Publishing, Wallingford, UK.

Chapter VII-19

Application and evaluation of entomopathogens for managing insects in stored products

Jeffrey C. Lord and James F. Campbell
USDA-ARS
Grain Marketing and Production
Research Center
Manhattan, KS 66502, USA

John D. Sedlacek
Community Research Service
Kentucky State University
Frankfort, KY 40601, USA

Patrick V. Vail
Deceased, formerly with USDA-ARS
San Joaquin Valley Agricultural
Sciences Center
Parlier, CA 93648, USA

1 Introduction

Insects are a major cause of post-harvest losses of stored foods and food products the world over. Stored-product insect infestation can occur beginning at harvest and continuing through bulk storage; conversion into processed commodities in food processing facilities; storage in warehouses; transportation in trucks, railcars, and ships; presentation on retail shelves; and ultimately storage in consumer pantries. At each of the locations in the processing and distribution channel there are differences in the nature of the environment, the pest species complex, and the management strategies and tactics used. Insects not only consume these commodities but can also reduce the 'value' of the commodity by contaminating it with insect fragments, whole bodies, feces, webbing, and foul smelling metabolic products. They may promote growth of a variety of microflora, some of which pose serious health hazards to humans and livestock, particularly those that create mycotoxins.

Lepidoptera have been the target of most of stored-products microbial control effort. This is due to their prominence in high value commodities and to the availability of the Lepidoptera-specific pathogens *Bacillus thuringiensis* (*Bt*) subsp. *kurstaki* and the Indianmeal moth granulovirus. Grains constitute by far the greatest quantity of stored agricultural products, but their low value precludes the use of relatively expensive microbial pesticides except in special situations. Beetles are the most important pest insects in stored grains, and there is a paucity of effective and inexpensive microbial insecticides for them. There are many naturally occurring pathogens of both Lepidoptera and Coleoptera in stored products, but there have been only isolated efforts to understand their impact.

L.A. Lacey and H.K. Kaya (eds.), *Field Manual of Techniques in Invertebrate Pathology*, 677–693.

A Major pest species

The pest insects most often associated with stored products are in the orders Coleoptera (600 species of beetles in 34 families (Hinton, 1945)), Lepidoptera (70 species in primarily four families (Cox and Bell, 1991; Sedlacek *et al.*, 1995), and Psocoptera. Stored-product insects can be grouped in different ways based on taxonomy, feeding preference, and life history traits. Some species feed on whole kernels, typically with one portion of the life cycle occurring within the kernel, and some feed on processed and damaged kernels (e.g., broken kernels, grain dust). Insects in the second group tend to be secondary pests in bulk stored grain, but become much more important as pests in processed food facilities.

1 Lepidoptera

The importance of Lepidoptera is much greater than is suggested by the number of species that infest stored products. In many commodities, only moths are monitored and treated. The almond moth (*Cadra cautella*), raisin moth (*C. figulilella*), Mediterranean flour moth (*Ephestia kuehniella*), tobacco moth (*E. elutella*), Indianmeal moth (*Plodia interpunctella*), rice moth (*Corcyra cephalonica*), and Angoumois grain moth (*Sitotroga cerealella*) are widely distributed and major pests of stored foods. It is the larval stage of stored-product moths that develops either in or on the commodity and causes damage. In addition to causing damage by feeding directly on stored commodities, they produce silk that binds kernels or food particles together and causes clogging and/or mechanical damage to various types of food industry machinery. Especially dense webbing may be produced by Indianmeal moth and almond moth larvae. Interestingly, the almond moth is second only to the red flour beetle, *Tribolium castaneum*, in the amount of damage caused to stored products on a world-wide scale (Cox and Bell, 1991). However, Indianmeal moth is the most widely distributed of all moths infesting stored foods. The most serious moth pest of stored grain in the New World and in Africa is the Angoumois grain moth, but modern grain storage practices and wide-spread use of the combine harvester have minimized the negative impacts of this species in most of the developed world.

Indianmeal moth, which is easily distinguished by the copper luster of its forewing markings, has been found in all temperate and tropical regions (Tzanakakis, 1959). It feeds on most dried fruits, nuts, grains, and grain products (Williams, 1964). The common name reflects its prevalence on corn. Indianmeal moth is considered a good colonizer because it is highly mobile (Campbell and Mullen, 2004) and has wide nutritional latitude. It can infest these commodities at any time during processing or storage and at any time while in marketing channels.

The Mediterranean flour moth is native to Europe but has spread through North America and Australia, and then to other parts of the world. Since the advent of the roller flour mill, it is a major pest of mills in temperate counties where larvae interfere with flour production by spinning webbing that clogs machines and by biting holes in silk screens. The webbing is produced by young larvae that spin tubes within which they remain until mature (Jacob and Cox, 1977).

Almond moth, which closely resembles the Mediterranean flour moth and tobacco moth, is a pest predominantly in the warmer regions of the world. However, it can be a pest in more temperate climates especially in heated storage facilities. Dried fruits and nuts are preferred foods, but the larvae may feed on many types of stored vegetable matter including flours, grains, cacao beans, and seeds (Cotton, 1950).

The Angoumois grain moth is small (1.5 cm in length), and the larvae are rarely seen because they feed strictly inside grain kernels which makes control difficult. The Angoumois grain moth is found globally in a variety of stored grains including corn, wheat, rice, sorghum, and millet. It has been found in northern latitudes, but is more common in temperate and tropical regions. It is particularly effective as a primary colonizer because it is highly mobile and has a flexible diet. Angoumois grain moth has been found in sites far removed from grain storage facilities and is capable of infesting grain both prior to and in storage. It is thought to compete with *Sitophilus* spp. for larval habitat and may dominate in drier habitats that are stressful for the weevils but be suppressed when conditions are favorable to the weevils, which destroy moth eggs (Chesnut and Douglas, 1971).

2 *Coleoptera*

The family Bostrichidae has two species that are important primary pests of grain. The lesser grain borer, *Rhyzopertha dominica*, is one of the most damaging insects found in grain, especially in wheat. The larger grain borer *Prostephanus truncatus* is a more serious pest of maize in warm tropical areas, especially where maize is stored on the cob. Adults of both species have functional flight wings and are strong flyers. The grub-like larvae develop primarily inside whole grain kernels and thus are rarely visible.

Female *R. dominica* lay up to 400 eggs, placing them among but not inside the grain kernels, sometimes in clusters of up to 45. Larvae chew into kernels or penetrate through cracks. Inside the kernels, the borers pass through three to five larval instars and a pupal stage to emerge as adults. In some instances, where broken kernels and much dust are present, larvae may feed and develop externally. Adult insects live an average of seven to eight months and often return to kernels from which insects have emerged to feed and more extensively damage the grain. Infestations are characterized by the presence of grain dust consisting of mostly fecal matter and a characteristic sweetish, musty odor. The lesser grain borer is better able to survive in drier grain than the weevils, which probably accounts for it being a more serious pest of farm-stored wheat in the central and southern plains of the United States.

Larger grain borer is similar to lesser grain borer in development and habits. It is a serious pest of corn, as well as dried cassava, and commonly occurs in Central and South America. It was introduced into Kenya in the 1970s and has rapidly spread across sub-Saharan Africa. The climate and the prevalence of farm storage of corn on the cob in parts of Africa are particularly favorable for larger grain borer infestations. Females bore into the attached kernels and lay eggs in side tunnels. Larvae feed mainly on dust, much of which is created by adult feeding.

Three Curculionidae are damaging primary pests of grain, the rice weevil, *Sitophilus oryzae*, the granary weevil, *S. granarius*, and the maize weevil, *S. zeamais*. They attack whole grains and typically do not inhabit fine materials such as flour but can infest formed cereal products such as pasta. Females chew small holes into grain kernels, and in some of these holes, they lay a single egg and seal the hole with a plug. Females can lay 300 to 400 eggs over a period of months. Development through adult emergence occurs within grain kernels, minimizing exposure to pathogens.

The rice weevil is found in all parts of the world where grain is used but is particularly abundant and damaging in warm climates where it breeds continuously. The rice weevil's habitat is almost identical to that of the granary weevil, one of the oldest known insect pests. Granary weevil cannot fly but has been carried from place to place in infested lots of grain and is now found in most parts of the world, especially in temperate climates. Maize weevils are capable of flight, and in warm areas, they commonly fly to fields and infest corn before it is harvested.

Grain beetles and flour beetles are economically important Coleoptera that mainly attack damaged kernels and grain products. The most notable pests among them are grain beetles of the genera *Cryptolestes* and *Oryzaephilus* and flour beetles of the genus *Tribolium*. Grain beetles are often the most abundant insects in stored grain. They belong to a guild of dorso-ventrally-flattened beetles that feed on damaged grain, dust and mold, but not undamaged grain. The development times are typically about a month under favorable conditions, and populations often build up rapidly, particularly in moist environments. The family Laemophloeidae includes eight species in the genus *Cryptolestes* that are secondary pests of cereals and cereal products. Two prominent species are the rusty grain beetle, *Cryptolestes ferrugineus*, a cold-tolerant species that is often the most abundant insect in stored grain in temperate climates, and the flat grain beetle, *C. pusillus*, a minute beetle of about 1.5 mm in adult length that is cosmopolitan and common in warm moist climates such as Southeast Asia. The members of the family Silvanidae are generally larger (2–4 mm adults) and more active than *Cryptolestes* species, but they lack their flying ability. The sawtoothed grain beetle, *Oryzaephilus surinamensis*, derives its common name from the tooth-like projections on the sides of its thorax. It is among the best-known cosmopolitan pests in stored grain and grain products (Howe, 1956). The merchant grain beetle, *O. mercator*, is often confused with the sawtoothed grain beetle, but it prefers oilseed products, including nuts (Loschiavo and Smith, 1970). Several other Silvanidae can infest cereals but are of minor importance.

Among the nine potential secondary pest species of the genus *Tribolium*, the red flour beetle, *T. castaneum*, and the confused flour beetle, *T. confusum*, are perhaps the most problematic due to their prevalence and ability to survive most control measures (Sokoloff, 1974). Adults of these species can live for over three years with females laying eggs for more than a year under laboratory conditions (Good, 1936). They are most damaging to flour and other milled products where eggs are laid directly in the food material. Where such material is abundant, flour beetles can develop with as little as 8% moisture content. Stressed or agitated flour beetles secrete quinones that have a pungent odor and can give a pink hue to heavily infested flour.

3 Psocids

A number of species in the insect order Psocoptera are associated with human structures and stored food. They are typically found in areas with high relative humidity that favor mold growth, which is a primary food source. Psocids typically occur at low densities and are not considered major pests in many parts of the world. However, in tropical countries and in sub-tropical zones with high temperature and humidity, psocids can build to large numbers and have an economic impact. They are also a problem in stored grain and animal feed warehouses in Australia, where large populations of *Liposcelis bostrychophila* can occur primarily as a secondary pest outbreak due to suppression of natural enemies by fumigations (Rees and Walker, 1990; Ho and Winks, 1995).

Additional information concerning the taxonomy, ecology and behavior of these stored-product pests, as well as others, can be found in Cotton (1963), Gorham (1987, 1991), Subramanyam and Hagstrum (1995a), and Rees (2004).

B Entomopathogens of stored-product pest insects

Two pathogens, a bacterium and a virus, have obtained US EPA registrations and labels for use in stored products. *Bacillus thuringiensis*, the most commercially successful of all insect pathogens, was described with *E. kuehniella* as its type host in 1915. *Bacillus thuringiensis* subsp. *kurstaki* is registered for use on stored grains, nuts, and dried fruits in the USA, primarily for Indianmeal moth and almond moth. Although several *Bt* isolates are effective against some leaf-feeding beetles, no *Bt* isolate has been found to have commercially practical efficacy for any of the major coleopteran pests of stored products (Moore *et al*., 2000).

Several baculoviruses, nucleopolyhedroviruses and granuloviruses, have been isolated from stored-product Lepidoptera, but they are not known to occur in beetles. The only virus to be extensively studied for the control of post-harvest pests is the granulovirus that infects the Indianmeal moth (PiGV). First described by Arnott and Smith (1968a, b), it was soon investigated for its potential as a microbial control agent. Because PiGV is highly infectious, autodissemination via traps can be used for its application (Vail *et al*., 1993). Two other baculoviruses have been isolated from the raisin moth (Kellen and Hoffman, 1983), which is also a cosmopolitan pest of stored dried fruit, but they have not been pursued for microbial control. With intensification of research on viruses as microbial control agents of stored-product pests in the 1970s, test commodities included grains (McGaughey, 1975; Kinsinger and McGaughey, 1976), peanuts, walnuts, almonds, and raisins (Hunter *et al*., 1973, 1977, 1979; Cowan *et al*., 1986; Vail *et al*., 1991). A number of lepidopterous stored-product pests have been challenged with PiGV, but none other than the type host was found to be susceptible, which attests to its specificity. Its efficacy having been demonstrated, a production and formulation process was developed and patented for PiGV (Vail, 1991), and it was registered for use by EPA in 2001.

While contaminating fungi can be a problem in commodities that are stored under unsuitable conditions and many stored-product pests can feed on fungi, those fungi that can contribute to insect control are not normally present in quantities that are lethal to insects. Two species, *Beauveria bassiana* and *Metarhizium anisopliae*, are under consideration for use as mycoinsectides for stored products (Moore *et al.*, 2000). Isolates of both are registered for uses other than stored products in the US, but no stored-product use has been labeled. While they have broad host ranges, their efficacies vary greatly for stored-product pests. Combination with desiccant dusts is one promising strategy to improve their performance (Lord, 2001, 2005)

A variety of protozoan and microsporidian pathogens infect stored-product Coleoptera. None of them is considered a candidate for development as a microbial insecticide, but they clearly have impacts as natural controls. Prominent among them are neogregarines, particularly *Mattesia* spp., and microsporidia of the genera *Nosema* and *Paranosema*. *Mattesia oryzaephili* was discovered in Europe and was recently found in the United States (Lord, 2001). It infects some of the most important pests of stored grain including *C. ferrugineus*, *O. surinamensis*, and *R. dominica*. *Mattesia trogodermae* infects several species of *Trogoderma* and has shown potential for population suppression by dissemination through pheromone traps (Shapas *et al.*, 1977). *Mattesia dispora* is primarily a pathogen of Lepidoptera, including *E. kuehniella* and *P. interpunctella*, but it also can infect some beetles (Lord, 2001).

The Microsporidia, which were formerly considered protozoa but were recently reclassified as fungi, can be found in most insect species. Several are well known in stored-product pests. The genus *Nosema* comprised all of these until the recent erection of *Paranosema*, which includes *P. whitei*, a well-studied pathogen of the red flour beetle (Milner, 1972; Sokolova *et al.*, 2005).

Although rarely associated with stored-product insects, nematodes of the commercially available genera, *Heterorhabditis* and *Steinernema*, have been shown to be pathogenic for important stored-product Coleoptera and Lepidoptera in laboratory assays (Mbata and Shapiro-Ilan, 2005; Ramos-Rodriguez *et al.*, 2006). Freedom from regulatory costs and restrictions greatly enhances their prospects for operational use under certain limited conditions. For more information on entomopathogens used in stored products, see Brower *et al.* (1996) and Moore *et al.* (2000).

2 Evaluation of microbial control agents

A Treatment preparation

Each microbial agent has its own characteristics that dictate the appropriate preparation and quantification. Commercial *Bt* is standardized in potency units while other pathogens may require potency determination by bioassay or viability testing. The commercial pathogens *Bt* and PiGV have established methods for preparation and handling, but less-developed agents require special considerations for stored products. *Beauveria bassiana* and *M. anisopliae* have often been suspended in water for laboratory tests with grain but would have a better residual effect and be more effective when used in combination with diatomaceous earth, a desiccating agent, if applied in dry form (Lord, 2005). This could be done with technical powder or a dry formulation. Limited ability to produce protozoan pathogens of stored-product pests restricts their application to inoculative releases rather than broadcast application, and inoculum preparation may be restricted to accumulation of infected host cadavers and perhaps mixing with bait.

Unlike *Bt*, baculoviruses are still in the early stages of development and do not have potency ratings. Baculoviruses must be produced either in living insects (Hunter *et al.*, 1973; Vail *et al.*, 1991) or insect cell cultures (Weiss and Vaughn, 1986). Thus, it is extremely important that candidate viruses be standardized prior to use because potency may vary between production batches. Without such standardization it is difficult to compare results among different lots of the same material or from investigator to investigator. Until the early 1970s, counts of infectious occlusion bodies (OB) per ml or g were made using a hemocytometer and a phase contrast microscope. Early efficacy studies of the baculoviruses of stored-product insects were conducted with "wet" formulations in which diseased larvae were homogenized in water (Hunter *et al.*, 1977), and rates were quantified by OBs per unit volume. There was inherent variation with that approach because the biological activity of the infectious material varied, and bioassays to determine virulence were initiated (Hunter, 1970; Cowan *et al.*, 1986; Vail *et al.*, 1991). Even the more recent freeze-dried preparations of OBs (Cowan *et al.*, 1986) with dose expressed as OBs or grams of preparation per gram of diet or commodity should have their potency confirmed (Johnson *et al.*, 1999). Bioassays of infectious activity have been conducted using whole grains in the case of Indianmeal moth or, more commonly, using a bran-based semi-synthetic diet composed of wheat bran, honey, glycerine, water, and brewer's yeast (Finney and Brinkman, 1967; Tebbets *et al.*, 1978). Usually, 0.24 liter untreated paper serving cups covered with clear lids and containing 20 g of the bran diet are used. Small holes may be punched in the lids to prevent water condensation. Prior to infesting the cups, 4 ml aliquots of serial dilutions of the virus (bacteria, protozoa, etc.) are uniformly mixed into the diet. Fifty neonate larvae are placed into each cup with a camel hair brush. At least 2 cups are used per dose, and adult emergence is recorded. The resulting data are subjected to probit analysis.

For preparations of fungal conidia, determination of percent germination as an index of viability is critical. Many methods have been used for this purpose. The important considerations are that conidia be spread well enough to examine individuals for germination and that the incubation time and temperature are selected to allow maximum germination without covering conidia with excessive mycelial growth. A simple method is to dip a cotton swap into a representative lot of conidia preparation and streak over the entire surface of a medium such as Sabouraud dextrose agar in a Petri dish to obtain areas of various conidia densities. Incubation of 18 hours at 26–28 °C is adequate for most preparations. The use of colony-forming units for viability is much less accurate because of the uncertainty of the number of conidia at the origin of each colony and much more time consuming because of the required incubation period.

A variety of commercial entomopathogenic nematode formulations have been developed. These fall into two categories: inert carrier formulations using materials such as sponge or vermiculite that allow free gas exchange and nematode movement and active carrier formulations using materials such as alginate gels or granular formulations that physically trap nematodes or reduce their mobility and metabolism by inducing anhydrobiosis (Grewal, 2002; Grewal and Peters, 2005). Nematodes can also be relatively easily produced in the laboratory for small-scale studies and stored in water for short periods of time before use (Kaya and Stock, 1997). Nematodes are typically applied in water and, depending on the formulation, may require a period of time between mixing and application.

Assessment of quality is very important before application, since there is considerable variation in commercial formulations (Gaugler *et al.*, 2000). The viability of infective juveniles can be tested using a dissecting microscope by looking at number of alive and dead individuals, but because viability and virulence are not always correlated, movement and/or infectivity assays are also needed. There has been considerable discussion of the best methods to evaluate nematode quality and there are a variety of methods available (Glazer and Lewis, 2000; Grewal and Peters, 2005). One-on-one bioassays are widely used to measure virulence using either filter paper (Converse and Miller, 1999) or sand (Grewal *et al.*, 1999). Behavioral bioassays may also be used to assess quality, but need to be selected based on the behavioral and ecological characteristics of the system (Glazer and Lewis, 2000). For example, nictation or standing behavior is correlated with infectivity of ambush foraging nematodes used against mobile hosts (Campbell and Gaugler, 1993).

B Sampling

There is an extensive body of literature that deals with all aspects of sampling of stored-product insects to evaluate population dynamics and impact of management tactics. Some of these methods can also be used to evaluate prevalence of pathogens in populations. Some general references on sampling techniques and the issues associated with sampling in stored-product environments include Subramanyam and Hagstrum (1995b), Subramanyam and Hagstrum (2000), and Campbell *et al.* (2004).

Monitoring strategies and tactics differ between bulk stored raw commodities and processed commodity facilities. Insect monitoring can involve sampling of the commodity itself using visual inspection or traps to determine if a patch of resource is infested, or indirect sampling of the insects dispersing among resource patches using tools such as pheromone traps. Bulk-stored commodity monitoring relies primarily on direct sampling of grain to determine density of insects or traps placed in the commodity or the headspace. For processed commodity facilities, direct sampling is difficult, so there is more reliance on indirect sampling using traps to capture dispersing individuals (Campbell *et al.*, 2004).

Traps are very useful for sampling adults, but are less efficient for insects of limited mobility, most importantly larvae. Because pathogens are most often transmitted to larvae, many of which do not survive to the adult stage, trapping alone may not be adequate to assess pathogen prevalence and impact. The larvae of some important grain pests, such as *Sitophilus* spp. weevils, lesser grain borer, and Angoumois grain moth, develop inside kernels making their detection more difficult. Detection in bulk grain is challenging because of the need to estimate the number of insects and/or pathogens present in a very large volume, but, because of the small volume of samples relative to the total volume of stored grain, low density and non-uniform distribution of insects, and the difficulty in taking samples from throughout the grain mass, extrapolation from the sample data is not highly accurate. When the pathogen prevalence is also low, extensive sampling is necessary to obtain adequate data. Insect distribution is uneven, and sampling may need to be concentrated in the areas of high density in order to obtain samples that are large enough to yield meaningful pathogen prevalence data. Unfortunately this approach introduces a bias because pathogen transmission rates are dependent on host density, and pathogen prevalence would be expected to correlate with density. Development of effective traps for stored-product insects has long been an area of major effort, resulting in a substantial body of literature covering all aspects. Attractant traps, reviewed by Burkholder and Ma (1985), Chambers (1990), and Subramanyam and Hagstrum (1995) are very useful for monitoring pest populations, but of limited value for pathogen monitoring. There are commercially available traps with either sex or aggregation pheromones and/or food attractants. Some trap types target flying adults, such as the sticky traps used for monitoring moths. Other trap types capture either individuals moving in bulk grain or walking on floors or walls in food facilities such as food processing plants, mills, or warehouses. Pheromone traps are effective primarily for adults, and the sex pheromone traps attract only adult males. Traps using food attractants, sometimes used in combination with pheromones, have some limited ability to capture larvae.

Pitfall and probe traps, reviewed by Cuperus *et al.* (1990), White *et al.* (1990), and Reed *et al.* (1991), are the best trap option for walking insects, including larval stages. Pitfall traps can be placed on floors in food facilities, and some traps can be placed on the surface of the grain. Since these traps can use pheromones and food attractants, as well as passively capturing individuals, they can be used for a wide range of species in a wide range of environments. Probe traps are used in bulk grain but require labor-intensive servicing, often in confined spaces that are subject to occupational hazard regulation. An automated system has been developed that counts insects as they fall through an infrared beam in the probe trap (Shuman *et al.*, 2005). It automatically sends counts and size dimensions to a computer where counts can be recorded and limited insect identification can be made. This technology has been commercially developed under the trade name StorMax Insector by OPI Systems of Calgary, Canada with the US Department of Agriculture. With both of these trap types, pathogens can be identified from the insects that accumulate in the wells of the traps. Since a sticky surface is not used in these types of traps, processing of captured insects for the detection of pathogens is easier than it would be for insects captured in sticky traps.

Adequate larval sampling of bulk commodities such as grain requires the processing of considerable volumes. Various sampling tools (*e.g.*, grain trier, pelican sampler, vacuum probe) are available for collecting grain samples depending on the volume of grain to sample and whether the grain is in a bin or being moved (Hagstrum, 1994;

Subramanyam and Hagstrum, 1995). Pneumatic grain samplers, such as Probe-A-Vac (Cargill, Minneapolis, MN) and Vac-A-Sample (Seedburo Equipment Co, Chicago, IL) are used to extract grain samples from a range of depths in stored grain. These samples are then sieved using either a hand sieve or an inclined sieve to remove external insects from the grain (White, 1983). A motorized incline sieve, such as the Insectomat (Samplex, Willow Park, UK), can be used to process large numbers of samples.

Because the larvae of many of the important bulk grain pests are internal feeders, they are difficult to detect using sieving. A range of techniques to detect internal feeding insects have been developed (e.g., staining, flotation, x-ray examination, sound detection, nuclear magnet resonance, ELISA, near-infrared reflectance spectroscopy), but most are still relatively labor and time intensive to perform on large amounts of grain (Pedersen, 1992; Dowell *et al.*, 1998). With some of these techniques, infested kernels can be identified and dissected to extract the insect. The presence of insects inside of kernels and the age structure of a population can also be measured by holding the grain samples and sieving at different time points to recover the emerged adults.

The patchy landscape structure and uneven distribution of insects in stored-product habitats must be taken into account in any sampling strategy (Campbell *et al.*, 2004). Spatial mapping has often been used to show how insects spread through habitats and where infested patches occur (Campbell *et al.*, 2002). These maps can be useful in selection of areas for concentrated sampling to determine pathogen prevalence.

C Experimental Protocols

1 Evaluation of bacteria in bulk stored grain

Application of microbial control agents to stored products, especially grains, is similar in technique to the use and application of synthetic chemical protectants (Brower *et al.*, 1996). The typical on-farm method of storing grains employs cylindrical, corrugated, galvanized steel grain bins. For testing bacteria efficacy, bin capacity can vary from 10–50 bushels (bu) or ca. 2.5–12.5 m^3 for pilot tests, but should not be less than 300 bu or ca. 10 m^3 for small on-farm or commercial storage. Ideally, there should be enough bins to replicate each treatment and control three to five times since there can be variation in insect density and species composition among bins. Frequently the number of available bins will not permit the desired replication, especially when there are several treatments. However, no fewer than three replicates per treatment should be performed. Treatments and controls are generally laid out in a randomized complete block design with blocking over locations or years. An ideal field research facility would have all grain bins in one location for synchronous replication of treatments. Aeration fans should be attached to the bins to maintain a uniform temperature in the grain mass and keep the temperature as low as practical. In general, grain temperature should not exceed 30 °C and moisture content should range from 10–14%, depending on grain type, geographical location, and storage time (Foster and Tuite, 1992). The experimental grain used should be from the current crop year. Studies can use natural infestations or bins can be tightly sealed and screened and known numbers of insects added to the bins to better standardize conditions (Flinn and Hagstrum, 2001).

Adult and larval stored-grain moth pests, unlike beetle pests, do not penetrate very deeply into the grain mass (McGaughey, 1978; Subramanyam and Cutkomp, 1985; Cox and Bell, 1991). Thus, it is not necessary to treat the entire grain mass for them. It is most effective and least costly to treat only the top 10 cm of the grain mass (McGaughey, 1978) with the quantity of treatment applied being proportional to area rather than volume. Dipel®2X (Valent USA) and Javelin WG (Certis USA), two *Bt* products registered for use on stored grain and peanuts in the US, are both labeled for application to the top 10 cm layer of grain.

An aqueous suspension of *Bt* should be applied to the grain surface and raked in to a depth of 10 cm, or it should be applied to the last 10 cm portion of the grain while it is being augered onto the top of the grain mass. Depth of treatment influences extent of control (McGaughey, 1978), and the depth of treated grain is more uniform using the latter method. The treated grain should then be leveled with a grain rake.

Sample Protocol

a Bin preparation

1. Two weeks before placing grain in previously used bins, remove all residual grain and debris by sweeping and vacuuming.
2. Disinfest bins by applying an appropriately labeled product such as Cyfluthrin (Tempo®). All bin surfaces, cracks, crevices, and the area beneath false floors should be treated by spraying directly onto/into each surface. Heat treatments are an alternative approach.

b Application of Bacillus thuringiensis

1. The amount of treated grain required to fill the experimental bins to a depth of 10 cm on top of the untreated grain should be predetermined and loaded into a grain wagon.
2. The time required to load this volume of grain into bins should be determined with a stop watch.
3. Similarly, the appropriate amount of water to treat the grain is predetermined based on the volume of grain in the top 10 cm of the grain mass.
4. Calibrate spray for synchrony with the duration of grain flow to empty the wagon so that all of the *Bt* suspension is applied as the grain flows past the sprayer nozzle.
5. To treat the grain, open the discharge chute at the bottom of the grain wagon allowing the grain to gravity feed from the wagon. Make sure grain flows freely down the chute at all times and spray the grain stream with the *Bt* suspension using a pump up hand sprayer as the grain falls into the auger hopper. The *Bt* suspension should be constantly agitated during this process.
6. Grain for untreated check bins is treated in an identical manner except it is sprayed with water only.

c Insect sampling (pre- and post-treatment)

1. Sample flying adults in bin head space using double-sided yellow sticky traps or baited or unbaited pheromone traps hung in the headspace, and/or glue boards placed on the sides of the bin.
2. Hang one trap in the center of the bin and one at each cardinal direction 0.5 m above the grain mass and 30 cm from the bin wall. Number of traps needed depends on the size of the bin.
3. Spool traps constructed of rolled corrugated cardboard strips measuring 2 by 50 cm are placed on the grain surface to enumerate wandering larvae, pupae, and empty cocoons in the grain mass.
4. Sample adult and larval populations weekly to determine average number present before the grain is treated.
5. If natural populations are inadequate, seed each grain mass monthly for 2–3 months with 1,000 eggs of Indianmeal moth and/or almond moth (*i.e.*, 200 eggs in the center and at each cardinal direction). Adjust timing to coincide with normal Indianmeal moth and almond moth seasonal phenology for specific region. Eggs can be counted directly or can be quantified using weight or volumetric estimates.
6. Sample bin head space and grain masses at 1 or 2 week intervals.
7. Wrap collected sticky cards in clear plastic kitchen wrap, label with date, bin number, and location and replace with a new sticky card during each sampling event. Pheromone traps should be replaced if trap captures or dust accumulation is high, but if trap captures are low, individual insects can be removed from traps on site. Pheromone lures should be replaced every two months. Spool traps should be placed in labeled plastic bags and returned to the laboratory along with sticky traps for identification and enumeration.

d Sampling for damage to grain

1. Remove one-liter grain samples from the center and 50 cm from the wall of each bin. Randomly select 100 g from each to examine and quantify feeding damage (Manis, 1992; Price *et al.*, 1999).
2. Sift grain with a sieve with openings that retain whole grains and quantify insect contaminants (*i.e.*, frass and body parts) from each sample numerically or by weight.
3. Estimate amount of webbing per bin by placing a metal frame of ca. 0.1 m^2 on the grain surface in three randomly selected locations and visually quantifying webbing covering the grain within each frame. Extrapolate to estimate total surface area webbed within each bin.

e Assessing persistence

1. At monthly intervals collect one 0.5 liter sample of treated grain from the treated zone in the center and at each cardinal direction from each treated and control grain bin. Homogenize the kernels from within each bin in a sealable plastic bag.
2. After removal of grain from the bins, place it in a freezer for several days to kill existing insects.
3. Place 150 g of grain in each of five 0.5 liter assay vessels and infest with 50 eggs or neonate larvae. The infested jars of grain should then be placed incubated at 27 °C and ≥ 60% RH and adult emergence quantified after 30 days.

f Environmental measurements

Temperature of the *Bt* treated portion of the grain masses (top 10 cm) at central and four peripheral locations in the grain bins should be taken with data loggers or at weekly intervals during the morning using thermocouples. Measure grain moisture content at sample periods with a meter or by drying samples in an oven at 100 °C for 3 days and measuring the percent weight loss.

2 Evaluation of baculoviruses on bulk-stored dried fruits and nuts

In general, most of our experience consists of using baculoviruses as protectants against reinfestation of moths after a suitable disinfestation treatment such as fumigation has been performed. It is envisioned that such applications would be utilized for protection during long-term storage or while the packaged commodities are in marketing channels prior to consumption.

The high price of dried fruits and nuts necessitates that initial studies be conducted in smaller containers such as 0.9, 3.8, 19 liter or larger containers, such as trash cans, depending on the size of the commodity unit (*e.g.*, raisins versus walnuts) (Hunter *et al.*, 1973; 1977; McGaughey, 1975; Kinsinger and McGaughey, 1976, Cowan *et al.*, 1986). These commodities may also be tested using packaged retail containers (*e.g.*, boxes, cellophane wrapped, etc.). Dried nuts may also be tested in 22.7 kg bulk shipping bags. If small scale pilot studies demonstrate efficacy, larger scale tests using larger quantities of materials may be conducted. For the pilot-scale tests, three to five replications are adequate to achieve sufficient sensitivity for statistical analyses. Neonate larvae or eggs are most commonly used for small scale commodity tests. There are approximately 50 Indianmeal moth eggs/mg, and hatch is generally about 95%. For small scale screening using 0.9 or 3.8 liter containers, 5 or 20 mg of eggs, respectively, should be placed on the commodity. Alternatively, neonate larvae can be used. Intermediate and larger scale tests typically are infested with eggs. Infesting with eggs is quite rapid when large numbers of larvae are required. Subsamples of eggs should be retained to quantify hatch and, thus, provide an estimate of the number of larvae placed on the commodity (Hunter *et al.*, 1973, 1977; Cowan *et al.*, 1986). In large scale semi-commercial tests, mated pairs of moths may be used instead to infest the commodities over a period of time with a suggested release rate of 10 pairs/week (Johnson *et al.*, 1999). Insects *in copula* are removed from colonies and placed on commodities. It is essential to know that females have mated so that infestation levels in different containers are uniform (Johnson *et al.*, 1999). In addition to larvae and moth counts and number of infested units, quantifying quality of representative units of the commodity in terms of damage by weight or volume or silk production is an excellent indicator of infestation and of control attained (Cowan *et al.*, 1986). Damage may be categorized in a number of ways such as undamaged, pin-hole, moderate, or severe. Alternatively, commercial grading systems can be used.

The commodity should be fumigated to disinfest before efficacy testing. Phosphine, sulfuryl flouride or, until its phase out, methyl bromide can be used depending on the commodity being tested. Freezing may also be used but may affect quality.

Aqueous suspensions or dry formulations of OBs should be applied as the commodity is placed in storage or when it is being packaged. Application should be made at a location in the handling procedure, such as on a conveyor belt or in an auger stream, that provides a thin layer, preferably a monolayer, on the commodity. There should be adequate agitation to insure uniform coverage. Aqueous sprays should be maintained at neutral or slightly acidic pH because baculoviruses are sensitive to high pH, which causes dissolution of OBs and considerable reduction in activity. Water added in liquid application to the commodity should not exceed 0.1% by weight, while dry formulations should not exceed 0.5% by weight (Cowan *et al.*, 1986). For aqueous applications, a wetting agent such as Triton X-100, Tween 80, Silwet L-77 should also be incorporated. As a starting point, the upper 95% confidence limit of the LD_{99} based on bioassays of the OBs/formulation as the application dose (expressed as mg/kg) of commodity can be used. Surface applications to the top 10 cm as used for grains have not been used for dried fruits and nuts. If OBs are to be diluted for dry application, ground (60 mesh) stabilized wheat germ can be utilized as a diluent. The dry OB technical material must be thoroughly mixed with the diluent. Similar procedures are used to treat commodities with OBs as with procedures described under *Bt*.

Sample Protocol

a Calculation of dosage and application procedure

1. Calculate doses by extrapolation from basic laboratory assays as mentioned above. The amount of PiGV powder to be used in an appropriate volume of water/dose is pre-determined and weighed off-site in the laboratory. Dry formulations may also be used.
2. Calibrate commodity treatment by adjusting commodity flow and OB delivery through hand held sprayer or larger pressurized sprayer while commodity moves through an auger or on a processing belt. Coverage can be monitored with either dry or wet applications by using fluorescent dyes or powders.
3. Treat a known volume or weight of commodity with OB by spraying a suspension or by applying a dry formulation. Aqueous applications should be made with cone-type nozzles that provide the minimum volume and coverage required to treat the commodity in the processing line. Coverage is extremely important as the viruses are only orally active.
4. Treat controls with the same volume of water or dry formulation without active ingredient.
5. Divide treated commodity into equal volumes and place into appropriately labeled containers and cover with muslin and a ventilated lid after seeding with eggs, larvae, or adults. Employ a randomized complete block design.

b Moth sampling methods

1. Seed commodity in each container with a counted, weighed or volumetrically determined number of eggs, neonate larvae, or mated pairs of adults.
2. Sample larval populations with spool traps to determine number present. Place 2–5 spool traps (depending on container size used and as described above) in the containers and replace and quantify larvae bi-weekly.
3. Quantify moth emergence by direct counts on the sides of containers/storages or by placing a pheromone baited sticky trap (Pherocon® II or Pherocon® 1C Trap, Trécé, Inc., Adair, OK) in the containers. Count moths weekly or check and replace traps bi-weekly for the duration of the test. Trapped moths may be quantified on-site or placed in clear plastic wrap and examined in the laboratory.

c Sampling methods to quantify damage to dried fruits and nuts

1. Remove a sample of dried fruit or nuts from each container. The amount will vary depending upon the size of the containers, but 100 units are typically selected at random.
2. Characterize and quantify damage caused by larvae as, undamaged, moderate if $< 33\%$ of the fruit or nuts is damaged, or severe if $> 33\%$ is damaged (Cowan *et al.*, 1986).
3. Weigh or measure volume of undamaged fruit or nuts.
4. Compare with control and/or industry standards.

d Assessing persistence

1. Collect enough units of the commodity to fill a 3.8 liter jar at monthly intervals throughout the duration of storage. Replicate samples should be taken.
2. Place the commodity sample in the freezer at −23 °C for one week to kill any live insects in the sample.
3. Place the commodity removed from the freezer into a 3.8 liter mason jar, allow it to come to

room temperature and infest with 20 mg of eggs or neonate larvae.

4. Place the jars containing commodity and eggs or larvae into an environmental chamber at 27 °C and ≥ 60% RH.
5. Quantify emerging adults after 30 or more days, depending on the commodity, until the microbial control agent is no longer effective. Damage may also be determined.

e Monitoring environmental conditions

Quantify temperature and moisture in the same manner as reported for *Bt*.

3 Evaluation of Protozoa on bulk grain

Protozoa field work in stored products will be restricted to assessment of their impact as naturally occurring and introduced chronic pathogens. Finding appropriate sites for assessment of long term pathogen impact is difficult but critical for success. An insect sampling history and an established relationship with a site's owner and workers are advisable. Initial surveys to determine the suitability of a site can be carried out by adult sampling with probe traps in grain storage bins. Computerized probe monitoring with the Insector system would minimize the sampling effort. Conventional sampling can also be done with triers and sticky traps. Because infections with most pathogens of interest are predominantly larval, adult samples serve primarily to quantify pest populations. Processing of large volumes of grain is necessary to obtain larval samples for determination of pathogen prevalence. This is best done with mechanized equipment. Samples can be collected with a pneumatic grain sampler such as a Probe-A-Vac and motorized incline sieves, such as Insectomat can be used to separate the insects. All insects of interest should be identified to species where practical or to genus as a minimum. The maximum practical number of locations will minimize the problems that arise in research that is conducted on working commercial facilities.

The predominant naturally occurring pathogens in pests of stored grain are PiGV, neogregarines, and Microsporidia (Moore *et al.*, 2000). Infections with these pathogens can be assessed microscopically, but doing so requires a great deal of tedious and uncomfortable labor and is prone to operator error. If the goal is extensive assessment of a specific organism, an enzyme-linked immunosorbent assay (ELISA) can be used to process more insects more rapidly and with greater reliability than be done with visual processing. The development of an ELISA requires a substantial commitment of time and money and is not practical for short-term work. For example, polyclonal antiserum for *Mattesia oryzaephili* cross-reacts with eugregarines and neogregarines, necessitating the procurement of monoclonal antibodies (Lord, 2007).

The following sample protocol is for *M. oryzaephili* and its principal host *C. ferrugineus*. The inoculum can be produced by exposing forth instar *O. surinamensis* to 10^6 oocysts/g of diet then incubating for 10 days before macerating larvae that develop fluorescence indicating heavy infection.

Grain bins on working farms or farm-scale bins can be used for studies of *M. oryzaephili* inoculative release for suppression of *C. ferrugineus* or monitoring population effects of natural occurrence. If working bins are used, then availability will dictate the physical parameters and insects present, making careful and complete recording of the conditions and fauna especially important. The following protocol is suggested for dedicated experimental bins.

Sample Protocol

1. Seal bins except for screened ports for aeration fans and roof ventilation. Clean thoroughly.
2. Fill with newly harvested grain.
3. To simulate natural insect immigration that occurs during the summer, add 40 *C. ferrugineus* adults per m^3 of grain to the bins each month from July to October.
4. Introduce the *M. oryzaephili* inoculum onto grain in four 0.01 m^3 cages with 16 mesh screening to allow passage of insects, but not grain. Apply in wheat flour with 10^6 oocysts/g at a rate of 1 g per cage.
5. Monitor temperatures continuously at the center of the bin and at a distance of 0.6 m from the side wall in each of the four cardinal directions using five monitors, such as HOBO H8 4-channel data loggers. The sensors in each bin are inserted into

the grain mass at depths of 0.15 m, 0.8m, 1.5m, and 2 m from the grain surface.

6. Assign two treatments (*M. oryzaephili* and control) randomly to six total bins.
7. Sample the grain in each of the bins for insects at monthly intervals starting in July using a pneumatic grain sampler. Three kg samples in each of three 0.8 m layers of wheat should be taken at three points 0.3 m from the bin center and at four points 0.6 m from the bin wall for a total of 21 samples from each bin.
8. All of the samples are processed with a motorized inclined sieve (Insectomat), to separate insects from the grain.
9. Insects are identified and subjected to the ELISA to determine *M. oryzaephili* prevalence.

4 *Evaluation of pathogens in processing, warehouse and retail environments*

Evaluation of the presence of pathogens in insect populations outside of bulk storage (e.g., in processing facilities, warehouses, retail stores, etc.) is more challenging due to the difficulties in sampling discussed above. Applications of pathogens to these types of environments is also difficult because of current limitations on using pathogens in commercial food facilities, the temporally and spatially patchy pest population distributions, the high levels of insect dispersal, and the challenges in evaluating pest population density. As a result, this type of field research will likely involve simulations of field conditions in which pest populations can be more readily sampled. Fortunately, given that targeted locations for field applications are in human created environments that are already relatively simplified, these simulations are more likely to be more realistic than in many other scenarios such as soil applications.

Pathogen applications are likely to be in the form of a treatment that is applied directly to flooring and wall surfaces and may be applied to cover surfaces or be limited to specific areas, such as cracks, crevices, wall baseboards, or as spot treatments to defined target areas. Treatment of spillage and residues outside of facilities is another potential target if pests emerging from these sources then move into food facilities. These types of experiments can be conducted using replicated locations in structures that are not necessarily food facilities, in replicated sheds or other small structures, or by using forms designed to replicate target locations such as cracks in a cement floor. Because each situation is unique and depends on the target insect, pathogen and environment, it is hard to develop a single specific procedure. There have also been relatively few studies of this type using pathogens. An example protocol from a study conducted to evaluate different pesticides application methods on pest populations under simulated warehouse sheds (Toews *et al.*, 2005) will be used to illustrate how these types of studies can be conducted. This design was used to evaluate the impact of treatments on pest populations in hidden structural refugia and involved both indirect sampling using pheromone traps and direct sampling of the pest populations in hidden areas that would not be easily assessable under field conditions. The basic approach is easily customizable to a particular research question.

Sample Protocol

1. Conduct research in small pilot-scale warehouses that are tightly sealed and ideally have temperature control. For example, shed buildings with interior dimensions of 2.8 m wide by 5.9 m long by 2 m tall with all interior surfaces sealed and coated with food grade epoxy. To reduce or prevent contamination of the building by treatments, the interior of the shed can be lined and surfaced to simulate field conditions. For example, the walls and floor of each warehouse can be covered with a contiguous piece of 0.15 mm thick polyethylene sheeting, and tape used to seal the edges of the sheeting. The entire floor can then be covered with 1.3 cm thick gypsum panels. The joints between gypsum panels should be filled with joint compound and sanded smooth, and the floor-wall junctions sealed with silicone sealant. To simulate a finished concrete floor in a commercial warehouse, the gypsum panel-covered floor can be coated with primer and sealer and top-coated with acrylic concrete floor sealer.
2. Environmental conditions can be controlled to correspond to conditions found in target location and conditions monitored by placing data loggers in each shed.

3. To simulate insect populations in hidden refugia, each warehouse is equipped with pilot scale shelving units to simulate large equipment or shelving units and provide structure and a place to conceal food patches. These may be custom made from a 2.5 cm square tubular steel frame, with a galvanized steel plate on the top and detachable kickplates along the sides with a small gap to permitted insect movement. Alternatively, portions of commercial shelving units, pallets or equipment can be used. Ideally, these units should be easily moveable so that insects in food patches under them can be easily sampled. They should also provide limited alternative refugia to facilitate sampling. Number of shelving units can be varied. We have used three in our experiments. Shelving units are positioned such that a sanitation buffer, a standard practice in commercial warehouses, is left between the interior warehouse walls and the shelving unit.
4. Food patches can be placed under shelving units. If testing efficacy against red flour beetle, *T. castaneum*, for example, wheat flour (10 g/patch) can be placed on 50 cm filter paper disks underneath each shelving unit. Each food patch can then be infested with 20 eggs, 20 small larvae (2nd–3rd instars), 20 large larvae (7th to 8th instars), 20 pupae, and 20 adults (1–2 wk old) to establish a population. Insect immigration can also be simulated by adding adult insects at a regular basis.
5. Treatments are applied in the pilot-scale warehouses 24 hours after insect release. Treatments can include pathogen type, concentration, application method and pattern, and corresponding untreated or insecticide treated controls. Hand-held sprayers can be used for applications. Applications can be targeted at different areas; for example the entire surface, along the inside perimeter of the shed, around the shelves, or under the shelves. After treatment, persons entering warehouses should wear disposable polyethylene overboots that are changed between each warehouse to avoid cross contamination.
6. A three step insect monitoring plan can be initiated 24 hours after application and continued on a weekly basis for as long as desired. First, dead insects can be collected from the floor of each warehouse, counted, and potentially assessed for the presence of the pathogen. Alternatively, they can be counted and left in the shed to facilitate transmission of pathogen. Second, relative estimates of insect populations can be obtained using pitfall traps (e.g., Dome traps, Trécé Inc., Adair, OK, baited with *T. castaneum* pheromone and food oil). Traps can be positioned in corners, along walls, in the middle, and under shelves. These first two types of monitoring provide the type of data that might be readily collected in the field. Third, to directly measure insect abundance in the food patches, subsamples of flour can be collected from each patch using a laboratory spatula. These patches may be pooled or processed separately. Flour samples should be taken to the laboratory, weighed, and sieved to determine the quantity of larvae, pupae, and adults present and whether they are alive or dead. Pathogen infection levels can also be tested using these collected individuals.
7. Treatments can be randomized among the sheds and blocked through time. Following each replication, the temporary floors are removed and discarded, warehouses are thoroughly vacuumed and cleaned, and new temporary floors are installed.
8. Number of dead adults and insects recovered in flour samples can be analyzed as a split-plot arrangement of a randomized complete block design with repeated measures. Individual warehouses are the main plot experimental unit for assessing effect of the applications. Week of the study can be treated as a repeated measures effect using a mixed model procedure to model the variance-covariance relationship of the response variables among weeks. Numbers of adults and larvae captured in pitfall traps can be analyzed as a split-split plot arrangement of a randomized complete block design with repeated measures. The main plot factor is the treatment, the subplot factor is the trap position, and the sub-subplot treatment factor is the week of study.

3 Summary

Development of entomopathogens effective against stored-product pests merits continuation because of the limited selection of synthetic chemical insecticides, insecticide resistance, issues of safety and public perception. In addition, potential resistance to *Bt* will affect its use in situations where chemical avoidance is desired. Successful, widespread use of *Bt* and baculoviruses will depend on our willingness to use reasonable resistance strategies. Some of those strategies can be addressed using multiple strain/toxin formulations, novel deployment strategies, and by continuing discovery phases of research aimed at identifying effective *Bt* or viral isolates for managing a broad spectrum of pests. For example, we have yet to identify effective *Bt* or viral isolates for managing stored-product beetle pests. There remain pathogens and strategies that have had only cursory evaluation for stored-product use. For example, *B. bassiana* can control some beetle pests such as the lesser grain borer and sawtoothed and rusty grain beetles in the laboratory particularly when used with desiccant (Lord, 2001, 2005) but has not been tested in field settings. Nematodes may be useful for sanitation treatment of empty grain bins. Introduction of protozoa pathogens may be used to exert suppression pressure on pest populations.

4 References

Arnott, H. J. and Smith, K. M. 1968a. An ultrastructural study of the development of a granulosis virus in the cells of the moth *Plodia interpunctella* (Hbn.). *J. Ultrastruct. Res.* **21**, 251–268.

Arnott, H. J. and Smith, K. M. 1968b. Ultrastructure and formation of abnormal capsules in a granulosis virus of the moth *Plodia interpunctella* (Hbn). *J. Ultrastruct. Res.* **22**, 136–158.

Brower, J. H., Smith, L., Vail, P. V., and Flinn, P. W. 1996. Biological control. *In* "Integrated Management of Insects in Stored Products" (Bh. Subramanyam and D. W. Hagstrum, Eds.), pp. 223–286. Marcel Dekker, Inc., New York.

Burkholder, W. E. and Ma, M. 1985. Pheromones for monitoring and control of stored-product insects. *Annu. Rev. Entomol.* **30**, 257–272.

Campbell, J. F. and Gaugler, R. 1993. Nictation behavior and its ecological implications in the host search strategies of entomopathogenic nematodes (Heterorhabditidae and Steinernematidae). *Behaviour* **126**, 155–170.

Campbell, J. F., Mullen, M. A., and Dowdy, A. K. 2002. Monitoring stored-product pests in food processing plants with pheromone trapping, contour mapping, and mark-recapture. *J. Econ. Entomol.* **95**, 1089–1101.

Campbell, J. F., Arthur, F. H., and Mullen, M. A. 2004. Insect management in food processing facilities. *In* "Advances in food and nutrition research" pp. 240–295. Elsevier, Inc., Amsterdam.

Campbell, J. F. and Mullen, M. A. 2004. Distribution and dispersal behavior of *Trogoderma variabile* and *Plodia interpunctella* outside a food processing plant. *J. Econ. Entomol.* **97**, 1455–1464.

Chambers, J. 1990. Overview of stored-product insect pheromones. *J. Kansas Entomol. Soc.* **63**, 490–499.

Chesnut, T. L. and Douglas, W. A. 1971. Competitive displacement between natural populations of the maize weevil and the Angoumois grain moth in Mississippi. *J. Econ. Entomol.* **64**, 864–868.

Converse, V. and Miller, R. W. 1999. Development of the one-on-one quality assessment assay for entomopathogenic nematodes. *J. Invertebr. Pathol.* **74**, 143–148.

Cotton, R. T. 1950. Notes on the almond moth. *J. Econ. Entomol.* **43**, 733.

Cotton, R. T. 1963. Pests of Stored Grain and Grain Products. Burgess Publishing Co., Minneapolis.

Cowan, D. K., Vail, P. V., Kok-Yokomi, M. L., and Schreiber, F. E. 1986. Formulation of a granulosis virus of *Plodia interpunctella* (Hübner) (Lepidoptera: Pyralidae): efficacy, persistence, and influence on oviposition and larval survival. *J. Econ. Entomol.* **79**, 1085–1090.

Cox, P. D. and Bell, C. H. 1991. Biology and ecology of moth pests of stored foods. *In* "Ecology and Management of Food-Industry Pests" FDA Tech. Bull. 4 (J. R. Gorham, Ed), pp. 181–193. Association of Official Analytical Chemists, Arlington, VA.

Cuperus, G. W., Fargo, W. S., Flinn, P. W., and Hagstrum, D. W. 1990. Variables affecting capture of stored-grain insects in probe traps. *J. Kans. Entomol. Soc.* **63**, 486–489.

Dowell, F. E., Throne, J. E., and Baker, J. E. 1998. Automated nondestructive detection of internal insect infestation of wheat kernels by near-infrared reflectance spectroscopy. *J. Econ. Entomol.* **91**, 899–904.

Finney, G. L. and Brinkman, D. 1967. Rearing the navel orangeworm in the laboratory. *J. Econ. Entomol.* **60**, 1109–1111.

Flinn, P. W. and Hagstrum, D. W. 2001. Augmentative releases of parasitoid wasps in stored wheat reduces insect fragments in flour. *J. Stored Prod. Res.* **37**, 179–186.

Foster, G. H. and Tuite, J. 1992. Aeration and stored grain management. *In* "Storage of Cereal Grains and Their Products" (D. B. Sauer, Ed.), pp. 219–247. American Association of Cereal Chemists, St. Paul, MN.

Gaugler, R., Grewal, P., Kaya, H. K., and Smith-Fiola, D. 2000. Quality assessment of commercially produced entomopathogenic nematodes. *Biol. Control* **17**, 100–109.

Glazer, I. and Lewis, E. E. 2000. Bioassays for entomopathogenic nematodes. *In* "Bioassays of Entomopathogenic Microbes and Nematodes" (A. Navon and K. R. S. Ascher Eds.), pp. 229–247. CAB International, Wallingford, UK.

Good, N. E. 1936. The flour beetles of the genus *Tribolium*. Tech. Bull. U.S. Dep. Agric. no. 498, 57pp.

Gorham, J. R. 1987. Insect and mite pests in food: an illustrated key. United States Department of Agriculture. Agriculture Handbook no. 655. vol. 1, 310pp.

Gorham, J. R. 1991. Insect and mite pests in food: an illustrated key. United States Department of Agriculture. Agriculture Handbook no. 655. vol. 2, 457pp.

Grewal, P. S., Converse, V., and Georgis, R. 1999. Influence of production and bioassay methods of infectivity of two ambush foragers (Nematoda: Steinernematidae). *J. Invertebr. Pathol.* **73**, 40–44.

Grewal, P. S. 2002. Formulation and application technology. *In* Entomopathogenic Nematodes (R. Gaugler, Ed.), pp. 265–287. CAB International, Wallingford, UK.

Grewal, P. S. and Peters, A. 2005. Formulation and Quality. *In* "Nematodes as biocontrol agents" (P. S. Grewal, R-U Ehlers, and D. I. Shapiro-Ilan, Eds.), pp. 79–89. CAB International, Wallingford, UK.

Hagstrum, D. W. 1994. Field monitoring and prediction of stored-grain insect populations. *Postharvest News Information*. **5**, 39–45.

Hagstrum, D. W. and Subramanyam, Bh. 2000. Monitoring and decision tools. *In* Alternatives to pesticides in stored-product IPM (Bh. Subramanyam and D. W. Hagstrum, Eds.), pp. 1–28. Kluwer Academic Publishers, Boston.

Hinton, H. E. 1945. A monograph of the beetles associated with stored products, vol. 1. British Museum (Natural History), London.

Ho, S. H. and Winks, R. G. 1995. The response of *Liposcelis bostrychophila* Badonnel and *L. entomophila* (Enderlein) (Psocoptera) to phosphine. *J. Stored Prod. Res.* **31**, 191–197.

Howe, R. W. 1956. The biology of the two common storage species of *Oryzaephilus* (Coleoptera, Cucujidae). *Ann. Appl. Biol.* **44**, 341–355.

Hunter, D. K. 1970. Pathogenicity of a granulosis virus of the Indian-meal moth. *J. Invertebr. Pathol.* **16**, 339–341.

Hunter, D. K., Hoffmann, D. F., and Collier, S. J. 1973. Effectiveness of a granulosis virus of the Indian meal moth as a protectant for stored inshell nuts: preliminary observations. *J. Invertebr. Pathol.* **22**, 481.

Hunter, D. K., Collier, S. S., and Hoffmann, D. F. 1977. Granulosis virus of the Indian meal moth as a protectant for stored inshell almonds. *J. Econ. Entomol.* **70**, 493–494.

Hunter, D. K., Collier, S. S., and Hoffmann, D. F. 1979. The effect of a granulosis virus on *Plodia interpunctella* (Hübner) (Lepidoptera: Pyralidae) infestations occurring in stored raisins. *J. Stored Prod. Res.* **15**, 65–69.

Jacob, T. A. and Cox, P. D. 1977. The influence of temperature and humidity on the life-cycle of *Ephestia kuehniella* Zeller (Lepidoptera: Pyralidae). *J. Stored Prod. Res.* **13**, 107–118.

Johnson, J. A., Vail, P. V., Soderstrom, E. L., Curtis, C. E., Brandl, D. G., Tebbets, J. S., and Valero, K. A. 1999. Integration of non-chemical treatments for control of navel orangeworm and Indian meal moth in postharvest walnuts. *J. Econ. Entomol.* **91**, 1437–1444.

Kaya, H. K. and Stock, S. P. 1997. Techniques in insect nematology. *In* "Manual of Techniques in Insect Pathology" (Lacey, L. A. Ed.), pp. 281–324. Academic Press, San Diego.

Kellen, W. R. and Hoffman, D. F. 1983. Longevity and fecundity of adult *Amyelois transitella* (Lepidoptera: Pyralidae) infected by two small RNA viruses. *Environ. Entomol.* **12**, 1542–1546.

Kinsinger, R. H. and McGaughey, W. H. 1976. Stability of *Bacillus thuringiensis* and a granulosis virus of *Plodia interpunctella* on stored wheat. *J. Econ. Entomol.* **69**, 149–154.

Lord, J. C. 2001. Desiccant dusts synergize the effect of *Beauveria bassiana* (Hyphomycetes: Moniliales) on stored-grain beetles. *J. Econ. Entomol.* **94**, 367–372.

Lord, J. C. 2005. Low humidity, moderate temperature, and desiccant dust favor efficacy of *Beauveria bassiana* (Hyphomycetes: Moniliales) for the lesser grain borer, *Rhyzopertha dominica* (Coleoptera: Bruchidae). *Biol. Control* **34**, 180–186.

Lord, J. C. 2007. Detection of *Mattesia oryzaephili* (Neogregarinorida: Lipotrophidae) in grain beetle laboratory colonies with an enzyme-linked immunosorbent assay. *J. Invertebr. Pathol.* **94**, 74–76.

Loschiavo, S. R. and Smith, L. B. 1970. Distribution of the merchant grain beetle, *Oryzaephilus mercator* (Silvanidae: Coleoptera) in Canada. *Can. Entomol.* **102**, 1041–1047.

Manis, J. M. 1992. Sampling, inspecting and grading. *In* "Storage of Cereal Grains and Their Products" (J. M. Manis, Ed.). pp. 563–588. Am. Assoc. Cereal Chemists, St. Paul, MN.

Mbata, G. N. and Shapiro-Ilan, D. I. 2005.Laboratory evaluation of virulence of heterorhabditid nematodes to *Plodia interpunctella* Hübner (Lepidoptera: Pyralidae). *Environ. Entomol.* **34**, 676–682.

McGaughey, W. H. 1975. Compatibility of *Bacillus thuringiensis* and granulosis virus treatments of stored grain with four grain fumigants. *J. Invertebr. Pathol.* **26**, 247–250.

McGaughey, W. H. 1978. Moth control in stored grain: efficacy of *Bacillus thuringiensis* on corn and method of evaluation using small bins. *J. Econ. Entomol.* **71**, 835–839.

Milner, R. J. 1972. *Nosema whitei*, a microsporidan pathogen of some species of *Tribolium*. II. effect on *Tribolium castaneum*. *J. Invertebr. Pathol.* **19**, 248–255.

Moore, D., Lord, J. C., and Smith, S. 2000. Pathogens *In* "Alternatives to Pesticides in Stored-Product IPM" (Bh. Subramanyam and D. W. Hagstrum eds.) pp. 193–227. Kluwer, Dortrecht, The Netherlands.

Pedersen, J. R. 1992. Insects: identification, damage, and detection. *In* "Storage of Cereal Grains and their

Products", 4th Edition, (Sauer, D. B. Ed.), pp. 435–489. American Association of Cereal Chemists, St. Paul.

Price, B. D., Sedlacek, J. D., and Weston, P. A. 1999. Pirimiphos-methyl and thiabendazole impact on the quality of on-farm stored shelled corn. *J. Entomol. Sci.* **34**, 225–238.

Ramos-Rodriguez, O., Campbell, J. F., and Ramaswamy, S. B. 2006. Pathogenicity of three species of entomopathogenic nematodes to some major stored-product insect pests. *J. Stored Prod. Res.* **42**, 241–252.

Reed, C. R., Wright, V. F., Mize, T. W., Pedersen, J. R., and Brockschmidt Evans, J. 1991. Pitfall traps and grain samples as indicators of insects in farm-stored wheat. *J. Econ. Entomol.* **84**, 1381–1387.

Rees, D. P. 1996. Coleoptera. *In* "Integrated Management of Insects in Stored Products" (Bh. Subramanyam and D. W. Hagstrum, Eds.), pp. 1–39. Marcel Dekker., Inc, New York.

Rees, D. P. 2004. Insects of stored products. Collingwood, Victoria.

Rees, D. P. and Walker, A. J. 1990. The effect of temperature and relative humidity on population growth of three *Liposcelis* species (Psocoptera: Liposcelidae) infesting stored products in tropical countries. *Bull. Entomol. Res.* **80**, 353–358.

Sedlacek, J. D., Weston, P. A., and Barney, R. J. 1995. Lepidoptera and Psocoptera. *In* "Integrated Management of Insects in Stored Products" (Bh. Subramanyam and D. W. Hagstrum, Eds.), pp. 41–70. Marcel Dekker., Inc, New York.

Shapas, T., Burkholdcr, W., and Boush, G. 1977. Population suppression of *Trogoderma glabrum* by using pheromone luring for protozoan pathogen dissemination. *J. Econ. Entomol.* **70**, 469–474.

Shuman, D., Weaver, D. K., and Larson, R. G. 2005. Performance of an analytical, dual infrared-beam, stored-product insect monitoring system. *J. Econ. Entomol.* **98**, 1723–1732.

Sokoloff, A. 1974. The biology of *Tribolium*. Vol. 2 Oxford University Press, Oxford, UK.

Sokolova, Y. Y., Issi, I. V., Morzhina, E. V., Tokarev, Y. S., and Vossbrinck, C. R. 2005. Ultrastructural analysis supports transferring *Nosema whitei* Weiser 1953 to the genus *Paranosema* and creation a new combination, *Paranosema whitei. J. Invertebr. Pathol.* **90**, 122–126.

Subramanyam, Bh. and Cutkomp, L. K. 1985. Moth control in stored grain and the role of *Bacillus thuringiensis*: an overview. *Res. Rev.* **94**, 1–47.

Subramanyam, B., Hagstrum, D. W., and Schenk, T. C. 1995. Sampling adult beetles (Coleoptera) associated with stored grain: comparing detection and mean trap catch efficiency of two types of probe traps. *Environ. Entomol.* **22**, 33–42.

Subramanyam, Bh. and Hagstrum, D. W. 1995a. "Integrated Management of Insects in Stored Products". Marcel Dekker, Inc, New York.

Subramanyam, B., and Hagstrum, D. W. 1995b. Sampling. *In* "Integrated management of insects in stored products" (B. Subramanyam and D. W. Hagstrum, eds.), Vol. 1995, pp. 135–193. Marcel Dekker, New York.

Subramanyam, B. and Hagstrum, D. W. 2000. Monitoring and decision making. *In* "Alternatives to Pesticides in Stored-Product IPM", (Bh. Subramanyan and D. W. Hagstrum Eds.), pp. 193–227. Kluwer, Dortrecht, The Netherlands.

Tebbets, J. S., Curtis, C. E., and Fries, R. D. 1978. Mortality of immature stages of the navel orangeworm stored at 3.5 °C. *J. Econ. Entomol.* **71**, 875–876.

Toews, M. D., Arthur, F. H., and Campbell, J. F. 2005. Role of food and structural complexity on capture of *Tribolium castaneum* (Herbst) (Coleoptera: Tenebrionidae) in simulated warehouses. *Environ. Entomol.* **34**, 164–169.

Tzanakakis, M. E. 1959. An ecological study of the Indian meal moth, *Plodia interpunctella*, with emphasis on diapause. *Hilgardia*. **29**, 205–246.

Vail, P. V. 1991. Novel virus composition to protect agricultural commodities. U.S. Patent No. 5,023,182.

Vail, P. V., Tebbets, J. S., Cowan, D. C., and Jenner, K.E. 1991. Efficacy and persistence of a granulosis virus against infestations of *Plodia interpunctella* (Hübner) (Lepidoptera: Pyralidae) on raisins. *J. Stored Prod. Res.* **27**, 103–107.

Vail, P. V., Hoffman, D. F., and Tebbets, J. S. 1993. Autodissemination of *Plodia interpunctella* granulosis virus by healthy adults. *J. Stored Product Res.* **29**, 71–74.

Weiss, S. A. and Vaughn, J. L. 1986. Cell culture methods for large-scale propagation of baculoviruses. *In* The Biology of Baculoviruses. Vol. II, (R. R. Granados and B. A. Federici, Eds.). pp 64–87. CRC Press, Inc. Boca Raton, FL.

White, G. G. 1983. A modified inclined sieve for separation of insects from wheat. *J. Stored Product Res.* **19**, 89–91.

White, N. D. G., Arbogast, R. T., Fields, P. G., Hillman, R. C., Loschiavo, S. R., Subramanyam, Bh., Throne, J. E., and Wright, V. F. 1990. The development and use of pitfall and probe traps for capturing insects in stored grain. *J. Kans. Entomol. Soc.* **63**, 506–525.

Williams, G. C. 1964. The life-history of the Indian meal-moth, *Plodia interpunctella*(Hübner) (Lep.: Phycitidae) in a warehouse in Britain and on different foods. *Ann. Appl. Biol.* **53**, 459–475.

Mention of trade names or commercial products in this publication is solely for the purpose of providing specific information and does not imply recommendation or endorsement by the U.S. Department of Agriculture

Chapter VII-20

Microbial control of urban pests – cockroaches, ants and termites

Richard J. Milner
CSIRO
Division of Entomology
Canberra, ACT 2601, Australia

Roberto M. Pereira
Urban Entomology Laboratory
Entomology & Nematology Dept.
University of Florida
Gainesville, FL 32611, USA

1 Introduction

Urban pests, represented in this chapter by cockroaches, ants and termites, present a unique challenge for microbial control. They are all known to detect and avoid both chemical and microbial pesticides, they have few natural enemies to integrate with microbial pesticides, the relatively constant environment is often inhibitory to infection by fungi, and few naturally occurring pathogens have been identified to develop. Add to this the expectation of householders to have quick, long-term control the pest and the difficulty of obtaining funding for research on the use of pathogens for urban pest control (Oi and Hinkle, 1997), and it becomes obvious why this is an under-researched area. However, there are incentives; an increasing proportion of householders and owners of buildings are becoming concerned about the use of synthetic chemicals because of allergies, other health concerns and, in the case of landlords, the threat of legal action. In addition, buildings offer protection from UV and extremes of temperature thus enhancing the persistence of microbial pesticides.

The scale of damage caused by these pests and the value of the urban pest control market has led a number of companies to consider possible investments to develop products for this market. Unfortunately, data on the efficacy of these products are rare in refereed journals so most of the methods described in this chapter have been developed by University or Government scientists.

Also needed are studies on the dynamics of insect pathogens when applied to urban environments, especially inside buildings. Humidity and temperature requirements during storage and after application must be determined for entomopathogens to be used in urban environments. Only a few papers report field-testing of pathogens and the methods used are similar to those developed for testing chemical pesticides. Pathogens offer the potential advantages of being more acceptable in a household situation, recycling within the pest population, more selective than chemicals and compatible with predators and parasites. As such they can be key components of an integrated pest managemen system.

L.A. Lacey and H.K. Kaya (eds.), Field Manual of Techniques in Invertebrate Pathology, 695–711.

2 Cockroaches

Cockroaches are one of the most familiar of all domestic pests, and they thrive in buildings provided there is a source of warmth, moisture and food. However, their pest status remains controversial. Cockroach problems have decreased drastically in recent years due to new active ingredients, formulations and applications techniques used in roach control (Robinson, 1999). As a consequence, research on biopesticides for roach control has also received less attention. Also, despite being capable of transmitting a variety of human pathogens, they have rarely been convincingly implicated in outbreaks of disease in the population (Roth and Willis, 1957; Baumholz *et al.*, 1997). Rather, their pest status is due to them being "disagreeable" that "their presence in an urban situation can be psychologically disturbing and cause considerable mental distress" (Piper and Frankie, 1978). They also cause allergies and this is of increasing concern in developed countries.

Of the 4,000 or so known species of cockroaches, only a dozen or so can be considered as pests. The German cockroach, *Blattella germanica*, is by far the most serious pest and occurs all over the temperate world. Generally, cockroaches are regarded as less serious pests in the tropics where the more open houses and warmer climates result in cockroaches flying in and out of houses; consequently, their populations do not build up to the size that cause such anguish in the temperate world. Two other species, the American cockroach, *Periplaneta americana*, and the oriental cockroach, *Blatta orientalis*, are also common pests. Other more minor pests will not be considered in this chapter.

A Biology of cockroaches

Cockroaches are hemimetabolous and among the most primitive of insects. *B. germanica* is a small cockroach growing up to 10–15 mm long. It favors temperatures around 30 °C, can complete its life-cycle in as little as 6 weeks, and each female can produce around 300 eggs. Thus, populations of German cockroaches can build up to high levels in a short time. *B. orientalis* is a larger species measuring 20–24 mm long. It is more tolerant of cool temperatures than the German cockroach. At 28 °C, it can complete its life-cycle in less than 1 year, while at cooler temperatures this time can be greater than 2 years. A female can produce over 150 eggs. The American cockroach is a very large insect measuring 28–44 mm and is the most serious cockroach pest of the tropics and subtropics. Warm conditions are needed with the optimum temperature range extending up to 33 °C. The life-cycle can be as short as 1 year. The adults often live for over a year, and females can produce 15 or more oothecae each containing about 18 eggs.

B Pathogens of cockroaches

Cockroaches are remarkably free of natural pathogens. The literature on cockroach pathogens was reviewed by Suiter (1997). The pathogenicity of many of the described "pathogens" is unknown and few have been field-tested. The most promising of these pathogens are fungi, such as *Cordyceps blattae*, which were shown by Roth and Willis (1960) as being highly pathogenic for German cockroaches. Various Microsporida and Haplosporida have also been reported but as with *Cordyceps* spp., these pathogens are difficult to mass produce and use as microbial insecticides. Lonc *et al.* (1997) reported that some strains of *Bacillus thuringiensis* subsp. *kurstaki* caused up to 45% mortality when fed at high concentrations to cockroaches (*B. orientalis, B. germanica* and *P. americana*). The only pathogens to have been effectively field tested are nematodes in the genus *Steinernema* (Appel *et al.*, 1993), while the fungus, *Metarhizium anisopliae*, was available for some years as BioPath™ an EcoScience product sold for control of cockroaches in the USA (Gunner *et al.*, 1990), but later withdrawn from the market. There is also a report of various isolates of *Beauveria bassiana* (Zukowski and Bajan, 1996) and *Paecilomyces fumosoroseus* and other fungi (Steenberg and Vagn-Jensen, 1998) being pathogenic to German cockroaches.

C Field testing of pathogens as microbial insecticides of cockroaches

Cockroaches live in environments that are usually too dry for either nematodes or fungi to recycle and there are no significant natural enemies. Therefore, to be commercially successful, a microbial insecticide has to provide a level of control equivalent, or close to equivalent, of a chemical. A slow kill may be acceptable, given the safety advantage of pathogens.

The two methods of field-testing chemicals are arena trials and tests under real-life conditions. The former allows reproducible results and can enable valid comparisons among products. However, arena trials may give misleadingly good results as inevitably they are an oversimplification of the real world. It is important that the virulence of the pathogen and the behavioral response of the cockroach to the formulation being used must be determined. Methods for testing virulence have been generally covered in *Manual of Techniques in Insect Pathology* (Goettel and Inglis, 1997). Testing for repellency is normally done using "Ebeling Chambers" (Ebeling *et al.*, 1966). No reports have been published on the repellency or acceptance of the BioPath chambers. For locusts, it has been shown that oil formulations of *M. flavoviride* enhance pathogenicity especially at low humidities (Bateman *et al.*, 1993). This led to EcoScience developing a modified BioPath chamber using an oil carrier (Prior, 1996); however, this version was never marketed. Interestingly oil-based formulations of chemical pesticides are more repellent than other formulations (Ali *et al.*, 1992). Thus, while the oil formulation may enhance the efficacy of the *Metarhizium* species, this benefit could be offset by a lack of acceptance by cockroaches.

Arena trials have been described by Le Patourel (1996) to assess the efficacy of a chemical pesticide pyrethroid WP and can be used for biologicals. The main features of this type of trial are:

1. Conduct tests under environmental conditions that are favorable for cockroaches such as a temperature of 28 °C, relative humidity at 50–60%, 12:12 light:dark cycle, with food and water provided *ad lib*.
2. Use wooden arena boxes, at least 50 × 120 × 13 cm, with the base and one end lined with filter paper, and the remaining three walls lined with glass plates. A harborage should be provided and the food and water placed about 5 cm away from the wall furthest from the harborage.
3. Introduce at least 50 adult German cockroaches into each arena and allow 3–4 days for them to acclimatize before applying treatment.
4. Apply candidate microbial pesticide either by spraying the plywood panels or by placing bait stations such BioPath chambers in various positions within the foraging areas of the arenas just prior to a dark phase to minimize disturbance.
5. Count the number of insects in the foraging area at intervals such as every 14 days and record the mortality.

Ying *et al.*, (1996) tested the EcoScience BioPath chambers, containing *M. anisopliae* conidia, against German cockroaches in an arena test and reported that with 2 chambers/m^2, the LT_{50} values were 4.1 and 11.9 days for male and female cockroaches, respectively. It is likely that relative humidity of the arena will affect the efficacy of the mycoinsecticide, and consequently, a relative humidity much higher than 50% may be preferred. Since there is transfer of the fungus from live cockroach to live cockroach (termed Horizontal Transfer™ by EcoScience), it may be important to continue the tests for at least 1 month to show the full effect. Kaakeh *et al.* (1996) reported that fungus-killed cockroaches were not cannibalized suggesting an avoidance by the healthy cockroaches. The use of combinations of pathogens with other products (Pachamuthu and Kamble, 2000; Zurek *et al.*, 2002; Lopes, 2005) may increase field efficiency of pathogens.

High levels of control are more difficult to achieve in real-life situations where there are a complex series of harborages, variable abundance of food and potential movement both within an individual apartment and other apartments in the same block, usually through plumbing ducts. Publicly owned apartment blocks are good sites for testing as they offer a number of similar apartments, enabling replication, and cockroaches are often a problem. Manweiler *et al.* (1993) tested the efficacy of *Steinernema carpocapsae* All strain for control of German cockroaches in apartments. While the nematodes gave comparable control to the chemical pesticide "Combat" (The Clorox Company, Pleasanton, CA, USA), both treatments left a significant residual populations and the authors concluded that improvements were needed such as the use of attractants in the nematode stations and improved placement.

The important steps in this type of trial are:

1. Select at least 30 similar apartments and place into three groups so that the cockroach populations within each group are as similar as possible.
2. Place 10 commercial sticky traps in specific locations between a vertical surface and an appliance. Traps should be left in place for 7 days and then returned to the laboratory for counting. The population density is assessed using these traps before treatment as well as at various intervals up to 12 weeks post-treatment.
3. For the microbial treatments, place 12 stations per apartment baited with nematodes or another pathogen with 11 in the kitchen and one in the bathroom. They should be replaced as required [Manweiler *et al.* (1993), replaced the nematode stations after 4 and 8 weeks]. Chemical insecticide stations should be used as a positive insecticide control and placed in a similar manner in 10 other apartments. These should be replaced in accordance with the manufacturer's instructions. At least six apartments should be used as untreated controls.
4. Assess the results by calculating the mean percentage change from the controls at different time intervals (for example, monthly) for both the microbial treatments and the chemical treatments.
5. Repeat the experiment at a different time of the year.

D Conclusions

Despite some limited commercial use in the USA, pathogens have not generally been shown to be effective against cockroaches. Both fungi and nematodes can control German cockroaches effectively in arena trials; however, under more realistic conditions it is difficult to achieve the high level of control demanded by the consumer. Low humidity is undoubtedly a factor which reduces the efficacy of fungal pathogens. Consequently, further evaluation of oil-based formulations is warranted.

3 Ants

A Pest species and their importance

Two large groups of pest ants can be recognized in the urban environment: (a) ants which live inside structures, and (b) ants which are a problem in urban landscapes, but outside human structures. Included in the first group are many ants that invade houses and cause no structural or other type of damage, but annoy homeowners. Also included in this group are ants that can damage the structural integrity of buildings (carpenter ants, acrobat ants), disseminate disease-causing organisms (pharaoh ants and others), or consume materials or otherwise cause financial or aesthetic damage. Included in the second group are ants that may attack humans or domestic animals, damage landscape plants and other materials outside human structures. These include fire ants which are major problems in the USA and some other countries such as Australia (Nattrass and Vanderwoude, 2001). A survey of ant problems in the USA showed that fire ants and other ants are major problems in 6% and 13% of the households, respectively (Whitmore *et al.*, 1992).

Because ants are social insects, special problems arise in controlling them, as well as in the evaluation of control levels, especially when entomopathogens and other biological control agents are considered (Pereira and Stimac, 1997). Due to a low genetic variability among individuals in an ant nest, effectiveness of pathogens can be compromised or enhanced (Sherman *et al.*, 1988). Nest hygiene, self-grooming and allogrooming (grooming between individuals in a social unit) can remove or spread entomopathogens (Oi and Pereira, 1993) and chemicals secreted by the ants can be harmful to the development of microorganisms (Schildknecht and Koob, 1971; Jouvenaz *et al.*, 1972; Storey *et al.*, 1991). However, in the nest, entomopathogens encounter high host densities, stable temperatures and humidities and no harmful radiation.

B Entomopathogens used against ants

In the past, urban ant pests have received minimum attention from researchers working with microbial control. Relatively few pathogens have been isolated from or tested against ants. Among tested microorganisms, most studies have targeted ant problems in agriculture, such as the leaf-cutting ants (Kermarrec *et al.*, 1986; Diehl-Fleig *et al.*, 1993). The ant known in the USA as the red imported fire ant, *Solenopsis invicta*, has been the focus of several attempts to isolate entomopathogens and use them as biological control agents (Williams *et al.*, 2003). Similar effort has been initiated with the ant *Myrmica rubra*, a recent invasive ant species in the USA (Groden *et al.*, 2005). The lack of recognized specific pathogens of ants prevents further development of microbial control against urban ants. Recently, new pathogen isolations from well-studied ant populations (Pereira *et al.*, 2002, Pereira, 2004) have demonstrated that ant pathogens are not as rare as once believed (Holldobler and Wilson, 1990). Nevertheless, there is a great need for studies on ant pathogens and their possible application in urban environments. Only fungi, nematodes, and microsporidia (now taxonomically placed with the fungi) have been seriously tested for use in the control of ant pests (Drees *et al.*, 1992, Briano and Williams, 1997, Pereira and Stimac, 1997). The fungi, *M. anisopliae* and *B. bassiana*, and the steinernematid nematode, *S. carpocapsae*, have been the focus of studies on microbial control of ants. Studies involving the field prevalence and release of the microsporidia, *Thelohania solenopsae* (Briano and Williams, 1997), and more recently *Vairimorpha invictae* (Oi *et al.*, 2005) may offer future alternative for control of fire ants. Efforts with bacteria have been limited to preliminary pathogenicity tests with mostly unsuccessful results (Miller and Brown, 1983; Jouvenaz, 1990). These studies used bacterial isolates usually obtained from ants and other insects.

Because several ant species have an efficient filtering system in the adult stage that prevents the ingestion of particulate material, entomopathogens which are acquired *per os* are usually effective only against some larval stages. Relatively little effort has been dedicated to the isolation and testing of pathogens from ant larvae. Viruses and bacteria, which are common pathogens in other members of the order Hymenoptera, are not well represented among the pathogens isolated from ant species (Oi and Pereira, 1993; Pereira and Stimac, 1997). However, recently a virus has been identified from fire ant populations (Valles *et al.*, 2004) and further studies may provide information on practical uses of this and other viruses against fire ants and other ant pests. Also, recent studies (Pereira *et al.*, 2002; Pereira, 2004; Pereira, unpublished) have shown that several pathogens that affect ant adults are actually acquired during the larval stages. This may indicate that the larvae are the most adequate targets for microbial applications against ants.

C Application methods

While most microbial control efforts against ants are aimed at developing biopesticides, classical biological strategies, with the introduction of the pathogens and reliance on natural spread of the disease organism, must be considered as well. Entomopathogens formulated as biopesticides have to be mass produced and applied in large amounts, whereas entomopathogens used in classical biological control do not require mass production and can be seeded into the ant population using small quantities of infectious material. In both cases however, there is an expectation of disease establishment in the ant population, with eventual elimination of the pest problem, ideally for several years. With the classical biological control approach, pest control is expected to be permanent with maintenance of the disease organism in the ant population. Biopesticides are designed to give short-term solutions to ant problems and control over a period of years is not usually a goal.

The location of an ant nest, whether indoors or outdoors, is an important factor in controlling ant pests with microorganisms. Ants found indoors may originate from outdoor nests and control measures applied inside buildings may not eliminate the problem. Outdoor nests may be easier to locate, but it is still often difficult to find ant nests and direct application of control agents to the ants or their nest may not be possible. Entomopathogens must be applied so they will reach the nest, and therefore spread among the entire ant population. Infection of the foraging ants alone is unlikely to provide lasting results since only a small proportion of the ant population engages in food gathering. Applications of entomopathogens in baits, which are carried into the nest by the ant workers, provide solutions in these cases. When ant nests can be located, direct application of entomopathogens to the ants and the nest environment is possible.

Indoor applications of entomopathogens must be considered in light of the possible exposure to humans and domestic animals. Although most entomopathogens are safe, care needs to be exercised in the application and formulation, for example, to avoid inhalation of fungal conidia. When applications are made outdoors, fewer restrictions on application methods and formulations are necessary but climatic factors may become more important in determining the effectiveness of formulations and application methods.

D *Testing entomopathogens for the control of ants indoors*

1 *Bait stations*

Bait stations containing chemical pesticides have become very common in the control of indoor cockroaches, ants and other insects. They consist of small structures with openings allowing entrance of the insects and contain an attractive bait and insecticide. This type of application is very easy for the homeowner to use and avoids exposure of humans and nontarget animals to the active ingredient. When chemical pesticides are used, the bait material usually contains a slow-acting toxicant. Because the toxicants do not kill the insect immediately, ants can transfer the active ingredient to nest mates, and greater mortality is obtained. Entomopathogens are naturally slow-acting, and therefore ideal for use in bait stations.

Bait stations containing the entomopathogenic fungus *B. bassiana* caused higher ant mortality than some commercial ant baits for control of several urban ant pests (Stimac and Pereira, 1997, 2001, 2006). The following protocol can be used for testing *B. bassiana* against several species of ant:

1. Coat the walls of small plastic boxes (17 × 12 × 5.5 cm or 19 × 14 × 9.5 cm) with liquid Fluon™ or Teflon™ to prevent ants escaping. These boxes will serve as the arenas where individual ant colonies will be kept.
2. Add nest cells (Petri dishes partially filled with either dental plaster or plaster of Paris, which is kept damp to maintain humidity, and with entrance holes on the lids or side walls) to the plastic boxes to provide shelter for the ants. Nest cells prepared with 60-mm diameter plastic Petri dishes will accommodate fire ants and carpenter ants, but a 35-mm can be used for pharaoh ants and other small species.
3. Prepare artificial ant colonies by separating small groups of individuals from larger colonies. The number of ants in each replicate will depend on the species and availability of ants. Typically, 0.5 g of fire ant workers is used per colony. For pharaoh ants, artificial colonies should consist of approximately 150 workers and 1 queen, and for carpenter ants 50 workers only. Ant colonies may be prepared 1–3 days before initiation of experiments to allow insects to adapt to the arena and nest cells.
4. Prepare the bait as a powder containing 35% ground roast peanuts, 50% cornstarch, 10% *B. bassiana* conidia and 5% drying agent (diatomaceous earth or a synthetic calcium silicate). Ensure the bait powder is free-flowing by passing it through a 60-mesh or finer sieve.
5. The bait stations, which may consist of a weigh dish, weighing paper, bait stations used with chemical pesticides, or any other similar structure, are filled with 0.5-2.0 g of the free-flowing bait powder and added to the plastic boxes.
6. Remove and count the dead ants after 3, 7, 14, 21 and 28 days. Finally, count the ants still surviving after 28 days.

Kelley-Tunis *et al.* (1995) conducted experiments with carpenter ants and entomopathogenic fungi using a bait station as the inoculation chamber. The chamber was similar to commercially available ant bait stations, and the fungal preparation was placed within the lid, above an ant-attractive material. Ants were inoculated when *B. bassiana* and *M. anisopliae* conidia were dislodged from the preparation and contacted them.

2 *Other applications*

Besides application from fixed stations, baits and other formulations containing entomopathogens can also be distributed along areas visited by ants and other insects inside buildings. Fungal pathogens, which need to contact the insect cuticle, are probably better used in powdered formulations that can readily be transferred onto the insect's body. Other pathogens, such as microsporidia, that need to be ingested by the host, must be formulated in a substrate that will be eaten by the target insects.

E *Testing entomopathogens for control of ants outdoors*

1 *Broadcast applications*

Both broadcast applications over large areas and the placement in piles of baits (10 g per pile, and 24 kg bait/ha) containing *B. bassiana* conidia have been used to control the southern fire ant *Solenopsis xyloni* (Pereira and Stimac, unpublished data). The placement of a bait into piles provides a concentrated source of attractive volatiles, but decreases chances of the bait being found randomly by a foraging ant. Non-attractive powders containing fungal conidia have also been used in broadcast application. A hydrophobic silica was used as the carrier for this formulation to prevent adherence of the material to wet soil. The hydrophobic nature of the carrier also promoted attachment of the fungal formulation to the insect cuticle as the ants moved over the treated area.

Alginate granules containing *B. bassiana* mycelium have been used in broadcast applications against fire ants (Bextine and Thorvilson, 2002). Granules coated with peanut oil as an ant attractant apparently caused some reduction in fire ant populations. Fungal infection was also detected in individual ants collected from the field. However, results of this research have not been successful enough to lead to commercial product development.

2 *Individual nest applications*

If an ant nest can be found, the application of entomopathogens in or on the nest, or the area immediately around the nest, will improve the chance of contact between the pathogen and the ant host. The application of granular bait containing chemical pesticides around nests is commonly recommended (Collins, 1992). Powder and granular baits containing *B. bassiana* have also been applied to *Solenopsis* spp. nests with variable results (Bextine and Thorvilson, 2002, Pereira and Stimac, unpublished). Infected caterpillars have also been used as baits to introduce *B. bassiana* into fire ant nests (Broome, 1974).

The use of injection devices to introduce pathogens into ant nests has produced good results (Oi *et al*., 1994). The procedures used in these experiments are as follows:

1. Mark the fire ant nests with flags and map their location in the study area. Assess the level of ant activity at each nest using an arbitrary rating system (see below).
2. Collect samples of ants (at least 100 individuals) by disturbing the nest and placing a spatula on top of it. Shake the ants climbing on the spatula into Fluon™-coated trays and transfer them to samples vials. In the laboratory, kill the ants by freezing and use them in infection determination as described below.
3. Inject 7–50 g of a hydrophobic powder formulation (90% hydrophobic fumed silica (TS-720, Cabot, Tuscola, IL, USA) and 10% *B. bassiana* conidia) into the fire ant nests. The use of a hydrophobic formulation prevents the adhesion of the powder to the wet wall of the nest galleries and ensures a better distribution of the applied material.
4. For step 3, use an injector consisting of a 2-liter plastic bottle, into which the powder formulation is added, and a discharge tube and rod, which was inserted into the ant nest. Compressed gas is then used to inject the material into the ant nests (Figure 1).
5. Deposit the powder formulation deep (10–60 cm) into the ant nest. Injection of the fungal formulation should be stopped when the formulation is observed emanating from openings at the surface of the nest. This guarantees a complete coverage of the nest galleries with biopesticide. Formulations containing $>$ 90% of fungal conidia mixed with diatomaceous earth do not produce good coverage because they do not disperse well within the ant nests. Such formulations remain at the injection site which will then be avoided by the ants. This avoidance is due to the repellent effect of high fungal content in formulations.

Drench and injection applications of the nematode, *S. carpocapsae*, in water suspensions

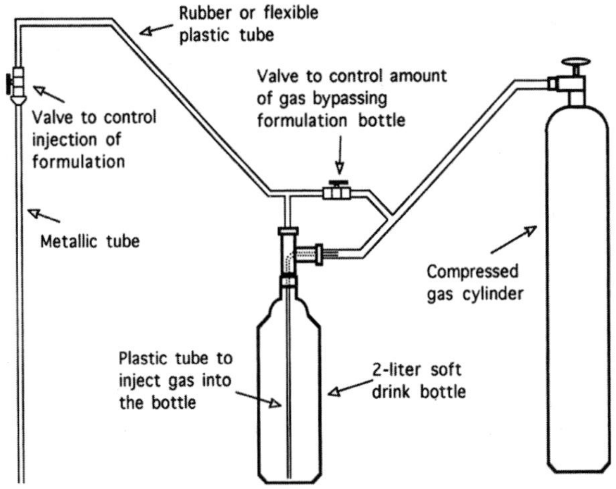

Figure 1. Injector used for application of hydrophobic powder formulation containing *Beauveria bassiana* conidia into fire ant nests. The formulation is added to the 2-liter plastic bottle. Compressed air from the tank will enter the bottle and suspend the formulation in air, which escapes through the tube at the top of the bottle. Some of the compressed air bypasses the formulation bottle and helps to siphon the formulation/air mixture from the formulation bottle. The formulation is injected into the fire ant nests through the metallic tube which is inserted into the soil. The formulation escapes the tube through holes at the tip of the metallic tube

have also been used against fire ants. Drenches ($2–5 \times 10^6$ nematodes in 3.8 liters of water per nest) were applied to the top and perimeter of fire ant mounds or injected (2×10^6 nematodes in 0.95 liter of water per nest) 0.5 m into mounds (Drees *et al.*, 1992). Significant reduction in the activity of treated nests was observed with these treatments, but new satellite mounds were found associated in 32–44 % of nematode-treated nests. This, and other applications that severely disturb the ants in the nests, tend to cause nest movement, limiting the efficacy of the entomopathogen.

F Evaluation methods

When ants and other social insects are targeted for control, the colony, rather than the individual, should be considered as the unit. Thus, the objective of control tactics must be the elimination of the reproductive potential in the nests. If accomplished, the colony will eventually die. The elimination of workers, may solve the immediate pest problem, but will not give long-term benefit. Quantification of percent mortality or infection of individuals from a nest may serve as an indicator of the effectiveness of microbial control measures, but percent nest mortality is the true measure of success.

Several indicators have been used in evaluating the effectiveness of entomopathogenic applications against urban ants:

1 Activity ratings

Activity ratings are usually based on visual observation of ant response to some disturbing stimulus. With fire ant colonies, the top of the nest is touched and disturbed, or air is blown into the nest, and the time it takes the ants to respond is evaluated. Ratings are assigned according to how the nest performs in relation to a normal response. Fire ant nests can be rated as active if > 30 ants respond within 20 seconds disturbance, and inactive otherwise (Oi *et al.*, 1994). Alternatively, a "progressive minimal mound disturbance technique" can be used and nests rated as active when > 50 ants respond to an increasing level of mound disturbance (Drees *et al.*, 1992). Depending on level of disturbance necessary to elicit the response, nests can be categorized into different levels of activity (*e.g.*, high, medium, low, inactive). Because insect activity can be affected by several factors including weather conditions, these ratings may not reflect the true population levels, or ability of the ants to cause damage.

2 Trapping and populations indexes

Two types of traps can be used to monitor ant populations: traps that prevent escape of the ants (pitfall traps, sticky cards), and traps that allow ants to leave and recruit more foragers. The former can be left in place for longer periods of time, but are usually less specific and many other insects may be caught. The second trap type is usually baited with attractive material that will be foraged by the ants. Because ants are free to leave, these are usually monitored or collected within minutes to hours of their deployment. Depending on how specific the attractant is, a limited number of species can be expected. Snap-cap tubes (8 cm × 2.5 cm diameter, 45-ml capacity) baited with several attractants including wieners, dog food, canned tuna fish, vegetable oil, honey water etc. can be used to monitor ants. Liquid attractants such as oils and honey water can be applied to absorbent paper (Kimwipe®) placed into the traps. Attractants are placed within the tubes which are laid sideways on the ground to allow free access of ants to the attractive material. Tubes can be placed next to ant nest entrances, near foraging trails, randomly in the field or in a grid pattern (*e.g.*, 10 × 10 m), depending on objective of study. Traps placed next to nest entrances or by foraging trails estimate populations of single nests whereas traps placed over an area, randomly or in a grid, estimate areawide populations. Snap-cap tubes can be left out for 10 to 240 minutes, then capped to trap ants and recovered. Ants are then freeze-killed and counted. Studies with fire ants have shown that an adequate estimate of field populations can be obtained by leaving baits consisting of wiener slices in the field for 30 to 60 minutes followed by evaluation of the percent of baits with ants (Pereira, 2003). If less than 30–40% of the baits are occupied by fire ants, the population is too low to justify application of control (Pereira, unpublished). Similar evaluation systems may be used for other ants.

A population index which takes into consideration the estimated size of the ant nest and the presence or absence of brood, has been used extensively for monitoring fire ant populations including those affected by an entomopathogen (Lofgren and Williams, 1983). Depending on the estimated number of ants in a nest, population indexes of 1 to 5 are assigned, and when worker brood is present, rates are multiplied by 2. In another evaluation system, nests without brood are rated between 1 and 5, whereas nests with brood receive rates of 6–10. Trained individuals are able to rate nests accurately and consistently, but this population index requires good familiarity with the ant species. Population indicators, such as size of nests, damage to structures and others, may also be used depending on the ant species. A good knowledge of the ant biology and behavior is necessary for ratings, which are good representations of the true population.

3 Percent infection

Whether ants are trapped or collected directed from nest or environment, they can be used in estimating the percent infection with fungal or bacterial pathogens using the following procedure (Pereira *et al.*, 1993):

1. Kill the ants by freezing. Allow enough time for complete kill since some ant species are quite resistant to freezing.
2. Surface-sterilize the ants by quickly dipping them in 95% ethanol (5–10 seconds).
3. Remove ants from ethanol and allow the ethanol to evaporate by placing the ants on a paper towel, if possible under a laminar flow hood or other sterile work area.
4. Place individual ants into empty wells on 96-well microtiter plates.
5. Cover each plate with moist paper towels or Kimwipes™.
6. Place plates in a plastic box with moist paper towel covering the bottom. Plates can be stacked in the box but the moist paper covering each plate should touch the moist box bottom. This allows the water to be drawn onto the paper to maintain a moist environment within each well on the plates. Alternatively, ants can be plated on 2% water agar plates.
7. Incubate plates at 25 °C.
8. Observe ants after 5–10 days and record as infected those showing signs of pathogen growth.

Microscopic observation of wet mounts or stained slides are used for observation of protozoans and nematodes. Staining and other relevant techniques are described in several chapters in Lacey (1997). Most ant diseases can only be detected in killed or dissected insects. However, fire ants infected with the protozoan-caused yellow head disease (Pereira *et al.*, 2002), with the fungus *Myrmicinosporidium durum* (Pereira, 2004), or with a yeast infection (Pereira, unpublished) can be diagnosed without being killed. Like these, other ant pathogens may be detectable in live insects.

Ants trapped or collected as they respond to a disturbing or attracting stimulus are not random samples of the ant population. Ants with advanced stages of a disease will likely not be present in these samples. Therefore, these techniques tend to underestimate the prevalence of disease in the ant population. Because of the social behavior of the ants, and behavior modifications caused by entomopathogenic infections (Oi and Pereira, 1993), other sample collection methods may not guarantee a random sample of the population either. Trapped ants usually provide a good indication of the prevalence of an entomopathogen in the ant population. However, depending on the trapping method and the ant behavior when affected by the entomopathogen, the true infection level may be quite different from the observed one.

4 Pathogen persistence in the environment

Plating of soil and other habitat materials onto selective medium [*e.g.*, for *B. bassiana*, Beilhartz *et al.* (1982)] can be used to monitor levels of some of the entomopathogens in the ant environment (Oi *et al.*, 1994). Obligate parasites will not be detected using media. *Galleria mellonella* (Zimmermann, 1986) or other suitable hosts can be used as pathogen bait for detection of fungi, nematodes and possible bacteria. Several methods described in Lacey (1997) can be used for detection of the different pathogens.

4 Termites

Termites belong to the order Isoptera (meaning equal wings) and some 2,500 species are recognized today. Although they are essentially tropical insects, they occur as far north as southern Canada and as far south as northern Argentina. Termites are thought to be closely related to cockroaches and to have evolved some 200 million years ago (Robinson, 1996). They are pests of buildings, forestry and crops over all of Africa, as well as much of Asia, Australia, South America, North America and southern Europe. It has been estimated that the 4 or 5 major termite pest species in the USA are responsible for over $1 billion damage to buildings and agriculture each year (Robinson, 1996). Termites are undoubtedly the most serious pest of buildings and human-made structures world-wide.

A Biology of termites

Termites are social insects living in colonies comprising of a king and queen together with numerous workers and soldiers. Their life cycle is hemimetabolous with the queen laying eggs that hatch into larvae which can develop into workers, soldiers or new reproductives. There are many variations on this simple pattern; colonies are always long-lived and primary queens often live for over 20 years and in some cases up to 50 years. Termites may be divided into "lower" and "higher" groups. Lower termites such as Mastotermitidae and Rhinotermitidae have cellulose-digesting flagellates in their guts, whereas the higher termites, or Termitidae, generally do not require specialized protozoans and have a more complex social organization. Some 80% of termite species belong to the higher termites and many cultivate fungi for food within the nest.

In the USA, the major urban pests are *Reticulotermes flavipes* and *Coptotermes formosanus*. Both species normally form large, diffuse, subterranean colonies comprising many millions of individuals and foraging over an area of 1000–2000 m. *C. formosanus* may also form nests in trees (Henderson and Forschler, 1997). Thus, a single colony is often responsible for damaging several houses and foraging distances of over 50 m are commonly recorded. *C. formosanus* is thought to have been accidentally introduced into the USA from Asia, possibly China, while *Reticulotermes* spp. are becoming increasingly serious pests in Europe especially in France.

In Australia, two other *Coptotermes* spp., *C. frenchi* and *C. acinaciformis*, are the main urban pests and both species form discrete nests of many millions of individuals often in trees. As with *C. formosanus* and *R. flavipes*, foraging distances can be 50 m or more. Mound builders such as *Nasutitermes exitiosus* are less serious pests. In other parts of the world, urban pests may be subterranean, mound builders or so-called dry-wood termites. Dry-wood termites are able to form nests away from the soil and so may even form nests high up in roof structures. These colonies can be small and numerous.

B Pathogens of termites

Fungi, such as *M. anisopliae* and *B. bassiana*, can be easily isolated from termites and from termite derived substances such as mound material (Zoberi and Grace, 1990; Zoberi, 1995). However, a large detailed survey in Australia concluded that *M. anisopliae* was not a significant natural pathogen (Milner *et al.*, 1998a). Colonies of termites must eventually decline and die; however, no pathogens have yet been shown to be the cause of colony mortality under natural conditions. Hanel (1982) listed a number of fungal pathogens recorded from termites and screened many of them for entomopathogenic activity. He reported that the most effective was *M. anisopliae*. Screening of over 100 isolates of this fungus has shown that most isolates are pathogenic to termites and that the most promising strains for biological control were originally isolated from termites or termite material (Milner *et al.*, 1998b). Recent field studies on Formosan termites have identified promising strains of *M. anisopliae* and *Paecilomyces fumosoroseus* which warrant detailed field testing (Wright *et al.*, 2005; Meikle *et al.*, 2005).

In Australia, a coelomic gregarine may be clearly visible as cysts in the hemocoel of workers of *C. lacteus*, but this protozoan has not been shown to be pathogenic (Milner, unpublished). Also, Gibbs *et al.* (1970) were able to isolate virus-like particles from *C. lacteus*, *N. exitiosus* and *Porotermes adamsoni* workers. Again pathogenicity has not been demonstrated. Nematodes such as *Steinernema* spp. are not very effective pathogens in the laboratory as the lethal concentration to kill 50% of worker termites is about 400 nematodes (Wang *et al.*, 2002; Mankowski *et al.*, 2005). In addition, they have not been reported as natural pathogens. Ochiel (1995) isolated the fungus *Cordycepioideus bisporus* (Pyrenomycetes: Clavicipitales) from alate adult *Macrotermes subhyalinus* in Kenya and was able to transmit the fungus to adults, but not to workers, by exposing them to soil mixed with *in vitro* produced ascospores and synnematal material.

C Field testing of pathogens as microbial insecticides against termites

There are very few reports of field-testing of pathogens against termites. Ideally treatment with a chemical or biological insecticide will result in colony mortality. In practice, this is difficult to achieve except where discrete colonies are formed in mounds or trees. More normally, treatment results in exclusion of termites by physical barrier or repellency, or treatment reduces the vigor of a colony for a limited time necessitating repeat treatment.

Entomopathogenic nematodes such as *Steinernema* spp. and *Heterorhabditis* spp. have provided some limited control of termites under field conditions, but have generally not proved successful for long-term suppression (Maudlin and Beal, 1989; Epsky and Capinera, 1988). Hanel and Watson (1983) were able to introduce *M. anisopliae* into nests of *N. exitiosus* by applying conidia to feeding sites and while some colonies were severely affected, others showed little effect. They concluded that there were unknown factors in the mound, which inhibited conidial germination and so rendered the pathogen dormant. The only successful direct treatment of urban termites is that reported by Milner *et al.* (1998b). The essential features of this method for treating nests with conidia of *M. anisopliae* are:

1. Locate the center of the nest by drilling into the mound or suspect tree. Ideally a temperature probe is used to ensure that the warmest part of the nest (the nursery) is detected.
2. Apply the fungus in a dust preparation of pure dry conidia, or conidia formulated in an inert carrier such as talc, by blowing the material into the center of the nest so as to ensure good distribution in the nursery area. For a large nest, 5 g of pure conidia has been found to be sufficient, assuming

a suitably virulent isolate of the fungus has been selected. For nests in trees which do not have an open system of galleries, it may be necessary to apply down several holes to ensure good distribution throughout the nursery area. A modified double-action pump sold for inflating air-beds is suitable for this type of application.

3. The effectiveness of the treatment can be evaluated using a data logger to measure temperature changes (Figure 2), or by sampling live termites periodically and incubating them to assess level of infection, or by destructive sampling of the nest and checking for survival of the royal pair as well as signs of cessation of reproduction. Infected termites often move down from the center of the colony and may be found, usually showing profuse sporulation, at the base or even under the base of the nest.

It is important to realize that live workers may continue to be found in the nest and surrounding feeding sites for many weeks after most of the colony has been killed. However, depending on the species of termites, nests can produce neotenic queens (*e.g.*, *Coptotermes* spp.) and recover despite the loss of the royal pair.

Often the source nest for an infestation in a building cannot be found or may be large and diffuse as is the case with *M. darwiniesis* and *R. flavipes*. In this case, the fungal dust preparation can be applied in a similar manner to that described for nests by blowing into damaged timber such as skirting boards. Small holes will need to be drilled at intervals of about 1 m to ensure good distribution of the conidia throughout the infested timber and the conidial preparation blown in through these holes using a plastic tube of similar diameter. It is important that a large number of termites be directly contacted by the conidia as it is likely that after this initial kill, termites will be repelled by the conidial deposit forcing them to feed elsewhere. BioBlast™, a *M. anisopliae* product that was sold by EcoScience in the USA for termite control, was formulated as a wettable powder. The product was mixed with water and then pumped into feeding galleries where termites were actively foraging. The aim of treatment was claimed to be to transfer the pathogen back to the nest; however, it may have acted mainly as a repellent (Grace, 1997; Pearce, 1997). Wooden power poles, often used for transmission of electricity in urban environments, may be attacked by termites. A recent report on a large field trial in New South Wales found that blowing conidia into infested poles was effective in controlling the termites. The fungus treatment reduced the infestation rate by 70% after 6 months which was similar to that achieved by the chemical insecticides, bifenthrin and triflumeron, but it was less effective than dazomet which was the most effective treatment giving 80% control after 12 months. The life of the fungus treatment was estimated as 22 months compared with 37 months for dazomet (Martin Horwood, pers comm., 2006). *Metarhizium* treatments may also be effective in non-urban environments.

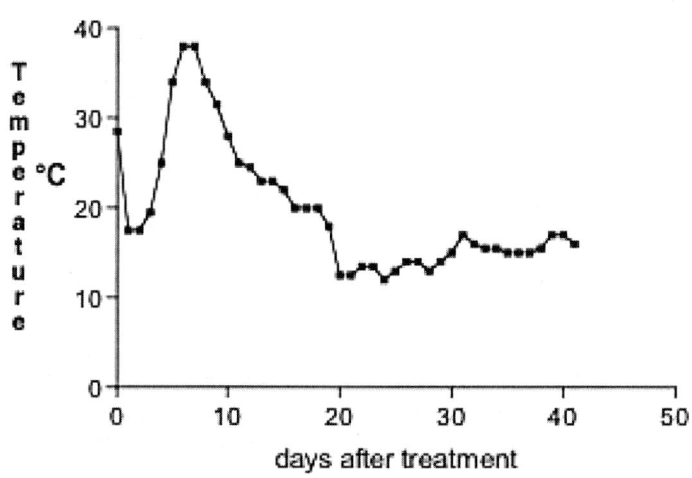

Figure 2. Temperature changes in a tree nest of *Coptotermes acinaciformis* following treatment with conidia of *Metarhizium anisopliae*

For example Maniania *et al.* (2001) reported that *M. anisopliae* applied to maize at planting reduced termite damage and increased the yield of grain. The effect was thought to be repellency though some termite mortality may also have occurred.

Methods for testing chemical insecticides as soil treatments for termite controls are similar in Australia and the USA and have been described by Lenz *et al.* (1990) and Kard *et al.* (1989). Similar methods were used by Milner *et al.* (1993) to show that *M. anisopliae* conidia can be applied to the soil surface, or mixed in with soil, or painted as a suspension in Tween 80 onto timber surfaces, to protect a susceptible piece of timber placed in the center. Soil surface treatment with 8×10^7 conidia/cm^2 of isolate FI-610 provided 100% protection for 24 weeks. Depending on prevailing environmental conditions this protection may last up to 3 years or more (Milner and Staples, unpublished).

Pathogens have not been rigorously tested for control of termites and so it is difficult to be prescriptive. As slow acting toxicants, fungi such as *M. anisopliae* are attractive for use in baiting strategies to kill colonies. Consequently, the methodology used by Su *et al.* (1996) to evaluate baiting methods, could also be used with pathogens. The essential features of this method are the use of mark-recapture to assess population density and to map the area of feeding by workers from a single colony. A large number of bait stations loaded with material such as susceptible timber are then deployed in the ground over the area. Once the termites start to feed at these baits, they are replaced with similar baits laced with a non-repellent formulation of a toxicant. These toxic baits are replaced as needed and the colony gradually declines. With hexaflumeron (a benzophenylurea) it may take in excess of 9 months to reduce the colony to below the detectable number of workers (Su *et al.*, 1996). Nile-blue is the most effective dye for labeling termites (Evans, 1997) but some of the assumptions (*e.g.*, random sampling) of standard mark-recapture formulae are violated with termites reducing the accuracy of this method for determining colony size (Forschler and Townsend, 1996; Thorne *et al.*, 1996).

A major problem with using *M. anisopliae* in baits containing conidia is the repellency of the infectious conidia. Isolates vary in their repellency and according to Begum and Jackson (1994) the repellency is independent of virulence. However, other workers have found that virulent isolates can also be repellent (Delante *et al.*, 1995; Rath and Tidbury, 1997) though this is dose-dependant (Milner and Staples, 1996). Promising ways of overcoming this repellency are to formulate the conidia in substrates such as agar (Delante *et al.*, 1995), organic amendments (Milner, 2000), and cellulose (Wang and Powell, 2004). These materials not only dilute the conidia, but also may mask the factors inducing repellency and are directly attractive for termites.

Ideally, candidate microbial products should be tested on infested houses or structures. However, as each example is different and it is not possible to have controls, the results may be impossible to interpret as the observed lack of termite activity may be due simply to the disturbance rather than the effect of the microbial treatment. Krueger *et al.* (1995) reported that of 101 structures treated with BioBlast™, 60% were still free of further termite activity 6–15 months later. Similarly, Milner *et al.* (1993) reported individual cases where *M. anisopliae* had apparently given good control of infestations in houses for a period of time. While these case histories and associated testimonials may be useful for marketing a product, they cannot constitute effective field testing, and therefore, should be in support of scientific testing of the types discussed in the preceding paragraphs.

D Conclusions

Urban termites present a unique challenge for testing pathogens as they often do not have a recognizable nest, have a complex behavior linked with their social structure, and are difficult to detect. Direct application to a nest will often give complete control with a single application, but this approach is not useful for termites such as *R. flavipes* where the nest is diffuse. If the nest is not affected, then the termites can continue to reproduce at a high rate and replace the individual termites killed by a localized application of a pathogen. Consequently, damage may continue or simply be transferred to another part of the building. Future acceptable control by pathogens may depend on either an integrated approach whereby the pathogen provides short-term cessation of damage while changes are made to the environment to make the building less attractive to termites, alternatively pathogens could be used in baits over a long period of time to eliminate the colony. These approaches are the subject of ongoing research.

5 References

Ali, F. A. F., El-Refai, A., Abdel-Rahman, A. and Abou-Donia, S. A. 1992. Effects of insecticides and environmental factors on the behaviour of cockroaches in Ebling choice-boxes. *Intern. Pest Control* **34**, 112–115.

Apel, A. G., Benson, E. P., Ellenberger, J. M. and Manweiler, S. A. 1993. Laboratory and field evaluations of an entomogenous nematode (Nematoda: Steinernematidae) for German cockroach (Dictyoptera: Blattellidae) control. *J. Econ. Entomol.* **86**, 777–784.

Bateman, R. P., Carey, M., Moore, D. and Prior, C. 1993. The enhanced infectivity of *Metarhizium flavoviride* in oil formulations to desert locusts at low humidities. *Ann. Appl. Biol.* **122**, 145–152.

Baumholz, M. A., Parish, L. C., Witkowski, J. H. A. and Nutting, W. B. 1997. The medical importance of cockroaches. *Intern. J. Dermatol.* **36**, 90–96.

Begum, H. A. and Jackson, C. W. 1994. Quantitative measurement of avoidance of entomopathogenic fungi by *Reticulitermes flavipes*. In "Vth International Colloquium on Invertebrate Pathology and Microbial Control – Abstracts", p. 325, Society for Invertebrate Pathology, Montpellier.

Beilhartz, V. C., Parbery, D. G. and Swart, H. J. 1982. Dodine: a selective agent for certain soil fungi. *Trans. Brit. Mycol. Soc.* **79**, 507–511.

Bextine, B. R. and Thorvilson, H. G. 2002. Field applications of bait-formulated *Beauveria bassiana* alginate pellets for biological control of the red imported fire ant (Hymenoptera: Formicidae). *Environ. Entomol.* **31**, 746–752.

Briano, J. A. and Williams, D. F. 1997. Effect of the microporidium *Thelohania solenopsae* (Microsporida: Thelohaniidae) on the longevity and survival of *Solenopsis richteri* (Hymenoptera: Formicidae) in the laboratory. *Fla. Entomol.* **80**, 366–376.

Broome, J. R. 1974. Microbial control of the imported fire ant, *Solenopsis richteri* Forel (Hymenoptera: Formicidae). MS thesis, Mississippi State University, Mississippi State, MS.

Collins, H. 1992. Control of imported fire ants: a review of current knowledge. USDA-APHIS Technical Bulletin No. 1807.

Delante, K. M., Grace, J. K. and Tome, C. H. M. 1995. Potential use of pathogenic fungi in baits to control Formosan subterranean termites (Isopt., Rhinotermitidae). *J. Appl. Entomol.* **119**, 429–433.

Diehl-Fleig, E., da Silva, M. E., Specht, A. and Valim-Labres, M. E. 1993. Efficiency of *Beauveria bassiana* for *Acromyrmex* spp. control (Hymenoptera: Formicidae). *An. Soc. Entomol. Brasil* **22**, 281–285.

Drees, B. M., Miller, R. W., Vinson, S. B. and Georgis, R. 1992. Susceptibility and behavioral response of red imported fire ant (Hymenoptera: Formicidae) to selected entomogenous nematodes (Rhabditida: Steinernematidae and Heterorhabditidae), *J. Econ. Entomol.* **85**, 365–370.

Ebeling, W., Wagner, R. E. and Reierson, D. A. 1966. Influence of repellency and the efficacy of blatticides. I. Learned modification of behavior of the German cockroach. *J. Econ. Entomol.* **59**, 1374–1388.

Epsky, N. D. and Capinera, J. L. 1988. Efficacy of the entomogenous nematode *Steinernema feltiae* against a subterranean termite, *Reticulitermes tibialis* (Isoptera: Rhinotermitidae). *J. Econ. Entomol.* **81**, 21–30.

Evans, T. A. 1997. Evaluation of markers for Australian subterranean termites (Isoptera: Rhinotermitidae and Termitidae). *Sociobiology* **29**, 277–292.

Forschler, B. T. and Townsend, M. L. 1996. Mark-release-recapture estimates of *Reticulitermes* spp. (Isoptera: Rhinotermitidae) colony foraging populations from Georgia, U.S.A. *Environ. Entomol.* **25**, 952–962.

Gibbs, A. J., Gay, F. J. and Wetherly, A. H. 1970. A possible paralysis virus of termites. *Virology* **40**, 1063–1065.

Goettel, M. S. and Inglis, G. D. 1997. Fungi: Hyphomycetes. *In* "Manual of Techniques in Insect Pathology" (L. A. Lacey, Ed.), pp. 213–249. Academic Press, London.

Grace, J. K. 1997. Biological control strategies for control of termites. *J. Agric. Entomol.* **14**, 281–289.

Groden, E., Drummond, F. A., Garnas, J. and Franceour, A. 2005. Distribution of an invasive ant, *Myrmica rubra* (Hymenoptera: Formicidae), in Maine. *J. Econ. Entomol.* **98**, 1774–1784.

Gunner, H. B., Agudelo-Silva, F. and Johnson, C. A. 1990. A method and device for the biological control of insects. US Patent PCT/US90/00274. 20 pp.

Hanel, H. 1982. Selection of a fungus species, suitable for biological control of the termite, *Nasutitermes exitiosus* (Hill). *Z. Angew. Entomol.* **94**, 237–245.

Hanel, H. and Watson, J. A. L. 1983. Preliminary field tests on the use of *Metarhizium anisopliae* for the control of *Nasutitermes exitiosus* (Hill) (Isoptera: Termitidae). *Bull. Entomol. Res.* **73**, 305–313.

Henderson, G. and Forschler, B. T. 1997. Termite bait tests. *Louisiana Agric.* **40**, 9–11.

Holldobler, B., and Wilson, E. 1990. "The Ants." Harvard University Press, Cambridge, MA.

Jouvenaz, D. P. 1990. Approaches to biological control of fire ants in the United States. *In*, "Applied Myrmecology:

A World Perspective." (R. K. Vander Meer, K. Jaffe, and A. Cedeno, Eds.), pp. 620–627. Westview Press, Boulder, CO.

Jouvenaz, D. P., Blum, M. S. and MacConnel, J. G. 1972. Antibacterial activity of venom alkaloids from the imported fire ant, *Solenopsis invicta* Buren. *Antimicrob. Agents Chemother.* **2**, 291–293.

Kaakeh, W., Reid, B. L. and Bennett, G. W. 1996. Horizontal transmission of the entomopathogenic fungus *Metarhizium anisopliae* (Imperfect Fungi: Hyphomycetes) and hydramethylnon among German cockroaches (Dictyoptera: Blattellidae). *J. Entomol. Sci.* **31**, 378–390.

Kard, B. M., Maudlin, J. K. and Jones, S. C. 1989. Evaluation of soil termiticides for control of subterranean termites (Isoptera). *Sociobiology* **15**, 285–297.

Kelley-Tunis, K. K., Reid, B. L. and Andis, M. 1995. Activity of entomopathogenic fungi in free-foraging workers of *Camponotus pennsylvanicus* (Hymenoptera: Formicidae). *J. Econ. Entomol.* **88**, 937–943.

Kermarrec, A., Febvay, G. and Decharme, M. 1986. Protection of leaf-cutting ants from biohazards: Is there a future for microbiological control? *In* "Fire Ants and Leaf-cutting Ants: Biology and Management." (C. S. Lofgren and R. K. Vander Meer, Eds.), pp. 339–356 Westview Press, Boulder, CO.

Krueger, S. R., Duan, H., Rath, A. C. and Rotramel, G. L. 1995. Development of Bio-Blast biological termiticide. *In* "28th Annual Meeting of the Society for Invertebrate Pathology," p.35. Cornell University, New York.

Lacey, L. A. (Ed.). 1997. "Manual of Techniques in Insect Pathology." Academic Press, London, UK.

Le Patourel, G. N. J. 1996. Forced contact and arena bioassays to assess the performance of a pyrethroid WP deposit against Oriental cockroach. *In* "Proceedings of the 2nd International Conference on Insect Pests in Urban Environments", pp. 303–308.

Lenz, M., Watson, J. A. L., Barrett, R. A. and Runko, S. 1990. The effectiveness of insecticidal soil barriers against subterranean termites in Australia. *Sociobiology* **17**, 9–36.

Lofgren, C. S. and Williams, D. F. 1983. Avermectin B_1 a highly potent inhibitor of reproduction by queens of the red imported fire ant (Hymenoptera: Formicidae). *J. Econ. Entomol.* **75**, 798–803.

Lonc, E., Lecadet, M.-M., Lachowicz, T. M. and Panek, E. 1997. Description of *Bacillus thuringiensis wratislaviensis* (H-41), a new serotype originating from Wroclaw (Poland), and other *Bt* soil isolates from the same area. *Lett. Appl. Microbiol.* **24**, 467–473.

Lopes, R. B. 2005. Controle de *Blatella germanica* com *Metarhizium anisopliae* e inseticidas reguladores de crescimento. Ph.D. Dissertation- Escola Superior de Agricultura Luiz de Queiroz, Univ. de São Paulo, Piracicaba, Brazil.

Maniania, N. K., Ekesi, S. and Songa, J. M. 2001. Managing termites in maize with the entomopathogenic fungus *Metarhizium anisopliae. Insect Sci. Appl.* **21**, 41–46.

Mankowski, M., Kaya, H. K., Grace, J. K. and Sipes, B. S. 2005. Differential susceptibility of subterranean termite castes to entomopathogenic nematodes. *Biocontrol Sci. Technol.* **15**, 367–377.

Manweiler, S., Appel, A. and Weber, T. 1993. Nematode-based biological control of German cockroaches. *In* "Proceedings of the 1st International Conference on Insect Pests in Urban Environments", pp. 173–180.

Maudlin, J. K. and Beal, R. H. 1989. Entomogenous nematodes for the control of subterranean termites *Reticulotermes* sp. (Isoptera: Rhinotermitidae). *J. Econ. Entomol.* **82**, 1638–1642.

Meikle, W. G., Mercadier, G., Rosengaus, R. B., Kirk, A. A., Derouane, F. and Quimby, P. C. 2005. Evaluation of an entomopathogenic fungus, *Paecilomyces fumosoroseus* (Wize) Brown and Smith (Deuteromycota: Hyphomyctes) obtained from Formosan subterranean termites (Isop., Rhinotermitidae). *J. Appl. Entomol.* **129**, 315–322.

Miller, S. R. and Brown, L. R. 1983. Studies on microbial fire ant pathogens. *Devel. Indust. Microbiol.* **24**, 443–445.

Milner, R. J. and Staples, J. A. 1996. Biological control of termites: results and experiences within a CSIRO project in Australia. *Biocontrol Sci. Technol.* **6**, 3–9.

Milner, R. J., Staples, J. A., Lenz, M., Lutton, G. G., McRae, C. M. and Watson, J. A. L. 1993. Insect Pest Control. US Patent 5,595,746; 16 pp.

Milner, R. J., Staples, J. A. and Lutton, G. G. 1998a. The selection of an isolate of the hyphomycete fungus, *Metarhizium anisopliae*, for control of termites in Australia. *Biol. Control* **11**, 240–247.

Milner, R. J., Staples, J. A., Lutton, G. G., Driver, F. and Watson, J. A. L. 1998b. Occurrence of *Metarhizium anisopliae* in nests and feeding sites of Australian termites. *Mycol. Res.* **102**, 216–220.

Milner, R. J. 2000. Improved formulations of *Metarhizium* for biological control of termites. Technical Report 86, CSIRO Entomology, Canberra, Australia. pp. 37.

Nattrass, R. and Vanderwoude, C. 2001. A preliminary investigation of the ecological effects of Red Imported Fire Ants (*Solenopsis invicta*) in Brisbane. *Ecol. Management Restoration* **2**, 220–223.

Ochiel, G. R. S. 1995. Biology and biocontrol potential of *Cordycepioideus bisporus* Stifler and *Paecilomyces fumosoroseus* (Wize) Brown and Smith on the higher termites *Macrotermes subhyalinus* Rambur in Kenya. Ph.D. Dissertation, Royal Veterinary and Agricultural University, Copenhagen, Denmark, 210 pp.

Oi, F. M. and Hinkle, N. C. 1997. Biorational approaches to urban pest management. *J. Agric. Entomol.* **14**, 227–229.

Oi, D. H. and Pereira, R. M. 1993. Ant behavior and microbial pathogens. *Fla. Entomol.* **76**, 63–74.

Oi, D. H., Pereira, R. M., Stimac, J. L. and Wood, L. A. 1994. Field applications of *Beauveria bassiana* for control of the red imported fire ant (Hymenoptera: Formicidae). *J. Econ. Entomol.* **87**, 623–630.

Oi D. H., Briano, J. A., Valles, S. M. and Williams, D. F. 2005. Transmission of *Vairimorpha invictae* (Microsporidia: Burenellidae) infections between red imported fire ant (Hymenoptera: Formicidae) colonies. *J. Invertebr. Pathol.* **88**, 108–115.

Pachamuthu, P. and Kamble, S. T. 2000. *In vivo* study on combined toxicity of *Metarhizium anisopliae* strain ESC-1 with sublethal doses of chlorpyrifos, propetamphos and cyfluthrin to German cockroaches. *J. Econ. Entomol.* **93**, 60–70.

Pearce, M. J. 1997. "Termites-Biology and Pest Management." CAB International, Wallingford, UK.

Pereira, R. M. 2003. Areawide suppression of fire ant populations in pastures: project update. *J. Agric. Urban Entomol.* **20**, 123–130.

Pereira, R. M. 2004. Occurrence of *Myrmicinosporidium durum* in *Solenopsis invicta* and other new host ants in Eastern United States. *J. Invertebr. Pathol.* **86**, 38–44.

Pereira, R. M. and Stimac, J. L. 1997. Biocontrol options for urban pest ants. *J. Agric. Entomol.* **14**, 231–248.

Pereira, R. M., Stimac, J. L. and Alves, S. B. 1993. Soil antagonism affecting the dose-response of workers of the red imported fire ant, *Solenopsis invicta*, to *Beauveria bassiana* conidia. *J. Invertebr. Pathol.* **61**, 156–161.

Pereira, R. M., Williams, D. F., Becnel, J. J. and Oi, D. H. 2002. Yellow head disease caused by a newly discovered *Mattesia* sp. in populations of the red imported fire ant, *Solenopsis invicta*. *J. Invertebr. Pathol.* **81**, 45–48.

Piper, G. L. and Frankie, G. W. 1978. Integrated management of urban cockroach populations. *In* "Perspectives in Urban Entomology" (G. W. Frankie and C. S. Koehler, Eds.), Academic Press, New York, pp. 417.

Prior, C. 1996. History, use and future of microbial pesticides in urban pest control. *In* "Proceedings of the 2nd international Conference on Insect Pests in the Urban Environment," pp. 31–41.

Rath, A. C. and Tidbury, C. A. 1997. Susceptibility of *Coptotermes acinaciformis* (Isoptera: Rhinotermitidae) and *Nasutitermes exitiosus* (Isoptera: Termitidae) to two commercial isolates of *Metarhizium anisopliae*. *Sociobiology* **28**, 67–72.

Robinson, W. H. 1996. "Urban Entomology." Chapman and Hall, London, UK.

Robinson, W. H. 1999. The changing pest status of the German cockroach in the urban environment. *Proc. 52nd New Zealand Plant Prot. Conf.* 1999: 16–21.

Roth, L. M. and Willis, E. R. 1957. The Medical and Veterinary Importance of Cockroaches. *Smithsonian Miscellaneous Collections*, No. 134. Lord Baltimore Press, Baltimore, MD.

Roth, L. M. and Willis, E. R. 1960. The biotic associations of cockroaches. *Smithsonian Miscellaneous Collections*, No. 141. Lord Baltimore Press, Baltimore, MD.

Schildknecht, H. and Koob, K. 1971. Myrmicacin, the first insect herbicide. *Angew. Chem. (Int. Edn.)* **10**, 124–125.

Sherman, P. W., Seeley, T. D. and Reeve, H. K. 1988. Parasites, pathogens, and polyandry in social Hymenoptera. *Am. Nat.* **131**, 602–610.

Steenberg, T. and Vagn-Jensen, K. M. 1998. Entomopathogenic fungi for control of German cockroach (*Blatella germanica*) and other synanthropic cockroaches. *Bull. OILB/SROP***21**, 145–150.

Stimac, J. L. and Pereira, R. M. 1997. Controlling cockroaches, carpenter ants, and pharaoh ants using strains of *Beauveria bassiana*. U.S. Patent 5,683,689, pp. 18.

Stimac, J. L. and Pereira, R. M. 2001. Methods and formulations for control of pests. U.S. Patent 6,254,864.

Stimac, J. L. and Pereira, R. M. 2006. Methods and formulations for control of pests. U.S. Patent RE38,958.

Storey, G. K., Vander Meer, R. K., Boucias, D. G. and McCoy, C. W. 1991. Effect of fire ant (*Solenopsis invicta*) venom alkaloids on the in vitro germination and development of selected entomopathogenic fungi. *J. Invertebr. Pathol.* **58**, 88–95.

Su, N.-Y., Scheffrahn, R. H. and Su, N. Y. 1996. Fate of subterranean termite colonies (Isoptera) after bait applications - an update and review. *Sociobiology* **27**, 253–275.

Suiter, D. R. 1997. Biological suppression of synanthropic cockroaches. *J. Agric. Entomol.* **14**, 259–270.

Thorne, B. L., Russek-Cohen, E., Forschler, B. T., Breisch, N. L. and Traniello, J. F. A. 1996. Evaluation of mark release recapture methods for estimating forager population size of subterranean termite (Isoptera: Rhinotermitidae) colonies. *Environ. Entomol.* **25**, 938–951.

Valles, S., Strong, C., Dang, P., Hunter, W., Pereira, R., Oi, D., Shapiro, A. and Williams, D. 2004. A picorna-like virus from the red imported fire ant, *Solenopsis invicta*: initial discovery, genome sequence, and characterization. *Virology* **328**, 151–157.

Wang, C. L. and Powell, J. E. 2004. Cellulose bait improves effectiveness of *Metarhizium anisopliae* as a microbial control of termites (Isoptera: Rhinotermitidae). *Biol. Control* **30** 523–529.

Wang, C., Powell, J. E. and Nguyen, K. 2002 Laboratory evaluations of four entomopathogenic nematodes for control of subterranean termites (Isoptera: Rhinotermitidae). *Environ. Entomol.* **31**, 381–387.

Whitmore, R. W., Kelly, J. E. and Reading, P. L. 1992. "National Home and Garden Pesticide Use Survey, Final Report, Volume I: Executive Summary, Results, and Recommendations". Research Triangle Institute/U.S. EPA, Research Triangle Park.

Williams, D. F., Oi, D. H., Porter, S. D., Pereira, R. M. and Briano, J. A.. 2003. Biological control of imported fire ants (Hymenoptera: Formicidae). *Am. Entomol.* **49**, 144–155.

Wright, M. S., Raina, A. K. and Lax, A. R. 2005. A strain of the fungus *Metarhizium anisopliae* for controlling subterranean termites. *J. Econ. Entomol.* **98**, 1451–1458.

Ying, Z. G., Wen, Y. H., Gang, H., Zhang, G. Y., Yang, H. W. and Huang, G. 1996. A preliminary study on the control of *Blatella germanica* (Dictyoptera: Blattellidae), with *Metarhizium anisopliae. Chinese J. Biol. Control* **12**, 100–102.

Zimmermann, G. 1986. The '*Galleria* bait method' for detection of entomopathogenic fungi in soil. *J. Appl. Entomol.* **102**, 213–215.

Zoberi, M. H. 1995. *Metarhizium anisopliae*, a fungal pathogen of *Reticulitermes flavipes* (Isoptera: Rhinotermitidae). *Mycologia* **87**, 354–359.

Zoberi, M. H. and Grace, J. K. 1990. Isolation of a pathogen *Beauveria bassiana* from *Reticulotermes flavipes* (Isoptera: Rhinotermitidae). *Sociobiology* **16**, 289–296.

Zukowski, K. and Bajan, C. 1996. The study of the usefulness of *Beauveria bassiana* in eradication of cockroaches (*Blatella germanica*). *Roczn. Pzh.* **47**, 343–349.

Zurek, L., Watson, D. W. and Schal, C. 2002. Synergism between the entomopathogenic fungus, *Metarhizium anisopliae*, and boric acid against the German cockroach, *Blattella germanica. Biol. Control* **23**, 296–302.

Chapter VII-21

Application and evaluation of entomopathogens for control of livestock and poultry pests

Dudley E. Pinnock

Microbial Products Pty. Ltd.
Kent Town
Adelaide
South Australia 5071, Australia

Bradley A. Mullens

Department of Entomology
University of California
Riverside, CA 92521, USA

1 Introduction

The control of livestock and poultry pests presents unique challenges. The hosts are warm-blooded vertebrate animals, capable of a wide range of behaviors and movement. Range cattle and sheep in arid environments, for example, may occupy sparse grassland regions of thousands of hectares and move freely over many kilometers in search of forage. In these situations, pest management strategies must be effective and sustainable with minimal human input or supervision. At the other extreme are intensive, confined animal operations that are growing in popularity, particularly in developed countries in western Europe, Australia, and North America. In these facilities, poultry, swine, cattle, and fur-bearers such as mink are held at high density. These confinement systems may exacerbate certain pest problems, such as muscoid flies, which develop in the accumulated manure, and facilitate the spread of some permanent ectoparasites such as lice or mites.

A fundamental difference between livestock and poultry pest management and similar efforts in plant crops relates to economic and damage thresholds. These fundamentals of integrated pest management can be much more difficult to define in animal systems. Certain production parameters, such as losses in weight gains, reduced milk yield or quality, or losses in feed utilization efficiency due to pests, can be measured. They can then be related to expected economic benefits when considering the costs and benefits of control efforts. Animals also feel discomfort and are subject to humanitarian concerns for their welfare, which may affect pest control decisions. Further, many pests associated with livestock and poultry, for example manure-breeding house flies, are less important as direct pests of the animals than they are of people nearby. This may result in very significant pest control costs to producers, who are compelled by public health authorities to control the problem or cease operations. These humanitarian and public health concerns are very difficult to define economically.

The microbial control of livestock and poultry pests, as in agricultural systems, can take the form of inoculative, augmentative, or inundative methods. There have been few inoculative

L.A. Lacey and H.K. Kaya (eds.), Field Manual of Techniques in Invertebrate Pathology, 713–734.

efforts in animal systems, but there is still potential for this approach as a part of integrated control programs. One example is the nematode parasite of house flies, *Paraiotonchium muscadomesticae* (Geden, 1997), which has significant sublethal effects at lower infection levels and causes direct mortality at higher levels. Inoculative releases, as in classical biological control, may assist in suppressing pest levels below a nuisance or damage threshold. Certain pathogens are relatively easy to produce in fairly large quantities, but also are found naturally at low prevalence; for example, *Beauveria bassiana*. Other pathogens may not be found naturally in the area of interest, but can be very effective when applied, such as certain strains of *Bacillus thuringiensis*. In most cases, it is hoped that inoculative releases may result in establishment, or at least allow the pathogen to persist and be retransmitted within a target population. Pathogens used in augmentative or inundative programs most often are biological insecticides and are not expected to persist.

The use of entomopathogens in livestock and poultry systems is still in its infancy. Thorough surveys for naturally occurring diseases remain to be done for many livestock and poultry pests in most areas. No doubt such surveys would reveal pathogens that might be manipulated either to achieve control by direct application, or to have their natural activity enhanced and more intentionally incorporated into integrated pest management programs. To date, most attention has focused either on microbial agents which cause obvious natural mortality, such as the mycosis of house flies by the *Entomophthora muscae* complex, or on agents that can be mass-produced, such as strains of *B. thuringiensis, B. bassiana*, or entomopathogenic nematodes.

In this chapter we describe first the microbial control of pests of livestock in extensive rangeland systems, with examples being the sheep blow fly, *Lucilia cuprina*, and sheep lice, *Bovicola (Damalinia) ovis*. Second, we describe the microbial control of pests in intensive, confined animal husbandry systems, such as the stable fly, *Stomoxys calcitrans*, house fly, *Musca domestica*, and the lesser mealworm, *Alphitobius diaperinus*.

2 Pests of extensive rangeland systems

A *Major pests of sheep*

1 *The "Australian" sheep blow fly,* Lucilia cuprina

The blow fly, *L. cuprina* (Diptera: Calliphoridae), is believed to have been accidentally introduced into Australia from Africa and/or Asia during the mid-19th century. In its indigenous regions, *L. cuprina* appears to have evolved as a species feeding on carrion during its larval stages. In Australia, where the many coprophagous species create intense competition for carrion, it has adapted to exploit live animals, mainly sheep, as a larval food source.

Female *L. cuprina* lay their egg masses on the soiled or wet wool of live sheep. The neonate larvae migrate to the skin of the animal where they mass together and scrape the host's skin with their mouth hooks, creating a cutaneous myiasis, a suppurating abrasion on which they feed. Each myiasis frequently is created by several thousands of larvae and can extend over the entire back or flank of the sheep. Unless treated without delay, the sheep will die within a few days, apparently of a syndrome similar to toxic shock; the larvae do not invade into deep tissues until after the death of the host. These myiases are called "flystrikes" by sheep farmers. It has been estimated that *L. cuprina* flystrikes cause sheep deaths and losses of wool production in Australia valued at over A$ 280 million every year (Australian Wool Innovation, 2007).

Previous flystrike control measures in general use were a combination of traditional sheep husbandry practices such as mulesing, pizzle dropping, docking and crutching, together with the topical application of insecticides by dipping or spraying ("jetting"). The insecticides used included the organochlorines dieldrin and aldrin, the carbamate butacarb, and the organophosphates diazinon, chlorfenvinphos, fenthion ethyl and dichlorfenthion (James, 1986).

The use of organochlorines was abandoned because of the risk of residues in sheep meat, lanolin and wool, and since the 1970s, the use of organophosphates has been greatly limited because of the development of widespread resistance of *L. cuprina* to these compounds. At the time of writing, diazinon, or diazinon combined with another toxicant, still is used as a local application directly onto an existing flystrike area to kill the *L. cuprina* larvae. Prophylactic treatment against flystrike with the triazine compound cyromazine, a chitin synthesis inhibitor, has been used widely since 1979 (James, 1986). Cyromazine does not control existing flystrikes.

2 *The sheep biting louse, Bovicola ovis*

The sheep biting louse, *Bovicola ovis* (Phthiraptera: Trichodectidae), was introduced into Australia with its host and now occurs in all sheep rearing areas. Merino sheep are especially prone to *B. ovis* infestations and untreated flocks may carry very high louse populations. The host's response to the severe irritation caused by lice is a reduction in wool production of up to 1 kg/sheep (Wilkinson, 1986), which represents a loss to the farmer of ca A$ 8.00/sheep at time of writing. Equally serious is that the sheep will rub against any solid structure available, such as rocks, fences, and gate posts. This rubbing results in derangement of the fleece and damage to the wool fibers, so that the wool is downgraded, further increasing the losses to the farmer.

Traditional control measures for sheep lice generally were based on insecticidal plunge dips using a range of toxicants, including arsenic compounds now banned. In 1981, backline or pour-on formulations of synthetic pyrethroids were introduced for lice control and their convenience of use and early efficacy gained them wide acceptance by Australian wool growers. In 1986, control failures of some of these backline treatments were reported and resistance of sheep lice to the synthetic pyrethroids was confirmed. By 1990, resistance factors to the synthetic pyrethroids in lice from some flocks were as high as 26.1 (Johnson *et al.*, 1990) and at time of writing, resistance factors very much higher have been recorded in lice from many commercial flocks throughout the Australian wool producing areas. Current control measures include the continued use of synthetic pyrethroids where sheep lice are still susceptible to these compounds, and the use of plunge or shower dips containing the chitin synthesis inhibitor diflubenzuron or a variety of organophosphate insecticides where synthetic pyrethroid resistance is prevalent.

B *Overview of the potential of microbial control*

1 *The "Australian" sheep blow fly*

For a microbial control method to be commercially successful, it must be cost-effective and offer distinct advantages over other, competing methods of control, or it must provide control of a pest where no other method is effective or available. For example, an increasing use of *B. thuringiensis* occurs in systems where the chemical insecticide resistance of the target pests is so high that an alternative means of control is an imperative, often as a component of an insecticide resistance management program.

There are at least two approaches to the microbial control of *L. cuprina* flystrike on sheep. The first approach is to introduce or apply a self-replicating, transmissible microbial agent which would greatly suppress the fertility of the field population of adult *L. cuprina*. Given time, this strategy would reduce the number of fertile egg masses that can be laid and so reduce the incidence of flystrike. The research with the microsporidium, *Octosporea* sp., has this objective and early results are encouraging (Smallridge *et al.*, 1995).

A second approach for control of *L. cuprina*, and which may be integrated with the first, is to deploy a microbial larvicide, for example, *B. thuringiensis*. This approach entails the solution of many new and difficult problems. The Australian sheep industry is based on flocks containing very large numbers of sheep ranging over vast tracts of land. Each farm (station) is managed with very low labor inputs. To be commercially successful in Australia, the microbial larvicide must provide reliable, cost-effective control or protection against flystrike for a period of at least 10, and preferably 12 weeks following a single application (Wilkinson, 1986). This extended period of performance must be maintained under ambient conditions of intense solar radiation and very high temperatures.

Readers familiar with the use of *B.thuringiensis* products for plant protection in agricultural systems will be aware of the extreme stringency of these requirements. Even with modern formulations, it is unusual for a

B. thuringiensis spray deposit on an agricultural crop to remain larvicidal for more than a few days following application.

The successful development of a microbial *B. thuringiensis* larvicide system that meets the stringent performance specifications described above required detailed knowledge of the microbial ecology of the sheep fleece. In this, the population dynamics of the competitive displacement strategy underlying the colonization of the fleece by the *B. thuringiensis* larvicide was crucial (Pinnock, 1994; Lyness *et al.*, 1994).

2 *Sheep lice*

If it is to be commercially successful, a microbial agent should control lice on off-shears or short wool sheep for a period of at least 20 weeks after a single application. This duration of control is an even more stringent requirement than that for control of sheep blow fly described above. Although the *B. thuringiensis* strains and mode of action for control of sheep lice are different from those for the control of sheep blow fly, the achievement of this duration of control by *B. thuringiensis* required similar knowledge of the microbial ecology of the fleece and of the population dynamics of the colonization of the fleece by the microbial agent (Pinnock *et al.*, 1994).

C *Field trial design*

In general, the purpose of microbial control field trials is to test or demonstrate the pest control efficacy or some other attribute of the microbial agent under realistic industry conditions. During the development stages of the agent, often many field trials may be conducted and the scale and design of these field trials will depend on the data being sought. Ultimately, the microbial agent must undergo field trials to demonstrate and satisfy the efficacy requirements necessary for registration of the agent as a product for sale to farmers.

In some countries, no prescribed design or scale of field trials is imposed; it is the responsibility of the applicant to provide sound efficacy data from properly designed and executed field trials. In other countries, such as Australia, the design and scale of registration trials are prescribed by the governmental registration authority. To provide the reader with examples of field trial scale and design, the Australian requirements for sheep blow fly and sheep lice product registration trials are described below.

1 *Pre-treatment procedures for sheep blow fly or sheep lice trials*

a *Pre-selection and assignment of animals to treatment groups*

Sheep are mustered and inspected and animals with abnormal physical or fleece conditions are rejected. Sheep to be used in the trial should be as uniform as possible in age, frame size, wrinkle, fleece type and wool length. As a general rule, weaners or young wethers are preferred because these are most susceptible to flystrike and lice and so provide the most sensitive test of the control agent. If a mixed commercial flock of ewes and mature wethers is to be used, these should be assigned so that there is no sexual bias in any of the dose or treatment groups. With this proviso, the sheep are assigned randomly to the different dose or treatment groups.

At this preliminary stage, it is a wise precaution to have larger numbers of sheep in the dose or treatment groups than will actually be used in the forthcoming field trial. This preserves the desired group size while allowing for any sheep which fail pretreatment inspection to be eliminated from the trial. The sheep in each dose or treatment group are then ear-tagged with plastic tags bearing an individual identification number for each animal. For ease of drafting after muster and throughout the trial, ear-tags of different colors may be used to indicate the different doses or treatments to be used.

b *Pre-treatment inspections*

For blow fly control trials, all sheep are given a detailed inspection immediately prior to the commencement of the trial to detect and eliminate animals that are scouring or wounded, have vaginal or urethral discharge or fleece abnormalities such as fleece rot or mycotic dermatitis, or have covert flystrikes.

For sheep lice trials, each animal is inspected as above. Sheep accepted for the lice trials are then restrained and a pre-treatment lice count made and recorded for each animal as described below in the scoring methods.

2 Design for control of sheep blow fly

In Australia, the registration and use of pesticides are regulated by the Australia Pesticides and Veterinary Medicines Authority (APVMA). The APVMA requires, among other data, the demonstration of efficacy of a pesticide, including a microbial control agent, before it can be registered as a saleable product. The demonstration of efficacy entails the execution of prescribed field trials. For control of the sheep blow fly, the demonstration of efficacy requirements entails compliance with the following field trial design.

A minimum of 5 to 10 field trials must be conducted in a range of different geographic zones, using sheep of a representative range of age, breed, sex and wool length. In most pastoral zones of Australia, the Merino or a Merino-cross is the predominant breed of sheep, and so it is expected that this breed also would be predominant in the field trials. If breeds other than Merino are to be included, it is recommended that weaners or unclassed hoggets be used because these are more susceptible to flystrike than are older sheep.

For each trial, a minimum of 100 sheep, inspected and known to be free of covert flystrikes, must be used for each dose level or treatment group. After application, (see below for methods), all treatment groups must be kept segregated until dry, and then managed as a single flock. A similarly sized, untreated control group may be included in the trial.

a Duration of sheep blow fly field trials

There is no prescribed duration for sheep blow fly field trials, but for practical purposes, the control agent should protect sheep from flystrike for at least 10, and preferably 12 weeks. This 12 week period determines the minimum duration of the field trial. During the development of *B. thuringiensis* as a control agent for sheep blow fly, the field trials were successfully run for 20 weeks.

b Scoring method of sheep blow fly field trials

All sheep in the trials undergo an individual, detailed bodily examination daily when intense blow fly pressure - "flywave" - conditions occur, and at least every 3 days at other times. The position, size, severity and date of all flystrikes are recorded and all struck sheep must be clipped, spot-treated and re-examined the following day to ensure that the spot-treatment has been effective. This protocol is a sensible animal ethics requirement intended to prevent deaths and minimize distress and suffering of the sheep in the trial.

As a general rule, an adequate level of blow fly pressure is deemed to have occurred, and therefore the control agent properly tested, if at least 15% of the untreated control sheep are struck during the trial.

3 Pen and field trial design for control of sheep lice

The prescribed trials of products for control of sheep lice are divided into two stages. The first stage comprises initial pen trials on only a few sheep per treatment or dose. At least two initial pen trials are required. These use low numbers of sheep - as few as 5 per dose or treatment group. The pen trials provide a preliminary examination of the efficacy of the control agent at different dose levels ("dosing trials") and of any adverse effects of the agent on the test animals.

The pen trials are followed by the second stage, which entails field trials on large numbers of sheep. There are two designs for the large field trials. The design to be adopted for any given field trial is determined by the length of wool on the sheep in that trial. For off-shears or short wool sheep, which are defined as sheep with up to 42 days' wool post shearing, in addition to the pen trials at least 5 field trials are required, each with at least 1,000 sheep per dose or treatment group. At least three of these field trials must be with fine wool Merino sheep. As indicated below, these field trials run for up to 20 weeks.

For sheep with more than 42 days' wool, at least 6 field trials are required in addition to the pen trials. In each trial, at least 500 sheep are required per dose or treatment group. These long wool field trials should be at fleece lengths equivalent to 3, 6 and 9 months' wool, with 2 field trials for each wool length.

a Duration of sheep lice field trials

If the control agent under trial is to be claimed to reduce lice numbers to undetectable levels, the field trials must run for 20 weeks or until the next shearing, whichever is the sooner. If the control agent is to have a lesser claim - for example, to reduce lice numbers by at least 95% over a period of 1 or 2 or 3 months, then the field trials must run for the claimed period plus at least 10 days, or until next shearing, whichever is the sooner.

b Scoring method of sheep lice field trials

The method of scoring sheep lice in the pen and field trials is to inspect and count lice on all sheep in the pen trials and on 25 of the sheep in each dose or treatment group in the field trials. These sheep are inspected and their lice counted before treatment, and at intervals post-treatment. The following sampling procedure is used.

The inspection and sampling procedure requires each sheep to be restrained. This may be achieved by placing the animal in a closed race, in a crutching cradle, or hog-tied on an inspection table. Once the sheep is restrained, all adult lice, nymphs and eggs are counted at 20 sites on each side - a total of 40 sites on each animal. The location of these sites has been determined by study of the distribution of lice on sheep, and the location of these sites is shown in Figure 1. A plastic mesh may be used as an aid to determining the location of the sampling sites, but with experience, the locations of these sites are learned and the mesh dispensed with. At each of the 40 sampling sites, the fleece is parted so that a line of skin 10 cm long is exposed. All lice, nymphs and eggs on this line of skin and on the exposed wool staples on either side of the parting, are counted. The efficacy of the control agent is determined by the reduction in lice numbers relative to the pre-treatment counts, as described in the preceding section.

D Application methods for sheep blow fly and sheep lice control

In general, the microbial control agent will be applied to the animal in a liquid suspension by one of three application methods in common use in the sheep industry. These methods are backline applications, spraying (jetting) applications and application by plunge or shower dips. These methods are described below.

1 Backline application methods for sheep blow fly and sheep lice control

Backline applications rely on the translocation of the control agent from a single or a few line or strip applications along the back of the sheep to effectively treat the entire animal. This method has, and is, being used for some ready-to-use chemical treatments such as the synthetic pyrethroids. The control agent is applied in a low volume of < 1 liter of suspension per sheep along the spine of the animal using a single nozzle, hand-operated syringe-type applicator.

Sheep to be treated are mustered into a permanent or temporary sheepyard, from where a sheepdog works the sheep so that they move in single file through a drafting race. As each sheep passes the operator, it receives its backline treatment. This application method has the advantages of high rate of throughput of sheep and low labor input (one person and a dog). A possible disadvantage of the backline application method is that the control agent forms a concentration gradient from dorsal to ventral around the animal, which may result in only low or discriminating doses being achieved in the axils or on the ventral side of the animal. Because of this effect, the use of low volume backline application for microbial control agents is still in the experimental phase.

An alternative backline application is the use of a pressure-fed, hand spraying ("jetting") wand (see below) to apply a medium volume of 1 to 2 liters/sheep. The jetting wand is drawn through the fleece along the spine of the animal, thoroughly wetting the skin. This method has been used successfully for the application of a *B. thuringiensis* formulation as a prophylactic treatment against flystrike if the application along the backline is extended to include the breech and pizzle areas.

2 Jetting application methods

Spraying ("jetting") applications use a system of fixed, flexible or hand-held nozzles or jets

Figure 1. Sites on the body of sheep where fleece partings should be made during examination for infestations of sheep lice in pen and field trials. Diagram reprinted by kind permission from "Guidelines for Ovine Lousicides Efficacy Submissions" 1996. National Registry Authority for Agricultural and Veterinary Chemicals, Australia

to apply the control agent, for example a *B. thuringiensis* formulation, over the entire animal as a suspension in water.

The jetting nozzles are supplied with the suspension by a hydraulic pump equipped with a pressure regulating valve and drawing the suspension from a reservoir tank. Generally, a delivery pressure of 700 to 800 kPa (101 to 116 psi) is selected, and some systems have a pressure reservoir to reduce pulsing or variations in delivery pressure and so maintain a more constant flow through the nozzles. Many jetting systems have a pressure relief valve which allows a recirculating or bypass line to take suspension back to the tank or reservoir, and so keep the contents thoroughly mixed. A diagram of a hand jetting wand is given above in Figure 2.

With an experienced operator, hand jetting is the most thorough jetting method. However, hand jetting is very labor intensive, and the need to reduce costs has led to the development of automatic jetting races, where the animals are passed rapidly through a set of fixed or flexible nozzles which apply the pressure spray automatically. An example of a jetting race is shown in Figures 3 and 4.

3 *Dip application methods*

Plunge dipping is the traditional method of treating sheep for ectoparasite control. A traditional sheep plunge dip is a concrete-lined trough approximately 0.75 m wide, 1.5 m deep and several meters long. Plunge dip capacities range from 7,000 to over 10,500 liters. Sheep fall into the dip from a single file race and are totally submerged for 2 to 3 sec. As they swim to the opposite end of the dip, they are submerged "dunked" twice again by operators wielding T-shaped poles. A ramp at the end of the dip allows the sheep easy exit on to a draining area where runoff is collected and returned to the dip. Research (Lund *et al.*, 1998) has shown that for effective coverage and penetration of the fleece, the dip swim length should be at least 9 m.

A recent development of the plunge dip is the mobile dip, which is a demountable, small plunge dip which can be towed behind a Land Rover® or similar four-wheel drive vehicle. These dips are operated by independent contractors and have the advantage of being able to travel to remote outback areas where the sheep are mustered. A typical mobile dip, the "Rippa Dippa S®" is shown in Figure 5. In this design, sheep from a single file race are lifted by conveyor belt and fall into an S - shaped plunge dip. This design has a swim length of approximately 12 m.

E *Assessment of persistence of microbial control agents*

Quantitative assessment is especially valuable during the developmental stages of the microbial agent, when different strains and/or different formulations are to be compared, and such assessments are essential if mathematical modeling is to be used as an aid to predicting dose delivery. Assessments of the persistence of a microbial agent for control of sheep ectoparasites may be made quantitatively by taking fleece samples

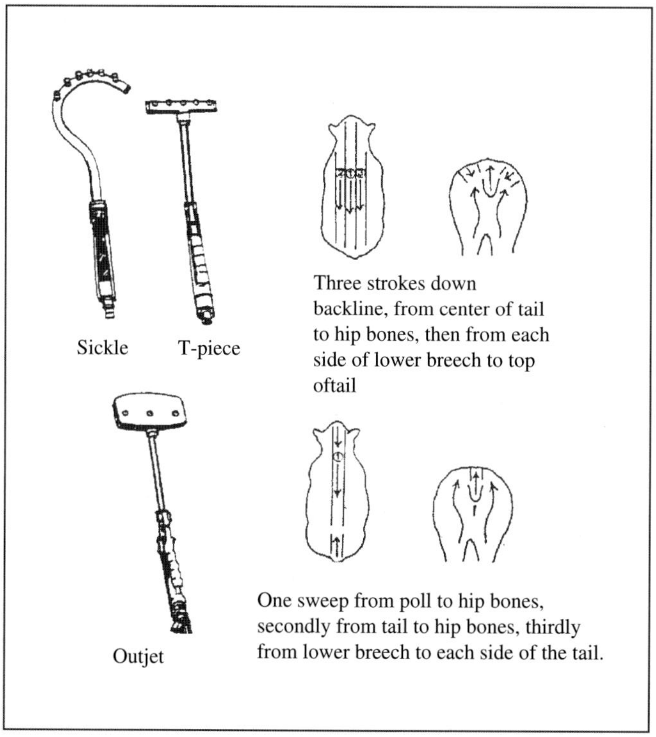

Figure 2. Hand jetting wands. Each wand consists of a hand piece with a cutoff valve and 3 to 6 cone nozzles. The wand is supplied with the treatment suspension by a hose and hydraulic pump – see text

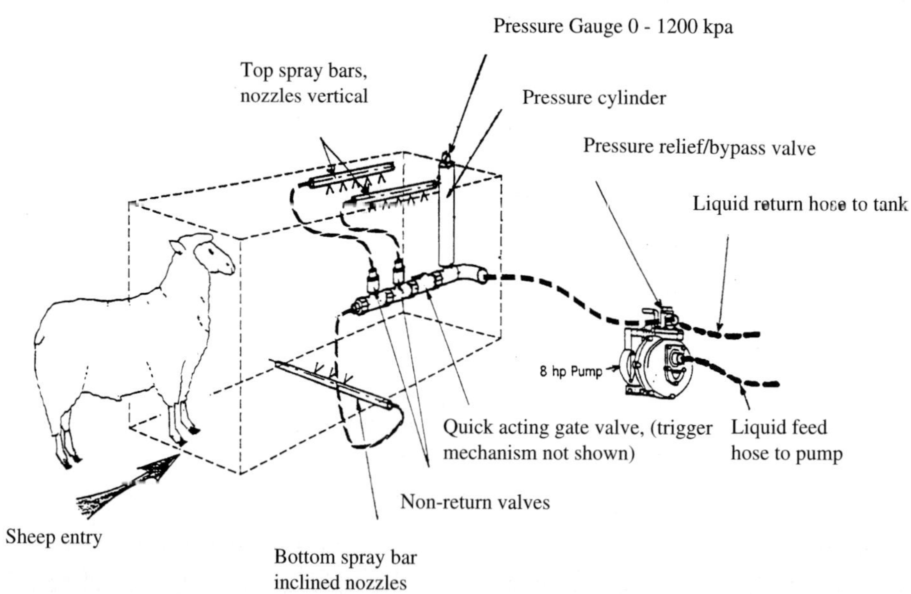

Figure 3. Automatic jetting race – general view, frame omitted to show the spray bars

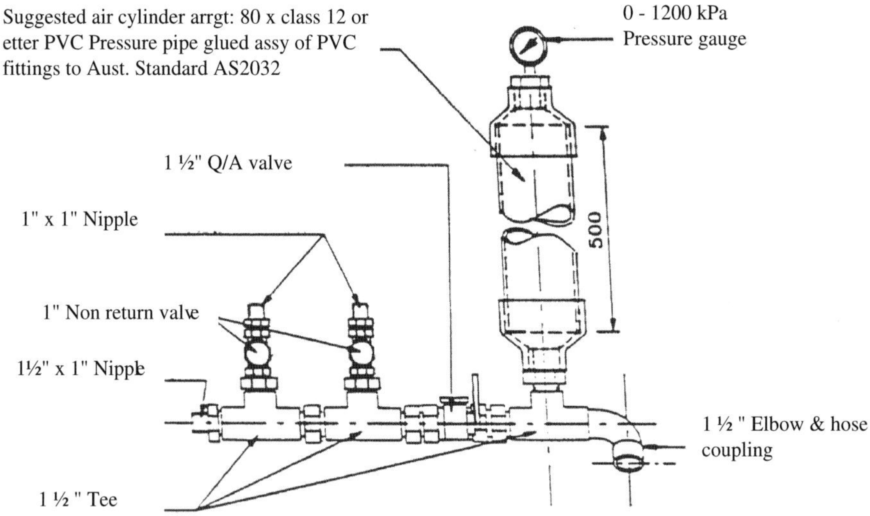

Figure 4. Automatic jetting race – detail of valving and pressure reservoir (1″ = 2.54 cm)

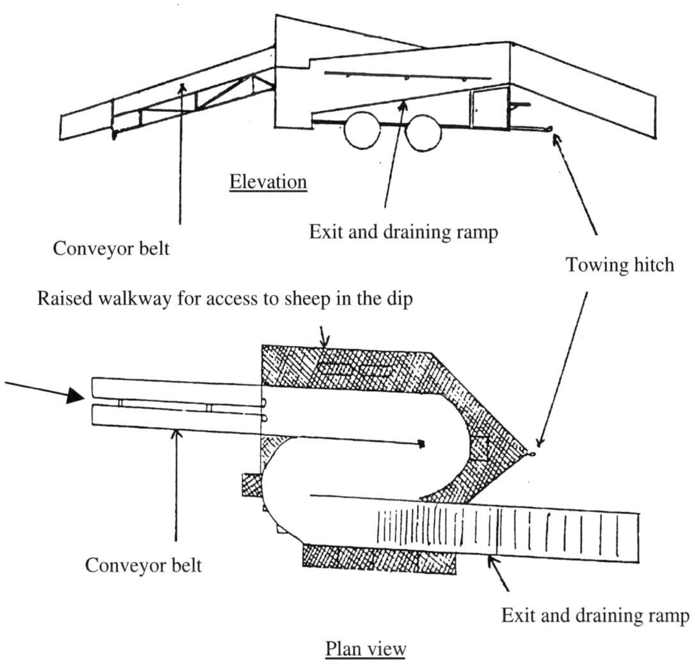

Figure 5. Mobile plunge dip- the "Rippa Dippa S"® in this design, sheep approach the dip through a single file race and are lifted and delivered into the dip by a V-shaped conveyor belt. The dip is folded to an S shape for compactness, and gives a swim length of over 11 m. The sheep exit the dip via a draining ramp. This dip can be operated by two persons and has a throughput of 700 to 800 sheep per hour

(Lyness *et al.*, 1994) and applying a laboratory assay method such as viable cell counts.

Qualitative assessment of persistence often is based on the duration of efficacy of the microbial agent, and this has a direct and practical utility. The duration of efficacy also is an essential element in the validation of predictive dose models. Estimation of the duration of efficacy may be made by continual or repeated, natural challenge of the treated animals or substrate by the pests in the field. For example, the sheep may be exposed to gravid *L. cuprina* or to untreated sheep with heavy lice infestations. Alternatively, the treated animals may be artificially challenged at intervals by an implant technique (Lyness *et al.*, 1994) using *L. cuprina* neonate larvae or by "seeding" the animals with a known number, usually 50, of young adult lice.

F Assessment of efficacy of microbial control agents

The assessment of efficacy of the control agent may be made by the scoring methods described above. To be regarded as efficacious, the control agent should reliably suppress the pest in the treated area or animal below a threshold density, or limit or prevent the pest's reinfestation or reproductive capacity for a certain period of time.

For sheep blow fly prophylactics, the control agent must protect the treated sheep from flystrike for at least 10 weeks post-treatment during periods of heavy blow fly pressure. For sheep lice control, the control agent either should reduce lice numbers to undetectable levels for at least 20 weeks post treatment, or reduce lice numbers by at least 95%, depending on the claim made for the product. The lice assessments are made using the 40 site scoring method described above.

3 Pests of intensive animal husbandry systems

A A Major pests of intensive animal husbandry

1 The common house fly, Musca domestica

The house fly is thought to have originated in eastern or central Africa, but has readily spread around the globe with human movements and is now cosmopolitan. It is regarded as a synanthropic fly, since it develops in large numbers in organic materials accumulated near people (*e.g.*, household garbage) or their livestock and poultry (West, 1950; Axtell and Arends, 1990). House flies readily develop in manure, especially from confined swine and poultry. They also may develop readily in other types of animal dung, but on dairies and cattle feedlots usually are most abundant in decaying foodstuffs or straw/hay contaminated with manure and urine. House flies prefer moisture levels above 60% for larval development, and cease development at temperatures below 12 °C (West, 1950; Stafford and Bay, 1987).

House flies disperse readily to nearby houses and are serious human nuisance pests, and most farmers will try to keep fly numbers in check for this reason. Proper manure management, particularly moisture control, is critical for both the house fly and stable fly. Many producers use insecticides for adult control (surface residual sprays, toxic baits). There is a growing awareness of the importance of biological control agents (predacious beetles and mites, parasitic wasps), and many producers now tend to avoid treating the larval habitat (accumulated manure) with broad-spectrum pesticides because of nontarget effects on beneficial arthropods. Economic losses from fly nuisance are difficult to quantify, but regulatory agencies may require significant and expensive control efforts such as intensive insecticide use or frequent manure removal. In extreme cases producers can be fined, sued or forced to cease operations. In addition to their direct nuisance potential, house flies defecate and regurgitate on surfaces they visit. These spots are unsightly. The behavior also contributes to the potential of flies to transmit or harbor pathogens such as trachoma (which may cause blindness), and particularly bacteria or viruses that cause enteric diseases of humans or animals (Greenberg, 1971). Recently, concern for fly transmission of such pathogens has increased in accordance with development of antibiotic resistance and particularly virulent strains such as *Escherichia coli* 0157:H7 (Szalanski *et al.*, 2004; Alam and Zurek, 2004). Other flies can be common and problematic in manure, garbage, and similar organic habitats, particularly *Fannia canicularis*. These are similar enough to the house fly for many of the same management principles to apply.

2 *The stable fly,* Stomoxys calcitrans

The biting stable fly is occasionally a human pest but is much more commonly found biting domestic livestock such as cattle or horses, and it can be a serious pest of dogs. The genus *Stomoxys* probably originated in Africa, where a number of species exist. Some species such as *S. nigra* are severe biting pests in their native range and areas where they have been introduced (*e.g.*, islands of Mauritius and Reunion), but *S. calcitrans* is the species most widely distributed in temperate zones, including Australia, Europe and North America (Zumpt, 1973).

S. calcitrans develops in some of the same habitats as house flies. Stable flies are not common in fresh manure, but may use it after it has decomposed somewhat. They prefer rotting vegetation or feedstuffs, often if those are polluted with livestock manure or urine. Rotting straw, wet hay, or silage waste are excellent habitats for stable flies. While the largest numbers of stable flies usually are found near the developmental sites, these flies are excellent dispersers and may travel several kilometers in search of hosts or oviposition sites. Under special conditions, particularly on weather fronts, stable flies can be transported up to hundreds of kilometers. This phenomenon occurs, for example, along the Florida panhandle in the USA, where down drafts deposit large numbers of stable flies from inland farms on beaches up to 225 km away (Broce, 1993).

Unlike *M. domestica*, which is mainly a human nuisance pest, both sexes of stable flies are obligate blood feeders and typically feed daily. The pain associated with their bites reduces weight gains, milk yield, and feed conversion efficiency in cattle, and stable fly economic impact on the USA cattle industry exceeds $400 million/year (Drummond *et al.*, 1988; Campbell, 1993). Economic injury levels on beef cattle are 2–3 flies/leg. Additionally, stable flies are capable of mechanical transmission of several livestock pathogens, such as Equine Infectious Anemia virus or *Trypanosoma evansi*, the causal agent of surra.

As is true for *M. domestica*, numbers of stable flies can be reduced substantially through proper sanitation, and natural biological control via predators and parasites is very important. Parasitic wasps which attack the pupal stage of the house fly and stable fly (particularly species of *Muscidifurax* and *Spalangia*) are reared commercially and sold to producers for release. Results of experimental releases have been inconsistent but may be beneficial, especially in habitats such as semi-enclosed dairy calf pens, where parasitoid releases can be focused. There also has been interest in sticky stable fly traps, which may reduce stable flies locally, provided fly numbers are not too high. At the farm level, however, it is most common for producers to try to protect their animals using insecticides such as pyrethroids applied directly to the host. Perhaps exacerbated by resistance, control is incomplete and quite temporary, usually only a few days at best.

3 *Manure flies*

Livestock and poultry pests can be divided into several groups, based on the degree of association between the pest and animal and the nature of the production systems. This affects our approach to sampling and potential application of a microbial agent. Permanent ectoparasites of either confined or range animals, such as lice, complete their entire life cycle of the host. In other cases only certain life stages are found on the host, but they remain there for substantial periods of time (days to weeks). This category includes adults of the horn fly and buffalo fly (*Haematobia* spp.) and to a lesser extent the face fly and bush fly (*M. autumnalis* and *M. vetustissima*). These flies spend much, but not all, of their adult life on the host, while immatures are found in dung. The common theme, however, is that both microbial control efforts and monitoring of pest and pathogen activity and prevalence often focus on the host animal.

A second category includes pests that periodically must contact the host, often to obtain blood, but do not spend much time there (a few minutes every 1–4 days). Adult stable flies would fall into this category. For these pests microbial control efforts must generally focus on immature habitats (decaying organic materials, older manure deposits), or on resting surfaces favored by these flies in the environment (*e.g.*, barn walls). An exception to this would be a pathogen capable of persisting on the hair or skin of a host or within a trap which utilizes host cues (*e.g.*, kairomones) to attract and infect the target pest fairly quickly. Because the direct association between pest and host is of primary interest, monitoring often focuses on pests coming to the host or associated cues (*e.g.*, CO_2-baited traps), while monitoring of the microbial agent often requires sampling the immatures in developmental habitats. Sampling the immatures is usually the more difficult task.

The third category includes pests closely associated with livestock or poultry, but which visit the animals only incidentally. Their economic impact is more indirect (pests of people nearby). Filth flies that develop in accumulated manure or animal feedstuffs, such as *M. domestica* or *Fannia* spp., are good examples. In these cases microbial control may be directed against either immature stages in developmental sites or adult stages in preferred resting or activity sites. Monitoring of the pathogens and target pests may require sampling adults, immatures in developmental sites, or both.

4 Lesser mealworm beetles

Confined poultry, such as caged egg-laying hens or birds raised for meat, are grown or held in buildings with variable levels of environmental control. Feed for these birds inevitably is spilled into the immediate area. This becomes a food source for larvae and adults of species such as the lesser mealworm beetle, *Alphitobius diaperinus* (Tenebrionidae). While other beetles, such as *Dermestes* spp., also can be common in the feed/feather/manure environments near birds, numbers of *A. diaperinus* often are the highest.

The lesser mealworm life cycle is approximately 50–70 days, but development is prolonged at lower temperatures, and adults may live for several months or longer (Axtell and Arends, 1990). Eggs are laid in cracks and crevices, and most of the developmental period is spent as a larva. The lesser mealworm causes serious damage to insulation, especially to polystyrene as a late instar larva burrows into it to pupate (Vaughan *et al.*, 1984). Perhaps even more significantly, the beetle is now appreciated as a persistent source/reservoir of several pathogens, including enteric bacteria such as *Campylobacter* spp. or viruses such as turkey coronavirus (Strother *et al.*, 2005; Watson *et al.*, 2000). The longevity of the beetles and their relative inaccessibility in the environment (adults or immatures secreted in cracks and crevices, or in pupation sites) make them a challenge to control and a particular concern for harboring pathogens longer-term or between flocks. Of biological control options, pathogens and nematodes have received the most attention for control of lesser mealworm beetles.

B Overview of potential microbial control agents for manure flies or beetles

1 Fungi

Many fungi have been reported, particularly from Diptera (*e.g.*, Steenberg *et al.*, 2001), but far fewer have been utilized for control. The *Entomophthora muscae* species complex (*E. muscae* and *E. schizophorae*) have been observed to cause epizootics in muscoid flies for many years (Mullens, 1990). Principal natural hosts include *M. domestica* and *Fannia* spp., and infection prevalence may exceed 70–80% during epizootics. A number of hosts may be infected by a given strain, although there is evidence for host specificity even within the Muscidae. More recently several attempts have been made to release infected flies and stimulate epizootics. These fungi can be maintained indefinitely *in vivo* in laboratory-reared *M. domestica* or grown (predominantly vegetatively) in tissue culture media such as Grace's medium. In the latter case, patent infections may be induced by injecting protoplasts into susceptible flies and then using cadaver-to-fly transmission to increase numbers of infected flies for release. While resting spores of *E. muscae* group fungi are reported in the literature, primary and secondary conidia are normally produced for expulsion from conidiophores; they are sticky, short-lived and almost impossible to handle or disseminate directly.

Natural *E. muscae* epizootics typically occur when host populations are high and temperatures are not too hot (maximum temperatures below 26–28 °C). Several researchers have managed to increase prevalence and stimulate earlier *M. domestica* epizootics on New York dairies through introduction of infected flies in two ways- deploying freshly killed cadavers with heads inserted into dishes of water agar in fly aggregation areas or release of laboratory-infected flies (Geden *et al.*, 1993; Steinkraus *et al.*, 1993). The level of infection was not sufficient to exert control, however. Short-term establishment of *E. schizophorae*, which is not naturally found in house flies on southern California dairies, was achieved by releasing critically ill flies in a covered calf-rearing facility. The pathogen did not persist, possibly due to behavioral fever (Six and Mullens, 1996). Enclosed habitats, such as poultry buildings, may offer more potential for control by *E. muscae* complex fungi (Kuramoto and Shimazu, 1997). Flies probably cannot bask to induce behavioral fever and are in fairly close proximity to wild flies. Kalsbeek *et al.* (2001), however, showed that house flies infected with *E. schizophorae* also exhibited behavioral fever indoors by resting on heat lamps provided for piglets in a swine barn.

The fungi *B. bassiana* and *Metarhizium anisopliae* have been screened in the laboratory against *M. domestica* and *S. calcitrans* (*e.g.*, Barson *et al.*, 1994; Watson *et al.*, 1995), and some field testing has been done with *M. domestica* (Watson *et al.* 1996). In general the effective concentrations of conidia in these tests have been high (> 10^6/cm^2). The tendency of house flies to rest inside structures such as calf hutches at night makes an inside surface application of conidia an attractive prospect for control. Natural *B. bassiana* prevalence in flies is usually < 1%, but prevalence was increased to 43–47% on treated farms (Watson *et al.*, 1996). The fungi, in general, have incubation periods of several days to a week or more, theoretically allowing flies to reproduce before death. Kaufman *et al.* (2005) recently tested *B. bassiana* sprayed as a coarse fog (mostly residual application to surfaces) and releases of pteromalid pupal parasitoids against house flies in caged layer houses. The results were compared with applications of pyrethrin insecticides plus the parasitoids. Fly numbers post-treatment were reduced by the fungus-parasitoid treatment. Renn *et al.* (1999) tested *M. anisopliae* against *M. domestica* in large enclosures (10 m^3) by using bait stations (sugar and muscalure) to attract/arrest flies and expose them to cultures of the fungus nearby. Flies did acquire many thousands of conidia, and almost all were killed in 10 days; the authors discussed how later fly-to-fly transmission could enhance the effectiveness of such an approach.

These same fungi have potential for control of *A. diaperinus*. For example, Steinkraus *et al.* (1991) tested *B. bassiana* against beetle larvae at doses of approximately 10^7–10^8 conidia per g of dust or ml of aqueous suspension, and achieved high infection rates (> 60%) for larvae treated directly, but low to moderate infection rates (< 30% for aqueous but higher for dust treatments) for beetle larvae distributed in poultry litter. Geden and Steinkraus (2003) achieved short-lived (2 weeks) but significant control (60–90% reduction in numbers of beetle larvae) using especially corn bait or granular formulations of *B. bassiana*. Because lesser mealworms feed on grains, there is potential for making more use of consumable baits for distribution of pathogens. Selection of high-pathogenic strains and appreciation of environmental variables such as temperature (Crawford *et al.*, 1998; Geden and Steinkraus, 2003; Alexandre *et al.*, 2006) are logical and necessary exercises to improve our ability to utilize fungi for beetle control.

2 *Nematodes*

Livestock and poultry pests have a number of parasitic nematodes, particularly Mermithidae and Tylenchida. The latter infect several important flies in the genus *Musca*, including *M. autumnalis*, *M. vetustissima*, and *M. domestica*. *Paraiotonchium (=Heterotylenchus) autumnalae* is a well-known associate of the face fly, *M. autumnalis*, and is typical of the group (see Kaya and Moon, 1978). Mated adult females in the larval habitat (in this case, cow dung) invade the host larva. They produce a parthenogenetic female generation whose progeny invade the ovaries of the adult host. Infected female flies then deposit gamogenetic nematodes, rather than eggs (parasitic castration). Recently a new member of the genus, *P. muscadomesticae*, was discovered infecting *M. domestica* in Brazil; the potential of this agent has yet to be assessed.

Like *P. autumnalis*, *P. muscadomesticae* causes parasitic castration and can substantially reduce fly survival (Geden, 1997).

Entomopathogenic nematodes in the genera *Steinernema* and *Heterorhabditis* can be readily mass-produced and have been tested in the laboratory, where they can infect several Diptera of veterinary importance. Field trials have been variable. Early experiments generally indicated these nematodes were ineffective and did not live very long or disperse well in poultry manure, especially wet manure where *M. domestica* was common (Georgis *et al.*, 1987; Renn, 1995). Somewhat drier poultry manure (< 60% moisture) appears to be more hospitable for the nematodes, particularly if they are encapsulated in alginate (Renn, 1995). Alginate encapsulation resulted in little infection and death of adult flies feeding on the capsules, but promising laboratory results were achieved when adult flies were exposed to nematode bait pads (Renn, 1994). This raised the potential of a bait formulation. The encapsulation may serve not only to protect the nematodes from desiccation, but from ammonia and similar chemicals in a hostile habitat such as poultry manure. Other types of animal manure also generally are not as nitrogen-rich as that of poultry, and the entomopathogenic nematodes may have better prospects there. If used in a bait station type approach, the porosity of substrate can have substantial effects on how well flies may be infected by such nematodes; cotton poplin was good, while Perlite was poor, for example (Renn and Wright, 2000).

Renn (1998) demonstrated suppression of house fly in swine barns (farrowing) using *S. feltiae* or *H. megidis* incorporated into baits, or manure applications of *S. feltiae* (encapsulated or not). Control was comparable with, or better than, methomyl bait. The larvicidal application was targeted to drier areas of the manure, and it is possible that the nematodes persisted well there and attacked larvae preparing to pupate in these drier areas. Taylor *et al.* (1998) screened a number of species and strains of *Steinernema* and *Heterorhabditis* for infectivity against larvae of *M. domestica* in bovine manure in the laboratory, and nematodes significantly reduced emergence of flies added to this substrate. Some nematodes remained infective in cattle manure (detected by *Galleria* sentinels) for up to 10 weeks after application. Nematodes similarly have been utilized against lesser mealworm beetles. Geden *et al.* (1987) applied 900,000 *S. feltiae* in 8 liters of water per 8.4 m^2 areas of floor soil (using a watering can) in broiler and turkey houses in North Carolina, USA. The target was particularly lesser mealworm pupae that might persist in soil between poultry flocks. Treated houses experienced slower and lower level increases in beetle populations after new flock introduction, but numbers were about equal after 10–13 weeks. They found evidence that nematodes could persist at least 7 weeks after application in the field, and one container bioassay showed persistence out to 6 months.

C Field trial design

1 Setting up plots for manure fly or beetle control

The following section refers primarily to filth flies such as *M. domestica*, *S. calcitrans*, or *Fannia* spp. The egg stage typically is short-lived, and it is assumed the primary target for control is the larvae or perhaps the pupae. Plot selection varies with the distribution of the target pest. In moist to wet substrates, the eggs of most flies hatch quickly (< 2 days), the larvae spend several days to a couple weeks maturing, and larvae move to a drier area nearby to pupate. Pupae emerge after several days to a week or two.

In confined animal systems, muscoid fly development occurs commonly at edges. The center of a feedlot, for example, is heavily disturbed by animal hooves, and flies cannot survive in numbers there. Rather, they are concentrated in relatively undisturbed manure under fencelines, or along feed bunks or water troughs. Even in undisturbed organic waste habitats, such as silo leakage or mounded manure stacks, flies are usually most common near the edges or near the surface. This may be due to fermentation heat or anoxia deeper in the manure mass. In certain situations, such as rows or areas of undisturbed manure accumulating beneath confined poultry in cages or dairy calves or swine on wooden slat floors, fly development can be extensive and relatively uniform, provided the moisture is consistently high (*e.g.*, 65–70%). Even here, however, it is much more common for immatures to be highly aggregated. Typically, moisture variation etc. results in areas where flies are concentrated, and these areas constitute a small proportion of total available habitat. Given limited time and resources, it thus is necessary to find the areas where flies are most abundant and focus attention on them. If time and resources permit, preliminary sampling should be done in a statistically sound way, and numbers of samples required can be adjusted according to the degree of fly aggregation and level of resolution desired (*e.g.*, Stafford and Bay, 1994).

Having thus stratified the habitat to focus sampling on suitable zones, multiples of these units are identified for treatment or control use. If at all possible, buffer (untreated) zones at least the size of treatment plots separate experimental plots. In manure, we often use sections in the range of 0.5–1 m^2 in surface area. These will accommodate removal of multiple small subsamples per plot over time. Pretreatment samples are taken, treatments are applied, and further samples are taken at pre-selected intervals after treatment.

If plots are indoors, such as an environmentally-controlled poultry or swine house, fly development may occur all year. Otherwise flies are quite seasonal (warm summer weather in temperate zones). In accumulated manure, they also occur in a succession. Flies are excellent colonizers and lay eggs as manure first begins to accumulate.

Relatively free of natural enemies, fly numbers are highest in the first few weeks after new manure accumulation begins. This needs to be considered in timing a control trial and in sampling. In a hen house which just had the manure cleaned out 4 weeks ago, for example, house flies would be expected to be numerous. However, a decline would be expected over the following 4–6 weeks. Good control plots or houses are critical to avoid attributing a natural decline in fly numbers to treatment effects.

As is true for fly immatures, both adult and immature beetles such as *A. diaperinus* also tend to be unevenly distributed in the environment, for example along water lines in a meat bird operation (Strother and Steelman, 2001). Determining their distribution prior to applying a biological control agent is therefore necessary both to optimize control and to allow accurate assessment of a control effort. Once relative distributions are known, experiments can be appropriately designed, using an approach rather similar to fly trials in many cases.

2 Sampling fly or beetle substrates

A volume for sampling must be selected. In many manure substrates with fairly high fly numbers, samples of 100–200 ml in size are good. Given the high degree of spatial heterogeneity, even after selective sampling, one may need to pool several individual smaller samples prior to processing. Manure can be removed using a hand trowel or coring device shown in Figure 6. Most fly immatures tend to be fairly close to the surface of a manure type substrate (the same tends to be true for beetles), but this can vary substantially with such things as fly species and moisture, so a core type sample which crosses depths often is preferred. Samples then can be handled in one of two basic ways.

If it is necessary to extract live larvae from the samples (for example to determine infection levels with a pathogen) samples can be spread on a screen in a Berlese funnel, where the larvae will crawl away from light and heat and separate themselves into a catch receptacle. There is not a good means of detecting dead larvae; if this is necessary one must search the medium visually.

If detailed information is not needed on numbers of immatures or infection levels, successful emergence of adults can be assessed by holding the samples. It usually is not possible to place emergence traps directly over the manure in the field, since traps may be buried by new manure deposits or disturbed/destroyed by animals. Emergence trap designs incorporate some type of one way funnel to retain adult flies which enter a collecting head (see Figure 7). A small amount of toxic fly bait may be added to the collecting head to incapacitate the adult flies before counting if desired (Mullens *et al.*, 1996). The collecting head is typically a clear plastic or glass container that admits light, so flies will move into it. Containers must have ventilation to prevent accumulation of toxic gases and oxygen depletion. Most fly species develop quickly enough that maintaining manure moisture in the emergence traps is not a critical issue. If possible, we try to hold manure samples

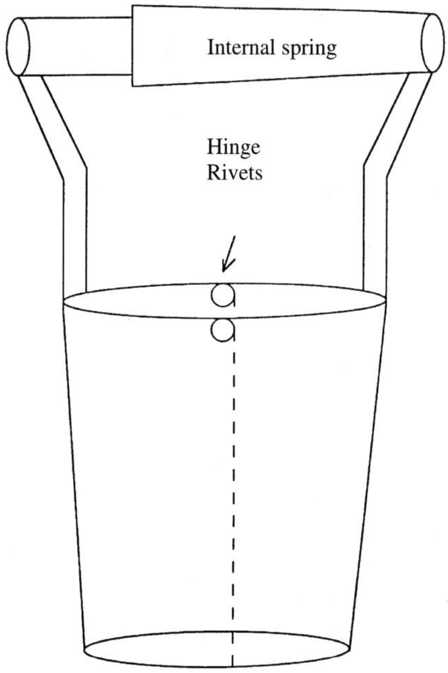

Figure 6. Bulb planter used to take core samples for sampling flies from manure substrates. By depressing the handle spring, the sample is released without compression or disturbing the surface

on-site for fly emergence, in a location which will not impede movement of people or equipment. This ensures manure temperatures are similar to ambient.

Beetles can also be sampled by removing portions of habitat and then extracting immatures or adults, either by visual search or using something such as a Berlese funnel. Beetles, however, are typically found in somewhat drier and looser substrates than are occupied by fly larvae. In a caged laying hen house with manure buildup beneath the hens, for example, fly larvae predominate in the upper (wetter) part of the manure accumulations, while

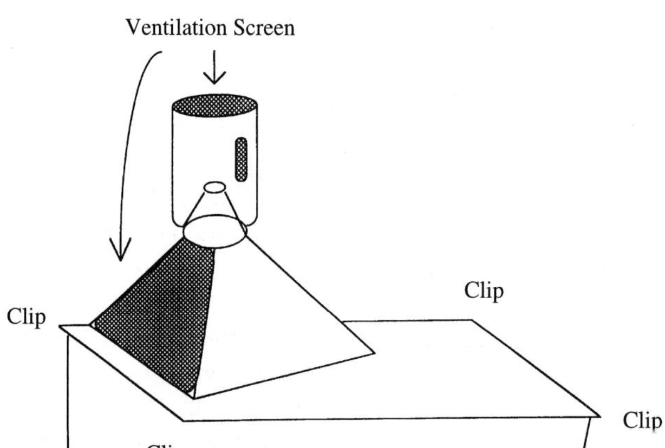

Figure 7. Trap used to collect flies emerging from manure samples placed into plastic dishpan. Top held on using spring clips. Small piece of screen glued on inside of collecting head may be wetted slightly to hold a small quantity of toxic fly bait, which will kill emerging flies prior to counting

beetles might be found more commonly in drier manure/feed mixtures near the base of the manure piles. In tests of pathogens, in fact, this direct search technique probably is underutilized. Direct searches and other extraction techniques might be uniquely useful in recovering pathogen-killed or moribund insects.

Due in part to relative ease of use (a major advantage, to be sure), the sampling tool most commonly used for studies of *A. diaperinus* field biology or control the past 20 years has been the tube trap (Safrit and Axtell, 1984), sometimes known as the Arends trap. This trap takes advantage of the beetles' marked tendency to occupy available cracks and crevices. It consists of a PVC plastic pipe, 4–5 cm in diameter and 24 cm in length. Corrugated cardboard is rolled lengthwise inside the tube such that the openings face toward the open ends of the PVC tube. The tube traps are easy to make and deploy, for example every 10–15 m along poultry house walls, manure accumulations, or water lines. They are left in place for a defined period (*e.g.*, a week) and then recovered and placed into individual plastic bags. Traps can be frozen prior to unrolling the cardboard and counting, categorizing and identifying the beetles, or the trap contents might also be processed alive if one wanted to determine infection levels. Trap counts are very useful in assessing relative activity (assumed to relate to density) of beetles from one time to another (*e.g.*, before vs. after treatment).

Plot selection and sampling for testing a microbial agent applied as a larvicide might proceed as follows:

1. Presample the habitat (*e.g.*, manure) to determine roughly where the larvae are. At this stage a simple, qualitative visual assessment is usually adequate.
2. Establish the plots. In many confined animal situations there might be rows or pits of manure that can be divided into sampling zones, with buffer areas in between. A randomized block design often is suitable here. Generally, one must take as many samples as is logistically feasible, if possible a minimum of 10.
3. Take the pretreatment samples. It often is advisable to take samples in pairs adjacent to each other in the habitat. One of these can be used for direct larval extraction on a Berlese funnel, while its pair is held for adult fly emergence (Mullens *et al.*, 1996).
4. Apply the microbial agent. Some of these can be applied directly to the surface of the substrate after mixing in water (*e.g.*, entomopathogenic nematodes). A low pressure hand garden sprayer (4–12 liters) is usually adequate for the purpose.
5. Take post-treatment samples. The sampling interval depends on the target pest, but, given the short life cycle of many flies, the initial post-treatment samples often are 2–5 days after treatment. Samples at 3, 7, 10, and 14 days would incorporate the expected duration of activity of many materials applied to a manure habitat which is receiving more manure daily. If at all possible, one should not depend solely on fly numbers to indicate "success" in control efforts, but take steps to determine if target insects have in fact been infected and/or killed by the pathogen. It is particularly useful to compare such infection levels to levels in the pretreatment period.

3 Sampling adult flies

Sampling adult flies to determine effectiveness of a microbial agent is challenging. For flies which frequent animals regularly, such as the horn fly *Haematobia irritans* or the stable fly *S. calcitrans*, counts on animals are frequently used. Exact estimates are very difficult, and quick visual estimates are more often used. With practice and comparison with more exact counts (*e.g.*, comparing one's estimate with a count from an enlarged photo taken at the same time), fairly accurate estimates are possible. Often only one side of an animal is counted, and this number used in a relative way to assess control (e.g. before vs. after treatment). Binoculars are an aid under field conditions with pastured or range animals. Visual counts are somewhat subjective and thus, it is important to have the same individual do the counts over time if at all possible. For some flies which visit a host less frequently, such as *Stomoxys*, attractive traps have been designed which yield a relative estimate of activity. Alsynite fiberglass traps covered with adhesive, for example, are very attractive to *Stomoxys* (Broce, 1988). A good general reference on sampling arthropod pests of livestock and poultry, especially Diptera, is provided by Lysyk and Moon (1994).

Pest flies such as *M. domestica* or *Fannia spp.* do not preferentially frequent animals, although they may rest on them sometimes. Such flies typically are sampled at resting sites in the environment near the manure or decaying feedstuffs which produced them. Several sampling methods have been used (*e.g.*, Lysyk and Axtell, 1986). It often is best to use two or more such methods, since each has biases. Methods 1, 2 and 4 below integrate numbers over time, while 3 and 5 depend heavily on weather conditions at a point in time.

1. Index cards (7.5 × 10 cm) can be tacked up in fly resting areas (often recognized by either large numbers of flies or fly regurgitation/fecal specks) within an enclosed animal house. Flies rest and leave specks behind which may be counted. If such cards are replaced at least weekly, they yield a relative estimate of fly activity. However, one is not certain what fly species left the speck, and this technique is useful only if one species predominates (*e.g., M. domestica*).
2. Sticky tapes also can be hung in fly aggregation areas, and both the number and identity of flies can be ascertained for relative comparisons of numbers over time. One also can walk through animal houses with such tapes and get an idea of fly numbers by how many adhere to the tape after a defined route has been walked.
3. Another sampling device is the Scudder grid, a device much like a picture frame which is thrown in designated areas over time to get a visual count of the number of flies which rest within the perimeter a few minutes later. Similarly, some workers have counted flies in designated resting areas (*e.g*,. the walls of a swine barn) visually. Such areas can be outlined on a wall using a marker or chalk. In either case, numbers should be counted at the same time of day, in the same location and by the same researcher.
4. Baited traps, using attractant materials such as the house fly pheromone, z-9-tricosene (Muscalure), with a toxicant also are useful in some settings. Such traps attract and kill flies which can be counted regularly (*e.g.*, weekly). A convenient trap can be fashioned from a 3.8 liter plastic milk jug (size is not critical, but such jugs are common in the USA). From 3 to 4 entrance holes 4–5 cm in diameter can be cut out in the top 1/3 of the jug and a small amount of toxic fly bait sprinkled in the bottom. Flies enter, feed on the bait and die there, where they can be counted later. Such traps are best used under cover (*e.g.*, hung from roof rafters) and care must be taken to keep animals and people from consuming the toxic bait accidentally.
5. Sweep net samples sometimes also are used to estimate relative fly numbers. If this is done, sweeps should be made in the same location for a set length of time or number of sweeps and at the same time of day to minimize the influence of fluctuating diurnal temperatures on fly activity.

Reducing adult fly numbers is usually the goal of control efforts, even though it is likely that application of a microbial agent might be directed at the immatures. The mobility of adult flies provides another major source of experimental error; adult flies taken in one area did not necessarily come from there. Further, it is dangerous to assume that fly numbers will be comparable even in enclosed poultry or swine houses of the same design and managed on the same schedule. Unanticipated factors may result in certain houses producing far higher numbers of adult flies than others. For these reasons it is desirable to sample the developmental substrate directly if possible, to attribute changes in numbers to a microbial treatment. This is particularly true if one is able to recover the microbial agent, or at least link it to dead hosts (*e.g.*, dissecting fly larvae to determine presence of nematodes). Such samples, coupled with adult sampling, allow one to attribute control unequivocally to the treatment. It is assumed that sampling adult flies would most likely be in conjunction with discrete confined animal housing such as swine or poultry. Keep in mind that the treatment unit here is the house; replicate houses may be quite limited, and it therefore is quite critical that good pretreatment assessment be done. Sampling adult flies in conjunction with a microbial treatment might proceed as follows.

1. Determine by observation where the adult flies rest, and where they are most active in daylight hours. Placement of sticky fly tapes may help determine the species composition and distribution. If the target is a single dominant species such as *M. domestica*, spot cards (white index cards) are useful. Place at least 4–6 of these in each house; the ends of the houses often have larger numbers of adult flies. It is good to have some sampling tool to establish whether the fly species composition shifts over the course

of the study. Sticky fly tapes placed near the spot cards, placed and changed on the same schedule, are useful.

2. If the pathogen affects adults (*e.g.*, many fungi), a sweep net can be used to capture a sample of at least 100 flies (pooled from different regions of the house) to hold for determining pretreatment pathogen prevalence (*e.g.*, Six and Mullens, 1996). Establish baseline levels of fly activity (pretreatment sampling) over several days to a week.
3. If possible, pair the treatment and control houses according to the fly activity. Take the 2 houses with the highest numbers of flies, and randomly assign one to the treatment and the other to control. Repeat with the next pair of houses, etc.
4. Apply the microbial agent.
5. Take post-treatment samples at intervals similar to those used for the substrate sampling discussed above. Again, if the pathogen affects adults, a net sample of flies can be held or dissected to determine pathogen prevalence directly. If one expects activity of the pathogen might extend beyond 2 weeks (*e.g.*, retransmission from initially infected flies or excellent persistence of the agent), sampling longer may well be warranted. If at all possible, one should not depend only on estimated numbers of flies to assess efficacy, since fly numbers can easily fluctuate for reasons unconnected to the treatment. While reduced fly numbers are of course the goal, cause and effect is more convincing if there is direct evidence of fly mortality caused by the pathogen in question.

D *Application methods for manure fly or beetle control*

1 *Application of nematodes*

Entomopathogenic nematodes can be mixed and prepared for application as a fly larvicide or against beetles to manure or organic substrates using standard techniques (Kaya and Stock, 1997). Some work against *M. domestica* in wet, high nitrogen poultry manure habitats suggests nematode survival and infectivity for fly larvae are very poor (*e.g.*, Georgis *et al.*, 1987), but some subsequent work in other manure substrates is more promising (Taylor *et al.*, 1998). As discussed by Renn (1998), targeting drier portions of the manure (< 60% moisture), where flies may move for pupation, may be a more effective strategy than trying to kill larvae directly in wetter manure. As a larvicidal application, this study had better efficacy using a direct nematode spray to swine manure compared with nematodes applied in calcium alginate capsules. However, the encapsulated nematodes also can be deployed in a bait station against adult flies. There is some laboratory (Renn, 1994) and field (Renn, 1998) evidence that such baits (design not yet published) can control flies in intensive swine units (or probably similar settings). Many confined animal facilities, especially poultry operations, are very dusty. This might be a serious problem in using a bait (*e.g.*, entomopathogenic nematodes) that must be delivered damp or wet.

Based on available literature it is difficult to recommend a specific nematode rate for larvicidal purposes. Much probably depends on the nematode strain selected and habitat characteristics. High application rates of 8–9 million/m^2 were insufficient to compensate for low infectivity on wet poultry manure (Mullens *et al.*, 1987), and such levels are economically prohibitive in most cases. Renn (1998) applied 0.5 million nematodes/sow stall (probably at least 2 m^2 of manure substrate below the pigs) twice weekly with results comparable to fly control using methomyl baits. Based on laboratory trials, Taylor *et al.* (1998) suggest that moderate numbers of *Steinernema* spp. also might be sufficient for fly control on bovine manure, but field trials remain to be done. Trials of nematodes against *A. diaperinus* have shown infectivity when applied to soil where beetles pupated, but control has been short-term (see Geden and Axtell, 1988).

2 *Application of fungi*

For purposes of laboratory infectivity screening, fungal conidia (Hypocreales) have been applied directly to adult or immature flies in aqueous or vegetable oil suspensions (Barson *et al.*, 1994). Some of these were highly infective at doses down to 10^4 or 10^5 conidia/ml. In selected cases, it might be possible to achieve some degree of control of a localized fly population with such an application. For example, one might spray adult flies in their overnight resting areas under the roof or eaves of confined animal houses and hope to hit a significant proportion of them directly. Indeed, the tendency of flies to rest in stereotypical areas on a farm provides some reasonable hope of placing them in proximity with a pathogen that can infect on contact. Conidia of these fungi can be prepared and applied using standard methods (Goettel and Inglis, 1997; Kaufman *et al.*, 2005).

In many confined animal settings, however, rough surfaces such as wood are more likely to exist and be used by flies as resting sites. Liquid applications of fungi are less suitable on such surfaces than are dust formulations (Geden *et al.*, 1995; Watson *et al.*, 1995). It is also possible that conidia of *B. bassiana* may be deployed as an attractant/feeding dry bait combined with sucrose (Geden *et al.*, 1995). In either case, experimental trials suggest effective concentrations are quite high ($10^7 - 10^8$ conidia/cm^2). In the field at a larger scale, application of such a conidial suspension still would require formulation to allow it to adhere well after spraying with standard equipment. While it also may be possible to infect some fly immatures within soil or manure substrates with fungi, it is likely that eventual use of fungi will be against adults.

For beetles, Geden and Steinkraus (2003) tested three fungi formulations. They included an "emusifiable concentrate" with conidia suspended in water at 2×10^{11} conidia/ml, a coarsely ground granular corn bait with 8.5×10^9 conidia/g, and residue from fungal propagation, a granular material with 6.5×10^8 conidia/g. For the field trials the liquid was applied using a calibrated sprayer, the bait and residue applications were applied using a restaurant grated cheese shaker.

4 Conclusion

Much research and development has been done on microbial control agents for use in broadacre agriculture and forestry, but there has been far less work on the microbial control of livestock pests. Yet, livestock industries worldwide face serious problems of increasing pesticide resistance of livestock pests and of greater stringency of regulations limiting chemical residues in animal products. Effective, non-chemical means of pest control would meet an urgent need and be welcomed. However, as we mentioned in the introduction to this chapter, the microbial control of livestock pests is, as yet, still in the pioneering stage, with various strategies showing promise but few fully operational programs. We hope that, in the years to come, microbial technologies for the control of pests will become an important part of livestock management.

5 References

Alam, M. J. and Zurek, L. 2004. Association of *Escherichia coli* O157: H7 with houseflies on a cattle farm. *Appl. Environ. Microbiol.* **70**, 7578–7580.

Alexandre, T. M., Alves L. F. A., Neves P. M. O. J, and Alves, S. B. 2006. Effect of temperature and poultry litter in *Beauveria bassiana* (Bals.) Vuill. and *Metarhizium anisopliae* (Metsch) virulence against the lesser mealworm *Alphitobius diaperinus* (Panzer) (Coleoptera: Tenebrionidae). *Neotrop. Entomol.* **35**, 75–82.

Axtell, R. C. and Arends, J. J. 1990. Ecology and management of arthropod pests of poultry. *Annu. Rev. Entomol.* **35**, 101–126.

Barson, G., Renn, N. and Bywater, A. F. 1994. Laboratory evaluation of six species of entomopathogenic fungi for the control of the house fly (*Musca domestica* L.), a pest of intensive animal units. *J. Invertebr. Pathol.* **64**, 107–113.

Broce, A. B. 1988. An improved alsynite trap for stable flies, *Stomoxys calcitrans* (Diptera: Muscidae). *J. Med. Entomol.* **25**, 406–409.

Broce, A. B. 1993. Dispersal of house flies and stable flies. *In* "Rural Flies in the Urban Environment." (G. D. Thomas and S. R. Skoda, Eds.). Res. Bull. 317, Univ. of Nebraska, Lincoln, NE.

Campbell, J. B. 1993. The economics of the fly problem. *In* "Rural Flies in the Urban Environment." (G. D. Thomas and S. R. Skoda, Eds.). Res. Bull. 317, Univ. of Nebraska, Lincoln, NE.

Crawford, P. J., Brooks, W. M., and Arends, J. J. 1998. Efficacy of field-isolated strains of *Beauveria bassiana* (Moniliales: Moniliaceae) as microbial control agents of the lesser mealworm (Coleoptera: Tenebrionidae). *J. Econ. Entomol.* **91**, 1295–1301.

Division of Entomology Biennial Report. 1988. p. 41. CSIRO, Canberra, Australia.

Drummond, R. O., George, J. E., and Kunz, S. E. 1988. "Control of Arthropod Pests of Livestock: A Review of Technology." CRC Press, Boca Raton, FL.

Geden, C. 1997. Evaluation of *Paraiotonchium muscadomesticae* (Nematoda, Tylenchida: Iotonchiidae), a potential biological control agent of the house fly (Diptera: Muscidae). *Biol. Control* **10**, 42–47.

Geden, C. J., Arends, J. J., and Axtell, R. C. 1987. Field trials of *Steinernema feltiae* (Nematoda: Steinernematidae) for control of *Alphitobius diaperinus* (Coleoptera: Tenebrionidae) in commercial broiler and turkey houses. *J. Econ. Entomol.* **80**, 136–141.

Geden, C. J. and Axtell, R. C. 1988. Effect of temperature on nematode (*Steinernema feltiae*) treatment of soil for control of lesser mealworm in turkey houses. *J. Econ. Entomol.* **81**, 800–803.

Geden, C. J., Steinkraus D. C., and Rutz., A. 1993. Evaluation of two methods of release of *Entomophthora muscae* (Entomophthorales: Entomophthoraceae) to infect house flies (Diptera: Muscidae) on dairy farms. *Environ. Entomol.* **20**, 1202–1208.

Geden, C. J., Rutz, D. A., and Steinkraus, D. C. 1995. Virulence of different isolates and formulations of *Beauveria bassiana* for house flies and the parasitoid *Muscidifurax raptor. Biol. Control* **5**, 615–621.

Geden C. J. and Steinkraus D. C. 2003. Evaluation of three formulations of *Beauveria bassiana* for control of lesser mealworm and hide beetle in Georgia poultry houses. *J. Econ. Entomol.* **96**, 1602–1607.

Georgis, R., Mullens, B. A., and Meyer, J. A. 1987. Survival and movement of insect parasitic nematodes in poultry manure and their infectivity against *Musca domestica. J. Nematol.* **19**, 292–295.

Goettel, M. S. and Inglis, G. D. 1997. Fungi: Hyphomycetes. *In* "Manual of Techniques in Insect Pathology." (L. A. Lacey, Ed.) pp. 213–249. Academic Press, New York.

Greenberg, B. 1971. "Flies and Disease." Princeton University Press, Princeton, NJ.

James, P. J. 1986. Current practices of blow fly control and future possibilities. *In* "Proceedings of the Sheep and Wool Refresher Course, Adelaide, February, 1986." South Australian Department of Agriculture and Australian Wool Corporation..

Johnson, P. W., Boray, J. C., and Dawson, R. L. 1990. Synthetic pyrethroid resistance *Damalinia ovis. In* "Resistance Management in Parasites of Sheep." (J. A. McKenzie, P. J. Martin and J. H. Arundal, Eds.) p. 21. Australian Wool Corporation, Melbourne, Australia.

Kalsbeek, V., Mullens, B. A., and Jespersen, J. B. 2001. Field studies of *Entomophthora* (Zygomycetes: Entomophthorales)-induced behavioral fever in *Musca domestica* (Diptera: Muscidae) in Denmark. *Biol. Control* **21**, 264–273.

Kaufman, P. E., Reasor, C., Rutz, D. A., Ketzis, J. K., and Arends, J. J. 2005. Evaluation of *Beauveria bassiana* applications against the house fly, *Musca domestica*, in commercial caged-layer poultry facilities in New York State. *Biol. Control* **33**, 360–367.

Kaya, H. K. and Moon, R. D. 1978. The nematode *Heterotylenchus autumnalis* and the face fly, *Musca autumnalis*: a field study in northern California. *J. Nematol.* **10**, 331–341.

Kaya, H. K. and Stock, S. P. 1997. Techniques in insect nematology. *In* "Manual of Techniques in Insect Pathology." (L. A. Lacey, Ed.) pp. 281–324. Academic Press, New York.

Kuramoto, H. and Shimazu, M. 1997. Control of house fly populations by *Entomophthora muscae* (Zygomycotina: Entomophthorales) in a poultry house. *Appl. Entomol. Zool.* **32**, 325–331.

Lund, R. D., Johnson, P. W., Gould, N. S. and van de Ven, R. J. 1998. "Improved Design and Use of Dipping Equipment for Sheep Lice Eradication." NSW Agriculture, Trangie, NSW, Australia

Lyness, E. W., Pinnock, D. E., and Cooper, D. J. 1994. Microbial ecology of sheep fleece. *Agric. Ecosys. Environ.* **49**, 103–112.

Lysyk, T. J. and Axtell, R. C. 1986. Field evaluation of three methods for monitoring populations of house flies *Musca domestica* (Diptera: Muscidae) and other filth flies in three types of poultry housing systems. *J. Econ. Entomol.* **79**, 144–151.

Lysyk, T. J. and Moon, R. D. 1994. Sampling arthropods in livestock management systems. In "Handbook of Sampling Methods for Arthropods in Agriculture." pp. 515–538, CRC Press, Boca Raton, FL.

Mullens, B. A. 1990. *Entomophthora muscae* (Entomophthorales: Entomophthoraceae) as a pathogen of filth flies. In "Biocontrol of Arthropods Affecting Livestock and Poultry." (D. A. Rutz and R. S. Patterson, Eds.) pp. 231–245. Westview Press, Boulder, CO.

Mullens, B. A., Meyer, J. A., and Georgis, R. 1987. Field tests of insect-parasitic nematodes (Rhabditida: Steinernematidae, Heterorhabditidae) against larvae of manure-breeding flies (Diptera: Muscidae) in caged-layer poultry facilities. *J. Econ. Entomol.* **80**, 438–442.

Mullens, B. A., Hinkle, N. C., and Szijj, C. E. 1996. Impact of alternating manure removal schedules on pest flies (Diptera: Muscidae) and associated predators (Coleoptera: Histeridae, Staphylinidae; Acarina: Macrochelidae) in caged-layer poultry manure in southern California. *J. Econ. Entomol.* **89**, 1406–1417.

Pinnock, D. E. 1994. The use of *Bacillus thuringiensis* for control of pests of livestock. *Agric. Ecosys. Environ.* **49**, 59–63.

Pinnock, D. E., Lyness, E. W., and Cooper, D. J. 1994 Microbial control by colonisation: the development of *Bacillus thuringiensis* for control of sheep ectoparasites. *In* "Proceedings of the Second Canberra Meeting on *Bacillus thuringiensis*." (R. J. Akhurst, Ed.) pp. 153–158. CSIRO, Canberra, Australia.

Renn, N. 1994. A comparison of baits containing different isolates of entomopathogenic nematodes for the control of *Musca domestica* using a large-scale laboratory facility. *Proc. Brighton Crop Prot. Conf.* pp. 1061–1066.

Renn, N. 1995. Mortality of immature houseflies (*Musca domestica* L.) in artificial diet and chicken manure after exposure to encapsulated entomopathogenic nematodes (Rhabditidae: Steinernematidae, Heterorhabditidae). *Biocontrol Sci. Technol.* **5**, 349–359.

Renn, N. 1998. The efficacy of entomopathogenic nematodes for controlling housefly infestations of intensive pig units. *Med. Vet. Entomol.* **12**, 46–51.

Renn, N., Bywater, A. F., and Barson, G. 1999. A bait formulated with *Metarhizium anisopliae* for the control of *Musca domestica* L. (Dipt., Muscidae) assessed in large-scale laboratory enclosures. *J. Appl. Entomol.* **123**, 309–314.

Renn, N. and Wright, E. 2000. The effect of artificial substrates on the pathogenicity of *Steinernema feltiae* (Rhabditida: Steinernematidae) to adult *Musca domestica. Nematology* **2**, 217–222.

Safrit, R. D. and Axtell, R. C. 1984. Evaluations of sampling methods for darkling beetles (Alphitobius diaperinus) in the litter of turkey and broiler houses. *Poultry Sci.* **63**, 2368–2375.

Six, D. L. and Mullens, B. A. 1996. Seasonal prevalence of *Entomophthora muscae* and introduction of *Entomophthora schizophorae* (Zygomycotina: Entomophthorales) in *Musca domestica* (Diptera: Muscidae) populations on California dairies. *Biol. Control* **6**, 315–323.

Smallridge, C. J., Cooper, D. J., and Pinnock, D. E. 1995. The effect of the microsporidian *Octosporea muscaedomesticae* on adult *Lucilia cuprina* (Diptera: Calliphoridae) *J. Invertebr.Pathol.* **66**, 196–197.

Stafford, K. C., III and Bay, D. E. 1987. Dispersion pattern and association of house fly, *Musca domestica* (Diptera: Muscidae), larvae and both sexes of *Macrocheles muscaedomesticae* (Acari:Macrochelidae) in response to poultry manure moisture temperature, and accumulation. *Environ. Entomol.* **16**, 159–164.

Stafford, K. C., III and Bay, D. E. 1994. Dispersion statistics and sample size estimates for house fly (Diptera: Muscidae) larvae and *Macrocheles muscaedomesticae* (Acari: Macrochelidae) in poultry manure. *J. Med. Entomol.* **31**, 732–737.

Steenberg, T., Jespersen, J. B., Jensen, K.-M. V., Nielsen, B. O., and Humber, R. A. 2001. Entomopathogenic fungi in flies associated with pastured cattle in Denmark. *J. Invertebr. Pathol.* **77**, 186–197.

Steinkraus, D. C., Geden, C. J., and Rutz, R. A. 1992. Susceptibility of lesser mealworm (Coleoptera: Tenebrionidae) to *Beauveria bassiana* (Moniliales: Moniliaceae): effects of host stage, substrate, formulation and host passage. *J. Med. Entomol.* **28**, 314–321.

Steinkraus, D. C., Geden, C. J., and Rutz, D. A. 1993. Prevalence of *Entomophthora muscae* (Cohn) Fresenius (Zygomycetes: Entomophthorales) in house flies (Diptera: Muscidae) on dairy farms in New York and induction of epizootics. *Biol.Control* **3**, 93–100.

Strother, K. O. and Steelman, C. D. 2001. Spatial analysis of *Alphitobius diaperinus* (Coleoptera: Tenebrionidae) in broiler production facilities. *Environ. Entomol.* **30**, 556–561.

Strother K. O., Steelman C. D., and Gbur E. E. 2005. Reservoir competence of lesser mealworm (Coleoptera: Tenebrionidae) for *Campylobacter jejuni* (Campylobacterales: Campylobacteraceae). *J. Med. Entomol.* **42**, 42–47.

Szalanski, A. L., Owens, C. B., Mckay, T., and Steelman, C. D. 2004. Detection of *Campylobacter* and *Escherichia coli* O157: H7 from filth flies by polymerase chain reaction *Med. Vet. Entomol.* **18**, 241–246.

Taylor, D. B., Szlanski, A. L, Adams, B. J., and Peterson, R. D. II. 1998. Susceptibility of house fly (Diptera: Muscidae) larvae to entomopathogenic nematodes (Rhabditida: Heterorhabditidae, Steinernematidae). *Environ. Entomol.* **27**, 1514–1519.

Vaughan, J. A., Turner, E. C., Jr., and Ruszler, P. L. 1984. Infestation and damage of poultry house insulation by the lesser mealworm, *Alphitobius diaperinus* (Panzer). *Poultry Sci.* **63**, 1094–1100.

Watson, D. W., Geden, C. J. Long, S. J., and Rutz, D. A. 1995. Efficacy of *Beauveria bassiana* for controlling the house fly and stable fly (Diptera: Muscidae). *Biol. Control* **5**, 405–411.

Watson, D. W., Rutz, D. A. and Long, S. J. 1996. *Beauveria bassiana* and sawdust bedding for the management of the house fly, *Musca domestica* (Diptera: Muscidae) in calf hutches. *Biol. Control* **7**, 221–227.

Watson, D. W., Guy, J. S., and Stringham, S. H. 2000. Limited transmission of turkey coronavirus in young turkeys by adult *Alphitobius diaperinus* (Coleoptera: Tenebrionidae). *J. Med. Entomol.* **37**, 480–483.

West, L. S. 1950. "The Housefly." Cornell Univ. Press, Ithaca, NY.

Wilkinson, F. C. 1986. Economics of sheep lice control. *In* "Proceedings of the Sheep and Wool Refresher Course, Adelaide, February, 1986". pp. 70–74 South Australian Department of Agriculture and The Australian Wool Corporation, Adelaide, Australia.

Zumpt, F. 1973. "The Stomoxyine Biting Flies of the World." Gustav Fisher Verlag, Stuttgart, Germany.

Chapter VII-22

Microbial control of mosquitoes and black flies

Ole Skovmand
Intelligent Insect Control
118 Chemin des Alouettes
34170 Castelnau le Lez France

James Kerwin
Analytic Core
School of Pharmacy
University of Southern California
Los Angeles, CA 90089, USA

Lawrence A. Lacey
USDA-ARS
Yakima Agricultural Research Laboratory
Wapato, WA 98951,USA

1 Introduction

Adult mosquitoes and black flies are responsible for the transmission of disease-causing agents of humans, and animals and they can lower the quality of life through their blood feeding activities. Mosquitoes transmit the causal agents of malaria, filariasis, and several arboviral diseases including yellow fever, dengue, West Nile fever and Japanese encephalitis. Malaria alone strikes 300–500 million people/year, killing 1.5–2.7 million. Black flies transmit the causal agents of human and bovine onchocerciasis as well as *Leucocytozoon* spp. in fowl. Most mosquito and black fly control programs are based predominantly on conventional broad-spectrum insecticides. Alternative interventions with biological control would enable abatement of pest and vector species without adverse environmental effects and concomitant hazards for humans and may supplement chemical control thus retarding or avoiding resistance problems.

In many abatement programs, larvae and adults of these dipterans are targeted for control efforts. The efficacy of control efforts is measured beginning shortly after application and continues for days or weeks to evaluate the residual effect, provided there is a continuous recruitment of larvae. This may not always be the case for multivoltine species and certainly not for univoltine species. Many mosquito breeding sites are only favorable for certain species for a limited time; for example, *Anopheles gambiae*, will only be found in rice fields before the plants grow to shade the water (Robert *et al.*, 1988) and *Aedes vexans* will often appear for just one generation after a flooding (Becker and Rettich, 1994). Tests for residual efficacy must, therefore, consider the ecology of larval habitat and species.

Assessing the effect of a larvicide application usually relies on monitoring larval population densities in treated and control sites. Since the nuisance or vector problem is related to the adult population only, the efficacy of a larvicide should also assess the impact on the adult population when feasible. Possible approaches to these problems are considered at the end of the chapter.

L.A. Lacey and H.K. Kaya (eds.), Field Manual of Techniques in Invertebrate Pathology, 735–750.

A number of entomopathogens have been reported from larval and adult mosquitoes and black flies including viruses, bacteria, rickettsia, protozoa, fungi (including Microsporidia) and nematodes (Chapman, 1974, 1985; Lacey and Undeen, 1986). Epizootics caused by one or more of these groups have occasionally been responsible for severe declines in larval populations. In order to optimize the control capability of entomopathogens and nematodes, reliable protocols for their mass production, formulation and application must be developed. A thorough understanding of the biology and ecology of targeted species is needed to optimize timing and frequency of applications. Relatively few species of entomopathogens have been exploited as microbial control agents (MCAs) of mosquitoes, and even fewer have been evaluated in the field against black flies. Those requiring *in vivo* production such as viruses, protozoa, certain fungi and mermithid nematodes are expensive to produce relative to those that can be mass produced on artificial media. In this chapter, we will emphasize those entomopathogens with the greatest control potential.

2 Microbial control of mosquito larvae

Bacillus thuringiensis subsp. *israelensis (Bti*; serotype H-14) and *B. sphaericus* (serotype 5a5b) are widely used for control of larval mosquitoes. They are safe for vertebrates and the vast majority of invertebrate nontarget organisms (Lacey and Mulla, 1990; Lacey and Siegel, 2000; Lacey and Merritt, 2003). An overview of developments in the past 20 years is presented by Lacey (2007) regarding the toxins of *Bti* and *B. sphaericus*, their modes of action, efficacy and factors that affect larvicidal activity, development of resistance, safety and their roles in integrated mosquito control. Due to their efficacy and relative specificity, both *Bti* and *B. sphaericus* can be ideal control agents in integrated programs especially where other biological control agents, environmental management, personal protection and the judicious use of insecticides are combined. Less commonly used are species in two fungal genera, *Lagenidium giganteum*[1] and *Coelomomyces* spp., and nematode species in the family Mermithidae, such as *Romanomermis culicivorax* and *R. iyengari.* An overview of mermithid biology, environmental limitations, and procedures for mass production of *R. culicivorax* is presented by Petersen (1985) and Platzer (2007). Kerwin and Petersen (1997) and Kerwin (2007) provide information on the biology, isolation, propagation and field testing of *L. giganteum.* The choice of the microbial organism to be used depends upon target species, availability of product, ecology of the target insect breeding site, formulation and application options. Target susceptibility is summarized for some species in Table 1.

A Formulations

Except for the bacterial products, the MCAs are mostly formulated as stabilized liquid suspensions of the agent. Bacterial pesticides are available in several forms: fluid concentrates, wettable powders, flowable granules, sand granules, pellets, and briquettes. Among the marketed products, only the briquettes of bacterial products have sustained released capabilities. Many more sophisticated formulations have been developed with the primary goal being extended residual effect, but none of these products are marketed, ostensibly due to high price. Researchers interested in such products should go through the patent literature.

B Experimental design

1 Selection of and setting up test plots

The selection of test plot should consider at least three aspects: (1) is the site representative for the target insect; (2) different field situations for the target insect in the area to be treated, and (3) compatibility of test site and product type. It is tempting to choose test plots based on ease of application, counting or access to the test site, but relying entirely on practical considerations often limits the general value of the study.

[1] The genus *Lagenidium* has now been placed in the Kingdom Chromista

Table 1. Some representative mosquito and black fly species, associated medical importance and their susceptibility to microbial control agents

Target Species	Medical importance	Susceptibility to selected microbial control agents[1]			
		Bacillus thuringiensis	*Bacillus sphaericus*[2]	*Lagenidium giganteum*	*Romanomermis culicivorax*
Aedes aegypti	Yellow fever Dengue	+++	–	+++	+
Ae. albopictus	Dengue	+++	–	+	+
Ae. vexans	Nuisance	+++	+	+	+
Anopheles gambiae	Malaria	++	+++	+	+++
An. stephensi	Malaria	++	++	+	+++
An. albimanus	Malaria	++	++	+	+++
Culex quinquefasciatus	Filariasis	+++	+++	+++	+
Psorophora spp.	Nuisance	+++	+++	+	+
Mansonia spp.	Encephalitis Filariasis *(Brugia)*	++	++	++	
Simulium damnosum	Onchocerciasis	+++	-	-	-
Simulium spp.	Nuisance	+++	-	-	-

[1]+moderately susceptible; +++ very susceptible; - not susceptible
[2]except in resistant populations (see Rao *et al.*, 1995; Nielsen-Leroux *et al.*, 1997, 2002; Mulla *et al.*, 2003).

Temperature, sun exposure and water chemistry may be different from site to site. If there is a range of water characteristics found within the experimental site, monitoring of water chemistry prior to application of the control agent will avoid treatment of sites in which there is little or no chance of larval infection. Such characteristics can also be used for stratifying the results of the treatment. Sites are matched in pairs according to important parameters with one/pair chosen randomly for treatment.

The number and distribution of sites will depend on many factors: purpose of the study, resources (time, personnel) available for treatment and sampling, the larval distribution (patchy, homogenous) and average density/site, the habitat and the life cycle of the target species. Monitoring treatment effects on the larval population by estimating reduction in average density/site may be insufficient where the core problem of control is to find the breeding sites. In such cases, monitoring the adult mosquito population should be included, and test and control areas must then be of such a size to reduce the effect of immigration and emigration.

a Container-breeding mosquitoes

The simplest habitats are those frequented by container breeding species such as *Aedes albopictus* and *Ae. aegypti*. These include containers used for household water supplies, plastic sheeting, flower vases, stock water tanks, jars, and tires (Hawley, 1988; Fukuda *et al.*, 1997). These discrete habitats are often clustered in high numbers in a given area, so numerous replicates for treated and control containers are available. Similar conditions are often encountered for small natural habitats such as treeholes and leaf axils serving as natural breeding sites of *Aedes* and other species (Laird, 1988; Frances *et al.*, 1989; Samarawickrema *et al.*, 1992). Environmental factors such as chemical characteristics of the water, exposure to sunlight, and temperature may be critical for small water volumes.

Estimates of mortality due to a microbial product are quite reliable in this habitat type since complications due to spatial distribution of immature mosquitoes are minimal. Special sampling methods are dealt with in section 2 C of this chapter.

b Mosquitoes breeding in temporary pools

A large number of mosquito species use temporary pools as larval habitats where the absence of predators and competitors can lead to a high larval density, developing more or less synchronously in a short period. These sites may dry up before the mosquitoes complete their life cycle. This category includes shallow snow pools, sporadically flooded drainage ditches, pastures subject to periodic inundation, soybean fields, or sites exposed to rivers reaching flood stage, and puddles of rain water. Species include univoltine floodwater and snow pool mosquitoes, primarily *Aedes* spp. (*e.g.*, *Ae. vexans, Ae. melanimon* and *Ae. stimulans*), but also species such as *Psorophora columbiae* in the southern and midwestern USA. In the rainy season in Subsahelian Africa, *An. gambiae* occupies transient, rain-filled puddles, with often just one generation/puddle even though the species is multivoltine in the area. The transient nature of the breeding habitats, and marked spatial heterogeneity (Skovmand and Sanogo, 1999), require careful choice of sites for field evaluations.

Some of these habitats may be transient and their numbers limited, making it impossible to select a large number of replicate plots. If the breeding habitats have dense, highly synchronized larval cohorts in very similar habitats, 6–8 replicates may be sufficient.

The habitat supporting mosquito breeding may consist of a single large body of water with discontinuous larval distribution. This situation would require treatment and control plots at a single site and perhaps the use of mechanical barriers (soil berms or wood/metal sheets) if larvae and microbial agent could move between treated and control areas. If barriers are not established, larvae from control areas should be collected and examined in the laboratory for presence of the applied agent.

c Large, permanent or semi-permanent habitats

The last major habitat type is exemplified by rice fields, fallow fields flooded to provide nesting sites and food for migratory wildfowl, and salt and freshwater marshes. Site selection for evaluation of microbial pest control agents in these habitats will be determined by the scale of the proposed tests, limitations of personnel, and time and cost constraints. For sufficiently large areas, it is possible to establish treatment and control plots within a single field, separated by several hundred meters or by natural barriers. Smaller test areas established within a single plot where water movement is minimal can be demarcated by wooden stakes, and adjacent fields can also be used. Careful considerations must be used in evaluating the effects of microbial applications that might create epizootics, since unexpected transfer of the applied material, either by wildlife or field operators monitoring mosquito populations is possible, and thereby creating problems in the control areas (Kerwin *et al.*, 1994).

For some of these habitats, the major problem is the very low density and sporadic spatial distribution of mosquitoes (Kerwin and Washino, 1985, 1986; Walker *et al.*, 1985). This requires much more intensive pre- and post-treatment sampling and extensive or exclusive reliance on sentinel mosquitoes (Kerwin and Washino, 1988). Larger plot sizes are often employed to overcome this problem, but this approach will likely be constrained by personnel and time limitations, and possibly by the amount of MCA available. Approaches to circumventing this problem are discussed in the succeeding section C.

Special statistical considerations have to be applied for such habitats, and one option is to regard each sample as either positive (with larvae) or negative (without larvae) and treat these data as binomial distributions.

C Sampling

Sampling of the target population is required prior to treatment of targeted species. Sampling methods before and after treatment will normally be the same to make results comparable. In all habitats other than small containers, design and interpretation of larval sampling become complex, especially if quantitative estimates of absolute numbers are required. Many attempts have been made to relate dip counts to absolute larval numbers in breeding sites of known size, but such efforts are difficult (Service, 1993). Factors that introduce bias into larval dip sampling include: immatures are only sampled if they are at the water surface; differences in submersion times of different species or different instars of the same species; time of day and weather conditions during sampling; variations in dipping technique between different samplers; type, amount and distribution of submerged and emergent vegetation and macro algae; and spatial dispersion and aggregation of immature mosquitoes.

Pre-sampling may extend to the season before the treatment if the ecology of the species is not known in the area, or if the study is a part of an epidemiological study to establish parameters such as percentage of female mosquitoes infected with parasite(s). Mosquito populations and the percentage of females infected with the disease-causing agent may fluctuate from year to year, and a control area with conditions similar to those of the intervention area must be included for both years. For shorter term studies, at least two pre-treatment counts are taken, one during identification and measuring of test sites for calculation of concentration/site, and one just before treatment in intervention and control areas.

Protocols to be used for sampling will be influenced by larval habitat and the initial design of the experimental plot. Pre-treatment sampling should include monitoring of indigenous parasites or pathogens that will affect larval survival in experimental plots. This establishes a baseline larval population against which relative or absolute population changes can be compared.

In spite of the bias associated with dipping (Service, 1993), the most common method for monitoring larval field populations is the use of a long-handled dipper. When the water surface is beyond reach, such as a deep well, a dipper cup without handle can be attached to a line. A variety of sizes have been used, ranging from 100 ml to one liter or more. Dippers with a white interior facilitate larval detection. Where microsporidian or some fungal infections are suspected, a dark colored dipper should be selected because these pathogens produce white coloration in larvae. If the number of larvae/dip is greater than 10, larvae are counted/dip. Otherwise, the contents are successively emptied into a white-bottomed container, and after a defined number of dips (10–50), the number of larvae is counted. In general, samples are returned to the habitat after counting. Subsamples may be brought to the laboratory for evaluation of delayed mortality, presence of pathogens, species determinations, etc. For habitats with low larval densities (*e.g.*, less than 2/dip), a variety of concentrating devices have been used, such as nylon mesh or muslin cloth attached to a solid support (Hagstrum, 1971; Fanara, 1973).

Where several dips are taken from a smaller site, suitable time intervals should pass between each dip since larvae and pupae will swim to the bottom when the habitat is disturbed. The number of dips taken will depend on the size of the test area and larval distribution. More dips are taken when the density is low, the distribution is heterogeneous or the habitat is large. A simple standard sampling protocol must be determined, (*e.g.*, 5 dips/pool or $3/m^2$). Fewer dips are taken/m in larger habitats. The preliminary protocol is used to determine optimum sampling regime.

Flood water species are usually highly aggregated and found at high densities, so 10–100 dips along linear transects are often sufficient to obtain reliable estimates of pre- and post-treatment populations. Habitats supporting lower densities will require more intensive sampling if indigenous populations are to be used to assess the efficacy of control agents.

Other methods for larval sampling include sweeps with aquatic nets in large, relatively deep larval habitats such as ponds and marshes (Hatfield *et al.*, 1985). Larger nets passing through the water will cause significant disturbance and avoidance behavior on the part of larvae and pupae, so successive sweeps should be some distance apart from one another. Another approach is to use both dip and net samples to provide complementary data (Meisch *et al.*, 1982; Tun-Lin *et al.*, 1994).

Area samplers are the third major type of sampling device. This involves the insertion of open-ended cylinders or rectangles made of metal, wood, plastic, Plexiglas, or PVC, as pioneered by Bates (1941). Depending upon the habitat and purpose of the experiment, larvae can then either be removed or sampled using dippers or nets. A watertight sampler with a small cylindrical constriction at the top is sunk into the substrate, with all but the constricted part submerged below water level. After an appropriate period of time (ca. 10–20 min.), larvae will have accumulated in the constriction in order to breathe, and can be counted or collected (Enfield and Pritchard, 1977; Service, 1993).

One last precaution applies specifically to *Coelomomyces* spp., a group of fungi that are obligate mosquito parasites. The life cycle of

Coelomomyces spp. involves obligate alternate hosts, usually a mosquito and a copepod or an ostracod (Whisler *et al.*, 1975; Apperson *et al.*, 1992). Therefore, the presence of the appropriate alternate host should be assured during pre-sampling.

Special considerations for three habitats are given below:

1 Container breeders

Sampling from containers and treeholes may be impossible using a standard dipper. Several devices have been developed for sampling from these types of habitats. Suction devices have been used to extract water containing larvae (Service, 1993), but these methods may completely change the distribution of the microbial product; for example, it may bring sedimented spores and toxins of *Bti* into suspension and thus back into the feeding zone. Simply pouring the content of a can or jar through a net to catch the larvae is of course also the end of that test site. If such destructive sampling techniques are used, each site (can, jar or tree-hole) can only be used once for monitoring and represents a sample of a population of breeding sites.

A passive, noninvasive, floating cup for sampling was tested by Undeen and Becnel (1994) in used tires colonized by *Ae. aegypti* and *C. quinquefasciatus*. Their study showed that a floating cup with a hole in the middle placed in tires accurately sampled larvae according to their density.

2 Temporary pools

The use of sentinel cages provides a means to obtain data using a known number of larvae (Case and Washino, 1979; Yap, 1991). Larvae of a certain instar are introduced into floating cages and their survival is followed at suitable intervals, both pre- and post-treatment.

Emerging adults may be trapped by setting the cage over the exit of a habitat, *e.g.* a cesspit, or they may be caught in a light trap or similar sampling device placed near the water surface. Otherwise, newly emerged mosquitoes are easily separated from females returning to deposit eggs by the color of their abdomen (*i.e.*, gravid females are darker with a whitish, swollen abdomen, while newly hatched are lighter in color and the abdomen is not swollen).

3 Large, more permanent habitats

A larger number of samples are often needed due to the low density and patchy distribution of larvae. Rice fields in Central California, for example, routinely average less than 10 larvae/100 dips using a standard one-pint (0.48 liter) dipper (Kerwin and Washino, 1986), so several hundred dips over hundreds of meters will be necessary to obtain sufficient numbers for generating statistically significant results.

D Application

1 Storage and handling

Microbial organisms have highly variable stability, depending not only on the species, but also on the stage of the species stored or applied and on the formulation. One of the major advantages of the *Bacillus* spp. products compared to most other microbial agents is their relative ease of storage and handling.

The spores of *Bti* and *B. sphaericus* are more resistant to abrasion and extremes in temperature than their vegetative cells. The spores and their associated toxins are active against larvae and are used in formulations for application. Dry formulations are more stable than liquid formulations. Liquid formulations should be stored at 5 °C for the long term (months or years), whereas powder formulations can be stored at room temperature for years without losing potency (Thiery and Hamon, 1998). For other microbial agents, the most durable stage is often not the most infective stage. Limited shelf-life is a major disadvantage of many fungal and nematode products, with a marked drop-off in efficacy observed within days or weeks after production. Preliminary laboratory or very limited field evaluations of the microbial agent should be conducted 24–48 h prior to field tests to ascertain viability and infectivity (for bioassay techniques, see Lacey, 1997).

Most zoospore-producing fungi and oomycetes that parasitize the aquatic stages of mosquitoes and black flies infect via motile spores. Motile spores lack cell walls and are dependent upon endogenous reserves for energy; therefore, it is not practical to store them or apply them in field trials. The (pre)sporangia and oospores may be applied and zoosporogenesis is induced *in situ* by dilution in water (Kerwin and Washino, 1986). Alginate formulation may improve shelf life of the zoospores, without losing infectivity (Guzman and Axtell, 1987).

Mermithid nematodes can be applied in the field using one of three stages: eggs, pre-parasites, or post-parasites. All of these stages are relatively susceptible to desiccation, but pre-parasites are especially sensitive, and lack the capability of eggs and post-parasites for prolonged survival in damp soil (Petersen *et al.*, 1978; Pailey and Balaraman, 1993).

Field application of *Romanomermis culicivorax* illustrates how application of a particular stage of a parasite can affect insect control. Most field trials of this mermithid have been with the pre-parasitic stage (Vladimorova *et al.*, 1990; Mijares and Pacheco, 1997) which is the infective stage of this parasite. The effects of pre-parasite applications can be assessed relatively quickly (ca. 4–7 days) following treatment. However, in rice fields, application of the pre-parasitic stage had minimal effect on indigenous mosquito larval populations (Zaim *et al.*, 1988), while treatment with the post-parasitic stage had a significant effect on both indigenous and sentinel larvae for weeks or months (Kerwin and Washino, 1985).

2 *Equipment*

The use of existing equipment familiar to operators in abatement districts with whom the researcher is working will enhance opportunities for incorporating a MCA into a control program (Bateman, 1997). Special application techniques have been developed for more fragile organisms (Kerwin and Washino, 1985). Equipment must be calibrated before use to ensure that the prescribed volume or weight of product is applied. Collecting at nozzles allows for volume or weight estimation. The number of droplets per unit area and droplet size can be roughly estimated with spray cards, or droplets may be caught on a silicone film on glass for microscopic examination. This may also allow for the monitoring of droplets with and without the microbial agent. These aspects are dealt with in more detail in Chapters I-2, III-1 and III-2.

When using equipment supplied by abatement agencies to apply the MCA, care must be taken that the equipment is clean and free of previous chemicals and adjuvants. The detergents used in some *Bt* formulations can loosen deposits in spray equipment, and the particles block the nozzles (Skovmand, unpublished). *Bt* formulations are compatible with most chemical insecticides and fungicides, but can be destroyed by fungicides containing copper. Most chemical insecticides are toxic to fungi and nematodes at low concentrations (ppm). Even when the active ingredient of an agrochemical or mosquito insecticide is compatible with the MCA, they may be formulated with detergents or solvents that degrade the MCA. Equipment for application of bacterial larvicides is covered in Chapters III-1 and III-2.

Equipment used for application of nematodes may range from beakers or buckets containing aqueous suspensions to Beecomist® or Micronair® atomizers for aerial field trials. Surprisingly, much of the ultra low volume (ULV) equipment developed for the application of chemical insecticides can be used with minor modifications for application of the larger, relatively fragile nematodes.

Filters and/or nozzle sizes of equipment (manual or mechanical, ground or air) are usually chosen to optimize application of liquid chemicals. The orifices may be too small for the MCA to exit, or the shearing forces generated by small diameter openings will cause high mortality in more fragile agents. Low pressure applicators can be used for application of even relatively fragile post-parasitic nematodes (Kerwin and Washino, 1985). Except for *Bt* application, removal of all filters is recommended especially when using velocity-driven aerial application systems such as Micronair atomizers. Appreciable shear will be generated even at the lowest velocities at which these systems can be operated. Small scale trials should be carried out initially when using these types of systems to determine possible undesirable effects.

Water suspensions of microbial agents will often sediment rapidly upon standing and the product should be agitated or pumped with a circulation pump to assure a homogeneous formulation. Care must be taken that these methods and in-tank agitators do not damage the MCA. Diluted products should be used the same day since the effect of preservatives is only assured at the concentration in the undiluted product, and activated stages may suffer from lack of oxygen.

E *Post-treatment sampling*

A combination of sampling of the indigenous population and monitoring the target insect held in sentinel cages will provide both a quantitative assessment of efficacy under relatively controlled conditions by the latter, and a relative estimate of MCA impact on the target population by the former. Where continuous recolonizing is not assured, water samples may be brought to the laboratory to measure residual effect. Recolonizing may be seen by the presence of egg rafts for *Culex* species or the presence of 1st instar larvae for all mosquito species since most MCA will not kill fast enough to prevent survival of newly oviposited eggs.

Use of the same protocols for pre- and post-treatment is usually feasible for short term monitoring, but for field trials involving multivoltine species, modifications may have to be made as densities change over the course of the breeding season. In some areas these changes lead to a succession of species, *e.g.* in rice fields (Robert *et al.*, 1988). Not all species have the same susceptibility to the MCA, so mosquito larvae must be determined to species and efficacy given/species. Mulla *et al.* (1971) suggested a simple model for compensating time-dependent development by using data from the control areas. The percentage reduction (R%) in a test field is corrected for the simultaneous change in the control field:

$$\mathbf{R\% = 100 - \left(\frac{C_1}{T_1} \times \frac{T_2}{C_2}\right) \times 100}$$

where C_1 and T_1 are the number of larvae in control and test plot prior to treatment, and C_2 and T_2 the number of larvae in the two plots post-treatment.

Post-treatment monitoring has to be adapted to the life cycle and possible recycling rate of the applied MCA. For many fungi and nematodes an initial high level of parasitism in the field, lasting for days or several weeks, is often followed by reduced activity that remains constant for the remainder of the breeding season. Therefore, intervals between sampling can often be extended later in the trial. Trials with *Bti* and *B. sphaericus* may provide extended effect also depending on concentration applied, formulation, and water depth (Lacey *et al.*, 1984). Sentinel larvae can be very useful for this purpose since younger instar larvae placed in the cages can often be left in the field for 4–7 days depending upon the species, growth rate, and environmental conditions.

A combination of larval and adult monitoring is ideal for field trials involving MCA applications in large areas since impact on the latter stage is the aim of all control programs. Service (1993) describes methods for adult mosquito collection in detail. The collection method of choice will be determined by the target species and the resources available. While a variety of techniques including animal and human baits, CO_2 traps, light traps, and red boxes are commonly used for adult sampling, the most direct assessment of the efficacy of MCA applications can be made using floating or submerged emergence traps (Davies, 1984). A variety of environmental conditions will influence sampling efficiency and adult emergence (Corbet, 1965; Kimerle and Anderson, 1967), so appropriate controls must be run concurrently.

F *Protocols for the evaluation of selected entomopathogens in specified habitats*

1 *Evaluation of* Bti *or* B. sphaericus *in permanent ponds, pools or cesspits*

a. Select at least 8 sites/treatment and control. Mark sites physically for ease of location and on a representative map of the area (topographic or field drawn). Mark the sites with code numbers and transfer the codes to a map. Measure key factors such as larval density, shading, degree of pollution, and water temperature. Where larval growth is rapid, exclude 1st and eventually 2nd instars in counting since the duration of their life may be shorter than the time it takes to obtain the total mortality effect of *Bti* and especially

B. sphaericus (2–3 days). Collect appropriate samples for laboratory examination to determine species composition and prevalence of parasitism.

b. Group sites according to key factors, including larval density and one or several of the above mentioned key factors. Within each group, plots are randomly assigned to treatments.
c. When possible, repeat sampling prior to treatment.
d. Calibrate application equipment to know the volume of output/sec. A higher dilution rate of product will allow for more application time and make it easier to dose on time. A 10 or 20 ml syringe with bent needle tip may serve as an accurate applicator in less than 3 m^2 areas if droplet size is not critical (Skovmand and Sanogo, 1999).
e. Sample 2 days after treatment and successively at day 4, 8, 14, etc. to measure residual effect. Collect a predetermined number of larvae from each site to determine species composition, delayed mortality rate, and prevalence of parasitism. In some cases, it may be necessary to bring water from the laboratory to transport the larvae, since the unavoidable disturbances under transport may create very unfavorable conditions. Remove predatory larvae.
f. Statistical treatment is based on repeated sampling from the same sites which reduce degrees of freedom. Several models exist, but some can only be used if all sites are observed throughout the period and groups are of the same size (*e.g.*, repeated measurement ANOVA tests), whereas others can tolerate some missing values (*e.g.*, general linear models).

2 *Evaluation of* Romanomermis culicivorax *in rice fields*

a. Sample larvae along linear transects, the length determined by the amount of nematodes available. Take several hundred dips due to the low larval density in this habitat. A 500 m transect each for treatment and control plots is sufficient if the treated area is limited by the quantity of post-parasite available for application. Weekly sampling in similar sites is used for comparative purposes.
b. Weigh out the post-parasitic stage of the nematode and suspend in distilled or clean tap water. Use Plexiglas tubes (1 m long, 4 cm diam.) sink 5–10 cm into the substrate (2.5–10 g/tube at 10 m intervals) for application, or a compressed-air backpack sprayer with an application rate of 20–100 g/100 m transect. Remove nozzles and filters to minimize damage to the fragile post-parasites. Spray continuously in a 1–5 m swath. Take aliquots from the sprayer periodically and examine samples in the laboratory for parasite viability.
c. Monitor parasite activity 3–4 weeks after application, the average time required for mating, oviposition and hatching of eggs. Use a combination of dip sampling and sentinel cages with 10–20 larvae of the targeted species/cage, to assess efficacy (Kerwin and Washino, 1985). Place 10–50 cages, 5–20 m apart in control and treated areas. Repeat this process weekly throughout the mosquito breeding season. Remove larvae 2–3 days after placement in the cages and monitor percent parasitism in the lab.
d. Monitor overwintering and migration of *R. culicivorax* by dip sampling and sentinel cages, with dip transects and cage placement at 0, 5, 10 and 20 m from the application transect.

3 *Evaluation of* Lagenidium giganteum *against univoltine and multivoltine species of mosquitoes*

a. When the asexual stage of the parasite is applied, start pre-treatment sampling of the indigenous larval population 1–2 days before application. For univoltine species as *Aedes* spp. and *Psorophora columbiae*, 25–50 dips/1000 m^2 is sufficient due to their high larval density. For multivoltine species, 50–100 dips/m^2 is required due to low densities, often 0.05–1 larva/dip. Bring larvae to the laboratory and monitor for the presence of *L. giganteum* or other parasites for 3–4 days. Sentinel cages can also be used to monitor for parasite activity prior to treatment. Depending upon the size of the plots, place 10–200 cages with 10–50 larvae in treatment and control plots 2–4 days prior to application. Remove cages 1–24 h prior to application. Monitor parasitic activity of larvae for 3–4 days. Add *L. giganteum* mycelium, laboratory reared larvae, and field collected larvae to samples of water brought to the laboratory. Monitor infectivity of both larval populations. Similar protocols are used if the sexual stage is applied, but subsequent steps in evaluating efficacy are modified as described below

b. Apply aqueous suspensions of *L. giganteum* at a rate of 0.5–15 × 10^9 cells/ha. Use higher application rates against floodwater species, in habitats with elevated levels of salinity and organic material, and in cold water (below 16 °C). Treat smaller habitats using backpack sprayers with nozzle and filters removed. For aerial applications use the same equipment as for *Bti* application, *e.g.* a Micronair atomizer or a Transland spray system. Remove filters to minimize shear and install large diameter nozzles. Minimize in-tank agitation to prevent mechanical cell disruption. Apply shortly after production due to limited shelf life. The sexual stage (oospores) can be applied in a habitat up to several months before mosquito populations are expected. This dormant stage is resistant to desiccation and mechanical abrasion and can remain viable for several years.
c. Post-treatment evaluations include a combination of dipping and sentinel cage placement as described in step 1 above. For univoltine species, monitor 2–3 days after treatment, or 4–6 days with slower developing larvae. In rice fields and marshes supporting multivoltine species and/or a succession of different species, monitor 2–3 days post-treatment, and then weekly for as long as breeding continues.
d. Since the oospore stage allows the parasite to overwinter, monitoring of treated plots can continue for 2 or more breeding seasons. If parasite activity is very low, sentinel cages can be left in plots for up to 6 days to minimize the manpower and maximize the chances of detecting recycling.

3 Potential of entomopathogens for control of black fly larvae

Black fly larval habitats are appreciably different than those of most mosquito species. The vast majority of simuliids are associated with moderate to relatively fast flowing, well-aerated water.

The discovery and development of *Bti* as a black fly larvicide have provided an effective and selective means of control (Undeen and Colbo, 1980; Molloy and Struble, 1989; Lacey and Merritt, 2003; Adler *et al.*, 2004). The majority of pest and vector simuliids obtain food from the water column by filter feeding (Currie and Craig, 1987). The parasporal inclusions containing the toxins responsible for larvicidal activity are within the size range of particles upon which black fly larvae feed. Despite the relatively brief exposure period because of the downstream movement of inoculum, the potency of *Bti* toxin and efficiency of larval filter feeding combine to increase efficacy.

Some factors that influence the efficacy of *Bti* in lotic systems are temperature, depth and volume of water, concentration of suspended organic matter, density of emergent vegetation, application strategy, formulation utilized (especially the effects of particle size), larval age and target species (Molloy *et al.*, 1981; Lacey *et al.*, 1982; Lacey and Undeen, 1984, 1986). In addition to efficacy just below the treatment point, one must also consider effective downstream carry, *i.e.* the distance below the treatment point where 80% or more of the larvae are killed.

A *Application and evaluation of* Bti *for black fly control*

1 *Experimental design*

The specifics of experimental designs for the evaluation of *Bti* against black fly larvae will depend on the objectives, *e.g.* efficacy of formulations, effect of environmental parameters on larvicidal activity, effect of target species, and effects on non-target organisms. Individual streams may be used for single tests, and replicate tests can be conducted over time by reusing the same stream, provided there is continuous oviposition. For univoltine populations it may be necessary to use several similar streams for replications within a fairly narrow time frame or to move upstream for subsequent tests. At least 3 replicate tests under similar stream and temperature conditions should be conducted for each research objective.

2 *Sampling*

As with evaluations of bacteria against mosquito larvae, pre- and post-treatment sampling of larvae is required to determine efficacy. Some factors that should be considered when estimating larval populations are presented by Colbo (1987). For the most part, field evaluation of *Bti* has employed 3 distinctly different techniques for determining larval mortality.

a. One method employs natural and/or artificial substrates to which simuliid larvae are attached. Larvae are sampled just before and 24 h following treatment. Black fly larvae will attach to many substrate types (Fredeen and Spurr, 1978; Walsh et al., 1981b; Undeen and Lacey, 1982). Artificial substrates can reduce sample variability by providing a defined surface area and uniform texture. Segments of braided nylon rope and soft plastic ribbons made of garbage bag are readily colonized by black fly larvae and have been successfully used as artificial substrates for *Bti* evaluation (Araujo-Coutinho and Lacey, 1990). Depending on population density, substrates may be colonized in as little as 24 h after being suspended in the stream.
b. A second method involves treatment of streams, followed by collection of larvae above (controls) and below the treatment point an hour after treatment, and holding them in containers in artificial current for 24 h at which time mortality is determined. Small scale trials may be carried out starting downstream with initial tests and progressively moving upstream in 100 m intervals. This enables replicated studies of a number of variables (application method, formulation, etc.) but does not allow studies on effective carry. Details of the method are described by Lacey and Undeen (1984). A variety of methods for maintaining field-collected larvae in the laboratory are presented by Lacey (1997).
c. A third method involves channeling stream water through small gutters, allowing collected larvae to attach to the surface of the gutter and then adding the candidate larvicide to the top of the gutter (Wilton and Travis, 1965; Lacey *et al.*, 1982; Guillet *et al.*, 1985). Although several formulations may be evaluated side by side, this method is not a true *in situ* evaluation and may not approximate conditions found in the stream.
d. For treatment programs in which the desired outcome is reduction of biting or pestiferous activity, sampling of adult black flies is the most effective method of determining the success of the program. Biting rates, numbers of flies attracted to CO_2 baited sticky traps, or bait animals are commonly employed for this purpose. Several methods for sampling adult black flies are reviewed by Service (1987).

3 *Protocol for evaluating* Bti *efficacy using artificial substrates*

a. Select a stream with confirmed black fly populations. Some key factors to consider: sampling points accessible at regular intervals; feasibility of measurement of stream discharge and feasibility of application of *Bti* to obtain even coverage of the stream. Measure and record salient features of the habitats (temperature, turbidity, presence of vegetation, species composition, larval density, etc.). If stream discharge fluctuates from day to day, measure velocity and discharge as close to the time of application of *Bti* as possible (see Appendices).
b. Place artificial substrates in the stream above and at regular intervals below the proposed treatment point. At least 10 artificial substrates at each sampling site are recommended. Generally, the greater the discharge of the stream, the greater will be the effective downstream carry of *Bti*. Very small creeks may afford only a dozen meters of carry (Undeen and Lacey, 1982), whereas very large rivers may enable several km of carry (Lacey *et al.*, 1982). The length of the stream for sampling depends on stream discharge. The intervals between sampling points depend on the desired degree of sampling precision. Small streams (500–25,000 liters/min discharge) enable 50–1,500 m of downstream carry (Undeen and Lacey, 1982).
c. Just before application of *Bti*, remove half of the substrates above and below the treatment point, and preserve samples from each one in 70% ethanol in separate containers labeled with collection site, date and distance from treatment point.
d. Mix the required amount of formulation in several liters of water. The product will mix more readily with stream water if it is first completely suspended in water. The amount of formulation will be determined by the total discharge of the stream during the treatment period and the desired concentration to be maintained during that interval. See Appendix 2 for an example.
e. Apply the water formulation mixture evenly across the stream over the desired time interval. It may be necessary to have more than one applicator if the width of the stream does not easily permit even coverage. In small streams, application can be made using a garden variety watering can.

Application from a bridge is ideal, otherwise standing in the stream may be necessary. In deeper streams, a pump type pressurized sprayer may be used. Removal of the nozzle tip enables squirting a stream of the *Bti*-water mixture for several meters. Care must be taken when applying the inoculum if applications are made while standing in the stream, since particulates distributed from the stream bed could interfere with larval feeding. If feeding is inhibited before or during exposure to *Bti*, fewer toxic inclusions will be consumed.

f. Post-treatment sampling: 24 h after application, remove remaining substrates at each site and place in individually labeled containers in 70% alcohol. *Bti*-killed larvae may still be attached to the substrates, but their elongated appearance is very distinct.
g. Count pre- and post-treatment samples to determine percent reduction due to treatment. Record the proportion of different instars and species from all samples. *Simulium* species have 7 instars. Neonate larvae are distinguished by the presence of an egg burster on the head capsule. Ultimate instars have well developed, melanized gill spots on the sides of the thorax. If there is a very high population density above the treatment point, downstream drift of larvae and colonization of substrates may be misinterpreted as treatment failure. This effect can be measured by placing clean substrates 10 m below the treatment point shortly after application, and collection of the substrates 24 h later to determine the amount of recolonization from upstream.

4 Special considerations

A decision as to whether adult sampling should be included as a component of a field trial must be made when designing pre-treatment sampling protocols. This will be difficult or impossible for most small scale trials since only the most isolated and discrete breeding sites will produce meaningful results on the impact of a MCA on adult populations of mosquitoes or black flies. Adult mosquitoes can travel several km in a 24 h period (Service, 1997), but in urban areas where *Culex pipiens* and *Ae. aegypti* are breeding, mosquitoes rarely move more than a few hundred m (Subra, 1972). Black flies have been documented routinely to be carried by the wind for 200–300 km, and more rarely even 400 km (Garms *et al.*, 1979; Walsh *et al.*, 1981a). Migration of adults from sites not treated by the MCA into the test site will render results from adult sampling useless for efficacy evaluations unless the treated area extends for tens or hundreds of hectares (or more for some species of black flies), or the targeted breeding sites are isolated from other breeding sites. Traps can be arranged to catch emerging mosquitoes but will not give an estimate of field efficacy which will also depend on the percentage of breeding sites identified and treated.

If the purpose of the experiment is to reduce or eliminate disease transmission, then efforts on monitoring the effect of MCA applications can be focused on reduced prevalence of disease, *e.g.*, dengue in a village or neighborhood of a large city. This approach should be taken if the entire focus of a project is the reduction of vector-borne diseases, and resources are so limited that pre- and post-treatment larval monitoring is not feasible. Success or failure to impact human or animal diseases may be due to factors completely unrelated to parasitism of the targeted vector. Transmission rates for some diseases can be indirectly monitored by morbidity or mortality among a population, but more direct assessments of attack rates involves serological surveys or other immunological screening. For political reasons, impact on disease transmission may nevertheless be monitored.

5 Acknowledgments

The authors wish to thank Don Hostetter and Joel Siegel for advice on the original manuscript and Bob Henszey for furnishing information on the measurement of hydrological parameters. Review of the revised chapter by Don Hostetter is gratefully appreciated.

6 Appendix 1: Simple method for measurement of stream discharge

A variety of meters are available to measure stream velocity. One of the most widely used is the Flo-Mate Portable flowmeter (Marsh-McBirney, Inc., Frederick, MD, USA).

Others include the Price AA meter and the Pygmy current meter. Fairly accurate measurement of stream discharge using current meters is possible by taking several depth and current readings and using the mid-section method as described by Brackensiek *et al.* (1979) and Harrelson *et al.* (1994). If use of a velocity meter is not possible, an orange fruit can be a useful float for velocity measurement. For small streams the following protocol will provide a simple method for determining discharge. Two people, a velocity meter or float, tape measure, stop watch, notebook and wooden stakes or survey tape will be needed.

1. Measure a 10 m segment of the stream and mark the beginning and end of the segment with wooden stakes or plastic survey ribbon tied to nearby vegetation. Try to select a section without excessive turbulence if possible. The more laminar the flow, the more accurate will be the determination of velocity.
2. Measure the width of the stream at 10 locations along the 10 m section of stream. Average the measurements.
3. Measure the depth of the stream at 5 locations across the stream at each of 5 locations every 2 m along the 10 m section of stream. Average the measurements.
4. If a velocity meter is available, measure the stream velocity at 5 locations across the stream in the middle of the 10 m section of stream. If one is not available, place the float (orange) a few meters above the upstream markers to allow it to attain the same velocity as the stream before beginning your measurement. Using a stopwatch or wristwatch, determine the time it takes for the float to pass between the two markers. Divide 10 m by the number of seconds it takes the float to traverse the segment. This gives you velocity in m/sec. Repeat this measurement at least 5 times. Average the velocities.
5. Calculate stream discharge by using the following formula: velocity (m/sec) x average depth (m) x average width (m)

For example, a stream with an average velocity of 0.65 m/sec, average depth of 0.45 m and average width of 5 m will have a discharge of 1.46 m^3/sec or 1,460 liters/sec. Undeen *et al.* (1984) and Undeen and Molloy (1996) offer an alternative to discharge for predicting rates of *Bti* based on stream width.

7 Appendix 2: Calculation of amount of *Bti* formulation required for treatment of streams

The amount of *Bti* formulation in parts per million (ppm = mg/liter or g/m^3) is calculated based on the desired concentration for a specified period of time. For example, if a concentration of 10 ppm (= 10 g/m^3) is desired for a one minute application in the stream described above, multiply the total discharge of the stream (in cubic meters) during one minute by 10 g ($1.46\,m^3/sec \times 60\,sec \times 10\,g/m^3$). The required amount of formulation will be 876 g of formulation.

8 References

Adler, P. H., Currie, C., and Wood, D. M. 2004. "The Black Flies (Simuliidae) of North America." Cornell University Press, Ithaca, NY.

Apperson, C. S., Federici, B. A., Stewart, W., and Tarver, F. R. 1992. Evidence for the copepods *Acanthocyclops robustus* and *Mesocyclops edax* as competent intermediate hosts for *Coelomomyces punctatus* during and epizootic in a larval population of the mosquito *Anopheles quadrimaculatus*. *J. Invertebr. Pathol.* **60**, 229–236.

Araujo-Coutinho, C. J. P. C. and Lacey, L. A. 1990. Field evaluation of *Bacillus thuringiensis* var. *israelensis* for control of black flies in the North Littoral Zone of Brazil's São Paulo State. *Bull. PAHO*. **24**, 307–313.

Bateman, R. 1997. Methods of application of microbial pesticide formulations for the control of grasshoppers and locusts. *Mem. Entomol. Soc. Canada* **171**, 69–81.

Bates, M. 1941. Field studies of the anopheline mosquitoes of Albania. *Proc. Entomol. Soc. Wash*. **43**, 37–58.

Becker, N. Rettich, F. 1994. Protocol for the introduction of new *Bacillus thuringiensis israelensis* products into the routine mosquito control program in Germany. *J. Am. Control Assoc*. **10**, 527–33.

Brakensiek, D. L., Osborn, H. B., and Rawls, W. L. 1979. "Field Manual for Research in Agricultural Hydrology". U.S. Department of Agriculture, Agricultural Handbook 224, 550 pp.

Case, T. J. and Washino, R. K. 1979. Flatworm control of mosquito larvae in rice fields. *Science* **206**, 1412–1414.

Chapman, H. C. 1974. Biological control of mosquito larvae. *Annu. Rev. Entomol.* **19**, 33–59.

Chapman, H. C. (Ed.) 1985. "Biological Control of Mosquitoes." *Am. Mosq. Control Assoc. Bull.* **6**, 219 pp.

Colbo, M. H. 1987. Problems in estimating black fly populations in their aquatic stages. *In* "Black Flies: Ecology,

Population Management, and Annotated World List" (R. Merritt and K. C. Kim, Eds.), pp. 77–89. Penn. State Univ. Press, University Park, PA.

Corbet, P. S. 1965. An insect emergence trap for quantitative studies in shallow ponds. *Can. Entomol.* **97**, 845–858.

Currie, D. C. and Craig, D. A. 1987. Feeding strategies of larval black flies. *In* "Black Flies: Ecology, Population Management and Annotated World List" (R. Merritt and K. C. Kim, Eds.), pp. 155–170. Penn. State Univ. Press, University Park, PA.

Davies, I. J. 1984. Sampling aquatic insect emergence, *In* "Methods for the Assessment of Secondary Productivity in Fresh Waters (2nd Edition)" (J. A. Downing and F. H. Rigler, Eds.), pp. 161–227. IBP Handbook, No. 17, Blackwell Scientific Publications, Oxford, UK.

Enfield, M. A. and Pritchard, G. 1977. Methods for sampling immature stages of *Aedes* spp. (Diptera: Culicidae) in temporary ponds. *Can. Entomol.* **109**, 1435–1444.

Fanara, D. M. 1973. An efficient aquatic sample concentrator. *Mosq. News* **33**, 603–604.

Frances, S. P., Sweeney, A. W., and Humber, R. A. 1989. *Crypticola clavulifera* gen. et sp. nov. and *Lagenidium giganteum*: Oomycetes pathogenic for dipterans infesting leaf axils in an Australian rain forest. *J. Invertebr. Pathol.* **54**, 103–111.

Fredeen, F. J. H. and Spurr, D. T. 1978. Collecting semi-quantitative samples of black fly larvae (Diptera: Simuliidae) and other aquatic insects from large rivers with the aid of artificial substrates. *Quaest. Entomol.* **14**, 411–431.

Fukuda, T., Willis, O. R., and Barnard, D. R. 1997. Parasites of the Asian tiger mosquito and other container-inhabiting mosquitoes (Diptera: Culicidae) in north central Florida. *J. Med. Entomol.* **34**, 226–233.

Garms, R., Walsh, J. F., and Davies, J. B. 1979. Studies on the reinvasion of the Onchocerciasis Control Programme in the Volta River Basin by *Simulium damnosum* s.l. in Savannen Guineas. *Z. Tropenmed. Parasitol.* **30**, 345–362.

Guillet, P., Hougard, J.-M., Doannio, J., Escaffre, H., and Duval, J. 1985. Evaluation de la sensibilité des larves du complexe *Simulium damnosum* à la toxine de *Bacillus thuringiensis* H 14. 1. Méthodologie. *Cah. ORSTOM, sér. Entomol. méd. Parasitol.* **23**, 241–250.

Guzman, D. R. and Axtell, R. C. 1987. Temperature and water quality effects in simulated woodland pools on the infection of *Culex* mosquito larvae by *Lagenidium giganteum* (Oomycetes: Lagenidiales) in North Carolina. *J. Am. Mosq. Control. Assoc.* **3**, 211–218.

Hagstrum, D. W. 1971. A model of the relationship between densities of larvae and adults in an isolated population of *Culex tarsalis* (Diptera: Culicidae). *Ann. Entomol. Soc. Am.* **64**, 1074–1077.

Harrelson, C. C., Rawlins, C. L., and Potyondo, J. P. 1994. "Stream Channel Reference Sites: An Illustrated guide to Field Technique." USDA Forest Service, Gen. Tech. Rep. RM-245, 61 pp.

Hatfield, L. D., Riner, J. L., and Norment, B. R. 1985. A dredge sampler for mosquito larvae. *J. Am. Mosq. Control Assoc.* **1**, 372–373.

Hawley, W. A. 1988. The biology of *Aedes albopictus. J. Am. Mosq. Control Assoc.* **4** (suppl), 1–39.

Kerwin, J. L., Dritz, D. A., and Washino, R. K. 1994. Pilot scale production and application in wildlife ponds of *Lagenidium giganteum* (Oomycetes: Lagenidiales). *J. Am. Mosq. Control Assoc.* **10**, 451–455.

Kerwin, J. L. 2007. *Lagenidium giganteum* (Oomycetes). *In* "Biorational Control of Mosquitoes" *Am. Mosq. Control Assoc. Bull.* **7**, *in press.*

Kerwin, J. L. and Petersen, E. E. 1997. Oomycetes and Chytridiomycetes, *In* "Manual of Techniques in Insect Pathology" (L. A. Lacey, Ed.), pp. 251–268. Academic Press, London.

Kerwin. J. L. and Washino, R. K. 1985. Recycling of *Romanomermis culicivorax* (Mermithidae: Nematoda) in rice fields in California, USA. *J. Med. Entomol.* **22**, 637–643.

Kerwin, J. L. and Washino, R. K. 1986. Ground and aerial application of the sexual and asexual stages of *Lagenidium giganteum* (Oomycetes: Lagenidiales) for mosquito control. *J. Am. Mosq. Control Assoc.* **2**, 182–189.

Kerwin, J. L. and Washino, R. K. 1988. Field evaluation of *Lagenidium giganteum* (Oomycetes:Lagenidiales) and description of a natural epizootic involving a new isolate of the fungus. *J. Med. Entomol.* **25**, 452–460.

Kimerle, R. A. and Anderson, N. H. 1967. Evaluation of aquatic insect emergence traps. *J. Econ. Entomol.* **60**, 1255–1259.

Lacey, L. A. 1997. Laboratory bioassay of bacteria against aquatic insects with emphasis on larvae of mosquitoes and black flies. *In* "Manual of Techniques in Insect Pathology" (L. A. Lacey, Ed.), pp. 79–90. Academic Press, London.

Lacey, L. A. 2007. *Bacillus thuringiensis* serovariety *israelensis* and *Bacillus sphaericus* for mosquito control. *In* "Biorational Control of Mosquitoes" *Am. Mosq. Control Assoc. Bull.* **7**, *in press.*

Lacey, L. A. and Merritt, R. W. 2003. The safety of bacterial microbial agents used for black fly and mosquito control in aquatic environments. *In*: "Environmental Impacts of Microbial Insecticides: Need and Methods for Risk Assessment" (H. M. T. Hokkanen and A. E. Hajek, eds.), pp. 151-168. Kluwer Academic Publishers Dordrecht.

Lacey, L. A. and Mulla, M. S. 1990. Safety of *Bacillus thuringiensis* (H-14) and *Bacillus sphaericus* to non-target organisms in the aquatic environment. *In* "Safety of Microbial Insecticides" (M. Laird, L. A. Lacey and E. W. Davidson, eds.). pp. 169–188. CRC Press, Boca Raton, FL.

Lacey, L. A. and Siegel, J. P. 2000. Safety and ecotoxicology of entomopathogenic bacteria. *In* "Entomopathogenic Bacteria: From Laboratory to Field Application" (J.-F. Charles, A. Delécluse, and C. Nielsen-LeRoux, Eds.), pp. 253-273. Kluwer Academic Publishers, Dordrecht, The Netherlands.

Lacey, L. A. and Undeen, A. H. 1984. Effect of formulation, concentration and application time on the efficacy of *Bacillus thuringiensis* (H-14) against black fly (Diptera: Simuliidae) larvae under natural conditions. *J. Econ. Entomol.* **77**, 412–418.

Lacey, L. A. and Undeen, A. H. 1986. Microbial control of black flies and mosquitoes. *Annu. Rev. Entomol.* **31**, 265–296.

Lacey, L. A., Escaffre, H., Philippon, B., Sékétéli, A. and Guillet, P. 1982. Large river treatment with *Bacillus thuringiensis* (H-14) for the control of *Simulium damnosum* in the Onchocerciasis Control Programme. *Z. Tropenmed. Parasitol.* **33**, 97–101.

Lacey, L. A.., Urbina, M. J., and Heitzman C. M. 1984. Sustained release formulations of *Bacillus sphaericis* and *Bacilus thuringiensis* (H-14) for control of container breeding *Culex quinquefasciatus. Mosq. News* **44**, 26–32.

Laird, M. 1988. "The Natural History of Larval Mosquito Habitats". Academic Press, New York.

Meisch, M. V., Anderson, A. L., Watson, R. L., and Olson, L. 1982. Mosquito species inhabiting ricefields in five growing regions of Arkansas. *Mosq. News* **42**, 341–346.

Mijares, A. S. and Pacheco, R. P. 1997. Reduction of mosquito larval densities in natural sites after introduction of *Romanomermis culicivorax* (Nematoda: Mermithidae) in Cuba. *J. Med. Entomol.* **34**, 1–4.

Molloy, D., Gaugler, R. and Jamnback, H. 1981. Factors influencing efficacy of *Bacillus thuringiensis* var. *israelensis* as a biological control agent of black fly larvae. *J. Econ. Entomol.* **74**, 61–64.

Molloy, D. and Struble, R. H. 1989. Investigation of the feasibility of the microbial control of black flies (Diptera: Simuliidae) with *Bacillus thuringiensis* var *israelensis* in the Adirondack Mountains of New York. *Bull. Soc. Vect. Ecol.* **14**, 266–276.

Mulla, M. S., Norland, R. L., Fanara, D. M., Darwazeh, H. A., and McKean, D. W. 1971. Control of chironomid midges in recreational lakes. *J. Econ. Entomol.* **64**, 300–307.

Mulla, M. S., Thavara, U., Tawatsin, A., Chomposri, J., and Su, T. Y. 2003. Emergence of resistance and resistance management in field populations of tropical *Culex quinquefasciatus* to the microbial control agent *Bacillus sphaericus. J. Amer. Mosq. Control Assoc.* **19**, 39–46.

Nielsen-Leroux, C., Pasquier, F., Charles, J. F., Sinegre, G., Gaven, B., and Pasteur, N. 1997. Resistance to *Bacillus sphaericus* involves different mechanisms in *Culex pipiens* (Diptera: Culicdae) larvae. *J. Med. Entomol.* **34**, 321–327.

Nielsen-LeRoux, C., Pasteur, N., Pretre, J., Charles, J.-F., Sheikh, H. B., and Chevillon, C. 2002. High resistance to *Bacillus sphaericus* binary toxin in *Culex pipiens* (Diptera: Culicidae): The complex situation of West Mediterranean countries. *J. Med. Entomol.* **39**, 729–735.

Pailey K. P. and Balaraman, K. 1993. Influence of soil moisture on survival and oviposition of *Romanomermis iyengari*, a mermithid parasite of mosquitoes. *Indian J. Malariol.* **30**, 221–225.

Petersen, J. J. 1985. Nematode parasites. *In* "Biological Control of Mosquitoes" (H. C. Chapman, Ed.). *Bull. Am. Mosq. Control Assoc.* **6**, 110–122.

Petersen, J. J., Chapman, H. C., Willis, O. R., and Fukuda, T. 1978. Release of *Romanomermis culicivorax* for the control of *Anopheles albimanus* in El Salvador II. Application of the nematode. *Am. J. Trop. Med. Hyg.* **27**, 1268–1273.

Platzer, E. G. 2007. Mermithid nematodes. *In* "Biorational Control of Mosquitoes" *Am. Mosq. Control Assoc. Bull.* **7**, *in press*.

Rao, D. R., Mani, T. R., Rajendran, R., Joseph, A. S., Gajanana, A., and Reuben, R. 1995. Development of a high level of resistance to *Bacillus sphaericus* in a field population of *Culex quinquefasciatus* from Kochi, India. *J. Am. Mosq. Control Assoc.* **11**, 1–5.

Robert, V. Ouari, B. Ouedraogo, V., and Carnevale, P. 1988. La succession des espéces anophéliennes et le cycle du riz; étude écologique des Culicidae adultes et larvaire dans la rizière de la Valle du Kou, Burkina Faso. *Acta Tropica.* **45**, 351–359.

Samarawickrema, W. A., Sone, F., Self, L. S., Cunnings, R. F., and Paulson, G. S. 1992. Distribution of breeding and control of the filariasis vector *Aedes samoanus* in leaf axils of Pandanus in Samoa. *Med. Vet. Entomol.* **6**, 367–370.

Service, M. W. 1987. Monitoring adult simuliid populations. *In* "Black Flies: Ecology, Population Management, and Annotated World List" (R. Merritt and K. C. Kim, Eds.), pp. 187–200. Penn. State Univ. Press, University Park, PA.

Service, M. W. 1993. "Mosquito Ecology-Field Sampling Methods" (2nd Edition). Elsevier Applied Science, London.

Service, M. W. 1997. Mosquito (Diptera: Culicidae) dispersal-the long and short of it. *J. Med. Entomol.* **34**, 579–588.

Skovmand, O. and Sanogo, E. 1999. Experimental formulations of *Bacillus sphaericus* and *Bacillus thuringiensis israelensis* against *Culex quinquefasciatus* and *Anopheles gambiae* (Dipetera: Culicidae) in Burkina Faso. *J. Med. Entomol.* **36**, 62–67.

Subra, R. 1972. Études écologiques sur *Culex pipiens fatigans* Wiedemann, 1828 (Diptera, Culicidae) dans une zone urbaine de savane soudanienne Ouest-Africaine.

Longevité et déplacements d'adultes marques avec des poudres fluorescentes. **10**, 3–36.

Thiery, I. and Hamon, S. 1998. Bacterial control of mosquito larvae: Investigation of stability of *Bacillus thuringiensis* var *israelensis* and *Bacillus sphaericus* standard powders. *J. Am. Mosq. Control Assoc.* **14**, 472–76.

Tun-Lin W., Kay, B. H., and Burkot, T. R. 1994. Quantitative sampling of immature *Aedes aegypti* in metal drums using sweep and dipping methods. *J. Am. Mosq. Control Assoc.* **10**, 390–396.

Undeen, A. H. and Becnel J. J. 1994. A device for monitoring populations of larval mosquitoes in container habitats. *J. Am. Mosq. Control Assoc.* **10**, 101–103.

Undeen, A. H. and Colbo, M. H. 1980. The efficacy of *Bacillus thuringiensis* var. *israelensis* against blackfly larvae (Diptera: Simuliidae) in their natural habitat. *Mosq. News* **40**, 181–184.

Undeen, A. H. and Lacey, L. A. 1982. Field procedures for the evaluation of *Bacillus thuringiensis* var. *israelensis* (serotype 14) against black flies (Simuliidae) and nontarget organisms in streams. *Misc. Pub. Entomol. Soc. Am.* **12**, 25–30.

Undeen, A. H., Lacey, L. A. and Avery, S. W. 1984. A system for recommending dosage of *Bacillus thuringiensis* (H-14) for control of simuliid larvae in small streams based upon stream width. *Mosq. News* **44**, 553–559.

Undeen, A. H. and Molloy, D. P. 1996. Use of stream width for determining the dosage rates of *Bacillus thuringiensis* var. *israelensis* for larval black fly (Diptera: Simuliidae) control. *J. Am. Mosq. Control Assoc.* **12**, 312–315.

Vladimirova, V. V., Pridantseva, E. P., Gafurov, A. K., and Muratova, M. E. 1990. A trial of *Romanomermis iyengari* and*R. culicivorax* as a means for controlling blood-sucking mosquitoes in the Tadzhik SSR. *Med. Parazitol. Parazit. Bolezni* pp. 42–45.

Walker, T. W., Meek, C. L., and Wright, V. L. 1985. Establishment and recycling of *Romanomermis culicivorax* (Nematoda: Mermithidae) in Louisiana ricelands. *J. Am. Mosq. Control Assoc.* **1**, 468–473.

Walsh, J. F., Davies, J. B., and Cliff, B. 1981a. World Health Organization Onchocerciasis Control Programme in the Volta River Basin, *In* "Blackflies-The Future for Biological Methods in Integrated Control" (M. Laird, Ed.), pp. 85–103. Academic Press, New York.

Walsh, D. J., Yeboah, D. Y., and Colbo, M. H. 1981b. A spherical sampling device for black fly larvae. *Mosq. News* **41**, 18–21.

Whisler, H. C., Zebold, S. L., and Shemanchuk, J. A. 1975. Life history of *Coelomomyces psorophorae. Proc. Natl. Acad. Sci.* USA **72**, 693–696.

Wilton, D. P. and Travis, B. V. 1965. An improved method for simulated stream tests of black fly larvicides. *Mosq. News* **25**, 118–123.

Yap, H. 1991. Field trials of *Bacillus sphaericus* for mosquito control. *In* "Bacterial control of mosquitoes and black flies" (H. de Barjac and D. J. Sutherland, Eds.), pp.307–320. Rutger Univ. Press, New Brunswick, NJ.

Zaim, M., Ladonni, H., Ershadi, M. R., Manoucheri, A. V., Sahabi, Z., Nazari, M., and Shahmohammadi, H. 1988. Field application *Romanomermis culicivorax* (Nematoda: Mermithidae) to control anopheline larvae in southern Iran. *J. Am. Mosq. Control Assoc.* **4**, 351–355.

Chapter VII-23

Terrestrial mollusc pests

Michael J. Wilson
Department of Plant and Soil Science
School of Biological Sciences
University of Aberdeen
Aberdeen AB24 3UU, Scotland, UK

1 Introduction

Many aquatic molluscs are pests; certain snails act as intermediate hosts of human and animal diseases, and zebra mussels foul water supply systems and some aquatic snails damage paddy rice. However, no biological pesticides are available for use against these pests. In contrast, one biological pesticide is commercially available for use against terrestrial mollusc pests and there are other candidate microorganisms with potential to be developed as biological pesticides.

Terrestrial molluscs are members of the class Gastropoda, mostly in the subclass Pulmonata with two families from the subclass Gymnomorpha (South, 1992). The common names slugs and snails are used to describe terrestrial molluscs, but these are not scientifically valid terms. The term slug is applied to animals that have no external shell or one that is small in relation to the size of their body, whereas snails have larger shells and are usually capable of withdrawing their entire body into the shell. However, intermediate forms occur and in all cases slugs have evolved from snail forms. This has happened on several occasions during the course of evolution such that many slugs that appear similar are quite distantly related, but have developed a similar morphology through convergent evolution (South, 1992).

Terrestrial molluscs have a relatively simple lifecycle in which they lay eggs that hatch to release juveniles. The juveniles resemble adults in shape, although their color is often different from the adults. Juveniles feed and develop into adults and reproduce either by hermaphroditism or amphimixis. Hermaphrodite species may be self-fertilizing (*e.g., Arion intermedius*), but more usually fertilization occurs through outcrossing (*e.g., Deroceras reticulatum*). Many species have distinct seasonal life cycles, which may be annual (*e.g.*, many *Arion* spp.) or as in the case of the giant African snail, *Achatina fulica*, over several years. Other species, such as the widespread pest *D. reticulatum*, are opportunistic breeders and can go through several generations in a year given optimum temperatures and adequate water availability. The biology of terrestrial molluscs has recently been reviewed (Barker, 2001).

Molluscs lack the rigid cuticle and water retentive capacity of most insects. In the absence of sufficient water, they tend to become inactive. Under dry conditions slugs move deep into the ground in search of moisture, rarely come to the surface, and do not cause crop damage. Snails, apart from very small species, are unable to move deep in the soil profile, and rely on the protective capacity of their shells to prevent desiccation. The animals withdraw into their shells and under prolonged dry conditions enter a state of aestivation in which a calcium pellicle may be formed over the mouth of the shell. Snails in this state can survive for many months. Damage forecasting systems for molluscs, therefore, need to include moisture components. This reason, combined with the difficulties in estimating population densities of many slug species, has

L.A. Lacey and H.K. Kaya (eds.), Field Manual of Techniques in Invertebrate Pathology, 751–765.

resulted in a scarcity of damage forecasting methods. However, the dependence on water for activity is beneficial from the point of biological control. Many biocontrol agents, particularly nematodes, cannot be applied under dry conditions. For mollusc control, this is unlikely to be a problem, since the pests are only damaging when moisture is plentiful.

2 Distribution and pest status

Molluscs have been reported as crop pests throughout every inhabited continent and their pest status has recently been reviewed (Barker, 2002a). However, because there are few specialists working on molluscs, data concerning the estimated monetary value of mollusc damage to different crops are scarce. In Europe, particularly in the northwestern countries with temperate climates and much rainfall, slugs are serious pests of arable crops (wheat, oilseed rape), vegetable crops (*e.g.*, lettuce, Brussels sprouts) and domestic gardens (Port and Port, 1986; Glen and Moens, 2002; Moens and Glen, 2002; Port and Ester, 2002). In North America, slugs are serious pests of field corn and soybeans, particularly in systems using reduced tillage (Hammond and Byers, 2002), and snails are pests of citrus production, particularly in California (Sakovich, 2002). In Central America, slugs are pests of beans (Rueda *et al.*, 2002). In Africa many species, but especially the giant African snail, *A. fulica*, are serious pests of a wide range of crops (Mead, 1961; Raut and Barker, 2002). In Australasia, slugs are a problem in pasture establishment in New Zealand (Barker, 2002b) and snails are pests in vine production in Northern Australia (Sanderson and Sirgel, 2002) and damage wheat in Southern Australia (Baker, 2002). In the latter case, snails do not damage the wheat by feeding, but in summer, climb to the top of the plants to aestivate. The sheer numbers of snails can damage harvesting machinery and lower the value of harvested wheat considerably.

3 Biological control options for molluscs

Very little is known about pathogens and parasites of terrestrial molluscs in comparison to the wealth of data on insect pathogens. This is clearly seen in the recent book reviewing natural enemies of molluscs (Barker, 2004) where the vast majority of chapters relate to predators of molluscs. There are no known viruses which infect slugs or snails and there are few known bacterial diseases (see review by Raut, 2004). There are no highly virulent pathogens with much potential to be developed commercially, but examples of bacteria which show a degree of molluscicidal activity include *Aerobacter* sp. (Mead, 1961), *Aeromonas hydrophila* (Dean *et al.*, 1970), *Vibrio parahaemolyticus* (Ducklow *et al.*, 1980) and *Bacillus brevis* (Singer *et al.*, 1997).

No fungi have been found that attack juvenile or adult molluscs, but fungal parasites of slug eggs have been described (Trevet and Esslemont, 1938; Arias and Crowell, 1963).

A promising protozoan biocontrol agent is the ciliate *Tetrahymena rostrata* (Brooks, 1968). This is lethal to certain slugs and can be mass-produced *in vivo*. However, while all the above have shown some degree of promise in laboratory studies, none has yet been developed as a commercial biological molluscicide.

To date, nematodes appear to have the greatest potential for biocontrol of slugs and snails and one nematode has been commercialized. Mengert (1953) studied nematodes associated with slugs and snails and concluded that nematodes are more likely to be associated with slugs than snails. This is because slugs generally live in the soil and are more likely to come into contact with nematodes, whereas snails tend to live on plants above the soil surface. Mengert (1953) found 29 species of nematodes associated with slugs and categorized them into four groups based on the degree of nematode adaptation to life within slugs. The first group contained 23 non-parasitic species that were transiently associated with slug slime or which survived ingestion by slugs and passage through the gut. The second group included the nematodes *Phasmarhabditis papillosa, P. neopapillosa* and *P. hermaphrodita*, (Mengert did not find *P. hermaphrodita* but grouped it with the other two species of *Phasmarhabditis*). These nematodes formed resistant dauer juveniles (third stage non-feeding juveniles with a retained second-stage cuticle) that showed a degree of adaptation to life within slugs, but were not parasitic. Mengert (1953) considered that dauer juveniles found their way into the body cavity of slugs and survived there until the slug died. Then they would feed on the cadaver. The third group contained one species of nematode, *Alloionema appendiculata*, which is a facultative and non-lethal parasite of the slug's foot. The fourth group contained three species of the genus *Angiostoma* that are obligate parasites of the slug's gut. There, adults produce eggs that are shed in the feces and later ingested by other slugs where the eggs hatch and develop into adults. In addition, there are reports of certain nematode species parasitizing the genital apparatus of molluscs (Morand and Petter, 1986; Morand and Hommay, 1990). The associations between terrestrial molluscs and nematodes have recently been reviewed (Grewal *et al.*, 2003; Morand *et al.*, 2004).

Other than the naturally occurring nematode parasites of slugs, two accounts of entomopathogenic nematodes killing molluscs have been published. Li *et al.* (1986) used *Steinernema glaseri* in laboratory assays to control the semi-aquatic snail, *Oncomelania hupensis*, and Jaworska (1993) reported pathogenicity of *Steinernema carpocapsae, S. feltiae*, and *Heterorhabditis bacteriophora* against the slug species, *Deroceras reticulatum* and *D. agreste*. To date, field efficacy of these nematodes has not been demonstrated, and in laboratory assays, Wilson *et al.* (1994c) could not demonstrate pathogenicity of *S. feltiae* or *H. megidis* against the slug *D. reticulatum*.

The only parasite that has been developed as a biocontrol agent is *P. hermaphrodita* although another promising candidate nematode has been tested in Australia, although this species is yet to be described (Charwat *et al.*, 2000). The majority of the research has been with *P. hermaphrodita* which is a lethal parasite capable of killing many species of pest slugs (Wilson *et al.*, 1993a; Tan and Grewal, 2001a). The nematodes form dauer juveniles that act as the infective stage by penetrating into the shell cavity of the slug *Deroceras reticulatum* through a natural opening connecting the shell cavity to the dorsal integumental pouch. Once inside, the juveniles develop into adults and reproduce. During this time, the slug develops a characteristic swelling in the mantle region (Figure 1). Within 7–21 days of infection, the slug dies, and nematodes spread from the mantle and feed on the entire slug cadaver. Eventually the food is depleted and the nematodes fail to develop into adults, once again forming dauer juveniles that leave in search of fresh hosts.

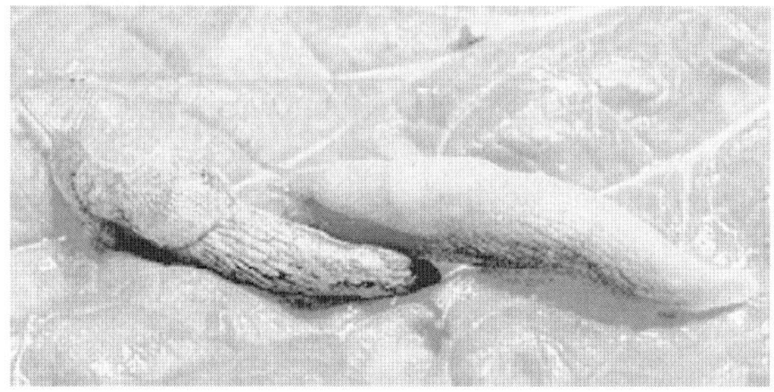

Figure 1. Healthy *Deroceras reticulatum* (background) and *D. reticulatum* infected by *Phasmarhabditis hermaphrodita* (foreground) showing characteristic swelling of the mantle

The nematode can be produced, and formulated using similar technology to that developed for entomopathogenic nematodes (Wilson and Grewal, 2005), and has been used successfully in several countries to control slug pests in a wide range of crops (Ester and Wilson, 2005). The nematode became commercially available to home gardeners in the United Kingdom in 1994 under the trade name Nemaslug® (Becker Underwood, Ames Iowa) and the product is now sold in fourteen European countries. While the nematode takes typically between 1 and 3 weeks to kill slugs, their feeding is strongly inhibited within a few days of infection, providing rapid crop protection (Glen *et al.*, 2000).

The mode of action of *P. hermaphrodita* is not known. Unlike the entomopathogenic nematodes, no symbiotically associated bacterium has yet been isolated. However, *P. hermaphrodita* is a bacterial feeder and when grown *in vitro* in monoxenic culture with different bacterial species (*e.g., Moraxella osloensis, Providencia rettgeri, Pseudomonas fluorescens*), growth and pathogenicity of nematodes are strongly influenced by bacterial species (Wilson *et al.*, 1995b, 1995c). It has been suggested that the virulence of the commercial product grown with *M. oslosensis* can be attributed to the action of the lipopolysacharide (endotoxin) present within the bacterium's cell wall (Tan and Grewal, 2001b; Tan *et al.*, 2002, 2003).

The vast majority of work on *P. hermaphrodita* has concentrated on control of slugs, and thus, the emphasis of this chapter will be on slug control. Since most snails live above the soil level, nematodes are not the ideal choice for their control. However, in the knowledge that certain snail species are susceptible to *P. hermaphrodita* (Wilson *et al.*, 1993a; Coupland, 1995; Wilson *et al.*, 2000), and that entomopathogenic nematodes have been used with some success in foliar and cryptic habitats (Begley, 1990), methods for studying snail control will also be included.

4 Field experiment techniques

A Nematode application

No published data on survival of *Phasmarhabditis* spp. nematodes through hydraulic or other types of sprayers are available, but there are data for application of entomopathogenic nematodes (see Chapter IV-5). Guidelines for handling entomopathogenic nematodes should be followed when using *P. hermaphrodita*. Entomopathogenic nematodes can be applied with conventional hydraulic spray equipment, providing that the filter's mesh size is not too small. Wilson *et al.* (1994a, 1996) applied nematodes to wheat fields using a knapsack sprayer fitted with a deflector nozzle at 1 bar (= 100 kPa) pressure and achieved slug control equivalent to the chemical standards. For smaller plots a watering can may be used, whereas for very large plots, tractor or helicopter mounted spraying equipment may be used (Ester and Wilson, 2005).

P. hermaphrodita is a soil organism, and care should be taken to apply the nematodes onto soil to ensure their survival. In many instances, when slugs damage seeds or seedlings, nematodes will be applied to bare or mostly bare soil. However, in some crops, (*e.g.*, potatoes), the soil will be largely covered by foliage. In such cases, the crop must be irrigated from overhead following nematode application. *P. hermaphrodita* may also be applied through irrigation lines. It has been applied through T-tape irrigation lines to control slugs in strawberries (Glen *et al.*, 1996).

P. hermaphrodita typically takes from 1 to 3 weeks to kill slugs. However, it inhibits slug feeding within 3 days of application (Glen *et al.*, 2000). Thus, nematode-applications are best made shortly before slug damage occurs. For example, in wheat or oilseed rape, nematodes should be applied at the time of seed drilling or shortly before. However, for potatoes, where damage occurs to the maturing tubers, nematodes should be applied about 4 to 6 weeks before harvest.

Commercial preparations of *P. hermaphrodita* (Nemaslug®) are formulated onto powdered vermiculite or clays, and this is simply mixed into the spray tank. No adjuvants are needed. There is little information available on compatibility of *P. hermaphrodita* with other crop-protection agents so tank mixing, at present, is not recommended. Nematodes are typically applied at a rate of 3×10^9/ha. As with entomopathogenic nematodes, it is best to apply *P. hermaphrodita* to damp soil and if available, apply up to 5 cm of irrigation afterwards. Soil water potential at the time of application should be determined. A filter paper method as described by Kaya and Stock (1997) is recommended. If irrigation is available, it is preferable to apply water as necessary to keep the soil surface moist at all times during the experiment to encourage slug activity.

The commercially available strain of *P. hermaphrodita* is sensitive to temperatures above 25 °C, and failure to control slugs in polyethylene tunnels has been attributed to high temperature inactivation (Glen, pers. comm). Thus, soil temperature at 5 cm depth should be recorded at the time of nematode application, and if possible, daily maximum, minimum and mean soil temperatures should also be recorded. In addition, precautions should be taken when handling and transporting the nematodes to avoid temperatures above 25 °C (*e.g.*, carry nematodes in a cooler and avoid exposure to direct sunlight).

B Mini-plots

Mini-plots can be extremely useful for testing biological molluscicides, especially when small amounts of inoculum are available (*i.e.*, for testing new strains). Mini-plots can be made specifically for use in slug experiments (*e.g.*, Wilson *et al.*, 1994b) or can be constructed in the field or in polyethylene tunnels used for vegetable production. Slugs tend to have a patchy distribution in fields, and thus, field experiments generally need to be done in large plots, and preferably several samples should be taken from each plot to overcome this problem. When mini-plots are used, an introduced slug population of uniform size may be added if slugs are not present in sufficient numbers at the experimental site. Rearing slugs in the laboratory can be labor intensive, and most workers rely on field collection for experimental animals. Slugs can be collected from grassland and cropped areas using baited refuge traps, as described in section D3. Because slugs have the ability to crawl up many vertical surfaces, barriers between plots need to be made in such a way that the slugs are contained. There are many ways to make mini-plots ranging from very simple to those that are more elaborate. Choice of construction is left to the individual worker based on available resources. Some examples of successfully used designs are given below. Whichever design is chosen, a minimum of four replicate mini-plots per treatment is recommended.

1 Packing crate miniplots

1. Miniplots can be prepared from standard plastic packing crates (80 cm × 60 cm (0.48 m^2) × 23 cm deep. Before filling the crates with soil drainage holes should be drilled in the bottom of the crates which should then be lined with plastic mesh to prevent slugs entering or leaving. A thin layer (1–2 cm) of coarse gravel should then be added to further aid drainage, and the plots are then filled with loam soil to a depth of approximately 20 cm.
2. Erect a 10 cm high fence of 0.8 mm woven copper mesh on top of the crates (Figure 2). Slugs are repelled by copper and will not cross such a barrier (Moens *et al.*, 1967). The copper mesh can be held in place on a metal or wooden frame. If a workshop is not available to make the frame, slugs can be contained within plots by painting polytetrafluoroethylene (Fluon, Whitford Plastics, Runcorn, UK) around the plastic crate immediately above the soil level. This substance has been used to prevent insect movement out of plots in many experiments and has been shown to form an effective barrier for slugs (Symondson, 1990).
3. Populate plots with slugs. Add between 10 and 20 adult *D. reticulatum* per plot, depending on availability of slugs. If the plots are being used with bare soil, *e.g.*, to test slug damage to seeds or young seedlings, it is advisable to put a small refuge in each plot for slugs to shelter from bright sunshine during the day (Figure 2).

Figure 2. A series of packing crate mini-plots with copper fences for use in biological control experiments. Small plant pot saucers are povided as resting sites for slugs during the day (photo, L. Simms)

2 Mini-plots made in situ *in protected vegetable production*

Plastic barriers coated with polytetrafluoroethylene can be erected in fields or glasshouses around individual groups of plants to act as miniplots. Wilson *et al.* (1995a) in two experiments using lettuce grown in polyethylene tunnels, used this technique.

1. Divide the lettuce crop into 2 × 1 m plots using barriers made of 16 cm high plastic lawn-edge material (available from most garden centers). Sink the lawn edge 5 cm into the ground.
2. Paint the protruding 11 cm of the barrier with polytetrafluoroethylene.

3. Add slugs to the plots to supplement the natural population if necessary (aim for at least 20 slugs/plot).

3 Use of commercial slug fences

Speiser and Andermatt (1996) used commercially available slug fences (supplier, Thomas Pfau, Würenlos, Switzerland) to construct plots measuring 2.5 × 2.5 m to assess the ability of *P. hermaphrodita* to protect lettuce, Chinese cabbage and kale. These fences have a lip at the top designed to prevent entrance of slugs and no additional material was needed to prevent slug movement. If these plots are constructed in fields with a high slug population, there is no need to add additional slugs.

4 Wooden barriers

A very simple design was used by Parrella *et al.* (1985).

1. Make individual plots 30 × 30 cm by sinking wooden paneling to a depth of 2.5 cm into the soil.
2. Coat barriers with a mollusc repellent material. Parrella *et al.* (1985) used a 2.5 cm band of rock salt glued around the top of the arena. Strips (1 cm wide or more) of copper can also be used at the tops of barriers to prevent movement. Fluon or copper mesh, as described above could also be used.

C Field plots

When using field plots with natural populations of slugs it is important to pick a suitable site with a large slug population, generally in excess of 100 slugs/m^2.

1. Select a favorable site for slugs. Slug populations tend to be highest in sites with clay soils, and sites that are shaded, known to retain moisture, or which drain poorly. The cropping history of the field is also likely to influence the slug population. If previous crops are ones with a dense canopy or ones that produce much leaf litter and provide ample refuges for slugs (*e.g.*, oilseed rape, corn, field beans), slug populations are more likely to be high.
2. Monitor site for slug presence using bait traps. Place ten refuge traps (see section D3) in a transect across the proposed site. Put a small heap (approximately 5 g) of molluscicidal bait pellets under each trap.
3. Leave traps out in the field for 3 days, and then count numbers of dead slugs under, and in the vicinity of the traps. If means of four or more slugs are recorded per trap, the site can be considered suitable.
4. Establish field plots of at least 6 × 6 m with a central 2 × 2m sampling zone and add treatments. If all data are recorded from within the central zone, buffers between plots are not needed (Glen *et al.*, 1991). However, buffers of at least 1 m wide will allow easy access to plots and help prevent spray drift. A minimum of four replicate plots per treatment should be used.
5. Sample plots for numbers of slugs and slug damage from within the 2 × 2 m sampling zone.

D Methods for estimating size of mollusc populations

Generally, population assessments of slugs should not be made until 3 weeks after nematode application. However, assessments made before this can be useful in determining the percentage of the slug population that has been infected. In such cases assessments may be made at weekly intervals after applying nematodes, and numbers of slugs showing characteristic swelling associated with *P. hermaphrodita* infection (Figure 1.) can be recorded.

Snails frequently live in plant foliage and under points of shelter (*e.g.*, stones, leaf litter) at the soil surface. While many individual slugs may be active on the soil surface, the majority of the population may be within the soil. Thus, the methods chosen for assessing mollusc populations will depend on whether slugs or snails are under investigation. Surface searching of defined areas will give good estimates of snail populations. This method has been used for slugs (Barry, 1969) but is likely to underestimate population size. The shells of snails can be painted which enables mark-recapture studies to be done. Accurate assessments of slug populations require taking soil samples of defined size and extracting slugs by one of several methods.

Sampling slugs can be difficult, and members of the small slug-research community have yet to settle on a standard protocol. Thus, it is important that if any of the methods described below are used directly, or modified, the method should be clearly explained when experiments are written up. Furthermore, the best methods require either much initial expenditure (*e.g.*, soil flooding, Defined Area Traps) or are very time consuming (soil washing), and extension workers or researchers with only a minor interest in mollusc pests may not be able to justify the time or expense. For snail pests, hand searching of the crop and plots provides a good estimate of numbers. For slugs we recommend using either soil washing or soil flooding, or Defined Area Traps for population assessments. For both slugs and snails, population assessments ought to be combined with an assessment of mollusc damage. If these methods of population assessment cannot be used, damage assessments should be the principal method for assessing molluscicide efficacy. Data from refuge traps may be used to give an indication of species present and efficacy, as long as caution is applied to their interpretation. Slugs and snails present should be identified to species, using keys. There is no one key for mollusc pests. Examples of keys useful in Europe and the USA include Chichester and Getz (1973), Cameron *et al.* (1983) and Kerney and Cameron (1979).

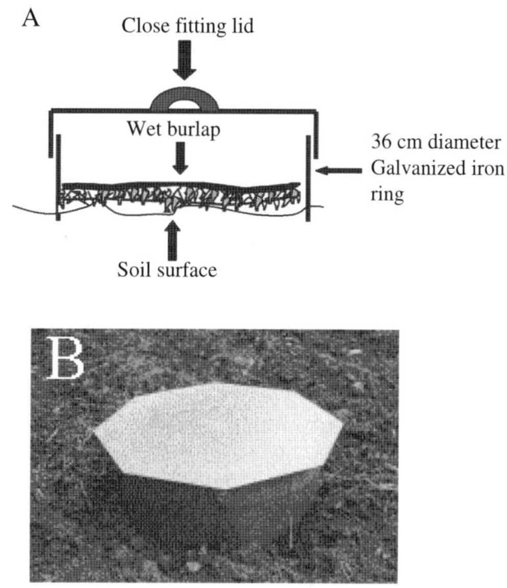

Figure 3. Defined Area Trap (DAT) A) vertical section through the trap showing component parts fully assembled and sunk into the ground for use and B) DAT trap in situ

1 Defined area traps

Defined Area Traps (DATs) were first described by Ferguson *et al.* (1989) then later by Ferguson and Hanks (1990), and have been further tested by Clements and Murray (1991). DATs were devised to provide a method that, unlike conventional refuge traps, can be used to gain an absolute estimate of slug population size per unit area. Unlike the soil washing or soil flooding techniques, the use of DATs does not necessitate the transport of large amounts of soil to the laboratory. Ferguson *et al.* (1989) compared DATs with soil flooding and found that the DATs had similar extraction efficiency to soil flooding but offered considerable savings in time and labor. In field experiments at least one DAT should be used per plot.

1. Make a ring of galvanized iron approximately 36 cm in diameter and 15 cm deep (Figure 3).
2. Add a close fitting steel lid or cover with wood and hammer the ring into the ground to a depth of 5 cm.
3. Trim the vegetation within the trap if it exceeds 5–6 cm in height to help prevent slug movement into or out of the traps during examination.
4. Place a piece of damp burlap sacking so that it covers all the ground within the trap.
5. Cover with a steel lid or a lid made of hardboard with the upper surface painted white to reflect heat. If a wooden lid is used, secure the lid to the ground using large pins.
6. Examine the DATs every 3 days, and remove and identify slugs. This process is repeated until no more slugs are found.

2 Soil sampling

a Hand sorting

Early attempts to assess slug populations relied on taking soil cores and hand sorting through the core to find slugs. However, these methods were found to give poor estimates of slug numbers with many smaller slugs being overlooked (Thomas, 1944). The problem is exacerbated because of the difficulty in hand sorting heavy clay soils, which often favor slugs. For people with only a casual interest in slug research and who have no soil washing or flooding facilities, this may be the only available option.

1. Take cores 10 cm diameter or wider to a depth of 10 cm.
2. Take at least four cores per plot.
3. Remove cores and carefully crumble and record numbers of slugs.

b *Flotation techniques*

These techniques are generally considered to be the most accurate method for assessing absolute population size of slugs. The ability to assess numbers of slug eggs is one of the key advantages of this method over soil flooding. Such techniques usually involve first breaking up the soil and passing it through sieves of various mesh sizes. The resulting soil samples are then placed into a solution with high specific gravity that causes slugs and eggs to float to the surface where they can be collected. South (1964) and Hunter (1968) used the following method with excellent results. However, the technique is very labor intensive.

1. Take at least one soil sample 30 × 30 × 10 cm deep per plot.
2. Break up soil samples using a water hose on a bank of three sieves (mesh size 8 mm, 2.5 mm and 0.8 mm). Slugs and eggs pass through the 8 mm sieve and collect on the lower two sieves.
3. Immerse each of the lower two sieves in a magnesium sulfate solution of relative density 1.17–1.20 (approximately 450 g $MgSO_4 \cdot 7H_2O$ dissolved in water and made up to 1 liter). Gently agitate the residues on the sieve. Organic matter including slugs and eggs float to the surface.
4. Wait 2 min and collect slugs and eggs floating on the surface.
5. Remove the sieve and examine the residues for remaining slugs and eggs.
6. Repeat steps 4 and 5 until no more slugs or eggs are found.

c *Cold water flooding*

South (1964) was the first to describe the cold water flooding technique. In this technique, soil samples of defined size are progressively flooded using a drip feed, over a period of up to 10 days. Slugs move up to the surface of the soil and can be collected on a daily basis. South's original method was devised for sampling slugs in turf and used samples of 30 × 30 × 10 cm deep. The samples were placed on their edge in plastic bins and flooded over a period of 5 days. Cold water flooding gives extraction efficiency similar to flotation techniques for juvenile and adult slugs (not eggs) but requires less labor. Thus, this technique has been modified by other workers and used extensively for arable soils (*e.g.*, Glen and Wiltshire, 1986) and woodland samples (Jennings and Barkham, 1975). However, the flooding procedure requires much space for the watertight troughs especially if many samples are to be taken. The following method of Glen and Wiltshire (1986) is recommended.

1. Take at least one soil sample 25 × 25 × 10 cm deep from each plot by sinking an aluminum frame into the soil and digging around the frame.
2. Transfer the soil into buckets with close fitting lids and holes at the base covered with 0.5 mm woven plastic mesh. This allows entrance of water but prevents the slugs from leaving.
3. Place the buckets in large watertight troughs and flood with a drip-feed consisting of a rubber hose with adjustable clamp, over a period of approximately 9 days.
4. Collect slugs forced to the surface as the water rises daily until samples are completely flooded.

3 *Refuge traps*

Refuge traps are commonly used in field studies as an indirect measure of population size because they are inexpensive and require little labor. Young *et al.* (1996) compared the relative merits of different trapping materials and trap baits. Refuge traps consist of a shelter for slugs that can be placed out in the field, and range from simple designs, to more complex ones. Traps may or may not contain some form of bait, and the baits may or may not be toxic. Rough cut bran or beer in a cup sunk into the ground can be used as bait to attract slugs to the trap. Alternatively, traps can contain toxic bait pellets (*e.g.*, methiocarb, thiodicarb metaldehyde or iron phosphate) which slugs eat and then die. The dead slugs can be counted under, and in the immediate vicinity of the trap. Simple slug traps can be made from roofing shingles (Byers and Bierlin, 1984) or inverted plant pot saucers (Glen and Wiltshire, 1986). More complex designs include the trap of Hommay and Briard (1988). These consist of moisture absorbing wadding sandwiched between two sheets of polyethylene, the upper one reinforced and colored silver (to reflect heat and keep the underside cool) and the lower one black and perforated to allow entrance of water. The traps are moistened first then pegged out in the field. A simple trapping method is as follows:

1. Place one inverted plant pot saucer in each plot without bait.
2. Examine the trap 3 days later and record slug numbers. This is best done early in the morning before the sun gets too hot and forces slugs underground.
3. Repeat assessments as needed by moving the trap to a different part of the plot for each assessment.

While trapping methods can be cheap and easy, there are drawbacks. Refuge traps do not give a measure of overall slug population. Numbers of slugs found under traps will be influenced by the amount of alternative shelter and food available, since slugs without food are more likely to travel further and thus, encounter the traps. If the traps contain bait, the attractiveness of the bait will depend on the availability of other foodstuffs. Refuge traps tend to selectively catch larger adults, under-representing numbers of juveniles in the population (Glen and Wiltshire, 1986). Furthermore, refuge traps seem particularly unsuitable for assessing slug numbers in experiments using the nematode parasite *P. hermaphrodita*. Wilson *et al.* (1994a) and Glen *et al.* (1994) found significantly more slugs under refuge traps in winter-wheat plots treated with *P. hermaphrodita* than in untreated plots. The percentage of infected slugs found under refuge traps in nematode-treated plots was significantly higher than that recorded in flooded soil samples from the same plots. Glen *et al.* (1994) suggested that this might result from a change in slug behavior following nematode infection and it is possible that slugs rest on the underside of traps to avoid contact with soil treated with *P. hermaphrodita* (Wilson *et al.*, 1999).

4 *Mark recapture*

Snails are easy to mark with enamel paints and mark recapture is an effective way of assessing populations. Slugs are more problematic. Slugs can be labeled by feeding them diets containing the radioisotopes ^{32}P or ^{131}I (Fretter, 1952; Francois *et al.*, 1965) or by inserting small radioactive wires made of ^{182}T (Moens *et al.*, 1965). More recently, methods by which slugs are freeze branded have been developed (Richter, 1976; Rollo *et al.*, 1983) and methods where dental injectors are used to inject dye (Alcian blue) into slugs (Hogan and Steele, 1986). These methods have not been compared with soil extraction techniques, and at present are not recommended.

E *Assessing crop damage*

Biological pesticides have many attributes different from those of chemical pesticides. In particular, they tend not to result in quick kills of the target pests. By assessing biological pesticides in the same way that we assess chemical pesticides, (*i.e.*, measuring pest population shortly after pesticide application), it is often possible to underestimate their efficacy. Pesticides are applied in order to protect crops and thus, the best measure of efficacy is the ability of the pesticide to reduce crop damage. Only in cases where the principal economic loss is by pest contamination (*e.g.*, some lettuce crops, Wilson *et al.*, 1995a) will assessing the residual slug population give an accurate measure of the pesticide's efficacy. In such cases several lettuce plants should be taken at random from each plot and examined thoroughly from the heart to the outer leaves and numbers of slugs recorded.

The type of damage caused by molluscs varies depending on the test crop and the mollusc species present and may include plant loss, plant damage and/or contamination. Care must be taken to distinguish mollusc damage from damage caused by insect or vertebrate pests.

1 *Lethal damage to seeds and seedlings*

If damage is predominantly to seeds (*e.g.*, wheat) or young seedlings (*e.g.*, oilseed rape) resulting in plant loss, counting numbers of established plants can assess slug damage.

1. Select two or three sites at random within the plot sampling zone.
2. Place a quadrat of known area (often $0.5m^2$) at each of the sites.
3. Count numbers of plants.

Alternatively, in row crops, rather than using quadrats it is possible to count numbers of established plants per unit length of drill row.

1. Throw a 0.5 m or longer cane at random into the plot.
2. Align the stick with the adjacent row.
3. Count numbers of plants along the length of the stick.
4. Repeat at least four times per plot.

2 Slug grazing on established plants

The method chosen depends largely on the extent of slug damage to the test crop.

1. If slug damage is slight, it is possible to record the proportion of plants that have suffered slug damage. Select plants (at least 20) at random in each plot and record presence or absence of slug damage.
2. If slug damage is more widespread, an assessment of total leaf area damage should be made. For most crops a five point scale of slug damage is suitable, $0 =$ no injury, $1 = 25\%$ eaten, $2 = 50\%$, $3 = 75\%$ eaten and $4 = 100\%$ eaten (Byers and Calvin, 1994). Alternatively estimate damage to the nearest 5 or 10%.
3. If damage is severe, measuring dry weight can be useful. Select at least 20 individual plants at random from each plot. If damage is largely confined to leaves, the roots can be removed from the plants prior to weighing so soil does not have to be cleaned from the roots. Oven-dry plants at 95 °C and weigh.

3 Combined measures of lethal and sublethal damage

Often it is convenient to assess overall slug damage by combining measures of lethal slug damage (plants destroyed) and grazing damage. Wilson *et al.* (1994a) in an experiment in wheat estimated slug damage in terms of undamaged plant equivalents.

1. Estimate mean numbers of plants per unit area as above – n
2. Estimate the mean proportion of leaf area damaged by slugs - a
3. Estimate the mean proportion of plants damaged by slugs – p
4. Calculate undamaged plant equivalents y with the formula $y = n(1 - ap)$.

4 Damage to fruit and tubers

In crops where fruit or tubers are damaged, (*e.g.*, strawberries or potatoes), assessments of presence or absence of slug damage should be recorded. Alternatively, numbers of individual holes caused by slug feeding can be recorded. Assessments should generally be made at the time of harvesting, and assessments should be made on several plants from each plot. It is difficult in such situations to measure the extent of area damaged or yield loss because extensive bacterial or fungal infection and subsequent decay often follow slug damage. If possible, it is best to supplement the above-described experimental assessments with assessments of economic relevance, (*e.g.*, crop yield or marketable value).

5 Acknowledgment

The author thanks Dr. Randy Gaugler for commenting on an earlier version of this manuscript.

6 Appendix

1. Method for monoxenization of *Phasmarhabditis hermaphrodita*

In order to mass-produce *P. hermaphrodita*, it is necessary to establish monoxenic cultures with a suitable growth promoting bacterium (*e.g.*, *Moraxella osloensis*).

1. Prepare sterile 3 cm diameter Petri dishes containing autoclaved lipid agar (10 g corn syrup, 5 g yeast extract, 25 g nutrient agar, 2.5 ml cod liver oil, 2 g $MgCl_2 6 \cdot H_2O$ in 1 liter distilled water).
2. Grow nutrient broth cultures of test bacterium for 24 h at 25 °C.

3. Spread one bacterial loop of the broth culture of one half of the 3 cm plates of lipid agar and incubate for 48 h at 25 °C.
4. Remove gravid hermaphrodites of *P. hermaphrodita* from an infected slug.
5. Using a mounted bristle to handle nematodes, change the nematodes through three changes of sterile Ringer's solution.
6. Place the nematodes in a sterile watch glass filled with a sterile solution of 0.02 % (w/v) sodium ethyl mercurithiosalicylate ("Thymerosal", Sigma, Poole, UK) and leave for 16 h at 10 °C.
7. Juveniles collected in this solution are transferred to 15 ml centrifuge tubes containing 10 ml of sterile Ringer's solution with 500 units/ml of benzyl penicillin and streptomycin sulfate and kept in this solution for 24 h at 10 °C.
8. Centrifuge at 90 × G for 5 min to concentrate the juveniles and pour off supernatant.
9. Re-suspend the juveniles in fresh sterile Ringer's solution without antibiotics.
10. Repeat 8 and 9 two more times to remove residual traces of antibiotics.
11. Place 10 of the juveniles on each 3 cm plate of lipid agar, close to the edge on the opposite side of the dish from the bacterial culture, such that nematodes have to cross 15 mm of sterile agar prior to reaching the test bacterium.
12. Incubate plates at 15 °C for approximately 3 weeks
13. After one week examine all plates for bacterial contamination introduced with the nematodes into the bacteria-free half of the plates. Discard any contaminated plates.

Once monoxenic cultures have been established in 3 cm Petri dishes they can be sub-cultured by cutting out a 1 cm^2 piece of agar containing nematodes and bacteria from the mature culture and placing it onto a 10 cm Petri dish of lipid agar. This should be done aseptically in a laminar flow hood to prevent contamination.

2. Mass cultivation method

Once monoxenic cultures have been established, it is necessary to bulk them up for field trials. Wilson *et al.* (1993b) used a modification of Bedding's (1984) method for rearing entomopathogenic nematodes.

1. Chop 650 g pig's kidney into small pieces, add 150 ml of water and homogenize in a Waring blender.
2. Melt 200 g of beef dripping in a large pan, add the kidney/water homogenate and stir over heat for 10 minutes.
3. Add 85 g of polyether-polyurethane foam chips no larger the 1 cm^3 into the warm medium and mix thoroughly
4. Fill 500 g conical flasks approximately 2/3 full of foam chips/medium mix, plug with non-absorbent cotton wool and autoclave for 25 min.
5. One day prior to addition of nematodes, add 10 ml of nutrient broth culture of the growth supporting bacterium to each flask. Add using a sterile pipette and distribute evenly over the foam surface, and incubate for 24 h at 25 °C.
6. Using sterile Ringer's solution, rinse all nematodes from a mature 10 cm diameter Petri dish culture into each flask using a laminar flow hood.
7. Incubate flasks at 15 °C until infective juvenile nematodes can be seen on the side of the flasks (approximately 3 weeks).
8. Transfer entire contents of each flask into a coarse 20 cm diameter soil sieve, place sieve in a 30 cm diameter plant pot saucer and fill with water such that the water just touches the bottom of the foam chips.
9. Leave for 24 h, after which time infective juvenile nematodes will have collected in the water.
10. Rinse nematodes from plant pot saucer through at 200 μm pore sieve, which will remove foam chips or spent medium from nematode suspension.
11. Pour nematodes onto a 50 μm pore sieve. Nematodes collect on the sieve and can be washed with fresh tap water to remove bacterial contaminants.
12. Wash nematodes off sieve into clean tap water and aerate at 10 °C until needed.
13. Each 500 ml flask will yield 1–2 million infective juveniles.

3. Assessing numbers of *P. hermaphrodita* within soil samples

Like entomopathogenic nematodes, *P. hermaphrodita* when used for biological control is applied as dauer larvae. Measuring persistence of *P. hermaphrodita* in soil is thus difficult as identifying dauer larvae to species within soil samples with a typical diverse nematode community requires specialized training and high quality microscopy. Unlike entomopathogenic nematodes, there is no equivalent to the *Galleria bait* method described in Chapter IV-5.

Recently a method has been devised to estimate numbers of *P. hermaphrodita* in soil samples using real time quantitative PCR (MacMillan *et al.*, 2006). While it is possible to extract DNA directly from soil samples, it is generally cheaper and more effective to extract the entire nematode community from the soil sample using standard methods, then extract the DNA from all nematodes. Numbers of *P. hermaphrodita* within the nematode sample can be estimated using qPCR run against a set of standards consisting of DNA extracted from known numbers of *P. hermaphrodita*. The method amplifies a fragment of the 18S ribosomal DNA genes that is know to be specific to *P. hermaphrodita* and if a dual labeled fluorescent probe is also used in the detection step, the morphologically indistinguishable *P. neopapillosa* will not be counted. Molecular biology protocols will vary between laboratories, and the guidelines below can be altered accordingly.

1. Take several soil samples (approx 10 g) from each plot.
2. Extract the entire nematode community using either Baermann funnels or a centrifugal flotation method (Hooper, 1986). If working in a nematology laboratory, use the labs' current protocol for nematode extraction.
3. Concentrate the nematodes by settling into a small volume (< 1 ml). The total genomic DNA can then be extracted from nematodes using proprietary kits (*e.g.*, QIAamp® DNA mini kits, Quiagen, Valencia, CA).
4. Prepare a set of *P. hermaphrodita* standards by adding 1, 30, 100, 300 and 1000 dauer larvae of *P. hermaphrodita* to micro-centrifuge tubes, and then extracting total genomic DNA as in step 3.
5. Prepare PCR tubes containing 12.6 μl proprietary PCR mix, (*e.g.*, Supermix®, Bio-Rad, Hercules, CA), 400nM forward primer (CGGGCTTAGTTTGTTGACT), 400 nM reverse primer (ACAACCATGATAGGC-CAATAGA), 200 nM dual labeled probe (FAM-TTCATCCGCTGAAGTCCGGAATTTT-TAMRA) and 2.5 μl of the DNA solution extracted above. It is advisable to add 400ng/μl bovine serum albumen to minimize PCR inhibition, particularly if DNA is extracted from soil directly. The entire reaction volume is then brought up to 25μl using sterile distilled water.
6. Load the test samples and the standards into the thermocycler and run the cycle as followed. Initial denaturation 95 °C for 3 minutes, then 30 cycles of 95 °C for 15 seconds, 60.5 for 1 minute. The cycle is completed with a final annealing/extension step of 56 °C for 10 minutes.

The threshold values when fluorescence is first detected are recorded during the cycle and test values can be run against the standard curve. This step is usually calculated by the thermocycler and will give an estimate of *P. hermaphrodita* genome equivalents in each reaction tube.

7 References

Arias, R. O. and Crowell, H. H. 1963. A contribution to the biology of the gray garden slug. *Bull. Southern Calif. Acad. Sci.* **62**, 83–97.

Baker, G. H. 2002. Helicidae and Hygromiidae as pests in cereal crops and pastures in southern Australia. *In* “Molluscs as Crop Pests” (G. M. Barker, Ed.), pp 193–216. CABI Publishing, Wallingford, UK.

Barker, G. M. 2001. “The Biology of Terrestrial Molluscs”, CABI Publishing, Wallingford, UK.

Barker, G. M. 2002a. “Molluscs as Crop Pests”, CABI Publishing, Wallingford, UK.

Barker, G. M. 2002b. Gastropods as pests in New Zealand pastoral agriculture, with emphasis on Agriolimacidae, Arionidae and Milacidae. *In* “Molluscs as Crop Pests” (G. M. Barker, Ed.), pp 361–424. CABI Publishing, Wallingford, UK.

Barker, G. M. 2004. “Natural Enemies of Terrestrial Molluscs”, CABI Publishing, Wallingford, UK.

Barry, B. D. 1969. Evaluation of chemicals for the control of slugs on field corn in Ohio. *J. Econ. Entomol.* **62**, 1277–1299.

Bedding, R. A., 1984. Large scale production, storage and transport of the insect parasitic nematodes *Neoaplectana* spp. and *Heterorhabditis* spp. *Ann. Appl. Biol.* **104**, 117–120.

Begley, J. W. 1990. Efficacy against insects in habitats other than soil. *In* “Entomopathogenic Nematodes in Biological Control” (R. Gaugler, and H. K. Kaya, Eds.), pp. 215–231. CRC Press, Boca Raton, FL.

Brooks, W. M. 1968. Tetrahymenid ciliates as parasites of the gray garden slug. *Hilgardia* **39**, 205–276.

Byers, R. A. and Bierlin, D. L. 1984. Continuous alfalfa: invertebrate pests during establishment. *J. Econ. Entomol.* **77**, 1500–1503.

Byers, R. A. and Calvin, D. D. 1994. Economic injury levels to field corn from slug (Stylommatophora: Agrolimacidae) feeding. *J. Econ. Entomol.* **87**, 1345–1350.

Cameron, R. A. D., Evesham, B. and Jackson, N. 1983. A field key to slugs of the British Isles. *Field Stud.* **5**, 807–824.

Charwat, S. M., Davies, K. A., and Hunt, C. H. 2000. Impact of a rhabditid nematode on survival and fecundity of *Cernuella virgata* (Mollusca: Helicidae). *Biocontrol Sci. Technol.* **10**, 147–155.

Chichester, L. F. and Getz, L. L. 1973. The terrestrial slugs of northeastern North America. *Sterkiana* **50**, 11–42.

Clements, R. O. and Murray, P. J. 1991. Comparison between defined-area slug traps and other methods of trapping slugs in cereal fields. *Crop Protect.* **10**, 152–154.

Coupland, J. B. 1995. Susceptibility of helicid snails to isolates of the nematode *Phasmarhabditis hermaphrodita* from Southern France. *J. Invertebr. Pathol.* **66**, 207–208.

Dean, W. W., Mead, A. R. and Northey, W. T. 1970. *Aeromonas liquefaciens* in the giant African snail *Achatina fulica. J. Invertebr. Pathol.* **16**, 346–351.

Ducklow, H. W., Tarraza, J. R., and Mitchell, R. 1980. Experimental pathogenicity of *Vibrio parahaemolyticus* for the schistosome-bearing snail *Biomphalaria glabrata. Can. J. Microbiol.* **26**, 503–506.

Ester, A. and Wilson, M. J. 2005. Application of slug-parasitic nematodes. *In* "Nematodes as Biocontrol Agents" (P. S. Grewal, R-U. Ehlers and D. I. Shapiro-Ilan, Eds.), pp 431–444. CABI Publishing, Wallingford, UK.

Ferguson, C. M., Barrat, B. I. P., and Jones, P. A. 1989. A new technique for estimating density of the field slug, *Deroceras reticulatum* (Muller). *In* "Slugs and Snails in World Agriculture" (I. F. Henderson, Ed.), pp. 331–337. British Crop Protection Council, Farnham, UK.

Ferguson, C. M. and Hanks, C. B. 1990. Evaluation of defined-area trapping for estimating the density of the field slug, *Deroceras reticulatum* (Muller). *Ann. Appl. Biol.* **117**, 451–454.

Francois, E., Riga, A. and Moens, R. 1965. Estimation des populations de *Agriolimax reticulatus* Muller au moyen de la technique de marquage au radiophosphore ^{32}P, et recapture. *Parasitica* **24**, 63–78.

Fretter, V. 1952. Experiments with ^{32}P and ^{131}I on species of *Helix, Arion* and *Agriolimax. Qtly. J. Microsc. Sci.* **93**, 133–146.

Glen, D. M. and Moens, R. 2002. Agriolimacidae, Arionidae and Milacidae as pests in West European cereals. *In* "Molluscs as Crop Pests" (G. M. Barker, Ed.), pp 271–300. CABI Publishing, Wallingford, UK.

Glen, D. M., Wilson, M. J., Brain, P., and Stroud, G. 2000. Feeding activity and survival of slugs, *Deroceras reticulatum* exposed to the rhabditid nematode, *Phasmarhabditis hermaphrodita*: a model of dose response. *Biol. Control* **17**, 73–81.

Glen, D. M., Wilson, M. J., Pearce, J. D., and Rodgers, P. B. 1994. Discovery and investigation of a novel nematode parasite for biological control of slugs. *In* "Proceedings of the 1994 Brighton Crop Protection Conference - Pests and Diseases", pp. 617–625. British Crop Protection Council, Farnham, UK.

Glen, D. M., Wilson, M. J., Hughes, L. A., Cargeeg, P., and Hajjar, A. 1996. Exploring and exploiting the potential of the rhabditid nematode *Phasmarhabditis hermaphrodita* as a biocontrol agent for slugs. *In* "Slug and Snail Pests in Agriculture" (I. F. Henderson, Ed.), pp. 271–280. British Crop Protection Council, Farnham, UK.

Glen, D. M. and Wiltshire, C. W. 1986. Estimating slug populations from bait-trap catches. *In* "Proceedings of the 1986 British Crop Protection Conference - Pests and Diseases", pp. 1151–1158. British Crop Protection Council, Farnham, UK.

Glen, D. M., Wiltshire, C. W., and Butler, R. C. 1991. Slug population changes following molluscicide treatment in relation to distance from edge of treated area. *Crop Protect.* **10**, 408–412.

Grewal, P. S., Grewal, S. K., Tan, L., and Adams, B. J. 2003. Parasitism of molluscs by nematodes. *J. Nematol.* **35**, 146–156.

Hammond, R. B. and Byers R. A. 2002. Agriolimacidae and Arionidae as pests in conservation tillage soybean and maize cropping in North America. *In* "Molluscs as Crop Pests" (G. M. Barker, Ed.), pp 301–314. CABI Publishing, Wallingford, UK.

Hogan, J. M. and Steele, G. R. 1986. Dye-marking slugs. *J. Mollusc. Stud.* **52**, 138–143.

Hommay, G. M. and Briard, P. 1988 Apport du piegeage dans le suivi des peuplements de limaces en grande culture. *Haliotes* **18**, 55–74.

Hunter, P. J. 1968. Studies on slugs of arable ground: I. sampling methods. *Malacologia* **6**, 370–389.

Jaworska, M. 1993. Laboratory infection of slugs (Gastropoda: Pulmonata) with entomopathogenic nematodes (Rhabditida: Nematoda). *J. Invertebr. Pathol.* **61**, 223–224.

Jennings, T. J. and Barkham, J. P. 1975. Slug populations in mixed deciduous woodland. *Oecologia* (Berlin) **20**, 279–286.

Kaya, H. K. and Stock, S. P. 1997. Techniques in insect nematology. *In* "Manual of Techniques in Insect Pathology" (L. Lacey, Ed.), pp. 281–324. Academic Press, San Diego, CA.

Kerney, M. P. and Cameron, R. A. D. 1979. "A Field Guide to the Land Snails of Britain and N. W. Europe." Collins, London.

Li, P. S., Deng, C. S., Zhang, S. G., and Yang, H. W. 1986. Preliminary tests on *Steinernema glaseri* infecting the snail, *Oncomelania hupensis*, an intermediate host of *Schistosoma japonica. Chinese J. Biol. Control* **2**, 50–52.

MacMillan, K., Blok, V., Young, I, Crawford, J. C., and Wilson, M. J. 2006. Quantification of the slug parasitic nematode *Phasmarhabditis hermaphrodita* from soil samples using real time qPCR. *Int. J. Parasitol.* **36**, 1453–1461.

Mead, A. R. 1961. "The Giant African Snail: A problem in Economic Malacology," Univ. of Chicago Press, Chicago, IL.

Mengert, H. 1953. Nematoden und Schnecken. *Z. Morphol. Oekol. Tiere*. **41**, 311–349.

Moens, R., François, E., Riga, A., and van den Bruel, W. E. 1967. A mechanical barrier against terrestrial gastropods. *Parasitica* **23**, 22–27.

Moens, R., François, E., Riga, A., and van den Bruel, W. E. 1965. Les radioisotopes en écologie animale. Première informations sur le comportement de *Agriolimax reticulatus* Muller. *Med. Landbouw. Opzoekings. Staat Gent* **3**, 1810–1823.

Moens, R. and Glen, D.,M. 2002. Agriolimacidae, Arionidae and Milacidae as pests in West European oilseed rape. *In* "Molluscs as Crop Pests" (G. M. Barker, Ed.), pp. 425–440. CABI Publishing, Wallingford, UK.

Morand, S. and Hommay, G. 1990. Redescription de *Agfa flexilis* Dujardin, 1845 (Nematoda: Agfidae) parasite de l'appareil génital de *Limax cinereoniger* Wolf, 1803 (Gastropoda, Limacidae). *Syst. Parasitol.* **15**, 127–132,

Morand, S. and Petter, A. J. 1986. *Nemhelix bakeri* n. gen. n. sp. (Nematoda, Cosmocercidae) parasite de l'appareil génital de *Helix aspersa* Müller (Gastropoda, Helicidae). *Can. J. Zool.* **64**, 2008–2011.

Morand, S., Wilson, M. J. and Glen, D. M. 2004. Nematodes (Nematoda) parasitic in terrestrial gastropods. *In* "Natural Enemies of Terrestrial Molluscs" (G. M. Barker, Ed.), pp. 525–558. CABI Publishing, Wallingford, UK.

Parrella, M. P., Robb, K. L., and Morishita, P. 1985. Snails and slugs in ornamentals. *Calif. Agric.*, Jan-Feb, 6–8.

Port, G. R. and Ester, A. 2002. Gastropods as pests in vegetable and ornamental crops in Western Europe. *In* "Molluscs as Crop Pests" (G. M. Barker, Ed.), pp. 337–352. CABI Publishing, Wallingford, UK.

Port, C. M. and Port, G. R. 1986. The biology and behavior of slugs in relation to crop damage and control. *Agric. Zool. Rev.* **1**, 255–299.

Raut, S. K. 2004. Bacterial and non-microbial diseases in terrestrial gastropods. *In* "Natural Enemies of Terrestrial Molluscs" (G. M. Barker, Ed.), pp. 525–558. CABI Publishing, Wallingford, UK.

Raut, S. K. and Barker, G. M. 2002. *Achatina fulica* Bowdich and other Achatinidae as pests in tropical agriculture. *In* "Molluscs as Crop Pests" (G. M. Barker, Ed.), pp. 55–114. CABI Publishing, Wallingford, UK.

Richter, K. O. 1976. A method for individually marking slugs. *J. Mollusc. Stud.* **42**, 146–151.

Rollo, C. D., Vertinsky, I. B., Wellington, W. G. and Kanetkar, V. K. 1983. Alternative risk-taking styles: The case of time-budgeting strategies of terrestrial gastropods. *Res. Pop. Ecol. (Kyoto)* **25**, 321–335.

Rueda, A., Caballero, R, Kaminsky, R., and Andrews, K. L. 2002. Vaginulidae in Central America, with emphasis on the bean slug *Sarasinula plebeia* (Fischer). *In* "Molluscs as Crop Pests" (G. M. Barker, Ed.), pp. 115–144. CABI Publishing, Wallingford, UK.

Sakovich, N. J. 2002. Integrated management of *Cantareus asperses* (Müller) (Helicidae) as a pest of citrus in California. *In* "Molluscs as Crop Pests" (G. M. Barker, Ed.), pp. 353–3360. CABI Publishing, Wallingford, UK.

Sanderson, G. and Sirgel, W. 2002. Helicidae as pests in Australian and South African grapevines *In* "Molluscs as Crop Pests" (G. M. Barker, Ed.), pp. 255–270. CABI Publishing, Wallingford, UK.

Singer, S., VanFleet, A., Viel, J. J., and Genevese, E. E. 1997. Biological control of the zebra mussel *Dreissena polymorpha* and the snail *Biomphalaria glabrata*, using Gramicidin S and D and molluscicidal strains of *Bacillus*. *J. Ind. Microbiol. Biotechnol.* **18**, 226–231.

South, A. 1964. Estimation of slug populations. *Ann. Appl. Biol.* **53**, 251–258.

South, A. 1992. "Terrestrial Slugs, Biology, Ecology and Control." Chapman and Hall, London, UK.

Speiser, B. and Andermatt, M. 1996. Field trials with *Phasmarhabditis hermaphrodita* in Switzerland. *In* "Slug and Snail Pests in Agriculture" (I. F. Henderson, Ed.), pp. 419–424. British Crop Protection Council, Farnham, UK.

Symondson, W. O. C. 1990. Chemical confinement of slugs; an alternative to electric fences. *J. Mollusc. Stud.* **59**, 259–261.

Tan, L. and Grewal, P. S., 2001a. Infection behaviour of the rhabditid nematode *Phasmarhabditis hermaphrodita* to the grey garden slug *Deroceras reticualtum*. *J. Parasitol.* **87**, 1349–1354.

Tan, L., and Grewal, P. S. 2001b. Pathogenicity of *Moraxella osloensis*, a bacterium associated with the nematode *Phasmarhabditis hermaphrodita*, to the slug *Deroceras reticulatum*. *Appl. Environ. Microbiol.* **67**, 5010–5016.

Tan, L. and Grewal, P. S. 2002. Endotoxin activity of *Moraxella osloensis* against the grey garden slug, Deroceras reticulatum. *Appl. Environ. Microbiol.* **68**, 3943–3947.

Tan, L. and Grewal, P. S. 2003. Characterization of the First Molluscicidal Lipopolysaccharide from *Moraxella osloensis*. *Appl. Environ. Microbiol.* **69**, 3646–3649.

Thomas, D. C. 1944. Discussion on slugs. II. Field sampling for slugs. *Ann. Appl. Biol.* **31**, 160–164.

Trevet, I. W. and Esslemont, J. M. 1938. A fungous parasite of the eggs of the gray field slug. *J. Quecket Microscopical Club* 4th series **1**, 1–3.

Wilson, M. J. and Grewal, P. S. 2005. Biology, production and formulation of slug-parasitic nematodes. *In* "Nematodes as Biocontrol Agents" (P. S. Grewal, R-U. Ehlers and D. I. Shapiro-Ilan, Eds.), pp. 421–430. CABI Publishing, Wallingford.

Wilson, M. J., Glen, D. M., and George, S. K. 1993a. The rhabditid nematode *Phasmarhabditis hermaphrodita* as a potential biological control agent for slugs. *Biocontrol Sci. Technol.* **3**, 503–511.

Wilson, M. J., Glen, D. M., George, S. K., and Butler, R. C. 1993b. Mass cultivation and storage of the rhabditid nematode *Phasmarhabditis hermaphrodita*, a biocontrol agent for slugs. *Biocontrol Sci. Technol.* **3**, 513–521.

Wilson, M. J., Glen, D. M., George, S. K., Pearce, J. D., and Wiltshire, C. W. 1994a. Biological control of slugs in winter wheat using the rhabditid nematode, *Phasmarhabditis hermaphrodita*. *Ann. Appl. Biol.* **125**, 377–390.

Wilson, M. J., Glen, D. M., George, S. K., and Wiltshire, C. W. 1994b. Mini-plot field experiments using the rhabditid nematode, *Phasmarhabditis hermaphrodita*, for biocontrol of slugs. *Biocontrol Sci. Technol.* **4**, 103–113.

Wilson, M. J., Glen, D. M., Hughes, L. A., Pearce, J. D., and Rodgers, P. B. 1994c. Laboratory tests of the potential of entomopathogenic nematodes for the control of field slugs (*Deroceras reticulatum*). *J. Invertebr. Pathol.* **64**, 182–187.

Wilson, M. J., Glen, D. M., George, S. K., and Hughes, L. A. 1995a. Biocontrol of slugs in protected lettuce using the rhabditid nematode *Phasmarhabditis hermaphrodita*. *Biocontrol Sci. Technol.* **5**, 233–242.

Wilson, M. J., Glen, D. M., George, S. K., and Pearce, J. D. 1995b. Selection of a bacterium for the mass production of *Phasmarhabditis hermaphrodita* as a biocontrol agent for slugs. *Fund. Appl. Nematol.* **18**, 419–425.

Wilson, M. J., Glen, D. M., Pearce, J. D., and Rodgers, P. B. 1995c. Monoxenic culture of the slug parasite *Phasmarhabditis hermaphrodita* with different bacteria in solid and liquid phase. *Fund. Appl. Nematol.* **18**, 159–166.

Wilson, M. J., Hughes, L. A., Hamacher, G. M., Barahona, L. D. and Glen, D. M. 1996. Effects of soil incorporation on the efficacy of the rhabditid nematode *Phasmarhabditis hermaphrodita* as a biological control agent for slugs. *Ann. Appl. Biol.* **128**, 117–126.

Wilson, M. J., Hughes, L. A., Jefferies, D., and Glen, D. M. 1999. Slugs, (*Deroceras reticulatum* and *Arion ater*. agg.) avoid soil treated with the rhabditid nematode *Phasmarhabditis hermaphrodita*. *Biol. Control* **16**, 170–176.

Wilson, M. J., Hughes, L. A., Hamacher, G. M., and Glen, D. M. 2000. Effects of *Phasmarhabditis hermaphrodita* on non-target molluscs. *Pest Man. Sci.* **56**, 711–716.

Young, A. G., Port, G. R., Craig, A. D., James, D. A., and Green, T. 1996. The use of refuge traps in assessing risk of slug damage: a comparison of trap material and bait. *In* "Slug and Snail Pests in Agriculture" (I. F. Henderson, Ed.), pp. 133–140. British Crop Protection Council, Farnham, UK.

SECTION VIII

TRANSGENIC PLANTS

Chapter VIII-1

Evaluating transgenic plants for suitability in pest and resistance management programs

Michael A. Caprio
Department of Entomology and Plant Pathology Mississippi State University Mississippi State, MS 39762, USA

Douglas V. Sumerford
USDA-ARS
Corn Insects and Crop Genetics Research Unit
Ames, IA 50010, USA

1 Introduction

The use of recombinant DNA techniques to develop plants that express exogenous proteins has the potential to revolutionize agriculture. Though transgenic plants will undoubtedly be important in the production of pharmaceuticals and the improvement of agronomic qualities of crops, this chapter will focus on traits that are intended to assist in management of insect populations. At present the only traits that have been successfully commercialized for pest population suppression have been proteins derived from the bacterium, *Bacillus thuringiensis (Bt)*, primarily for the control of a range of lepidopteran species such as European corn borer *(Ostrinia nubilalis)*, tobacco budworm *(Heliothis virescens)* and cotton bollworm *(Helicoverpa zea)* (Höfte and Whiteley, 1989; Fischhoff, 1996). In addition, the Colorado potato beetle (*Leptinotarsa decemlineata*) has been targeted by one product. These transgenic crops represent a paradigm shift for insect pest control. In theory, the expression or synthesis of these toxic proteins can be custom tailored to specific agroecosystems, and the expression can be limited to certain time periods, specific plant tissue, perhaps even to specific types of feeding damage (Gasser and Fraley, 1989). Expression levels can be altered through the use of promoters and screening of insertion events to develop plants that express high or low levels of toxin. Multiple toxins/traits can be stacked or pyramided into the same plant, either increasing the level of control of targeted pests or increasing the range of beneficial traits combined into single varieties. The utilization of these plants can significantly reduce the use of conventional pesticides in agricultural systems such as cotton in the mid-southern USA, which in turn can increase the number of generalist biological control agents present in transgenic fields (compared to conventionally treated fields) (Luttrell *et al.*, 1995; Mascarenhas and Luttrell, 1997, Shelton *et al.*, 2002). We focus primarily on the plant and on operational factors in this chapter. Questions regarding insect biology itself such as fitness costs associated with resistance or the dominance of those costs, while important in assessing risk, are not covered here.

While transgenic plants offer many unique opportunities for the management of pest populations, they also present new challenges. Perhaps the greatest ecological challenge is the potential for the rapid evolution of resistance in pest populations to the toxins

L.A. Lacey and H.K. Kaya (eds.), Field Manual of Techniques in Invertebrate Pathology, 769–789.

(McGaughey, 1985; Shelton *et al.*, 1993; Whalon and McGaughey, 1993; Caprio, 1994, 1998; Roush, 1994; Tabashnik, 1994; Alstad and Andow, 1995; Bauer, 1995; Gould *et al.*, 1995; Kennedy and Whalon, 1995). Because many strains of transgenic plants express toxins in all plant parts throughout the season (constitutive expression), all individuals in local pest populations are exposed to selection. Simulation models suggest that pest populations could evolve resistance in as little as 1–2 years under such worst case conditions (ILSI-HESI, 1998; http://www.msstate.edu/Entomology/PGjava/ILSI model.html, Gould *et al.*, 1997; Roush, 1997; Tabashnik, 1997). Thus, transgenic plants must be evaluated not only on their capabilities for managing pest populations, but also on their compatibility with proactive resistance management programs (ILSI HESI, 1998; FIFRA Scientific Advisory Panel Subpanel on *Bacillus thuringiensis (Bt)* Plant-Pesticides and Resistance Management, 1998).

Developing appropriate resistance management programs for transgenic plants is complicated because the perceived benefits of the transgenic crop and the costs of the evolution of resistance to that crop vary with different stakeholders (conventional and organic growers, seed and technology companies, environmentalists, etc.). The balance between immediate use of transgenic plant technology and long term sustainability cannot be established solely on a scientific basis, but must include sociological inputs (which it receives from the scientific community, general public, environmental groups; all mediated by, in a regulatory manner, the EPA). Certainly one goal of resistance management should be to improve the efficiency of strategies, for example by ecological manipulations, thereby reducing the magnitude of this division. As the efficiency of refuges is increased, the expected rate of resistance evolution decreases while the economic burden placed on growers in terms of refuge size may also decrease. Whatever the resistance management strategy chosen, successful implementation of that strategy will depend on our ability to effectively evaluate and monitor both the performance of transgenic crops (in particular the levels of protein expression) and the frequencies and genetic characteristics of resistance alleles present within field pest populations. Efficient bioassay techniques will be required to monitor these parameters across regions where transgenic crops are grown (Bolin *et al.*, 1998; Andow and Hutchinson, 1998).

2 Resistance management

Resistance management strategies must be tailored for each agroecosystem, and the pests that move among the crops that comprise that system. Resistance management strategies that are appropriate in one ecosystem dominated by highly dispersive insects may be inappropriate in other ecosystems where the targeted insects are less mobile.

A *Resistance management alternatives*

One resistance management strategy for transgenic plants emphasizes limiting the rate of selection by utilizing the lowest level of expression possible while maintaining adequate pest control. The other primary resistance management option is to develop plants that express high levels of toxin (commonly referred to as a high dose), and combine these plants with a refuge of non-expressing plants.

The "low dose" strategy relies on low selection coefficients to maintain susceptibility to the toxin for as long as possible while maintaining economically acceptable levels of control (Georghiou and Taylor, 1977; Tabashnik and Croft, 1982; Roush, 1994; Caprio, 1994; Gould, 1998). While intuitively easy to understand, this strategy may be quite complicated to implement. When multiple pest species are present, a low dose for one insect may be high for another, resulting in either rapid resistance evolution or inadequate control of at least one targeted insect. In the case of *Bt*-transgenic plants, a low dose strategy was not chosen by industry in part because of sublethal effects on target insects and in part because of unacceptable and/or unpredictable control in crops. Low doses of *Bt*-endotoxins retard growth rates of susceptible genotypes of many targeted pests (*e.g.*, Sims *et al.*, 1996; though see Liu *et al.*, 1995). This can result in a temporal asynchrony developing between the emergence of susceptible and resistant adults. Developmental delays induced by sublethal effects can hasten resistance development or retard resistance when overwintering also influences insect phenology (Peck *et al.*, 1999). Interactions between sublethal effects and natural enemies can also hasten resistance evolution (Gould *et al.*, 1991; Johnson and Gould, 1992).

The primary alternative to the low dose strategy is to combine a high dose of the toxin in transgenic plants with a refuge of plants that express no toxin. This strategy assumes that the relatively large number of susceptible pests produced in refuges will mate with the few resistant homozygotes that emerge in the transgenic crop. The heterozygotes produced are presumably killed by the high dose expressed in the transgenic plants. An essential element of this strategy is that resistance is functionally recessive and the dose is high enough so that fewer than 5% of heterozygotes survive (Roush, 1994, 1997). As with the low dose alternative, the implementation of this strategy is complicated when there is a complex of pests with varying levels of toxin susceptibility feeding on a single transgenic crop (Gould, 1994). The high dose in Monsanto's BollGard® *Bt*-cotton that kills > 99% of susceptible *H. virescens*, a key pest of cotton in the mid-southern USA, may kill less than 95% of susceptible *H. zea*, another major pest in the same region (Benedict *et al.*, 1993). If resistant larvae feeding on transgenic plants have delayed development compared to susceptible larvae on non-transgenic refuge plants, the temporal asynchrony and assortative mating that results may, as with the low dose strategy, influence the rate of resistance evolution (Peck *et al.*, 1999).

B Determining if an event is a high dose event

A central issue of the high dose strategy is determining what constitutes a sufficiently high dose. The 1998 Environmental Protection Agency Scientific Advisory Panel (EPA-SAP) subpanel report on *Bacillus thuringiensis* plant-pesticides and resistance management (FIFRA Scientific Advisory Panel Subpanel on *Bacillus thuringiensis (Bt)* Plant-Pesticides and Resistance Management, 1998) recommended that a high dose should be one that exceeds, by 25-fold, the toxin concentration needed to kill 99% of susceptible individuals at an identical stage. The subpanel recognized that this number must be subject to modification as additional data become available, and that it was possible that an allele might be found conferring sufficiently high fitness to heterozygotes to allow survival at the 25-fold dose. Current empirical data suggest that this level of toxin expression is not consistently high enough to cause high mortality among heterozygotes with known *Bt* resistance alleles (Figure 1). We recommend that a higher value, 50-fold, be adopted. At the 25-fold level, 23% of the currently reported fitness values for heterozygotes exceed the standard.

If the standard were raised so that a high dose product would produce a titer of toxin 50-fold greater than the LC_{50} of the susceptible strain, only a single population (5.8%) would exceed the critical value. This analysis assumes that the slope of the dose mortality curves for the susceptible and F_1 populations are equal so that the ratios calculated from LC_{50}s are identical to the ratios from the LD_{99}s.

How can it be determined if a theoretical "high dose" [*i.e.*, a dose that is 25-fold greater than the LC_{99} of the homozygous susceptible genotype (SS)] is expressed under ideal conditions? If this dose is not expressed under field conditions, what are the alternatives? Is the LC_{99} (lethal concentration that kills 99% of the population) or the EC_{99} (the concentration that reduces weight of the treated sample to 1% that of the untreated control), a suitable alternative? The EPA-SAP recommended five different methods for determining if a cultivar meets this high dose requirement. Because each method has practical difficulties, it is worthwhile to briefly discuss them separately.

1. Conducting serial dilution bioassay with artificial diet overlaid with lyophilized plant proteins. This method has utility in situations where plant expression of the transgenic insecticidal protein is high and sensitivity of the target insect to the protein is also high. When target insect sensitivity is lower (such as that of *H. zea* toward commercial transgenic cotton cultivars expressing Cry1Ac protein at maximum levels of 10–20 μg/g (fresh tissue weight), then it is difficult to add the required amount of tissue into, or on top of, the diet

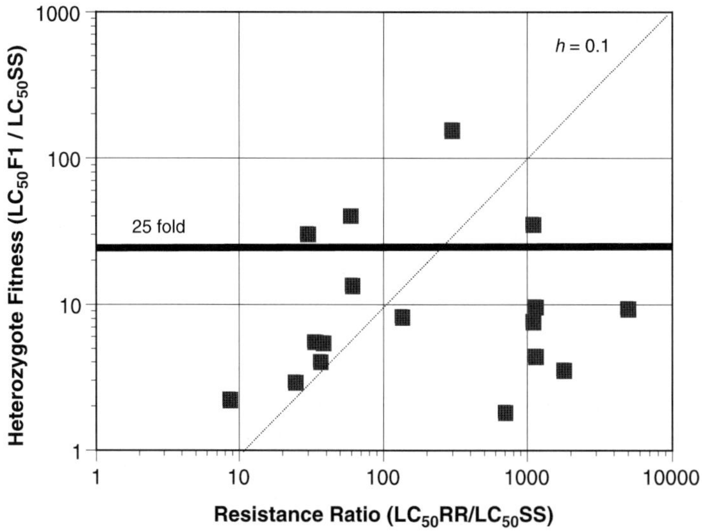

Figure 1. The ratio of heterozygote fitness to susceptible fitness as a function of the overall level of resistance. The dark horizontal line is the 25-fold ratio recommended by the EPA-SAP as the minimal level to determine if the titer produced by a crop is high enough to qualify the product as a high dose product. All points to the left of the dashed line have dominance values that exceed 0.1 (on a 0–1 scale). (Data compiled from: McGaughey and Beeman, 1988; Gould *et al.*, 1992; Tabashnik *et al.*, 1992; Gould *et al.*,, 1995; Rahardja and Whalon, 1995; Chaufaux *et al.*, 1997; Liu and Tabashnik, 1997; Tang *et al.*, 1997; Huang *et al.*, 1999)

to achieve an LC_{50} concentration. Excessively high amounts of lyophilized control (non-transgenic) tissue added to diet usually result in unacceptable stunting and/or mortality due to effects from secondary plant compounds and allelochemicals (Lukefahr and Martin, 1966; Shaver and Lukefahr, 1969).

Another potential problem is the assumption that the LC_{50} dose of a transgenic protein in the whole plant matrix is exactly equivalent to the LC_{50} dose of the same protein mixed into an artificial diet. This is almost certainly untrue. Sublethal doses of transgenic proteins in artificial diet assays stunt larvae but mortality, much of which occurs among neonate larvae, is minimized because the relative humidity in diet cups or wells approaches 100% thereby reducing larval death from desiccation. Mortality of larvae under the stress of *Bt* protein exposure is probably synergized by the effects of desiccation (S. R. Sims, unpublished observations).

2. Assaying plant cultivars with 25-fold lower levels of endotoxin expression. This proposed procedure is complicated because transgenic protein expression in both the commercial plant event and lower expression event is dynamic, varying with plant structure and growth stage. The approach has the best chance of working if foliage from young (pre-reproductive) plants is used and the plants are "isolines" or plant lines having the same genetic background.
3. Sampling large numbers of the commercial cultivar in the field to determine that fewer than 1 in 10,000 larvae derived from natural oviposition survive. This approach is confounded by a lack of information on the size of the initial cohort (eggs) placed on the commercial cultivar. In a season with extremely high populations of the target insect, huge numbers of ova might be placed on the commercial plants resulting in a relatively large number of surviving larvae despite little change in the *Bt* susceptibility of the population. Therefore, a large initial cohort might result in a false inference regarding less than expected "high dose" expression.
4. Sampling large numbers of the commercial cultivar in the field to determine that fewer than 1 in 10,000 artificially infested larvae survive. This approach might have utility only if a sufficiently diverse sample of larvae can be obtained. This means, at a minimum, that there must be a collection of a large number of females and a sample of progeny from each female should be tested on the commercial transgenic cultivar.

5. Assaying with older instar larvae of the target insect with a 25-fold greater tolerance to the toxin than neonate larvae. The details of this assay make it problematical. First, it is unlikely that a later instar of any specific target species will be 25-fold less susceptible to the toxin or even close to that value. Second, "tolerance" is not clarified as being dependent or independent of larval body weight. Larger larvae of some susceptible insects do not show decreased tolerance to *Bt* insecticidal proteins if increased body weight is considered (Ferro and Lyon, 1991).

C Spatial aspects of refuge placement

The high dose/refuge strategy can be adapted to many different agroecosystems by varying the spatial distribution of the refuge. In agroecosystems where larval movement among plants is rare (and adults are not affected by the toxin), a seed-mixture may be an option. A percentage of the transgenic seed sold in this case is non-transgenic, so that each field would be a fine-grained mosaic of transgenic and non-transgenic plants. This strategy minimizes the spatial segregation between susceptible adults emerging from refuge plants and resistant adults emerging from transgenic plants, reducing the risks associated with assortative mating. If the insect stages that are susceptible to the toxin move between plants or between tissues of plants that vary in expression, the effective dominance of alleles that confer resistance to the toxin and the risk of resistance evolution may be increased (Mallet and Porter, 1992). Heterozygous larvae are presumably able to move from non-transgenic plants onto transgenic plants and survive at higher rates earlier in development than susceptible larvae, and this movement results in a "selection window" that can significantly increase the rate of resistance evolution. Shelton *et al.* (1998) also suggested that larval movement in combination with high dose plants may act to decrease the effective size of the refuge. As larvae age, they generally become increasingly tolerant of *Bt* endotoxins (though exceptions exist, see Liu *et al.*, 1995). If empirical evidence suggests that larvae do move between plants, then the seed-mixture strategy should be avoided and care should be taken to reduce the interface across which larvae could move between isolated refuges and transgenic fields.

Semi-isolated refuges in various spatial patterns are an alternative to seed mixtures (Caprio, 2001). The segregation of transgenic and non-transgenic plants reduces the dominance shifts associated with larval movement, but introduces new problems. If the distance between transgenic and nontransgenic plants becomes too great, the probability of matings between resistant individuals emerging from transgenic crops increases. When multiple species feed on a transgenic crop, the optimal distance between refuges and transgenic crops for one insect may be suboptimal for another pest. A potential sociological problem with the isolated refuge design is that non-transgenic plants can be easily identified and growers might reduce the effectiveness of this strategy by treating refuges with chemical pesticides.

If the proportion of refuge habitat is small and the semi-isolated refuges are distributed in a fine-grained pattern, most of the eggs from females emerging in the refuges will be oviposited on transgenic plants, where they presumably have little chance of surviving. For example, diamondback moth females, *Plutella xylostella*, were found to oviposit at random in a mixture of transgenic and non-transgenic broccoli (Tang *et al.*, 1999). In this situation, the transgenic crop acts as a sink for the refuge population. Here the refuge strategy can fail because of the ovipositional patterns of adults. Although it decreases the proportion of the population that mates at random, increasing the isolation of the refuges may increase their overall effectiveness (Caprio *et al.*, 2004). Caprio and Suckling (1995) found that isolated refuges in a simulated transgenic apple-clover system were much more effective than the internal understory refuges, though adult light brown apple moth from the latter were assumed to mate at random with adults emerging in the transgenic apples. Models suggest that there may be an optimal level of isolation (Figure 2). Not all models will identify this intermediate level (Peck *et al.*, 1999) unless there is independent population regulation in each patch and sufficiently broad disperal parameters are examined (which may exceed reasonable expectations for specific species). Resistance evolution can be accelerated by source-sink dynamics if

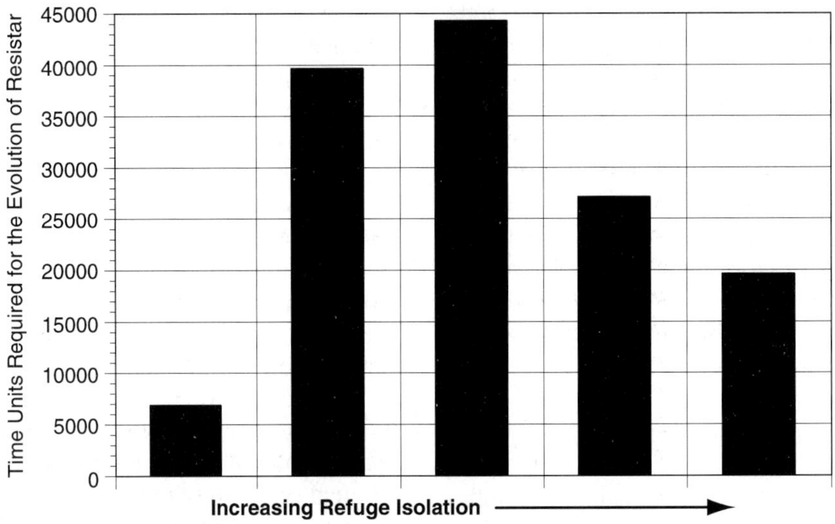

Figure 2. The time required for resistance evolution with various levels of isolation between refuges and transgenic fields. The results are based on a spatially explicit model of resistance in *Helicoverpa zea* (Caprio, 1998)

the habitat is too fine-grained, or by a high degree of non-random mating if the habitat is too course-grained. The proportion of the habitat allocated to an untreated refuge is dependent on the time frame over which one wishes to utilize a particular toxin and upon the risk associated with the loss of the toxin if efficacy declines due to resistance.

D *Policy and the initial deployment of transgenic crops*

These questions depend in part on sociological factors, and a strictly scientific answer is not possible. Andow and Hutchinson (1998) proposed two alternative approaches for developing refuge size recommendations. The first is a proactive resistance management policy, in which a conservative interpretation of available data is used to develop a resistance management strategy. This approach would deploy a sufficiently large refuge from the initial release of the transgenic plants to make it unlikely that resistance would evolve within the desired time horizon. This policy would require that most testing of transgenic plants, and evaluations of insect ecology, be complete before release of the toxins. These evaluations would likely emphasize quality and thoroughness rather than speed.

A second approach would be a relatively rapid introduction of transgenic plants combined with an adaptive resistance management policy. This policy would adopt a less conservative interpretation of available data but utilize constant monitoring and evaluation of test results to alter resistance management as necessary. As resistance alleles are identified in the field, refuge size requirements might be adjusted, depending upon the genetic characteristics and the frequency of the observed alleles. The adaptive resistance management policy is dependent on our ability to identify rare resistance alleles while still at low frequencies. The technology and monitoring procedures required to accomplish this task have not been demonstrated in the field. Though estimates of rare allele frequencies have been accomplished in several insects (Gould *et al.*, 1997; Andow *et al.*, 1998), the process is laborious and expensive and a cautious initial approach should probably be adopted. Because of population substructure and genetic variation between populations, the frequency of resistance alleles may be unevenly distributed across populations, and multiple sites across the area where the transgenic crop is grown would require monitoring. The exact pattern would be dependent on population substructure and gene flow. Random events, in particular local extinction events of resistance alleles, make it more difficult to predict where resistance will initially evolve. This complicates resistance monitoring because it limits our ability to concentrate monitoring efforts on areas of perceived high risk. In general, because many populations must be monitored, this policy would emphasize the use of rapid screening techniques. These techniques should be sufficiently sensitive to detect alleles at low frequencies.

E Choosing an appropriate resistance management strategy

The choice of a resistance management strategy will depend on many factors, including characteristics of, (1) the toxin itself (toxicity spectrum, sublethal effects), (2) the expression pattern of the toxin in a plant (including effects from insertion events, promoters and genetic manipulation of the gene), (3) the ecology of the targeted pests (larval movement, adult mating patterns, population substructure, alternative hosts), (4) expected utilization patterns of the gene (both in the crop immediately being considered, but also other crops if insects susceptible to the toxin move between crops), and (5) the expected characteristics of resistance (based on laboratory selection experiments and prior field experience with other insects or resistance to similar toxins).

1 The toxin

The toxicity spectrum of the toxin and the number of different insects affected by the toxin can indirectly impact resistance management options by increasing the number of targeted insects. A targeted insect is currently susceptible to the toxin but one in which the evolution of resistance would have a negative impact to growers or other stakeholders (ILSI-HESI, 1998). Targeted insects may include a broader range of species than those the crop is actually labeled to control. If certain parameters are required to implement the strategy (*e.g.*, high mortality of susceptible insects), then this goal must be met for each targeted insect or the strategy must be adapted to delay resistance in the insect with the highest risk of resistance.

2 Sublethal effects

Sublethal effects may also limit resistance management options for a toxin. For example, low doses of *Bt* proteins may slow the growth of susceptible larvae. A temporal discontinuity may then occur in the development of susceptible and resistant individuals resulting in assortative mating between these genotypes. For partially recessive traits, this non-random mating can increase the risk of resistance evolution (Peck *et al.*, 1999). If susceptible larvae develop more slowly than resistant larvae, they may spend more time in stages attacked by natural enemies, and this differential survivorship could also alter the risk of resistance evolution (Johnson and Gould, 1992).

3 Pattern of expression

Both protein toxin and non-transgenic protein expression generally decline with plant age, particularly as plants senesce following bloom. Pests that attack plants early in the season will be exposed to higher doses than other pests (or later generations) that infest the same plants later in the season. If a high level of expression is required to control a pest population, late season pests may not be adequately managed. The issue is further complicated when expression in different tissues of the same plant may vary at different rates through the season. If there is tissue specific expression of the toxin, the impact of larval movement among tissues on resistance evolution must also be evaluated (Mallet and Porter, 1992).

4 Pest ecology

Pest ecology must be considered in evaluating the utility of resistance management programs for specific transgenic events. As previously noted, larval movement between expressing and non-expressing plants can alter the effective dominance of recessive resistance traits and increase the risk of resistance evolution via such traits (Mallet and Porter, 1992).

5 Movement and mating patterns of adults

Movement and mating behavior can also be important ecological considerations. If adults mate before dispersing, an increase in assortative mating between genotypes that predominate in refuges and transgenic fields could alter the effective dominance of resistance alleles. Models suggest that the impact of non-random mating is greatest when the functional dominance of the resistance allele is recessive ($h < 0.1$, ILSI 1998, http://www.msstate.edu/Entomology/PGjava/ILSImodel.html). When a high dose/refuge strategy is utilized, oviposition patterns may also be important because of their impact on population dynamics. If a small proportion of the habitat is maintained as a refuge, and the mixture of refuges and transgenic fields is too fine grained (*i.e.*, each adult experiences both habitats in proportion to their frequency), most of the eggs from adults emerging in refuges would be laid in transgenic fields where they would have little chance of surviving. The transgenic fields in this case act as sinks for the refuge populations, and, if the habitat is too fine-grained, can lead to collapse of the refuge populations and rapid resistance evolution.

Once an appropriate resistance management strategy has been chosen given the unique characteristics of the agroecosystem, techniques must be developed and implemented to monitor resistance allele frequencies in all targeted pests.

3 Risk assessment modeling

Ultimately it is recommended that some modeling effort be used to evaluate the combined effects of the transgenic event in the plant, the insect biology and the landscape/agroecosystem in which it is planned to utilize the transgenic plant. There is considerable controversy regarding such models, and their complexity can range from simple, several variable models (*e.g.*, May and Dobson, 1986) to complex, multi habitat, multi-species models (Caprio and Hoy, 1994, 1995). The goals of such models can vary from understanding the evolutionary process that leads to resistance to the assessment of risk of the potential for the development of resistance within given time horizons. It is the latter type of model that is most useful in developing policy with regard to effective deployment of transgenic plants. These models must not only identify mean time to resistance, but also estimate the variance about that estimate to provide an adequate estimate of risk. The variance may include demographic stochasticity (unless the population size of resistance alleles is large enough so that these effects are unimportant), as well as variance generated by uncertainty in parameter values used in the model. The relative importance of these two variance components will vary with insect biology, the ecosystem, and our knowledge. Note that the latter can be changed and lead to changes in risk assessments.

Stochastic variance is primarily related to the population size of resistance alleles, which is in turn related to overall population size and resistance allele frequencies. Other factors, such as dispersal rates and migrant pool size in metapopulations may also be important (Whitlock and McCauley, 1990). When a risk assessment model is run repeatedly without parameter uncertainty, an estimate of the variance due to stochastic effects on resistance evolution is obtained (Figure 3A). Note that this is the minimal amount of variance that can be obtained without changing the model. The distribution that results is the best estimate of time to resistance that can be

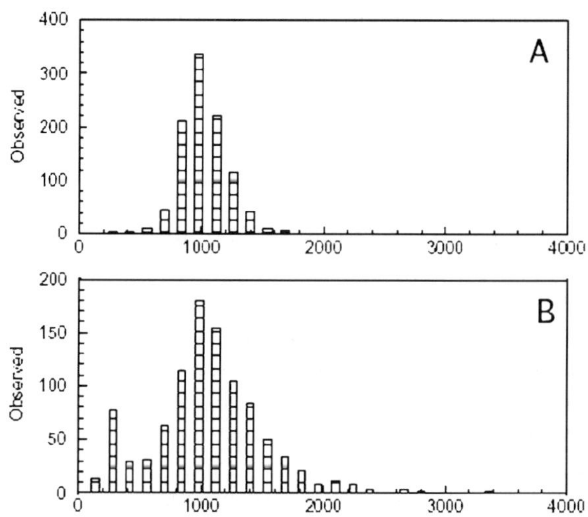

Figure 3. Distributions of time until product failure for simulations of western corn rootworm resistance to adulticidal sprays with A) stochastic variability alone and B) with stochastic variability and 10% uncertainty in each of 19 different parameters (Caprio *et al.*, 2006)

obtained, as we assume in this model that we know all the parameters perfectly. This figure is a good example of why the distribution of time to resistance can be important. While the mean is a good estimate of the overall performance of the model (and is a good measure for understanding the effects of different factors on the evolution of resistance), it provides no indication of the likelihood that resistance will evolved in shorter times. Using the distribution in Figure 3A, it is easy to set minimal time horizons over which you would like susceptibility to products to be maintained, and estimate the probability of product failure in shorter time frames.

Parameter uncertainty is more difficult to incorporate into risk assessment models. Conventionally this was accomplished by developing a series of scenarios, ranging from worst case to best case scenarios. The disadvantage of this approach is that little assistance is given in determining the likelihood of each scenario. If the worst case scenario uses the worst likely values for all parameters, how likely is it that all parameters will have those values in the field? The first step is to develop descriptions of the uncertainty we have in each parameter to be included in the analysis (Vose, 2000). These distributions may be any appropriate shape that adequately describes our uncertainty. They can range from a uniform distribution (implying no prior knowledge), to normal or triangular distributions, or even user supplied functions (Evans and Olson, 2002). The model is then run a number of times (usually 1000–5000 times), and new parameter values chosen from the appropriate distributions for each instance or iteration. The result is a distribution of times to resistance that is a reflection of both stochastic events and parameter uncertainty (Figure 3B). These distributions can then be used to determine the likelihood of undesirable events and guide adoption of resistance management strategies to limit the risk of such events to acceptable levels.

This risk assessment technique has advantages and disadvantages over conventional scenario modeling. It can be time consuming (weeks are often required to obtain a suitable number of simulation instances), it requires a model that is a reasonable representation of reality and is capable of handling the range of simulations without error, and humans tend to underestimate risk and make the distributions of parameter uncertainty too narrow (Evans and Olson, 2002). The technique does, however, put scenarios into a reasonable probabilistic framework. The equivalent of the worst case scenario has some probability of occurring in the 1000 runs or instances of the model, but its likelihood will be adjusted by our uncertainty distributions. In other words, the worst case scenario is put into the probabilistic context of other likely outcomes. Risk assessment modeling can also produce a type of sensitivity analysis, though it differs in several important ways from a conventional sensitivity analysis. Sensitivity analysis is normally done by altering one parameter while holding all other parameters constant and determining the effect on model results. In contrast, after completing the simulations, a risk assessment will have data on the individual values of the parameters and the observed outcomes, but against a background of changes at all other parameters. The data can be analyzed using regression, principal components, Bayesian model averaging or other techniques to identify critical parameters. The techniques allow estimation of interactions between parameters, which is not possible with conventional sensitivity analysis. In contrast to conventional sensitivity analysis, the results are also conditioned on our uncertainty. If we are absolutely certain in parameter's value there will be no variance in that parameter and it will not enter into the sensitivity analysis. A conventional sensitivity analysis examines structural relationships in the model, while a risk assessment will condition those relationships on our uncertainty in parameter values. The latter is more relevant to identifying areas of biological systems that require additional research.

Risk assessment models offer the opportunity to formalize our lack of knowledge about agricultural ecosystems and to determine the impact of that uncertainty on our assessment of the risk of undesirable events occurring in unacceptable time horizons. These assessments are not infallible. They are dependent on having good models that incorporate the important biological components. The uncertainty distributions may be subjective and tend to underestimate risk. The assessments do, however, incorporate all available knowledge into a cohesive framework. They allow for incorporation of input from many different sources and stakeholders. As new knowledge becomes available, it can be integrated into the uncertainty distributions or the model itself and the assessment revised. In policy development, where decisions often have to be made immediately, on the basis of incomplete knowledge, risk assessment models can improve our estimates of the potential hazards or advantages of releasing transgenic varieties.

4 Monitoring insect populations for tolerance to transgenic plants

We know that the evolution of insect resistance to transgenic plants is a threat, and that appropriate resistance management strategies can mitigate this threat. The rate of evolution to toxins produced by transgenic plants is dependent on the genetic architecture of the resistance trait (relative resistance of homozygotes, dominance, number of loci, interactions between loci) and the initial allelic frequencies present within insect populations (Comins, 1979; Caprio and Tabashnik, 1992; Tabashnik, 1994; Gould *et al.*, 1997; Andow and Alstad, 1998; Gould, 1998). One approach to delay the evolution of resistance is to adapt the resistance management strategy to the current genetic characteristics of the insect population (Andow and Hutchinson, 1998; Andow and Alstad, 1998). To do this, it is necessary to track the evolutionary changes of resistance alleles in targeted pest populations through monitoring efforts.

In order for resistance management strategies to be optimally effective, the frequency of resistance alleles must be very low (Comins, 1979; Roush and Miller, 1986; Tabashnik, 1994; Sims *et al.*, 1996; Gould, 1998). Furthermore, some resistance management approaches work best when the gene action is mostly recessive (*e. g.*, high dose strategy with refuge). If dominant alleles are detected, changes in the resistance management strategy may be warranted.

The goal of resistance monitoring is to aid researchers in evaluating resistance management strategies by determining baseline levels of susceptibility to transgenic insecticides, detecting changes in the frequencies of resistance alleles, detecting the presence of dominant alleles, and documenting control failures due to resistant insects. During the past two decades, entomologists have documented many cases of resistance in field populations of economic pests (Wolfenbarger *et al.*, 1982; Dennehy and Granett, 1984; Gunning *et al.*, 1984; Staetz, 1985). Most of these monitoring efforts were begun to detect the prevalence of resistance alleles before control failures occurred. To accomplish this goal, dose-response studies were (and still remain) the major method used to monitor changes in resistance levels (Robertson *et al.*, 1984; Roush and Miller, 1986; ffrench-Constant and Roush, 1990). However, dose-response studies are not sensitive enough to detect resistance alleles that are rare (Roush and Miller, 1986). By the time significant changes in the LC_{50}/LC_{95} have occurred in wild populations, resistance has essentially evolved (Figure 4, Roush and Miller, 1986). Roush and Miller (1986) concluded that monitoring with a diagnostic dose (single dose that can distinguish between susceptible and resistant phenotypes) was more efficient and effective than dose-response studies. For the same sampling effort from field populations, it is more informative to test all collected larvae on one diagnostic dose instead of dividing them among many doses. With regard to the use of a diagnostic-dose assay, Roush and Miller (1986) found that larger samples of larvae were required for dose-mortality

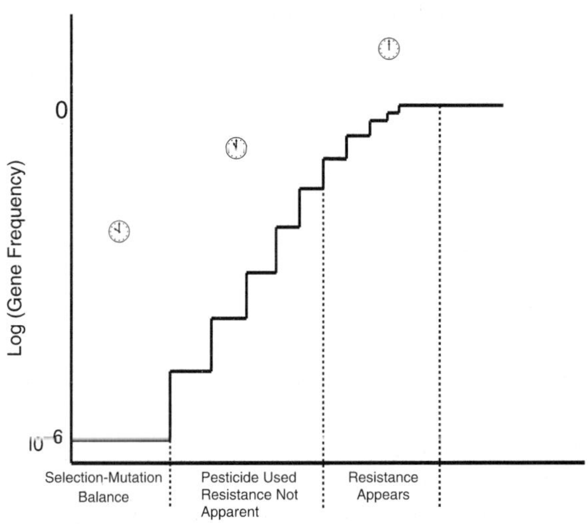

Figure 4. The three major phases in the evolution of resistance. The current focus in resistance monitoring is the development of technologies to detect changes in resistance allele frequencies earlier in phase 2

studies in order to obtain the same probability of detecting a resistant individual.

Even when a single diagnostic dose is used to monitor resistance alleles, the probability of false negatives at low allelic frequencies may be unacceptable, especially if the resistance trait is recessive (Roush and Miller, 1986; Andow and Alstad, 1998; Gould and Tabashnik, 1998). For a single-locus trait that is completely recessive, the probability of finding a resistant individual will be one in a million when the initial allelic frequency is 10^{-3}. If 1000 larvae were tested at the diagnostic dose, the researcher would only have a 0.1% chance of finding at least one resistant individual. For a completely dominant trait, the researcher's odds would improve (13.5% chance of finding no resistant larvae, N = 1000). Even though the probability of sampling a resistant larva is much better when the trait is dominant, the expectation of a false negative is still great at low allelic frequencies.

Because of the limitations of both the dose-mortality analyses and diagnostic dose studies, other approaches to monitor rare resistance alleles have been evaluated (Gould *et al.*, 1997; Andow and Alstad, 1998; Andow and Hutchinson, 1998; Andow *et al.*, 1998). Debate about the merits of resistance-monitoring methods primarily focus on the difficulties of detecting resistance alleles at low initial frequencies, the cost of monitoring, and its logistical plausibility (Roush and Miller, 1986; Gould *et al.*, 1997; Gould and Tabashnik, 1998; Andow and Alstad, 1998). Below we describe the relative merits and problems with these alternative methods.

5 Methods to monitor resistance alleles

A *Recovery of resistance alleles*

When colonies of highly resistant insects are available, they can be used to screen wild individuals for resistance alleles by the "allele-recovery method" (Gould *et al.*, 1997). Gould *et al.*, (1995) selected for high levels of tolerance to the Cry1Ac toxin in a field-collected colony of *H. virescens*. The YHD2 strain LC_{50} was >2000X greater than the susceptible strain LC_{50}. In a classical genetic analysis, Gould *et al.*, (1995) demonstrated that a major gene with mostly recessive inheritance controlled a major portion of the resistance trait. Heckel *et al.* (1997) used an allozyme-linkage map to locate quantitative-trait loci affecting Cry 1Ac tolerance in YHD2. They found that one linkage group accounted for approximately 80% of the variance in log weights of backcross individuals, and the tolerance locus/loci associated with this linkage group had a mostly recessive mode of inheritance.

A series of crosses were set up to screen wild individuals for the presence of resistance by mating them to YHD2. The allele-recovery method uses a diagnostic-dose assay, paired with single-family analyses, to improve the chances of detecting and isolating recessive alleles occurring at low frequency. The crosses take advantage of the mostly recessive, single locus nature of Cry 1Ac tolerance in YHD2. Gould *et al.* (1997) used a diagnostic dose that could distinguish heterozygous from homozygous resistant individuals. The mating design is illustrated in Figure 5. The major assumptions of this technique are that: (1) the same locus identified in the selected strain will govern resistance in wild populations of *H. virescens*, (2) resistance alleles from the laboratory strain and field individuals will have similar expression, and (3) that the genetic characteristics of the resistance allele (R′) from the wild males will be similar to the resistance allele from YHD2 females. When the allelic frequency is low in the field, heterozygous individuals with the resistance allele greatly outnumber resistant

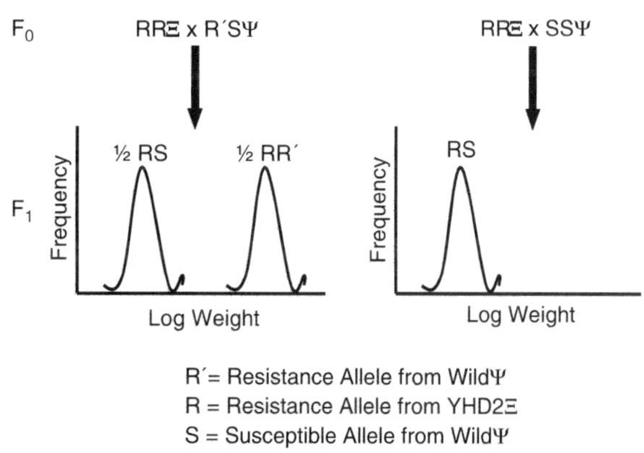

Figure 5. The mating design used by Gould *et al.*, (1995) to screen wild individuals of *H. virescens* for the presence of resistance by mating them to a resistant (YHD2) strain

homozygotes. To detect one heterozygous individual from the field would require at least 1000 matings when the frequency of the resistance allele is 10^{-3}.

Gould *et al.* (1997) screened $\sim$ 50 larvae each from 1025 single-pair matings. Four of these families produced larvae that were larger (weighed more) than the remaining families. These "outlier" families were sib mated to produce F_2 progeny. If the wild males that produced these outlier families possessed a resistance allele R, their larger F_1 offspring would be homozygous resistant (RR) while their small offspring would be heterozygotes (RS). Gould *et al.* (1997) tested this prediction by crossing sibs based on their growth rate on diet containing a diagnostic dose of the toxin (1/3 largest larvae (LL) and 1/3 smallest larvae (SL)). Single-pair matings confirmed that offspring from LL × LL crosses grew significantly faster on diet containing Cry1Ac than the larvae from SL × SL crosses.

The technique of Gould *et al.* (1997) improves the chance of detecting resistance traits encoded by recessive or partially recessive alleles by isolating resistance alleles within inbred families. In order to employ this technique, however, it is necessary to have a resistant laboratory strain. If such strains are available, it would be possible to estimate the initial frequency of a resistance allele and track changes in its frequency as the transgenic plants are used by growers (Gould *et al.*, 1997). Resistance alleles at other loci could also be detected if their expression were dominant. The detection of dominant traits, however, would only make the estimate of the allelic frequency based on a recessive model more conservative. The development of new *Bt* proteins and non-*Bt* insecticidal proteins will make it more difficult to isolate and select strains of insects (such as Gould's YHD2 *H. virescens* strain) that are highly resistant to single or multiple proteins. The allele-recovery method requires large numbers of families and a diagnostic dose that clearly distinguishes heterozygous from homozygous-resistant individuals. In the Gould *et al.* (1997) study, > 50, 000 larvae were sampled. The resources required to maintain and study > 1000 families will place severe economic and logistical limits on the utility of this procedure.

B F2 screen

Resistant strains are seldom available and the genetic mechanism(s) of resistance traits is usually unknown prior to starting a resistance-monitoring plan. Andow and Alstad (1998) designed a method, the "F_2 screen", to avoid these problems. The purpose of the F_2 screen is to maintain genetic variation among female isolines and to concentrate homozygous genotypes so that they may be detected using plant tissue or a diagnostic-dose assay (Andow and Alstad, 1998). Mated females possess four haploid genomes; two from their own genetic material and two from the male with whom they mated. Therefore, mated females collected from the wild represent a random sample of four alleles for a single-locus resistance trait. With the F_2 screen, the offspring (F_1 generation) of each mated female are sib-mated to produce F_2-isofemale families. Approximately 1/16 of the F_2 larvae will be homozygous for each of the grandparental alleles. For families beginning with one resistance allele from one of the grandparents, at least 1/16 of the F_2 larvae should be homozygous resistant. The F_2 screen, therefore, allows the detection of recessive and partially recessive resistance alleles within and among F_2 families. If many families are screened, the allelic frequency can be estimated from the number of F_2 families where a resistance allele is detected (see Andow and Alstad (1998) for details about statistical analysis).

Andow *et al.* (1998) used the F_2 screen to detect alleles affecting resistance to Cry1Ab protein in a population of *O. nubilalis*. In a sample of 91 female isolines, they determined (with 95% confidence) that the frequency of the resistance alleles was < 0.013. A larger number of isolines would have improved the precision of their estimate and this precision would be required to adaptively manage resistance to the Cry1Ab toxin.

As with the allele-recovery method, the F_2 screen utilizes information from family analyses. The strengths of the F_2 screen include. (1) the ability to detect recessive and partially recessive resistance alleles, (2) the collection of genetic information from wild populations before resistant lab colonies are created, and (3) the ability to apply rigorous statistical confidence intervals to estimates of allelic frequencies. In addition, families that produce resistant larvae may be saved for further analyses. Further, analyses could also determine if more than one resistance mechanism is operating within populations. The disadvantage of the F_2 screen is that many isolines must be screened. Andow *et al.* (1998) report that if no resistance alleles are found among 1200 family lines, it can be concluded with 95% confidence that the initial frequency of the resistance allele is less than 10^{-3}. Therefore, the F_2 screen requires significantly more effort per insect than the diagnostic-dose assay (Andow *et al.*, 1998). However, Andow *et al.* (1998) conclude that improvements in collecting, rearing, and handling methods could reduce the cost and labor involved in screening F_2 families of *O. nubilalis*. Schneider (1999) noted that the confidence limits calculated by Andow *et al.* (1998) were overly conservative, and that only 750 lines need to be analyzed for that same level of confidence.

The F_2 screen does not make assumptions about the genetic similarities between resistant alleles in laboratory strains and field populations. In fact, it does not require a resistant strain of the target insect. It does, however, assume that field-collected females have only mated once and that inbreeding depression does not interfere with the accuracy and precision of the allele-frequency estimate. While this assay is a very interesting approach for evaluating initial gene frequency, it is clearly impractical for routine monitoring. For example, the 5 authors of the research report using this procedure could only test a total of 91 female lines of *O. nubilalis* (Andow *et al.*, 1998). One advantage of this assay is that it could readily be used for isolating resistant strains that could potentially be used for monitoring using the allele recovery method (Gould *et al.*, 1997).

C Monitoring sentinel fields

A less labor-intensive approach to monitoring is to plant small plots of susceptible and transgenic plants. These plots can be easily sampled for resistant individuals. In conjunction with a laboratory bioassay using a diagnostic dose, collections of insects from the two plant types can also be compared for their relative tolerances to the toxin. However, when resistance is controlled by a single locus with recessive inheritance, monitoring sentinel fields is limited in the same way as the diagnostic-dose assay; *i.e.*, by the time resistant individuals are detected, the resistance allele may already be at a high frequency. An additional problem is "off-types", plants from the transgenic seed stock that do not express the toxin, generating false positives (Andow and Hutchinson, 1998). Accidents or mix-ups during seed packaging can also lead to the inclusion of non-transgenic seed in bags of transgenic seed.

D Monitoring using larval growth inhibition assays

Sims *et al.* (1996) presented a larval growth inhibition assay approach for monitoring the sensitivity of Lepidoptera to *Bt* insecticidal proteins. *H. virescens* and *H. zea*, major lepidopteran pests targeted for control by transgenic cotton, were used to compare traditional dose-response mortality assays with a growth inhibition assay. Larval growth inhibition assays using sublethal Cry1Ac protein concentrations were considerably more sensitive than dose-response mortality assays. EC_{99} values were 0.058 μg/ml (0.030–0.086) for *H. virescens* and 28.8 μg/ml (−7.4 − 65.1) for *H. zea*. These estimates were considerably lower (114-fold less for *H. virescens*, 463-fold less for *H. zea*) than the corresponding LC_{99} estimates for the Cry1Ac protein.

Growth inhibition assays were easy to set-up and read and could readily deliver a diagnostic dose allowing for visual discrimination of resistant (generally 2nd or 3rd instars after 7 days) from susceptible (living 1st instar or dead larvae) phenotypes. The ability of a larval growth assay, combined with a diagnostic dose, to unambiguously separate resistant from susceptible insects was validated using a Cry1Ac protein resistant strain of *H. virescens* (YHD2) and F_1 hybrids (YHD2 × susceptible strain).

E Larval feeding disruption monitoring schemes

Bailey *et al.* (1998) used a larval feeding disruption assay to accurately identify *H. virescens* larvae resistant to the Cry1Ac protein. The assay distinguished both the highly resistant YHD2 *H. virescens* strain (Gould *et al*.,, 1995) as well as F_1 hybrids resulting from crossing YHD2 with a susceptible *H. virescens* strain. The assay was scored by measuring the number of fecal pellets produced by larvae exposed to diagnostic doses of Cry1Ac incorporated into a blue-colored indicator diet. Susceptible larvae exposed to diagnostic doses of CylAc protein in the diet fed very little and produced few or no blue fecal pellets compared to heterozygous or homozygous larvae that produced many blue fecal pellets. The assay can also be used to distinguish between the nearly identical larvae of *H. zea* and *H. virescens* since *H. zea* is less susceptible to Cry1Ac.

F Field monitoring using transgenic events with lower expression levels of the transgenic protein

This approach has significant practical and functional potential. In a very real sense, the transgenic plant administers the diagnostic dose. At least two plant lines might be used; one line producing protein at a relatively low level and the other line at a considerably higher level. Low protein levels are appropriate for insects sensitive to *Bt* insecticidal protein such as *H. virescens* in cotton or *O. nubilalis* in corn. Low protein expression could make reduced susceptibility, inherited as a recessive or partially recessive trait, functionally dominant and therefore detectable. Higher protein levels are useful for locations where the pest complex includes less susceptible species. In the USA, the best known of the less susceptible species is *H. zea*, that readily infests both cotton and corn. Many non-commercialized transgenic lines are available with insecticidal protein levels exceeding those of commercial varieties. These high expressing transgenic lines often have only minor agronomic shortcomings that would not be expected to compromise their potential utility as monitoring tools. Infestation of the plants could be achieved from either natural oviposition or artificial inoculation Surviving 2nd or 3rd instar larvae would then be collected and susceptibility confirmed using a feeding disruption assay (Bailey *et al*., 1998). Development of larvae is completed on standard artificial diet, and progeny from the emerging adults are available for determining the resistance level compared to a susceptible strain.

An alternative to this approach is to grow and infest the transgenic monitoring plants under more controlled conditions of a greenhouse or growth chamber. Although this is less natural than a field plot, it would be easier to maintain cohorts of plants of consistent growth stage and insecticidal protein concentration. Infestations would require field collection of eggs or gravid females.

Some potential shortcomings to this approach should be detailed. First, the entire process is dependent on the cooperation and support of industry. Monitoring will require rather large quantities of seed from appropriate transgenic plant lines that possess precisely known levels of insecticidal proteins. Because protein levels can vary in different plant tissues and at different plant ages (Greenplate *et al*., 1998; J. T. Greenplate, personal communication) the details of this variation will need to be analyzed and understood. Quantitation of insecticidal protein levels will require the use of validated ELISA or bioassay (Sims and Berberich, 1996). In addition, seeds from the chosen plant lines must be increased in a systematic way during the growing season or in a winter nursery. This seed will then be used for monitoring by the company of origin or sent to cooperators for monitoring in the field. Considerable quality control regarding seed and plant quality and uniform production of insecticidal proteins within and between growth seasons is required. Insecticidal protein production typically decreases at about the time plants reach the reproductive stage of development (Greenplate *et al*., 1998). This decrease can have significant implications for monitoring because a large ($\geq$ 3rd instar) larva found on reproductive and post-reproductive stage plants could easily represent a false positive resistance response. A decrease in insecticidal protein concentration with advanced plant development requires that monitoring be done prior to the end of the reproductive stage.

G Considerations in designing a resistance monitoring protocol

The most important issue in resistance monitoring is whether resistance can be detected before it has essentially evolved. In considering the approaches above, we focused on the sensitivity of the techniques to detect alleles at low frequencies, especially when the resistance trait is recessive or partially recessive. Without massive sampling effort, dose-mortality and diagnostic-dose assays are not adequately sensitive to changes in allelic frequencies until resistance is prevalent. The F_2 screen and the allele recovery techniques use a diagnostic dose to detect recessive alleles at low frequencies, but build on the strength of family analyses to add statistical confidence to estimates of allelic frequencies. However, family analyses require considerable capital, space, organization, and large sample sizes (both in terms of numbers of families, and to a lesser extent numbers of individuals per family) to adequately detect resistance alleles with statistical confidence. In addition, the biology and genetics of the pest(s) and cropping system(s) must be factored into the monitoring program.

1 Movement of the target pest

If monitoring is to be used to adaptively manage resistance, the population structure of the pest will influence the scale at which populations should be monitored. There are two related ways to look at this issue: (1) How many populations and at what geographic scale does one need to monitor the pest species to assess the geographic structure of genetic variation for tolerance, or (2) To what geographic scale can one extend information collected from a single population? Information about the scale of tolerance could be used to determine if some regions are more likely to be resistance "hot spots". Hot spots could be present due to prior use of microbial pesticide products or simply due to differences in the initial gene frequencies of resistance alleles. At present there are no methods to adequately address the geographic scale of genetic variation of susceptibility (Andow and Hutchinson, 1998). Regional monitoring using multiple populations and the F_2 screen/allele recovery is unrealistic. To adequately determine the genetic variation within and between populations would require large numbers of families from each population sampled. General information about the movement of adults would be a more useful tool than intensively monitoring populations at extremely fine scales.

2 Genetic basis of resistance

The genetic basis of resistance will dictate the most efficient monitoring strategy. When resistance is recessive, a diagnostic-dose assay will be inefficient at detecting low frequencies of resistant alleles. However, if resistance is functionally dominant or at high frequencies, diagnostic doses may be a reasonable alternative to the family-based methods. Dominant alleles in the pest may be more readily detected if the toxin titer in the plant is low to moderate, or if the concentration is temporally and/or spatially variable within the plant. When no prior knowledge is available, the F_2 screen may offer some insights into the mode of inheritance of the resistance trait.

Most of our discussion about monitoring resistance alleles has focused on monogenic models of inheritance. If more than one locus governs resistance, it will be difficult to estimate changes in the allelic frequencies at these loci. Polygenic inheritance may be more amenable to quantitative genetic analyses (Tabashnik, 1994; Falconer and Mackay, 1996; Gould and Tabashnik, 1998). Such analyses include estimates of heritability, measures of response to selection, and other single-family mating designs. Because the heritability of a trait is dependent on the additive genetic variation present in a population, which is influenced by the frequencies of alleles affecting the trait, monitoring changes in the heritability offers a crude approximation to changes in allelic frequencies. In a similar way, the response to selection is dependent on the heritability of a trait. However, quantitative-genetic analyses have their own problems. Heritabilities generally have large errors associated with their estimates. Large numbers of families are therefore necessary to provide statistical confidence in a heritability estimate and also to detect changes in heritability. Furthermore, the amount of additive genetic variation present will be low when the resistance alleles are rare, greatest at intermediate allele frequencies, and low again at higher allelic frequencies. The relationship between additive genetic variation and allele frequencies will also be dependent on the dominance of the trait. If adaptive resistance management is planned, quantitative-genetic tools are not as informative since it may be very difficult to transform heritability (or genetic variance) estimates into allele frequencies. However, quantitative genetic analyses may offer a good starting point to qualitatively demonstrate that genetic variation for resistance is present in populations.

In addition, by examining the distribution of tolerances both within and between families, sib analyses may offer insight into the general genetic architecture of the resistance trait. The F_2 screen could be used in this way with species more tolerant of the toxin.

3 *Multiple pests*

For some cropping systems, different resistance monitoring procedures may be required for individual pest species. Some pests may be more tolerant of the transgenic toxin than others, have different dispersal behavior, or differ with respect to larval movement. In the eastern portion of the USA's Cotton Belt, both *H. virescens* and *H. zea* are serious pests of cotton. Resistance in both of these pests would threaten the long-term use of *Bt* cotton transformed with the cry1Ac transgene. *H. zea* is much more tolerant (10–40% survival relative to non-*Bt* cotton) of Cry1Ac protein than *H. virescens* (< 1% survival), and it is possible that resistance in *H. zea* will be polygenic (Gould and Tabashnik, 1998). Gould *et al.* (1997) used the allele recovery technique to estimate the frequency of one resistance factor in *H. virescens*. However, quantitative-genetic analyses may be a more efficient approach for initial investigations of *H. zea* populations. Sumerford (unpubl. data) found that *H. zea* larvae that grew large on artificial diet containing the Cry1Ac protein produced offspring that grew significantly larger on this toxin relative to the offspring of less tolerant individuals (no difference in larval growth on control diets). These preliminary results suggest that *H. zea* can respond to modest selection. Luttrell *et al.* (1999) selected field collected populations of *H. zea* with a number of *Bt* formulations as well as HD73 (purified protein from early *Bt*-cotton). After 7 generations of selection, the resistance ratio of the selected colony exceeded 118 fold, and by the 8th generation of selection, mortality at the highest dose was limited to 18%. Responses to formulated *Bt* products (Javelin™, Dipel ES™) exceeded 6000 and 1×10^9 fold, respectively, after 8 generations of selection (Wan, 1995).

Estimates of heritability, family analyses or diagnostic-dose assays followed by selection, may be more efficient to initially characterize the tolerance of populations of *H. zea* to Cry1Ac than methods such as the F_2 screen.

H *General conclusion and recommendations (monitoring)*

To effectively monitor the frequency of resistance alleles in wild populations of insects, researchers must balance the concerns of statistical precision at low allelic frequencies, costs of sampling, and the organization and labor required to intensively sample many individuals or families. Much of industry's interest in monitoring for changes in insect susceptibility to insecticidal proteins stems from specific requirements attached to the EPA conditional registration of these plant products. It is unreasonable to expect that industry monitoring efforts will significantly exceed these requirements. However, it is not clear if the EPA requirements are adequate for monitoring changes in susceptibility over the huge acreage on which transgenic crops such as corn and cotton are now, and will be, planted in the USA. In addition, EPA monitoring mandates do not even begin to address the challenge of resistance monitoring of corn, cotton, and other transgenic crops in other countries throughout the world.

The techniques that offer the most to adaptive resistance management are the F_2 screen and the allele-recovery method, since they offer the best estimates of allelic frequencies. However, both of these methods require large commitments in labor, space, and capital. Can such an effort be continued on an annual basis to monitor changes in allelic frequencies? The time and labor involved with these two techniques make them unlikely choices for industry support. It must be stressed that industry is typically not equipped or adequately staffed to conduct this type of research. These assays are also poor choices in situations where transgenic plants are somehow damaged by insect attack during the growing season and the possibility of reduced insect susceptibility requires rapid investigation.

Andow *et al.* (1998) were limited to screening 91 F_2 isolines for resistance alleles in a single population of *O. nubilalis*. Even with this effort, Andow *et al.* (1998) were only able to estimate the allele frequency as < 0.013. The effort necessary to sample many populations over time may not be feasible.

Larval growth inhibition and feeding disruption share a similar concept – reduction in larval growth or the by-products (frass) of growth – and are more promising assays for short term monitoring. They are reasonably simple, realistic assays that can be executed directly on collected insects or their progeny. Turnaround time is rapid (one week or less) and the assays are easy to score visually. Neither assay makes assumptions about the genetic basis for reduced susceptibility. These assays are sufficiently sensitive to detect very slight changes in protein susceptibility present at very low frequency in a population. As noted previously, a diagnostic dose coupled with these assays appears to offer the most promise as an efficient procedure to test for resistance.

The use of living transgenic plants with carefully characterized insecticidal protein expression levels could be the most useful of all monitoring methods discussed here. However, substantial logistical and industry cooperation issues require resolution before this option can be tested. Survival and continued development of larvae on transgenic plants is the true test of resistance and eminent potential for a product control failure. Field selection for resistance involves insect exposure to the insecticidal protein in a living plant matrix containing lignins, tannins, and other allelochemicals. These plant constituents differ greatly from the ingredients used within artificial insect diets for laboratory resistance selection. Laboratory-selected strains resistant to insecticidal proteins often fail to survive on transgenic plants (Whalon *et al.*, 1993; Huang *et al.*, 1997) suggesting that resistant laboratory strains may not always represent the genotypes capable of causing field control failures.

The Food Quality Protection Act of 1996 required revision of the U.S. EPA processes for evaluating and registering pesticides. One revision that EPA enacted related to transgenic crops that produce insecticidal proteins. This revision required that a satisfactory resistance management program be in place prior to registration. This program includes monitoring for changes in pest susceptibility to the transgenic proteins. Susceptibility monitoring is presently conducted, in some way, on an individual basis by each company that sells transgenic crops. As far as we know, these monitoring data are shared neither among competitors nor with the academic community. A strong argument could be made for employing an independent party to conduct the overall monitoring effort. Efficient and impartial monitoring of insect susceptibility requires centralized testing facilities operated by organizations such as University Extension, USDA, or contract testing firms. The test organization(s) chosen could be responsible for all aspects of monitoring (sample collection, rearing, testing, confirmation, reporting) or it could rely on industry or other groups to do the sampling.

The monitoring organization would use assay methodology generally recognized by academic consultants as appropriate for the insect(s) involved. If multiple labs are involved to ensure that at least one lab is operating at all times during the sampling and monitoring season, then all should use the same assay(s) for consistency and comparison of results. The optimum amount of sampling effort required during each growing season remains unclear. However, one simple procedure would be to relate the number of sites sampled to the total acreage of the transgenic crop. These sample sites should be reasonably well distributed throughout the crop range. When more than one company markets a transgenic crop product, sampling responsibilities are allocated as a function of number of hectares sold during that particular year. Similarly, financial responsibility for support of a testing facility could be related to market share for the particular insect control transgenic crop. Reporting of the monitoring test results should be simultaneously made both to the individual companies and to the EPA.

All the monitoring methods discussed have limitations, particularly if dynamic changes in resistance management strategies are dependent on rapid, accurate estimates of resistance allele frequencies. Estimates of initial allelic frequency in conjunction with population movement research may be the best approach until better monitoring strategies are developed. The potential of DNA technologies to help map resistance factors in insect genomes may offer some new monitoring strategies in the future.

6 Remedial action or mitigation of resistance

Relatively little attention has been given to action plans in the event that evidence for a change in susceptibility is detected in a pest population. This evidence could range from relatively minor (a few individuals sampled from one location are able to grow on a diagnostic dose of protein) to major (a control failure occurs and most insects tested prove to be resistant). In the early stages of resistance, action plans could range from moderate (*e.g.*, an increase in the size of the untreated, non-transgenic refuge) to severe (*e.g.*, a reduction or elimination of transgenic crops from a large "quarantine" buffer zone around the area where resistance was detected).

7 Conclusions

The rational use of genetically modified crops should require the development of three different plans. A resistance management plan should be developed to minimize the risk of rapid resistance evolution. Though it must be assumed that resistance will always evolve, these strategies should be designed to minimize that risk within the constraints of the various stake holders. The second plan to be developed should focus on effective methods to monitor pest adaptation to the transgenic plants. Such monitoring is difficult because one can never be certain how or by what mechanism pests will adapt, but prior experience and newer techniques such as the F_2 screen may be informative here. Finally, some type of remediation plan must be in place soon after implementation of the transgenic crop. Each of these plans will require input from a diverse range of experts and will require considerable capital input to implement. Nonetheless, the potential for rapid resistance evolution if such crops are utilized carelessly suggests that the development of these plans should be carefully considered prior to release of new transgenic crop varieties.

8 References

Alstad, D. N. and Andow, D. A. 1995. Managing the evolution of insect resistance to transgenic plants. *Science* **268**, 1894–1896.

Andow, D. A. and Alstad, D. N. 1998. F_2 screen for rare resistance alleles. *J. Econ. Entomol.* **91**, 572–578.

Andow, D. A., Alstad, D. N., Pang, Y. H., Bolin, P. C., and Hutchinson, W. D. 1998. Using an F_2 screen to search for resistance alleles to *Bacillus thuringiensis* toxin in European corn borer (Lepidoptera: Crambidae). *J. Econ. Entomol.* **91**, 579–584.

Andow, D. A. and Hutchinson, W. D. 1998. *Bt*-corn resistance management. *In* "Now or Never: Serious New Plans to Save a Natural Pest Control." (M. Mellon and J. Rissler, Eds.), pp. 19–66. Union of Concerned Scientists, Washington, DC.

Bauer, L. S. 1995. Resistance: a threat to the insecticidal crystal proteins of *Bacillus thuringiensis*. *Fla. Entomol.* **78**, 414–443.

Bailey, W. D., Zhao G., Carter, L. M., Gould, F., Kennedy, G. G., and Roe, R. M. 1998. Feeding disruption bioassay for species and *Bacillus thuringiensis* resistance diagnosis for *Heliothis virescens* and *Helicoverpa zea* in cotton (Lepidoptera: Noctuidae). *Crop Prot.* **17**.

Benedict, J. H., Sachs. E. S., Altman, D. W., Ring, D. R., Stone, T. B., and Sims, S. R. 1993. Impact of delta-endotoxin-producing transgenic cotton on insect-plant interactions with *Heliothis virescens* and *Helicoverpa zea* (Lepidoptera: Noctuidae). *Environ. Entomol.* **22**, 1–9.

Bolin, P. C., Hutchison, W. D., Andow, D. A., and Ostlie, K. R. 1998. Monitoring for European corn borer (Lepidoptera: Crambidae) resistance to *Bacillus thuringiensis*: Logistical considerations when sampling larvae - Note. *J. Agric. Entomol.* **15**, 231–238.

Caprio, M. A. 1994. *Bacillus thuringiensis* gene deployment and resistance management in single- and multi-tactic environments. *Biocontrol Sci. Technol.* **4**, 487–497.

Caprio, M. A. 1998. Evaluating resistance management strategies for multiple toxins in the presence of external refuges. *J. Econ. Entomol.* **91**, 1021–1031.

Caprio, M. A. 2001. Source-sink dynamics between transgenic and non-transgenic habitats and their role in the evolution of resistance. *J. Econ. Entomol.* **94**, 698–705.

Caprio, M. A., Faver, M. K. and Hankins, G. 2004. Evaluating the impacts of refuge width on source-sink dynamics between transgenic and non-transgenic habitats. *J. Insect Science*, **4**, 3. insectscience.org/4.3

Caprio, M. A. and Hoy, M. A. 1994. Metapopulation dynamics affect resistance development in a predatory mite. *J. Econ. Entomol.* **87**, 525–534.

Caprio, M. A. and Hoy, M. A. 1995. Premating isolation generates frequency-dependent selection which alters the rate at which pesticide resistance strains of arthropod natural enemies can be established. *J. Econ. Entomol.* **88**, 205–212.

Caprio, M. A. and Suckling, D. M. 1995. Mating disruption reduces the risk of resistance development to transgenic apple orchards: simulations of the light brown apple moth. *In*: Proceedings of the 48th N. Z. Weed and Pest Control Conference, pp. 52–58.

Caprio, M. A. and Tabashnik, B. E. 1992. Gene flow accelerates local adaptation among finite populations: simulating the evolution of insecticide resistance. *J. Econ. Entomol.* **85**, 611–620.

Caprio, M. A., Nowatzki, T., Siegfried, B., Meinke, L. J., Wright, R. J., and Chandler, L. D. 2006. Assessing the risk of resistance to aerial applications of methyl-parathion in the western corn rootworm. *J. Econ. Entomol.* **99**, 483–493.

Chaufaux, J., Muller-Cohn, J., Buisson, C., Sanchis, V., Lereclus, D., and Pasteur, N. 1997. Inheritance of resistance to the *Bacillus thuringiensis* CryIC toxin in *Spodoptera littoralis* (Lepidoptera: Noctuidae). *J. Econ. Entomol.* **90**, 873–878.

Comins, H. N. 1979. The management of pesticide resistance: models. *In* "Genetics In Relation to Insect Management" (M. A. Hoy and J. J. McKelvey, Eds.), pp. 54–69. The Rockefeller Foundation.

Dennehy, T. J. and Granett, J. 1984. Monitoring dicofol-resistant spider mites (Acari: Tetranychidae) in California cotton. *J. Econ. Entomol.* **77**, 1386–1392.

Evans, J. R. and D. L. Olson. 2002. Introduction to Simulation and Risk Analysis. Prentice-Hall, Inc. Upper Saddle River, NJ, USA. 392 pp.

Falconer, D. and Mackay T. F. C. 1996. "Introduction to Quantitative Genetics". Longman, Essex, England.

Ferro, D. N. and Lyon, S. M. 1991. Colorado potato beetle (Coleoptera: Chrysomelidae) larval mortality: Operative effects of *Bacillus thuringiensis* subsp. *san diego*. *J. Econ. Entomol.* **84**, 806–809.

ffrench-Constant, R. H. and Roush, R. T. 1990. Resistance detection and documentation. *In* "The Relative Roles of Pesticidal and Biochemical Assays. Pesticide Resistance in Arthropods" (R. T. Roush and B. E. Tabashnik, Eds.), pp. 4–38. Chapman and Hall, New York.

FIFRA Scientific Advisory Panel, Subpanel on *Bacillus thuringiensis (Bt)* Plant-Pesticides and Resistance Management, February 9 and 10, 1998. Transmittal of the final report. US EPA Docket Number OPP #00231.

Fischhoff, D. A. 1996. Insect-resistant crop plants. *In* "Biotechnology and Integrated Pest Management." (G. J. Persley, Ed.), pp. 214–227. CAB International, Wallingford, UK.

Gasser, C. S. and Fraley, R. T. 1989. Isolation of tissue-specific cDNAs from tomato pistils. *Plant Cell Rep.* **1**, 15–24.

Georghiou, G. P. and Taylor, C. E. 1977. Operational influences in the evolution of insecticide resistance. *J. Econ. Entomol.* **70**, 653–658.

Gould, F. 1994. Potential and problems with high-dose strategies for pesticidal engineered crops. *Biocontrol Sci. Technol.* **4**, 451–461.

Gould, F. 1998. Sustainability of transgenic insecticidal cultivars: integrating pest genetics and ecology. *Annu. Rev. Entomol.* **43**, 701–736.

Gould, F, Anderson, A., Reynolds, A., Bumgarner, L., and Moar, W. 1995. Selection and genetic analysis of a *Heliothis virescens* (Lepidoptera: Noctuidae) strain with high levels of resistance to *Bacillus thuringiensis* toxins. *J. Econ. Entomol.* **88**, 1545.

Gould, F., Anderson, A., Jones, A., Sumerford, D., Heckel, D. G., Lopez, J., Micinski, S., Leonard, R., and Laster, M., 1997. Initial frequency of alleles for resistance to *Bacillus thuringiensis* toxins in field populations of *Heliothis virescens*. *Proc. Natl. Acad. Sci. U S A* **94**, 3519–3523.

Gould, F., Kennedy, G. G., and Johnson, M. T. 1991. Effects of natural enemies on the rate of herbivore adaptation to resistant host plants. *Entomol. Exp. Appl.* **58**, 1–14.

Gould, F., Martinez-Ramirez, A., Anderson, A., Ferre, J., Silva, F. J., and Moar, W. J. 1992. Broad-spectrum resistance to *Bacillus thuringiensis* toxins in *Heliothis virescens*. *Proc. Natl. Acad. Sci. USA* **89**, 7986–7990.

Gould, F. and Tabashnik, B. 1998. *Bt*-cotton resistance management, *In* "Now or Never Serious New Plans to Save a Natural Pest Control" (M. Mellon and J. Rissler, Eds.), pp. 67–105. Union of Concerned Scientists, Washington, DC.

Greenplate, J. T., Head, G. P., Penn, S. R., and Kabuye, V. T. 1998. Factors potentially influencing the survival of *Helicoverpa zea* on Bollgard cotton. *In* "Proceedings, 1998 Beltwide Cotton Production and Research Conference". National Cotton Council, Memphis, TN.

Gunning, R. V., Easton, C. S., Greenup, L. R., and Edge, V. E. 1984. Pyrethroid resistance in *Heliothis armigera* (Hübner) (Lepidoptera: Noctuidae) in Australia. *J. Econ. Entomol.* **77**, 1283–1287.

Heckel, D. G., Gahan, L. C., Gould, F., and Anderson, A. 1997. Identification of a linkage group with a major effect on resistance to *Bacillus thuringiensis* CryIAc endotoxin in the tobacco budworm (Lepidoptera: Noctuidae). *J. Econ. Entomol.* **90**, 75–86.

Höfte, H. and Whitely, H. R. 1989. Insecticidal crystal proteins of *Bacillus thuringiensis*. *Microbiol. Rev.* **53**, 242–255.

Huang, F., Buschman, L., Higgins, R. A., and McGaughey, W. H. 1999. Inheritance of resistance to

Bacillus thuringiensis toxin (Dipel ES) in the European corn borer. *Science* **284**, 965–967.

Huang, F., Higgins, R. A., and Buschman, L. L. 1997. Baseline susceptibility and changes in susceptibility to *Bacillus thuringiensis* subsp. *kurstaki* under selection pressure in European corn borer (Lepidoptera: Pyralidae). *J. Econ. Entomol.* **90**, 1137–1143.

ILSI Health and Environmental Sciences Institute (ILSI HESI). 1998. An evaluation of insect resistance management in *Bt* field corn: A science-based framework for risk assessment and risk management. Report of an expert panel. ILSI Press, Washington DC.

Johnson, M. T. and Gould, F. 1992. Interaction of genetically engineered host plant resistance and natural enemies of *Heliothis virescens* (Lepidoptera: Noctuidae) in tobacco. *Environ. Entomol.* **21**, 586–597.

Kennedy, G. G. and Whalon, M. E. 1995. Managing pest resistance to *Bacillus thuringiensis* endotoxins: constraints and incentives to implementation. *J. Econ. Entomol.* **88**, 454–460.

Liu, Y. B. and Tabashnik, B. E. 1997. Inheritance of resistance to the *Bacillus thuringiensis* toxin Cry1C in the diamondback moth. *Appl. Environ. Microbiol.* **63**, 2218–2223.

Liu, Y. B., Tabashnik, B. E., and Johnson, M. W. 1995. Larval age affects resistance to *Bacillus thuringiensis* in diamondback moth (Lepidoptera: Plutellidae). *J. Econ. Entomol.* **88**, 788–792.

Lukefahr, M. J. and Martin, D. F. 1966. Cotton plant pigments as a source of resistance to the bollworm and tobacco budworm. *J. Econ. Entomol.* **59**, 176–179.

Luttrell, R. G., Mascarenhas, V. J., Schneider, J. C., Parker, C. D., and Bullock, P. D. 1995. Effect of transgenic cotton expressing endotoxin protein on arthropod populations in Mississippi cotton. *In* "Proceedings, 1995 Beltwide Cotton Production and Research Conference" pp. 760–763. National Cotton Council, Memphis, TN.

Luttrell, R. G., Wan L., and Knighten, K. 1999. Variation in susceptibility of noctuid (Lepidoptera) larvae attacking cotton and soybean to purified endotoxin proteins and commercial formulations if *Bacillus thuringiensis. J. sEcon. Entomol.* **92**, 21–32.

Mallet, J. and Porter, P. 1992. Preventing insect adaptation to insect-resistance crops: are seed mixtures or refugia the best strategy? *Proc. R. Soc. Lond. B* **250**, 165–169.

Mascarenhas, V. J. and Luttrell, R. G. 1997. Combined effect of sublethal exposure to cotton expressing the endotoxin protein of *Bacillus thuringiensis* and natural enemies on survival of bollworm (Lepidoptera: Noctuidae) larvae. *Environ. Entomol.* **26**, 939–945.

May, R. M. and A. P. Dobson. 1986. Population dynamics and the rate of evolution of pesticide resistance. pp. 258–266. In "Pesticide Resistance: Strategies and Tactics for Management" (E. H. Glass *et al.*, Eds). National Academy Press, Washington, DC, 1986.

McGaughey, W. H. 1985. Insect resistance to the biological insecticide *Bacillus thuringiensis. Science* **229**, 193–195.

McGaughey, W. H. and Beeman, R. W. 1988. Resistance to *Bacillus thuringiensis* in colonies of Indianmeal moth and almond moth (Lepidoptera: Pyralidae). *J. Econ. Entomol.* **81**, 28–33.

Peck, S. L., Gould, F., and Ellner, S. P. 1999. Spread of resistance in spatially extended regions of transgenic cotton: implications for management of *Heliothis virescens* (Lepidoptera: Noctuidae). *J. Econ. Entomol.* **92**, 1–16.

Rahardja, U. and Whalon, M. E. 1995. Inheritance of resistance to *Bacillus thuringiensis* subsp. *tenebrionis* CryIIIA delta-endotoxin in Colorado potato beetle (Coleoptera: Chrysomelidae). *J. Econ. Entomol.* **88**, 21–26.

Robertson, J. L., Smith, K. C., Sevin, N. E., and Lavigne, R. J. 1984. Effects of dose and sample size on the precision of lethal dose estimates in dose-mortality regression. *J. Econ. Entomol.* **77**, 833–837.

Roush, R. T. 1994. Managing pests and their resistance to *Bacillus thuringiensis*: Can transgenic crops be better than sprays? *Biocontrol Sci. Technol.* **4**, 501–516.

Roush, R. T. 1997. Managing resistance to transgenic crops. *In* "Advances In Insect Control: The Role of Transgenic Plants" (N. Carozzi and M. Kozeil, Eds.), pp. 271–294. Taylor and Francis, London.

Roush, R. T. and Miller, G. L. 1986. Considerations for design of insecticide resistance monitoring programs. *J. Econ. Entomol.* **79**, 293–298.

Schneider, J. C. 1999. Confidence interval for Baysean estimates of resistance allele frequencies. *J. Econ. Entomol.* **92**, 755.

Shaver, T. N. and Lukefahr, M. J. 1969. Effect of flavonoid pigments and gossypol on growth and development of the bollworm, tobacco budworm, and pink bollworm. *J. Econ. Entomol.* **62**, 643–646.

Shelton, A. M., Robertson, J. L., Tang, J. D., Perez, C., Eigenbrode, S. D., Preisler, H. K., Wilsey, W. T., and Cooley, R. J. 1993. Resistance of diamondback moth (Lepidoptera: Plutellidae) to *Bacillus thuringiensis* subspecies in the field. *J. Econ. Entomol.* **86**, 697–705.

Shelton, A. M., Tang, J. D., Earle, E. D. and Roush, R. T. 1998. Can we manage resistance to Bt-engineered plants? Results of greenhouse and field tests. pp. 258–266. *In* "Pest Management-Future Challenges." (M. P. Zalucki, R. A. I. Drew and G. G. White, Eds.), Sixth Australasian Applied Entomological Research Conference. The University of Queensland Printery, Brisbane, Australia.

Shelton, A. M., Zhao, J. Z. and Roush, R. T. 2002. Economic, ecological, food safety, and social consequences of the deployment of Bt transgenic plants. *Annu. Rev. Entomol.* **47**, 845–881.

Sims, S. R. and, Berberich, S. A. 1996. *Bacillus thuringiensis* CryIA protein levels in raw and processed seed of transgenic cotton: determination using insect bioassay and ELISA. *J. Econ. Entomol.* **89**, 247–251.

Sims, S. B., Greenplate, J. T., Stone, T. B., Caprio, M. A., and Gould, F. L. 1996. Monitoring strategies for early detection of Lepidoptera resistance to *Bacillus thuringiensis* insecticidal proteins,. *In* "Molecular Genetics and Ecology of Pesticide Resistance" (T. M. Brown, Ed.), pp. 229–242. *Am. Chem. Soc. Symp*. Series. No 645.

Staetz, C. A. 1985. Susceptibility of *Heliothis virescens* (F.) (Lepidoptera: Noctuidae) to permethrin from across the cotton belt: a five-year study. *J. Econ. Entomol*. **78**. 505–510.

Tabashnik, B. E. 1994. Evolution of resistance to *Bacillus thuringiensis. Annu. Rev. Entomol*. **39**, 47–79.

Tabashnik, B. E. 1997. Seeking the root of insect resistance to transgenic plants. *Proc. Nat. Acad. Science*. **94**, 3488–3490

Tabashnik, B. E. and Croft, B. A. 1982. Managing pesticide resistance in crop-arthropod complexes: interactions between biological and operational factors. *Environ. Entomol*. **11**, 1137–1144.

Tabashnik, B. E., Schwartz J. M., Finson, N., and Johnson, M. W. 1992. Inheritance of resistance to *Bacillus thuringiensis* in diamondback moth (Lepidoptera: Plutellidae). *J. Econ. Entomol*. **85**, 1046–1055.

Tang, J. D., Gilboa, S., Roush, R. T., and Shelton, A. M.. 1997. Inheritance, stability and lack-of-fitness costs of field-selected resistance to *Bacillus thuringiensis* in diamondback moth (Lepidoptera: Plutellidae) from Florida. *J. Econ. Entomol*. **90**,732–741.

Tang, J. D., Collins, H. L., Roush, R. T., Metz, T. D., Earle, E. D., and Shelton, A. M. 1999. Survival, weight gain, and oviposition of resistant and susceptible *Plutella xylostella* (Lepidoptera: Plutellidae) on broccoli expressing Cry1Ac toxin of *Bacillus thuringiensis. J. Econ. Entomol* **92**, 47–55.

Vose, D. 2000. Risk analysis: a quantitative guide. Wiley & Sons, West Sussex, England. 418pp.

Wan, L. 1995. Susceptibility and avoidance behavior to *Bacillus thuringiensis* attacking cotton and soybean. M. S. Thesis, Mississippi State University, pp. 137.

Wolfenbarger, D. A., Raulston, J. R., Bartlett A. C., Donaldson, G. E., and Lopez, P. P. 1982. Tobacco budworm: selection for resistance to methyl parathion from a field-collected strain. *J. Econ. Entomol*. **75**, 211–215.

Whalon, M. E. and McGaughey, W. H. 1993. Insect resistance to *Bacillus thuringiensis. In* "Advanced Engineered Pesticides" (L. Kim, Ed.), pp. 215–231. Marcel Dekker, Inc., New York.

Whalon, M. E., Miller, D. L., Hollingworth, R. M., Grafius, E. J., and Miller, J. R. 1993. Selection of a Colorado potato beetle (Coleoptera: Chrysomelidae) strain resistant to *Bacillus thuringiensis. J. Econ. Entomol*. **86**, 226–233.

Whitlock, M. C. and McCauley, D. E. 1990. Some population genetic consequences of colony formation and extinction: genetic correlations within founding groups. *Evolution* **44**, 1717–1724.

SECTION IX

RESISTANCE

Chapter IX-1

Resistance to insect pathogens and strategies to manage resistance: An update

A. M. Shelton, P. Wang and J.-Z. Zhao
Department of Entomology
Cornell University
New York State
Agricultural Experiment Station
Geneva, NY 14456, USA

R. T. Roush
UC Statewide IPM Program
Department of Entomology
University of California
Davis, CA 95616, USA

1 Introduction

Resistance to traditional synthetic pesticides has become one of the major driving forces altering the development of integrated pest management (IPM) programs worldwide. Early definitions of resistance focused on the "development of strains capable of surviving a dose lethal to a majority of individuals in a normal population" (cited by ffrench-Constant and Roush, 1990). Sawicki (1987) proposed a definition that "resistance (is) a genetic change in response to selection by toxicants that may impair control in the field." Others have suggested that genetic changes in behavior, rather than physiology, may effect resistance and these also need to be taken into account (Lockwood *et al.*, 1984; Sparks *et al.*, 1989; but see also Roush and Daly, 1990, for a more skeptical view). In the most recently published survey, there were over 500 species of arthropods that had developed strains resistant to one or more of the five principal classes of insecticides (Georghiou and Lagunes-Tejeda, 1991). Interestingly, the survey did not include insect pathogens as one of the principal classes, although it listed the Japanese beetle, *Popilla japonica*, and the oriental beetle, *Anomala* (*Exomala*) *orientalis*, as having developed resistance to *Paenibacillus* (*Bacillus*) *popilliae*. Because of resistance to these other classes of insecticides as well as concern about some of the deleterious effects of traditional insecticides, there has been increased interest in using pathogens for insect control. Will their increased use result in a higher incidence of resistance to these control tactics as well?

Insect diseases caused by fungi (including microsporidia), viruses, nematodes, protozoa and bacteria are relatively common and widespread in nature. They can be very important natural control factors. Widespread high-level prevalence of naturally occurring diseases (epizootics) frequently reduces pest populations below damaging levels (Roberts *et al.*, 1991). More than 100 years ago, entomologists proposed using insect pathogens for pest control and until recently many entomologists believed that resistance to pathogens would be unlikely. This unfounded optimism rested on the belief that because pathogens were natural, had short persistence, and had coexisted with insects over eons, resistance

L.A. Lacey and H.K. Kaya (eds.), Field Manual of Techniques in Invertebrate Pathology, 793–811.

would not occur (Roush, 1998). As in many other instances, insects have proven us wrong. Indeed, the long evolutionary associations between insects and pathogens suggest that where genes for resistance exist, their frequencies prior to widespread commercial use of the pathogens may be higher than was the case for chemical insecticides (an example will be given below). While some insects are or never were susceptible to certain pathogens (and therefore would not be considered resistant in our sense), others have already developed resistance in the laboratory and/or field. With increasing use of pathogens and/or their products, it is likely that we will see increasing cases of resistance. The Resistant Arthropods Database (http://whalonlab.msu.edu/rpmnews/) should be a valuable repository for such information and readers are encouraged to enter information into it.

2 Documented cases of resistance

A Bacteria

1 Overview

Bacteria which are applied for insect pest management are primarily in the genera *Bacillus* and *Paenibacillus*. The milky disease bacteria (*Paenibacillus (=Bacillus) popilliae* and *P. lentimorbus*) are primarily used against the Japanese beetle, while subspecies of *B. thuringiensis* (*Bt*) *aizawai* and *kurstaki* are targeted for Lepidoptera, *Bt israelensis* (*Bti*) for mosquitoes and *Bt tenebrionis* for Colorado potato beetle, a pest which has developed resistance to most other alternative insecticides. By far the most widely used of these strains is *Bt* subsp. *kurstaki* (*Btk*) which has been applied for more than four decades to control lepidopteran pests of horticultural crops, ornamentals and forest systems. Especially for *Btk* and *Bt tenebrionis* (*Btt*) and in contrast to the other microbial pesticides we will discuss, it is not the living organism that is most important in control, but bacterial products, the δ-endotoxins, that are the main source of mortality (Höfte and Whiteley, 1989). This is well illustrated by the increasingly dominant use of toxins derived from *Bt* in transgenic crops (Chapter VIII-1) and, to a lesser extent, in recombinant bacteria such as *Pseudomonas* for crop sprays. Although the most extensive use of *Btk* sprays is probably for control of various forest and shade trees pests such as gypsy moth, spruce budworm, tent caterpillars and the like, perhaps the most intensive use has been against lepidopterous pests of cruciferous crops and especially the diamondback moth , *Plutella xylostella*, the most important insect pest of crucifers worldwide (Talekar and Shelton, 1993).

2 Cases of resistance

The diamondback moth has a long history of developing resistance to insecticides used against it. Since it evolved resistance to the first modern synthetic insecticide (DDT) and has developed resistance to most other available insecticides within less than a decade after their wide-spread use, it was not surprising that *P. xylostella* was the first insect to be detected with resistance to *Btk* in the field. Different strains of *Bt* have different δ-endotoxins (Höfte and Whiteley, 1989), and only some of the toxins are active against any particular insect. Extensive use of sprays of *Btk* to control *P. xylostella* has led to several cases of resistance to Cry1A and related *Bt* toxins in Australia (Ahmad *et al.*, 1998), Florida (Shelton *et al.*, 1993), Hawaii (Tabashnik *et al.*, 1990), Japan (Hama *et al.*, 1992), the Philippines (Ferré *et al.*, 1991), Central America (Perez and Shelton, 1997), and China (Zhao *et al.*, 1993). Resistance in one field-collected population was high enough that insects could survive on plants engineered to express the Cry1A(c) toxin at high levels (Metz *et al.*, 1995). A colony established from field-collected insects from Hawaii (Liu *et al.*, 1996) displayed about 20-fold resistance to Cry1C toxins contained in *Bt aizawai* (*Bta*), and a population collected in South Carolina with an initial 30-fold resistance level has further been selected on Cry1C-producing plants to > 100-fold resistance to the Cry1C protein (Cao *et al.*, 1999). In addition to *P. xylostella*, *Bt*-resistant populations of the cabbage looper, *Trichoplusia ni*, have been recently identified in commercial greenhouses in Canada, where *Bt* had been extensively applied (Janmaat and Myers, 2003). In these greenhouses, *T. ni* populations showed a level of resistance to the *Btk* formulation, Dipel, as high as 160-fold. The resistance in these *T. ni* larvae was primarily to the toxin Cry1Ac and the resistant individuals can survive on Cry1Ac transgenic broccoli plants (Kain *et al.*, 2004).

Prior to the discovery of field resistance, the ability of an insect species to evolve resistance to *Bt* in the laboratory was demonstrated with the Indianmeal moth (*Plodia interpunctella*, a pest of stored grain) in the mid-1980s (McGaughey, 1985). In these tests, populations eventually developed resistance levels of 140-fold. In total, laboratory populations of at least 10 species of Lepidoptera, 2 species of Coleoptera and 4 species of Diptera have been selected for resistance and 9 of the 16 have evolved resistance greater than 10-fold (Tabashnik, 1994; Ferré and Van Rie, 2002). The most recent example of resistance to *Bt* was in *Culex pipiens* from New York. A strain collected from Syracuse displayed 33-fold resistance to a preparation of *Bt israelensis* (Paul *et al.*, 2005). This is by far the highest level of resistance reported to date and may portent future problems in mosquito control.

3 Resistance mechanism and genetics

The pathogenesis of *Bt* toxicity in insects involves multiple steps: (a) dissolution of the parasporal crystals in the insect midgut lumen; (b) proteolytic cleavage of the protoxin by midgut digestive proteinases, resulting in activation of the protoxin to an active toxin; (c) passage of the toxin through the midgut peritrophic membrane to reach the midgut epithelium; and (d) binding of the toxin to the midgut brush border receptors, followed by insertion of the toxin into the midgut cell membrane or by activation of a cellular signaling pathway, leading to cell lysis and insect death (Schnepf *et al.*, 1998; Bravo *et al.*, 2004; Zhang *et al.*, 2005). Alteration of any of these steps may affect the toxicity of *Bt* toxins in insects and can be potentially involved in *Bt* resistance. Therefore, various mechanisms have been suggested in *Bt*-resistant insects (Ferré and Van Rie, 2002; Griffitts and Aroian, 2005).

Solubilization of Cry protein crystals in the midgut is a factor determining the toxicity in insects (Aronson *et al.*, 1991). Therefore, reduced solubilization could be a potential mechanism for *Bt* resistance in insects (Schnepf *et al.*, 1998). Midgut digestive proteinases are critically involved in both activation of the protoxins into their active forms and inactivation (degradation) of the activated toxins in the midgut (Oppert *et al.*, 1994, 1997; Forcada *et al.*, 1996; Shao *et al.*, 1998; Oppert, 1999). Excessive degradation of *Bt* toxin by the midgut proteinases could potentially contribute to low toxicity of *Bt* toxins in insensitive or *Bt*-resistant insect hosts (Forcada *et al.*, 1996; Shao *et al.*, 1998). Similarly, insufficient activation of *Bt* toxins by midgut serine proteinases can also be a mechanism for *Bt* resistance (Oppert *et al.*, 1997). Alteration of midgut trypsin activities has been observed in *Bt*-resistant strains of *P. interpunctella* and *O. nubilalis* (Oppert *et al.*, 1997; Li *et al.*, 2004).

The target site of *Bt* toxins is the insect midgut epithelium. Active *Bt* toxins must penetrate through the midgut peritrophic membrane (PM) to reach the target site. Therefore, the PM can be an important factor for the toxicity of *Bt* in insects (Rees *et al.*, 2002). Recently, trapping of *Bt* toxin Cry1Ac in the PM was observed in a Cry1Ac-resistant strain of *Bombyx mori* (Hayakawa *et al.*, 2004). These observations suggest that the PM may be involved in *Bt* resistance. Upon contact with the midgut epithelium, the *Bt* toxin binds to the midgut brush border. The specific binding of *Bt* toxins to the midgut receptors is a critical event for the toxicity of *Bt* toxins. Studies on interactions of *Bt* toxins with several Lepidoptera species demonstrated that reduced binding of *Bt* toxins to the midgut brush border membranes could result in reduced toxin activities in insects (Lee *et al.*, 1999) and is a primary mechanism for *Bt* resistance (Ferré *et al.*, 1991; Tabashnik *et al.*, 1994; Tang *et al.*, 1996). Currently, identified midgut proteins that may serve as the receptor for a *Bt* toxin include midgut cadherin-like proteins, aminopeptidases N (APNs) and membrane-bound alkaline phosphatase (Sangadala *et al.*, 1994; Knight *et al.*, 1994; Gill *et al.*, 1995; Valaitis *et al.*, 1995; Vadlamudi *et al.*, 1995; Francis and Bullar, 1997; Keeton and Bullar, 1997; Gahan *et al.*, 2001; Jenkins and Dean, 2001; Jurat-Fuentes and Adang, 2004; Herrero *et al.*, 2005). More recently, midgut glycolipids have also been reported to serve as receptors for *Bt* toxins (Griffitts *et al.*, 2005). In *B. mori* midgut, a 252 kD protein has been recently identified as a novel *Bt* toxin binding protein (Hossain *et al.*, 2004). Resistance to Cry toxins in several Lepidoptera species has been attributed to mutations or lack of expression of a gene coding for the *Bt* receptor in the midgut (Gahan *et al.*, 2001; Morin *et al.*, 2003; Xu *et al.*, 2005; Herrero *et al.*, 2005; Jurat-Fuentes and Adang, 2004). Recessive mutations in a cadherin gene are associated with resistance to Cry1A in at least three species, *i.e.*, *Heliothis virescens, Pectinophora gossypiella*, and *Helicoverpa armigera* (Gahan *et al.*, 2001; Morin *et al.*, 2003; Xu *et al.*, 2005).

In addition, other mechanisms involved in *Bt* resistance have been reported to include aggregation of *Bt* toxin proteins by the midgut esterase (Gunning *et al.*, 2005), elevated melanization activity of larval hemolymph and the midgut (Rahman *et al.*, 2004; Ma *et al.*, 2005) and possibly increased antioxidation activities in *Bt*-resistant insects (Candas *et al.*, 2003). Clearly, mechanisms for *Bt* resistance in insects may be multifaceted.

In Lepidoptera the most common type of resistance to Cry1A toxins is known as "Mode 1" resistance, which is characterized by a high level of resistance to one or more Cry1A toxins, recessive inheritance, reduced binding of one or more Cry1A toxins to the midgut brush border membrane and little or no cross-resistance to Cry1C toxin (Tabashnik *et al.*, 1998). "Mode 1" resistance results from an alteration of the midgut binding sites, which leads to reduced binding of Cry1A toxins to the brush border membranes in homozygous resistant individuals, but has little or no effect on the binding in heterozygous individuals (Tabashnik, 1994; Ferré and Van Rie, 2002). Specific binding of a Cry toxin to the midgut brush border membrane is an essential event for the toxicity of Cry toxins in insects and critically determines the toxicity (Schnepf *et al.*, 1998). Domain II, particularly the loop regions of domain II, in the Cry toxins are involved in the binding of the toxins to the midgut epithelium. Therefore, individual toxins within the same class (*e.g.*, Cry1A with its three subclasses of Cry1Aa, Cry1Ab and Cry1Ac), which are grouped by their protein sequence similarity, likely share the same target site, while toxins in different classes (*e.g.*, Cry1A vs. Cry1C) may have a different target site in the insect (Crickmore *et al.*, 1998; Ferré and Van Rie, 2002). Comparisons of resistance among *P. xylostella, H. virescens, P. gossypiella* and *Plodia interpunctella* show that at least one strain of each of these species has a 'mode 1' type of resistance. On the other hand, at least one strain of each of these three species does not fit the mode 1 pattern. Such differences between and within species may be important as more sophisticated management practices are developed.

In populations of *P. xylostella* that have evolved resistance to Cry 1A toxins, the primary mechanism of resistance generally appears to be a reduced binding of the toxin to the epithelial layer of the brush border membrane of the midgut (Tabashnik *et al.*, 1997). Formulated *Btk* products also contain the HD-1 spore. Tests with a population of *P. xylostella* collected from Florida, which had developed > 1500-fold resistance in the field, showed high levels of resistance to the HD-1 spore as well as all Cry1A protoxins (Tang *et al.*, 1996). When the HD-1 spore was combined with a toxin, it synergized toxicity to some toxins but not those to which the insects had developed resistance.

This same Florida population was used to study the inheritance, stability and fitness cost of resistance to *Btk* (Tang *et al.*, 1997). As with most other strains (Tabashnik *et al.*, 1998), genetic analysis of F_1 and backcross larvae indicated that resistance was an incompletely recessive, autosomal trait probably controlled by a single allele. The instability of resistance has figured prominently in resistance management debates, although fitness costs seem unlikely to have a major impact on resistance evolution unless they are very high for individuals carrying just one resistance allele (*i.e.*, the heterozygotes) (Roush, 1997). Reports to date indicate that there is considerable variation in resistance stability and fitness costs for *Bt* among populations of diamondback moth, with resistance in some populations appearing to have high fitness costs at least to resistant homozygotes (Tabashnik *et al.*, 1994), and at least until stabilized by continued laboratory selection (Tabashnik *et al.*, 1995). However, the major portion of resistance (probably the major gene) in the Florida population has generally seemed to be stable. From the initial > 1,500-fold resistance, resistance fell to about 300-fold within 3 generations in the absence of selection. Unlike previous cases of resistance to *Bt* (Tabashnik *et al.*, 1994), resistance in this Florida colony remained stable at about 300-fold (Tang *et al.*, 1997), and could be restored to > 1,000-fold after only a single application of *Bt*. Further, resistance did not perceptibly decline in populations with a 50% initial resistance allele frequency.

T. ni is the second insect species that has evolved resistance to *Bt* under agricultural conditions, in commercial vegetable greenhouses. Inheritance analysis of the resistance to *Btk* in the greenhouse-evolved *T. ni* showed a polygenic inheritance pattern and the inheritance was autosomal and incompletely recessive (Janmaat *et al.*, 2004). Further analysis of the greenhouse-evolved resistance to the major Cry toxin in *Btk*, Cry1Ac, showed that the resistance was controlled by a single genetic locus. Similarly, the resistance gene to Cry1Ac was autosomal and incompletely recessive (Kain *et al.*, 2004). The resistance of *T. ni* to *Bt* is associated with a fitness cost and the magnitude of fitness cost negatively correlates with the suitability of the host plants on which *T. ni* feeds (Janmaat and Myers, 2005).

The frequency of resistance alleles for *Bt* has been studied in populations that had not yet been exposed to significant commercial use of Cry toxins. Gould *et al.* (1997) estimated that this "initial frequency" of resistance alleles for Cry 1A is about 10^{-3} for *H. virescens*. Although this estimate may be high (Roush and Shelton, 1997; Roush, 1998), it does seem likely on the basis of successful selection experiments and studies with genetic markers (cited in Gould *et al.*, 1997) that resistance alleles for Cry1A may be on the order of 10^{-4}, which is much higher than generally supposed for chemical insecticides (less than 10^{-6}). Perhaps, resistance to *Bt* toxins in *H. virescens* is relatively more common because this species is often exposed to *Bt* in nature, a speculation that may apply to other species and other naturally occurring agents.

B Viruses

1 Overview

Baculoviruses have been the dominant entomopathogenic virus group used in insect pest management. A list of viruses "successfully" used for insect pest management was supplied by Roberts *et al.* (1991), although they noted that the degree of success is difficult to establish from the literature. In a recent review, Moscardi (1999) provides an assessment of the application of baculoviruses for control of Lepidoptera and discusses how they are used in various crops throughout the world.

Viruses have not been cost effective compared to synthetic insecticides. Their narrow host range and relatively slow rate at which they kill insects place them at a disadvantage in many crops with low damage thresholds, but they have been used successfully in some field crops with higher thresholds. Genetically engineered viruses may overcome some of the limitations of naturally-occurring viruses. If this happens and viruses become more widely and intensively used, resistance to them may become problematic. Given the history of resistance to insecticides that act hormonally or on the nervous system, including the sodium channel (Roush and Tabashnik, 1990; Denholm *et al.*, 1998), resistance may evolve specifically to some of the genes being introduced to the viruses (Roush, 1999).

2 Cases of resistance

Although there are reports of insects becoming increasingly "resistant" to baculovirus infections as they age, this is not really resistance in the sense we use, but rather simply changes in developmental susceptibility. Still, some of these can be rather dramatic, such as the 34,000-fold difference from the first to the fifth instar of *Mamestra brassicae* (Evans, 1981). True resistance (as we have defined it), however, has been developed through laboratory selections. In an early summary of studies, Roberts *et al.* (1991) noted that of six insect species selected over several generations for resistance, four became less susceptible to the virus against which they were challenged although resistance levels were < 10-fold in two of the species and resistance was unstable in at least one of them. Included in this early work were high levels of resistance (140-fold) in the potato tubermoth, *Phthorimaea operculella*, when it was challenged by a granulovirus for only 6 generations (Briese and Mende, 1983). Other important agricultural pests that have developed resistance to viruses through laboratory selection are the cabbage looper, *Trichoplusia ni* (Milks and Theilmann, 2000), the Indianmeal moth, *Plodia interpunctella* (Boots and Begon, 1993) and the fall armyworm, *Spodoptera frugiperda* (Fuxa *et al.*, 1988). Also worth noting is the important work on virus resistance in the silkworm, *Bombyx mori*. Resistance of the silkworm to a cypovirus (formerly cytoplasmic polyhedrosis virus) has been clearly demonstrated and the impact on the silk industry of this virus can be substantial (for an overview of the early work on virus resistance in the silkworm, see Watanabe, 1971).

Still, as noted by Fuxa (1993), there were no documented cases of resistance through the early 1990s in the field to a virus employed as a microbial control agent. However, this situation changed with *Anticarsia gemmatalis*, a pest of soybeans. Populations collected from Brazil were assayed with the *A. gemmatalis* multiple-embedded nucleopolyhedrovirus (*Ag*MNPV) obtained in Brazil in 1979. Although the Brazilian populations did not differ in susceptibility (based on non-overlap of the LC_{50} values), there was a positive correlation with the number of years that field sites had been sprayed with the virus (Abot *et al.*, 1995). In a follow-up study, Abot *et al.* (1996) collected colonies from two sites in Brazil and one site in the USA. One of the Brazilian colonies was initiated from a site that had been treated for four seasons with *Ag*MNPV. When the colonies were challenged for 3–4 generations, the Brazilian populations developed resistance ratios of > 1,000 while the USA population increased only 5-fold. A more recent example is resistance to a product containing the codling moth (*Cydia pomonella*) granulovirus (CpGV). When this product was used over several years, some populations in Germany and France developed resistance (Fritsch *et al.*, 2005; Eberle and Jehle, 2006; Sauphanor *et al.*, 2006). These examples may predict things to come.

3 Resistance mechanism and genetics

The genetics of resistance to viruses has been studied in a few species. Naturally occurring resistance in *Heliothis subflexa* to *Baculovirus heliothis* appeared to be controlled by a single major gene, as revealed in crosses with the cotton pest, *H. virescens* (Ignoffo *et al.*, 1985). In *Spodoptera frugiperda* when challenged with a single NPV, Reichelderfer and Benton (1974) reported that "resistance was due to a single gene or genes lacking dominance." The opposite appears to be the case with *C. pomonella* since resistance appears to be highly dominant (Sauphanor *et al.*, 2006). In a study of resistance in *B. mori* to viral infections, Watanabe (1986) noted that complete protection against the development of epizootics can be ensured by rearing silkworm varieties which are homozygous for a recessive non-susceptibility gene. In the case of *A. gemmatalis* (Abot *et al.*, 1996), the large differences in resistance evolution between the populations from Brazil and the USA could have been due to the occurrence of different genes or gene frequencies resulting from different exposure levels since *Ag*MNPV does not naturally occur in the USA. In a recent study in which a laboratory population of *A. gemmatalis* was challenged by *Ag*MNPV, the population readily regained susceptibility to the virus. There were indications that resistant individuals were less fit since they produced fewer eggs and had a lower hatch and longer development times (Fuxa and Richter, 1998). Further tests have indicated that optical brighteners, when combined into diet with *Ag*MNPV, enhanced mortality of *Ag*MNPV-resistant insects. Although the mechanism remains unclear (Morales *et al.*, 2001), studies have shown optical brighteners disrupt the peritrophic membrane (Wang and Granados, 2000, Okuno *et al.*, 2003).

Studies with *Drosophila* have elucidated resistance to the sigma rhabdovirus (Wyers *et al.*, 1995). In this case it appears that the *ref (2)P* gene of *Drosophila melanogaster* interferes with viral replication. This gene is highly variable and the different alleles are considered permissive or restrictive for viral replication. Similar studies with agricultural or medical pests are needed for a clearer understanding of how resistance to viruses can evolve and be managed.

A morphological approach to understanding the potential mechanisms of resistance was undertaken in a study of Indianmeal moth. Larvae were inoculated with a granulovirus, dissected over time, and tissues examined with an electron microscope (Begon *et al.*, 1993). Four possible resistance mechanisms were suggested: the peritrophic membrane, which formed a barrier; the basal lamina of the hemocoel, which impeded the secondary spread of the infection; hemocytes, which consumed viral particles; and the midgut cells, which passed viral particles into the midgut lumen in the later stages of infection.

C Fungi

1 Overview

Virtually all insect orders are susceptible to some fungal disease. Worldwide there are more than 700 species of entomopathogenic fungi in approximately 100 genera, but only 10 species have been or are currently being developed for insect control (Hajek and St. Leger, 1994). As in the case of viruses, fungi can be introduced into a population in several ways. Control strategies that have been successful include permanent introduction and establishment, augmentative releases, and environmental manipulation or conservation (Hajek and St. Leger, 1994). However, the majority of control programs that rely on fungi as the principal management tool use fungi essentially as a frequently applied biological insecticide. Such mycoinsecticides have resulted in successful control of pests in both greenhouse and field situations. There is particular interest in using fungi to control whiteflies which have become increasingly problematic in recent years (Faria and Wraight, 2001), especially in cases where resistance to other insecticides has evolved. In China, one species of fungi, *Beauveria bassiana*, has been used on over 1,000,000 ha for control of the pine caterpillar (primarily *Dendrolimus punctatus*) and 300,000 ha of corn for control of *Ostrinia furnacalis* (Roberts *et al.*, 1991). In the USA, commercial formulations of this fungus are registered for control of whiteflies, grasshoppers, thrips and aphids on several important crops. Entomopathogenic fungi with enhanced efficacies may contribute in a significant and sustainable manner to control vector-borne diseases such as malaria, dengue and filariasis (Scholte *et al.*, 2004).

2 Cases of resistance

Cases of true resistance are unknown, although this may be due to the relatively limited use of fungi to date or a lack of follow-up studies. As with bacteria and viruses, there is considerable intraspecific variability of response to fungi which results in differences in pathogenicity. Likewise, there is simultaneous variability in the host. Successful laboratory selection for resistance does not appear to have taken place, and there are no documented cases of resistance in the field. However, there are instances in the field which indicate that real resistance could occur due to genetic changes in the host. For example, aphid clones susceptible and resistant to *Pandora neoaphidis* coexist in Australia (Milner, 1982, 1985) and this would be due to genetic variation in host susceptibility. The increased use of fungi, coupled with improved follow-up, may detect resistance in some of the 26 locations where there has been permanent introduction of one of 19 different host-pathogen systems (Roberts and Hajek, 1992). In the case of fungi used to control mosquito-borne diseases, no cases of resistance have been observed but such cases may arise in the future because effective control programs require repeated rather than single applications during mosquito breeding periods (Scholte *et al.*, 2004).

3 Potential resistance mechanisms and genetics

Pathogenesis is a complex process involving many steps. A number of potential mutations in insects along this pathway may lead to changes in susceptibility. For example, the cuticle serves as multi-layered barrier to fungi. The failure of a fungus to penetrate this barrier could be caused by the presence of inhibitory compounds in the cuticle, the lack of factors needed for recognition on the cuticle, or lack of proper nutrients in the cuticle. Once the fungus has entered the insect's body, the host may be killed by some combination of mechanical damage caused by the fungus, depletion of the host's nutrients or toxins produced by the fungus. Again, changes in the insect could alter the insect-pathogen interaction at any one of these steps and result in resistance.

D Microsporidia

1 Overview

Microsporidia are now placed with the fungi. They are transmitted orally by ingestion of spores and some are also transmitted transovarially via the eggs or by parasitoids (Roberts *et al.*, 1991). Infection by microsporidia usually results only in somewhat reduced pest population densities through lowered fecundity and other such sublethal effects, so their place in pest management has been limited. However, two species that might have potential use as microbial insecticides are *Paranosema (=Nosema) locustae* and *Varimorpha necatrix*. The former is the only microsporidia registered by the EPA and has been used in a bait for grasshopper control in situations such as rangeland where the economic threshold is high. Due to their limitations in mass production and efficacy, Roberts *et al.* (1991) suggested that it might be more realistic to consider microsporidia as the "microbial counterparts to parasitoids and predators rather than as candidates for development as microbial insecticides."

2 Cases of resistance

A careful search of the literature has found no cases of resistance to microsporidia.

3 Potential resistance mechanisms and genetics

Like fungi, infection by microsporidia appears to be a complex process of interactions between the pathogen and its host. Too little is known about this process, and much less is known about genetic plasticity to provide insight into the development of resistance to particular microsporidia.

E Nematodes

1 Overview

Although nearly 40 nematode families are associated with insects, species in three families (Mermithidae, Steinernematidae and Heterorhabditidae) are the predominant nematodes used for insect control. Steinernematidae and Heterorhabditidae are characterized by their mutualistic relationship with bacterial pathogens in the genera *Xenorhabdus* and *Photorhabdus*, respectively (see Chapter IV-5), whereas mermithids are more like insect parasitoids. In the case of steinernematids and heterorhabditids, the infective nematodes enter host insects and release the bacterium that multiplies, causes septicemia, and kills the insect within 48 hours. The primary success of these two nematode families as biological control agents has been in sites where the nematodes have been sheltered from environmental extremes. The introduction of nematodes into plant stems or soil environments where they can then use volatiles from the insect to locate their hosts has proven to be successful for such pests as the black vine weevil, citrus weevil, mole crickets and Japanese beetle. For mermithids their most effective use has been against aquatic insects, especially mosquito larvae.

2 Cases of resistance

As with other pathogens, host susceptibility varies with different nematode species and strains. Although not true resistance, host susceptibility to nematodes may also change with age of the host. In the case of *Simulium vittatum* infection by *Steinernema (=Neoaplectana) carpocapsae*, the basis for 'resistance' was the physical exclusion of the comparatively large nematodes during the early instars of the host (Gaugler and Molloy, 1981). Because many soil insects have co-evolved with steinernematids and heterorhabditids, they are often 'resistant' to these nematodes. For example, scarab grubs have sieve plates over their spiracles and elaterid larvae have an oral filter that prevents nematodes from entering and infecting the larvae (see Chapter IV-5). In a series of studies with nematodes against Japanese beetle larvae, the infection process by the nematodes and the defensive mechanisms by the beetle have been studied in detail. In the case of *Steinernema glaseri* and *Heterorhabditis bacteriophora*, each used different strategies for killing the larvae but caused similar mortality. *S. glaseri* tolerated the gut fluid and avoided the host immune system, while *H. bacteriophora* had poor tolerance to the host gut fluid but overcame this by releasing its bacteria during or soon after penetrating its host (Wang *et al.*, 1995). In a later study Wang and Gaugler (1999) found evidence of a surface coat protein from *S. glaseri* that suppresses the immune response of beetle larvae. In the case of the beetle's defensive mechanisms, it displayed aggressive behaviors such as brushing with the legs and rubbing with its raster to remove attacking nematodes or simply make evasive behaviors in the presence of nematodes (Gaugler *et al.*, 1994). While these studies do not document cases of resistance due to selection pressure over time, they do suggest potential mechanisms of resistance. However, the mosquito, *Culex quinquefasciatus*, developed *bona fide* behavioral resistance to the mermithid, *Romanomermis culicivorax*, under laboratory selection after 300 generations (Petersen, 1978) and *Anopheles quadrimaculatus* developed resistance to *Diximermis peterseni* after 4 years of exposure (Woodward and Fukuda, 1977). No cases of resistance have been reported in the field.

3 Resistance mechanisms and genetics

As with fungi and microsporidia, the complex interactions between nematodes and their hosts involve multiple steps, any one of which may be altered by changes in the genome of the host. A prime limitation of nematodes has been their susceptibility to environmental factors and hence lack of persistence. Selection and/or development of strains that have increased persistence may lead to increased use and perhaps potential resistance problems. Insect immunity to nematodes might result from encapsulation, intracellular melanization, changes in behavior, or resistance to the bacterium *Xenorhabdus*. Perhaps these factors could also occur through genetic changes in previously susceptible insects and thereby create truly resistant populations. In the case of *A. quadrimaculatus*, resistance appears to have occurred because the resistant host was much more active and attempted to remove attached preparasites. The resistance mechanism of *C. p. quinquefasciatus* is not known.

3 Characteristics of pathogens influencing resistance

The most spectacular documented case of resistance in the field to any of the pathogens listed above is with *P. xylostella* and *Bt*. We speculate that this is due not to some inherent characteristic of this system, but rather to the heavy use of this bacterially produced pesticide against this specific pest. In this case, use was increased because other control tactics had failed (Shelton *et al.*, 1993). In the case of field resistance in *A. gemmatalis* to *Ag*MNPV in Brazil, resistance occurred only after several years of wide-spread use, but such use may have been due to economic and environmental reasons rather than to the failure of other tactics.

However, one can speculate that the more complex the infection process, the more likely that the host population may carry rare resistance alleles to combat a critical step in the process. For example, when conidia of *Metarhizium anisopliae* attach to a host, they produce appressoria on the cuticle, infection pegs in the epicuticle, hyphae that penetrate plates in the procuticle and yeast-like hyphal bodies within the hemocoel, which eventually lead to the death of the insect (Hajek and St. Leger, 1994). At various stages, genetic variants could physiologically challenge the fungal invasion.

Perhaps, the factors which would most influence the development of resistance are application, use pattern and formulation, especially in terms of persistence of exposure of the control agent. As noted above, in Florida use of *Bt* increased when synthetic products failed, and we estimate that high levels of resistance to *Btk* in *P. xylostella* occurred within a 4 year period (Shelton *et al.*, 1993). The more a microbial insecticide is used, the more likely resistance to it will evolve. In the case of *Bt* toxins that are engineered into plants and thereby significantly increase persistence, it is especially appropriate that resistant management strategies be developed and utilized when the plants are first deployed (Chapter VIII-1). Projections indicate that *Bt* transgenic plants have the potential to capture one-third of the insecticide market (Krattiger, 1997). This will likely lead to further selection for *Bt* resistance to foliar applications as well as when expressed in plants, if for no other reason than that *Bt* toxins will find greatly increased use.

Although plants which have been engineered to express a microbial toxin effectively make it more persistent and may increase the likelihood for resistance, transgenic plants may also produce such a consistently high dose as to delay resistance more effectively than sprays in the presence of a "refuge" of untreated individuals (Roush, 1994). Companies have produced plants which express *Bt* toxins and utilized a high dose strategy which calls for high expression of the toxin ($> LC_{95}$ of heterozygous RS insects) in combination with a refuge of non-transformed plant (Chapter VIII-1). The benefits of utilizing a refuge have been proved in greenhouse studies (Shelton *et al.*, 1998, Tang *et al.*, 2001). In field studies with *P. xylostella* when applied as a spray, development of resistance to *Btk* has been slower with lower rates, but this also resulted in increased damage to the crop (Perez *et al.*, 1997b). This indicates the need to balance rate with efficacy in an overall resistance management program. Stability of resistance can also be influenced by another aspect of formulation besides application. In the case of *Bt*, the two major strains that have been used against *P. xylostella* are *Bta* and *Btk*. While they share some of the same toxins, *e.g.*, Cry1A, *Bta* has an additional toxin, Cry1C, which has a different receptor site and consequently can control insects that have developed resistance only to *Btk*. However, using a *Bta* product against a population of *P. xylostella* can maintain or increase resistance to Cry1A, which might otherwise have declined if a pure Cry 1C product was used (Tang *et al.*, 1995). In contrast to sprays, however, models suggest that pyramiding two toxins into the same cultivar may greatly delay resistance (Roush, 1994, 1997, 1998), and laboratory selection tests on *H. armigera* (Zhao *et al.*, 1999) and greenhouse selection experiments on *P. xylostella* (Zhao *et al.*, 2003) have provided direct evidence. Thus, one needs to be concerned not only about the method of delivering *Bt*, whether it be as a spray or incorporated into plants, but also about the concentration and the composition of the toxins.

4 Detecting resistance

As noted previously, not all insects within a population are equally susceptible to all pathogens and, for those that are similarly susceptible, there may be large differences in susceptibility depending on the stage. It is within this framework that entomologists must use techniques that detect true resistance. Tremendous improvements have been made in insecticide resistance detection methods (ffrench-Constant and Roush, 1990) and our current understanding of biochemistry and ecological genetics of resistance to chemical insecticides (Roush and Tabashnik, 1990; Denholm *et al.*, 1998). Outside of *P. xylostella* and *Bt* toxins and the *ref (2)P* gene of *D. melanogaster* which interferes with virus replication, however, we have little knowledge of fundamental aspects that can be used to detect resistance of pathogens or their products and must rely more on standardizing methods for monitoring resistance through dose-response and, more practically, diagnostic dose methods (for a more thorough treatment see ffrench-Constant and Roush, 1990; Halliday and Burnham, 1990; Chapter VIII-1).

Testing of microbials should be done in a fashion similar to how the insect would become exposed in the field to the pathogen. The leaf-dip bioassay technique has been used most frequently to assess *P. xylostella* resistance to *Bt* (Tabashnik *et al.*, 1990; Shelton *et al.*, 1993) and the LC_{50} and the corresponding 95% CL have also been the most frequent criterion used to compare the susceptibility of two or more populations. Field-collected larvae can be brought back to the laboratory, but it is usually necessary to rear them for one or more generations so that enough insects are available for testing. Perez *et al.* (1997a) compared *P. xylostella* in leaf-dip bioassays with the mortality of larvae caused by residues of field applications of *Btk* and mortality of larvae in three diagnostic concentrations of a *Btk* product incorporated in an artificial diet. They concluded that LC_{50}s of a *B. thuringiensis* product (Javelin) > 0.6 mg (AI)/liter in leaf dip bioassays can be associated with low levels of mortality in field applications and that the diagnostic concentration of 20.5 mg/ml artificial diet can be used as an on-farm assay for monitoring *P. xylostella* resistance to *Btk*. In those cases where toxins of *Bt* have been incorporated into a plant, one can use the plant as an assay substrate and simply put the larvae on the plant (Chapter VIII-1).

Realistically, the sample sizes needed to reliably detect resistance across a wide geographic area are so large that resistance management must be seen as a preventative rather than reactive activity. Even with excellent diagnostic doses, one needs a sample size of at least 3000 individuals per location to detect resistance phenotypes even when they are at a frequency of 0.1% (Roush and Miller, 1986), a frequency so high that failures due to resistance may occur in just a few more generations (Roush, 1994). Even reliance on a sample of 3000 is optimistic, because perfect diagnostic doses rarely exist given the inconsistencies and variations typical in bioassays (Halliday and Burnham, 1990). Even if the diagnostic dose is coupled to the F_2 screen as discussed by Caprio *et al.* (Chapter VIII-1), the labor requirements would be enormous. It was proved that the diagnostic diet assay was a better F_2 screen method to detect alleles in *P. xylostella* than using transgenic plants expressing a high level of a *Bt* toxin (Zhao *et al.*, 2002). The estimated probabilities of false positives and false negatives were 33% and 1%, respectively, for detecting Cry1Ac resistance at the allele frequency of 0.012 using the diagnostic diet assay. Careful validation of the screening method for each insect-crop system is necessary before the F_2 screen can be used to detect rare *Bt* resistance alleles in field populations. Given that resistance can often evolve on very local scales (*e.g.*, Tabashnik *et al.*, 1987; Roush *et al.*, 1990), many locations would have to be monitored to assure the detection of resistance before it became widespread.

A DNA-based method was recently developed that could detect the mutation in genomic DNA of each of the three *Bt* resistant alleles (r1, r2, r3) of a cadherin gene in *P. gossypiella* (Morin *et al.*, 2004)

5 Management of resistance

A Characteristics of insects influencing resistance

The proportion of a population that is exposed to a selection agent is among the most important factors influencing the evolution of resistance, particularly when coupled with long range dispersal to ensure mating between selected and unselected individuals. In many comparisons of similar or closely related pests (even with the same species in different habitats), resistance evolves more quickly where a high proportion of the population is exposed each generation (Tabashnik and Croft, 1985; Roush and Daly, 1990), *i.e.*, when there are only small "refuges" of untreated individuals. In the presence of refuges, larger scale insect migrations further assist the delay of resistance by allowing an influx of potentially susceptible alleles into an area where selection pressure is occurring, and thereby help dilute resistant alleles. In Brazil, where one million ha of soybeans were treated annually with *Ag*MNPV, this represented only 10% of the total soybean acreage. Fuxa (1993) suggests quite sensibly that one factor preventing wide-spread development of resistance is the migration of adults from soybean areas not treated with *Ag*MNPV to areas where the pathogen has been extensively used.

In contrast, many other aspects of pest biology that are commonly thought to influence the rate of resistance evolution have much less clear or consistent influences. A large number of generations per year do not consistently accelerate resistance (Rosenheim and Tabashnik, 1990, 1991); some species with few annual generations, such as the Colorado potato beetle (Roush *et al.*, 1990; Zhao *et al.*, 2000), evolve resistance very quickly, apparently due to intense selection within each generation. The resistance genes themselves can have a strong influence on the rate of evolution. Resistance will evolve faster with dominant alleles (*i.e.*, where the heterozygous carriers of one resistance allele and one susceptible allele survive) (Tabashnik and Croft, 1982; Roush, 1994, 1997, 1998). However, as noted above, fitness costs to resistance do not seem to have a significant effect on slowing resistance unless the costs are large and affect heterozygotes (Roush, 1997), which seems generally not to be the case (Roush and Daly, 1990). However, an alternative view about the value of fitness costs in slowing resistance evolution was suggested in *P. gossypiella* (Carrieré *et al.*, 2001a, b; Tabashnik *et al.*, 2005).

B Promote refuges: apply agent less often and only on "hot spots"

Given that one of the most obvious and consistent influences on the rate at which resistance evolves is the proportion of the population in "refuges" each generation, it seems clear that a key tactic for the management of resistance is to provide refuges where possible. The best way to increase refuges for pests with a history of resistance evolution would be to reduce the percentage of the pest population that is sprayed within the crop.

While this advice may at first seem impractical, it is completely consistent with the long-term thrust of IPM: reduce the frequency of sprays or portion of the crop covered, at least where sprays fail to provide an economic return (*i.e.*, pests exceed economic or action threshold densities). In particular, there are many examples in which "spot treatment" is often sufficient to control pests without treating the entire population. For example, Colorado potato beetles are often concentrated along field margins (Roush and Tingey, 1992). Those areas internal to the field serve as refuges for insects that are not sufficiently dense to cause loss but which can dilute resistance.

C Avoid persistent formulations

Closely related to the notion of spraying less often is to avoid highly persistent formulations (Roush, 1989, 1999). Users want enough persistence to control the pest, but excessive persistence can continue to select for resistance long after the pest has been suppressed below damaging numbers, just as would calendar spraying. Most biopesticides probably will have short persistence, which would tend to avoid resistance, but it may be technically feasible to use them in slow release devices. Such delivery systems may be counterproductive in the long term by creating persistent selection.

D Avoid high application rates

Contrary to popular belief, there is no general advantage to applying high rates in sprays, either in theory or experiments (Tabashnik and Croft, 1982; Roush 1989, 1998). There may be an advantage for transgenic crops (Roush, 1994, 1997, 1998) or closed systems like granaries (Roush, 1989) where there can be tight control of the application rate even if the exposure is persistent, but sprays will inherently suffer from less than thorough coverage and decay, such that the 95% mortality of heterozygotes needed for a "high dose strategy" (Chapter VIII-1) is simply not possible (Tabashnik and Croft, 1982; Roush, 1994, 1998). Further, excessively good control can eliminate prey for natural enemies such that they starve even if not killed by the spray itself, which has the potential for inducing a pesticide treadmill that increases dependence on the sprays and faster resistance. On the other hand, there is also no clear evidence or theory that a lower rate of application will in itself significantly slow resistance, at least not within the range of rates that will still control the pests, although low rates may be more easily integrated with the survival and use of predators or parasitoids (Roush, 1989, 1994).

E Avoid pesticide mixtures

Contrary to another popular belief, it is not necessarily true that mixtures of pesticides will delay resistance. Experimental studies have failed to consistently find any advantage to mixtures over sequential or rotational use of the same insecticides (Tabashnik, 1989; Immaraju *et al.*, 1990), and theoretical models showed that mixtures will significantly delay resistance only when several conditions are met (Gould, 1986; Roush, 1989, 1997, 1998). To be most effective, mixtures must meet several requirements, especially that there must be high mortality from each of the agents when they are sprayed alone. Almost all individuals resistant and exposed to one pesticide must be killed by the other for mixtures to be highly effective (Roush, 1989, 1998).

These conditions will probably be rarely met for sprays of microbial agents. For example, experiments with mixtures of *Bt* serotypes, applied at concentrations that did not provide high levels of control when used individually, failed to delay resistance in Indianmeal moth (McGaughey and Johnson, 1992). Not only is this experiment a good model for the field (where control with microbials probably rarely exceeds 90%), the results are just as would be predicted from the modeling papers listed above. We have previously mentioned problems of dual selection with mixtures of Cry 1A and Cry 1C toxins as found in *Bta*.

F Use rotations rather than mosaics

Rotating the use of pesticides over an entire area in a so-called "window strategy" based on a seasonal calendar has proven to be a very effective resistance management tactic both in terms of adoption and efficacy (Roush, 1989; Forrester *et al.*, 1993). The use of different compounds at roughly the same time in neighboring fields creates a mosaic of treatment patterns and should be avoided, as should very short-term rotations within a generation (Roush, 1989). The problem for both is that you have simultaneous selection with several pesticides, resulting in much lower pesticide durability (as illustrated by Roush, 1993).

To make sense of this, consider a case where selection for resistance is so strong that resistance occurs in a single generation. If you have two pesticides and use the first pesticide, you still have the second pesticide for the next generation, giving a total of two generations of control. If instead, you split the population in half, roughly half of the population will be resistant to each of the pesticides in the next generation (depending on what assumptions you make about the dominance of resistance), which would most likely result in failure to control the population in the field after one generation. Long term rotations, over at least a month in length and in some cases for 2 years, have proved to be extremely successful for cotton insecticides and acaricides in Australia and Zimbabwe (Roush, 1989; Forrester *et al.*, 1993). Recently we evaluated the development of resistance in *P. xylostella* to three insecticides (*Bt*, spinosad, and indoxacarb) when these products are used in a rotation strategy compared to when the products are used in a mosaic fashion in which all products are used simultaneously in greenhouse cage experiments. After 9 generations of selection, the overall results showed that rotation of each insecticide was better than a mosaic of all three insecticides for resistance management (Zhao *et al.*, unpublished data).

6 Conclusions

It has been the general experience of entomologists that resistance will evolve to any control agent that is intensively applied on populations with small refuges. Even if we exclude *Bt* as a unique case, resistance will almost certainly occur to at least some insect pathogens, as has already been found for viruses and, to a lesser extent, nematodes. Determining whether or not resistance has historically been a problem for the pest in question is a simple test of the need for a resistance management strategy.

Although resistance management is often perceived as a complex problem, the list of potential practical tactics is generally so short that choosing which would be useful is not difficult with a minimal amount of information about pest biology (Roush, 1989). Most of these tactics are also complementary and are most effective when adopted before selection commences. Thus, even though resistance management can always be improved later with additional data, one should aim to adopt a resistance management plan at the first introduction of the product to keep resistance frequencies as low as is reasonably practical.

The most common problem in resistance management is not one of assembling the obvious tactics, but of gaining implementation of the strategy. The history of resistance management implies quite strongly that for problematic pests, it is at least in the interests of the pesticide user and the local community to find a path to implementation.

7 Acknowledgments

The authors thank H. Collins, J. Curtis, R. Gaugler, R. Granados, J. Grant, H. Kaya, R. Milner and M. Villani for their help on this chapter.

8 References

Abot, A. R., Moscardi, F., Fuxa, J. R., Sosa-Gomez, D. R., and Richter, A. R. 1995. Susceptibility of populations of *Anticarsia gemmatalis* from Brazil and the United States to a nuclear polyhedrosis virus. *J. Entomol. Sci.* **30**, 62–69.

Abot, A. R., Moscardi, F., Fuxa, J. R., Sosa-Gomez, D. R., and Richter, A. R. 1996. Development and resistance by *Anticarsia gemmatalis* from Brazil and the United States to a nuclear polyhedrosis virus under laboratory selection pressure. *Biol. Control* **7**, 126–130.

Ahmad, M., Tang, J. D., Shelton, A. M., and Roush, R. 1998. Survey for resistance to *B. thuringiensis* in field populations of *Plutella xylostella* in Australia. *Proc. 29th Annu. Australian Conf.*, Brisbane.

Aronson, A. I., Han, E.-S., McGaughey, W., and Johnson, D. 1991. The solubility of inclusion proteins from *Bacillus thuringiensis* is dependent upon protoxin composition and is a factor in toxicity to insects. *Appl. Environ. Microbiol.* **57**, 981–986.

Begon, M., Daud, K., Young, P., and Howells, R. 1993. The invasion and replication of a granulosis virus in the Indian meal moth, *Plodia interpunctella*: an electron microscope study. *J. Invertebr. Pathol.* **61**, 281–295.

Boots, M. and Begon, M. 1993. Trade-offs with resistance to a granulosis virus in the Indian meal moth, examined by a laboratory evolution experiment. *Functional Ecol.* **7**, 528–534.

Bravo, A., Gomez, I., Conde, J., Munoz-Garay, C., Sanchez, J., Miranda, R., Zhuang, M., Gill, S. S., and Soberon, M. 2004. Oligomerization triggers binding of a *Bacillus thuringiensis* Cry1Ab pore-forming toxin to aminopeptidase N receptor leading to insertion into membrane microdomains. *Biochem. Biophys. Acta.* **1667**, 38–46.

Briese, D. T. and Mende, H. A. 1983. Selection for increased resistance to a granulosis virus in the potato moth, *Phthorimaea operculella. Bull. Entomol. Res.* **73**, 1–9.

Candas, M., Loseva, O., Oppert, B., Kosaraju, P., and Bulla, L. A. Jr. 2003. Insect resistance to *Bacillus thuringiensis*, alterations in the Indianmeal moth larval gut proteome. *Mol. Cell. Prot.* **2**, 19–28.

Cao, J., Tang, J. D., Strizhov, N., Shelton, A. M., and Earle, E. D. 1999. Transgenic broccoli with high levels of *Bacillus thuringiensis* Cry1C protein control diamondback moth larvae resistant to Cry1A or Cry1C broccoli. *Mol. Breeding* **5**, 131–141.

Carrieré, Y., Ellers-Kirk, C., Patin, A. L., Sims, M. A., Meyer, S., Liu, Y.-B., Dennehy, T., and Tabashnik, B. E. 2001a. Overwintering cost associated with resistance to transgenic cotton in the pink bollworm (Lepidoptera: Gelechiidae). *J. Econ. Entomol.* **94**, 935–941.

Carrieré, Y., Ellers-Kirk, C., Liu, Y.-B., Sims, M. A., Patin, A. L., Dennehy, T. J., and Tabashnik, B. E. 2001b. Fitness costs and maternal effects associated with resistance to transgenic cotton in the pink bollworm (Lepidoptera: Gelechiidae) *J. Econ. Entomol.* **94**, 1571–1576.

Crickmore, N., Zeigler, D. R., Feitelson, J., Schnepf, E,. Van Rie, J., Lereclus, D., Baum, J., and Dean, D. H. 1998. Revision of the nomenclature for the *Bacillus thuringiensis* pesticidal crystal proteins. *Microbiol. Mol. Biol. Rev.* **62**, 807–813.

Denholm, I., Pickett, J. A., and Devonshire, A. L. 1998. Insecticide resistance: from mechanisms to management (a discussion meeting issue). *Phil. Trans. R. Soc. London B* **353**, 1673–1795.

Eberle, K. E. and Jehle, J. A. 2006. Field resistance of codling moth against *Cydia pomonella* granulovirus (CpGV) is autosomal and incompletely dominant inherited. *J. Invertebr. Pathol.* **93**, 201–206.

Evans, H. F. 1981. Quantitative assessment of the relationship between dosage and response of the nuclear polyhedrosis virus of *Mamestra brassicae. J. Invertebr. Pathol.* **37**, 101–109.

Faria, M. and Wraight, S. P. 2001. Biological control of *Bemisia tabaci* with fungi. *Crop Protection* **20**, 767–778.

Ferré, J., Real, M. D., Van Rie, J., Jansens, S., and Peferoen, M. 1991. Resistance to the *Bacillus thuringiensis* bioinsecticide in a field population of *Plutella xylostella* is due to a change in a midgut membrane receptor. *Proc. Natl. Acad. Sci. USA* **88**, 5119–5123.

Ferré, J. and Van Rie, J. 2002. Biochemistry and genetics of insect resistance to *Bacillus thuringiensis. Annu. Rev. Entomol.* **47**, 501–533.

ffrench-Constant, R. H. and Roush, R. T. 1990. Resistance detection and documentation; the relative roles of pesticidal and biochemical assays. *In* "Pesticide Resistance to Arthropods" (R. T. Roush and B. E. Tabashnik, Eds.), pp. 4–38. Chapman and Hall, New York.

Forcada, C., Alcacer, E., Garcera, M. D., and Martinez, R. 1996. Differences in the midgut proteolytic activity of two *Heliothis virescens* strains, one susceptible and one resistant to *Bacillus thuringiensis* toxins. *Arch. Insect Biochem. Physiol.* **31**, 257–272.

Forrester, N. W., Cahill, M., Bird, L. J., and Layland, J. K. 1993. Management of pyrethroid and endosulfan resistance in *Helicoverpa armigera* (Lepidoptera: Noctuidae). *In* Australia. Bull. Entomol. Res, Suppl 1.

Francis, B. R. and Bullar, L. A. Jr. 1997. Further characterization of BT-R1, the cadherin-like receptor for Cry1Ab toxin in tobacco hornworm (*Manduca sexta*) midguts. *Insect Biochem. Mol. Biol.* **27**, 541–550.

Fritsch, E., Undorf-Spahn, K., Kienzle, J., Zebitz, C. P. W., and Huber, J. 2005. Apfelwickler-granulovirus: erste hinmeise auf unterschiede in der empfindlichkeit lokaler apfelwickler-populationen. *Nachrichtenbl. Dtsch. Pflanzenschutzd.* **57**, 29–34.

Fuxa, J. R. 1993. Insect resistance to viruses. *In* "Parasites and Pathogens of Insects, Vol. 2, Pathogens" (N. E. Beckage, S. N. Thompson and B. A. Federici, Eds), pp. 197–209. Academic Press, San Diego, CA.

Fuxa, J. R., Mitchell, F. L., and Richter, A. R. 1988. Resistance of *Spodoptera frugiperda* to a nuclear polyhedrosis virus in the field and laboratory. *Entomophaga* **33**, 55–63.

Fuxa, J. R. and Richter, A. R. 1998. Repeated reversion of resistance to nucleopolyhedrovirus by *Anticarsia gemmatalis*. *J. Invertebr. Pathol.* **71**, 159–164.

Gahan, L. J., Gould, F., and Heckel, D. G. 2001. Identification of a gene associated with Bt resistance in *Heliothis virescens*. *Science* **293**, 857–860.

Gaugler, R. and Molloy, D. 1981. Instar susceptibility of *Simulium vittatum* to the entomogenous nematode *Neoaplectana carpocapsae*. *J. Nematol.* **13**, 1–5.

Gaugler, R., Wang, Y., and Campbell, J. F. 1994. Aggressive and evasive behaviors in *Popillia japonica* larvae: defenses against entomopathogenic nematode attack. *J. Invertebr. Pathol.* **64**, 193–199.

Georghiou, G. P. and Lagunes-Tejeda, A. 1991. The Occurrence of Resistance to Pesticides in Arthropods. *Food and Agriculture Organization*. Rome, Italy.

Gill, S. S., Cowles, E. A., and Francis, V. 1995. Identification, isolation and cloning of a *Bacillus thuringiensis* Cry1Ac toxin-binding protein from the midgut of the lepidopteran insect *Heliothis virescens*. *J. Biol. Chem.* **270**, 27277–27282.

Gould, F. 1986. Simulation models for predicting durability of insect-resistant germplasm: a deterministic diploid, two locus model. *Environ. Entomol.* **15**, 1–10.

Gould, F., Anderson, A., Jones, A., Sumerford, D., Heckel, D. G., Lopez, J., Micinski, S., Leonard, R., and Laster, M. 1997. Initial frequency of alleles for resistance to *Bacillus thuringiensis* toxins in field populations of *Heliothis virescens*. *Proc. Natl. Acad. Sci. USA* **94**, 3519–3523.

Griffitts, J. S. and Aroian, R. V. 2005. Many roads to resistance: how invertebrates adapt to Bt toxins. *BioEssays* **27**, 614–624.

Griffitts, J. S., Haslam, S. M., Yang, T., Garczynski, S. F., Mulloy, B., Morris, H., Cremer, P. S., Dell, A., Adang, M. J., and Aroian, R. V. 2005. Glycolipids as receptors for *Bacillus thuringiensis* crystal toxin. *Science* **307**, 922–925.

Gunning, R. V., Dang, H. T., Kemp, F. C., Nicholson, I. C., and Moores, G. D. 2005. New resistance mechanism in *Helicoverpa armigera* threatens transgenic crops expressing *Bacillus thuringiensis* Cry1Ac toxin. *Appl. Environ. Microbiol.* **71**, 2558–2563.

Hajek, A. E. and St. Leger, R. J. 1994. Interactions between fungal pathogens and insect hosts. *Annu. Rev. Entomol.* **39**, 293–322.

Halliday, W. R. and Burnham, K. P. 1990. Choosing the optimal diagnostic dose for monitoring insecticide resistance. *J. Econ. Entomol.* **83**, 1151–1159.

Hama, H., Suzuki, K., and Tanaka, H. 1992. Inheritance and stability of resistance to *Bacillus thuringiensis* formulations in the diamondback moth, *Plutella xylostella*. *Appl. Entomol. Zool.* **27**, 355–362.

Hayakawa, T., Shitomi, Y., Miyamoto, K., and Hori, H. 2004. GalNAc pretreatment inhibits trapping of *Bacillus thuringiensis* Cry1Ac on the peritrophic membrane of *Bombyx mori*. *FEBS Lett.* **576**, 331–335.

Herrero, S., Gechev, T., Bakker, P. L., Moar, W. J., and de Maagd, R. A. 2005. *Bacillus thuringiensis* Cry1Ca-resistant *Spodoptera exigua* lacks expression of one of four aminopeptidase N genes. *BMC Genomics* **6**, 96.

Höfte, H. and Whiteley, H. R. 1989. Insecticidal crystal proteins of *Bacillus thuringiensis*. *Microbiol. Rev.* **53**, 242–255.

Hossain, D. M., Shitomi, Y., Moriyama, K., Higuchi, M., Hayakawa, T., Mitsui, T., Sato, R., and Hori, H. 2004. Characterization of a novel plasma membrane protein, expressed in the midgut epithelia of *Bombyx mori*, that binds to Cry1A toxins. *Appl. Environ. Microbiol.* **70**, 4604–4612.

Ignoffo, C. M., Huettel, M. D., McIntosh, A. H., Garcia, C., and Wilkening, P. 1985. Genetics of resistance of *Heliothis subflexa* to *Baculovirus heliothis*. *Ann. Entomol. Soc. Am.* **78**, 468–473.

Immaraju, J. A., Morse, J. G., and Hobza, R. F. 1990. Field evaluation of insecticide rotation and mixtures as strategies for citrus thrips resistance management in California. *J. Econ. Entomol.* **83**, 306–14.

Janmaat, A. F. and Myers, J. 2003. Rapid evolution and the cost of resistance to *Bacillus thuringiensis* in greenhouse populations of cabbage loopers, *Trichoplusia ni*. *Proc. R. Soc. Lond. B. Biol. Sci.* **270**, 2263–2270.

Janmaat, A. F., Wang, P., Kain, W., Zhao, J. -Z., and Myers, J. 2004. Inheritance of resistance to *Bacillus thuringiensis* subsp. kurstaki in *Trichoplusia ni*. *Appl. Environ. Microbiol.* **70**, 5859–5867.

Janmaat, A. F. and Myers, J. H. 2005. The cost of resistance to *Bacillus thuringiensis* varies with the host plant of *Trichoplusia ni*. *Proc Biol. Sci.* **272**,1031–1038.

Jenkins, J. L. and Dean, D. H. 2001. Binding specificity of *Bacillus thuringiensis* Cry1Aa for purified, native *Bombyx mori* aminopeptidase N and cadherin-like receptors. *BMC Biochem.* **2**, 12.

Jurat-Fuentes, J. L. and Adang M. J. 2004. Characterization of a Cry1Ac-receptor alkaline phosphatase in susceptible and resistant *Heliothis virescens* larvae. *Eur. J. Biochem.* **271**, 3127–3135.

Kain, W. C., Zhao, J.-Z., Janmaat, A. F., Myers, J. Shelton, A. M. and Wang, P. 2004. Inheritance of resistance to *Bacillus thuringiensis* Cry1Ac toxin in a greenhouse-derived strain of cabbage looper (Lepidoptera: Noctuidae). *J. Econ. Entomol.* **97**, 2073–2078.

Keeton, T. P. and Bullar, L. A. Jr. 1997. Ligand specificity and affinity of BT-R, the *Bacillus thuringiensis* Cry1A toxin receptor from *Manduca sexta*, expressed in mammalian and insect cell cultures. *Appl. Environ. Microbiol.* **63**, 3419–3425.

Knight, P. J. K., Crickmore, N. and Ellar, D. J. 1994. The receptor for *Bacillus thuringiensis* Cry1a(C) delta-endotoxin in the brush border membrane of the lepidopteran *Manduca sexta* is aminopeptidase N. *Mol. Microbiol.* **11**, 429–436.

Krattiger, A. F. 1997. Insect resistance in crops: A case study of *Bacillus thuringiensis* (Bt) and its transfer to developing countries. *International Service for the Acquisition of Agribiotech Applications. ISAAA Briefs No. 2*. ISAAA: Ithaca, NY.

Lee, M. K., You, T. H., Gould, F. L. and Dean, D. H. 1999. Identification of residues in domain III of *Bacillus thuringiensis* Cry1Ac toxin that affect binding and toxicity. *Appl. Environ. Microbiol.* **65**, 4513–4520.

Li, H., Oppert, B., Higgins, R. A., Huang, F., Zhu, K. Y., and Buschman, L. L. 2004. Comparative analysis of proteinase activities of *Bacillus thuringiensis*-resistant and -susceptible *Ostrinia nubilalis* (Lepidoptera: Crambidae). *Insect Biochem. Mol. Biol.* **34**, 753–762.

Liu, Y. B., Tabashnik, B. E., and Pusztai-Carey, M. 1996. Field-evolved resistance to *Bacillus thuringiensis* toxin Cry1C in diamondback moth. *J. Econ. Entomol.* **89**, 798–804.

Lockwood, J. A., Sparks, T. C., and Story, R. N. 1984. Evolution of insect resistance to insecticides: A reevaluation of the roles of physiology and behavior. *Bull. Entomol. Soc. Am.* **30**, 41–51.

Ma, G., Roberts, H., Sarjan, M., Featherstone, N., Lahnstein, J., Akhurst R., and Schmidt, O. 2005. Is the mature endotoxin Cry1Ac from *Bacillus thuringiensis* inactivated by a coagulation reaction in the gut lumen of resistant *Helicoverpa armigera* larvae? *Insect Biochem. Mol. Biol.* **35**, 729–739.

McGaughey, W. H. 1985. Insect resistance to the biological insecticide *Bacillus thuringiensis. Science* **229**, 193–195.

McGaughey, W. H. and Johnson, D. E. 1992. Toxicity of different serotypes and toxins of *Bacillus thuringiensis* to resistant and susceptible Indian meal moth. *J. Econ. Entomol.* **80**, 1122–1126.

Metz, T. M., Roush, R. T., Tang, J. D., Shelton, A. M., and Earle, E. D. 1995. Transgenic broccoli expressing a *Bacillus thuringiensis* insecticidal crystal protein: implications for pest resistance management strategies. *Mol. Breeding* **1**, 309–317.

Milks, M. L., and Theilmann. 2000. Serial selection for resistance to a wild-type and a genetically modified nucleopolyhedrosis virus in *Trichoplusia ni. Biol. Control* **19**, 283–289.

Milner, R. J. 1982. On the occurrence of pea aphids, *Acyrthosiphon pisum*, resistant to isolates of the fungal pathogen *Erynia neoaphidis. Entomol. Exp. Appl.* **32**, 23–27.

Milner, R. J. 1985. Distribution in time and space of resistance to the pathogenic fungus, *Erynia neoaphidis*, in the pea aphid, *Acyrthosiphon pisum. Entomol. Exp. Appl.* **37**, 235–240.

Morales, L., Moscardi, F., Sosa-Goméz, D. R. Paro, F. E., and Soldorio, I. 2001. Fluorescent brightners improve *Anticarsia gemmatalis* nucleopolyhedrosis (*Ag*MNPV) activity on *Ag*MNPV-susceptible and resistant strains of the insect. *Biol. Control* **20**, 247–253.

Morin, S., Biggs, R. W., Sisterson, M. S., Shriver, L., Ellers-Kirk, C., Higginson, D., Holley, D., Gahan, L. J., Heckel, D. G., Carrieré, Y., Brown, J. K., and Tabashnik, B. E. 2003. Three cadherin alleles associated with resistance to *Bacillus thuringiensis* in pink bollworm. *Proc. Natl. Acad. Sci. USA* **100**, 5004–5009.

Morin, S., Henderson, S., Fabrick, J. A, Carrieré, Y,, Dennehy, T.J,, Brown, J. K., and Tabashnik, B. E. 2004. DNA-based detection of Bt resistance alleles in pink bollworm. *Insect Biochem Mol Biol.* **34**, 1225–33.

Moscardi, F. 1999. Assessment of the application of baculoviruses for control of Lepidopotera. *Annu. Rev. Entomol.* **44**, 257–289.

Oppert, B., Kramer, K. J., Johnson, D. E., MacIntosh, S. C., and McGaughey, W. H. 1994. Altered protoxin activation by midgut enzymes from a *Bacillus thuringiensis* resistant strain of *Plodia interpunctella. Biochem. Biophys. Res. Comm.* **198**, 940–947.

Oppert, B., Kramer, K. J., Beeman, R. W., Johnston, D., and McGaughey, W. H. 1997. Proteinase-mediated insect resistance to *Bacillus thuringiensis* toxins. *J. Biol. Chem.* **272**, 23473–23476.

Oppert, B. 1999. Proteases interactions with *Bacillus thuringiensis* insecticidal toxins. *Arch. Insect Biochem. Physiol.* **42**, 1–12.

Paul, A., Harrington, L. C., Zhang, L., and Scott, J. G. 2005. Insecticide resistance in *Culex pipiens* from New York. *J. Am. Mosquito Control Assoc.* **21**, 305–309.

Perez, C. P. and Shelton, A. M. 1997. Resistance of *Plutella xylostella* to *Bacillus thuringiensis* Berliner in Central America. *J. Econ. Entomol.* **90**, 87–93.

Perez, C. J., Tang, J. D., and Shelton, A. M. 1997a. Comparison of leaf-dip and diet bioassays for monitoring *Bacillus thuringiensis* resistance in field populations of diamondback moth. *J. Econ. Entomol.* **90**, 94–101.

Perez, C. J., Shelton, A. M., and Roush, R. T. 1997b. Managing diamondback moth resistance to foliar applications of *Bacillus thuringiensis*: testing strategies in field cages. *J. Econ. Entomol.* **90**, 1462–1470.

Petersen, J. J. 1978. Development of resistance by the southern house mosquito to the parasitic nematode *Romanomermis culicivorax. Environ. Entomol.* **7**, 518–520.

Rahman, M. M., Roberts, H. L., Sarjan, M., Asgari, S., and Schmidt, O. 2004. Induction and transmission of *Bacillus thuringiensis* tolerance in the flour moth *Ephestia kuehniella. Proc. Natl. Acad. Sci. USA.* **101**, 2696–9.

Rees, J. S., Jarrett, P., and Ellar, D. J. 2002. The peritrophic membrane plays an additional role in protecting lepidopteran insects from *Bacillus thuringiensis* Cry δ-endotoxins. *J. Insect Sci.* **2**, 47–48.

Reichelderfer, C. F. and C. V. Benton. 1974. Some genetic aspects of resistance of *Spodoptera frugiperda* to a nuclear polyhedrosis virus. *J. Invertebr. Pathol.* **23**, 378–382.

Roberts, D. W., Fuxa, J. R., Gaugler, R., Goettel, M., Jaques, R., and Maddox, J. 1991. Use of pathogens in insect control. *In* "CRC Handbook on Pest Management in Agriculture, Vol 2" (D. Pimentel, Ed.), pp. 243–278. CRC Press, Boca Raton, FL.

Roberts, D. W. and Hajek, A. E. 1992. Entomopathogenic fungi as bioinsecticides. *In* "Frontiers in Industrial Mycology" (G. F. Leatham, Ed.), pp. 144–159. Chapman and Hall, New York.

Rosenheim, J. A. and Tabashnik, B. E. 1990. Evolution of pesticide resistance: interactions between generation time and genetic, ecological, and operational factors. *J. Econ. Entomol.* **83**, 1184–1193.

Rosenheim, J. A. and Tabashnik, B. E. 1991. Influence of generation time on the rate of response to selection. *Am. Natur.* **137**, 527–541.

Roush, R. T. 1989. Designing resistance management programs: How can you choose? *Pest. Sci.* **26**, 423–441.

Roush, R. T. 1993. Occurrence, genetics and management of insecticide resistance. *Parasitol. Today* **9**, 174–179.

Roush, R. T. 1994. Managing pests and their resistance to *Bacillus thuringiensis*: Can transgenic crops be better than sprays? *Biocontrol Sci. Technol.* **4**, 501–516.

Roush, R. T. 1997. Managing resistance to transgenic crops. *In* "Advances in Insect Control: The Role of Transgenic Plants" (N. Carozzi and M. Koziel, Eds.), pp 271–294. Taylor and Francis, London.

Roush, R. T. 1998. Two toxin strategies for management of insecticidal transgenic crops: Can pyramiding succeed where pesticide mixtures have not? *Phil. Trans. R. Soc. London, Series B*, **353**, 1777–1786.

Roush, R. T. 1999. Strategies for resistance management. *In* "Methods in Biotechnology Series #5, Biopesticides: Use and Delivery" (F. R. Hall and J. J. Menn, Eds.) pp. 575–593. Humana Press, Totawa, NJ.

Roush, R. T. and Daly, J. C. 1990. The role of population genetics in resistance research and management. *In* "Pesticide Resistance in Arthropods" (R. T. Roush and B. E. Tabashnik, Eds.), pp. 97–152. Chapman and Hall, New York.

Roush, R. T. and Miller, G. L. 1986. Considerations for design of insecticide resistance monitoring programs. *J. Econ. Entomol.* **79**, 293–298.

Roush, R. T. and Shelton, A. M. 1997. Assessing the odds: The emergence of resistance to Bt transgenic plants. *Nature Biotech.* **15**, 816–817.

Roush, R. T. and Tabashnik, B. E. (Eds.). 1990. "Pesticide Resistance in Arthropods", Chapman and Hall, New York.

Roush, R. T. and Tingey, W. M. 1992. Evolution and management of resistance in the Colorado potato beetle, *Leptinotarsa decemlineata*. *In* "Resistance '91: Achievements and Developments in Combating Pesticide Resistance" (I. Denholm, A. L. Devonshire, and D. W. Holloman, Eds.) pp. 61–74. Elsevier Applied Science, Essex, UK.

Roush, R. T., Hoy, C. W., Ferro, D. N., and Tingey, W. M. 1990. Insecticide resistance in the Colorado potato beetle: influence of crop rotation and insecticide use. *J. Econ. Entomol.* **83**, 315–319.

Sangadala, S., Walters, F. S., English, L. H., and Adang, M. J. 1994. A mixture of *Manduca sexta* aminopeptidase and phosphatase enhances *Bacillus thuringiensis* insecticidal Cry1Ac toxin binding and $^{86}Rb^+ - K^+$ efflux in vitro. *J. Biol. Chem.* **269**, 10088–10092.

Sauphanor, B., Berling, M., Toubon, J.-F., Reyes, M., and Delnatte, J. 2006. Carpocapse des pommes: cas de résistance aux virus de la granulose dans le Sud-Est. *Phytoma* **590**, 24–27.

Sawicki, R. M. 1987. Definition, detection and documentation of insecticide resistance. *In* "Combating Resistance to Xenobiotics: Biological and Chemical Approaches" (M. G. Ford, D. W. Holloman, B. P. S. Khambay and R. M. Sawicki, Eds.) pp. 105–117. Ellis Horwood, Chichester, UK.

Schnepf, E., Crickmore, N., Van Rie, J., Lereclus, D., Baum, J., Feitelson, J., Zeigler, D. R., and Dean, D. H. 1998. *Bacillus thuringiensis* and its pesticidal crystal proteins. *Microbiol. Mol. Biol. Rev.* **62**, 775–806.

Scholte, E. J., Knols, B. G. J., Samson, R. A., and Takken, W. 2004. Entomopathogenic fungi for mosquito control: a review. *J. Insect Sci.* **4**, (19): 1–24.

Shao, Z., Cui, Y., Liu, X., Yi, H., Jig, J., and Yu, Z. 1998. Processing of delta-endotoxin of *Bacillus thuringiensis* subsp. *kurstaki* HD-1 in *Heliothis armigera* midgut juice and the effects of protease inhibitors. *J. Invertebr. Pathol.* **72**, 73–81.

Shelton, A. M., Robertson, J. L., Tang, J. D., Perez, C., Eigenbrode, S. D., Preisler, H. K., Wilsey, W. T., and Cooley, R. J. 1993. Resistance of diamondback moth to *Bacillus thuringiensis* subspecies in the field. *J. Econ. Entomol.* **86**, 697–705.

Shelton, A. M., Tang, J. D., Roush, R. T., and Earle, E. D. 1998. Can we manage resistance to Bt-engineered plants? Results of greenhouse and field tests. *Proc. Sixth Australian Appl. Entomol. Res. Conf.*, Brisbane, Australia, October 1998. pp. 258–266.

Shohei, O., Takatsuka, J., Nakai, M., Ototake, S., Masui, A., and Kunimi, Y. 2003. Viral-enhancing activity of various stilbene-derived brighteners for a *Spodoptera litura* nucleopolyhedrovirus. *Biol. Control* **26**, 146–152.

Sparks, T. C., Lockwood, J. A., Byford, R. L., Graves, J. B., and Leonard, B. R. 1989. The role of behavior in insecticide resistance. *Pest. Sci.* **26**, 383–399.

Tabashnik, B. E. 1989. Managing resistance with multiple pesticide tactics: theory, evidence, and recommendations. *J. Econ. Entomol.* **82**, 1263–69.

Tabashnik, B. E. 1994. Evolution of resistance to *Bacillus thuringiensis. Annu. Rev. Entomol.* **39**, 47–79.

Tabashnik, B. E. and Croft, B. A. 1982. Managing pesticide resistance in crop-arthropod complexes: interactions between biological and operational factors. *Environ. Entomol.* **11**, 1137–1144.

Tabashnik, B. E. and Croft, B. A. 1985. Evolution of pesticide resistance in apple pests and their natural enemies. *Entomophaga* **30**, 37–49.

Tabashnik, B. E., Cushing, N. L., and Johnson, M. W. 1987. Diamondback moth resistance to insecticides in Hawaii: intra-island variation and cross-resistance. *J. Econ. Entomol.* **80**, 1091–1099.

Tabashnik, B. E., Cushing, N. L., Finson, N., and Johnson, M. W. 1990. Field development of resistance to *Bacillus thuringiensis* in diamondback moth. *J. Econ. Entomol.* **83**, 1671–1676.

Tabashnik, B. E., Finson, N., Groeters, F. R., Moar, W. J., Johnson, M. W., Luo, K., and Adang, M. J. 1994. Reversal of resistance to *Bacillus thuringiensis* in *Plutella xylostella. Proc. Natl. Acad. Sci. USA* **91**, 4120–4124.

Tabashnik, B. E., Finson, N., Johnson, M. W., and Heckel, D. G. 1995. Prolonged selection affects stability of resistance to *Bacillus thuringiensis* in diamondback moth. *J. Econ. Entomol.* **88**, 219–224.

Tabashnik, B. E., Liu, Y.-B., Malvar, T., Heckel, D. G., Masson, L., Ballester, V., Granero, F., Ménsua, J. L., and Ferré, J. 1997. Global variation in the genetic and biochemical basis of diamondback moth resistance to *Bacillus thuringiensis. Proc. Natl. Acad. Sci. USA* **94**, 12780–12785.

Tabashnik, B. E., Liu, Y-B., Malvar, T., Heckel, D. G., Masson, L., and Ferré, J. 1998. Insect resistance to *Bacillus thuringiensis*: uniform or diverse? *Phil. Trans. R. Soc. London B* **353**, 1751–1756.

Tabashnik, B. E., Dennehy, T. J., and Carrieré, Y. 2005. Delayed resistance to transgenic cotton in pink bollworm. *Proc. Natl. Acad. Sci. USA* **102**, 15389–15393.

Talekar, N. T. and Shelton, A. M. 1993. Biology, ecology and management of the diamondback moth. *Annu. Rev. Entomol.* **38**, 275–301.

Tang, J. T., Shelton, A. M., Moar, W. J., and Roush, R. T. 1995. Consequences of shared toxins in strains of *Bacillus thuringiensis* for resistance in diamondback moth. *Resistant Pest Manag. Newsletter*, **7(1)**, 5–7.

Tang, J. T., Shelton, A. M., Van Rie, J., De Roeck, S., Moar, W. J., Roush, R. T., and Peferoen, M. 1996. Toxicity of *Bacillus thuringiensis* spore and crystal protein to resistant diamondback moth. *Appl. Environ. Micro.* **62**, 564–569.

Tang, J. T., Roush, R. T., Gilboa, S., and Shelton, A. M. 1997. Inheritance, stability and lack-of-fitness costs of field-selected resistance to *Bacillus thuringiensis* in diamondback moth from Florida. *J. Econ. Entomol.* 90, 732:741.

Tang, J. D., Collins, H. L., Metz, T. D., Earle, E. D., Zhao, J. Z., Roush, R. T., and Shelton, A. M. 2001. Greenhouse tests on resistance management of Bt transgenic plants using refuge strategies. *J. Econ. Entomol.* **94**, 240–247.

Vadlamudi, R. K., Weber, E., Li, T. H., and Bulla, L. A. 1995. Cloning and expression of a receptor for an insecticidal toxin of *Bacillus thuringiensis. J. Biol. Chem.* **270**, 5490–5494.

Valaitis, A. P., Lee, M. K., Rajamohan, F., and Dean, D. H. 1995. Brush border membrane aminopeptidase-N in the midgut of the gypsy moth serves as the receptor for the Cry1A(c) delta-endotoxin of *Bacillus thuringiensis. Insect Biochem. Mol. Biol.* **25**, 1143–1151.

Wang, Y. and Gaugler, R. 1999. *Steinernema glaseri* surface coat protein suppresses the immune response of *Popillia japonica* larvae. *Biol. Control* **14**, 45–50.

Wang, P. and Granados, R. R. Calcofluor disrupts the midgut defense system in insects. 2000. *Insect Biochem. Mol. Biol.* **30**, 135–143.

Wang, Y., Campbell, J. F., and Gaugler, R. 1995. Infection of entomopathogenic nematodes *Steinernema glaseri* and *Heterorhabditis bacteriophora* against *Popillia japonica* larvae. *J. Invertebr. Pathol.* **66**, 178–184.

Watanabe, H. 1971. Resistance of the silkworm to cytoplasmic-polyhedrosis virus. *In* "The Cytoplasmic-polyhedrosis Virus of the Silkworm" (H. Aruga and Y. Tanada, Eds.), pp. 169–184. University of Tokyo Press, Japan.

Watanabe, H. 1986. Resistance of the silkworm, *Bombyx mori*, to viral infections. *Agric. Ecosys. Environ.* **15**, 131–139.

Woodward, D. B. and Fukuda, T. 1977. Laboratory resistance of the mosquito *Anopheles quadrimaculatus* to the mermithid nematode *Diximermis peterseni. Mosq. News* **37**, 192–195.

Wyers, F., Petitjean, A. M., Dru, P., Gay, P., and Contamine, D. 1995. Localization of domains within Drosophila Ref(2) protein involved in the intracellular control of sigma rhabdovirus multiplication. *J. Virol.* **69**, 4463–4470.

Xu, X., Yu, L., and Wu, Y. 2005. Disruption of a cadherin gene associated with resistance to Cry1Ac-endotoxin of *Bacillus thuringiensis* in *Helicoverpa armigera. Appl. Environ. Microbiol.* **71**, 948–954.

Zhang, X., Candas, M., Griko, N. B., Rose-Young, L., and Bulla, L. A. Jr. 2005. Cytotoxicity of *Bacillus thuringiensis* Cry1Ab toxin depends on specific binding of the toxin to the cadherin receptor BT-R1 expressed in insect cells. *Cell Death Differ.* **12**, 1407–1416.

Zhao, J.-Z., Zhu, G., Ju, Z. and Wang, W. 1993. Resistance of the diamondback moth to *Bacillus thuringiensis* in China. *Resistant Pest Manag. Newsletter* 5, 11–12.

Zhao, J.-Z., Fan, Y., Fan, X., Shi, X., and Lu, M. 1999. Evaluation on the effect of delaying resistant adaptation of *Helicoverpa armigera* by transgenic tobacco co-expressing Bt and CpTI genes. *Chinese Bull. Sci.* 44, 1871–1873.

Zhao, J.-Z., Bishop, B. A., and Grafius, E. 2000. Inheritance and synergism of resistance to imidacloprid in the Colorado potato beetle (Coleoptera: Chrysomelidae). *J. Econ. Entomol.* **93**, 1508–1514.

Zhao J.-Z, Li, Y. X., Collins, H. L., and Shelton, A. M. 2002. Examination of the F_2 screen for rare resistance alleles to *Bacillus thuringiensis* toxins in the diamondback Moth. *J. Econ. Entomol.* **95**, 14–21.

Zhao, J.-Z., Cao, J., Li, Y. X., Collins, H. L., Roush, R. T., Earle, E. D., and A. M. Shelton. 2003. Transgenic plants expressing two *Bacillus thuringiensis* toxins delay insect resistance evolution. *Nature Biotechnol.* **21**, 1493–1497.

SECTION X

NON-TARGET ORGANISMS

Chapter X-1

Guidelines for evaluating effects of entomopathogens on non-target organisms

Ann E. Hajek

Department of Entomology
Cornell University
Ithaca, NY 14853, USA

Mark S. Goettel

Agriculture & Agri-Food Canada
Lethbridge Research Centre
Lethbridge, Alberta T1J 4B1, Canada

1 Introduction

The term "non-target" organism has been adopted to indicate all living things aside from pests being treated for control. As such, this term can be extremely broad and unwieldy. However, studies of the effects of entomopathogens on non-target organisms become more focused because of pathogen specificity. By definition, entomopathogens live by parasitizing insects and it is uncommon that these pathogens also attack organisms outside of Insecta. Moreover, for each specific instance, only those species occurring in the area of interest would normally be considered as non-targets.

There has always been interest in the effects of entomopathogens on commercially important non-target organisms such as bees and silkworms. With the increasing interest in use of natural enemies of insects for pest control, the potential effects of entomopathogens on non-targets important for pest control must also be seriously taken into consideration. More recently, interest has broadened to focus on the direct and indirect effects of entomopathogens on all members of ecosystems and on the ecosystem itself. This, of course, makes an evaluation of the effects of entomopathogens on non-targets much more difficult.

The North American Microbial Biocontrol Working Group identified four potential hazards associated with the use of microorganisms for the biological control of plant pests and diseases: (1) displacement of non-target microorganisms, (2) allergenicity of humans and other animals, (3) toxigenicity to non-target organisms, and (4) pathogenicity to non-target organisms (Cook *et al.*, 1996). These unintended non-target effects are independent of whether the pathogen in question is indigenous or non-indigenous, naturally-occurring or modified. However, the potential risks, and our ability to evaluate them, will very much depend on the origin and biology of the pathogen and its potential use.

This chapter is not an extensive review of the literature documenting non-target effects of entomopathogens (see Flexner *et al.*, 1986; Laird *et al.*, 1990; Hokkanen and Hajek, 2003) but is specifically intended to serve as a broad overview of methods for evaluating the effects of entomopathogens on non-targets. Some similar considerations can be found in discussions of methods for testing the effects of synthetic chemical pesticides on non-target invertebrates (Jepson, 1989), *e.g.*, conducting studies in the field as well as the standard laboratory studies (*e.g.*, Hassan, 1989). Protocols for testing the effects of

L.A. Lacey and H.K. Kaya (eds.), Field Manual of Techniques in Invertebrate Pathology, 815–833.

pathogens on non-targets have been suggested for a few specific entomopathogen/host systems (*e.g.*, James and Lighthart, 1992; Grace, 1994; Goettel *et al.* 1996; Hokkanen *et al.*, 2003) but, at present, protocols for most systems are nonexistent and standardized protocols have not been developed. In this chapter, we will discuss the range of non-targets that might be considered during studies of potential non-target impacts and will predominantly focus our discussion on non-target invertebrates. We will emphasize field methods, but will also describe laboratory methods as necessary, providing examples where possible. Finally, we will discuss how present regulations attempt to address safety for non-target organisms. Readers interested in testing the effects of entomopathogens on vertebrates should refer to guidelines presented by Siegel (1997) and other authors in Laird *et al.* (1990).

2 Defining host range

The effects of entomopathogens on non-target organisms are not only determined by infectivity but also can be influenced by the ecological niches occupied by the pathogen and potential hosts. The range of species that an entomopathogen is able to infect, usually under optimized conditions, (*e.g.*, high relative humidity and no ultraviolet radiation, high pathogen doses, etc.), is often referred to as its "physiological host range." Physiological host range is determined in the laboratory and demonstrates which species the pathogen can directly infect. Typical effects measured include some combination of mortality, infection, and/or pathogen reproduction. On the other hand, "ecological host range," defined as the host range manifested by pathogens under field conditions, is of primary concern regarding non-target impacts. In essence, the ecological host range of a pathogen can only be tested after a pathogen species is present in the environment.

By knowing the physiological host range of a pathogen, we can attempt to predict its ecological host range. However, it is fairly typical that hosts can be infected in the laboratory that are never found infected under field conditions (Hajek *et al.*, 1996; Hajek and Butler, 1999). This is thought to be the result of the complex biotic and abiotic interactions that occur in the field and which are difficult to duplicate in the laboratory. Consequently, great caution must be exercised when attempting to extrapolate laboratory results to the field situation.

Solter and Maddox (1998) have proposed the terms "indigenous host," an insect species normally found infected in the field, "natural host," an insect species from which the pathogen has been isolated, and "laboratory host," an insect species that develops normal infections under laboratory conditions. In practice, testing host specificity is generally an exercise in identifying laboratory hosts but can also include identifying those species that are marginally acceptable to pathogens (= "marginal hosts"). The impacts of pathogens on some non-target organisms that are marginal hosts can be complex and difficult to determine because interactions are atypical. For example, some pathogens may be able to kill a marginal host, but not reproduce within it. Due to the unusual nature of infections in suboptimal hosts, careful attention must be paid to evaluation of non-targets being tested in the laboratory in order to correctly identify effects on non-targets in the field.

As a separate consideration, for some organisms, pathogens do not kill but instead cause sublethal effects, such as reduced food consumption, reduced adult longevity, and decreased fecundity. The occurrence of sublethal effects in target hosts could influence organisms in the ecosystem whose life cycles depend on the target host, *e.g.*, beneficials (see Flexner *et al.*, 1986; Goettel and Hajek, 2001). However, non-targets exposed to a pathogen could themselves be affected sublethally; in such cases, detailed observations are necessary to detect these effects. For more discussion on the intricacies and difficulties in determining host range of candidate pathogens, the readers are referred to Hokkanen and Hajek (2003) and chapters therein.

3 Non-target concerns differ by application strategy

There have always been primary concerns about the direct impact of entomopathogens on insects being reared as a product. This would principally include honey bees and silkworms although other insects are reared for food, dyes, etc., according to regional practices. With the increased use of parasitoids and predators for biological control of insects, knowledge of the impact of entomopathogens on these beneficial agents is critical to Integrated Pest Management (IPM) programs.

In recent years, there has also been growing interest in the effects of entomopathogens on the total native invertebrate fauna in the ecosystem, especially for aquatic and forested ecosystems that are inhabited by indigenous arthropod populations. In particular, questions regarding the direct impact of entomopathogens on sensitive populations of threatened and endangered insects must be addressed. While threatened and endangered species that might be present in the area of interest should be evaluated, this can be very difficult due to the rarity of these individuals. Even when it would be possible to rear individuals from a limited number of females for testing, permission to conduct studies with threatened or endangered insects can be difficult to obtain (L. S. Bauer, pers. comm.).

Direct effects are more easily documented than indirect effects although both can occur. As a direct effect, one can readily understand how entomopathogens applied in the field would contact free-ranging bees or parasitoids, causing an effect if these non-targets were susceptible. In contrast, indirect effects of pathogens can be highly variable. Insects in rearing facilities (*e.g.*, silkworms) could be affected by pathogens if they are being fed field-collected foliage on which the pathogen occurs or if rearing practices are not meticulous. An indirect effect could also occur if decreasing populations of one or more target species precipitated a change in a food web, especially if numerous and specialized predators and parasitoids depended on the target species. In an excellent example of an indirect effect from application of an entomopathogen, applications of *Bacillus thuringiensis* (*Bt*) against *Lymantria dispar* caused sublethal effects in target larvae that were not killed outright (Weseloh and Andreadis, 1982; Weseloh, 1984). The slower development of these *L. dispar* larvae through the early instars resulted in a greater period of time that larvae were susceptible to *Cotesia melanoscelus*, a parasitoid attacking only early instars (Weseloh *et al.*, 1983). *Bt* did not directly impact the parasitoid but the sublethal effects on hosts indirectly increased parasitoid populations in *Bt*-treated plots compared with untreated controls.

Studies of non-target effects on the indigenous invertebrate fauna can be enormous and may require extensive efforts. However, while the possibilities for non-targets that should potentially be evaluated can be extremely broad, studies of effects on non-targets can be focused, depending on the control method and type of pathogen used.

A Pathogen introductions for establishment

Classical biological control is aimed at permanent establishment of an exotic agent in a new area (see Chapter VI-1). Once an exotic pathogen is introduced, it has the potential to persist in the environment and spread from the sites of release. Consequently, before release of a pathogen for classical biological control, knowledge of its potential effects on non-targets is critical. Emphasis should be placed on those organisms that live in the same habitats within an ecosystem as the target pest (Maddox *et al.*, 1992; Hajek *et al.*, 2003; Hokkanen *et al.*, 2003). Host-specific pathogens are generally released in small quantities and are expected to propagate in the environment within the host populations, responding to control increasing populations of hosts. Studies of non-target effects should thus address the potential of the pathogen to become established and persist by infecting non-target populations.

Much useful information can be gained by studying the epizootiology of the pathogen in its area of endemicity. One can go to the country of origin of the pathogen and evaluate which hosts are infected in the field, how well the pathogen persists, what non-target organisms are normally exposed to the inoculum, and any additional effects that are apparent in the area where the pathogen already occurs. These data could give an indication of the types of hosts infected in the field, and would lend information regarding taxonomy of those groups that are infected as well as the ecological niches of insects infected in the field. Unfortunately, such studies can be challenging because levels of both the host and pathogen often are very low in the area of origin.

Alternatively, data on the ecological specificity of closely related pathogens could be used to gain some idea about the potential host range in the field.

In addition to learning about the direct effects on non-target organisms (*i.e.*, pathogenicity and establishment in non-target populations), it is important to evaluate the potential impact of depletion of the target host populations on non-target populations, an indirect effect. Studies of both direct and indirect effects are of special interest if the target population is endemic (= a new association release strategy; *i.e.*, see Lockwood, 1993a, b) and therefore part of an established food web. Studies of indirect effects are especially difficult and require a thorough knowledge of the target organism's ecology and role in food chains.

In summary, Solter and Becnel (2003) suggest the following steps be taken to evaluate the safety of microsporidia to non-target organisms. These steps should be applicable to most entomopathogens being considered for introduction for classical biological control.

1. Conduct a literature review to determine what is known of the host range of the candidate or related species.
2. Choose ecologically relevant potential non-target hosts to test for susceptibility.
3. Look for non-target host responses that do not occur in the natural host.
4. Conduct horizontal transmission bioassays with important non-target species. Is the pathogen well adapted to these potential new hosts?
5. Conduct vertical transmission bioassays with important non-target hosts. Can the pathogen complete its life cycle within its new host?
6. Evaluate the ability of the pathogen to utilize a host cell. This is most pertinent relative to intracellular pathogens such as viruses, rickettsiae (*e.g.*, *Wolbachia*) and microsporidia.
7. Survey ecologically relevant non-target hosts in the area of origin of the candidate pathogen.

B Inundative use of indigenous pathogens

For indigenous pathogens applied as microbial insecticides, all organisms in the area of application could potentially come into contact with concentrations of the pathogen that are much higher than would normally be encountered. Pathogens applied inundatively, such as *Bt*, often have relatively broader host ranges and require large concentrations to cause disease. Because dose is clearly important in determining susceptibility of hosts, inundative applications could directly affect species not normally influenced by the pathogen. Therefore, studies of potential effects of inundative releases on non-targets are warranted even if indigenous pathogens are being used. However, when an indigenous pathogen is being used, the potential that the pathogen would become newly established within a non-target population, and spread within this population outside of the immediate spray area, is quite remote. Once a pathogen has been applied, inoculum levels usually rapidly return to pre-spray, background levels. For inundative applications of pathogens, it is only the non-target organisms coming in direct contact with the microbial insecticide that are at risk, and therefore risk is essentially very similar to chemical insecticide applications.

Consequently, with inundative use of indigenous pathogens, emphasis should be placed on the pathogen's persistence, because it is persistence that would determine more significant, long-term effects. If negative effects on non-target organisms are demonstrated in preliminary field studies or after operational use, then the use of the pathogen can be altered or stopped altogether. After application, any affected non-target populations should gradually return to initial levels because pathogen concentrations would decline, depending on the persistence level of the pathogen in question (*i.e.*, any negative effect should be fully reversible if the pathogen does not persist well in the environment) and transmissibility of the pathogen from environmental reservoirs.

Methods for non-target evaluation can be essentially the same as those used for evaluation of applications against target organisms. Inundative applications frequently are used for highly managed agricultural fields. In such cases, choice of non-target organisms to monitor would normally include those species that are essential or beneficial in the immediate area of pathogen application.

C Inundative use of non-indigenous pathogen species or pathotypes

Since non-indigenous pathogens used inundatively could become established, in essence, the potential threat to non-target organisms could be the same as for pathogens that would be used for classical biological control. Most pathogens being developed as microbial insecticides are essentially ubiquitous, and therefore are indigenous worldwide. However, molecular techniques often demonstrate differences among isolates of the same species from different locations and these different pathotypes can have remarkably different host ranges. Therefore, attention must be paid to potential differences among strains before an agent should be considered for introduction. As with pathogens intended for use in classical biological control, much information on potential risks to non-target invertebrates can be gained by studying the effects and persistence of the pathogen in the areas where it is native or in other areas where it is being used inundatively.

D Genetically-engineered pathogens

Genetically-altered pathogens pose unique difficulties because they have never before existed in nature in their altered form and may contain genes from non-related taxa. Assessment of the potential effects resulting from the release of genetically-engineered pathogens can be a difficult process, as microbial ecology in general is still very poorly understood. Consequently, the potential effects on non-target organisms can only be definitively determined once the pathogen has been released. Pertinent potential risks could include passage of genes to other organisms, increased host range, irreversible depletion of the target host and competitive displacement of the parent strain.

Much information can be gained from the biology and ecology of the parent strain and predictions can also be made based on the exact nature of the engineered traits. Laboratory and microcosm studies could be conducted to test many of the potential risks. If the genetic modification does not alter the host range of the pathogen, then the ecological host range of the unmodified pathogen could be used to predict the host specificity of the modified pathogen. If the genetic modification alters the host range, then extensive laboratory bioassays need to be conducted. In addition, studies to determine whether the ecological attributes of the pathogen (*e.g.*, sensitivity to ultraviolet radiation, temperature range) have been altered could help in understanding whether there is any indication that the modified pathogen would persist in new niches or infect non-targets not previously encountered.

A detailed framework for assessment of the potential risks of genetically modified organisms is provided by the Scientists' Working Group on Biosafety (1998a, b).

4 Methods for non-target evaluation

Due to the disparity between ecological and physiological host ranges, efforts must be made to investigate both the idealized host range in the laboratory, preferably using hosts that occur in the area of concern, and the actual host range that would be expected under field conditions. Investigating the ecological host range poses many challenges and direct studies are not always possible, as with both exotic pathogens and genetically modified pathogens because neither exist in the environment prior to release. By careful design of laboratory, semi-field and, if possible, field studies, information for risk assessment pertinent to non-target organisms can be obtained.

In developing protocols for evaluating effects of pathogens on non-target organisms, regardless of whether the testing is to take place in the laboratory, semi-field or field, the inclusion of both positive (inoculated, susceptible hosts) and negative (non-inoculated, susceptible hosts) controls is critical for comparisons. Negative controls are important to include because one often has not previously worked with non-target species and their responses to testing conditions can differ from target species (*e.g.*, *Danaus plexippus* larvae often cannot survive the same high humidities under which *L. dispar* larvae are maintained when challenged with entomopathogenic fungi (A.E.H., unpubl. data)). Little useful information will be gained if results demonstrate lack of susceptibility of a non-target organism, without at the same time demonstrating a positive effect on the target host under the same conditions.

Methods for rapid and definitive identification of pathogens, diagnosis of disease, and isolation of pathogens are essential for laboratory, semi-field and field evaluations. In addition, consideration must be given to methods for quantification of pathogen impact. Typically, numbers of hosts killed are compared with negative controls. However, other parameters that might directly negatively affect non-target populations should also be considered. Such parameters may include the capability of the pathogen to replicate within the non-target host, the possibility of latent infections or sub-lethal infections causing reductions in longevity and fecundity. It must be remembered that non-target infections may be abnormal if the species in question is a marginal host [*e.g.*, reproduction of microsporidia occurs in different tissues in marginal hosts compared with fully susceptible hosts (Solter and Maddox, 1998)]. Non-target studies by scientists with experience working with the pathogens in question could add significantly to our meager knowledge of marginal infections.

A Laboratory

Studies of potential effects of entomopathogens on non-target organisms generally begin with laboratory assays of host specificity. Bioassay methods for different groups of pathogens are described in detail in Lacey (1997), Navon and Ascher (1999), and Hokkanen and Hajek (2003). Here, we will discuss issues specific to tests of host specificity for delineating the potential ecological host range.

The types of hosts to evaluate are discussed above but we once again stress that the species potentially present in the field should be tested rather than only testing readily available organisms from laboratory colonies. In fact, even if laboratory colonies of non-targets that occur in the release area are available, one should consider whether responses by members of a laboratory colony would be indicative of responses by wild populations. Working with wild populations or their progeny is possible in many instances, but frequently requires more planning. Rearing conditions may be more complicated or unknown and the numbers of insects available for testing may be low.

Unfortunately, in almost all instances, we know little about the factors that determine host specificity of a pathogen. Perhaps in the future, an understanding of the physiological basis of host recognition by pathogens and the conditions necessary for pathogens to infect and survive within susceptible hosts but not within non-hosts will assist in the prediction of ecological host ranges. This knowledge would be very useful because it is impossible to test all species present in the field.

One major difficulty in testing specificity in the laboratory as a preparation for understanding non-target impacts in the field is determining an appropriate dose. This is problematic because we do not know precisely what doses organisms are exposed to in the field. For exposures to pathogens used inundatively as microbial insecticides, the dose can be estimated with more precision. Field applications can be quantified by use of spray cards, selective media, and other methods (see Lacey, 1997 and other chapters in this book). However, special attention should be paid to placing spray cards in the microhabitat of the organisms of concern or collecting soil for analysis from the area inhabited by the insects of concern. Nevertheless, it is difficult to estimate the dose that some non-target organisms may acquire secondarily, *e.g.*, while walking on an inoculated surface.

For pathogens being introduced for establishment and cycling in the environment, the dose to test in the laboratory is more difficult to determine. For such pathogens, the concentrations to be encountered in the field can swing from extremely low levels during enzootic phases to high concentrations during epizootics. In these systems, it is more difficult to determine ecologically meaningful doses for laboratory testing. The more conservative approach would be to test higher doses that are based on estimates of the pathogen load in the environment or on doses known to infect susceptible species. Where possible, a series of doses should be tested.

Attempts should be made to mimic as much as possible the parameters expected in the field. Consequently, parameters such as dose, inoculation method and temperature and humidity should be as close as possible to those that the non-target organisms are expected to encounter under field situations. Little useful information will be gained if laboratory assays demonstrate susceptibility of non-target organisms under unrealistic conditions never expected to occur in the field.

B Semi-field

In an effort to duplicate natural conditions as much as possible, semi-field assays can be carried out. Semi-field studies range from reproducing field conditions in the laboratory in microcosms, to caging test organisms in the field, to collecting and rearing field-collected insects. However it must be realized that results from such assays must still not be used as conclusive evidence of the effects that would occur in free-living non-target insects. Nevertheless, in many instances semi-field assays can provide information that is far superior to the results obtained from laboratory studies, as they can mimic many of the complex environmental parameters that are difficult, or next to impossible, to duplicate in the laboratory. For exotic or genetically altered organisms, it is sometimes possible to carry out semi-field evaluations, not only in microcosms but in the field under strict containment.

1 Microcosms

A microcosm is "an intact, minimally disturbed piece of an ecosystem brought into the laboratory for study... that behaves ecologically like its counterpart in the actual field" (NRC, 1989 as cited by Krimsky *et al.*, 1995). It allows for standardization and replication of experiments to a degree not possible in the environment, while at the same time maintaining many of the complexities of the environment, all under stringent containment.

A major consideration in the design of a microcosm is the calibration or comparison of the microcosm with the field. The majority of microcosm studies have neglected to make the vital link between the effects within the microcosm and within the field (Frederickson *et al.*, 1990). Scale and component inclusions are also major considerations. In microcosms, the scales under which processes occur are frequently maximally limited to mm. A microcosm essentially requires the exclusion of higher trophic levels because they cannot be naturally maintained in such miniature systems. Consequently, microcosms are best suited for studies of microorganisms in the soil, which is the ultimate sink for most microorganisms, although studies with small insects and plants are also possible. Microcosms are most useful in the study of genetically-altered or exotic pathogens, since pathogens in microcosms are studied in complete confinement. Microcosms have been used to study fate and persistence, gene transfer, movement and ecological effects of a pathogen, all factors that may be important in risk assessment of potential effects on non-target organisms.

During an elegant example of the use of microcosms, Kruetzweiser *et al.* (1996) studied the functional effects of *Bt* on respiration and decomposition activity of microbial communities on leaf surfaces. Microcosms consisted of circular Plexiglass chambers, 6.5 cm in diam. and 7 cm high, mounted in a water bath that maintained constant temperature. Magnetic stir bars in each chamber provided constant water circulation. Leaf disks (23 mm diam.) were cut from birch trees and placed in a laboratory river water flow-through trough for 4 days to allow colonization by natural aquatic microbial communities. Respiration of microbial communities on the leaf disks was measured as oxygen depletion in darkened chambers in the microcosm tank. Four replicate chambers served as controls, 4 were treated with *Bt* and 4 chambers were used to monitor changes in dissolved oxygen in the river water alone. For microbial decomposition determination, leaf disks were randomly allocated to 12 groups of 9 disks each. Five disks in each group were weighed to determine pretreatment weights. The remaining disks were used for measuring protozoan and bacterial density. Experiments were conducted sequentially at different test concentrations of *Bt*. The control and treated water was replaced every 2 days to discard accumulated metabolic wastes. After 9 days, the pre-weighed disks were rinsed, air-dried and weighed to determine loss of mass of the leaf material. Microbial densities were quantified by scraping the surfaces of the disks, washing and fixing the residue, and then counting microbes under the microscope. Study results compared to microbial decomposition in natural stream channels suggested that contamination of water-courses with *Bt* is unlikely to significantly impact microbial communities in terms of decomposition of detritus.

2 *Field cages and containment methods*

Exposing non-target organisms in cages in the field provides information on susceptibility under more realistic conditions. Researchers can include the required numbers of wild or laboratory colony non-targets in cages and place cages in appropriate locations. Particular attention must be paid to the type of cage and its placement in order to attempt to eliminate cage effects as much as possible. Use of data loggers to continuously measure parameters such as insolation, temperature and humidity within and outside of cages is recommended.

Caged trials in the field are not possible for exotic or engineered pathogens that cannot be confined within cages. The potential for movement from cages should be clearly understood, especially for pathogens that disperse well on their own (*e.g.*, Entomophthorales, some entomopathogenic nematodes). In these cases, precautions should be incorporated into studies to prevent movement, *e.g.*, provision of structures preventing wind exposure or preventing escape into the surrounding soil.

Semi-field tests of genetically-altered organisms include strict requirements for containment and post-experimental decontamination. For general issues and considerations regarding choice of project sites and design of barriers, the reader is referred to the Manual prepared by the Scientists' Working Group on Biosafety (1998a). As an example, we describe the protocols from the first release of a genetically-altered baculovirus in the United Kingdom (J. Cory, pers. comm.). Although these protocols were used for testing target insects, as an alternative they could be adapted for field tests against non-targets prior to pathogen release. Cages of fine nylon mesh that did not degrade when exposed to UV were constructed within larger cages that excluded larger wildlife. The insects being tested were epigeal lepidopterans, so the bottom edges of the mesh were buried in the soil to contain the test insects. Before use, all mobile organisms were excluded from the mesh cages using pitfall traps, with herbicide applications around the plots to reduce the chances of any intruders. Of course, strict precautions to prevent spread of the genetically-engineered virus were taken; all personnel wore special suits, foot covers, and gloves that were regularly changed and autoclaved. Everything carried to or from the site was double-bagged and closely accounted for.

In another example of containment during field trials with engineered pathogens, an entomopathogenic nematode genetically engineered to enhance thermotolerance was evaluated at a field site bordered on three sides by buildings (Gaugler *et al.*, 1997). Releases were conducted within 30 cm diam. PVC cylinders buried in the soil to a depth of 14 cm. A 10 cm wide strip of grass was cleared around each cylinder and the 1 cm of cylinder protruding from the soil was coated with a sticky substance to prevent any potential dispersal of nematodes via phoretic transport on soil invertebrates.

3 *Laboratory incubation of field-collected organisms*

A popular method for assessment of target as well as non-target effects of field applications of pathogens is to periodically collect insects from the experimental plots, incubate them under laboratory conditions and then assess mortality. However, unless extreme care is taken, such assessments can provide misleading information. Such assays can measure exposure of the non-target organisms to the inoculum under field conditions and not necessarily infection and disease progression that would occur in the field. As a caveat, using these techniques disease initiation may take place only under the stress of transfer to laboratory conditions that favor the pathogen; therefore, results would overestimate infection levels occurring under field conditions. Of course, the converse is also possible with decreased disease after transfer to the laboratory (in comparison with leaving these organisms in the field) if laboratory conditions in some way do not favor the pathogen.

In addition, it can be very difficult to rear field-collected insects under controlled conditions for any length of time without incurring extensive mortality. Therefore, if possible, after collection non-targets should be maintained for only a limited time period in order to eliminate stress that might begin if rearing conditions are suboptimal or begin to deteriorate. Stressful rearing conditions or provision of non-optimal or unnatural foods can alter susceptibility to pathogens. Attempts to overcome such artifacts are sometimes made by surface sterilizing field-collected insects prior to laboratory incubation. However, it is generally difficult to ensure complete sterilization of the external cuticle and a significant amount of inoculum may also be contained within the gut.

This method of assessment must usually be classified as a semi-field method when field conditions are not exactly simulated after field collection. In such cases, a more accurate method of assessment would be to immediately diagnose presence of disease in field-collected insects (see section C2a below).

As another type of semi-field testing strategy, instead of rearing field-collected insects, the pathogen can be collected in the field and used in the laboratory to challenge insects. The susceptibility of non-target organisms to the levels of inoculum present in the field can be tested by exposing test insects to field-collected substrates, preferably in the field or under simulated field conditions in the greenhouse or laboratory. This technique differs from microcosm studies in that attempts are not necessarily made to duplicate all aspects of the field habitat and these studies are not especially conducted in microcosms. Thus, leaves from the field that could bear pathogen propagules (*i.e.*, spores, infective juvenile nematodes, viral occlusion bodies) could be placed in laboratory cages with herbivorous insects. Soil-dwelling or wandering hosts could be placed in cages in the laboratory containing field-collected soils bearing pathogen inoculum. For example, during studies of resting spores of the Entomophthorales, soil is brought from the field and placed in microcosm boxes in the laboratory at appropriate spring-time temperatures and soil moisture levels (*e.g.*, Hajek *et al.*, 2000). Insects are exposed to resting spore-laden soil but exposure is limited so that the test insects do not die in the boxes and thereby add to the pathogen propagules in the soil. To gain more information from such studies, if possible, pathogen propagules in or on the field-collected substrate should be quantified prior to exposure (see Lacey, 1997; Hajek *et al.*, 2007, and other chapters in this book for methods of quantification).

C Field

Naturally, the most useful and definitive information on the effects of pathogens on non-targets is gained from field studies. Unfortunately, the numbers of individuals of each species collected in the field are often low, especially for indigenous non-targets. Therefore, collection across numerous sites can be very important in order to have a more representative sample of less common species. In some instances, sampling across years can be important for two reasons. First, different species are differentially abundant during different years, *e.g.*, endemic North American lymantriid populations are frequently at low levels, although individual species will increase in abundance occasionally. More importantly, sampling across years can provide an idea of changes in non-target populations over several years after applications of pathogens (*e.g.*, Miller, 1990; Sample *et al.*, 1996).

1 Prerequisites for field studies

a Basic knowledge of host specificity

For any evaluation of non-target organisms in the field, an initial understanding of host specificity is essential. Many entomopathogens are specialists attacking a fairly narrow spectrum of hosts (Federici and Maddox, 1996). Even for pathogens attacking a broader range of hosts, individual strains within a species can differ in host specificity. It is important to initially define the range of potential hosts in order to focus more closely on specific non-targets. In situations where an exotic strain of an indigenous pathogen is being investigated for potential use, host specificity tests are important to determine if the pathotype in question is similar in host range to indigenous strains. If it differs in specificity from indigenous strains, then introductions of the pathotype should be considered more similar to an introduction of an exotic species.

b Methods for pathogen identification

It is not uncommon that researchers do not know the identities of indigenous pathogens in an area of interest prior to testing a pathogen isolate for potential release. Because pathogen species range in distribution from cosmopolitan to those occurring over restricted ranges, it is possible that a pathogen of interest for control or a close relative could already occur in a release area. In addition, the pathogens normally infecting indigenous species that would potentially be evaluated as non-targets are frequently not known. Therefore, to evaluate the impact of pathogens on non-targets in the field, it is necessary to be able to recognize the strain of pathogen that is being released; this would especially be important for pathogens that cause epizootics on their own and possibly is less important for species such as *Bt* that are not known to cause significant levels of mortality without manipulation. In most instances, simple morphological criteria are not adequate for identification and molecular techniques are necessary (Lacey, 1997). Such techniques should be as precise as possible because other strains of the species of pathogen of interest could already be present in the area of release; in this case, molecular techniques that differentiate strains within a species are required. For example, a polyclonal enzyme-linked immunosorbent assay (ELISA) could not differentiate well between the closely related Asian *Entomophaga maimaiga* and North American strains of *Entomophaga aulicae* (Hajek *et al.*, 1991). In contrast, species specific DNA probes used as dot blots readily differentiated between these sister species (Walsh, 1996).

Use of molecular methods for detection and quantification has grown substantially in recent years. For example, field studies of the fate of a released strain of *Beauveria bassiana* used molecular markers developed through comparisons of many isolates with the strain being released. After release, this fungus was isolated from insect cadavers (Wang *et al.*, 2004) and environmental samples (Castrillo *et al.*, 2003), using selective media. Next, specific molecular markers were used to quantify how many samples contained the released strain. Another way to detect a pathogen strain after release has been genetic transformation prior to release, so that the pathogen is easily recognizable (*e.g.*, with green fluorescent protein (GFP); Hu and St. Leger, 2002). However, this latter method brings with it potential regulatory issues due to releasing genetically modified organisms in the field. Studies have also been conducted to evaluate whether genetic exchange between strains of fungi is occurring in the field after commercially available fungal strains are released; such exchange would alter the genetic structure of the indigenous *B. bassiana* populations. *B. bassiana* was isolated from soil samples and both AFLP and RAPD were used to detect recombination between the commercial strain and indigenous strains (P. K. Mishra *et al.*, unpubl. data).

Because field sampling can yield extremely large numbers of samples and infection of non-targets could be infrequent, techniques for processing large numbers of samples, such as dot blots, are useful.

c Methods for disease diagnosis

Historically, signs and symptoms of infected insects or signs from cadavers have often been used first to diagnose disease. However, infections must frequently be well advanced for the appearance of diagnostic signs and symptoms and, for many diseases, there are no externally visible indications to aid in identification of the pathogen in question. In addition, non-target hosts can have marginal infections that result in non-standard signs and symptoms. In any case, the appearance alone of the infected insect or cadaver is usually not adequate for pathogen identification.

The next step in disease diagnosis usually is to observe smears of body tissues under the microscope. However, this also often is not adequate for identification due to various factors, frequently including the lack of morphological characters. Immediate assessment of infection is optimal so that saprophytes and/or rearing conditions do not obscure the pathogen that infected in the field. However, if the infection is very early, the pathogen might not be detected during microscopic examination because adequate time has not elapsed for development of structures by the pathogen that aid in visual identification.

Optimally, visual examination should be followed with use of molecular methods for detection and identification of specific pathogens. Especially in the case of marginal infections, infections can be latent or difficult to detect visually and would, therefore, be missed without molecular methods. In one example, recombinant nucleopolyhedroviruses (NPVs) were developed containing reporter genes allowing easy detection (Huang *et al.*, 1997). A variety of insects were infected with these NPVs and subsequent tests confirmed the presence of the pathogens in numerous hosts. While latent infections were rare, it was only due to these techniques that such marginal infections could be detected at all. It is important to consider such marginal infections in evaluations of risk to non-target organisms because a heterologous virus may trigger a homologous virus in a host (Doyle *et al.*, 1990). Unfortunately, little is known about latent and sublethal infections for most pathogen groups.

d Methods for pathogen quantification

Much useful information on the potential effects of an entomopathogen on non-target organisms can be gained through knowledge of pathogen persistence in areas where the pathogen will be applied. For persistence studies, accurate methods for pathogen retrieval or quantification are necessary. Methods include use of sentinel insects, selective media, and soil extraction and are detailed in Lacey (1997), Hajek *et al.* (2007), and other chapters in this book.

As an example, in a comprehensive set of studies aimed at quantifying a pathogen in the environment, the presence and persistence of *B. bassiana* released for control of the pinewood nematode vector *Monochamus alternatus* was studied due to concerns regarding potential effects on silkworm production if populations of this fungus persisted and increased after release. Selective media was used to isolate this fungus from the air and soil and densities of colonies in treated sites were compared with control sites (Shimazu *et al.*, 2002b). *B. bassiana* was also released in soil and, over time, different media were used to isolate and quantify general fungal populations, actinomycetes, and bacteria as well as *B. bassiana* to compare treated versus control sites (Shimazu *et al.*, 2002a). In cases where strains of released pathogens are of specific interest, molecular markers could be used to identify the released versus background indigenous strains.

2 Estimating effects on non-target host populations

The definitive evaluation of the efficacy of an entomopathogen attacking a pest is often the determination of the reduction of pest damage (*e.g.*, crop yields). However, with few exceptions (*e.g.*, honey yields), this usually is not applicable to non-target organisms, many of which are not pests. Consequently, parameters such as disease prevalence, host population density, and biodiversity are more appropriate to evaluate.

The scope of non-target species to collect and evaluate can be narrowed through knowledge of the pathogen host range determined during laboratory studies. For pathogens being applied as microbial insecticides, all potential species in the environment could be evaluated because pathogen propagules will have been released at unusually high titers throughout the area. Alternatively, for pathogens being released with the goal of establishment in the environment, the range of species to sample is more restricted. Geographic range, macrohabitat and microhabitat have been suggested as major factors determining the niche of a parasite (Adamson and Caira, 1994; Rohde, 1994). Therefore, it is important to understand where transmission occurs in the environment in order to evaluate those organisms that occur in these microhabitats.

As a warning, for studies evaluating a diversity of non-targets, identification of many invertebrates collected in the field as immature stages can be very difficult or impossible; in cases where the organisms collected die as immatures, identification of some groups to species is not possible. As a result, numerous studies report data from large-scale field efforts as the number of individuals for each insect order or suborder. Perhaps maintaining a reference collection of immatures of identified species can assist with identifications of some groups, but with others, *e.g.*, many microlepidopteran larvae, at present there is often no possibility for species identification using larvae. Of course, molecular techniques to identify each individual insect species could be developed but this is clearly not feasible for large-scale studies including many insect species.

a Disease prevalence

In general, system-specific techniques for sampling non-targets of interest should be used to estimate disease prevalence. As examples, diverse non-target aquatic invertebrates were collected after applications of *Bt* and *Bacillus sphaericus* (see Lacey and Mulla, 1990; Lacey and Merritt, 2003) and diverse insects in cabbage fields treated with *Metarhizium anisopliae* were pitfall-trapped (Hu and St. Leger, 2002). Collected invertebrates were evaluated to detect pathogens after collection. Optimally, collected organisms should be evaluated immediately for the presence of pathogens (as previously discussed). Specific methods for disease diagnosis are presented in Lacey (1997) and other chapters in this book.

For example, quantification of fungal prevalence within the bodies of individual insects using selective media can assess the direct effects of entomopathogenic Hypocreales on non-target insects under field conditions. Goettel *et al.* (1996) used an oatmeal dodine selective medium to quantify prevalence of *B. bassiana* in field-collected, non-target arthropods following fungal application to rangeland and alfalfa. Non-target invertebrates collected in sweep nets or from pitfall traps were immediately killed by freezing, surface disinfected in 70% ethanol, rinsed in water, homogenized and plated onto the selective medium. Arthropods were collected immediately before conidial application and then at 24-h intervals for up to 10 days. Temporary increases in *B. bassiana* prevalence were observed in coccinellid beetles and harvestmen. Subsequent laboratory studies (M.S.G., unpubl. data) indicated that these increases in numbers of colony-forming units represented ingested conidia within alimentary tracts rather than blastospores within the hemocoels of infected insects. Only leaf-cutting bees, collected 10 days post-application, exhibited greater than 1000 colony-forming units per individual, indicating that they were infected under field conditions (Goettel *et al.*, 1996). These results suggest that field collections should preferably be carried out over longer periods, because disease initiation can be prolonged under field conditions.

With care, for specific types of host/pathogen systems one could consider rearing field-collected insects under controlled conditions closely simulating the field to determine disease prevalence. When rearing conditions do not closely mimic the field, we consider such a technique as less realistic, *i.e.*, a semi-field method; (see section 4B3). To use this technique, the pathogen in question should not be a facultative pathogen that could readily cause increased infection and mortality if hosts are stressed. The non-target rearing conditions after collection should closely replicate field conditions, exerting minimal stress on the pathogen or the potential hosts. As an example of a relatively stress-free rearing strategy, some researchers have collected insects from the field and reared them in sleeves on host plants in the field. In this instance, the potential hosts were thus readily accessible for observation during rearing. Some techniques that have been adopted are using field-collected food for rearing (although food surfaces should be sterilized to kill any unwanted pathogens) or rearing insects in a field insectary so that insects are reared at the correct ambient temperatures for that time of year. Field-collected insects should be reared for only a relatively brief time, as an indication of the mortality that would occur during that period of time in the field. Then, the population in the field should be resampled to collect more individuals for rearing. If such techniques are used, a method for calculating the season-long infection level based on these "snapshots" of infection has been proposed by Elkinton *et al.* (1992) (see Chapter V-2).

b Population density

Determining the effects of an entomopathogen on the target host population density is usually an arduous task in itself. Consequently, it is next to impossible to include every non-target organism exposed to the pathogen in such evaluations. However, at times it may be warranted to investigate population densities of specific non-target organisms. Such situations could arise if there is evidence from laboratory assays that an important non-target organism may be affected under certain conditions or if non-targets known to be specifically associated with the target pest are thought to be impacted, *e.g.*, direct or indirect effects on beneficials (see Flexner *et al.*, 1986). Alternately, in some situations, it may be prudent to monitor population densities of either certain sentinel beneficial organisms known to be essential in IPM programs or non-targets that are important members of the invertebrate community.

Experimental design is essentially the same as that used to evaluate the effects of the pathogen on the target host population. Methods for monitoring non-target organisms in particular must be carefully chosen to ensure that population density assessments are valid. It is imperative that the population densities of target organisms be assessed concurrently; if changes in target host population densities due to the pathogen are not apparent, then information on lack of effects on the non-target populations are not of much use. Replicated treatments should consist of pathogen- treated populations and non-treated controls. If possible, carrier and chemical treatments should be included. Carrier-only treatments will help determine if any detrimental effects are due to the carrier itself and not the active control agent (*i.e.*, pathogen). The chemical treatment can help demonstrate the sensitivity of the population monitoring methods employed.

c *Native biodiversity*

Pathogens can have indirect effects on non-target organisms and consequently impacts could also occur at the community level. In addition, such community level influences may occur across time and space. To investigate non-target effects on communities, standard ecological measurements, such as diversity indices would be appropriate (Magurran, 2004). Such indices would be especially suitable for broad-based studies carried out over numerous years comparing numerous treated versus untreated sites.

For such studies, plot size and study design will depend on many factors including non-target species being investigated, number of treatments, habitat, etc. Plots should be established that would be large enough to provide adequate buffers from cross-contamination from immigrating and emigrating invertebrates. Methods for sampling should be used that are suitable for the insects to be collected, *e.g.*, pitfall traps for some Coleoptera. Numbers of traps should be adjusted according to plot size, to ensure that depletion due to trapping does not occur. Optimally, samples should be collected in advance of application, on the day of application and at intervals at least up to 20 days after application, depending on the characteristics of the pathogen in question. The longer the pre-treatment sampling, the more pertinent the information on possible differences between plots after treatment.

D *A special case: honey bees*

Virtually all trials of the effects of microbial control agents on non-targets require tests of honey bees, *Apis mellifera*. Assessment of hazards to honey bees presents unique difficulties because these are social insects. Laboratory assays typically use caged bees isolated from their queen. This disrupts the social aspects of bee behavior as well as maintains bees at lower temperatures than those typically maintained within the hive. Laboratory assays therefore result in stressful situations for the worker bees, which undoubtedly reduces longevity and renders them more susceptible to pathogens than under normal conditions. The longevity of worker bees is typically 4–5 weeks (Free and Spencer-Booth, 1959) whereas high mortalities in caged worker bees can occur within 10 days (Vandenberg, 1990). Consequently, outdoor nucleus hives should be used whenever possible to evaluate safety of pathogens to honey bees.

The following methods to assess effects of pathogens on outdoor honey bee colonies are adapted from Cantwell *et al.* (1966), Morton and Moffett (1972) and Bromenshenk *et al.* (1996).

1 Establish colonies

Establish nine mite- and disease-free honey bee nucleus colonies for each pathogen to be tested. Place a thermocouple in the brood area and connect it to a data recorder to monitor the daily maximum and minimum temperatures within the hive. On the day before treatment, identify and delineate an area of brood cells in each colony that contains mostly eggs and some young larvae. Take a cell by cell census and mark the area.

2 Assess hygienic behavior

Because of the variation in hygienic behavior among hives and its importance in the epizootiology of bee diseases (Rothenbuhler, 1964), the hygienic behavior of each colony must be assessed first. One method is to remove a brood frame from each colony, place a metal cylinder on part of the capped brood, and rotate and gently push the cylinder so that it cuts through the comb, down to the plastic foundation sheet, thereby forming a diked area on the comb (J. Bromenshenk, pers. comm.). Pour ca. 80 ml of liquid nitrogen into the cylinder and wait until it evaporates. Once the cylinder has thawed, remove it. The brood cells exposed to the liquid nitrogen will have been killed. Place an acetate sheet directly over the frozen brood cells and mark the position of 20 cells that look undamaged within the frozen area. Mark the frame and acetate sheet positions so that the sheet can later be placed in the same position. Return the brood frame to the colony with the frozen side placed in the colony's center. After a minimum of 15 days, remove the frame, position the marked acetate sheet over the frozen area and note the numbers of capped, emptied or otherwise altered cells. The percent of altered cells as compared to the original 20 frozen cells serves as an index of hygienic behavior (*i.e.*, the efficiency with which dead brood is removed from a colony).

Based on the results of the hygienic behavior tests, treatments should be assigned to groups of three colonies, attempting to assure that hygienic behaviors are equally distributed among treatments.

3 Treatment

Prepare inoculum and decide on a method of dose administration. Cantwell *et al.* (1966) administered pathogens (viruses and bacteria) by feeding. Bromenshenk *et al.* (1996) argued that spraying adult honey bees is more realistic than feeding, since the most direct exposure route of commercially applied microbial control agents to honey bees would be direct contact, and not through the ingestion of pollen or nectar. A third method of bee exposure would be to apply the pathogen directly to a crop field, as in an operational application against a target host. Regardless of inoculation method, treatments should consist of the active pathogen, inactivated pathogen (either killed by gamma or UV-irradiation or heat) and a control. At the time of administration, the propagules are enumerated and the viability assessed according to established protocols for the pathogen in question (see Lacey, 1997). If required, treatments can be re-administered at intervals based upon the expected use pattern of the pathogen against its target hosts.

a Inoculation by feeding (after Cantwell et al., *1966)*

Place the pathogen in 200 ml of sterile sucrose solution (1:1 sugar to water, w/w) and place in a feeder. Close the hive entrance for 8 hours or until all of the inoculum has been consumed. After re-opening the hive, provide a clean sucrose solution for the remainder of the test. It is important to determine if the sucrose solution has any detrimental effect on the activity of the pathogen prior to using this method. Dosages of the pathogen should be at least equal to those causing high mortality in the target hosts.

b Topical inoculation (after Bromenshenk et al., *1996)*

Open and treat colonies in succession, beginning with the three control hives, followed by the inactivated pathogen, then by the active pathogen. Open each hive and tranquilize the bees with smoke. Remove ca. one third of the bees by gently brushing them into a lidded pre-weighed plastic container. Determine the net weight of bees and estimate the number of bees within the container (estimating that each bee weighs an average of 0.092 g). Anaesthetize the bees by placing them for 15 to 20 min in a cooler containing dry ice. Cooling time needs to be adjusted according to the prevailing weather conditions. Spread each group of immobilized bees evenly on a spray platform inside a disposable spray container and spray evenly using an airbrush positioned overhead, or another suitable application device. Precise methods of spraying and monitoring of the dose delivered to each spray group will vary according to the nature of the pathogen (see Lacey, 1997). Where possible, remove samples of sprayed bees and enumerate the number of propagules that each bee receives.

Return treated bees to the appropriate colony prior to their reviving. The entire procedure for each colony should require about 30 min. Appropriate procedures must be followed to ensure that cross contamination does not occur. Concentrations chosen should at least approximate pathogen levels expected during operational use of the pathogen against target hosts.

c Inoculation through operational application

Choose replicate treatment plots with adequate buffer zones to ensure that mixing of bees between treatments does not occur. Hives can be placed into the plots prior to or immediately after treatment, depending on the expected exposure during operational use of the pathogen. For a worst case scenario, treatments should take place during conditions of peak bee foraging.

4 Evaluation of effects

a Collection of dead workers and larvae at the hive entrance

Following treatment, place dead bee traps (Atkins *et al.*, 1970) at each hive entrance (Figure 1). These devices trap 90 to 95% of dead bees and larvae being removed from the hive by the worker bees. Collect dead bees daily and process for pathogen diagnoses according to methods appropriate for the pathogen in question (see Lacey, 1997).

b Survival and development of the brood

After an interval appropriate to the development of the pathogen being tested, open the colonies and reexamine the marked area of the brood, paying special attention to empty cells or cells with cadavers or larvae showing disease symptoms. Mark new brood and take new censuses if treatments are to be repeated. The number of frames having eggs, larvae, pupae or emerging adults, and the areas of sealed and unsealed brood can also be used to measure brood production (Morton and Moffett, 1972).

c Monitoring adult bees

At regular intervals, remove samples of living adult bees from within each colony and process for presence of pathogens or disease according to protocols appropriate for the pathogen in question (see Lacey, 1997).

d Monitoring hive temperature

Analyze the daily hive temperature profiles to determine if there are significant differences between treatment and control hives. Temperature anomalies in treated hives are a sign of a stressed or diseased hive (J. Bromenshenk, pers. comm.).

If at all possible, assessments of the effects on the target population should be carried out at the same time that effects on honey bees are being determined. This is especially important when pathogen application is done on an operational scale.

Recent developments enable continuous electronic monitoring of hives, including tracking individual bees and enumerating the numbers of bees flying into and out of each hive (J. Bromenschenk, pers. comm.). This technology will greatly improve the precision of estimating detrimental effects of pathogens on honey bee colonies.

5 Regulations

Most protocols for determining potential effects of pathogens on non-target organisms for release or registration are governed by regulations

Figure 1. Dead bee trap used to collect dead bees emerging from nucleus hives. A) Dead bee trap which is set on a two-story hive. Note the data logger within the plastic cup on the right side of the trap. The temperature probe is inserted into the hive, adjacent to the brood. B) The key to the dead bee trap: two staggered perforated plates with holes just large enough for a worker to pass through. This arrangement forces workers to drop dead bees and larvae as they exit the hive. C) Dead bees are dropped on the trap's bottom, which can be pulled out allowing removal of cadavers and bee frass without opening the colony. Photos courtesy of Stefan Jaronski, Sidney, MT.

that vary among countries (Jaronski *et al.*, 2003). Although there are efforts under way to harmonize these regulations, there is still great disparity among the regulations between different countries. Many regulations include detailed protocols on non-target testing that need to be followed. Others allow the researchers to adapt and develop protocols for their special cases. Regulations can differ by type of pathogen, *e.g.*, in the United States, viruses, bacteria, fungi, and protozoa are regulated in ways more similar to chemical pesticides (toxicology studies are required although these are not as extensive as requirements for chemical pesticides), while nematodes are not regulated. Unfortunately, most regulations principally only require detailed laboratory testing, even for indigenous pathogens. Although at present laboratory assessments are economical and efficacious, they may provide little useful information regarding impact on non-target organisms under field conditions (*e.g.*, Hajek and Butler, 1999; Jaronski *et al.*, 2003).

As new field diagnostic methods are being devised, it is becoming possible to evaluate effects of entomopathogens on non-target organisms under field conditions. Although at present most decisions regarding non-target risks are still based on data from laboratory evaluations, we feel that in the future, it will be possible to obtain more meaningful risk assessments through studies using microcosms, semi-field and field assessments.

6 Acknowledgments

We thank L. Castrillo and J. Vandenberg, USDA, ARS, J. Bromenshenk, Montana State University, and J. Cory, Algoma Univ. Coll., Sault Ste. Marie, Canada for graciously providing their unpublished information and manuscripts, and S. Jaronski, USDA, ARS for providing the photos. L. Solter provided helpful additions to this chapter.

7 References

Adamson, M. L. and Caira, J. N. 1994. Evolutionary factors influencing the nature of parasite specificity. *Parasitology* **109**, S85–S95.

Atkins, E. L., Todd, F. E., and Anderson, L. D. 1970. Honey bee field research aided by Todd dead bee hive entrance trap. *Calif. Agric.* **24**, 12–13.

Bromenshenk, J. J., Jaronski, S. T. and Loeser, M. R. 1996. Multiple endpoint, holistic assessment of the effects of Mycotrol WP (*Beauveria bassiana* Strain GHA) on outdoor *Apis mellifera* L. colonies, 69 pp. Tech. Rep. Mycotech 95–05.

Cantwell, G. E., Knox, D. A., Lehnert, T., and Michael, A. S. 1966. Mortality of the honey bee, *Apis mellifera*, in colonies treated with certain biological insecticides. *J. Invertebr. Pathol.* **8**, 228–233.

Castrillo, L. A., Vandenberg, J. D., and Wraight, S. P. 2003. Strain-specific detection of introduced *Beauveria bassiana* in agricultural fields by use of sequence-characterized amplified region markers. *J. Invertebr. Pathol.* **82**, 75–83.

Cook, R. J., Bruckart, W. L., Coulson, J. R., Goettel, M. S., Humber, R. A., Lumsden, R. D., Maddox, J. V., McManus, M. L., Moore, L., Meyer, S. F., Quimby, P. C., Stack, J. P., and Vaughn, J. L. 1996. Safety of microorganisms intended for pest control and plant disease control: A framework for scientific evaluation. *Biol. Control* **7**, 333–351.

Doyle, C. J., Hirst, M. L., Cory, J. S., and Entwistle, P. F. 1990. Risk assessment studies: Detailed host range testing of wild-type cabbage moth, *Mamestra brassicae* (Lepidoptera: Noctuidae), nuclear polyhedrosis virus. *Appl. Environ. Microbiol.* **56**, 2704–2710.

Elkinton, J. S., Buonaccorsi, J. P., Bellows T. S., Jr., and Van Driesche, R. G. 1992. Marginal attack rate, *k*-values and density dependence in the analysis of contemporaneous mortality factors. *Res. Popul. Ecol.* **34**, 29–44.

Federici, B. A. and Maddox, J. V. 1996. Host specificity in microbe-insect interactions. *BioScience* **46**, 410–421.

Flexner, J. L., Lighthart, B. and Croft, B. A. 1986. The effects of microbial pesticides on non-target, beneficial arthropods. *Agric. Ecosys. Environ.* **16**, 203–254.

Frederickson, J. K., Van Voris, P., Bentjen, S. A., and Bolton, H., Jr. 1990. Terrestrial microcosms for evaluating the environmental fate and risks associated with the release of chemicals or genetically engineered microorganisms to the environment. *Hazard Assessment Chem.* **7**, 157–202.

Free, J. B. and Spencer-Booth, Y. 1959. The longevity of worker honey bees (*Apis mellifera*). *Proc. R. Entomol. Soc. Lond.* (*A*) **34**, 141–150.

Gaugler, R., Wilson, M. and Shearer, P. 1997. Field release and environmental fate of a transgenic entomopathogenic nematode. *Biol. Control* **9**, 75–80.

Goettel,. M. S, and Hajek, A. E. 2001. Evaluation of nontarget effects of pathogens used for management of arthropods *In* "Evaluating Indirect Ecological Effects of Biological Control" (E. Wajnberg, J. K. Scott and P. C. Quimby, Eds.), pp. 81–97. CABI Press, Wallingford, U.K.

Goettel, M. S., Inglis, G. D., Duke, G. M. and Lord, J. C. 1996. Effect of *Beauveria bassiana* (Mycotech strain GHA) on invertebrates in rangeland and alfalfa ecosystems. Final Report for the Untited States Environmental Protection Agency, Washington, D.C.

Grace, J. K. 1994. Protocol for testing effects of microbial pest control agents on non-target subterranean termites (Isoptera: Rhinotermitidae). *J. Econ. Entomol.* **87**, 269–274.

Hajek, A. E. and Butler, L. 1999. Predicting the host range of entomopathogenic fungi. *In* "Non-target Effects of Biological Control" (P. A. Follett & J. J. Duan, Eds.), pp. 263–276. Kluwer Acad. Publ., Dordrecht, Netherlands.

Hajek, A. E., Butler, L., Walsh, S. R. A., Silver, J. C., Hain, F. P., Hastings, F. L., ODell, T. M., and Smitley, D. R. 1996. Host range of the gypsy moth (Lepidoptera: Lymantriidae) pathogen *Entomophaga maimaiga* (Zygomycetes: Entomophthorales) in the field versus the laboratory. *Environ. Entomol.* **25**, 709–721.

Hajek, A. E., Butler, L., Liebherr, J. K., and Wheeler, M. M. 2000. Risk of infection by the fungal pathogen *Entomophaga maimaiga* among Lepidoptera on the forest floor. *Environ. Entomol.* **29**, 645–650.

Hajek, A. E., Butt, T. M., Stelow, L. I., and Gray, S. M. 1991. Detection of *Entomophaga maimaiga* (Zygomycetes: Entomophthorales) using enzyme linked immunosorbent assay (ELISA). *J. Invertebr. Pathol.* **58**, 1–9.

Hajek, A. E., Delalibera Jr., I., and Butler, L. 2003. Entomopathogenic fungi as classical biological control agents. *In* "Environmental Impacts of Microbial Insecticides" (H. M. T. Hokkanen and A. E. Hajek. Eds.), pp. 15–34. Kluwer Academic Publishers, Dordrecht, The Netherlands.

Hajek, A. E., Fuxa, J. R., and Kunimi, Y. 2007. Microorganisms associated with soil arthropods. *In* "Manual of Environmental Microbiology, 3rd ed." (C. J. Hurst, R. L. Crawford, J. L. Garland, D. A. Lipson, A. L. Mills, and L. D. Stetzenbach, Eds.). ASM Press, Washington, D.C. (in press).

Hassan, S. A. 1989. Testing methodology and the concept of the IOBC/WPRS working group. *In* "Pesticides and Non-target Invertebrates" (P. D. Jepson, Ed.), pp. 1–18. Intercept, Wimborne, Dorset, UK.

Hokkanen, H. M. T., Bigler, F., Burgio, G., van Lenteren, J. C., and Thomas, M. B. 2003. Ecological risk assessment framework for biological control agents. *In* "Environmental Impacts of Microbial Insecticides" (H. M. T. Hokkanen and A. E. Hajek, Eds.), pp 1–14. Kluwer Academic Publishers, Dordrecht, The Netherlands.

Hokkanen, H. M. T., and Hajek, A. E. (Eds.) 2003. "Environmental Impacts of Microbial Insecticides." Kluwer Academic Publishers, Dordrecht, Netherlands.

Hokkanen, H. M. T., Zeng, Q-Q., and Menzler-Hokkanen, I. 2003. Assessing the impacts of *Metarhizium* and *Beauveria* on bumblebees. *In* "Environmental Impacts of Microbial Insecticides" (H. M. T. Hokkanen and

A. E. Hajek, Eds.), pp. 63–72. Kluwer Academic Publishers, Dordrecht, The Netherlands.

Hu, G. and St. Leger, R. J. 2002. Field studies using a recombinant mycoinsecticide (*Metarhizium anisopliae*) reveal that it is rhizosphere competent. *Appl. Environ. Microbiol.* **68**, 6383–6387.

Huang, X.-P., Davis, T. R., Hughes, P., and Wood, A. 1997. Potential replication of recombinant baculoviruses in non-target insects species: Reporter gene products as indicators of infection. *J. Invertebr. Pathol.* **69**, 234–245.

James, R. R. and Lighthart, B. 1992. Protocol for testing the effects of fungal pesticides on non-target beetles using *Hippodamia convergens* (Coleoptera: Coccinellidae). EPA/600/R-92/160. Natl. Tech. Info. Serv. Publ. PB92 217 488. Springfield, VA.

Jaronski, S. T., Goettel, M. S., and Lomer, C. L. 2003 Regulatory requirements for ecotoxicological assessments of microbial insecticides – how relevant are they? *In* "Environmental Impacts of Microbial Insecticides" (H. M. T. Hokkanen and A. E. Hajek. Eds.), pp 237–260. Kluwer Academic Publishers, Dordrecht, The Netherlands.

Jepson, P. C. (Ed.) 1989. "Pesticides and Non-target Invertebrates." Intercept, Wimborne, Dorset, UK.

Krimsky, S., Wrubel, R. P., Naess, I. G., Levy, S. B., Wetzler, R. E., and Marshall, B. 1995. Standardized microcosms in microbial risk assessment. *BioScience* **45**, 590–599.

Kruetzweiser, D. P., Gringorten, J. L., Thomas, D. R. and Butcher, J. T. 1996. Functional effects of the bacterial insecticide *Bacillus thuringiensis* var. *kurstaki* on aquatic microbial communities. *Ecotoxicol. Environ. Safety* **33**, 271–280.

Lacey, L. A. (Ed.) 1997. "Manual of Techniques in Insect Pathology." Academic Press, London.

Lacey, L. A. and Merritt, R. W. 2003. The safety of bacterial microbial agents used for black fly and mosquito control in aquatic environments. *In* "Environmental Impacts of Microbial Insecticides (H. M. T. Hokkanen and A. E. Hajek, eds.), pp. 151–168. Kluwer Academic Publishers, Dordrecht, The Netherlands.

Lacey, L. A. and Mulla, M. S. 1990. Safety of *Bacillus thuringiensis* ssp. *israelensis* and *Bacillus sphaericus* to nontarget organisms in the aquatic environment. *In* "Safety of Microbial Insecticides" (M. Laird, L. A. Lacey and E. W. Davidson, Eds.), pp. 169–188. CRC Press, Boca Raton, FL.

Laird, M., Lacey, L. A., and Davidson, E. W. (Eds.). 1990. "Safety of Microbial Insecticides." CRC Press, Boca Raton, FL.

Lockwood, J. 1993a. Environmental issues involved in biological control of rangeland grasshoppers (Orthoptera: Acrididae) with exotic agents. *Environ. Entomol.* **22**, 503–518.

Lockwood, J. 1993b. Benefits and costs of controlling rangeland grasshoppers (Orthoptera: Acrididae) with exotic agents. *Environ. Entomol.* **22**, 904–913.

Maddox, J. V., McManus, M. L., Jeffords, M. R., and Webb, R. E. 1992. Exotic insect pathogens as classical biological control agents with an emphasis on regulatory considerations. *In* "Selection Criteria and Ecological Consequences of Importing Natural Enemies" (W. C. Kauffmann and J. E. Nechols, Eds.), pp. 27–39. Entomol. Soc. Amer., Lanham, MD.

Magurran, A. E. 2004. "Measuring Biological Diversity". Blackwell Publ., Malden, MA.

Miller, J. C. 1990. Field assessment of the effects of a microbial pest control agent on non-target Lepidoptera. *Amer. Entomol.* **36**, 135–139.

Morton, H. L. and Moffett, J. O. 1972. Ovicidal and larvicidal effects of certain herbicides on honey bees. *Environ. Entomol.* **1**, 611–614.

Navon, A. and Ascher, K. R. S. (Eds). 1999. "Bioassays of Entomopathogenic Microbes and Nematodes." CABI Publishing, Wallingford, UK.

[NRC] National Research Council. 1989. Field testing genetically modified organisms: framework for decisions. National Academy Press, Washington.

Rohde, K. 1994. Niche restriction in parasites: proximate and ultimate causes. *Parasitology* **109**, S69–S84.

Rothenbuhler, W. C. 1964. Behavior genetics of nest cleaning in honey bees. IV. Responses of F1 and backcross generations to disease-killed brood. *Am. Zool.* **4**, 111–123.

Sample, B. E., Butler, L., Zivkovich, C., Whitmore, R. C., and Reardon, R. 1996. Effects of *Bacillus thuringiensis* Berliner var. *kurstaki* and defoliation by the gypsy moth (*Lymantria dispar* (L.) (Lepidoptera: Lymantriidae)) on native arthropods in West Virginia. *Can. Entomol.* **128**, 573–592.

Scientists' Working Group on Biosafety. 1998a. Manual for Assessing Ecological and Human Health Effects of Genetically Engineered Organisms. Part One: Introductory Materials and Supporting Text for Flowcharts. The Edmonds Institute, Washington.

Scientists' Working Group on Biosafety. 1998b. Manual for Assessing Ecological and Human Health Effects of Genetically Engineered Organisms. Part Two: Flowcharts and Worksheets. The Edmonds Institute, Washington.

Shimazu, M., Maehara, N., and Sato, H. 2002a. Density dynamics of the entomopathogenic fungus, *Beauveria bassiana* Vuillemin (Deuteromycotina: Hyphomycetes) introduced into forest soil, and its influence on other soil microorganisms. *Appl. Entomol. Zool.* **37**, 268–269.

Shimazu, M., Sato, H., and Maehara, N. 2002b. Density of the entomopathogenic fungus, *Beauveria bassiana* Vuillemin (Deuteromycotina: Hyphomycetes) in forest air and soil. *Appl. Entomol. Zool.* **37**, 19–26.

Siegel, J. P. 1997. Testing the pathogenicity and infectivity of entomopathogens to mammals. *In* "Manual of Techniques in Insect Pathology" (L. A. Lacey, Ed.), pp. 325–336. Academic Press, London.

Solter, L. F., and Becnel, J. J. 2003. Environmental safety of microsporidia. *In* "Environmental Impacts of Microbial Insecticides" (H. M. T. Hokkanen and A. E. Hajek, Eds.), pp. 93–118. Kluwer Academic Publishers, Dordrecht, The Netherlands.

Solter, L. F., and Maddox, J. V. 1998. Physiological host specificity of microsporidia as an indicator of ecological host specificity. *J. Invertebr. Pathol.* **71**, 207–216.

Vandenberg, J. D. 1990. Safety of four entomopathogens for caged adult honey bees (Hymenoptera: Apidae). *J. Econ. Entomol.* **83**, 755–759.

Walsh, S. R. A. 1996. Development of molecular markers for the detection and differentiation of *Entomophaga* strains pathogenic for insects. Ph.D. dissertation, Univ. Toronto, Canada.

Wang, C., Fan, M., Li, Z., and Butt, T. M. 2004. Molecular monitoring and evaluation of the applicationof the insect-pathogenic fungus *Beauveria bassiana* in southeast China. *J. Appl. Microbiol.* **96**, 861–870.

Weseloh, R. M. 1984. Effect of the feeding inhibitor Plictran and low *Bacillus thuringiensis* Berliner doses on *Lymantria dispar* (L.) (Lepidoptera: Lymantriidae): Implications for *Cotesia melanoscelus* (Hymenoptera: Braconidae). *Environ. Entomol.* **13**, 1371–1376.

Weseloh, R. M. and Andreadis, T. G. 1982. Possible mechanisms for synergism between *Bacillus thuringiensis* and the gypsy moth (Lepidoptera: Lyjmantriidae) parasitoid, *Apanteles melanoscelus* (Hymenoptera: Braconidae). *Ann. Entomol. Soc. Amer.* **75**, 435–438.

Weseloh, R. M., Andreadis, T. G., Moore, R. E., Anderson, J. R., Dubois, N. R., and Lewis, F. B. 1983. Field confirmation of a mechanism causing synergism between *Bacillus thuringiensis* and the gypsy moth parasitoid, *Apanteles melanoscelus*. *J. Invertebr. Pathol.* **41**, 99–103.

Subject Index

C

M

O

Q

R

S

T